DIRECTORY OF GEOSCIENCE DEPARTMENTS

46th Edition

Christopher M. Keane and
Cynthia Martinez, Editors

American Geological Institute
Alexandria, Virginia

Cover

These cinder cones are associated with a volcanic vent that formed about 70,000 to 75,000 years ago north of present day Flagstaff at Arizona's SP Crater. Photo copyright Michael Collier. AGI Imagebank ID# ih0c3z.

AGI Staff

Christopher M. Keane	Editor
Cindy Martinez	Editor
John Rasanen	Advertising

Typeset in Times New Roman and Arial using Adobe InDesign CS3.

Printed in the United States of America.
Published by the American Geological Institute.

ISBN 0-922152-80-2
ISSN 0364-7811

American Geological Institute
4220 King Street, Alexandria, VA 22302-1502
(703) 379-2480 FAX (703) 379-7563
cmm@agiweb.org
http://www.agiweb.org

Introduction

The 46th Edition of the Directory of Geoscience Departments brings some new features. This year we have listings of the leadership of the major geoscience professional and scientific societies in the United States. Also, by popular demand, we are expanding the statistics section derived from the data gathered each year to compile this publication. The selected statistical analyses are based on the most requested data for 2008.

The listings for non-US departments has expanded dramatically, representing increased participation in the surveys used to compile this edition. The data used to compile the directory is provided by the individuals and departments listed. AGI only edits the information for consistency and format. Any errors or omissions most likely reflect the entries of the individuals or the primary department contact. If you identify issues, please consider contacting Cindy Martinez at cmm@agiweb.org with information about a current responsible contact for the department in question, so that we may further encourage updates to the next edition.

The American Geological Institute does not warrant the accuracy of any of the self-reported information, but we do encourgage individuals and departments to update material they believe is out-of-date.

For the 46th edition of the Directory of Geoscience Departments, **412** departments updated their departmental records. The 2008 directory lists **1,132** geoscience departments, surveys, and agencies. This includes **1,008** geoscience departments, with **724** being in the United States and **284** outside of the United States. Of the U.S. geoscience departments listed, **724** are degree-granting.

AGI gratefully acknowledges the continued cooperation of geoscience department heads in providing information about their departments.

Christopher M. Keane
Editor

Notations used in the 46th edition of the Directory of Geoscience Departments:

- 2008 AGI Academic Associates in bold
- Year of most recent department updates in parentheses
- Most recent enrollments/degrees
- Selected non-U.S. colleges and universities in 45 countries
- Federal agencies that employ or are of professional interest to geoscientists
- A new individual specialty notation system

DIRECTORY OF GEOSCIENCE DEPARTMENTS
46th Edition 2008

Table of Contents

University/Agency Table of Contents v
Directory of Geoscience Departments
U.S. Section 1
Non-U.S. Section 178
U.S. State Geological Surveys 217
U.S. Federal Agencies and National Laboratories 235
Geoscience Societies. 249
Statistical Compendium 2007 259
List of Faculty Specialty Codes 263
Faculty Index 265

University/Agency Table of Contents

Alabama

Alabama A&M University 1
Auburn University 1
University of Alabama 2
University of Alabama, Dauphin Island 2
University of Montevallo 2
University of South Alabama 2
University of West Alabama 2

Alaska

University of Alaska, Anchorage 3
University of Alaska, Fairbanks 3

Arizona

Arizona State University 3
Arizona Western College 4
Central Arizona College 4
Glendale Community College 4
Mesa Community College 4
Northern Arizona University 4
Phoenix College 5
Prescott College 5
University of Arizona 5
Yavapai College 8

Arkansas

Arkansas Tech University 8
College of the Ozarks 8
Southern Arkansas University 8
University of Arkansas, Fayetteville 8
University of Arkansas, Little Rock 9
University of Arkansas, Pine Bluff 9

California

American River College 9
Antelope Valley College 9
Bakersfield College 9
Barstow Community College 9
Cabrillo College 9
California Institute of Technology 9
California Lutheran University 10
California Polytechnic State University 10
California State Polytechnic Univ, Pomona 10
California State University, Bakersfield 11
California State University, Chico 11
California State University, Dominguez Hills 11
California State University, East Bay 11
California State University, Fresno 11
California State University, Fullerton 12
California State University, Long Beach 12
California State University, Los Angeles 12
California State University, Northridge 12
California State University, Sacramento 13
California State University, San Bernardino 13
California State University, Stanislaus 13
Cerritos College 13
Chaffey College 13
Chapman University 14
City College of San Francisco 14
College of the Desert 14
College of the Redwoods 14
College of the Siskiyous 14
Cosumnes River College 14
Crafton Hills College 14
Cypress College 14
De Anza College 14
Diablo Valley College 14
East Los Angeles College 15
El Camino College 15
Foothill College 15
Fresno City College 15
Fullerton College 15
Glendale Community College 15
Golden West College 15
Grossmont College 15
Humboldt State University 15
Irvine Valley College 16
Las Positas College 16
Loma Linda University 16
Long Beach City College 16
Los Angeles City College 16
Los Angeles County Mus of Natural History 16
Los Angeles Harbor College 16
Los Angeles Valley College 16
Marin Community College 17
Mendocino College 17
Merced College 17
MiraCosta College 17
Modesto Junior College 17
Moss Landing Marine Laboratories 17
Mount San Antonio College 17
Napa Valley College 17
Naval Postgraduate School 18
Occidental College 18
Ohlone College 18
Orange Coast College 18
Palomar College 18
Pasadena City College 18
Pomona College 19
Rio Hondo College 19
Riverside Community College 19
Saddleback Community College 19
San Bernardino County Museum 19
San Bernardino Valley College 19
San Diego State University 19
San Francisco State University 20
San Jose City College 20
San Jose State University 20
Santa Ana College 20
Santa Barbara City College 20
Santiago Canyon College 20
Shasta Community College 20
Sonoma State University 21
Southwestern College 21
Stanford University 21
University of California, Berkeley 22
University of California, Davis 23
University of California, Irvine 24
University of California, Los Angeles 25
University of California, Riverside 26
University of California, San Diego 26
University of California, Santa Barbara 27
University of California, Santa Cruz 27

University/Agency Table of Contents

University of San Diego ... 29
University of Southern California ... 29
University of the Pacific ... 29
Ventura College ... 29
West Valley College ... 29
Whittier College ... 30
Yuba Community College ... 30

Colorado

Adams State College ... 30
Colorado College ... 30
Colorado Northwestern Community College ... 30
Colorado School of Mines ... 30
Colorado State University ... 31
Denver Museum of Nature & Science ... 31
Fort Lewis College ... 32
Mesa State College ... 32
Metropolitan State College of Denver ... 32
Pikes Peak Community College ... 32
United States Air Force Academy ... 32
University of Colorado ... 32
University of Colorado, Denver ... 33
University of Denver ... 34
University of Northern Colorado ... 34
Western State College of Colorado ... 34

Connecticut

Central Connecticut State University ... 34
Eastern Connecticut State University ... 34
Southern Connecticut State University ... 35
University of Connecticut ... 35
University of New Haven ... 35
Wesleyan University ... 35
Western Connecticut State University ... 36
Yale University ... 36

Delaware

University of Delaware ... 36

District of Columbia

Carnegie Institution of Washington ... 37
George Washington University ... 38
Nat Academy of Sci/National Res Council ... 38
Smithsonian Inst / Nat Air & Space Mus ... 38
Smithsonian Inst / Nat Mus of Natural History ... 38
University of the District of Columbia ... 39

Florida

Broward Community Coll, Central Campus ... 39
Eckerd College ... 39
Florida Atlantic University ... 39
Florida Institute of Technology ... 39
Florida International University ... 40
Florida Keys Community College ... 41
Florida State University ... 41
Gulf Coast Community College ... 42
Jacksonville University ... 42
Miami-Dade Community Coll (North Campus) ... 42
Miami-Dade Comm Coll (Wolfson Campus) ... 42
Nova Southeastern University ... 42
Okaloosa-Walton College ... 43
Pensacola Junior College ... 43
Saint Petersburg College, Clearwater ... 43
Saint Petersburg Junior College ... 43
Tallahassee Community College ... 43
University of Central Florida ... 43
University of Florida ... 43
University of Miami ... 44
University of South Florida, Saint Petersburg ... 45
University of South Florida, Tampa ... 46
University of West Florida ... 46

Georgia

Columbus State University ... 46
East Georgia College ... 46
Emory University ... 46
Emory University-Oxford College ... 47
Fort Valley State University ... 47
Gainesville College ... 47
Georgia Institute of Technology ... 47
Georgia Perimeter College ... 47
Georgia Southern University ... 48
Georgia Southwestern State University ... 48
Georgia State University ... 48
Gordon College ... 49
Mercer University ... 49
Shorter College ... 49
University of Georgia ... 49
University of West Georgia ... 49
Valdosta State University ... 50

Guam

University of Guam ... 50

Hawaii

University of Hawai'i at Hilo ... 50
University of Hawai'i, Manoa ... 50

Idaho

Boise State University ... 52
Brigham Young University, Idaho ... 52
College of Southern Idaho ... 52
Idaho State University ... 52
Lewis-Clark State College ... 53
University of Idaho ... 53

Illinois

Augustana College ... 53
Black Hawk College ... 53
Bradley University ... 53
College of Lake County ... 54
Columbia College ... 54
Concordia University ... 54
Eastern Illinois University ... 54
Elgin Community College ... 54
Field Museum of Natural History ... 54
Illinois State University ... 54
Illinois Valley Community College ... 54
Lincoln Land Community College ... 55
Northeastern Illinois University ... 55
Northern Illinois University ... 55
Northwestern University ... 55
Olivet Nazarene University ... 55
Principia College ... 56
Southern Illinois University Carbondale ... 56
University of Chicago ... 56
University of Illinois, Chicago ... 56
University of Illinois, Urbana-Champaign ... 57

University/Agency Table of Contents

Western Illinois University 57
Wheaton College 58
William Rainey Harper College 58

Indiana
Ball State University 58
DePauw University 58
Earlham College 58
Hanover College 58
Holy Cross College 58
Indiana State University 58
Indiana Univ/Purdue University, Fort Wayne 59
Indiana Univ/Purdue University, Indianapolis 59
Indiana University Northwest 59
Indiana University, Bloomington 59
Indiana University, Indianapolis 60
Purdue University 60
Tri-State University 61
University of Notre Dame 61
University of Southern Indiana 61
Vincennes University Junior College 61

Iowa
Cornell College 61
Iowa State University of Science & Tech 61
University of Iowa 62
University of Northern Iowa 62

Kansas
Cowley County Community College 63
Emporia State University 63
Fort Hays State University 63
Kansas State University 63
University of Kansas 64
Wichita State University 64

Kentucky
Eastern Kentucky University 64
Morehead State University 65
Murray State University 65
Northern Kentucky University 65
University of Kentucky 65
University of Louisville 66
Western Kentucky University 66

Louisiana
Centenary College of Louisiana 66
Louisiana State University 66
Louisiana Tech University 67
McNeese State University 67
Northwestern State University 67
Tulane University 68
University of Louisiana, Lafayette 68
University of Louisiana, Monroe 68
University of New Orleans 68

Maine
Bates College 68
Bowdoin College 69
Colby College 69
University of Maine 69
University of Maine, Farmington 70
University of Maine, Presque Isle 70
University of Southern Maine 70

Maryland
Comm Coll of Baltimore County, Catonsville 70
Frostburg State University 70
Montgomery College 70
The Johns Hopkins University 70
Towson University 71
United States Naval Academy 71
University of Maryland 71

Massachusetts
Amherst College 72
Bentley College 73
Boston College 73
Boston University 73
Bridgewater State College 74
Fitchburg State College 74
Greenfield Community College 74
Hampshire College 74
Harvard University 74
Massachusetts Institute of Technology 75
Mount Holyoke College 75
Nichols College 75
North Adams State College 75
Northeastern University 75
Salem State College 76
Smith College 76
Tufts University 76
University of Massachusetts 76
Univ of Massachusetts, Boston/Harbor 77
University of Massachusetts/Lowell 77
Wellesley College 77
Williams College 77
Woods Hole Oceanographic Institution 77
Worcester State College 79

Michigan
Adrian College 79
Albion College 79
Calvin College 79
Central Michigan University 79
Charles Stewart Mott Community College 80
Concordia University 80
Delta College 80
Eastern Michigan University 80
Ferris State University 80
Gogebic Community College 80
Grand Valley State University 80
Hope College 80
Jackson Community College 81
Lake Superior State University 81
Macomb Comm College, Center Campus 81
Michigan State University 81
Michigan Technological University 81
Muskegon Community College 82
Northern Michigan University 82
Saint Clair County Community College 82
University of Michigan 82
University of Michigan, Dearborn 83
Washtenaw Community College 83
Wayne State University 83
Western Michigan University 84

University/Agency Table of Contents

Minnesota

Anoka-Ramsey Community College.................. 84
Bemidji State University 84
Carleton College.. 84
Concordia College... 84
Gustavus Adolphus College 84
Itasca Community College 85
Macalester College... 85
Minnesota State University 85
Saint Cloud State University 85
University of Minnesota, Duluth....................... 85
University of Minnesota, Morris........................ 85
University of Minnesota, Twin Cities 86
University of Saint Thomas............................... 87
Winona State University................................... 87

Mississippi

Millsaps College... 87
Mississippi State University.............................. 87
University of Mississippi.................................. 88
University of Southern Mississippi 88

Missouri

Central Methodist University 88
Meramec Community College 88
Northwest Missouri State University 89
St. Louis Comm College, Florissant Valley 89
Saint Louis University....................................... 89
Southwest Missouri State University 89
Truman State University................................... 90
University of Central Missouri.......................... 90
University of Missouri...................................... 90
University of Missouri, Columbia..................... 90
University of Missouri, Rolla............................ 90
University of Missouri-Kansas City.................. 91
Washington University..................................... 91
William Jewell College 92

Montana

Flathead Valley Community College 92
Montana State University................................. 92
Montana State University, Billings 92
Montana Tech of the University of Montana 92
Rocky Mountain College.................................. 93
University of Montana...................................... 93
University of Montana, Western....................... 93

Nebraska

Bellevue University.. 93
Chadron State College...................................... 93
Creighton University.. 94
University of Nebraska at Omaha 94
University of Nebraska, Kearney 94
University of Nebraska, Lincoln 94

Nevada

Desert Research Institute.................................. 95
University of Nevada, Las Vegas 95
University of Nevada, Reno 95

New Hampshire

Dartmouth College 98
Keene State College 98
Plymouth State College..................................... 98
University of New Hampshire........................... 98

New Jersey

Fairleigh Dickinson University 99
Kean University.. 99
Montclair State University 99
New Jersey City University.............................. 99
Princeton University... 99
Richard Stockton College of New Jersey.......... 101
Rider University 101
Rutgers, The State University of New Jersey.... 101
Rutgers, The State Univ of New Jersey, Newark 101
Stockton State College 101
Trenton State College...................................... 102
Union County College..................................... 102
William Paterson College................................ 102

New Mexico

College of Santa Fe 102
Dine College, Shiprock 102
Eastern New Mexico University 102
New Mexico Highlands University.................. 102
New Mexico Institute of Mining and Tech........ 102
New Mexico State University, Las Cruces........ 103
University of New Mexico 104
Western New Mexico University 104

New York

Adelphi University 104
Adirondack Community College..................... 105
Alfred University... 105
American Museum of Natural History............. 105
Binghamton University 105
Brooklyn College (CUNY) 106
Broome Community College........................... 106
Buffalo State College 106
City College (CUNY)...................................... 106
Colgate University... 106
College of Staten Island 106
Columbia University 107
Cornell University... 108
Dowling College.. 109
Graduate School of the City Univ of New York 109
Hamilton College 110
Hartwick College... 110
Hobart & William Smith Colleges 110
Hofstra University... 110
Hudson Valley Community College................ 110
Hunter College (CUNY) 110
Jefferson Community College......................... 111
Lehman College (CUNY) 111
Long Island University, Brooklyn Campus....... 111
Long Island Univ, C.W. Post Campus.............. 111
Molloy College.. 111
Nassau Community College............................ 111
Onondaga Community College....................... 111
Orange County Community College............... 111
Pace University, New York Campus 112
Paleontological Research Institution................ 112
Plattsburgh State University (SUNY) 112
Queens College (CUNY)................................. 112
Rensselaer Polytechnic Institute...................... 112
Skidmore College.. 112
St. Lawrence University.................................. 113

University/Agency Table of Contents

Suffolk County Community College ... 113
SUNY College at Brockport ... 113
SUNY Maritime College ... 113
SUNY New Paltz ... 113
SUNY Ulster County Community College ... 113
SUNY, Albany ... 113
SUNY, Buffalo ... 114
SUNY, Cortland ... 114
SUNY, Fredonia ... 114
SUNY, Geneseo ... 115
SUNY, Oneonta ... 115
SUNY, Oswego ... 115
SUNY, Potsdam ... 115
SUNY, Purchase ... 115
SUNY, Stony Brook ... 116
Syracuse University ... 116
The College of Saint Rose ... 117
Union College ... 117
United States Military Academy ... 117
University of Rochester ... 117
Utica College ... 117
Vassar College ... 117
York College (CUNY) ... 118

North Carolina

Appalachian State University ... 118
Brevard College ... 118
Cape Fear Community College ... 118
Duke University ... 118
East Carolina University ... 119
Elizabeth City State University ... 119
Guilford College ... 119
North Carolina Agricultural & Tech State Univ 119
North Carolina Central University ... 119
North Carolina State University ... 120
University of North Carolina Wilmington ... 120
University of North Carolina, Asheville ... 121
University of North Carolina, Chapel Hill ... 121
University of North Carolina, Charlotte ... 121
University of North Carolina, Pembroke ... 121
Western Carolina University ... 122

North Dakota

Dickinson State University ... 122
Minot State University ... 122
North Dakota State University ... 122
University of North Dakota ... 122

Ohio

Ashland University ... 123
Bowling Green State University ... 123
Case Western Reserve University ... 123
Central State University ... 123
Cincinnati Museum Center ... 123
Cleveland Museum of Natural History ... 123
Cleveland State University ... 124
Denison University ... 124
Kent State University ... 124
Kent State University, Stark Campus ... 124
Kent State University, Tuscarawas ... 124
Marietta College ... 124
Miami University ... 124
Mount Union College ... 125
Muskingum College ... 125
Oberlin College ... 125
Ohio State University ... 125
Ohio State University at Newark ... 126
Ohio State University, Lima ... 126
Ohio University, Lancaster ... 127
Ohio Wesleyan University ... 127
Shawnee State University ... 127
The College of Wooster ... 127
University of Akron ... 127
University of Cincinnati ... 127
University of Dayton ... 128
University of Toledo ... 128
Wittenberg University ... 128
Wright State University ... 128
Youngstown State University ... 128

Oklahoma

Northeastern Oklahoma A&M College ... 129
Northeastern State University ... 129
Oklahoma State University ... 129
University of Oklahoma ... 129
University of Tulsa ... 130

Oregon

Central Oregon Community College ... 130
Clatsop Community College ... 130
Eastern Oregon University ... 130
Oregon State University ... 130
Portland State University ... 132
Southern Oregon University ... 132
Southwestern Oregon Comm College ... 132
University of Oregon ... 133
Western Oregon University ... 133
Willamette University ... 133

Pennsylvania

Allegheny College ... 133
Bloomsburg University ... 133
Bryn Mawr College ... 134
Bucknell University ... 134
California University of Pennsylvania ... 134
Carnegie Museum of Natural History ... 134
Clarion University ... 134
Dickinson College ... 135
Drexel University ... 135
Edinboro University of Pennsylvania ... 135
Elizabethtown College ... 135
Franklin and Marshall College ... 135
Gannon University ... 136
Indiana University of Pennsylvania ... 136
Juniata College ... 136
Kutztown University of Pennsylvania ... 136
La Salle University ... 136
Lafayette College ... 136
Lehigh University ... 136
Lock Haven University ... 137
Mansfield University ... 137
Mercyhurst College ... 137
Millersville University ... 137
Montgomery County Community College ... 137
Moravian College ... 137
Pennsylvania State University, Altoona ... 137

University/Agency Table of Contents

Pennsylvania State University, Erie 138
Pennsylvania State University, Hazleton........... 138
Pennsylvania State University, Monaca 138
Pennsylvania State Univ, University Park......... 138
Point Park College.. 139
Shippensburg University 139
Slippery Rock University 140
Susquehanna University................................... 140
Temple University 140
The State Museum of Pennsylvania.................. 140
Thiel College.. 140
University of Pennsylvania 140
University of Pittsburgh 140
University of Pittsburgh at Johnstown 141
University of Pittsburgh, Bradford.................... 141
West Chester University................................... 141
Wilkes University... 141
York College of Pennsylvania.......................... 141

Puerto Rico

University of Puerto Rico................................. 142

Rhode Island

Brown University 142
Community College of Rhode Island................ 142
Roger Williams University............................... 142
University of Rhode Island............................... 142

South Carolina

Clemson University.. 143
Coker College... 143
College of Charleston....................................... 144
Furman University.. 144
University of South Carolina............................ 144
Winthrop University... 145
Wofford College... 145

South Dakota

Black Hills State University............................. 145
South Dakota School of Mines & Tech............ 145
South Dakota State University 145
University of South Dakota.............................. 146

Tennessee

Austin Peay State University............................ 146
Rhodes College... 146
Tennessee Technological University................ 146
University of Memphis..................................... 146
University of Tennessee, Chattanooga 147
University of Tennessee, Knoxville 147
University of Tennessee, Martin....................... 148
University of the South..................................... 148
Vanderbilt University 148

Texas

Amarillo College.. 148
Angelo State University 148
Austin Community College.............................. 148
Baylor University 148
Brookhaven College.. 149
Del Mar College.. 149
Hardin-Simmons University............................. 149
Houston Community College System.............. 149
Kilgore College 149
Lamar University.. 149
Laredo Community College............................. 149
Lee College... 149
Midwestern State University............................ 150
North Harris College 150
Odessa College.. 150
Rice University... 150
Saint Mary's University.................................... 151
Sam Houston State University 151
San Antonio College... 151
South Plains College 151
Southern Methodist University 151
Stephen F. Austin State University.................. 151
Sul Ross State University................................. 152
Tarleton State University.................................. 152
Tarrant County Junior College 152
Texas A&M University 152
Texas A&M University, Commerce 154
Texas A&M University, Corpus Christi 154
Texas A&M University, Kingsville.................. 154
Texas Christian University 155
Texas Tech University 155
The University of Texas at San Antonio........... 155
Trinity University 155
University of Houston 155
University of Houston Downtown 156
University of North Texas 156
University of Texas at Austin 156
University of Texas at Austin, Jackson School of Geosciences 156
University of Texas, Arlington 158
University of Texas, Dallas 158
University of Texas, El Paso 158
University of Texas, Pan American.................. 159
University of Texas, Permian Basin 159
University of Texas, San Antonio..................... 159
West Texas A&M University............................ 159

Utah

Brigham Young University............................... 159
Snow College 160
Southern Utah University................................. 160
University of Utah.. 160
Utah State University 161
Weber State University.................................... 162

Vermont

Castleton State College 162
Johnson State College 162
Lyndon State College 162
Middlebury College.. 162
Norwich University.. 162
University of Vermont...................................... 162

Virginia

College of William & Mary............................. 163
Emory & Henry College.................................. 163
George Mason University 163
Hampton University 163
J. Sargeant Reynolds Community College........ 164
James Madison University 164
Lynchburg College 164
Mary Washington College................................ 164
Northern Virginia Comm Coll-Loudoun........... 164

University/Agency Table of Contents

Old Dominion University ... 164
Radford University ... 164
Randolph-Macon College ... 165
Tidewater Community College ... 165
University of Mary Washington ... 165
University of Virginia ... 165
University of Virginia College, Wise ... 166
Virginia Polytechnic Inst & State Univ ... 166
Virginia Western Community College ... 167
Washington & Lee University ... 167

Washington

Central Washington University ... 167
Eastern Washington University ... 167
Grays Harbor College ... 168
Green River Community College ... 168
Lower Columbia College ... 168
Olympic College ... 168
Pacific Lutheran University ... 168
Seattle Central Community College ... 168
Shoreline Community College ... 168
Tacoma Community College ... 168
University of Puget Sound ... 168
University of Washington ... 169
Washington State University ... 171
Washington State University, Vancouver ... 172
Western Washington University ... 172
Whitman College ... 172
Whitworth College ... 172
Yakima Valley College ... 173

West Virginia

Concord University ... 173
Glenville State College ... 173
Marshall University ... 173
Potomac State College ... 173
West Virginia University ... 173
West Virginia University, Parkersburg ... 173

Wisconsin

Beloit College ... 173
Lawrence University ... 174
Milwaukee Area Technical College ... 174
Milwaukee Public Museum ... 174
Northland College ... 174
Saint Norbert College ... 174
University of Wisconsin Colleges ... 174
University of Wisconsin, Eau Claire ... 174
University of Wisconsin, Extension ... 174
University of Wisconsin, Green Bay ... 175
University of Wisconsin, Madison ... 175
University of Wisconsin, Milwaukee ... 176
University of Wisconsin, Oshkosh ... 176
University of Wisconsin, Parkside ... 177
University of Wisconsin, Platteville ... 177
University of Wisconsin, River Falls ... 177
University of Wisconsin, Stevens Point ... 177
University of Wisconsin, Superior ... 177
University of Wisconsin, Whitewater ... 177

Wyoming

Casper College ... 177
University of Wyoming ... 178

Australia

Australian National University ... 178
Curtin University of Technology ... 178
Flinders University ... 179
James Cook University ... 179
Latrobe University ... 179
Macquarie University ... 179
Monash University ... 179
Queensland University of Technology ... 179
Southern Cross University ... 179
The Flinders University of South Australia ... 180
University of Adelaide ... 180
University of Canberra ... 180
University of Melbourne ... 180
University of New England ... 180
University of New South Wales ... 180
University of Newcastle ... 180
University of Queensland ... 180
University of Sydney ... 180
University of Tasmania ... 180
University of Western Australia ... 181
University of Wollongong ... 181

Austria

Univ. Leoben ... 181
Univ. of Vienna ... 181
University of Innsbruck ... 181

Azerbaijan

Baku State University ... 181

Belgium

K.U. Leuven ... 181
Universite de Liege ... 181
Vrije University Brussel ... 181

Brazil

Federal University of Bahia Geophysics ... 181
Universidade De Brasilia ... 181
University of Campinas Geoscience Institute ... 181

Bulgaria

Bulgarian Academy of Sciences (BAS) ... 181
Sofia University St. Kliment Ohridski ... 182
Univ of Mining and Geology, St. Ivan Rilski ... 182

Canada

Acadia University ... 182
Athabasca University ... 182
Brandon University ... 182
Brock University ... 182
Cambrian College of Applied Arts & Tech ... 182
Capilano College ... 183
Carleton University ... 183
Dalhousie University ... 183
Douglas College ... 184
Ecole Polytechnique ... 184
Lakehead University ... 185
Laurentian University, Sudbury ... 185
McGill University ... 185
McMaster University ... 186
Memorial University of Newfoundland ... 186
Mount Royal College ... 187
Northern Alberta Institute of Tech ... 187
Queen's University ... 187

University/Agency Table of Contents

Royal Ontario Museum 187
Royal Tyrrell Museum of Palaeontology 188
Saint Francis Xavier University 188
Simon Fraser University 188
Sir Sandford Fleming College 188
Sir Wilfred Grenfell College 188
The Manitoba Museum 189
Universite de Montreal 189
Universite du Quebec 189
Universite du Quebec a Chicoutimi 189
Universite du Quebec a Montreal 190
Universite du Quebec a Rimouski 190
Universite Laval 190
University College of Cape Breton 190
University College of the Cariboo 190
University of Alberta 191
University of British Columbia 191
University of Calgary 193
University of Guelph 193
University of Manitoba 194
University of New Brunswick 194
University of New Brunswick, Saint John 194
University of Ottawa 194
University of Regina 195
University of Saskatchewan 195
University of Toronto 195
University of Victoria 196
University of Waterloo 196
University of Western Ontario 197
University of Windsor 198
York University 198

Chile

Universidad de Chile 198

China

China University of Geosciences 198
University of Hong Kong 198

Czech Republic

Charles University 198
Masaryk University 199

Denmark

Aarhus University 199
Technical Univ. of Denmark 199
University of Copenhagen 199

Estonia

University of Tartu 199

Fiji

University of the South Pacific 199

Finland

Helsinki University of Technology 199
University of Helsinki 199
University of Oulu 199

France

Ecole des Mines de Paris 200
Ecole Nationale Supérieure de Géologie (ENSG) 200
Ecole Nat Superieure des Mines de Nancy 200
Ecole Normale Superieure de Lyon 200
Universite de Bourgogne 200
Universite de Pau 200
Universite Pierre et Marie Curie (Paris VI) 200
University Claude Bernard 200
University of Lille 200
University of Nice-Sopia Antipollis-1 200
University of Pierre & Marie Curie (Paris VI) .. 200
University of Rennes I 200

Germany

Aachen University of Technology 200
Bonn Universitat 201
Freien Universitat Berlin 201
Georg-August University of Gottingen 201
Karlsruhe University 201
Koln University 201
LMU - University of Munich 201
Ludwig-Maximilians-Universitat Munich 202
Martin-Luther-Universitaet Halle-Wittenberg... 202
Technical University Clausthal 202
Univ. of Bochum 202
Univ. of Erlangen-Nuremberg 202
Univ. of Freiburg 202
Universitaet Muenster 202
Universitat Bonn 202
Universitat Hannover 202
Universitat Leipzig 202
Universitat Wurzburg 202
University of Cologne 202
University of Erlangen 202
University of Hamburg 202
University of Mainz 202
Universität Kiel 203
Universität Tübingen 203

Greece

National Technical University of Athlens 203
University of Athens 203

Hungary

Eotvos Lorand University 203
Eotvos University 203
Univesity of Szeged 203

India

Presidency College 203
Roorkee University 203

Ireland

Trinity College 203
University College Cork 203
University College Dublin 203

Israel

Ben Gurion University of the Negev 204
Hebrew University of Jerusalem 204
Tel Aviv University 204

Italy

Messina Universita 204
Universita de Bologna 204
Universita de Firenze 204
Universita de Pavia 204
Universita Degli Studi 204
Universita di Bari 204
Universita di Cagliari 204
Universita di Calabria 204
Universita di Milano 204
Università di Modena e Reggio Emilia 204

University/Agency Table of Contents

Universita di Palermo..204
Universita di Siena..204
Universita di Torino..204
Universita di Udine...204
Universita Pisa...204
Universita' Di Napoli..205
University of Parma..205
Univsita di Padova...205

Japan

Ehime University...205
Hirosaki University..205
Hiroshima University...205
Kagoshima University..205
Kanazawa University...205
Kobe University...205
Kumamoto University..205
Miyazaki University...205
Nagoya University...205
Nihon University..205
Shimane University..205
Shizuoka University...205
Tohoku University..206
Toyama University...206
Tsukuba University..206
University of Tokyo...206
Yokohama National University........................206

Korea, South

Seoul National University................................206

Kuwait

Kuwait University..206

Latvia

University of Latvia...206

Lithuania

Vilnius University..206

Malaysia

Universiti Putra Malaysia..................................206
Universiti Tecnologi Malaysia..........................206
University of Malaya...206

Mexico

Centro de Estudios Sup del Estado Sonora.......206
Centro de Inv Científica y de Ed Sup de Ensenada..207
Facultad de Minas, Metalurgia y Minas............207
Mexico Universidad Nacional Autonoma.........207
Univ Autonoma de Baja California Sur............207
Universidad Autonoma de Chihuahua...............207
Univ Autonoma de San Luis Potosi..................208
Univ Nacional Autonoma de Mexico................208

Netherlands

Universiteit Utrecht...208
VU University Amsterdam................................209

New Zealand

University of Auckland.....................................209
University of Canterbury..................................209
University of Otago...209
Victoria Univ of Wellington..............................209

Norway

University of Bergen...209
University of Tromso...209

Poland

Jagiellonian University......................................209
Silesian University...209

Portugal

Oporto University..209
Universidade de Coimbra..................................209
Universidade de Lisboa.....................................210
Universidade de Tras-os-Montes.......................210
Universidade do Minho......................................210

Qatar

University of Qatar..210

Romania

Babes-Bolyai University....................................210
Iasi University Al.I. Cuza.................................210

Russia

Moscow State University..................................210
Perm State University..210
Rostov State University.....................................210
Saint-Petersburg State University.....................210

Slovakia

Comenius University...210

Slovenia

Ljubljana University..210

South Africa

Rhodes University..210
University of Cape Town..................................211
University of Durban-Westville........................211
University of Port Elizabeth.............................211

Spain

Coruna University...211
Univ Complutense de Madrid...........................211
Universidad de Cadiz..211
Universidad de Granada....................................211
Universidad de Jaen..211
Universidad de Oviedo......................................211
Universidad de Zaragoza...................................211
Universitat de Barcelona...................................211
Universitat de les Illes Balears.........................212

Sweden

Chalmers University of Technology.................212
Karlstads University..212
University of Stockholm..................................212
Uppsala Universitet...212

Switzerland

Basel University..212
Bern University...212
ETH Zurich..212
Univ de Lausanne..212
University of Basel..212
University of Bern...212
University of Fribourg......................................212
University of Geneva..212
University of Neuchatel....................................212

Turkey

Ankara University...212
Hacettepe Universitesi.......................................212
Hacettepe University...213
Middle East Technical University....................213

University/Agency Table of Contents

Ukraine
State Mining University of Ukraine 213
United Kingdom
Bristol University 213
Brunel University 213
Cardiff University 213
Durham University 213
Edinburgh University 214
Imperial College 214
Keele University 214
Kingston University 214
Leicester University 214
Open University 214
Oxford Brooks University 214
Staffordshire Univeristy 214
The University of Liverpool 214
University of Aberdeen 214
University of Birmingham 214
University of Derby 215
University of Exeter 215
University of Glasgow 215
University of Keele 215
University of Leeds 215
University of Newcastle Upon Tyne 215
University of Nottingham 215
University of Oxford 215
University of Portsmouth 215
University of Reading 215
University of Sheffield 215
University of Wales, Aberystwyth 215
Zimbabwe
University of Zimbabwe 215
Alabama
Geological Survey of Alabama 217
Alaska
Div of Geological & Geophysical Surveys 217
Arizona
Arizona Geological Survey 217
Arkansas
Arkansas Geological Commission 217
California
California Geological Survey 217
Colorado
Colorado Geological Survey 218
Connecticut
Geological and Natural History Survey 218
Delaware
University of Delaware 218
Florida
Florida Geological Survey 219
Georgia
Georgia Dept of Natural Resources 219
Hawaii
Hawai'i Geological Survey 219
Idaho
University of Idaho 219
Illinois
Illinois State Geological Survey 220
Illinois State Water Survey 221
Indiana
Indiana Geological Survey 221
Iowa
Iowa Dept of Natural Resources 222
Kansas
University of Kansas 222
Kentucky
University of Kentucky 222
Louisiana
Louisiana State University 223
Maine
Dept of Conservation 223
Maryland
Maryland Dept of Natural Resources 223
Maryland Geological Survey 223
Massachusetts
Office of the State Geologist 224
Michigan
Michigan Dept of Environmental Quality 224
Minnesota
University of Minnesota, Twin Cities 224
Mississippi
Mississippi Office of Geology 224
Missouri
Missouri Dept of Natural Resources 225
Montana
Montana Tech - University of Montana 226
Nebraska
Unversity of Nebraska - Lincoln 226
Nevada
University of Nevada 227
New Hampshire
New Hampshire Geological Survey 227
New Jersey
New Jersey Geological Survey 227
New Mexico
New Mexico Institute of Mining & Tech 228
New York
New York State Geological Survey 228
North Carolina
North Carolina Geological Survey 228
North Dakota
North Dakota Geological Survey 229
Ohio
Ohio Dept of Natural Resources 229
Oklahoma
University of Oklahoma 229
Oregon
Dept of Geology & Mineral Industries 229
Pennsylvania
Bureau of Topo & Geologic Survey 230
Puerto Rico
Puerto Rico Bureau of Geology 231
Rhode Island
University of Rhode Island 231

University/Agency Table of Contents

South Carolina
South Carolina Geological Survey 231
South Dakota
Dept of Env & Natural Resources 231
Tennessee
Tennessee Division of Geology 231
Texas
University of Texas at Austin 231
Utah
Utah Geological Survey 232
Vermont
Dept of Environmental Conservation 233
Virginia
Division of Mineral Resources 233
Washington
Division of Geology & Earth Resources 233
West Virginia
Geological & Economic Survey 234
Wisconsin
University of Wisconsin - Extension 234
Wyoming
Wyoming State Geological Survey 234
Federal Agencies
Argonne National Laboratory 235
Dept of Agriculture 235
Dept of Commerce 236
Dept of Defense 236
Dept of Energy 238
Dept of Health & Human Services 238
Dept of Justice 238
Dept of Labor 239
Dept of State 239
Dept of the Interior 239
Dept of the Treasury 240
Dept of Transportation 240
Environmental Protection Agency 240
Federal Emergency Management Agency 241
Jet Propulsion Laboratory 241
Lawrence Berkeley National Laboratory 241
Lawrence Livermore National Laboratory 241
Lockheed Martin Idaho Technologies 241
Los Alamos National Laboratory 242
National Aeronautics & Space Administration 244
National Science Foundation 244
National Transportation Safety Board 244
Nuclear Regulatory Commission 244
Oak Ridge National Laboratory 244
Pacific Northwest National Laboratory 245
Smithsonian Institution 246
Societies
American Association of Petroleum Geologists 249
American Association of Stratigraphic Palynologists 249
American Geological Institute 249
American Geophysical Union 249
American Institute of Hydrology 249
American Institute of Professional Geologists 249
American Rock Mechanics Association 250
American Society of Limnology and Oceanography 250
Association for Women Geoscientists 250
Association of American State Geologists 250
Association of Earth Science Editors 250
Association of Environmental and Engineering Geologists 251
The Clay Minerals Society 251
Council on Undergraduate Research, Geosciences Div. 251
Environmental and Engineering Geophysical Society 251
Friends of Mineralogy 252
Geo-Institute of ASCE 252
Geological Society of America 252
Geological Society of London 252
Geoscience Information Society 252
History of Earth Sciences Society 252
International Association of Hydrogeologists/ U.S. National Chapter 253
International Basement Tectonics Association 253
Mineralogical Society of America 253
National Association of Black Geologists and Geophysicists 253
National Association of Geoscience Teachers 253
National Association of State Boards of Geology 254
National Earth Science Teachers Association 254
National Speleological Society 254
North American Commission on Stratigraphic Nomenclature 254
Paleobotanical Section of the Botanical Society of America 254
Paleontological Research Institution 254
Paleontological Society 254
Petroleum History Institute 255
Seismological Society of America 255
SEPM (Society for Sedimentary Geology) 255
Society for Mining, Metallurgy, and Exploration 255
The Society for Organic Petrology 256
Society of Economic Geologists 256
Society of Exploration Geophysicists 256
Society of Independent Professional Earth Scientists 256
Society of Mineral Museum Professionals 256
Society of Vertebrate Paleontology 256
Soil Science Society of America 257
U.S. Permafrost Association 257

DEPARTMENTS AND FACULTIES

Universities and colleges in the United States are arranged alphabetically by state. Non-U.S. institutions are listed alphabetically by country following the U.S. section. Departments whose names appear in bold are Academic Associates of the American Geological Institute (AGI) as of January 17, 2008. A complete description of these departments — including contact information, admissions procedures, degree requirements, financial information, research facilities and support, full description of faculty teaching and research specialties, department geoscience specialties, and historical enrollments and degrees-granted data — may be seen in AGI's online Guide to Geoscience Departments: (guide.agiweb.org). For information about AGI's Academic Associates program, visit www.agiweb.org/associates/academic or contact Cindy Martinez: (703) 379-2480 ext 227; (cmm@agiweb.org).

The letters in parentheses after the department names indicate the degree(s) offered ("A" - Associate's, "B" - Bachelor's, "M" - Master's, "D" - Doctoral). "[FC]" appears if a department offers a field camp, field course, or both. Basic enrollment and degrees-granted data for the most recent year reported since 1995 are shown below department information: degree level, enrollment, and degrees granted (in parentheses).

The letter code following individual faculty listings indicates their research specialties. Captial letters indicate general area, lowercase letters indicate subspecialty. Two or more consecutive lowercase letters refer to multiple subspecialties in the same major focus area refered to by the preceding capital letter. The list of specialty codes appears immediately before the faculty index.

Alabama

Alabama A&M University

Dept of Plant & Soil Science (A,B,M,D) (2006)
P.O. Box 1208
Normal, AL 35762
p. (256) 372-4185 f. (256) 372-5429
monday.mbila@email.aamu.edu
http://www.aamu.edu
Department Secretary: Catharine Strother
Enrollment (1996): B: 6 (0) M: 9 (0) D: 6 (1)

Chair:
Govind Sharma, (D), Kansas State, 1970, Oo
Professor:
Tommy L. Coleman, (D), Iowa State, 1980, So
James W. Shuford, (D), Penn State, 1975, So
Robert W. Taylor, (D), Michigan State, 1977, Sc
Associate Professor:
Zachary Senwo, (D), Iowa State, 1995, Sc
Teferi D. Tsegaye, (D), Maryland, 1994, Sp
Associate Scientist:
Andrew Manu, (D), Iowa State, 1984, Sd
Assistant Professor:
Monday O. Mbila, (D), Iowa State, 2000, Sd
Adjunct Professor:
George S. Da Bai, (D), Colorado Mines, 1976, Nm

Auburn University

Dept of Agronomy & Soils (M,D) (2007)
201 Funchess Hall
Auburn University, AL 36849
p. (334) 844-4100 f. (334) 844-3945
touchjt@auburn.edu
Administrative Assistant: Linda Bankston
Enrollment (2006): M: 8 (2) D: 6 (3)

Head:
Joseph T. Touchton, (D), Illinois, 1977, Sc
Extension Agronomist-Soils:
Charles C. Mitchell, (D), Florida, 1980, Sc
Professor:
Jake H. Dane, (D), Colorado State, 1976, Sp
Elizabeth A. Guertal, (D), Oklahoma State, 1993, Sc
Joey N. Shaw, (D), Georgia, 1997, Sd
Charles W. Wood, (D), Colorado State, 1990, Sc
Associate Professor:
Yucheng Feng, (D), Penn State, 1995, Sb
John W. Odom, (D), Purdue, 1977, Sc
Assistant Professor:
Julie Howe, Wisconsin, 2005, Sc

Dept of Geology & Geography (B,M) (2007)
210 Petrie Hall
Auburn University, AL 36849-5305
p. (334) 844-4282 f. (334) 844-4486
lillyet@auburn.edu
http://www.auburn.edu/academic/science_math/geology/docs
Administrative Assistant: Eva Lilly
Enrollment (1999): B: 46 (11) M: 16 (5)

Chair:
Charles E. Savrda, (D), S California, 1986, Gd
Professor:
Cyrus B. Dawsey, (D), Florida, 1975, Oy
Willis E. Hames, (D), Virginia Tech, 1990, Gp
David T. King, Jr., (D), Missouri, 1980, Gr
Ming-Kuo Lee, (D), Illinois, 1993, Hw
James A. Saunders, (D), Colorado Mines, 1986, Cl
Mark G. Steltenpohl, (D), North Carolina, 1985, Gt
Lorraine W. Wolf, (D), Alaska (Fairbanks), 1989, Ys
Associate Professor:
Phil L. Chaney, (D), Louisiana State, 1999, Oy

Ronald D. Lewis, (D), Texas, 1982, Pi
Luke J. Marzen, (D), Kansas State, 2001, Oy
Ashraf Uddin, (D), Florida State, 1996, Gt
Assistant Professor:
Toni A. Alexander, (D), Louisiana State, 2004, Oy
Visiting Professor:
Dennis R. Ruez, Jr., (M), Florida, 1999, Pe
Emeritus:
Thomas J. Carrington, (D), Virginia Tech, 1965, Gr
Robert B. Cook, (D), Georgia, 1971, Eg
Tom L. Martinson, (D), Kansas, 1969, Oy

University of Alabama

Dept of Geological Sciences (B,M,D) (2007)
Box 870338
201 7th Avenue
Room 202 Bevill Bldg.
Tuscaloosa, AL 35487-0338
p. (205) 348-5095 f. (205) 348-0818
dcapps@geo.ua.edu
http://www.geo.ua.edu
Enrollment (2007): M: 11 (11) D: 2 (2)

Chair:
Harold H. Stowell, (D), Princeton, 1987, Gp
Professor:
Paul Aharon, (D), Australian National, 1980, Ge
Donald J. Benson, (D), Cincinnati, 1976, Gd
Rona J. Donahoe, (D), Stanford, 1984, Cl
Ernest A. Mancini, (D), Texas A&M, 1974, Pe
Carl W. Stock, (D), North Carolina, 1977, Pi
Chunmiao Zheng, (D), Wisconsin, 1988, Hw
Assistant Professor:
C. Fred T. Andrus, (D), Georgia, 2000, Ge
Andrew Goodliffe, (D), Hawaii, 1998, Yg
Timothy L. Masterllark, (D), Wisconsin, 2000, GqvYd
Delores Robinson, (D), Arizona, 2001, Gc
Joshua J. Schwartz, (D), Wyoming, 2007, GiCc
Geoffrey Tick, (D), Arizona, 2003, Hw
Amy L. Weislogel, (D), Stanford, 2006, Gs
Adjunct Assistant Professor:
Michael G. Bersch, (D), New Mexico, 1990, Gi
Emeritus:
Richard H. Groshong, Jr., (D), Brown, 1971, Gc
W. Gary Hooks, (D), North Carolina, 1961, Gm
Geochemical Research Lab Manager:
Elizabeth Y. Graham, (B), Alabama, 1975, Ca

University of Alabama, Dauphin Island

Marine Science Program (A,B,M,D) (1999)
P.O. Box 369
101 Bienville Boulevard
Dauphin Island, AL 36528
p. (334) 861-7528 f. (334) 861-7540
gcrozier@disl.org
Department Secretary: Carolyn F. Wood
Enrollment: No data reported since 1995

Director:
William W. Schroeder, (D), Texas A&M, 1971, Og
Professor:
Thomas S. Hopkins, (D), California (San Diego), 1967, Ob
Associate Professor:
Jonathan R. Pennock, (D), Delaware, 1983, Oc
Cooperating Faculty:
George F. Crozier, (D), California (San Diego), 1966, Oo
Geology Librarian:
Connie Mallon, Oo

University of Montevallo

Dept of Geology (2006)
Station 6480
Montevallo, AL 35115
p. (205) 665-6462 f. (205) 665-6477
brandeb@um.montevallo.edu

Head:
Barbara L. Brande, (M), SUNY (Stony Brook), 1979, Ys

University of South Alabama

Dept of Earth Sciences (B) (2007)
307 University Blvd N.
LSCB 136
Mobile, AL 36688-0002
p. (251) 460-6381 f. (251) 461-1487
gsebasti@jaguar1.usouthal.edu
http://www.southalabama.edu/geography
Department Secretary: Lisa Roberts
Enrollment (2006): B: 60 (0)

Chair:
Glenn R. Sebastian, (D), N Colorado, 1967, Oy
Professor:
Wayne C. Isphording, (D), Rutgers, 1966, Ca
Associate Professor:
David T. Allison, (D), Florida State, 1992, Gc
Murlene W. Clark, (D), Florida State, 1983, Pi
Miriam L. Fearn, (D), Louisiana State, 1995, Pe
Douglas W. Haywick, (D), James Cook, 1990, Gs
Instructor:
Lee S. Yokel, (M), Auburn, 1996, Ge
Lecturer:
Carolyn Y. Davis, (M), Houston, 1971, Oo
Adjunct Professor:
Erich M. Mueller, (D), Miami, 1983, Ob
Emeritus:
George M. Lamb, (D), Colorado, 1964, Gs

Dept of Marine Sciences (M,D) (2002)
307 University Drive
Mobile, AL 36688-0002
p. (334) 460-7136 f. (334) 460-7357
nmerrill@usouthal.edu
Department Secretary: Amada Gonzalez
Enrollment (1996): M: 22 (1) D: 11 (0)

Chair:
Robert L. Shipp, (D), Florida State, 1971, Ob
Professor:
Ronald P. Kiene, (D), SUNY (Stony Brook), 1986, Oc
Associate Professor:
Richard B. Aronson, (D), Harvard, 1985, Ob
Adjunct Professor:
Douglas W. Haywick, (D), James Cook, 1990, Gs
Erich M. Mueller, (D), Miami, 1983, Ob

University of West Alabama

College of Natural Sciences & Mathematics (B) (2007)
UWA Station 7
Livingston, AL 35470
p. (205) 652-3400 (Ext. 3414) f. 205 652 3831
arindsberg@uwa.edu
Enrollment: No data reported since 1995

Director:
Julia E. Massey, (D), Alabama, 1993, Oo
Chair:
Douglas Wymer, (D)
Associate Professor:
Andrew K. Rindsberg, (D), Colorado Mines, 1986, Gg

Adjunct Professor:
John C. Hall, (D), Alabama, 1985, Og

Alaska

University of Alaska, Anchorage

Dept of Geological Sciences (B) (2007)
3211 Providence Drive
Anchorage, AK 99508-4670
p. (907) 786-6840 f. (907) 786-6850
aftm@uaa.alaska.edu
http://geology.uaa.alaska.edu/index.html
Administrative Assistant: Kathy Mills
Enrollment (2006): B: 90 (0)

Associate Professor:
Kristine J. Crossen, (D), Washington, 1997, Gl
Terry R. Naumann, (D), Idaho, 1998, Gi
Assistant Professor:
Patricia A. Heiser, (D), Alaska (Fairbanks), 1997, Pe
LeeAnn Munk, (D), Ohio State, 2001, Ge
Pete Stelling, (D), Alaska (Fairbanks), 2003, Gv
Emeritus:
Anne D. Pasch, (M), Alaska Methodist, 1972, Pg

University of Alaska, Fairbanks

Alaska Quaternary Center (2005)
907 Yukon Drive
Fairbanks, AK 99775-1200
p. (907) 474-7758 f. (907) 474-5469
ffsjf@aurora.alaska.edu
Administrative Assistant: Leicha Welton
Enrollment: No data reported since 1995

Director:
Mary E. Edwards, (D), Cambridge, 1980, Pe
Professor:
W. Scott Ambruster, (D), California (Davis), 1981, Oo
Bruce P. Finney, (D), Oregon State, 1987, Ou
David M. Hopkins, (D), Harvard, 1955, Gg
Roger W. Powers, (D), Wisconsin, 1973, Oo
David K. Swanson, (D), Sd
Dan L. Wetzel, (B), Alaska (Fairbanks), 1975, Oo
Research Associate:
Wendy H. Arundale, (D), Michigan State, 1976, Oo
Peter Bowers, (M), Washington State, 1980, Ga
Owen Mason, (D), Alaska, 1990, Ga
Adjunct Professor:
Daniel H. Mann, (D), Washington, 1983, Pe
Geology Librarian:
Judy Triplehorn, Oo

Dept of Geology & Geophysics (B,M,D) (2007)
900 Yukon Drive
P.O. Box 755780
Fairbanks, AK 99775-5780
p. (907) 474-7565 f. (907) 474-5163
geology@uaf.edu
http://www.uaf.edu/geology/
Enrollment (2006): B: 46 (13) M: 46 (11) D: 26 (2)

Chair:
Michael T. Whalen, (D), Syracuse, 1993, Gs
Associate Dean, CNSM:
Paul Layer, (D), Stanford, 1986, Ym
Professor:
James E. Beget, (D), Washington, 1981, Gm
Douglas Christensen, (D), Michigan, 1987, Ys
Jeffrey T. Freymueller, (D), South Carolina, 1991, Yd
Rainer J. Newberry, (D), Stanford, 1980, Em
Vladimir Romanovsky, (D), Alaska (Fairbanks), 1996, Yg
Virgil Sharpton, (D), Brown, 1984, Xg
Wesley K. Wallace, (D), Washington, 1981, Gc
Associate Professor:
Bernard Coakley, (D), Columbia, 1991, Yg
Hajo Eicken, (D), Bremen (Germany), 1990, Og
Sarah Fowell, (D), Columbia, 1994, Pl
Mary J. Keskinen, (D), Stanford, 1979, Gx
Paul McCarthy, (D), Guelph, 1995, Gs
Anupma Prakash, (D), Roorkee (India), 1966, Ye
Assistant Professor:
Patrick S. Druckenmiller, (D), Pv
Emeritus:
Keith Echelmeyer, (D), Caltech, 1983, Gl
David B. Stone, (D), Newcastle, 1963, Ym
Donald M. Triplehorn, (D), Illinois, 1961, Gs
Laboratory Director:
Kenneth P. Severin, (D), California (Davis), 1987, Oo
Related Staff:
William Witte, (D), Columbia, 1991, Gg

Dept of Mining & Geological Engineering (B,M,D) (1999)
P.O. Box 755800
Fairbanks, AK 99775-5800
p. (907) 474-7388 f. (907) 474-6635
fnedw@uaf.edu
http://137.229.20.100/
Administrative Assistant: Carol Leakey
Enrollment (1999): B: 54 (10) M: 7 (3) D: 1 (1)

Head:
Sukumar Bandopadhyay, (D), Penn State, 1982, Nm
Professor:
Gang Chen, (D), Virginia Tech, 1989, Nm
Scott L. Huang, (D), Missouri (Rolla), 1981, Nr
Hsing K. Lin, (D), Utah, 1985, Oo
Associate Professor:
Rajive Ganguli, (D), Kentucky, 1999, Nm
Paul A. Metz, (D), Imperial Coll, 1991, Em
Daniel E. Walsh, (M), Alaska (Fairbanks), 1984, Oo
Assistant Professor:
Edwin E. Kroeger, (D), Alaska (Fairbanks), 1998, Ng
Terril E. Wilson, (D), Kentucky, 1998, Nm
Emeritus:
Pemmasani D. Rao, (D), Penn State, 1961, Oo

Museum (1999)
Box 756960
Fairbanks, AK 99775-6960
p. (907) 474-7505 f. (907) 474-5469
ffrag@aurora.alaska.edu
Administrative Assistant: Marie D. Ward

Curator:
Roland A. Gangloff, (D), California (Berkeley), 1975, Pi
Associate Scientist:
Paul E. Mattheus, (D), Alaska, 1997, Po
Adjunct Professor:
William A. Clemens, (D), California (Berkeley), 1960, Pv
Dale Guthrie, (D), Chicago, 1963, Pv
Donald M. Triplehorn, (D), Illinois, 1961, Gs

Arizona

Arizona State University

Dept of Geological Sciences (B,M,D) (2004)
Box 871404
Tempe, AZ 85287-1404
p. (480) 965-5081 f. (480) 965-8102
geology@asu.edu

http://geology.asu.edu
Enrollment (1999): B: 66 (10) M: 32 (10) D: 24 (3)

Director:
Laurie Leshin, (D), Caltech, 1994, Xc
Chair:
James A. Tyburczy, (D), Oregon, 1983, Yx
Director, Astrobiology Program:
Jack D. Farmer, (D), California (Davis), 1978, Py
Professor:
Donald M. Burt, (D), Harvard, 1972, Eg
Peter R. Buseck, (D), Columbia, 1962, Cg
Philip R. Christensen, (D), California (Los Angeles), 1981, Xg
Jonathan H. Fink, (D), Stanford, 1979, Gv
Ronald Greeley, (D), Missouri, 1966, Xg
Richard Hervig
John R. Holloway, (D), Penn State, 1970, Cp
L. Paul Knauth, (D), Caltech, 1973, Cs
Stephen J. Reynolds, (D), Arizona, 1982, Gt
Everett Shock, (D), Oo
Edmund Stump, (D), Ohio, 1976, Gc
Stanley N. Williams, (D), Dartmouth, 1983, Gv
Senior Scientist:
Gary Huss, (D), Minnesota, 1987, Xc
Associate Professor:
Ramon Arrowsmith, (D), Stanford, 1994, Gc
Thomas G. Sharp, (D), Arizona State, 1990, Gz
Assistant Professor:
Amanda Clarke, (D), Gv
Matthew J. Fouch, (D), Brown, 1999, Yg
Edward Garnero, (D), Caltech, 1994, Ys
Hilairy Hartnett
Steven Semken
Research Associate:
Mikhail Y. Zolotov, (D), Vernadsky Inst (Russia), 1990, Cg
Emeritus:
David H. Krinsley, (D), Chicago, 1956, Gs
John Larimer, Lehigh, 1966, Gg
Robert F. Lundin, (D), Illinois, 1962, Pm
Carleton B. Moore, (D), Caltech, 1960, Cg
Donal M. Ragan, (D), Washington, 1961, Gc
Business Operations Manager:
Cammy Cecil, (B), Arizona State, 2002, Oo

Arizona Western College

Dept of Geology (A) (2007)
2020 S. Avenue 8E
Yuma, AZ 85365
p. (928) 317-6000 f. (928) 344-7730
fred.croxen@azwestern.edu
Enrollment (2007): A: 3 (3)

Professor:
Fred W. Croxen III, (M), N Arizona, 1977, HwPvOi
Adjunct Professor:
Earl E. Burnett, (B), Arizona, 1965, HwGc
F Michael Conway, (D), Michigan Tech, 1995, Gv
Maureen Garrett, (M), Gg
John Wharton, (B), Oi

Central Arizona College

Science / Mathematics Div (2004)
Coolidge, AZ 85228
p. (602) 426-4444
maria_wise@centralaz.edu

Instructor:
Allan E. Morton, (M), Brigham Young, 1975, Oe

Glendale Community College

Dept of Geology & Applied Sciences (A) (2004)
6000 West Olive
Glendale, AZ 85302
p. (623) 845-3847 f. (623) 845-3846
leland@glendale.edu
http://www.gc.maricopa.edu/appliedscience/ge-oweb/geohome.html
Enrollment: No data reported since 1995

Professor:
Gary Calderone, (D), Arizona, 1988, Ym
Stan Celestian, (M), N Arizona, 1976, Oe
Julia Johnson, (M), Arizona State, 2000, Hy
Steve Kadel, (M), Arizona State, 1993, Xg
Pamela J. Nelson, (M), Arizona State, 1997, Gv
Emeritus:
J. Robert Thompson, (M), N Arizona, 1968, Ge

Mesa Community College

Dept of Physical Science (A) (2005)
1833 West Southern Ave
Mesa, AZ 85202
p. (480) 461-7015 f. (480) 461-7234
rleighty@mail.mc.maricopa.edu
Administrative Assistant: Kelli Santistevan
Enrollment: No data reported since 1995

Professor:
Donna M. Benson, (M), Arizona State, 1996, Gg
Raymond W. Grant, (D), Harvard, 1968, Gg
Robert S. Leighty, (D), Arizona State, 1997, Og
Armand J. Lombard, (M), Arizona, 1977, Gg

Northern Arizona University

Dept of Geography, Planning, and Recreation (B,M) (2007)
P.O. Box 15016
Flagstaff, AZ 86011
p. (928) 523-2650 f. (928) 523-2275
deborah.martin@nau.edu
http://www.geog.nau.edu
Administrative Assistant: Deborah Martin

Professor and Chair:
Pamela Foti, (D), Wisconsin, 1988, Oo
Professor:
Graydon L. Berlin, (D), Tennessee, 1970, Or
Leland R. Dexter, (D), Colorado, 1986, Oy
Christina B. Kennedy, (D), Arizona, 1989, Oo
Alan A. Lew, (D), Oregon, 1986
Associate Professor:
Samantha Arundel, (D), Arizona State, 1999, Oiy
Carolyn M. Daugherty, (D), Arizona State, 1987, Oo
Charles Hammersley, (D), New Mexico, 1988
Rebecca D. Hawley, (D), Arizona State, 1994, Oo
Assistant Professor:
Aaron Divine, (M), N Arizona, 2003
Ruihong Huang, (D), Wisconsin (Milwaukee), 2003
Senior Lecturer:
Stephen R. Dodd, (M), Slippery Rock, 1981, Oo
Judith Hammersley, (M), New Mexico, 1985
Emeritus:
Robert O. Clark, (D), Denver, 1970, Oo
Stanley W. Swarts, (D), California (Los Angeles), 1975, Oy
George A. Van Otten, (D), Oregon State, 1977, Oo

Dept of Geology (B,M) (2007)
P O Box 4099
Building 12, Room 100
Knoles Drive

Flagstaff, AZ 86011-4099
p. (928) 523-4561 f. (928) 523-9220
larry.middleton@nau.edu
http://www4.nau.edu/geology/
Enrollment (2006): B: 87 (0) M: 30 (0)

Chair:
Roderic A. Parnell, (D), Dartmouth, 1982, Cl
Acting Chair:
Larry T. Middleton, (D), Wyoming, 1979, Gs
Director, Quaternary Sci Program:
James I. Mead, (D), Arizona, 1983, Po
Professor:
Ronald C. Blakey, (D), Iowa, 1973, Gr
David S. Brumbaugh, (D), Indiana, 1972, Ys
Ernest M. Duebendorfer, (D), Wyoming, 1986, Gc
David K. Elliott, (D), Bristol (UK), 1979, Pv
Thomas D. Hoisch, (D), S California, 1985, Gp
Paul Morgan, (D), London, 1973, Yh
Associate Professor:
Darrell S. Kaufman, (D), Colorado, 1991, Gm
Michael H. Ort, (D), California (Santa Barbara), 1991, Gv
Mary R. Reid, (D), MIT, 1987, Gi
Nancy Riggs, (D), California (Santa Barbara), 1991, Gv
Abraham E. Springer, (D), Ohio State, 1994, Hw
Paul J. Umhoefer, (D), Washington, 1989, Gt
Assistant Professor:
James C. Sample, (D), California (Santa Cruz), 1986, Cl
Research Associate:
Joseph E. Hazel, Jr., (M), N Arizona, 1991, Gs
Matthew A. Kaplinski, (M), N Arizona, 1990, Yg
Adjunct Associate Professor:
Darrell S. Kaufman, (D), Colorado, 1991, Gm
Adjunct Professor:
Wendell A. Duffield, (D), Stanford, 1967, Gv
John H. Sass, (D), Australian National, 1965, Yg
Wesley Ward, (D), Washington, 1978, Gs
Emeritus:
Charles "Chuck" W. Barnes, (D), Wisconsin, 1965, Gc
Stanley S. Beus, (D), California (Los Angeles), 1963, Pi
Augustus S. Cotera, Jr., (D), Texas, 1962, Gg
Richard F. Holm, (D), Washington, 1969, Gv
Jack D. Nations, (D), California (Berkeley), 1969, Pm
Laboratory Director:
James Wittke, (D), Texas, 1984, Gi
Research Professor:
David D. Gillette, (D), S Methodist, 1974, Pv
Office Manager:
Susan M. Sabala-Foreman, (A), N Arizona, 1981, Oo

Phoenix College

Physics Dept (A) (2006)
1202 West Thomas Road
Phoenix, AZ 85013
p. (602) 285-7244
d.speed@pcmail.maricopa.edu

Chair:
Don Speed, (M), Arizona, 1980, Gg
Instructor:
Richard Cups, (M), Arizona State, 1983

Prescott College

Dept of Environmental Studies (B) (2005)
220 Grove Avenue
Prescott, AZ 86301
p. (928) 350-2201 f. (928) 776-5137
labbott@prescott.edu
http://www.prescott.edu

Professor:
Lon Abbott, (D), California (Santa Cruz), 1993, Gt
Tim Crews, (D), Cornell, 1993, Sb
John Herring, (D), Washington, 1994, Oa

University of Arizona

Dept of Geography & Regional Dev (B,M,D) (2007)
Harvill Building
PO Box 210076
Tucson, AZ 85721
p. (520) 621-1652 f. (520) 621-2889
jpjones@email.arizona.edu
http://geog.arizona.edu/
Enrollment (2006): B: 10 (2) M: 4 (2) D: 6 (1)

Professor:
Michael E. Bonine, (D), Texas, 1975, Ou
Andrew Comrie, (D), Pennsylvania, 1992, Oy
Lay J. Gibson, (D), California (Los Angeles), 1968, Ou
John Paul Jones, III, (D)
Stuart E. Marsh, (D), Stanford, 1978, Or
Sallie A. Marston, (D), Colorado, 1986, Ou
David A. Plane, (D), Pennsylvania, 1981, Oo
Associate Professor:
Carl J. Bauer, (D), California (Berkeley), 1995, Hs
Beth A. Mitchneck, (D), Columbia, 1990, Oo
Paul Robbins, (D), Clark, 1996, Oo
Marvin Waterstone, (D), Rutgers, 1983, Oo
Connie Woodhouse, (D), Arizona, 1995, Pe
Stephen R. Yool, (D), California (Santa Barbara), 1985, Or
Assistant Professor:
Keiron D. Bailey, (D), Kentucky, 2002
Christopher A. Scott, (D), Cornell, 1998, Hw
Adjunct Professor:
Barbara Becker, (D), Texas, 1989, Oo
Julio Betancourt, (D), Arizona, 1989, Oo
Gary Christopherson, (D), Oi
Katherine Hirschboeck, (D), Arizona, 1985, Oa
Vance Holliday, (D), Colorado, 1982, Ga
Laura Huntoon, (D), Pennsylvania, 1991, Oo
Charles F. Hutchinson, (D), California (Riverside), 1978, Or
Miranda Joseph, (D), Stanford, 1995, Oo
Janice J. Monk, (D), Illinois, 1972, Oo
Barbara Morehouse, (D), Arizona, 1993, Oo
Elizabeth Oglesby, (D), California (Berkeley), 2002, Oo
Vera K. Pavlakovich, (D), Kent State, 1986, Oo
Thomas W. Swetnam, (D), Arizona, 1987, Oo
Margaret Wilder, (D), Arizona, 2002, Oo

Dept of Geosciences (B,M,D) (2007)
Gould-Simpson Building
1040 E. Fourth Street
Tucson, AZ 85721
p. (520) 621-6000 f. (520) 621-2672
enedina@email.arizona.edu
http://www.geo.arizona.edu
Administrative Assistant: Enedina Cervantes
Department Business Manager: Gayle Zizzo
Enrollment (2006): B: 70 (0) M: 23 (0) D: 52 (0)

Chair:
Susan L. Beck, (D), Michigan, 1987, Ys
Professor:
Victor R. Baker, (D), Colorado, 1971, Gm
Mark D. Barton, (D), Chicago, 1981, Eg
Clement G. Chase, (D), California (San Diego), 1970, Yg
Andrew S. Cohen, (D), California (Davis), 1982, Ps
George H. Davis, (D), Michigan, 1971, Gc
Owen K. Davis, (D), Minnesota, 1981, Pl
Peter G. DeCelles, (D), Indiana, 1984, Gd

Karl W. Flessa, (D), Brown, 1973, Po
George E. Gehrels, (D), Caltech, 1986, Gt
Vance Holliday, (D), Colorado, 1982, Oo
Roy A. Johnson, (D), Wyoming, 1984, Ys
Jonathan T. Overpeck, (D), Brown, 1985, Oa
P. Jonathan Patchett, (D), Edinburgh, 1976, Cc
Jay Quade, (D), Utah, 1990, Sc
Randall M. Richardson, (D), MIT, 1978, Yg
Joaquin Ruiz, (D), Michigan, 1983, Cg
Spencer R. Titley, (D), Arizona, 1958, Eg
George Zandt, (D), MIT, 1978, Ys

Staff Scientist:
Christopher J. Eastoe, (D), Tasmania, 1979, Cs

Research Scientist:
Warren Beck, (D), Oo
David Dettman, (D), Michigan, 1994, Cs

Senior Scientist:
A. J. Timothy Jull, (D), Bristol, 1976, Cg

Associate Professor:
Julia E. Cole, (D), Columbia, 1992, Pe
Robert T. Downs, (D), Virginia Tech, 1992, Gi
Mihai N. Ducea, (D), Caltech, 1998, Gt
Eric Seedorff, (D), Stanford, 1987, Eg

Associate Staff Scientist:
Lanny R. McHargue, (D), Arizona, 1994, Cg

Associate Research Scientist:
Nathaniel A. Lifton, (D), Arizona, 1997, Cg

Assistant Professor:
Paul Kapp, (D), California (Los Angeles), 2001, Oo
Jon D. Pelletier, (D), Cornell, 1997, Gm

Senior Research Associate:
John J. Chesley, (D), Michigan, 1993, Cg

Research Associate:
Hersh Gilbert, (D), Colorado, 2001, Oo
David Johnson, (D), Arizona, 2000, Eg
Allister Rees, (D), London (UK), 1990

Senior Lecturer:
Peter L. Kresan, (M), Arizona, 1975, Gg

Adjunct Lecturer:
Floyd Gray, (M), Massachusetts, 1982

Adjunct Associate Professor:
Thomas R. Van Devender, (D), Arizona, 1973, Gg
Robert Webb, (D), Arizona, 1985

Adjunct Professor:
Anna K. Behrensmeyer, (D), Harvard, 1973, Pe
Julio Betancourt, (D), Arizona, 1989, Oo
John Dohrenwend, (D), Stanford, 1975, Gm
Eric R. Force, (D), Lehigh, 1970, Eg
Richard Hay, (D), Princeton, 1952, Gd
Robert J. Kamilli, (D), Harvard, 1976, Eg
Charles F. Kluth, (D), Arizona, 1982, Gc
Larry Mayer, (D), Arizona, 1982, Gm
J. Dale Nations, (D), California (Berkeley), 1969, Pm
Fernando Ortega-Gutierrez, (D), Leeds, 1975, Gt
Charles Prewitt, (D), MIT, 1969, Oo
Thierry Sempere, (D), Ecole des Mines de Paris (France), 1981, Gg
John H. Stewart, (D), Stanford, 1961, Gt
Steven R. Taylor, (D), MIT, 1980, Ys
Raymond M. Turner, (D), Washington State, 1954, Po
Alexander T. Wilson, (D), California (Berkeley), 1954, Cs
Jack A. Wolfe, (D), California (Berkeley), 1960, Pb

Visiting Professor:
James F. Hays, (D), Harvard, 1966, Cp

Emeritus:
William B. Bull, (D), Stanford, 1960, Gm
William R. Dickinson, (D), Stanford, 1958, Gt
John M. Guilbert, (D), Wisconsin, 1962, Em
DeVerle P. Harris, (D), Penn State, 1965, Eg
C. Vance Haynes, Jr., (D), Arizona, 1965, Ga
Everett H. Lindsay, (D), California (Berkeley), 1967, Pv
Austin Long, (D), Arizona, 1966, Cs
Paul S. Martin, (D), Michigan, 1956, Pe
Edgar J. McCullough, Jr., (D), Arizona, 1963, Ge
Joseph F. Schreiber, Jr., (D), Utah, 1958, Gs

Dept of Hydrology & Water Resources (B,M,D) (2006)
226 Harshbarger Building #11
PO Box 210011
Tucson, AZ 85721-0011
p. (520) 621-5082 f. (520) 621-1422
programs@hwr.arizona.edu
http://www.hwr.arizona.edu
Enrollment (2006): B: 13 (2) M: 27 (13) D: 34 (7)

Head:
Thomas Maddock, III, (D), Harvard, 1972, Hq

Regents Professor:
Victor R. Baker, (D), Colorado, 1971, Gm
Shlomo P. Neuman, (D), California (Berkeley), 1968, Hq

Director, SAHRA Center:
W. James Shuttleworth, (D), Manchester (UK), 1971, Hs

Associate Director, SAHRA Center:
Hoshin V. Gupta, (D), Case Western, 1984, Hs

Professor:
Donald R. Davis, (D), Arizona, 1971, Hq
Peter A. Troch, (D), Ghent, 1993, Hs
Tian-Chyi Jim Yeh, (D), New Mexico Tech, 1983, Hq

Associate Head:
Michael D. Bradley, (D), Michigan, 1971, Oo

Associate Professor:
Paul A. Ferre, (D), Waterloo, 1997, Hw
Thomas Meixner, (D), Arizona, 1999, Cg
Marek G. Zreda, (D), New Mexico Tech, 1994, Hw

Assistant Professor:
Paul D. Brooks, (D), Colorado, 1995, Cg
Jennifer C. McIntosh, (D), Michigan, 2004, Cg

Associate Director, SAHRA Center:
Gary C. Woodard, (D), Michigan, 1981, Oo

Adjunct Professor:
Roger C. Bales, (D), Caltech, 1984, Cg
Robert A. Clark, (D), Texas A&M, 1964, Hs
Leo S. Leonhart, (D), Arizona, 1978, Hw
Robert MacNish, (D), Michigan, 1966, Hw
Soroosh Sorooshian, (D), California (Los Angeles), 1978, Hq
Donald W. Young, (D), Arizona, 1994, Oo

Adjunct Associate Professor:
David C. Goodrich, (D), Arizona, 1990, Hs
Robert H. Webb, (D), Arizona, 1985, Gm

Adjunct Assistant Professor:
James F. Hogan, (D), Dartmouth, 2001, Cg
James E. Smith, (D), Waterloo, 1995, Hw
James C. Washburne, (D), Arizona, 1994, Hs

Emeritus:
Nathan Buras, (D), California (Los Angeles), 1962, Hq
Stanley N. Davis, (D), Yale, 1955, Hw
Lucien Duckstein, (D), Colorado State, 1962, Hq
Simon Ince, (D), Iowa, 1953, Hs
Austin Long, (D), Arizona State, 1966, Cl
Ernest T. Smerdon, (D), Missouri State, 1959, Hs
Peter J. Wierenga, (D), California (Davis), 1968, Sp
Lorne G. Wilson, (D), California (Davis), 1962, Hw

Related Staff:
Erma Santander, Oo
Terrie Thompson, Oo

Cooperating Faculty:
Mark L. Brusseau, (D), Florida, 1989, Hw

Jonathan D. Chorover, (D), California (Berkeley), 1993, Cg
Bonnie C. Colby, (D), Wisconsin, 1983, Oo
R. B. Hawkins, (D), Colorado State, 1968, Hs
Katherine K. Hirschboeck, (D), Arizona, 1985, Pe
Katharine L. Jacobs, (M), California (Berkeley), 1981, Hg
Kevin E. Lansey, (D), Texas, 1987, Hs
Sharon B. Megdal, (D), Princeton, 1981, Hg
Steven L. Mullen, (D), Washington, 1985, Oa
Ferenc Szidarovszky, (D), K Marx Univ of Econ, 1977, Hq
Juan B. Valdes, (D), MIT, 1976, Hs
Robert G. Varady, (D), Arizona, 1981, Oo
Arthur W. Warrick, (D), Iowa State, 1967, Sp
Marvin Waterstone, (D), Rutgers, 1983, Oo

Dept of Mining & Geological Engineering (B,M,D) (2007)
1235 E. James E. Rogers Way
Mines Building, Room 229
PO Box 210012
Tucson, AZ 85721-0012
p. (520) 621-6063 f. (520) 621-8330
mgedept@email.arizona.edu
http://www.mge.arizona.edu
Enrollment (2007): B: 71 (6) M: 22 (0) D: 12 (0)

Head:
Mary M. Poulton, (D), Arizona, 1990, Ng
Director, Lab for Advanced Subsurface Imaging:
Ben K. Sternberg, (D), Wisconsin, 1977, Ye
Professor:
William G. Davenport, (D)
J. Brent Hiskey, (D), Utah, 1973, Nx
Pinnaduwa H. S. W. Kulatilake, (D), Ohio State, 1981, Nr
Project Manager:
Robert R. Casavant, (D), Arizona, 2001, Gm
Associate Professor:
John M. Kemeny, (D), California (Berkeley), 1986, Nr
Moe Momayez, (D), Nm
Assistant Professor:
Sean Dessureault, (D), British Columbia, 2002, Nx
Jinhong Zhang, (D)
Adjunct Professor:
John R (Ros) Hill, (M), Colorado Mines, Nm
Ihor Kunasz, (D)
Emeritus:
Charles E. Glass, (D), Ne
William C. Peters, (D), Colorado, 1957, Ng
Business Manager:
Olga Ortiz

Dept of Planetary Sciences (M,D) (2005)
1629 E. University Boulevard
Tucson, AZ 85721-0092
p. (520) 621-6963 f. (520) 621-4933
acad_info@lpl.arizona.edu
http://www.lpl.arizona.edu
Administrative Assistant: Pamela S. Streett
Enrollment (2001): D: 33 (6)

Head:
Michael J. Drake, (D), Oregon, 1972, Xc
Research Professor:
Martin G. Tomasko, (D), Princeton, 1969, Oa
Professor:
Victor R. Baker, (D), Colorado, 1971, Xg
William V. Boynton, (D), Carnegie Mellon, 1971, Xc
Robert H. Brown, (D), Hawaii, 1982, Xc
Uwe Fink, (D), Penn State, 1965, Oa
Tom Gehrels, (D), Chicago, 1956, Oo
Richard J. Greenberg, (D), MIT, 1972, Xy
William B. Hubbard, (D), California (Berkeley), 1967, Xy
Donald M. Hunten, (D), McGill, 1950, Oa
Jack R. Jokipii, (D), Caltech, 1965, Xy
Harold P. Larson, (D), Purdue, 1967, Oo
John S. Lewis, (D), California (San Diego), 1968, Xc
Jonathan I. Lunine, (D), Caltech, 1985, Xc
Renu Malhotra, (D), Cornell, 1988, Oo
Alfred McEwen, (D), Arizona State, 1988, Xg
Henry J. Melosh, (D), Caltech, 1972, Gt
George H. Rieke, (D), Harvard, 1969, Oo
Timothy D. Swindle, (D), Washington (St. Louis), 1986, Xm
Roger Yelle, (D), Wisconsin, 1984, Oo
Senior Scientist:
Jozsef Kota, (D), Roland Eotvos, 1980
Associate Professor:
Caitlin Griffith, (D), SUNY (Stony Brook), 1991, Oo
David Kring, (D), Harvard, 1989, Xc
Robert Kursinski, (D), Caltech, 1997, Oa
Assistant Professor:
Joe Giacalone, (D), Kansas, 1991, Xy
Dante Lauretta, (D), Washington (St. Louis), 1997, Oo
Adam Showman, (D), Caltech, 1998, Oo
Senior Research Scientist:
Lyle A. Broadfoot, (D), Saskatchewan, 1963, Or
Jay B. Holberg, (D), California (Berkeley), 1974, Oa
Lonnie L. Hood, (D), California (Los Angeles), 1979, Xy
Larry A. Lebofsky, (D), MIT, 1974, Og
Bill R. Sandel, (D), Rice, 1972, Oa
Research Associate:
Alexander Dessler, (D), Duke, 1956, Oo
Ralph Lorenz, (D), Kent (UK), 1994
Elizabeth Turtle, (D), Arizona, 1998
Associate Research Scientist:
Robert S. McMillan, (D), Texas, 1977, Oo
Peter Smith, (M), Arizona, 1977, Or
Emeritus:
Elizabeth Roemer, (D), California (Berkeley), 1955, Oo
Charles P. Sonett, (D), California (Los Angeles), 1954, Yg
Robert G. Strom, (M), Stanford, 1957, Xg
Manager, Academic Affairs:
Joan M. Weinberg, (B), Hunter, 1963, Oo

Dept of Soil, Water & Environmental Science (B,M,D) (2004)
PO Box 210038
429 Shantz Building, #38
Tucson, AZ 85721-0038
p. (520) 621-1646 f. (520) 621-1647
sw@ag.arizona.edu
http://ag.arizona.edu/swes
Enrollment (2001): B: 49 (18) M: 17 (3) D: 42 (15)

Head:
Jeffrey C. Silvertooth, (D), Oklahoma State, 1986, So
Professor:
Mark L. Brusseau, (D), Florida, 1989, Hw
Charles P. Gerba, (D), Miami, 1973, Sb
Ed Glenn, (D), Hawaii, 1978, Oo
David M. Hendricks, (D), California, 1966, Sc
Alfredo R. Huete, (D), Arizona, 1984, Or
Raina M. Maier, (D), Rutgers, 1988, Oo
Allan D. Matthias, (D), Iowa State, 1979, Oa
Ian L. Pepper, (D), Ohio State, 1975, Oo
Donald F. Post, (D), Purdue, 1967, Sd
Arthur W. Warrick, (D), Iowa State, 1967, Sp
Peter J. Wierenga, (D), California, 1968, Sp
Senior Scientist:
James Riley, (D), Arizona, 1968, Oo
Associate Professor:
Joan E. Curry, (D), California (Davis), 1992, Sc
Robert Frye, (D), Arizona, 1983, Oo

Christopher C. Rensing, (D), Free (Berlin), 1996, Sb
Associate Professor & Extension Specialist:
Kevin Fitzsimmons, (D), Arizona, 1999, Oo
Associate Scientist:
Charles Sanchez, (D), Iowa State, 1986, Oo
Assistant Professor:
Thomas L. Thompson, (D), Iowa State, 1991, So
Research Associate:
James Walworth, (D), Georgia, 1985, So
Associate Research Scientist:
Janick F. Artiola, (D), Arizona, 1980, Sc
Adjunct Senior Research Scientist:
Stephen Nelson, So
Adjunct Associate Professor:
Dean Martens, So
Jim Yeh, Sp
Adjunct Assistant Professor:
Paul (Ty) Ferre, So
Christopher Straub, So
Bryan Williams, So
Adjunct Professor:
Floyd Adamsen, So
Extension Specialist:
Paul W. Brown, (D), Wisconsin, 1981, Ow

Laboratory of Tree Ring Research (2004)
West Stadium, Room 105
Tucson, AZ 85721
p. (520) 621-2191 f. (520) 621-8229
pgress@ltrr.arizona.edu
Administrative Assistant: Pamela Smith
Administrative Associate: Phyllis Gress
Director:
Malcolm K. Hughes, (D), Durham (UK), 1970, Pe
Professor:
Jeffrey S. Dean, (D), Arizona, 1967, Oo
Steven W. Leavitt, (D), Arizona, 1982, Cs
Charles W. Stockton, (D), Arizona, 1971, Hq
Associate Professor:
Lisa J. Graumlich, (D), Washington, 1985, Pe
Katherine K. Hirschboeck, (D), Arizona, 1985, Oa
Thomas W. Swetnam, (D), Arizona, 1987, Pe
Research Associate:
Thomas P. Harlan, (M), Arizona, 1962, Oo
David M. Meko, (D), Arizona, 1981, Hq
Richard L. Warren, (B), Arizona, 1962, Oo
Emeritus:
Bryant Bannister, (D), Arizona, 1960, Oo
Harold C. Fritts, (D), Ohio State, 1956, Pe

Yavapai College
Geology Dept (A) (2004)
1100 E. Sheldon Street
Prescott, AZ 86301
p. (928) 776-2331 f. (928) 717-7698
beth_boyd@yc.edu
http://www.yc.edu/geology.nsf
Enrollment: No data reported since 1995
Chair:
Beth Nichols Boyd, (M), Arizona, 1983, Cg

Arkansas

Arkansas Tech University
Dept of Physical Sciences-Geology (B) (2007)
1701 N. Boulder Ave
McEver Science Building
Russellville, AR 72801
p. (479) 9640816 f. (479) 964-0837
rcohoon@atu.edu
http://www.atu.edu
Enrollment (2007): B: 24 (9)
Professor:
Richard R. Cohoon, (D), Oklahoma State, 1974, Ge
Stephen W. Kline, (D), Georgia, 1984, Gx
Associate Professor:
Cathy Baker, (D), Iowa, 1986, Gg

College of the Ozarks
Div of Science & Mathematics (1999)
415 College Avenue
Clarksville, AR 72830
p. (501) 754-8261
fknight@ozarks.edu
Professor:
George Shellenberger, (M), Emporia State, 1973, Oe

Southern Arkansas University
Dept of Physical Sciences (Geology) (B,M) (2001)
Magnolia, AR 71753
p. (501) 235-4296 f. (501) 235-4936
lcdavis@saumag.edu
Administrative Assistant: Ann Collier
Enrollment: No data reported since 1995
Chair:
Leo C. Davis, (D), Iowa, 1975, Pv

University of Arkansas, Fayetteville
Arkansas Water Resources Center (2004)
113 Ozark Hall
Fayetteville, AR 72701
p. 501-575-4403 f. 501-575-3177
ralphd@mail.uark.edu
http://www.uark.edu/depts/awrc
Administrative Assistant: Theresa J. Russell
Director:
Kenneth F. Steele, (D), North Carolina, 1971, Cg
Research Associate:
Terry E. Nichols, (D), Chicago, 1973, Oo
Laboratory Director:
Marc A. Nelson, (D), Arkansas, 1992, Cl

Dept of Geosciences (B,M) (2006)
Divisions of Geography and Geology
113 Ozark Hall
Fayetteville, AR 72701
p. (479) 575-3355 f. (479) 575-3469
tcenter@comp.uark.edu
http://www.uark.edu/depts/geology
Administrative Assistant: Teresa J. Center
Enrollment (2006): B: 71 (27) M: 61 (22)
Chair:
Pamela E. Jansma, (D), Northwestern, 1988, Gc
Professor:
John V. Brahana, (D), Missouri, 1973, Hw
John C. Dixon, (D), Colorado, 1983, Gm
Margaret J. Guccione, (D), Colorado, 1982, Gm
John G. Hehr, (D), Michigan State, 1971, Ow
Ronald H. Konig, (D), Cornell, 1959, Eg
Walter L. Manger, (D), Iowa, 1971, Ps
Glen S. Mattioli, (D), Northwestern, 1987, Gv
Thomas R. Paradise, (D), Arizona State, 1993, Oi
David W. Stahle, (D), Arizona State, 1990, Pb
Kenneth F. Steele, (D), North Carolina, 1971, Cg
Doy L. Zachry, (D), Texas, 1969, Gr

Associate Professor:
Stephen K. Boss, (D), North Carolina, 1994, Yr
Fiona M. Davidson, (D), Nebraska, 1991, Oo
Ralph K. Davis, (D), Nebraska, 1992, Hw
Thomas O. Graff, (D), Kansas, 1973, Ou
Assistant Professor:
Jackson D. Cothren, (D), Ohio State
Jason A. Tullis, (D), South Carolina, 2001, Or
Adjunct Professor:
Phillip D. Hays, (D), Texas A&M, 1996, Cs
Patricia A. Whalen, (D), Texas (Dallas), 1985, Ps
Emeritus:
Malcolm K. Cleaveland, (D), Arizona, 1983, Oi
Kern C. Jackson, (D), Wisconsin, 1951, Gi
Harold C. MacDonald, (D), Kansas, 1969, Or
Richard S. Smith, (D), N Colorado, 1974, Oy
George H. Wagner, (D), Iowa, 1941, Cg

University of Arkansas, Little Rock

Dept of Earth Sciences (B) (2006)
2801 South University Avenue
Little Rock, AR 72204-1099
p. (501) 569-3546 f. (501) 569-3271
saleslie@ualr.edu
http://www.ualr.edu/ersc/
Administrative Assistant: Tammy D. Starks
Enrollment (2003): B: 26 (3)

Chair:
Stephen A. Leslie, (D), Ohio State, 1995, Po
Professor:
Jeffrey B. Connelly, (D), Tennessee, 1992, Gc
Michael T. Ledbetter, (D), Rhode Island, 1978, Og
Assistant Professor:
Margaret E. McMillan, (D), Wyoming, 2003, Gm
Instructor:
Don R. Owens, (M), Arkansas, 1967, Em
William J. Sims, (M), Kentucky, 1997, Gs
Emeritus:
Philip L. Kehler, (D), S Methodist, 1970, Gr

University of Arkansas, Pine Bluff

School of Agriculture, Fisheries & Human Sci (B,M) (2004)
1200 North University Drive
PO Box 4990
Pine Bluff, AR 71611
p. (870) 543-8529 f. (501) 543-8033
mccray_j@vx4500.uapb.edu
Department Secretary: Margaret L. Taylor
Enrollment (1998): B: 15 (1)

Head:
Owen A. Porter, (D), North Dakota State, 1975, Sc
Chair:
Leslie J. Glover, (D), Massachusetts, 1978, Sc
Associate Professor:
Robert A. Fitz, (M), Tuskeegee, 1969, Oo

California

American River College

Dept of Earth Science (A) (2008)
4700 College Oak Drive
Sacramento, CA 95841
p. (916) 484-8107 f. (916) 484-8725
ARCEarthScience@losrios.edu
http://arc.losrios.edu/~earthsci/
Enrollment (1996): B: 27 (0)

Chair:
Melissa H. Levy, (M), E Tennessee State, 1994, GgOgy
Professor:
John E. Aubert, (M), California (Davis), 1994, Oy
Charles E. Thomsen, (M), Cal State (Chico), 1994, Oy
GIS Coordinator:
Hugh H. Howard, (D), Kansas, 2003, Oi
Assistant Professor:
Glenn Jaecks, (D), California (Davis), 2002, GgOgPg
Adjunct:
Paul M. Veisze, (M), California (Berkeley), 1985, Oi
Adjunct Professor:
Terry J. Boroughs, (M), Ohio State, 1992, GgXgCg
Richard L. Oldham, (M), Nevada, 1972, Gg

Antelope Valley College

Science Div - Geology Dept (2004)
Lancaster, CA 93536
p. (805) 943-3241 (Ext. 280)
rbalogh@avc.edu

Professor:
Richard S. Balogh, (M), California (Los Angeles), 1976, Gl
Warren B. Houghton, (M), California (Davis), 1957, Gg

Bakersfield College

Physical Science Dept (A) (2004)
1801 Panorama Drive
Bakersfield, CA 93305
p. (661) 395-4391 f. (661) 395-4328
rschiffm@bc.cc.ca.us

Acting Chair:
Robert A. Schiffman, (M), California (Santa Barbara), 1971, Ga
Professor:
John C. Lyman, (B), Cal State (San Diego), 1962, Gg
Instructor:
Robert Lewy, (M), Cal State (Bakersfield), 1992, Gx
Michael Oldershaw, (M), California (Davis), 1987, Eo

Barstow Community College

Geology Dept (A) (2004)
2700 Barstow Road
Barstow, CA 92311
p. (760) 252-2411 f. (760) 252-1875
sbulkley@bcconline.com

Chair:
Zhenying Jiang, (D), Utah, 1991, Oo
Instructor:
Scott Bulkley, (M), N Arizona, 1995, Oo

Cabrillo College

Dept of Geology (A) (2004)
6500 Soquel Drive
Aptos, CA 95003
p. (408) 479-6495

Instructor:
David Schwartz, (M), San Jose State, 1983, Gu

California Institute of Technology

Div of Geological & Planetary Sciences (B,M,D) (2007)
1200 East California Boulevard
MC 170-25
Pasadena, CA 91125
p. (626) 395-6111 f. (626) 568-0935
farley@gps.caltech.edu
http://www.gps.caltech.edu

Enrollment (2007): B: 19 (6) D: 102 (8)

Chair:
Kenneth A. Farley, (D), California (San Diego), 1991, Cg
Professor:
Geoffrey A. Blake, (D), Caltech, 1985, Xc
Michael E. Brown, (D), California (Berkeley), 1994, Xy
Robert W. Clayton, (D), Stanford, 1981, Ye
John M. Eiler, (D), Wisconsin, 1994, Cg
Charles Elachi, (D), Caltech, 1971
Michael C. Gurnis, (D), Australian National, 1987, Ys
Thomas H. Heaton, (D), Caltech, 1979, Ne
Donald V. Helmberger, (D), California (San Diego), 1967, Ys
Andrew P. Ingersoll, (D), Harvard, 1966, Oa
Joseph L. Kirschvink, (D), Princeton, 1979, Pg
Shrinivas R. Kulkarni, (D), California (Berkeley), 1983, Xy
Dianne K. Newman, (D), MIT, 1997, Py
George R. Rossman, (D), Caltech, 1971, Gz
Jason B. Saleeby, (D), California (Santa Barbara), 1975, Gc
Kerry E. Sieh, (D), Stanford, 1977, Gt
Mark Simons, (D), MIT, 1995, Ys
David J. Stevenson, (D), Cornell, 1976, Xy
Joann M. Stock, (D), MIT, 1988, Gt
Edward M. Stolper, (D), Harvard, 1979, Cp
Jeroen Tromp, (D), Princeton, 1992, Ys
Paul O. Wennberg, (D), Harvard, 1994, Oa
Brian P. Wernicke, (D), MIT, 1982, Gc
Yuk L. Yung, (D), Harvard, 1974, Oa
Associate Professor:
Jess F. Adkins, (D), MIT, 1997, Cm
Paul D. Asimow, (D), Caltech, 1997, Cp
Assistant Professor:
Oded Aharonson, (D), MIT, 2002
Jennifer M. Jackson, (D), U Illinois, 2005
Swaminathan Krishnan, (D), Caltech, 2003
Nadia Lapusta, (D), Harvard, 2001
Victoria Orphan, (D), California (Santa Barbara), 2001, Py
Mark Richardson, (D), California (Los Angeles), 1999
Tapio Schneider, (D), Princeton, 2001
Senior Research Associate:
Hermann Engelhardt, (D), Tech (Munich), 1964, Gl
Research Associate:
Egill Hauksson, (D), Columbia, 1981, Ys
Emeritus:
Thomas J. Ahrens, (D), RPI, 1962, Yx
Arden L. Albee, (D), Harvard, 1957
Clarence R. Allen, (D), Caltech, 1954
Don L. Anderson, (D), Caltech, 1962, Ys
Donald S. Burnett, (D), California (Berkeley), 1963, Xc
Peter M. Goldreich, (D), Cornell, 1963, Xy
David G. Harkrider, (D), Caltech, 1963, Ys
Barclay Kamb, (D), Caltech, 1956, Gl
Hiroo Kanamori, (D), Tokyo, 1964, Ys
Duane O. Muhleman, (D), Harvard, 1963, Xy
Bruce C. Murray, (D), MIT, 1955, Xg
Leon T. Silver, (D), Caltech, 1955, Cc
Hugh P. Taylor, (D), Caltech, 1959, Cs
Gerald J. Wasserburg, (D), Chicago, 1954, Cc
Peter J. Wyllie, (D), St. Andrews, 1958, Cp
Geology Librarian:
Jim O'Donnell, Oo

California Lutheran University

Dept of Geology (B) (2007)
60 West Olsen Road #3700
Thousand Oaks, CA 91360-2787
p. (805) 493-3264 f. (805) 493-3392
bilodeau@clunet.edu
Enrollment (2007): B: 5 (3)

Chair:
William L. Bilodeau, (D), Stanford, 1979, GcsGg
Professor:
Linda A. Ritterbush, (D), California (Santa Barbara), 1990, Pi

California Polytechnic State University

Earth & Soil Sciences Dept (B,M) (2004)
One Grand Avenue
San Luis Obispo, CA 93407
p. (805) 756-2261 f. (805) 756-5412
jstevens@calpoly.edu
http://www.earth.soils.calpoly.edu
Enrollment (2002): B: 126 (19) M: 3 (0)

Chair:
Thomas J. Rice, (D), North Carolina State, 1981, Sd
Professor:
Delmar D. Dingus, (D), Oregon State, 1975, So
Brent G. Hallock, (D), California (Davis), 1979, Sf
William L. Preston, (D), Oregon, 1979, Oe
Thomas A. Ruehr, (D), Colorado State, 1975, Sb
Terry L. Smith, (D), Nebraska, 1980, So
Associate Professor:
Antonio Garcia, (D), Gg
Assistant Professor:
Christopher (Chip) S. Appel, (D), Florida, 2001, Sc
Lynn E. Moody, (D), California (Riverside), 1992, Sd
Emeritus:
Ronald D. Taskey, (D), Oregon State, 1977, Sf
Administrative Coordinator:
Joan M. Stevens, Oo
Instructional Tech:
Craig Stubler, (B), California Polytech, 1996, So

Physics Dept (2007)
San Luis Obispo, CA 93407
p. (805) 756-2448 f. (805) 756-2435
rsaenz@calpoly.edu
http://www.calpoly.edu/~phys/
Administrative Assistant: Alice Gold

Professor:
David H. Chipping, (D), Stanford, 1970, Ng
Assistant Professor:
Antonio F. Garcia, (D), California (Santa Barbara), Gm
Emeritus:
Kenneth A. Hoffman, (D), California (Berkeley), 1973, Ym

California State Polytechnic Univ, Pomona

Dept of Geological Sciences (B) (2007)
3801 West Temple Avenue
Pomona, CA 91768
p. (909) 869-3454 f. (909) 869-2920
jaklasik@csupomona.edu
http://geology.csupomona.edu
Administrative Assistant: Monica Giannini
Enrollment (2007): B: 45 (5)

Chair:
John A. Klasik, (D), Louisiana State, 1976, Gu
Professor:
David R. Berry, (D), California (Los Angeles), 1983, Pg
David R. Jessey, (D), Missouri (Rolla), 1981, Eg
Associate Professor:
Jonathan A. Nourse, (D), Caltech, 1989, Gt
Assistant Professor:
Jeffrey S. Marshall, (D), Penn State, 2000, Gm
Jascha Polet, (D), Caltech, 1999, Ysg
Emeritus:
Lawrence Herber, (D), Nevada (Reno), 1968, Ng
Donald W. Tarman, (D), Arizona, 1975, Gc

California State University, Bakersfield

Dept of Physics and Geology (B,M) (2007)
9001 Stockdale Hwy
Bakersfield, CA 93311
p. (661) 654-3027 f. (661) 654-2040
dbaron@csub.edu
http://www.csub.edu/Geology/
Administrative Assistant: Alicia Parks
Enrollment (2007): B: 25 (5) M: 12 (3)

Chair:
Robert M. Negrini, (D), California (Davis), 1986, Ym
Professor:
Dirk Baron, (D), Oregon, 1995, Cl
Janice Gillespie, (D), Wyoming, 1991, Go
Robert A. Horton, Jr., (D), Colorado Mines, 1985, Gd
Thomas Meyer, (D), California (Los Angeles), 1976, Oo
Steven W. Mitchell, (D), George Washington, 1980, Ps
Jorge Talamantes, (D), California (Riverside), 1990, Oo
Assistant Professor:
Staci Loewy, (D), Texas, 2002, Gc
Lecturer:
Patrick Mickler, (D), Texas, 2004, Gg
Emeritus:
John R. Coash, (D), Yale, 1954, Gg
Daniel P. Detwiler, (D), Yale, 1952, Oo
Edward A. Laskowski, (D), California (Los Angeles), 1966, So

California State University, Chico

Dept of Geological and Environmental Sci (B,M) (2006)
Chico, CA 95929-0205
p. (530) 898-5262 f. (530) 898-5234
gjohnston@csuchico.edu
http://www.csuchico.edu/geos
Administrative Assistant: Julie Garnett
Enrollment (2006): B: 114 (0) M: 26 (0)

Chair:
K. R. Gina Johnston, (D), Florida, 1970, Hs
Professor:
David L. Brown, (D), California (Berkeley), 1995, Hg
Terence T. Kato, (D), California (Los Angeles), 1976, Gt
Gregory R. Taylor, (D), Washington, 1987, Oa
Associate Professor:
Ann Bykerk-Kauffman, (D), Arizona, 1990, Gc
Karin A. Hoover, (D), Johns Hopkins, 1998, Hw
William M. Murphy, (D), California (Berkeley), 1985, Cl
Tal M. Nahir, (D), North Carolina, 1993, Oa
Assistant Professor:
Jeff T. Price, (D), Wayne State, 1995, Oa
Randy S. Senock, (D), Kansas State, 1994, Sf
Russell Shapiro, (D), California (Santa Barbara), 1998, Ps
Rachel Teasdale, (D), Idaho, 2001, Gv
Lecturer:
John L. McMurtry, (M), Chico State, 1991, Hw
Emeritus:
Jerold J. Behnke, (D), Nevada (Reno), 1968, Hw
Victor A. Fisher, (D), Florida State, 1968, Gr
Richard A. Flory, (D), Oregon State, 1975, Ps
Rolland K. Hauser, (D), Chicago, 1967, Oa
James L. Regas, (D), Harvard, 1969, Xg
Howard L. Stensrud, (D), Washington, 1970, Em

California State University, Dominguez Hills

Earth Sciences (B) (2007)
1000 E. Victoria Street
Carson, CA 90747
p. (310) 243-3377 f. (310) 243-2593
vknauss@csudh.edu
http://www.nbs.csudh.edu/earth/
Administrative Assistant: Virginia L. Knauss
Enrollment (2007): B: 22 (2)

Chair:
Ralph Saunders, (D), Arizona, 1996, Oo
Associate Dean, Natural & Behavioral Sciences:
Rodrick Hay, (D), Arizona, 1996, Or
Professor:
Brendan A. McNulty, (D), California (Santa Cruz), 1996, Gc
Assistant Professor:
John Keyantash, (D), California (Los Angeles), 2001, Hg
Aisish Sinha, (D), USC, 1997, Gg
Lecturer:
Judith A. King-Rundel, (M), California (Los Angeles), 2003, Oe

California State University, East Bay

Dept of Earth & Environmental Sciences (B,M) (2007)
25800 Carlos Bee Boulevard
Hayward, CA 94542-3088
p. (510) 885-3486 f. (510) 885-2526
geology@csueastbay.edu
http://www.sci.csueastbay.edu/earth
Administrative Assistant: Julie M. Mielke
Enrollment (2007): B: 54 (12) M: 10 (6)

Chair:
Jeffery C. Seitz, (D), Virginia Tech, 1994, Cp
Associate Professor:
Mitchell Craig, (D), Georgia Tech, 1990, Yg
Luther M. Strayer, (D), Minnesota, 1998, Gc
Adjunct Professor:
Carlota Escutia, (D), Polytech (Barcelona), 1992, Ou
Emeritus:
Elwood R. Brooks, (D), Wisconsin, 1964, Gv
Sue E. Hirschfeld, (D), California (Berkeley), 1971, Gt
Alexis N. Moiseyev, (D), Stanford, 1966, Cg
Detlef A. Warnke, (D), S California, 1965, Ou
Michael M. Wood, (D), Arizona, 1969, Gz
Related Staff:
James Allen, (M)
Cooperating Faculty:
Joy Andrews, (D), California (Berkeley), Cl
Michael Lee, (D), London, 1990, Ou
Susan Opp, (D), Massachusetts, 1988, Oo

Dept of Geography & Environmental Studies (B,M) (2005)
25800 Carlos Bee Blvd
Hayward, CA 94542
p. (510) 885-3193 f. (510) 885-2353
david.larson@csueastbay.edu
http://class.csueastbay.edu/geography
Enrollment (2005): B: 0 (22) M: 14 (2)

Chair:
David J. Larson, (D), California (Berkeley), 1994, Ou
Professor:
Scott W. Stine, (D), California (Berkeley), 1987, Gm
Associate Professor:
Karina Garbesi, (D), California (Berkeley), 1994, Cg
Michael D. Lee, (D), London Schl of Econ, 1990, Hg
Gary Li, (D), SUNY (Buffalo), 1997, Oi
David Woo, (D), California (Santa Barbara), 1991, Or

California State University, Fresno

Dept of Earth & Environmental Sciences (B,M) (2004)
2345 E. San Ramon Avenue
MH24
Fresno, CA 93740-8031
p. (559) 278-3086 f. (559) 278-5980
frakah@csufresno.edu

Administrative Support Coord: Evangelina Balli
Enrollment (2001): B: 26 (7) M: 9 (1)

Chair:
C. John Suen, (D), MIT, 1978, Hw
Professor:
Arthur H. Barabas, (D), Yale, 1977, Em
Bruce A. Blackerby, (D), California (Los Angeles), 1965, Gi
Roland H. Brady, (D), California (Davis), 1986, Ng
Fraka J. M. Harmsen, (D), Victoria (NZ), 1984, Gs
Robert D. Merrill, (D), Texas, 1974, Gm
Associate Professor:
Stephen D. Lewis, (D), Columbia, 1982, Yg
Assistant Professor:
P. Graham Mortyn, (D), Scripps, 2000, Ou
Keith D. Putirka, (D), Columbia, 1997
Zhi (Luke) Wang, (D), Leuven (Belgium), 1997, Ge
Lecturer:
Robert G. Dundas, (D), California (Berkeley), 1994, Pv
Adjunct Professor:
Donald Hunsaker, Jr., (D), California (Los Angeles), 1980, Ge
Emeritus:
Jon C. Avent, (D), Washington, 1965, Gr
Seymour Mack, (D), Syracuse, 1957, Gi

California State University, Fullerton

Dept of Geological Sciences (B,M) (2007)
800 N. State College Boulevard
P.O. Box 6850
Fullerton, CA 92834-6850
p. (714) 278-3882 f. (714) 278-7266
geology@fullerton.edu
http://geology.fullerton.edu
Enrollment (2006): B: 72 (0) M: 20 (0)

Chair:
David D. Bowman, (D), S California, 1999, Ys
Professor:
Gaylen R. Carlson, (D), Iowa, 1973, Oe
Diane Clemens-Knott, (D), Caltech, 1992, Ca
John H. Foster, (D), California (Riverside), 1980, Ng
General Education Coordinator:
Jeffrey R. Knott, (D), California (Riverside), 1998, Ge
Associate Professor:
Phillip A. Armstrong, (D), Utah, 1996, Yg
W. Richard Laton, (D), W Michigan, 1997, Hw
Brady P. Rhodes, (D), Montana, 1984, Gc
Assistant Professor:
Nicole Bonuso, (D), S California, 2005, Pie
Brandon Browne, (D), Alaska, 2005, Gv
Matthew E. Kirby, (D), Syracuse, 2001, Gs
Adam D. Woods, (D), S California, 1998, Gs
Instructor:
Hassan Amini, (D), Colorado, 1983, Ge
Kelly R. Bovard, (M), California (Riverside), 2001, Gm
Patricia M. Butcher, (M), Utah, 1993, Hw
Otto Figueroa, (M), Cal State (Fullerton), 2004, Hw
Wayne G. Henderson, (M), George Washington, 1997, Pg
James L. Ruhle, (M), Massachusetts, 1960, Pg
Stephen I. Wareham, (M), Loma Linda, 1986, Go
Kristin Weaver-Bowman, (M), S California, 1999, Ys
Emeritus:
Gerald F. Brem, (D), California (Riverside), 1977, Cg
Neil J. Maloney, (D), Oregon State, 1965, Og
Prem K. Saint, (D), Minnesota, 1973, Hw
Margaret S. Woyski, (D), Minnesota, 1946, Gi

California State University, Long Beach

Dept of Geological Sciences (B,M) (2007)
1250 Bellflower Boulevard
Long Beach, CA 90840-3902
p. (562) 985-4809 f. (562) 985-8638
scfinney@csulb.edu
http://www.cnsm.csulb.edu/departments/geology/
Administrative Assistant: Margaret Costello
Enrollment (2007): B: 71 (8) M: 19 (6)

Chair:
Stanley C. Finney, (D), Ohio State, 1977, Ps
Professor:
Richard J. Behl, (D), California (Santa Cruz), 1992, Gs
Robert D. Francis, (D), California (San Diego), 1980, Co
Roswitha B. Grannell, (D), California (Riverside), 1969, Yv
Jack Green, (D), Columbia, 1953, Gv
Associate Professor:
Gregory J. Holk, (D), Caltech, 1998, Cs
Assistant Professor:
Thomas Kelty, (D), California (Los Angeles), 1998, Gc
Nate Onderdonk, (D), California (Santa Barbara), 2003, Gtm
Lora R. Stevens (Landon), (D), Minnesota, 1997, Gn
Adjunct Professor:
Elihu Goldish, (D), Caltech, 1956, Gz
Emeritus:
Kwan M. Chan, (D), Liverpool, 1966, Oc
Paul J. Fritts, (D), Colorado, 1969, Pm
Charles T. Walker, (D), Leeds, 1952, Cl
Robert E. Winchell, (D), Ohio State, 1963, Gz

California State University, Los Angeles

Dept of Geological Sciences (B,M) (2006)
5151 State University Drive
Los Angeles, CA 90032
p. (323) 343-2400 f. (323) 343-2435
geology@calstatela.edu
http://www.calstatela.edu/dept/geology/
Enrollment (2006): B: 23 (4) M: 23 (4)

Chair:
Pedro C. Ramirez, (D), California (Santa Cruz), 1990, Gs
Professor:
Kim Bishop, (D), S California, 1994, Gc
Richard W. Hurst, (D), California (Los Angeles), 1975, Cc
Gary A. Novak, (D), Virginia Tech, 1970, Gz
Associate Professor:
Barry Hibbs, (D), Texas, 1993, Hw
Assistant Professor:
David P. Mayo, (D), S California, 1996, Oe
Nathan Onderdonk, (D), California (Santa Barbara), 2003, Gt
Emeritus:
Ivan P. Colburn, (D), Stanford, 1961, Gs
Alan A. Colville, (D), Indiana, 1961, Gz
Terry E. Davis, (D), California (Santa Barbara), 1969, Cc
Robert J. Stull, (D), Washington, 1969, Gi

California State University, Northridge

Dept of Geography (B,M) (2008)
18111 Nordhoff Street
Northridge, CA 91330-8249
p. (818) 677-3532 f. (818) 677-2723
geography@csun.edu
www.csun.edu/geography

Professor:
Julie E. Laity, (D), California (Los Angeles), 1982, Oy
Gong-yuh Lin, (D), Hawaii, 1974, Oa
Amalie Orme, (D), California (Los Angeles), 1983, Gm
Eugene Turner, (D), Washington, 1977, Or
Associate Professor:
Helen M. Cox, (D), California (Los Angeles), 1998, Oa
Yifei Sun, (D), SUNY (Buffalo), 2000, Oi

Assistant Professor:
Shawna J. Dark, (D), California (Los Angeles), 2003, Oy

Dept of Geological Sciences (B,M) (2006)
18111 Nordhoff Street
MD: 8266
Northridge, CA 91330-8266
p. (818) 677-3541 f. (818) 677-2820
geology@csun.edu
http://geology.csun.edu
Administrative Assistant: Mari C. Flores
Enrollment (2006): B: 32 (8) M: 33 (4)

Professor:
Vicki A. Pedone, (D), SUNY (Stony Brook), 1991, Gd
Gerald W. Simila, (D), California (Berkeley), 1979, Yg
Richard L. Squires, (D), Caltech, 1973, Pg
M. Ali Tabidian, (D), Nebraska, 1987, Hw

Associate Professor:
Kathleen M. Marsaglia, (D), California (Los Angeles), 1989, Gs
Jon R. Sloan, (D), California (Davis), 1980, Pm
J. Douglas Yule, (D), Caltech, 1996, Gt

Assistant Professor:
Elena A. Miranda, (D), Wyoming, 2006, Gc
Elizabeth A. Nagy-Shadman, (D), Caltech, 1997, Cc
Jorge A. Vazquez, (D), California (Los Angeles), 2004, Gi
Dayanthie Weeraratne, (D), Carnegie Inst, 2005, Yg

Emeritus:
Herbert G. Adams, (D), California (Los Angeles), 1971, Ng
Lorence G. Collins, (D), Illinois, 1959, Gz
George C. Dunne, (D), Rice, 1972, Gc
Peter J. Fischer, (D), S California, 1972, Gu
A. Eugene Fritsche, (D), California (Los Angeles), 1969, Gd
Allan M. Gutstadt, (D), Northwestern, 1954, Gr

California State University, Sacramento

Dept of Geology (B,M) (2007)
6000 J Street
Placer Hall
Sacramento, CA 95819-6043
p. (916) 278-6337 f. (916) 278-4650
dave_evans@csus.edu
http://www.asnet.csus.edu/geol/
Administrative Support Coordinator: Rosemary C. Burnham
Enrollment (2006): B: 63 (0) M: 15 (0)

Professor:
Diane H. Carlson, (D), Washington State, 1984, Gc
Brian Hausback, (D), California (Berkeley), 1984, Gv
Susan Clark Slaymaker, (D), North Carolina, 1973, Yg

Associate Professor:
Kevin J. Cornwell, (D), Nebraska, 1994, Gm
David G. Evans, (D), Louisiana State, 1989, Hw
Tim C. Horner, (D), Ohio State, 1992, Hw

Assistant Professor:
Judith E. Kusnick, (D), California (Davis), 1996, Oe

Lecturer:
Mark D. Boryta, (D), New Mexico Tech, 1997, Cg
Nancy Buening, (D), California (Davis), 1997, Gg
Karl A. Mertz, (D), California (Santa Cruz), 1984, Gr
Barbara J. Munn, (D), Virginia Tech, 1997, Em
Marlon A. Nance, (M), California (Davis), 1987, Gg
Maria C. Tobia, (M), Cal State (Chico), 1997, Gg

Emeritus:
Norman C. Janke, (D), California (Los Angeles), 1964, Gd
David F. McGeary, (D), California (San Diego), 1969, Gd
Charles C. Plummer, (D), Washington, 1969, Gp

Instructional Support Tech:
Steven W. Rounds, (B), Cal State (Sacramento), 1990, Gg

California State University, San Bernardino

Dept of Geological Sciences (B,M) (2006)
5500 University Parkway
San Bernardino, CA 92407-2397
p. (909) 537-5336 f. (909) 537-5112
alsmith@csusb.edu
www.geology.csusb.edu
Administrative Assistant: Christina Palmer
Enrollment (2006): B: 30 (3) M: 4 (0)

Chair:
Alan L. Smith, (D), California (Berkeley), 1969, Gv

Professor:
Bonnie J. Brunkhorst, (D), Iowa, 1987, Oe
Louis A. Fernandez, (D), Syracuse, 1969, Gi
Joan E. Fryxell, (D), North Carolina, 1984, Gc
Sally F. McGill, (D), Caltech, 1992, Gt

Associate Professor:
W. Britt Leatham, (D), Ohio State, 1987, Ps
Erik Melchiorre, (D), Washington, 1997, Cs

Cooperating Faculty:
Stuart S. Sumida, (D), California (Los Angeles), 1987, Pv

California State University, Stanislaus

Dept of Geology (B) (2007)
One University Circle
Turlock, CA 95382
p. (209) 667-3466 f. (209) 667-3099
hferriz@geology.csustan.edu
http://geology.csustan.edu
Administrative Assistant: Dawn McCulley
Enrollment (2007): B: 21 (4)

Professor:
Mario J. Giaramita, (D), California (Davis), 1989, Gp

Acting Chair:
Horacio Ferriz, (D), Stanford, 1985, Ng

Associate Professor:
Julia Sankey, (D), Louisiana State, 1998, Pe

Assistant Professor:
Robert D. Rogers, (D), Texas, 2003, GcmGt

Lecturer:
Garry F. Hayes, (M), Nevada (Reno), 1988, Gg

Emeritus:
Abbas Kimyai, (D), New York, 1964, Pl
Vladmir S. Tuman, (D), Stanford, 1964, Yv

Cerritos College

Earth Science Dept (A) (2007)
11110 E. Alondra Boulevard
Norwalk, CA 90650
p. (562) 860-2451 (Ext. 2668)
rkreger@cerritos.edu
http://www.cerritos.edu/earth-science/
Enrollment (2007): A: 4 (0)

Chair:
Robert D. Kreger, (D), Illinois, 1988, Oy

Assistant Professor:
Donald E. Hallinger, (M), Indiana, 1963, Hy
Tor Lacy, (M), Cal State (Long Beach), 2005, Gg
Crystal LoVetere, (M), Cal State (Fullerton), Oy
Megan McQuarrie, (M), Duke, Gu

Instructor:
Dan DeKraker, (M), Cal State (Fullerton), 2004, Oe
Gary D. Johnpeer, (M), Arizona State, 1977, Ng

Chaffey College

Geology Dept (A) (2006)
5885 Haven Avenue

Rancho Cucamonga, CA 91737-3002
p. (909) 941-2367 f. (909) 941-2783
terry.giugni@chaffey.edu
Enrollment: No data reported since 1995

Head:
Judy A. Lowman, (M), Cal State (Los Angeles), 1989, Gv
Professor:
Jane Warger, (M), Columbia, 1990, Cg
Adjunct Professor:
Lori Clark, (M), SUNY (Stony Brook), 2000, Cm
Leigh Dudash, (M), Texas, 1988, Yg
Jeffrey Johnston, (M), Cal State (Long Beach), 1985, Gg
John Maas, (D), California (Riverside), 1976, Yg
Franky Telles, (M), Cal State (Northridge), 2005, Gn

Chapman University

Dept of Natural Sciences (B) (2004)
Orange, CA 92668
p. (714) 997-6831 f. (714) 532-6048
science@chapman.edu
Enrollment (1997): B: 23 (4)

Emeritus:
Arthur E. Flint, (D), Chicago, 1954, Eg

City College of San Francisco

Dept of Earth Sciences (A) (2006)
Box S-138
50 Phelan Avenue
San Francisco, CA 94112
p. (415) 239-3104 f. (415) 239-3314
dhess@ccsf.edu
http://www.ccsf.edu/Earth
Enrollment: No data reported since 1995

Chair:
Darrel E. Hess, (M), California (Los Angeles), 1990, Oy
Instructor:
Joyce Lucas-Clark, (D), Stanford, Pg
Wanda Simpson-Baczek, Gz

College of the Desert

Div of Science & Mathematics (A) (2004)
Palm Desert, CA 92260
p. (619) 773-2573 f. (619) 341-8678
rortiz@collegeofthedesert.edu
Administrative Assistant: Valorie Stiff
Enrollment: No data reported since 1995

Chair:
Brett Romer, (M), New Mexico Highlands, 1964, Oo
Professor:
George L. Meyer, (M), California (Santa Barbara), 1967, Gg
Adjunct Professor:
Harry M. Quinn, (M), S California, 1968, Ps

College of the Redwoods

Dept of Science (Earth Science) (A) (2005)
7351 Tompkins Hill Road
Eureka, CA 95501
p. (707) 476-4224 f. (707) 476-4424
dave-bazard@redwoods.edu
http://online.redwoods.cc.ca.us/DEPTS/science/earth/
Enrollment: No data reported since 1995

Professor:
David R. Bazard, (D), Arizona, 1991, Gg
Associate Instructor:
Bob McPherson, (M), Humboldt State, 1989, Ys

College of the Siskiyous

Dept of Natural Sciences (A) (2006)
800 College Avenue
Weed, CA 96094
p. (530) 938-5255 f. (530) 938-5227
hirt@siskiyous.edu
http://www.siskiyous.edu/class/geology/
Enrollment (2006): A: 1 (0)

Instructor:
William H. Hirt, (D), California (Santa Barbara), 1989, Gi

Cosumnes River College

Dept of Science, Mathematics & Engineering (A) (2007)
8401 Center Parkway
Sacramento, CA 95823
p. (916) 691-7210 f. (916) 691-7142
montanrl@CRC.losrios.edu
Department Secretary: Sue McCoy

Chair:
Debra Sharkey, (M), California (Davis), 1994, Oy
Assistant Professor:
Hiram Jackson, (M), California (Davis), Gg
Adjunct Professor:
Gerry Drobny, (M), Washington State, 1981, Gg

Crafton Hills College

Geology Dept (A) (2007)
11711 Sand Canyon Road
Yucaipa, CA 92399
p. (714) 794-2161 f. (714) 794-0423
rihughes@craftonhills.edu
Enrollment (2007): A: 9 (4)

Assistant Professor:
Richard O. Hughes III, (M), Ohio, 1994, Gl

Cypress College

Physical Sciences Dept (A) (2005)
9200 Valley View Street
Cypress, CA 90630
p. (714) 484-7153 f. (714) 527-8238
vcastle@cypresscollege.edu
Enrollment (1996): B: 11 (0)

Professor:
Russell L. Flynn, (M), San Diego State, 1971, Op
Adjunct Professor:
Hank Wadleigh, (B), Cal State (Long Beach), 1957, Gg
Visiting Professor:
Curtis J. Williams, (M), Cal State (Los Angeles), 1996, Gg
Emeritus:
Keith E. Green, (M), S California, 1958, Pg
Altus Simpson, (M), S California, 1958, Go

De Anza College

Physical Science & Mathematics Div (A) (2004)
21250 Stevens Creek Boulevard
Cupertino, CA 95014
p. (408) 864-8664
ljb6159@fhda.edu
Enrollment (1995): B: 6 (0)

Professor:
Marek A. Cichanski, (D), S California, 1995, Gg

Diablo Valley College

Dept of Physical Sciences (A) (2005)
321 Golf Club Road
Pleasant Hill, CA 94523

p. ((92) 5) –685 (Ext. 1230 x2461)
BSinha@dvc.edu
Administrative Assistant: Nicole Gomes

Chair:
Jean Hetherington, (M), Washington, 1983, Gg

Instructor:
Binita Sinha, (D), Oi
Ted Wieded, (M), Oo

East Los Angeles College

Dept of Anthropology/Geography/Geology (A) (2005)
Monterey Park, CA 91754
p. (213) 265-8838
grimmejc@elac.edu

Chair:
Gerald R. Licari, (D), California (Los Angeles), 1971, Pm

Professor:
John Grant, (M), North Carolina, 1958, Oo
Domenick Miretti, (D), California (Los Angeles), 1979, Oy
James D. Sullivan, (M), California (Riverside), 1966, Oy

El Camino College

Dept of Earth Sciences (A) (2004)
Torrance, CA 90506
p. (310) 660-3371 f. (310) 660-3409
jholliday@elcamino.edu
Administrative Assistant: Carol Quiroz
Enrollment: No data reported since 1995

Associate Professor:
Joseph W. Holliday, (M), Oregon State, 1982, Gg

Instructor:
Suzi Alwash, (D), S California, 1988, Gg

Emeritus:
Ronald H. Arntson, (M), California (Los Angeles), 1956, Gg
Evelyn J. Baldwin, (D), Arizona, 1971, Gg
Bruce Fitzpatrick, (D), Nova, 1979, Gg
Burton S. Glass, (M), Hofstra, 1961, Gg

Foothill College

Geology Dept (A) (2004)
12345 El Monte Road
Los Altos Hills, CA 94022
p. (650) 949-7491
dileonardo@foothill.edu
Enrollment: No data reported since 1995

Head:
Christopher G. DiLeonardo, (D), California (Santa Cruz), 2001, Gc

Fresno City College

Earth/Physical Science Dept (2007)
Math, Science & Engineering Div
1101 E. University Avenue
Fresno, CA 93741
p. (559) 442-4600 f. (559) 497-8814
craig.poole@fresnocitycollege.edu
Enrollment (2006): A: 8 (0)

Instructor:
Brandy Anglen, (D), Indiana, 2005, Gg
T. Craig Poole, (M), Cal State (Fresno), 1987, Gg

Fullerton College

Div of Natural Sciences (Geology) (A) (2004)
Fullerton, CA 92634
p. (714) 992-7105 (Ext. 275) f. (714) 447-4097
rsimpson@fullcoll.edu
Enrollment (1997): B: 83 (0)

Director:
Allen G. Brown, (D), California (Berkeley), 1972, Oo

Instructor:
Richard P. Lozinsky, (D), New Mexico Tech, 1988, Gr
Tom Norton, Gg

Glendale Community College

Dept of Geology (A) (2004)
Glendale, CA 91208
p. (818) 240-1000 (Ext. 517)
leland@glendale.cc.ca.us
Enrollment: No data reported since 1995

Professor:
Poorna Pal, (D), Ym

Instructor:
John F. Leland, (D), California (Los Angeles), 1999, Cc

Distinguished Visiting Professor:
Kate Hutton, (D), Gg

Golden West College

Inst of Science, Health & Physical Education (A) (2006)
Huntington Beach, CA 92647
p. (714) 892-7711 (Ext. 494)
rgibson@gwc.cccd.edu
Enrollment (2005): A: 2 (0)

Professor:
Ronald C. Gibson, (M), California (Riverside), 1964, Gc
Bernard J. Gilpin, (M), California (Riverside), 1976, Ys

Grossmont College

Dept of Earth Sciences (A) (2006)
8800 Grossmont College Drive
El Cajon, CA 92020
p. (714) 644-7887
Gary.Jacobson@gcccd.net
http://www.grossmont.net
Enrollment (2004): A: 15 (0)

Professor:
Chris Hill, (D), Gg

Instructor:
Gary L. Jacobson, (M), San Diego State, 1982, Gu

Humboldt State University

Dept of Geology (B,M) (2007)
1 Harpst Street
Arcata, CA 95521-8299
p. (707) 826-3931 f. (707) 826-5241
arizzi@humboldt.edu
http://www.humboldt.edu/~geology/
Enrollment (2006): B: 64 (0) M: 7 (0)

Professor:
Kenneth R. Aalto, (D), Wisconsin, 1970, Gr
Raymond M. Burke, (D), Colorado, 1976, Gm
Susan M. Cashman, (D), Washington, 1977, Gc
Lorinda Dengler, (D), California (Berkeley), 1978, Yg
Andre K. Lehre, (D), California (Berkeley), 1982, Gm
William C. Miller, (D), Tulane, 1984, Pi

Associate Professor:
Mark Hemphill-Haley, (D), Oregon, 2000, Gt

Assistant Professor:
Brandon E. Schwab, (D), Oregon, 2000, Gi

Research Associate:
Robert C. McPherson, (M), Humboldt State, 1989, Gt

Adjunct Professor:
Eileen Hemphill-Haley, (D), California (Santa Cruz), 1991, Pm
Angela S. Jayko, (D), California (Santa Cruz), Gt
Harvey M. Kelsey, (D), California (Santa Cruz), 1976, Gm

Thomas Lisle, (D), California (Berkeley), 1976, Hs
Mary Ann Madej, (D), Washington, Gm
Jon R. Pedicino, (D), Arizona, 1996
Silvio K. Pezzopane, (D), Oregon, 1993, Gt
Carol S. Prentice, (D), Caltech, Gt
Carol Prentice, (D), 1989
Leslie Reid, (D), Washington, 1989, Gm
William E. Weaver, (D), 1986

Emeritus:
Gary A. Carver, (D), Washington, 1972, Gmt
John D. Longshore, (D), Rice, 1966, Gi
Alistair W. McCrone, (D), Kansas, 1961, Gs

Irvine Valley College

Dept of Geological Sciences (A) (2007)
5500 Irvine Center Drive
Irvine, CA 92618
p. (949) 451-5561
astinson@ivc.edu
http://www.ivc.edu
Administrative Assistant: Liz Nichols
Enrollment (2007): A: 20 (5)

Professor:
George Brogan, (M), San Diego State, Gc
Amy L. Stinson, (M), San Diego State, 1990, Gc

Adjunct Professor:
Mark Bordelon, (M)
Donna McKeown, (M)

Las Positas College

Dept of Geology: Science & Technology (2004)
3033 Collier Canyon Road
Livermore, CA 94550
p. (510) 373-4919

Instructor:
Ruth L. Hanna, (M), California (Davis), 1988, Gg

Loma Linda University

Dept of Natural Sciences, Geology Program (B,M) (2005)
Griggs Hall, Room 138
Loma Linda, CA 92350
p. (909) 824-4530 f. (909) 824-4859
pbuchheim@univ.llu.edu
Department Secretary: Judy Cushman
Enrollment (1999): B: 1 (0) M: 8 (0)

Director:
H. Paul Buchheim, (D), Wyoming, 1978, Gn

Professor:
Leonard R. Brand, (D), Cornell, 1970, Pv
Clyde L. Webster, (D), Colorado State, 1969, Ct

Assistant Professor:
Ben Clausen, (D), Colorado, 1987, Yg
Robert A. Cushman, (D), Colorado Mines, 1994, Ps

Long Beach City College

Dept of Physical Science (A) (2004)
4901 East Carson Street
Long Beach, CA 90808
p. (562) 938-4448 f. (562) 938-4425
jtomson@lbcc.cc.ca.us

Head:
Janice H. Tomson, (M), San Jose, 1985, Gs

Assistant Professor:
Charles M. Curtis, (M), S California, 1966, Gs
Yuet-Ling O'Connor, (D), S California, 1998, Gi

Los Angeles City College

Chemistry & Geophysical Sciences Dept (A) (2006)
855 N. Vermont Ave.
Los Angeles, CA 90029
p. (323) 953-4000 (Ext. 2606) f. 323-953-4013
brownad@lacitycollege.edu
Enrollment (2006): A: 4 (0)

Acting Chair:
Aaron Brown, (D), Utah State, 1985, Cg

Associate Professor:
Lawrence Miyaki, (M), Oy

Assistant Professor:
Christina O'Boyle, (M), Gg

Instructor:
Marla Conti, (M), Oy
Mary Prichard, (M), Oy

Los Angeles County Mus of Natural History

Research and Collections Branch (2004)
900 Exposition Boulevard
Los Angeles, CA 90007
p. (213) 763-3361 f. (213) 747-0204
kjohnson@nhm.org

Chief Curator:
John M. Harris, (D), Bristol, 1970, Pv

Curator:
Lawrence G. Barnes, (D), California (Berkeley), 1972, Pv
Kenneth E. Campbell, (D), Florida, 1973, Pv
Anthony R. Kampf, (D), Chicago, 1976, Gz

Associate Curator:
Luis M. Chiappe, (D), Buenos Aires, 1992, Pv

Assistant Curator:
Joe D. Stewart, (D), Kansas, 1984, Pv

Emeritus:
Lou Ella R. Saul, (M), California (Los Angeles), 1959, Pi
David P. Whistler, (D), California (Berkeley), 1969, Pv

Curatorial Assistant:
Shelley M. Cox, (B), California (Los Angeles), 1972, Pv
Cathy McNassor, (M), Cal State (Los Angeles), 1981, Pv

Collections Manager:
Dorothy Ettensohn, (M), Illinois, 1989, Gz
Samuel A. McLeod, (D), California (Berkeley), 1981, Pv
Christopher A. Shaw, (M), Cal State (Long Beach), 1981, Pv

Los Angeles Harbor College

Dept of Earth Science (A) (2006)
1111 Figueroa Place
Wilmington, CA 90744
p. (310) 522-8226 f. 310 233 4223
munasit@lahc.edu
http://www.lahc.cc.ca.us/

Instructor:
Patricia Kellner, (D)
John Mack
Tissa Munasinghe, (D)
Melanie Renfrew, (D)
Susan White

Los Angeles Valley College

Dept of Earth Science and Anthropology (A) (2006)
5800 Fulton Ave.
Valley Glen, CA 91401
p. (818) 778-5514 f. (818) 947-2620
gauthidj@lavc.edu
Enrollment (2006): A: 20 (6)

Chair:
Donald J. Gauthier, (M), California (Los Angeles), 1993, Oi

Associate Professor:
Angus A. MacDonald, (M), Cal State (Northridge), 1964, Gm
Assistant Professor:
Bob S. Barlow, (M), California (Los Angeles), 1967, Og
Jacqueline Hams, (M), Og
Adjunct Professor:
Meredith L. Leonard, (M), Cal State (Northridge), Ow
Emeritus:
Richard M. Raskoff, (M), Cal State (Northridge), 1966, Oy

Marin Community College

Geology Dept (A) (2006)
835 College Ave
Kentfield, CA 94904
p. (415) 485-9525 f. (415) 485-9543
jim@marin.cc.ca.us
http://www.marin.cc.ca.us/~geology
Enrollment: No data reported since 1995

Professor:
Donald J. Foss, (M), Boise State, 1980, Og
Emeritus:
James L. Locke, (M), San Jose State, 1971, Og

Mendocino College

Earth Science Dept (A) (2007)
1000 Hensley Creek Road
Ukiah, CA 95482
p. (707) 468-3002 f. (707) 468-3120
Enrollment: No data reported since 1995

Professor:
Steve Cardimona, (D), Texas, 1992, Ys

Merced College

Science, Mathematics & Engineering Div (A) (2005)
3600 M Street
Merced, CA 95348
p. (209) 384-6293 f. (209) 384-6338
eyrich.t@mccd.edu
Enrollment: No data reported since 1995

Chair:
Terry L. Eyrich, (D), California (Davis), 1973, Oo
Adjunct Professor:
Shirley Styhl, (M)

MiraCosta College

Dept of Physical Sciences (A) (2004)
1 Barnard Drive
Oceanside, CA 92056
p. (619) 757-2121 (Ext. 394)

Professor:
Keith H. Meldahl, (D), Arizona, 1990, Gg
Christopher V. Metzler, (D), California (San Diego), 1987, Gg

---

Dept of Social Sciences (A) (2004)
1 Barnard Drive
Oceanside, CA 92056
p. (619) 757-2121 (Ext. 233)

Chair:
Herschel I. Stern, (D), Oregon, 1988, Oy

Modesto Junior College

Dept of Engineering, Math & Physical Sciences (A) (2004)
435 College Avenue
Modesto, CA 95350
p. (209) 575-6172 f. (209) 575-6666
Enrollment (1999): A: 9 (5)

Instructor:
Donald C. Ahrens, (D), N Colorado, Og
Garry F. Hayes, (M), Nevada (Reno), Gg

Moss Landing Marine Laboratories

Moss Landing Marine Laboratories (B,M) (2006)
8272 Moss Landing Road
Moss Landing, CA 95039
p. (831) 771-4400 f. (831) 632-4403
coale@mlml.calstate.edu
www.mlml.calstate.edu
Enrollment: No data reported since 1995

Director:
John H. Martin, (D), Rhode Island, 1966, Ob
Professor:
Gregor M. Cailliet, (D), California (Santa Barbara), 1972, Ob
Kenneth S. Johnson, (D), Oregon State, 1979, Oc
Nicholas A. Welschmeyer, (D), Washington, 1982, Ob
Assistant Professor:
Ivano Aiello, (D), Florence (Italy), Gu
Michael Graham, (D), Scripps, Ob
James T. Harvey, (D), Oregon, 1987, Ob
Erika McPhee-Shaw, (D), Washington, Op
Adjunct Professor:
John S. Oliver, (D), California (San Diego), 1980, Ob

Mount San Antonio College

Dept of Earth Sciences and Astronomy (A) (2007)
1100 North Grand Avenue
Walnut, CA 91789
p. (909) 594-5611 (Ext. 4439) f. (909) 594-7661
mvcaputo@earthlink.net
Enrollment (2006): A: 17 (0)

Chair:
Julie Ali-Bray, (M), S California, 1999, Oo
Professor:
Mark Boryta, (D), New Mexico Tech, 1997, Cg
Karen Bridges, (M), Massachusetts, 1999, Og
Mario V. Caputo, (D), Cincinnati, 1988, Gg
Tim Castellano, (D), California (Santa Cruz), 2001, Oo
Micol Christopher, (D), Caltech, 2007, Oo
Sara Difiori, (M), California (Los Angeles), 2007, Gc
Terri Long, (M), Southern California, 1986, Og
Becca Walker, (M), Arizona, 2005, Og
Craig A. Webb, (M), Duke, 1996, Og
Part-time Instructor:
David Amiel, (D), 2006, Oc
Linda Chan, (D), Oo
Barbara Grubb, (M), Cal State (Long Beach), 1991, Ps
Gary Johnpeer, (M), Arizona State, 1977, Ng
Larry Mendenhall, (D), Oregon State, 1961, Ow
Charles Roberts, (M), Ohio, 1972, Pm
Annie Scott, (M), Og
Alex Wagner, (M), Og
Curtis Williams, (M), Og
Emeritus:
Hallock J. Bender, (D), San Gabriel, 1959, Gg
John W. Burns, (M), Pittsburgh, 1960, Oo
Damon P. Day, (M), Michigan Tech, 1965, Gg
Ron N. Hartman, (M), Cal State (Los Angeles), 1973, Xm
Kazimierz M. Pohopien, (M), McGill, 1951, Ge
Harold V. Thurman, (M), Cal State (Los Angeles), 1966, Og
Geotechnician:
Mark Koestel, (B), Arizona, 1978, Em

Napa Valley College

Dept of Earth & Environmental Sciences (A) (2004)

2277 Napa-Vallejo Hwy.
Napa, CA 94558
p. (707) 253-3161 f. 888-456-5545
rdellavalle@napavalley.edu
http://www.napavalley.edu
Administrative Assistant: Sue Borer
Department Secretary: Karen Clark
Enrollment: No data reported since 1995

Chair:
Richard S. Della Valle, (D), New Mexico, 1980, Cg
Adjunct Professor:
W. Darwin Myers, (D), Wisconsin, 1971, Gc
Jeffery Schaffer, (M), California (Berkeley), 1969, Oy
Tom Williams, (M), Colorado State, 1982, Ge

Naval Postgraduate School

Dept of Meteorology (M,D) (2004)
589 Dyer Road, Room 254
Monterey, CA 93943-5114
p. 831-656-2516 f. 831-656-3061
wash@nps.navy.mil

Dept of Oceanography (M,D) (2007)
833 Dyer Road, Room 328
Monterey, CA 93943-5122
p. (408) 656-2673 f. (408) 656-2712
bourke@oc.nps.navy.mil
Department Secretary: Hoda T. Salib
Enrollment (2006): M: 40 (0) D: 4 (0)

Chair:
Mary L. Batteen, (D), Oregon State, 1984, Op
Research Professor:
Steven R. Ramp, (D), Rhode Island, 1986, Op
Professor:
Robert H. Bourke, (D), Oregon State, 1972, Op
Ching-Sang Chiu, (D), MIT/Woods Hole, 1985, Op
Peter C. Chu, (D), Chicago, 1985, Op
Curtis A. Collins, (D), Oregon State, 1967, Op
Roland W. Garwood, (D), Washington, 1976, Op
Albert J. Semtner, (D), Princeton, 1973, Op
Edward B. Thornton, (D), Florida, 1970, On
Associate Professor:
Thomas H. Herbers, (D), California (San Diego), 1990, Op
Wieslaw Maslowski, (D), Alaska (Fairbanks), 1994, Op
Jeffrey D. Paduan, (D), Oregon State, 1987, Op
Leslie K. Rosenfeld, (D), MIT/Woods Hole, 1987, Op
Timothy P. Stanton, (M), Auckland (NZ), 1978, Op
Assistant Professor:
Pierre-Marie R.G. Poulain, (D), California (San Diego), 1989, Op
Lecturer:
Henry Jones, (M), Naval Postgrad Sch, 1986, Op
Adjunct Professor:
James R. Clynch, (D), Brown, 1974, Yd
Newell Garfield, (D), Rhode Island, 1990, Op
Ly N. Le, (D), State Hydromet Inst (St. Petersburg), 1977, Op
Julie L. McClean, (D), Old Dominion, 1993, Op
Robin T. Tokmakian, (D), Naval Postgrad Sch, 1997, Op
James H. Wilson, (D), Penn State, 1974, Op
Yuxia Zhang, (D), Texas A&M, 1994, Op
Emeritus:
Eugene C. Haderlie, (D), California (Berkeley), 1950, Ob
Glenn H. Jung, (D), Texas A&M, 1955, Op
Robert G. Paquette, (D), Washington, 1941, Op
Warren C. Thompson, (D), Texas A&M, 1953, Op
Eugene D. Traganza, (D), Miami, 1966, Oc
Stevens P. Tucker, (D), Oregon State, 1972, Og
Joseph J. Von Schwind, (D), Texas A&M, 1968, Op
Jacob B. Wickham, (M), Scripps, 1949, Op

Occidental College

Dept of Geology (B) (2006)
1600 Campus Road
Los Angeles, CA 90041
p. (323) 259-2823 f. (323) 341-4858
jangarcia@oxy.edu
http://www.oxy.edu/
Administrative Assistant: Jan Garcia
Enrollment (2005): B: 9 (7)

Professor:
Scott W. Bogue, (D), California (Santa Cruz), 1982, Ym
Donald R. Prothero, (D), Columbia, 1982, Pv
Margaret E. Rusmore, (D), Washington, 1985, Gc
James L. Sadd, (D), South Carolina, 1987, Oi
Associate Professor:
James A. Woodhead, (D), Princeton, 1977, Gz

Ohlone College

Dept of Geology (A) (2004)
43600 Mission Boulevard
Fremont, CA 94539
p. (415) 659-6225
pbelasky@ohlone.edu

Acting Head:
George P. Rodgers, (M), San Francisco, 1990, Gg
Associate Professor:
Paul Belasky, (D), California (Los Angeles), 1994, Pq

Orange Coast College

Div of Mathematics & Science (A) (2007)
2701 Fairview Road
Box 5005
Costa Mesa, CA 92626-5005
p. (714) 432-5647 f. (714) 432-5960
ebender@occ.cccd.edu
Enrollment (2004): A: 18 (0)

Chair:
Edward E. Bender, (D), Southern California, 1994, Giz
Professor:
Jan R. Yett, (M), Washington, 1978, Pm
Instructor:
Beverly J. Berekian, (M), Cal State (Fullerton), 2007, Gzi
Gregory V. Bishop, (M), Cal State (Los Angeles), 2000, OsHwOy

Palomar College

Dept of Earth Sciences (A) (2006)
1140 West Mission Road
San Marcos, CA 92069
p. (760) 744-1150 (Ext. 2512) f. (760) 744-8123
pdeen@palomar.edu
http://www.palomar.edu/earthscience/
Department Secretary: Brenda Morris

Professor:
Steven G. Spear, (D), California (Riverside), 1986, Gg
Associate Professor:
Patricia A. Deen, (M), San Diego State, 1984, Og
Lisa Dubois, (D), Brown, 1996, Og
Alan P. Trujillo, (M), N Arizona, 1984, Og

Pasadena City College

Dept of Physical Science (Geology) (A) (2005)
1570 E. Colorado Boulevard
Pasadena, CA 91106
p. (626) 585-7138 f. (626) 585-7160
dacantarero@paccd.cc.ca.us
Enrollment (1999): A: 7 (0)

Chair:
David N. Douglass, (D), Dartmouth, 1987, Ym
Instructor:
Martha House, (D), MIT, Gg
Gerald L. Lewis, (M), Cal State (Long Beach), 1965, Ps
Yuet-Ling O'Connor, (D), S California, Gg
Geology Lab Tech:
Debra A. Cantarero, (A), Pasadena City Coll, 2000, Oo

Pomona College

Geology Dept (B) (2006)
609 N. College Avenue
Claremont, CA 91711-6339
p. (909) 621-8675 f. (909) 621-8552
rhazlett@pomona.edu
http://www.geology.pomona.edu
Administrative Assistant: Lori Keala
Enrollment (2004): B: 15 (6)

Chair:
Linda A. Reinen, (D), Brown, 1993, Gc
Professor:
Richard W. Hazlett, (D), S California, 1986, Gv
Associate Professor:
Eric B. Grosfils, (D), Brown, 1995, Xg
Assistant Professor:
Robert R. Gaines, (D), California (Riverside), 2003, Gs
Emeritus:
Donald H. Zenger, (D), Cornell, 1962, Gr
Geology Librarian:
Brian Ebersole, (N), Oo

Rio Hondo College

Dept of Geology (2007)
3600 Workman Mill Rd.
Whittier, CA 90601-1699
p. (310) 692-0921 (Ext. 3564) f. (310) 699-0480
mforrest@riohondo.edu
Enrollment: No data reported since 1995

Chair:
Michael R. Forrest, (D), S California, 1997, Ys

Riverside Community College

Dept of Physical Science (Geological Sciences) (A) (2005)
4800 Magnolia Avenue
Riverside, CA 92506-1299
p. (909) 222-8350
sblair@rccd.cc.ca.us

Chair:
James M. Mehegan, (D), Dalhousie, 1988, Gi

Saddleback Community College

Dept of Astronomy-Geosciences-Physics (A) (2004)
28000 Marguerite Parkway
Mission Viejo, CA 92692
p. (714) 582-4820 f. (949) 347-1936
mamelotte@saddleback.cc.ca.us
Enrollment: No data reported since 1995

Professor:
Peter E. Borella, (D), S California, 1975, Gu
Associate Professor:
James L. Repka, (D), California (Santa Cruz), 1998, Gm

San Bernardino County Museum

Geological Sciences Div (2005)
2024 Orange Tree Lane
Redlands, CA 92374
p. (909) 307-2669 f. (909) 307-0539
jsagebiel@sbcm.sbcounty.gov

Senior Curator:
Kathleen B. Springer, (M), California (Riverside), Og
Curator:
J. Chris Sagebiel, (M), Texas, 1998, Pv

San Bernardino Valley College

Dept of Geology (A) (2005)
San Bernardino, CA 92410
p. (909) 888-6511 (Ext. 1453) f. (909) 888-1267
swilson@craftonhills.edu

Head:
Jerry D. Horne, (M), Houston, 1961, Ge
Adjunct Professor:
Donald G. Buchanan, (M), Naval Postgrad Sch, 1975, Og
Emeritus:
Floyd J. Williams, (D), Columbia, 1958, Gc

San Diego State University

Dept of Geological Sciences (B,M) (2007)
5500 Campanile Drive
San Diego, CA 92182-1020
p. (619) 594-5586 f. (619) 594-4372
department.office@geology.sdsu.edu
http://www.geology.sdsu.edu
Administrative Assistant: Pia Parrish
Enrollment (2007): B: 51 (12) M: 30 (10)

Professor:
Steven M. Day, (D), California (San Diego), 1977, Ys
Gary H. Girty, (D), Columbia, 1983, Gc
David L. Kimbrough, (D), California (Santa Barbara), 1982, Cc
Gary L. Peterson, (D), Washington, 1963, Gr
Thomas K. Rockwell, (D), California (Santa Barbara), 1983, Gm
Senior Scientist:
Barry B. Hanan, (D), Virginia Tech, 1980, Cc
Robert J. Mellors, (D), Indiana, 1995, Ye
Co-Director, Ctr IT & Infrastructure:
Eric G. Frost, (D), S California, 1983, Gc
Associate Professor:
Lindsey Leighton, (D), Michigan, 1999, Pe
Kim B. Olsen, (D), Utah, 1994, Ye
Kathryn W. Thorbjarnarson,(D),California (Los Angeles),1990, Hq
Assistant Professor:
Jared R. Morrow, (D), Colorado, 1997, Gs
Aaron Pietruszka, (D), Hawaii, 2003, Ct
Stephen Schellenberg, (D), S California, 2000
Lecturer:
Victor E. Camp, (D), Washington State, 1976, Gv
Damon P. DeYoung, (M), California (Riverside), 2005, Gg
Kevin Robinson, (M), San Diego State, 1996, Gc
Isabelle Sacramentogrilo, (M), San Diego State, 1999, Gg
Adjunct Professor:
Harold W. Magistrale, (D), Caltech, 1990, Ys
Emeritus:
Patrick L. Abbott, (D), Texas, 1973, Gs
Richard W. Berry, (D), Washington (St. Louis), 1963, Gz
Kathe K. Bertine, (D), Yale, 1970, Cl
Clive E. Dorman, (D), Oregon State, 1974, Op
R. Gordon Gastil, (D), California (Berkeley), 1954, Gt
David Huntley, (D), Colorado Mines, 1976, Hw
George R. Jiracek, (D), California (Berkeley), 1972, Ye
J. Philip Kern, (D), California (Los Angeles), 1968, Pg
Daniel Krummenacher, (D), Geneva, 1959, Cc
Claude Monte Marshall, (D), Stanford, 1971, Ym
Richard H. Miller, (D), California (Los Angeles), 1975, Ps
Anton D. Ptacek, (D), Washington (St. Louis), 1965, Gx

Richard L. Threet, (D), Washington, 1952, Gm
Michael J. Walawender, (D), Penn State, 1972, Gi

San Francisco State University

Dept of Geosciences (B,M) (2004)
1600 Holloway Avenue
San Francisco, CA 94132
p. (415) 338-2061 f. (415) 338-7705
geosci@sfsu.edu
http://tornado.sfsu.edu
Enrollment (2004): B: 37 (3) M: 20 (6)

Chair:
Oswaldo Garcia, (D), SUNY (Albany), 1976, Oa
Professor:
Charles E. Bickel, (D), Harvard, 1971, Gx
David P. Dempsey, (D), Washington, 1985, Oa
Karen Grove, (D), Stanford, 1989, Gs
John P. Monteverdi, (D), California (Berkeley), 1977, Oa
David A. Mustart, (D), Stanford, 1972, Gg
Raymond Pestrong, (D), Stanford, 1965, Ng
Lisa D. White, (D), California (Santa Cruz), 1989, Pg
Associate Professor:
John Caskey, (D), Nevada (Reno), 1996, Gc
Newell (Toby) Garfield, (D), Rhode Island, 1990, Op
Assistant Professor:
Matthew J. La Force, (D), Idaho, 1999, Hg
Leonard Sklar, (D), California (Berkeley), 2003, Gm
Adjunct Professor:
E. Jan Null, (D), California (Davis), 1974, Oa
Ted Roush, (D), Hawaii, 1987, Yg
Emeritus:
York T. Mandra, (D), Stanford, 1958, Pg
Erwin Seibel, (D), Michigan, 1972, On
Raymond Sullivan, (D), Glasgow, 1960, Go
Geosciences Tech:
Bruce Davies, Oo

San Jose City College

Dept of Physical Science (2005)
San Jose, CA 95128
p. (408) 288-3716 f. (408) 275-9386
leandra.martin@sjcc.edu
Department Secretary: Marlene Underwood

Director:
D. James Samuelson, (M), California (Santa Barbara), 1969, Oo
Instructor:
Judy Mariant, (M), San Jose State, 1994, Gg
Emeritus:
John W. Martin, (M), San Jose State, 1959, Og

San Jose State University

Dept of Geology (B,M) (2007)
One Washington Square
San Jose, CA 95192-0102
p. (408) 924-5050 f. (408) 924-5053
andersen@geosun.sjsu.edu
http://www.geosun.sjsu.edu/
Enrollment (2007): B: 29 (10) M: 26 (6)

Chair:
Richard L. Sedlock, (D), Stanford, 1988, GtOe
Professor:
David W. Andersen, (D), Utah, 1973, Gs
Ellen P. Metzger, (D), Syracuse, 1984, Gp
Robert B. Miller, (D), Washington, 1980, Gc
June A. Oberdorfer, (D), Hawaii, 1983, Hw
Donald L. Reed, (D), California (San Diego), 1985, Gu
John W. Williams, (D), Stanford, 1970, Ng
Associate Professor:
Paula Messina, (D), CUNY, 1998, Oe
Jonathan S. Miller, (D), North Carolina, 1994, Gi
Assistant Professor:
Emmanuel Gabet, (D), California (Santa Barbara), 2002, Gm
Emeritus:
Calvin H. Stevens, (D), S California, 1963, Ps

Santa Ana College

Dept of Physical Science (A) (2004)
Santa Ana, CA 92706
p. (714) 564-6610 f. (714) 564-6379
hughes_phil@sac.edu
Enrollment: No data reported since 1995

Chair:
Philip Hughes, (M), Ohio State, 1986, Gs
Instructor:
Richard Greenwood, (D), S California, Gg
William Phelps, (D), California (Riverside), 2004, Pg
Stephen I. Wareham, (M), Loma Linda, 1986, Go
Adjunct Professor:
Jerusha Dahm, (M), Colorado State, Gg
Thomas Norton, (N), Go

Santa Barbara City College

Dept of Earth & Planetary Sciences (A) (2008)
721 Cliff Drive
Santa Barbara, CA 93109
p. (805) 965-0581 (Ext. 2315)
schultz@sbcc.edu
Enrollment (2000): A: 47 (12)

Chair:
Robert S. Gray, (D), Arizona, 1965, Gs
Professor:
Frederic Marschak, (M), California (Santa Barbara), 1978, Xg
Jan Schultz, (M), California (Santa Barbara), 1991, Gge
Associate Professor:
Carl Sundbeck, (M), California (Santa Barbara), 1982, Oy
Lab Supervisor:
William H. Harz, (B), Whittier, 1972, Hg
Associate Scientist:
Jan Dependahl, (B), California (Santa Barbara), 1977, Oe
Assistant Professor:
Jeff Meyer, (D), California (Santa Barbara), 1992, Gg
Erin O'Connor, (M), Xg
Related Staff:
Naomi E. Sullwold, Oe

Santiago Canyon College

Dept of Physical Sciences (A) (2004)
8045 East Chapman Avenue
Orange, CA 92869-4512
p. (714) 564-4788

Chair:
Debra A. Brooks, (M), Texas A&M, 1989, Yg
Adjunct Professor:
Gail F. Montwill, (M), Cal State (Los Angeles), 1988, Gg
Lizanne V. Simmons, (B), Brigham Young, 1979, Gr
Amy L. Stinson, (M), San Diego State, 1990, Gc
Adam D. Woods, (D), S California, 1998, Gs

Shasta Community College

Science, Industry & Natural Resources (A) (2006)
11555 Old Oregon Trail
PO Box 496006
Redding, CA 96049-6006
p. (530) 225-4660 f. (530) 225-4823

rreed@shastacollege.edu
Enrollment: No data reported since 1995
Instructor:
Randy Reed, (M), Geology, Ge

Sonoma State University

Dept of Geology (B) (2006)
1801 East Cotati Avenue
Rohnert Park, CA 94928
p. (707) 664-2334 f. (707) 664-3012
matt.james@sonoma.edu
http://www.sonoma.edu/geology/
Administrative Assistant: Gayle Walker
Enrollment (2006): B: 35 (12)

Chair:
Matthew J. James, (D), California (Berkeley), 1987, Pi
Professor:
Thomas B. Anderson, (D), Colorado, 1969, Gs
Walt Vennum, (D), Stanford, 1971, Gi
Associate Professor:
Daniel B. Karner, (D), California (Berkeley), 1997, Cc
Assistant Professor:
Matty Mookerjee, (D), Rochester, 2005, Gc
Michael E. Smith, (D), Wisconsin, 2006, Gd
Emeritus:
Rolfe C. Erickson, (D), Arizona, 1968, Gx
William H. Wright, (D), Illinois, 1970, Gc
Equipment Tech:
Sue Nosker, (M), Nevada (Reno), 1981, Oo

Southwestern College

Math/Science Div (2004)
Chula Vista, CA 92010
p. (619) 421-6700 (Ext. 290) f. (619) 421-1189
ktyner@swc.cc.ca.us

Professor:
Wilbur J. Switzer, (M), Cal State (Long Beach), 1961, Og

Stanford University

Dept of Geological and Environmental Sci (B,M,D) (2007)
Stanford, CA 94305-2115
p. (650) 723-0847 f. (650) 725-0979
ges-dept@pangea.stanford.edu
http://pangea.stanford.edu/GES/GES.html
Administrative Assistant: Elaine Andersen
Enrollment (2006): B: 21 (0) M: 27 (0) D: 40 (0)

Chair:
Jonathan F. Stebbins, (D), California (Berkeley), 1983, Cg
Research Professor:
J. Michael Moldowan, (D), Michigan, 1972, Co
Professor:
Atilla Aydin, (D), Stanford, 1978, Gc
Dennis K. Bird, (D), California (Berkeley), 1978, Cg
Gordon E. Brown, Jr., (D), Virginia Tech, 1970, Gz
C. Page Chamberlain, Yr
Robert B. Dunbar, (D), California (San Diego), 1981, On
Marco T. Einaudi, (D), Harvard, 1969, Em
W. Gary Ernst, (D), Johns Hopkins, 1959, Cp
Steven M. Gorelick, (D), Stanford, 1981, Hw
Stephan A. Graham, (D), Stanford, 1976, Go
James C. Ingle, Jr., (D), S California, 1966, Pm
Andre G. Journel, (D), Nancy (France), 1977, Gq
Juhn G. Liou, (D), California (Los Angeles), 1970, Gp
Keith Loague, (D), British Columbia, 1986, Hq
Donald R. Lowe, (D), Illinois, 1967, Gd
Gail A. Mahood, (D), California (Berkeley), 1979, Gi
Pamela A. Matson, (D), Oregon State, 1983, Sb
Michael O. McWilliams, (D), Australian National, 1978, Ym
Elizabeth L. Miller, (D), Rice, 1977, Gc
David D. Pollard, (D), Stanford, 1969, Gc
Paul Switzer, (D), Harvard, 1965, Gq
Associate Professor:
Christopher F. Chyba, (D), Cornell, 1991, Gh
Assistant Professor:
Scott E. Fendorf, (D), Delaware, 1992, Sc
Adina Paytan, (D), California (San Diego), 1996, Oc
Consulting Professor:
Richard Bernknopf, (D), George Washington, 1980, Oo
H. Edward Clifton, Yr
Alan K. Cooper, Yr
Francois Farges, Yr
Paul Hsieh, Yr
Consulting Associate Professor:
Trevor Dumitru, (D), Melbourne (Australia), 1990, Gc
Thomas Holzer, (D), Stanford, 1970, Ng
Joseph W. Ruetz, (M), Stanford, 1977, Gg
Joseph Wooden, (D), North Carolina, 1975, Cg
Lecturer:
Karen Seto, Yr
Courtesy Professor:
Peter G. Brewer, (D), Liverpool, 1967, Oc
David A. Clague, (D), California (San Diego), 1974, Ob
Peter K. Kitanidis, Yr
James O. Leckie, (D), Harvard, 1970, Cl
Stephen G. Monismith, Yr
Charles K. Paul, Yr
Courtesy Associate Professor:
James P. Barry, Yr
Ronaldo I. Borja, Yr
Francisco P. Chavez, Yr
David L. Freyberg, Yr
Simon L. Klemperer, (D), Cornell, 1985, Yr
Anders Nilsson, Yr
Alfred Spormann, Yr
Debra S. Stakes, (D), Oregon State, 1978, Ob
Courtesy Assistant Professor:
Kevin Arrigo, (D), S California, 1992, Og
Visiting Associate Professor:
Chunmiao Zheng, Yr
Visiting Professor:
Peter Barrett, Yr
Dennis L. Corwin, Yr
Shige Maruyama, Yr
Emeritus:
Robert G. Coleman, (D), Stanford, 1957, Gi
Robert R. Compton, (D), Stanford, 1949, Gx
William R. Evitt, Yr
John W. Harbaugh, (D), Wisconsin, 1955, Gq
Konrad B. Krauskopf, Yr
Ronald J. P. Lyon, (D), California (Berkeley), 1954, Or
George A. Parks, (D), MIT, 1960, Cl
Irwin Remson, (D), Columbia, 1954, Hw
Tjeerd H. Van Andel, (D), Groningen (Neth), 1950, Gu

Dept of Geophysics (B,M,D) (2007)
Panama Street & Lomita Mall
Stanford, CA 94305-2215
p. (650) 723-4746 f. (650) 725-7344
lmnelson@stanford.edu
Administrative Assistant: Lauren Nelson
Enrollment (2006): M: 20 (0) D: 21 (0)

Chair:
Rosemary J. Knight, (D), Stanford, 1985, Hy
Professor:
Gregory C. Beroza, (D), MIT, 1989, Ys

Jon F. Claerbout, (D), MIT, 1967, Ye
Steven Gorelick, Yr
Jerry M. Harris, (D), Caltech, 1980, Ys
Simon L. Klemperer, (D), Cornell, 1985, Gt
Gerald M. Mavko, (D), Stanford, 1977, Yg
Marcia K. McNutt, (D), California (San Diego), 1978, Yr
Amos M. Nur, (D), MIT, 1969, Yg
Joan Roughgarden, (D), Harvard, 1971, Ob
Paul Segall, (D), Stanford, 1981, Yg
Norman H. Sleep, (D), MIT, 1972, Yg
Mark D. Zoback, (D), Stanford, 1975, Yg

Senior Scientist:
Jack Dvorkin, (D), Moscow, 1980, Yg

Associate Professor:
Kevin Arrigo, (D), S California, 1992, Yr
Biondo L. Biondi, (D), Stanford, 1990, Ye
Azadeh Tabazadeh, (D), California (Los Angeles), 1994, Oa
Howard A. Zebker, (D), Stanford, 1984, Yx

Assistant Professor:
Jesse F. Lawrence, (D), St. Louis, 2004, Oo

Consulting Faculty:
James G. Berryman, (D), Wisconsin, 1975, Ye
David Scholl, (D), Stanford, 1962, Yr

Adjunct Professor:
William Ellsworth, (D), MIT, 1977, Yg
Stewart A. Levin, (D), Stanford, 1987, Yg
Walter Mooney, (D), Wisconsin, 1978, Ys

Emeritus:
Antony C. Fraser-Smith, (D), Auckland, 1966, Ym
Robert L. Kovach, (D), Caltech, 1962, Ys
George A. Thompson, (D), Stanford, 1949, Yg

Geology Librarian:
Charlotte Derksen, Oo

University of California, Berkeley

Dept of Earth & Planetary Science (B,M,D) (2007)
307 McCone
Berkeley, CA 94720-4767
p. (510) 642-3993 f. (510) 643-9980
wwwgeo@eps.berkeley.edu
http://eps.berkeley.edu
Enrollment (2002): B: 43 (18) M: 1 (1) D: 6 (6)

Chair:
Hans-Rudolf Wenk, (D), Zurich, 1965, Gz

Professor:
Walter Alvarez, (D), Princeton, 1967, Grh
Jillian Banfield, (D), Johns Hopkins, 1990, CoGz
William B. N. Berry, (D), Yale, 1957, OuGe
James K. Bishop, (D), MIT/Woods Hole, 1977, Oc
George H. Brimhall, (D), California (Berkeley), 1972, Eg
Roland Burgmann, (D), Stanford, 1993, Gt
Donald J. DePaolo, (D), Caltech, 1978, Cc
William E. Dietrich, (D), Washington, 1982, Gm
Inez Fung, (D), MIT, 1977, Oa
B. Lynn Ingram, (D), Stanford, 1992, Cs
Raymond Jeanloz, (D), Caltech, 1979, Gy
James W. Kirchner, (D), California (Berkeley), 1990, Ge
Michael Manga, (D), Harvard, 1994, GvXy
Mark A. Richards, (D), Caltech, 1986, Yg
Barbara A. Romanowicz, (D), Paris, 1979, Ys

Associate Professor:
Kristie Boering, (D), Stanford, 1991, Oa
Douglas S. Dreger, (D), Caltech, 1992, Ys

Assistant Professor:
Richard M. Allen, (D), Princeton, 2001, Ys
Burkhard Militzer, (D), Illinois (Champaign-Urbana), 2000, Gy

Adjunct Associate Professor:
David L. Alumbaugh, (D), California (Berkeley), 1993, Ye
Paul Renne, (D), California (Berkeley), 1987, Cc

Adjunct Professor:
Steven Pride, (D), Texas A&M, 1991, Ys

Visiting Professor:
William D. Collins, (D), Chicago, 1988, Oa

Emeritus:
Mark S. Bukowinski, (D), California (Los Angeles), 1975, Gy
Ian S. Carmichael, (D), London, 1958, Gi
Garniss H. Curtis, (D), California (Berkeley), 1951, Cc
Lane R. Johnson, (D), Caltech, 1966, Ys
Chi-Yuen Wang, (D), Harvard, 1964, Yg

Related Staff:
Doris Sloan, (D), California (Berkeley), 1981, Ge

Cooperating Faculty:
Fatemah Van Buren, Oo

Dept of Engineering Geoscience (B,M,D) (2004)
Berkeley, CA 94720
p. (510) 642-3157 f. (510) 642-3805
hfmorrison@berkeley.edu
Enrollment (1997): M: 5 (2) D: 7 (0)

Professor:
Alex Becker, (D), McGill, 1964, Ye
Huntly Frank Morrison, (D), California (Berkeley), 1967, Ye

Senior Scientist:
Ki Ha Lee, (D), California (Berkeley), 1978, Ye

Assistant Professor:
James W. Rector, (D), Stanford, 1990, Ye

Dept of Integrative Biology (B,M,D) (2004)
Berkeley, CA 94720
p. (510) 642-4958 f. (510) 643-6264
Administrative Assistant: Derek Maskell
Administrative Assistant: Mona Radice
Enrollment: No data reported since 1995

Vice Chair:
Robert J. Full, (D), SUNY (Buffalo), 1984, Pi

Professor:
Roy L. Caldwell, (D), Iowa, 1969, Pi
F. Stuart Chapin, (D), Stanford, 1973, Pe
William A. Clemens, (D), California (Berkeley), 1960, Pv
Harry W. Green, (D), Tennessee, 1977, Pv
Carole S. Hickman, (D), Stanford, 1975, Po
Ned K. Johnson, (D), California (Berkeley), 1961, Pv
Mimi A. R. Koehl, (D), Duke, 1976, Po
David R. Lindberg, (D), California (Santa Cruz), Pi
Jere H. Lipps, (D), California (Los Angeles), 1966, Po
Brent Mishler, (D), Harvard, Pe
James L. Patton, (D), Arizona, 1968, Pv
Thomas M. Powell, (D), California (Berkeley), 1970, Op
Montgomery Slatkin, (D), Harvard, 1970, Po
Glennis Thompson, (D), Melbourne, 1974, Oo
David B. Wake, (D), S California, 1964, Po
Marvalee H. Wake, (D), S California, 1968, Po

Associate Professor:
Anthony D. Barnosky, (D), Washington, 1983, Pv
Thomas Duncan, (D), Michigan, 1976, Pb
Kevin Padian, (D), Yale, 1980, Pv
Mary E. Power, (D), Washington, 1981, Pe
Wayne P. Sousa, (D), California (Santa Barbara), 1977, Ob

Assistant Professor:
Nan Arens, (D), Harvard, Pb
Carla D'Antonio, (D), California (Santa Barbara), 1990, Pe
Deborah L. Penry, (D), Washington, 1988, Ob

Adjunct Professor:
Donald Weston, (D), William & Mary, 1983, Ob

Emeritus:
George W. Barlow, (D), California (Los Angeles), 1958, Ob

Joseph Gregory, California (Berkeley), 1949, Pv
Howard J. Hutchison, (D), California (Berkeley), Pv
William Z. Lidicker, Jr., (D), Illinois, 1957, Pv
Robert Ornduff, (D), California (Berkeley), 1961, Pb
Thelma Rowell, (D), Cambridge, 1959, Pv
Donald Savage, California (Berkeley), 1938, Pv
James W. Valentine, (D), California (Los Angeles), 1958, Po

Librarian:
Vivienne Roumani-Denn, Oo

Dept of Materials Sci & Mineral Eng (Earth Res) (B,M,D) (2004)
577 Evans Hall
Berkeley, CA 94720-1760
p. (510) 642-3801 f. (510) 643-5792
hfmorrison@berkeley.edu
Enrollment: No data reported since 1995

Professor:
Alex Becker, (D), McGill, 1963, Yg
George A. Cooper, (D), Cambridge, 1967, Np
Didier deFontaine, (D), Northwestern, 1967, Om
Lutgard DeJonghe, (D), California (Berkeley), 1970, Om
Thomas M. Devine, (D), MIT, 1974, Om
Fiona M. Doyle, (D), Imperial Coll, 1983, Nx
James W. Evans, (D), SUNY (Buffalo), 1970, Nx
Douglas W. Fuerstenau, (D), MIT, 1953, Nx
Andreas Glaeser, (D), MIT, 1981, Om
Ronald Gronsky, (D), California (Berkeley), 1977, Om
Eugene Haller, (D), Basel, 1970, Om
J. W. Morris, Jr., (D), MIT, 1969, Om
H. Frank Morrison, (D), California (Berkeley), 1967, Yg
T. N. Narasimhan, (D), California (Berkeley), 1975, Hw
Robert O. Ritchie, (D), Cambridge, 1973, Om
Timothy Sands, (D), California (Berkeley), 1984, Om
Kalanadh V. S. Sastry, (D), California (Berkeley), 1970, Nx
Eicke Weber, (D), Cologne, 1976, Om

Associate Professor:
Tad W. Patzek, (D), Silesian Tech, 1979, Np
James W. Rector, III, (D), Stanford, 1990, Ys

Assistant Professor:
Daryl Chrzan, (D), California (Berkeley), 1989, Om
Mauro Ferrari, (D), California (Berkeley), 1989, Om

Museum of Paleontology (2006)
1101 Valley Life Sciences Bldg
Berkeley, CA 94720
p. (510) 642-1821 f. (510) 642-1822
rlcaldwell@berkeley.edu
http://www.ucmp.berkeley.edu
Enrollment (1995): B: 2 (2) D: 4 (4)

Director:
Roy Caldwell, (D), Iowa, Po

Curator:
Walter Alvarez, (D), Princeton, 1967, Gr
William B. Berry, (D), Yale, 1957, Ps
Roger Byrne, (D), Wisconsin, 1972, Pl
Carole S. Hickman, (D), Stanford, 1975, Po
Lynn Ingram, (D), Stanford, Po
David R. Lindberg, (D), California (Santa Cruz), 1983, Po
Jere H. Lipps, (D), California (Los Angeles), 1966, Po
Kevin Padian, (D), Yale, 1980, Pv
Tim White, (D), Michigan, Pv

Principal Museum Scientist:
Mark B. Goodwin, (M), California (Berkeley), 1988, Pv

Museum Scientist:
Diane Erwin, (D), Alberta, 1990, Pb
Pat Holroyd, (D), Duke, 1994, Pv

Education & Public Programs:
Judith Scotchmoor, (B), California (Berkeley), Pg

Associate Curator:
Anthony D. Barnosky, (D), Washington, 1983, Pv

Curator:
William A. Clemens, (D), California (Berkeley), 1960, Pv

Emeritus:
Zach M. Arnold, (D), California, 1953, Po
Joseph T. Gregory, (D), California, 1938, Pv
Clarke Howell, (D), Chicago, Oo
J. Howard Hutchison, (D), California (Berkeley), 1976, Pv
Howard E. Schorn, (M), California (Berkeley), 1966, Pb
James W. Valentine, (D), California (Los Angeles), 1958, Po

Preparator:
Jane Mason, Oo

Cooperating Faculty:
Bruce Tiffney, (D), Harvard, 1977, Pb
Gareet Vermeij, (D), Yale, 1971, Po

University of California, Davis

Dept of Geology (B,M,D) (2007)
174 PHY/GEO Bldg
One Shields Ave.
Davis, CA 95616-5270
p. (530) 752-0350 f. (530) 752-0951
geology@ucdavis.edu
http://www.geology.ucdavis.edu
Enrollment (2006): B: 53 (15) M: 14 (4) D: 30 (4)

Chair:
Louise H. Kellogg, (D), Cornell, 1988, Yg

Professor:
Sandra J. Carlson, (D), Michigan, 1986, Po
Howard W. Day, (D), Brown, 1971, Gp
Charles E. Lesher, (D), Harvard, 1985, Gi
James S. McClain, (D), Washington, 1979, Yr
Isabel P. Montañez, (D), Virginia Tech, 1990, Gd
Jeffrey F. Mount, (D), California (Santa Cruz), 1980, Gm
John Rundle, (D), California (Los Angeles), 1976, Yd
James R. Rustad, (D), Minnesota, 1992, Cl
Peter Schiffman, (D), Stanford, 1978, Gp
Howard J. Spero, (D), California (Santa Barbara), 1986, Pe
Dawn Y. Sumner, (D), MIT, 1995, Gs
Donald L. Turcotte, (D), Caltech, 1958, Yg
Geerat J. Vermeij, (D), Yale, 1971, Pe
Kenneth L. Verosub, (D), Stanford, 1973, Ym
Robert A. Zierenberg, (D), Wisconsin, 1983, Cs

Associate Professor:
Ryosuke Motani, (D), Toronto, 1997, Pv

Assistant Professor:
Magali I. Billen, (D), Caltech, 2001, Yg
Kari M. Cooper, (D), California (Los Angeles), 2001, Gi
Eric S. Cowgill, (D), California (Los Angeles), 2001, Gc
Tessa M. Hill, (D), California (Santa Barbara), 2004, Pe
Qing-zhu Yin, (D), Max Planck, 1995, Cc

Research Associate:
Gary Acton, (D), Northwestern, 1990, Ym
Irina Delusina, (D), Tallinn Inst of Geology (Estonia), 1989, Pl
Joan L. Florsheim, (D), California (Santa Barbara), 1988, Gm
Sarah M. Roeske, (D), California (Santa Cruz), 1988, Gc
Ann D. Russell, (D), Washington, 1994, Cm
Peter Thy, (D), Aarhus (Denmark), 1982, Gx
Jeffrey R. Unruh, (D), California (Davis), 1990, Gc

Lecturer:
David A. Osleger, (D), Virginia Tech, 1990, Gr

Emeritus:
Richard Cowen, (D), Cambridge, 1966, Po
John Dewey, (D), London (UK), 1960, Gt
Charles G. Higgins, (D), California (Berkeley), 1950, Gm
Eldridge M. Moores, (D), Princeton, 63, Gt
Robert J. Twiss, (D), Princeton, 1970, Gc

Cooperating Faculty:
William H. Casey, (D), Penn State, 1985, Cl
James A. Doyle, (D), Harvard, 1970, Pb
Graham E. Fogg, (D), Texas, 1986, Hw
Alexandra Navrotsky, (D), Chicago, 1967, Om
Librarian:
Carol J. La Russa, (M), California (Los Angeles), 1978, Oo

Dept of Land, Air & Water Resources (B,M,D) (2004)
1110 Plant and Environmental Sciences building
One Shields Avenue
Davis, CA 95616
p. (916) 752-1406 f. (916) 752-1552
jwhopmans@ucdavis.edu
http://lawr.ucdavis.edu/
Department Manager: Sandra Filby
Enrollment (1999): B: 158 (19) M: 30 (17) D: 58 (5)
Chair:
Michael J. Singer, (D), Minnesota, 1972, So
Professor:
Caroline S. Bledsoe, (D), Colorado State, 1970, So
William H. Casey, (D), Penn State, 1985, Cl
Randy A. Dahlgren, (D), Washington, 1987, So
Graham E. Fogg, (D), Texas, 1986, Hw
Mark E. Grismer, (D), Colorado State, 1971, Hw
James H. Richards, (D), Alberta, 1981, So
Dennis E. Rolston, (D), California (Davis), 1971, So
Randal J. Southard, (D), North Carolina State, 1983, So
Associate Professor:
William R. Horwath, (D), Michigan State, 1993, So
Kate Scow, (D), Cornell, 1988, So
Susan Ustin, (D), California (Davis), 1983, Or
Emeritus:
James W. Biggar, (D), Utah State, 1956, So
Richard G. Burau, (D), Minnesota, 1963, So
Kenneth K. Tanji, (D), Kyoto, 1997, So

Dept of Land, Air, & Water Resources - Hydrology Prog (B,M,D) (2004)
113 Veihmeyer Hall
One Shields Avenue
Davis, CA 95616-8628
p. (530) 752-3060 f. (530) 752-5262
jwhopmans@ucdavis.edu
http://lawr.ucdavis.edu/
Chair:
Jan W. Hopmans, (D), Auburn, 1985, Sp
Professor:
William H. Casey, (D), Penn State, 1985, Sc
Graham E. Fogg, (D), Texas, 1986, Hy
Mark E. Grismer, (D), Colorado State, 1984, Sp
Theodore C. Hsiao, (D), Illinois, 1963, Sb
Miguel A. Marino, (D), California (Los Angeles), 1972, Hw
Susan L. Ustin, (D), California (Davis), 1983, Or
Wesley W. Wallender, (D), Utah State, 1982, Hg
Water Management Specialist:
Terry L. Prichard, (D), California (Davis), 1975, Oo
Plant-Water Relations Specialist:
Stephen R. Grattan, (D), California (Riverside), 1984, Oo
Irrigation Specialist:
Larry J. Schwankl, (D), California (Davis), 1991, Oo
Irrigation & Soil Specialist:
David A. Goldhamer, (D), California (Davis), 1980, Oo
Irrigation & Drainage Specialist:
Blaine R. Hanson, (D), Colorado State, 1977, Hg
Groundwater Hydrology Specialist:
Thomas L. Harter, (D), Arizona, 1993, Hw

Associate Professor:
Gregory B. Pasternack, (D), Johns Hopkins, 1998, Sp
Carlos E. Puente, (D), MIT, 1984, Hq
Emeritus:
James W. Biggar, (D), Utah State, 1956, Sc
Robert M. Hagan, California (Davis), 1948, Hg
Allen W. Knight, (D), Utah, 1965, Hs
Donald R. Nielsen, (D), Iowa State, 1958, Sp
William O. Pruitt, Washington State, 1951, Ow
Verne H. Scott, Colorado State, 1959, Hs
Kenneth K. Tanji, (D), California (Davis), 1961, Sc

Graduate Group in Hydrologic Sciences (B,M,D) (2004)
113 Veihmeyer Hall
Davis, CA 95616
p. (530) 752-0453 f. (530) 752-5262
jwhopmans@ucdavis.edu
http://lawr.ucdavis.edu/hyd/
Administrative Assistant: Diane Swindall
Enrollment (1995): M: 13 (13) D: 1 (1)
Chair:
Graham E. Fogg, (D), Texas, 1986, Hy
Professor:
William H. Casey, (D), Pennsylvania, 1985, Cl
Jeannie Darby, (D), Texas, 1988, Hg
Mark E. Grismer, (D), Colorado State, 1984, Hw
Jan W. Hopmans, (D), Auburn, 1985, So
Theodore C. Hsiao, (D), Illinois, 1964, Oo
B. E. Larock, (D), Stanford, 1966, Oo
Jay R. Lund, (D), Washington, 1986, Hs
M. A. Marino, (D), California (Los Angeles), 1972, Hw
James S. McClain, (D), Washington, 1979, Yr
Eldridge M. Moores, (D), Princeton, 1963, Gt
Jeffrey F. Mount, (D), California (Santa Cruz), 1980, Gs
Dennis E. Rolston, (D), California (Davis), 1971, So
K. K. Tanji, (M), California (Davis), 1961, Oo
B. C. Weare, (D), SUNY (Buffalo), 1974, Oa
Associate Professor:
Bruce Kutter, (D), Cambridge, 1983, Ne
Carlos E. Puente, (D), MIT, 1984, Hg
Emeritus:
Robert A. Matthews, (B), California (Berkeley), 1953, Ge

University of California, Irvine

Dept of Earth System Science (B,D) (2007)
School of Physical Sciences
Irvine, CA 92697-3100
p. (949) 824-8794 f. (949) 824-3874
ejford@uci.edu
http://www.ess.uci.edu
Enrollment (2006): B: 67 (0) D: 31 (0)
Chair:
Susan E. Trumbore, (D), Columbia, 1989, Sc
Advance Professor:
Ellen R. M. Druffel, (D), California (San Diego), 1980, Oc
Professor:
Michael L. Goulden, (D), Stanford, 1992, Oo
Gudrun Magnusdottir, (D), Colorado State, 1989, Oa
Michael J. Prather, (D), Yale, 1975, Oa
William S. Reeburgh, (D), Johns Hopkins, 1967, Oo
Eric S. Saltzman, (D), Rosenstiel, 1986, Oa
Soroosh Sorooshian, (D), California (Los Angeles), 1978, Hq
Assistant Professor:
James T. Randerson, (D), Stanford, 1998, Oo
Associate Researcher:
Stanley C. Tyler, (D), California (Irvine), 1983, Oa

University of California, Los Angeles

Dept of Atmospheric and Oceanic Sciences (B,M,D) (2006)
7127 Math Sciences
Box 951565
Los Angeles, CA 90095
p. (310) 825-1217 f. (310) 206-5219
deptinfo@atmos.ucla.edu
http://www.atmos.ucla.edu

Chair:
Lawrence Lyons, (D), California (Los Angeles), 1972, Oa
Professor:
Michael Ghil, (D), Courant Inst, 1975, Oa
Kuo Nan Liou, (D), New York, 1970, Oa
James McWilliams, (D), Harvard, 1971, Oa
Carlos R. Mechoso, (D), Princeton, 1978, Oa
J. David Neelin, (D), Princeton, 1987, Oa
Richard M. Thorne, (D), MIT, 1968, Oa
Richard Turco, (D), Illinois, 1971, Oa
Associate Professor:
Robert Fovell, (D), Illinois, 1988, Oa
Assistant Professor:
Suzanne Paulson, (D), Caltech, 1991, Oa
Bjorn B. Stevens, (D), Colorado State, 1996, Oa
Jochen P. Stutz, (D), Heidelberg, 1996, Oa
Lecturer:
Jeffrey Lew, (D), California (Los Angeles), 1985, Oa
Emeritus:
Akio Arakawa, (D), Tokyo, 1961, Oa
Michio Yanai, (D), Tokyo, 1961, Oa

Dept of Earth & Space Sciences (B,M,D) (2007)
595 Charles Young Drive East
3806 Geology Building
Los Angeles, CA 90095-1567
p. (310) 825-3880 f. (310) 825-2779
chair@ess.ucla.edu
http://www.ess.ucla.edu
Enrollment (2001): B: 71 (11) M: 6 (3) D: 51 (9)

Chair:
David D. Jackson, (D), MIT, 1969, YsdYg
Director, Institute of Geophysics and Planetary Physics:
T. Mark Harrison, (D), Australian National, 1981, Cc
Director, Ctr Study Evolution & Origin of Life:
J. William Schopf, (D), Harvard, 1968, Po
Professor:
G. Peter Bird, (D), MIT, 1976, Yg
Paul M. Davis, (D), Queensland (Australia), 1974, YsmYg
Raymond V. Ingersoll, (D), Stanford, 1976, Gt
Vladimir Keilis-Borok, (D), Russian Acad of Sci, 1948, Ys
Margaret G. Kivelson, (D), Radcliffe, 1957, Xy
Craig E. Manning, (D), Stanford, 1989, Gp
Kevin D. McKeegan, (D), Washington (St. Louis), 1987, Xc
Robert L. McPherron, (D), California (Berkeley), 1968, Xy
Mark Moldwin, (D), Boston, 1993, Xy
William I. Newman, (D), Cornell, 1979, Xy
David A. Paige, (D), Caltech, 1985, Xy
Gilles Peltzer, (D), Paris, 1987, Gt
Bruce N. Runnegar, (D), Queensland (Australia), 1967, Po
Christopher T. Russell, (D), California (Los Angeles), 1968, Yg
Ray Walker, (D), California (Los Angeles), 1973, Xy
John T. Wasson, (D), MIT, 1958, Xm
An Yin, (D), S California, 1988, Gt
Edward Young, (D), S California, 1990, Cs
Associate Professor:
Vassilis Angelopoulos, (D), California (Los Angeles), 1993, XyOa
Jonathan M. Aurnou, (D), Johns Hopkins, 1999, XyYmg
Assistant Professor:
Caroline Beghein, (D), Utrecht, 2003, Ysg
Abby Kavner, (D), California (Berkeley), 1997, Yg
Edwin A. Schauble, (D), Caltech, 2002, Cg
Adjunct Professor:
Paul M. Merifield, (D), Colorado, 1963, Ng
Emeritus:
Orson L. Anderson, (D), Utah, 1951, Yx
Donald Carlisle, (D), Wisconsin, 1950, Eg
Paul J. Coleman, (D), California (Los Angeles), 1966, Xy
Wayne A. Dollase, (D), MIT, 1966, Gz
Clarence A. Hall, (D), Stanford, 1956, Gt
Isaac R. Kaplan, (D), S California, 1962, Cg
Art Montana, (D), Penn State, 1966, Cp
Gerhard Oertel, (D), Bonn, 1945, Gc
Walter E. Reed, (D), California (Berkeley), 1972, Gs
John L. Rosenfeld, (D), Harvard, 1954, Gp
Ronald Shreve, (D), Caltech, 1958, Gm

Inst of Geophysics & Planetary Physics (2007)
3845 Slichter Hall
Los Angeles, CA 90095-1567
p. (310)825-1664 f. (310)206-3051
kolwin@igpp.ucla.edu
www.igpp.ucla.edu

Director:
Michael Ghil, (D), New York, 1975, Oa
Professor:
Vassilis Angelopoulos
Maha Ashour-Abdalla, (D), Imperial Coll (UK), 1971, Oo
Friedrich H. Busse, (D), Munich, 1967, Oo
Richard E. Dickerson, (D), Minnesota, 1957, Oo
T. Mark Harrison, (D), Australian National, 1980, Cc
Charles F. Kennel, (D), Princeton, 1964, Oo
Margaret G. Kivelson, (D), Radcliffe, 1957, Xy
Robert L. McPherron, (D), California (Berkeley), 1968, Xy
James C. McWilliams, (D), Harvard, 1971, Oa
Paul H. Roberts, (D), Cambridge, 1967, Ym
Bruce Runnegar, (D), Queensland, 1967, Po
Christopher T. Russell, (D), California (Los Angeles), 1968, Yg
J. William Schopf, (D), Harvard, 1968, Po
Gerald Schubert, (D), California (Berkeley), 1964, Yg
Karl O. Stetter, (D), Tech (Munich), 1973, Po
Richard Turco, (D), Illinois, 1971, Oa
Raymond J. Walker, (D), California (Los Angeles), 1973, Oo
John T. Wasson, (D), MIT, 1958, Xm
Edward Young
Senior Scientist:
Stanislav I. Braginsky, (D), Moscow Inst, 1948, Ym
Robert J. Strangeway, (D), London, 1978, Oo
M I. Venkatesan, (D), Madras, 1973, Co
Paul Warren, (D), California (Los Angeles), 1979, Xm
Associate Professor:
J. David Neelin, (D), Princeton, 1987, Oa
Associate Research Scientist:
Jean Berchem, (D), California (Los Angeles), 1986, Oo
Gregory Kallemeyn, (D), California (Los Angeles), 1982, Cg
Frank T. Kyte, (D), California (Los Angeles), 1983, Ct
Alan E. Rubin, (D), New Mexico, 1982, Xm
David Schriver, (D), California (Los Angeles), 1988, Yg
Fred Schwab, (B), California (Los Angeles), 1960, Ys
Assistant Research Scientist:
Kayo Ide, (D), Caltech, 1990, Og
Krishan Khurana, (D), Durham, 1984, Oo
Robert Richard, (D), California (Los Angeles), 1988, Yg
William Smythe, (D), California (Los Angeles), 1979, Ys
Ferenc D. Varadi, (D), California (Los Angeles), 1989, Oo
Emeritus:
Orson L. Anderson, (D), Utah, 1951, Yx
Paul J. Coleman, Jr., (D), California (Los Angeles), 1966, Gc

California

Isaac R. Kaplan, (D), S California, 1962, Cg
Leon Knopoff, (D), Caltech, 1949, Ys
Ronald L. Shreve, (D), Caltech, 1958, Gm
Department Manager:
James Nakatsuka, Oo

University of California, Riverside

Dept of Earth Sciences (B,M,D) (2006)
University of California Riverside
Earth Science Department
Riverside, CA 92521
p. (951) 827-3182 f. (951) 827-3434
joslync@ucr.edu
http://earthscience.ucr.edu
Enrollment (2002): B: 36 (9) M: 4 (3) D: 19 (1)

Chair:
Mary L. Droser, (D), S California, 1987, Po
Professor:
James H. Dieterich, (D), Yale, 1968, Yg
Harry W. Green, II, (D), California (Los Angeles), 1968, Yx
Nigel C. Hughes, (D), Bristol (UK), 1990, Po
Tien-Chang Lee, (D), S California, 1973, Yh
Richard A. Minnich, (D), California (Los Angeles), 1978, Oy
Stephen K. Park, (D), MIT, 1984, Ym
Peter M. Sadler, (D), Bristol (UK), 1973, Gd
Associate Professor:
Martin J. Kennedy, (D), Adelaide, 1994, Gr
Michael A. McKibben, (D), Penn State, 1984, Cg
David D. Oglesby, (D), California (Santa Barbara), 1999, Ys
Alan E. Williams, (D), Brown, 1980, Cs
Assistant Professor:
Elizabeth Cochran, (D), California (Los Angeles), 2005, Ys
Gordon Love, (D), Strathclyde, 1995, Co
Adjunct Assistant Professor:
Katherine J. Kendrick, (D), California (Riverside), 1999, Gm
Thomas A. Scott, (D), California (Berkeley), 1987, Oy
Adjunct Professor:
Larissa F. Dobrzhinetskaya, (D), Inst of Physics of Earth (Moscow), 1978, Gx
Douglas M. Morton, (D), California (Los Angeles), 1966, Gp
Emeritus:
Wilfred A. Elders, (D), Durham, 1961, Gi
Michael A. Murphy, (D), California (Los Angeles), 1954, Ps
Michael O. Woodburne, (D), California (Berkeley), 1966, Pv
Senior Museum Scientist:
Marilyn A. Kooser, (D), California (Riverside), 1980, Gr

University of California, San Diego

Scripps Institution of Oceanography (D) (2006)
Graduate Department
10550 North Torrey Pines Rd.
La Jolla, CA 92037
p. (858) 534-3206
siodept@sio.ucsd.edu
http://sio.ucsd.edu/
Enrollment (2000): D: 29 (0)

Chair:
Myrl C. Hendershott, (D), Harvard, 1966, Op
Director, Inst of Geophys & Planetary Phys:
J. Bernard H. Minster, (D), Caltech, 1974, Ys
Professor:
Duncan C. Agnew, (D), California (San Diego), 1979, Ys
Laurence Armi, (D), California (Berkeley), 1975, Op
Gustaf Arrhenius, (D), Stockholm, 1952, Cg
Farooq Azam, (D), Czech Acad of Sci, 1968, Ob
Jeffrey L. Bada, (D), California (San Diego), 1968, Co
Wolfgang H. Berger, (D), California (San Diego), 1968, Gs
Michael J. Buckingham, (D), Reading, 1971, Op
Ronald S. Burton, (D), Stanford, 1981, Ob
Steven C. Cande, (D), Columbia, 1976, Yr
Catherine G. Constable, (D), California (San Diego), 1987, Ym
Steven C. Constable, (D), Australian National, 1983, Yr
Paul J. Crutzen, (D), Stockholm, 1973, Oa
Paul K. Dayton, (D), Washington, 1970, Ob
Leroy M. Dorman, (D), Wisconsin, 1970, Yg
D. John Faulkner, (D), Imperial Coll (UK), 1965, Oc
Horst Felbeck, (D), Muenster, 1979, Ob
William H. Fenical, (D), California (Riverside), 1968, Oc
Peter J. S. Franks, (D), MIT/Woods Hole, 1990, Ob
Carl H. Gibson, (D), Stanford, 1962, Oo
Joris M. Gieskes, (D), Manitoba, 1965, Oc
Robert T. Guza, (D), California (San Diego), 1974, On
James W. Hawkins, (D), Washington, 1963, Gp
John A. Hildebrand, (D), Stanford, 1983, Yr
William S. Hodgkiss, Jr., (D), Duke, 1975, Oo
Nicholas D. Holland, (D), Stanford, 1965, Ob
Glenn R. Ierley, (D), MIT, 1982, Op
Jeremy B. C. Jackson, (D), Yale, 1971, Po
Miriam Kastner, (D), Harvard, 1970, Cl
Charles D. Keeling, (D), Northwestern, 1954, Co
Charles F. Kennel, (D), Princeton, 1964, Oo
Nancy Knowlton, (D), California (Berkeley), 1978, Ob
William A. Kuperman, (D), Maryland, 1972, Op
Devendra x. Lal, (D), Bombay, 1958, Xc
Lisa A. Levin, (D), California (San Diego), 1982, Ob
Peter F. Lonsdale, (D), California (San Diego), 1974, Gu
J. Douglas Macdougall, (D), California (San Diego), 1972, Cm
T. Guy Masters, (D), Cambridge, 1979, Ys
W. Kendall Melville, (D), Southampton (UK), 1974, Op
William A. Newman, (D), California (Berkeley), 1962, Ob
Pearn P. Niiler, (D), Brown, 1964, Op
Mark D. Ohman, (D), Washington, 1983, Ob
John A. Orcutt, (D), California (San Diego), 1976, Yr
Robert L. Parker, (D), Cambridge, 1966, Ym
Robert Pinkel, (D), California (San Diego), 1974, Op
V. Ramanathan, (D), SUNY (Stony Brook), 1973, Oa
Dean H. Roemmich, (D), MIT/Woods Hole, 1980, Op
Daniel L. Rudnick, (D), California (San Diego), 1987, Op
Richard L. Salmon, (D), California (San Diego), 1976, Op
David T. Sandwell, (D), California (Los Angeles), 1981, Yr
John G. Sclater, (D), Cambridge, 1966, Yh
Robert E. Shadwick, (D), British Columbia, 1983, Ob
Peter M. Shearer, (D), California (San Diego), 1986, Ys
Richard C. Somerville, (D), New York, 1966, Oa
Dariusz Stramski, (D), Gdansk (Poland), 1985, Op
George Sugihara, (D), Princeton, 1983, Ob
Lynne D. Talley, (D), MIT/Woods Hole, 1982, Op
Lisa Tauxe, (D), Columbia, 1983, Ym
Victor D. Vacquier, (D), California (Berkeley), 1968, Ob
Martin Wahlen, (D), Bern, 1969, Cs
Ray F. Weiss, (D), California (San Diego), 1970, Cm
Clinton D. Winant, (D), S California, 1972, On
William R. Young, (D), MIT/Woods Hole, 1981, Op
Associate Professor:
Douglas H. Bartlett, (D), Illinois, 1985, Ob
Kevin M. Brown, (D), Durham (UK), 1987, Gu
Paterno R. Castillo, (D), Washington (St. Louis), 1987, Gi
Paola Cessi, (D), MIT, 1987, Op
Christopher D. Charles, (D), Columbia, 1991, Pe
David M. Checkley, (D), California (San Diego), 1978, Ob
Andrew G. Dickson, (D), Liverpool (UK), 1978, Cm
Margo G. Haygood, (D), California (San Diego), 1984, Ob
David R. Hilton, (D), Cambridge, 1985, Cs
Ralph F. Keeling, (D), Harvard, 1988, Oa
Brian Palenik, (D), MIT/Woods Hole, 1989, Ob
Jeffrey P. Severinghaus, (D), Columbia, 1995, Pe

Detlef B. Stammer, (D), Kiel, 1992, Ow
Bradley T. Werner, (D), Caltech, 1987, Gm
Assistant Professor:
Katherine A. Barbeau, (D), MIT/Woods Hole, 1998, Oc
Adjunct Professor:
Jay P. Barlow, (D), California (San Diego), 1982, Ob
Richard B. Deriso, (D), Washington, 1978, Ob
John R. Hunter, (D), Wisconsin, 1962, Ob
Michael S. Longuet-Higgins, (D), Cambridge, 1951, Op
William F. Perrin, (D), California (Los Angeles), 1972, Ob
Paul E. Smith, (D), Iowa, 1962, Ob
Research Professor:
John A. McGowan, (D), California (San Diego), 1960, Ob
Emeritus:
George E. Backus, (D), Chicago, 1956, Yg
Andrew A. Benson, (D), Caltech, 1942, Ob
Charles S. Cox, (D), Scripps, 1954, Op
Harmon Craig, (D), Chicago, 1951, Cs
Joseph R. Curray, (D), Scripps, 1959, Gu
James T. Enright, (D), California (Los Angeles), 1961, Ob
Edward A. Frieman, (D), Brooklyn Polytech, 1951, Oo
J. Freeman Gilbert, (D), MIT, 1956, Ys
Edward D. Goldberg, (D), Chicago, 1949, Oc
Harold T. Hammel, (D), Cornell, 1953, Ob
Richard A. Haubrich, (D), Wisconsin, 1958, Ys
Francis T. Haxo, (D), Stanford, 1947, Ob
Robert R. Hessler, (D), Chicago, 1960, Ob
Douglas L. Inman, (D), Scripps, 1953, On
Gerald L. Kooyman, (D), Arizona, 1966, Ob
Ralph A. Lewin, (D), Yale, 1950, Ob
Walter H. Munk, (D), Scripps, 1947, Op
Joseph L. Reid, (M), Scripps, 1950, Op
Richard H. Rosenblatt, (D), California (Los Angeles), 1959, Ob
George G. Shor, (D), Caltech, 1954, Yr
Fred N. Spiess, (D), California (Berkeley), 1951, Yr
Victor Vacquier, (M), Wisconsin, 1928, Yh
Kenneth M. Watson, (D), Iowa, 1948, Op
Edward L Winterer, (D), California (Los Angeles), 1954, Gd
A. Aristides Yayanos, (D), Penn State, 1967, Ob

University of California, Santa Barbara

Dept of Geological Sciences (B,M,D) (2007)
Room 1006
Webb Hall
Santa Barbara, CA 93106-9630
p. (805) 893-4831 f. (805) 893-2314
mattinson@geol.ucsb.edu
http://www.geol.ucsb.edu
Undergraduate Advisor: Christine S. Choi
Enrollment (2006): B: 96 (0) M: 9 (0) D: 33 (0)
Chair:
Bruce P. Luyendyk, (D), California (San Diego), 1969, Yr
Professor:
Ralph J. Archuleta, (D), California (San Diego), 1976, Ys
Tanya M. Atwater, (D), California (San Diego), 1972, Yr
Stanley M. Awramik, (D), Harvard, 1973, Po
James R. Boles, (D), Otago (NZ), 1972, Gd
Cathy J. Busby, (D), Princeton, 1983, Gs
Michael J. DeNiro, (D), Caltech, 1977, Cs
Rachel M. Haymon, (D), California (San Diego), 1982, Cm
Edward A. Keller, (D), Purdue, 1973, Gm
James P. Kennett, (D), Victoria (NZ), 1965, Gu
Ken C. Macdonald, (D), MIT, 1975, Yr
James M. Mattinson, (D), California (Santa Barbara), 1970, Cc
William A. Prothero, (D), California (San Diego), 1967, Ys
Frank J. Spera, (D), California (Berkeley), 1977, Gi
Arthur G. Sylvester, (D), California (Los Angeles), 1966, Gc
Toshiro Tanimoto, (D), California (Berkeley), 1982, Ys
Bruce H. Tiffney, (D), Harvard, 1977, Pb
Associate Professor:
Phillip B. Gans, (D), Stanford, 1987, Gt
Bradley R. Hacker, (D), California (Los Angeles), 1998, Gp
David W. Lea, (D), MIT, 1990, Oc
Andre R. Wyss, (D), Columbia, 1989, Pv
Assistant Professor:
Jordan Clark, Hg
Researcher:
Michael D. Fuller, (D), Cambridge, 1961, Ym
Emeritus:
John C. Crowell, (D), California (Los Angeles), 1947, Gt
Richard V. Fisher, (D), Washington, 1957, Gr
Clifford A. Hopson, (D), Johns Hopkins, 1955, Gi
Robert M. Norris, (D), California (San Diego), 1951, Gm
George R. Tilton, (D), Chicago, 1951, Cc
Donald W. Weaver, (D), California (Berkeley), 1960, Ps
William S. Wise, (D), Johns Hopkins, 1961, Gz

Inst for Crustal Studies (2004)
Mail Code 1100
Santa Barbara, CA 93106-1100
p. (805) 893-8231 f. (805) 893-8649
admin@quake.crustal.ucsb.edu
http://www.crustal.ucsb.edu
Management Services Officer: Giulia F. Brofferio
Director:
Douglas W. Burbank, (D), Dartmouth, 1982, Gm
Associate Director:
Bradley R. Hacker, (D), California (Los Angeles), 1988, Gx
Professor:
Ralph J. Archuleta, (D), California (San Diego), 1976, Ys
Tanya M. Atwater, (D), California (San Diego), 1972, Yr
Cathy Busby, (D), Princeton, 1983, Gs
Edward A. Keller, (D), Purdue, 1973, Gt
Bruce P. Luyendyk, (D), California (San Diego), 1969, Yr
William A. Prothero, (D), California (San Diego), 1967, Ys
Frank J. Spera, (D), California (Berkeley), 1977, Cp
Arthur G. Sylvester, (D), California (Los Angeles), 1966, Gt
Toshiro Tanimoto, (D), California (Berkeley), 1982, Ys
Senior Scientist:
Lorne G. Everett, (D), Arizona, 1972, Hw
Associate Professor:
Phillip B. Gans, (D), Stanford, 1987, Gt
Hugo A. Loaiciga, (D), California (Davis), 1986, Hg
Associate Research Seismologist:
Jamison H. Steidl, (D), California (Santa Barbara), 1995, Ys
Assistant Research Scientist:
Marc J. Kamerling, (D), California (Santa Barbara), 1994, Yr
Peng Cheng Liu, (D), Dalian (China), 1990, Ys
Christopher C. Sorlien, (D), California (Santa Barbara), 1994, Yr
Associate Scientist:
Stephen J. Cullen, (M), Montana State, 1981, Hw
Kim B. Olsen, (D), Utah, 1994, Ys
Assistant Professor:
Jordan Clark, (D), Columbia, 1995, Cg
Research Associate:
Peter W. Rodgers, (D), California (Berkeley), 1965, Ys
Emeritus:
John C. Crowell, (D), California (Los Angeles), 1947, Gc

University of California, Santa Cruz

Center for the Dynamics & Evolution of the Land-Sea Interface (2005)
1156 High Street
Room A234, Earth & Marine Sciences Building
Santa Cruz, CA 95064
p. (831) 459-4089 f. (831) 459-3074
acr@es.ucsc.edu

Professor:
Robert S. Anderson, (D), Washington, 1986, Gm
Kenneth W. Bruland, (D), California (San Diego), 1974, Oc
Daniel P. Costa, (D), California (Santa Cruz), Ob
Margaret L. Delaney, (D), MIT/Woods Hole, 1983, Oc
Andrew T. Fisher, (D), Miami, 1989, Hw
Russell Flegal, (D), Oregon State, Cg
Laurel R. Fox, (D), California (Santa Barbara), Oo
Dianne Gifford-Gonzalez, (D), California (Berkeley), Oo
Paul L. Koch, (D), Michigan, 1989, Py
Mark S. Mangel, (D), British Columbia, 1978, Oo
Donald C. Potts, (D), California (Santa Barbara), Ob
Mary W. Silver, (D), California (San Diego), 1971, Ob
Lisa C. Sloan, (D), Penn State, 1990, Pe
James C. Zachos, (D), Rhode Island, 1988, Cm
Jonathan P. Zehr, (D), California (Davis), 1985, Ob
Associate Professor:
Mark Carr, (D), California (Santa Barbara), Ob
Brent Haddad, (D), California (Berkeley), 1996, Oo
Karen D. Holl, (D), Virginia Tech, 1994, Ou
Christina Ravelo, (D), Columbia, 1991, Cs
Donald R. Smith, (D), California (Santa Cruz), Oo
Assistant Professor:
Patrick Y. Chuang, (D), Caltech, 1999, Oa
Don Croll, (D), California (Santa Cruz), Ob
Raphael M. Kudela, (D), S California, 1995, Or
Margaret A. McManus, (D), Old Dominion, 1996, On
Slawek Tulaczyk, (D), Caltech, 1998, Gl

Center for the Origin, Dynamics, & Evolution of Planets (2005)
1156 High Street
Room A234, Earth & Marine Sciences Building
Santa Cruz, CA 95064
p. (831) 459-4089 f. (831) 459-3074
kory@es.ucsc.edu

Professor:
Robert S. Anderson, (D), Washington, 1986, Gm
Peter Bodenheimer, (D), California (Berkeley), Xc
Frank Bridges, (D), California (San Diego), 1968, Yx
Robert S. Coe, (D), California (Berkeley), 1966, Ym
Gary A. Glatzmaier, (D), Colorado, 1980, Xy
Elise Knittle, (D), California (Berkeley), 1988, Gy
Thorne Lay, (D), Caltech, 1983, Ys
Douglas Lin, (D), Cambridge, Xc
Steven Vogt, (D), Texas, 1978, Xc
Quentin C. Williams, (D), California (Berkeley), 1988, Gy
Associate Professor:
Erik Asphaug, (D), Arizona, 1993, Xg
Other:
Don Korycansky, Xc

Center for the Study of Imaging & Dynamics of the Earth (2006)
1156 High Street
Room A234, Earth & Marine Sciences Building
Santa Cruz, CA 95064
p. (831) 459-4089 f. (831) 459-3074
sschwartz@es.ucsc.edu

Professor:
Robert S. Coe, (D), California (Berkeley), 1966, Ym
Andrew T. Fisher, (D), Miami, 1989, Hw
James B. Gill, (D), Australian National, Gi
Gary A. Glatzmaier, (D), Colorado, 1980, Yg
Elise Knittle, (D), California (Berkeley), 1988, Gy
Thorne Lay, (D), Caltech, 1983, Ys
Marcia McNutt, (D), California (San Diego), 1978, Yr
J. Casey Moore, (D), Princeton, 1971, Gc
Susan Y. Schwartz, (D), Michigan, 1988, Ys
Eli A. Silver, (D), California (San Diego), 1969, Yr
Quentin C. Williams, (D), California (Berkeley), 1988, Gy
Research Geophysicist:
Steven N. Ward, (D), Princeton, 1978, Ys
Ru-shan Wu, (D), MIT, 1984, Ys
Associate Professor:
Erik Asphaug, (D), Arizona, 1993, Xg
Associate Research Geophysicist:
Xiao-bi Xie, (D), Chinese Acad of Sci, 1988, Ys
Xixi Zhao, (D), California (Santa Cruz), 1987, Ym
Assistant Professor:
Jeremy K. Hourigan, (D), Stanford, Gt
Slawek Tulaczyk, (D), Caltech, 1988, Gl

Earth & Planetary Sciences Dept (B,M,D) (2008)
1156 High Street
Earth & Marine Sciences Building, Room A234
Santa Cruz, CA 95064
p. (831) 459-4089 f. (831) 459-3074
pkoch@pmc.ucsc.edu
http://www.es.ucsc.edu
Administrative Assistant: Mary Nosse
Enrollment (2006): B: 144 (41) M: 17 (10) D: 45 (4)

Chair:
Paul L. Koch, (D), Michigan, 1989, Pv
Professor:
Erik Asphaug, (D), Arizona, 1993, Xg
Robert S. Coe, (D), California (Berkeley), 1966, Ym
Andrew T. Fisher, (D), Miami, 1989, Hw
James B. Gill, (D), Australian National, 1972, Gi
Gary A. Glatzmaier, (D), Colorado, 1980, Xy
Gary B. Griggs, (D), Oregon State, 1968, On
Elise Knittle, (D), California (Berkeley), 1988, Gy
Thorne Lay, (D), Caltech, 1983, Ys
Susan Y. Schwartz, (D), Michigan, 1988, Ys
Eli A. Silver, (D), California (San Diego), 1969, Yr
Lisa C. Sloan, (D), Penn State, 1990, Pe
Quentin C. Williams, (D), California (Berkeley), 1988, Gy
James C. Zachos, (D), Rhode Island, 1988, Cm
Associate Professor:
Emily Brodsky, (D), Caltech, Ys
Francis Nimmo, (D), Cambridge, Xy
Slawek Tulaczyk, (D), Caltech, 1998, Gl
Assistant Professor:
Patrick Y. Chuang, (D), Caltech, 1999, Oa
Matthew E. Clapham, (D), Southern California, 2006, Pe
Jeremy Hourigan, (D), Stanford, Gc
Lecturer:
Hilde Schwartz, (D), California (Santa Cruz), 1983, Pv
Emeritus:
Kenneth L. Cameron, (D), Virginia Tech, 1971, Gx
Robert E. Garrison, (D), Princeton, 1964, Gs
Leo F. Laporte, (D), Columbia, 1960, Pg
Karen C. McNally, (D), California (Berkeley), 1976, Ys
J. Casey Moore, (D), Princeton, 1971, Gc
Othmar T. Tobisch, (D), London (UK), 1963, Gc
Gerald E. Weber, (D), California (Santa Cruz), 1980, Ng
On Leave:
Marcia McNutt, (D), California (San Diego), 309, Yr
Grad/Undergrad Program Advisor:
Cathy Smith, (B), California (Santa Cruz), Oo
Front Office Manager:
Jennifer M. Fish, (B), California (Santa Cruz), Oo
Department Manager:
Judy Van Leuven, Oo

Inst of Geophysics & Planetary Physics (2007)
1156 High Street
Earth & Marine Sciences Building, Room A234

Santa Cruz, CA 95064
p. (831) 459-4089 f. (831) 459-3074
tlay@es.ucsc.edu
http://natsci.ucsc.udu/igpp
Enrollment (1997): B: 121 (45) M: 23 (12) D: 61 (6)
Director:
Thorne Lay, (D), Caltech, 1983, Ys

University of San Diego

Dept of Marine Science & Env Studies (B,M) (2005)
Alcala Park
San Diego, CA 92110
p. (619) 260-4600 (Ext. 4418) f. (619) 260-6874
adkinsk@sandiego.edu
Enrollment (1998): B: 74 (14)
Director:
Hugh I. Ellis, (D), Florida, 1976, Ob
Associate Professor:
Ronald S. Kaufmann, (D), California (San Diego), 1992, Ob
Anne A. Sturz, (D), California (San Diego), 1991, Cm
Assistant Professor:
Michel A. Boudrias, (D), California (San Diego), 1992, Ob
Richard Gonzalez, (D), Penn State, Oo
Sarah Gray, (D), California (Santa Cruz), Oo
Adjunct Professor:
Ann B. Bowles, (D), California (San Diego), Oo
Scott Eckert, (D), Georgia, Oo
Joseph R. Jehl, (D), Michigan, 1967, Ob
Donald B. Kent, (M), San Diego State, 1980, Ob
Brent S. Stewart, (D), California (Los Angeles), 1988, Ob
Pam Yochem, (D), California (Davis), Oo
Cooperating Faculty:
Gerald N. Estberg, (D), Cornell, 1966, Ow
Mary Sue Lowery, (D), California (San Diego), 1987, Ob
LeeAnn Otto, (D), British Columbia, 1981, Oo
Marie Simovich, (D), California (Riverside), 1985, Oo

University of Southern California

Dept of Earth Sciences (B,M,D) (2007)
3651 Trousdale Parkway ,ZHS117
Los Angeles, CA 90089-0740
p. (213) 740-6106 f. (213) 740-8801
earthsci@usc.edu
http://www.usc.edu/dept/earth
Administrative Assistant: Vardui Ter-Simonian
Student Services Coordinator: Cynthia Waite
Enrollment (1996): B: 31 (7) M: 18 (2) D: 40 (2)
Chair:
David J. Bottjer, (D), Indiana, 1978, Pe
Research Professor:
David A. Okaya, (D), Stanford, 1985, Ys
Professor:
J. Lawford Anderson, (D), Wisconsin, 1975, Gi
Yehuda Ben-Zion, (D), S California, 1990, Ys
Douglas E. Hammond, (D), Columbia, 1975, Cm
Thomas H. Jordan, (D), Caltech, 1972, Ys
Steven P. Lund, (D), Minnesota, 1981, Ym
Jean Morrison, (D), Wisconsin, 1988, Gp
Kenneth Nealson, (D), Chicago, 1969, Py
Scott R. Paterson, (D), California (Santa Cruz), 1986, Gc
John Platt, (D), California (Santa Barbara), 1973, Gc
Charles G. Sammis, (D), Caltech, 1971, Yg
Sergio Sanudo, (D), Py
Lowell D. Stott, (D), Rhode Island, 1989, Pm
Ta-liang Teng, (D), Caltech, 1966, Ys
Associate Professor:
William M. Berelson, (D), S California, 1986, Cm
Frank A. Corsetti, (D), California (Santa Barbara), 1998, Gs
James F. Dolan, (D), California (Santa Cruz), 1988, Gt
Katrina Edwards, (D), Py
Research Associate Professor:
Yong-Gang Li, (D), S California, 1988, Ys
Ellen S. Platzman, (D), ETH, 1990
Research Assistant Professor:
Ronald L. Biegel, (D), S California, 1989, Yg
Ann E. Blythe, (D), Cornell, 1992, Gt
Andrea Donnellan, (D), Caltech, 1991, Yx
Rob Rye, (D), Harvard, 1999, Py
Danijel Schorlemmer, (D), ETH Zurich, 2004, Ys
Assistant Professor:
Thorsten W. Becker, (D), Harvard, 2002, Ys
Adjunct Professor:
Robert G. Douglas, (D), California (Los Angeles), 1966, Pe
Emeritus:
Alfred G. Fischer, (D), Columbia, 1950, Gs
Donn S. Gorsline, (D), S California, 1958, Gu
Thomas L. Henyey, (D), Caltech, 1968, Yg
Teh-Lung Ku, (D), Columbia, 1966, Cg
Bernard W. Pipkin, (D), Arizona, 1964, Ng

University of the Pacific

Dept of Geosciences (B) (2007)
3601 Pacific Avenue
Stockton, CA 95211
p. (209) 946-2482 f. (209) 946-2362
lkfox@pacific.edu
http://www.pacific.edu/college/geosciences
Enrollment (2007): B: 15 (5)
Chair:
Lydia K. Fox, (D), California (Santa Barbara), 1989, GizOe
Professor:
Eugene F. Pearson, (D), Wyoming, 1972, GsPgOg
Assistant Professor:
Kurtis C. Burmeister, (D), Illinois, 2005, GctEg
Laura K. Rademacher, (D), California (Santa Barbara), 2002, ClHyOs
Emeritus:
Roger T. Barnett, (D), California (Berkeley), 1973, Oy
J. Curtis Kramer, (D), California (Davis), 1976, Gg

Ventura College

Geosciences Dept (A) (2005)
4667 Telegraph Road
Ventura, CA 93003
p. (805) 654-6400 (Ext. 1265) f. (805)654-6374
lhall@vcccd.net
Chair:
Luke D. Hall, (M), W Kentucky, 1975, Oy
Assistant Professor:
Steve D. Palladino, (M), California (Santa Barbara), 1994, Oi
Instructor:
William Budke, (M), Os

West Valley College

Dept of Geology (A) (2004)
14000 Fruitvale Avenue
Saratoga, CA 95070-5698
p. (408) 741-2437
blopez@es.ucsc.edu
Acting Head:
Robert Lopez, (D), California (Santa Cruz), 1997, Gg
Instructor:
H. Shade, Gg
Emeritus:
Theodore C. Herman, (M), Michigan, 1961, Oe

Whittier College
Dept of Earth and Environmental Science (B) (2004)
Whittier, CA 90608
p. (562) 907-4220 f. (562) 693-6117
jvermilye@whittier.edu
Enrollment (1999): B: 23 (3)
Visiting Professor:
Andrew H. Wulff, (D), Massachusetts, 1998, Gi
Emeritus:
William B. Wadsworth, (D), Northwestern, 1966, Gi

Yuba Community College
Geology Dept (2004)
Marysville, CA 95901
p. (916) 741-6940
Instructor:
Lee Mitchell, (M), California (Davis), 1957, Gg

Colorado

Adams State College
Biology and Earth Sciences (B) (2007)
208 Edgemont Blvd
Alamosa, CO 81102
baybarro@adams.edu
http://www.adams.edu/geology
Enrollment (2007): B: 20 (4)
Professor:
Brent Ybarrondo, (D), Vermont, 1993
Associate Professor:
Robert G. Benson, (D), Colorado Mines, 1997, Gg
Assistant Professor:
Jared M. Beeton, (D), Kansas, 2007, OySd

Colorado College
Geology Dept (B) (2007)
14 E Cache La Poudre
Colorado Springs, CO 80903
p. (719) 389-6621 f. (719) 389-6910
geology@coloradocollege.edu
http://www.coloradocollege.edu/dept/gy
Administrative Assistant: Catherine M. Bailie
Enrollment (2007): B: 29 (12)
Chair:
Paul M. Myrow, (D), Memorial, 1987, Gs
Associate Dean:
Jeffrey B. Noblett, (D), Stanford, 1980, Gx
Professor:
Eric M. Leonard, (D), Colorado, 1981, Gl
Assistant Chair:
Christine S. Siddoway, (D), California (Santa Barbara), 1993, Gc
Associate Professor:
Henry C. Fricke, (D), Michigan, 1997, Cs
Visiting Assistant Professor:
Steve Quane, (D), British Columbia, 2004, Gv
Emeritus:
William A. Fischer, (D), Colorado, 719, Pg
Eiler L. Henrickson, (D), Minnesota, 1956, Eg
Technical Director:
Stephen G. Weaver, (D), Colorado Mines, 1988, Gx

Colorado Northwestern Community College
Dept of Geology (A) (2006)
500 Kennedy Drive
Rangely, CO 81648
p. (303) 675-3261
todd.ward@cncc.edu
Enrollment: No data reported since 1995
Chair:
Todd W. Ward, (D), Colorado State, 2001
Instructor:
Robert A. Horntvedt, (B), W Michigan, 1980, Eo

Colorado School of Mines
Dept of Chemistry & Geochemistry (M,D) (2004)
Golden, CO 80401
p. (303) 273-3610 f. (303) 273-3629
sdaniel@mines.edu
http://www.mines.edu
Enrollment (2000): M: 4 (0) D: 5 (2)
Head:
Stephen R. Daniel, (D), Colorado Mines, 1970, Oo
Professor:
Dean W. Dickerhoof, (D), Illinois, 1961, Oo
Ronald W. Klusman, (D), Indiana, 1969, Ca
Donald L. Macalady, (D), Wisconsin, 1969, Cl
Patrick MacCarthy, (D), Cincinnati, 1975, Co
Michael J. Pavelich, (D), SUNY (Buffalo), 1970, Oo
Kent J. Voorhees, (D), Utah State, 1970, Co
Thomas R. Wildeman, (D), Wisconsin, 1967, Ct
Associate Professor:
Scott W. Cowley, (D), S Illinois, 1975, Oo
Mark E. Eberhart, (D), MIT, 1983, Om
Craig Simmons, (D), SUNY (Stony Brook), 1976, Cp
Kim R. Williams, (D), Michigan State, 1986, Ca
David T. Wu, (D), California (Berkeley), 1991, Oo
Associate Scientist:
Martin B. Goldhaber, (D), California (Los Angeles), 1968, Ce
Assistant Professor:
Jeffrey Harlan, (D), Texas (Arlington), 1992, Om
Daniel M. Knauss, (D), Virginia Tech, 1994, Oo
Kevin W. Mandernack, (D), California (San Diego), 1992, Py

Dept of Geology & Geological Engineering (B,M,D) (2007)
1516 Illinois Street
Golden, CO 80401-1887
p. (303) 273-3800 f. (303) 273-3859
mhitzman@mines.edu
http://www.mines.edu
Administrative Assistant: Marilyn Schwinger
Enrollment (2006): B: 85 (0) M: 83 (0) D: 32 (0)
Head:
Murray W. Hitzman, (D), Stanford, 1983, Em
Professor:
Wendy J. Harrison, (D), Manchester, 1979, Cl
Neil F. Hurley, (D), Michigan, 1986, Go
Eileen P. Poeter, (D), Washington State, 1980, Ng
Samuel B. Romberger, (D), Penn State, 1968, Em
A. Keith Turner, (D), Purdue, 1969, Ng
Richard F. Wendlandt, (D), Penn State, 1978, Gx
Associate Professor:
David A. Benson, (D), Nevada (Reno), 1998, Hq
L. Graham Closs, (D), Queen's, 1973, Ce
John B. Curtis, (D), Ohio State, 1989, Go
Jerry D. Higgins, (D), Missouri (Rolla), 1980, Ng
Gregory S. Holden, (D), Wyoming, 1978, Gx
John D. Humphrey, (D), Brown, 1987, Gd
Kevin Mandernack, (D), California (San Diego), 1992, Co
John E. McCray, (D), Arizona, 1998, Hw
Eric P. Nelson, (D), Columbia, 1981, Gc
Bruce D. Trudgill, (D), Imperial Coll (UK), 1989, Gc
Assistant Professor:
Michael N. Gooseff, (D), Colorado, 2001, Hs
Adjunct Professor:
Fred F. Meissner, (M), Colorado Mines, 1954, Go

Emeritus:
L. Trowbridge Grose, (D), Stanford, 1955, Gt
John D. Haun, (D), Wyoming, 1953, Go
Richard W. Hutchinson, (D), Wisconsin, 1954, Em
Keenan Lee, (D), Stanford, 1969, Or
John E. Warme, (D), California (Los Angeles), 1966, Pi
Robert J. Weimer, (D), Stanford, 1953, Gr

Dept of Geophysics (B,M,D) (2006)
1500 Illinois Street
Golden, CO 80401
p. (303) 273-3450 f. (303) 273-3478
ssummers@mines.edu
http://www.mines.edu/academic/geophysics/
Enrollment (2006): B: 60 (0) M: 48 (0) D: 12 (0)

Professor:
Michael L. Batzle, (D), MIT, 1978, Yg
Thomas L. Davis, (D), Colorado Mines, 1974, Ye
Dave Hale, (D), Stanford, 1997, Ye
Gary R. Olhoeft, (D), Toronto, 1975, Yg
Roel Snieder, (D), Utrecht (Neth), 1987, Ye
Ilya D. Tsvankin, (D), Moscow State, 1982, Ye
Senior Scientist:
Warren Hamilton, (D), California (Los Angeles), 1951, Ys
Misac Nabighian, (D), Columbia, 1967, Nm
Associate Research Professor:
Robert D. Benson, (M), Colorado Mines, 1984, Ye
Associate Professor:
Yaoguo Li, (D), British Columbia, 1992, Yv
Assistant Professor:
Paul Sava, (D), Stanford, 2005, Ye
Emeritus:
Kenneth L. Larner, (D), MIT, 1970, Ye
Research Professor:
Norman Bleistein, (D), Courant Inst, 1965, Yg

Dept of Mining Engineering (B,M,D) (2007)
1600 Illinois
Golden, CO 80401
p. (303) 273-3701 f. (303) 273-3719
smann@mines.edu
Administrative Assistant: Shannon Mann
Enrollment (2006): B: 108 (0) M: 16 (0) D: 23 (0)

Head:
Tibor G. Rozgonyi, (D), Tech (Hungary), 1974, Nm
Professor:
Kadri Dagdelen, (D), Colorado Mines, 1985, Nm
M. Ugor Ozbay, (D), Witwatersrand, 1988, Nr
Levent Ozdemir, (D), Colorado Mines, 1977, Nr
Associate Professor:
Mark Kuchta, (D), Lulea Univ of Tech, 1990, Nm
Hugh Miller, (D), Colorado Mines, 1996, Nm
Masami Nakagawa, (D), Cornell, 1988, Or
Assistant Professor:
Scott Kieffer, (A), California (Berkeley), 1998, Nr
Research Assistant Professor:
Vilem Petr, (D), Colorado Mines, 2000, Nm
Emeritus:
Miklos D. G. Salamon, (D), Durham, 1962, Nr
Baki Yarar, (D), Imperial Coll (UK), Nx
Cooperating Faculty:
Paul Bartos, Oo

Colorado State University

Dept of Atmospheric Science (M,D) (2006)
Fort Collins, CO 80523-1371
p. (970) 491-8360 f. (970) 491-8449
jen@atmos.colostate.edu
http://www.atmos.colostate.edu
Enrollment (2006): M: 47 (23) D: 54 (10)

Head:
Steven A. Rutledge, (D), Washington, 1983, Oa
Professor:
Jeffrey L. Collett, Jr., (D), Caltech, 1989, Oa
William R. Cotton, (D), Penn State, 1970, Oa
Richard H. Johnson, (D), Washington, 1975, Oa
Sonia M. Kreidenweis, (D), Caltech, 1989, Oa
David A. Randall, (D), California (Los Angeles), 1976, Oa
Wayne H. Schubert, (D), California (Los Angeles), 1973, Oa
Graeme L. Stephens, (D), Melbourne (Australia), 1977, Oa
Thomas H. Vonder Haar, (D), Wisconsin, 1968, Oa
Associate Professor:
A. Scott Denning, (D), Colorado State, 1994, Oa
Christian D. Kummerow, (D), 1987, Yr
Assistant Professor:
David W. J. Thompson, (D), Washington, 2000, Yr

Dept of Geosciences (B,M,D) (2007)
Fort Collins, CO 80523-1482
p. (970) 491-5661 f. (970) 491-6307
sallys@warnercnr.colostate.edu
http://www.cnr.colostate.edu/GEO/
Administrative Assistant: Barbara J. Holtz
Enrollment (2003): B: 49 (19) M: 24 (5) D: 13 (3)

Head:
Judith L. Hannah, (D), California (Davis), 1980, Gi
Professor:
Eric A. Erslev, (D), Harvard, 1981, Gc
Ellen E. Wohl, (D), Arizona, 1988, Gm
Senior Scientist:
Holly J. Stein, (D), North Carolina, 1985, Em
Associate Professor:
Dennis L. Harry, (D), Texas (Dallas), 1989, Yg
Jerry F. Magloughlin, (D), Minnesota, 1993, Gp
John R. Ridley, (D), Edinburgh, 1982, Eg
William E. Sanford, (D), Cornell, 1992, Hw
Sally J. Sutton, (D), Cincinnati, 1987, Gd
Assistant Professor:
Sven O. Egenhoff, (D), Tech, 2000, Gs
Sara L. Rathburn, (D), Colorado State, 2001, Gg
Research Associate:
James R. Chappell, (B), Colorado State, 2000, Ge
Ronald J. Karpilo, Jr, (M), Denver, 2004, Oy
Richard J. Markey, (M), N Illinois, 1990, Cc
Stephanie O'Meara, (M), Colorado State, 1997, Gc
Heather I. Stanton, (M), Colorado State, 2002, Gg
Trista L. Thornberry-Ehrlich, (M), Colorado State, 2001, Gg
Emeritus:
Donald O. Doehring, (D), Wyoming, 1968, Ge
Frank G. Ethridge, (D), Texas A&M, 1970, Gs
Robert B. Johnson, (D), Illinois, 1954, Ng
Malcolm E. McCallum, (D), Wyoming, 1964, Gz
Stanley A. Schumm, (D), Columbia, 1955, Gm
Tommy B. Thompson, (D), New Mexico, 1966, Em
James P. Waltz, (D), Stanford, 1967, Hw
Joseph L. Weitz, (D), Yale, 1954, Gc

Denver Museum of Nature & Science

Dept of Earth Sciences (2004)
DMNS 2001 Colorado Boulevard
Denver, CO 80205-5798
p. (303) 370-6473 f. (303) 331-6492
http://www.dmns.org

Chair:
Kirk R. Johnson, (D), Yale, 1989, Pb

Chief Preparator, Curator of Paleontology:
Ken Carpenter, (D), Colorado, 1996, Pv
Preparator:
Bryan Small, (M), Texas Tech, 1987, Pv
Curator, Paleontology:
Richard Stucky, (D), Colorado, 1982, Pv
Curator, Geology:
Jack Murphy, (D), Colorado (Denver), 2000, Gz
Collections Manager:
Logan D. Ivy, (D), Colorado, 1993, Pv
Chief Curator, Curator of Paleontology:
Russell Graham, (D), Texas, 1976, Pv
Office Manager:
Paula A. Hentz, (B), W Carolina, 1989, Oo

Fort Lewis College

Geosciences Dept (B) (2007)
Durango, CO 81301
p. (970) -247-7278 f. (970) 247-6053
collier_j@fortlewis.edu
http://geo.fortlewis.edu
Enrollment (2005): B: 69 (10)

Professor:
James D. Collier, (D), Colorado Mines, 1982, Cg
Ray Kenny, (D), Arizona State, 1991, Gm
Associate Professor:
Gary Gianniny, (D), Wisconsin, 1995, Gs
David A. Gonzales, (D), Kansas, 1997, Gx
Kimberly Hannula, (D), Stanford, 1993, Gc
Scott White, (D), Utah, 2001, Oy
Instructor:
Jeffrey Cary, (M), W Washington, 1990, Eg
Adjunct Professor:
Charles Burnham, (D), MIT, 1961, Gz
Mary L. Gillam, (D), Colorado, 1998, Gm
Laboratory Director:
Lauren K. Heerschap, (M), Colorado, 2003, Gt

Mesa State College

Dept of Physical & Environmental Sciences (A,B) (2005)
1100 North Avenue
Grand Junction, CO 81501-3122
p. (970) 248-1599 f. (970) 248-1700
rcole@mesastate.edu
http://www.mesastate.edu/snsm/geology
Enrollment (2005): B: 47 (5)

Program Coordinator:
Rex D. Cole, (D), Utah, 1975, Gs
Professor:
Verner C. Johnson, (D), Tennessee, 1975, Yg
Associate Professor:
Andres Aslan, (D), Colorado, 1994, Gm
Richard F. Livaccari, (D), New Mexico, 1994, Gc
Assistant Professor:
Gigi Richard, (D), Colorado State, 2001, Hs
Lecturer:
Kovschak Anthony, (M), Texas (Arlington), 1973, Gg
Harold W. Hase, (M), Michigan Tech, 1973, Em
Lawrence Jones, (D), Wyoming, 1996, Gs
Adjunct Professor:
William C. Hood, (D), Montana, 1964, Gg
Dave Wolny, (B), Mesa State, 1992, Ys
Emeritus:
Dell R. Foutz, (D), Washington State, 1966, Go
James B. Johnson, (D), Colorado, 1979, Gl
Jack E. Roadifer, (D), Arizona, 1966, Gx

Metropolitan State College of Denver

Earth & Atmospheric Sciences Dept (B) (2004)
P.O. Box 173362, Campus Box 22
Denver, CO 80217-3362
p. (303) 556-3143 f. (303) 556-4436
cronoblj@mscd.edu
Administrative Assistant: Diane Hollenbeck
Enrollment: No data reported since 1995

Chair:
James M. Cronoble, (D), Colorado Mines, 1977, Gg
Professor:
John R. Kilcoyne, (D), Washington, 1973, Oy
Anthony A. Rockwood, (M), Colorado State, 1976, Ow
Roberta A. Smilnak, (D), Clark, 1973, Og
Associate Professor:
Thomas J. Corona, (M), Colorado State, 1978, Ow
Assistant Professor:
Robert E. Leitz, (M), California (Berkeley), 1974, Gz
Rafael Moreno, (D), Colorado State, 1992, Og
Emeritus:
James MacLachlan, (D), Princeton, Gg
H. Dixon Smith, (D), Minnesota, 1960, Og

Pikes Peak Community College

Health, Env, Natural and Physical Sciences (A) (2006)
5675 South Academy Boulevard
Box CC13
Colorado Springs, CO 80906
p. (719) 540-7393 f. (719) 579-3030
Mark.Izold@ppcc.edu
Enrollment (2006): A: 5 (5)

Assistant Professor:
Mark D. Izold, (M), Ohio State, 1993, Pm

United States Air Force Academy

Dept of Economics & Geography (B) (2005)
HQ USAFA/DFEG
2354 Fairchild Drive, Suite 6K110
USAF Academy, CO 80840-5701
p. (719) 333-3080 f. (719) 333-7137
steve.gordon@usafa.af.mil
http://www.usafa.af.mil/dfeg
Enrollment (2005): B: 16 (10)

Professor:
Terry W. Haverluk, (D), Minnesota, 1993, Og
Associate Professor:
Steven J. Gordon, (D), Arizona State, 1999, Gm
Assistant Professor:
Barbra J. Rusnak, (M), Og
Instructor:
Troy Carlson, (M), Colorado (Colorado Springs), 2004, Or
Gareth Johnson, (M)
Brett J. Machovina, (M), Ohio State, Oi
Visiting Professor:
Taylor C. Barnes, (D), Illinois, 1984, Eg
Related Staff:
Danny Portillo, (B), Oi

University of Colorado

Dept of Geography (B,M,D) (2006)
Campus Box 260
Boulder, CO 80309-0260
p. (303) 492-8312 f. (303) 492-7501
signer@colorado.edu
Administrative Assistant: Marcia Signer
Enrollment (2006): B: 170 (80) M: 31 (11) D: 40 (6)

Chair:
Susan W. Beatty, (D), Cornell, 1981, So
Professor:
Roger G. Barry, (D), Southampton, 1965, Oa
T. Nelson Caine, (D), Australian National, 1966, Gm
Konrad Steffem, (D), ETH (Switzerland), 1983, Or
Thomas T. Veblen, (D), California (Berkeley), 1975, Og
Mark W. Williams, (D), California (Santa Barbara), 1990, Og
Associate Professor:
John Pitlick, (D), Colorado State, 1988, Gm
Assistant Professor:
Suzanne Anderson, (D), California (Berkeley), 1995, Gm
Geology Librarian:
Suzanne Larsen, Oo

Dept of Geological Sciences (B,M,D) (2007)
Campus Box 399
Boulder, CO 80309-0399
p. 303 492-8141 f. 303 492-2606
Mary.Kraus@Colorado.edu
http://www.colorado.edu/geolsci/
Program Assistant: B. Beth Hanson
Enrollment (2007): B: 105 (15) M: 16 (5) D: 55 (12)
Chair:
Mary J. Kraus, (D), Colorado, 1983, GsSa
Professor:
Robert Anderson, (D), Washington, 1986, Gm
Roger G. Bilham, (D), Cambridge, 1970, Yg
David A. Budd, (D), Texas, 1984, Gd
G. Lang Farmer, (D), California (Los Angeles), 1983, Cg
Shemin Ge, (D), Johns Hopkins, 1990, Hw
Bruce M. Jakosky, (D), Caltech, 1982, Xg
Gifford H. Miller, (D), Colorado, 1975, Cc
Peter Molnar, (D), Columbia, 1970, Gt
Anne F. Sheehan, (D), MIT, 1991, Ys
Joseph R. Smyth, (D), Chicago, 1970, Gz
Charles R. Stern, (D), Chicago, 1973, Gi
James P. Syvitski, (D), British Columbia, 1978, Gs
Paul Weimer, (D), Texas, 1989, Go
James W. C. White, (D), Columbia, 1983, Cs
Associate Professor:
Craig H. Jones, (D), MIT, 1987, Ys
Scott J. Lehman, (D), Colorado, 1989, Og
Stephen J. Mojzsis, (D), California (San Diego), 1997, Xc
Karl J. Mueller, (D), Wyoming, 1992, Gc
Eric E. Small, (D), California (Santa Cruz), 1998, Hw
Greg E. Tucker, (D), Penn State, 1996, Gm
Assistant Professor:
Karen Chin, (D), California (Santa Barbara), 1996, Po
Jaelyn J. Eberle, (D), Wyoming, 1996, Pv
Rebecca Flowers, (D), MIT, 2005, CgcGt
Brian M. Hynek, (D), Washington, 2003, Xg
Thomas M. Marchitto, (D), MIT, 1999, Oc
Jason Neff, (D), Stanford, 1999, Ge
Matthew J. Pranter, (D), Colorado Mines, 1999, Go
Dena M. Smith, (D), Arizona, 2000, Pg
Alexis Templeton, (D), Stanford, 2002, ClGe
Research Associate:
Kevin H. Mahan, (D), Massachusetts, 2005, Gp
Instructor:
Lon Abbott, (D), California (Santa Cruz), 1993, Gg
Adjunct Professor:
David M. Anderson, (D), Brown, 1991, Og
Richard J. Goldfarb, (D), Colorado, 1989, Oo
Anne E. Jennings, (D), Colorado, 1989, Pe
Roy M. Kligfield, (D), Columbia, 1978, Gc
Darrell K. Nordstrom, (D), Stanford, 1977, Cl
Douglas S. Robertson, (D), MIT, 1975, Oo
Emeritus:
John T. Andrews, (D), Nottingham (UK), 1965, Gl
William W. Atkinson, Jr., (D), Harvard, 1973, Em
Peter W. Birkeland, (D), Stanford, 1961, Sd
William C. Bradley, (D), Stanford, 1956, Gm
Bruce F. Curtis, (D), Harvard, 1949, Hw
Don L. Eicher, (D), Yale, 1958, Pm
Alexander Goetz, (D), Caltech, 1967, Or
Carl Kisslinger, (D), St. Louis, 1952, Ys
Edwin E. Larson, (D), Colorado, 1965, Ym
James L. Munoz, (D), Johns Hopkins, 1966, Cp
Peter Robinson, (D), Yale, 1960, Pv
Donald D. Runnells, (D), Harvard, 1964, Cl
Hartmut A. Spetzler, (D), Caltech, 1969, Ys
Theodore R. Walker, (D), Wisconsin, 1952, Gd
Earth Sciences Librarian:
Suzanne T. Larsen, (M), Denver, 1973, Oo
Laboratory Director:
John W. Drexler, (D), Michigan Tech, 1982, Ca

Univ of Colorado Museum (2006)
Campus Box 265
Boulder, CO 80309-265
p. (303) 492-6165 f. 303-735-2347
linda.cordell@colorado.edu
http://cumuseum.colorado.edu/
Assistant Professor:
Karen Chin, (D), California (Santa Barbara), 1996, Po
Jaelyn J. Eberle, (D), Wyoming, 1996, Pv
Dena M. Smith, (D), Arizona, 2000, Po
Adjunct Curator:
Kenneth Carpenter, Pv
Mary Dawson, Pv
Trihn Dzanh, Pv
Emmett Evanoff, Pi
Sherri Gabbert, Pg
Jeff Indeck, Pv
Jonathan Marcot, Pv
Greg McDonald, Pv
Malcolm McKenna, Pv
Herbert Meyer, Pb
Karen Sears, Pv
Paola Villa, Pg
Emeritus:
Judith A. Harris, (D), Cambridge, 1972, Pv
Peter Robinson, (D), Yale, 1960, Pv
Museum Associate:
Emily Bray, Pg
Donna Engard, Pg
Frank Fisher, Pg
David Kohls, Pg
Pat Monaco, Pg
Steve Wallace, Pg
Collection Manager:
Amy P. Moe-Hoffman, (M), Colorado, 2002, Pg
Tonia Superchi-Culver, (M), South Dakota Mines, 2001, Pg
Associate Curator, Nanofossils:
William Hay, Pm
Associate Curator, Micropaleontology:
Donald Eicher, Pm
Associate Curator, Fossil Primates:
Herbert Covert, Pv

University of Colorado, Denver

Dept of Geography, & Environmental Science (B) (2004)
CB 172
P.O. Box 173364
Denver, CO 80217-3364

p. (303) 556-2276 f. (303) 556-6197
frederick.chambers@cudenver.edu
Administrative Assistant: Nancy Coca
Enrollment (1997): B: 52 (6)

Chair:
John W. Wyckoff, (D), Utah, 1980, Oy
Professor:
Wesley E. LeMasurier, (D), Stanford, 1965, Gv
Martin G. Lockley, (D), Birmingham (UK), 1977, Pi
Edward Nuhfer, (D), New Mexico, 1980, Ng
John G. Weihaupt, (D), Wisconsin (Milwaukee), 1973, Ou
Assistant Professor:
Jeffrey G. Pittman, (D), Texas, 1992, Pv
Adjunct Professor:
James Jehn, (M), Hw
Mort D. Turner, (D), Eg

University of Denver

Dept of Geography (B,M,D) (2004)
2050 E. Iliff Avenue
Boettcher Center West, Room 120
Denver, CO 80208
p. (303) 871-2654 f. (303) 871-2201
kescobar@du.edu
http://www.du.geography.edu/
Enrollment (2000): B: 25 (7) M: 9 (2) D: 4 (0)

Chair:
David B. Longbrake, (D), Iowa, 1972, Og
Professor:
Andrew R. Goetz, (D), Og
Michael W. Kerwin, (D), Gg
Terrence J. Toy, (D), Denver, 1973, Gm
Associate Professor:
Michael J. Keables, (D), Wisconsin, 1986, Oa
Donald G. Sullivan, (D), California (Berkeley), 1988, Og
Assistant Professor:
Sharolyn Anderson, (D), Oi
Paul C. Sutton, (D), California (Santa Barbara), 1999, Oi
Matthew Taylor, (D), Og
Adjunct Professor:
William Emmett, (N), Oy
Richard F. Hadley, (M), Minnesota, 1950, Hg
Reuben F. Miller, (M), Colorado State, 1958, So
Martha A. Narey, (D), Denver, 1999, Og
Waite Osterkamp, Oy
Emeritus:
Laurance C. Herold, (D), Southampton (UK), 1968, Og

University of Northern Colorado

Earth Sciences (B,M) (2007)
Campus Box 100
School of Chem., ESCI, Physics
Greeley, CO 80639
p. (970) 351-2647 f. (970) 351-4197
william.hoyt@unco.edu
http://esci.unco.edu
Administrative Assistant: V. Ouellette
Enrollment (2007): B: 124 (27) M: 9 (4)

Director:
William H. Hoyt, (D), Delaware, 1982, Gu
Director, MAST Inst.:
Steve Anderson, (D), Arizona, 1990, GvOe
Professor:
Richard D. Dietz, (D), Colorado, 1965, Xg
Assistant Professor:
Graham Baird, (D), Minnesota, 2006, Gcz
Emmett Evanoff, (D), Colorado, 1990, Grs
Gary R. Huffines, (D), Texas A&M, 1999, Oa
Paul Nutter, (D), Oklahoma, 2003, Oaw
Lucinda Shellito, (D), California (Santa Cruz), 2004, Oa
Lecturer:
Rita F. Leafgren, (M), N Colorado, 1965, Gg
Libby Prueher, (D), Michigan, 1999, Gg
Adjunct Research Faculty:
Maynard Slaughter, (D), Pittsburgh, 1962, Cg
Adjunct Professor:
Walter A. Lyons, (D), Chicago, 1970, Oa
Victor J. Mayer, (D), Colorado, 1966, Oe
Jared Morrow, (D), Colorado, 1997, Gds
Emeritus:
Kenneth D. Hopkins, (D), Washington, 1976, Gm
William D. Nesse, (D), Colorado, 1977, Gx
K. Lee Shropshire, (D), Colorado, 1974, Pg
Leslie W. Trowbridge, (D), Michigan, 1961, Oe
Laboratory Director:
Patricia Mills, (M), OooOo

Western State College of Colorado

Dept of Geology (B) (2007)
Gunnison, CO 81231
p. (303) 943-2015 f. (303) 943-7120
astork@western.edu
http://www.western.edu/geology
Enrollment (2007): B: 53 (11)

Rady Chair:
James Coogan, (D), Wyoming, 1992, Gc
Moncrief Chair:
Ronald J. Hill, (D), California (Los Angeles), 1994, GoCos
Professor:
Robert P. Fillmore, (D), Kansas, 1994, Gs
Allen L. Stork, (D), California (Santa Cruz), 1984, Giv
Assistant Professor:
David W. Marchetti, (D), Utah, 2006, GmCl
Emeritus:
Bruce L. Bartleson, (D), Colorado, 1968, Gs
Thomas L. Prather, (D), Colorado, 1964, Gc

Connecticut

Central Connecticut State University

Dept of Physics & Earth Sciences (B,M) (2004)
1615 Stanley Street
New Britain, CT 06050-4010
p. (860) 832-2930 f. (860) 832-2946
antar@CCSU.EDU
Department Secretary: Sandra O'Day
Enrollment (1997): B: 29 (8)

Chair:
Ali A. Antar, (D), Connecticut, Oo
Professor:
Charles A. Baskerville, (D), New York, 1965, Ng
Sandra Burns, (D), Connecticut, 1972, Oe
Steven B. Newman, (D), SUNY (Albany), Oa
Associate Professor:
Marsha Bednarski, (D), Connecticut, 1997, Oe
Kristine Larsen, (D), Connecticut, 1988, Xy
Emeritus:
Charles W. Dimmick, (D), Tulane, 1969, Ge
Related Staff:
R. Craig Robinson, (B), Millersville, 1971, Xy

Eastern Connecticut State University

Environmental Earth Science Dept (B,M) (2006)
83 Windham Street
Willimantic, CT 06226

p. (860) 465-4317 f. (860) 465-5213
CarlquistZ@easternct.edu
http://www.easternct.edu/depts/eearthsci/index.htm
Enrollment (2000): B: 79 (12)

Chair:
Roy R. Wilson, (D), Oregon State, 1984, Oi
Professor:
Sherman M. Clebnik, (D), Massachusetts, 1975, Gl
Associate Professor:
Catherine A. Carlson, (D), Michigan State, 1994, Hw
James Andrew Hyatt, (D), Queen's, 1993, Gm
Assistant Professor:
Peter A. Drzewiecki, (D), Wisconsin, 1996, Gs
Lecturer:
Susan Bruening, (M), E Connecticut, 1988, Og
Vishnu R. Khade, (D), Cincinnati, 1987, Oo
James Motyka, (M), E Connecticut, 1987, Og
Wesley Winterbottom, (M), Connecticut, 1988, Oo
Emeritus:
Henry I. Snider, (D), New Mexico, 1966, Ge

Southern Connecticut State University

Dept of Earth Sciences (B) (2006)
501 Crescent Street
New Haven, CT 06515
p. (203) 392-5835 f. (203) 392-6885
coronc1@southernct.edu
http://www.southernct.edu/departments/earth-science/
Enrollment (2006): B: 74 (8)

Professor:
Cynthia R. Coron, (D), Toronto, 1982, Em
Associate Professor:
Thomas H. Fleming, (D), Ohio State, 1995, Gi
James W. Fullmer, (D), MIT, 1979, Ow
Adjunct Professor:
Susan Butts, (D), Idaho, Pi
Mihali Felipe, (D), Yale, Yg
Emeritus:
John W. Drobnyk, (D), Rutgers, 1962, Gg

University of Connecticut

Center for Integrative Geosciences (B,M,D) (2006)
345 Mansfield Road
U-2045
Storrs, CT 06269-2045
p. (860) 486-4435 f. (860) 486-1383
geology@uconn.edu
http://www.geosciences.uconn.edu
Administrative Assistant: Abigail W. Howe
Enrollment (2006): B: 4 (12) M: 2 (3) D: 8 (1)

Program Director:
Pieter Visscher, (D), Groningen, 1991, Co
Graduate Director:
Zoe Cardon, (D), Stanford, 1994, Py
Professor:
Vernon F. Cormier, (D), Columbia, 1976, Ys
William F. Fitzgerald, (D), MIT/Woods Hole, 1970, Oc
Raymond Joesten, (D), Caltech, 1970, Gp
Gary A. Robbins, (D), Texas A&M, 1983, Hw
Robert M. Thorson, (D), Washington, 1970, Gm
Thomas Torgersen, (D), Columbia, 1977, Cm
Associate Professor:
Timothy Byrne, (D), California (Santa Cruz), 1981, Gc
Jean M. Crespi, (D), Colorado, 1985, Gc
Lanbo Liu, (D), Stanford, 1993, Yg
Assistant Professor:
Andrew M. Bush, (D), Harvard, 2005, Pg
Adjunct Professor:
J. Gregory McHone, (D), North Carolina, 1978, Gi
Emeritus:
Larry Frankel, (D), Nebraska, 1956, Pm
Alfred J. Frueh, (D), MIT, 1949, Gz
Norman H. Gray, (D), McGill, 1971, Gx
Homer C. Liese, (D), Utah, 1962, Ct
Anthony R. Philpotts, (D), Cambridge, 1963, Gi

Dept of Marine Sciences (M,D) (2004)
1080 Shennecossett Road
Groton, CT 06340
p. (860)405-9151 f. (860)405-9153
mscadm03@uconnvm.uconn.edu
Enrollment: No data reported since 1995

Head:
Robert B. Whitlatch, (D), Chicago, 1976, Ob
Professor:
Walter F. Bohlen, (D), MIT/Woods Hole, 1969, Op
Richard A. Cooper, (D), Rhode Island, 1963, Ob
Hans G. Dam, (D), SUNY (Stony Brook), 1989, Ob
William F. Fitzgerald, (D), MIT/Woods Hole, 1970, Oc
James N. Kremer, (D), Rhode Island, 1975, Ob
Edward C. Monahan, (D), MIT, 1966, Oa
Thomas Torgersen, (D), Columbia, 1977, Cl
Associate Professor:
George B. McManus, (D), SUNY (Stony Brook), 1986, Ob
James O'Donnell, (D), Delaware, 1986, Op
Assistant Professor:
Pieter T. Visscher, (D), Groningen, 1991, Co
Contractor:
Mary DiGiacomo-Cohen, (B), Rochester, 1985, Gl

University of New Haven

Dept of Biology & Environmental Sciences (B,M) (2004)
300 Orange Avenue
West Haven, CT 06516
p. (203) 932-7101 f. (203) 931-6097
rzajac@newhaven.edu
http://www.newhaven.edu
Administrative Assistant: Alexis Zuhuski
Enrollment (1998): B: 17 (3) M: 68 (12)

Undergraduate Director:
R. Laurence Davis, (D), Rochester, 1980, Ge
Graduate Director:
Roman N. Zajac, (D), Connecticut, 1985, Oo
Professor:
Henry E. Voegeli, (D), Rhode Island, 1970, Oo
Assistant Professor:
David T. Osgood, (D), Virginia, 1996, Cl
Adjunct Professor:
Steven P. Broker, (M), Yale, 1983, Oo
Norman L. Richards, (D), Rhode Island, 1969, Ou
Sandra L. Selenskas, (D), SUNY (Buffalo), 1987, Oo
Robert H. Wardwell, (M), Maryland, 1974, Oo
Michael H. Ziskin, (M), Polytechnic (New York), 1985, Oo
Practitioner-in-Residence:
Daniel DePodesta, (M), New Haven, 1979, Oy
Cooperating Faculty:
George L. Wheeler, (D), Maryland, 1978, Cl

Wesleyan University

Dept of Earth & Environmental Sciences (B,M) (2007)
265 Church Street
Room 455
Middletown, CT 06459-0139
p. (860) 685-2244 f. (860) 685-3651
vharris@wesleyan.edu

http://www.wesleyan.edu/ees
Administrative Assistant: Virginia M. Harris
Enrollment (2006): B: 29 (15) M: 5 (2)

Chair:
Johan C. Varekamp, (D), Utrecht (Neth), 1979, CgGzv
Professor:
Barry Chernoff, (D), Michigan, 1983, Ge
Peter C. Patton, (D), Texas, 1976, Gm
Associate Professor:
Martha S. Gilmore, (D), Brown, 1997, XgGmOr
Suzanne B. O'Connell, (D), Columbia, 1986, GsuGe
Assistant Professor:
Timothy C.W. Ku, (D), Michigan, 2001, Cl
Phillip G. Resor, (D), Stanford, 2003, Gct
Dana Royer, (D), Yale, 2002, PeoPb
Research Professor:
James P. Greenwood, (D), Brown, 1997, Xc
Ellen Thomas, (D), Utrecht (Neth), 1979, Pm
Emeritus:
Jelle Z. De Boer, (D), Utrecht (Neth), 1963, YmGta
James T. Gutmann, (D), Stanford, 1972, Gv
Facilities Manager:
Joel LaBella, (B), S Connecticut, 1987, Oo

Western Connecticut State University

Dept of Biological & Environmental Sciences (B,M) (2004)
Danbury, CT 06810
p. (203) 837-8791 f. (203) 837-8320
halliburtonr@wcsu.edu
Administrative Assistant: Mildred Jackson
Enrollment: No data reported since 1995

Professor:
Ray Castillo, (D), SUNY (Albany), 1979, Oa
Ronald Chriss, (D), Connecticut, 1976, Cm
Melvin Goldstein, (D), New York, 1976, Ow
Kal Kanungo, (D), Ohio State, 1963, Ob
Phillip K. Lu, (D), Columbia, 1977, Oo
Howard Russock, (D), West Virginia, 1975, Ob
Yves M. Tourre, (D), Virginia, 1979, Op
Associate Professor:
Dennis W. Dawson, (D), New Mexico State, 1977, Oo
Assistant Professor:
Tom Philbrick, (D), Connecticut, 1980, Oo
Adjunct Professor:
Lloyd H. Burckle, (D), New York, 1971, Gu
James DeAngelis, (M), McGill, 1984, Ng
Brian R. Spies, (D), MacQuarie, 1980, Ye
Robert Stewart, (D), W Ontario, 1982, Hw
Emeritus:
Chin Chen, (D), Boston, 1962, Gu

Yale University

Dept of Geology & Geophysics (B,D) (2006)
210 Whitney Avenue
P.O. Box 208109
New Haven, CT 06520-8109
p. (203) 432-3114 f. (203) 432-3134
ernestine.jones@yale.edu
http://geology.yale.edu/
Enrollment (2004): B: 10 (4) D: 33 (4)

Professor:
Jay J. Ague, (D), California (Berkeley), 1987, Gx
David Bercovici, (D), California (Los Angeles), 1989
Robert A. Berner, (D), Harvard, 1962, Cl
Mark T. Brandon, (D), Washington, 1984, Gc
Derek E G Briggs, (D), Cambridge, 1976
Jacques Gauthier, (D), California (Berkeley), 1984, Pv
Robert B. Gordon, (D), Yale, 1955, Nr
Leo J. Hickey, (D), Princeton, 1967, Pb
Shun-ichiro Karato, (D), Tokyo, 1977
Jeffrey J. Park, (D), California (San Diego), 1985, Ys
Danny M. Rye, (D), Minnesota, 1972, Cs
Brian J. Skinner, (D), Harvard, 1955, Cg
Ronald B. Smith, (D), Johns Hopkins, 1975, Oa
Karl K. Turekian, (D), Columbia, 1955, Xm
George Veronis, (D), Brown, 1954, Op
Elisabeth S. Vrba, (D), Cape Town, 1974, Pv
John S. Wettlaufer, (D), Washington, 1991
Senior Research Scientist:
Edward W. Bolton, (D), California (Los Angeles), 1985, Gq
Senior Scientist:
Ellen Thomas, (D), Utrecht (Neth), 1979, Pm
Associate Professor:
Ruth E. Blake, (D), Michigan, 1997, Cg
David A D Evans, (D), Caltech, 1998
Steven C. Sherwood, (D), Scripps, 1995
Assistant Professor:
Jun Korenaga, (D), MIT, 2000
Mark Pagani, (D), Penn State, 1998
Adjunct Professor:
Leo W. Buss, (D), Johns Hopkins, 1979, Po
Thomas E. Graedel, (D), Michigan, 1969, Cl
Visiting Professor:
Adolf Seilacher, (D), Tubingen, 1951, Pi
Research Affiliate:
William C. Graustein, (D), Yale, 1981, Cl
Related Staff:
Elizabeth K. Berner, (D), Radcliffe, 1959, Cl
Laurent Bonneau, (M), Connecticut, 1997, Oa
James O. Eckert, (D), Texas A&M, 1988, Gp
Jason Evans, (D), Australian National, 2001
Roland Geerken, (D), Karlsruhe, 1992, Or
Zhenting Jiang, (D), Gottingen, 1998
Dorothy M. Koch, (D), Yale, 1993, Yg
Gerard A. Olack, (D), Purdue, 1990
Francis J. Robinson, (D), Hong Kong Sci & Tech, 1999
David E. Seidemann, (D), Yale, 1976, Cg
H. Catherine W. Skinner, (D), Adelaide, 1959, Gz
Stuart N. Thomson, (D), Univ Coll (London), 1992

Peabody Museum of Natural History (2004)
170 Whitney Avenue
Box 208118
New Haven, CT 06520-8118
p. (203) 432-3750 f. (203) 432-9816
http://www.peabody.yale.edu/

Curator:
Jacques Gauthier, (D), California (Berkeley), 1984, Pv
Leo Hickey, (D), Princeton, 1967, Pb
Karl K. Turekian, (D), Columbia, 1955, Xm
Elisabeth S. Vrba, (D), Cape Town, 1974, Pv
Emeritus:
John H. Ostrom, (D), Columbia, 1960, Pv

Delaware

University of Delaware

Dept of Geological Sciences (B,M,D) (2006)
Newark, DE 19716
p. (302) 831-2569 f. (302) 831-4158
smcgeary@udel.edu
http://www.geosci.udel.edu
Administrative Assistant: Cheryl Doherty
Enrollment (2006): B: 45 (10) M: 9 (5) D: 8 (3)

Acting Chair:
Susan McGeary, (D), Stanford, 1984, Yg
Professor:
Ronald E. Martin, (D), California (Berkeley), 1981, Pm
James E. Pizzuto, (D), Minnesota, 1982, Gm
John F. Wehmiller, (D), Columbia, 1971, Cl
Associate Professor:
John A. Madsen, (D), Rhode Island, 1987, Yr
Assistant Professor:
Arthur C. Trembanis, (D), William & Mary, 2004, On
Emeritus:
Billy P. Glass, (D), Columbia, 1968, Xm
Robert R. Jordan, (D), Bryn Mawr, 1964, Gr
John C. Kraft, (D), Minnesota, 1955, Gs
Peter B. Leavens, (D), Harvard, 1967, Gz
Frederick M. Swain, (D), Kansas, 1943, Ps
Allan M. Thompson, (D), Brown, 1968, Gx
Cooperating Faculty:
A. Scott Andres, (M), Hy
Katharina Billups, (D), California (Santa Cruz), 1998, Ou
Shreeram Inamdar, (D), Virginia Tech, Hg
Thomas E. McKenna, (D), Texas, 1997, Hy
Peter P. McLaughlin, Jr., (D), Gr
Michael ONeal, (D), Washington, 2005, Gm
Kelvin W. Ramsey, (D), Delaware, Gs
William S. Schenck, (M), Delaware, Gg
Christopher K. Sommerfield, (D), SUNY (Stony Brook),1997, On
John H. Talley, (M), Delaware, Eg
William J. Ullman, (D), Chicago, 1982, Oc
Kyungsoo Yoo, (D), California (Berkeley), Os

Oceanography Program (M,D) (2006)
700 Pilottown Road
Lewes, DE 19958
p. (302) 645-4279 f. (302) 645-4007
cms-academics@udel.edu
http://www.ocean.udel.edu
Enrollment (2000): M: 21 (2) D: 16 (5)
Professor:
Thomas M. Church, (D), California (San Diego), 1970, Cm
Victor Klemas, (D), Braunschweig (Germany), 1965, Or
George W. Luther, III, (D), Pittsburgh, 1972, Cm
Jonathan H. Sharp, (D), Dalhousie, 1972, Oc
William J. Ullman, (D), Chicago, 1982, Cm
Ferris Webster, (D), MIT, 1961, Op
Xiao-Hai Yan, (D), SUNY (Stony Brook), 1989, Or
Associate Professor:
Katharina Billups, (D), California (Santa Cruz), 1998, Pe
Douglas C. Miller, (D), Washington, 1985, Ob
Associate Scientist:
Charles H. Culberson, (D), Oregon State, 1972, Oc
Richard T. Field, (D), Delaware, 1994
Assistant Professor:
David A. Hutchins, (D), California (Santa Cruz), 1994, Ob
Christopher K. Sommerfield, (D), SUNY (Stony Brook), 1997, Gs
Adjunct Professor:
Richard B. Coffin, (D), Delaware, 1986, Oo
James Crease, (D), Cambridge, 1951, Op
Marilyn L. Fogel, (D), Texas, 1977, Oc
Norden E. Huang, (D), Johns Hopkins, 1967, Oo
David E. Krantz, (D), South Carolina, Cs
Kamlesh Lulla, (D), Indiana State, 1963, Or
Donald B. Nuzzio, (D), Rutgers, 1982, Oc
Manmohan Sarin, (D), Gujarat (India), 1984, Oo
Alain J. Veron, (D), Paris, 1988, Oc
John F. Wehmiller, (D), Columbia, Gu

Emeritus:
Jin Wu, (D), Iowa, 1964, Op

District of Columbia

Carnegie Institution of Washington

Dept of Terrestrial Magnetism (2006)
5241 Broad Branch Road, N.W.
Washington, DC 20015-1305
p. (202) 478-8820 f. (202) 478-8821
dept@dtm.ciw.edu
http://www.dtm.ciw.edu
Fiscal Officer: Terry L. Stahl
Director:
Sean C. Solomon, (D), MIT, 1971, Yg
Senior Scientist:
Conel M. O'D Alexander, (D), Essex (UK), 1987, Xc
Alan P. Boss, (D), California (Santa Barbara), 1979, Oo
R. Paul Butler, (D), Maryland, 1993, Oo
Richard W. Carlson, (D), California (San Diego), 1980, Cc
John E. Chambers, (D), Manchester, 1994, Oo
Erik H. Hauri, (D), MIT/Woods Hole, 1992, Cc
David E. James, (D), Stanford, 1967, Ys
Alan T. Linde, (D), Queensland (Australia), 1972, Ys
Larry R. Nittler, (D), Washington (St. Louis), 1996, Xc
I. Selwyn Sacks, (D), Witwatersrand, 1961, Ys
Sara Seager, (D), Harvard, 1999, Oo
Steven B. Shirey, (D), SUNY (Stony Brook), 1984, Cc
Paul G. Silver, (D), California (San Diego), 1982, Ys
Alycia J. Weinberger, (D), Caltech, 1998, Oo
Senior Fellow:
Vera C. Rubin, (D), Georgetown, 1954, Oo
Research Associate:
Alceste Z. Bonanos, (D), Harvard, 2005, Oo
Henner Busemann, (D), ETH (Zurich), 1998, Ca
Fred Ciesla, (D), Arizona, 2003, Xm
John H. Debes, (D), Penn State, 2005, Oo
Hannah Jang-Condell, (D), Harvard, 2004, Oo
Mercedes López-Morales, (D), North Carolina, 2004, Oo
Isamu Matsuyama, (D), Toronto, 2005, Oo
Brian Savage, (D), Caltech, 2004, Ys
Ivan P. Savov, (D), S Florida (Tampa), 2004, Cg
Scott S. Sheppard, (D), Hawaii, 2004, Oo
Taka'aki Taira, (D), Hokkaido, 2004, Ys
Margaret Turnbull, (D), Arizona, 2004, Oo
Lara S. Wagner, (D), Arizona, 2005, Oo
Linda M. Warren, (D), California (San Diego), 2003, Ys
Dayanthie S. Weeraratne, (D), Brown, 2004, Ys
Emeritus:
L. Thomas Aldrich, (D), Minnesota, 1948, Yg
John A. Graham, (D), Australian National, 1964, Oo
Fouad Tera, (D), Vienna, 1962, Cc
George W. Wetherill, (D), Chicago, 1953, Og
Cooperating Faculty:
Shaun J. Hardy, Oo

Geophysical Laboratory (2005)
5251 Broad Branch Road, N.W.
Washington, DC 20015-1305
p. (202) 478-8900 f. (202) 478-8901
w.huntress@gl.ciw.edu
http://www.gl.ciw.edu/
Director:
Wesley T. Huntress, Jr., (D), Stanford, 1968, Xc
Senior Scientist:
George D. Cody, (D), Penn State, 1992, Co
Ronald E. Cohen, (D), Harvard, 1985, Gy
Yingwei Fei, (D), Brooklyn, 1989, Cp

Marilyn L. Fogel, (D), Texas, 1977, Co
Alexander Goncharov, (D), TMRRAS, 1983, Gy
Russell J. Hemley, (D), Harvard, 1983, Yx
Ho-kwang Mao, (D), Rochester, 1968, Yx
Bjorn O. Mysen, (D), Penn State, 1974, Cp
Douglas Rumble, III, (D), Harvard, 1969, Gp
Andrew Steele, (D), Portsmouth, 1996, Xy
Viktor Struzhkin, (D), Moscow Inst of Physics & Tech, 1991, Gy

Summer Education Coordinator:
Stephen A. Gramsch, (D), Chicago, 1994, Gy

Staff Associate:
Burkhard Militzer, (D), Illinois, 2000, Yx
Przemyslaw Dera, (D), Mickiewicz (Poland), 1999, Gy
James H. Scott, (D), Wisconsin (Milwaukee), 1999, Co

Research Tech:
Jinfu Shu, (M), Wuhan, 1981, Yx

Research Scientist:
Nabil Boctor, (D), Purdue, 1976, Cg

Beamline Scientist:
Jingzhu Hu, (D), Chinese Acad of Sci, 1965, Yx
Zhen-xian Liu, (D), Jilian (China), 1990, Gy
Maddury Somayazulu, (D), Bombay, 1992, Yx

Adjunct Senior Research Scientist:
Peter M. Bell, (D), Harvard, 1963, Yx
Robert M. Hazen, (D), Harvard, 1975, Gz

Emeritus:
T. Neil Irvine, (D), Caltech, 1959, Gi

Cooperating Faculty:
Shaun J. Hardy, Oo
Gotthard Saghi Szabo, (D), Maryland, 1995, Gy
John Straub, (M), Oo

George Washington University

Dept of Geography (B,M) (2004)
619 21st Street, NW
Washington, DC 20052
p. (202) 994-6185 f. (202) 994-0458
mprice@gwu.edu
Department Secretary: Karin Johnston

Chair:
Dorn C. McGrath, Jr., (M), Harvard, 1959, Oy

Professor:
John C. Lowe, (D), Clark, 1969, Og

Assistant Professor:
Elizabeth Chacko, (D), California (Los Angeles), 1997, Oy
Doult O. Fuller, (D), Maryland, 1994, Oy
Marie D. Price, (D), Syracuse, 1990, Oy

Visiting Professor:
Micheal Brewer, (D), Delaware, 1997, Or
Thomas Foltltin, (D), Montana, 1980, Oy

Nat Academy of Sci/National Res Council

Board on Earth Sciences and Resources (2007)
The National Academies Keck Center, 618
500 5th Street, NW
Washington, DC 20001
p. (202) 334-2744 f. (202) 334-1377
adesouza@nas.edu

Director:
Anthony R. deSouza, (D), Reading (UK), 1968, Og

Senior Scientist:
David A. Feary, (D), Australian National, 1984
Anne M. Linn, (D), California (Los Angeles), 1991, Gs

Smithsonian Inst / Nat Air & Space Mus

Center for Earth & Planetary Studies (2007)
MRC 315, P.O. Box 37012
6th and Independence Ave., SW
Washington, DC 20013-7012
p. (202) 633-2470 f. (202) 786-2566
grantj@si.edu
http://nasm.si.edu
Administrative Assistant: Donna Slattery
Enrollment (1998): B: 3 (0)

Chair:
James R. Zimbelman, (D), Arizona State, 1984, Xg

Geologist:
Robert A. Craddock, (D), Virginia, 1999, Xg
Ted A. Maxwell, (D), Utah, 1977, Xg
Priscilla L. Strain, (B), Smith, 1974, Or
Thomas R. Watters, (D), George Washington, 1985, Xg

Senior Scientist:
Bruce A. Campbell, (D), Hawaii, 1991, Xy
John A. Grant, (D), Brown, 1990, Xg

Research Associate:
Andrew K. Johnston, (B), George Washington, 1991, Oy

Photo Librarian:
Rosemary Steinat, Oo

Smithsonian Inst / Nat Mus of Natural History

Dept of Mineral Sciences (2007)
NHB MRC 119
10th & Constitution Avenue, NW
PO Box 37012
Washington, DC 20013-7012
p. (202) 633-1860 f. (202) 357-2476
gvp@si.edu
http://mineralsciences.si.edu/

Chair:
Sorena S. Sorensen, (D), California (Los Angeles), 1984, Gp

Research Geologist:
Jonathan M. Castro, (D), Oregon, 1999, Gvi
Elizabeth Cottrell, (D), Columbia, 2004, Cp
Glenn J. MacPherson, (D), Princeton, 1981, XcGi
Timothy J. McCoy, (D), Hawaii, 1994, Xmg
Jeffrey E. Post, (D), Arizona State, 1981, Gz
Edward P. Vicenzi, (D), Rensselaer Polytechnic, 1991, CaXc
Michael A. Wise, (D), Manitoba, 1987, Gz

Collection Manager:
Russell Feather, (B), George Mason, 1982, Gz
Leslie Hale, (B), Maryland, 1989, Gg
Stacia Largen, (B), George Mason, 2002, Oo
LuElla Speakman, (M), Missouri-Columbia, 2006, Gg
Linda Welzenbach, (M), Bowling Green, 1992, Xm

Associate Scientist:
Cathe Brown, (M), Maryland, 1996, Gz
Timothy Gooding, (B), Hampshire, 1990, Gg
Paul Kimberly, (M), Michigan Tech, 1995, GvOi
Amelia Logan, (D), La Plata (Argentina), 1991, Ca
Timothy O'Hearn, (M), George Washington, 1984, GvCa
Timothy Rose, (M), Maryland, 1991, CaGv
Lee Siebert, (M), George Washington, 1983, Gv
Edward Venzke, (M), Minnesota - Duluth, 1993, Gv
Richard Wunderman, (D), Michigan Tech, 1988, Gv

Post Doctoral Fellow:
Emma Bullock, (D), Open Univ (England), 2006, CaXm
Eloise Gaillou, (D), Institut des Matériaux Jean Rouxel - Nantes (France), 2006, Gz
Mariek Schmidt, (D), Oregon State, 2005, XgGi

Contract Geologist:
Carter Hearn, (D), Johns Hopkins, 1959, Gi
Marc Lipella, (B), Maryland, 2004, Cg
Sally K. Sennert, (M), Pittsburgh, 2003, GvOr

Emeritus:
Roy S. Clarke, Jr., (D), George Washington, 1976, Xm

Richard S. Fiske, (D), Johns Hopkins, 1960, Gv
Brian H. Mason, (D), Stockholm, 1943, Xm
William G. Melson, (D), Princeton, 1964, GvtGa
Tom Simkin, (D), Princeton, 1965, Gv
Management Support:
Phyllis McKenzie, Oo
Ellen Thurnau, (M), Duke, 1981, Oo

Dept of Paleobiology (2006)
MRC121
NMNH, Smithsonian Institution
P.O. Box 37012
Washington, DC 20013-7012
p. (202) 633-1320 f. (202) 786-2832
paleodept@si.edu
http://www.nmnh.si.edu/paleo/
Curator:
Scott L. Wing, (D), Yale, 1981, Pb
Curator:
Anna K. Behrensmeyer, (D), Harvard, 1973, Pv
William A. DiMichele, (D), Illinois, 1979, Pb
Brian T. Huber, (D), Ohio State, 1988, Pm
Conrad C. Labandeira, (D), Chicago, 1991, Pi
Ian G. Macintyre, (D), McGill, 1967, Gs
Curator:
Martin A. Buzas, (D), Yale, 1963, Pm
Douglas H. Erwin, (D), California (Santa Barbara), 1985, Pi
Daniel J. Stanley, (D), Grenoble, 1964, Ou
Curator:
Matthew T. Carrano, (D), Chicago, 1998, Pv
Curator:
Gene Hunt, (D), Chicago, 2003, Pm
Geologist:
Thomas Dutro, (D), Pi
Bevan French, (D), Xm
Sergius Mamay, (D), Pb
John Pojeta, (D), Pi
Neuman Robert, (D), Pi
Curator:
Alan H. Cheetham, (D), Columbia, 1959, Po
Robert Emry, (D), Columbia, 1970, Pv
Clayton E. Ray, (D), Harvard, 1962, Pv
Thomas R. Waller, (D), Columbia, 1966, Pi
Associate Chair & Collections Manager:
Jann W. M. Thompson, (B), George Washington, 1970, Pg

University of the District of Columbia

Dept of Biological & Environmental Science (A,B) (2005)
ATTN: Geoscience Program Advisor
MS 44-04
4200 Connnecticut Avenue, NW
Washington, DC 20008
p. (202) 274-5886 f. (202) 274-5932
magraham@yahoo.com
Enrollment: No data reported since 1995
Professor:
A. Jose Jones, (D), Georgetown, 1973, Ob
James Preer, (D), Caltech, 1970, Hg
Associate Professor:
Mohamed Elhelu, (D), Howard, 1976, Og
Victoria Guerrero, (D), Howard, 1975, Oo
Thomas Kakovitch, (M), Maryland, 1972, Oa
Harkewal Sekhon, (D), Wyoming, 1971, Og
Allen Wilkens, (M), Catholic, 1973, So
Instructor:
Ahmed Abdi, (D), Hood, 1988, Sd

Florida

Broward Community Coll, Central Campus

Dept of Physical Sci(Oceanography & Geology) (A) (2006)
3501 SW Davie Road
Fort Lauderdale, FL 33314
p. (954) 201-6771 f. (954) 201-6847
jmuza@broward.edu
Administrative Assistant: Nicki Pickett
Enrollment (2005): A: 10 (10)
Professor:
Jay P. Muza, (D), Florida State, 1996, Ou
Laura Precedo, (D), Emory, 1993, Oc
Angel Rodriguez, (M), California (San Diego), 1989, Ob
Assistant Professor:
Lewis Fox, (D), Delaware, 1981, Oc
Adjunct Professor:
Kristi Foster, (M), Nova Southeastern, 2004, Ob
Susan Frandsen, (M), Florida Atlantic, 2001, Og
Lon-Clay Hill, (D), Iowa, 1989, Xm
Stephen Vaos, (D), Florida State, 1980, Gu
Planetarium Director:
Susan Barnett, (B), Yale, Gg
Lab Manager:
Lynn A. Curtis, (B), Florida Atlantic, 1997, Gg

Eckerd College

Dept of Marine Science (B) (2004)
4200 54th Avenue South
St. Petersburg, FL 33711
p. (727) 864-8200 f. (727) 864-7964
natsci@eckerd.edu
http://www.eckerd.edu/academics/nas/msn/
Enrollment (1997): B: 149 (37)
Professor:
Gregg R. Brooks, (D), S Florida, 1986, Gu
Associate Professor:
Joel B. Thompson, (D), Syracuse, 1989, Py
Assistant Professor:
Laura R. Wetzel, (D), Washington (St. Louis), 1997, Yr

Florida Atlantic University

Dept of Geography & Geology (B,M) (2006)
Physical Sciences, 336
777 Glades Road
Boca Raton, FL 33431
p. (561) 297-3250 f. (561 297-2745
ivy@fau.edu
http://www.geosciences.fau.edu
Enrollment (2005): B: 88 (14) M: 25 (11)
Professor:
Edward J. Petuch, (D), Miami, 1980, Pg
Jorge I. Restrepo, (D), Colorado State, 1987, Hq
Associate Professor:
Russell L. Ivy, (D), Florida, 1992, Oo
Anton Oleinik, (D), Purdue, 1998, Pe
Charles E. Roberts, (D), Penn State, 1991, Or
David L. Warburton, (D), Chicago, 1978, Cg
Research Associate:
Tobin Hindle, (D), Florida Atlantic, 2006, Hw
Related Staff:
Thomas Corcoran, (M), Florida Atlantic, 1998, Oo

Florida Institute of Technology

Dept of Marine & Env Systems - Environmental Science Program (B,M,D) (2005)
150 West University Boulevard

Melbourne, FL 32901
p. (321) 674-8096 f. (321) 674-7212
dmes@fit.edu
http://www.fit.edu/dmes
Staff Assistant: Carmen Serrano
Enrollment (1998): B: 82 (22) M: 68 (8) D: 22 (0)
Head:
George A. Maul, (D), Miami, 1974, Op
Chair:
John G. Windsor, Jr., (D), William & Mary, 1976, Oc
Professor:
Thomas V. Belanger, (D), Florida, 1979, Oo
John H. Trefry, (D), Texas A&M, 1977, Cm
Associate Professor:
Charles R. Bostater, (D), Delaware, 1990, Op
Assistant Professor:
Steven Lazarus, (D), Oklahoma, Oa
Adjunct Professor:
Joseph A. Angelo, (D), Arizona, 1976, Og
Diane Barile, (M), Florida Inst of Tech, Ou
Michael F. Helmletter, (D), Old Dominion, Oc
J. J. Keaffaber, (D), Florida, Oc
Carlton R. Parks, (M), Oklahoma, 1981, Ow
D. R. Resio, (D), Virginia, Op
Ned P. Smith, (D), Wisconsin, 1972, Op
Geology Librarian:
Celine Alvey, (D), Oo

Dept of Marine & Env Systems-Ocean Eng Prog (B,M,D) (2004)
150 West University Boulevard
Melbourne, FL 32901
p. 321-674-8096 f. 321-674-7212
dmes@fit.edu
http://www.fit.edu/dmes
Staff Assistant: Carmen Serrano
Enrollment: No data reported since 1995
Head:
George A. Maul, (D), Miami, 1974, Op
Chair:
Andrew Zborowski, (D), Gdansk Univ of Tech, 1968, Oo
Professor:
Geoffry Swain, (D), Southampton, 1981, Oo
Associate Professor:
Lee E. Harris, (D), Florida Atlantic, 1995, Oo
Assistant Professor:
Eric D. Thosteson, (D), Florida, 1997, Oo
Stephen L. Wood, (D), Oregon State, 1995, Oo
Adjunct Professor:
W. A. Cleary, (M), RI Naval War Coll, 1951, Oo
Emeritus:
John C. Sainsbury, (D), Southampton, 1966, Oo

Dept of Marine & Env Systems-Oceanography Prog (B,M,D) (2004)
150 West University Boulevard
Melbourne, FL 32901
p. (321) 674-8096 f. (321) 674-7212
dmes@fit.edu
http://www.fit.edu/dmes
Staff Assistant: Carmen Serrano
Enrollment: No data reported since 1995
Head:
George A. Maul, (D), Miami, 1974, Op
Acting Chair:
John G. Windsor, (D), William & Mary, 1976, Oc
Professor:
Thomas V. Belanger, (D), Florida, Hw
Iver W. Duedall, (D), Dalhousie, 1973, Oc
Geoffrey W.J Swain, (D), Southampton, 1981, Ob
John H. Trefry, (D), Texas A&M, 1977, Cm
Gary A. Zarillo, (D), Georgia, 1979, Ou
Associate Professor:
Charles R. Bostater, (D), Delaware, 1990, Op
Lee E. Harris, (D), Florida Atlantic, 1995, On
Assistant Professor:
Elizabeth A. Irlandi, (D), North Carolina, 1993, Ob
Eric D. Thosteson, (D), Florida, On
Research Associate:
Simone Metz, (D), Florida Inst of Tech, 1986, Cm
Chih-Shin Shieh, (D), Florida Inst of Tech, 1988, Oc
Robert P. Trocine, (M), Florida Inst of Tech, 1979, Cm
Adjunct Professor:
Diane D. Barile, (M), Florida Inst of Tech, 1975, Ou
Michael F. Helmstetter, (D), Oc
J. J. Keaffaber, (D), Florida, Oc
D. R. Resio, (D), Virginia, Op
Ned P. Smith, (D), Wisconsin, 1972, Op
Emeritus:
Dean R. Norris, (D), Texas A&M, 1969, Ob

Meteorology Program (B,M) (2004)
150 West University Boulevard
Melbourne, FL 32901
p. 321-674-8096 f. 321-674-7212
dmes@fit.edu
http://www.fit.edu/dmes
Head:
George A. Maul, (D), Miami, 1974, Op
Professor:
Thomas V. Belanger, (D), Florida, Hw
John G. Windsor, Jr., (D), William & Mary, Oc
Associate Professor:
Joseph R. Dwyer, (D), Chicago, 1994, Oa
Tom Utley, (D), Florida Inst of Tech, 1996, Ow
Assistant Professor:
Steven M. Lazarus, (D), Oklahoma, Oa

Florida International University

Center for the Study of Matter at Extreme Conditions (M,D) (2004)
VH-150
Miami, FL 33199
p. 305-348-3030 f. 305-348-3070
saxenas@fiu.edu
http://cesmec.fiu.edu
Enrollment (2001): M: 2 (0) D: 2 (4)
Director:
Surendra K. Saxena, (D), Uppsala, 1967, Gy

Dept of Earth Sciences (B,M,D) (2007)
Miami, FL 33199
p. (305) 348-2365 f. (305) 348-3877
geology@fiu.edu
http://www.fiu.edu/orgs/geology
Enrollment (2007): B: 30 (7) M: 11 (5) D: 20 (5)
Chair:
Bradford M. Clement, (D), Columbia, 1985, Ym
Professor:
Grenville Draper, (D), West Indies, 1979, Gc
Rosemary Hickey-Vargas, (D), MIT, 1983, Cg
Jose F. Longoria, (D), Texas (Dallas), 1972, Ps
Florentin J-M. R. Maurrasse, (D), Columbia, 1973, Ps
Gautam Sen, (D), Texas (Dallas), 1981, Gi
Associate Professor:
William T. Anderson, (D), ETH (Switzerland), 2000, Cs

Laurel S. Collins, (D), Yale, 1988, Po
Michael Gross, (D), Penn State, 1993, Gc
Andrew W. Macfarlane, (D), Harvard, 1989, Em
Dean Whitman, (D), Cornell, 1993, Yg
Assistant Professor:
Rene Price, (D), Miami, 2001, Hw
Michael Sukop, (D), Kentucky, 2001, Hq
Ping Zhu, (D), Miami, 2002, Oa
Distinguished Research Professor:
Hugh E. Willoughby, (D), Miami, 1977, Oa
Instructor:
Deborah Arnold, (B), Florida Intl, Gg
Lecturer:
Neptune Srimal, (D), Rochester, 1986, Gt
Adjunct Professor:
Jose Antonio Barros, (D), Miami, 1995, Or
Michael Wacker, (M), Florida Intl, Gg
Research Scientist:
Edward Robinson, (D), London, 1969, Pm
Distinguished Research Professor:
Stephen E. Haggerty, (D), London, 1968, Gz
Research Lab Manager:
Diane H. Pirie, (B), Florida Intl, 1980, Gg
Cooperating Faculty:
Gabriel Guitierrez-Alonso, (D), Oviedo, 1992, Gc

Florida Keys Community College

Dept of Physical Science (2004)
Key West, FL 33040
p. (305) 296-9081 (Ext. 214)
gragg_b@firn.edu
Instructor:
Dennis Loehr, (D), California, 1980, Cg
Jeff Sunquist, (M), Wisconsin, 1980, Gg

Florida State University

Dept of Geological Sciences (B,M,D) (2005)
108 Carraway Building
Tallahassee, FL 32306-4100
p. (850) 644-5861 f. (850) 644-4214
karl@gly.fsu.edu
http://www.gly.fsu.edu
Program Assistant: Tami S. Karl
Enrollment (1998): B: 23 (8) M: 21 (5) D: 18 (3)
Chair:
LeRoy A. Odom, (D), North Carolina, 1971, Cg
Professor:
James B. Cowart, (D), Florida State, 1974, Cc
James F. Tull, (D), Rice, 1973, Gc
Sherwood W. Wise, Jr, (D), Illinois, 1970, Pm
Associate Professor:
Anthony J. Arnold, (D), Harvard, 1983, Pm
Joseph F. Donoghue, (D), S California, 1981, Gu
Xiaolong Hu, (D)
Stephen A. Kish, (D), North Carolina, 1983, Eg
Humayan Munir, (N)
William C. Parker, (D), Chicago, 1980, Pi
Vincent J. Salters, (D), MIT, 1989, Cg
Associate Scientist:
Yang Wang, (D), Utah, 1992, Cg
Assistant Professor:
Jennifer Georgen, (D)
Emeritus:
George W. Devore, (D), Chicago, 1952, Gp
John K. Osmond, (D), Wisconsin, 1954, Cc
Paul C. Ragland, (D), Rice, 1962, Ca

Dept of Meteorology (A,B,M,D) (2007)
404 Love Building
1017 Academic Way
P.O. Box 3064520
Tallahassee, FL 32306-4520
p. (850) 644-6205 f. (850) 644-9642
bobe@met.fsu.edu
http://www.met.fsu.edu
Enrollment (2007): B: 155 (34) M: 61 (23) D: 27 (5)
Chair:
Robert E. Ellingson, (D), Florida State, 1972, Ow
Professor:
Henry E. Fuelberg, (D), Texas A&M, 1976, Ow
T. N. Krishnamurti, (D), Chicago, 1959, Ow
Sharon E. Nicholson, (D), Wisconsin, 1976, Ow
Peter S. Ray, (D), Florida State, 1973, Ow
Xiaolei Zou, (D), Inst of Atm Physics, 1988, Ow
Associate Chair:
Paul H. Ruscher, (D), Oregon State, 1988, Ow
Associate Professor:
Jon E. Ahlquist, (D), Wisconsin, 1981, Ow
Ming Cai, (D), Illinois, 1990, Ow
Carol Ann Clayson, (D), Colorado, 1995, Ow
Philip Cunningham, (D), SUNY (Albany), 2000, Ow
Gousheng Liu, (D), Nagoya, 1990, Ow
Assistant Professor:
Mark A. Bourassa, (D), Purdue, 1993, Ow
Robert Hart, (D), Penn State, 2001, Ow
Paul Reasor, (D), Colorado State, 2000, Ow
Emeritus:
Albert I. Barcilon, (D), Harvard, 1965, Ow
Noel E. LaSeur, (D), Chicago, 1953, Ow
James J. O'Brien, (D), Texas A&M, 1966, Ow
Richard L. Pfeffer, (D), MIT, 1957, Ow

Dept of Oceanography (M,D) (2004)
Tallahassee, FL 32306-4320
p. 850 644-6700 f. 850 644-2581
sbrimm@ocean.fsu.edu
http://ocean.fsu.edu
Administrative Assistant: Stephenie Brimm
Academic Coordinator: Michaela Lupiani
Coord Administrative Services: Linda D. Carter
Enrollment (2000): M: 46 (3) D: 65 (8)
Chair:
Nancy H. Marcus, (D), Yale, 1976, Ob
David Thistle, (D), California (San Diego), 1977, Ob
Professor:
William C. Burnett, (D), Hawaii, 1974, Oc
Jeffery P. Chanton, (D), North Carolina, 1985, Cm
Allan J. Clarke, (D), Cambridge, 1976, Op
William K. Dewar, (D), MIT/Woods Hole, 1983, Op
Ya Hsueh, (D), Johns Hopkins, 1965, Op
Richard L. Iverson, (D), Oregon State, 1972, Ob
Ruby Krishnamurti, (D), California (Los Angeles), 1967, Op
William M. Landing, (D), California (Santa Cruz), 1983, Oc
Doron Nof, (D), Wisconsin, 1976, Op
Kevin G. Speer, (D), MIT/Woods Hole, 1988, Op
Melvin E. Stern, (D), MIT, 1956, Op
Wilton Sturges, (D), Johns Hopkins, 1966, Op
Georges L. Weatherly, (D), Nova, 1971, Op
Associate Professor:
Joel Kostka, (D), Delaware, 1993, Ob
Assistant Professor:
Lita M. Proctor, (D), SUNY (Stony Brook), 1991, Ob
Emeritus:
John W. Winchester, (D), MIT, 1955, Oc

Florida

Gulf Coast Community College

Dept of Physical Science (A) (2005)
5230 West U.S. Highway 98
Panama City, FL 32401-1058
p. (850) 769-1551 (Ext. 2858) f. (850) 873-3529
jwebb@gulfcoast.edu
http://www.gulfcoast.edu
Enrollment: No data reported since 1995

Professor:
John Mix, (D), Florida State, 1975, Oe
Assistant Professor:
Jamieson D. Webb, (M), Auburn, 2001, Gg

Jacksonville University

Dept of Biology & Marine Science (B) (2004)
2800 University Boulevard North
Jacksonville, FL 32211
p. (904) 256-7302 f. (904) 256-7573
lclemen@ju.edu
Enrollment (2001): B: 39 (6)

Professor:
Ted T. Allen, (D), Oo
Lee Ann J. Clements, (D), South Carolina, 1988, Ob
Kenneth D. Hoover, (D), Oo
Assistant Professor:
Karen E. Jackson, (D), Oo
Daniel McCarthy, (N)
Curtis Small, (D)

Miami-Dade Community Coll (North Campus)

Dept of Chemistry - Earth Science (A) (2007)
11380 NW 27th Avenue
Miami, FL 33167
p. (305) 237-1084 f. (305) 237-1529
sorbon@mdc.edu (Department Secretary
http://www.mdc.edu/kendall/chmphy/
Enrollment (1996): A: 16 (9)

Miami-Dade Comm Coll (Wolfson Campus)

Dept of Natural Sciences, Health & Wellness (A) (2005)
300 NE 2nd Avenue
Miami, FL 33132
p. (305) 237-3025 f. (305) 237-3645
jsalinas@mdc.edu
Department Secretary: Ileana Baldizon
Enrollment (2000): A: 4 (0)

Professor:
Michael Kaldor, (M), SUNY (Buffalo), 1969, Gg
Assistant Professor:
Tony Barros, (D), Miami, 1995, Gc

Nova Southeastern University

Inst of Marine & Coastal Studies (M,D) (2006)
Oceanographic Center
8000 North Ocean Drive
Dania Beach, FL 33004
p. (954) 262-3600 f. (954) 262-4098
imcs@nsu.nova.edu
http://www.nova.edu/ocean
Administrative Assistant: Nichole Dean
Administrative Assistant: Greg Jacoski
Administrative Assistant: Peggy Oellrich
Enrollment (2006): M: 300 (47) D: 10 (2)

Director:
Richard C. Spieler, (D), Louisiana State, 1975, Ob
Dean:
Richard E. Dodge, (D), Yale, 1978, Py
Professor:
Charles Messing, (D), Miami, 1981, Ob
James D. Thomas, (D), Florida Inst of Tech, 1991, Ob
Associate Professor:
Patricia L. Blackwelder, (D), South Carolina, 1976, Py
Curtis Burney, (D), Rhode Island, 1981, Oc
Joshua Feingold, (D), Miami, 1994, Ob
Edward Keith, (D), California (Santa Cruz), 1984, Ob
Bernhard M. Riegl, (D), Cape Town, 1993, Ob
Mahmood S. Shivji, (D), Washington, 1990, Ob
Alexander V. Soloviev, (D), Shirshov Inst of Ocean, 1992, Op
Assistant Professor:
David Gilliam, (D), Nova Southeastern, 1999, Ob
Sean Kennan, (D), Hawaii, 1997, Og
Sam Purkis, (D), Vrije (Amsterdam), 2004, Or
Jennifer Regage, (D), Kentucky, 2003, Oo
Research Associate:
Demian D. Chapman, (M), Nova Southeastern, 2004, Ob
Dan Fahy, (M), Nova Southeastern, 2005, Oo
Greg Foster, (M), Nova Southeastern, 2005, Or
Kristi Foster, (M), Nova Southeastern, 2005, Or
Shaun Gill, (M), Nova Southeastern, 2006, Oo
Kevin P. Helmle, (M), Nova Southeastern, 1998, Ob
Erin Hodel, (M), Nova Southeastern, Oo
Lance Jordan, (M), Nova Southeastern, 2003, Ob
Lindsey Klink, (M), Nova Southeastern, Oo
Alison Moulding, (D), Miami, 2005, Oo
Melissa Phillips, (M), Nova Southeastern, Oo
Pat Quinn, (M), Nova Southeastern, Ob
Abigail Renegar, (M), Nova Southeastern, 2003, Oo
Vince Richards, (M), Nova Southeastern, 2005, Ob
Nicole Stephens, (M), Nova Southeastern, Oo
Maureen Trnka, (M), Nova Southeastern, Oo
Joanna Walczak, (M), Nova Southeastern, Oo
Brian K. Walker, (M), Nova Southeastern, 2002, Oo
Asst Professor Law School:
Brion Blackwelder, (D), Nova Southeastern
Adjunct Professor:
Paul Arena, (D), Nova Southeastern, 2005, Oo
Barry Barker, (D), Syssex Coll of Tech, Oo
Gregory Booten, (D), Ohio State, Oo
Mensun Bound, (M), Rutgers, Oo
Barbara Brunnick, (D), Union Inst, Oo
Nancy Craig, (D), SUNY (Stony Brook), Oo
O P. Dwivedi, (D), Queens, Oo
Randy Edwards, (D), Miami, Oo
Mark Farber, (D), Miami, 1990
Jay Fleisher, (D), New York, Oo
Nancy Gassman, (D), Miami, Oo
William Hamlett, (D), Clemson, Oo
Patrick Hardigan, (D), Wyoming, Oo
David Kerstetter, (D), Virginia Inst of Marine Sci, Oo
Vladimir Kosmynin, (D), Moscow State, Gm
Philip Light, (D), James Cook, Oo
R. Aidan Martin, (M), Queensland, Oo
Frank Mazzotti, (D), Penn State, Oo
Donald McCorquodale, (D), Nova, 1987, Oc
Peter Moeller, (D), California (Berkeley), Oo
Stacy Myers, (M), Nova Southeastern, Oi
Carol Palmer, (D), Hawaii, Oo
Esther Peters, (D), Rhode Island, Oo
Brian Polkinghorn, (D), Syracuse, Oo
Keith Ronald, (D), McGill, Oo
Scott Schatz, (D), New England Coll of Opt, Oo
Steffen Schmidt, (D), Columbia, Oo
Michael Stanhope, (D), Simon Fraser, Oo
Theodore Switzer, (D), Louisiana State, Oo
Willam Venezia, (D), Clemson, Oo

Marianne Walch, (D), Harvard, Oo
Alan Watson, (M), Guelph, Oo
Brad Weatherbee, (D), Hawaii, Oo
Scott White, (D), Bristol, Oo

Senior Programmer:
Kevin Kohler, (M), Miami, Oo

Project Manager:
Stefanie Ouellette, (B), Oo

Geology Librarian:
Kathleen Maxson, (M), Florida Intl, 2002, Oo

Distance Education Coordinator:
Jane Dougan, (M), Oo

Administrative Coordinator:
Melissa L. Dore, (B), Wells Coll, 1990, Oo

Administrative Accountant:
Jane Naraine, (M), Nova Southeastern, 2004

Chief Diving Officer, Harbour Master:
Lance Robinson, (U), Oo

Okaloosa-Walton College

Science Dept (A) (2007)
100 College Blvd.
Niceville, FL 32578
p. (850) 729-5376 f. (850) 729-5215
science@owc.edu
Enrollment (1995): B: 80 (0)

M.K. Environmental Institute:
Allison Beauregard, Delaware, 2006, Ge

Professor:
Jonathan R. Bryan, (D), Tennessee, 1991, Pi

Pensacola Junior College

Dept of Physical Sciences (2004)
Pensacola, FL 32504-8998
p. (904) 484-1150
estout@pjc.edu

Professor:
Kenith G. Exum, (D), S Mississippi, 1971, Gg
Brooke L. Towery, (M), Ball State, 1969, Gg

Associate Professor:
Fredrick L. Gamble, (D), N Colorado, 1983, Og

Saint Petersburg College, Clearwater

Dept of Natural Sciences (A) (2007)
2465 Drew Street
Clearwater, FL 33765
p. (727) 791-2542 f. (727) 791-2620
hemme.bill@spcollege.edu

Professor:
Carl Opper, (M), Florida, 1982, Hw

Saint Petersburg Junior College

Dept of Natural Science (A,B) (2006)
6605 5th Avenue North
St Petersburg, FL 33710
p. (727) 341-4381 f. (727) 341-4306
culverw@spcollege.edu
http://www.spjc.cc.fl.us/spg/science/dns/index.htm
Administrative Assistant: Lana Turner
Enrollment (2000): A: 7 (0)

Chair:
Robert N. Davie, Jr., (D), Florida, 1975, Yg

Professor:
William C. Culver, (M), 1973, Gg

Assistant Professor:
Ross I. Brightman, (D), S Florida, 1993, Ob
John V. Gartner, Jr., (D), S Florida, 1990, Ob
Thomas M. Lancraft, (M), California, 1982, Ob

Tallahassee Community College

Div of Science & Mathematics (A) (2004)
Tallahassee, FL 32304
p. (850) 922-8152 f. (850) 921-1261
brownf@tcc.fl.edu
Administrative Assistant: Monika Bedard
Enrollment (1996): B: 6 (0)

Instructor:
Howard J. Fischer, (D), Idaho, 1982, Gd
Peter H. Greaney, (M), Ohio State, 1972, Gs
Nancy A. Perrin, (D), North Dakota, 1987, Gs

University of Central Florida

Dept of Chemistry (Geology) (2005)
Orlando, FL 32816
p. (407) 823-2246 f. (407) 823-2252
kbelfiel@mail.ucf.edu
Office Manager: Shirley Sheppard

Associate Professor:
Frank B. Kujawa, (D), Johns Hopkins, 1968, Cp

Cooperating Faculty:
Clarence M. Head, (D), Georgia, 1975, Gm
Shiou-San Kuo, (D), Michigan State, 1972, Ng

University of Florida

Dept of Geological Sciences (B,M,D) (2007)
P.O. Box 112120
241 Williamson Hall
Gainesville, FL 32611-2120
p. (352) 392-2231 f. (352) 392-9294
adavis@geology.ufl.edu
http://web.geology.ufl.edu/
Enrollment (2007): M: 20 (6) D: 12 (5)

Chair:
Michael R. Perfit, (D), Columbia, 1977, Giu

Professor:
James T. Channell, (D), Newcastle upon Tyne, 1975, YmGt
David A. Foster, (D), SUNY (Albany), 1989, GtCc
David A. Hodell, (D), Rhode Island, 1986, GuPe
Jonathan B. Martin, (D), California (San Diego), 1993, ClHg
Guerry H. McClellan, (D), Illinois, 1964, En
Paul A. Mueller, (D), Rice, 1971, Cc
Neil D. Opdyke, (D), Durham (UK), 1958, Ym
Elizabeth Screaton, (D), Lehigh, 1995, Hg

Associate Professor:
Mark Brenner, (D), Florida, 1983, Gn
Paul F. Ciesielski, (D), Florida State, 1978, Pm
John M. Jaeger, (D), SUNY (Stony Brook), 1998, Gs
Ellen E. Martin, (D), California (San Diego), 1993, Cl
Joseph G. Meert, (D), 1993, YgGt

Assistant Professor:
Peter N. Adams, (D), California (Santa Cruz), 2004, Gm
Kyoungwon Min, (D), California (Berkeley), 2002, Cc
Philip S. Neuhoff, (D), Stanford, 2000, Cl
Raymond Russo, (D), Yg
Andrew Zimmerman, (D), William and Mary/VIMS, 2000, Co

Research Associate:
Jason H. Curtis, (D), Pe
Ann L. Heatherington, (D), Washington (St. Louis), 1988, Cc
Kainian Huang, (D), Acad Sinica, 1986, Ym
Ray G. Thomas, (B), Florida, 1990, Og

Lecturer:
Matthew Smith, (D), Gi
Jim Vogl, (D), 2000, Gct

Courtesy Professor:
Paul A. Colinvaux, Yr
Emeritus:
Frank N. Blanchard, (D), Michigan, 1960, Gz
Earl C. Pirkle, (D), Cincinnati, 1956, Eg
Anthony F. Randazzo, (D), North Carolina, 1968, Gd
Douglas L. Smith, (D), Minnesota, 1972, Yh
Daniel P. Spangler, (D), Arizona, 1969, Hw
Program Assistant:
Alison Davis, (B), Florida, 2003
Assistant Curator:
Jonathan Bloch, (D), Pv
Affiliate Faculty:
Micheal W. Binford, (D), Indiana, 1980, Or
Joann Mossa, (D), Louisiana State, 1990, Gm
Cooperating Faculty:
David L. Dilcher, (D), Yale, 1964, Pb
Douglas S. Jones, (D), Princeton, 1980, Pg
Steven R. Manchester, (D), Indiana, 1981, Pb
Bruce J. McFadden, (D), Columbia, 1976, Pv
Claire L. Schelske, (D), Michigan, 1961, Pe

Florida Museum of Natural History (2008)
PO Box 117800
Gainesville, FL 32611-7800
p. ((35) 2) -392- (Ext. 21, ext. f. (352) 392-8783
bmacfadd@flmnh.ufl.edu
http://www.flmnh.ufl.edu
Administrative Assistant: Pam Dennis
Director:
Douglas S. Jones, (D), Princeton, 1980, Pg
Associate Director:
Graig D. Shaak, (D), Pittsburgh, 1972, Pe
Associate Curator:
Bruce J. MacFadden, (D), Columbia, 1976, Pv
Steven R. Manchester, (D), Indiana, 1981, Pb
David W. Steadman, (D), Arizona, 1982, Pq
Research Associate:
David M. Jarzen, (D), Toronto, 1973, Pb
Graduate Research Professor:
David L. Dilcher, (D), Yale, 1964, Pb
Research Associate:
Ann S. Cordell, (M), Florida, 1983, Gx
Roger W. Portell, (B), Florida, 1985, Pi
Sylvia J. Scudder, (M), Florida, 1993, Pg

Soil & Water Science Dept (B,M,D) (2004)
PO Box 110510
106 Newell Hall
Gainesville, FL 32611-0510
p. (352) 392-1803 Ext. 340 f. (352) 392-3399
krr@ufl.edu
http://soils.ifas.ufl.edu
Enrollment (2001): B: 34 (6) M: 28 (5) D: 26 (1)
Chair:
K Ramesh Reddy, (D), Louisiana State, 1976, Sf
Distinguished Professor:
George H. Snyder, (D), Ohio State, 1967, Sc
Professor:
David V. Calvert, (D), Iowa State, 1962, Sc
Mary E. Collins, (D), Iowa State, 1980, Sd
Nicholas B. Comerford, (D), SUNY (Syracuse), 1980, Sf
Donald A. Graetz, (D), Wisconsin, 1970, Sc
James H. Graham, Jr., (D), Oregon State, 1980, Sb
Edward A. Hanlon, Jr., (D), Oklahoma State, 1983, So
Willie G. Harris, Jr., (D), Virginia Tech, 1984, Sc
Arthur G. Hornsby, (D), Oklahoma State, 1972, Sp
Lena Q. Ma, (D), Colorado State, 1991, Sc
Robert S. Mansell, (D), Iowa State, 1968, Sp
Peter Nkedi-Kizza, (D), California, 1979, Sp
George A. O'Connor, (D), Colorado State, 1970, Sc
Hugh L. Popenoe, (D), Florida, 1960, So
John E. Rechcigl, (D), Virginia Tech, 1986, So
R Dean Rhue, (D), Oregon State, 1975, Sc
Jerry B. Sartain, (D), North Carolina, 1973, So
Craig D. Stanley, (D), Iowa State, 1978, Sp
Associate Professor:
Yuncong Li, (D), Maryland, 1993, Sc
Thomas A. Obreza, (D), Florida, 1983, So
Andrew V. Ogram, (D), Tennessee, 1988, Sb
Scientist:
Li-Tse Ou, (D), Rochester, 1972, Sb
Assistant Professor:
Samira H. Daroub, (D), Michigan State, 1994, So
William F. Debusk, (D), Florida, 1996, Sf
Sabine Grunwald, (D), Giessen, 1996, So
James W. Jawitz, (D), Florida, 1999, Sp
S Rao Mylavarapu, (D), Clemson, 1996, Sc
Arnold W. Schumann, (D), Georgia, 1997, So
P. Christopher Wilson, (D), Clemson, 1999, So
Research Assistant Professor:
Mark W. Clark, (D), Florida, 2000, Sf
Vimala D. Nair, (D), Gottingen, 1978, Sc
Joseph P. Prenger, (D), Missouri, 1987, So
John R. White, (D), Florida, 1999, Sf
Ann C. Wilkie, (D), Univ Coll (Ireland), 1984, Sb
Emeritus:
Randall B. Brown, (D), Oregon State, 1980, Sd
Gerald Kidder, (D), Oklahoma State, 1969, So
Brian L. McNeal, (D), California (Riverside), 1965, Sc

University of Miami

Dept of Geological Sciences (B,M,D) (2007)
1301 Memorial Drive
43 Cox Science Building
Coral Gables, FL 33124
p. (305) 284-4253 f. (305) 284-4258
geology@umiami.ir.miami.edu
Administrative Assistant: Ruth Goodin
Enrollment (2007): B: 28 (0)
Sr. Associate Dean:
Jackie E. Dixon, (D), Caltech, 1992, GipGv
Dean:
Larry C. Peterson, (D), Brown, 1984, Gr
Professor:
David E. Fisher, (D), Florida, 1958, Cs
Harold R. Wanless, (D), Johns Hopkins, 1971, Gs
Senior Lecturer:
Teresa A. Hood, (D), Miami, 1991, Gg
Associate Professor:
John R. Southam, (D), Illinois, 1974, Op
Assistant Professor:
James S. Klaus, (D), Illinois, 2005, Gg
Karrie-Ann H. Kubatko, (D), Notre Dame, 2005, Ge
Scientist:
Donald F. McNeill, (D), Miami, 1989, Gg

Div of Marine Geology & Geophysics (M,D) (2007)
RSMAS
4600 Rickenbacker Causeway
Miami, FL 33149-1098
p. (305) 421-4663 f. (305) 421-4632
mgg@rsmas.miami.edu
http://rsmas.miami.edu/divs/mgg/
Enrollment (2007): M: 5 (1) D: 26 (6)
Professor:
Keir Becker, (D), Scripps, 1981, Yr

Jacqueline E. Dixon, (D), Caltech, 1991, Cp
Timothy Dixon, (D), California (San Diego), 1979, Or
Gregor P. Eberli, (D), ETH (Switzerland), 1985, Gr
Robert N. Ginsburg, (D), Chicago, 1953, Gs
Christopher G.A. Harrison, (D), 1964, Yr
James Natland, (D), Scripps, 1975, Gi
Larry C. Peterson, (D), Brown, 1984, Gu
Peter K. Swart, (D), King's Coll (London), 1980, Cg

Research Associate Professor:
Ruth P. Reid, (D), Miami, 1979, Gs

Associate Professor:
Mark Grasmueck, (D), ETH (Switzerland), 1995, Yr
Shimon Wdowinski, Harvard, 1990, Yd

Assistant Professor:
Falk Amelung, (D), Lous Pasteur, 1996, Oo
Eugene Rankey, (D), Kansas, 1996, Oo

Cooperating Faculty:
Patricia L. Blackwelder, (D), South Carolina, 1976, Po

University of South Florida, Saint Petersburg

Center for Ocean Technology (2006)
140 7th Avenue South, MSL 119
St. Petersburg, FL 33701-5016
p. (727) 553-1280 f. (727) 553-3967
flournoy@marine.usf.edu
http://cot.marine.usf.edu/
Administrative Assistant: Antoinette Flournoy

Director:
Larry C. Langebrake, (M), S Florida, 1990, Oo

Research Assistant:
David P. Fries, (M), S Florida, 1999, Oo
Eric A. Kaltenbacher, (M), Dayton, 1996, Oo
Chad E. Lembke, (B), S Florida, 1995, Oo
James T. Patten, (B), West Florida, 1981, Oo
D. Randy Russell, (M), S Florida, 1990, Oo
Scott A. Samson, (D), Iowa, 1994, Oo
R. Timothy Short, (D), Tennessee, 1987, Oo

Mechanical Engineer:
Michael Hall, (B), North Carolina State, 1996, Oo

Field Support Engineer:
Graham Tilbury, (B), Natal (Durban), 1996, Oo

Research Associate:
Andres Cardenas, (D), S Florida, Oc

Business Development Manager:
Carol S. Steele, (D), Fielding Inst, 2000, Oo

Dept of Marine Science (M,D) (2004)
140 7th Avenue South
St. Petersburg, FL 33701
p. (727) 553-1130 f. (727) 553-1189
npiehl@marine.usf.edu
http://www.marine.usf.edu
Administrative Officer: Robert Jolley
Enrollment (1999): M: 63 (9) D: 48 (6)

Dean, College of Marine Science:
Peter R. Betzer, (D), Rhode Island, 1970, Cm

Professor:
Norman J. Blake, (D), Rhode Island, 1972, Ob
Robert H. Byrne, (D), Rhode Island, 1974, Cm
Larry J. Doyle, (D), S California, 1972, Gs
Pamela Hallock-Muller, (D), Hawaii, 1977, Pm
Albert C. Hine, (D), South Carolina, 1975, Gs
Thomas L. Hopkins, (D), Florida State, 1964, Ob
John H. Paul, (D), Miami, 1980, Ob
Joan B. Rose, (D), Arizona, 1985, Oo
Joseph J. Torres, (D), California, 1980, Ob
Edward S. Van Vleet, (D), Rhode Island, 1978, Co
John J. Walsh, (D), Miami, 1969, Ob
Robert H. Weisberg, (D), Rhode Island, 1975, Op

Associate Professor:
Paula G. Coble, (D), MIT/Woods Hole, 1992, Cm
Boris Galperin, (D), Israel Inst of Tech, 1982, Op
Mark E. Luther, (D), North Carolina, 1982, Op
Gary T. Mitchum, (D), Florida State, 1984, Op
Frank E. Muller-Karger, (D), Maryland, 1988, Or
Terrence M. Quinn, (D), Brown, 1989, Cm
Gabriel A. Vargo, (D), Rhode Island, 1976, Ob

Assistant Professor:
Peter Howd, (D), Oregon State, 1991, Ou
Michael W. Howell, (D), South Carolina, 1988, Ou
Sarah F. Tebbens, (D), Columbia, 1994, Yr

Research Associate:
Robert Chen, (D), Washington, 1986, Oo
Rick Cole, (B), Florida Inst of Tech, 1983, Op
David K. Costello, (M), S Florida, 1988, Op
Dwight A. Dieterle, (M), San Jose, 1979, Ob
Joseph Donnelly, (M), S Florida, 1986, Ob
Jeff C. Donovan, (B), Florida, 1985, Op
Meredith A. Haines, (D), Dalhousie, 1993, Og
Weilin Hou, (D), S Florida, 1997, Op
Debra E. Huffman, (D), S Florida, 1994, Oo
Zhongping Lee, (D), S Florida, 1994, Op
Stanley D. Locker, (D), Rhode Island, 1989, Ou
Yu-Ran Luo, (D), Dalhousie, 1991, Oc
David J. Mallinson, (D), S Florida, 1995, Gu
Phillip N. Reinersman, (D), S Florida, 1988, Op
Chunzai Wang, (D), S Florida, 1995, Op
Huijun Yang, (D), Florida State, 1991, Op
Wensheng Yao, (D), Miami, 1995, Oc

Research Assistant:
David C. Burwell, (M), Tennessee, 1991, Op
David C. English, (M), Washington, 1983, Op
Robert G. Steward, (M), S Florida, 1980, Op
Rodger T. Stokes, (M), S Florida, 1996, Ob

Adjunct Professor:
Thomas G. Bailey, (D), California, 1984, Ob
Ronald C. Baird, (D), Harvard, 1969, Ob
Christopher Barton, (D), Yale, 1983, Gg
Theresa M. Bert, (D), Yale, 1985, Ob
William P. Bissett, (D), S Florida, 1997, Ob
Greg R. Brooks, (D), S Florida, 1986, Ou
Roy E. Crabtree, (D), William & Mary, 1984, Ob
Richard A. Davis, Jr., (D), Illinois, 1964, Gs
Michael J. Durako, (D), S Florida, 1991, Og
William W. Falls, (D), S Mississippi, 1983, Ob
John V. Gartner, (D), S Florida, 1990, Og
Robert B. Halley, (D), SUNY (Stony Brook), 1974, Gu
Evamaria W. Koch, (D), S Florida, 1993, Og
Lynn A. Leonard, (D), S Florida, 1994, Oo
Gary W. Litman, (D), Minnesota, 1972, Oo
Dan Marelli, (D), Florida State, 1987, Oo
Richard E. Matheson, (D), Texas A&M, 1983, Ob
Anne Meylan, (D), Florida, 1984, Ob
Robert G. Muller, (D), Hawaii, 1976, Ob
Esther C. Peters, (D), Rhode Island, 1985, Ob
John E. Reynolds, III, (D), Miami, 1980, Ob
Gary E. Rodrick, (D), Oklahoma, 1971, Ob
Egil Sakshaug, (D), Oslo, 1967, Oo
Abby H. Sallenger, (D), Virginia, 1975, On
Eugene Shinn, (B), Miami, 1957, Gg
Albert C. Smith, (D), California (Irvine), 1967, Ob
Thomas J. Smith, III, (D), Virginia, 1984, Oo
Karen A. Steidinger, (D), S Florida, 1979, Ob
Rick Stumpf, (D), Delaware, 1984, Og
Yves Tardy, (D), Louis Pasteur, 1969, Cg
Carmelo R. Tomas, (D), Rhode Island, 1977, Ob

Marguerite A. Toscano, (D), S Florida, 1996, Gu
Randy Wells, (D), California (Santa Cruz), 1986, Ob
Roland F. Wollast, (D), Libre de Bruxelles, 1960, Cg
Scott Wright, (D), Connecticut, 1987, Oo
Emeritus:
John C. Briggs, (D), Stanford, 1951, Ob
Harold J. Humm, (D), Duke, 1945, Ob

University of South Florida, Tampa

Dept of Geology (B,M,D) (2004)
SCA 528
4202 East Fowler Avenue
Tampa, FL 33620-5201
p. (813) 974-2236 f. (813) 974-2654
mhaney@chuma1.cas.usf.edu
Administrative Assistant: Mary Haney
Enrollment (2003): B: 54 (15) M: 31 (9) D: 11 (2)

Chair:
Charles B. Connor, (D), Dartmouth, 1987, Gv
Professor:
Richard A. Davis, Jr., (D), Illinois, 1964, Gs
Mark T. Stewart, (D), Wisconsin, 1976, Hw
H. Leonard Vacher, (D), Northwestern, 1971, Hw
Associate Professor:
Peter J. Harries, (D), Colorado, 1993, Pg
Eric A. Oches, (D), Massachusetts, 1994, Gm
Jeffrey G. Ryan, (D), Columbia, 1989, Ct
Assistant Professor:
Sarah E. Kruse, (D), MIT, 1989, Yg
Thomas Pichler, (D), Ottawa, 1998, Cl
Mark C. Rains, (D), California (Davis), 2002, Hw
Ping Wang, (D), S Florida, 1995, Gs
Instructor:
Thomas C. Juster, (D), S Florida, 1995, Hw
Eleanour Snow, (D), Brown, 1987, Gz
Assistant Curator:
Raymond Denicourt, Gz
Cooperating Faculty:
Jeffrey Heikoop, Cl
Lisa Robbins, Py
James E. Sorauf, Pg
Barbara W. Leyden, (D), Indiana, 1982, Pl
Robert Morton, (D), West Virginia, 1972, Gs
Thomas M. Scott, (D), Florida State, Gr
Sam B. Upchurch, (D), Northwestern, 1970, Hw

University of West Florida

Dept of Environmental Studies (B) (2006)
11000 University Parkway
Pensacola, FL 32514
p. 850-474-2746 f. 850-857-6036
environmental@uwf.edu
http://uwf.edu/environmental/
Enrollment (2006): B: 91 (23)

Professor:
Klaus J. Meyer-Arendt, (D), Louisiana State, 1987, On
Associate Professor:
Johan Liebens, (D), Michigan State, 1995, So
Assistant Professor:
Christopher Houser, (D), Toronto, 2003, On
Zhiyong Hu, (D), Georgia, 2003, Or
Matthew Schwartz, (D), Delaware, 2002, Hy
John Waldron, (D), Texas A&M, 2002, Oy
Instructor:
Tom Banjanin, (D), George Washington, 1975, Ou
John D. Barksdale, (M), Florida State, 1988, Ge
Taylor Kirschenfeld, (M), West Florida, 1988, Ob
Hilde Snoeckx, (D), Michigan, 1995, Ou
Lecturer:
Wilbur G. Hugli, (D), West Florida, 2001, Oa
GIS Coordinator:
Stuart Hamilton, (M), SUNY (Buffalo), 2003, Oi

Georgia

Columbus State University

Dept of Chemistry & Geology (B) (2006)
4225 University Avenue
Columbus, GA 31907-5645
p. (706) 568-2075 f. (706) 569-3133
hanley_tom@colstate.edu
http://chemgeo.colstate.edu
Enrollment (2002): B: 16 (1)

Acting Chair:
Thomas B. Hanley, (D), Indiana, 1975, Gc
Professor:
William J. Frazier, (D), North Carolina, 1973, Gs
David R. Schwimmer, (D), SUNY (Stony Brook), 1973, Pv
Research Associate:
G. Dent Williams, (B), Auburn, 1965, Pv

East Georgia College

Science & Mathematics (A) (2004)
131 College Circle
Swainsboro, GA 30401
p. (912) 289-2073 f. (912) 289-2080
edbolton@ega.edu
http://www.ega.peachnet.edu/stracher.html
Enrollment (1999): A: 3 (2)

Associate Professor:
Glenn B. Stracher, (D), Nebraska, 1989, Om

Emory University

Dept of Environmental Studies (B) (2004)
Mathematics & Science Center
400 Dowman Drive
Atlanta, GA 30322
p. (404) 727-4216 f. (404) 727-4448
hpayne@learnlink.emory.edu
http://www.emory.edu/COLLEGE/ENVS
Enrollment (2000): D: 1 (0)

Chair:
Lance Gunderson, (D), Florida, 1992, Sf
Associate Professor:
William B. Size, (D), Illinois, 1971, Gi
Assistant Professor:
Sonia Altizer, (D), Minnesota, 1998, Sf
Lowell Pritchard, (D), Florida, 1990, Sf
Lore Ruttan, (D), California (Davis), 1999, Ob
Tracy Yandle, (D), Indiana, 2001, Ob
Instructor:
Anne M. Hall, (M), Georgia Tech, 1985, Gz
Senior Lecturer:
C Woodbridge Hickcox, (D), Rice, 1971, Gg
Campus Env Officer:
John Wegner, (D), Carleton, 1995, Sf
Lecturer:
Anthony J. Martin, (D), Georgia, 1991, Pe
Adjunct Professor:
Pamela J. W. Gore, (D), George Washington, 1983, Gs
Emeritus:
Howard R. Cramer, (D), Northwestern, 1954, Pg
Willard H. Grant, (D), Johns Hopkins, 1955, Cg

Emory University-Oxford College

Dept of Geology (A) (2004)
Oxford, GA 30054
p. (770) 784-8345 f. (770) 784-8423
shender@emory.edu
http://www.emory.edu/OXFORD/ScienceMath/Geology/
Enrollment (2003): A: 3 (1)

Associate Professor:
Stephen W. Henderson, (D), Georgia, 1984, Pe

Fort Valley State University

Cooperative Developmental Energy Program (2004)
Box 5800 FVSU
1005 State University Drive
Fort Valley, GA 31030
p. (912) 825-6454 f. (912) 825-6618
crumblyi@fvsu.edu
http://www.fac.fvsu.edu/as/cd/index.htm
Enrollment (1995): B: 37 (0)

Director:
Isaac J. Crumbly, (D), North Dakota State, 1970, Oo
Assistant Director:
Jackie Hodges, (B), Georgia Southern, 1983, Oo
Associate Professor:
Aditya Kar, (D), Oklahoma, 1997, Ct
Related Staff:
James Brown, (M), Fort Valley State, 1997, Oo

Gainesville College

Div of Natural Science (A) (2005)
Gainesville, GA 30501
p. (770) 718-3666 f. (770) 718-3770
lrogers@gc.peachnet.edu
Administrative Assistant: Mary McDade
Enrollment: No data reported since 1995

Chair:
Lewis F. Rogers, (D), Georgia, 1979, Ge
Research Associate:
Christopher J. Semerjian, (B), Georgia, 1994, Ge

Georgia Institute of Technology

School of Earth & Atmospheric Sciences (B,M,D) (2004)
221 Bobby Dodd Way
Atlanta, GA 30332-0340
p. (404) 894-3893 f. (404) 894-5638
rita.bryan@eas.gatech.edu
Enrollment (1999): B: 59 (21) M: 25 (13) D: 51 (10)

Professor:
Glen R. Cass, (D), Caltech, 1978, Oa
William L. Chameides, (D), Yale, 1974, Oa
George Chimonas, (D), Sussex (UK), 1965, Oa
Derek M. Cunnold, (D), Cornell, 1965, Oa
Douglas D. Davis, (D), Florida, 1966, Oa
Robert E. Dickinson, (D), MIT, 1966, Oa
Philip N. Froelich, (D), Rhode Island, 1979, Cm
Chia S. Kiang, (D), Georgia Tech, 1970, Oa
Shaw C. Liu, (D), Pittsburgh, 1972, Oa
Leland T. Long, (D), Oregon State, 1968, Ys
Robert P. Lowell, (D), Oregon State, 1972, Yr
Edward Michael Perdue, (D), Georgia Tech, 1973, Co
Robert G. Roper, (D), Adelaide, 1963, Oa
Paul H. Wine, (D), Florida State, 1974, Oa
Senior Research Engineer:
Karsten Baumann, (D), Stuttgart, 1992, Oa
Research Scientist:
Mian Chin, (D), Georgia Tech, 1992, Oa
Hann-wen Guan, (D), Nebraska, 1994, Oa
Chao Luo, (D), Peking, 1990, Oa
Robert Stickel, (D), Rice, 1979, Oa
Principal Research Scientist:
Fred L. Eisele, (D), Vermont, 1975, Oa
Senior Scientist:
Carlos Cardelino, (D), MIT, 1978, Oa
Michael E. Chang, (D), Georgia Tech, 1997, Oa
Gao Chen, (D), Georgia Tech, 1995, Oa
Scott T. Sandholm, (D), Michigan State, 1982, Oa
James C. St. John, (D), Georgia Tech, 1997, Oa
Hsiang-Jui Wang, (D), Georgia Tech, 1995, Oa
Jing-Zhi Zhao, (D), Hertfordshire, 1983, Oa
Associate Professor:
Robert X. Black, (D), MIT, 1990, Oa
Rong Fu, (D), Columbia, 1991, Oa
Gregory L. Huey, (D), Wisconsin, 1992, Oa
Ellery D. Ingall, (D), Yale, 1991, Cm
J. Marion Wampler, (D), Columbia, 1963, Cc
Yuhang Wang, (D), Harvard, 1997, Oa
Assistant Professor:
Michael H. Bergin, (D), Carnegie Mellon, 1995, Oa
James B. Gaherty, (D), MIT, 1995, Yv
Daniel Lizarralde, (D), MIT/Woods Hole, 1997, Yv
Carolyn D. Ruppel, (D), MIT, 1992, Yv
Martial Taillefert, (D), Northwestern, 1997, Cm
Rodney J. Weber, (D), Minnesota, 1995, Oa
Adjunct Associate Professor:
Carmen Nappo, (D), Georgia Tech, 1989, Oa
Adjunct Assistant Professor:
Thomas D. Christina, (D), Caltech, 1989, Cl
Kavita Philip, (D), Cornell, 1996, Co
Adjunct Professor:
Clark R. Alexander, (D), North Carolina State, 1990, Cl
Jackson O. Blanton, (D), Oregon State, 1968, Cl
Paul Curtzen, (D), Stockholm, 1968, Oa
James L. Harding, (D), Texas A&M, 1964, Oa
Richard Johnke, (D), Washington, 1976, Cl
Demetrius P. Lalas, (D), Cornell, 1968, Oa
Hiram Levy, II, (D), Harvard, 1966, Oa
Richard S. Lindzen, (D), Harvard, 1964, Oa
David W. Menzel, (D), Michigan, 1958, Ob
Leonard Newman, (D), MIT, 1956, Oa
Wolfgang Seiler, (D), Mainz, 1970, Oa
Stuart G. Wakeham, (D), Washington, 1976, Co
Herbert L. Windom, (D), California (San Diego), 1968, Cm

Georgia Perimeter College

Geology Dept, Clarkston Campus (A) (2007)
555 North Indian Creek Drive
Clarkston, GA 30021
p. 678-891-3754 f. 678-891-3747
pgore@gpc.edu
http://gpc.edu/~mokafor/gpcsci/sci_team.html
Enrollment (2007): A: 3 (0)

Professor:
Pamela J. W. Gore, (D), George Washington, 1983, GsOeGn
Associate Professor:
E. Lynn Zeigler, (M), Emory, 1989, Gd
Adjunct Professor:
Edward F. Albin, (D), Georgia, 1997, Xg
Don A. Lundy, (M), Wyoming, 1978, Hy
Debia F. McCulloch, (M), Emory, 1987, Gg
Roy A. (Rick) Nixon III, (D), Emory, 1981, Gm

Geology Dept, Dunwoody Campus (A) (2007)
2101 Womack Road
Dunwoody, GA 30338-4497
p. (770) 274-5050 f. (770) 274-5565

janderso@gpc.edu
http://www.gpc.edu/~dunsci/depthp.htm
Administrative Assistant: Elizabeth Harris
Enrollment (2007): A: 7 (3)

Chair:
John R. Anderson, Jr., (D), Pittsburgh, 1986, Ps
Associate Professor:
Dion Stewart, (D), Penn State, 1980, Gi
Instructor:
Jill Ghelerter, (M), Florida Atlantic, 2005, Hw
Kimberly Schulte, (M), Tulsa, 1997, Ge
Adjunct Professor:
Carol Gelbaum, (M), Brooklyn, 1974, Gg
Nicole Vermillion, (M), Georgia State, 2005, Gg

Geology Dept, Lawrenceville (A) (2004)
1000 University Center Lane
Lawrenceville, GA 30043
p. (678)407-5217
ezoeckle@gpc.edu
Administrative Assistant: Polly A. Bouker

Instructor:
Deniz Altin, (M), Gz

Georgia Southern University

Applied Coastal Research Laboratory (2004)
10 Ocean Science Circle
Savannah, GA 31411
p. (912) 598-2463 f. (912) 598-2366
henry@skio.peachnet.edu

Professor:
Jonathan Copeland, (D), SUNY (Stony Brook), 1975, Ob
Vernon James Henry, (D), Texas A&M, 1961, Ou
Research Associate:
Anthony M. Foyle, (D), Old Dominion, 1994, Gu
Adjunct Professor:
Clark R. Alexander, (D), North Carolina State, 1990, Ou

Dept of Geology and Geography (B) (2007)
1100 Herty Drive
Statesboro, GA 30460-8149
p. (912) 681-5361 f. (912) 681-0668
DRhodes@GeorgiaSouthern.edu
http://cost.georgiasouthern.edu/geo/
Administrative Assistant: Sarita N. Warren
Enrollment (2007): B: 98 (6)

Chair:
Dallas D. Rhodes, (D), Syracuse, 1973, Gm
Professor:
Fredrick J. Rich, (D), Penn State, 1979, Pl
Associate Professor:
Pranoti M. Asher, (D), Connecticut, 1995, Gi
Jonathan H. Geisler, (D), Columbia, 2001, Pv
James S. Reichard, (D), Purdue, 1995, Hw
Charles H. Trupe, (D), North Carolina, 1997, Gc
R. Kelly Vance, (D), New Mexico Tech, 1989, Eg
Mark R. Welford, (D), Illinois, 1993, Oy
Assistant Professor:
Eleanor J. Camann, (D), North Carolina, 2005, On
Thomas E. Chapman, (D), Florida State, 2007, OioOo
Wei Tu, (D), Texas A&M, 2004, Oi
Robert A. Yarbrought, (D), Georgia, 2006, Oo
Adjunct Professor:
Clark R. Alexander, Jr., (D), North Carolina State, 1990, Ou
Vernon J. Henry, (D), Texas A&M, 1961, Ou
Emeritus:
Gale A. Bishop, (D), Texas, 1971, Pi
James H. Darrell, (D), Louisiana State, 1973, Pl

Daniel B. Good, (D), Tennessee, 1973, Oo

Georgia Southwestern State University

Dept of Geology & Physics (B) (2005)
800 Wheatley Street
Americus, GA 31709
p. (229) 931-2353 f. (229) 931-2770
daskren@canes.gsw.edu
Administrative Assistant: Debbie Standridge
Enrollment (2004): B: 15 (2)

Chair:
Daniel R. Askren, (D), Georgia, 1992, Gz
Professor:
Burchard D. Carter, (D), West Virginia, 1981, Ps
Thomas J. Weiland, (D), North Carolina, 1988, Gi
Associate Professor:
Samuel T. Peavy, (D), Virginia Tech, 1997, Yg
Assistant Professor:
Svilen Kostov, (D), CUNY, 1992, Oo
Emeritus:
Daniel D. Arden, (D), California (Berkeley), 1961, Ps
Harland E. Cofer, (D), Illinois, 1957, Gz
John P. Manker, (D), Rice, 1975, Gs

Georgia State University

Dept of Geosciences (B,M,D) (2007)
340 Kell Hall
PO Box 4105
33 Gilmer St. SE
Atlanta, GA 30302-4105
p. (404) 413-5750 f. (404) 413-5768
wcelliott@gsu.edu
http://monarch.gsu.edu/geosciences/
Administrative Assistant: Tyra Hines
Enrollment (2005): B: 79 (19) M: 28 (17) D: 4 (0)

Chair:
W. Crawford Elliott, (D), Case Western, 1988, Cl
Professor:
William J. Fritz, (D), Montana, 1980, Gs
Associate Professor:
Hassan A. Babaie, (D), Northwestern, 1984, Gc
Jeremy W. Crampton, (D), Penn State, 1994, Oi
Jeremy E. Diem, (D), Arizona, 2000, Oa
Eirik J. Krogstad, (D), SUNY (Stony Brook), 1988, Cc
Timothy E. La Tour, (D), W Ontario, 1979, Gp
Seth E. Rose, (D), Arizona, 1987, Hw
Dona J. Stewart, (D), Florida, 1994, Oo
Assistant Professor:
Jordan A. Clayton, (D), Colorado, 2005, Oo
Daniel Deocampo, Rutgers
Lawrence M. Kiage, (D), Louisiana State, 2007, Oo
Senior Lecturer:
Kenneth J. Terrell, (M), Cincinnati, 1980, Pv
Adjunct Professor:
John R. Anderson, (D), Pittsburgh, 1986, Pg
Pamela C. Burnley, (D), California (Davis), 1990, Gy
Beth A. Christensen, (D), South Carolina, 1997, Gs
John Costello, (D), South Carolina, Gc
Aditya Kar, (D), Oklahoma, 1997, Gi
Hermann Lebit, (D), ETH (Zurich), 1995, Gc
Catalina M. Luneburg, (D), ETH (Zurich), 1995, Gc
J. Marion Wampler, (D), Columbia, 1963, Cc
Visiting Professor:
Anil K.K Gangadharan, (D), Cochin Univ of Sci & Tech, 1986, Oa
Emeritus:
Sanford H. Bederman, (D), Minnesota, 1973, Oo
Truman A. Hartshorn, (D), Iowa, 1968, Oo
Vernon J. Henry, Jr., (D), Texas A&M, 1961, Gu

Richard Pillsbury, (D), Penn State, 1968, Oo
W. Robert Power, (D), Johns Hopkins, 1959, En
Senior Lecturer:
John Allensworth, (D), Kent State, 1982, Oy

Gordon College

Div of Natural Sciences & Nursing (A) (2004)
Barnesville, GA 30204
p. (404) 358-5000 (Ext. 5197) f. (404) 358-3031
Administrative Assistant: Virginia Martin
Enrollment (1995): B: 3 (0)
Chair:
Judith M. Malachowski, (D), Virginia, 1993, Oo

Mercer University

Dept of Physics & Earth Sciences (B,M) (2004)
1400 Coleman Avenue
Macon, GA 31207
p. (912) 744-2597 f. (912) 752-2802
dod_bd@mercer.edu
Department Secretary: Brenda Walraven
Enrollment (1997): B: 7 (4)
Professor:
Bruce D. Dod, (D), S Mississippi, 1973, Og
Dan R. Quisenberry, (D), World Open, 1980, Or
Assistant Professor:
Geoffrey W. Hayden, (D), Pennsylvania, 1987, Om
Robert L. Huffman, (D), Massachusetts, 1985, Oo
Research Associate:
Paul P. Sipiera, (D), Otago (NZ), 1985, Xm
Instructor:
Barbara D. Henley, (M), Mercer, 1978, Gg

Shorter College

Div of Natural Sciences (B) (2004)
Rome, GA 30161
p. (404) 291-2121 (Ext. 263)
callee@shorter.edu
Enrollment: No data reported since 1995
Associate Professor:
J. B. Campbell, (D), Georgia, 1975, Oo
Instructor:
Andrew C. Kemp, (M), Emory, 1987, Oo

University of Georgia

Dept of Crop & Soil Science (B,M,D) (2004)
311 Plant Sciences Building
Athens, GA 30602-7272
p. (706) 542-2461 f. (706) 542-0914
Enrollment (1998): A: 0 (7) B: 19 (2) M: 18 (5) D: 32 (3)
Professor:
Domy C. Adriano, (D), Kansas State, 1970, Sc
Paul M. Bertsch, (D), Kentucky, 1983, Sc
Gary J. Gascho, (D), Michigan State, 1968, Sc
James E. Hook, (D), Penn State, 1975, Sp
Edward T. Kanemasu, (D), Wisconsin, 1969, Sp
David E. Kissel, (D), Kentucky, 1969, Sc
William P. Miller, (D), Virginia Tech, 1981, Sc
David E. Radcliffe, (D), Kentucky, 1984, Sp
William P. Segars, (D), Clemson, 1972, Sc
Larry M. Shuman, (D), Penn State, 1970, Sc
Associate Professor:
Miguel L. Cabrera, (D), Kansas State, 1986, So
Peter G. Hartel, (D), Oregon State, 1984, Sb
Owen C. Plank, (D), Virginia Tech, 1973, Sc
Larry T. West, (D), Texas A&M, 1986, Sd
Assistant Professor:
Glendon H. Harris, (D), Michigan State, 1993, Sb

Dept of Geology (B,M,D) (2006)
210 Field Street
Room 308
Athens, GA 30602
p. (706) 542-2652 f. (706) 542-2425
beatrice@gly.uga.edu
http://www.gly.uga.edu/
Administrative Assistant: Beatriz R. Stephens
Enrollment (2000): M: 1 (0)
Head:
Michael F. Roden, (D), MIT, 1982, Gi
Associate Head:
Paul A. Schroeder, (D), Yale, 1992, Gz
Professor:
Douglas E. Crowe, (D), Wisconsin, 1990, Em
R. David Dallmeyer, (D), SUNY (Stony Brook), 1972, Gt
Ervan G. Garrison, (D), Missouri, 1979, Ga
Susan T. Goldstein, (D), California (Berkeley), 1984, Pm
Steven M. Holland, (D), Chicago, 1990, Gr
John E. Noakes, (D), Texas A&M, 1962, Cc
Valentine A. Nzengung, (D), Georgia Tech, 1993, Cl
Alberto E. Patino-Douce, (D), Oregon, 1990, Gp
L. Bruce Railsback, (D), Illinois, 1989, Gd
Samuel E. Swanson, (D), Stanford, 1974, Gi
James (Jim) E. Wright, (D), California (Santa Barbara), 1981, Gt
Associate Professor:
Raymond Freeman-Lynde, (D), Columbia, 1980, Gu
Robert B. Hawman, (D), Princeton, 1988, Ys
Christopher S. Romanek, (D), Texas A&M, 1991, Cs
Sally E. Walker, (D), California (Berkeley), 1988, Pm
David B. Wenner, (D), Caltech, 1971, Cs
Sandra J. Wyld, (D), Stanford, 1991, Gc
Assistant Professor:
John F. Dowd, (D), Yale, 1984, Hy
Senior Lecturer:
Patino Douce Marta, (D), Buenos Aires, 1990, Gi
Adjunct Assistant Professor:
Roger A. Burke, (D), S Florida, 1985, Cl
Adjunct Professor:
Elizabeth J. Reitz, (D), Florida, 1979, Ga
Sandra Whitney, (D), Georgia, 1992, Ga
Emeritus:
Gilles O. Allard, (D), Johns Hopkins, 1956, Eg
Norman Herz, (D), Johns Hopkins, 1950, Gx
Mark Rich, (D), Illinois, 1959, Ps
James A. Whitney, (D), Stanford, 1972, Gi

University of West Georgia

Dept of Geosciences (B) (2007)
1601 Maple Street
Callaway Building, Room 101
Carrollton, GA 30118
p. (678) 839-4050 f. (678) 839-4071
chollaba@westga.edu
http://www.westga.edu/~geosci/
Administrative Assistant: Anita M. Cline
Enrollment (2007): B: 70 (14)
Chair:
Curtis L. Hollabaugh, (D), Washington State, 1980, CgGz
Professor:
Julie Bartley, (D), California (Los Angeles), 1994, CsGtOe
David M. Bush, (D), Duke, 1991, Ou
Timothy M. Chowns, (D), Newcastle, 1968, Gs
Randal L. Kath, (D), SD Sch of Mines & Tech, 1990, Gc

Director, GLOBE Teacher Training Partnership:
Rebecca L. Dodge, (D), Colorado Mines, 1982, Oy
Associate Professor:
James R. Mayer, (D), Texas, 1995, Hw
Jeong C. Seong, (D), Georgia, 1999, Oi
Assistant Professor:
Christopher A. Berg, (D), Texas, 2007, Gpi
Georgina G. DeWeese, (D), Tennessee, 2007, Oy
Hannes Gerhardt, (D), Arizona, 2007, Oo
Philip M. Novack-Gottshall, (D), Duke, 2004, Pg
Nathan A. Walter, (D), Florida State, 2005, Og
Research Associate:
Randa R. Harris, (M), Tennessee, 2001, Hs
Emeritus:
Thomas J. Crawford, (M), Emory, 1957, Eg
Richard P. Sanders, (D), Illinois, 1971, Gp
Lab Coordinator:
John D. Congleton, (M), S Methodist, 1990, Pv

Valdosta State University

Dept of Physics, Astronomy & Geosciences (B) (2006)
1500 North Patterson Street
Valdosta, GA 31698-0055
p. (229) 333-5752 f. (229) 219-1201
echatela@valdosta.edu
http://www.valdosta.edu/phy/
Enrollment (2005): B: 58 (4)
Head:
Edward E. Chatelain, (D), Iowa, 1984, Pi
Professor:
Frank A. Flaherty, (D), Fordham, 1983, Oo
Martha A. Leake, (D), Arizona, 1982, Xg
Kenneth S. Rumstay, (D), Ohio State, 1984, Oo
Associate Professor:
Cecilia S. Barnbaum, (D), California (Los Angeles), 1992, Oo
Eric C. Brevik, (D), Iowa State, 2001, Sd
Mary A. Fares, (D), Tennessee Tech, 1992, Oo
Michael G. Noll, (D), Kansas, 2000, Oo
William B. Pollard, (D), MIT, 1979, Om
Pre-Engineering Director:
Barry Hojjatie, (D), Florida, 1990, Om
Assistant Professor:
Brian H. Bossak, (D), Florida State, 2003, Ow
Can Denizman, (D), Florida, 1998, Hw
Judy Grable, (D), Tennessee, 2001, Hs
Mark S. Groszos, (D), Florida State, 1996, Gc
Paul Vincent, (D), Texas A&M, 2001, Oi
Instructor:
Perry A. Baskin, (M), Valdosta State, 1972, Oo
Emeritus:
Calvin T. Harrington, (D), Florida State, 1975, Oo
Dennis W. Marks, (D), Michigan, 1970, Oo
Arnold E. Somers, Jr., (D), Virginia Tech, 1975, Oo

Guam

University of Guam

Div of Natural Sciences (B) (2004)
Mangilao, GU 96923
p. (671) 734-9305
Enrollment (1998): M: 80 (2)
Professor:
Frank H. Kilmer, (D), California (Berkeley), 1963, Ps

Water & Energy Research Inst (M) (2004)
Mangilao, GU 96923
p. (671) 735-2685 f. (671) 734-8890
khosrow@uog.edu
Administrative Assistant: Delores Goolsby
Enrollment (1996): M: 18 (0)
Director:
Henry G. Siegrist, Jr., (D), Penn State, 1961, Ge
Professor:
Leroy F. Heitz, (D), Idaho, 1981, Hs
Shahram Khosrowpanah, (D), Colorado State, 1984, Hs
Associate Professor:
Gary R. W. Denton, (D), London, 1974, Ob
John W. Jenson, (D), Oregon State, 1993, Hy
Mark A. Lander, (D), Hawaii, 1990, Ow
Brenna Lorenz, (D), Memorial, 1985, Gi
Assistant Professor:
Peter P. Motaralli, (D), Cornell, 1989, So
Research Associate:
John M.U. Jocson, (M), Guam, 1998, Hw
Instructor:
Charles Guard, (M), Hawaii, 1974, Ow
Lab Manager:
Harold R. Wood, (B), Fort Hays State, 1969, Oo

Hawaii

University of Hawai'i at Hilo

Dept of Geology (B) (2006)
200 W. Kawili Street
Hilo, HI 96720-4091
p. (808) 933-3383 f. (808) 933-3677
kenhon@hawaii.edu
Enrollment (1995): A: 0 (1) B: 33 (6)
Chair:
Ken Hon, (D), Colorado, 1987, Gv
Associate Professor:
James L. Anderson, (D), S California, 1987, Gc
Jene D. Michaud, (D), Arizona, 1992, Hq
Visiting Professor:
Steven P. Lundblad, (D), North Carolina, 1994, Gr
Emeritus:
Joseph B. Halbig, (D), Penn State, 1969, Ca
Affiliate Faculty:
Walther M. Barnard, (D), Penn State, 1965, Gz
Robert W. Decker, (D), Colorado Mines, 1953, Oo
Cheryl A. Gansecki, (D), Stanford, 1998, Cc
Christina C. Heliker, (M), W Washington, 1984, Gv
John P. Lockwood, (D), Princeton, 1966, Gv

University of Hawai'i, Manoa

Dept of Geology & Geophysics (B,M,D) (2005)
1680 East-West Road, POST 701C
Honolulu, HI 96822
p. (808) 956-8763 f. (808) 956-5512
lanthony@soest.hawaii.edu
http://www.soest.hawaii.edu/gg
Enrollment (2004): B: 45 (12) M: 25 (8) D: 34 (2)
Chair:
Paul Wessel, (D), Columbia, 1989, Yr
Interim Associate Dean:
Patricia A. Cooper, (D), Hawaii, 1985, Ys
Professor:
Frederick K. Duennebier, (D), Hawaii, 1972, Yr
Charles H. Fletcher, (D), Delaware, 1986, Gs
L. Neil Frazer, (D), Princeton, 1978, Ys
Michael O. Garcia, (D), California (Los Angeles), 1976, Gi
Craig R. Glenn, (D), Rhode Island, 1987, Gs
Bruce F. Houghton, (D), Otago (NZ), 1977, Gv
John Mahoney, (D), California (San Diego), 1984, Ca
Gregory F. Moore, (D), Cornell, 1977, Gt

Brian N. Popp, (D), Illinois, 1986, Cs
C. Barry Raleigh, (D), California (Los Angeles), 1963, Nr
John M. Sinton, (D), Otago (NZ), 1976, Gi
Brian Taylor, (D), Columbia, 1982, Gt
Associate Professor:
Janet M. Becker, (D), California (San Diego), 1989, Oo
Aly I. El-Kadi, (D), Cornell, 1983, Hw
Kenneth H. Rubin, (D), California (San Diego), 1991, Cc
Assistant Professor:
F. Scott Anderson, (D), Arizona State, 1998, Xy
Benjamin Brooks, (D), Cornell, 1999, Gt
Robert A. Dunn, (D), Oregon, 1999, Yr
Sarah Fagents, (D), Lancaster (UK), 1994, Gv
Eric J. Gaidos, (D), MIT, 1996, Yr
Victoria E. Hamilton, (D), Arizona State, 1998, Xy
Julia E. Hammer, (D), Oregon, 1998, Gv
Garrett T. Ito, (D), MIT/Woods Hole, 1996, Yr
Stephen J. Martel, (D), Stanford, 1987, Ng
Douglas Pyle, (D), Oregon, 1994, Cg
Gregory Ravizza, (D), Yale, 1991, Cm
Kathleen Ruttenberg, (D), Yale, 1990, Cg
Research Associate:
Alexander N. Krot, (D), Moscow State (Russia), 1989, Xm
Instructor:
Patricia D. Lee, (M), Hawaii, 1985, Gg
Researcher:
Johanna M. Resig, (D), Kiel, 1965, Pm
Emeritus:
Pow-Foong Fan, (D), California (Los Angeles), 1965, Gs
Ralph Moberly, (D), Princeton, 1956, Gu
Assistant Specialist:
Daniel Kent Ross, (D), Houston, 1994, Gi
Student Academic Advisor:
Leona M. Anthony, (B), 1986, Oo
Cooperating Faculty:
T. Bruce Appelgate, (D), Hawaii, 1995, Yr
Jeffrey F. Bell, (D), Hawaii, 1984, Xg
Keith A. W. Crook, (D), New England, 1959, Gs
Eric H. De Carlo, (D), Hawaii, 1982, Ca
Margo H. Edwards, (D), Columbia, 1992, Gu
Luke P. Flynn, (D), Hawaii, 1992, Or
Gerard J. Fryer, (D), Hawaii, 1980, Ys
Patricia B. Fryer, (D), Hawaii, 1981, Gu
Michael Fuller, (D), Cambridge, 1961, Ym
Milton A. Garces, (D), California (San Diego), 1995, Ys
Andrew J. Harris, (D), Open, 1996, Gv
B. Ray Hawke, (D), Brown, 1978, Xg
Emilio Herrero-Bervera, (D), Hawaii, 1984, Ym
Richard N. Hey, (D), Princeton, 1975, Yr
Barbara H. Keating, (D), Texas (Dallas), 1976, Ym
Klaus Keil, (D), Johannes-Gutenberg (Germany), 1961, Xm
Loren W. Kroenke, (D), Hawaii, 1972, Gu
Barry R. Lienert, (D), Texas (Dallas), 1976, Yg
Paul G. Lucey, (D), Hawaii, 1986, Xy
Fred T. Mackenzie, (D), Lehigh, 1962, Cm
Murli H. Manghnani, (D), Montana State, 1962, Yx
Fernando Martinez, (D), Columbia, 1988, Yr
Floyd W. McCoy, Jr., (D), Harvard, 1974, Gu
Peter J. Mouginis-Mark, (D), Lancaster (UK), 1977, Xg
Scott K. Rowland, (D), Hawaii, 1987, Oo
Jane E. Schoonmaker, (D), Northwestern, 1981, Cl
Edward R. Scott, (D), Cambridge (UK), 1971, Xm
Shiv K. Sharma, (D), Indian Inst of Tech, 1973, Gx
G. Jeffrey Taylor, (D), Rice, 1970, Xm
Donald M. Thomas, (D), Hawaii, 1977, Ca
Affiliate Graduate Faculty:
David A. Clague, (D), California (San Diego), 1974, Gv
Allen L. Clark, (D), Idaho, 1967, Eg
Jonathan C. Gradie, (D), Arizona, 1978, Xg
Kevin T. M. Johnson, (D), MIT, 1990, Gi
David M. Johnston, (D), Massey, 1998, Gv
James P. Kauahikaua, (D), Hawaii, 1982, Gv
Laura S.L. Kong, (D), MIT, 1990, Ys
Paul G. Okubo, (D), MIT, 1986, Ys
Donald A. Swanson, (D), Johns Hopkins, 1964, Gv
Carl Thornber, (D), Colorado, 1992, Cg
Gordon Tribble, (D), Hawaii, 1990, Og

Dept of Meteorology (B,M,D) (2004)
2525 Correa Road, HIG 331
Honolulu, HI 96822
p. (808) 956-8775 f. (808) 956-2877
http://lumahai.soest.hawaii.edu/
Enrollment (1996): B: 13 (1) M: 12 (2) D: 15 (2)
Chair:
Kevin Hamilton, (D), Princeton, 1981
Professor:
Gary M. Barnes, (D), Virginia, 1980, Ow
Steven Businger, (D), Washington, 1986, Ow
Yi-Leng Chen, (D), Illinois, 1980, Ow
Pao-Shin Chu, (D), Wisconsin
Tim Li, (D), Hawaii, 1993
Duane E. Stevens, (D), Harvard, 1977, Ow
Bin Wang, (D), Florida State, 1984, Ow
Shang-Ping Xie, (D), Hokkaido, 1991
Associate Professor:
P. Anders Daniels, (D), Michigan, 1971, Ow
Thomas A. Schroeder, (D), Purdue, 1974, Ow
Yuqing Wang, (D), Monash, 1995
Cooperating Faculty:
Antony D. Clarke, (D), Washington, 1983, Oc

Dept of Oceanography (B,M,D) (2006)
1000 Pope Road
Honolulu, HI 96822
p. (808) 956-7633 f. (808) 956-9225
ocean@soest.hawaii.edu
http://www.soest.hawaii.edu/oceanography/
Student Services Specialist: Nancy Koike
Enrollment (2001): B: 56 (0) M: 41 (3) D: 26 (0)
Chair:
Edward A. Laws, (D), Harvard, 1972, Cm
Undergraduate Chair:
Jane E. Schoonmaker, (D), Northwestern, 1981, Cm
Research Professor:
James P. Cowen, (D), California (Santa Cruz), 1983, Cm
Professor:
Robert R. Bidigare, (D), Texas A&M, 1981, Ob
Barry J. Huebert, (D), Northwestern, 1970, Oc
David M. Karl, (D), California (San Diego), 1978, Ob
Michael R. Landry, (D), Washington, 1976, Ob
Yuan-Hui Li, (D), Columbia, 1967, Cm
Roger Lukas, (D), Hawaii, 1981, Op
Douglas S. Luther, (D), MIT, 1980, Op
Fred T. Mackenzie, (D), Lehigh, 1962, Cm
Lorenz Magaard, (D), Kiel, 1963, Op
Alexander Malahoff, (D), Hawaii, 1965, Ou
Julian P. McCreary, (D), California (San Diego), 1977, Op
Christopher Measures, (D), Southampton, 1978, Oc
Peter Muller, (D), Hamburg, 1974, Op
Kelvin J. Richards, (D), Southampton, 1978, Og
Kathleen C. Ruttenberg, (D), Yale, Cg
Francis J. Sansone, (D), North Carolina, 1980, Co
Craig R. Smith, (D), California (San Diego), 1983, Ob
Richard E. Young, (D), Miami, 1968, Ob

Senior Scientist:
Patricia Fryer, (D), Hawaii, 1981, Gu
Richard W. Grigg, (D), California (San Diego), 1969, Ob
Associate Professor:
Eric Firing, (D), MIT, 1978, Op
Pierre J. Flament, (D), California (San Diego), 1986, Op
Gary M. McMurtry, (D), Hawaii, 1979, Cm
Brian N. Popp, (D), Illinois, 1986, Cm
Bo Qiu, (D), Kyoto, 1990, Op
Niklas Schneider, (D), Hawaii, 1992, Og
Associate Scientist:
Antony D. Clarke, (D), Washington, 1983, Oa
Eric H. De Carlo, (D), Hawaii, 1982, Cm
Assistant Professor:
Mark A. Merrifield, (D), California (San Diego), 1989, Oo
Grieg F. Steward, (D), California (San Diego), 1996, Ob
Emeritus:
Klaus Wyrtki, (D), Kiel, 1950, Op
Investigator:
Valerie M. Franck, (D), California (Santa Barbara), 2002
Cooperating Faculty:
Shannon K. Atkinson, (D), Murdoch, 1985, Oo
Thomas A. Clarke, (D), California (San Diego), 1968, Ob
Keith A. W. Crook, (D), New England, 1959, Gt
Walter Dudley, (D), Hawaii, 1976, Ou
Louis M. Herman, (D), Penn State, Oo
Rudolf C. Kloosterziel, (D), Utrecht (Neth), 1990, Op
John R. Sibert, (D), Columbia, 1968, Oo
Luis M. Tupas, (D), Tokyo, 1991, Ob
John C. Wiltshire, (D), Hawaii, 1983, Gu
Christopher D. Winn, (D), Hawaii, 1984, Ob
Affiliate Grad Faculty:
Whitlow W. L. Au, (D), Washington State, 1970, Oo
John E. Bardach, (D), Wisconsin, 1949, Oo
Paul K. Bienfang, (D), Hawaii, 1977, Oo
Richard N. Hey, (D), Princeton, 1975, Gu
Gary T. Mitchum, (D), Florida State, 1984, Op
Jeffrey J. Polovina, (D), California (Berkeley), 1974, Oo
Lewis M. Rothstein, (D), Hawaii, 1983, Oo
Eric W. Vetter, (D), California (San Diego), 1995, Oo

Idaho

Boise State University

Dept of Geosciences (B,M,D) (2004)
1910 University Drive
Mail Stop 1535
Boise, ID 83725
p. (208) 426-1581 f. (208) 4126-4061
cjnorth@boisestate.edu
http://earth.boisestate.edu/
Administrative Assistant: Liz Johansen
Enrollment: No data reported since 1995
Chair:
Clyde J. Northrup, (D), MIT, 1996, Gc
Professor:
Paul R. Donaldson, (D), Colorado Mines, 1974, Ye
John R. Pelton, (D), Utah, 1979, Ys
Walter S. Snyder, (D), Stanford, 1977, Gr
Craig M. White, (D), Oregon, 1980, Gi
Research Professor:
Vladimir I. Davydov, (D), St. Petersburg, 1982, Pg
Mitchell W. Lyle, (D), Oregon State, 1978, Ou
Associate Professor:
James P. McNamara, (D), Alaska (Fairbanks), 1997, Hq
Paul Michaels, (D), Utah, 1993, Ne
Assistant Professor:
Partha Routh, (D), British Columbia, 1999, Cg
David E. Wilkins, (D), Utah, 1997, Oy
Research Associate:
Warren Barrash, (D), Idaho, 1986, Hw
William Clement, (D), Wyoming, 1995, Yg
Thomas M. Clemo, (D), British Columbia, Yg
Michael D. Knoll, (D), British Columbia, 1996, Yg
Lee M. Liberty, (M), Wyoming, 1992, Ys
Mark D. Schmitz, (D), MIT, 2001, Cc
Karen Viskupic, (D), MIT, 2002, Gg
James Zollweg, (M), St. Louis, 1982, Ys
Adjunct Professor:
Virginia S. Gillerman, (D), California (Berkeley), 1982, Eg
Debra Matthews, (B), Gg
Jean Parker, (M), Gg
Michele Stoklosa, (D), Wisconsin, 2002, Gs
Emeritus:
Elton B. Bentley, (D), Oregon, 1974, Oy
Kenneth M. Hollenbaugh, (D), Idaho, 1968, Eg
Claude Spinosa, (D), Iowa, 1968, Pi
Charles J. Waag, (D), Arizona, 1968, Gc
Mont M. Warner, (D), Iowa, 1963, Gs
Monte D. Wilson, (D), Idaho, 1969, Gm
Spencer H. Wood, (D), Caltech, 1975, Ge

Brigham Young University, Idaho

Dept of Geology (B) (2007)
525 South Center Street
Rexburg, ID 83460-0510
p. (208) 496-1905 f. (208) 496-1950
moored@byui.edu
http://www.byui.edu/Geology
Enrollment (2007): B: 79 (51)
Chair:
Daniel K. Moore, (D), Rensselaer, 1997, GxiCp
Professor:
Robert W. Clayton, (D), Southern California, 1993, GcYgNg
Forest J. Gahn, (D), Michigan, 2004, PogGs
Benjamin R. Jordan, (D), Vermont, 2004, GvOuGx
William W. Little, (D), Colorado, 1995, GsdGm
Mark D. Lovell, (M), Idaho, 1998, HyOie
Julie B. Willis, (D), Utah, 2007, GtOri
Emeritus:
Glenn F. Embree, (D), Idaho, 1976, Gx
Roger D. Hoggan, (D), 1971, PeGg

College of Southern Idaho

Dept of Physical Science (A) (2007)
Box 1238
315 Falls Avenue
Twin Falls, ID 83301
p. (208) 732-6400 f. (208) 736-2136
swillsey@csi.edu
Enrollment (2007): A: 24 (1)
Assistant Professor:
Shawn P. Willsey, (M), N Arizona, 2000, GgcGv

Idaho State University

Dept of Geosciences (A,B,M) (2007)
STOP 8072
Physical Science Building #3
921 S. 8th Ave
Pocatello, ID 83209-8072
p. (208) 282-3365 f. (208) 282-4414
geology@isu.edu
http://www.isu.edu/departments/geology/
Administrative Assistant: Melissa Neiers
Enrollment (2005): B: 100 (14) M: 31 (19)

Chair:
Scott S. Hughes, (D), Oregon State, 1983, Gx
Professor:
Paul K. Link, (D), California (Santa Barbara), 1982, Gs
Michael O. McCurry, (D), California (Los Angeles), 1985, Gz
David W. Rodgers, (D), Stanford, 1987, Gc
Glenn D. Thackray, (D), Oregon State, 1989, Gl
Associate Professor:
Nancy F. Glenn, (D), Nevada (Reno), 2000, Or
Assistant Professor:
Daniel P. Ames, (D), Utah, 2002, Oi
Benjamin T. Crosby, (D), MIT, 2006, GmtHs
Leif Tapanila, (D), Utah, 2005, Pi
Instructor:
Lori Tapanila, (M), Utah, 2006, Gg
Emeritus:
David E. Fortsch, (M), S California, 1973, Oy
Office Specialist:
Michele Hughes, Oo
GIS Director:
Keith Weber, (M), Montana, 19, Oi
Cooperating Faculty:
James D. Stephens, (D), Utah, 1960, Gz
Affiliate Faculty:
John A. Welhan, (D), California, 1981, Hw

Lewis-Clark State College

Earth Sciences (A,B) (2007)
Div of Natural Science & Mathematics
500 8th Ave
Lewiston, ID 83501
p. (208) 790-2283 f. (208) 790-2064
klschmidt@lcsc.edu
http://www.lcsc.edu/geology/
Administrative Assistant: Connie Hallen
Enrollment (2007): A: 1 (0) B: 15 (2)
Associate Professor:
Keegan L. Schmidt, (D), S California, 2000, Gc

University of Idaho

Dept of Geological Sciences (B,M,D) (2007)
7th & Line Street
PO Box 443022
Moscow, ID 83844-3022
p. (208) 885-6192 f. (208) 885-5724
geology@uidaho.edu
http://www.geology.uidaho.edu
Administrative Assistant: Debbie K. Jensen
Enrollment (2007): B: 74 (11) M: 34 (8) D: 3 (1)
Chair:
Dennis J. Geist, (D), Oregon, 1985, Gi
Professor:
Mickey E. Gunter, (D), Virginia Tech, 1987, Gz
Peter E. Isaacson, (D), Oregon State, 1974, Pi
John S. Oldow, (D), Northwestern, 1978, Gc
James L. Osiensky, (D), Idaho, 1983, Hw
Judith Totman Parrish, (D), California (Santa Cruz), 1979, Pe
Kenneth F. Sprenke, (D), Alberta, 1983, Ye
Scott A. Wood, (D), Princeton, 1985, Cg
Associate Professor:
Jerry P. Fairley, (D), California (Berkeley), 2000, Hw
Gary S. Johnson, (D), 1991, Hw
Simon A. Kattenhorn, (D), Stanford, 1998, Gc
William C. McClelland, (D), Arizona, 1990, Gt
Assistant Professor:
Susan E. Childers, (D), 1997, Gg

Soil Science Div (B,M,D) (2004)
Dept of Plant, Soil, & Entomological Sciences
Moscow, ID 83844-2339
p. (208) 885-7012 f. (208) 885-7760
pmcdaniel@uidaho.edu
Enrollment (1995): B: 8 (4) M: 14 (2) D: 5 (2)
Professor:
Guy Knudsen, (D), Sb
Robert L. Mahler, (D), North Carolina State, 1980, Oo
Associate Professor:
Bradford D. Brown, (D), Utah State, 1985, Oo
Matthew J. Morra, (D), Ohio State, 1986, Sb
Assistant Professor:
Jodi Johnson-Maynard, (D), California (Riverside), 1999, Sd
Paul A. McDaniel, (D), North Carolina State, 1989, Sd
Daniel G. Strawn, (D), Delaware, 1999, Sc

Illinois

Augustana College

Dept of Geology (B) (2006)
639 38th St.
Rock Island, IL 61201
p. (309) 794-7318 f. (309) 794-7564
michaelwolf@augustana.edu
http://www.augustana.edu/academ/geology
Administrative Assistant: Kathy Nelson
Enrollment (2003): B: 16 (2)
Chair:
Michael B. Wolf, (D), Caltech, 1992, Gi
Professor:
William R. Hammer, (D), Wayne State, 1979, Pv
Assistant Professor:
Jeffrey C. Strasser, (D), Lehigh, 1996, Gm
Emeritus:
Richard C. Anderson, (D), Chicago, 1955, Gm
Museum Educational Programs Coordinator:
Susan Kornreich Wolf, (B), Hamilton, 1987, Oe

Black Hawk College

Dept of Natural Science & Engineering (A) (2007)
6600 34th Avenue
Moline, IL 61265
p. (309) 796-5000 f. (309) 792-3230
harwoodr@bhc.edu
http://facweb.bhc.edu/academics/science
Administrative Assistant: Tara Carey
Chair:
Brian Glaser, (N)
Professor:
Richard D. Harwood, (M), N Arizona, 1989, Gv

Bradley University

Dept of Geological Sciences (B) (2004)
1501 W. Bradley Avenue
Peoria, IL 61625
p. (309) 677-2345 f. (309) 677-2330
fossil@hilltop.bradley.edu
Enrollment (1998): B: 22 (7)
Chair:
Merrill W. Foster, (D), Harvard, 1970, Pg
Professor:
Donald R. Gorman, (D), Illinois, 1962, Gs
Associate Professor:
Henry L. Helenek, (D), Brown, 1971, Gt

College of Lake County
Geology Dept (A) (2004)
Grayslake, IL 60030-1198
p. (847) 543-2491 f. (847) 223-1017
riepe@clcillinois.edu
Enrollment (1997): B: 9 (0)
Professor:
Ronald A. Riepe, (M), Indiana, 1970, Gr
Douglas B. Sherman, (M), S California, 1963, Gu

Columbia College
Dept of Science & Mathematics (2007)
600 South Michigan Avenue
Chicago, IL 60605
p. (312) 344-7396 f. (312) 344-8075
crasinariu@popmail.colum.edu
http://www.colum.edu
Professor:
Gerald E. Adams, (D), Northwestern, 1985, Gi

Concordia University
Dept of Natural Science and Geography (B) (2007)
7400 Augusta
River Forest, IL 60305
p. (708) 771-8300 f. (708) 209-3176
Anita.Briedis@cuchicago.edu
Enrollment (2007): B: 2 (1)
Professor:
Anita Briedis, (D), Illinois (Chicago), OooOo
William H. Kammrath, (D), Illinois, 1971, Oy
Patricia Rose, (D), Chicago, 1991, Oy

Eastern Illinois University
Dept of Geology/Geography (B) (2004)
600 Lincoln Avenue
Charleston, IL 61920-3099
p. (217) 581-2626 f. (217) 581-6613
cfab@ux1.eiu.edu
http://www.eiu.edu/~geoscience/
Administrative Assistant: Susan Kile
Enrollment (2002): B: 19 (7)
Chair:
Alan Baharlou, (D), Tulsa, 1973, Cg
Professor:
Craig A. Chesner, (D), Michigan Tech, 1988, Gx
Vincent P. Gutowski, (D), Cal State, 1977, Gm
Robert B. Jorstad, (D), Idaho, 1983, Gr
Godson C. Obia, (D), Oklahoma, 1986, Oo
James F. Stratton, (D), Indiana, 1975, Pg
Associate Professor:
Kathleen M. Bower, (D), New Mexico, 1993, Ng
Belayet H. Khan, (D), Pittsburgh, 1985, Oy
Betty E. Smith, (D), SUNY (Buffalo), 1994, Oi
Assistant Professor:
Michael W. Cornebise, (D), Tennessee, 2003, Og
James A. Davis, (D), Kansas State, 2001, Og
John P. Stimac, (D), Oregon, 1996, Gc
Instructor:
Robert Cataneo, (M), E Illinois, 2003, Oa
Lester B. Stoner, (B), E Illinois, 1983, Og
Emeritus:
Raymond N. Pheifer, (D), Indiana, 1979, Ec
Related Staff:
Steven DiNaso, (B), E Illinois, 1995, Oi

Elgin Community College
Dept of Math, Sci, Tech & Health Professions (2004)
1700 Spartan Drive
Elgin, IL 60123
p. 847/214-7359 f. 847/214-7965
gdimoplon@elgin.edu
http://elgin.edu
Instructor:
Timothy M. Millen, (M), N Illinois, 1982, Cg

Field Museum of Natural History
Dept of Geology (2005)
1400 S. Lake Shore Drive
Chicago, IL 60605-2496
p. (312) 665-7621 f. (312) 665-7641
klawson@fieldmuseum.org
http://www.fieldmuseum.org
Administrative Assistant: Karsten L. Lawson
Administrative Assistant: Elaine Zeiger
Chair:
Olivier C. Rieppel, (D), Basel, 1978, Pv
Curator, Meteoritics:
Meenakshi Wadhwa, (D), Washington (St. Louis), 1994, Xm
Curator, Fossil Fishes:
Lance Grande, (D), CUNY, 1983, Pv
Curator, Fossil Amphibians & Reptiles:
John R. Bolt, (D), Chicago, 1968, Pv
Associate Curator, Fossil Invertebrates:
Scott H. Lidgard, (D), Johns Hopkins, 1984, Pi
Emeritus:
Matthew H. Nitecki, (D), Chicago, 1968, Pi
William D. Turnbull, (D), Chicago, 1967, Pv
Bertram G. Woodland, (D), Chicago, 1962, Gp
Associate Curator, Fossil Invertebrates:
Peter J. Wagner, (D), Chicago, 1995, Pi

Illinois State University
Dept of Geography-Geology (B,M) (2007)
206 Felmley Hall of Science
Campus Box 4400
Normal, IL 61790-4400
p. (309) 438-7649 f. (309) 438-5310
geo@ilstu.edu
http://www.geo.ilstu.edu/contact_us/
Enrollment (2007): B: 49 (6) M: 18 (5)
Chair:
David H. Malone, (D), Wisconsin, 1994, Gc
Professor:
James E. Day, (D), Iowa, 1988, Pi
Associate Professor:
Dagmar Budikova, (D), Calgary, 2001, Ge
Elizabeth M. King, (D), Wisconsin, 2001, Gx
Robert S. Nelson, (D), Iowa, 1970, Gm
Stephen Van der Hoven, (D), Utah, 2000, Hg
Assistant Professor:
Amy M. Bloom, (D), Utah, 2005, Gn
Eric W. Peterson, (D), Missouri, 2002, Hg
Adjunct Professor:
Toby J. Dogwiler, (D), Missouri (Colombia), 2002, Gm
Walton R. Kelly, (D), Virginia, 1993, HgCg
Edward Mehnert, (D), Illinois, 1998, Hg
Andre J. Pugin, (D), Geneva (Switzerland), 1989, Gl
William W. Shilts, (D), Tulane, 1970, Gg
Geographic Tech:
William Shields, (M), Illinois State, 2001, Gg

Illinois Valley Community College
Dept of Geology (A) (2007)
Natural Sciences & Business Division

815 North Orlando Smith Avenue
Oglesby, IL 61348-9692
p. (815)224-0394
mike_phillips@ivcc.edu
http://www.ivcc.edu/phillips/geology/geology.html
Enrollment (2007): A: 4 (2)

Professor:
Michael Phillips, (M), S Illinois, 1990, GgeGm

Lincoln Land Community College

Math and Sciences (A) (2007)
Springfield, IL 62794-9256
p. (217) 786-4923
dean.butzow@llcc.edu
Enrollment (2006): A: 7 (0)

Northeastern Illinois University

Dept of Earth Science (B,M) (2008)
5500 N. St. Louis Avenue
Chicago, IL 60625
p. (773) 442-6050 f. (773) 442-5710
K-Bartels@neiu.edu
http://www.neiu.edu/~deptesci/
Administrative Assistant: Kate Ekman
Enrollment (2007): B: 32 (7) M: 7 (0)

Head:
Karen S. Bartels, (D), MIT, 1991, Gi

Professor:
Laura L. Sanders, (D), Kent State, 1986, HwGe
Hans D. Upadhyay, (D), Memorial, 1973, Gc

Instructor:
Cary Easterday, (M), Ohio State, 2004, Pg
Mohammad Fariduddin, (D), N Illinois, 1999, Pe
Jean M. Hemzacek Laukant, (M), N Illinois, 1986, Sc

Laboratory Director:
Peter Fortsas, (B), NE Illinois, 1984, Oo

Northern Illinois University

Dept of Geology and Env Geosciences (B,M,D) (2007)
Davis Hall 312, Normal Rd.
De Kalb, IL 60115-2854
p. (815) 753-1943 f. (815) 753-1945
geology@geol.niu.edu
http://jove.geol.niu.edu
Administrative Assistant: Amy Polzin
Enrollment (2006): B: 70 (17) M: 17 (2) D: 16 (5)

Chair:
Jonathan H. Berg, (D), Massachusetts, 1976, Gp

Assistant Chair:
Mark P. Fischer, (D), Penn State, 1994, Gc

Professor:
Colin J. Booth, (D), Penn State, 1984, Hw
Phillip J. Carpenter, (D), New Mexico Tech, 1984, Yg
Paul Loubere, (D), Oregon State, 1980, Pe
Eugene C. Perry, (D), MIT, 1963, Cs
Ross D. Powell, (D), Ohio State, 1980, Gs
Reed P. Scherer, (D), Ohio State, 1991, Pm
James A. Walker, (D), Rutgers, 1982, Gi

Associate Professor:
Melissa E. Lenczewski, (D), Tennessee, 2001, Co
Paul R. Stoddard, (D), Northwestern, 1989, Gt
Jay A. Stravers, (D), Colorado, 1986, Gl

Teacher Certification Coordinator:
B. Kathy Kitts, (D), Washington, 2002, Xc

Assistant Professor:
Mark R. Frank, (D), Maryland, 2001, Cp

Research Associate:
Monica Carroll, (D), Georgia, 2006, Cl
D. Neil Dickey, Jr., (D), N Illinois, 1984, Gi
Stefan W. Vogel, (D), California (Santa Cruz), 2004, Gl

Adjunct Professor:
Jean E. Bogner, (D), N Illinois, 1996, Cl
Luis E. Marin, (D), N Illinois, 1990, Hw
Lyle D. McGinnis, (D), Illinois, 1965, Yg
Michael G. Mudrey, Jr., (D), Minnesota, 1973, Eg

Emeritus:
C. Patrick Ervin, (D), Wisconsin, 1972, Yv
Hsin-Yi Ling, (D), Washington (St. Louis), 1963, Pm
Carla W. Montgomery, (D), MIT, 1977, Cc

Northwestern University

Earth and Planetary Sciences (B,M,D) (2007)
1850 Campus Drive
Locy Hall
Evanston, IL 60208-2150
p. (847) 491-3238 f. (847) 491-8060
eps@earth.northwestern.edu
http://www.earth.northwestern.edu/
Administrative Assistant: Charlotte M. Hayes
Enrollment (2007): B: 7 (1) M: 0 (1) D: 0 (3)

Chair:
Bradley B. Sageman, (D), Colorado, 1991, Ps

Professor:
Craig R. Bina, (D), Northwestern, 1987, Yg
Neal E. Blair, (D), Stanford, 1980, Co
Donna M. Jurdy, (D), Michigan, 1974, Yg
Abraham Lerman, (D), Harvard, 1964, Cg
Emile A. Okal, (D), Caltech, 1978, Ys
Seth Stein, (D), Caltech, 1978, Ys

Joint Faculty:
Jean-Francois Gaillard, (D), Paris 7, 1987, Cl

Assistant Professor:
Matthew T. Hurtgen, (D), Penn State, 2003, Gs
Steven D. Jacobsen, (D), Colorado, 2001, Gy
Andrew D. Jacobson, (D), Michigan, 2001, Cl
Francesca A. Smith, (D), Chicago, 2002
Suzan van der Lee, (D), Princeton, 1996, Ys

Research Associate:
Michael Stefanick, (D), Princeton, 1980, Yg

Adjunct Professor:
Jennifer C. McElwain, (D), London (UK), 1997, Pb

Adjunct Professor Emeritus:
Johannes Weertman, (D), Carnegie Inst, 1951, Om

Emeritus:
G. Edward Birchfield, (D), Chicago, 1962, Pe

Librarian:
Cunera Buys, Oo

Olivet Nazarene University

Dept of Physical Sciences (B) (2007)
One University Avenue
Bourbonnais, IL 60914
p. (815) 939-5394 f. (815) 939-5071
mreams@olivet.edu
http://geology.olivet.edu/
Enrollment (2007): B: 12 (4)

Chair:
Max W. Reams, (D), Washington (St. Louis), 1968, Gsm

Professor:
Kevin E. Brewer, (D), Nevada (Reno), 1994, HwOiYg

Director, Strickler Planetarium & Observatory:
Brock C. Schroeder, (D), Walden, 2006, Xg

Associate Professor:
Charles W. Carrigan, (D), Michigan, 2005, CcGxc

Instructor:
Priscilla Skalac, (M), Olivet Nazarene, 2005, Oe

Principia College

Dept of Geology (2007)
1 Maybeck Place
Elsah, IL 62028
p. (618) 374-5294 f. (618) 374-5122
janis.treworgy@prin.edu
http://prin.edu/mammoth
Enrollment: No data reported since 1995

Chair:
Janis D. Treworgy, (D), Illinois (Urbana), 1985, GgPv

Southern Illinois University Carbondale

Dept of Geology (B,M,D) (2006)
1259 Lincoln Drive, Mailcode 4324
Parkinson 102
Carbondale, IL 62901
p. (618) 453-3351 f. (618) 453-7393
geology@geo.siu.edu
http://www.science.siu.edu/geology/
Administrative Assistant: Sonja S. Bailey
Administrative Assistant: Sharon J. Simons
Enrollment (2005): B: 32 (9) M: 14 (5) D: 8 (0)

Chair:
Steven P. Esling, (D), Iowa, 1983, Hw

Professor:
Nicholas Pinter, (D), California (Santa Barbara), 1992, Gm
Dhananjay Ravat, (D), Purdue, 1989, Ye
John L. Sexton, (D), Indiana, 1974, Ye

Associate Professor:
Ken B. Anderson, (D), Melbourne, 1989, Co
Richard H. Fifarek, (D), Oregon State, 1985, Em
Scott E. Ishman, (D), Ohio, 1990, Pe
John E. Marzolf, (D), California (Los Angeles), 1970, Gs

Assistant Professor:
Eric C. Ferre, (D), Toulouse, 1989, Gc

Research Associate:
Harvey Henson, (M), S Illinois, 1989, Yg
William Huggett, (B), S Illinois, 1981, Ec

Lecturer:
Susann B. Pinter, (M), Gottingen, 1991, Gg

Leader, Coal Characterization Lab:
John C. Crelling, (D), Penn State, 1973, Ec

Emeritus:
Russell R. Dutcher, (D), Penn State, 1960, Ec
Charles O. Frank, (D), Syracuse, 1973, Gx
Stanley E. Harris, Jr., (D), Iowa, 1947, Ge
Paul D. Robinson, (M), S Illinois, 1963, Gz
John E. Utgaard, (D), Indiana, 1963, Pi
Jay Zimmerman, Jr., (D), Princeton, 1968, Gc

University of Chicago

Dept of Geophysical Sciences (B,M,D) (2004)
5734 S. Ellis Avenue
Chicago, IL 60637
p. (773) 702-8101 f. (773) 702-9505
info@geosci.uchicago.edu
http://geosci.uchicago.edu
Administrative Assistant: David J. Leslie
Enrollment (1999): B: 27 (6) D: 50 (3)

Chair:
David B. Rowley, (D), SUNY (Albany), 1983, Gc

Professor:
Alfred T. Anderson, Jr., (D), Princeton, 1963, Gv
Victor Barcilon, (D), Harvard, 1963, Or
Peter R. Crane, (D), Reading, 1981, Po
John E. Frederick, (D), Colorado, 1975, Oa
Lawrence Grossman, (D), Yale, 1972, Xc
David Jablonski, (D), Yale, 1979, Pi
Susan M. Kidwell, (D), Yale, 1982, Gs
Michael C. LaBarbera, (D), Duke, 1976, Po
Douglas R. MacAyeal, (D), Princeton, 1983, Yr
Raymond T. Pierrehumbert, (D), MIT, 1980, Oa
Frank M. Richter, (D), Chicago, 1972, Gt
Joseph V. Smith, (D), Cambridge, 1951, Gz
Ramesh C. Srivastava, (D), McGill, 1964, Oa
Alfred M. Ziegler, (D), Oxford, 1964, Gg

Associate Professor:
David Archer, (D), Washington, 1990, Yr
Michael J. Foote, (D), Chicago, 1989, Pg
Dion L. Heinz, (D), California (Berkeley), 1986, Yx

Assistant Professor:
Munir Humayun, (D), Chicago, 1994, Cg
Noboru Nakamura, (D), Princeton, 1989, Oa

Lecturer:
Meenakshi Wadhwa, (D), Washington, 1994, Og

Emeritus:
Robert N. Clayton, (D), Caltech, 1955, Cs
Dave Fultz, (D), Chicago, 1947, Oa
Ole J. Kleppa, (D), Norwegian Inst of Tech, 1956, Cp
Hsiao-Lan Kuo, (D), Chicago, 1948, Oa
George W. Platzman, (D), Chicago, 1948, Oa
David M. Raup, (D), Harvard, 1957, Po

University of Illinois, Chicago

Dept of Earth and Environmental Sciences (B,M,D) (2007)
845 W. Taylor Street
2440 SES
MC 186
Chicago, IL 60607-7059
p. (312) 996-3155 f. (312) 413-2279
msting@uic.edu
http://www.uic.edu/depts/geos/
Enrollment (2000): M: 5 (0)

Head:
Neil C. Sturchio, (D), Washington (St. Louis), 1983, Cg

Professor:
Martin F. J. Flower, (D), Manchester, 1971, Gi
Steven L. Forman, (D), Colorado, 1986, Gl
Stephen J. Guggenheim, (D), Wisconsin, 1976, Gz
Kathryn L. Nagy, (D), Texas A&M, 1988, Cg
Roy E. Plotnick, (D), Chicago, 1983, Pi
Carol A. Stein, (D), Columbia, 1984, Yg

Associate Professor:
Peter T. Doran, (D), Nevada (Reno), 1996, Hw
Fabien Kenig, (D), Orleans (France), 1991, Ge

Adjunct Associate Professor:
Jean Bogner, (D), N Illinois, 1996, Hw
John R. Bolt, (D), Chicago, 1968, Pv
Mohammed Sultan, (D), Washington (St. Louis), 1984, Gx

Adjunct Professor:
Martin Perlmutter, (D), Miami, 1982, Gs

Emeritus:
August F. Koster Van Groos, (D), Leiden (Neth), 1966, Cp
Kelvin S. Rodolfo, (D), S California, 1967, Gu

Secretary:
Minnie O. Jones, Oo

Research Specialist:
Jeaneth Gomez, (B), Illinois (Chicago)

Electron Microprobe Analyst:
Gene Harris, (M), 1973, Oo

Business Manager:
Mary Louise Schick, (B), Cumberland Coll, 1982, Oo
Administrative Secretary:
Marilyn Cartlidge
Related Staff:
James Pierson, (M)

University of Illinois, Urbana-Champaign

Dept of Atmospheric Sciences (M,D) (2007)
105 S. Gregory Street
Urbana, IL 61801-3070
p. (217) 333-2046 f. (217) 244-4393
dept@atmos.uiuc.edu
http://www.atmos.uiuc.edu
Enrollment (2006): M: 19 (9) D: 18 (3)

Acting Head:
Robert M. Rauber, (D), Colorado State, 1985, Oa
Professor:
Mankin Mak, (D), MIT, 1968, Oa
Walter A. Robinson, (D), Columbia, 1985, Oa
Michael E. Schlesinger, (D), California (Los Angeles), 1976, Oa
Robert B. Wilhelmson, (D), Illinois, 1972, Oa
Donald J. Wuebbles, (D), California (Davis), 1983, Oa
Associate Professor:
Larry Di Girolamo, (D), McGill, 1996, Or
Atul K. Jain, (D), India, 1988, Oa
Greg M. McFarquhar, (D), Toronto, 1992, Or
Assistant Professor:
Somnath Baidya Roy, (D), Rutgers, 2001, Oa
Stephen W. Nesbitt, (D), Utah, 2003, Oa
Peter K. Snyder, (D), Wisconsin, 2004, Oa
Tracy E. Twine, (D), Wisconsin, 2004, Oa
Instructor:
Eric R. Snodgrass, (M), Illinois, 2006, Oa
Lecturer:
Donna J. Charlevoix, (M), California (Davis), 1996, Oa
Emeritus:
Kenneth V. Beard, (D), California (Los Angeles), 1970, Oa
John E. Walsh, (D), MIT, 1974, Oa

Dept of Geology (B,M,D) (2007)
1301 West Green St.
245 Natural History Bldg.
Urbana, IL 61801
p. (217) 333-3542 f. (217) 244-4996
mkt@uiuc.edu
http://www.geology.uiuc.edu
Enrollment (2006): B: 64 (19) M: 13 (4) D: 22 (1)

Head:
Stephen Marshak, (D), Columbia, 1983, Gc
Professor:
Jay D. Bass, (D), SUNY (Stony Brook), 1982, Gy
James L. Best, (D), London, 1985, Gs
Craig M. Bethke, (D), Illinois, 1984, Hw
Wang-Ping Chen, (D), MIT, 1979, Ys
Marcelo Garcia, (D), Minnesota, 1989, Gm
Susan W. Kieffer, (D), Caltech, 1971, Gv
Gary Parker, (D), Minnesota, 1974, Gm
Associate Professor:
Stephen P. Altaner, (D), Illinois, 1985, Gz
Chu-Yung Chen, (D), MIT, 1983, Cg
Bruce W. Fouke, (D), SUNY (Stony Brook), 1993, Gs
Feng-Sheng Hu, (D), Washington, 1994, Pe
Thomas M. Johnson, (D), California (Berkeley), 1995, Hw
Craig C. Lundstrom, (D), California (Santa Cruz), 1996, Cp
Xiaodong Song, (D), Caltech, 1994, Ys
Assistant Professor:
Alison M. Anders, (D), Washington, 2005, Gm
Jie (Jackie) Li, (D), Harvard, 1998, Gy
Lecturer:
Michael A. Stewart, (D), Duke, 2000, Gi
Adjunct Professor:
Robert J. Finley, (D), South Carolina, 1975, Go
Leon R. Follmer, (D), Illinois, 1970, Sa
Hannes E. Leetaru, (D), Illinois, 1997, Go
Morris W. Leighton, (D), Chicago, 1951, Eo
William W. Shilts, (D), Syracuse, 1970, Gs
M. Scott Wilkerson, (D), Illinois, 1991, Gc
Emeritus:
Thomas F. Anderson, (D), Cs
Daniel B. Blake, (D), California (Berkeley), 1966, Pi
Albert V. Carozzi, (D), Geneva, 1948, Gd
Donald L. Graf, (D), Columbia, 1950, Cl
Albert T. Hsui, (D), Cornell, 1972, Yg
George deV. Klein, (D), Yale, 1960, Gs
Ralph L. Langenheim, Jr., (D), Minnesota, 1951, Gr
C. John Mann, (D), Wisconsin, 1961, Gq
Alberto S. Nieto, (D), Illinois, 1974, Ng
Philip A. Sandberg, (D), Stockholm, 1965, Gd
Teaching Specialist:
Eileen A. Herrstrom, (D), Iowa, 1995, Gi
Ann D. Long, (M), Leeds, 1981, Gm
Senior Research Scientist:
Robert A. Sanford, (D), Michigan State, 1996, Py
Research Scientist:
Jonathan H. Tomkin, (D), Australian National, 2002, Gm
Research Programmer:
Stephen D. Hurst, (D), California (Davis), 1991, Gc
Geochemistry Specialist:
Justin Glessner, (B), Washington, 1998, Cs

Dept of Natural Resources & Env Sciences (B,M,D) (2004)
W-503 Turner Hall
1102 South Goodwin Avenue
Urbana, IL 61801-4798
p. (217) 333-2770 f. (217) 244-3219
g-rolfe@uiuc.edu
Enrollment (1998): B: 270 (71) M: 73 (36) D: 62 (18)

Professor:
Charles W. Boast, (D), Iowa State, 1970, Sp
Mark B. David, (D), New York, 1983, Sf
John J. Hassett, (D), Utah State, 1970, Sc
Robert L. Jones, (D), Illinois, 1962, Sc
Richard L. Mulvaney, (D), Illinois, 1983, Sc
Theodore R. Peck, (D), Wisconsin, 1962, Sc
Joseph W. Stucki, (D), Purdue, 1975, Sc
Associate Professor:
Robert G. Darmody, (D), Maryland, 1980, Sd
Timothy R. Ellsworth, (D), California, 1989, Sp
Kenneth R. Olson, (D), Cornell, 1983, Sc
F. William Simmons, (D), North Carolina State, 1987, Sp
Assistant Professor:
Robert J. M. Hudson, (D), MIT, 1989, Cg
Gregory F. McIsaac, (D), Illinois, 1994, Hs

Western Illinois University

Dept of Geology (B) (2007)
1 University Circle
Macomb, IL 61455
p. (309) 298-1151 f. (309) 298-3399
pl-calengas@wiu.edu
http://www.wiu.edu/geology/
Administrative Assistant: Kristie Parkins

Enrollment (2007): B: 20 (0)

Chair:
Peter L. Calengas, (D), Indiana, 1977, En
Professor:
Jack B. Bailey, (D), Illinois, 1975, Pg
Leslie A. Melim, (D), S Methodist, 1991, Gd
Associate Professor:
Steven W. Bennett, (D), Indiana, 1994, Hw
Kyle R. Mayborn, (D), California (Davis), 2000, Gp
Assistant Professor:
Amy L. Brock, (D), UNLV, 2007, GmSa
Cooperating Faculty:
Robert E. Johnson, (B), W Illinois, 1978, Oo

Wheaton College

Dept of Geology & Environmental Science (B) (2006)
501 E. College Ave.
Wheaton, IL 60187
p. (630) 752-5063 f. (630) 752-5996
geology@wheaton.edu
Enrollment (2006): B: 15 (7)

Chair:
Stephen O. Moshier, (D), Louisiana State, 1986, Gd
Professor:
James A. Clark, (D), Colorado, 1977, Gm
Jeffrey K. Greenberg, (D), North Carolina, 1978, Gt
Emeritus:
Gerald H. Haddock, (D), Oregon, 1967, Gi

William Rainey Harper College

Technology, Mathematics & Science Div (2005)
1200 West Algonquin Road
Palatine, IL 60067-7398
p. (847) 925-6374 f. (847) 925-6049
sgriffit@harpercollege.edu

Professor:
Paul P. Sipiera, (M), NE Illinois, 1975, Xm
Associate Professor:
Larry Knight, (M), S Illinois, 1968, Gg

Indiana

Ball State University

Dept of Geology (A,B,M) (2007)
Muncie, IN 47306
p. (765) 285-8270 f. (765) 285-8265
asamuels@bsu.edu
http://www.bsu.edu/geology
Administrative Assistant: M. Joan Gray
Enrollment (2006): B: 34 (0) M: 7 (0)

Chair:
Alan C. Samuelson, (D), Penn State, 1972, Hw
Professor:
Richard H. Fluegeman, (D), Cincinnati, 1987, Pm
R. Scott Rice-Snow, (D), Penn State, 1983, Gm
Associate Professor:
Jeffry D. Grigsby, (D), Cincinnati, 1989, Gs
Assistant Professor:
Klaus Neumann, (D), Alabama, 1999, Cl
Kirsten N. Nicholson, (D), Joseph Fourier, 1999, Gi
Emeritus:
Henry E. Kane, (D), California, 1965, Gm
Harlan H. Roepke, (D), Texas, 1970, Gs
Geological Tech:
Michael Kutis, (B), Wisconsin (Platteville), 1989, Gg

DePauw University

Dept of Geosciences (B) (2007)
602 South College Avenue
Julian Science Center
Greencastle, IN 46135
p. (765) 658-4654 f. (765) 658-4732
mswilke@depauw.edu
http://www.depauw.edu/acad/geosciences/
Administrative Assistant: Bonnie L. Bryan
Enrollment (2007): B: 27 (7)

Professor:
James G. Mills, (D), Michigan State, 1991, Gi
Frederick M. Soster, (D), Case Western, 1984, Gs
Associate Professor:
M. Scott Wilkerson, (D), Illinois, 1991, Gct
Assistant Professor:
Tim D. Cope, (D), Stanford, 2003, GsOi
Jeanette K. Pope, (D), Virginia Tech, 2002, Ge
Emeritus:
Robert D. Loring, (M), Indiana, 1948, Oy
James A. Madison, (D), Washington (St. Louis), 1968, Gz

Earlham College

Geosciences Dept (B) (2006)
801 National Road West
Richmond, IN 47374-4095
p. (765) 983-1429 f. (765) 983-1691
streeme@earlham.edu
http://www.earlham.edu/geos
Administrative Assistant: Jayne H. Arnold
Enrollment (2006): B: 18 (4)

Assistant Professor:
Ronald L. Parker, (M), Vermont, 1986, Hy
Meg Streepey, (D), Michigan, 2001, Gt
Emeritus:
Jon W. Branstrator, (D), Cincinnati, 1975, Pg
Charles W. Martin, (D), Wisconsin, 1962, Gg

Hanover College

Dept of Geology (B) (2007)
PO Box 890
Hanover, IN 47243-0890
p. (812) 866-7306 f. (812) 866-72164
worcestr@hanover.edu
Enrollment (2007): B: 12 (5)

Chair:
Heyo Van Iten, (D), Michigan, 1989, Pg
Professor:
Peter A. Worcester, (D), Miami (Ohio), 1976, Gi
Associate Professor:
Kenneth A. Bevis, (D), Oregon State, 1998, Gl
Emeritus:
Stanley M. Totten, (D), Illinois, 1962, Gl

Holy Cross College

Dept of Geology (A,B) (2006)
P.O. Box 308
Notre Dame, IN 46556
p. (219) 239-8417 f. (219) 239-8371
wcaponigri@hcc-nd.edu
Enrollment (1996): A: 0 (59) B: 78 (0)

Emeritus:
Winifred F. Caponigri, (M), Notre Dame, 1971, Gg

Indiana State University

Dept of Geography, Geology, & Anthropology (B,M,D) (2006)

159 Science Building
Terre Haute, IN 47809
p. (812) 237-2444 f. (812) 237-8029
sberta@isugw.indstate.edu
http://www.indstate.edu/gga/
Enrollment (2005): B: 67 (15) M: 28 (13) D: 21 (4)

Chair:
Susan M. Berta, (D), Oklahoma, 1986, Gm
Professor:
Prodip K. Dutta, (D), Indiana, 1983, Gs
Basil Gomez, (D), Southampton, 1980, Gm
Steven Pontius, (D), Minnesota, 1977, Oo
C. Russell Stafford, (D), Arizona State, 1981, Ga
Associate Professor:
Gregory Bierly, (D), Michigan State, 1996, Oa
Sandra S. Brake, (D), Colorado Mines, 1989, Ge
Brian Ceh, (D), W Ontario, 1994, Oi
Kathleen M. Heath, (D), Utah, 2001, Ga
Ryan Jensen, (D), Florida, 2000, Or
Robert C. Larson, (D), Indiana, 1975, Oo
Nancy J. Obermeyer, (D), Chicago, 1987, Oi
Mohameden Ould-Mey, (D), Kentucky, 1995, Oo
Qihao Weng, (D), Georgia, 1999, Or
Assistant Professor:
Meredith Beilfuss, (D), Indiana, 2004, Oe
Jay Gatrell, (D), West Virginia, 1999, Oi
Jennifer Latimer, (D), Indiana, Cm
Shawn Phillips, (D), New York, 2001, Oo
Anthony Rathburn, (D), Duke, 1992, Ob
James Speer, (D), Tennessee, 2001, Pe
Research Associate:
Nelson Dias, (D), Indiana State, 2002, Or
Karla Hansen-Speer, (D), Washington, 2006, Ga
Elena Perez, (D), California (San Diego), 2001, Pm
Instructor:
Leigh Ann Lawson, (M), Indiana State, 2003, Gg
Emeritus:
William A. Dando, (D), Minnesota, 1969, Oo
Paul W. Mausel, (D), North Carolina, 1965, Or
John R. McGregor, (D), Illinois, 1965, Oo
John E. Oliver, (D), Columbia, 1970, Oa
Assistant Anthropology Lab Director:
Mark Cantin, (M), Indiana State, 2000, Ga
Other:
David Holycross, (M), Indiana State, 2004, Ga

Indiana Univ/Purdue University, Fort Wayne

Dept of Geosciences (B) (2007)
2101 East Coliseum Boulevard
Fort Wayne, IN 46805-1499
p. (260) 481-6249 f. (260) 481-6880
isiorho@ipfw.edu
http://www.geosci.ipfw.edu/
Administrative Assistant: Linda D. Whitlock
Enrollment (2007): B: 16 (3)

Chair:
Solomon A. Isiorho, (D), Case Western, 1987, HwGeOy
Professor:
Anne S. Argast, (D), SUNY (Binghamton), 1986, Gdz
Carl N. Drummond, (D), Michigan, 1994, Gs
James O. Farlow, (D), Yale, 1980, PvoPe
Assistant Professor:
Benjamin F. Dattilo, (D), Cincinnati, 1994, GrPiGs
Emeritus:
Dipak K. Chowdhury, (D), Texas A&M, 1961, Ys
Jack A. Sunderman, (D), Indiana, 1963, Gr
Related Staff:
Clarence Tennis

Indiana Univ/Purdue University, Indianapolis

Dept of Geology (B,M) (2005)
723 West Michigan Street
Indianapolis, IN 46202-5132
p. (317) 274-7484 f. (317) 274-7966
ibsz100@iupui.edu
http://www.geology.iupui.edu
Enrollment (1998): B: 16 (2)

Chair:
Andrew P. Barth, (D), S California, 1989, Gi
Professor:
Gabriel M. Filippelli, (D), California (Santa Cruz), 1994, Cm
Robert D. Hall, (D), Indiana, 1973, Gl
Associate Professor:
Pascal P. de Caprariis, (D), RPI, 1973, Hq
Joseph F. Pachut, Jr., (D), Michigan State, 1977, Po
Gary D. Rosenberg, (D), California (Los Angeles), 1972, Po
Lenore P. Tedesco, (D), Miami, 1991, Gs
Assistant Professor:
Eliot A. Atekwana, (D), W Michigan, 1996, Hw
Kathy J. Licht, (D), Colorado, 1999, Gl
R. Jeffrey Swope, (D), Colorado, 1997, Gz
Adjunct Professor:
Timothy S. Brothers, (D), California (Los Angeles), 1985, Oy
Michael R. Cohen, (D), Cornell, 1968, Oe
Timothy G. Fisher, (D), Calgary, 1993, Gl
Swapan K. Ghosh, (D), Syracuse, 1975, Cl
Hendrik M. Haitjema, (D), Minnesota, 1982, Hy
Frederick W. Kleinhans, (D), Ohio State, 1971, Xg
Gregory H. Lindsey, (D), Johns Hopkins, 1992, Oo
Catherine J. Souch, (D), British Columbia, 1990, Pe
Emeritus:
Arthur Mirsky, (D), Ohio State, 1960, Ge

Indiana University Northwest

Dept of Geosciences (A,B) (2005)
3400 Broadway
Gary, IN 46408
p. (219) 980-6738 f. (219) 980-6673
eargyila@iun.edu
http://www.iun.edu/~geos/
Enrollment (1997): B: 18 (1)

Chair:
Timothy G. Fisher, (D), Calgary, 1993, Gl
Assistant Professor:
Zoran Kilibarda, (D), Gr
Visiting Assistant Professor:
Kristin Huysken, (D), Michigan State, 1996, Gi
Visiting Professor:
Leslie Hasbargen, Gs

Indiana University, Bloomington

Dept of Geological Sciences (B,M,D) (2007)
1001 E. 10th Street
Bloomington, IN 47405
p. (812) 855-5582 f. (812) 855-7899
geochair@indiana.edu
http://www.indiana.edu/~geosci/
Administrative Assistant: Lindsay Bugher
Enrollment (2005): B: 54 (7) M: 19 (20) D: 24 (8)

Chair:
Simon C. Brassell, (D), Bristol, 1980, Co
Professor:
David L. Bish, (D), Penn State, 1977, Gz
Jeremy D. Dunning, (D), North Carolina, 1978, Gc
Michael W. Hamburger, (D), Cornell, 1986, Ys
Gary L. Pavlis, (D), Washington, 1982, Ys

Mark Person, (D), Johns Hopkins, 1990, Hw
Lisa M. Pratt, (D), Princeton, 1981, Co
Edward M. Ripley, (D), Penn State, 1976, Em
Robert P. Wintsch, (D), Brown, 1975, Gp
Senior Scientist:
Arndt Schimmelmann, (D), California (Los Angeles), 1985, Cs
Associate Professor:
James G. Brophy, (D), Johns Hopkins, 1984, Gi
Claudia C. Johnson, (D), Colorado, 1993, Pe
Greg A. Olyphant, (D), Iowa, 1979, Hw
Juergen Schieber, (D), Oregon, 1985, Gs
Chen Zhu, (D), Johns Hopkins, 1992, Oo
Associate Scientist:
Chusi Li, (D), Toronto, 1993, Ce
Research Associate:
Peter Sauer, (D), Colorado, 1997, Pe
Senior Lecturer:
Bruce Douglas, (D), Princeton, 1983, Gc
Adjunct Professor:
David Dilcher, (D), Yale, 1964, Pb
Hendrick M. Haitjema, (D), Minnesota, 1982, Hw
John M. Hayes, (D), MIT, 1966, Co
Brian D. Keith, (D), RPI, 1974, Gs
Christopher G. Maples, (D), Indiana, 1985, Pg
Maria D. Mastalerz, (D), Poland, 1988, Ec
Peter J. Ortoleva, (D), Cornell, 1970, Cl
Carl B. Rexroad, (D), Iowa, 1955, Pi
Jeffrey White, (D), Syracuse, 1984, Hs
Charles E. Wier, (D), Indiana, 1955, Ec
Emeritus:
Robert F. Blakely, (D), Indiana, 1973, Ys
J. Robert Dodd, (D), Caltech, 1961, Po
John B. Droste, (D), Illinois, 1956, Gs
Donald E. Hattin, (D), Kansas, 1954, Gr
Erle G. Kauffman, (D), Michigan, 1961, Po
Noel C. Krothe, (D), Penn State, 1976, Hw
Judson Mead, (D), MIT, 1949, Yg
Enrique Merino, (D), California (Berkeley), 1973, Cl
Haydn H. Murray, (D), Illinois, 1951, En
Albert J. Rudman, (D), Indiana, 1963, Ys
Lee J. Suttner, (D), Wisconsin, 1966, Gd
David G. Towell, (D), MIT, 1963, Cg
Laboratory Director:
Erika R. Elswick, (D), Cincinnati, 1998, Cl

School of Public & Environmental Affairs (2004)
SPEA Building
Bloomington, IN 47408
p. (812) 855-7485 f. (812) 855-7802
Department Secretary: Charlie Johnson
Chair:
Hendrik M. Haitjema, (D), Minnesota, 1982, Hw
Professor:
Ronald A. Hites, (D), MIT, 1968, Co
David F. Parkhurst, (D), Wisconsin, 1970, Oo
James C. Randolph, (D), Carleton, 1972, Sf
Jeffrey R. White, (D), Syracuse, 1984, Hg
Senior Scientist:
William W. Jones, (M), Wisconsin, 1975, Hs
Associate Professor:
Debera A. Backhus, (D), MIT, 1990, Cl
Diane S. Henshel, (D), Washington, 1987, Oo
Philip S. Stevens, (D), Harvard, 1990, Oa
Assistant Professor:
Vicky J. Meretsky, (D), Arizona, 1995, Sf
Flynn W. Picardal, (D), Arizona, 1992, So
Geology Librarian:
Mike Parrish, Oo

Indiana University, Indianapolis

Dept of Geography (A,B) (2004)
213 Cavanaugh Hall
Indianapolis, IN 46202
p. (317) 274-8877 f. (317) 274-2347
tbrother@iupui.edu
http://www.iupui.edu/~geogdept/
Enrollment (1995): B: 43 (14)
Chair:
Timothy S. Brothers, (D), California (Los Angeles), 1985, Oy
Professor:
F. L. Bein, (D), Florida, 1974, Oy
Associate Professor:
Catherine J. Souch, (D), British Columbia, 1990, Gm
Assistant Professor:
Jeffey S. Wilson, (D), Indiana State, 1998, Oy

Purdue University

Dept of Earth & Atmospheric Sci (B,M,D) (2007)
550 Stadium Mall Drive
West Lafayette, IN 47907-2051
p. (765) 494-3258 f. (765) 496-1210
eas-info@purdue.edu
http://www.purdue.edu/eas
Enrollment (2007): B: 114 (22) M: 15 (8) D: 47 (13)
Head:
Lawrence W. Braile, (D), Utah, 1973, Ys
Distinguished Professor:
John Cushman, (D), Iowa State, 1978, Sp
Joseph S. Francisco, (D), MIT, 1983, Oa
Associate Head:
Ernest M. Agee, (D), Missouri, 1968, Oa
Professor:
Eric Calais, (D), Nice (France), 1991, Yg
Maarten de Hoop, (D), Delft (Neth), 1992, Gq
Andrei Gabrielov, (D), Moscow State, 1973, Ys
Harshvardhan, (D), SUNY (Stony Brook), 1976, Oa
Gerald H. Krockover, (D), Iowa, 1970, Oe
Robert L. Nowack, (D), MIT, 1985, Ys
James G. Ogg, (D), California (San Diego), 1981, Gs
Kenneth D. Ridgway, (D), Rochester, 1992, Gs
Daniel Shepardson, (D), Iowa, Oe
Paul B. Shepson, (D), Penn State, 1982, Oa
Yuch-Ning Shieh, (D), Caltech, 1968, Cs
Phillip J. Smith, (D), Wisconsin, 1967, Oa
Wen-Yih Sun, (D), Chicago, 1975, Oa
Terry R. West, (D), Purdue, 1966, Ng
William J. Zinsmeister, (D), California (Riverside), 1974, Ps
Associate Head:
Timothy R. Filley, (D), Penn State, 1997, Co
Associate Professor:
Indrajeet Chaubey, (D), Oklahoma State, 1997, Hg
Steven J. Fritz, (D), North Carolina, 1976, Hw
Alexander Gluhovsky, (D), USSR Acad of Sci, Oa
Darryl E. Granger, (D), California (Berkeley), 1996, Oa
Matthew Huber, (D), California (Santa Cruz), 2001, Pe
Eric Riggs, (D), California (Riverside), 2000, Oe
Thomas M. Tharp, (D), Wisconsin, 1978, Nm
R. Jeff Trapp, (D), Oklahoma, 1994, Oa
Indiana State Climatologist:
Dev Niyogi, (D), North Carolina State, 2000, Oa
Assistant Professor:
Michael Baldwin, (D), Oklahoma, 2003, Oa
Brenda Beitler Bowen, (D), Utah, 2005, Gs
Gabriel Bowen, (D), California (Santa Cruz), 2003, Cs
Noah S. Diffenbaugh, (D), California (Santa Cruz), 2004, Pe
Lucy M. Flesch, (D), Carnegie Inst, 2005, Yg

Andrew M. Freed, (D), Arizona, 1998, Gg
Hersh Gilbert, (D), Colorado, 2001, Ys
Kevin Gurney, (D), Colorado State, 2004, Oa
Jennifer S. Haase, (D), Scripps, 1992, Yd
Sonia Lasher-Trapp, (D), Oklahoma, 2000, Oa
Greg Michalski, (D), California (San Diego), 2003, Cs
Wen-wen Tung, (D), California (Los Angeles), 2002, Oa
Dongbin Xiu, (D), Brown, 2003, Gq
Qianlai Zhuang, (D), Alaska (Fairbanks), Oo

Lecturer:
Barbara C. Cooper, (D), Purdue, 1994, Hw

Visiting Professor:
Saad Haq, (D), Gt

Emeritus:
William J. Hinze, (D), Wisconsin, 1957, Ye
Arvid M. Johnson, (D), Penn State, 1965, Gc
Darrell I. Leap, (D), Penn State, 1974, Hy
Dayton G. Vincent, (D), MIT, 1969, Oa

Cooperating Faculty:
Carolyn J. Laffoon, (M), Indiana, 1986, Oo
Chris Miller, (M), Iowa, 2002, Oo

Tri-State University

Science Dept (B) (2004)
300 South Darling Street
Angola, IN 46703
p. (219) 665-4528 f. (219) 665-4292
bensona@tristate.edu
Department Secretary: Mindy Donnelly

University of Notre Dame

Dept of Civil Engineering & Geological Sciences (B,M,D) (2004)
156 Fitzpatrick Hall
Notre Dame, IN 46556
p. (219) 631-5380 f. (219) 631-9236
cegeos@nd.edu
http://www.nd.edu/~cegeos
Administrative Assistant: Tammy Youngs
Enrollment (2001): B: 12 (11) M: 1 (1) D: 16 (5)

Chair:
Peter C. Burns, (D), Manitoba, 1994, Gz

Professor:
Jeremy B. Fein, (D), Northwestern, 1989, Cl
William G. Gray, (D), Princeton, 1974, Hw
Stephen E. Silliman, (D), Arizona, 1986, Hw

Associate Professor:
Patricia A. Maurice, (D), Ge
Clive R. Neal, (D), Leeds (UK), 1985, Gi
J. Keith Rigby, Jr., (D), Columbia, 1976, Pv
Joannes J. Westerink, (D), MIT, 1984, On

Emeritus:
William M. Fairley, (D), Johns Hopkins, 1962, Gp
Michael J. Murphy, C.S.C., Rev., (M), California, 1953, Gg
Erhard M. Winkler, (D), Vienna, 1945, Ge

University of Southern Indiana

Dept of Geology & Physics (B) (2007)
8600 University Boulevard
Evansville, IN 47712
p. (812) 464-1794 f. (812) 465-1052
nking@usi.edu
http://www.usi.edu/science/geology/
Administrative Assistant: Kim E. Schauss
Enrollment (2007): B: 40 (0)

Chair:
Norman R. King, (D), Indiana, 1973, Gr

Associate Professor:
Joseph A. DiPietro, (D), Oregon State, 1990, Gc
Paul K. Doss, (D), N Illinois, 1991, Hw
James Durbin, (D), Nebraska, 1999, Gm

Assistant Professor:
Tony Maria, (D), Rhode Island, 2000, Gv

Instructor:
Carrie L. Wright, (M), Wright State, 2006, Oe

Vincennes University Junior College

Dept of Earth Science & Agriculture (A) (2006)
Vincennes, IN 47591
p. (812) 885-4235
Enrollment: No data reported since 1995

Chair:
James D. Stewart, (M), Ball State, 1970, Oe

Professor:
John A. Parsons, (M), Ball State, 1972, Oe
George W. Stephenson, (M), Ball State, 1966, Xg

Associate Professor:
William T. Bandy, (M), Indiana, 1981, Gr
Neal E. Catt, (M), Indiana State, 1970, Oy

Assistant Professor:
Charles Mansfield, (D), Florida, 1990, So

Iowa

Cornell College

Dept of Geology (B) (2007)
600 First Street SW
Mount Vernon, IA 52314
p. (319) 895-4307 f. (319) 895-5667
bgreenstein@cornellcollege.edu
http://www.cornellcollege.edu/geology
Enrollment (2007): B: 11 (3)

Chair:
Benjamin J. Greenstein, (D), Cincinnati, 1990, Gu

Associate Professor:
Rhawn F. Denniston, (D), Iowa, 2000, Ge

Assistant Professor:
Emily O. Walsh, (D), California (Santa Barbara), 2003, Gp

Instructor:
Jill R. Hopper, (B), Cornell, 2007

Emeritus:
Paul L. Garvin, (D), Colorado, 1969, Gz

Iowa State University of Science & Tech

Dept of Agronomy (B,M,D) (2004)
2101 Agronomy Hall
Ames, IA 50011-1010
p. (515) 294-1360 f. (515) 294-3163
slf@iastate.edu
http://www.agron.iastate.edu/
Department Secretary: Pam Hinderaker
Enrollment (1999): B: 35 (9) M: 31 (10) D: 16 (5)

Director:
Dennis R. Keeney, (D), Iowa, 1965, So

Acting Chair:
Thomas E. Loynachan, (D), North Carolina State, 1975, So

Associate Dean, College of Agriculture:
Gerald A. Miller, (D), Iowa State, 1974, So

Professor:
Raymond W. Arritt, (D), Colorado State, 1985, Ow
Alfred M. Blackmer, (D), Iowa State, 1977, Sc
Richard E. Carlson, (D), Iowa State, 1971, Ow
Richard M. Cruse, (D), Minnesota, 1978, So
V. P. (Bill) Evangelou, (D), California (Davis), 1981, Sc

William J. Gutowski, (D), MIT, 1984, Ow
Robert Horton, (D), New Mexico State, 1981, Sp
Douglas L. Karlen, (D), Kansas State, 1978, So
Randy J. Killorn, (D), Idaho, 1983, So
Thomas B. Mooreman, (D), Washington State, 1983, So
Jonathan A. Sandor, (D), California (Berkeley), 1983, Sd
John W. Schafer, Jr., (D), Michigan State, 1968, So
Eugene S. Takle, (D), Iowa State, 1971, Oa
M. Ali Tatabatai, (D), Iowa State, 1965, Sc
Elwynn Taylor, (D), Washington (St. Louis), 1970, Oa
Regis D. Voss, (D), Iowa State, 1962, So
Douglas N. Yarger, (D), Arizona, 1967, Oa

Associate Professor:
Cynthia Cambardella, (D), Colorado, 1991, So
Dan B. Jaynes, (D), Penn State, 1983, Sp
Thomas A. Kaspar, (D), Iowa State, 1982, So
David A. Laird, (D), Iowa State, 1987, Sc
Sally D. Logsdon, (D), Virginia Tech, 1985, Sp
Antonio W. Mallarino, (D), Iowa State, 1991, So
Michael L. Thompson, (D), Ohio State, 1980, Sd

Assistant Professor:
Lee Burras, (D), Ohio State, 1992, Sd
Larry Halverson, (D), Wisconsin, 1991, So
Stanley J. Henning, (D), Oregon State, 1975, So
Thomas A. Polito, (D), Iowa State, 1987, So

Emeritus:
Jerry K. Radke, (D), Wisconsin, 1965, Sp

Geology Librarian:
Peter A. Peterson, (D), Illinois, 1953, Oo

Dept of Geological & Atmospheric Sci (B,M,D) (2008)
253 Science I
Ames, IA 50011-3212
p. (515) 294-4477 f. (515) 294-6049
geology@iastate.edu
http://www.ge-at.iastate.edu/
Administrative Assistant: DeAnn Frisk
Enrollment (2007): B: 32 (3) M: 16 (6) D: 6 (1)

Chair:
Carl E. Jacobson, (D), California (Los Angeles), 1980, Gc

Professor:
Igor A. Beresnev, (D), USSR Acad of Sci, 1986, Yg
Tsing-Chang Chen, (D), Michigan, 1975, Oa
William A. Gallus, (D), Colorado State, 1993, Oa
William J. Gutowski, (D), MIT, 1984, Oa
Neal R. Iverson, (D), Minnesota, 1989, Gl
William W. Simpkins, (D), Wisconsin, 1989, Hw
Paul G. Spry, (D), Toronto, 1984, Em
Eugene S. Takle, (D), Iowa State, 1971, Oa

Associate Professor:
Cinzia C. Cervato, (D), ETH (Switzerland), 1990, Gg
German Mora, (D), Indiana, 2000, Cs
Kenneth E. Windom, (D), Penn State, 1976, Cp
Xiaoqing Wu, (D), California (Los Angeles), 1992, Oa

Assistant Professor:
Jiasong Fang, (D), Texas A&M, 1993, Ge
Kristie Franz, (D), California (Irvine), 2006, Hs
Chris Harding, (D), Houston, 2001, Gq

Senior Lecturer:
Jane P. Dawson, (D), New Mexico, 1995, Gp

Lecturer:
David M. Flory, (M), Iowa State, 2003, Ow

Adjunct Professor:
Michael R. Burkart, (D), Iowa, 1976, Hy

Emeritus:
Robert D. Cody, (D), Colorado, 1968, Gs
Karl E. Seifert, (D), Wisconsin, 1963, Ct
Carl F. Vondra, (D), Nebraska, 1963, Gr

Douglas N. Yarger, (D), Arizona, 1967, Oa

Teaching Lab Coordinator:
Mark E. Mathison, (M), Iowa State, 2000, Oe

University of Iowa

Dept of Geoscience (B,M,D) (2008)
121 Trowbridge Hall
Iowa City, IA 52242
p. (319) 335-1818 f. (319) 335-1821
geology@uiowa.edu
http://www.uiowa.edu/~geology/
Enrollment (2007): B: 51 (18) M: 22 (7) D: 19 (1)

Chair:
Charles T. Foster Jr., (D), Johns Hopkins, 1975, Gp

Collections Manager, Paleontology:
Tiffany S. Adrain, (M), Iowa, 2003, Oo

Professor:
Ann F. Budd, (D), Johns Hopkins, 1978, Po
Philip H. Heckel, (D), Rice, 1966, Gd
Mark K. Reagan, (D), California (Santa Cruz), 1987, Gi
You-Kuan Zhang, (D), Arizona, 1990, Hw

Associate Professor:
Jonathan M. Adrain, (D), Alberta, 1993, Pi
E. Arthur Bettis III, (D), Iowa, 1995, Oo
Christopher A. Brochu, (D), Texas, 1997, Pv
Jane A. Gilotti, (D), Johns Hopkins, 1987, Gc
Walter A. Illman, (D), Arizona, 1999, Hg
Frank H. Weirich, (D), Toronto, 1982, Gm

Assistant Professor:
Jeffrey A. Dorale, (D), Minnesota, 2001, Cs
David W. Peate, (D), Open, 1989, Cg
Ingrid A. Peate, (D), Royal Holloway, 2003, Gv
Hallie J. Sims, (D), Chicago, 2000, Pe

Post-doc Research Scholar:
Michael Rowe, (D), Oregon State, 2006, Gi

Adjunct Associate Professor:
Scott J. Carpenter, (D), Michigan, 1991, Cs
Gregory A. Ludvigson, (D), Iowa, 1988, Gd
Brian J. Witzke, (D), Iowa, 1981, Ps

Adjunct Assistant Professor:
Raymond R. Anderson, (D), Iowa, 1992, Gr
Rhawn F. Denniston, (D), Iowa, 2000, Cs
James K. Huber, (D), Minnesota, 2001, Pl
Douglas J. Schnoebelen, (D), Indiana, 1999, Hg

Adjunct Professor:
David L. Campbell, (D), California (Berkeley), 1969, Yg

Visiting Assistant Professor:
Qing Zhang, (D), Minnesota, 2006, Gc

Emeritus:
Richard G. Baker, (D), Colorado, 1969, Pl
Robert S. Carmichael, (D), Pittsburgh, 1967, Ye
Lon D. Drake, (D), Ohio State, 1968, Hy
Brian F. Glenister, (D), Iowa, 1956, Pi
Richard A. Hoppin, (D), Caltech, 1951, Gc
Gilbert Klapper, (D), Iowa, 1962, Pm
George R. McCormick, (D), Ohio State, 1964, Gz
Holmes A. Semken, Jr., (D), Michigan, 1965, Pv
Keene Swett, (D), Edinburgh, 1965, Gs

Rock Lab Manager:
Matthew J. Wortel, (M), Iowa, 2007, Oo

Office Manager:
Georgie Greiner, Oo

Geoscience Librarian:
Leo Clougherty, Oo

University of Northern Iowa

Dept of Earth Science (B) (2007)
121 Latham Hall

Cedar Falls, IA 50614-0335
p. (319) 273-2759 f. (319) 273-7124
james.walters@uni.edu
http://www.earth.uni.edu/
Administrative Assistant: Lois Jerke
Enrollment (2007): B: 72 (20)

Head:
James C. Walters, (D), Rutgers, 1975, Gm

Professor:
Timothy M. Cooney, (D), N Colorado, 1976, Oe
Alan C. Czarnetzki, (D), Wisconsin, 1992, Ow
Thomas A. Hockey, (D), New Mexico State, 1988, Oo
Mohammad Z. Iqbal, (D), Indiana, 1994, Hw
Siobahn M. Morgan, (D), Washington, 1991, Oo

Associate Professor:
Kenneth J. De Nault, (D), Stanford, 1974, Gz
John R. Groves, (D), Iowa, 1983, Pg

Assistant Professor:
Chad E. Heinzel, (D), Northern Illinois, 2005, Gag

Adjunct Professor:
Lee S. Potter, (D), Texas, 1996, Gz

Emeritus:
Wayne I. Anderson, (D), Iowa, 1964, Pg
Lynn A. Brant, (D), Penn State, 1980, Gn
Walter E. De Kock, (D), Ohio State, 1972, Oe

Lab Tech:
Steven J. Smith, (M), N Iowa, 2000, Og

Kansas

Cowley County Community College

Dept of Natural Sciences (Geology) (A) (2004)
125 South 2nd
PO Box 1147
Arkansas City, KS 67005-1147
p. (620) 441-5297 f. (620) 441-5359
cales@cowley.edu
http://cowleycollege.com
Administrative Assistant: Marcia Cales

Chair:
Michelle Schoon, Oo

Instructor:
Ernie Morrison, Oo

Emporia State University

Earth Science Dept (B,M) (2005)
1200 Commercial
Emporia, KS 66801-5087
p. (620) 341-5330 f. (620) 341-6055
sleezerr@emporia.edu
www.emporia.edu/s/www/earthsci/earthsci.htm
Enrollment (2005): B: 29 (5) M: 19 (4)

Professor:
James S. Aber, (D), Kansas, 1978, Gl

Associate Professor:
Michael Morales, (D), California (Berkeley), 1987, Pv
Richard O. Sleezer, (D), Kansas, 2001, Os
Kenneth W. Thompson, (D), Iowa State, 1991, Oe

Assistant Professor:
Marcia K. Schulmeister, (D), Kansas, 2000, Ge

Lecturer:
Susan W. Aber, (D), Emporia State, 2005, Gz

Fort Hays State University

Dept of Geosciences (B,M) (2007)
600 Park Street
Tomanek Hall
Hays, KS 67601-4099
p. (785) 628-5389 f. (785) 628-4096
geosciences@fhsu.edu
http://www.fhsu.edu/geo
Administrative Assistant: Patricia Duffey
Enrollment (2007): B: 122 (18) M: 33 (6)

Chair:
John F. Heinrichs, (D), Colorado, 1996, OyrOl

Professor:
Kenneth R. Neuhauser, (D), South Carolina, 1973, Gc
Paul Phillips, (D), Kansas, 1977, Oe
Richard J. Zakrzewski, (D), Michigan, 1968, Pv

Associate Professor:
Richard Lisichenko, (D), Kansas State, 2000, Oi
Tom Schafer, (D), Kansas State, 2000, Oi

Assistant Professor:
Ron Schott, (D)

Instructor:
Kara Kuntz, (M), Oo
Mark Tschudi, (M), Or

Adjunct Professor:
Bruce Basye, (M), Fort Hays State, 1996, Ge
William H. Heimann, (M), Fort Hays State, 1987, Ge
Case Morris, (B)

Cooperating Faculty:
Joseph R. Thomasson, (D), Iowa State, 1976, Pb

Kansas State University

Dept of Agronomy (B) (2004)
2004 Throckmorton
Manhattan, KS 66506
p. (913) 532-6101 f. (613) 532-6094
Department Secretary: Dana Minihan
Enrollment (1996): B: 63 (9) M: 18 (5) D: 19 (5)

Head:
Gerry L. Posler, (D), Sf

Research Scientist:
George H. Liang, (D), So

Professor:
John L. Havlin, (D), Sc
MaryBeth Kirkham, (D), Sp
Clenton E. Owensby, (D), Sf
Gary M. Paulsen, (D), Sp
Mickey D. Ransom, (D), Sd
Bill T. Schapaugh, (D), So
Paul A. Schwab, (D), Sc
Rollie Sears, (D), So
Loyd Stone, (D), Sp
Steve J. Thien, (D), Sc
Richard L. Vanderlip, (D), So
Steve M. Welch, (D), Sp

Associate Professor:
Paula J. Bramel-Cox, (D), So
Stanley W. Ehler, (D), Sd
Walter H. Fick, (D), Sf
Jay M. Ham, (D), Sp
Gerard J. Kluitenberg, (D), Sp
Gary M. Pierzynski, (D), Sc
Chuck W. Rice, (D), Sb

Assistant Professor:
John O. Fritz, (D), Sf
Michael J. Horak, (D), Sb

Instructor:
Sylvia Bodolay, (B), So
Lance R. Gibson, (M), Sd

Librarian:
Judy Adams, Oo

Dept of Geology (B,M) (2008)
Thompson Hall
Manhattan, KS 66506-3201
p. (785) 532-6724 f. (785) 532-5159
lindasd@ksu.edu
http://www.ksu.edu/geology/
Administrative Assistant: Linda Clough
Enrollment (2007): B: 36 (7) M: 10 (7)

Interim Head:
George R. Clark, II, (D), Caltech, 1969, Poy
Professor:
Sambhudas Chaudhuri, (D), Ohio State, 1966, Cc
Charles G. Oviatt, (D), Utah, 1984, Gm
Associate Professor:
Allen W. Archer, (D), Indiana, 1983, Gr
Matthew W. Totten, (D), Oklahoma, 1992, Gso
Assistant Professor:
Matthew E. Brueseke, (D), Miami (Ohio), 2006, GivGt
Keith B. Miller, (D), Rochester, 1988, Pe
Joel Q.G Spencer, (D), Glasgow (UK), 1996, CcGs
Iris M. Totten, (D), New Orleans, 2003, Oe
Emeritus:
Robert L. Cullers, (D), Wisconsin, 1971, Ct
Ronald R. West, (D), Oklahoma, 1970, Po

University of Kansas

Dept of Geology (B,M,D) (2005)
1475 Jayhawk Boulevard
Room 120
Lindley Hall
Lawrence, KS 66045-7613
p. (785) 864-4974 f. (785) 864-5276
egravatt@ku.edu
http://www.ku.edu/~geology/
Enrollment (2005): B: 54 (15) M: 41 (9) D: 21 (2)

Chair:
Robert H. Goldstein, (D), Wisconsin, 1986, Gs
Professor:
Roger L. Kaesler, (D), Kansas, 1965, Pm
Carl D. McElwee, (D), Kansas, 1970, Hg
Donald W. Steeples, (D), Stanford, 1975, Ys
W. Randall Van Schmus, (D), California (Los Angeles), 1964, Cc
J. Douglas Walker, (D), MIT, 1985, Gc
Senior Scientist:
Gisela Dreschoff, (D), Tech (Braunschweig), 1972, Ce
Rolfe Mandel, Ga
Associate Professor:
Ross A. Black, (D), Wyoming, 1989, Ye
J. Rick Devlin, (D), Waterloo, 1994, Hw
Luis A. Gonzalez, (D), Michigan, 1989, Cs
John Gosse, (D), Lehigh, 1994, Gm
Stephen T. Hasiotis, (D), Colorado, 1997, Pe
Diane Kamola, (D), Georgia, 1989, Gs
Bruce S. Lieberman, (D), Columbia, 1994, Po
Gwen L. Macpherson, (D), Texas, 1989, Hw
Beth McClellan, (D), Tennessee, 1993, Gp
Anthony W. Walton, (D), Texas, 1972, Gs
Associate Scientist:
Susan E. Nissen
Assistant Professor:
Kirsten P. Nicolaysen, Gt
Jennifer A. Roberts, (D), Texas, 2000, Co
Daniel F. Stockli, (D), Stanford, 1999, Gc
Michael H. Taylor, (D), California (Los Angeles), 2004, Gt
George P. Tsoflias, (D), Texas, 1999, Yg
Adjunct Assistant Professor:
James M. Butler, (D), Stanford, 1986, Hw
Adjunct Professor:
Timothy R. Carr, (D), Wisconsin, 1981, Go
John H. Doveton, (D), Edinburgh, 1969, Gq
Evan K. Franseen, (D), Wisconsin, 1989, Gd
Lee C. Gerhard, (D), Kansas, 1964, Gs
Leonard Krishtalka, (D), Kansas, 1976, Pv
Larry D. Martin, (D), Kansas, 1973, Pv
Michael T. Meyer, Hg
Richard D. Miller, (M), Kansas, 1983, Yx
Marios Sophocleous, (D), Alberta, 1978, Hw
Edith Taylor, (D), Ohio State, 1983, Pb
Thomas N. Taylor, (D), Illinois, 1964, Pb
W. Lynn Watney, (D), Kansas, 1985, Gr
Donald O. Whittemore, (D), Penn State, 1973, Cl
Emeritus:
Ernest E. Angino, (D), Kansas, 1961, Cl
Louis F. Dellwig, (D), Michigan, 1954, Or
Wakefield Dort, Jr., (D), Stanford, 1955, Gm
Paul Enos, (D), Yale, 1965, Gs
William W. Hambleton, (D), Kansas, 1951, Oo
John Harbaugh, (D)
William M. Merrill, (D), Ohio State, 1950, Gr
Richard A. Robison, (D), Texas, 1962, Pg
Albert J. Rowell, (D), Leeds, 1953, Pi

Wichita State University

Dept of Geology (B,M) (2004)
1845 Fairmount
Wichita, KS 67260-0027
p. (316) 978-3140 f. (316) 978-7229
john.gries@wichita.edu
Administrative Assistant: K. L. Smith
Enrollment (1997): B: 41 (6) M: 28 (3)

Chair:
John C. Gries, (D), Texas, 1970, Gt
Professor:
William D. Bischoff, (D), Northwestern, 1985, Cl
Salvatore J. Mazzullo, (D), RPI, 1974, Go
Associate Professor:
Collette D. Burke, (D), Wisconsin (Milwaukee), 1983, Pm
Assistant Professor:
Hongsheng Cao, (D), Florida State, 2001, Cl
William C. Parcell, (D), Alabama, 2000, Gr
Wan Yang, (D), Texas, 1999, Gs
Lecturer:
Toni K. Jackman, (M), Wichita State, 1984, Ge
David L. Schaffer, (B), Utah, 1982, Ow
Emeritus:
James N. Gundersen, (D), Minnesota, 1958, Ga
Daniel F. Merriam, (D), Kansas, 1961, Gr
Peter G. Sutterlin, (D), Northwestern, 1958, Gd

Kentucky

Eastern Kentucky University

Dept of Earth Sciences (B,M) (2006)
521 Lancaster Avenue
Roark 103
Richmond, KY 40475-3102
p. (859) 622-1273 f. (859) 622-1451
melissa.dieckmann@eku.edu
http://www.earthscience.eku.edu
Enrollment (2005): B: 28 (5) M: 14 (4)

Acting Chair:
Melissa S. Dieckmann, (D), Notre Dame, 1995, Hs
Professor:
Ralph O. Ewers, (D), McMaster, 1982, Hw
Stewart S. Farrar, (D), SUNY (Binghamton), 1976, Gc

Malcolm P. Frisbie, (D), Penn State, 1987, Oo
Bruce R. MacLaren, (D), Wisconsin, 1978, Oo
Jon R. Maki, (D), Michigan State, 1969, Oo
Associate Professor:
Claude S. Dean, (D), Harvard, 1976, Gc
Assistant Professor:
Walter S. Borowski, (D), North Carolina, 1998, Gu
Danita M. LaSage, (D), Kentucky, 2004, Hw
Tom Lierman, (D), George Washington, 1995, Gd
John C. White, (D), Baylor, 2002, Gp
Associate Dean:
Gary L. Kuhnhenn, (D), Illinois, 1976, Gd

Dept of Geography & Geology (B,M) (2008)
Roark 103
521 Lancaster Avenue
Richmond, KY 40475-3102
p. (859) 622-1273 f. (859) 622-1451
richard.sambrook@eku.edu
http://www.geoscience@eku.edu
Enrollment (1999): B: 193 (0)
Chair:
Bruce Davis, (D), California (Los Angeles), 1980, Oi
Professor:
David N. Zurick, (D), Hawaii, 1986, Oo
Instructor:
Glenn A. Campbell, Oo

Morehead State University

Dept of Physical Sciences (B) (2005)
123 Lappin Hall
Morehead, KY 40351
p. (606) 783-2914 f. (606) 783-5002
a.carnevali@moreheadstate.edu
http://www.moreheadstate.edu/physsci/
Administrative Assistant: Regina Kissick
Enrollment (2005): B: 14 (5)
Associate Professor:
Marshall Chapman, (D), Massachusetts, 1996, Gv
Eric Jerde, (D), California (Los Angeles), 1991, Cg
Charles E. Mason, (M), George Washington, 1981, Pi
Steven K. Reid, (D), Texas A&M, 1991, Gs
Joan M. Whitworth, (D), Colorado, 1995, Oe
Instructor:
Jennifer O'Keefe, (M), Texas A&M, 2001, Gn

Murray State University

Dept of Geosciences (B,M) (2007)
104A Wilson Hall
Murray, KY 42071-3311
p. (270) 809-2591 f. (270) 809-3276
tom.kind@murraystate.edu
http://www.murraystate.edu/geosciences
Administrative Assistant: Carruth Kitrell
Enrollment (2006): B: 49 (5) M: 12 (7)
Chair:
Thomas C. Kind, (D), Indiana State, 1976, Or
Dean, College of Sci, Eng & Tech:
Neil V. Weber, (D), Indiana State, 1972, Oy
Professor:
Ken Carstens, (D), Washington (St. Louis), 1978, Oo
George W. Kipphut, (D), Columbia, 1978, Op
Kit W. Wesler, (D), North Carolina, 1981, Oo
Associate Professor:
Haluk Cetin, (D), Purdue, 1993, Oi
Burl I. Naugle, (D), Kentucky, 1979, Hq
Assistant Professor:
Lara Homsey, (D), Pittsburgh, 2004, Ga
Qiaofeng (Robin) Zhang, (D), W Ontario, 2002, Oi
Lecturer:
V. Lynn Leasure, (M), Murray State, 1982, Gg
Adjunct Professor:
Michael R. Busby, (M), Murray State, 1996, Oo
Richard Grainger, (M), Murray State, 1973, Oy
Emeritus:
Peter W. Whaley, (D), Louisiana State, 1969, Gr
Related Staff:
Jane L. Benson, (M), Murray State, 1986, Oi

Northern Kentucky University

Dept of Physics and Geology (2005)
204H Natural Sciences Center
Highland Heights, KY 41099
p. (859) 572-5307
rockawayj@nku.edu
Administrative Assistant: Linda Gee
Enrollment: No data reported since 1995
Assistant Professor:
Janet Bertog, (D), Cincinnati, 2002, Gd
Samuel Boateng, (D), Missouri (Rolla), 1996, Hw
Lecturer:
Reuben G. Bullard, (M), Cincinnati, 2000, Gd
Sarah E. Johnson, (M), Purdue, 1997, Ng
Visiting Professor:
John D. Rockaway, (D), Purdue, 1968, Ng
Related Staff:
Dan Spence, Oo

University of Kentucky

Dept of Earth and Environmental Sci (B,M,D) (2007)
101 Slone Research Building
Lexington, KY 40506-0053
p. (859) 257-3758 f. (859) 323-1938
geowat@uky.edu
http://www.uky.edu/AS/Geology
Administrative Assistant: Pam K. Stephens
Enrollment (2007): B: 13 (0) M: 12 (0) D: 2 (0)
Chair:
William A. Thomas, (D), Virginia Tech, 1960, Gt
Professor:
Richard I. Barnhisel, (D), Virginia Tech, 1964, Gz
Frank R. Ettensohn, (D), Illinois, 1975, Ps
Dhananjay Ravat, (D), Purdue, 1989
Associate Professor:
Alan E. Fryar, (D), Alberta, 1992, Hw
David P. Moecher, (D), Michigan, 1988, Gx
Kieran D. O'Hara, (D), Brown, 1984, Gc
Susan M. Rimmer, (D), Penn State, 1984, Ec
Edward W. Woolery, (D), Kentucky, 1998, Ys
Assistant Professor:
Harold D. Rowe, (D), Stanford, 2002, Cs
Lecturer:
J. Richard Bowersox, (D), South Florida, 2006, Oo
Kent Ratajeski, (D), North Carolina, 1999, Oo
Adjunct Professor:
Donald R. Chestnut, (D), Kentucky, 1988, Ec
James C. Cobb, (D), Illinois, 1981, Ec
James S. Dinger, (D), Nevada (Reno), 1977, Hw
James A. Drahovzal, (D), Iowa, 1966, Eo
Cortland F. Eble, (D), West Virginia, 1988, Pl
Stephen R. Fisher, (D), Texas, 1982, Cl
Stephen F. Greb, (D), Kentucky, 1992, Ec
Christopher Groves, (D), Virginia, 1993, Hg
James C. Hower, (D), Penn State, 1978, Ec
John D. Kiefer, (D), Illinois, 1970, Ng
Thomas L. Robl, (D), Kentucky, 1977, Cg

Zhenming Wang, (D), Kentucky, 1998, Ys
Gerald A. Weisenfluh, (D), South Carolina, 1982, Ec
Emeritus:
William H. Dennen, (D), MIT, 1949, Cg
Irving S. Fisher, (D), Harvard, 1952, Gz
Bruce R. Moore, (D), Melbourne, 1967, Gd
Lyle V. A. Sendlein, (D), Iowa State, 1964, Hw
Ronald L. Street, (D), St. Louis, 1975, Ys
John Thrailkill, (D), Princeton, 1965, Cl
Cooperating Faculty:
Mary Spencer, Oo

Dept of Mining Engineering (B,M,D) (2007)
230 Mining & Mineral Resources Building
Lexington, KY 40506-0107
p. (606) 257-8026 f. (606) 323-1962
rsweigar@engr.uky.edu
Administrative Assistant: Tish Carter
Administrative Assistant: Angela C. Stotz
Enrollment (1997): B: 53 (6) M: 9 (4) D: 15 (1)
Chair:
Richard J. Sweigard, (D), Penn State, 1984, Nm
Professor:
Joseph W. Leonard, (M), Penn State, 1958, Nm
Richard S. Mateer, (D), Pittsburgh, 1950, Nm
Kot F. Unrug, (D), Poland, 1966, Nr
Associate Professor:
Gene T. Lineberry, (D), West Virginia, 1982, Nm
Andrew M. Wala, (D), Poland, 1972, Nm
Jon C. Yingling, (D), Pittsburgh, 1988, Nm
Assistant Professor:
Joseph Sottile, Jr., (D), Penn State, 1991, Nm
Adjunct Professor:
John G. Groppo, (D), Kentucky, 1992, Nm
Bhupendra K. Parekh, (D), Penn State, 1979, Nm

University of Louisville
Dept of Geography & Geosciences (A,B,M) (2004)
NAB 205
Louisville, KY 40292
p. (502) 852-6844 f. (502) 852-4560
caleut01@gwise.louisville.edu
Administrative Assistant: Sharon Vance
Enrollment (1996): B: 39 (9)
Chair:
Clara A. Leuthart, (D), Louisville, Hs
Professor:
Don E. Bierman, (D), Michigan State, 1970, Oo
Anthony O. Clarke, (D), California (Riverside), 1977, Gm
James E. Conkin, (D), Cincinnati, 1960, Pg
Arthur W. Dakan, (D), California (Los Angeles), 1974, Oo
K. L. Gauri, (D), Bonn, 1964, Ge
David A. Howarth, (D), Ohio State, 1979, Oa
Graham H. Hunt, (D), Alberta, 1961, Gg
George A. Lager, (D), British Columbia, 1976, Gz
Associate Professor:
Jafar Hadizadeh, (D), London, 1980, Gc
Dennis L. Spetz, (D), Indiana, 1971, Oo
Assistant Professor:
John L. Anderson, (D), Kentucky, 1974, Oo
Lecturer:
Robert W. Forbes, Oo
Visiting Professor:
Keith R. Mountain, (D), Ohio State, 1990, Oa

Western Kentucky University
Dept of Geography & Geology (B,M) (2007)
1906 College Heights Blvd
#31066
Bowling Green, KY 42101-1066
p. (270) 745-4555 f. (270) 745-6410
david.keeling@wku.edu
http://www.wku.edu/geoweb
Enrollment (2007): B: 207 (41) M: 31 (9)
Head:
David Keeling, (D), Oregon, 1992, Og
Professor:
Stuart Foster, (D), Ohio State, 1988, Oa
Christopher Groves, (D), Virginia, 1993, Hg
Kenneth W. Kuehn, (D), Penn State, 1982, Ec
Michael May, (D), Indiana, 1993, Ge
Louis M. Trapasso, (D), Indiana State, 1978, Oa
Associate Professor:
Catherine Algeo, (D), Louisiana State, 1997, Oi
Rezaul Mahmood, (D), Oklahoma, 2000, Oa
Fredrick D. Siewers, (D), Illinois, 1995, Pg
Assistant Professor:
John All, (D), Arizona, 2001, Oy
Aaron Celestian, (D), Stoney Brook, 2005, GzYgGg
Stephen Kenworthy, (D), Johns Hopkins, 2002, Gm
Yanmei Li, (D), The Ohio State, 2006, Ouo
Andrew Wulff, (D), Massachusetts, 1993, Gv
Jun Yan, (D), Buffalo, 2004, Oi
Instructor:
William Blackburn, (M), W Kentucky, 2002
Kevin Cary, (M), W Kentucky, 2000, Oi
Margaret Crowder, (M), Florida, 1998, Oe
Scott Dobler, (M), Bowling Green, 1996, Oa
Debra D. Kreitzer, (M), W Kentucky, 1998, Oi
Emeritus:
James M. Bingham, (M), Memphis State, 1965, Og
Nicholas Crawford, (D), Clark, 1977, Hg

Louisiana

Centenary College of Louisiana
Dept of Geography & Geology (B) (2007)
2911 Centenary Boulevard
Shreveport, LA 71104
p. (318) 869-5055 f. (318) 869-5795
svetter@centenary.edu
Enrollment (2007): B: 15 (6)
Professor:
Mary L. Barrett, (D), Johns Hopkins, 1987, Gs
Scott K. Vetter, (D), South Carolina, 1989, Gi
Associate Professor:
David B. Bieler, (D), Illinois, 1983, Gt

Louisiana State University
Dept of Agronomy (B,M,D) (2006)
104 M. B. Sturgis Hall
Baton Rouge, LA 70803
p. (225) 578-2110 f. (225) 578-1403
http://www.agronomy.lsu.edu
Enrollment (1997): B: 214 (55) M: 8 (6) D: 5 (6)
Head:
Freddie A. Martin, (D), Cornell, 1970, Oo
Professor:
Gary A. Breitenbeck, (D), Iowa State, 1984, Sb
Hussein M. Selim, (D), Iowa State, 1971, Sp
Assistant Professor:
Lewis A. Gaston, (D), Florida, 1987, Sc
Maud Walsh, (D), Louisiana State, Ge
Jim Wang, (D), Iowa State, Sc
Instructor:
Michal Lindsey, (M), Louisiana State, 1998, So

Dept of Geography & Anthropology (B,M,D) (2006)
227 Howe-Russell Geoscience Complex
Baton Rouge, LA 70803
p. (225) 578-5942 f. (225) 578-2912
gachair@lsu.edu
www.ga.lsu.edu
Enrollment (2006): B: 54 (23) M: 12 (6) D: 22 (7)

Professor:
Patrick A. Hesp, (D), Sydney, 1981, Gm
Richard H. Kesel, (D), Maryland, 1971, Hg
Anthony J. Lewis, (D), Kansas, 1971, Or

Associate Professor:
Steven Namikas, (D), S California, 1999, Gm

Dept of Geology & Geophysics (B,M,D) (2006)
E235 Howe-Russell Geoscience Complex
Baton Rouge, LA 70803-4101
p. (225) 578-3353 f. (225) 578-2302
thibod@lsu.edu
http://www.geol.lsu.edu
Enrollment (2006): B: 57 (8) M: 25 (9) D: 17 (1)

Chair:
Laurie C. Anderson, (D), Wisconsin, 1991, Pi

Professor:
Ajoy K. Baksi, (D), Toronto, 1970, Cc
William J. Blanford, (D), Arizona, 2000, Hw Gary R.
Michael Blum, (D), Texas, 1991, Gs
Byerly, (D), Michigan State, 1974, Gi
Lui-Heung Chan, (D), Harvard, 1966, Oc
Barbara L. Dutrow, (D), S Methodist, 1985, Gz
Brooks B. Ellwood, (D), Rhode Island, 1977, Ym
Ray E. Ferrell, Jr., (D), Illinois, 1966, Cl
Jeffrey S. Hanor, (D), Harvard, 1967, Cl
Darrell J. Henry, (D), Wisconsin, 1981, Gp
Jeffrey A. Nunn, (D), Northwestern, 1981, Yg

Associate Professor:
Philip J. Bart, (D), Rice, 1998, Gr
Juan M. Lorenzo, (D), Columbia, 1991, Ys

Assistant Professor:
Huiming Bao, (D), Princeton, 1998, Cs
Brian Darby, (D), S California, 2003, Gt
Annette Summers Engel, (D), Texas, 2004, Py
Jonathan Tomkin, (D), Australian National, 2000, Gm

Associate Curator, LSU Museum of Natural Science:
Judith A. Schiebout, (D), Texas, 1973, Pv

Emeritus:
Arnold H. Bouma, (D), Utrecht (Neth), 1961, Gu
George Hart, Oo
Donald Kupfer, Oo
Clyde Moore, Oo
James E. Roche, (D), Illinois, 1969, Gg
Barun K. Sen Gupta, (D), Indian Inst of Tech, 1963, Pm

Dept of Oceanography & Coastal Sciences (M,D) (2007)
1002 Energy Coast and Environment Building
Baton Rouge, LA 70803
p. (225) 578-6308 f. (225) 578-6307
ocean@lsu.edu
http://www.ocean.lsu.edu
Administrative Assistant: Gaynell Gibbs
Enrollment (2006): M: 26 (0) D: 32 (0)

Chair:
Charles A. Wilson, (D), South Carolina, 1984, Ob

Professor:
John W. Day, Jr., (D), North Carolina State, 1971, Ob
Robert P. Gambrell, (D), North Carolina State, 1974, Cg
Shih A. Hsu, (D), Texas, 1969, Oa
Oscar K. Huh, (D), Penn State, 1968, Or
Paul A. LaRock, (D), RPI, 1968, Ob
Irving A. Mendelssohn, (D), North Carolina State, 1974, Ob
Samuel P. Meyers, (D), Columbia, 1957, Ob
Richard F. Shaw, (D), Maine, 1981, Ob
Robert E. Turner, (D), Georgia, 1974, Ob
William J. Wiseman, Jr., (D), Johns Hopkins, 1969, Op

Associate Professor:
Donald M. Baltz, (D), California (Davis), 1980, Ob
Robert S. Carney, (D), Oregon State, 1977, Ob
Richard E. Condrey, (D), Washington, 1981, Ob
Masamachi Inoue, (D), Texas A&M, 1982, Op
Lawrence J. Rouse, Jr., (D), Louisiana State, 1969, Op
Greg W. Stone, (D), Maryland, 1990, Og

Assistant Professor:
Mark C. Benfield, (D), Texas A&M, 1991, Ob
Samuel J. Bentley, (D), SUNY (Stony Brook), 1998, Ob
Jaye E. Cable, (D), Florida State, 1996, Ob
Stephen P. Faulkner, (D), Duke, 1994, Cg

Adjunct Professor:
Michael J. Dagg, (D), Washington, 1975, Ob
Quay Dortch, (D), Washington, 1980, Ob
James P. Geaghan, (D), North Carolina, 1980, Ob
Dubravko Justic, (D), Zagreb, 1989, Op
Brent A. McKee, (D), North Carolina, 1986, Cg
Karen L. McKee, (D), Louisiana State, 1993, Ob
Ralph J. Portier, (D), Louisiana State, 1982, Ob
Nancy N. Rabalais, (D), Texas, 1983, Ob
Harry H. Roberts, (D), Louisiana State, 1969, Gs
Paul W. Sammarco, (D), SUNY, 1978, Oo
Nan D. Walker, (D), Cape Town, 1989, Og

Emeritus:
James M. Coleman, (D), Louisiana State, 1966, Gs

Louisiana Tech University

Geosciences Program (B,M) (2004)
600 W. Arizona Street
Ruston, LA 71272
p. (318) 257-3972 f. (318) 257-2562
mm@engr.latech.edu
Department Secretary: Connie McKenzie
Enrollment (1997): B: 23 (1)

Associate Professor:
Maureen McCurdy, (D), Wisconsin, 1990, Hw
Gary S. Zumwalt, (D), California (Davis), 1976, Pi

Emeritus:
Leo A. Herrmann, (D), Johns Hopkins, 1951, Go

McNeese State University

Dept of Chemistry, Geology Section (B) (2004)
Lake Charles, LA 70609
p. (318) 475-5771 f. (318) 475-5249
rdarbeau@mail.mcneese.edu
Enrollment (1998): B: 3 (1)

Director:
Robert C. Rettke, (D), Case Western, 1976, Gd

Professor:
Robert F. Mahfoud, (D), Brigham Young, 1971, Em

Adjunct Professor:
James W. Batchelor, (M), Miami (Ohio), 1949, Pm
Walter Morse, (M), McNeese State, 1972, Ge

Northwestern State University

Dept of Chemistry and Physics (B,M) (2004)
Natchitoches, LA 71497
p. (318) 357-5501 f. (318) 357-5599
withey@nsula.edu

Enrollment: No data reported since 1995

Director:

Kelly Knowlton, (D), Texas A&M, 1991, Gg

Tulane University

Dept of Earth and Environmental Sci (B,M,D) (2006)
6823 St. Charles Ave.
120 Dinwiddie Hall
New Orleans, LA 70118
p. (504) 865-5198 f. (504) 865-5199
mreine@tulane.edu
http://www.tulane.edu/~eens/
Enrollment (2006): B: 16 (0)

Chair:

Stephen A. Nelson, (D), California (Berkeley), 1979, Gi

Professor:

Ronald L. Parsley, (D), Cincinnati, 1969, Po

Associate Professor:

Mead A. Allison, (D), SUNY (Stony Brook), 1993, Gs
George C. Flowers, (D), California (Berkeley), 1979, Cg
Torbjörn E. Törnqvist, (D), Utrecht (Neth), 1993, Gs

Assistant Professor:

Nancye H. Dawers, (D), Columbia, 1997, Gc
Suzanne Leclair, (D), Binghamton, 2000, Gs

University of Louisiana, Lafayette

Dept of Geology (B,M) (2007)
BOX 44530
Lafayette, LA 70504-4530
p. (337) 482-6468 f. (337) 482-5723
geology@louisiana.edu
http://geology.louisiana.edu
Administrative Assistant: Ramona Viguerie
Enrollment (2007): B: 47 (2) M: 19 (6)

Professor:

Gary L. Kinsland, (D), Rochester, 1974, Yg
Brian E. Lock, (D), Cambridge (UK), 1969, Gs
Herman H. Rieke, (D), S California, 1970, Np

Associate Professor:

Timothy W. Duex, (D), Texas, 1983, Hg
Carl Richter, (D), Tubingen, 1990, Ym

Assistant Professor:

Victoria C. Hover, (D), Michigan, 1996, Cg
Jack Stalnaker, (D), Texas A&M, 2004, Yg

Instructor:

Cathy Bishop, (M), Louisiana, Gg

Emeritus:

Walter P. Kessinger, (D), Louisiana State, 1974, Pm

Dept of Renewable Resources (B) (2004)
P.O. Box 44650
Lafayette, LA 70504-4650
p. (318) 482-6647
Enrollment: No data reported since 1995

Director:

Lynn J. Dessene, (D), Iowa State, 1967, Sf

Assistant Professor:

Lee Burras, (D), Ohio State, 1992, Sd

Adjunct Professor:

Arville Touchet, (B), SW Louisiana, 1959, Sd

University of Louisiana, Monroe

Dept of Geosciences (B,M) (2007)
700 University Avenue
Monroe, LA 71209-0550
p. (318) 342-1878 f. (318) 342-1755
camille@ulm.edu
http://hurricane.geos.ulm.edu/~geos
Department Secretary: Ruthie Sampson
Enrollment (2006): B: 45 (0)

Dean:

Mervin Kontrovitz, (D), Tulane, 1971, Pm

Head:

Eric A. Pani, (D), Texas Tech, 1987, Oa

Professor:

Rene A. De Hon, (D), Texas Tech, 1970, Gz
Lloyd N. Glawe, (D), Louisiana State, 1966, Pm
Lynn L. LeBlanc, (D), Texas A&M, 1970, Oa
Leonard M. Young, (D), Texas, 1968, Gs

Associate Professor:

Gary L. Stringer, (D), S Mississippi, 1990, Oe

Assistant Professor:

Michael A. Camille, (D), Texas A&M, 1991, Oy
Darrell P. Kruger, (D), Louisiana State, 1992, Oy
Samuel W. Mitcham, (D), Tennessee, 1984, Oo
Paul A. Washington, (D), Connecticut, 1985, Hw

Adjunct Professor:

Robert P. Connolly, (D), Illinois, 1996, Ga
Joe Saunders, (D), Texas A&M, 1985, Ga

Emeritus:

Marcus B. Mapp, (M), Mississippi State, 1950, Gg

University of New Orleans

Dept of Earth and Environmental Sciences (B,M,D) (2005)
2000 Lakeshore Drive
New Orleans, LA 70148
p. (504) 280-6325 f. (504) 280-7396
spenland@uno.edu
http://www.uno.edu/geology/
Enrollment (2005): B: 106 (11) M: 35 (6) D: 4 (0)

Chair:

Shea Penland, (D), Louisiana State, 1990, Gm

Professor:

William H. Busch, (D), Oregon State, 1981, Ou
Terry L. Pavlis, (D), Utah, 1982, Gt
Denise J. Reed, (D), Cambridge, 1986, Gm
A. K. Mostofa Sarwar, (D), Indiana, 1983, Ye
Laura F. Serpa, (D), Cornell, 1986, Ys
William B. Simmons, (D), Michigan, 1973, Gz
Ronald K. Stoessell, (D), California (Berkeley), 1977, Cl

Associate Professor:

Kraig L. Derstler, (D), California (Davis), 1985, Pv
Frank R. Hall, (D), Rhode Island, 1991, Ym

Assistant Professor:

Mark A. Kulp, (D), Kentucky, 2000, Gr
Christopher D. Parkinson, (D), London (UK), 1991, Gp

Adjunct Professor:

Miles O. Hayes, (D), Texas, 1965, Gs
Jacqueline Michel, (D), South Carolina, 1980, Cg
Karen L. Webber, (D), Rice, 1988, Gv
Michael A. Wise, (D), Manitoba, 1987, Gz

Emeritus:

Gary C. Allen, (D), North Carolina, 1968, Gp

Maine

Bates College

Dept of Geology (B) (2007)
Carnegie Science Center
44 Campus Avenue
Lewiston, ME 04240-6084
p. (207) 786-6490 f. (207) 786-8334
jcreasy@bates.edu

http://www.bates.edu/
Administrative Assistant: Laurie McConnell
Enrollment (2007): B: 10 (6)

Professor:
John W. Creasy, (D), Harvard, 1974, Gx
J. Dykstra Eusden, Jr., (D), Dartmouth, 1988, Gc
Michael J. Retelle, (D), Massachusetts, 1985, Gl
Associate Professor:
Beverly J. Johnson, (D), Colorado, 1998, Cl
Assistant Instructor:
Marita Bryant, (M), Free (Berlin), 1984, Gg
Lecturer:
Gene A. Clough, (D), Caltech, 1978, Ym
Emeritus:
Roy L. Farnsworth, (D), Boston, 1961, Gl
Related Staff:
Charlotte Lehmann, (M), Maine (Orono), 1991, Cl

Bowdoin College

Dept of Geology (B) (2007)
6800 College Station
Brunswick, ME 04011
p. (207) 725-3628 f. (207) 798-7037
mparker@bowdoin.edu
http://academic.bowdoin.edu/geology/
Administrative Assistant: Marjorie Parker
Enrollment (2007): B: 19 (8)

Chair:
Rachel J. Beane, (D), Stanford, 1997, Gp
Associate Professor:
Edward P. Laine, (D), MIT, 1977, Gu
Peter D. Lea, (D), Colorado, 1989, Gl
Service Learning Coord/Lab Instr:
Cathryn K. Field, (M), Smith, 2000, Ge
Lab Instructor:
Joanne Urquhart, (M), Dartmouth, 1987, Ge
Emeritus:
Arthur M. Hussey, II, (D), Illinois, 1961, Gc

Colby College

Dept of Geology (B) (2007)
5800 Mayflower Hill
Waterville, ME 04901-8858
p. (207)859-5800 f. (207)859-5868
ragastal@colby.edu
http://www.colby.edu/geology/
Administrative Assistant: Alice M. Ridky
Enrollment (2007): B: 32 (4)

Chair:
Robert A. Gastaldo, (D), S Illinois, 1978, PbGs
Professor:
Robert E. Nelson, (D), Washington, 1982, Pe
Visiting Professor:
Bruce F. Rueger, (D), Colorado, 2002, Pl
Lara Storm, (D), RPI, 2005, Gpz
Walter "Bill" Sullivan, (D), Wyoming, 2007, Gct
Emeritus:
Donald B. Allen, (D), Illinois, 1972, Eg
Donaldson Koons, (D), Columbia, 1945, Gm
Harold R. Pestana, (D), Iowa, 1965, Pi
Cooperating Faculty:
S. Cole, Oo

University of Maine

Dept of Earth Sciences (B,M,D) (2007)
111 Bryand Global Sciences Center
Orono, ME 04469-5790
p. (207) 581-2152 f. (207) 581-2202
dianne.perro@umit.maine.edu
hhttp://www.geology.um.maine.edu/
Administrative Assistant: Dianne Perro
Enrollment (2007): B: 33 (9) M: 11 (8) D: 14 (3)

Chair:
Joseph T. Kelley, (D), Lehigh, 1979, Gu
Research Professor:
Edward S. Grew, (D), Harvard, 1972, Gp
Roger L. Hooke, (D), Caltech, 1965, Gm
Professor:
Daniel F. Belknap, (D), Delaware, 1979, Gu
Joseph V. Chernosky, Jr., (D), MIT, 1973, Gx
George H. Denton, (D), Yale, 1965, Gl
Terence J. Hughes, (D), Northwestern, 1968, Gl
Scott E. Johnson, (D), James Cook, 1989, Gc
Thomas B. Kellogg, (D), Columbia, 1973, Pm
Peter O. Koons, (D), ETH (Switzerland), 1982, Gt
Daniel R. Lux, (D), Ohio State, 1981, Gi
Paul A. Mayewski, (D), Ohio State, 1973, Pe
Stephen A. Norton, (D), Harvard, 1967, Cl
Associate Professor:
Kirk A. Maasch, (D), Yale, 1989, Oa
Andrew S. Reeve, (D), Syracuse, 1996, Hw
Assistant Professor:
Christopher C. Gerbi, (D), Maine, 2005, Gz
Karl J. Kreutz, (D), New Hampshire, 1998, Cs
Research Assistant Professor:
Brenda L. Hall, (D), Maine, 1997, Gl
Lab Coordinator:
Alice R. Kelley, (M), Lehigh, 1981, Ga
Instructor:
Martin G. Yates, (D), Indiana, 1988, Em
Faculty Associate:
Marc C. Loiselle, (D), MIT, 1979, Hy
Robert G. Marvinney, (D), Syracuse, 1986, Gc
Thomas K. Weddle, (D), Boston, 1991, Hy
Adjunct Professor:
Henry N. Berry, (D), Massachusetts, 1989, Gg
Stephen M. Dickson, (D), Maine, 1999, Gu
Woodrow B. Thompson, (D), Ohio State, 1975, Gl
Elizabeth A. Wilson, (D), Delaware, 1978, Eo
Emeritus:
Harold W. Borns, Jr., (D), Boston, 1959, Gl
Edward R. Decker, (D), Harvard, 1966, Gc
Bradford A. Hall, (D), Yale, 1964, Gr
Philip H. Osberg, (D), Harvard, 1952, Gc
Cooperating Faculty:
Gordon S. Hamilton, (D), Cambridge, 1992, Gl
Geology Librarian:
James Bird, Oo

Dept of Oceanography (M,D) (2006)
5741 Libby Hall
Orono, ME 04469-5741
p. (207) 581-4381 f. (207) 581-4388
davidt@maine.edu
www.ume.edu/~marine/gradoceanography.html
Enrollment: No data reported since 1995

Chair:
Detmar F. Schnitker, (D), Illinois, 1967, Ou
Professor:
Lawrence M. Mayer, (D), Dartmouth, 1976, Oc
James D. McCleave, (D), Montana State, 1967, Ob
Leslie D. Watling, (D), Delaware, 1974, Ob
Associate Professor:
L. Kenneth Fink, Jr., (D), Miami, 1969, Ou
Bernard J. McAlice, (D), Rhode Island, 1969, Ob
Cynthia H. Pilskaln, (D), Harvard, 1985, Oc

Robert S. Steneck, (D), Johns Hopkins, 1982, Ob
Research Associate:
David W. Townsend, (D), Maine, 1981, Ob
Adjunct Professor:
Neal R. Pettigrew, (D), MIT/Woods Hole, 1981, Op
Cooperating Faculty:
Daniel F. Belknap, (D), Delaware, 1979, Gu
Ian R. Davison, (D), Dundee, 1983, Ob
Joseph T. Kelley, (D), Lehigh, 1980, Gu
Gary M. King, (D), Georgia, 1978, Oc
Vijay Panchang, (D), Maine, 1985, Oo
Bryan R. Pearce, (D), Florida, 1972, Oo
Michael L. Prentice, (D), Brown, 1988, Gu
J. Malcolm Shick, (D), Texas A&M, 1974, Ob
Robert L. Vadas, (D), Washington, 1968, Ob

University of Maine, Farmington

Natural Sciences Dept (Geology Div) (B) (2007)
173 High Street
Farmington, ME 04938
p. (207) 778-8151 f. (207) 778-7365
eastler@maine.edu
http://sciences.umf.maine.edu
Administrative Assistant: Nancy Barker
Enrollment (2007): B: 15 (5)
Professor:
Thomas E. Eastler, (D), Columbia, 1970, GeOrGg
Associate Professor:
David Gibson, (D), Queens (Belfast), 1984, GigGz
Douglas N. Reusch, (D), Maine, 1998, GtCgOu
Assistant Professor:
Julia F. Daly, (D), Maine, 2002, GmlGu

University of Maine, Presque Isle

Div of Mathematics & Science (B) (2004)
181 Main Street
Presque Isle, ME 04769
p. (207) 768-9482 f. (207) 764-0311 (Ext. 215)
mccartney@polaris.umpi.maine.edu
Department Secretary: Connie Leveque
Enrollment: No data reported since 1995
Professor:
Kevin McCartney, (D), Florida State, 1988, Pm

University of Southern Maine

Dept of Geosciences (B) (2005)
37 College Avenue
Gorham, ME 04038
p. (207) 780-5350 f. (207) 228-8361
glidden@usm.maine.edu
http://www.usm.maine.edu/~geos/
Administrative Assistant: Karen I. Glidden
Enrollment (1998): B: 20 (5)
Chair:
Irwin D. Novak, (D), Cornell, 1971, Gm
Professor:
Stephen G. Pollock, (D), Rutgers, 1975, Gx
Mark T. Swanson, (D), SUNY (Albany), 1982, Gc
Associate Professor:
Charles R. Fitts, (D), Minnesota, 1990, Hw
Lab Associate:
Margaret E. Vose, (M), S Maine, 2003, Og

Maryland

Comm Coll of Baltimore County, Catonsville

School of Mathematics & Science (A) (2007)
800 S. Rolling Road
Catonsville, MD 21228
p. (410) 455-4516 f. (410) 455-6938
DLudwikoski@ccbcmd.edu
www.ccbcmd.edu
Department Secretary: Suzanne Mrozinski
Assistant Professor:
David J. Ludwikoski, (M), Toledo, 1993, Oe

Frostburg State University

Dept of Geography (B) (2007)
101 Braddock Rd.
Frostburg, MD 21532
p. (301) 687-4369 f. (301) 687-4495
ccaupp@frostburg.edu
http://www.frostburg.edu/dept/geog/index.htm
Administrative Assistant: Gale Yutzy
Enrollment (1998): B: 21 (7)
Chair:
Craig L. Caupp, (D), Utah State, 1986, OoHs
Professor:
Henry W. Bullamore, (D), Iowa, 1978, Oo
Francis L. Precht, (D), Georgia, 1989, Oy
Associate Professor:
Fritz Kessler, (D), Kansas, 1999, Or
James C. Saku, (D), Saskatchewan, 1995, Oo
George W. White, (D), Oregon, 1994, Oo
Assistant Professor:
David L. Arnold, (D), Indiana, 1994, Oa
Matthew E. Ramspott, (D), Kansas, 2006, Or
Daniel L. Welsch, (D), Virginia, 2003, HgCgSb

Montgomery College

Dept of Physics, Engineering & Geosciences (A) (2005)
51 Mannakee Street
Rockville, MD 20850
p. (301) 279-5230
kenneth.beem@montgomerycollege.edu
Administrative Assistant: Mary Deep
Professor:
Kenneth A. Beem, (D), Cincinnati, 1973, Po
Associate Professor:
Terry L. Dyroff, (M), S Illinois, 1972, Gm

The Johns Hopkins University

Dept of Geography & Environmental Eng (B,M,D) (2007)
313 Ames Hall
34th & Charles Streets
Baltimore, MD 21218-2681
p. (410) 516-7092 f. (410) 516-8996
bouwer@jhu.edu
http://www.jhu.edu/dogee
Enrollment (1997): M: 2 (4) D: 24 (2)
Chair:
Edward John Bouwer, (D), Stanford, 1982, Oo
Professor:
William P. Ball, (D), Stanford, 1989, Oo
Grace S. Brush, (D), Harvard, 1956, Pl
Hugh Ellis, (D), Waterloo, 1984, Oo
Steve H. Hanke, (D), Colorado, 1969, Oo
Benjamin F. Hobbs, (D), Cornell, 1983, Oo
Charles R. O'Melia, (D), Michigan, 1963, Oo
A. Lynn Roberts, (D), MIT, 1991, Oo
Erica J. Schoenberger, (D), California (Berkeley), 1984, Oo
Alan T. Stone, (D), Caltech, 1983, Cl
Peter W. Wilcock, (D), MIT, 1987, Gm
M. Gordon Wolman, (D), Harvard, 1953, Gm

Assistant Professor:
Kai Loon Chen, (D), Yale, 2008, Oo
Seth Guikema, (D), Stanford, 2003, Oo
Markus Hilpert, (D), Oo
Catherine Norman, (D), Oo
Emeritus:
John J. Boland, (D), Johns Hopkins, 1973, Oo
Eugene D. Shchukin, (D), Moscow State, 1958, Oo
Related Staff:
Hedy Alavi, (D), Ohio State, 1983, Oo
Christine Kavanagh, (M), Oo

The Morton K. Blaustein Dept of Earth & Planetary Sciences (B,M,D) (2005)
3400 N Charles Street
301 Olin Hall
Baltimore, MD 21218
p. (410) 516-7135 f. (410) 516-7933
rrouse@jhu.edu
http://www.jhu.edu/~eps/
Enrollment (2003): M: 5 (5) D: 25 (2)
Chair:
Peter L. Olson, (D), California (Berkeley), 1977, Yg
Professor:
John M. Ferry, (D), Harvard, 1975, Gp
George W. Fisher, (D), Johns Hopkins, 1963, Gp
Grant Garven, (D), British Columbia, 1982, Hw
Lawrence A. Hardie, (D), Johns Hopkins, 1965, Gs
Bruce D. Marsh, (D), California (Berkeley), 1974, Gi
Thomas R. Osborn, (D), California (San Diego), 1969, Op
Steven M. Stanley, (D), Yale, 1968, Po
Darrell F. Strobel, (D), Harvard, 1969, Oa
Dimitri A. Sverjensky, (D), Yale, 1980, Cl
David R. Veblen, (D), Harvard, 1976, Gz
Associate Professor:
Thomas W.N. Haine, (D), Southampton, 1993, Op
A. Hope Jahren, (D), California (Berkeley), 1996, Pe
Darryn W. Waugh, (D), Cambridge, 1991, Oa
Senior Lecturer:
Linda A. Hinnov, (D), Johns Hopkins, 1994, Gs
Katalin Szlavecz, (D), Eotvos (Hungary), 1981, Pi
Lecturer:
Sakiko N. Olsen, (D), Johns Hopkins, 1972, Gp
Emeritus:
Owen M. Phillips, (D), Cambridge, 1955, Op
Principal Research Scientist:
Albert Arking, (D), Cornell, 1959, Oa
Associate Research Scientist:
David C. Elbert, (U), Gz
Linda Veblen, (D), Johns Hopkins, 1984, Gi
Associate Research Scholar:
Edward Urban, (D), Delaware, 1989, Og

Towson University

Dept of Physics, Astronomy & Geosciences (B) (2007)
8000 York Road
Towson, MD 21252-0001
p. (410) 704-3020 f. (410) 704-3511
dvanko@towson.edu
http://wwwnew.towson.edu/physics/geosciences/
Enrollment (2007): B: 12 (4)
Chair:
David A. Vanko, (D), Northwestern, 1982, Gx
Professor:
Rachel J. Burks, (D), Texas, 1985, Gc
Associate Professor:
Jonathan K. Filer, (D), North Carolina, 1992, Gr
Steven M. Lev, (D), SUNY (Stony Brook), 1998, Cl

United States Naval Academy

Dept of Oceanography (B) (2006)
572C Holloway Road
Annapolis, MD 21402-5026
p. (410) 293-6550 f. (410) 293-2137
drsmith@usna.edu
http://www.nadn.navy.mil/Oceanography/
Administrative Assistant: Helen T. Foggo
Enrollment (2005): B: 88 (48)
Professor:
Peter L. Guth, (D), MIT, 1980, Gc
Associate Professor:
David R. Smith, (D), Texas A&M, 1979, Ow
Mario E. C. Vieira, (D), Johns Hopkins, 1983, Op
Assistant Professor:
Henry Jones, (D), Naval Postgrad Sch, 2001, Op
Emil Petruncio, (D), Naval Postgrad Sch, 1996, Or
Cecily N. Steppe, (D), Delaware, 2001, Ob
Michelle K. Whisenhant, (D), Naval Postgrad Sch, 2003, Ow
Instructor:
Carl A. Hager, Naval Postgrad Sch, 1997, Op
Michael Cornelius, (M), Naval Postgrad Sch, 2004, Og
Brad J. Storey, Naval Postgrad Sch, 2003, Og
Karen M. Wingeart, (M), Naval Postgrad Sch, 2001, Ow
Oceanography Librarian:
Barbara Yoakum, Oo

University of Maryland

Dept of Geology (B,M,D) (2007)
Geology Building (#237), Room 1118
College Park, MD 20742-4211
p. (301)405-4082 f. (301)314-9661
geology@geol.umd.edu
http://www.geol.umd.edu
Administrative Assistant: Dorothy Brown
Enrollment (2005): B: 35 (8) M: 16 (9) D: 12 (2)
Chair:
Michael Brown, (D), Keele (UK), 1975, Gx
Professor:
Philip A. Candela, (D), Harvard, 1982, Cg
William F. McDonough, (D), Australian National, 1988, Cg
Roberta L. Rudnick, (D), Australian National, 1988, Gx
Richard J. Walker, (D), SUNY (Stony Brook), 1984, Cg
Ann G. Wylie, (D), Columbia, 1972, Gz
Associate Research Scientist:
Philip M. Piccoli, (D), Maryland, 1992, Cg
Associate Professor:
James Farquhar, (D), Alberta, 1995, Cs
Alan J. Kaufman, (D), Indiana, 1990, Cg
Karen L. Prestegaard, (D), California (Berkeley), 1982, Hg
Assistant Professor:
Andrew Campbell, (D), Chicago, 1993, Gz
Saswata Hier-Majumder, (D), Minnesota, 2004, Yg
Aaron Martin, (D), Arizona, 2005, Gt
Research Associate:
Sonja Aulbach, (D), Macquarie, 2005, Cg
Ralf Halama, (D), Tuebingen, 2003, Cg
Sarah Penniston-Dorland, (D), Johns Hopkins, 2004, Gp
Tetsuya Yokoyama, (D), Okayama, 1999, Cg
Director, Earth, Life & Time Program:
Thomas R. Holtz, (D), Yale, 1992, Pv
Lecturer:
John W. Merck, Jr., (D), Texas, 1997, Pv
Affiliate Professor:
Antonio Busalacchi, (D), Florida State, 1982, Og
Adjunct Associate Professor:
John Bohlke, (D), California (Berkeley), 1986, Cg

Adjunct Assistant Professor:
Julia Baldwin, (D), MIT, 2003, Gx
Boswell A. Wing, (M), Johns Hopkins, 1998, Cg
Adjunct Professor:
Yingwei Fei, (D), CUNY, 1989, Cg
Sorena Sorensen, (D), California (Los Angeles), 1984, Gp
E-An Zen, (D), Harvard, 1955, Og
Visiting Research Associate:
Timothy E. Johnson, (D), Derby, 1999, Gx
Jacqueline Mann, (D), Maryland, 2005, Cg
Renato Moraes, (D), Brasilia, 1997, Gx
Shuhei Ono, (D), Penn State, 1991, Cg
Visiting Assistant Research Scientist:
William Minarik, (D), RPI, 1993, Cg
Paul Tomascak, (D), Maryland, 1995, Cg
Visiting Professor:
Harry Becker, (D), Mainz, 1994, Cc
Marilyn Fogel, (D), Texas, 1977, Gu
Jeffrey Ryan, (D), Columbia, 1989, Cg
Professor Emeritus:
Luke L Y. Chang, (D), Chicago, 1963, Cp
Associate Professor Emeritus:
Peter B. Stifel, (D), Utah, 1964, Pg
Research Professor:
George Helz, (D), Penn State, 1970, Cg
Lecturer:
Elizabeth Brabson, (B), Maryland, 2002, Cg
Tracey Centorbi, (B), Maryland, 2003, Cg
Faculty Research Assistant:
Richard Ash, (D), Open, 1990, Ca
Nicholas Collins, (B), Manchester, 1998, Cg
Daniel Earnest, (M), Maryland, 2002, Gz
Todd Karwoski, (B), Maryland, 2005, Cg
Seung Ryeol Lee, (D), Seoul National, 2000, Cg
Assistant Research Scientist:
Igor Puchtel, (D), Russian Acad of Sci, 1992, Cg
Cooperating Faculty:
Bruce James, (D), Vermont, 1981, Sc
Daniel Lathrop, (D), Texas, 1991, Yg

Dept of Nat Resource Sci & Landscape Arch (B,M,D) (2004)
2102 Plant Sciences Building
College Park, MD 20742-4432
p. (301) 405-4356 f. (301) 314-9308
fgouin@alsc.umd.edu
Administrative Assistant: Pat Dolecek
Department Secretary: Kathy Hunt
Enrollment (1995): B: 132 (0) M: 38 (6) D: 22 (5)
Acting Chair:
Francis R. Gouin, (D), Maryland, 1969, Oo
Professor:
Timothy J. Ng, (D), Purdue, 1976, Oo
Bruno Quebedeaux, (D), Cornell, 1968, Oo
Theophanes Solomos, (D), Cambridge, 1962, Oo
Christopher Walsh, (D), Cornell, 1980, Oo
Robert C. Wiley, (D), Oregon State, 1953, Oo
Associate Professor:
Edward Beste, (D), Purdue, 1971, Oo
John C. Bouwkamp, (D), Michigan State, 1969, Oo
Gerald F. Deitzer, (D), Georgia, 1971, Oo
Charles A. McClurg, (D), Penn State, 1970, Oo
Madis Pihlak, (M), California (Berkeley), 1983, Ou
Franklin D. Schales, (D), Cornell, 1963, Oo
Harry J. Swartz, (D), Cornell, 1980, Oo
Assistant Professor:
Gary D. Coleman, (D), Nebraska, 1989, Oo
Margarita Hill, (M), California (Davis), 1989, Ou
Elmina J. Hilsenrath, (M), Massachusetts, 1973, Ou
Jack B. Sullivan, (M), Virginia, 1980, Ou
Lecturer:
Henry G. Mityga, (D), Maryland, 1976, Oo

Marine-Estuarine-Env Sci Graduate Program (M,D) (2004)
2113 Agriculture/Life Sciences Surge Building
College Park, MD 20742
p. (301) 405-6938 f. (301) 314-4139
meesgrad@deans.umd.edu
http;//www.mees.umd.edu
Enrollment (1999): M: 60 (16) D: 109 (15)
Director:
Kennedy T. Paynter, (D), Iowa State, 1985, Oo
Professor:
Joel E. Baker, (D), Minnesota, 1988, Co
Shenn-Yu Chao, (D), North Carolina State, 1979, Op
Thomas R. Fisher, Jr., (D), Duke, 1975, Ob
Robert H. Gardner, (D), North Carolina State, 1975, Ou
Patricia M. Gilbert, (D), Harvard, 1982, Ob
Rodger Harvey, (D), Georgia, 1985, Co
George Heiz, (D), Penn State, 1971, Cm
Micheal R. Roman, (D), New Hampshire, 1976, Ob
Lawrence P. Sanford, (D), MIT, 1984, Op
Kenneth P. Sebens, (D), Washington, 1977, Ob
Diane Stoecker, (D), SUNY (Stony Brook), 1979, Ob
Raymond R. Weil, (D), Virginia Tech, 1977, Sb
Richard Weismiller, (D), Michigan State, 1969, Or
Associate Professor:
William Boicourt, (D), Johns Hopkins, 1973, Op
James Carton, (D), Princeton, 1983, Op
Keith N. Eshleman, (D), MIT, 1985, Hg
Robert L. Hill, (D), Iowa State, 1984, Sb
Micheal S. Kearney, (D), Ontario, 1981, Gu
Robert P. Mason, (D), Connecticut, 1991, Ct
Karen L. Prestegaard, (D), California (Berkeley), 1982, Hw
Jennifer E. Purcell, (D), California (Santa Barbara), 1981, Ob
Marjorie Reaka-Kudla, (D), California (Berkeley), 1975, Ob
Philip A. Townsend, (D), North Carolina, 1997, Or
Leonard J. Walstad, (D), Harvard, 1987, Ou
Research Associate Professor:
Jeffery C. Cornwell, (D), Alaska, 1983, Oc
Assistant Professor:
Mark S. Castro, (D), Virginia, 1991, Oa
William Currie, (D), New Hampshire, 1995, Cg
Raleigh Hood, (D), California (San Diego), 1990, Ob
Glenn E. Moglen, (D), MIT, 1995, Hw
Ronald L. Siefert, (D), Caltech, 1996, Ca
Alba Torrents, (D), Johns Hopkins, 1992, Sb
Research Professor:
Lawrence W. Harding, (D), Stanford, 1978, Ob
Research Associate Professor:
Todd M. Kana, (D), Harvard, 1982, Ob

Massachusetts

Amherst College

Dept of Geology (B) (2004)
P.O. Box 2238
Amherst, MA 01002-5000
p. (413) 542-2233 f. (413) 542-2713
geology@amherst.edu
http://www.amherst.edu/~geology
Enrollment (2001): B: 18 (8)
Chair:
John T. Cheney, (D), Wisconsin, 1975, Gi
Professor:
Edward S. Belt, (D), Yale, 1963, Gs
Peter D. Crowley, (D), MIT, 1985, Gc

Tekla A. Harms, (D), Arizona, 1986, Gc
Assistant Professor:
Anna M. Martini, (D), Michigan, 1997, Cl
Adjunct Professor:
Margery C. Coombs, (D), Columbia, 1971, Pv

Bentley College

Dept of Natural and Applied Sciences (2007)
175 Forest Street
Waltham, MA 02452-4705
p. (781) 891-2980 f. (781) 891-2838
pdavis@bentley.edu
Administrative Assistant: Martha E. Keating
Enrollment (1996): B: 7 (0)

Professor:
P. Thompson Davis, (D), Colorado, 1980, Gl
Adjunct Professor:
Janette Gartner, (M), Massachusetts, 2000, Hy
Liane M. Stevens, (M), Massachusetts, 2002, Gp
Laboratory Director:
Anna K. Tary, (M), Boston, 1999, Gl

Boston College

Dept of Geology & Geophysics (B,M,D) (2004)
140 Commonwealth Avenue
213 Devlin Hall
Chestnut Hill, MA 02467-3809
p. (617) 552-3640 f. (617) 552-2462
dipietro@bc.edu
http://www.bc.edu/geology
Administrative Assistant: Angelina DiPietro
Enrollment (1999): B: 2 (0) M: 4 (0)

Professor:
Emanuel G. Bombolakis, (D), MIT, 1963, Gc
John E. Ebel, (D), Caltech, 1981, Ys
J. Christopher Hepburn, (D), Harvard, 1972, Gp
David C. Roy, (D), MIT, 1970, Gr
Associate Professor:
Rudolph Hon, (D), MIT, 1976, Gi
Alan L. Kafka, (D), SUNY (Stony Brook), 1980, Yg
Alan L. Kafka, (D), SUNY (Stony Brook), 1980, Yg
Assistant Professor:
Kevin G. Harrison, (D), Columbia, 1994, Cg
Gail C. Kineke, (D), Washington, 1993, On
David P. Lesmes, (D), Texas A&M, 1993, Yg
Lecturer:
Judith Hepburn, (D), Harvard, 1976, Ge
David Krauss, (D), Massachusetts, 1997, Po
Dale S. Weiss, (M), Boston Coll, 1983, Hw
Research Professor:
Paul K. Strother, (D), Harvard, 1980, Pb
Adjunct Professor:
Andrew R. Lazarewicz, (D), Hawaii, 1979, Oa
Randolph J. Martin, III, (D), MIT, 1968, Gc
Alfredo Urzua, (D), MIT, 1981, Hq
Emeritus:
John F. Devane, S.J., (M), Fordham, 1949, Ym
James W. Skehan, S.J., (D), Harvard, 1953, Gc
Lab Coordinator:
Kenneth G. Galli, (D), Massachusetts, 2003, Gg

Weston Observatory (2004)
381 Concord Road
Weston, MA 02493
p. (617) 552-8300 f. (617) 552-8388
ebel@bc.edu
http://www.bc.edu/westonobservatory

Director:
John E. Ebel, (D), Caltech, 1981, Yg
Chair:
Alan Kafka, (D), SUNY (Stony Brook), 1980, Ys
Associate Scientist:
Michael Hagerty, (D), California (Santa Cruz), 1998, Ys
Anastasia Moulis, (M), Boston Coll, 2003, Ys
Dina Smith, (M), Boston Coll, 2001, Ys
Research Associate:
John Beck, (D), Boston, 1998, Pb
Vincent Murphy, (M), Boston Coll, 1957, Yg
Adjunct Professor:
Paul Strother, (D), Harvard, 1980, Pb
Alfredo Urzua, (D), MIT, 1981, Hq
Emeritus:
James W. Skehan, (D), Harvard, 1953, Gc

Boston University

Center for Remote Sensing (2006)
685 Commonwealth Avenue
Room 433
Boston, MA 02215
p. (617) 353-9709 f. (617) 353-3200
crsadmin@bu.edu
Administrative Assistant: Krissa D. Nichols
Enrollment: No data reported since 1995

Director:
Farouk El-Baz, (D), Missouri (Rolla), 1964, Or
Professor:
Sucharita Gopal, (D), California (Santa Barbara), 1988, Oi
Alan Strahler, (D), Johns Hopkins, 1969, Or
Curtis E. Woodcock, (D), California (Santa Barbara), 1986, Or
Associate Professor:
Mark A. Friedl, (D), California (Santa Barbara), 1994, Or
Kenneth L. Kvamme, (D), California (Santa Barbara), 1983, Ga
Guido D. Salvucci, (D), MIT, 1994, Hq
Research Associate Professor:
Magaly Koch, (D), Boston, 1993, Ga
Cordula Robinson, (D), Univ Coll (UK), 1991, Gm
Crystal Schaaf, (D), Boston, 1994, Or
Research Associate:
Eman Ghoneim, (D), Southampton, 2002, Gm
John Hodges, (M), Boston, 1998
Geology Librarian:
Nasim Momen, Oo

Dept of Earth Sciences (B,M,D) (2007)
675 Commonwealth Avenue
Boston, MA 02215
p. (617) 353-2532 f. (617) 353-3290
earth@bu.edu
http://www.bu.edu/es/
Enrollment (2007): B: 15 (15) M: 1 (1) D: 4 (4)

Chair:
Guido D. Salvucci, (D), MIT, 1994, Hq
Professor:
Geoffrey A. Abers, (D), MIT, 1989, Yg
Richard Murray, (D), California (Berkeley), 1991, Pg
Terry A. Plank, (D), Columbia, 1992, Cg
Associate Professor:
Rachel Abercrombie, (D), Reading, 1991, Gt
Ulrich Faul, (D), Oregon, 1994, Gc
Duncan M. FitzGerald, (D), South Carolina, 1977, On
Andrew Kurtz, (D), Cornell, 2000, Clg
Amala Mahadevan, (D), Stanford, 1995, Op
David R. Marchant, (D), Edinburgh, 1993, Gm

Maureen E. Raymo, (D), Columbia, 1989, Pe
Assistant Professor:
Ethan F. Baxter, (D), California (Berkeley), 2000, Cp
Colleen Dalton, (D), Harvard, 2007, Ysg
Sergio Fagherazzi, (D), Padua, 1999, Gm
Paul Hall, (D), Rhode Island, 2003, Yg

Dept of Geography (B,M,D) (2005)
675 Commonwealth Avenue
4th Floor
Boston, MA 02215-1401
p. (617) 353-2525 f. (617) 353-8399
geography@bu.edu
http://geography.bu.edu/
Administrative Assistant: Michael J.M. Holmes
Enrollment (2004): M: 5 (0) D: 39 (0)
Chair:
Mark Friedl, (D), California (Santa Barbara), 1993, Or
Research Professor:
Yuri Knyazikhin, (D), Georgia, 1985, Or
Professor:
William Anderson, (D), Boston, 1984, Oo
Lata R. Chatterjee, (D), Johns Hopkins, 1973, Ou
Cutler J. Cleveland, (D), Illinois, 1988, Oo
Robert K. Kaufmann, (D), Pennsylvania, Oo
T. R. Lakshmanan, (D), Ohio State, 1965, Oo
Ranga Myneni, (D), Or
Alan H. Strahler, (D), Johns Hopkins, 1969, Or
Curtis E. Woodcock, (D), California (Santa Barbara), 1986, Or
Associate Professor:
Sucharita Gopal, (D), California (Santa Barbara), 1988, Oo
Guido Salvucci, (D), MIT, 1994, Sp
Associate Scientist:
Ian Sue Wing, (D), MIT
Assistant Professor:
Bruce Anderson, (D), California (San Diego), 1998, Oy
Anthony Patt, (D), Harvard, 2000, Oo
Nathan Phillips, (D), Duke, 1997, Oo
Joan Walker, (D), MIT, 2001, Oo
Research Professor:
Xiaowen Li, (D), California (Santa Barbara), Or
Research Associate Professor:
Crystal Schaaf, (D), Boston, 1994, Or
Research Assistant Professor:
Feng Gao, (D), Beijing Normal, 1997, Or
Research Associate:
John Hodges, (M), Boston, 1998, Or
Tigga Kingston, (N)
Nikolay Shabanov, (N), shab
Elena Tsvetsinskaya, (N)
Master Lecturer:
Syed Hasnath, (D), Boston, 1989, Oo
Lecturer:
Graham Walker, (N), Oe

Bridgewater State College

Dept of Earth Sciences (B,M) (2007)
Conant Science Building
Bridgewater, MA 02325
p. (508) 697-1390 f. (508) 697-1785
jsulanowski@bridgew.edu
http://www.bridgew.edu/EarthSciences/
Administrative Assistant: Brenda Crowley
Enrollment (2007): B: 68 (15)
Chair:
Jacek K. Sulanowski, (D), Chicago, 1978, Pe
Professor:
Richard L. Enright, (D), Rutgers, 1969, Hg
Peter J. Saccocia, (D), Minnesota, 1991, Cm
Associate Professor:
Robert Cicerone, (D), MIT, 1991, Ys
Michael Krol, (D), Lehigh, 1996, Gt
Visiting Professor:
Joseph Doyle, (M), New Hampshire, Gg
Peter Fletcher, (M), Massachusetts, Sd
Gerard Martin, (M), W Michigan, 1984, Hw
Emeritus:
Robert F. Boutilier, (D), Boston, 1963, Gz
Ira E. Furlong, (D), Boston, 1960, Gm

Fitchburg State College

Dept of Geophysical Sciences (B) (2005)
160 Pearl Street
Fitchburg, MA 01420-2697
p. (978) 665-3378 f. (978) 665-3081
lguth@fsc.edu
Administrative Assistant: Anne Burrill
Enrollment (2005): B: 2 (6)
Associate Professor:
Lawrence R. Guth, (D), Rice, 1991, Gc

Greenfield Community College

Dept of Earth Science (2006)
1 College Drive
Greenfield, MA 01301
rdlittle2000@aol.com
Enrollment (1996): B: 2 (0)
Emeritus:
Richard D. Little, (M), S California, 1971, Gg

Hampshire College

School of Natural Science (B) (2004)
Amherst, MA 01002
p. (413) 582-5373 f. (413) 582-5448
cdavanzo@hampshire.edu
Administrative Assistant: Joan Barrett
Professor:
John B. Reid, Jr., (D), MIT, 1970, Gi
Geology Librarian:
Helaine Selin, Oo

Harvard University

Dept of Earth and Planetary Sciences (B,D) (2007)
Hoffman Laboratory
20 Oxford Street
Cambridge, MA 02138-2902
p. (617) 495-2351 f. (617) 495-8839
lester@eps.harvard.edu
http://www.eps.harvard.edu
Enrollment (2003): D: 58 (0)
Chair:
Jeremy Bloxham, (D), Cambridge, 1985, Yg
Professor:
James G. Anderson, (D), Colorado, 1970, Oa
Adam M. Dziewonski, (D), Polish Acad of Sci, 1965, Ys
Goran A. Ekstrom, (D), Harvard, 1987, Ys
Brian F. Farrell, (D), Harvard, 1981, Ow
Paul F. Hoffman, (D), Johns Hopkins, 1970, Gc
Daniel J. Jacob, (D), Caltech, 1985, Oa
Stein B. Jacobsen, (D), Caltech, 1980, Cc
Andrew H. Knoll, (D), Harvard, 1977, Pb
Charles H. Langmuir, (D), Cg
Charles R. Marshall, (D), Pg
James J. McCarthy, (D), California (San Diego), 1971, Ob
Michael B. McElroy, (D), Queen's, 1962, Oa

Richard J. O'Connell, (D), Caltech, 1969, Yg
James R. Rice, (D), Lehigh, 1964, Yg
Daniel P. Schrag, (D), California (Berkeley), 1993, Cg
John Shaw, (D), Princeton, 1993, Gt
Eli Tziperman, (D), Og
Steven C. Wofsy, (D), Harvard, 1971, Oa

Assistant Professor:
Sujoy Mukhopadhyay, (D), Cg
Ann Pearson, (D), Cm
Sarah T. Stewart-Mukhopadhyay, (D), Xg

Research Professor:
Allan R. Robinson, (D), Harvard, 1959, Op

Emeritus:
Charles W. Burnham, (D), MIT, 1961, Gz
James N. Butler, (D), Harvard, 1959
Cornelius S. Hurlbut, Jr., (D), Harvard, 1933, Gz
Ulrich Petersen, (D), Harvard, 1963, Eg
Raymond Siever, (D), Chicago, 1950, Gd
James B. Thompson, Jr., (D), MIT, 1950, Gp

Related Staff:
Carl A. Francis, (D), Virginia Tech, 1980, Gz

Massachusetts Institute of Technology

Dept of Earth, Atm, & Planetary Sci (B,M,D) (2006)
77 Massachusetts Avenue, 54-918
Cambridge, MA 02139
p. (617) 253-2127 f. (617) 253-8298
eapsinfo@mit.edu
Enrollment (2005): B: 35 (6) M: 2 (7) D: 155 (15)

Head:
Maria T. Zuber, (D), Brown, 1986, Xy

Professor:
Richard P. Binzel, (D), Texas, 1986, Xm
Samuel A. Bowring, (D), Kansas, 1985, Gg
Edward A. Boyle, (D), MIT, 1976, Oc
Rafael L. Bras, (D), MIT, 1975, Hg
B. Clark Burchfiel, (D), Yale, 1961, Gc
James L. Elliot, (D), Harvard, 1972, Oa
Kerry A. Emanuel, (D), MIT, 1978, Ow
J. Brian Evans, (D), MIT, 1978, Yx
Glenn R. Flierl, (D), Harvard, 1975, Op
Frederick A. Frey, (D), Wisconsin, 1967, Ct
Timothy L. Grove, (D), Harvard, 1976, Gi
Bradford H. Hager, (D), Harvard, 1978, Ys
Thomas A. Herring, (D), MIT, 1983, Yd
Richard S. Lindzen, (D), Harvard, 1964, Ow
Paola M. Malanotte-Rizzoli, (D), California (San Diego), 1978, Op
John C. Marshall, (D), Imperial Coll (UK), 1980, Op
F Dale Morgan, (D), MIT, 1981, Yg
Raymond A. Plumb, (D), Manchester, 1972, Ow
Ronald G. Prinn, (D), MIT, 1971, Oa
Daniel H. Rothman, (D), Stanford, 1986, Yg
Leigh H. Royden, (D), MIT, 1982, Gt
Roger Summons, (D), New South Wales, Py
M Nafi Toksoz, (D), Caltech, 1963, Ys
Robert van der Hilst, (D), Utrecht (Neth), 1990, Ys
Jack Wisdom, (D), Caltech, 1981, Og
Carl I. Wunsch, (D), MIT, 1967, Op

Principal Research Scientist:
Robert W. King, (D), MIT, 1975, Yg
Robert E. Reilinger, (D), Cornell, 1979, Yg

Senior Scientist:
Michael Follows, (D), 1990, Op
Luisa T. Molina, Oa
William Rodi, (D), 1989, Yg
Chien Wang, (D), Oa

Associate Professor:
Dara Entekhabi, (D), MIT, 1990, Hg
Raffaele Ferrari, (D), Scripps, 2001, Op
Sara Seager, (D), Harvard, 1999, Xy

Assistant Professor:
Tanja Bosak, (D), Caltech, 2004, Py
Alison Malcolm, (D), Colorado Mines, 2005, Yx
Stephane Rondenay, (D), British Columbia, 2001, Ys
Sang-Heon (Dan) Shim, (D), Princeton, 2001, Yx
Benjamin Weiss, (D), Caltech, 2003, Ym

Senior Lecturer:
Lodovica Illari, (D), Imperial Coll (UK), 1982, Ow

Emeritus:
William F. Brace, (D), Gg
Charles C. Counselman, III, (D)
Edward N. Lorenz, (D), MIT, 1948, Ow
Theodore R. Madden, (D), MIT, 1961, Yg
Gordon H. Pettengill, (D), California (Berkeley), 1955, Og
William F. Pinson, Jr, Gg
Frederick A. Sanders, (D), Ow
M. Gene Simmons, (D), Harvard, 1962, Yg
John B. Southard, (D), Harvard, 1966, Gs
Peter H. Stone, (D), Harvard, 1964, Oa

Mount Holyoke College

Dept of Geology and Geography (B) (2007)
50 College Street
Clapp Laboratory #320
South Hadley, MA 01075-6419
p. (413) 538-2278 f. (413) 538-2239
cvasquez@mtholyoke.edu
http://www.mtholyoke.edu/acad/earth
Enrollment (2007): B: 28 (10)

Professor:
Steven R. Dunn, (D), Wisconsin, 1989, Gp
Girma Kebbede, (D), Syracuse, 1981, Oy
Mark McMenamin, (D), California (Santa Barbara), 1984, PgGst
Alan Werner, (D), Colorado, 1988, GleGm

Associate Professor:
M. Darby Dyar, (D), MIT, 1985, Gz
Michelle J. Markley, (D), Minnesota, 1998, GcgGt
Thomas L. Millette, (D), Clark, 1989, OriGm

Emeritus:
Martha M. Godchaux, (D), Oregon, 1969, Gv

Laboratory Director:
Penny M. Taylor, (M), SUNY (Oneonta), 2000, Gg

Nichols College

Dept of Environmental Science (A,B) (2006)
PO Box 5000
Dudley, MA 01571
p. (508) 213-2168 f. (508) 213-2225
peltoms@nichols.edu
http://www.nichols.edu/departments/glacier/

Professor:
Mauri S. Pelto, (D), Maine, 1989, Gl

North Adams State College

Dept of Geography (2004)
North Adams, MA 01247
p. (413) 664-4511 (Ext. 358) f. (413) 663-3033
fjonessn@mcla.edu

Assistant Professor:
Donald A. Hyers, (D), Arizona State, Oy
Sharon M. Thomas, (D), Arizona State, 1980, Oy

Northeastern University

Dept of Earth and Environmental Sciences (B) (2007)
14 Holmes Hall

Massachusetts

Boston, MA 02115
p. (617) 373-3176 f. (617) 373-4378
environment@neu.edu
http://ees.neu.edu/
Administrative Assistant: Leah Ben-Ami
Enrollment (2006): B: 90 (16)

Professor:
Richard H. Bailey, (D), North Carolina, 1973, Pi
Acting Chair:
Malcolm Hill, (D), California (Santa Cruz), 1979, OiGz
Associate Professor:
Peter S. Rosen, (D), William & Mary, 1976, Onu
Martin E. Ross, (D), Idaho, 1978, Gi
Instructor:
Jennifer S. Cole, (D), Syracuse, 1999, GeHw
Visiting Professor:
Daniel Douglass, (D), Wisconsin, 2005, Gl
Laboratory Director:
Ryan Hill, (B), Northeastern, 2001, Gg

Salem State College

Geological Sciences Dept (B) (2005)
352 Lafayette Street
Salem, MA 01970
p. (978) 542-6282 f. (978) 542-6132
Enrollment (2005): B: 33 (8)

Chair:
Peter Sablock, (D), Idaho, 1991, Gc
Professor:
James L. Cullen, (D), Brown, 1984, Gs
Lindley S. Hanson, (D), Boston, 1988, Gm
Associate Professor:
Jeanette M. Sablock, (D), Idaho, 1991, Gx
Jack Sweeney, (D), Buffalo, 1975, Yv
Assistant Professor:
Douglas Allen, (D), Minnesota, 2003, Cl
Lecturer:
Louise Bolge, (D), Rutgers, 2005, Ca

Smith College

Dept of Geology (B) (2007)
Clark Science Center
44 College Lane
Northampton, MA 01063
p. (413) 585-3805 f. (413) 585-3786
krichard@email.smith.edu
www.science.smith.edu/departments/Geology/
Administrative Assistant: Kathy D. Richardson
Enrollment (2007): B: 28 (9)

Professor:
John B. Brady, (D), Harvard, 1975, Gx
H. Robert Burger, (D), Indiana, 1966, Gc
Lawrence D. Meinert, (D), Stanford, 1980, Em
Robert M. Newton, (D), Massachusetts, 1978, Gm
Associate Professor:
Bosiljka Glumac, Chair, (D), Tennessee, 1997, Gs
Amy L. Rhodes, (D), Dartmouth, 1996, Cg
Assistant Professor:
Sara B. Pruss, (D), Southern California, 2004, Pg
Instructor:
Steve Gaurin, (M), Columbia, 2003, Ou
Lecturer:
Mark E. Brandriss, (D), Stanford, 1994, Gx
Emeritus:
H. Allen Curran, (D), North Carolina, 1968, Pg
Related Staff:
Tony Caldanaro, (M), Tennessee, 1994

Tufts University

Dept of Geology (B) (2007)
Lane Hall
Medford, MA 02155
p. (617) 627-3494 f. (617) 627-3584
anne.gardulski@tufts.edu
http://ase.tufts.edu/geology/
Administrative Assistant: Janet Silvano
Enrollment (2007): B: 14 (6)

Professor:
Grant Garven, (D), British Columbia, 1982, Hwy
John C. Ridge, (D), Syracuse, 1985, Gln
Chair:
Anne F. Gardulski, (D), Syracuse, 1987, Grs
Associate Professor:
Robert L. Reuss, (D), Michigan, 1970, Gzo
Lecturer:
Jacob Benner, (M), Utah, 2002, Pge
Emeritus:
Charles E. Stearns, (D), Harvard, 1950, Gm

University of Massachusetts

Dept of Geosciences (B,M,D) (2006)
233 Morrill Science Center
611 North Pleasant St.
Amherst, MA 01003-9297
p. (413) 545-2286 f. (413) 545-1200
head@geo.umass.edu
http://www.geo.umass.edu/
Enrollment (2006): B: 90 (0) M: 30 (0) D: 30 (0)

Head:
Michael L. Williams, (D), New Mexico, 1987, Gc
Professor:
Raymond S. Bradley, (D), Colorado, 1974, Oa
Julie Brigham-Grette, (D), Colorado, 1985, Gl
Laurie Brown, (D), Oregon State, 1974, Ym
Julie Graham, (D), Clark, 1984, Oo
James A. Hafner, (D), Michigan, 1970, Oo
R. Mark Leckie, (D), Colorado, 1984, Pm
Lynn Margulis, (D), California, 1965, Py
Rutherford H. Platt, (D), Chicago, 1971, Ou
J. Michael Rhodes, (D), Australian National, 1970, Ca
Richard W. Wilkie, (D), Washington, 1968, Oo
Richard F. Yuretich, (D), Princeton, 1976, Cl
Associate Professor:
Stephen J. Burns, (D), Duke, 1987, Cs
Christopher D. Condit, (D), New Mexico, 1984, Gi
Michele L. Cooke-Andresen, (D), Stanford, 1996, Gc
Robert DeConto, (D), Colorado, 1996, Oa
Piper Gaubatz, (D), California (Berkeley), 1968, Oo
Michael J. Jercinovic, (D), New Mexico, 1988, Gx
William D. McCoy, (D), Colorado, 1981, Oy
Sheila J. Seaman, (D), New Mexico, 1988, Gi
Stan Stevens, (D), California (Berkeley), 1983, Oo
Assistant Professor:
David Boutt, (D), New Mexico Tech, Hw
Steven Petsch, (D), Yale, 2000, Co
Qian Yu, (D), California (Berkeley), 2005, Oi
Manager, Climate System Research Center:
Francis T. Keimig, (M), Rutgers, 1992, Ow
Research Associate Professor:
Mathias Vuille, (D), Bern, 1995, Ow
Adjunct Professor:
Margery C. Coombs, (D), Columbia, 1973, Oo
Joseph I. Goldstein, (D), MIT, 1964, Oo
Douglas R. Hardy, (D), Massachusetts, 1995, Hy
Peter T. Panish, (D), Massachusetts, 1989, Gp

Emeritus:
John Hubert, (D), Harvard, Gs
George E. McGill, (D), Princeton, 1958, Gc
Stearns A. Morse, (D), McGill, 1962, Gi
Peter Robinson, (D), Harvard, 1963, Gc

Univ of Massachusetts, Boston/Harbor

Dept of Geography & Earth Science (A,B) (2004)
100 Morrissey Boulevard
Boston, MA 02125
p. (617) 287-5280 f. (617) 287-3929
michael.kenny@umb.edu
Administrative Assistant: Helen Buchanan
Enrollment (1997): B: 72 (18)

Chair:
Janet E. Duncan, (D), Boston, 1971, Ou
Professor:
John F. Looney, Jr., (D), Boston, 1972, Oe
Associate Professor:
Andrew N. Genes, (D), Syracuse, 1973, Gl
Zong-Guo Xia, (D), Cincinnati, 1993, Or
Assistant Professor:
Richard Gelpke, (D), Northeastern, 1998, Oy
Alberto Giordano, Syracuse, 1999, Or
David A. Nellis, (D), Boston, 1973, Em
Lecturer:
Yang Feng, Or
Michael Trust, Or
Adjunct Professor:
Glen Hazelton, (D), North Carolina State, 1987, Oo
Visiting Professor:
Thomas Brewer, Hw
Emeritus:
Michael G. Mensoian, (D), Connecticut, 1962, Oy
Systems Analyst & Network Administrator:
William Mahoney, Oo

University of Massachusetts/Lowell

Dept of Env, Earth, & Atmospheric Sci (B,M) (2004)
Lowell, MA 01854
p. (978) 934-3900 f. (978) 934-3069
eeas@uml.edu
Enrollment (1997): B: 136 (0)

Chair:
Arnold L. O'Brien, (D), Boston, 1973, Hw
Professor:
Frank P. Colby, (D), MIT, 1983, Ow
G. Nelson Eby, (D), Boston, 1971, Cg
Robert R. Gamache, (D), Massachusetts, 1978, Oa
Bodo W. Reinisch, (D), Massachusetts, 1970, Oa
Associate Professor:
Dan S. Golomb, (D), Hebrew, 1958, Oa
Richard Z. Gore, (D), Boston, 1973, Gz
Paul Song, (D), California (Los Angeles), 1991, Oa
Emeritus:
Robert C. Curtis, (D), Penn State, 1958, Ow
Wen Tang, (D), New York, 1960, Ow

Wellesley College

Dept of Geosciences (B) (2006)
106 Central Street
Wellesley, MA 02481-8203
p. (781) 283-3151 f. (781) 283-3642
jbesancon@wellesley.edu
http://www.wellesley.edu/Geosciences/
Department Secretary: Maria Raffi
Enrollment (2006): B: 7 (0)

Chair:
James R. Besancon, (D), MIT, 1975, Gz
Professor:
Margaret D. Thompson, (D), Harvard, 1976, Gc
Assistant Professor:
Britt A. Argow, (D), Boston, 2006, Gs
Daniel J. Brabander, (D), Brown, 1997, Cl
Lab Instructor:
Rebecca Mattison, (D), Massachusetts, 1999, Pv
Instructor:
Kathleen W. Gilbert, (M), Miami, 1995, Cs

Williams College

Dept of Geosciences (B) (2008)
947 Main Street
Williamstown, MA 01267
p. (413) 597-2221 f. (413) 597-4116
patricia.e.acosta@williams.edu
http://www.williams.edu/Geoscience
Administrative Assistant: Patricia E. Acosta
Enrollment (2007): B: 10 (10)

Chair:
Paul Karabinos, (D), Johns Hopkins, 1981, Gc
Professor:
David P. Dethier, (D), Washington, 1977, Gm
Markes E. Johnson, (D), Chicago, 1977, Ps
Reinhard A. Wobus, (D), Stanford, 1966, Gi
Associate Professor:
Ronadh Cox, (D), Stanford, 1993, Gs
Assistant Professor:
Heather M. Stoll, (D), Princeton, 1998, Cl
Research Associate:
Gudveig Baarli, (D), Oslo, 1989, Ps
David H. Backus, (D), Washington, 1998, Ps
Mark E. Brandriss, (D), Stanford, 1993, Gi
Emeritus:
William T. Fox, (D), Northwestern, 1962, Gs
Science Librarian:
Helena Warburg, Oo

Woods Hole Oceanographic Institution

Dept of Geology & Geophysics (D) (2007)
Woods Hole, MA 02543-1541
p. (508) 289-2388 f. (508) 457-2187
etulka@whoi.edu
http://www.whoi.edu/science/GG/dept/
Administrative Assistant: Maryanne F. Ferreira
Enrollment (2006): D: 23 (0)

Chair:
Susan E. Humphris, (D), MIT/Woods Hole, 1977, Gu
Senior Scientist:
William B. Curry, (D), Brown, 1980, Pm
Robert S. Detrick, (D), MIT/Woods Hole, 1978, Yr
Henry J B. Dick, (D), Yale, 1975, Gi
Daniel J. Fornari, (D), Columbia, 1978, Gu
Chris German, (D), Cambridge, 1988, Gt
Lloyd D. Keigwin, (D), Rhode Island, 1979, Cs
Jian Lin, (D), Brown, 1988, Gt
Delia W. Oppo, (D), Columbia, 1989, Pm
Hans Schouten, (D), Utrecht (Neth), 1970, Yr
Nobumichi Shimizu, (D), Tokyo, 1968, Ca
Deborah K. Smith, (D), California, 1985, Yr
Ralph A. Stephen, (D), Cambridge, 1978, Ys
Brian E. Tucholke, (D), MIT/Woods Hole, 1973, Gu
Associate Scientist:
Joan M. Bernhard, (D), California (San Diego), 1990, Cg
Karen L. Bice, (D), Penn State, 1997, Pe
Juan Pablo Canales Cisneros, (D), Barcelona, 1997, Yr

Massachusetts

Jeffrey Donnelly, (D), Brown, 2000, Gu
Robert L. Evans, (D), Cambridge, 1991, Yr
Glenn A. Gaetani, (D), MIT, 1996, Gi
Daniel Lizarralde, (D), MIT/Woods Hole, 1997, Ys
Olivier Marchal, (D), Paris, 1996, Pe
Daniel C. McCorkle, (D), Washington, 1987, Cm
Jeffrey J. McGuire, (D), MIT, 2000, Ys
Jerry F. McManus, (D), Columbia, 1997, Pe
Robert A. Reves-Sohn, (D), California (San Diego), 1996, Yr
Kenneth W. Sims, (D), California (Berkeley), 1995, Gi
Maurice A. Tivey, (D), Washington, 1988, Gt

Assistant Scientist:
Mark D. Behn, (D), MIT/Woods Hole, 2002, Yr
Ilya Buynevich, (D), Boston, 2001, Yr
Sarah B. Das, (D), Penn State, 2003, Pm
Liviu Giosan, (D), SUNY (Stony Brook), 2001, Gu

Assistant Professor:
S. Adam Soule, (D), Oregon, 2003, Gu

Senior Research Specialist:
Graham D. Layne, (D), Toronto, 1988, Ca
Ann P. McNichol, (D), MIT, 1986, Oc
Robert J. Schneider, (D), Tufts, 1968, Ct
Karl F. Von Reden, (D), Hamburg, 1983, Ct

Research Specialist:
James E. Broda, (B), Penn State, 1970, Gs
John A. Collins, (D), MIT/Woods Hole, 1989, Ys
Alan R. Gagnon, (B), New Hampshire, 1983, Cs
Steven J. Manganini, (B), Nasson, 1974, Ou
Dorinda R. Ostermann, (B), Pomona, 1978, Pe
Mark L. Roberts, (D), Duke, 1988, Yg
Stephen A. Swift, (D), MIT/Woods Hole, 1986, Yr
Frank B. Wooding, (B), Harvard, 1965, Yr
Li Xu, (D), Xiamen, 1992, Yr

Information Systems Associate:
S. Thompson Bolmer, (B), Colby, 1974, Yr
Richard J. Healy, (B), St. Lawrence, 1987, Oo

Engineer:
Peter B. Landry, (B), NEIT, 1990, Ca

Research Associate:
Jurek Blusztajn, (D), Polish Acad of Sci, 1985, Cc
Mary R. Carman, (M), Iowa, 1984, Gu
Anne L. Cohen, (D), Cape Town, 1993, Pe
Virginia Edgcomb, (M), Delaware, 1997, Cg
Kathryn L. Elder, (B), Massachusetts, 1985, Cs
Peter C. Lemmond, (B), Lehigh, 1978, Oo
Bret Longworth, Ou
Dana R. Stuart-Gerlach, (B), Michigan, 1993, Ou

Adjunct Scientist:
Johnson R. Cann, (D), Cambridge, 1963, Ys
Wayne Crawford, (D), California (San Diego), 1994, Ou
Javier Escartin, (D), MIT, 1996, Gg
Kai-Uwe Hinrichs, (D), Oldenburg, 1997, Cg
Barbara E. John, (D), California (Santa Barbara), 1987, Gc
Peter B. Kelemen, (D), Washington, 1987, Gi
Charles Langmuir, (D), SUNY (Stony Brook), 1980, Cg
Larry Mayer, (D), Scripps, 1979, Gg
Peter Meyer, (D), Rhode Island, 1984, Ou
Maureen Raymo, (D), Columbia, 1986, Pe
Roger C. Searle, (D), Newcastle upon Tyne, 1969, Yg
Jonathan E. Snow, (D), MIT/Woods Hole, 1993, Cg
Uri S. ten Brink, (D), Columbia, 1986, Yg

Retired:
Wilfred B. Bryan, (D), Wisconsin, 1959, Gu

Emeritus:
William A. Berggren, (D), Stockholm, 1962, Ps
Carl O. Bowin, (D), Princeton, 1960, Yr
Graham S. Giese, (D), Chicago, 1966, On
Stanley R. Hart, (D), MIT, 1960, Cg
John M. Hayes, (D), MIT, 1966, Ct
Susumu Honjo, (D), Hokkaido, 1961, Ou
George P. Lohmann, (D), Brown, 1972, Pe
David A. Ross, (D), California (San Diego), 1965, Gu
Elazar Uchupi, (D), S California, 1962, Ou
Richard P. Von Herzen, (D), California (San Diego), 1960, Yh

Visiting Investigator:
Laurent Montesi, (D), MIT, 2002, Yg
Wen-Lu Zhu, (D), SUNY (Buffalo), 1996, Yx

Dept of Marine Chemistry & Geochemistry (M,D) (2007)
360 Woods Hole Road
MS 25
Woods Hole, MA 02543-1541
p. (508) 289-2328 f. (508) 457-2193
mcg@whoi.edu
http://www.whoi.edu/page.do?pid=7146
Administrative Assistant: Sheila A. Clifford
Administrative Assistant: Lauren Ledwell
Administrative Assistant: Donna Mortimer
Administrative Assistant: Mary Zawoysky
Department Administrator: Susan A. Casso
Enrollment (2006): D: 26 (0)

Chair:
Timothy I. Eglinton, (D), Newcastle, 1988, Co

Senior Scientist:
Ken O. Buesseler, (D), MIT/Woods Hole, 1986, Oc
Scott C. Doney, (D), MIT, 1991, Cm
William J. Jenkins, (D), McMaster, 1974, Oc
Mark D. Kurz, (D), MIT/Woods Hole, 1982, Cc
Daniel J. Repeta, (D), MIT/Woods Hole, 1982, Co
Edward R. Sholkovitz, (D), California (San Diego), 1972, Oc
James A. Yoder, (D), Rhode Island, 1979, Cm

Associate Scientist:
Matthew A. Charette, (D), Rhode Island, 1998, Cm
Konrad A. Hughen, (D), Colorado, 1997, Cc
William R. Martin, (D), MIT/Woods Hole, 1985, Oc
Bernhard Peucker-Ehrenbrink, (D), Max Planck, 1994, Cm
Christopher M. Reddy, (D), Rhode Island, 1997, Co
Mak A. Saito, (D), MIT/Woods Hole, 2001, Oc
Jeffrey S. Seewald, (D), Minnesota, 1990, Cp
Margaret K. Tivey, (D), Washington, 1989, Cm

Assistant Scientist:
Karen L. Casciotti, (D), Princeton, 2002, Oc
Marco J.L. Coolen, (D), Oldenburg, 2000, Cm
Elizabeth B. Kujawinski, (D), MIT/Woods Hole, 2000, Oc
Phoebe Lam, (D), California (Berkeley), 2005, Cm
Carl Lamborg, (D), Connecticut, 2003, Oc
Laura Robinson, (D), Oxford, 2002, Cm
Olivier Rouxel, (D), Inst Nat Poly de Lorraine, 2002, Cm
Dierdre Toole, (D), California (Santa Barbara), 2003, Cm
Benjamin Van Mooy, (D), Washington, 2003, Oc

Research Associate:
Joshua M. Curtice, (B), Massachusetts, 1992, Cc
Helen Fredricks, (D), Plymouth (UK), 2000, Oc
Meagan Gonneea, Cm
Matt McIlvin, (D), Massachusetts, 2004, Oc
Daniel Montlucon, (M), SUNY (Stony Brook), 1998, Co
Steven M. Pike, (M), Rhode Island, 1998, Ca
Sean Sylva, Oc

Information Systems Associate:
Cyndy Chandler, (B), SUNY (Geneseo), 1975, Og
Ivan D. Lima, (D), Miami, 1999, Oc

Adjunct Professor:
Wolfgang Bach, Cm
Minhan Dai, Cm
Michael D. DeGrandpre, (D), Washington, 1990, Ca
Geoffrey Eglinton, (D), Manchester, 1966, Co

Roger Francois, Cm
Markus Keinast, Cm
Thomas Trull, Cm
Senior Research Specialist:
Nelson M. Frew, (D), Washington, 1971, Ca
Dempsey E. Lott, (M), Florida State, 1973, Oc
Scientist Emeritus:
Michael P. Bacon, (D), MIT/Woods Hole, 1976, Oc
Werner G. Deuser, (D), Penn State, 1963, Cs
John W. Farrington, (D), Oc
Frederick L. Sayles, (D), Manchester, 1968, Oc
Geoffrey Thompson, (D), Manchester, 1965, Cm
Oliver C. Zafiriou, (D), Johns Hopkins, 1966, Oc
Research Associate:
David P. Schneider, (B), Penn State, 1965, Ct
Oceanographer Emeritus:
Jean K. Whelan, (D), MIT, 1965, Co
Research Specialist:
David M. Glover, (D), Alaska, 1985, Oc
Carl G. Johnson, (M), Maine, 1983, Ca
Robert K. Nelson, (B), C Connecticut, 1980, Co
Guest Investigator:
Lary A. Ball, (B), Maine, 1973, Oc
Senior Research Assistant:
Scot P. Birdwhistell
Joanne Goudreau, Cm
Margaret Sulanowska, Cm
Research Assistant:
Tracy Atwood, Cm
Kevin Cahill, Cm
Paul Henderson, Cm

Worcester State College

Dept of Natural & Earth Sciences (B) (2006)
Worcester, MA 01602-2597
p. (508) 929-8641 f. (508) 793-8191
pbenjamin@worcester.edu
Enrollment (2006): B: 79 (0)

Chair:
Patricia A. Benjamin, (D), Clark, 2002, Oy
Associate Professor:
William A. Belanger, (M), Worcester Polytech Inst, 1962, Og
Richard G. Hunt, (M), S Connecticut, 1968, Oy
Francisco Lamelas, (D), Michigan, 1990, Yg
James V. Lyons, (D), Clark, 1993, Og
Assistant Professor:
William J. Hansen, (D), CUNY, 2002, Oi
Sudha Swaminathan, (D), Michigan, Yg

Michigan

Adrian College

Dept of Earth Sciences (B) (2007)
110 S. Madison St.
Adrian, MI 49221
p. (517) 265-5161 f. (517) 264-3331
slhanson@adrian.edu
http://www.adrian.edu/academics/ERTH/
Enrollment (2007): B: 10 (4)

Chair:
Sarah L. Hanson, (D), Utah, 1995, Gi
Professor:
Forest E. Haines, (D), Missouri (Rolla), 1977, Gd

Albion College

Dept of Geological Sciences (B) (2005)
611 E. Porter St.
Albion, MI 49224
p. (517) 629-0313 f. (517) 629-0264
wbartels@albion.edu
http://www.albion.edu/geology/
Enrollment (2005): B: 30 (6)

Professor:
William S. Bartels, (D), Michigan, 1986, Pv
Russell G. Clark, (D), Dartmouth, 1972, Gi
Beth Z. Lincoln, (D), California (Los Angeles), 1985, Gc
Timothy N. Lincoln, (D), California (Los Angeles), 1978, Em
Associate Professor:
Thomas I. Wilch, (D), New Mexico Tech, 1997, Gl
Assistant Professor:
Christopher M. Van de Ven, (D), Stanford, 2003, Oi
Emeritus:
John Wm. Parker, (D), Texas, 1967, Gr
Lawrence D. Taylor, (D), Ohio State, 1962, Gl

Calvin College

Dept of Geology, Geography, & Env Studies (B) (2007)
3201 Burton SE
Grand Rapids, MI 49546
p. (616) 526-8415 f. (616) 526-6501
aay@calvin.edu
http://www.calvin.edu/academic/geology/
Enrollment (2007): B: 26 (10)

Chair:
Ralph F. Stearley, (D), Michigan, 1990, PvOgGh
Professor:
Henry Aay, (D), Clark, 1978, Oo
Johnathan Bascom, (D), Iowa, 1989, Oo
Janel M. Curry, (D), Minnesota, 1985, Oo
Gerald K. Van Kooten, (D), California (Santa Barbara), 1980, EgCePi
Associate Professor:
Deanna van Dijk, (D), Waterloo, 1998, GmOny
Assistant Professor:
Kenneth A. Bergwerff, (M), Grand Valley State, 1988, Oe
Jason VanHorn, (D), Ohio State, 2007, Oi
Emeritus:
Clarence Menninga, (D), Purdue, 1966, CcGg
Davis A. Young, (D), Brown, 1969, Giz

Central Michigan University

Dept of Geology (B,M) (2006)
314 Brooks Hall
Mount Pleasant, MI 48859
p. (989) 774-3179 f. (989) 774-2142
matty1dj@cmich.edu
http://www.cst.cmich.edu/units/gel/
Administrative Assistant: Lorraine K. Thompson
Enrollment (2006): B: 18 (13)

Professor:
David J. Matty, (D), Rice, 1984, Gi
Jane M. Matty, (D), Michigan State, 1992, Cl
Reed Wicander, (D), California (Los Angeles), 1973, Pl
Associate Professor:
Kathleen C. Benison, (D), Kansas, 1997, Gs
Sven S. Morgan, (D), Virginia Tech, 1998, Gc
Assistant Professor:
Maria Mercedes Gonzalez, (D), Nacional del Sur (Argentina), 1997, Em
Patrick Kinnicutt, (D), MIT, 1995, Gq
Mona-Liza C. Sirbescu, (D), Missouri, 2002, Gz
James J. Student, (D), Virginia Tech, 2002, Gg
Research Associate:
Brenda B. Bowen, (D), Utah, 2005, Gs

Instructor:
Lisa Boettcher, (M), SD Sch of Mines & Tech, 2004, Gg
Keith Johnson, (B), Winona, 1987, Ye
Emeritus:
Richard V. Dietrich, (D), Yale, 1951, Gx
James S. Monroe, (D), Montana, 1976, Gs
Wayne E. Moore, (D), Cornell, 1950, Eo
Geology Librarian:
David Ginsberg, (M), Chicago, 1972, Gp

Charles Stewart Mott Community College

Dept of Geology (A) (2006)
Flint, MI 48503
p. (313) 232-6291
scimath@mcc.edu
Professor:
David Maguire, (M), E Michigan, 1981, Xg
Chester H. Wilson, (D), Michigan State, 1967, Ye

Concordia University

Div of Natural Sciences & Mathematics (B) (2005)
4090 Geddes Road
Ann Arbor, MI 48105
p. (734) 995-7300 f. (734) 995-4610
allenr@cuaa.edu
Enrollment: No data reported since 1995
Professor:
Heather M. Czartoski, (M), Cleveland State, 1003, Ge

Delta College

Dept of Geology (A) (2007)
1961 Delta Road
University Center, MI 48710-0002
p. (989) 686-9252 f. (989) 686-8736
tlclarey@delta.edu
Department Secretary: Barb Jurmanovich
Enrollment (2007): A: 4 (1)
Department Chair:
Timothy L. Clarey, (D), W Michigan, 1996, HwGc
Lecturer:
Mary C. Gorte, (M), RPI, 1983, Gs
Emeritus:
Barry A. Carlson, (D), Michigan State, 1974, Yg
Paul A. Catacosinos, (D), Michigan State, 1972, Gr
Cooperating Faculty:
Kevin T. Dehne, (M), E Michigan, 1992, Og

Eastern Michigan University

Dept of Geography & Geology (B) (2005)
Ypsilanti, MI 48197
p. (743) 487-8589 f. (734) 487-6979
yulanda.woods@EMICH.EDU
Enrollment (2001): B: 135 (0)
Head:
Michael Kasenow, (D), W Michigan, 1994, Hg
Professor:
Eugene Jaworski, (D), Louisiana State, 1971, Or
Carl F. Ojala, (D), Georgia, 1972, Oa
Constantine N. Raphael, (D), Louisiana State, 1967, Gu
Associate Professor:
Steven T. LoDuca, (D), Rochester, 1990, Pi
Assistant Professor:
Kevin Blake, (D), Wisconsin, 1998, Gl
Michael Bradley, (D), Utah, 1988, Gc
Maria-Serena Poli, (D), Padova, 1995, Pe

Ferris State University

Dept of Physical Sciences (2004)
ASC-3021
Big Rapids, MI 49307
p. (616) 592-2580 f. (616) 592-2545
heckf@ferris.edu
Administrative Assistant: Leona Royer
Professor:
Frederick R. Heck, (D), Northwestern, 1987, Gg

Gogebic Community College

Math-Science Div (2005)
E4946 Jackson Rd.
Ironwood, MI 49938
p. (906) 932-4231
marlak@adm.gogebic.edu
http://www.gogebic.edu/
Instructor:
Robert Seasor, (B), Michigan Tech, 1964, Gg

Grand Valley State University

Dept of Geology (B) (2004)
1 Campus Drive
125 Padnos Hall
Allendale, MI 49401
p. (616) 895-3728 f. (616) 895-3740
bramerm@gvsu.edu
http://www.gvsu.edu/
Administrative Assistant: Mary Ann Bramer
Enrollment (2002): B: 77 (17)
Professor:
William J. Neal, (D), Missouri, 1968, Gs
Patricia E. Videtich, (D), Brown, 1982, Gd
Associate Professor:
Kevin C. Cole, (D), Arizona, 1990, Gz
Stephen R. Mattox, (D), N Illinois, 1992, Oe
Virginia L. Peterson, (D), Massachusetts, 1992, Gc
Peter E. Riemersma, (D), Wisconsin, 1997, Hw
John C. Weber, (D), Northwestern, 1995, Gc
Assistant Professor:
Patrick Colgan, (D), 1996, Gl
Angela M. Hessler, (D), Stanford, 2001, Oe
Figen A. Mekik, (D), N Illinois, 1999, Pe
Peter Wampler, (D), Oregon State, 2004, Gm
Research Associate:
Tari N. Mattox, (M), N Illinois, 1991, Gv
Affiliate Instructor:
Larry Fegel, (M), Michigan State, 1979, Gg
Adjunct Professor:
Kelly L. Heid, (B), Missouri (Kansas City), 1985
Emeritus:
Thomas E. Hendrix, (D), Wisconsin, 1960, Gc
Richard H. Lefebvre, (D), Northwestern, 1966, Gx
Norman W. Ten Brink, (D), Washington, 1971, Gm
Lab Supervisor:
Greg C. Wilson, (M), S Illinois, 1987, Gm

Hope College

Dept of Geological & Environmental Sciences (B) (2006)
35 E 12th Street
P.O. Box 9000
Holland, MI 49422-9000
p. (616) 395-7540 f. (616) 395-7118
peterson@hope.edu
Administrative Assistant: Lois Roelofs
Enrollment (2006): B: 20 (0)

Chair:
Jon W. Peterson, (D), Chicago, 1989, Ge
Professor:
Edward C. Hansen, (D), Chicago, 1983, Gm
Associate Professor:
Brian E. Bodenbender, (D), Michigan, 1994, Pi
Graham F. Peaslee, (D), SUNY (Stony Brook), 1987, Oo
Emeritus:
J. Cotter Tharin, (D), Illinois, 1960, Gs

Jackson Community College

Dept of Geology & Geography (A) (2007)
2111 Emmons Road
Jackson, MI 49201
p. (517) 787-0800 (Ext. 157) f. (517) 789-1632
econgeorged@jccmi.edu
http://www.jackson.cc.mi.us
Enrollment: No data reported since 1995
Professor:
George D. Econ, (M), S Illinois, 1975, Ge

Lake Superior State University

Geology/Physics (B) (2007)
650 W. Easterday Avenue
Sault Ste. Marie, MI 49783
p. (906) 635-2267 f. (906) 635-2266
lbrown@lssu.edu
http://geology.lssu.edu/
Administrative Assistant: Donna White
Enrollment (2007): B: 26 (8)
Chair:
Lewis M. Brown, (D), New Mexico, 1973, Pi
Professor:
Paul R. Kelso, (D), Minnesota, 1993, Ym
Assistant Professor:
Matt Spencer, (D), Penn State, 2005, Gl
Adjunct Professor:
Gerry Bennett, Eg
Manfred Engel, (M), Germany, Eg
Science Lab Manager:
Mary St. Antoine, (B), Lake Superior State, 1990, Cg

Macomb Comm College, Center Campus

Dept of Science (Geology) (A) (2005)
44575 Garfield Road
Clinton Township, MI 48038-1139
p. (586) 286-2154 f. (586) 286-2170
schaferc@macomb.edu
Enrollment (2005): A: 3 (0)
Professor:
Carl M. Schafer, (M), Montana, 1998, Gg

Michigan State University

Dept of Geological Sciences (B,M,D) (2007)
206 Natural Science Building
East Lansing, MI 48824-1115
p. (517) 355-4626 f. (517) 353-8787
geosci@msu.edu
http://www.geology.msu.edu
Administrative Assistant: Michelle Torres
Enrollment (2004): B: 101 (24) M: 18 (5) D: 10 (1)
Chair:
Ralph E. Taggart, (D), Michigan State, 1971, Pb
Professor:
Robert L. Anstey, (D), Indiana, 1970, Pi
Kazuya Fujita, (D), Northwestern, 1979, Ys
Grahame J. Larson, (D), Ohio State, 1976, Gl
David T. Long, (D), Kansas, 1977, Cl
Randall Schaetzl, (D), Illinois, 1987, Gm
Duncan F. Sibley, (D), Oklahoma, 1975, Gs
Michael A. Velbel, (D), Yale, 1984, Gd
Associate Professor:
Danita S. Brandt, (D), Yale, 1985, Pi
Michael D. Gottfried, (D), Kansas, 1991, Pv
David W. Hyndman, (D), Stanford, 1995, Hy
Phanikumar S. Mantha, (D), Indian Inst of Sci, 1990, Hw
Roger B. Wallace, (D), Michigan, 1981, Hq
Assistant Professor:
Brian A. Hampton, (D), Purdue, 2006, Gs
Julie C. Libarkin, (D), Arizona, 1999, Oe
Tyrone Rooney, (D), Pennsylvania State, 2006, Gi
Associate Adjunct Professor:
Stephen K. Hamilton, (D), California (Santa Barbara), 1994, Cg
David B. Westjohn, (D), Michigan State, 1990, Hy
Assistant Adjunct Professor:
Sheridan K. Haack, (D), Michigan State, 1992, Sb
Adjunct Professor:
Lina C. Patino, (D), Rutgers, 1997, Ca
Visiting Research Associate:
Kevin G. Mackey, (D), Michigan State, 1999, Ys
Emeritus:
Hugh F. Bennett, (D), Wisconsin, 1967, Ys
F William Cambray, (D), London, 1964, Gc
Aureal T. Cross, (D), Cincinnati, 1943, Pb
James H. Fisher, (D), Illinois, 1953, Go
George P. Merk, (D), Michigan State, 1972, Gs
Max M. Mortland, (D), Illinois, 1951, Gz
Harold B. Stonehouse, (D), Toronto, 1952, Gz
James W. Trow, (D), Chicago, 1948, Eg
Thomas A. Vogel, (D), Wisconsin, 1963, Gi

Michigan Technological University

A. E. Seaman Mineral Museum (2006)
1400 Townsend Drive
Houghton, MI 49931-1295
p. (906) 487-2572 f. (906) 487-3027
tjb@mtu.edu
http://www.museum.mtu.edu
Director, Advancement & Planning:
Stanley J. Dyl, II, (M), Michigan Tech, 1979, Gz
Director:
Theodore J. Bornhorst, (D)
Curator:
George W. Robinson, (D), Queen's, 1979, Gz

Dept of Geological & Mining Eng & Sci (B,M,D) (2007)
1400 Townsend Drive
Houghton, MI 49931-1295
p. (906) 487-2531 f. (906) 487-3371
geo@mtu.edu
http://www.geo.mtu.edu
Administrative Assistant: Kelly McLean
Enrollment (2007): B: 48 (7) M: 28 (5) D: 14 (1)
Chair:
Wayne D. Pennington, (D), Wisconsin, 1979, Ys
Dean, Graduate School:
Jacqueline E. Huntoon, (D), Penn State, 1990, Gs
Curator:
George W. Robinson, (D), Queen's, 1979, Gz
Professor:
Suzanne Beske-Diehl, (D), Wyoming, 1977, Ym
Theodore J. Bornhorst, (D), New Mexico, 1980, Eg
Jimmy F. Diehl, (D), Wyoming, 1977, Ym
Alex S. Mayer, (D), North Carolina, 1992, Hy

William I. Rose, (D), Dartmouth, 1970, Gv
James R. Wood, (D), Johns Hopkins, 1973, Cl
Associate Professor:
John S. Gierke, (D), Michigan Tech, 1990, Hy
William J. Gregg, (D), SUNY (Albany), 1979, Gc
Assistant Professor:
Aleksey K. Smirnov, (D), Rochester, 2002, Yv
Gregory P. Waite, (D), Utah, 2004, Ys
Research Professor:
Roger M. Turpening, (D), Michigan, 1966, Ye
Research Associate Professor:
Judith W. Budd, (D), Michigan Tech, 1997, Or
James W. Vallance, (D), Michigan Tech, 1994, Gv
Adjunct Associate Professor:
Iain Matthew Watson, (D), Cambridge, 2000, Gv
Adjunct Assistant Professor:
Jeremy Shannon, (D), Michigan Tech, PhD, Gv
Albert Wylie, (D), Michigan Tech, 2002, Go
Adjunct Professor:
William F. Griffin, (D), Wyoming, 1991, Hw
Robert S. Regis, (D), Michigan Tech, 1997, Gl
Stanley J. Vitton, (D), Michigan, 1990, Ng
Emeritus:
Allan M. Johnson, (D), Michigan Tech, 1971, Nm
S. Douglas McDowell, (D), Caltech, 1966, Gx
Charles T. Young, (D), Wisconsin, 1977, Ye
On Leave:
Gregg J. S. Bluth, (D), Penn State, 1990, Or
Department Facilities Manager:
Robert Barron, (B), Michigan Tech, 1979, Og
Related Staff:
Amie S. Ledgerwood, Oo

Muskegon Community College

Dept of Mathematics & Physical Sciences (A) (2007)
221 S. Quarterline Road
Muskegon, MI 49442
p. (231) 777-0289 f. (231) 777-0487
john.bartley@muskegoncc.edu
http://www.muskegoncc.edu/pages/652.asp
Administrative Assistant: Paula Halloran
Enrollment: No data reported since 1995
Professor:
John W. Bartley, (D), Michigan State, 1988, GgPi

Northern Michigan University

Dept of Geography (A,B) (2004)
3001New Science Building
Marquette, MI 49855-5342
p. (906) 227-1176 f. (906) 227-1621
mbroadwa@nmu.edu
Department Secretary: Beverly Boyer
Enrollment (1997): B: 19 (2)
Head:
Michael J. Broadway, (D), Illinois, 1983, Oo
Director, Env Science Program:
Ronald C. Sundell, (D), Northwestern, 1991, Oo
Associate Professor:
Stephen R. DeGoosh, (D), Indiana State, 1990, Oo
Martha J. Ferrill, (D), Nebraska, 1989, Oo
Assistant Professor:
John B. Anderton, (D), Wisconsin, 1995, Ga

Saint Clair County Community College

Geography Dept (A) (2005)
Port Huron, MI 48060
p. (313) 984-3881 (Ext. 268) f. (313) 984-2852
rhunckler@sc4.edu
http://www.sc4.edu/geography
Department Secretary: Geri Gibson
Enrollment: No data reported since 1995
Professor:
Robert Hunckler, (M), Michigan State, 1996, Og

University of Michigan

Dept of Atmospheric, Oceanic, & Space Sciences (B,M,D) (2004)
2455 Hayward
Ann Arbor, MI 48109-2143
p. (313) 764-3335 f. (313) 764-4585
tamas@umich.edu
Department Manager: S. Due
Enrollment (1995): B: 19 (6) M: 23 (10) D: 71 (13)
Chair:
Lennard A. Fisk, (D), California (San Diego), 1969, Oa
Professor:
Sushil K. Atreya, (D), Michigan, 1973, Oa
John R. Barker, (D), Carnegie Mellon, 1979, Oa
John P. Boyd, (D), Harvard, 1976, Op
S. Roland Drayson, (D), Michigan, 1967, Oa
Anthony W. England, (D), MIT, 1970, Oo
Tamas I. Gombosi, (D), Hungarian Acad of Sci, 1983, Oa
Paul B. Hays, (D), Michigan, 1964, Oa
Stanley J. Jacobs, (D), Harvard, 1963, Op
Timothy L. Killeen, (D), Univ Coll (London), 1975, Oa
William R. Kuhn, (D), Colorado, 1966, Oa
Andrew F. Nagy, (D), Michigan, 1963, Oa
Joyce Penner, (D), Oo
Perry J. Samson, (D), Wisconsin, 1979, Oa
John F. Vesecky, (D), Stanford, 1967, Or
James C. G. Walker, (D), Columbia, 1964, Oa
Research Scientist:
Sanford Sillman, (D), Harvard, 1987, Ow
Senior Scientist:
Larry H. Brace, (B), Michigan, 1958, Gg
George R. Carignan, Oo
John T. Clarke, (D), Johns Hopkins, 1980, Oa
C. Robert Clauer, (D), California (Los Angeles), 1980, Og
Michael R. Combi, (D), Toledo, 1977, Oa
William E. Sharp, (D), Colorado, 1970, Oa
Tong-Wha Shyn, (D), Michigan, 1969, Oo
Associate Professor:
Dennis G. Baker, (D), MIT, 1970, Oa
Mary Anne Carroll, (D), MIT, 1983, Oa
Guy A. Meadows, (D), Purdue, 1977, Op
Associate Scientist:
Gilda E. Ballester, (D), Johns Hopkins, 1989, Oa
Alan G. Burns, (D), Canterbury (NZ), 1986, Og
Mark D. Burrage, (D), Leicester (UK), 1988, Og
William B. Cook, (D), Florida State, 1992, Oa
Jason M. Daida, (D), Stanford, 1991, Oa
Walter M. Harris, (D), Michigan, 1992, Oa
Roberta M. Johnson, (D), California (Los Angeles), 1987, Oa
George V. Khazanov, (D), Irkutsk State (Russia), 1974, Og
Janet U. Kozyra, (D), Michigan, 1986, Oa
Ruth S. Lieberman, (D), Washington, 1992, Oa
Ronald H. Miller, (D), MIT, 1988, Og
Richard J. Niciejewski, (D), W Ontario, 1987, Og
Craig E. Rasmussen, (D), Utah State, 1984, Og
Wilbert R. Skinner, (D), Michigan, 1984, Oa
Qian Wu, (D), Maryland, 1989, Og
Assistant Professor:
Brian E. Gilchrist, (D), Stanford, 1991, Oo
Gerald J. Keeler, (D), Michigan, 1987, Og
Peter J. Sousounis, (D), Penn State, 1990, Oa

Lecturer:
Lee H. Somers, (D), Michigan, 1969, Oo
Adjunct Professor:
Vincent J. Abreu, (D), Texas (Dallas), 1974, Og
Hal G. Marshall, (D), Iowa State, 1983, Oo
Emeritus:
Frederick L. Bartman, (D), Michigan, 1967, Oa
A. Nelson Dingle, (D), MIT, 1947, Oa
Thomas Donahue, (D), Johns Hopkins, 1947, Oa
Ernest G. Fontheim, (D), Lehigh, 1960, Oo
Donald J. Portman, (D), Johns Hopkins, 1955, Oa

Dept of Geological Sciences (B,M,D) (2007)
2534 C.C. Little Building
1100 North University Avenue
Ann Arbor, MI 48109-1005
p. (734) 764-1435 f. (734) 763-4690
robpatte@umich.edu
http://www.lsa.umich.edu/geo/
Administrative Assistant: Robert J. Patterer
Enrollment (1999): B: 29 (8) M: 27 (5) D: 37 (12)
Chair:
Rodney C. Ewing, (D), Stanford, 1974, Gz
Professor:
Tomasz K. Baumiller, (D), Chicago, 1990, Pg
Joel D. Blum, (D), Caltech, 1990, Cc
Maria Clara Castro, (D), Paris, 1995, Hy
Eric J. Essene, (D), California (Berkeley), 1967, Gp
Daniel C. Fisher, (D), Harvard, 1975, Pi
Philip D. Gingerich, (D), Yale, 1974, Pv
Gerald Keeler, (D), Michigan, 1987, Oo
Stephen E. Kesler, (D), Stanford, 1966, Eg
Rebecca A. Lange, (D), California (Berkeley), 1989, Gi
Kyger C. Lohmann, (D), SUNY (Stony Brook), 1977, Cs
Samuel B. Mukasa, (D), California (Santa Barbara), 1984, Cc
Robert M. Owen, (D), Wisconsin, 1975, Oc
Larry J. Ruff, (D), Caltech, 1981, Ys
Gerald R. Smith, (D), Michigan, 1965, Ps
Lars P. Stixrude, (D), California (Berkeley), 1991, Ys
Ben A. van der Pluijm, (D), New Brunswick, 1984, Gc
Rob van der Voo, (D), Utrecht (Neth), 1969, Ym
Lynn M. Walter, (D), Miami, 1983, Cl
Youxue Zhang, (D), Columbia, 1989, Cp
Associate Professor:
Udo Becker, (D), Virginia Tech, 1995, Oo
Robyn J. Burnham, (D), Washington, 1987, Pb
Todd A. Ehlers, (D), Utah, 2001, Yg
Carolina R. Lithgow-Bertelloni, (D), California (Berkeley), 1994, Yg
Christopher J. Poulsen, (D), Penn State, 1999, Pe
Jeroen Ritsema, (D), California (Santa Cruz), 1995, Ys
Peter J. van Keken, (D), Utrecht (Neth), 1989, Yg
Assistant Professor:
Marin Clark, (D), MIT, 2003, Gm
Ingrid Hendy, (D), California (Santa Barbara), 2000, Py
Nathan Niemi, (D), Caltech, 2001, Gc
Jeffrey A. Wilson, (D), Chicago, 1999, Gg
Research Scientist:
Jeffrey C. Alt, (D), Miami, 1984, Ou
Catherine E. Badgley, (D), Yale, 1982, Pv
Associate Research Scientist:
Chris Hall, (D), Toronto, 1982, Cc
Shaopeng Huang, (D), Acad Sinica, 1990, Yh
Josep M. Pares, (D), Barcelona, 1988, Ym
Assistant Research Scientist:
Yuhui Ai, (D), Harbin Eng, 1998
James D. Gleason, (D), Arizona, 1994
Jie Lian, (D), Michigan, 2003, Gz
Shanan Peters, (D), Chicago, 2003, Pi
Mirjam Schaller, (D), Bern, 2001, Gm
Adjunct Assistant Research Scientist:
Roland C. Rouse, (D), Michigan, 1972, Gz
Adjunct Assistant Professor:
Karen L. Webber, (D), Rice, 1988, Gv
Adjunct Professor:
John W. Geissman, (D), Michigan, 1980, Ym
William B. Simmons, (D), Michigan, 1973, Gz
Emeritus:
Charles Beck, (D), Cornell, 1955, Pg
William Farrand, (D), Michigan, 1960
William C. Kelly, (D), Columbia, 1954, Eg
Philip A. Meyers, (D), Rhode Island, 1972, Co
Theodore C. Moore, (D), California (San Diego), 1968, Ou
James O'Neil, (D), Chicago, 1963, Cs
Donald R. Peacor, (D), MIT, 1962, Gz
Henry N. Pollack, (D), Michigan, 1963, Yh
David K. Rea, (D), Oregon State, 1974, Gu
Bruce H. Wilkinson, (D), Texas, 1973, Gs

University of Michigan, Dearborn

Dept of Natural Sciences (B) (2007)
4901 Evergreen Road
Dearborn, MI 48128
p. (313) 593-5277 f. (313) 593-4937
vmnaik@umd.umich.edu
Department Secretary: Tina Barlow
Acting Chair:
Kent S. Murray, (D), California (Davis), 1981, Hw
Professor:
Don Bord, (D), Dartmouth, 1976, Xy
Associate Professor:
John Riebesell, (D), Chicago, 1975, Ou
Lecturer:
David Matzke, (M), Michigan, 1975, Xy
Gordon Parker, (D), Wayne State, 1966, Cg
Adjunct Professor:
Michael Favor, (B), Michigan, 1985, Oo
Mark Uhen, (D), Michigan, 1996, Pg
Emeritus:
Charles Olson, (D), Wisconsin, 1952, Or
Geology Librarian:
Donna Hellman, Oo

Washtenaw Community College

Dept of Geology (A) (2007)
Ann Arbor, MI 48106
p. (313) 973-3582 f. (313) 677-5414
salbach@wccnet.edu
Administrative Assistant: Irene Oliver
Enrollment (1996): B: 4 (0)
Instructor:
David P. Thomas, (M), E Michigan, 1970, Oe
Geology Librarian:
Del Blaine, Oo

Wayne State University

Geology Dept (B,M) (2006)
0224 Old Main Building
4831 Cass Avenue
Detroit, MI 48201
p. (313) 577-2506 f. (313) 577-0517
baskaran@wayne.edu
http://sun2.science.wayne.edu/~geology/
Enrollment (2006): B: 26 (10) M: 6 (1)

Acting Head:
Mark M. Baskaran, (D), Phy Res Lab (India), 1985, Cc
Associate Professor:
Jeffrey L. Howard, (D), California (Santa Barbara), 1987, Gs
Assistant Professor:
Larry D. Lemke, (D), Michigan, 2003, Ge
Edmond H. van Hees, (D), Michigan, 2000, Em
Lecturer:
Charles F. Barker, (M), Boston, 1988, Gg
Ann Purdy, (D), Michigan State, 1995, Gg
Grazyna Sledzinski, (M), Wayne State, 1994, Gg
Art Weinle, (M), Ohio State, 1973, Gg
John M. Zawiskie, (M), Wayne State, 1979, Gg
Academic Services Officer:
David J. Lowrie, (B), Wayne State, 1964, Gg

Western Michigan University

Dept of Geosciences (B,M,D) (2007)
1187 Rood Hall
Kalamazoo, MI 49008-5241
p. (269) 387-5486 f. (269) 387-5513
mohamed.sultan@wmich.edu
http://www.wmich.edu/geology/
Administrative Assistant: Kathyrn Wright
Enrollment (2007): B: 82 (0) M: 27 (0) D: 7 (0)

Chair:
Mohamed Sultan, (D), Washington, 1984, Or
Professor:
Ronald B. Chase, (D), Montana, 1968, Gp
Alan E. Kehew, (D), Idaho, 1977, Hw
Michelle A. Kominz, (D), Columbia, 1986, Yr
R. V. Krishnamurthy, (D), Phy Res Lab (India), 1984, Cs
Christopher J. Schmidt, (D), Indiana, 1975, Gc
Associate Professor:
David Barnes, (D), California, 1982, Gd
Daniel P. Cassidy, (D), Notre Dame, 1995, Hw
G Michael Grammer, (D), Miami, 1991, Gs
Johnson R. Haas, (D), Washington, 1993, Cl
Duane R. Hampton, (D), Colorado State, 1989, Hw
Carla M. Koretsky, (D), Johns Hopkins, 1997, Cl
William A. Sauck, (D), Arizona, 1972, Yg
Assistant Professor:
Heather L. Petcovic, (D), Oregon State, June, Gi

Minnesota

Anoka-Ramsey Community College

Dept of Natural Science (Geology) (2006)
11200 Mississippi Boulevard
Coon Rapids, MN 55433
p. (763) 433-1174 f. (763) 433-1521
ed.wehling@anokaramsey.edu
http://www.ar.cc.mn.us/wehling/nat_sci.htm

Other:
Ed Wehling, (M), Minnesota, 1993, Oa

Bemidji State University

Center for Environmental, Earth, & Space Studies (A,B,M) (2007)
1500 Birchmont Drive NE
Bemidji, MN 56601
p. (218) 755-2783 f. (218) 755-4107
tkroeger@bemidjistate.edu
http://www.bemidjistate.edu/ceess/
Administrative Assistant: Stephanie Hendricks
Enrollment (2006): B: 8 (0) M: 5 (0)

Professor:
Dragoljub D. Bilanovic, (D), Technion Israel Inst Tech, 1990, GePyCa
Timothy J. Kroeger, (D), North Dakota, 1995, PlGsHw
Patrick G. Welle, (D), Wisconsin, unkn, Oo
Assistant Professor:
Jason M. Dahl, (M), Brown, unkn, Xgm

Carleton College

Dept of Geology (B) (2008)
One North College Street
Northfield, MN 55057
p. (507) 646-4407 f. (507) 646-4400
ehaberot@carleton.edu
http://www.carleton.edu/departments/geol
Administrative Assistant: Ellen T. Haberoth
Enrollment (2006): B: 59 (11)

Professor:
Mary E. Savina, (D), California (Berkeley), 1982, Gm
Associate Professor:
Clinton A. Cowan, (D), Queen's, 1992, Gs
Cameron Davidson, (D), Princeton, 1991, Gx
Bereket Haileab, (D), Utah, 1994, Gx
Assistant Professor:
Sarah J. Titus, (D), Wisconsin, 2006, Gc
Director, Science Ed Resource Center:
Cathryn A. Manduca, (D), Caltech, 1988, Gi
Emeritus:
Caryl E. Buchwald, (D), Kansas, 1966, Gg
Technical Director:
Timothy D. Vick, (M), Wisconsin (River Falls), 1974, Oe
Geology Librarian:
Charles Priore, Oo

Concordia College

Dept of Environmental Studies (B) (2006)
St. Paul, MN 55104
p. (612) 641-8497
Enrollment: No data reported since 1995

Professor:
Carroll Peter, (D), Minnesota, 1980, Og
Assistant Professor:
Dale M. Trapp, (M), Michigan, 1972, Og

Gustavus Adolphus College

Dept of Geology (B) (2005)
800 West College Avenue
St Peter, MN 56082
p. (507) 933-7335 f. (507) 933-6285
welsh@gac.edu
Administrative Assistant: Kathy Scholl
Enrollment (2005): B: 10 (7)

Chair:
James L. Welsh, (D), Wisconsin, 1982, Gx
Assistant Professor:
Benjamin J.C. Laabs, (D), Wisconsin, 2004, Gl
Research Associate:
Mark D. Johnson, (D), Wisconsin, 1984, Gl
Julie A. Maxson, (D), Minnesota, 1996, Gs
Emeritus:
Keith J. Carlson, (D), Chicago, 1966, Pv
Chester O. Johnson, (M), Chicago, 1970, Gm
Cooperating Faculty:
Daniel Mollner, Oo

Itasca Community College

Dept of Mathematics & Science (A) (2004)
Grand Rapids, MN 55744
p. (218) 327-4227 f. (218) 327-4350
dcarlson@it.cc.mn.us
Administrative Assistant: Janice Padden
Enrollment: No data reported since 1995

Related Staff:
Robert J. Schwob, (M), Bemidji State, 1968, Oe

Macalester College

Geology Dept (B) (2007)
1600 Grand Avenue
St Paul, MN 55105
p. (651) 696-6105 f. (651) 696-6122
rogers@macalester.edu
http://www.macalester.edu/geology/
Department Secretary: Barb Ekeberg
Enrollment (2007): B: 16 (5)

Chair:
Raymond R. Rogers, (D), Chicago, 1995, Gs

Professor:
John P. Craddock, (D), Michigan, 1988, Gc

Associate Professor:
Karl R. Wirth, (D), Cornell, 1991, Gi

Assistant Professor:
Kristina A. Curry Rogers, (D), SUNY (Stony Brook), 2001, Pv
Kelly MacGregor, (D), California (Santa Cruz), 2002, Gm

Emeritus:
Gerald F. Webers, (D), Minnesota, 1964, Pg

Geology Lab Supervisor:
Jeffrey T. Thole, (M), Washington State, 1991, Gg

Minnesota State University

Earth Science Program (B,M) (2006)
Department of Geography
7 Armstrong Hall
Mankato, MN 56001
p. (507) 389-2617 f. (507) 389-2980
donald.friend@mnsu.edu
http://www.mnsu.edu/geog/
Enrollment (2004): B: 50 (10)

Director:
Donald A. Friend, (D), Arizona State, 1997, Gm

Professor:
Daryl Adams, (D), Georgia, 1987, Oe
Paul Brown, (D), Colorado, 1980, Ga
Cecil S. Keen, (D), Wisconsin (Milwaukee), 1987, Oa
Martin D. Mitchell, (D), Illinois, 1993, Ou
James Pierce, (D), Iowa State, 1980, Xc
Beth Proctor, (D), SUNY (Buffalo), 1987, Hs

Associate Professor:
Bryce W. Hoppie, (D), California (Santa Cruz), 1996, Ge
Brock McMillan, (D), Kansas State, 1999, So

Assistant Professor:
Changjoo Kim, (D), Ohio State, 2004, Oi
Dean Moosavi, (D), New Hampshire, 1998, Oe
Forrest Wilkerson, (D), Texas State, 2004, Oy
Fei Yuan, (D), Minnesota, 2003, Or

Instructor:
Ginger Schmid, (D), Texas State, 2004, Sd

Saint Cloud State University

Dept of Earth & Atmospheric Sciences (B) (2004)
720 4th Avenue South
St Cloud, MN 56301-4498
p. (320) 308-3260 f. (320) 308-4262
arhansen@stcloudstate.edu
bulletin.stcloudstate.edu/departments/esci.asp
Administrative Assistant: Jan Maslonkowski
Enrollment (1998): B: 103 (22)

Chair:
Gregory D. Nastrom, (D), Minnesota, 1980, Oa

Professor:
Anthony R. Hansen, (D), Iowa State, 1981, Ow
Charles L. Nelson, (D), Chicago, 1974, On
Robert A. Weisman, (D), SUNY (Albany), 1988, Ow

Associate Professor:
Jean L. Hoff, (D), North Dakota, 1989, Gs
Alfred H. Pekarek, (D), Wyoming, 1974, Gg

Assistant Professor:
Rodney Kubesh, (D), Illinois, 1991, Ow
Kate Pound, (D), Otago (NZ), 1993, Gt

University of Minnesota, Duluth

Dept of Geological Sciences (B,M,D) (2007)
229 Heller Hall
1114 Kirby Drive
Duluth, MN 55812
p. (218) 726-8385 f. (218) 726-8275
geol@d.umn.edu
http://www.d.umn.edu/geology
Administrative Assistant: Cathy A. Dziuk
Administrative Assistant: Claudia J. Rock
Enrollment (2007): M: 17 (7) D: 6 (0)

Professor:
Erik T. Brown, (D), MIT, 1990, Og
Steven M. Colman, (D), Colorado, 1977, Gn
John W. Goodge, (D), California (Los Angeles), 1987, Gp
Vicki L. Hansen, (D), California (Los Angeles), 1987, Gc
Thomas C. Johnson, (D), California (San Diego), 1975, Gn
Howard Mooers, (D), Minnesota, 1988, Gl
Ronald L. Morton, (D), Carleton, 1976, Eg

Associate Professor:
Christina D. Gallup, (D), Minnesota, 1997, Cc
Penelope Morton, (D), Carleton, 1982, Eg
John B. Swenson, (D), Minnesota, 2000, Gr
Nigel J. Wattrus, (D), Minnesota, 1984, Yr

Assistant Professor:
Karen Gran, (D), Washington, 2005, Gms

Emeritus:
James A. Grant, (D), Caltech, 1964, Gp
John C. Green, (D), Harvard, 1960, Gi
Charles L. Matsch, (D), Wisconsin, 1971, Gl
Richard W. Ojakangas, (D), Stanford, 1964, Gd
George R. Rapp, (D), Penn State, 1960, Ga

University of Minnesota, Morris

Div of Science & Mathematics (B) (2004)
Geology Discipline
Science Building
600 East 4th Street
Morris, MN 56267
p. (320) 589-6300 f. (320) 589-6371
geol@mrs.umn.edu
http://www.mrs.umn.edu
Enrollment (2002): B: 26 (9)

Coordinator:
James F. Cotter, (D), Lehigh, 1984, Gl

Professor:
James B. Van Alstine, (D), North Dakota, 1980, Pg

Associate Professor:
Keith A. Brugger, (D), Minnesota, 1992, Gl

Assistant Professor:
Peter M. Whelan, (D), California (Santa Cruz), 1988, Gx

University of Minnesota, Twin Cities

Dept of Civil Engineering (B,M,D) (2005)
500 Pillsbury Drive SE
Minneapolis, MN 55455
p. (612) 625-5522 f. (612) 626-7750
cive@umn.edu
http://www.ce.umn.edu/
Enrollment (2005): B: 17 (4) M: 5 (0) D: 10 (1)

Head:
John S. Gulliver, (D), Minnesota, 1980, Hs
Professor:
Emmanuel M. Detournay, (D), Minnesota, 1983, Nr
Andrew Drescher, (D), Inst of Fund Tech Res (Poland), 1968, So
Efi Foufoula-Georgiou, (D), Florida, 1985, Hg
Joseph F. Labuz, (D), Northwestern, 1985, Nr
Otto D. Strack, (D), Delft (Neth), 1973, Hw
Vaughan R. Voller, (D), Sunderland, 1980, Om
Associate Professor:
Randal J. Barnes, (D), Colorado Mines, 1985, Oo
Bojan B. Guzina, (D), Colorado, 1996, Nr
Karl A. Smith, (D), Minnesota, 1980, Nx
Research Associate:
Sonia Mogilevskaya, (D), Russian Acad of Sci, 1987
Adjunct Professor:
Peter A. Cundall, (D), Nr

Dept of Geology & Geophysics (B,M,D) (2006)
108 Pillsbury Hall
310 Pillsbury Drive SE
Minneapolis, MN 55455-0219
p. (612) 624-1333 f. (612) 625-3819
geology@umn.edu
http://www.geo.umn.edu
Student Support Services: Sharon J. Kressler
Enrollment (2006): B: 51 (11) M: 10 (5) D: 46 (8)

Head:
David L. Kohlstedt, (D), Illinois, 1970, Yx
Distinguished Professor:
R. Lawrence Edwards, (D), Caltech, 1988, Cc
Director, Minnesota Geological Survey:
Harvey Thorleifson, (D), Colorado, 1989, Gg
Professor:
E. Calvin Alexander, Jr., (D), Missouri (Rolla), 1970, Cc
Subir K. Banerjee, (D), Cambridge, 1963, Ym
Peter J. Hudleston, (D), Imperial Coll (UK), 1969, Gc
Emi Ito, (D), Chicago, 1979, Cs
Sally G. Kohlstedt, (D), Illinois, 1972, Oo
Bruce M. Moskowitz, (D), Minnesota, 1980, Ym
Christopher Paola, (D), MIT/Woods Hole, 1983, Gs
Hans Olaf Pfannkuch, (D), Paris, 1962, Hw
Justin Revenaugh, (D), MIT, 1989, Ys
William E. Seyfried, Jr., (D), S California, 1977, Cl
James H. Stout, (D), Harvard, 1970, Gp
Christian P. Teyssier, (D), Monash, 1985, Gc
Donna L. Whitney, (D), Washington, 1991, Gp
David A. Yuen, (D), California (Los Angeles), 1978, Yg
Associate Professor:
Marc M. Hirschmann, (D), Washington, 1992, Gi
Karen L. Kleinspehn, (D), Princeton, 1982, Gt
Assistant Professor:
David L. Fox, (D), Michigan, 1999, Pg
Katsumi Matsumoto, (D), Columbia, 2000, Oc
Lesley A. Perg, (D), California (Santa Cruz), 2001, Gm
Martin O. Saar, (D), California (Berkeley), 2003, Hg
Research Associate:
Thelma Berquó, (D), Ym
Randy Calcote, (D), Minnesota, 2000, Py
Hai Cheng, (D), Nanjing, 1988, Cc
Kang Ding, (D), Acad Sinica, 1987, Cg
Paul H. Glaser, (D), Minnesota, 1978, Gn
Andrey Kosterov, (D), Ym
Ivanka Stefanova, (D), Sofia, 1991, Pl
Tony Withers, (D), Bristol, 1997, Gi
Mark Zimmerman, (D), Minnesota, 1999, Yx
Adjunct Professor:
James E. Almendinger, (D), Minnesota, 1988, Pe
Val W. Chandler, (D), Purdue, 1977, Ye
Mark B. Edlund, (D), Michigan, 1998, Pe
Daniel R. Engstrom, (D), Minnesota, 1983, Pe
Michael Jackson, (D), Michigan, 1986, Ym
Carrie E. Jennings (formerly Patterson), (D), Minnesota, 1996, Gl
Robert G. Johnson, (D), Iowa State, 1952, Pe
Thomas C. Johnson, (D), California (San Diego), 1975, Pe
Kent C. Kirkby, (D), Wisconsin, 1994, Gg
James D. Miller, (D), Gx
Anthony Runkel, (D), Texas, 1988, Gs
Visiting Professor:
Patrice Rey, Gt
Emeritus:
Frederick M. Swain, (D), Kansas, 1943, Ps
Paul W. Weiblen, (D), Minnesota, 1965, Gi
Herbert E. Wright, Jr., (D), Harvard, 1943, Gl
Other:
Annia K. Fayon, (D), Arizona State, 1997, Gc
Anders Noren, Gn
Kathy Ohler, Oo
Zhong Zhang, Cl
Geology Librarian:
Janice M. Jaguszewski, (M), Illinois (Urbana), 1990, Oo
Related Staff:
Ellery Frahm, Oo

Dept of Soil, Water, & Climate (B,M,D) (2004)
1991 Upper Buford Circle
St. Paul, MN 55108-6028
p. (612) 625-9734 f. (612) 625-2208
enater@soils.umn.edu
http://www.soils.umn.edu
Administrative Director: Marjorie J. Bonse
Executive Secretary: Karen J. Mellem
Enrollment (2002): M: 21 (4) D: 14 (1)

Head:
Edward A. Nater, (D), California (Davis), 1987, Sd
Soil Scientist:
Donald C. Reicosky, (D), Illinois, 1969, So
Professor:
Deborah L. Allan, (D), California (Riverside), 1987, Sc
James L. Anderson, (D), Wisconsin, 1976, Sd
James C. Bell, (D), Penn State, 1990, Sd
Paul R. Bloom, (D), Cornell, 1978, Sc
Terence H. Cooper, (D), Michigan State, 1975, Sd
Peter H. Graham, (D), W Australia, 1964, Sb
Satish C. Gupta, (D), Utah State, 1972, Sp
Thomas R. Halbach, (M), Wisconsin, 1974, Sb
John A. Lamb, (D), Nebraska, 1984, Sc
Gary L. Malzer, (D), Purdue, 1973, Sc
Jean A. Molina, (D), Cornell, 1967, Sb
John F. Moncrief, (D), Wisconsin, 1981, Sp
David J. Mulla, (D), Purdue, 1983, Sp
Gyles W. Randall, (D), Wisconsin, 1972, So
George W. Rehm, (D), Minnesota, 1969, Sc
Pierre C. Robert, (D), Minnesota, 1982, Sd
Carl J. Rosen, (D), California (Davis), 1983, Sc
Michael J. Sadowsky, (D), Hawaii, 1983, Sb
Michael A. Schmitt, (D), Illinois, 1985, Sc

Mark W. Seeley, (D), Nebraska, 1977, Oa
Associate Professor:
Albert L. Sims, (D), North Carolina State, 1992, Sc
Timothy Wagar, (M), Minnesota, 1984, Sc
Assistant Professor:
Timothy J. Griffis, (D), McMaster, 2000
Neil C. Hansen, (D), Minnesota, 1998, Sp
Jane M F Johnson, (D), Minnesota, 1995
Jennifer Y. King, (D), California (Irvine), 1999
Randall K. Kolka, (D), Minnesota, 1996
Pamela J. Rice, (D), Iowa State, 1996
Jeffrey S. Strock, (D), North Carolina State, 1999, So
Dong Wang, (D), Wisconsin, 1993, Sp
Research Associate:
Alan E. Olness, (D), Minnesota, 1973, So
Adjunct Professor:
John M. Baker, (D), Texas A&M, 1987, Sp
Charles E. Clapp, (D), Cornell, 1957, Sb
William C. Koskinen, (D), Washington State, 1980, Sc
Robert B. Mabagala, (D), Michigan State, 1991
Michael P. Russelle, (D), Nebraska, 1982, Sc
Rodney T. Venterea, (D), California (Davis), 2000
Emeritus:
H. H. Cheng, (D), Illinois, 1961, Sb
Robert H. Dowdy, (D), Michigan State, 1966, Sc

University of Saint Thomas

Dept of Geology (B) (2007)
OWS 153
2115 Summit Avenue
St Paul, MN 55105
p. (651) 962-5241 f. 651-962-5209
tahickson@stthomas.edu
Enrollment (2006): B: 18 (0)
Chair:
Melissa A. Lamb, (D), Stanford, 1998, Yr
Associate Professor:
Thomas A. Hickson, (D), Stanford, 1999, Yr

Winona State University

Dept of Geoscience (B) (2008)
P.O. Box 5838
Winona, MN 55987
p. (507) 457-5260 f. (507) 457-5681
geoscience@winona.edu
http://www.winona.edu/geology
Enrollment (2007): B: 65 (15)
Associate Professor & Director Southeastern Minnesota Water Resources Center:
Toby Dogwiler, (D), Missouri, 2002, GmHqOw
Professor:
James H. Meyers, (D), Indiana, 1971, Gs
Catherine L. Summa, (D), MIT, 1993, Gt
Associate Professor:
Stephen T. Allard, (D), Wyoming, 2003, Gc
Assistant Professor:
Jennifer L.B. Anderson, (D), Brown, 2004, Yg
Emeritus:
John F. Donovan, (D), Cornell, 1963, Eg
Dennis N. Nielsen, (D), North Dakota, 1973, Gm
Associate VP, Academic Affairs:
Nancy O. Jannik, (D), New Mexico Tech, 1989, Hw

Mississippi

Millsaps College

Dept of Geology (B) (2007)
Box 150308
1701 North State Street
Jackson, MS 39210
p. (601) 974-1343 f. (601) 974-1345
harrijb@millsaps.edu
http://www.millsaps.edu/geology/index.shtml
Enrollment (2007): B: 10 (3)
Chair:
James B. Harris, (D), Kentucky, 1992, Ye
Assistant Professor:
Stan Galicki, (D), Mississippi, 2002, Ge
Zachary A. Musselman, (D), Kentucky, 2006, Gm
Emeritus:
Delbert E. Gann, (D), Missouri Sch of Mines, 1976, Gz

Mississippi State University

Dept of Geosciences (B,M) (2007)
P. O. Box 5448
108 Hilbun Hall
East Lee Blvd.
Mississippi State, MS 39762
p. (662) 325-3915 f. (662) 325-9423
schmitz@ra.msstate.edu
www.msstate.edu/dept/geosciences/4site/home.htm
Administrative Assistant: Kim Chewe
Administrative Assistant: Mary A. Dean
Administrative Assistant: Patricia Wilson
Enrollment (2007): B: 155 (33) M: 58 (20)
Head:
Darrel W. Schmitz, (D), Texas A&M, 1991, Hw
Outreach & Cont Ed Director:
Mark S. Binkley, (D), Indiana State, 1989, Ow
Professor:
John E. Mylroie, (D), RPI, 1977, Gm
Charles L. Wax, (D), Louisiana State, 1977, Oa
Associate Professor:
Michael E. Brown, (D), North Carolina, 1999, Ow
Christopher P. Dewey, (D), Memorial, 1983, Pm
F. Leo Lynch, (D), Texas, 1994, Gz
John C. Rodgers, (D), Georgia, 1999, Oy
Assistant Professor:
Shrinidhi Ambinakudige, (D), Florida State, 2006, Oiy
Jinmu Choi, (D), Georgia, 2004, Oi
Renee M. Clary, (D), Louisiana State, 2003, Oe
William H. Cooke, III, (D), Mississippi State, 1997, Or
Grady Dixon, (D), Arizona State, 2005, Ow
Jamie Dyer, (D), Georgia, 2005, Ow
Brenda L. Kirkland, (D), Louisiana State, 1992, Gd
Karen S. McNeal, (D), Texas A & M, 2007, Co
Kathleen M. Sherman-Morris, (D), Florida State, 2006, Ow
Research Associate:
Katarzynz Grala, (M), Iowa State, 2004, Oi
Distance Learning Director:
Douglas M. Gillham, (M), Mississippi State, 1999, Ow
Instructor:
Christa M. Haney, (M), Mississippi State, 1999, Ow
Jonathan G. Harris, (M), Mississippi State, 1995, Gg
Justyn D. Jackson, (M), Mississippi State, 2006, Ow
Dalton W. Miller, (M), Ole Miss, 1981, Og
Joan R. Mylroie, (M), Murray State, 1979, Oe
Nel Ruffin, (M), Alabama, 1996, Oy
Reynold J. Vandewege, (M), Mississippi State, 2006, Ow
Tim Wallace, (M), Mississippi State, 1994, Ow
Adjunct Professor:
Paul J. Croft, (D), Rutgers, 1991, Ow
Patrick J. Fitzpatrick, (D), Colorado State, 1995, Ow
James May, (D), Texas A&M, 1988, Hw
Gueorgui V. Mostovoi, (D), Moscow State, 1984, Ow

Charles G. O'Hara, (D), Ole Miss, 1994, Oir
Jack C. Pashin, (D), Kentucky, 1990, GoEcGs
Scott A. Sampson, (D), Nebraska, 1988, OirOs
Janet E. Simms, (D), Texas A& M, 1991, Yg
Jayaram Veeramony, (D), Delaware, 1999, Og
Emeritus:
John M. Kaye, (D), Louisiana State, 1974, Gg
Troy J. Laswell, (D), Missouri, 1953, Gm
Ernest E. Russell, (D), Tennessee, 1965, Gr
Academic Coordinator:
Cindy Wasson, (B), Mississippi State, 1985

University of Mississippi

Dept of Geology & Geological Eng (B,M,D) (2006)
School of Engineering
P.O. Box 1848
118 Carrier
University, MS 38677-1848
p. (662) 915-7498 f. (662) 915-5998
geology@olemiss.edu
http://www.geo.olemiss.edu/
Enrollment (2006): B: 76 (8) M: 12 (2) D: 9 (2)
Chair:
Gregory L. Easson, (D), Missouri (Rolla), 1996, Or
Research Professor:
James R. Woolsey, (D), Georgia, 1977, Oo
Professor:
R. P. Major, (D), Brown, 1984, Gd
Research Associate Professor:
Thomas M. McGee, (D), Utrecht (Neth), 1991, Ys
Associate Professor:
Gregg R. Davidson, (D), Arizona, 1995, Hw
Robert M. Holt, (D), New Mexico Tech, 2000, Hq
Joel S. Kuszmaul, (D), California (Berkeley), 1993, Nr
Research Assistant Professor:
Robin C. Buchannon, (D), Mississippi, 1995, Oo
Craig J. Hickey, (D), Alberta, 1994, Ys
Assistant Professor:
Terry L. Panhorst, (D), Nevada (Reno), 1996, Em
Emeritus:
Nolan B. Aughenbaugh, (D), Purdue, 1963, Nr
George D. Brunton, (D), Indiana, 1957, Ng
William R. Reynolds, (D), Florida State, 1966, Gs
Alphonse C. Van Besien, (D), Missouri (Rolla), 1985, Ng
Coordinator, Academic & Admin Affairs:
Cathy A. Grace, (M), Mississippi, 1996, Gg

University of Southern Mississippi

Dept of Geography and Geology (B,M,D) (2007)
118 College Drive, Box 5051
Walker Science Building, Room 127
Hattiesburg, MS 39406
p. (601) 266-4526 f. (601) 266-6219
gail.russell@usm.edu
http://gg.usm.edu/dept/
Administrative Assistant: Pam Watts
Enrollment (2003): B: 32 (7) M: 12 (3)
Chair:
Gail S. Russell, (D), Florida State, 1978, Cc
Professor:
Maurice A. Meylan, (D), Hawaii, 1978, Gu
David M. Patrick, (D), Oklahoma, 1972, Ng
Assistant Professor:
Jerry "Joby" Bass, (D)
David F. Ufnar, (D), Iowa, 2002, Gs
Instructor:
Lin F. Pope, (M), S Mississippi, 1983, Gz
Adjunct Professor:
Thomas H. Orsi, (D), Texas A&M, 1994, Yr

Dept of Marine Science (B,M,D) (2007)
Stennis Space Center Teaching Site
1020 Balch Boulevard
Stennis Space Center, MS 39529
p. (800) 644-4207 f. (228) 688-1121
marine.science@usm.edu
http://www.usm.edu/marine
Enrollment (2007): M: 14 (3) D: 9 (1)
Chair:
Steven E. Lohrenz, (D), MIT/Woods Hole, 1985, Ob
Professor:
Vernon L. Asper, (D), MIT/Woods Hole, 1986, Ou
Charlotte A. Brunner, (D), Rhode Island, 1978, Ou
Vladimir D. Kamenkovich, (D), Shirshov Inst, 1969, Op
Donald G. Redalje, (D), Hawaii, 1980, Ob
Alan M. Shiller, (D), California (San Diego), 1982, Oc
Associate Professor:
Laodong Guo, (D), Texas A&M, 1995, Oc
Stephan Howden, (D), Rhode Island, 1996, Op
Dmitri Nechaev, (D), Shirshov Inst, 1986, Op
Assistant Professor:
David W. Dodd, (D), New Brunswick, 1994, Oi
Scott Milroy, (D), South Florida, 2007, Ob
Karen Orcutt, (D), Wales (Swansea), 1999, Ob
Jerry Wiggert, (D), Southern California, 1995, Og
Kevin Yeager, (D), Texas A&M, 2002, On
Instructor:
Charles Meador, (M), S Mississippi, 2003, Oo
Business Manager:
Bridgett May, (M), Oo

Gulf Coast Research Laboratory (2006)
Department of Coastal Sciences
Geology Section
Ocean Springs, MS 39566-7000
p. (601) 872-4200 f. (601) 872-4204
angelia.bone@usm.edu
http://www.coms.usm.edu
Administrative Assistant: Angelia Bone
Head:
Ervin G. Otvos, (D), Massachusetts, 1964, On

Missouri

Central Methodist University

Div of Science (2006)
411 CMU Square
Fayette, MO 65248
p. (816) 248-3391 (Ext. 57370) f. (660)248-1634
delliott@centralmethodist.edu
Department Secretary: Nancy Symes
Enrollment (1998): B: 8 (0)
Head:
Larry J. Peery, (D), Oklahoma State, Yg
Professor:
Dan R. Elliott, (D), Missouri, 1981, Gg

Meramec Community College

Dept of Physical Sciences (Geology) (A) (2004)
11333 Big Bend Boulevard
St Louis, MO 63122-5799
p. (314) 984-7724 f. (314) 984-7112
pbhattacharyya@stlcc.edu
www.stlcc.cc.mo.us/mc/dept/physci/geology.htm
Enrollment (2000): A: 17 (0)

Chair:
Robert J. Monaghan, (M), S Illinois, 1970, Oe
Professor:
Peter H. Kellams, (M), Indiana, 1968, Gg
Associate Professor:
Paul D. Nelson, (M), Kansas State, 1952, Oe
Instructor:
Lisa A. Williams, (M), Missouri (Rolla), 1997, Og

Northwest Missouri State University

Dept of Geology-Geography (B,M) (2006)
800 University Drive
Maryville, MO 64468
p. (660) 562-1723 f. (660) 562-1055
geosci@.nwmissouri.edu
Enrollment (2003): B: 176 (53)

Chair:
Gregory D. Haddock, (D), Idaho, 1996, Oi
Associate Professor:
Mark W. Corson, (D), South Carolina, 1997, Oy
Theodore L. Goudge, (D), Oklahoma State, 1984, Oy
Assistant Professor:
Patricia L. Drews, (D), South Carolina, 1999, Oi
Richard M. Felton, (M), Missouri, 1979, Pg
James Hickey, (D), Dartmouth, 2006, Ge
Ming-Chih Hung, (D), Utah, 2003, Or
Yanfen Le, (D), Georgia, 2005, Oi
Leah D. Manos, (M), Tennessee, 1997, Oy
John P. Pope, (D), Iowa, 2006, Ps
C. Renee Rohs, (D), Kansas, 2000, Cg
Yi-Hwa Wu, (D), Utah, 2003, Oi
Instructor:
Jeffrey Bradley, (M), Oklahoma State, 1991, Og

St. Louis Comm College, Florissant Valley

Dept of Geology (A) (2005)
St. Louis, MO 63135
p. (314) 595-2380
bstinchcomb@fv.stlcc.edu
Enrollment (2001): A: 3 (0)

Chair:
Bruce L. Stinchcomb, (D), Missouri (Rolla), 1978, Pg
Instructor:
Richard L. Cameron, (D), Ohio State, 1963, Gl
Michael F. Fix, (M), Washington (St. Louis), 1968, Pg

Saint Louis University

Dept of Earth & Atmospheric Sciences (B,M,D) (2006)
3642 Lindell Ave, ONeil Hall
Room 205
St. Louis, MO 63108
p. (314) 977-3131 f. (314) 977-3117
dannevik@eas.slu.edu
http://www.eas.slu.edu
Administrative Assistant: Laurie Hausmann
Administrative Assistant: Cyndi Wise
Enrollment (2001): B: 43 (9) M: 13 (6) D: 11 (2)

Chair:
William P. Dannevik, (D), St. Louis, 1984, Oa
Professor:
Benjamin F. Abell, (M), St. Louis, Ow
David J. Crossley, (D), British Columbia, 1973, Yg
Robert B. Herrmann, (D), St. Louis, 1974, Ys
Associate Professor:
John Encarnacion, (D), Michigan, 1994, Gx
Charles E. Graves, (D), Iowa State, 1988, Oa
Timothy M. Kusky, (D), Johns Hopkins, 1990, Gg
Robert W. Pasken, (D), St. Louis, 1981, Ow
Assistant Professor:
Benjamin de Foy, (D), Cambridge, 1998, Oa
David Kirschner, (D), Minnesota, 1992, Cg
Keith D. Koper, (D), Washington (St. Louis), 1998, Ys
Zaitao Pan, (D), Iowa State, 1996, Oa
Lupei Zhu, (D), Caltech, 1998, Ys
Emeritus:
Kenneth G. Brill, (D), Oo
Brian J. Mitchell, (D), S Methodist, 1970, Ys
Albert J. Pallmann, (D), Cologne, 1958, Oa

Southwest Missouri State University

Dept of Geography, Geology & Planning (B,M) (2004)
Box 87, 901 S. National
Springfield, MO 65804-0089
p. (417) 836-5800 f. (417) 836-6006
geology@smsu.edu
http://geosciences.smsu.edu
Administrative Assistant: Joanne Keller
Department Secretary: Deana Gibson
Enrollment (1996): B: 43 (5)

Head:
James Skinner, (D), Florida, Oo
Professor:
William H. Cheek, (D), Michigan State, 1976, Oo
William Corcoran, (D), Michigan State, 1981, Oa
Elias Johnson, (D), Oklahoma, 1977, Oy
Erwin J. Mantei, (D), Missouri (Rolla), 1965, Ct
Kevin L. Mickus, (D), Texas (El Paso), 1989, Yg
James F. Miller, (D), Wisconsin, 1970, Ps
Paul A. Rollinson, (D), Illinois, 1988, Oo
Burl E. Self, (D), Oklahoma State, 1978, Oo
Kenneth C. Thomson, (D), Utah, 1970, Em
Associate Professor:
Rex Cammack, (D), South Carolina, 1995, Or
Stanley C. Fagerlin, (D), Missouri, 1980, Pm
Melida Gutierrez, (D), Texas (El Paso), 1992, Hy
Dimitri Ioannides, (D), Rutgers, 1994, Ou
Rajinder S. Jutla, (D), Virginia Tech, 1995, Ou
Judith Meyer, (D), Wisconsin, 1994, Oo
Thomas D. Moeglin, (D), Nebraska, 1978, Ng
Robert T. Pavlowsky, (D), Wisconsin, 1995, Gm
Thomas G. Plymate, (D), Minnesota, 1986, Gx
Charles W. Rovey, (D), Wisconsin (Milwaukee), 1990, Hy
Yongwei Zhang, (D), S Illinois, 1990, Ou
Assistant Professor:
Mario Daoust, (D), McGill, Oa
Diane M. May, (M), S Illinois (Edwardsville), 1974, Oo
Lecturer:
Jill Black, (M), SW Missouri State, Oo
Deborah Corcoran, (M), Michigan State, 1980, Oo
Adjunct Professor:
Thomas J. Aley, (M), California (Berkeley), 1962, Hy
Visiting Professor:
Jim Davis, (D), Kansas State, 2001, Oo
Emeritus:
David A. Castillon, (D), Michigan State, 1972, Gm
Charles D. Collins, (M), Arkansas, 1971, Oy
Russel L. Gerlach, (D), Nebraska, 1974, Oo
Vincent E. Kurtz, (D), Oklahoma, 1960, Ps
Donald H. McInnis, (D), Missouri, 1970, Oa
Milton D. Rafferty, (D), Nebraska, 1970, Oy
Other:
John C. Catau, (D), Michigan State, 1973, Oo
Cooperating Faculty:
Frederick J. May, (M), S Illinois, 1974, Oo
John S. Petersen, (M), SW Missouri State, 1986, Oo

Truman State University

Div of Science (2004)
Kirksville, MO 63501
p. (816) 785-4597 f. (816) 785-4045
sellis@truman.edu
Administrative Assistant: Marsha Redmon

Assistant Professor:
Jeffrey Osborn, (D), Ohio State, 1992, Pb

University of Central Missouri

Dept of Earth Science (B) (2007)
Warrensburg, MO 64093
p. (660) 543-4933 f. (660) 543-4355
nold@ucmo.edu
Enrollment (2006): B: 40 (0)

Professor:
James D. Loch, (D), Missouri, 1992, Ps
John L. Nold, (D), Montana, 1968, Em

Assistant Professor:
Adel E. Haj, (D), Iowa, 2007, Gm
Sarah D. Zellers, (D), Texas, 1995, Gg

Instructor:
Mark A. Dudley, (M), Missouri (Rolla), 1998, Gg
Kuppusamy Panneerselvam, (D), Florida Intl, 2001, Cg

University of Missouri

Dept of Geological Sciences (B,M,D) (2007)
101 Geology Building
Columbia, MO 65211
p. (573) 882-6785 f. (573) 882-5458
BennettKA@missouri.edu
http://web.missouri.edu/~geolwww/
Administrative Assistant: Marsha Huckabey
Enrollment (2007): B: 46 (4) M: 13 (6) D: 10 (1)

Chair:
Kevin L. Shelton, (D), Yale, 1982, Em

Professor:
Mian Liu, (D), Arizona, 1989, Yg
Peter I. Nabelek, (D), SUNY (Stony Brook), 1983, Ct
Michael B. Underwood, (D), Cornell, 1982, Gs
Carol M. Wicks, (D), Virginia, 1992, Hw

Associate Professor:
Robert L. Bauer, (D), Minnesota, 1981, Gc
Cheryl A. Kelley, (D), North Carolina, 1993, Co
Kenneth A. MacLeod, (D), Washington, 1992, Pg
Eric A. Sandvol, (D), New Mexico State, 1995, Ys

Assistant Professor:
Martin Appold, (D), Johns Hopkins, 1998, Hw
Marie-Helene Cormier, (D), California (Santa Barbara), 1994, Yr
Francisco Gomez, (D), Cornell, 1999, Gt
Mitchell D. Schulte, (D), Washington, 1997, Cg
Alan Whittington, (D), Gv

Adjunct Professor:
Chris C. Sorlien, (D), California (Santa Barbara), 1994, Yg
Angela Speck, (D), Univ Coll (London), 1998, Xc
Samson Tesfaye, (D), Colorado, 1999, GtcOr

Emeritus:
Raymond L. Ethington, (D), Iowa, 1958, Pm
Thomas J. Freeman, (D), Texas, 1962, Gd
Glen R. Himmelberg, (D), Minnesota, 1965, Gp
William D. Johns, (D), Illinois, 1952, Gz

Research Assistant Professor:
Youqing Yang, (D), State Seis Bureau (China), 1995, Yg

Cooperating Faculty:
Stephen Stanton, Oo

University of Missouri, Columbia

Dept of Soil, Environmental & Atmospheric Sciences (B,M,D) (2007)
302 Anheuser-Busch Natural Resources Building
Columbia, MO 65211-7250
p. (573) 882-6301 f. (573) 884-5070
lupoa@missouri.edu
http://www.snr.missouri.edu/seas/
Enrollment (2006): B: 15 (15) M: 8 (8) D: 1 (1)

Professor:
Stephen H. Anderson, (D), North Carolina State, 1985, Sp

Associate Professor:
Clark J. Gantzer, (D), Minnesota, 1980, Os
Milon F. George, (D), Minnesota, 1975, Os
Anthony R. Lupo, (D), Purdue, Oa
Patrick S. Market, (D), St. Louis, 1999, Oa
Randall J. Miles, (D), Texas A&M, 1981, Sd
Peter P. Motavalli, (D), Cornell, 1989, Os

Assistant Professor:
Neil I. Fox, (D), Salford (UK), 1998, Oa
Keith W. Goyne, (D), Penn State, 2003, Sc

Adjunct Professor:
Robert J. Kremer, (D), Mississippi State, 1981, Os
E. John Sadler, (D), Texas A&M, Sp

Emeritus:
Robert W. Blanchar, (D), Minnesota, 1964, Sc
James R. Brown, (D), Iowa State, 1963, Os
Ernest C. Kung, (D), Wisconsin, 1963, Oa
Stephen E. Mudrick, (D), MIT, 1973, Oa

Other:
Michael T. Aide, (D), Mississippi State, Sc
E Eugene Alberts, (D), Purdue, 1979, Hw
Claire Baffaut, (D), Purdue, Os
J Glenn Davis, (D), Iowa State, Os
Frieda Eivazi, (D), Iowa State, 1980, Sc
Newell R. Kitchen, (D), Colorado State, 1990, Sp
Robert N. Lerch, (D), Colorado State, 1990, Os
Ronald Rinehart, (D), Colorado State, Oa
W. Gene Stevens, (D), Mississippi State, Os
Christopher K. Wikle, (D), Iowa State, Oa
John Yang, (D), Missouri, Os
Fred J. Young, (D), Missouri, Os

University of Missouri, Rolla

Dept of Geological Sciences & Eng (B,M,D) (2006)
129 McNutt Hall
Rolla, MO 65409-0410
p. (573) 341-4616 f. (573) 341-6935
rocks@umr.edu
Enrollment (2006): B: 45 (18) M: 25 (6) D: 8 (1)

Chair:
Robert C. Laudon, (D), Texas, 1975, Go

Program Head:
Norbert H. Maerz, (D), Waterloo, 1990, Nr
Daopu Numbere, (D), Oklahoma, Np
Francisca E. Oboh-Ikuenobe, (D), Cambridge (UK), 1990, Pl

Professor:
Neil L. Anderson, (D), Calgary, 1986, Ye
Jeff Cawlfield, (D), California (Berkeley), 1987, Hy

Associate Professor:
Mohamed Abdelsalam, (D), Texas (Dallas), 1993, Or
Shari Dunn-Norman, (D), Heriot-Watt, Np
Stephen Shangxing Gao, (D), California (Los Angeles), 1995, Ys
John P. Hogan, (D), Virginia Tech, 1990, Gi
Kelly Hong Liu, (D), California (Los Angeles), 1998, Ys
J David Rogers, (D), California (Berkeley), Ne
Thomas M. Whitworth, (D), Purdue, 1993, Hy
David J. Wronkiewicz, (D), New Mexico Tech, 1989, Ce

Assistant Professor:
Baojun Bai, (D), New Mexico Tech, 2005, Np
A Curt Elmore, (D), Arizona, Ge
Leslie Gertsch, (D), Colorado Mines, 1989, Nm
Lecturer:
Cheryl Seeger, (D), Missouri (Rolla), Eg
Adjunct Professor:
James Vandike, (M), South Dakota Mines, Hw
Emeritus:
David J. Barr, (D), Purdue, 1968, Or
Richard D. Hagni, (D), Missouri, Eg
Leonard Koederitz, (D), Missouri (Rolla), Np
Gerald B. Rupert, (D), Missouri (Rolla), Yg
Alfred c. Spreng, (D), Wisconsin, Pg
Don L. Warner, (D), California (Berkeley), 1964, Hw

University of Missouri-Kansas City

Dept of Geosciences (B,M,D) (2007)
5100 Rockhill Road
Room 420, Robert H. Flarsheim Hall
Kansas City, MO 64110-2499
p. (816) 235-1334 f. (816) 235-5535
geosciences@umkc.edu
http://cas.umkc.edu/geo/
Enrollment (2007): B: 85 (19) M: 13 (2) D: 11 (1)
Chair:
Syed E. Hasan, (D), Purdue, 1978, Ng
Professor:
Raymond M. Coveney, (D), Michigan, 1972, Em
Steven L. Driever, (D), Georgia, 1977, Oo
Wei Ji, (D), Connecticut, 1991, Oy
Associate Professor:
Jimmy Adegoke, (D), Penn State, 2000, Oa
Daniel P. Hopkins, (D), Louisiana State, 1987, Oo
James B. Murowchick, (D), Penn State, 1984, Cg
Tina M. Niemi, (D), Stanford, 1992, Gt
Assistant Professor:
Caroline P. Davies, (D), Arizona State, 2000, Pl
Jejung Lee, (D), Northwestern, 2001, Gq
Emeritus:
Richard J. Gentile, (D), Missouri, 1965, Gr
Edwin D. Goebel, (D), Kansas, 1966, Gg
Paul L. Hilpman, (D), Kansas, 1969, Ge
Eldon J. Parizek, (D), Iowa, 1949, Gm

Washington University

Dept of Earth & Planetary Sciences (B,M,D) (2006)
Campus Box 1169
1 Brookings Drive
(EPSC Building)
St Louis, MO 63130-4899
p. (314) 935-5610 f. (314) 935-7361
heuer@levee.wustl.edu
http://epsc.wustl.edu
Administrative Assistant: Gail Kremser
Administrative Officer: Richard Heuermann
Enrollment (2005): B: 18 (0) M: 6 (0) D: 29 (0)
Chair:
Raymond E. Arvidson, (D), Brown, 1974, Or
Director, McDonnell Center for the Space Sciences:
Roger J. Phillips, (D), California (Berkeley), 1968, Xy
Professor:
Robert E. Criss, (D), Caltech, 1981, Cs
Robert F. Dymek, (D), Caltech, 1977, Gp
M. Bruce Fegley, (D), MIT, 1980, Xc
William B. McKinnon, (D), Caltech, 1981, Xg
Jill D. Pasteris, (D), Yale, 1980, Em
Frank A. Podosek, (D), California (Berkeley), 1969, Cc
William H. Smith, (D), Princeton, 1966, Xc
Douglas A. Wiens, (D), Northwestern, 1984, Ys
Senior Scientist:
Edward Guinness, (D), Washington (St. Louis), 1980, Or
Alian Wang, (D), Sci/Tech (France), 1987, Ca
Brigitte Wopenka, (D), Tech (Vienna), 1982, Ca
Associate Professor:
Jan P. Amend, (D), California (Berkeley), 1995, Co
Viatcheslav S. Solomatov, (D), Moscow Inst of Physics & Tech, 1990, Yg
Robert D. Tucker, (D), Yale, 1985, Cc
Michael E. Wysession, (D), Northwestern, 1991, Ys
Assistant Professor:
Carrine E. Blank, (D), California (Berkeley), 2002, Oo
Jennifer R. Smith, (D), Pennsylvania, 2001, Ga
Research Scientist:
Ghassan Al-Eqabi, (D), St. Louis, 1994, Ys
Keith Bennett, (M), Washington, 1985
Joyce C. Brannon, (D), Washington, 1984, Cc
James Conder, (D), Brown, 2001, Ys
Rex A. Couture, (D), California (San Diego), 1977, Ca
John Freeman
Patrick McGuire
Zhan Peng, (D), Hawaii, 1998, Cc
Christopher Reese
Patrick J. Shore, (D), S Methodist, 1985, Gc
Rigobert Tibi, (D), GeoForschungsZentr (Potsdam), 2000, Oo
Emeritus:
Ghislaine Crozaz, (D), Libre de Bruxelles, 1967, Xm
Harold L. Levin, (D), Washington (St. Louis), 1956, Pi
Research Professor:
Anne M. Hofmeister, (D), Caltech, 1984, Gz
Ernst K. Zinner, (D), Washington (St. Louis), 1972, Xm
Research Associate Professor:
Bradley L. Jolliff, (D), SD Sch of Mines & Tech, 1987, Gx
Randy L. Korotev, (D), Wisconsin, 1976, Ct
Katharina Lodders-Fegley, (D), Max Planck, 1991, Xc
Julie Morris, (D), MIT, 1984, Cc
Postdoc:
Ryan Zeigler, (D), Washington, 2004
Earth & Planetary Sciences Librarian:
Clara P. McLeod, (M), 1965, Oo

Environmental Studies Program (B) (2007)
Box 1169
One Brookings Drive
St. Louis, MO 63130-4899
p. (314) 935-7047 f. (314) 935-7361
enstadmin@levee.wustl.edu
http://www.enst.wustl.edu
Administrative Assistant: Barbara Winston
Enrollment (2007): B: 85 (37)
Director:
Jan P. Amend, (D), California (Berkeley), Co
Professor:
Raymond E. Arvidson, (D), Brown, 1974, Oo
Pratim Briswas, (D), Caltech
Robert E. Criss, (D), Caltech, Cs
Willem H. Dickhoff, (D), Free Univ (Amsterdam)
Mike Dudukovic, (D), Illinois Inst Tech
Robert F. Dymek, (D), Caltech, Gi
Claude Evans, (D), SUNY (Stony Brook)
Bruce Fegley, (D), MIT, 1980, Xc
T.R. Kidder, (D), Harvard
Maxine I. Lipeles, (D), Harvard
William R. Lowry, (D), Stanford, 1988
Edward Macias, (D), MIT, 1970, Oa
Jill D. Pasteris, Yale, Oo

Bruce Petersen, (D), Harvard
Robert Pollak, (D), MIT
Tab Rasmussen, (D), Duke
Barbara Schaal, (D), Yale, 1974
Glenn D. Stone, (D), Arizona, 1988
Robert W. Sussman, (D), Duke, 1972
Alan R. Templeton, (D), Michigan, 1972

Associate Professor:
Richard Axelbaum, (D), California (Davis)
Jon M. Chase, (D), Chicago
Clare Palmer, (D)
Jay Turner, (D), Washington (St. Louis)

Assistant Professor:
Lars Angenent, (D), Iowa State
Geoff Childs, (D), Indiana
Ellen Damschen, (D), North Carolina State
Daniel Giammar, (D)
Tiffany Knight, (D), Pittsburgh
Ken Olsen, (D), Washington (St. Louis)
John Orrock, (D), Iowa State
Jen R. Smith, (D), Pennsylvania, Ga
Jane Wolff, (D), Harvard

Engineering & Science Director:
Beth Martin, (M), Washington (St. Louis), 1996

Emeritus:
Owen J. Sexton, Michigan, Ooo

Geology Librarian:
Clara McLeod

William Jewell College

Dept of Biology (2004)
Liberty, MO 64068
p. (816) 781-3806 (Ext. 230)
heruthd@william.jewell.edu

Associate Professor:
Charles F. J. Newlon, (M), Missouri, 1962, Og

Montana

Flathead Valley Community College

Math and Science Div (2004)
777 Grandview Drive
Kalispell, MT 59901
p. (406) 752-5222 (Ext. 290) f. (406) 755-4044
pmartino@fvcc.edu

Chair:
William A. Blood, (D), Colorado Mines, 1968, Em

Instructor:
Janette Y. Black, (M), Montana State, 1984, Gg
Cara J. Hodge, (M), Colorado State, 1982, Eg

Montana State University

Dept of Earth Sciences (B,M,D) (2008)
P.O. Box 173480
Bozeman, MT 59717-3480
p. (406) 994-3331 f. (406) 994-6923
scuster@montana.edu
http://www.montana.edu/wwwes/
Enrollment (2007): B: 210 (24) M: 48 (7) D: 12 (2)

Head:
David R. Lageson, (D), Wyoming, 1980, Gc

Professor:
Katherine J. Hansen, (D), Colorado, 1981, Oa
William W. Locke, (D), Colorado, 1980, Gm
David W. Mogk, (D), Washington, 1984, Gp
Cathy Whitlock, (D), Washington, 1983, Pey
William K. Wyckoff, (D), Syracuse, 1982, Oo

Associate Professor:
Joseph M. Ashley, Jr., (D), Colorado, 1971, Oy
Stephan G. Custer, (D), Montana, 1976, Hg
Michael Gardner, (D)
Jian-yi Liu, (D), Minnesota, 1992, Oo
James G. Schmitt, (D), Wyoming, 1982, Gr

Assistant Professor:
Todd C. Feeley, (D), California (Los Angeles), 1993, Gi
Mark L. Skidmore, (D), Alberta, 2001, ClPyOy
David J. Varricchio, (D), Montana State University, 1995, Pv

Adjunct Assistant Professor:
Stuart Challender, (M), Utah State University, 1986, Oi

Adjunct Assistant Professor:
Frankie Jackson, (D), Montana State University, 2007, Pv
Colin Shaw, (D), New Mexico, 2001, GcEmGp

Adjunct Professor:
David Bowen, (D), Colorado, 2001, Gd
John R. Horner, Pv

Emeritus:
John Montagne, (D), Wyoming, 1955, Gl

Dept of Land Resources & Env Sciences (B,M,D) (2004)
334 Leon Johnson Hall
P.O. Box 173120
Bozeman, MT 59717-3120
p. (406) 994-7060 f. (406) 994-3933
jefj@montana.edu
Department Secretary: Peggy Humphrey
Enrollment (2000): B: 93 (13) M: 25 (2) D: 13 (0)

Professor:
James W. Bauder, (D), Utah State, 1974, Sf
Lisa J. Graumlich, (D), Washington, 1985, Ou
William P. Inskeep, (D), Minnesota, 1985, Sc
Jeffrey S. Jacobsen, (D), Oklahoma State, 1985, So
Gerald A. Nielsen, (D), Wisconsin, 1963, Sd
David M. Ward, (D), Wisconsin, 1975, Oo

Senior Scientist:
Dennis R. Neuman, (M), Montana State, 1972, Oo

Associate Professor:
Richard E. Engel, (D), Minnesota, 1983, Sc
Cliff Montagne, (D), Montana State, 1976, Sd
Jon M. Wraith, (D), Utah State, 1989, Sp

Assistant Professor:
Paul B. Hook, (D), Colorado State, 1992, Sf
Rick L. Lawrence, (D), Oregon State, 1998, Or

Adjunct Professor:
Douglas J. Dollhopf, (D), Montana State, 1975, Oo

Montana State University, Billings

Dept of Biological & Physical Sciences (B) (2006)
1500 N. 30th Street
Billings, MT 59101
p. (406) 657-2341 (Ext. 2028)
tzwick@msubillings.edu
Enrollment: No data reported since 1995

Dean:
Tasneem Khaleel, (D), Bangalore (India), 1970, Po

Professor:
Matt Benacquista, (D), Montana State, 1989, Xy
Thomas T. Zwick, (D), N Colorado, 1977, Oe

Montana Tech of the University of Montana

Dept of Chemistry & Geochemistry (B,M) (2004)
1300 West Park Street
Butte, MT 59701-8997
p. (406) 496-4207 f. (406) 496-4135
dhobbs@mtech.edu
Department Secretary: Wilma Immonen

Enrollment (1998): M: 6 (4)
Chair:
Douglas A. Coe, (D), Oregon State, 1974, Cg
Professor:
Douglas A. Drew, (D), Wyoming, 1971, Cg
Donald Stierle, (D), California (Riverside), 1979, Co
Assistant Professor:
Douglas Cameron, (D), Purdue, 1979, Ca
John D. Hobbs, (D), New Mexico, 1991, Oo
Research Associate:
Wayne Olmsted, (B), Montana State, 1962, Ca
Andrea Stierle, (D), Montana State, Oo
Instructor:
Stephen R. Parker, (M), Indiana, 1972, Oo
Emeritus:
Frank E. Diebold, (D), Colorado Mines, 1967, Cl
Alexis Volborth, (D), Helsinki, 1954, Ca

Dept of Geological Engineering (B,M) (2007)
Montana School of Mines & Engineering
1300 West Park Street
Butte, MT 59701
p. (406) 496-4262 f. (406) 496-4260
dwolfgram@mtech.edu
http://www.mtech.edu/geo_eng
Enrollment (2004): B: 33 (0) M: 12 (0)
Professor:
Diane Wolfgram, (D), California (Berkeley), 1977, Emo
Professor:
Christopher H. Gammons, (D), Penn State, 1988, Cg
Mary M. MacLaughlin, (D), California (Berkeley), 1997, Nr
Mark A. Sholes, (D), Texas, 1978, Ec
Willis D. Weight, (D), Wyoming, 1989, HwGq
Adjunct Assistant Professor:
Colleen G. Elliott, (D), New Brunswick, 1988, Gc

Geophysical Engineering (B,M) (2007)
1300 West Park Street
Butte, MT 59701
p. (406) 496-4401 f. (406) 496-4704
mspeece@mtech.edu
http://www.mtech.edu/mines/geophysical
Enrollment (2006): B: 25 (2) M: 7 (6)
Professor:
Curtis A. Link, (D), Houston, 1993, Ye
Marvin A. Speece, (D), Wyoming, 1992, Ye
Assistant Professor:
Leif Cox, (D), Utah, 2007, Ye
Xiaobing Zhou, (D), Alaska (Fairbanks), 2002, Or

Rocky Mountain College
Dept of Geology (B) (2006)
1511 Poly Drive
Billings, MT 59102
p. (406) 657-1101 f. (406) 259-9751
kalakayt@rocky.edu
http://geology.rocky.edu
Enrollment (2005): B: 11 (2)
Chair:
Thomas J. Kalakay, (D), Wyoming, 2001, Gc
Assistant Professor:
Lawrence S. Jones, (D), Wyoming, 1996, Gs

University of Montana
Dept of Geology (B,M,D) (2005)
32 Campus Drive #1296
Missoula, MT 59812-1296
p. (406) 243-2341 f. (406) 243-4028
gl_sds@selway.umt.edu
http://www.umt.edu/geology/default.htm
Administrative Assistant: Christine Foster
Administrative Assistant: Loreene Skeel
Enrollment (2005): B: 60 (15) M: 30 (10) D: 10 (0)
Chair:
Steven D. Sheriff, (D), Wyoming, 1981, Ym
Professor:
Johnnie N. Moore, (D), California (Los Angeles), 1976, Gs
James W. Sears, (D), Queen's, 1979, Gt
George D. Stanley, Jr., (D), Kansas, 1977, Pi
William W. Woessner, (D), Wisconsin, 1978, Hw
Associate Professor:
Marc S. Hendrix, (D), Stanford, 1992, Gs
Nancy W. Hinman, (D), California (San Diego), 1987, Co
Assistant Professor:
Julia Baldwin, (D), MIT, Gp
Rebecca Bendick, (D), Colorado, Yg
Manny Gabet, (D), California (Santa Barbara), Gm
Joel Harper, (D), Wyoming, Gl
Cooperating Faculty:
Christine M. Brick, (D), Montana, 1996, Cl

University of Montana, Western
Dept of Environmental Sciences (B) (2006)
710 South Atlantic Street
Dillon, MT 59725-3598
p. (406) 683-7017 f. (406) 683-7493
s_roberts@umwestern.edu
http://www.umwestern.edu/envirosci/
Administrative Assistant: Thomas G. Satterly
Enrollment (2002): B: 144 (0)
Chair:
Sheila M. Roberts, (D), Calgary, 1996, Ge
Professor:
R. Stephen Mock, (D), Montana State, 1989, Ca
Robert C. Thomas, (D), Washington, 1993, Gs
Craig E. Zaspel, (D), Montana State, 1975, Yg
Associate Professor:
Andrea L. Easter-Pilcher, (D), Montana State, 1993, Py
Delena Norris-Tull, (D), Texas, 1990, Oe
Assistant Professor:
Richard E. Clark, (D), Brigham Young, 1975, Oo
Eric G. Dyreson, (D), Arizona, 1997, Gq
Instructor:
Nils Troedsson, (M), Montana, 1960, Oo
Director, Math/Science Upward Bound:
Brian K. Pilcher, (M), Louisiana State, 1981, Sf
Adjunct Professor:
Peter Bengeyfield, (M), West Virginia, 1969, Hg
Bryce A. Bohn, (M), Utah State, 1998, Sf
Emily M. Geraghty, (M), Washington State, 2002, Gc
Thomas J. Kalakay, (D), Wyoming, Gc
Marvin E. Kauffman, (D), Princeton, 1960, Gr

Nebraska

Bellevue University
Dept of Natural Sciences (B) (2004)
1000 Galvin Road South
Bellevue, NE 68005-3098
p. (402) 293-3790 f. (402) 293-2023
jackie@bellevue.edu
Enrollment (1997): B: 25 (1)

Chadron State College
Dept of Geociences (B) (2005)

Chadron, NE 69337
p. (308) 432-6377 f. (308) 432-6434
mleite@csc.edu
http://www.csc.edu/geoscience
Administrative Assistant: Carly Schmitt
Enrollment (2005): B: 6 (4)

Professor:
Michael B. Leite, (D), Wyoming, 1992, Gg

Creighton University

Dept of Atmospheric Sciences (B,M) (2006)
2500 California Plaza
Omaha, NE 68178
p. (402) 280-2641 f. (402) 280-1731
sonora@creighton.edu
Enrollment (2006): B: 24 (5) M: 5 (2)

Chair:
Arthur V. Douglas, (D), Arizona, 1976, Og

Assistant Professor:
Jason T. Martinelli, (D), St. Louis, 2004, Ow
Jon M. Schrage, (D), Purdue, 1998, Ow

Adjunct Professor:
Steven Byrd, (M), St. Thomas, Yg
John Eylander, (M), South Dakota Mines, Ow
Bryon C. Miller, (M), Creighton, 1997, Oa
Richard Ritz, (M), Texas A&M, Ow

University of Nebraska at Omaha

Dept of Geography and Geology (B,M) (2006)
6001 Dodge Street
DSC 260
Omaha, NE 68182-0199
p. (402) 554-2662 f. (402) 554-3518
jpeake@mail.unomaha.edu
http://maps.unomaha.edu/geo/
Administrative Assistant: Brenda Todd
Enrollment (2006): B: 31 (4)

Professor:
Michael P. Bishop, (D), Indiana State, 1987, Or
George F. Engelmann, (D), Columbia, 1978, Pv
Harmon D. Maher, Jr., (D), Wisconsin, 1984, Gc
John F. Shroder, Jr., (D), Utah, 1967, Gm

Associate Professor:
Jeffrey S. Peake, (D), Louisiana State, 1977, Oy
Robert D. Shuster, (D), Kansas, 1985, Gi

University of Nebraska, Kearney

Dept of Geography & Earth Science (B) (2004)
Kearney, NE 68849
p. (308) 865-8355 f. (308) 865-8980
darts@unk.edu
Enrollment (2002): B: 165 (0)

Chair:
Gordon E. Bennett, (M), San Diego State, 1972, Oy

Professor:
Marvin Stone, (D), Kansas, 1978, Oe

Associate Professor:
Steele Becker, (M), Indiana, 1965, Oy
Stan Dart, (M), Rutgers, 1969, Oy
Robert D. Lind, (M), N Colorado, 1968, Oo

Assistant Professor:
Laurence C. Becker, (D), London, 1995, Oo

University of Nebraska, Lincoln

Dept of Geosciences (B,M,D) (2007)
214 Bessey Hall
Lincoln, NE 68588-0340
p. (402) 472-2663 f. (402) 472-4917
dwatkins1@unl.edu
http://www.geosciences.unl.edu
Staff Assistant: Janelle Bird
Staff Assistant: Toni Penton
Enrollment (1997): B: 280 (59) M: 15 (15) D: 33 (5)

Chair:
David K. Watkins, (D), Florida State, 1984, Pm

Professor:
Christopher R. Fielding, (D), Durham (UK), 1982, Gs
Sherilyn C. Fritz, (D), Minnesota, 1985, Gn
Ronald J. Goble, (D), Queens, 1977, GzCc
Priscilla C. Grew, (D), California (Berkeley), 1967, Gg
David M. Harwood, (D), Ohio State, 1986, Pm
Robert M. Hunt, (D), Columbia, 1971, Pv
Merlin P. Lawson, (D), Clark, 1973, Oy
Nancy Lindsley-Griffin, (D), California (Davis), 1982, Gt
David B. Loope, (D), Wyoming, 1981, Gs
Robert J. Oglesby, (D), Yale, 1990, Oa
Darryll T. Pederson, (D), North Dakota, 1971, Hw
Norman D. Smith, (D), Brown, 1967, Gs
James B. Swinehart, (M), Nebraska, 1979, Gr
Vitaly A. Zlotnik, (D), Moscow, 1979, Hq

Associate Professor:
Mark R. Anderson, (D), Colorado, 1985, Oa
Tracy D. Frank, (D), Michigan, 1996, CsGdCm
Qi S. Hu, (D), Colorado State, 1992, Oa
R. Matt Joeckel, (D), Iowa, 1993, Gr
Richard M. Kettler, (D), Michigan, 1990, Cl
John D. Lenters, (D), Cornell, 1997, Oa
Frank Rack, (D), Texas A&M, 1992, Ou
Clinton M. Rowe, (D), Delaware, 1988, Oa

Assistant Professor:
Adam L. Houston, (D), Illinois, 2004, Oa
Erkan Istanbulluoglu, (D), Utah State, 2003, Hq
Durelle T. Scott, (D), Colorado, 2001, Cl
Jun Wang, (D), Alabama (Huntsville), 2006, Oa

Research Associate Professor:
Mary Anne Holmes, (D), Florida State, 1989, Gs

Adjunct Professor:
John R. Griffin, (D), California, 1973, Eo

Emeritus:
Samuel B. Treves, (D), Ohio State, 1959, Gi
Michael R. Voorhies, (D), Wyoming, 1966, Pv
William Wayne, (D), Indiana, 1943, Gm

School of Natural Resource Sciences (M,D) (2004)
236 L.W. Chase Hall
Lincoln, NE 68583-0728
p. (402) 472-3679 f. (402) 472-6614
msailors2@unl.edu
Department Secretary: Elsie E. Connelly
Enrollment (1995): M: 8 (2) D: 16 (2)

Director:
Blaine L. Blad, (D), Minnesota, 1970, Oa

Professor:
Kenneth G. Hubbard, (D), Utah State, 1981, Oa
Shashi B. Verma, (D), Colorado State, 1971, Oa
Albert Weiss, (D), Cornell, 1975, Oa
Donald A. Wilhite, (D), Nebraska, 1975, Oa

Associate Professor:
Elizabeth A. Walter-Shea, (D), Nebraska, 1987, Oa

Assistant Professor:
Michael J. Hayes, (D), Missouri, 1994, Oa
Steven J. Meyer, (D), Nebraska, 1990, Oa
David E. Stooksbury, (D), Virginia, 1992, Oa

Nevada

Desert Research Institute

Earth & Ecosystems Science Div (2007)
2215 Raggio Parkway
Reno, NV 89512-1095
p. (775) 673-7300 f. (775) 673-7485
bj@dri.edu

Executive Director:
Michael Auerbach, (D), Florida State, 1982, Py

Professor:
John Arnone, (D), Yale, 1988, Py
Colleen M. Beck, (D), California (Berkeley), 1979, Sa
Nicholas Lancaster, (D), Cambridge, 1977, Gm
Eric McDonald, (D), New Mexico, 1994, Sd
David E. Rhode, (D), Washington, 1987, Pe

Staff Remote Sensing GIS Scientist:
Christopher Kratt, (M), Nevada (Reno), 2005, Or

Staff Geomorphologist:
Sophie Baker, (M), Dalhousie, 2005, Gm
Graham Dalldorf, (M), Oregon, 2003, Gma

Assistant Research Soil Scientist:
Todd G. Caldwell, (M), Nevada (Reno), 1999, Hw

Assistant Research Geomorphologist:
Steven N. Bacon, (M), Humboldt State, 2003, Gm

Associate Professor:
Kenneth D. Adams, (D), Nevada (Reno), 1997, Gm
Mary Cablk, (D), Oregon State, 1997, Or
Lynn Fenstermaker, (D), Nevada, 2003, Or
Christian H. Fritsen, (D), S California, 1996, Py
Giles Marion, (D), California (Berkeley), 1974, Sc
Kenneth C. McGwire, (D), California (Santa Barbara), 1992, Oy
David A. Mouat, (D), Oregon State, 1974, Ge
Alison E. Murray, (D), California (Santa Barbara), 1998, Py
Paul S. J. Verburg, (D), Wageningen Ag, 1998, Sb

Assistant Professor:
Thomas F. Bullard, (D), New Mexico, 1995, Gm
Donald E. Sabol, Jr, (D), Washington, 1991, Or
Stephen F. Zitzer, (D), Illinois, 1990, Sf

GIS/Remote Sensing Scientist:
Timothy B. Minor, (M), California (Santa Barbara), 1982, Or

University of Nevada, Las Vegas

Geoscience Dept (B,M,D) (2007)
4505 Maryland Parkway
Box 454010
Las Vegas, NV 89154-4010
p. (702) 895-3262 f. (702) 895-4064
geodept@unlv.edu
geoscience.unlv.edu
Administrative Assistant: Linda Robison
Enrollment (2006): B: 88 (0) M: 30 (0) D: 20 (0)

Chair:
Frederick W. Bachhuber, (D), New Mexico, 1971, Pl

Professor:
Jean S. Cline, (D), Virginia Tech, 1990, Em
David K. Kreamer, (D), Arizona, 1982, Hw
Margaret N. Rees, (D), Kansas, 1984, Gs
Stephen Rowland, (D), California (Santa Cruz), 1978, Pi
Eugene I. Smith, (D), New Mexico, 1970, Gi
David L. Weide, (D), California (Los Angeles), 1974, Gm

Senior Scientist:
Clay Crow, (D), New Mexico Tech, 1988, Ca

Associate Professor:
Rodney V. Metcalf, (D), New Mexico, 1990, Gp
Wanda J. Taylor, (D), Utah, 1989, Gc
Michael L. Wells, (D), Cornell, 1991, Gc

Assistant Professor:
Brenda J. Buck, (D), New Mexico State, 1996, Sd
Andrew D. Hanson, (D), Stanford, 1998, Co
Elizabeth Jacobson, (D), Arizona, 1984, Hg
Richard L. Orndorff, (D), Kent State, 1994, Gq
Terry L. Spell, (D), SUNY (Albany), 1991, Cc

Research Associate:
Nicholas Wilson, (D), Dalhousie, 1998, Og
Kathleen Zanetti, (M), Idaho, 1997, Gg

Visiting Professor:
Karem Azmy, (D), Ottawa, 1997, Gd

Cartographer:
Nathan Stout, (B), Nevada, 1982, Oo

University of Nevada, Reno

Mackay School of Earth Sci and Eng Director's Office (2008)
Mail Stop 0168
Reno, NV 89557-0168
p. (775) 784-6987 f. (775) 784-1766
ewaday@unr.edu
http://www.mines.unr.edu/Mackay/
Administrative Assistant: Melodie Gander
Administrative Assistant: Erika Waday

Director:
James V. Taranik, (D), Colorado Mines, 1975, Ye

Coordinator, Mackay Student Services:
Elizabeth Ball, (M), Nevada (Reno), 1997, Oo
Leslie K. Rumph, (M), Nevada (Reno), 1997, Oo

Administrator, W.M. Keck Museum:
Rachel Dolbier, (M), Nevada (Reno), 2001, Gg

Dept of Geography (B,M,D) (2007)
Mail Stop 0154
Reno, NV 89557-0154
p. (775) 784-6995 f. (775) 784-1058
smensing@unr.edu
http://www.unr.edu/geography/
Administrative Assistant: Shari Baughman
Enrollment: No data reported since 1995

Chair:
Scott A. Mensing, (D), California (Berkeley), 1993, Oy

Professor:
Gary J. Hausladen, (D), Syracuse, 1983, Oy
Paul F. Starrs, (D), California (Berkeley), 1989, Oy

Associate Professor:
Franco Biondi, (D), Arizona, 1994, Sf

Assistant Professor:
Scott Bassett, (D), Harvard, 2001, Ou
Kate Berry, (D), Colorado, 1993, Oy
Jill Heaton, (D), Oregon State, 2001, Oi
Jeffrey Underwood, (D), Georgia, 1999, Ow

Research Associate:
Scotty Strachan, (M), Nevada (Reno), 2001

Instructor:
Patrick F. Guiberson, (M), Nebraska, 1997, Oy

Adjunct Professor:
Douglas Duer, (D), Louisiana State, 2000, Oo
Richard A. Eigenheer, (D), California (Davis), 1976
Jordan Hastings, (D), California (Santa Barbara)
Anne Macfarlane, (M), Nevada (Reno), 2001, Ou
Victoria Randlett, (D), California (Berkeley), 1999
Christopher Ryan, (M), Nevada (Reno), 1998, Oy
Peter Wigand, (D), Washington State, 1985, Pe

Dept of Geological Sci and Eng (B,M,D) (2007)
Mail Stop 0172
Reno, NV 89557-0172
p. (775) 784-6050 f. (775) 784-1833

Nevada

gina@hydro.unr.edu
http://mines.unr.edu/geology/
Administrative Assistant: Cheryl Breland
Administrative Assistant: Ingrid Ramos
Enrollment: No data reported since 1995

Chair:
Regina Tempel, (D), Colorado Mines, 1993, Cl
Research Professor:
Patricia H. Cashman, (D), Southern California, 1979, Gc
Regent Professor:
James V. Taranik, (D), Colorado Mines, 1975, Ye
Professor:
John Anderson, (D), Columbia, 1976, Ys
Greg B. Arehart, (D), Michigan, 1992, Eg
James R. Carr, (D), Arizona, 1983, Ng
Robert Karlin, (D), Oregon State, 1984, Ou
John N. Louie, (D), Caltech, 1987, Ye
Richard A. Schultz, (D), Purdue, 1987, Nr
Richard A. Schweickert, (D), Stanford, 1972, Gt
Tommy B. Thompson, (D), New Mexico, 1966, Eg
James H. Trexler, (D), Washington, 1984, Gs
Robert J. Watters, (D), Imperial Coll (UK), 1972, Nr
Steven G. Wesnousky, (D), Columbia, 1982, Ys
Stephen W. Wheatcraft, (D), Hawaii, 1979, Hq
Research Associate Professor:
Gary Oppliger, (D), California (Berkeley), 1982, Ye
Simon R. Poulson, (D), Penn State, 1990, Cs
Associate Professor:
Wendy M. Calvin, (D), Colorado, 1991, Or
Paula J. Noble, (D), Texas, 1993, Pm
Emeritus:
Frank Dickson, (D), California (Los Angeles), 1955, Gg
Malcolm J. Hibbard, (D), Washington, 1962, Gi
Liang-Chi Hsu, (D), California (Los Angeles), 1966, Cp
Lawrence T. Larson, (D), Wisconsin, 1962, Eg
Joseph Lintz, (D)
Donald C. Noble, (D), California (Berkeley), 1953, Ge
Post-Doctoral Scholar:
Christine Hatch, (D), California (Santa Cruz), 2007, Gg
Manoj Menon, (D), Swiss Fed Inst Tech, 2006, Gg
Yasuhiro Oba, (D), Okayama, 2007
Manager, Electron Microbeam Facility:
John K. McCormack, (D), Nevada (Reno), 1997, Gz
Cooperating Faculty:
John W. Bell, (M), Arizona State, 1974, Ng
Glenn W. Berger, (D), Toronto, 1973, Gg
Geoffrey Blewitt, (D), Caltech, 1986, Yd
Tom Bullard, (D), New Mexico, 1995, Gm
James E. Faulds, (D), New Mexico, 1989, Gc
Nick Lancaster, (D), Chamberlain, 1977, Gm
Greg Pohll, (D), Nevada (Reno), Ca
Jonathan G. Price, (D), California (Berkeley), 1977, Eg
Lisa A. Shevenell, (D), Nevada (Reno), 1990, Hw
Lisa Stillings, (D), Penn State, 1994, Hw
Scott W. Tyler, (D), Nevada (Reno), 1990, Hg
Alan R. Wallace, (D), Oregon State, 1983, Eg
Peter Wigand, (D), Washington State, 1985, Pl

Department of Mining Engineering (B,M,D) (2008)
Mail Stop 0173
Reno, NV 89557-0173
p. (775) 784-6961 f. (775) 784-1833
cscott@unr.edu
http://www.unr.edu/mines/mine-eng
Administrative Assistant: Carla Scott
Enrollment (2007): B: 4 (2) M: 1 (1)

Acting Chair:
Danny L. Taylor, (D), Colorado Mines, 1980, Nm
Professor:
Jaak Daemen, (D), Minnesota, 1975, Nr
George Danko, (D), Budapest, 1985, Nm
Pierre Mousset-Jones, (D), London, 1988, Nm
Associate Professor:
Kathleen A. Altman, (D), Nevada (Reno), 1999, Nx
Development Technician:
John D. Leland, (M), Stanford, 1983

Nevada Bureau of Mines and Geology (2006)
Mail Stop 178
Reno, NV 89557-0088
p. (775) 784-6691 f. (775) 784-1709
jprice@unr.edu
http://www.nbmg.unr.edu/
Administrative Assistant: Terri Garside

Director:
Jonathan G. Price, (D), California (Berkeley), 1977, Gg
Senior Scientist:
John W. Bell, (M), Arizona State, 1974, Ng
Geoffrey Blewitt, (D), Caltech, 1996, Yd
Stephen B. Castor, (D), Nevada (Reno), 1972, Eg
Craig M. de Polo, (D), Nevada (Reno), 1998, Ng
James E. Faulds, (D), New Mexico, 1989, Gc
Larry J. Garside, (M), Nevada (Reno), 1968, Gg
Christopher D. Henry, (D), Texas, 1975, Gg
P. Kyle House, (D), Arizona, 1996, Gm
Paul J. Lechler, (D), Nevada (Reno), 1995, Cg
John Muntean, (D), Stanford, 1998
Hans-Peter Plag, (D), Free (Berlin), 1988, Yd
Alan R. Ramelli, (M), Nevada (Reno), 1988, Ng
Lisa Shevenell, (D), Nevada (Reno), 1990, Hw
Research Assistant Professor:
William Hammond, (D), Oregon (Eugene), 2000, Yd
Cornelis Kreemer, (D), SUNY (Stony Brook), 2001, Yd
GIS Supervisor:
Ronald H. Hess, (M), Nevada (Reno), 1998, Oy
Geologic Information Specialist:
David A. Davis, (M), Nevada (Reno), 1990, Gg
Assistant Geochemist:
Mario O. Desilets, (M), Nevada (Reno), 1992, Cg
Research Associate:
Daphne D. LaPointe, (M), Montana, 1977, Gg
Publications Manager:
Susan L. Tingley, (B), California (Los Angeles), 1966, Oy
Emeritus:
Harold F. Bonham, Jr., (M), Nevada (Reno), 1963, Em
John W. Erwin, (M), Yg
Liang-Chi Hsu, (D), California (Los Angeles), 1966, Cp
Joseph V. Tingley, (M), Nevada (Reno), 1963, Em
Publications Manager:
Elizabeth Crouse, (B), Kansas, 1996, Gg

Nevada Seismological Lab (2007)
Mail Stop 0174
Reno, NV 89557-0174
p. (775) 784-4975 f. (775) 784-4165
mainofc@seismo.unr.edu
http://www.seismo.unr.edu
Administrative Assistant: Lori McClelland
Administrative Assistant: Erik Williams

Director:
John G. Anderson, (D), Columbia, 1976, Ys
Associate Director:
John N. Louie, (D), Caltech, 1987, Ye
Professor:
James N. Brune, (D), Columbia, 1961, Ys
Steve Wesnousky, (D), Columbia, 1982, Ys

Seismic Network Manager & Development Director:
Ken D. Smith, (D), Nevada (Reno), 1991, Ys
Research Associate Professor:
Rasool Anooshehpoor, (D), California (San Diego), 1988, Ys
Glenn P. Biasi, (D), Oregon, 1994, Ys
Ileana Tibuleac, (D), Southern Methodist, 1999, Ys
Research Assistant Professor:
Matthew Purvance, (D), Nevada (Reno), 2005, Ys
Adjunct Research Assistant Professor:
Satish Pullammanappallil
Adjunct Research Associate:
Bill Honjas
Emeritus Seismic Network Manager:
David von Seggern, (D), Penn State, 1982, Ys
Emeritus Research Design Engineer:
Wally Nicks, (B), Nevada (Reno), 1983
Post-Doctoral Scholar:
Aasha Pancha, (D), Nevada (Reno), 2007, Ys
Leiph Preston, (D), Washington, 2003, Ys
James Scott
Seismic Systems Analyst:
David Slater, (B), Calgary, 1990
Seismic Systems Administrator:
Carol Freinkel, (M), Nevada (Reno), 2003
Seismic Records Technician:
Arturo Aburto
Tom Rennie
Network Seismologist:
Diane dePolo, (M), Nevada (Reno), 1989, Ys
GIS Specialist & Webmaster:
Yui Miyata, (B), Nevada (Reno), 2003
Development Technician:
Nathan Edwards
Austin Wilson
Associate Engineer:
John Torrisi

Center for Neotectonic Studies (2004)
MS 169
Reno, NV 89557-0135
p. (702) 784-1382 f. (702) 784-1382
Director:
Steven G. Wesnousky, (D), Columbia, 1982, Ys

Center for Research in Economic Geology (2008)
Mail Stop 0169
Reno, NV 89557-0169
p. (775) 784-1382 f. (775) 682-7241
shelley@mines.unr.edu
Administrative Assistant: Shelley Harvey
Director:
Tommy Thompson, (D), New Mexico, 1966, Eg

Great Basin Center for Geothermal Energy (2008)
Mail Stop 0172
Reno, NV 89557-0172
p. (775) 784-7018 f. (775) 327-5801
geothermal@mines.unr.edu
http://www.unr.edu/geothermal/index.html
Director:
Lisa A. Shevenell, (D), Nevada (Reno), 1990, Hw
Associate Director:
Kwang Kim, (D), Arizona State, 1992, Oo
Research Professor:
Geoffrey Blewitt, (D), Caltech, 1986, Yd
Research Geologist:
James Faulds, (D), New Mexico, 1989, Gc
Chief Geochemist:
Paul Lechler, (D), Nevada (Reno), 1995, Cg
Research Associate Professor:
Gary Oppliger, (D), California (Berkeley), 1982, Ye
Associate Professor:
Wendy Calvin, (D), Colorado, 1991, Or
John Louie, (D), Caltech, 1987, Ye
Research Assistant Professor:
Mark Coolbaugh, (D), Nevada (Reno), 2003, Gg
Cornelis Kreemer, SUNY (Stony Brook), 2001, Yd
Research Scientist/GIS:
Richard Zehner, (M), Montana, 1987, Gg
Staff Research Associate:
Robin Penfield
Chris Sladek, (B), Colorado State, 1994, Gg
Program Officer:
Peggy Brown

Graduate Program of Hydrologic Sciences (M,D) (2008)
MS 0175
Reno, NV 89557-0175
p. (775) 784-6469 f. (775) 784-1953
hydro@unr.edu
http://www.hydro.unr.edu
Administrative Assistant: Pamela Love
Enrollment (1997): M: 43 (12) D: 25 (5)
Director:
John Warwick, (D), Penn State, 1983, Hs
Professor:
John C. Guitjens, (D), California (Davis), 1968, Hq
Dale W. Johnson, (D), Washington, 1975, Sb
Glenn C. Miller, (D), California (Davis), 1977, Hg
Wally Miller, (D), California (Riverside), 1973, Hg
Rangesan Narayanan, (D), Utah State, 1976, Hg
Scott Tyler, (D), Nevada (Reno), 1990, Hg
Steve Wheatcraft, (D), Hawaii, 1979, Hw
Research Hydrologist:
Brian Andraski
Associate Research Professor:
Ahmed Hassan, (D), Purdue, 1996, Oo
Associate Professor:
George Danko, (D), Hungarian Acad of Sci, 1985, Hq
George C. J. Fernandez, (D), Texas A&M, 1985, Hg
Robert Karlin, (D), Oregon State, 1984, Hg
John Louie, (D), Caltech, 1987, Yg
Gary Norris, (D), California (Berkeley), 1977, Hs
Raj Siddarthan, (D), British Columbia, 1984, Hs
Sherman Swanson, (D), Oregon State, 1983, Hs
Gina Tempel, (D), Colorado Mines, 1993, Cl
Associate Scientist:
Sue Donaldson, (D), Nevada (Reno), 1994, Cg
Lisa Shevenell, (D), Nevada (Reno), 1990, Hw
Assistant Research Professor:
Kumud Acharya, (D), Saitama University, Oo
Kenneth Adams, (D), Nevada (Reno), 1997, Gam
Assistant Professor:
Keith E. Dennett, (D), Georgia Tech, 1995, Hs
Kyle House, (D), Arizona, 1996, Gm
Roger Jacobson, (D), Penn State, 1973, Cg
Eric McDonald, (D), New Mexico, 1994, Sp
Greg Pohll, (D), Nevada (Reno), 1996, Hg
Jonathan G. Price, (D), California (Berkeley), 1977, Cg
Robert G. Qualls, (D), Georgia, 1989, Oo
Mark Walker, (D), Cornell, 1998, Oo
Steve G. Wells, (D), Cincinnati, 1976, Oo
Research Associate:
Dean Adams, (D), Utah State, 1972, Oo
Charalambos Papelis, (D), Stanford, 1992, Cl

Lisa Stillings, (D), Penn State, 1994, Cm
Assistant Research Professor:
John J. Arnone, (D), Yale, 1988, Sp
Tom Bullard, (D), New Mexico, 1995, Gm
Gayle L. Dana, (D), Nevada (Reno), 1997, Or
Jennifer Duan, (D), Mississippi, 1998, Hs
Christian H. Fritsen, (D), S California, 1995, Og
Joseph McConnell, (D), Arizona, 1997, Oo
Michael Young, (D), Arizona, 1995, Hg
Adjunct Professor:
Jeanne Chambers, (D), Utah State, 1987, Py
Jim Thomas, (D), Nevada (Reno), 1996, Cg
Assistant Director, Biological Resources Research Center:
Mary Peacock, (D), Arizona State, 1995, Oo
Cooperating Faculty:
Nick Lancaster, (D), Cambridge, 1977, Gm

New Hampshire

Dartmouth College

Dept of Earth Sciences (B,M,D) (2007)
228 Fairchild Bldg, HB 6105
Hanover, NH 03755
p. (603) 646-2373 f. (603) 646-3922
earth.sciences@dartmouth.edu
http://www.dartmouth.edu/~earthsci
Department Administrator: Jodie Davi
Enrollment (2007): B: 31 (19) M: 11 (6) D: 10 (4)

Chair:
Carl E. Renshaw, (D), Stanford, 1993, Hg
Professor:
James L. Aronson, (D), Caltech, 1966, Cc
Xiahong Feng, (D), Case Western, 1991, Cs
Associate Professor:
William Brian Dade, (D)
Mukul Sharma, (D), Rochester, Ge
Leslie J. Sonder, (D), Harvard, 1986, Gq
Assistant Professor:
Benjamin C. Bostick, (D), Stanford, 2001
Robert L. Hawley, (D), Washington, 2005, Gl
James H. Scott, (D), Wisconsin (Milwaukee), 1999, Py
Adjunct Professor:
Gary D. Johnson, (D), Iowa State, 1971, Gr
Emeritus:
Richard W. Birnie, (D), Harvard, 1975, Gz
Research Professor:
Brian P. Jackson, (D), Georgia, 1998, Ca

Keene State College

Dept of Geology (B) (2006)
Mail Stop 2001
229 Main Street
Keene, NH 03435-2001
p. (603) 358-2553 f. (603) 358-2257
pnielsen@keene.edu
http://kilburn.keene.edu/
Enrollment (2006): B: 20 (3)

Chair:
Peter A. Nielsen, (D), Alberta, 1977, Gx
Professor:
Timothy T. Allen, (D), Dartmouth, 1992, Hw
J. Russell Harkay, (D), Vermont, 1979, Yg
Jerry P. Jasinski, (D), Wyoming, 1977, Gz
Associate Professor:
Steven D. Bill, (D), Case Western, 1982, Pg
Adjunct:
Charles M. Kerwin, (D), New Hampshire, 2006, Gg
Carol Leger, (B), Keene State Coll, 2005, Og

Plymouth State College

Natural Science Dept (B) (2004)
17 High Street
Plymouth, NH 03264
p. (603) 536-5000 (Ext. 2325) f. (603) 535-2723
wwf@mail.plymouth.edu
http://oz.plymouth.edu/~natsci/
Enrollment (1996): B: 48 (0)

Chair:
Wavell W. Fogleman, (D), Tulane, 1968, Oo
Professor:
James P. Koermer, (D), Utah, 1980, Ow
Larry T. Spencer, (D), Colorado State, 1968, Gg
Joseph Zabransky, Jr., (D), Wisconsin, 1976, Ow
Associate Professor:
Mark P. Turski, (D), Texas, 1994, Og
Assistant Professor:
Eric Hoffman, (D), SUNY (Albany), 2000, Ow

University of New Hampshire

Dept of Earth Sciences (B,M,D) (2007)
121 James Hall
56 College Road
Durham, NH 03824-3589
p. (603) 862-1718 f. (603) 862-2649
earth.sciences@unh.edu
http://www.unh.edu/esci/index.html
Administrative Assistant: Susan E. Clark
Enrollment (2007): B: 51 (3) M: 45 (5) D: 21 (2)

Chair:
William C. Clyde, (D), Michigan, 1997, Pg
Professor:
Francis S. Birch, (D), Princeton, 1969, Yg
Wallace A. Bothner, (D), Wyoming, 1967, Gc
Janet W. Campbell, (D), Virginia Tech, 1973, Oa
Larry A. Mayer, (D), California (San Diego), 1979, Gu
Robert W. Talbot, (D), Wisconsin, 1981, Oa
Karen L. Von Damm, (D), MIT/Woods Hole, 1984, Cl
Charles J. Vorosmarty, (D), New Hampshire, 1991, Hq
Associate Professor:
J. Matthew Davis, (D), New Mexico Tech, 1994, Hw
Jack E. Dibb, (D), SUNY (Binghamton), 1988, Oa
Stephen E. Frolking, (D), New Hampshire, 1993, Oa
Robert J. Griffin, (D), Caltech, 2000, Oa
Jo Laird, (D), Caltech, 1977, Gp
James M. Pringle, (D), MIT/Woods Hole, 1998, Op
Cameron P. Wake, (D), New Hampshire, 1993, Gl
Larry G. Ward, (D), South Carolina, 1978, Gu
Assistant Professor:
Julia G. Bryce, (D), California (Santa Barbara), 1998, Cg
Joel E. Johnson, (D), Oregon State, 2004, Gu
Linda Kalnejais, (D), MIT/Woods Hole, 2005, Oc
Joseph M. Licciardi, (D), Oregon State, 2000, Gl
Affiliate Professor:
P. Thompson Davis, (D), Colorado, 1980, Gl
James Gardner, (D), Columbia, 1973, Gu
Dork L. Sahagian, (D), Chicago, 1987, Gm
Peter J. Thompson, (D), Massachusetts, 1985, Gc
David R. Wunsch, (D), Kentucky, 1992, Cl
Affiliate Associate Professor:
Michael L. Prentice, Brown, 1988, Pe
Douglas C. Vandemark, (M), New Hampshire, 2005, Or
Affiliate Assistant Professor:
David P. Brown, (D), Arizona, 2004, Oa
Jeffrey B. Johnson, (D), Washington, 2000, Gv
J. Ru Morrison, (D), Wales (Bangor), 1999
Ruth K. Varner, (D), New Hampshire, 2000

Emeritus:
Franz E. Anderson, (D), Washington, 1967, Yr
S. Lawrence Dingman, (D), Harvard, 1970, Hq
Henri E. Gaudette, (D), Illinois, 1963, Cc
Francis R. Hall, (D), Stanford, 1961, Hw
Theodore C. Loder, (D), Alaska, 1971, Oc
Cecil J. Schneer, (D), Cornell, 1954, Gz
Herbert Tischler, (D), Michigan, 1961, Ps

Dept of Natural Resources (B,M,D) (2004)
215 James Hall
56 College Rd
Durham, NH 03824-3589
p. (603) 862-1020 f. (603) 862-4976
Administrative Assistant: Linda Scogin
Enrollment (1998): B: 60 (23) M: 3 (0)
Chair:
Theodore E. Howard, (D), Oregon State, 1982, Oo
Professor:
Robert D. Harter, (D), Purdue, 1966, Sc
Assistant Professor:
Elizabeth A. Rochette, (D), Washington State, 1994, Sc

New Jersey

Fairleigh Dickinson University

Dept of Chemistry & Geology (B) (2004)
M-SB-1-01
285 Madison Avenue
Madison, NJ 07940-1099
p. (201) 593-8790 f. (201) 593-8795
boyer@fdu.edu
http://www.fdu.edu
Enrollment: No data reported since 1995
Chair:
William D. Fordham, (D), Yale, Cg
Professor:
Paul S. Boyer, (D), Rice, 1970, Pi
Geology Librarian:
James Fraser, Oo

Kean University

Dept of Geology & Meteorology (B) (2007)
1000 Morris Avenue
Union, NJ 07083-0411
p. (908) 737-3690 f. (908) 737-3699
cmurphy@kean.edu
Department Secretary: Christine Pacia
Enrollment (2006): B: 160 (0)
Chair:
Robert Metz, (D), RPI, 1967, Gr
Professor:
Donald B. Krall, (D), Rutgers, 1972, Gl
Richard L. Kroll, (D), Syracuse, 1972, Gx
Charles H. Murphy, (D), SUNY (Albany), 1973, Ow
Shing Yoh, (D), Drexel, 1989, Ow
Constantine S. Zois, (D), Rutgers, 1980, Ow
Associate Professor:
Carrie M. Manfrino, (D), Miami, 1995, Gu
Assistant Professor:
John Dobosiewic, (D), Rutgers, 2003, Hg
Paul J. Croft, (D), Rutgers, 1991, Ow
Kikombo Ngoy, (D), Oregon, 1996, Oy
Laboratory Director:
William C. Heyniger, (B), Kean, 1995, Oo

Montclair State University

Dept of Earth & Environmental Studies (B,M) (2004)
1 Normal Avenue
Upper Montclair, NJ 07043
p. (973) 655-4448 f. (973) 655-4072
debeusb@mail.montclair.edu
http://www.csam.montclair.edu/earth/eesweb
Administrative Assistant: Barbara De Beus
Enrollment (2002): B: 48 (12) M: 40 (19)
Chair:
Jonathan M. Lincoln, (D), Northwestern, 1990, Gr
Associate Dean, Science & Mathematics:
Michael A. Kruge, (D), California (Berkeley), 1985, Co
Professor:
Harbans Singh, (D), Rutgers, 1973, Oo
William Solecki, (D), Rutgers, 1990, Ou
Rolf Sternberg, (D), Syracuse, 1971, Oo
Robert W. Taylor, (D), St. Louis, 1971, Oo
John V. Thiruvathukal, (D), Oregon State, 1968, Yg
Associate Professor:
Zhaodong Feng, (D), Kansas, 1992, Oi
Matthew L. Gorring, (D), Cornell, 1997, Gi
Duke U. Ophori, (D), Alberta, 1986, Hw
Gregory A. Pope, (D), Arizona State, 1994, Gm
Assistant Professor:
Mark J. Chopping, (D), Nottingham, 1998, Or
Huan E. Feng, (D), SUNY (Stony Brook), 1997, Cm
Adjunct Professor:
Kathryn Black, (M), Oklahoma, 1966, Oy
Matthew S. Tomaso, (M), Texas, 1995, Ga
Christine Valenti, (M), Montclair State, 1997, Gg
Laboratory Director:
Yoko Sato, (M), Montclair State, 2000, Gg

New Jersey City University

Dept of Geoscience/Geography (B) (2006)
Rossey Hall - Room 608
2039 Kennedy Boulevard
Jersey City, NJ 07305-1597
p. (201) 200-3161 f. (201) 200-2298
jmarchisin@njcu.edu
http://www.njcu.edu/Geo/geo.html
Enrollment (2006): B: 37 (19)
Professor:
Martin Abend, (D), Syracuse, 1955, Oy
Associate Professor:
Debra Freile, (D), Ge
John Marchisin, (M), Montclair State, 1965, Gg
Research Associate:
John M. O'Brien, (D), California (Santa Barbara), 1973, Gs
Lecturer:
William W. Montgomery, (D), W Michigan, 1998, Hw
Adjunct Professor:
George Papcun, (M), Oe
Howard Zlotkin, (M), Oe
Related Staff:
Nicolai Kaplun, Oo

Princeton University

Dept of Civil Eng/Env Eng & Water Resources Prg (B,M,D) (2004)
E-220 Engineering Quad
Princeton, NJ 08544
p. (609) 258-4655 f. (609) 258-2799
celia@Princeton.EDU
Department Secretary: Maryann Rothberg
Enrollment (1998): B: 40 (18) M: 1 (1) D: 20 (5)

Director:
Michael A. Celia, (D), Princeton, 1983, Hw
Professor:
Peter R. Jaffe, (D), Vanderbilt, 1981, Hw
Ignacio Rodriguez-Iturbe, (D), Colorado State, 1967, Hg
James A. Smith, (D), Johns Hopkins, 1981, Hg
Eric F. Wood, (D), MIT, 1974, Hg
Assistant Professor:
Catherine A. Peters, (D), Carnegie Mellon, 1992, Hw

Dept of Geosciences (B,D) (2005)
113 Guyot Hall
Princeton, NJ 08544-1003
p. (609) 258-4100 f. (609) 258-1274
bbw@princeton.edu
http://www.geoweb.princeton.edu
Enrollment (2005): B: 16 (11) D: 27 (2)
Chair:
F. Anthony Dahlen, (D), California (San Diego), 1969, Ys
Associate Chair:
Guust M. Nolet, (D), Utrecht (Neth), 1976, Ys
Professor:
Michael L. Bender, (D), Columbia, 1970, Cg
Lincoln S. Hollister, (D), Caltech, 1966, Gp
Gerta Keller, (D), Stanford, 1978, Pm
Francois M M. Morel, (D), Caltech, 1971, Cl
Tullis C. Onstott, (D), Princeton, 1980, Py
Michael Oppenheimer, (D), Oa
S. George H. Philander, (D), Harvard, 1970, Op
Allan M. Rubin, (D), Stanford, 1988, Yg
Jorge L. Sarmiento, (D), Columbia, 1978, Oc
John Suppe, (D), Yale, 1969, Gc
Bess B. Ward, (D), Washington, 1982, Ob
Research Staff:
Amal Jayakumar, (D), GOA, 1999, Cm
Fuming Jiang, (D), Tsukuba, 2001, Gy
Atsushi Kubo, (D), Gakushin, 1999, Gy
Adam B. Kustka, (D), SUNY, 2002, Ct
Pascale F. Poussart, (D), Harvard, 2004, Co
Associate Professor:
Thomas Duffy, (D), Caltech, 1992, Gy
Satish C B. Myneni, (D), Ohio State, 1995, Sc
Assistant Professor:
Adam C. Maloof, (D), Harvard, 2004, Ym
Nadine S. McQuarrie, (D), Arizona, 2001, Gt
Daniel M. Sigman, (D), MIT, 1997, Cg
Hess Fellow:
Andrew Leier, (D), Arizona, 2005, Gs
Maria G. Prokopenko, (D), S California, 2004, Cg
Research Associate:
Nicolas Cassar, (D), Hawaii, 2004, Ob
John D. Kessler, (D), California (Irvine), 2005, Cm
Patrick J. McGinn, (D), Australian National, 2003, Cm
Gregory D. O'Mullan, (D), Princeton, 2005, Ob
Jeremy J. Rich, (D), Oregon State, 2003, Sb
Jeffra K. Schaefer, (D), Rutgers, 2005, Ct
Caroline B. Tuit, (D), MIT/Woods Hole, 2003, Ob
Visiting Research Fellow:
Jenny L. Baeseman, (D), Colorado, 2004, Ob
Emeritus:
William E. Bonini, (D), Wisconsin, 1957, Yg
Kenneth S. Deffeyes, (D), Princeton, 1959, Eg
Robert A. Phinney, (D), Caltech, 1961, Ys
Franklyn B. Van Houten, (D), Princeton, 1941, Gs
Geosciences Librarian:
Patricia Gaspari-Bridges, (M), Oo
Undergraduate/Graduate Administrator:
Sheryl A. Rickwell, Oo

Technical Staff:
Bruce A. Barnett, (M), Alaska, 1984, Yr
Robert J. Mika, (M), Drexel, 1981, Cg
Frank Yi Wang, (D), NENU China, 1996, Cg
Department Manager:
Deborah A. Smith, (A), Rider, 1983, Oo
Academic Lab Manager:
Laurel P. Goodell, (M), Princeton, 1983, Oe
Danielle M. Schmitt, (M), W Michigan, 1999, Oe

Geological Engineering Program (B,M) (2005)
219 Guyot Hall
Princeton, NJ 08544
p. (609) 258-3598 f. (609) 258-1274
fad@princeton.edu
Department Manager: Mary Ann Nicoletti
Undergraduate & Graduate Secretary: Nancy A. Janos
Enrollment (1998): B: 13 (0)
Professor:
Francois M.M. Morel, (D), Caltech, 1971, Cm
Guust Nolet, (D), Utrecht (Neth), 1976, Om
S. George Philander, (D), Harvard, 1970, Op
Jean-Herve Prevost, (D), Stanford, 1974, Ng
Associate Professor:
Michael A. Celia, (D), Princeton, 1983, Hw
Peter R. Jaffe, (D), Vanderbilt, 1981, Hg
Jim A. Smith, (D), Johns Hopkins, Sp
Geology Librarian:
Patricia Gaspari-Bridges, Oo

Program in Atmospheric & Oceanic Sci (B,M,D) (2005)
Sayre Hall
Princeton, NJ 08544-0710
p. (609) 258-6571 f. (609) 258-1274
jvcallan@splash.princeton.edu
http://www.aos.princeton.edu/
Enrollment (2004): B: 8 (8) D: 27 (6)
Chair:
F. Anthony Dahlen, (D), California (San Diego), 1969, Ys
Professor:
Michael L. Bender, (D), Columbia, 1970, Cg
Lincoln S. Hollister, (D), Caltech, 1966, Gp
Gerta Keller, (D), Stanford, 1978, Pm
Francois M M Morel, (D), Caltech, 1971, Cl
Guust Nolet, (D), Utrecht (Neth), 1976, Ys
Michael Oppenheimer, (D), Gg
S. George Philander, (D), Harvard, 1970, Op
Robert A. Phinney, (D), Caltech, 1961, Ys
Jorge L. Sarmiento, (D), Columbia, 1978, Oc
John Suppe, (D), Yale, 1969, Gc
Gregory van der Vink, (D), Gg
Bess B. Ward, (D), Washington, 1982, Ob
Associate Professor:
Satish C. B. Myneni, (D), Ohio State, 1995, Sc
Tullis Onstott, (D), Princeton, 1980, Yg
Allan M. Rubin, (D), Stanford, 1988, Gc
Assistant Professor:
Thomas Duffy, (D), Caltech, 1992, Yx
Nadine McQuarrie, (D), Gg
Daniel M. Sigman, (D), MIT, 1997, Cg
Emeritus:
William E. Bonini, (D), Wisconsin, 1957, Yg
Kenneth S. Deffeyes, (D), Princeton, 1959, Eg
W. Jason Morgan, (D), Princeton, 1964, Yg
Franklyn B. Van Houten, (D), Princeton, 1941, Gs
Related Staff:
John V. Fahey, (B), Gg

Laurel P. Goodell, (M), Princeton, 1983, Yr
Danielle M. Schmitt, (M), W Michigan, 1999, Yr
Debbie Smith, (B), Gg

Richard Stockton College of New Jersey

Environmental Studies & Geology Program (B) (2007)
Division of Natural Science and Mathematics
PO Box 0195
Pomona, NJ 08240-0195
p. (609) 652-4620 f. (609) 748-5515
mgeller@stockton.edu
http://talon.stockton.edu/eyos/page.cfm?siteID=14&pageID=17
Enrollment (2005): B: 122 (31)

Director, Coastal Research Center:
Stewart Farrell, (D), Massachusetts, 1972, On
Professor:
Claude M. Epstein, (D), Brown, 1971, Hw
Raymond G. Mueller, (D), Kansas, 1981, So
Roger C. Wood, (D), Harvard, 1971, Pv
Program Coordinator:
Michael D. Geller, (D), SUNY (Binghamton), 1979, Oo
Michael J. Hozik, (D), Massachusetts, 1981, Ym
Associate Professor:
William J. Cromartie, (D), Cornell, 1974, Oo
Weihong Fan, (D), Colorado State, 1993, Oi
Lynn F. Stiles, (D), Cornell, 1970, Oo
George Zimmermann, (D), Rutgers, 1982, Oo
Assistant Professor:
Tait Chirenje, (D), Florida, Sc

Rider University

Dept of Geological, Environmental, and Marine Sciences (GEMS) (B) (2007)
2083 Lawrenceville Road
Lawrenceville, NJ 08648-3099
p. (609) 896-5092 f. (609) 895-5782
husch@rider.edu
Enrollment (2007): B: 93 (0)

Chair:
Jonathan M. Husch, (D), Princeton, 1982, GieGz
Assistant Provost:
Kathleen M. Browne, (D), Miami, 1993, Oge
Associate Professor:
Reed A. Schwimmer, (D), Delaware, 1999, GsOgn
Hongbing Sun, (D), Florida State, 1995, HwCgSc
Assistant Professor:
Gabriela W. Smalley, (D), Maryland, 2002, ObcOp
Emeritus:
Mary Jo Hall, (D), Lehigh, 1981, Gs
Reginald Shagam, (D), Princeton, 1956, Gt

Rutgers, The State University of New Jersey

Dept of Geological Sciences (B,M,D) (2007)
Wright Lab
610 Taylor Road
Piscataway, NJ 08854-8066
p. (732) 445-2044 f. (732) 445-3374
kgm@rci.rutgers.edu
http://geology.rutgers.edu/
Enrollment (2006): B: 25 (0)

Chair:
Kenneth G. Miller, (D), MIT/Woods Hole, 1982, Gu
Vice Chair:
Carl C. Swisher III, (D), California (Berkeley), 1992, Cc
Professor:
Gail M. Ashley, (D), British Columbia, 1977, Gs
Marie-Pierre Aubry, (D), Pierre & Marie Curie, 1983, Yr
Michael J. Carr, (D), Dartmouth, 1974, Gv
Paul G. Falkowski, (D), British Columbia, 1975, Ob
Mark D. Feigenson, (D), Princeton, 1982, Cg
Claude T. Herzberg, (D), Edinburgh, 1975, Cp
Roger H. Hewins, (D), Toronto, 1971, Xm
Dennis V. Kent, (D), Columbia, 1974, Ym
George R. McGhee, Jr., (D), Rochester, 1978, Po
Gregory S. Mountain, (D), Columbia, 1981, Gu
Peter A. Rona, (D), Yale, 1967, Yr
Martha O. Withjack, (D), Brown, 1977, Gc
Associate Professor:
Craig S. Feibel, (D), Utah, 1988, Ga
Roy W. Schlische, (D), Columbia, 1990, Gt
Robert M. Sherrell, (D), MIT/Woods Hole, 1991, Cm
James D. Wright, (D), Columbia, 1991, Cs
Assistant Research Professor:
Vadim Levin, (D), Columbia, 1996, Ys
Assistant Professor:
Ying Fan Reinfelder, (D), Utah State, 1992, Hw
Nathan Yee, (D), Notre Dame, 2001, Py
Research Associate:
James V. Browning, (D), Rutgers, 1996, Oo
Jeremy S. Delaney, (D), Belfast, 1978, Gx
Miriam E. Katz, (D), Rutgers, 2001, Pm
Bridget Wade, (D), Edinburgh, 2002, Pm
Adjunct Professor:
Peter P. Sugarman, (D), Rutgers, 1995, Gr
Brigitte M. Zanda, (D), Paris, 1988, Xm
Distinguished Visiting Profesor:
William A. Berggren, (D), Stockholm, 1962, Pm
Emeritus:
Richard K. Olsson, (D), Princeton, 1958, Pm
Robert E. Sheridan, (D), Columbia, 1968, Yr
Cooperating Faculty:
Colomban de Vargas, (D), Geneva, 2000, Pm
John R. Reinfelder, (D), SUNY (Stony Brook), 1993, Cl

Rutgers, The State Univ of New Jersey, Newark

Dept of Earth & Environmental Sciences (B,M,D) (2007)
101 Warren Street
Smith Hall, room 135
Newark, NJ 07102
p. (973) 353-5100 f. (973) 353-1965
morrin@andromeda.rutgers.edu
http://geology.newark.rutgers.edu
Administrative Assistant: M. Elizabeth Morrin
Enrollment (2006): B: 12 (3) M: 8 (3) D: 12 (1)

Professor:
Alexander E. Gates, (D), Virginia Tech, 1986, Gc
Associate Professor:
Andrew E. Kasper, (D), Connecticut, 1970, Pb
Lee S. Slater, (D), Lancaster (UK), 1997, Yg
Assistant Professor:
Yuan Gao, (D), Rhode Island, 1994, Oa
Emeritus:
Sam L. Agron, (D), Johns Hopkins, 1949, Gc
Hessle F. Garner, (D), Iowa State, 1953, Gm
Warren Manspeizer, (D), Rutgers, 1964, Gr
John H. Puffer, (D), Stanford, 1969, Gi
George Theokritoff, (D), London, 1961, Pi
Andreas H. Vassiliou, (D), Columbia, 1969, Gz

Stockton State College

Marine Science Program (B) (2004)
Pomona, NJ 08240
p. (609) 652-1776 (Ext. 4548) f. (609) 748-5515
iaprod152@stockton.edu

Administrative Assistant: Charlaine Claxton
Enrollment: No data reported since 1995
Professor:
Stewart C. Farrell, (D), Massachusetts, 1972, On
Matthew Landau, (D), Florida Inst of Tech, 1983, Og
Yitzhak Y. Sharon, (D), Princeton, 1966, Yr
Associate Professor:
Richard P. Hager, (D), New Hampshire, 1974, Oo
Michael J. Hozik, (D), Massachusetts, 1976, Gm
Carol Slocum, (D), West Indies, 1988, Oo
Assistant Professor:
Gordan Grguric, (D), Florida Inst of Tech, 1993, Oc

Trenton State College

Physics Dept (2004)
North American Geological Alliance
Trenton, NJ 08650-4700
p. (609) 771-1855 (Ext. 2569) f. (609) 771-3228
goldstef@tcnj.edu
Director:
Fredric R. Goldstein, (D), Rutgers, 1974, Oe
Adjunct Professor:
Thomas Gillespie, (M), Rutgers, 1986, Gc
John M. O'Brien, (D), California (Santa Barbara), 1973, Gs

Union County College

Dept of Biology (A) (2005)
Cranford, NJ 07016
p. (201) 709-7196
ombrello@hawk.ucc.edu
Administrative Assistant: Helen Gmitro
Enrollment: No data reported since 1995
Associate Professor:
Raymond J. Daly, (M), Rutgers, 1975, Pv

William Paterson College

Dept of Environmental Science & Geography (B) (2005)
Science Hall
Wayne, NJ 07470
p. (201) 595-2721 f. (201) 595-2338
pardir@wpunj.edu
Administrative Assistant: Beverly Middleton
Enrollment (2001): B: 40 (40)
Chair:
Richard R. Pardi, (D), Pennsylvania, 1983, Cc
Professor:
Sharon Hanks, (D), Rutgers, 1974, Oo
Michael Sebetich, (D), Rutgers, 1973, Oo
Associate Professor:
Robert J. McCallum, (D), Minnesota, 1970, Oo
Lance Risley, (D), Georgia, 1987, Oo
Philip Thiuri, (D), Syracuse, 1976, Oo
Assistant Professor:
Sheridan L. Dodge, (D), Michigan State, Oo
Karen Swanson, (D), Penn State, 1989, Cc
Steve Vail, (D), California (Davis), 1990, Oo
Librarian:
Jackie Hill, Oo

New Mexico

College of Santa Fe

Dept of Sciences and Conservation Studies (B) (2005)
1600 St. Michael's Drive
Santa Fe, NM 87505-7634
p. (505) 473-6300 f. (505) 473-6399
http://www.csf.edu
Enrollment (1998): B: 9 (2)
Professor:
David Johnson, (D), Colorado, 1983, Oo
William O. Sayre, (D), Southampton, 1980, Hg
Associate Professor:
Thomas Antonio, (D), Oo
Assistant Professor:
Carmen Burch, (D), Oo
Lenny Gannes, (D), Oo
Timothy Meehan, (D), Oo

Dine College, Shiprock

Div of Mathematics, Science, & Technology (A) (2005)
P.O. Box 580
Yucca Street
Shiprock, Navajo Nation, NM 87420-0580
p. (505) 368-3630 f. (505) 368-3636
sagould@dinecollege.edu
http://www.dinecollege.edu
Enrollment (2002): A: 18 (0)
Chair:
Steven C. Semken, (D), MIT, 1989, Ge
Instructor:
Douglas C. Isley, (D), Wyoming, 1979, Oi
Coordinator:
Doris Bahee-Cody, (A), Oo

Eastern New Mexico University

Dept of Physical Sciences/Geology (B) (2004)
Portales, NM 88130
p. (505) 562-2174 f. (505) 562-2192
robert.pierce@enmu.edu
http://www.enmu.edu/~piercer/geol/welcome.htm
Enrollment (1997): B: 15 (0)
Chair:
Robert W. Pierce, (D), Illinois, 1969, Gr
Associate Professor:
James T. Constantopoulos, (D), Idaho, 1989, Em

New Mexico Highlands University

Environmental Geology Program (B,M) (2007)
Box 9000
Las Vegas, NM 87701
p. (505) 454-3000 f. (505) 454-3063
lindlinej@nmhu.edu
Enrollment (2007): B: 15 (2) M: 1 (0)
Associate Professor:
Jennifer Lindline, (D), Bryn Mawr, 1997, GipGe
Assistant Professor:
Michael S. Petronis, (D), New Mexico, 2005, YmGcv

New Mexico Institute of Mining and Tech

Dept of Earth & Environmental Science (B,M,D) (2007)
801 Leroy
Socorro, NM 87801
p. (505) 835-5634 f. (505) 835-6436
geos@nmt.edu
http://www.ees.nmt.edu/
Enrollment (2003): M: 81 (19) D: 21 (2)
Chair:
Robert S. Bowman, (D), New Mexico State, 1982, So
Professor:
Richard C. Aster, (D), California (San Diego), 1991, Ys
Andrew R. Campbell, (D), Harvard, 1984, Cs
Kent C. Condie, (D), California (San Diego), 1965, Ct
Jan M. H. Hendrickx, (D), New Mexico State, 1984, Hw
Thomas L. Kieft, (D), New Mexico, 1983

Philip R. Kyle, (D), Victoria (NZ), 1976, Gi
David I. Norman, (D), Minnesota, 1977, Em
Fred M. Phillips, (D), Arizona, 1981, Hw
John L. Wilson, (D), MIT, 1974, Hw

Associate Professor:
Gary Axen, (D), Harvard, Gc
Bruce I. Harrison, (D), New Mexico, 1991, Gm
David B. Johnson, (D), Iowa, 1978, Ps
Peter S. Mozley, (D), California (Santa Barbara), 1988, Gs

Assistant Professor:
Susan L. Bilek, (D), California (Santa Cruz), 2003, Ys
William C. McIntosh, (D), New Mexico Tech, 1989, Cc
Glenn Spinelli, (D), California (Santa Cruz), 2002, Hw
Harold J. Tobin, (D), California (Davis), 1995, Ys
Enrique R. Vivoni, (D), MIT, 2003, Hs

Research Associate:
Penelope Boston, (D), Colorado
Brent Newman, (D), Cs

Director, Nat Cave & Karst Research Inst:
Louise D. Hose, (D), Louisiana State, 1990, Gs

Adjunct Professor:
Robert Andres, (D), Michigan Tech, Gv
George S. Austin, (D), Iowa, 1971, En
James M. Barker, (M), California (Santa Barbara), 1972, En
Paul W. Bauer, (D), New Mexico Tech, 1988, Gc
Ronald F. Broadhead, (M), Cincinnati, 1979, Eo
Jill S. Buckley, (D), Heriot-Watt (UK), 1996, Np
Steven M. Cather, (D), Texas, 1986, Gc
Richard Chamberlin, (D), Colorado Mines, Gr
Charles E. Chapin, (D), Colorado Mines, 1965, Gv
Sean D. Connell, (M), California (Riverside), 1995, Gm
Nelia W. Dunbar, (D), New Mexico Tech, 1989, Cg
Laurel B. Goodwin, (D), California (Berkeley), Gc
John W. Hawley, (D), Illinois, 1962, Ge
Matthew T. Heizler, (D), California (Los Angeles), 1993, Cc
Lawrence H. Jaksha, (B), N Colorado, 1959, Ys
Shari A. Kelley, (D), S Methodist, 1984, Gt
Lewis Land, North Carolina, Hy
David W. Love, (D), New Mexico, 1980, Ge
James T. McCord, (D), New Mexico Tech, 1988, Hq
Virginia T. McLemore, (D), Texas (El Paso), 1993, Eg
Mark Murray, MIT, Gc
Michael J. Pullin, Kent State, 1999
Marshall A. Reiter, (D), Virginia Tech, 1969, Yh
Peter A. Scholle, (D), Princeton, 1970, Gs
Daniel B. Stephens, (D), Arizona, 1979, Hw
Vincent C. Tidwell, (D), New Mexico Tech, 1999, Hw
Dana S. Ulmer-Scholle, (D), S Methodist, 1992, Gs

Emeritus:
Gerardo W. Gross, (D), Penn State, 1959, Yx
Lawrence H. Lattman, (D), Cincinnati, 1953, Gm
Allan R. Sanford, (D), Caltech, 1958, Ys
John W. Schlue, (D), California (Los Angeles), 1975, Ys

Chief Gofer:
Patricia Mills, (B), New Mexico Tech, 1998, Oo

Dept of Mineral Engineering (B,M,D) (2006)
Campus Station
Socorro, NM 87801-9990
p. (505) 835-5345 f. (505) 835-5252
mojtabai@nmt.edu
Department Secretary: Lucero Joanna
Enrollment: No data reported since 1995

Chair:
Navid Mojtabai, Arizona, 1990, Nr

Professor:
William X. Chavez, Jr., (D), California (Berkeley), 1984, Em

Associate Professor:
Cathrine T. Aimone-Martin, (D), Northwestern, 1982, Ng

Assistant Professor:
Baolin Deng, (D), Johns Hopkins, 1995, Cg
Randal S. Martin, (D), Washington State, 1992, Oa

Adjunct Professor:
William Haneberg, (D), Cincinnati, 1989, Ng
Per-Anders Persson, (D), Cambridge, 1960, Nr
Ingar F. Walder, (D), New Mexico Tech, 1991, Cl

Emeritus:
George B. Griswold, (D), Arizona, Nx
Kalman I. Oravecz, (D), Witwatersrand, 1967, Nr

New Mexico State University, Las Cruces

Dept of Agronomy & Horticulture (A,B,M,D) (2004)
Box 30003
Dept. 3Q
Las Cruces, NM 88003-0003
p. (505) 646-3405 f. (505) 646-6041
Department Secretary: Paula Ross
Enrollment: No data reported since 1995

Head:
LeRoy A. Daugherty, (D), Cornell, 1975, Sd

Professor:
William C. Lindemann, (D), Minnesota, 1978, Sb
Bobby D. McCaslin, (D), Minnesota, 1974, Sc
Theodore W. Sammis, (D), Arizona, 1974, Sp

Assistant Professor:
Dean Heil, (D), California (Berkeley), 1991, Sc
Tim L. Jones, (D), Washington State, 1989, Sp
H. C. Monger, (D), New Mexico State, 1990, Sa

Dept of Geological Sciences (B,M) (2007)
MSC 3AB, P.O. Box 30001
Breland Hall, Room 129
Corner of Breland & Stewart Street
Las Cruces, NM 88003
p. (575) 646-2708 f. (575) 646-1056
geology@nmsu.edu
http://www.nmsu.edu/~geology/
Administrative Assistant: Lee Hubbard
Enrollment (2007): B: 41 (3) M: 23 (9)

Head:
Nancy J. McMillan, (D), S Methodist, 1986, Gi

Professor:
Katherine A. Giles, (D), Arizona, 1991, Gs
Timothy F. Lawton, (D), Arizona, 1983, Gr
Gregory H. Mack, (D), Indiana, 1977, Gs

Assistant Professor:
Jeffrey M. Amato, (D), Stanford, 1995, Gc

Instructor:
Marilyn C. Huff, (M), Georgia Tech, 1985, Gg

Adjunct Professor:
Glenn F. Huff, (D), Louisiana State, 1993, Hw

Emeritus:
Thomas H. Giordano, (D), Penn State, 1978, Cl
William E. King, (D), Wisconsin, 1959, Pi
William R. Seager, (D), Arizona, 1966, Gc

Related Staff:
Nila Bourland

Dept of Physics (B,M,D) (2005)
Graduate Program in Geophysics
Las Cruces, NM 88003
p. (505) 646-3831 f. (505) 646-1934
jni@nmsu.edu
http://geophysics.nmsu.edu
Enrollment (1998): D: 8 (0)

Director:
James F. Ni, (D), Cornell, 1984, Ys
Head:
Gary Kyle, (D), Minnesota, 1978, Yg
Professor:
Robert L. Armstrong, (D), Johns Hopkins, 1970, Oa
George H. Goedecke, (D), RPI, 1961, Oa
Associate Professor:
Thomas M. Hearn, (D), Caltech, 1985, Ys
Assistant Professor:
Viatcheslav S. Solomatov, (D), Moscow Inst of Physics & Tech, 1990, Yg

University of New Mexico

Dept of Earth & Planetary Sciences (B,M,D) (2007)
Northrop Hall, Room 141
MSC03 2040 , 1 University of New Mexico
Albuquerque, NM 87131-0001
p. (505) 277-4204 f. (505) 277-8843
epsdept@unm.edu
http://epswww.unm.edu/
Enrollment (2007): B: 101 (20) M: 22 (9) D: 26 (4)

Chair:
John W. Geissman, (D), Michigan, 1980, Ym
Professor:
Carl A. Agee, (D), Columbia, Oe
Yemane Asmerom, (D), Arizona, 1988, Cc
Adrian J. Brearley, (D), Manchester (UK), 1984, Gz
Laura J. Crossey, (D), Wyoming, 1985, Cl
David J. Gutzler, (D), MIT, 1986, Oa
Karl E. Karlstrom, (D), Wyoming, 1980, Gt
Barry S. Kues, (D), Indiana, 1974, Pi
Leslie M. McFadden, (D), Arizona, 1982, Sd
James J. Papike, (D), Minnesota, 1964, Ca
Jane E. Selverstone, (D), MIT, 1985, Gp
Zachary D. Sharp, (D), Michigan, 1987, Cs
Gary A. Smith, (D), Oregon State, 1986, Gs
Senior Scientist:
Victor J. Polyak, (D), Texas Tech, 1998, Cg
Associate Professor:
Maya Elrick, (D), Virginia Tech, 1990, Gs
Peter J. Fawcett, (D), Penn State, 1994, Pe
Tobias Fischer, (D), Arizona State, 1999, Gv
Rhian Jones, (D), Manchester (UK), 1986, Gz
Grant A. Meyer, Gm
Mousumi Roy, (D), MIT, 1997, Yg
Louis Scuderi, (D), California (Los Angeles), Pe
Gary Weissmann, (D), California (Davis), Hw
Assistant Professor:
Joseph Galewsky, (D), California (Santa Cruz), Ow
Research Associate:
Nieu-Viorel Atudorei, (D), Lausanne (Switzerland), 1998, Cs
Frans J.M. Rietmeijer, (D), Utrecht (Neth), 1979, Gp
Timothy Wawrzyniec, (D), New Mexico, Gt
Lecturer:
Matthew Nyman, (D), Virginia Tech, 1992, Gp
Adjunct Professor:
Bruce Allen, (D), New Mexico, Gn
Sidney Ash, (D), Reading (England), 1966, Pb
W. Scott Baldridge, (D), Caltech, 1979, Gi
M. Susan Barger, (D), Penn State, 1982, Om
John Bloch, (D), Calgary
David Coblentz, (D), Arizona
Fraser Goff, (D), California (Santa Cruz), 1977, Cg
Spencer Lucas, (D), Yale, 1983, Pv
Sean McKenna, (D), Colorado Mines
Duane M. Moore, (D), Illinois
V. Rama Murthy, (D), Yale
Aurora Pun, (D), New Mexico, 1996, Ca
Thomas E. Williamson, (D), New Mexico, 1993, Pv
Kenneth H. Wohletz, (D), Arizona State, 1980, Gv
Emeritus:
Roger Y. Anderson, (D), Stanford, 1960, Gn
Michael E. Campana, (D), Arizona, 1975, Hw
Wolfgang E. Elston, (D), Columbia, 1953, Eg
Rodney E. Ewing, (D), Stanford, 1974, Gz
Stephen P. Huestis, (D), California (San Diego), 1976, Yg
Cornelis Klein, (D), Harvard, 1965, Gz
Lee A. Woodward, (D), Washington, 1962, Gc

Inst of Meteoritics (2007)
MSC03 2050
1 University of New Mexico
Albuquerque, NM 87131
p. (505) 277-1644 f. (505) 277-3577
iom@unm.edu
Administrative Assistant: Shannon Clark

Professor:
Carl B. Agee, (D), Columbia, 1988, XcGyCp
Research Professor:
V. Rama Murthy, (D), Yale, 1957, XcCc
James J. Papike, (D), Minnesota, 1964, Gz
Robert C. Reedy, (D), Columbia, 1969, Xcy
David C. Rubie, (D), Leicester, 1972
Senior Scientist III:
David S. Draper, (D), Oregon, 1991, GiCpXc
Penelope L. King, (D), Arizona State, 1999, GiXcCp
David T. Lescinsky, (D), Arizona State, 1999, Gvi
Horton E. Newsom, (D), Arizona, 1981, Xc
Charles K. Shearer, Jr., (D), Massachusetts, 1983, Gi
Senior Scientist I:
James M. Karner, (D), New Mexico, 2003, Gz
Research Assistant Professor:
Barbara A. Cohen, (D), Arizona, 2000, Xcg
Research Specialist:
Paul V. Burger, (M), New Mexico, 2005, Gz
Research Scientist III:
Michael N. Spilde, (M), SD Sch of Mines & Tech, 1987, Gz

Water Resources Program (2004)
1915 Roma NE, Room 1044
Albuquerque, NM 87131-1217
p. 505 277 5249 f. 505 277 5226
aquadoc@unm.edu
http://www.unm.edu/~wrp/

Director:
Michael E. Campana, Oo

Western New Mexico University

Dept of Natural Sciences (2007)
P.O. Box 680
1000 West College Avenue
Silver City, NM 88062
p. (505) 538-6227 f. (505) 538-6228
dowsem@wnmu.edu
http://www.wnmu.edu/academic/nspages
Enrollment: No data reported since 1995

Professor:
Mary E. Dowse, (D), West Virginia, 1980, Og

New York

Adelphi University

Environmental Studies Program (B,M) (2007)
South Avenue

Garden City, NY 11530
p. (516) 877-4170 f. (516) 877-4209
schlosse@adelphi.edu
Enrollment (2007): B: 39 (0) M: 19 (0)

Professor:
Anthony E. Cok, (D), Dalhousie, 1970, OuGeu

Assistant Professor:
Beth A. Christensen, (D), South Carolina, 1997, PmeGs

Adirondack Community College

Science Div (A) (2004)
640 Bay Road
Queensbury, NY 12804
p. (518) 743-2325 f. (518) 745-1433
minkeld@acc.sunyacc.edu
http://www.sunyacc.edu
Enrollment (1997): B: 5 (0)

Professor:
Donald H. Minkel, (D), RPI, 1991, Gc

Alfred University

Dept of Geology (B) (2004)
Saxon Drive
Alfred, NY 14802
p. (607) 871-2208 f. (607) 871-2746
fmuller@alfred.edu
Department Secretary: Ann Murray
Enrollment: No data reported since 1995

Professor:
Michele M. Hluchy, (D), Dartmouth, 1988, Cl
Otto H. Muller, (D), Rochester, 1974, Gc

Emeritus:
Daniel B. Sass, (D), Cincinnati, 1959, Pi
J. Scott Weaver, (D), Rochester, 1971, Gz

American Museum of Natural History

Dept of Earth & Planetary Sciences (2004)
Central Park West at 79th Street
New York, NY 10024-5192
p. (212)769-5390 f. (212)769-5339
kateh@amnh.org
http://research.amnh.org/earthplan
Administrative Assistant: Kate V. Hazel

Chair:
James D. Webster, (D), Arizona State, 1987, Eg

Curator:
George E. Harlow, (D), Princeton, 1977, Ga
Edmond A. Mathez, (D), Washington, 1981, Gx

Associate Curator:
Denton Ebel, (D), Purdue, 1993, Xm

Senior Scientist:
Alfred Duba, (D), Chicago, 1971, Gg

Scientific Assistant:
Jamie Newman, (M), Brooklyn Coll, Gg
Christine M. Tappen, (M), North Carolina (Wilmington), 1998, Gi

Research Scientist:
Robert A. Fogel, (D), Brown, 1989, Cp
Charles W. Mandeville, (D), Rhode Island, 1995, Gv

Research Associate:
Hans G. Ave Lallamant, (D), Gt
Nehru E. Cherukupalli, (D), Gi
Harold Connolly, Jr., (D), Xm
John Delaney, (D), Gg
Rosamond Kinzler, (D), Gi
Arthur Langer, (D), Gg
John Longhi, (D), Gi
Hanna Nevkasil, (D), Gi
Wayne G. Powell, (D), Eg
Federica Raia, (D), Gv
Alan Rice, (D), Yg
George Rossman, (D), Gz
Virginia B. Sisson, Gp
Heather Sloan, (D), Yr
Sorena Sorenson, (D), Gp
David Walker, (D), Gi
Michael K. Weisberg, (D), CUNY, 1991, Xm

Senior Scientific Assistant:
Joseph Boesenberg, (M), Rutgers, 1995, Xm

Postdoc:
Rondi M. Davies, (D), Macquarie, 1998, Gz

Other:
M. Francesca Sintoni, (D), Firenze (Italy), 2002, Gi

Div of Paleontology (D) (2007)
Central Park West at 79th Street
New York, NY 10024
p. (212) 769-5810 f. (212) 769-5842
jflynn@amnh.org
http://paleo.amnh.org/
Administrative Assistant: Alejandra Lora
Enrollment (2007): D: 7 (2)

Dean, and Frick Curator:
John J. Flynn, (D), Columbia, 1983, Pv

Provost:
Michael J. Novacek, (D), California (Berkeley), 1978, Pv

Curator:
Niles Eldredge, (D), Columbia, 1969, Pi
Neil H. Landman, (D), Yale, 1982, Pi
John G. Maisey, (D), London, 1974, Pv
Jin Meng, (D), Columbia, 1991, Pv
Mark A. Norell, (D), Yale, 1989, Pv

Frick Curator Emeritus:
Malcolm C. McKenna, (D), California (Berkeley), 1958, Pv
Richard H. Tedford, (D), California (Berkeley), 1960, Pv

Curator Emeritus:
Roger L. Batten, (D), Columbia, 1956, Pi
Eugene S. Gaffney, (D), Columbia, 1969, Pv

Binghamton University

Dept of Geological Sci and Env Studies (B,M,D) (2007)
Binghamton, NY 13902-6000
p. (607) 777-2264 f. (607) 777-2288
dmjenks@binghamton.edu
http://geology.binghamton.edu
Administrative Assistant: Carol Slavetskas
Enrollment (2005): B: 40 (12) M: 20 (8) D: 10 (2)

Chair:
Tim K. Lowenstein, (D), Johns Hopkins, 1982, Cl

Professor:
John S. Bridge, (D), St. Andrews, 1974, Gs
Robert V. Demicco, (D), Johns Hopkins, 1981, Gs
Steven R. Dickman, (D), California (Berkeley), 1977, Yg
David M. Jenkins, (D), Chicago, 1980, Gx
William D. MacDonald, (D), Princeton, 1965, Gc
H. Richard Naslund, (D), Oregon, 1980, Gi
Herman E. Roberson, (D), Illinois, 1959, Gs
Francis T. Wu, (D), Caltech, 1966, Ys

Associate Professor:
Richard E. Andrus, (D), SUNY (Syracuse), 1974, Py
Jeffrey S. Barker, (D), Penn State, 1984, Ys
Joseph R. Graney, (D), Michigan, 1994, Cl
Peter L. K. Knuepfer, (D), Arizona, 1984, Gt

Assistant Professor:
Karen M. Salvage, (D), Penn State, 1998, Hw

Research Associate:
Michael Hubenthal, (B), Slippery Rock, 1996, Oe
Alan Jones, (D), Purdue, 1964, Ys
E. Jay Sarton, (M), S California, 1971, Oe
Michael N. Timofeeff, (D), Binghamton, 2001, Cl
Emeritus:
James R. Beerbower, (D), Chicago, 1954, Pg
Donald R. Coates, (D), Columbia, 1956, Gm
Thomas W. Donnelly, (D), Princeton, 1959, Gg
James E. Sorauf, (D), Kansas, 1962, Pi

Brooklyn College (CUNY)

Dept of Geology (B,M,D) (2006)
2900 Bedford Avenue
Brooklyn, NY 11210
p. (718) 951-5416 f. (718) 951-4753
wpowell@brooklyn.cuny.edu
http://depthome.brooklyn.cuny.edu/geology/
Administrative Assistant: Joan Kanakakis
Enrollment (2006): B: 17 (0) M: 14 (0) D: 4 (0)
Chair:
Wayne G. Powell, (D), Queen's, 1994, Gp
Professor:
John A. Chamberlain, (D), Rochester, 1971, Po
Nehru E. Cherukupalli, (D), Madras (India), 1963, Gi
David J. Leveson, (D), Columbia, 1960, Gh
E. Lynn Savage, (D), Rutgers, 1967, Gs
David E. Seidemann, (D), Yale, 1975, Cc
Associate Professor:
Stephen U. Aja, (D), Washington State, 1989, Cl
Constantin Cranganu, (D), Oklahoma, 1997, Go
Assistant Professor:
Zhongqi Cheng, (D), Ohio State, 2001, Ca
Emeritus:
Somdev Bhattacharji, (D), Chicago, 1959, Gt
Gerald M. Friedman, (D), Columbia, 1952, Gs
William H. Harris, (D), Brown, 1971, Cm
Laboratory Director:
Guillermo Rocha, (M), CUNY, 1994, Gg

Broome Community College

Dept of Eng Science/Physics/Physical Sci (A) (2007)
Upper Front Street
Box 1017
Binghamton, NY 13902
p. (607) 778-5000 (Ext. 5114) f. (607) 771-5345
goodman_k@sunybroome.edu
Professor:
Karen J. Goodman, (M), Cornell, 1975, Og
Bruce K. Oldfield, (M), SUNY (Binghamton), 1988, Gg
Adjunct Professor:
Clare Gibbs, (M), Tennessee, 1962, Og

Buffalo State College

Dept of Earth Sciences (B,M) (2007)
1300 Elmwood Avenue
Buffalo, NY 14222
p. (716) 878-6731 f. (716) 878-4524
solargs@buffalostate.edu
http://www.buffalostate.edu/earthsciences
Administrative Assistant: Nancy Williams
Enrollment (2007): B: 88 (18) M: 11 (4)
Chair:
Gary S. Solar, (D), Maryland, 1999, GcpGt
Professor:
Jill K. Singer, (D), Rice, 1986, GsOp
Associate Professor:
Richard J. Batt, (D), Colorado, 1987, PiGr
John E. Mack, (D), Fordham, 1971, Xg
Assistant Professor:
Elisa T. Bergslien, (D), SUNY (Buffalo), 2002, ClHw
Kevin K. Williams, (D), Johns Hopkins, 2002, GmXg
Emeritus:
Richard J. Foster, (M), Case Western, 1965, Gm
James R. Orgren, (D), SUNY (Buffalo), 1973, Og
Carl K. Seyfert, (D), Stanford, 1967, Gc
Irving Tesmer, (D), Syracuse, Pg
Director, Field Studies:
Bettina Martinez-Hackert, (D), SUNY (Buffalo), 2006, Gvm
Planetarium Director:
Arthur W. Gielow, (M), Buffalo State, 1992, Oo

City College (CUNY)

Dept of Earth & Atmospheric Sciences (B,M) (2004)
New York, NY 10031
p. (212) 650-6984 f. (212) 650-6482
steiner@scisun.sci.ccny.cuny.edu
Enrollment: No data reported since 1995
Chair:
Jeffrey Steiner, (D), Stanford, 1970, Cp
Professor:
Stanley Gedzelman, (D), MIT, 1970, Oa
Edward E. Hindman, (D), Washington, 1975, Oa
Margaret A. Winslow, (D), Columbia, 1979, Gc
Associate Professor:
Patricia M. Kenyon, (D), Cornell, 1986, Yg
Anthony R. Prave, (D), Penn State, 1986, Gs
Adjunct Professor:
Willard J. Pierson, (D), New York, 1949, Op

Colgate University

Dept of Geology (B) (2007)
13 Oak Drive
Hamilton, NY 13346
p. (315) 228-7201 f. (315) 228-7187
jmcnamara@mail.colgate.edu
http://departments.colgate.edu/geology
Administrative Assistant: Jodi McNamara
Enrollment (1999): B: 54 (28)
Chair:
Constance M. Soja, (D), Oregon, 1985, Pi
Professor:
Richard April, (D), Massachusetts, 1978, Cg
Paul Pinet, (D), Rhode Island, 1972, Ou
Bruce Selleck, (D), Rochester, 1975, Gs
Associate Professor:
Karen Harpp, (D), Cornell, 1994, Cg
Amy Leventer, (D), Rice, 1988, Ou
William H. Peck, (D), Wisconsin, 2000, Gp
Assistant Professor:
Martin Wong, (D), California (Santa Barbara), 2005, Gtc
Senior Lecturer:
Dianne M. Keller, (M), Colgate, 1988, Cg
Emeritus:
James McLelland, (D), Chicago, 1961, Gp

College of Staten Island

Engineering Science & Physics (2004)
2800 Victory Boulevard
Staten Island, NY 10314
p. (718) 982-2827 f. (718) 982-2830
robbins@postbox.csi.cuny.edu

http://www.library.csi.cuny.edu/dept/as/geo/geo.html
Department Secretary: H. Marchek
Assistant Professor:
Anderson A. Ohan, (M), New York, 1965, Gc
Adjunct Professor:
Alan I. Benimoff, (D), Lehigh, 1984, Gi
Peter A. Rigotti, (D), Pittsburgh, 1976, Yg

Columbia University

Dept of Earth & Environmental Engineering (B,M,D) (2004)
Henry Krumb School of Mines
500 West 120 Street
918 Mudd Bldg
New York, NY 10027
p. (212) 894-2905 f. (212) 854-3054
pwb1@columbia.edu
http://www.seas.columbia.edu/columbia/departments/krumb
Administrative Assistant: Co'Quesie Gilbert
Department Administrator: Barbara Algin
Enrollment (1999): B: 6 (0) M: 13 (2) D: 3 (3)
Acting Chair:
Nickolas J. Themelis, (D), McGill, 1961, Nx
Professor:
Paul F. Duby, (D), Columbia, 1962, Nx
Peter Schlosser, (D), Heidelberg, 1985, Cg
Ponisseril Somasundaran, (D), California (Berkeley), 1964, Nx
Tuncel M. Yegulalp, (D), Columbia, 1968, Nm
Associate Professor:
Ross Bagtzoglou, (D), California (Berkeley), 1990, Hw
Senior Research Scientist:
Roelof Versteeg, Yg
Adjunct Professor:
Vasilis M. Fthenakis, (D), New York, 1991, Oa
Emeritus:
Stefan H. Boshkov, (M), Columbia, 1942, Nm
John T. Kuo, (D), Stanford, 1958, Yg
Malcolm T. Wane, (M), Columbia, 1954, Nm

Dept of Earth & Environmental Sci (B,M,D) (2006)
P.O. Box 1000
Route 9W
Palisades, NY 10964
p. (845) 365-8550 f. (845) 365-8163
mia@ldeo.columbia.edu
http://eesc.columbia.edu
Administrative Assistant: J. Mia Leo
Enrollment (1999): B: 77 (13) M: 16 (10) D: 71 (9)
Chair:
Nicholas Christie-Blick, (D), California (Santa Barbara), 1979, Gs
Professor:
Wallace S. Broecker, (D), Columbia, 1958, Oc
Mark A. Cane, (D), MIT, 1975, Op
Joel E. Cohen, (D), Harvard, 1970, Oo
Peter M. Eisenberger, (D), Harvard, 1967, Oo
Richard G. Fairbanks, (D), Brown, 1977, Pe
Arnold L. Gordon, (D), Columbia, 1965, Op
Dennis E. Hayes, (D), Columbia, 1966, Yr
James D. Hays, (D), Columbia, 1964, Pe
Peter B. Kelemen, (D), Washington, 1988, Gi
William H. Menke, (D), Columbia, 1981, Ys
John C. Mutter, (D), Columbia, 1982, Yr
Paul E. Olsen, (D), Yale, 1983, Pv
Lorenzo M. Polvani, (D), MIT, 1988, Ow
G. Michael Purdy, (D), Cambridge, 1974, Yr
Paul G. Richards, (D), Caltech, 1970, Ys
Peter Schlosser, (D), Heidelberg, 1985, Hw
Christopher H. Scholz, (D), MIT, 1967, Yx
H. James Simpson, Jr., (D), Columbia, 1970, Cm
Lynn R. Sykes, (D), Columbia, 1965, Ys
David Walker, (D), Harvard, 1972, Cp
Associate Professor:
Mark H. Anders, (D), California (Berkeley), 1989, Gc
Peter B. de Menocal, (D), Columbia, 1991, Pe
Steven L. Goldstein, (D), Columbia, 1986, Cs
Kevin L. Griffin, (D), 1994, Oo
Sidney R. Hemming, (D), SUNY (Stony Brook), 1994, Cs
Patrick Louchouarn, (D), Quebec (Montreal), 1997, Co
Stephanie L. Pfirman, (D), MIT, 1985, Gv
Adam H. Sobel, (D), MIT, 1998, Oa
Marc W. Spiegelman, (D), Cambridge, 1989, Yg
Assistant Professor:
Tony L. Burgess, (D), Arizona, 1988, Oo
Andrea K. Gerlak, (D), Arizona, 1997, Oo
Anupma Gupta, (D), Columbia, 2001, Gt
Lecturer:
DelWayne R. Bohnenstiehl, (D), Columbia, 2002, Gu
James R. Cochran, (D), Columbia, 1977, Yv
David T. Ho, (D), Columbia, 2001, Hg
Andrew A. Lacis, (D), Iowa, 1970, Oa
Dorothy M. Peteet, (D), New York, 1983, Pl
Drew T. Shindell, (D), SUNY (Stony Brook), 1995, Oa
Christopher Small, (D), California (San Diego), 1993, Yr
Bruno Tremblay, (D), McGill, 1996, Op
Jeffrey K. Weissel, (D), New South Wales, 1982, Yr
Adjunct Professor:
Robert F. Anderson, (D), MIT, 1981, Oc
Roger N. Anderson, (D), California (San Diego), 1973, Yh
W. Roger Buck, IV, (D), MIT, 1984, Yg
Anthony D. Del Genio, (D), California (Los Angeles), 1978, Oa
James E. Hansen, (D), Iowa, 1967, Oa
Kim A. Kastens, (D), California (San Diego), 1981, Gu
Arthur L. Lerner-Lam, (D), California (San Diego), 1982, Ys
Douglas G. Martinson, (D), Columbia, 1982, Op
Hsien Wang Ou, (D), MIT, 1979, Op
Walter C. Pitman, III, (D), Columbia, 1967, Yr
David H. Rind, (D), Columbia, 1976, Oa
William B. F. Ryan, (D), Columbia, 1971, Gu
Taro Takahashi, (D), Columbia, 1957, Oc
Spahr C. Webb, (D), California (San Diego), 1984, Yr
Emeritus:
Roger L. Batten, (D), Columbia, 1955, Pi
Rhodes W. Fairbridge, (D), W Australia, 1947, Gm
Malcolm C. McKenna, (D), California (Berkeley), 1958, Pv
Norman D. Newell, (D), Yale, 1933, Pi
Cooperating Faculty:
Lisa Fish, (M), Oo

Lamont-Doherty Earth Observatory (M,D) (2007)
P.O. Box 1000
61 Route 9W
Palisades, NY 10964
p. (845) 359-2900 f. (845) 359-2931
director@ldeo.columbia.edu
http://www.ldeo.columbia.edu
Enrollment (2006): M: 40 (0) D: 83 (6)
Director:
G. Michael Purdy, (D), Cambridge, 1974, Yr
Chair:
Mark A. Cane, (D), MIT, 1975, Op
Professor:
Wallace S. Broecker, (D), Columbia, 1957, Cg
Nicholas Christie-Blick, (D), California (Santa Barbara), 1979, Gs
Peter Eisenberger, (D), Harvard, 1967, Oo
Goran Ekstrom, (D), Harvard, 1987, Yg

Richard Fairbanks, (D), Brown, 1977, Cg
Steven Goldstein, (D), Columbia, 1986, Cg
Arnold L. Gordon, (D), Columbia, 1965, Op
Dennis E. Hayes, (D), Columbia, 1966, Yr
Peter Kelemen, (D), Washington, 1987, Gi
William H. Menke, (D), Columbia, 1981, Ys
John C. Mutter, (D), Columbia, 1982, Yr
Paul E. Olsen, (D), Yale, 1983, Gm
Paul G. Richards, (D), Caltech, 1970, Ys
Peter Schlosser, (D), Heidelberg, 1985, Hw
Christopher H. Scholz, (D), MIT, 1967, Ys
H. James Simpson, (D), Columbia, 1970, Cg
David Walker, (D), Harvard, 1972, Gx

Senior Scholar:
Robert F. Anderson, (D), MIT, 1981, Cg
Roger N. Anderson, (D), California (San Diego), 1973, Yr
Edward R. Cook, (D), Arizona, 1985, Hw
William B. F. Ryan, (D), Columbia, 1961, Yr
Taro Takahashi, (D), Columbia, 1957, Cm

Senior Research Scientist:
Robin E. Bell, (D), Columbia, 1989, Yr
Roger W. Buck, (D), MIT, 1984, Yr
Suzanne Carbotte, (D), California (Santa Barbara), 1992, Yr
Dake Chen, (D), SUNY (Stony Brook), 1989, Op
Steven Chillrud, (D), Columbia, 1995, Cg
James R. Cochran, (D), Columbia, 1977, Yr
David S. Goldberg, (D), Columbia, 1985, Yr
Stanley Jacobs, (B), MIT, 1962, Op
Kim Kastens, (D), California (San Diego), 1981, Gu
Won-Young Kim, (D), Uppsala (Sweden), 1986, Ys
Yochanan Kushnir, (D), Oregon State, 1985, Oa
Arthur L. Lerner-Lam, (D), California (San Diego), 1982, Ys
John Longhi, (D), Harvard, 1976, Gi
Douglas G. Martinson, (D), Columbia, 1982, Op
Hsien W. Ou, (D), MIT/Woods Hole, 1979, Op
Richard Seager, (D), Columbia, 1990, Ow
Leonardo Seeber, (B), Columbia, 1964, Ys
Christopher Small, (D), California (San Diego), 1993, Or
William Smethie, (D), Washington, 1979, Cg
Michael Steckler, (D), Columbia, 1980, Yv
Mingfang Ting, (D), Princeton, 1990, Oa
Alexander Van Geen, (D), MIT/Woods Hole, 1989, Cg

Associate Professor:
Mark H. Anders, (D), California (Berkeley), 1989, Gc
Peter deMenocal, (D), Columbia, 1991, Pe
Kevin Griffin, (D), Duke, 1994, Oo
Sidney Hemmings, (D), SUNY (Stony Brook), 1994, Cg
Marc Spiegelman, (D), Cambridge, 1989, Ys

Associate Research Scientist:
Andrew Barclay, (D), Oregon, 1998, Ys
Andrew Goodwillie, (D), Oxford, 1993, Yr
Gilles Guerin, (D), Columbia, 2000, Gu
Huei-Ping Huang, (D), Illinois, 1997
Andrew Juhl, (D), Scripps, 2000, Ob
Timothy Kenna, (D), MIT/Woods Hole, 2002, Cg
Samar Khatiwala, (D), Columbia, 2000, Op
Juerg Matter, (D), Swiss Fed Inst of Tech, 2001, Cg
Robert Newton, (D), Columbia, 2001, Ct
Frank Nitsche, (D), Alfred Wegener Inst (Germany), 1997, Gu
Joerg Schaefer, (D), Swiss Fed Inst, 2000, Pe
David Schaff, (D), Stanford, 2001, Ys
Susanne Straub, (D), Kiel (Germany), 1991, Gv
Ajit Subramaniam, (D), SUNY (Stony Brook), 1995, Or
Andreas Thurnherr, (D), Southampton, 1999, Op
Tina van de Flierdt, (D), ETH (Switzerland), 2003, Cg
Trevor Williams, (D), Edinburgh, 1994, Gu
Gisela Winckler, (D), Heidelberg, 1998, Cg
William Wright, (D), Arizona, 2001
Christopher Zappa, (D), Washington, 1999, Op

Special Research Scientist:
Pierre E. Biscaye, (D), Yale, 1964, Cm
Enrico Bonatti, (D), Pisa, 1967, Yr
Kenneth Hunkins, (D), Stanford, 1960, Yg
Klaus H. Jacob, (D), Goethe (Frankfurt), 1968, Ys
Gordon Jacoby, (D), Columbia, 1971, Hw
George J. Kukla, (D), Charles (Prague), 1953, Pe
Walter Pitman, (D), Columbia, 1967, Yr
Robert D. Stoll, (D), Columbia, 1962, Ye

Senior Scientist:
Spahr Webb, (D), California (San Diego), 1984, Yr

Senior Research Scientist:
John Diebold, (D), Columbia, 1980, Yr
Alberto Malinverno, (D), Columbia, 1989, Gu

Research Scientist:
Brendon Buckley, (D), Tasmania, 1997, Oo
Connie Class, (D), Karlsruhe, 1994, Cg
James Gaherty, (D), MIT, 1995, Ys
David Ho, (D), Columbia, 2001, Cg
Geraldo Iturrino, (D), Miami, 1995, Yr
Alexey Kaplan, (D), Gubkin Inst of Tech (Moscow), 1990, Op
Mikhail Kogan, (D), Inst of Physics (Moscow), 1977, Yd
Beate Liepert, (D), Ludwig Maximilians, 1995, Ow
Wade McGillis, (D), California (Berkeley), 1993, Cm
Naomi Naik, (D), Wisconsin, 1987
Raymond N. Sambrotto, (D), Alaska, 1983, Ob
Bruce Shaw, (D), Chicago, 1989, Ys
Colin Stark, (D), Leeds (UK), 1991, Gg
Michael Studinger, (D), Alfred Wegener Inst (Germany), 1998, Yr
Maria Tolstoy, (D), California (San Diego), 1994, Yr
Felix Waldhauser, (D), ETH (Switzerland), 1996, Ys
Xiaojun Yuan, (D), California (San Diego), 1994, Op

Cornell University

Dept of Earth & Atmospheric Sciences, (B,M,D) (2004)
Colleges of Engineering, CALS & Arts & Sci
2122 Snee Hall
Ithaca, NY 14853-1504
p. (607) 255-5267 f. (607) 254-4780
cugeosci@geology.cornell.edu
http://www.eas.cornell.edu/
Enrollment (1997): B: 47 (18) M: 2 (4) D: 41 (6)

Chair:
Teresa A. Jordan, (D), Stanford, 1979, Gr

Professor:
Richard W. Allmendinger, (D), Stanford, 1979, Gc
Muawia Barazangi, (D), Columbia, 1971, Ys
Larry D. Brown, (D), Cornell, 1976, Ye
Lawrence M. Cathles, (D), Princeton, 1968, Eg
John L. Cisne, (D), Chicago, 1973, Pg
Steven Colucci, (D), SUNY (Albany), 1982, Oa
Kerry H. Cook, (D), North Carolina State, 1984, Oa
Bryan L. Isacks, (D), Columbia, 1965, Ys
Robert W. Kay, (D), Columbia, 1970, Gi
Suzanne M. Kay, (D), Brown, 1975, Gi
Susan Riha, (D), Washington, 1980, Sf
Tammo S. Steenhuis, (D), Wisconsin, 1977, So
William M. White, (D), Rhode Island, 1977, Cc
Daniel Wilks, (D), Oregon, 1986, Oa

Associate Professor:
Arthur DeGaetano, (D), Rutgers, 1989, Oa
Louis A. Derry, (D), Harvard, 1989, Cl
Charles H. Greene, (D), Washington, 1985, Ob
David Hysell, (D), Cornell, 1992, Yg

Senior Research Associate:
Michelle F. Goman, (D), California (Berkeley), 1996, Oo

Research Associate:
Francisco Gomez, (D), Cornell, 1999, Gt
Steven Losh, (D), Yale, 1985, Eg
Bruce Monger, (D), Hawaii, 1993, Ob
Alexandra Moore, (D), Harvard, 1993, Oe
Maura S. Weathers, (D), Cornell, 1978, Oo
Michael Wizevich, (D), Eg
Senior Lecturer:
Mark Wysocki, (M), Cornell, 1988, Oa
Adjunct Associate Professor:
Warren D. Allmon, (D), Harvard, 1988, Pg
James Knapp, (D), MIT, 1983, Gc
Adjunct Professor:
Dogan Seber, (D), Cornell, 1996, Ys
Visiting Scientist:
Kodjopa Attoh, (D), Eg
John Gephart, (D), Eg
Visiting Lecturer:
William Brice, (D), Gg
Emeritus:
William A. Bassett, (D), Columbia, 1959, Gz
John M. Bird, (D), RPI, 1962, Gt
Arthur L. Bloom, (D), Yale, 1959, Gm
Daniel E. Karig, (D), California (San Diego), 1970, Yr
Warren Knapp, (D), Wisconsin, 1968, Oa
Frank H. T. Rhodes, (D), Birmingham, 1950, Pi
Donald L. Turcotte, (D), Caltech, 1958, Yg
Cooperating Faculty:
Peter Gierasch, (D), Oo
Michael Kelley, (D), California (Berkeley), 1970, Oa
Fred H. Kulhawy, (D), California (Berkeley), 1969, Ng
Thomas O'Rourke, (D), Illinois, 1975, Ng
Jean-Yves Parlange, (D), Brown, 1962, Hg
Andy L. Ruina, (D), Brown, 1981, Yx
A. L. Ruoff, (D), Utah, 1955, Cp
Steven W. Squyres, (D), Cornell, 1981, Xg
Joseph Veverka, (D), Harvard, 1969, Xg

Inst for the Study of the Continents (2005)
3122 Snee Hall
Ithaca, NY 14853-1504
p. (607) 255-3474 f. (607) 254-4780
easinfo@cornell.edu
http://www.eas.cornell.edu/
Enrollment (1998): B: 39 (8) M: 3 (3) D: 42 (6)
Director:
Larry D. Brown, (D), Cornell, 1976, Ye
Associate Director:
Muawia Barazangi, (D), Columbia, 1971, Ys
Professor:
Lawrence M. Cathles, (D), Princeton, 1968, Eg
John L. Cisne, (D), Chicago, 1973, Pg
Bryan L. Isacks, (D), Columbia, 1965, Ys
Teresa A. Jordan, (D), Stanford, 1979, Gr
Robert W. Kay, (D), Columbia, 1970, Gi
Suzanne M. Kay, (D), Brown, 1975, Gp
Donald L. Turcotte, (D), Caltech, 1958, Yg
Research Staff:
Dogan Seber, (D), Cornell, 1995, Ys
Adjunct Associate Professor:
James H. Knapp, (D), MIT, 1989, Oo
Adjunct Professor:
K. Douglas Nelson, (D), SUNY (Albany), 1979, Gt
Visiting Scientist:
John W. Gephart, (D), Brown, 1986, Gc

Dowling College
Dept of Earth and Marine Sciences (1999)
KSC 116
Oakdale, NY 11769-1999
p. (631) 244-3394 f. (631) 589-6644
Tanacre@dowling.edu

Graduate School of the City Univ of New York
PhD Program in Earth & Environmental Sci (D) (2004)
365 Fifth Avenue
New York, NY 10016
p. (212) 817-8240 f. (212) 817-1513
josleeb@gc.cuny.edu
Administrative Assistant: Lina C. McClain
Enrollment (2004): D: 40 (4)
Executive Officer:
Jeffrey P. Osleeb, (D), SUNY (Buffalo), 1974, Oo
Distinguished Professor:
Gerald M. Friedman, (D), Columbia, 1952, Gs
Professor:
Hannes K. Brueckner, (D), Yale, 1968, Cc
John A. Chamberlain, (D), Rochester, 1971, Po
Nicholas K. Coch, (D), Yale, 1965, Gs
Kathleen Crane, (D), California (San Diego), 1977, Yh
Eric Delson, (D), Columbia, 1973, Pv
Robert M. Finks, (D), Columbia, 1959, Pi
Stanley D. Gedzelman, (D), MIT, 1970, Ow
Victor Goldsmith, (D), Massachusetts, 1972, On
Daniel Habib, (D), Penn State, 1965, Pl
Charles A. Heatwole, (D), Michigan State, 1974, Oo
Edward E. Hindman, (D), Washington, 1975, Ow
Reza M. Khanbilvardi, (D), Penn State, 1983, Hy
Arthur M. Langer, (D), Columbia, 1965, Gz
Irene S. Leung, (D), California (Berkeley), 1969, Gz
David J. Leveson, (D), Columbia, 1960, Oo
David C. Locke, (D), Kansas State, 1965, Ca
Allan Ludman, (D), Pennsylvania, 1969, Gg
Sara L. McLafferty, (D), Iowa, 1979, Oo
Cherukupalli E. Nehru, (D), Madras (India), 1963, Gi
Robert S. Prezant, (D), Delaware (Lewes), 1981, Py
David E. Seidemann, (D), Yale, 1976, Cc
David H. Speidel, (D), Penn State, 1964, Cg
Dennis Weiss, (D), New York, 1971, Pg
Margaret S. Winslow, (D), Columbia, 1979, Gc
Associate Professor:
Sean C. Ahearn, (D), Wisconsin, 1986, Or
Stephen U. Aja, (D), Washington State, 1989, Cl
Patrick W. G. Brock, (D), Leeds, 1963, Gg
Lin A. Ferrand, (D), Princeton, 1988, Hq
Patricia M. Kenyon, (D), Cornell, 1986, Yg
Cecilia M. McHugh, (D), Columbia, 1993, Ou
Inez Miyares, (D), Arizona State, 1994, Oo
Jeffrey Steiner, (D), Stanford, 1970, Gx
Assistant Professor:
Charles R. Ehlschlaeger, (D), California (Santa Barbara), 1998, Oo
Mohamed B. Ibrahim, (D), Alberta, 1985, Oy
Robert P. Nolan, (D), CUNY, 1986, Co
Wayne G. Powell, (D), Queen's, 1994, Gp
William G. Wallace, (D), SUNY (Stony Brook), 1996, Og
Yan Zheng, (D), Columbia, 1999, Cm
Adjunct Professor:
Niles Eldredge, (D), Columbia, 1969, Pg
George E. Harlow, (D), Princeton, 1977, Gz
Neil H. Landman, (D), Yale, 1982, Ps
Edmond A. Mathez, (D), Washington, 1981, Gi
Jin Meng, (D), Columbia, 1991, Pv
Martin Prinz, (D), Columbia, 1961, Gi
Karl H. Szekielda, (D), Aix (France), 1967, Or

Emeritus:
Somdev Bhattacharji, (D), Chicago, 1959, Gt
Saul B. Cohen, (D), Harvard, 1955, Oy
Otto L. Franke, (D), Karlsruhe, 1962, Gc
William H. Harris, (D), Brown, 1971, Hw
Richard S. Liebling, (D), Columbia, 1963, Gz
Peter H. Mattson, (D), Princeton, 1957, Gc
Andrew McIntyre, (D), Columbia, 1967, Pe
Joaquin Rodriguez, (D), Indiana, 1960, Pi
Surenda K. Saxena, (D), Uppsala (Sweden), 1964, Cg
B. Charlotte Schrieber, (D), RPI, 1974, Gd
Frederick C. Shaw, (D), Harvard, 1965, Pi
David L. Thurber, (D), Columbia, 1964, Cl

Hamilton College

Dept of Geology (B) (2007)
198 College Hill Road
Clinton, NY 13323
p. (315) 859-4367 f. (315) 859-4807
trayne@hamilton.edu
http://www.hamilton.edu/academics/
Enrollment (2007): B: 27 (10)

Chair:
Todd W. Rayne, (D), Wisconsin, 1993, Hy
Professor:
Cynthia R. Domack, (D), Rice, 1985, Pg
Eugene W. Domack, (D), Rice, 1982, Gs
Barbara J. Tewksbury, (D), Colorado, 1981, Gc
Associate Professor:
David G. Bailey, (D), Washington State, 1990, Gi
Assistant Professor:
Michael L. McCormick, (D), Michigan, 2002, Py
Emeritus:
Donald B. Potter, (D), Caltech, 1954, Gg

Hartwick College

Dept of Geological and Environmental Sci (B) (2007)
Johnstone Science Center
Oneonta, NY 13820
p. (607) 431-4658 f. (607) 431-4374
Johnsone@hartwick.edu
http://www.hartwick.edu/x13179.xml
Administrative Assistant: Nancy Heffeman
Enrollment (2007): B: 30 (6)

Chair:
Eric L. Johnson, (D), SUNY (Binghamton), 1990, Gp
Professor Emeritus:
David Hutchison, (D), West Virginia, 1968, GxgGi
Professor:
Robert C. Titus, (D), Boston, 1974, Ps
Assistant Professor:
David H. Griffing, (D), Binghamton, 1994, Gd

Hobart & William Smith Colleges

Dept of Geoscience (B) (2007)
Lansing Hall
Geneva, NY 14456
p. (315) 781-3586 f. (315) 781-3860
geoscience@hws.edu
academica.hws.edu/geoscience
Enrollment (2007): B: 19 (9)

Chair:
Nan Crystal Arens, (D), Harvard, 1993, Po
Professor:
John D. Halfman, (D), Duke, 1987, Gn
D. Brooks McKinney, (D), Johns Hopkins, 1985, Gx
Assistant Professor:
Tara M. Curtin, (D), Arizona, Pe
Neil Laird, (D), Illinois, Ow
Adjunct Professor:
David C. Kendrick, (D), Harvard, 1997, Po
Technician:
Barbara Halfman, Oo

Hofstra University

Geology Dept (B) (2007)
114 Hofstra University
Hempstead, NY 11549
p. (516) 463-5564 f. (516) 463-5120
geocmm@hofstra.edu
http://www.hofstra.edu/Academics/Colleges/HCLAS/GEOL/index.html
Administrative Assistant: Lena Hiller
Enrollment (2007): B: 35 (5)

Chair:
Charles M. Merguerian, (D), Columbia, 1985, Gc
Professor:
Dennis Radcliffe, (D), Queen's, 1966, Gz
Associate Professor:
J Bret Bennington, (D), Virginia Tech, 1994, Pi
Assistant Professor:
Emma Christa Farmer, (D), Columbia, 2005, Ou
Adjunct Professor:
Nehru Cherukupalli, (D), Madras, 1963, Gg
Lillian Hess Tanguay, (D), CUNY, 1993, Gg
Richard Liebling, (D), Columbia, 1963, Gg
Steven C. Okulewicz, (M), CUNY (Brooklyn), 1979, Gg

Hudson Valley Community College

Biology, Chemistry, Physics Dept (2007)
80 Vandenburgh Ave.
Troy, NY 12180
p. (518) 629-7453 f. (518)629-7799
schaepet@hvcc.edu

Assistant Professor:
Ruth H. Major, (M), Syracuse, 1989, Ggh

Hunter College (CUNY)

Dept of Geography (B,M) (2006)
695 Park Avenue
Room 1006 North Building
New York, NY 10021
p. (212) 772-5265 f. (212) 772-5268
office@geo.hunter.cuny.edu
http://www.geo.hunter.cuny.edu
Administrative Assistant: Dana G. Reimer
Enrollment (2006): B: 135 (30) M: 40 (8)

Chair:
William Solecki, (D), Rutgers, 1990, Ou
Professor:
Sean C. Ahearn, (D), Wisconsin, 1986, Oi
Philip Gersmehl, (D), Georgia, 1970, Oy
Charles A. Heatwole, (D), Michigan State, 1974, Oo
Ines Miyares, (D), Arizona State, 1994, Oo
Senior Scientist:
Karl H. Szekielda, (D), Marseille, 1967, Or
Associate Professor:
Jochen Albrecht, (D), Vechta (Germany), 1995, Oi
Allan Frei, (D), Rutgers, 1997, Hs
Rupal Oza, (D), Rutgers, 1999, Oo
Marianna Pavlovskaya, (D), Clark, 1998, Oi
Haydee Salmun, (D), Johns Hopkins, 1989, Yg
Assistant Professor:
Frank Buonaiuto, (D), SUNY (Stony Brook), 2003, On

Hongmian Gong, (D), Georgia, 1997, Oo
Mohamed Ibrahim, (D), Alberta, 1985, Oo
Wenge Ni-Meister, (D), Boston, 1997, Or
Randye L. Rutberg, (D), Columbia, 2000, Cc
Director, SPARs Lab:
Thomas Walter, (M), Miami (Ohio), 1984, Oo
Research Associate:
Carol Gersmehl, (M), Georgia, 1970, Oe
Adjunct Professor:
Jack Eichenbaum, (D), Michigan, 1972, Oy
Anthony Grande, (M), CUNY (Baruch), 1999, Oo
Edward Linky, (D), Duquesne Law, 1973, Oo
Teodosia Manecan, (D), Bucharest, 1985, Gp
Faye Melas, (D), CUNY, 1989, Gs
Douglas Williamson, (D), CUNY, 2003, Oi
GeoScience Lab Tech:
Amy Jeu, (M), Minnesota, 2002, Or

Jefferson Community College

Science Div (A) (2007)
1220 Coffeen Street
Watertown, NY 13601
p. (315) 786-2320 f. (315) 788-0716
fflorence@sunyjefferson.edu
Enrollment (2006): A: 2 (0)
Associate Professor:
Frank P. Florence, (D), RPI, 1991, Gg

Lehman College (CUNY)

Environmental, Geographic, and Geological Sci (B) (2007)
250 Bedford Park Boulevard West
Bronx, NY 10468-1589
p. (718) 960-8660 f. (718) 960-8584
heather.sloan@lehman.cuny.edu
http://www.lehman.edu/deannss/geography/
Enrollment (2004): B: 9 (4)
Professor:
Eric Delson, (D), Pv
Irene S. Leung, (D), California (Berkeley), 1969, Gz
Zong-Guo Xia, (D), CUNY, 1993, Oi
Associate Professor:
Juliana Maantay, (D), Oi
Assistant Professor:
Hari Pant, (D), Cg
Heather Sloan, (D), Paris 6, 1993, Yr
Emeritus:
Edward F. Bergman, (D), Washington, 1971, Oo
Frederick C. Shaw, (D), Harvard, 1965, Ps

Long Island University, Brooklyn Campus

Dept of Physics (A) (2004)
Brooklyn, NY 11201
p. (718) 488-1054 f. (718) 488-1465
Professor:
Richard Macomber, (D), Iowa, 1963, Pg
Adjunct Professor:
Richard A. Jackson, (D), Massachusetts, 1980, Gc
Alan Siegelberg, (M), Brooklyn, 1977, Gg
Emeritus:
Samuel R. Kamhi, (D), Columbia, 1963, Gz

Long Island Univ, C.W. Post Campus

Dept of Earth & Environmental Sciences (B,M) (2005)
720 Northern Boulevard
Brookville, NY 11548-1300
p. (516) 299-2318 f. (516) 299-3945
maboorst@liu.edu
Administrative Assistant: Beth Rondot
Enrollment (1999): B: 1 (0)
Chair:
Margaret F. Boorstein, (D), Columbia, 1977, Og
Associate Professor:
Victor DiVenere, (D), Columbia, 1995, Ym
Lillian Hess-Tanguay, (D), CUNY, 1993, Gs
E Mark Pires, (D), Michigan State, 1998, Og
Emeritus:
Robert S. Harrison, (D), Cambridge, 1965, Oy
Heinrich Toots, (D), Wyoming, 1965, Pg

Molloy College

Dept of Chemistry, Physics, Earth Sciences (A) (2004)
Rockville Centre, NY 11570
p. (516) 678-5000 (Ext. 263)
jwinter@molloy.edu
Enrollment: No data reported since 1995
Chair:
John H. Winter, (D), Columbia, 1989, Ga

Nassau Community College

Dept of Physical Sciences (Geology) (A) (2007)
1 Education Drive
Garden City, NY 11530
p. (516) 572-7278 f. (516) 572-7750
riganol@ncc.edu
http://www.sunynassau.edu/dptpages/physci
Department Secretary: Grace Miller
Enrollment (2007): A: 5 (0)
Chair:
Lance S. Rigano, (M), Adelphi, 1988, Hw
Professor:
Theodore Benitt, (M), CUNY (Brooklyn), 1970, Gz
Michael J. Sichko, Jr., (M), CUNY (Brooklyn), 1965, Gi
Associate Professor:
Norman Harris, (M), Adelphi, 1976, Gg
Instructor:
David A. Stern, (D), CUNY, 2001, Ge
Joann Thissen, (M), SUNY (Stony Brook), Gg

Onondaga Community College

Chemistry & Physical Science Dept (A) (2004)
Syracuse, NY 13215
p. (315) 469-7741 (Ext. 2432)
giganted@sunyocc.edu
Associate Professor:
George C. Kelley, (D), Syracuse, 1975, Og
Assistant Professor:
Margaret J. Harris, (M), SUNY (Cortland), 1987, Gg
Adjunct Professor:
Thomas Hauf, (B), Florida State, 1984, Ow
Barbara M. Hill, (M), Cal State (Long Beach), 1989, Ge
Renaldo Mercado, (M), Ge
Barbara K. Nissen, (M), Syracuse, 1990, Ge
Edward S. Ponto, (M), Syracuse, 1975, Og
Peggy Vance, (B), California (Santa Cruz), Ge

Orange County Community College

Dept of Science & Engineering (A) (2007)
Middletown, NY 10940
p. (845) 341-4570 f. (845) 341-4594
lobrien@sunyorange.edu
Enrollment (2007): A: 1 (0)
Professor:
Lawrence E. O'Brien, (M), Michigan, 1972, Gg

New York

Pace University, New York Campus

Dept of Chemistry & Physical Sciences (2004)
1 Pace Plaza
New York, NY 10038
p. (212) 346-1502 f. (212) 346-1256
eweiser@fsmail.pace.edu
Department Secretary: Pat Calegari

Assistant Professor:
Stephen T. Lofthouse, (M), Hunter (CUNY), 1974, Og

Adjunct Professor:
Anatole Dolgoff, (M), Miami (Ohio), 1960, Og
William Hansen, (M), Hunter, 1991, Oa
John Marchisin, (M), Rutgers, 1965, Og
Nathan Reiss, (D), New York, 1973, Oa

Paleontological Research Institution

Paleontological Research Inst (2005)
1259 Trumansburg Road
Ithaca, NY 14850
p. (607) 273-6623 f. (607) 273-6620
allmon@museumoftheearth.org
http://www.priweb.org
Administrative Assistant: Starr Hays

Director:
Warren D. Allmon, (D), Harvard, 1988, Pe

Education Director:
Robert M. Ross, (D), Harvard, 1990, Po

Plattsburgh State University (SUNY)

Center for Earth & Environmental Science (B) (2007)
101 Broad Street
102 Hudson Hall
Plattsburgh, NY 12901
p. (518) 564-2028 f. (518) 564-5267
cees@plattsburgh.edu
http://www.plattsburgh.edu/cees
Enrollment (2006): B: 58 (0)

Distinguished Service Professor:
James C. Dawson, (D), Wisconsin, 1970, Ge

Professor:
Donald D. Adams, (D), Dalhousie, 1973, Cl
Donald J. Bogucki, (D), Tennessee, 1970, Or
David A. Franzi, (D), Syracuse, 1984, Gl
Thomas H. Wolosz, (D), SUNY (Stony Brook), 1984, Pe

Assistant Professor:
Edwin A. Romanowicz, (D), Syracuse, 1993, Hw

Adjunct Professor:
Carol Treadwell-Steitz, (D), New Mexico, 1996, Gg

Queens College (CUNY)

School of Earth & Environmental Sciences (B,M,D) (2006)
65-30 Kissena Boulevard
Flushing, NY 11367
p. (718) 997-3300 f. (718) 997-3299
geology@qc.edu
http://www.qc.edu/EES
Department Secretary: Laura Mancia
Enrollment (2000): B: 76 (13) M: 7 (1)

Chair:
Daniel Habib, (D), Penn State, 1965, Pl

Professor:
Hannes K. Brueckner, (D), Yale, 1968, Cg
Nicholas K. Coch, (D), Yale, 1965, Gs
Robert M. Finks, (D), Columbia, 1959, Pi
Allan Ludman, (D), Pennsylvania, 1969, Gg
David H. Speidel, (D), Penn State, 1964, Cg

Associate Professor:
Patrick W. G. Brock, (D), Leeds, 1963, Gg
Cecilia McHugh, (D), Columbia, 1993, Ou

Assistant Professor:
Yan Zhen, (D), Columbia, 1998, Cl

Research Associate:
Pamela Brock, (D), CUNY, 1993, Gg
Mohamad El Tabakah, (D), CUNY, 1994, Cl

Emeritus:
Eugene A. Alexandrov, (D), Columbia, 1959, Eg
Peter H. Mattson, (D), Princeton, 1957, Gc
Andrew McIntyre, (D), Columbia, 1967, Pe
B. Charlotte Schreiber, (D), RPI, 1974, Gd
David L. Thurber, (D), Columbia, 1964, Cl

Senior Lab Tech:
Daniel J. Klinger, (B), Queen's, 1998, Gm

Rensselaer Polytechnic Institute

Dept of Earth & Environmental Sciences (B,M,D) (2007)
Science Center 1W19
110 8th Street
Troy, NY 12180-3590
p. (518) 276-6474 f. (518) 276-2012
ees@rpi.edu
http://www.rpi.edu/dept/geo
Administrative Assistant: Kimberly Watson
Enrollment (2006): B: 14 (8) M: 5 (1) D: 5 (2)

Chair:
Frank S. Spear, (D), California (Los Angeles), 1976, Gp

Professor:
Teofilo A. Abrajano, Jr., (D), Washington, 1984, Cs
Robert McCaffrey, (D), California (Santa Cruz), 1981, Yg
Steven W. Roecker, (D), MIT, 1981, Yg
E. Bruce Watson, (D), MIT, 1976, Cp

Associate Professor:
Richard F. Bopp, (D), Columbia, 1979, Co

Senior Research Scientist:
Charles A. Williams, (D), Arizona, 1990, Yg

Research Associate Professor:
Daniele J. Cherniak, (D), SUNY (Albany), 1990, Gx
David A. Wark, (D), Texas, 1989, Gi

Research Associate:
Joseph M. Pyle, (D), RPI, 2001, Gp
Jay Thomas, (D), Ce

Emeritus:
M. Brian Bayly, (D), Chicago, 1962, Gc
Gerald Friedman, (D), Gs
Samuel Katz, (D), Columbia, 1955, Yg
Robert G. La Fleur, (D), RPI, 1961, Hw
Donald S. Miller, (D), Columbia, 1960, Cc

Director, Teaching & Lab Facilities:
Jonathan D. Price, (D), Oklahoma, 1998, Cp

Skidmore College

Dept of Geosciences (B) (2007)
815 North Broadway
Saratoga Springs, NY 12866
p. (518) 580-5190 f. (518) 580-5199
kcartwri@skidmore.edu
Administrative Assistant: Eleanor Hutchins
Enrollment (2007): B: 11 (2)

Associate Professor:
Richard H. Lindemann, (D), RPI, 1980, Pi

Assistant Professor:
Kyle K. Nichols, (D), Vermont, 2002, Gm

St. Lawrence University
Dept of Geology (B) (2005)
Brown Hall
Canton, NY 13617-1475
p. (315) 229-5851 f. (315) 229-5804
jbursnall@stlawu.edu
Administrative Assistant: Sherrie Kelly
Enrollment (2005): B: 30 (14)
Chair:
John T. Bursnall, (D), Cambridge, 1975, Gx
Professor:
J. Mark Erickson, (D), North Dakota, 1971, Po
Associate Professor:
Catherine H. Shrady, (D), Massachusetts, 1990, Gc
Assistant Professor:
Stephen D. Robinson, (D), McGill, 2000, Hg
Technician:
Matt VanBrocklin, (M), Oo

Suffolk County Community College
School of Mathematics and Natural Sciences (A) (2004)
Selden, NY 11784
p. (516) 451-4315 f. (516) 451-4433
gaillaa@sunysuffolk.edu
Head:
Robert L. Warasila, (D), SUNY (Stony Brook), 1976, Xm
Professor:
Scott Mandia, (M), Penn State, 1990, Ow

SUNY College at Brockport
Dept of the Earth Sciences (B) (2005)
350 New Campus Drive
Brockport, NY 14420-2936
p. (585) 395-2636 f. (585) 395-2416
earthsci@esc.brockport.edu
http://www.esc.brockport.edu
Administrative Assistant: Lauri A. Mulley
Enrollment (2005): B: 100 (20)
Chair:
James A. Zollweg, (D), Cornell, 1994, Hs
Associate Professor:
Whitney J. Autin, (D), Louisiana State, 1989, Gs
Jose A. Maliekal, (D), Hawaii, 1989, Oa
Judy A. Massare, (D), Johns Hopkins, 1983, Pv
Mark R. Noll, (D), Delaware, 1989, Cl
Assistant Professor:
Paul L. Richards, (D), Penn State, 1999, Hg
Scott M. Rochette, (D), St. Louis, 1998, Ow
Stephen W. Weinbeck, (D), Texas Tech, 2005, Ow
Adjunct Professor:
David A. Boehm, (M), SUNY (Buffalo), 2003, Gg
Christine Crafts, (B), SUNY (Brockport), 2000, Ow
Jutta S. Dudley, (D), SUNY (Buffalo), 1998, Gg
Eugene Gartland, (M), Rochester, 1973, Gg
William G. Glynn, (M), Texas A&M, 1984, Gg
Paul J. Mroz, (D), SUNY (Buffalo), 1991, Ow
Linda J. Schaffer, (M), SUNY (Brockport), 1988, Oe
Emeritus:
Robert W. Adams, (D), Johns Hopkins, 1964, Gs
Robert M. Cassie, (D), Wisconsin, 1965, Gc
John E. Hubbard, (D), Colorado State, 1968, Hg
Richard M. Liebe, (D), Iowa, 1962, Ps
John M. Williams, (B), Goshen, Oa
On Leave:
Robert S. Weinbeck, (D), Iowa State, 1980, Oa
Systems Administrator:
Thomas M. McDermott, Oo

SUNY Maritime College
Science Dept (B) (2006)
6 Pennyfield Avenue
Bronx, NY 10465
p. (718) 409-7380
mdeangelis@sunymaritime.edu
Enrollment (2005): B: 89 (21)
Chair:
Kathy Olszewski, (D)
Associate Professor:
Marie deAngelis, (D), Oc
Assistant Professor:
Anthony Manzi, (M), Ow

SUNY New Paltz
Geological Sciences (B,M) (2007)
1 Hawk Drive
New Paltz, NY 12561
p. (845) 257-3760 f. (845) 257-3755
vollmerf@newpaltz.edu
http://www.newpaltz.edu/geology/
Enrollment (2007): B: 73 (14) M: 3 (1)
Chair:
Frederick W. Vollmer, (D), Minnesota, 1985, Gct
Professor:
Martin S. Rutstein, (D), Brown, 1969, Gz
Associate Professor:
Alvin S. Konigsberg, (D), Syracuse, 1969, Oa
Assistant Professor:
Alexander J. Bartholomew, (D), Cincinnati, 2006, GrPi
Shafiul H. Chowdhury, (D), W Michigan, 1999, Hw
John A. Rayburn, (D), Binghamton, 2004, Gml
Emeritus:
Gilbert J. Brenner, (D), Penn State, 1962, Pl
Constantine Manos, (D), Illinois, 1963, Gs
Russell H. Waines, (D), California (Berkeley), 1965, Pg
Related Staff:
Donald R. Hodder, (B), SUNY (Geneseo), 1985, Gg

SUNY Ulster County Community College
Mathematics, Physical Sciences, & Eng (A) (2006)
Burroughs 107
Cottekill Road
Stone Ridge, NY 12484
p. 845-687-5230 f. 845-687-5083
schimmrs@sunyulster.edu
http://people.sunyulster.edu/esc
Enrollment: No data reported since 1995
Assistant Professor:
Steven Schimmrich, (M), SUNY (Albany), 1991, Og
Instructor:
Karen Helgers, (M), SUNY (New Paltz), 1983, Og

SUNY, Albany
Dept of Earth & Atmospheric Sciences (B,M,D) (2007)
1400 Washington Avenue
Albany, NY 12222
p. (518) 442-4466 f. (518) 442-5825
chair@atmos.albany.edu
http://www.atmos.albany.edu
Administrative Assistant: Sharon Baumgardner
Administrative Assistant: Lynn M. Hughes
Enrollment (2004): B: 81 (27) M: 8 (5) D: 6 (4)
Chair:
Vincent P. Idone, (D), SUNY (Albany), 1982, Oa
Professor:
Lance F. Bosart, (D), MIT, 1969, Oa

New York

John W. Delano, (D), SUNY (Stony Brook), 1977, Cp
Kenneth L. Demerjian, (D), Ohio State, 1973, Oa
Daniel Keyser, (D), Penn State, 1981, Oa
William S. Kidd, (D), Cambridge, 1974, Gt
Arthur Z. Loesch, (D), Chicago, 1973, Oa
John E. Molinari, (D), Florida State, 1979, Oa
Senior Research Professor:
David R. Fitzjarrald, (D), Virginia, 1980, Oa
Lee C. Harrison, (D), Washington, 1982, Oa
Garland Lala, (D), SUNY (Albany), 1972, Oa
Richard R. Perez, (D), SUNY (Albany), 1983, Oa
James J. Schwab, (D), Harvard, 1983, Oa
Christopher J. Walcek, (D), California (Los Angeles), 1983, Oa
Wei-Chyung Wang, (D), Columbia, 1973, Oa
Associate Professor:
Robert G. Keesee, (D), Colorado, 1979, Oa
Andrei Lapenis, (D), St. Petersburg, 1986, Cg
Braddock K. Linsley, (D), New Mexico, 1990, Cs
Christopher D. Thorncroft, (D), Reading, 1988, Oa
Assistant Professor:
John G. Arnason, (D), Stanford, 1995, Gx
Karen I. Mohr, (D), Texas, 2000, Hy
Research Associate:
Stephen S. Howe, (M), Penn State, 1981, Cs
David Knight, (D), Washington, 1987, Oa
Lecturer:
Michael G. Landin, (M), SUNY (Albany), 1982, Ow
Visiting Professor:
Michael Reeder, (D), Monash
James Steenburgh, (D), Washington
Morris Weisman, (D), Penn State, 1990
Emeritus:
Gregory D. Harper, (D), California (Berkeley), 1980, Gt
Winthrop D. Means, (D), California (Berkeley), 1960, Gc
Akiho Miyashiro, (D), Tokyo, 1953, Gp
Volker A. Mohnen, (D), Munich, 1966, Oa
George W. Putman, (D), Penn State, 1961, Ge
Jon T. Scott, (D), Wisconsin, 1963, Oa

SUNY, Buffalo

Dept of Geology (B,M,D) (2007)
876 Natural Sciences Complex
Buffalo, NY 14260
p. (716) 645-6800 (Ext. 6100) f. (716) 645-3999
geology@buffalo.edu
http://www.geology.buffalo.edu
Administrative Assistant: Alison A. Lagowski
Enrollment (2007): B: 51 (13) M: 42 (7) D: 15 (2)
Chair:
Charles E. Mitchell, (D), Harvard, 1983, Po
Professor:
Richelle Allen-King, (D), Waterloo, 1991, Cg
Marcus I. Bursik, (D), Caltech, 1988, Gv
Mary Alice Coffroth, (D), Miami, 1988, Gu
Rossman F. Giese, (D), Columbia, 1962, Gz
Robert D. Jacobi, (D), Columbia, 1979, Gr
Howard R. Lasker, (D), Chicago, 1978, Gu
Research Professor:
Anton Schenk, (D), Switzerland, 1972
Gregory Valentine, (D), Santa Barbara, 1988
Associate Professor:
Matthew W. Becker, (D), Texas, 1996, Hw
Tracy K. P. Gregg, (D), Arizona State, 1995, Gv
Research Associate Professor:
Gregory S. Baker, (D), Kansas, 1999, Yg
Research Assistant Professor:
Thomas E. Jordan, (D), Buffalo, 2003, Yg
William Korth, (D), Pittsburgh, 1981, Pv
Gerald J. Smith, (D), SUNY (Buffalo), 1997, Gs
Assistant Professor:
Tracy L. Bank, (D), Virginia Tech, 2003, Cg
Jason P. Briner, (D), Colorado, 2003, Gl
Eliza Calder, (D), Bristol, 1999, Gv
Beata Csatho, Miskolc (Hungary), 1993, Yg
Zhangshuan Hou, (D), Berkeley, 2006, Yg
Research Assistant Professor:
Silvio DeAngelis, (D), Bologna, 2004
Research Associate:
Yarrow Axford, (D), Boulder, 2007
Adjunct Associate Professor:
H. David Sheets, (D), SUNY (Buffalo), Oo
Adjunct Assistant Professor:
Richard S. Laub, (D), Cincinnati, 1973
Gary S. Solar, (D), Maryland, 1999, Gc
Adjunct Professor:
Gordon C. Baird, (D), Rochester, 1975, Pg
Gary S. Lash, (D), Lehigh, 1980, Gc
Carel J. van Oss, (D), Paris, 1955, Gy
Visiting Professor:
Jorg Maletz, (D), Tech (Berlin), 1992, Pg
Distinguished Professor:
Michael F. Sheridan, (D), Stanford, 1965, Gv
Emeritus:
Parker Calkin, (D), Ohio, 1963
John S. King, (D), Wyoming, 1963, Gc
Paul H. Reitan, (D), Oslo, 1959, Gp

SUNY, Cortland

Geology Dept (B,M) (2007)
342 Bowers Hall
PO Box 2000
Cortland, NY 13045
p. (607) 753-2815 f. (607) 753-2927
GeologyDept@cortland.edu
http://www.cortland.edu/geology/
Administrative Assistant: Susan K. Nevins
Enrollment (2007): B: 48 (15) M: 8 (6)
Professor & Chair:
Christopher P. Cirmo, (D), Syracuse, 1994, Hg
Professor:
Robert S. Darling, (D), Syracuse, 1992, Gx
Associate Professor:
David J. Barclay, (D), SUNY (Buffalo), 1998, Gl
Christopher A. McRoberts, (D), Syracuse, 1994, Pg
Assistant Professor:
Gayle C. Gleason, (D), Brown, 1993, Gc
Lecturer:
Julie L. Barclay, (M), SUNY (Buffalo), 1998, Gg
Laboratory Director:
John R. Driscoll, (A), Northwest Electronic, 1974, Oo

SUNY, Fredonia

Dept of Geosciences (B) (2007)
Fredonia, NY 14063-1198
p. (716) 673-3303 f. (716) 673-3347
berkley@fredonia.edu
http://www.fredonia.edu/department/geosciences/
Enrollment (2007): B: 35 (3)
Chair:
John L. Berkley, (D), New Mexico, 1977, GxXgGe
Professor:
Gordon C. Baird, (D), Rochester, 1975, Ps
Walther M. Barnard, (D), Penn State, 1965, Gz
Gary G. Lash, (D), Lehigh, 1980, Gr
Michael P. Wilson, (D), Syracuse, 1981, GmHgYg

Associate Professor:

Ann K. Deakin, (D), SUNY (Buffalo), 1996, OyiGg

Instructor:

Kimberly Weborg-Benson, (M), Illinois, 1991, GgOaPy

SUNY, Geneseo

Dept of Geological Sciences (B) (2008)
1 College Circle
Geneseo, NY 14454
p. (585) 245-5291 f. (585) 245-5116
loretta@geneseo.edu
http://www.geneseo.edu/~gsci/
Administrative Assistant: Loretta A. Stratton
Enrollment (2007): B: 48 (12)

Chair:

D. Jeffrey Over, (D), Texas Tech, 1990, Ps

Professor:

Richard A. Young, (D), Washington (St. Louis), 1966, Gm

Assistant Professor:

Dori J. Farthing, (D), Johns Hopkins, 2001, Gz
Scott D. Giorgis, (D), Wisconsin, 2003, Gc
Benjamin J.C. Laabs, (D), Wisconsin, 2004, GmlGe
Amy L. Sheldon, (D), Utah, 2002, Hw

Emeritus:

Phillip D. Boger, (D), Ohio State, 1976, Cg
William J. Brennan, (D), Colorado, 1968, Gc
Richard B. Hatheway, (D), Cornell, 1969, Gx
James W. Scatterday, (D), Ohio State, 1963, Po

SUNY, Oneonta

Dept of Earth Sciences (B,M) (2006)
Ravine Parkway
209 Sci. #1
Oneonta, NY 13820-4015
p. (607) 436-3707 f. (607) 436-3547
blechmjb@oneonta.edu
http://www.oneonta.edu/academics/earths
Enrollment (2006): M: 5 (2)

Professor:

Jerome B. Blechman, (D), Wisconsin, 1979, Oa
James R. Ebert, (D), SUNY (Binghamton), 1984, Gr
P. Jay Fleisher, (D), Washington State, 1967, Gl

Associate Professor:

James R. Albanese, (D), RPI, 1979, Pg
Brent K. Dugolinsky, (D), Hawaii, 1976, Ou
Peter D. Muller, (D), SUNY (Binghamton), 1980, Gc

Assistant Professor:

Devin Castendyk, (D), Auckland (NZ), 2004, Hw
Richard Grimaldi, (D), SD Sch of Mines & Tech, 2003, Oa
Tatiana Vislova, (D), Minnesota, 2003, Gx

Emeritus:

Stephen Berman, (D), Wisconsin, 1972, Oa
Arthur N. Palmer, (D), Indiana, 1969, Hg
Robert T. Simmonds, (D), Illinois, 1961, Pg
Philo C. Wilson, (D), Washington State, 1954, Og
Duane D. Wohlford, (D), Colorado, 1965, Gp

SUNY, Oswego

Dept of Earth Sciences (B) (2007)
Route 104
Oswego, NY 13126
p. (315) 312-3065 f. (315) 312-2789
dvalenti@Oswego.EDU
http://www.oswego.edu/academics/colleges_and_departments/departments/earth_sciences/
Enrollment (2007): B: 40 (0)

Professor:

Alfred J. Stamm, (D), Wisconsin, 1976, Ow
David W. Valentino, (D), Virginia Tech, 1993, GtcGp

Associate Professor:

Robert J. Ballentine, (D), SUNY (Albany), 1979, Ow
Scott W. Roby, (D), Texas, 1987, Oo

Assistant Professor:

Steven T. Skubis, (D), SUNY (Albany), 1994, Ow
Scott Steiger, (D), Texas A&M, 2005, Ow
Paul B. Tomascak, (D), Maryland, 1995, Cg

Visiting Instructor:

Karen R. Valentino, (M), Virginia Tech, 1993, GgOm

Visiting Professor:

David L. Rodland, (D), Virginia Tech, 2004, Pi

Emeritus:

Ronald S. Chaldu, (D), Indiana, 1976, Oo
Anthony Del Prete Jr., (D), RPI, 1972, Og
Robert E. Maurer, (D), Utah, 1970, Gz
Raymond H. Schneider, (D), Tennessee, 1973, Ge
Paul M. Shaver, (D), Syracuse, 1969, Oe
David J. Thomas, (D), SUNY (Binghamton), 1972, Pi

Librarian:

Chris Hebblethwaite, Oo

SUNY, Potsdam

Dept of Geology (B) (2007)
220 Timerman Hall
44 Pierrepont Avenue
Potsdam, NY 13676
p. (315) 267-2286 f. (315) 267-2695
badgerrl@potsdam.edu
Administrative Assistant: Roberta Greene
Enrollment (2007): B: 42 (4)

Chair:

Robert L. Badger, (D), Virginia Tech, 1989, Gi

Professor:

Frank A. Revetta, (D), Rochester, 1970, Yg

Assistant Professor:

Lisa M. Amati, (D), Oklahoma, 2004, Pi
Christopher R. Kelson, (D), Georgia, 2006, Gz
Michael C. Rygel, (D), Dalhousie, 2005, Gs

Emeritus:

James D. Carl, (D), Illinois, 1962, Gz
William T. Kirchgasser, (D), Cornell, 1967, Ps
Neal R. O'Brien, (D), Illinois, 1963, Gs
Bradford B. Van Diver, (D), Washington, 1964, Gp

SUNY, Purchase

Program In Environmental Science (B) (2004)
Purchase, NY 10577
p. (914) 251-6646 f. (914) 251-6635
naturalsciences@purchase.edu
Administrative Assistant: Donna Shaw
Enrollment (1996): B: 44 (23)

Professor:

Gerard M. Capriulo, (D), SUNY (Stony Brook), 1982, Ob
Paul L. Steineck, (D), Louisiana State, 1973, Pm
Joel Tenenbaum, (D), Harvard, 1968, Oa

Associate Professor:

Jon C. Cooper, (D), Wisconsin, 1974, Oo
Maryann McEnroe, (D), California (Davis), 1981, Oc
James M. Utter, (D), Rutgers, 1971, Oo

Assistant Professor:

Barbara L. Dexter, (D), Oregon State, 1982, Ob
Emanuel D. Ottu, (D), McMaster, 1981, Oo

Geology Librarian:

Richard Arsenty, Oo

New York

SUNY, Stony Brook

Dept of Geosciences (B,M,D) (2006)
Nicolls Road
Stony Brook, NY 11794-2100
p. (631) 632-8200 f. (631) 632-8240
claire.ondrovic@sunysb.edu
http://www.geosciences.stonybrook.edu
Administrative Assistant: Claire A. Ondrovic
Enrollment (2005): B: 70 (15) M: 3 (4) D: 41 (3)

Professor:
Daniel M. Davis, (D), MIT, 1983, Yg
Gilbert N. Hanson, (D), Minnesota, 1964, Cc
William E. Holt, (D), Arizona, 1989, Ys
Robert C. Lieberman, (D), Columbia, 1969, Gy
Scott M. McLennan, (D), Australian National, 1981, Cg
Hanna Nekvasil, (D), Penn State, 1986, Cp
John B. Parise, (D), James Cook, 1980, Gz
Richard J. Reeder, (D), California (Berkeley), 1980, Cl
Martin A. Schoonen, (D), Penn State, 1989, Cl
Donald J. Weidner, (D), MIT, 1972, Gy
Teng-fong Wong, (D), MIT, 1981, Yx
Associate Professor:
Lianxing Wen, (D), Caltech, 1998, Ys
Assistant Professor:
Brian L. Phillips, (D), Illinois, 1990, Gz
E. Troy Rasbury, (D), SUNY (Stony Brook), 1998, Cc
Research Associate:
Liping Wang, (D), Gy
Curator:
Stephen C. Englebright, (M), SUNY (Stony Brook), 1975, Oe
Lecturer:
Christiane Wilson Stidham, (D), Univ, Yg
Adjunct Professor:
Robert C. Aller, (D), Yale, 1977, Cm
Henry J. Bokuniewicz, (D), Yale, 1976, Yr
Forster A. Catherine, (D), Pennsylvania, 1990, Pv
Jiuhua Chen, (D), Inst Materials Str Sci, 1994, Gy
J. Kirk Cochran, (D), Yale, 1979, Oc
Roger D. Flood, (D), MIT, 1978, Gu
David W. Krause, (D), Michigan, 1982, Pv
Baosheng Li, (D), SUNY (Stony Brook), 1996, Gy
Michael T. Vaughan, (D), SUNY (Stony Brook), 1979, Yx
Emeritus:
Garman Harbottle, (D), Columbia, 1949, Cc
Donald H. Lindsley, (D), Johns Hopkins, 1961, Cp
Laboratory Director:
Owen C. Evans, (D), SUNY (Stony Brook), 1994, Cg
Other:
Maureen O'Leary, (D), Johns Hopkins, 1997, Pv
Robert Rapp, (D), Gy

Marine Sciences Research Center (B,M,D) (2006)
Stony Brook, NY 11794-5000
p. (516) 632-8700 f. (516) 632-8820
dconover@notes.cc.sunysb.edu
http://www.msrc.sunysb.edu
Enrollment (1998): M: 72 (14) D: 96 (12)

Director:
David O. Conover, (D), Massachusetts, 1982, Ob
Distinguished Professor:
Robert C. Aller, (D), Yale, 1977, Cm
Nicholas S. Fisher, (D), SUNY (Stony Brook), 1974, Oc
Cindy Lee, (D), California (San Diego), 1975, Co
Professor:
Henry J. Bokuniewicz, (D), Yale, 1976, Yr
Malcolm J. Bowman, (D), Saskatchewan, 1970, Op
J. Kirk Cochran, (D), Yale, 1979, Oc
Roger D. Flood, (D), MIT/Woods Hole, 1978, Ou
Marvin A. Geller, (D), MIT, 1969, Oa
Sultan Hameed, (D), Manchester, 1968, Oa
Glenn R. Lopez, (D), SUNY (Stony Brook), 1976, Ob
Mary I. Scranton, (D), MIT/Woods Hole, 1977, Oc
Gordon T. Taylor, (D), S California, 1983, Ob
Prasad Varanasi, (D), California (San Diego), 1967, Oa
Dong-Ping Wang, (D), Miami, 1975, Op
Minghua Zhang, (D), Inst of Atm Physics, 1987, Oa
Senior Scientist:
Josephine Y. Aller, (D), S California, 1975, Ob
Peter M. J. Woodhead, (B), Durham (UK), 1953, Ob
Research Staff:
Robert A. Armstrong, (D), Minnesota, 1975, Oc
Associate Professor:
Robert Armstrong, (D), Ob
Bruce J. Brownawell, (D), MIT/Woods Hole, 1986, Oc
Robert M. Cerrato, (D), Yale, 1980, Ob
Edmund Chang, (D), Princeton, 1993, Oa
Brian Colle, (D), Washington, 1997, Oa
Christopher Gobler, (D), SUNY (Stony Brook), 1999, Ob
Paul F. Kemp, (D), Oregon State, 1985, Ob
Darcy J. Lonsdale, (D), Maryland, 1979, Ob
Kamazima M. Lwiza, (D), Wales, 1991, Oc
John E. Mak, (D), California (San Diego), 1992, Oa
Anne McElroy, (D), MIT/Woods Hole, 1985, Oc
Sergey Piontkovski, (D), Inst of Biology of the S Seas, 1978, Ob
Sergio A. Sanudo-Wilhelmy, (D), California (Santa Cruz), 1993, Oc
Assistant Professor:
Allam Bassem, (D), W Brittany (France), 1998, Ob
David Black, (D), Ou
Jackie Collier, (D), Stanford, 1994, Ob
Michael Frisk, (D), Ob
Daniel Knopf, (D), Swiss Fed Inst of Tech, 2003, Oa
Bradley Peterson, (D), S Alabama, 1998, Ob
Nicole Riemer, (D), Karlsruhe (Germany), 2002, Oa
Joseph Warren, (D), MIT, 2001, Ob
Research Professor:
Charles Flagg, (D), MIT, 1977, Op
Research Associate:
Frank J. Roethel, (D), SUNY (Stony Brook), 1981, Oc
Director, Waste Reduction & Management Inst:
R. L. Swanson, (D), Oregon State, 1971, Og
Adjunct Professor:
Douglas Hill, (D), Columbia, 1977, Oo
Lee K. Koppelman, (D), New York, 1968, On
Joel S. O'Connor, (D), Rhode Island, 1965, Og
Douglas W. R. Wallace, (D), Dalhousie, 1985, Oc
Distinguished Professor Emeritus:
Robert D. Cess, (D), Pittsburgh, 1959, Oa
Emeritus:
Peter K. Weyl, (D), Chicago, 1953, Og
Charles F. Wurster, (D), Stanford, 1957, Ob
Cooperating Faculty:
Theodore D. Goldfarb, (D), California (Berkeley), 1959, Oo
Stewart M. Harris, (D), Northwestern, 1964, Om
Herbert Herman, (D), Northwestern, 1961, Oo
Manuel Lerdau, (D), Stanford, 1992, Ob
Jeffrey Levinton, (D), Yale, 1971, Ob
Jack S. Mattice, (D), Syracuse, 1971, Ob
Sheldon Reaven, (D), California (Berkeley), 1975, Oo
Lawrence B. Slobodkin, (D), Yale, 1951, Ob
Duane E. Waliser, (D), California (San Diego), 1992, Oa

Syracuse University

Dept of Earth Sciences (B,M,D) (2004)
204 William B. Heroy Geology Laboratory
Syracuse, NY 13244-1070
p. (315) 443-2672 f. (315) 443-3363

kdnelson@syr.edu
http://earthsciences.syr.edu
Enrollment (2001): B: 22 (8) M: 12 (5) D: 8 (0)

Chair:
Scott D. Samson, (D), Arizona, 1990, Cc
Professor:
Henry T. Mullins, (D), North Carolina, 1978, Gu
Cathryn R. Newton, (D), California (Santa Cruz), 1983, Po
Donald I. Siegel, (D), Minnesota, 1981, Hw
Alumni Associate Professor:
Geoffrey O. Seltzer, (D), Minnesota, 1992, Gl
Associate Professor:
Suzanne L. Baldwin, (D), SUNY (Albany), 1988, Cc
Paul G. Fitzgerald, (D), Melbourne, 1988, Gt
Christopher A. Scholz, (D), Duke, 1989, Gs
Assistant Professor:
Linda C. Ivany, (D), Harvard, 1997, Pie
Adjunct Professor:
Charles T. Driscoll, Jr., (D), Cornell, 1980, Hy
Emeritus:
M. E. Bickford, (D), Illinois, 1960, Cc
Gary M. Boone, (D), Yale, 1959, Gp
James C. Brower, (D), Wisconsin, 1964, Po
Dirk De Waard, (D), Utrecht (Neth), 1947, Gp
Bryce M. Hand, (D), Penn State, 1964, Gs
Ernest H. Muller, (D), Illinois, 1952, Gm
John James Prucha, (D), Princeton, 1950, Gc
Joseph E. Robinson, (D), Alberta, 1968, Eo

The College of Saint Rose

Dept of Physical and Biological Sciences (B) (2007)
432 Western Ave.
Albany, NY 12203
p. 518-454-5140 f. 518-458-5446
smithj@strose.edu
http://www.strose.edu/Future_Students/Academics/School_of_Math_and_Science/science_ug.asp
Enrollment (2007): B: 7 (0)

Assistant Professor:
Eric Eslinger, (D), Case Western, 1971, EoGzHy
Stephanie Maes, (D), Wisconsin, 2006, GicEg
Jacqueline A. Smith, (D), Syracuse, 2005, GlmGe

Union College

Geology Dept (B) (2007)
807 Nott Street
Schenectady, NY 12308-3107
p. (518) 388-6770 f. (518) 388-6417
geology@union.edu
http://www.union.edu/PUBLIC/GEODEPT/
Administrative Assistant: Jean Conley
Enrollment (2007): B: 14 (6)

Chair:
John I. Garver, (D), Washington, 1989, GtCcGr
Research Professor:
Robert L. Fleischer, (D), Harvard, 1956, Cc
Professor:
Kurt T. Hollocher, (D), Massachusetts, 1985, GxCga
Donald T. Rodbell, (D), Colorado, 1991, GlmGe
George Shaw, (D), Washington, 1971, Yx
Assistant Professor:
Holli M. Frey, (D), Michigan, 2005, GvCag
Research Associate:
Robert Finks, (D), Columbia, 1959, Pg
Visiting Professor:
Jaclyn Cockburn, (M), Queens, 2004, HgGeHs
Related Staff:
William S. Neubeck, (M), SUNY (Binghamton), 1980, Gm

United States Military Academy

Dept of Geography & Environmental Eng (B) (2007)
West Point, NY 10996
p. (914) 938-2300 f. (914) 938-3339
bw0384@usma.edu
http://www.dean.usma.edu/geo/gene.htm
Department Secretary: Jean Keller
Enrollment (2006): B: 268 (0)

Head:
Wendell C. King, (D), Tennessee, 1988, Hs
Academy Professor:
Joe D. Manous, (D), Minnesota, 2000, Hs
Associate Professor:
John A. Brockhaus, (D), Idaho, 1987, Yd
Marie C. Johnson, (D), Brown, 1990, Gi
Michael B. Kelley, (D), RPI, 1996, Hs
Assistant Professor:
Jon C. Malinowski, (D), North Carolina, 1995, Oo
Instructor:
Christopher A. Gellasch, (M), Indiana, 1994, Gr

University of Rochester

Dept of Earth & Environmental Sciences (B,M,D) (2007)
227 Hutchison Hall
RC Box 270221
Rochester, NY 14627-0221
p. (585) 275-5713 f. (585) 244-5689
ees@earth.rochester.edu
http://www.earth.rochester.edu
Administrative Assistant: Marjorie Goodison
Enrollment (1998): M: 8 (5) D: 16 (1)

Chair:
Udo Fehn, (D), Munich, 1973, Eg
Professor:
Asish R. Basu, (D), California (Davis), 1975, Gx
Cynthia J. Ebinger, (D), MIT/Woods Hole, 1988, Gt
Gautam Mitra, (D), Johns Hopkins, 1977, Gc
Robert J. Poreda, (D), California (San Diego), 1983, Cg
John A. Tarduno, (D), Stanford, 1987, Ym
Associate Professor:
Carmala N. Garzione, (D), Arizona, 2000, Gs
Adjunct Assistant Professor:
William Chaisson, (D), Massachusetts, 1996, Pi
Emeritus:
Lawrence W. Lundgren, (D), Yale, 1958, Ge
Geology Librarian:
Isabel Kaplan, Oo
Other:
Rory D. Cottrell, (D), Rochester, 2000, Ym
Pennilyn Higgins, (D), Wyoming, 2000, Cs

Utica College

Dept of Geology (B) (2007)
1600 Burrstone Road
Utica, NY 13502
p. (315) 792-3134 f. (315) 792-3831
skanfoush@utica.edu
Enrollment: No data reported since 1995

Chair:
Sharon L. Kanfoush, (D), Florida, 2002, GgsOg

Vassar College

Dept of Geology & Geography (B) (2006)
Box 735
124 Raymond Avenue
Poughkeepsie, NY 12604-0735

p. (845) 437-5540 f. (845) 437-7577
geo@vassar.edu
http://geologyandgeography.vassar.edu/index.html
Administrative Assistant: Lois Horst
Enrollment (2005): B: 12 (6)

Chair:
Jill S. Schneiderman, (D), Harvard, 1987, Gs
Professor:
Brian J. Godfrey, (D), California (Berkeley), 1984, Ou
Associate Professor:
Jeffrey R. Walker, (D), Dartmouth, 1987, Sc
Yu Zhou, (D), Minnesota, 1995, Ou
Assistant Professor:
Mary A. Cunningham, (D), Minnesota, 2001, Oi
Brian G. McAdoo, (D), California (Santa Cruz), 1999, Yr
Kirsten M. Menking, (D), California (Santa Cruz), 1995, Gm
Joseph Nevins, Ou
Technical Staff:
Meg E. Stewart, (M), Nevada, 1994, Oi

York College (CUNY)

Dept of Natural Sciences: Geology Discipline (B) (2004)
Jamaica, NY 11451
p. (718) 262-2079 f. (718) 262-2652
lwj@york.cuny.edu
Enrollment (1999): B: 10 (2)

Head:
Arthur P. Loring, (D), New York, 1966, Gm
Professor:
Stephen Lakatos, (D), RPI, 1971, Cc
Assistant Professor:
Stanley Schleifer, (D), CUNY, 1996, Ge
Related Staff:
Michael Ashby, (B), York, 1994, Ca

North Carolina

Appalachian State University

Dept of Geography & Planning (B,M) (2007)
323 Rankin Science West
ASU Box 32066
Boone, NC 28608-2066
p. (828) 262-3000 f. (828) 262-3067
youngje@appstate.edu
http://www.geo.appstate.edu
Enrollment (2001): B: 82 (46) M: 15 (7)

Chair:
James E. Young, (D), Minnesota, 1994, Oe
Professor:
Garry V. Cooper, (D), Texas A&M, 1979, Oo
Michael W. Mayfield, (D), Tennessee, 1984, Hs
Peter T. Soule, (D), Georgia, 1989, Oa
Roger A. Winsor, (D), Illinois, 1975, Oo
Associate Professor:
Jeff Colby, (D), Colorado, 1995, Oi
Richard J. Crepeau, (D), California (Irvine), 1995, Oo
Kathleen Schroeder, (D), Minnesota, 1995, Oo
Assistant Professor:
Chris Badurek, (D), Buffalo, 2005, Oi
Rob Brown, (D), Louisiana State, 2001, Oo
Jana Carp, (D), Illinois (Chicago), 1999, Oo
Gabrielle Katz, (D), Colorado, 2001, Hg
Baker Perry, (D), North Carolina, 2006, Og
Instructor:
Arthur B. Rex, (M), Appalachian State, 1980, Oi
Emeritus:
Ole Gade, (D), Michigan State, 1972, Oy
Neal G. Lineback, (D), Tennessee, 1970, Oy

Dept of Geology (B) (2005)
PO Box 32067
195 Rankin Science Building
Boone, NC 28608-2067
p. (828) 262-3049 f. (828) 262-6503
watersja@appstate.edu
http://www.geology.appstate.edu/
Enrollment (1999): B: 38 (2)

Chair:
Johnny A. Waters, (D), Indiana, 1976, Pi
Professor:
Richard N. Abbott, (D), Harvard, 1977, Gz
Ellen A. Cowan, (D), N Illinois, 1988, Gm
Associate Professor:
Steven J. Hageman, (D), Illinois, 1992, Pi
Assistant Professor:
William P. Anderson, (D), North Carolina State, 1999, Hw
Research Associate:
Keith C. Seramur, (M), N Illinois, 1988, Hw
Instructor:
Lucio D'Alberto, (M), Ferrara (Italy), 1992, Gc
Lecturer:
Neil E. Johnson, (D), Virginia Tech, 1986, Gz
Laura Mallard, (M), Vermont, 2000, Gt
Adjunct Faculty:
Marg J. McKinney, (M), North Carolina, 1968, Pg
Adjunct Professor:
Mark G. Adams, (D), North Carolina, 1995, Gp
B. Charlotte Schreiber, (D), RPI, 1974, Gs
Emeritus:
John E. Callahan, (D), Queen's, 1973, Eg
Frank K. McKinney, (D), North Carolina, 1970, Pi
Loren A. Raymond, (D), California (Davis), 1973, Gx
Fred Webb, (D), Virginia Tech, 1965, Gr

Brevard College

Geology Program (2006)
1 Brevard College Lane
Brevard, NC 28712
p. 828-884-8377 f. (828) 884-3790
reynoljh@brevard.edu
http://www2.brevard.edu/reynoljh/
Administrative Assistant: Denise Poole

Geology Minor Coordinator:
James H. Reynolds, (D), Dartmouth, 1987, Gr

Cape Fear Community College

Dept of Math/Science (2006)
411 N. Front Street
Wilmington, NC 28401
khufham@cfcc.edu
Administrative Assistant: Sandra Moore

Chair:
Kenneth Hufham

Duke University

Div of Earth & Ocean Sciences (B,M,D) (2007)
Nicholas School of the Environment and Earth Sciences
Box 90227
Durham, NC 27708-0227
p. (919) 684-5847 f. (919) 684-5833
eosoffice@duke.edu
http://www.nicholas.duke.edu/eos
Enrollment (2007): B: 25 (6) M: 5 (3) D: 21 (7)

Chair:
M. Susan Lozier, (D), Washington, 1989, Op
Professor:
Paul A. Baker, (D), California (San Diego), 1981, Cl
Alan E. Boudreau, (D), Washington, 1986, Gi
Bruce H. Corliss, (D), Rhode Island, 1978, Pm
Peter K. Haff, (D), Virginia, 1970, Oo
Robert B. Jackson, (D), Utah State, 1992, Oo
Emily M. Klein, (D), Columbia, 1988, Gi
Associate Professor:
A. Bradshaw Murray, (D), Minnesota, 1995, Ou
Richard G. Newell, (D), Harvard, 1997, Eg
Lincoln F. Pratson, (D), Columbia, 1993, Gs
Avner Vengosh, (D), Australian National, 1990, Hw
Associate Scientist:
Gary S. Dwyer, (D), Duke, 1996, Gs
Eylon Shalev, (D), Yale, 1993, Ys
Instructor:
Alex Glass, (D), Illinois, 2006, Ggu
Adjunct Professor:
Fred K. Boadu, (D), Georgia Tech, 1994, Yg
James S. Clark, (D), Minnesota, 1988, Oo
David J. Erickson, (D), Rhode Island, 1987
Mark N. Feinglos, (D), McGill, 1973, Gz
Richard F. Kay, (D), Yale, 1973, Oo
Peter E. Malin, (D), Princeton, 1978, Ys
Bruce F. Molnia, (D), South Carolina, 1972, Or
Daniel D. Richter, (D), Duke, 1980, Os
William H. Schlesinger, (D), Cornell, 1976, Oo
Emeritus:
Richard T. Barber, (D), Stanford, 1967, Ob
Duncan Heron, (D), North Carolina, 1958, Gr
Daniel A. Livingstone, (D), Yale, 1953, Pe
Ronald D. Perkins, (D), Indiana, 1962, Gd
Orrin H. Pilkey, Jr., (D), Florida State, 1962, Ou

East Carolina University

Dept of Geological Sciences (B,M) (2006)
101 Graham Building
Greenville, NC 27858-4353
p. (252) 328-6360 f. (252) 328-4391
culvers@ecu.edu
http://www.geology.ecu.edu
Administrative Assistant: Dare Merritt
Enrollment (2006): B: 24 (10) M: 25 (5)
Chair:
Stephen J. Culver, (D), Wales, 1976, Pm
Professor:
Richard L. Mauger, (D), Arizona, 1966, Gi
Distinguished Research Professor:
Stanley R. Riggs, (D), Montana, 1967, Gu
Associate Professor:
David P. Lawrence, (D), SUNY (Binghamton), 1975, Gc
Donald W. Neal, (D), West Virginia, 1979, Gr
Catherine A. Rigsby, (D), California (Santa Cruz), 1989, Gs
Richard K. Spruill, (D), North Carolina, 1981, Hw
Terri L. Woods, (D), S Florida, 1988, Cl
Assistant Professor:
David Reide Corbett, (D), Florida State, 1999, Oc
David Mallinson, (D), S Florida, 1995, Gu
Visiting Assistant Professor:
Stephen B. Harper, (D), Georgia, 1996, Gm
Emeritus:
Charles Q. Brown, (D), Virginia Tech, 1959, Gs

Elizabeth City State University

Dept of Geological, Env & Marine Sci (B) (2004)
1704 Weeksville Road
Elizabeth City, NC 27909
p. (252) 335-3375 f. (252) 335-3465
gems@mail.ecsu.edu
Enrollment (2003): B: 35 (2)
Chair:
Francisco C. San Juan, (D), Florida State, 1982, Eg
Professor:
William A. Porter, (D), Maryland, 1986, Oy
Associate Professor:
Thomas J. Rossbach, (D), North Carolina, 1992, Ps
Assistant Professor:
Kathleen M. Fischer, (D), Oregon State, 1984, Cm
Env Scientist:
Elizabeth B. Noble, (M), North Carolina State, 1986, Op
Instructor:
B Marie Jenkins, (B), Elizabeth City State, 1995, Ge
Adjunct Instructor:
Jerome Leete, (M), East Carolina, 1981, Ge

Guilford College

Dept of Geology & Earth Science (B) (2004)
5800 West Friendly Avenue
Greensboro, NC 27410-4173
p. (336) 316-2263 f. (336) 316-2951
calmy@guilford.edu
http://www.guilford.edu/academic/geology
Enrollment (2001): B: 20 (3)
Professor:
Charles C. Almy, (D), Rice, 1965, Gg
Marlene McCauley, (D), California (Los Angeles), 1986, Cp
Associate Professor:
David M. Dobson, (D), Michigan, 1997, Gu
Emeritus:
Cyril H. Harvey, (D), Nebraska, 1960, Gr

North Carolina Agricultural & Tech State Univ

Dept of Natural Resources and Env Design (B,M) (2006)
E. Market St
Carver Hall
Greensboro, NC 27411
p. (919) 334-7543 f. (919) 334-7844
reddyg@ncat.edu
Enrollment (2005): B: 14 (55)
Chair:
Gudigopuram B. Reddy, (D), Georgia, 1974, So
Professor:
M. R. Reddy, (D), Georgia, 1973, So
Godfrey A. Uzochukwu, (D), Nebraska, 1983, Gz
Associate Professor:
Charles Raczkowski, (D), North Carolina State, So
Manuel Reyes, (D), Louisiana State, 1992, Hw
Assistant Professor:
Arona Diouf, (D), Gg

North Carolina Central University

Dept of Geography (B) (2004)
Box 19765
Durham, NC 27707-19765
p. (919) 560-6233 f. (919) 560-3324
abarnett@nccu.edu
Department Office Assistant: Margaret B. Guy
Enrollment: No data reported since 1995
Chair:
Albert P. Barnett, (D), Arizona, 1980, Or
Professor:
Woodrow W. Nichols, Jr., (D), California (Los Angeles), 1973, Oy

Associate Professor:
Vinston Burton, Jr., (D), California (Los Angeles), 1974, Gm
Dennis P. Enberg, (D), North Carolina, 1975, Oo
Jasper L. Harris, (D), North Carolina, 1983, Oy
Harris Williams, (D), Arizona, 1979, Oy
Geography Librarian:
Dennis P. Enberg, Oo

North Carolina State University

Dept of Marine, Earth & Atmospheric Sci (B,M,D) (2007)
P.O. Box 8208
Raleigh, NC 27695-8208
p. (919) 515-3711 f. (919) 515-7802
fountain@ncsu.edu
http://www.meas.ncsu.edu
Enrollment (2007): B: 227 (50) M: 59 (17) D: 38 (13)
Head:
John C. Fountain, (D), California (Santa Barbara), 1975, Hw
Director, Graduate Programs:
Gerald S. Janowitz, (D), Johns Hopkins, 1967, Op
Professor:
Viney P. Aneja, (D), North Carolina State, 1977, Oa
Satyapal S. Arya, (D), Colorado State, 1968, Oa
David J. DeMaster, (D), Yale, 1979, Oc
David B. Eggleston, (D), William & Mary, 1991, Ob
Ronald V. Fodor, (D), New Mexico, 1972, Gi
David P. Genereux, (D), MIT, 1991, Hw
James P. Hibbard, (D), Cornell, 1988, Gc
Daniel Kamykowski, (D), California (San Diego), 1973, Ob
Elana L. Leithold, (D), Washington, 1987, Gs
Leonard J. Pietrafesa, (D), Washington, 1973, Op
Sethu S. Raman, (D), Colorado, 1972, Oa
Fred H. M. Semazzi, (D), Nairobi, 1983, Oa
Thomas G. Wolcott, (D), California (Berkeley), 1971, Ob
Lian Xie, (D), Miami, 1992, Oa
Associate Professor:
Ruoying He, (D), S Florida, 2002, Op
Michael M. Kimberley, (D), Princeton, 1974, Ce
Mary H. Schweitzer, (D), Montana State, 1995, Po
Ping-Tung Shaw, (D), MIT/Woods Hole, 1982, Op
William J. Showers, (D), Hawaii, 1982, Cs
Yang Zhang, (D), Iowa, 1994, Oa
Associate Scientist:
Sandra Yuter, (D), Washington, 1996, Oa
Assistant Professor:
Anantha Aiyyer, (D), SUNY Albany, 2003, Oa
Del Bohnenstiehl, (D), Columbia, 2003, Yr
Julia A. Clarke, (D), Yale, 2002, Pv
Cynthia N. Cudaback, (D), Washington, 1998, Gu
Gary M. Lackmann, (D), SUNY (Albany), 1995, Oa
Jingpu P. Liu, (D), William & Mary, 2001, Gu
Nicholas Meskhidze, (D), Georgia Tech, 2003, Oa
Matthew Parker, (D), Colorado State, 2002, Ow
Research Associate:
Jing Lin, (D), William & Mary, 2001, Hs
Senior Research Professor:
S. T. Rao, (D), SUNY (Albany), 1973, Oa
Adjunct Professor:
Neal E. Blair, (D), Stanford, 1980, Co
Visiting Professor:
Earle N. Buckley, (D), North Carolina, 1980, Og
Emeritus:
Charles E. Knowles, (D), Texas A&M, 1970, Op
Thomas F. Malone, (D), MIT, 1946, Oa
Allen J. Riordan, (D), Wisconsin, 1977, Oa
Dale A. Russell, (D), Columbia, 1964, Pv
Edward F. Stoddard, (D), California (Los Angeles), 1976, Gp
Gerald F. Watson, (D), Florida State, 1971, Oa
Charles W. Welby, (D), MIT, 1952, Hw
Donna L. Wolcott, (D), California (Berkeley), 1972, Ob

Dept of Soil Science (B,M,D) (2005)
Box 7619
Raleigh, NC 27695-7619
p. (919) 515-2655 f. (919) 515-2167
keith_cassel@ncsu.edu
Administrative Assistant: Ashru Shah
Enrollment (1997): B: 74 (12) M: 32 (2) D: 38 (1)
Professor:
Aziz Amoozegar, (D), Arizona, 1977, Sp
Stephen W. Broome, (D), North Carolina, 1973, Sf
Donald K. Cassel, (D), California (Davis), 1968, Sp
John L. Havlin, (D), Colorado State, 1983, Sc
Dean L. Hesterberg, (D), California (Riverside), 1988, Sc
Michael T. Hoover, (D), Penn State, 1983, Sd
Greg D. Hoyt, (D), Georgia, 1981, Sb
Daniel W. Israel, (D), Oregon State, 1973, Sb
Harold J. Kleiss, Illinois, 1972, Sd
Deanna L. Osmond, (D), Cornell, 1991, Sb
Wayne P. Robarge, (D), Wisconsin, 1975, Sc
Thomas J. Smyth, (D), North Carolina State, 1981, Sc
Michael J. Vepraskas, (D), Texas A&M, 1980, Sd
Michael G. Wagger, (D), Kansas State, 1983, Sb
Associate Professor:
David A. Crouse, (D), North Carolina State, 1996, Sc
Carl Crozier, (D), North Carolina State, 1992, Sb
David Lindbo, (D), Massachusetts, 1990, Sd
Richard A. McLaughlin, (D), Purdue, 1985, Sc
Assistant Professor:
Alexandria Graves, (D), Virginia Tech, 2003, Sb
Rory O. Maguire, (D), Sb
Wei Shi, (D), Purdue, Sb
Jeffrey G. White, (D), Cornell, 1988, Or
Emeritus:
James W. Gilliam, (D), Mississippi State, 1965, Sb

University of North Carolina Wilmington

Dept of Geography and Geology (B,M) (2007)
601 South College Road
Wilmington, NC 28403-5944
p. (910) 962-3490 f. (910) 962-7077
laws@uncw.edu
http://www.uncw.edu/earsci/
Administrative Assistant: Catherine Morris
Administrative Assistant: Anne Sutter
Enrollment (2007): B: 78 (21) M: 32 (5)
Chair:
Richard A. Laws, (D), California (Berkeley), 1983, Pm
Professor:
William F. Ainsley, (D), North Carolina, 1977, Og
William J. Cleary, (D), South Carolina, 1972, Gu
James A. Dockal, (D), Iowa, 1980, Gd
Nancy R. Grindlay, (D), Rhode Island, 1991, Yg
W. Burleigh Harris, (D), North Carolina, 1975, Gs
Patricia H. Kelley, (D), Harvard, 1979, Pg
Lynn A. Leonard, (D), S Florida, 1993, Gu
Paul A. Thayer, (D), North Carolina, 1967, Gd
Associate Professor:
Lewis J. Abrams, (D), Rhode Island, 1992, Gu
Robert T. Argenbright, (D), California (Berkeley), 1990, Oo
Michael M. Benedetti, (D), Wisconsin, 2000, Gm
David E. Blake, (D), Washington State, 1991, Gp
Douglas W. Gamble, (D), Georgia, 2000, Oy
Joanne N. Halls, (D), South Carolina, 1996, Oi

Eric J. Henry, (D), Arizona, 2001, Hw
Mary E. Hines, (D), Louisiana State, 1992, Ou
John R. Huntsman, (D), Bryn Mawr, 1978, Gc
Michael S. Smith, (D), Washington (St. Louis), 1990, Gz
Assistant Professor:
Craig R. Tobias, (D), William & Mary, 1999, Cs
Lecturer:
Roger D. Shew, (M), North Carolina, 1979, Eo
Laboratory Director:
Yvonne Marsan, (B), North Carolina (Wilmington), Oi

University of North Carolina, Asheville

Dept of Atmospheric Sciences (B) (2007)
CPO 2450
One University Heights
Asheville, NC 28804-3299
p. (828) 251-6149 f. (828) 232-5046
ahuang@unca.edu
http://www.atms.unca.edu
Department Secretary: Linda Allman
Enrollment (2006): B: 55 (0)
Chair:
Alex Huang, (D), Purdue, 1984, Oa
Professor:
Edward A. Brotak, (D), Yale, 1977, Oa
Assistant Professor:
Leigh Orf, (D), Wisconsin, 1997, Oa
Lecturer:
Frank Quinlan, (M), Penn State, 1984, Oa

Dept of Environmental Studies (B) (2004)
CPO 2330
Asheville, NC 28804-8511
p. (828) 251-6441 f. (828) 251-6041
drobbins@unca.edu
http://bulldog.unca.edu/envr_studies/
Administrative Assistant: Debra C. Robbins
Enrollment (2001): B: 123 (40)
Director:
Richard P. Maas, (D), North Carolina, 1981, Oo
Professor:
Gary L. Miller, (D), North Carolina, 1968, Oo
J William Miller, (D), Georgia, 1985, Ge
Associate Professor:
Kevin K. Moorhead, (D), Florida, 1986, Sb
Assistant Professor:
Delores M. Eggers, (D), North Carolina, 1999, Oo
Barbara C. Reynolds, (D), Georgia, 2000, Oo
Irene M. Rossell, (D), SUNY (Syracuse), 1995, Oo
Geology Librarian:
Ellie E. Marsh, (M), Emporia State, 1995, Oo

University of North Carolina, Chapel Hill

Dept of Geological Sciences (B,M,D) (2006)
CB 3315, Mitchell Hall
Chapel Hill, NC 27599-3315
p. (919) 966-4516 f. (919) 966-4519
lbenning@email.unc.edu
http://www.geosci.unc.edu
Enrollment (2006): B: 39 (4) M: 40 (2) D: 18 (2)
Professor:
John G. Bane, Jr., (D), Florida State, 1975, Op
Larry K. Benninger, (D), Yale, 1976, Cl
Joseph G. Carter, (D), Yale, 1976, Po
Paul D. Fullagar, (D), Illinois, 1963, Cc
Allen F. Glazner, (D), California (Los Angeles), 1981, Gi
Christopher S. Martens, (D), Florida State, 1972, Oc
Jose A. Rial, (D), Caltech, 1979, Ys
Associate Professor:
Louis R. Bartek, III, (D), Rice, 1989, Gu
Jonathan M. Lees, (D), Washington, 1989, Yg
Kevin G. Stewart, (D), California (Berkeley), 1987, Gc
Assistant Professor:
Drew S. Coleman, (D), Kansas, 1991, Cg
Stephen R. Meyers, (D), Northwestern, 2003, Pe
Michael E. Oskin, (D), Caltech, 2002, Gm
Donna M. Surge, (D), Michigan, 2001, Pe
Emeritus:
John M. Dennison, (D), Wisconsin, 1960, Gr
A. Conrad Neumann, (D), Lehigh, 1963, Gu
John J. Rogers, (D), Caltech, 1955, Gg
Joseph St. Jean, (D), Indiana, 1956, Pm
Daniel A. Textoris, (D), Illinois, 1963, Gd
Geological Sciences Librarian:
Miriam Kennard, (M), Illinois, 1979

University of North Carolina, Charlotte

Dept of Geography & Earth Sciences (B,M,D) (2007)
9201 University City Boulevard
Charlotte, NC 28223-0001
p. (704) 687-5995 f. (704) 687- 5966
jfbender@email.uncc.edu
www.uncc.edu/colleges/arts_and_sciences/geoearth/
Enrollment (2007): B: 130 (45) M: 24 (7) D: 4 (0)
Professor:
John F. Bender, (D), SUNY (Stony Brook), 1980, GiCtOu
John A. Diemer, (D), SUNY (Binghamton), 1985, Gs
Associate Professor:
Craig J. Allan, (D), York, 1992, Hg
Andy R. Bobyarchick, (D), SUNY (Albany), 1983, Gc
Scott P. Hippensteel, (D), Delaware, 2000, Gr
Walter E. Martin, (D), Tennessee, 1984, Oga
Ross Meentemeyer, (D), North Carolina, 2000, Oi
Assistant Professor:
John Chadwick, (D), Florida, 2003, Or
Matt Eastin, (D), Colorado State, 2003, Ow
Martha Cary Eppes, (D), New Mexico, 2002, Os
Brian J. Etherton, (D), Penn State, Ow
Anne Jefferson, (D), Oregon State, 2006, HgGm
Lecturer:
Jake Armour, (M), New Mexico, Gg
Laboratory Director:
William Garcia, (M), Cincinnati, Pv

University of North Carolina, Pembroke

Geology & Geography Dept (2007)
PO Box 1510
Pembroke, NC 28372-1510
p. (910) 775-4024 f. (910) 521-6606
geo@uncp.edu
http://www.uncp.edu/geo/
Enrollment (1995): B: 3 (2)
Chair:
Martin B. Farley, (D), Penn State, 1987, PlOe
Professor:
Thomas E. Ross, (D), Tennessee, 1977, Oy
Associate Professor:
Dennis J. Edgell, (D), Kent State, 1992, Oa
Assistant Professor:
P. Lee Phillips, (D), Iowa, 2004, Gd
Lecturer:
Nathan E. Phillippi, (M), South Dakota State, 2004, Oy
Erin E. Poliakon, (M), Illinois (Chicago), 2005, Og

Western Carolina University

Dept of Geosciences & Natural Resources Management (B) (2007)
Stillwell Building Room 349
Cullowhee, NC 28723-9047
p. (828) 227-7367 f. (828) 227-7647
lkolenb@wcu.edu
Enrollment (1998): B: 15 (2)

Head:
Lawrence G. Kolenbrander, (D), Colorado State, 1981, Oo
Professor:
Jerry R. Miller, (D), S Illinois, 1990, Gm
Steven P. Yurkovich, (D), Brown, 1972, Gx
Associate Professor:
Paul G. White, (D), Denver, 1973, Gm
Robert S. Young, (D), Duke, 1995, Gs
Assistant Professor:
Mark L. Lord, (D), Hw

North Dakota

Dickinson State University

Dept of Natural Science (B) (2004)
Dickinson, ND 58601
p. (701) 227-2114 f. (701) 227-2006
Michael.Hastings@dickinsonstate.edu
Enrollment: No data reported since 1995

Associate Professor:
Larry D. League, (M), Kansas, 1971, Oy

Minot State University

Dept of Geoscience (B) (2007)
500 University Avenue West
Minot, ND 58707
p. (701) 858-3873 f. (701) 858-3163
webster@minotstateu.edu
Enrollment (2006): B: 18 (0)

Chair:
Allen J. Kihm, (D), Colorado, 1984, Pv
John R. Webster, (D), Indiana, 1992, Gi
Professor:
Clark Markell, (D), Ohio State, 1971, Oe

North Dakota State University

Dept of Geosciences (B,M,D) (2008)
Stevens Hall
Fargo, ND 58105-5517
p. (701) 231-8455 f. (701) 231-7149
bernhardt.saini-eidukat@ndsu.edu
http://www.ndsu.edu/geosci
Administrative Assistant: Suzy Schmoll
Enrollment (2007): B: 31 (5) M: 10 (1) D: 0 (1)

Chair:
Bernhardt Saini-Eidukat, (D), Minnesota, 1991, Gi
Professor:
Allan C. Ashworth, (D), Birmingham, 1969, Pe
Donald P. Schwert, (D), Waterloo, 1978, Pe
Assistant Professor:
Ken Lepper, (D), Oklahoma State, 2001, Cc
Peter Oduor, (D), Missouri (Rolla), 2004, Oi
Instructor:
Elaine C. Hatzenbuhler, (B), Kansas State, 1971, Oe
Adjunct Professor:
Gregory J. McCarthy, (D), Penn State, 1969, Gz
Douglas C. Munski, (D), Illinois, 1978, Oo

Dept of Soil Science (2004)
Walster Hall
Fargo, ND 58105
p. (701) 231-8901 f. (701) 231-7861
Department Secretary: Pam Loose
Enrollment (1995): B: 12 (2) M: 4 (7) D: 11 (3)

Director:
Lynn J. Brun, (D), Kansas State, 1972, Oa
Professor:
Edward J. Deibert, (D), Nebraska, 1976, Og
John W. Enz, (D), Minnesota, 1976, Oa
Robert J. Goos, (D), Colorado, 1980, Sc
John T. Moraghan, (D), Iowa State, 1961, Sc
Lyle D. Prunty, (D), Iowa State, 1978, Sp
Jimmie L. Richardson, (D), Iowa State, 1974, Sf
Assistant Professor:
Larry J. Cihacek, (D), Iowa State, 1976, Og
David W. Franzen, (D), Illinois, 1993, Sc
Joseph F. Giles, (D), Colorado, 1974, Sc
David G. Hopkins, (D), North Dakota State, 1997, Sd
Mohamed Khan, (D), Clemson, 1998, Oa
Raymond E. Knighton, (D), Cornell, 1985, Sp

University of North Dakota

Dept of Geography (B,M,D) (2004)
Box 8274 University Station
Grand Forks, ND 58202
p. (701) 777-4591

Chair:
Douglas C. Munski, (D), Illinois, 1978, Oe
Associate Professor:
Lowell Goodman, (D), Illinois, 1972, Ou
Mohammad Hemmasi, (D), Indiana, 1971, Oo
Assistant Professor:
John L. Hammen, (D), Indiana State, 1988, Oy
Robert Seidel, (D), North Dakota, 1986, Oy
Paul Todhunter, (D), California (Los Angeles), 1986, Oy

Dept of Geology & Geological Eng (B,M,D) (2004)
Box 8358
Grand Forks, ND 58202
p. (701) 777-2811 f. (701) 777-4449
geology_ge@mail.und.nodak.edu
http://www.geology.und.nodak.edu
Administrative Assistant: Gina Eastman
Enrollment (1999): B: 64 (11) M: 8 (4) D: 3 (0)

Chair:
Richard D. LeFever, (D), California (Los Angeles), 1979, Gs
Interim Director:
William D. Gosnold, (D), S Methodist, 1976, Yh
Professor:
Frank R. Karner, (D), Illinois, 1963, Gx
Dexter Perkins, III, (D), Michigan, 1979, Gp
Associate Professor:
Philip J. Gerla, (D), Arizona, 1983, Hw
Joseph H. Hartman, (D), Minnesota, 1984, Pi
Scott F. Korom, (D), Utah State, 1991, Hw
Ronald K. Matheney, (D), Arizona State, 1989, Cs
Assistant Professor:
Frank W. Beaver, (D), North Dakota, 1986, Hw
Nels F. Forsman, (D), North Dakota, 1985, Eg
Ahmad Ghassemi, (D), Oklahoma, 1996, Nr
Richard L. Josephs, (D), Iowa, 2000, Ga
Adjunct Professor:
John P. Bluemle, (D), North Dakota, 1972, Gm
Gerald H. Groenewold, (D), North Dakota, 1972, Hw
Emeritus:
Sidney B. Anderson, (D), North Dakota, 1951, Eo

Alan M. Cvancara, (D), Michigan, 1965, Pi
F. D. Holland, Jr., (D), Cincinnati, 1958, Pi
John R. Reid, (D), Michigan, 1961, Gm
Geology Librarian:
Kathleen Spencer, Oo

Ohio

Ashland University

Dept of Chemistry/Geology/Physics (B) (2006)
401 College Avenue
Ashland, OH 44805
p. (419) 289-6268 f. (419) 289-5333
mhudson@ashland.edu
www3.ashland.edu/academics/arts_sci/geology
Enrollment (2005): B: 9 (4)
Associate Professor:
Michael R. Hudson, (D), Miami (Ohio), 1994, Gp
Assistant Professor:
Nigel Brush, (D), California (Los Angeles), 1992, Gx

Bowling Green State University

Dept of Geology (B,M) (2008)
190 Overman Hall
Bowling Green, OH 43403
p. (419) 372-2886 f. (419) 372-7205
geology@bgsu.edu
http://geology.bgsu.edu
Administrative Assistant: Pat A. Wilhelm
Enrollment (2007): B: 31 (2) M: 25 (9)
Professor:
James E. Evans, (D), Washington, 1988, Gs
Charles M. Onasch, (D), Penn State, 1977, Gc
Don C. Steinker, (D), California (Berkeley), 1969, Po
Robert K. Vincent, (D), Michigan, 1973, Yg
Chair:
Sheila J. Roberts, (D), Arizona, 1992, CgHw
Associate Professor:
John R. Farver, (D), Brown, 1988, Gy
Joseph P. Frizado, (D), Northwestern, 1980, Oi
Kurt S. Panter, (D), New Mexico Tech, 1995, Gv
Jeffrey A. Snyder, (D), Ohio State, 1996, Gm
Peg M. Yacobucci, (D), Harvard, 1999, Po
Assistant Professor:
Enrique Gomezdelcampo, (D), Tennessee, 2003, Ge
Peter Gorsevski, (D), Idaho, 2002, OirGm
Emeritus:
Richard D. Hoare, (D), Missouri, 1957, Pi
John A. Howe, (D), Nebraska, 1961, Pv
Charles F. Kahle, (D), Kansas, 1962, Gs
Joseph J. Mancuso, (D), Michigan State, 1960, Em
Charles C. Rich, (D), Harvard, 1960, Gm

Case Western Reserve University

Dept of Geological Sciences (B,M,D) (2008)
10900 Euclid Avenue
A.W. Smith #112
Cleveland, OH 44106-7216
p. (216) 368-3690 f. (216) 368-3691
lmd3@case.edu
http://geology.case.edu
Enrollment (2007): B: 14 (3) M: 2 (1) D: 2 (0)
Chair:
Gerald Matisoff, (D), Johns Hopkins, 1978, Cl
Professor:
Peter L. McCall, (D), Yale, 1975, Po
Samuel M. Savin, (D), Caltech, 1967, Cs
Associate Professor:
Ralph P. Harvey, (D), Pittsburgh, 1990, Xm
Beverly Z. Saylor, (D), MIT, 1996, Gs
Peter J. Whiting, (D), California (Berkeley), 1990, Gm
Assistant Professor:
Steven A. Hauck, II, (D), Washington (St. Louis), 2001, Yg
Adjunct Professor:
Philip O. Banks, (D), Caltech, 1963, Cc
Enriqueta Barrera, (D), Case Western, Yr
Joseph T. Hannibal, (D), Kent State, 1990, Pi
Michael Ketterer, (D), Colorado, 1985, Ca
Richard C. Schmidt, (D), McGill, 1956, Em

Central State University

Intl Center for Water Resources Management (B) (2004)
C.J. McLin Bldg
Wilberforce, OH 45384
p. (937) 376-6212 f. (937) 376-6257
Department Administrative Coordinator: Phyllis A. Chilton
Enrollment: No data reported since 1995
Director:
Subramania I. Sritharan, (D), Colorado State, 1984, Hg
Professor:
Samuel Okunade, (D), Kent State, 1986, Gm
Associate Professor:
Sam Laki, (D), Michigan State, 1992, Hg

Cincinnati Museum Center

Museum of Natural History & Science (2006)
1301 Western Avenue
Cincinnati, OH 45203
p. (513) 287-7000 f. (513) 287-7095
storrsg@cincymuseum.org
http://www.cincymuseum.org
Assistant Vice President:
Glenn W. Storrs, (D), Yale, 1986, Pv
Curator:
Brenda Hanke, (D), California (Riverside), 2004, Pi
Research Associate:
Carlton E. Brett, (D), Michigan, 1978, Po
Nigel C. Hughes, (D), Bristol, 1990, Pi
David L. Meyer, (D), Yale, 1971, Pi
Arnold I. Miller, (D), Chicago, 1986, Pq

Cleveland Museum of Natural History

Dept of Invertebrate Paleontology (2007)
1 Wade Oval Drive
Cleveland, OH 44106-1767
p. (216) 231-4600 (Ext. 3233) f. (216) 231-5919
hannibal@cmnh.org
Head:
Joseph T. Hannibal, (D), Kent State, 1990, Pi
Assistant:
Douglas Dunn, (A), Lakeland Comm Coll, 1974, Pg

Dept of Mineralogy (2007)
1 Wade Oval Drive
University Circle
Cleveland, OH 44106-1767
p. (216) 231-4600 (Ext. 3229) f. (216) 231-5919
dsaja@cmnh.org
http://www.cmnh.org

Dept of Paleobotany (2004)
1 Wade Oval Drive
University Circle

Cleveland, OH 44106-1767
p. (216) 231-4600 f. (216) 231-5919
schitale@cmnh.org
Head:
Shyamala Chitaley, (D), Reading (UK), 1955, Pb

Dept of Vertebrate Paleontology (2006)
1 Wade Oval Drive
University Circle
Cleveland, OH 44106-1767
p. (216) 231-4600
mryan@cmnh.org
Head:
Michael J. Ryan, (D)
VP Tech:
Gary Jackson
Casting Technician:
David Chapman

Cleveland State University

Dept of Biological, Geological, & Env Sci (B) (2007)
2121 Euclid Avenue
Cleveland, OH 44115-2414
p. (216) 687-2440 f. (216) 687-6972
bges@csuohio.edu
http://bgesweb.artscipub.csuohio.edu
Enrollment (2006): B: 15 (0)
Professor:
Wentworth B. Clapham, Jr., (D), Chicago, 1968, Ge
Howard H. Lo, (D), Washington (St. Louis), 1970, Gi
Michael J. S. Tevesz, (D), Yale, 1974, Pi
Associate Professor:
Abbed Babaei, (D), Missouri, 1985, Gc
Instructor:
John Fleshin, (M), Cleveland State, Gg
Heather Gallacher, (M), Toledo, Gg

Denison University

Dept of Geosciences (B) (2007)
F.W. Olin Science Hall
Granville, OH 43023
p. (740) 587-6217 f. (740) 587-6774
hawkins@denison.edu
www.denison.edu/academics/departments/geosciences/
Administrative Assistant: Jude Hall
Enrollment (2006): B: 20 (4)
Chair:
David P. Hawkins, (D), MIT, 1996, GxCc
Professor:
Tod A. Frolking, (D), Wisconsin, 1985, Oy
Associate Professor:
David C. Greene, (D), Nevada (Reno), 1995, Gc
Assistant Professor:
David H. Goodwin, (D), Arizona, 2002, Pi
Johanna M. Kieniewicz, (D), Washington, 2007, Pe
Emeritus:
Kennard B. Bork, (D), Indiana, 1967, Pi
Robert J. Malcuit, (D), Michigan State, 1973, Gx

Kent State University

Dept of Geology (B,M,D) (2007)
221 McGilvrey Hall
Kent, OH 44242
p. (330) 672-2680 f. (330) 672-7949
geology@kent.edu
http://dept.kent.edu/geology/
Administrative Assistant: Karen E. Smith
Enrollment (2000): B: 8 (0)
Chair:
Donald F. Palmer, (D), Princeton, 1968, Yg
Professor:
Peter S. Dahl, (D), Indiana, 1977, Gp
Yoram Eckstein, (D), Hebrew (Israel), 1977, Hw
Richard A. Heimlich, (D), Yale, 1959, Gi
Daniel K. Holm, (D), Harvard, 1992, Gc
Abdul Shakoor, (D), Purdue, 1982, Ng
Alison J. Smith, (D), Brown, 1991, Gn
Neil A. Wells, (D), Michigan, 1984, Gs
Associate Professor:
Ernest H. Carlson, (D), McGill, 1966, Ce
Assistant Professor:
David B. Hacker, (D), Kent State, 1998, Gc
Joseph D. Ortiz, (D), Oregon State, 1995, Gs
Carrie E. Schweitzer, (D), Kent State, 2000, Pi
Donna L. Witter, (D), Oregon State, 1995, Op
Emeritus:
Rodney M. Feldmann, (D), North Dakota, 1967, Pi

Kent State University, Stark Campus

Dept of Geology (B) (2007)
6000 Frank Avenue NW
North Canton, OH 44720
p. (330) 499-9600 (Ext. 53303) f. (330) 494-6121
cschweit@kent.edu
http://www.personal.kent.edu/~cschweit/Stark/
Department Secretary: Sharon Schreffler
Enrollment (1996): B: 4 (0)
Associate Professor:
Carrie E. Schweitzer, (D), Kent State, 2000, Pi

Kent State University, Tuscarawas

Dept of Geology (A) (2007)
University Drive NE
New Philadelphia, OH 44663
p. (330) 339-3391 f. (330) 339-3321
hcunningham@tusc.kent.edu
Enrollment (1997): B: 16 (0)
Instructor:
John F. Beltz, (M), Akron, 1992, Ge

Marietta College

Dept of Petroleum Engineering & Geology (B) (2006)
215 Fifth Street
Marietta, OH 45750
p. (740) 376-4776 f. (740) 376-4777
chaser@marietta.edu
http://www.marietta.edu
Administrative Assistant: Susan P. Martin
Enrollment (2006): B: 14 (3)
Professor:
Robert W. Chase, (D), Penn State, 1980, Np
Associate Professor:
Frederick R. Voner, (D), Miami, 1985, Gx
Assistant Professor:
David L. Jeffery, (D), Texas A&M, 2003, Go
Instructor:
Wendy Bartlett, (M), Texas A&M, Ge
Veronica Freeman, (M), Texas, 1993, Pg

Miami University

Dept of Geology (B,M,D) (2007)
114 Shideler Hall
Oxford, OH 45056
p. (513) 529-3216 f. (513) 529-1542

edwardca@muohio.edu
http://www.muohio.edu/geology
Administrative Assistant: Cathy Edwards
Enrollment (2007): B: 29 (7) M: 12 (4) D: 13 (4)

Chair:
William K. Hart, (D), Case Western, 1982, Gi
Director, Inst for Env Science:
Mark R. Boardman, (D), North Carolina, 1978, On
Professor:
Yildirm Dilek, (D), California (Davis), 1989, Gt
Associate Professor:
Brian S. Currie, (D), Arizona, 1998, Gs
Hailiang Dong, (D), 1997, Gz
Jonathan Levy, (D), Wisconsin, 1993, Hw
John F. Rakovan, (D), SUNY (Stony Brook), 1996, Cl
Elisabeth Widom, (D), California (Santa Cruz), 1991, Cc
Assistant Professor:
Michael Brudzinski, (D), Illinois, 2002, Ys
Jason Rech, (D), Arizona, 2001, Gm
Instructor:
Janelle Sikorski, (M), N Arizona, 2004, Gm
Adjunct Professor:
Cindy K. Carney, (D), West Virginia, 1987, Oo
Carol A. Eddy-Dilek, (D), California (Davis), 1997, Hw
David C. Kuentz, (M), Texas (Arlington), 1986, Ca
Arnold I. Miller, (D), Chicago, 1986, Pi
Monica Tsang, (M), Delaware, 1992, Hw
Visiting Instructor:
Paul E. Holm, (D), Illinois, 1980, Gc
Emeritus:
A. Dwight Baldwin, Jr., (D), Stanford, 1966, Hw
John M. Hughes, (D), Dartmouth, 1981, Gz
Wayne D. Martin, (D), Cincinnati, 1955, Gs
Robert G. McWilliams, (D), Washington, 1968, Ps
John K. Pope, (D), Cincinnati, 1966, Pi
David M. Scotford, (D), Johns Hopkins, 1950, Gp
David P. Stewart, (D), Syracuse, 1954, Gl
Cooperating Faculty:
R. Hays Cummins, (D), Texas A&M, 1984, Oo

Mount Union College

Dept of Geology (B) (2004)
Alliance, OH 44601
p. (330) 823-3672 f. (330) 823-8531
graylm@muc.edu
http://www.muc.edu/gy
Enrollment (1997): B: 7 (2)

Chair:
Lee M. Gray, (D), Rochester, 1985, Po
Adjunct Professor:
Leonard G. Epp, (D), Penn State, 1970, Ob
Visiting Assistant Professor:
Mark A. McNaught, (D), Rochester, 1991, Gc
Emeritus:
Robert G. Wiese, Jr., (D), Harvard, 1961, Eg

Muskingum College

Dept of Geology (B) (2004)
163 Stormont Street
New Concord, OH 43762
p. (740) 826-8242 f. (740) 826-8229
ericlaw@muskingum.edu
http://muskingum.edu/~geology/index.html
Enrollment (2001): B: 13 (2)

Chair:
Eric W. Law, (D), Case Western, 1982, Gp
Professor:
Jack Kovach, (D), Ohio State, 1974, Cl
Assistant Professor:
Stephen R. Van Horn, (D), Connecticut, 1996, Ge

Oberlin College

Dept of Geology (B) (2004)
52 West Lorain Street
Oberlin, OH 44074-1044
p. (440) 775-8350 f. (440) 775-8038
steven.wojtal@oberlin.edu
http://www.oberlin.edu/~Geopage/
Enrollment (2001): B: 24 (8)

Professor:
Bruce M. Simonson, (D), Johns Hopkins, 1982, Gs
William R. Skinner, (D), Columbia, 1966, Gx
Steven F. Wojtal, (D), Johns Hopkins, 1982, Gc
Assistant Professor:
Jonathan M. Castro, (D), Oregon, 1999, Gv
Karla M. Parsons-Hubbard, (D), Rochester, 1993, Po
Research Associate:
Roy W. Simonson, (D), Wisconsin, 1938, Sd
Visiting Professor:
Dyanna M. Czeck, (D), Minnesota, 2001, Gc
Dennis K. Hubbard, (D), South Carolina, 1977, On
Science Librarian:
Alison Ricker, Oo

Ohio State University

Atmospheric Sciences Program (B,M,D) (2004)
103 Bricker Hall
190 North Oval Mall
Columbus, OH 43210-1361
p. (614) 292-2514 f. (614) 292-6213

Director:
John N. Rayner, (D), Canterbury (NZ), 1965, Oa
Professor:
Jeffery C. Rogers, (D), Colorado, 1979, Oa

Dept of Civil & Environmental Eng (B,M,D) (2004)
Rm 470
2070 Neil Avenue
Columbus, OH 43210
p. (612) 292-6753 f. (614) 292-2957
mcgee.1@osu.edu
Enrollment: No data reported since 1995

Chair:
Keith W. Bedford, (D), Cornell, 1974, Oo
Professor:
Christopher Jekeli, (D), Ohio State, 1981, Yd
Joel L. Morrison, (D), Wisconsin, 1968, Oo
Anton F. Schenk, (D), Swiss Fed Inst, 1972, Or
Associate Professor:
Rongxing Li, (D), Tech (Berlin), 1990, Oo
Burkhard Schaffrin, (D), Bonn, 1982, Yd
C. K. Shum, (D), Texas, 1982, Oo
Assistant Professor:
Ayman F. Habib, (D), Ohio State, 1994, Oo
Alan J. Saalfeld, (D), Maryland, 1993, Oo
Lecturer:
Lois Brunty, (B), Ohio State, 1989, Oo
Dean Ringle, (B), Ohio State, 1981, Oo
Emeritus:
John D. Bossler, (D), Ohio State, 1972, Oo
Dean C. Merchant, (D), Ohio State, 1967, Or
Ivan I. Mueller, (D), Ohio State, 1960, Yd
Richard H. Rapp, (D), Ohio State, 1964, Yv
Urho A. Uotila, (D), Ohio State, 1959, Yd

Dept of Geography (B,M,D) (2007)

1036 Derby Hall
154 North Oval Mall
Columbus, OH 43210-1361
p. (614) 292-2514 f. (614) 292-6213
okelly.1@osu.edu

Professor:
David H. Bromwich, (D), Wisconsin, 1979, Oa
Ellen E. Mosley-Thompson, (D), Ohio State, 1979, Oa
Jeffrey C. Rogers, (D), Colorado, 1979, Oa
Associate Professor:
Jay S. Holgood, (D), Ohio State, 1984, Oa
Assistant Professor:
Jason E. Box, (D), Colorado, 2002, Oa
Bryan G. Mark, (D), Syracuse, 2001, Oa
David F. Porinchu, (D), California (Los Angeles), 2002, Oa
Emeritus:
A. John Arnfield, (D), McMaster, 1973, Oa
John N. Rayner, (D), Canterbury (NZ), 1965, Oa

School of Earth Sciences (B,M,D) (2007)

275 Mendenhall Lab
125 South Oval Mall
Columbus, OH 43210-1398
p. (614) 292-2721 f. (614) 292-7688
geology@osu.edu
http://www.geology.ohio-state.edu
Administrative Assistant: Susie L. Shipley

Enrollment (2005): B: 70 (5) M: 34 (10) D: 20 (0)

Ohio Eminent Scholar:
Frank W. Schwartz, (D), Illinois, 1972, Hg
Ohio Eminent Scholar:
Michael G. Bevis, (D), Cornell, 1982, Yx
Director, Byrd Polar Research Center:
William B. Lyons, (D), Connecticut, 1979, Cl
Associate Dean:
Jeffrey J. Daniels, (D), Colorado Mines, 1974, Yg
Professor:
William I. Ausich, (D), Indiana, 1978, Po
Loren E. Babcock, (D), Kansas, 1990, Po
Edwin S. Bair, (D), Penn State, 1980, Hw
Michael Barton, (D), Manchester, 1975, Gx
Yu-Ping Chin, (D), Michigan, 1988, Hg
Kenneth A. Foland, (D), Brown, 1972, Cc
Christopher Jekeli, (D), Ohio State, 1981, Yd
Kenneth C. Jezek, (D), Wisconsin, 1980, Or
Hallan C. Noltimier, (D), Newcastle upon Tyne, 1965, Ym
CK Shum, (D), Texas, 1982, Yd
Lonnie G. Thompson, (D), Ohio State, 1976, Gl
Ralph R. B. von Frese, (D), Purdue, 1980, Ye
Peter N. Webb, (D), Utrecht (Neth), 1966, Pm
Senior Scientist:
John W. Olesik, (D), Wisconsin, 1982, Ca
Associate Director, Graduate Studies:
Motomu Ibaraki, (D), Waterloo, 1994, Hg
Associate Director, Administration:
Lawrence A. Krissek, (D), Oregon State, 1982, Gs
Associate Dean:
Anne E. Carey, (D), Nevada (Reno), 1995, Hq
Associate Professor:
Garry D. McKenzie, (D), Ohio State, 1968, Gl
Douglas E. Pride, (D), Illinois, 1969, Em
Alan J. Saalfeld, (D), Maryland, 1993, Yd
Burkhard A. Schaffrin, (D), Bonn, 1983, Yd
Mary W. Scott, (M), Pittsburgh, 1989, Oo
Terry J. Wilson, (D), Columbia, 1983, Gc
Assistant Professor:
Douglas E. Alsdorf, (D), Cornell, 1996, Hg
Andrea G. Grottoli, (D), Houston, 1998
Steven K. Lower, (D), Virginia Tech, 2001, Py
Wendy R. Panero, (D), California (Berkeley), 2001, Yg
Matthew R. Saltzman, (D), California (Los Angeles), 1996, Gs
Lindsay M. Schoenbohm, (D), MIT, 2004, Gt
Adjunct Professor:
Michael C. Hansen, (D), Ohio State, 1986, Pg
Thomas G. Naymik, (D), Ohio State, 1978, Hw
Emeritus:
Stig M. Bergstrom, (D), Lund (Sweden), 1961, Ps
Colin B. Bull, (D), Birmingham (UK), 1951, Gl
James W. Collinson, (D), Stanford, 1966, Ps
Charles E. Corbato, (D), California (Los Angeles), 1960, Ye
James W. Downs, (D), Virginia Tech, 1983, Gz
David H. Elliot, (D), Birmingham, 1965, Gi
Gunter Faure, (D), MIT, 1961, Cc
George E. Moore, (D), Harvard, 1947, Gc
Charles H. Summerson, (D), Illinois, 1942, Gs
Walter C. Sweet, (D), Iowa, 1954, Ps
Thomas N. Taylor, (D), Illinois, 1964, Pb
Rodney T. Tettenhorst, (D), Illinois, 1960, Gz
Russell O. Utgard, (D), Indiana, 1969, Ge
Sidney E. White, (D), Syracuse, 1952, Gm

School of Env and Natural Resources (B,M,D) (2007)

2021 Coffey Road
Room 210 Koffman Hall
Columbus, OH 43210
p. (614) 292-2265 f. (614) 292-7432
director@snr.osu.edu
http://snr.osu.edu/
Administrative Assistant: Mary Capoccia

Enrollment (1998): B: 119 (43) M: 17 (3) D: 17 (3)

Eminent Scholar:
Richard P. Dick, (D), Iowa State, 1986, Sb
Associate Director:
Donald J. Eckert, (D), Ohio State, 1978, Sc
Professor:
Jerry M. Bigham, (D), North Carolina State, 1977, Sd
Frank G. Calhoun, (D), Florida, 1971, Sd
Warren A. Dick, (D), Iowa State, 555, Sb
Rattan Lal, (D), Ohio State, Sp
Associate Professor:
Nicholas T. Basta, (D), Iowa State, 1989, Sc
Edward L. McCoy, (D), Oregon State, 1984, Sp
Assistant Professor:
Dawn Ferris, (D), Minnesota, 1997, Sf
Robert Mullen, (D), Oklahoma State, 2003, So
Brian K. Slater, (D), Wisconsin, Sd

Ohio State University at Newark

School of Earth Sciences (2008)

1179 University Drive
Newark, OH 43055
stjohn.2@osu.edu

Assistant Professor:
Daniel Leavell, (D), Massachussets, 1983
Instructor:
James St. John, (M), Ohio State University, 1994

Ohio State University, Lima

Dept of Geological Sciences (A) (2004)

Lima, OH 45804-3576
p. (419) 995-8208 f. (419) 995-8091
kleffner.1@osu.edu

Enrollment: No data reported since 1995

Associate Professor:
Mark A. Kleffner, (D), Ohio State, 1988, Ps

Ohio University, Lancaster

Geological Sciences (2007)
1570 Granville Pike
Lancaster, OH 43130
p. (740) 654-6711 (Ext. 231) f. (740) 654-6711
geological_sciences@ohio.edu
http://www.lancaster.ohiou.edu/
Enrollment (2006): B: 40 (10) M: 16 (8)

Chair:
Gregory C. Nadon, (D), Toronto, 1991, Gs
Professor:
R. Damian Nance, (D), Cambridge, 1978, Gt
Associate Professor:
Elizabeth H. Gierlowski-Kordesch, (D), Case Western, 1985, Gn
Douglas H. Green, (D), Wisconsin, 1989, Yg
David L. Kidder, (D), California (Santa Barbara), 1987, Gs
Dina L. Lopez, (D), Louisiana State, 1992, Cl
Mary W. Stoertz, (D), Wisconsin, 1989, Hw
Assistant Professor:
Keith A. Milam, (D), Tennessee, 2006, Xg
Gregory S. Springer, (D), Colorado State, 2002, Gm
Alycia L. Stigall, (D), Kansas, 2004, Po
Adjunct Professor:
Daniel Hembree, (D), Kansas, 2005, Pi
Visiting Professor:
Ahmad Abdel-Fattah, (D), Texas (El Paso), 2005, Hw
Emeritus:
Gene W. Heien, (M), Indiana, 1962, Gz
Royal H. Mapes, (D), Iowa, 1977, Pi
Thomas R. Worsley, (D), Illinois, 1970, Gu
Technician & Information Tech:
Timothy A. Grubb, (A), Washington State Comm Coll, 2002
Administrative Associate:
Jayde S. Gater, (A), Hocking Coll, 1998

Ohio Wesleyan University

Dept of Geology & Geography (B) (2007)
61 S. Sandusky Street
Delaware, OH 43015
p. (740) 368-3615 f. (740) 368-3999
khfryer@owu.edu
http://geology.owu.edu
Administrative Assistant: Dorothy Carina
Enrollment (2006): B: 11 (2)

Chair:
Karen H. Fryer, (D), Illinois, 1986, Gc
Director, Urban Studies:
Richard D. Fusch, (D), Oregon, 1972, Og
Director, Env Studies:
David H. Hickcox, (D), Oregon, 1979, Oy
Professor:
Barton S. Martin, (D), Massachusetts, 1991, Gx
Associate Professor:
John B. Krygier, (D), Penn State, 1995, Oi
Keith O. Mann, (D), Iowa, 1987, Pi
Visiting Professor:
Ronald J. McChesney, (D), Ohio State, 2006, Og

Shawnee State University

Dept of Natural Sciences (B) (2007)
940 Second Street
Portsmouth, OH 45662
p. (740) 351-3456 f. (740) 351-3596
jbauer@shawnee.edu
http://www.shawnee.edu
Administrative Assistant: Sharon Messer
Enrollment (2007): B: 6 (0)

Professor:
Jeffrey A. Bauer, (D), Ohio State, 1987, Ps
Assistant Professor:
Kurt A. Shoemaker, (D), Miami, 2004, Gi

The College of Wooster

Dept of Geology (B) (2007)
Scovel Hall
944 College Mall
Wooster, OH 44691-2363
p. (330) 263-2380 f. (330) 263-2249
preeder@wooster.edu
http://www.wooster.edu/geology
Administrative Assistant: Patrice Reeder
Enrollment (2007): B: 20 (13)

Chair:
Robert J. Varga, (D), California (Davis), 1980, Gc
Professor:
Mark A. Wilson, (D), California (Berkeley), 1982, Pi
Associate Professor:
Gregory C. Wiles, (D), SUNY (Buffalo), 1992, Gl
Visiting Professor:
Johannes Koch, (D), Simon Fraser, 2006, Gl
Matthew "Rocky" Severs, (D), Virginia Tech, 2007, Gx

University of Akron

Dept of Geology & Environmental Sci (B,M) (2007)
Akron, OH 44325-4101
p. (330) 972-7630 f. (330) 972-7611
geology@uakron.edu
http://www.uakron.edu/geology/
Administrative Assistant: Elaine Butcher
Enrollment (2006): B: 61 (7) M: 15 (6)

Chair:
John P. Szabo, (D), Iowa, 1975, Gl
Professor:
Lindgren L. Chyi, (D), McMaster, 1972, Eg
Annabelle M. Foos, (D), Texas (Dallas), 1984, Cg
David A. McConnell, (D), Texas A&M, 1987, Gc
Lisa E. Park, (D), Arizona, 1995, Pi
Ira D. Sasowsky, (D), Penn State, 1992, Hw
Associate Professor:
La Verne M. Friberg, (D), Indiana, 1976, Gp
John A. Peck, (D), Rhode Island, 1995, Gs
David N. Steer, (D), Cornell, 1996, Ys
Research Associate:
Thomas J. Quick, (M), Akron, 1983, Ca
Instructor:
John F. Beltz, (M), Akron, 1992, Pm
Adjunct Professor:
Linda R. Barrett, (D), Michigan State, 1995, Sd
Timothy Matney, (D), Pennsylvania, 1993, Ga
Emeritus:
A. W. Gerhard Kunze, (D), Penn State, 1973, Yg

University of Cincinnati

Dept of Geology (B,M,D) (2007)
500 Geology/Physics Building
P. O. Box 210013
Cincinnati, OH 45221-0013
p. (513) 556-3732 f. (513) 556-6931
cannelss@uc.edu
http://www.uc.edu/geology
Enrollment (2006): B: 40 (0) M: 8 (0) D: 24 (0)

Head:
Arnold I. Miller, (D), Chicago, 1986, Pi

Professor:
Carlton E. Brett, (D), Michigan, 1978, Pe
Madeleine Briskin, (D), Brown, 1973, Pe
John E. Grover, (D), Yale, 1972, Gz
Warren D. Huff, (D), Cincinnati, 1963, Gz
Attila I. Kilinc, (D), Penn State, 1969, Cp
Thomas V. Lowell, (D), SUNY (Buffalo), 1986, Gl
J. Barry Maynard, (D), Harvard, 1972, Cl
David L. Meyer, (D), Yale, 1971, Pi
David Nash, (D), Michigan, 1977, Gm
Associate Professor:
Thomas J. Algeo, (D), Michigan, 1989, Gs
Kees A. De Jong, (D), Utrecht (Neth), 1969, Gt
Craig Dietsch, (D), Yale, 1985, Gp
Lewis Owen, (D), Leicester (UK), 1988, Gl
Adjunct Professor:
Brenda R. Hanke, (D), California (Riverside), 2004, Pi
Glenn W. Storrs, (D), Yale, 1986, Pv
Emeritus:
Paul E. Potter, (D), Chicago, 1952, Gs

University of Dayton

Dept of Geology (B) (2007)
300 College Park
Dayton, OH 45469-2364
p. (937) 229-3432 f. (937) 229-2889
Donna.Hess@notes.udayton.edu
http://artssciences.udayton.edu/geology
Administrative Assistant: Donna Hess
Enrollment (2007): B: 28 (10)

Chair:
Allen J. McGrew, (D), Wyoming, 1992, Gc
Professor:
Donald Pair, (D), Syracuse, 1991, Gl
Michael R. Sandy, (D), London, 1984, Pi
Associate Professor:
Daniel Goldman, (D), SUNY (Buffalo), 1993, Po
Andrea M. Koziol, (D), Chicago, 1988, Cp
Shuang-Ye Wu, (D), Cambridge, 2000, Oy
Instructor:
Katherine Schoenenberger, (M), Cincinnati, 2001, Gl
Adjunct Professor:
Stacey Hundley, (D), Ohio State, 2007, Gg
Sue Klosterman, (M), Wright State, 2005, Gg
Heidi McGrew, (M), Stanford, Yg
Steve Roberts, (B), Wright State, Gg
Emeritus:
Charles J. Ritter, (D), Michigan, 1971, Ct

University of Toledo

Dept of Earth, Ecological & Env Sci (B,M,D) (2004)
2801 W. Bancroft Street
Toledo, OH 43606-3390
p. (419) 530-2009 f. (419) 530-4421
eees@utnet.utoledo.edu
http://www.eeescience.utoledo.edu/
Department Secretary: Stephany Mikols
Enrollment (2000): B: 17 (0) M: 14 (0)

Associate Chair:
Elliot J. Tramer, (D), Georgia, 1969, Oo
Professor:
Jiquan Chen, (D), Washington, 1991, Oo
Stephen L. Goldman, (D), Missouri, 1971, Oo
James A. Harrell, (D), Cincinnati, 1983, Ga
Michael W. Phillips, (D), Virginia Polytechnic, 1972, Gz
Associate Professor:
Vernon M. Brown, (D), Missouri (Rolla), 1983, Gi
Mark J. Camp, (D), Ohio State, 1974, Pi
Daryl F. Dwyer, (D), Michigan State, 1986, Oo
Timothy G. Fisher, (D), Calgary, 1993, Gl
Johan F. Gottgens, (D), Florida, 1992, Oo
David E. Krantz, (D), South Carolina, 1988, Gs
James Martin-Hayden, (D), Connecticut, 1994, Hq
Daryl L. Moorhead, (D), Tennessee, 1985, Oo
Deborah A. Neher, (D), California (Davis), 1990, Sb
Alison L. Spongberg, (D), Texas A&M, 1994, Co
Donald J. Stierman, (D), Stanford, 1977, Yg
Assistant Professor:
Scott A. Heckathorn, (D), Illinois, 1995, Oo
Christine M. Mayer, (D), Illinois, 1998, Oo
William V. Sigler, (D), Purdue, 1999, Oo

Wittenberg University

Dept of Geology (B) (2005)
P.O. Box 720
Springfield, OH 45501-0720
p. (937) 327-7335 f. (937) 327-6340
jritter@wittenberg.edu
http://www4.wittenberg.edu/academics/geol/
Enrollment (2005): B: 9 (1)

Professor:
Kenneth W. Bladh, (D), Arizona, 1978, Gz
John B. Ritter, (D), Penn State, 1990, Gm
Associate Professor:
Katherine L. Bladh, (D), Arizona, 1976, Gi
Michael J. Zaleha, (D), SUNY (Binghamton), 1994, Gs
Emeritus:
Robert W. Morris, (D), Columbia, 1969, Pi

Wright State University

Dept of Geological Sciences (B,M) (2004)
3640 Colonel Glenn Highway
260 Brehm Lab
Dayton, OH 45435
p. (937) 775-3455 f. (937) 775-3462
deborah.cowles@wright.edu
http://www.wright.edu/geology
Administrative Manager: Deborah Cowles
Enrollment (2002): B: 25 (6) M: 70 (27)

Chair:
Paul J. Wolfe, (D), Case Western, 1966, Ye
Professor:
C. B. Gregor, (D), Utrecht (Neth), 1967, Gs
Byron Kulander, (D), West Virginia, 1966, Gc
Associate Professor:
Abinash Agrawal, (D), North Carolina, 1990, Co
Cindy K. Carney, (D), West Virginia, 1987, Gd
Songlin Cheng, (D), Arizona, 1984, Hw
David F. Dominic, (D), West Virginia, 1988, Gs
Ernest C. Hauser, (D), Wisconsin, 1982, Ye
Robert W. Ritzi, Jr., (D), Arizona, 1989, Hw
William Slattery, (D), CUNY, 1994, Gr
Assistant Professor:
Roderic I. Brame, (D), Virginia Tech, 2001, Oe
Doyle Watts, (D), Michigan, 1979, Or
Adjunct Professor:
Carol A. Winhusen, (M), Wright State, 1971, Og
Emeritus:
Benjamin H. Richard, (D), Indiana, 1966, Ye

Youngstown State University

Dept of Geological & Environmental Sci (B,M) (2007)
One University Plaza
2120 Moser Hall
Youngstown, OH 44555

p. (330) 941-3612 f. (330) 941-1754
amjacobs@ysu.edu
http://www.as.ysu.edu/~geology/
Enrollment (2007): B: 48 (4) M: 21 (1)

Chair:
Alan M. Jacobs, (D), Indiana, 1967, Ge
Professor:
Raymond E. Beiersdorfer, (D), California (Davis), 1992, Cg
Director, Env Studies Program:
Isam E. Amin, (D), Nevada (Reno), 1987, Hw
Associate Professor:
Jeffrey C. Dick, (D), Kent State, 1992, Ng
Assistant Professor:
Joseph E. Andrew, (D), Kansas, 2002, Gc
Felicia P. Armstrong, (D), Oklahoma State, 2003, Sb
Shane V. Smith, (D), Washington State, 2005, Gs
Instructor:
Harry Bircher, (M), Youngstown State, 1995, Ge
Brian M. Greene, (D), Kent State, 2001, Ng
Lawrence P. Gurlea, (M), Pennsylvania, 1971, Cg
Engineering Adjunct:
Scott C. Martin, (D), Clarkson, 1984, Ge
Douglas M. Price, (D), Notre Dame, 1988, Cg
Emeritus:
Ann G. Harris, (M), Miami (Ohio), 1958, Ge
Ikram U. Khawaja, (D), Indiana, 1969, Ec
Charles R. Singler, (D), Nebraska, 1969, Gs
Cooperating Faculty:
Thomas P. Diggins, (D), SUNY, 1997, Py
Carl G. Johnston, (D), Cincinnati, 1992, Py

Oklahoma

Northeastern Oklahoma A&M College

Dept of Physical Science (A,B) (2004)
200 I Street, NE
Miami, OK 74354
p. (918) 540-6266 f. (918) 540-6488
swells@neoam.edu
http://www.neoam.edu
Department Secretary: Billy Minson
Enrollment (1996): B: 1 (0)

Associate Professor:
Randy M. Jones, (D), Pittsburgh State, 1990, Oo
Mary R. Kirkpatrick, (M), Pittsburgh State, 1989, Oo
Instructor:
Matthew G. Mayfield, (M), Missouri, 1987, Oo
Sam H. Wells, (B), E New Mexico, 1963, Gg

Northeastern State University

Dept of Geology (A,B) (2007)
Science Building
Tahlequah, OK 74464
p. (918) 456-5511 (Ext. 3817) f. (918) 458-2193
simmsj@cherokee.nsuok.edu
http://nsuok.edu
Enrollment (2006): A: 15 (6) B: 13 (4)

Assistant Professor:
John J. Simms, (M), Kansas, 1975, Gg
Instructor:
Dorothy F. Simms, (M), Kansas, 1973, Pg

Oklahoma State University

School of Geology (B,M) (2005)
105 Noble Research Center
Stillwater, OK 74078-3031
p. (405) 744-6358 f. (405) 744-7841
sandrae@okstate.edu
http://geology.okstate.edu
Administrative Assistant: Nancy Dryden
Enrollment (2005): B: 73 (4) M: 27 (4)

Head:
Jay M. Gregg, (D), Michigan State, 1982, Gs
Professor:
Ibrahim Cemen, (D), Penn State, 1983, Gc
Associate Professor:
Darwin Boardman, II, (D), Texas Tech, 1990, Ps
Surinder K. Sahai, (D), Wyoming, 1974, Yg
Assistant Professor:
Elizabeth J. Catlos, (D), California (Los Angeles), 2000, Gz
Anna Cruse, (D), MIT, 2003, Co
Todd Halihan, (D), Texas, 2000, Hw
Stanley T. Paxton, (D), Penn State, 1983, Go
James Puckette, (D), Oklahoma State, 1996, Go
Alex Simms, (D), Rice, 2005, Gs
Emeritus:
Arthur Hounslow, (D), Carleton, 1968, Cl
Douglas Kent, (D), Iowa State, 1969, Hw
Wayne Pettyjohn, (D), Boston, 1964, Hw
Vernon Scott, (D), Utah, 1975, Oe
Gary Stewart, (D), Kansas, 1973, Go
John D. Vitek, (D), Iowa, 1973, Gm
Related Staff:
Sandy Earls, Oo

University of Oklahoma

Mewbourne School of Petroleum & Geological Eng (B,M,D) (2004)
100 East Boyd Street
Energy Center T301
Norman, OK 73019
p. (405) 325-2921 f. (405) 325-7477
mpge@ou.edu
http://mpge.ou.edu/
Enrollment (2004): B: 161 (21) M: 75 (35) D: 18 (4)

Director:
Dean S. Oliver, (D), Washington, 1980, Np
Professor:
Younane Abousleiman, (D), Delaware, 1991, Nr
Faruk Civan, (D), Oklahoma, 1978, Np
Roy M. Knapp, (D), Kansas, 1973, Np
Chandra S. Rai, (D), Hawaii, 1977, Np
Jean-Claude Roegiers, (D), Minnesota, 1974, Nr
Subhash N. Shah, (D), New Mexico, 1974, Np
Carl H. Sondergeld, (D), Cornell, 1977, Np
Djebbar Tiab, (D), Oklahoma, 1976, Np
Michael Wiggins, (D), Texas A&M, 1991, Np
Don Zhang, (D), Arizona, 1994, Np
Associate Professor:
Samuel O. Osisanya, (D), Texas, 1991, Np
Assistant Professor:
Richard G. Hughes, (D), Stanford, 1998, Np
Related Staff:
Deborah Bradley, Oo

School of Geology & Geophysics (B,M,D) (2007)
100 East Boyd
810 Energy Center
Norman, OK 73019-0628
p. (405) 325-3253 f. (405) 325-3140
sggdir@hoth.gcn.ou.edu
http://geology.ou.edu
Geology Librarian: Claren Kidd
Enrollment (2006): B: 87 (0) M: 40 (0) D: 24 (0)

Oregon

Director:

Roger M. Slatt, (D), Alaska, 1970, Go

Professor:

John P. Castagna, (D), Texas, 1983, Ye
R. Douglas Elmore, (D), Michigan, 1981, Ym
Michael H. Engel, (D), Arizona, 1980, Co
James M. Forgotson, Jr., (D), Northwestern, 1956, Go
M. Charles Gilbert, (D), California (Los Angeles), 1965, Cp
Charles W. Harper, Jr., (D), Caltech, 1964, Pi
David London, (D), Arizona State, 1981, Cp
Charles J. Mankin, (D), Texas, 1958, Gd
Shankar Mitra, (D), Johns Hopkins, 1977, Gc
R. Paul Philp, (D), Sydney, 1972, Co
Stephen R. Westrop, (D), Toronto, 1984, Pi
Alan J. Witten, (D), Rochester, 1975, Ye

Associate Professor:

Judson L. Ahern, (D), Cornell, 1980, Yg
David Deming, (D), Utah, 1988, Yh
Thomas A. Dewers, (D), Indiana, 1990, Cl
John D. Pigott, (D), Northwestern, 1981, Ye
Barry L. Weaver, (D), Birmingham (UK), 1980, Ct
Roger A. Young, (D), Toronto, 1979, Ye

Assistant Professor:

Richard Lupia, (D), Chicago, 1997, Pm
Kevin J. Smart, (D), Tennessee, 1996, Gc
Gerilyn S. Soreghan, (D), Arizona, 1992, Gs

Research Assistant Professor:

Thurman E. Scott, Jr., (D), Texas (Dallas), 1989, Gc

Adjunct Professor:

Richard L. Cifelli, (D), Columbia, 1983, Pv

Emeritus:

Robert L. Dubois, (D), Washington, 1954, Ym
David W. Stearns, (D), Texas A&M, 1969, Gc

Related Staff:

Daniel O'Meara, (D), Princeton, 1979, Np

Cooperating Faculty:

Jean-Claude Roegiers, (D), Minnesota, 1974, Nr

Electron Microprobe Operator:

George B. Morgan, (D), Oklahoma, 1988, Gi

Teaching Associate:

Michael J. Soreghan, (D), Arizona, 1994, Gs

University of Tulsa

Dept of Geosciences (B,M,D) (2006)
600 South College Avenue
Tulsa, OK 74104-3189
p. (918) 631-2517 f. (918) 631-2091
jbt@utulsa.edu
http://www.geo.utulsa.edu
Administrative Assistant: Carol Holly
Enrollment (2006): B: 38 (7) M: 10 (3) D: 3 (1)

Chair:

J. Bryan Tapp, (D), Oklahoma, 1983, Gc

Assoc VP for Research & Dean of Graduate School:

Janet Haggerty, (D), Hawaii, 1982, Gu

Professor:

Peter J. Michael, (D), Columbia, 1983, Gi
Kerry Sublette, (D), Tulsa, 1985, Ge

Senior Scientist:

Winton Cornell, (D), Rhode Island, 1987, Gi

Dean, Eng & Natural Sciences:

Steven J. Bellovich, (D), Nebraska, 1974, Oy

Associate Professor:

Dennis R. Kerr, (D), Wisconsin, 1989, Gs

Assistant Professor:

Berton Fisher, (D), Case Western, 1979, Cg
Kumar Ramachandran, (D), Victoria, 2001, Ye

Research Associate:

Eleanor Jennings, (D), Oklahoma, 2005, Sb
Robert W. Scott, (D), Kansas, 1967, Gs

Emeritus:

Colin Barker, (D), Oxford, 1965, Cg

Oregon

Central Oregon Community College

Dept of Science (A) (2006)
Bend, OR 97701
p. (503) 382-6112 (Ext. 263) f. (503) 385-5978
breynolds@cocc.edu
Enrollment (1997): B: 8 (0)

Associate Professor:

Robert W. Reynolds, (D), Idaho, 1994, Gv

Clatsop Community College

Dept of Geology (A) (2004)
Astoria, OR 97103
p. (503) 325-0910 (Ext. 2240)
pkeefe@clatsopcc.edu
Enrollment: No data reported since 1995

Instructor:

Paul D. See, (M), Portland State, 1964, Gg

Eastern Oregon University

Science Dept (2007)
Badgley Hall
One University Boulevard
La Grande, OR 97850-2807
p. (541) 962-3351 f. (541) 962-3873
jvantass@eou.edu
http://www.eou.edu/~geology/

Professor:

Jay Van Tassell, (D), Duke, 1979, GsrPg

Adjunct Professor:

Mark Ferns, (D), Oregon, 1979, GvcEg

Visiting Professor:

Henrietta Laustsen, (D), Colorado, 1997, OiGfz

Oregon State University

College of Oceanic & Atmospheric Sci (M,D) (2007)
104 COAS Admin Building
Corvallis, OR 97331-5503
p. (541) 737-3504 f. (541) 737-2064
student_adviser@coas.oregonstate.edu
http://www.coas.oregonstate.edu
Administrative Assistant: Kay Yates
Enrollment (2007): M: 61 (26) D: 43 (12)

Dean:

Mark R. Abbott, (D), California (Davis), 1978, Obr

Associate Dean:

Robert A. Duncan, (D), Australian National, 1976, Ou

Senior Research:

Harold P. Batchelder, (D), Oregon State, 1986, Ob
Curtiss O. Davis, (D), Washington, 1973, Ob

Distinguished Professor:

Dudley B. Chelton, (D), California (San Diego), 1980, Opr
Patricia A. Wheeler, (D), California (Irvine), 1976, Ob

Director:

Michael Harte, (D), Victoria, 1994, Og

Professor:

Jeffery R. Barnes, (D), Washington, 1983, Oa
Jack A. Barth, (D), MIT/Woods Hole, 1987, Op
James A. Coakley, (D), California (Berkeley), 1972, Oar

Robert W. Collier, (D), MIT, 1978, Cm
Frederick Colwell, (D), Virginia Tech, 1986, GuPy
Timothy J. Cowles, (D), Duke, 1977, Ob
Gary D. Egbert, (D), Washington, 1987, Yrm
Kelly K. Falkner, (D), MIT/Woods Hole, 1989, Oc
Martin R. Fisk, (D), Rhode Island, 1978, Ou
Michael Freilich, (D), Scripps Coll, 1981, Op
David W. Graham, (D), MIT/Woods Hole, 1987, Ou
Robert A. Holman, (D), Dalhousie, 1979, On
Gary P. Klinkhammer, (D), Rhode Island, 1979, OuCm
P. Michael Kosro, (D), California (San Diego), 1984, Op
Ricardo M. Letelier, (D), Hawaii, 1994, Ob
Murray D. Levine, (D), Washington, 1976, Op
Jim McManus, (D), Oregon State, 1992, GuCm
Robert N. Miller, (D), California (Berkeley), 1976, Op
Alan C. Mix, (D), Columbia, 1985, Ou
James N. Moum, (D), British Columbia, 1984, Op
Nicklas G. Pisias, (D), Rhode Island, 1978, Ou
Frederick G. Prahl, (D), Washington, 1982, Co
Clare E. Reimers, (D), Oregon State, 1982, Cm
Roger Samelson, (D), Oregon State, 1987, Op
Adam Schultz, (D), Washington, 1986, Yr
Barry F. Sherr, (D), Georgia, 1977, Ob
Evelyn B. Sherr, (D), Duke, 1974, Ob
Eric Skyllingstad, (D), Wisconsin, 1986, Oa
P. Ted Strub, (D), California (Davis), 1983, Op
Marta Torres, (D), Oregon State, 1988, OcCa
Anne M. Trehu, (D), MIT/Woods Hole, 1982, Ygs
Michael H. Unsworth, (D), Edinburgh, 1968, Oa
Robert A. Wheatcroft, (D), Washington, 1990, Ob

Senior Research:
W. Scott Pegau, (D), Oregon State, 1996, Op
Frank Tepley, (D), California (Los Angeles), 1999, CmGi
Cynthia Twohy, (D), Washington, 1992, Oa
Hemantha Wijesekera, (D), Oregon State, 1992, Op

Associate Professor:
Edward Dever, (D), MIT, 1995, Op
Chris Goldfinger, (D), Oregon State, 1994, Ou
Miguel Goni, (D), Washington, 1992, Oc
Burke Hales, (D), Washington, 1995, Oc
Robert N. Harris, (D), Utah, 1996, Yr
Anthony Koppers, (D), Vrije, 1998, Cmc
James Lerczak, (D), Scripps, 2000, Op
Ricardo P. Matano, (D), Princeton, 1989, Op
John L. Nabelek, (D), MIT, 1984, Ys
Jonathan Nash, (D), Oregon State, 2000, Op
James G. Richman, (D), MIT/Woods Hole, 1977, Op
William D. Smyth, (D), Toronto, 1990, Op
Yvette H. Spitz, (D), Old Dominion, 1995, Ob
Paul Vincent, (D), Colorado, 1998, Gu
Richard J. Vong, (D), Washington, 1985, Oa

Assistant Professor:
Kelly Benoit-Bird, (D), Hawaii, 2003, Ob
Zanna Chase, (D), New York (Columbia), 2001, OcCm
Lorenzo Ciannelli, (D), Washington, 2002, Ob
Anthony D'Andrea, (D), SUNY (Stony Brook), 1999, Ob
Mick Haller, (D), Deleware, 1999, Oo
Alexandre Kurapov, (D), St. Petersburg Marine Tech (Russia), 1994, Op
H. Tuba Ozkan-Haller, (D), Delaware, 1998, On
Andreas Schmittner, (D), Bern, 1999, Yr
Jeffrey Shaman, (D), Columbia, 2003, Oa
R. Kipp Shearman, (D), Oregon State, 1999, Op
Karen Shell, (D), Scripps, 2004, Oa
Joseph Stoner, (D), Quebec, 1995, Yrm
Peter Strutton, (D), Flinders, 1998, Ob

Postdoc:
Leo Berline, (D), Joseph Fourier, 2006, Ob
Brandy Cervantes, (D), Oregon State, 2004, Op
Jeff Hatten, (D), Oc
Anna Kelbert, (D), Cardiff, 2006, Yr
Sangil Kim, (D), Arizona, 2005, Op
Andrey Koch, (D), Op
Prasanta Patro, (D), Yr
David Rivas Camargo, (D), Op
Tiago Silva, (D), Gu
Qingtao Song, (D), Rhode Island, 2006, Op
Kush Tandon, (D), Louisiana, 1998, Yr
Jacqueline Tweddle, (D), Ob
Dan Tyler, (D), Oregon State, 2004, Oa
Angelique White, (D), Oregon State, 2006, Ob

Research Associate:
Tim Boyd, (D), California (San Diego), 1989, Op
Russell H. Desiderio, (D), Stanford, 1984, Op
Scott M. Durski, (D), Rutgers, 2000, Op
Anatoli Erofeev, (D), Shirshov Inst of Oceanology, 1990, Op
Svetlana Erofeeva, (D), Moscow Engr Physics Inst, 1988, Yg
Corinne James, (D), Oregon State, 1987, Op
Alexander Perlin, (D), Tel Aviv, 2000, Op
Natalie Perlin, (D), Tel Aviv, 2000, Oa
Stephen D. Pierce, (D), Oregon State, 1998, Op
Scott Springer, (D), Washington, 1994, Op
William R. Tahnk, (D), Oregon State, 2001, Oa
Barry Vanhoff, (D), Washington State, 1996, Op

Emeritus:
John S. Allen, (D), Princeton, 1968, Op
Andrew F. Bennett, (D), Harvard, 1971, Op
Douglas R. Caldwell, (D), Chicago, 1963, Op
Andrew G. Carey, Jr., (D), Yale, 1962, Ob
G. Brent Dalrymple, (D), California (Berkeley), 1963, Ou
Roland de Szoeke, (D), Nova, 1973, Op
Thomas M. Dillon, (D), California (Davis), 1974, Op
Steven K. Esbensen, (D), California (Los Angeles), 1976, Oa
James W. Good, (D), Oregon State, 1992, Og
Louis I. Gordon, (D), Oregon State, 1973, Oc
Adriana J. Huyer, (D), Oregon State, 1974, Op
George H. Keller, (D), Illinois, 1966, Ou
Paul D. Komar, (D), California (San Diego), 1969, On
LaVerne D. Kulm, (D), Oregon State, 1965, Ou
Shaul Levi, (D), Washington, 1974, Yg
Larry Mahrt, (D), Wisconsin, 1972, Oa
Charles B. Miller, (D), California (San Diego), 1969, Ob
David M. Nelson, (D), Alaska, 1975, Ob
Stephen J. Neshyba, (D), Texas A&M, 1965, Op
Clayton A. Paulson, (D), Washington, 1967, Op
William G. Pearcy, (D), Yale, 1960, Ob
Bernd R. Simoneit, (D), Bristol, 1976, Oc
Lawrence F. Small, (D), Iowa State, 1961, Ob
Robert L. Smith, (D), Oregon State, 1964, Op
J. Ronald V. Zaneveld, (D), Oregon State, 1971, Op

Dept of Geosciences (B,M,D) (2007)
104 Wilkinson Hall
Corvallis, OR 97331-5506
p. (541) 737-1201 f. (541) 737-1200
geo-info@geo.oregonstate.edu
http://www.geo.oregonstate.edu
Office Manager: Melinda Peterson
Enrollment (2006): B: 99 (26) M: 56 (33) D: 24 (8)

Director:
Michael E. Campana, (D), Arizona, 1975, Hg

Chair:
Roger L. Nielsen, (D), S Methodist, 1983, Gi

Professor:
Sherman H. Bloomer, (D), California (San Diego), 1982, Gi
Peter U. Clark, (D), Colorado, 1984, Gl

Shanika de Silva, (D), Open (UK), 1987, Gv
John H. Dilles, (D), Stanford, 1984, Em
Anita L. Grunder, (D), Stanford, 1986, Gi
Julia A. Jones, (D), Johns Hopkins, 1983, So
A. Jon Kimerling, (D), Wisconsin, 1976, Or
Robert J. Lillie, (D), Cornell, 1984, Ye
Aaron T. Wolf, (D), Wisconsin, 1992, Hg
Dawn J. Wright, (D), California (Santa Barbara), 1994, Ou

Senior Research:
Christopher Daly, (D), Oregon State, 1994, Ow
Randall Keller, (D), Oregon State, Gg
Adam J. R. Kent, (D), Australian National, 1995, Ca
Peter Ruggiero, (D), Oregon State, 1997, On

Associate Professor:
Laurence Becker, (D), California (Berkeley), 1989, Oo
Edward J. Brook, (D), MIT, 1993, Ou
Ronald E. Doel, (D), Princeton, 1990, Gh
Roy D. Haggerty, (D), Stanford, 1995, Hw
Andrew J. Meigs, (D), S California, 1995, Gc
Anne Nolin, (D), California (Santa Barbara), 1993, Or
Alan J. Woods, (D), California (Los Angeles), 1978, Gm

Assistant Research Professor:
Mary V. Santelmann, (D), Minnesota, 1988, Oy

Assistant Professor:
Hannah Gosnell, (D), Colorado, 2000, Ou
Stephen Lancaster, (D), MIT, 1999, Hs

Research Associate:
Ellen J. Moore, (M), Oregon, 1950, Pi

Senior Instructor:
Steve Cook, (D), Florida, 1995, Oo

Emeritus:
Cyrus W. Field, (D), Yale, 1961, Em
Robert E. Frenkel, (D), California (Berkeley), 1967, Oy
Philip L. Jackson, (D), Kansas, 1977, Oy
Robert D. Lawrence, (D), Stanford, 1968, Gc
Gordon E. Matzke, (D), Syracuse, 1975, Og
Keith W. Muckleston, (D), Washington, 1970, Oo
Alan R. Niem, (D), Wisconsin, 1971, Gd
Charles L. Rosenfeld, (D), Pittsburgh, 1974, Gm
William H. Taubeneck, (D), Columbia, 1955, Gx
Edward M. Taylor, (D), Washington State, 1967, Gv
Robert S. Yeats, (D), Washington, 1958, Gt

Program Coordinator:
Kaplan Yalcin, (D), New Hampshire, 2005, Og

Cooperating Faculty:
Joseph R. Glasmann, (D), Oregon State, 1982, Gs
Gordon E. Grant, (D), Johns Hopkins, 1986, Gm
Munther J. Haddadin, (D), Washington, 1969, Hw
Steven W. Hostetler, (D), Oregon, 1984, Oa
John M. Logan, (D), Oklahoma, 1965, Gt
George W. Moore, (D), Yale, 1960, Gt
Frederick J. Swanson, (D), Oregon, 1972, Gm
R. Denis White, (M), Boston, 1984

Portland State University

Dept of Geology (B,M,D) (2007)
P.O. Box 751
Portland, OR 97207
p. (503) 725-3022 f. (503) 725-3025
andrew@pdx.edu
http://www.geol.pdx.edu
Administrative Assistant: Nancy Eriksson
Enrollment (2007): B: 30 (11) M: 15 (4) D: 6 (0)

Professor:
Scott F. Burns, (D), Colorado, 1980, Ng
Michael L. Cummings, (D), Wisconsin, 1978, Gg
Andrew G. Fountain, (D), Washington, 1992, Gl
Curt D. Peterson, (D), Oregon State, 1983, Gs

Associate Professor:
Sherry L. Cady, (D), California (Berkeley), 1994, Py
Kenneth M. Cruikshank, (D), Purdue, 1991, Gc
Christina L. Hulbe, (D), Chicago, 1998, Gl
Martin J. Streck, (D), Oregon State, 1994, Gi

Assistant Professor:
Georg H. Grathoff, (D), Illinois, 1996, Gz
Robert Benjamin Perkins, (D), Portland State, 2000, Hq
Alexander (Alex) M. Ruzicka, (D), Arizona, 1996, Xm

Research Associate:
Tomas Hode, (D), Stockholm, 2005, Xg
Richard Hugo, (D), Washington State, 1998, Gy

Instructor:
Thomas C. Lindsay, (M), California (Davis), 1982, Pe

Adjunct Professor:
Elizabeth Carter, (D), Lausanne, 1993, Pm
Michael L. Feves, (D), MIT, 1977, Nr
Cynthia Gardner, (D), Stanford, 1984, Gi
Richard Iverson, (D), Stanford, 1984, Ng
Roger Nielsen, (D), S Methodist, 1983, Gi
James O'Connor, (D), Arizona, 1990, Hs
Thomas Pierson, (D), Washington, 1977, Gm
Frank Reckendorf, (D), Oregon State, 1973, Gm
Evelyn Roeloffs, (D), Wisconsin, 1982, Ng
Joseph Walder, (D), Stanford, 1984, Gl

Emeritus:
G. Thomas Benson, (D), Yale, 1963, Gc
Paul E. Hammond, (D), Washington, 1963, Gi
Ansel G. Johnson, (D), Stanford, 1973, Ye
Richard E. Thoms, (D), California, 1965, Ps

Research Assistant:
David Percy, (B), Portland State, 1999, Oi

Southern Oregon University

Dept of Geology (B) (2006)
1250 Siskiyou Boulevard
Ashland, OR 97520
p. (541) 552-6479 f. (541) 552-6415
lane@sou.edu
Administrative Assistant: Susan Koralek
Enrollment (2006): B: 45 (5)

Chair:
Charles L. Lane, (D), California (Los Angeles), 1987, Hg

Dean, Science:
Joseph L. Graf, Jr., (D), Yale, 1975, Em

Professor:
Jad A. D'Allura, (D), California (Davis), 1977, Gc

Associate Professor:
Eric Dittmer, (M), San Jose State, 1972, Gg

Assistant Professor:
William S. Elliott, (D), Indiana, 2002, Gs

Adjunct Professor:
Vernon J. Crawford, (M), Oregon, 1970, Gg
Harry W. Smedes, (D), Washington, 1959, Gi
Richard Ugland, (M), Utah, 1974, Gg

Emeritus:
Monty A. Elliott, (D), Oregon State, 1971, Gr
William B. Purdom, (D), Arizona, 1960, Gx

Southwestern Oregon Comm College

Dept of Geology (A) (2007)
1988 Newmark
Coos Bay, OR 97420-2912
p. (541) 888-7216 f. (541) 888-7285
rmetzger@socc.edu
Enrollment (2007): A: 4 (0)

Professor:
Ronald A. Metzger, (D), Iowa, 1991, Pms

University of Oregon

Dept of Geography (B,M,D) (2007)
107 Condon Hall
Eugene, OR 97403-1251
p. (541) 346-4555 f. (541) 346-2067
uogeog@uoregon.edu
http://geography.uoregon.edu
Enrollment (2007): B: 102 (60) M: 18 (15) D: 19 (2)

Professor:
Patrick J. Bartlein, (D), Wisconsin, 1978, Pe
Patricia F. McDowell, (D), Wisconsin, 1980, Gm

Associate Professor:
Shaul E. Cohen, (D), Chicago, 1991, Oo
Susan W. Hardwick, (D), California (Davis), 2000, Oo
Alexander B. Murphy, (D), Chicago, 1987, Oo

Assistant Professor:
Dan Gavin, (D), Washington, 2000, Oy
Derrick Hindery, (D), California (Los Angeles), 2003, Oo
Amy Lobben, (D), Michigan State, 1999, OioOo
Lise Nelson, (D), Washington, 2000, Oo
Gary H. Searl, (M), Oregon, 1966, Oo
Xiaobo Su, (D), Singapore, 2007, Oo
Peter A. Walker, (M), Oregon, 1966, Oo

Adjunct Professor:
Nicholas Kohler, (D), Oregon, 2004, Oi

Emeritus:
Stanton A. Cook, (D), California (Berkeley), 1960, Oo
Carl L. Johannessen, (D), California (Berkeley), 1959, Oo
Clyde P. Patton, (D), California (Berkeley), 1953, Oo
Edward T. Price, (D), California (Berkeley), 1950, Oo
Everett G. Smith, Jr., (D), Minnesota, 1962, Oo
Alvin W. Urquhart, California (Berkeley), 1962, Oo
Ronald Wixman, (D), Columbia, 1978, Oo

Dept of Geological Sciences (B,M,D) (2005)
1272 University of Oregon
Eugene, OR 97403-1272
p. (541) 346-4573 f. (541) 346-4692
arbeiter@uoregon.edu
http://darkwing.uoregon.edu/~dogsci/
Administrative Assistant: Vicki E. Arbeiter
Enrollment (1997): B: 66 (10) M: 14 (3) D: 26 (4)

Head:
A. Dana Johnston, (D), Minnesota, 1983, Cp

Professor:
Katharine V. Cashman, (D), Johns Hopkins, 1987, Gv
Eugene D. Humphreys, (D), Caltech, 1982, Yg
Mark H. Reed, (D), California (Berkeley), 1977, Em
Gregory J. Retallack, (D), New England, 1978, Oo
Jack M. Rice, (D), Washington, 1972, Gp
Douglas R. Toomey, (D), MIT/Woods Hole, 1987, Ys
Harve S. Waff, (D), Oregon, 1970, Yg
Ray J. Weldon, (D), Caltech, 1986, Gc

Associate Professor:
Rebecca J. Dorsey, (D), Princeton, 1989, Gr

Assistant Professor:
Ilya N. Bindeman, (D), Chicago, 1998, Cs
Alan W. Rempel, (D), Cambridge, 2001
Joshua J. Roering, (D), California (Berkeley), 2000, Gm
David A. Schmidt, (D), California (Berkeley), 2002, Yd
Paul Wallace, (D), California (Berkeley), 1991, Gv

Research Associate:
Emilie E. Hooft, (D), MIT, 1996, Yr
C. Patrick Ryan, (M), Oregon, 1985, Yg

Instructor:
Marli G. Miller, (D), Washington, 1997, Gc
Elise Mezger Weldon, (M), S California, 1986, Gl

Lecturer:
Claudia Owen, (D), Washington, 1988, Gp

Courtesy Professor:
David H. Krinsley, (D), Chicago, 1956, Gd

Emeritus:
Ewart M. Baldwin, (D), Cornell, 1943, Gr
Sam Boggs, Jr., (D), Colorado, 1964, Gd
M. Allan Kays, (D), Washington (St. Louis), 1960, Gp
Alexander R. McBirney, (D), California (Berkeley), 1961, Gv
William N. Orr, (D), Michigan State, 1967, Pm
Norman M. Savage, (D), Sydney, 1968, Pi

Cooperating Faculty:
Arthur J. Boucot, (D), Harvard, 1953, Pi
John M. Logan, (D), Oklahoma, 1965, Nr
Cathy Whitlock, (D), Washington, 1983, Pe

Western Oregon University

Earth and Physical Sciences Dept (B) (2004)
345 N. Monmouth Ave.
Monmouth, OR 97361
p. 503-838-8398 f. 503-838-8072
taylors@wou.edu
www.wou.edu/las/physci/geology/earthsci.html
Enrollment (2004): B: 23 (7)

Associate Professor:
Jeffrey A. Myers, (D), Santa Barbara, 1998, Gs
Stephen B. Taylor, (D), West Virginia, 1999, Gm
Jeffrey H. Templeton, (D), Oregon State, 1998, Gv

Willamette University

Environmental and Earth Sciences Dept (2007)
900 State Street
Salem, OR 97301
p. (503) 370-6587
spike@willamette.edu
http://www.willamette.edu/cla/ees/

Chair:
Karen Arabas, (D), Penn State, 1997, Oo

Professor:
H. Peter Eilers, (D), Oregon State, 1974, On

Endowed Dempsey Chair:
Joe Bowersox, (D), Wisconsin, 1995, Oo

Assistant Professor:
Scott Pike, (D), Georgia, 2000, Gg

Pennsylvania

Allegheny College

Dept of Geology (B) (2006)
520 North Main Street
Meadville, PA 16335
p. (814) 332-2350 f. (814) 332-2981
bschwart@allegheny.edu
http://merlin2.alleg.edu/dept/geology/
Enrollment: No data reported since 1995

Professor:
Robert K. Schwartz, (D), Indiana, 1972, Gs

Associate Professor:
Ron Cole, (D), Rochester, 1993, Gt

Assistant Professor:
Rachel O'Brien, (D), Washington State, 2000, Hw

Bloomsburg University

Dept of Geography and Geosciences (B) (2007)
400 East Second Street
Bloomsburg, PA 17815-1301
p. (570) 389-4108 f. (570) 389-3028

dbraun@bloomu.edu
http://departments.bloomu.edu/geo/
Enrollment (2007): B: 45 (12)

Professor:
John E. Bodenman, (D), Penn State, 1995, Oo
Duane D. Braun, (D), Johns Hopkins, 1976, GmHwGl
Sandra J. Kehoe-Forutan, (D), Queensland, 1991, Oo
Michael K. Shepard, (D), Washington (St. Louis), 1994, XgYgOr
Dale A. Springer, (D), Virginia Tech, 1982, PivOe
Karen M. Trifonoff, (D), Kansas, 1994, Oi

Associate Professor:
Shahalam M. N. Amin, (D), Kent State, 1991, SoOyn
Patricia J. Beyer, (D), Arizona State, 1997, HsOyGm
Cynthia Venn, (D), Pittsburgh, 1996, OgClOc

Assistant Professor:
Jeffrey C. Brunskill, (D), SUNY - Buffalo, 2005, OiwOy
Joseph C. Hill, (M), Tennessee, 1999, GcpGg
John G. Hintz, (D), Kentucky, 2005, Oio
Brett T. McLaurin, (D), Wyoming, 2000, GrsGd

Bryn Mawr College

Dept of Geology (B) (2007)
101 North Merion Avenue
Bryn Mawr, PA 19010-2899
p. (610) 526-5115 f. (610) 526-5086
aweil@brynmawr.edu
http://www.brynmawr.edu/geology/
Enrollment (2007): B: 32 (14)

Chair:
Arlo B. Weil, (D), Michigan, 2001, Gc

Professor:
W. Bruce Saunders, (D), Iowa, 1970, Pi

Associate Professor:
Donald C. Barber, (D), 2000, Gs

Emeritus:
Maria Luisa B. Crawford, (D), California (Berkeley), 1965, Gx
William A. Crawford, (D), California (Berkeley), 1965, Cg
Lucian B. Platt, (D), Yale, 1960, Gc

Associate Curator:
Juliet C. Reed, (M), Bryn Mawr, Gz

Map Curator:
Betsy Reese, (U), Yr

Department Secretary:
Barbara Edwards, (N), Oo

Geology Librarian:
Terri Freedman, (U), Oo

Bucknell University

Dept of Geology (B) (2007)
231 O'Leary Center
Lewisburg, PA 17837
p. (570) 577-1382 f. (570) 577-3031
lmertz@bucknell.edu
http://www.departments.bucknell.edu/geology/
Enrollment (2007): B: 14 (7)

Chair:
Christopher G. Daniel, (D), RPI, 1998, Gp

Professor:
R. Craig Kochel, (D), Texas, 1980, Gm

Associate Professor:
Mary Beth Gray, (D), Rochester, 1991, Gc
Carl S. Kirby, (D), Virginia Tech, 1993, Cl
Jeffrey M. Trop, (D), Purdue, 2000, Gs

Assistant Professor:
Ellen K. Herman, (D), Penn State, 2006, Hw

Laboratory Director:
Brad C. Jordan, (M), Rhode Island, 1983, Gi

California University of Pennsylvania

Dept of Earth Sciences (B,M) (2005)
250 University Avenue
California, PA 15419
p. (724) 938-4180 f. (724) 938-5780
wickham@cup.edu
http://www.cup.edu/sci & tech/earth_sci/default.htm
Administrative Assistant: Pamela Higinbotham
Enrollment (2005): B: 10 (0) M: 7 (0)

Chair:
Thomas Wickham, (D), Penn State, 2000, Oy

Professor:
Lawrence L. Moses, (D), Pittsburgh, 1984, Oy

Associate Professor:
Thomas Mueller, (D), Illinois, 1999, Oi

Assistant Professor:
Swarndeep S. Gill, (D), Wyoming, 2002, Oa
Chad Kauffman, (D), Nebraska, 2000, Oa
Jamie Mitchem, (D), South Carolina, 2004, Oa
Daria L. Nikitina, (D), Delaware, 2000, Ge
Susan Ryan, (D), 2005, Og
G Wayne Smith, (D), Wilfrid Laurier, 2005, Og

Carnegie Museum of Natural History

Div of Paleontology (2005)
4400 Forbes Avenue
Pittsburgh, PA 15213
p. (412) 622-3246 f. (412) 622-8837
beardc@CarnegieMNH.org
Enrollment (1995): B: 1 (0)

Section Chair:
Mary R. Dawson, (D), Kansas, 1957, Pv

Curator:
David S. Berman, (D), California (Los Angeles), 1969, Pv

Associate Curator:
K. Christopher Beard, (D), Johns Hopkins, 1989, Pv
Zhexi Luo, (D), California (Berkeley), 1987, Pv

Preparator:
Amy C. Henrici, (M), Pittsburgh, 1990, Pv
Alan R. Tabrum, (M), SD Sch of Mines & Tech, 1981, Pv
Norman Wuerthele, (B), Pittsburgh, 1966, Pv

Collections Manager:
Elizabeth A. Hill, Pv
Albert Kollar, (B), Southhampton, 1974, Pi

Clarion University

Dept of Anthropology, Geography & Earth Sci (B) (2004)
Clarion, PA 16214
p. (814) 393-2317 f. (814) 393-2004
MCGREEVY@clarion.edu
http://clarion.edu/artsci/ages
Department Secretary: Earla L. McCall
Enrollment: No data reported since 1995

Chair:
Patrick V. McGreevy, (D), Minnesota, 1984, Oo

Professor:
David W. Howes, (D), Wisconsin (Milwaukee), 1983, Or
Robert R. McKay, (D), Oklahoma, 1982, Oo
Frank J. Vento, (D), Pittsburgh, 1985, Ga
Craig E. Zamzow, (D), Texas, 1983, Gi

Associate Professor:
John J. Ernissee, (D), South Carolina, 1982, Pi
Susan C. Prezzano, (D), SUNY (Binghamton), 1993, Oo
Paul T. Ryberg, (D), Arizona, 1984, Gs
Gerald A. Thomas, (D), Kentucky, 1993, Oo
Anthony J. Vega, (D), Louisiana State, 1994, Oo

Assistant Professor:
Yasser M. Ayad, (D), Montreal, 2000, Oi
Stephen J. Shulik, (D), Pittsburgh, 1979, Ym

Dickinson College

Dept of Geology (B) (2007)
P.O. Box 1773
Carlisle, PA 17013-2896
p. (717) 245-1355 f. (717) 245-1971
niemitz@dickinson.edu
http://www.dickinson.edu/departments/geol/
Administrative Assistant: Mary Orr
Enrollment (2007): B: 18 (8)

Professor:
Marcus M. Key, Jr., (D), Yale, 1988, Pi
Jeffery W. Niemitz, (D), S California, 1978, ClOuGn
Associate Professor:
Benjamin R. Edwards, (D), British Columbia, 1997, Gi
Assistant Professor:
Meagen Pollock, (D), Duke, 2007, Gi
Peter B. Sak, (D), Penn State, 2002, Gc
Emeritus:
Noel Potter, Jr., (D), Minnesota, 1969, Gm
William W. Vernon, (D), Lehigh, 1964, Gz
Technician:
Robert Dean, (M), Texas (El Paso), 2004, Gi

Drexel University

Paleontology and Geology Program (B,M,D) (2007)
4-270A Alumni Hall
Philadelphia, PA 19104
p. (215) 895-2344 f. (215) 895-1363
lacovara@drexel.edu
Administrative Assistant: Sharon Stokes
Enrollment (2006): B: 3 (0) M: 5 (0)

Director:
Edward L. Doheny, PG, (D), Indiana, 1967, Ng
Professor:
George Harris, PE, (D), Cornell, 1967, Ne
Robert M. Koerner, PE, (D), Duke, 1968, Sp
J. Richard Weggel, PE, (D), Illinois, 1968, On
Associate Professor:
Aspasia Zerva, (D), Illinois, 1985, Ne
Instructor:
Chad H. Freed, PG, (M), Drexel, 1993, Ng
Kathleen F. Stetser, (M), Delaware, 1992, Ps
Adjunct Professor:
Craig R. Calabria, PE, (D), Salford (UK), 1996, Sp
Gary P. Crawford, PG, (M), Drexel, 1984, Ng
George E. Duda, PE, (M), Drexel, 1977, Nr
Roy E. Hunt, PG, PE, (M), Uppsala, 1952, Ng
Jack C. Hwang, PE, (D), Minnesota, 1972, Hq
David M. Kargbo, (D), Nebraska, 1984, Sc
Kenneth Lacovara, (D), Delaware, 1995, Ps
Matthew A. Lahvis, (D), Drexel, 1993, Hw
Carl T. Mastropaolo, (M), Drexel, 1992, Hq
Dennis J. Mohan, PE, (M), Drexel, 1972, Sp
Michael J. Monteleone, PE, (M), Drexel, 1986, Ng
J. Mark Nielson, (M), Princeton, 1988, Hw
Margaret E. Passmore, (M), Drexel, 1990, Cg
Gary N. Paulachok, PG, (M), Drexel, 1985, Hw
Gary W. Snyder, PE; PG, (M), Drexel, 1992, Yg
Vincent W. Uhl, (M), Arizona, 1976, Hw
Geology Librarian:
Edward L. Doheny, PG, Ng

Edinboro University of Pennsylvania

Dept of Geosciences (B) (2007)
214 Cooper Hall
Edinboro, PA 16444-0001
p. (814) 732-2529 f. (814) 732-2593
kmoyer@edinboro.edu
www.edinboro.edu/cwis/geosci/htdocs/geohome.html
Enrollment (2007): B: 39 (13)

Chair:
Kerry A. Moyer, (D), Penn State, 1993, Oa
Professor:
Baher A. Ghosheh, (D), SUNY (Buffalo), 1988, Oo
David Hurd, (D), Cleveland State, 1997, Og
Henry Lawrence, (D), Oregon, 1985, Ou
Dale Tshudy, (D), Kent State, 1993, Pi
Brian S. Zimmerman, (D), Washington State, 1991, Em
Associate Professor:
Jeffrey J. Gryta, (D), Penn State, 1987, Gm
Laurie A. Parendes, (D), Oregon State, 1997, Oo
Joseph F. Reese, (D), Texas, 1995, Gc
Eric Straffin, (D), Nebraska, 2000, Gm
Assistant Professor:
Richard Deal, (D), South Carolina, 2000, Oi
Karen Eisenhart, (D), Colorado, 2004, Oy
Tadesse Kidane-Mariam, (D), Iowa, 2001, Oo
Susan Lucas, (D), Wilfrid Laurier, 1999, Oo

Elizabethtown College

Dept of Physics (2006)
Elizabethtown, PA 17022-2298
p. (717) 361-1323 (Ext. 323) f. (717) 361-1207
degoedek@etown.edu
Department Secretary: Rebecca Weis

Associate Professor:
Michael A. Scanlin, (D), Penn State, Ye
Lecturer:
David Ferruzza, (M), MIT, 1967, Oa

Franklin and Marshall College

Dept of Earth and Environment (B) (2008)
PO Box 3003
Lancaster, PA 17604-3003
p. (717) 291-4133 f. (717)291-4186
roger.thomas@fandm.edu
http://www.fandm.edu/earthandenvironment.xml
Administrative Assistant: Sarah K. Esh
Enrollment (2007): B: 52 (5)

Chair:
Roger D. K. Thomas, (D), Harvard, 1970, PoGh
Professor:
Carol B. de Wet, (D), Cambridge (UK), 1989, Gs
Dorothy J. Merritts, (D), Arizona, 1987, Gm
Stanley A. Mertzman, (D), Case Western, 1971, Gi
Robert S. Sternberg, (D), Arizona, 1982, Ym
Associate Professor:
Andrew P. deWet, (D), Cambridge (UK), 1989, Ge
Robert C. Walter, (D), Case Western, 1981, Cc
Assistant Professor:
Zeshan Ismat, (D), Rochester, 2002, Gc
Keely B. Maxwell, (D), Yale, 2004, Oy
James E. Strick, (D), Princeton, 1997, Ghe
Christopher J. Williams, (D), Pennsylvania, 2002, Pe
Research Associate:
Steven Sylvester, (M), Franklin & Marshall, 1971, Ca
Director of Public Policy:
Richard V. Pepino, (M), Villanova, 1970, Ou

Emeritus:
Edward C. Beutner, (D), Penn State, 1968, Gt
Robert A. Wiebe, (D), Stanford, 1966, Gi

Gannon University

Earth Science Program (B) (2007)
109 University Square
PMB 3183
Erie, PA 16541
p. (814) 871-7453 f. (814) 871-5565
olanrewa001@gannon.edu
Enrollment (1996): B: 16 (2)

Assistant Professor:
Johnson Olanrewaju, (D), Penn State, 2002, Cg

Indiana University of Pennsylvania

Dept of Geoscience (B) (2007)
111 Walsh Hall
Indiana, PA 15705
p. (724) 357-2379 f. (724) 357-6208
geoscience-info@iup.edu
www.iup.edu/geoscience
Enrollment (2006): B: 66 (0)

Chair:
Steven A. Hovan, (D), Michigan, 1993, Ou
Professor:
Karen Rose Cercone, (D), Michigan, 1984, Cl
John F. Taylor, (D), Missouri, 1984, Ps
Associate Professor:
Michael A. Poage, (D), Dartmouth, Cs
Assistant Professor:
Kenneth S. Coles, (D), Columbia, Oe
Jon C. Lewis, (D), Conn, Gc
Instructor:
Yvonne K. Branan, (D), Michigan Tech, ABD, Gv
Emeritus:
Joseph C. Clark, (D), Stanford, 1966, Gr
Frank W. Hall, (D), Montana, 1969, Gc
Darlene S. Richardson, (D), Columbia, 1974, Gd
Connie J. Sutton, (M), Indiana (Penn), 1968, Oe
Department Secretary:
April Mazur, Oo

Juniata College

Dept of Geology (B) (2004)
Huntingdon, PA 16652
p. (814) 641-3601 f. (814) 641-3685
mutti@juniata.edu
http://departments.juniata.edu/geology
Enrollment (2001): B: 24 (6)

Chair:
Laurence J. Mutti, (D), Harvard, 1978, Gx
Associate Professor:
David Lehmann, (D), Rochester, 1993, Gr
Assistant Professor:
Ryan Mathur, (D), Arizona, 2000, Gc
Emeritus:
J. Peter Trexler, (D), Michigan, 1964, Ps
Robert H. Washburn, (D), Columbia, 1966, Gs

Kutztown University of Pennsylvania

Dept of Physical Science (B) (2007)
Boehm Science Building Room 135
Kutztown, PA 19530
p. (610) 683-4447 f. (610) 683-1352
simpson@kutztown.edu
http://www.kutztown.edu/acad/geology
Department Secretary: Donna Moore
Enrollment (2001): B: 138 (22)

Chair:
Edward L. Simpson, (D), Virginia Tech, 1987, Gs
Professor:
Bruce F. Rowell, (D), Nebraska, 1974, Po
Madan M. Varma, (D), Indiana, 1975, Yg
Assistant Professor:
Kurt Friehauf, (D), Stanford, 1998, Em
Sarah E. Tindall, (D), Arizona, 2000, Gc

La Salle University

Dept of Geology & Environmental Science (B) (2006)
20th and Olney Avenue
Philadelphia, PA 19141
p. (215) 951-1268 f. (215) 951-1488
bart@lasalle.edu
http://www.lasalle.edu/academ/geo_env
Enrollment (2006): B: 30 (5)

Chair:
Henry A. Bart, (D), Nebraska, 1974, Gs
Professor:
Alice L. Hoersch, (D), Johns Hopkins, 1977, Gp
Adjunct Professor:
Natalie Flynn, (M), Temple, 1999, Gz

Lafayette College

Dept of Geology & Env Geosciences (B) (2006)
Easton, PA 18042
p. (610) 330-5193 f. (610) 330-5717
geology@lafayette.edu
http://www.lafayette.edu/~geology/
Department Secretary: Rohana Meyerson
Enrollment (2006): B: 23 (6)

Professor:
Dru Germanoski, (D), Colorado State, 1989, Gm
Guy L. Hovis, (D), Harvard, 1971, Gy
Associate Professor:
Lawrence L. Malinconico, (D), Dartmouth, 1982, Yg
Assistant Professor:
Kira Lawrence, (D), Brown, 2005, Pe
David Sunderlin, (D), Chicago, 2005, GsPo
Research Associate:
Mary Ann Malinconico, (D), Columbia, 2002, Eo
Emeritus:
Richard W. Faas, (D), Iowa State, 1964, Gs
Laboratory Director:
John R. Wilson, (M), Virginia Tech, 2001, Oi

Lehigh University

Dept of Earth & Environmental Sci (B,M,D) (2007)
31 Williams Drive
Bethlehem, PA 18015-3126
p. (610) 758-3660 f. (610) 758-3677
ljc0@lehigh.edu
http://www.ees.lehigh.edu
Administrative Assistant: Laura J. Cambiotti
Enrollment (2007): M: 15 (0) D: 12 (0)

Chair:
Frank J. Pazzaglia, (D), Penn State, 1993, Gm
Director, South Mountain College:
Peter K. Zeitler, (D), Dartmouth, 1983, Cc
Director, Env Initiative:
Dork Sahagian, (D), Chicago, 1987
Dean, College of Arts and Sciences:
Anne S. Meltzer, (D), Rice, 1988, Ys

Professor:
David J. Anastasio, (D), Johns Hopkins, 1988, Gc
Gray E. Bebout, (D), California (Los Angeles), 1989, Gp
Edward B. Evenson, (D), Michigan, 1972, Gl
Kenneth P. Kodama, (D), Stanford, 1978, Ym
Deputy Provost:
Carl O. Moses, (D), Virginia, 1987, Cl
Associate Professor:
Bruce R. Hargreaves, (D), California (Berkeley), 1977, Ob
Donald P. Morris, (D), Colorado, 1990, Oo
Zicheng Yu, (D), Toronto, 1997, Pe
Assistant Professor:
Robert K. Booth, (D), Wyoming, 2003, Oo
Stephen C. Peters, (D), Michigan, 2001, Cl
Joan Ramage, (D), Cornell, 2001, Or
Research Associate:
Bruce D. Idleman, (D), SUNY (Albany), 1990, Cc
Emeritus:
Bobb Carson, (D), Washington, 1971, Gu
Paul B. Myers, (D), Lehigh, 1960, Hw
Hayden N. Pritchard, (D), Lehigh, 1963, Ob
Dale R. Simpson, (D), Caltech, 1960, Gz

Lock Haven University

Dept of Geology & Physics (B) (2007)
401 N. Fairview Street
Lock Haven, PA 17745-2390
p. (570) 484-2048 f. (570) 484-2407
mkhalequ@lhup.edu
Administrative Assistant: Barbara Greene
Enrollment (2007): B: 30 (5)
Chair:
Md. Khalequzzaman, (D), 1998, HwOni
Assistant Professor:
Loretta D. Dickson, (D), Connecticut, 2006, GicGe
Thomas C. Wynn, (D), 2004, GsPiEo

Mansfield University

Dept of Geography & Geology (B) (2006)
Belknap Hall
Mansfield, PA 16933
p. (570) 662-4614 f. (570) 662-4126
rdodson@mansfield.edu
http://geoggeol.mansfield.edu/index.html
Enrollment (2006): B: 55 (17)
Chair:
Russell L. Dodson, (D), Michigan State, 1984, Gm
Associate Professor:
Hobart M. King, (D), West Virginia, 1982, Ge
Kathryn F. Thorne, (D), Michigan State, 1992, Oo
Assistant Professor:
Jennifer Demchak, (D), West Virginia, 2005, Hs

Mercyhurst College

Dept of Geology (B,M) (2007)
501 East 38th Street
Erie, PA 16546
p. (814) 824-2581 f. (814) 824-2594
adovasio@mercyhurst.edu
http://mai.mercyhurst.edu
Enrollment (2006): B: 8 (4)
Director:
James M. Adovasio, (D), Utah, 1970, Ga
Associate Professor:
M. Raymond Buyce, (D), RPI, 1975, Gs
Assistant Professor:
Andrew Breckenridge, (D), Minnesota, 2005, Gl

Millersville University

Dept of Earth Sciences (B) (2007)
PO Box 1002
Millersville, PA 17551
p. (717) 872-3289 f. (717) 871-4725
esci@millersville.edu
http://www.millersville.edu/~esci
Enrollment (2007): B: 29 (9)
Chair:
Richard D. Clark, (D), Wyoming, 1987, Oa
Professor:
Sepideh Yalda, (D), St. Louis, 1997, Oa
Associate Professor:
Alex J. DeCaria, (D), Maryland, 2000, Oa
L. Lynn Marquez, (D), Northwestern, 1998, Cg
R. Ramananantoandro, (D), Washington, 1971, Ng
Assistant Professor:
Ajoy Kumar, (D), Old Dominion, 1996, Op
Jason R. Price, (D), Michigan State, 2003, Gs
Todd D. Sikora, (D), Penn State, 1996, Oa
Instructor:
Joseph Calhoun, (B), Penn State, Oa
G. Robert Ganis, (D), Leicester, 2004, Gg
Mary Ann Schlegel, (M), Og
Professor:
Robert S. Ross, (D), Florida State, 1977, Oa
Emeritus:
William M. Jordan, (D), Wisconsin, 1965, Gs
Bernard L. Oostdam, (D), Delaware, 1971, Ou
Charles K. Scharnberger, (D), Washington (St. Louis), 1971, Gc
Yin S. Soong, (D), Florida State, 1977, Op

Montgomery County Community College

Dept of Science (A) (2006)
Blue Bell, PA 19422
p. (215) 641-6446
rkuhlman@mc3.edu
Professor:
Robert Kuhlman, (M), Bryn Mawr, 1975, Gg
Instructor:
George Buchanan, (M), Drexel, 1992, Ng
Adjunct Professor:
Kelly Corrado, (M), Drexel, 2003
Laurie Martin-Vermilyea, (D), South Carolina, 1992, Oe
Frank Roberts, (D), Bryn Mawr, 1969, Gp
Anthony Stevens, (M), Florida, 1981, Ge

Moravian College

Dept of Physics & Earth Science (B) (2007)
1200 Main Street
Bethlehem, PA 18018-6650
p. (610) 861-1440 f. (610) 625-7918
gerencher@moravian.edu
http://www.physics.moravian.edu
Department Secretary: Lou Ann Vlahovic
Enrollment (2005): B: 1 (0)
Professor:
Joseph J. Gerencher, Jr., (D), Penn State, 1982, Gg
Librarian:
Rita Berk, Oo

Pennsylvania State University, Altoona

Environmental Earth Systems Sciences (B) (2004)
Altoona, PA 16601
p. (814) 949-5272 f. (814) 949-5011
pdz1@psu.edu

Acting Head:
Brian B. Tormey, (D), Penn State, 1980, Gm
Assistant Professor:
Marieta P. Staneva, (D), Sofia, 1985, Oy
Senior Associate Registrar:
Garry L. Burkle, (M), Illinois State, 1970, Gg
Lecturer:
Nathaniel S. Winstead, (M), Penn State, 1995, Ow
Paul D. Zell, (M), Pittsburgh, 1985, Pi
Geology Librarian:
Linda R. Musser, Oo

Pennsylvania State University, Erie

Geoscience Dept (2004)
Erie, PA 16510
p. (814) 898-6277
amf11@psu.edu
Assistant Professor:
Eva Tucker, (M), Cincinnati, 1962, Gg

Pennsylvania State University, Hazleton

Science Dept (2004)
76 University Drive
Hazleton, PA 18202
p. (570) 450-3000 f. (570) 450-3182
http://www.hn.psu.edu
Assistant Professor:
George Tseo, (D), Adelaide, 1986, Oo
Instructor:
Daniel H. Vice, (D), Penn State, 1996, Eg

Pennsylvania State University, Monaca

Dept of Geosciences (B) (2004)
100 University Drive
Monaca, PA 15061
p. (412) 773-3867 f. (412) 773-3557
jac7@psu.edu
Enrollment (1998): B: 15 (0)
Assistant Professor:
John A. Ciciarelli, (D), Penn State, 1971, Gg

Pennsylvania State Univ, University Park

Dept of Crop and Soil Sciences (B,M,D) (2007)
116 Agricultural Sciences & Industries Building
University Park, PA 16802
p. (814) 865-6541 f. (814) 863-7043
dms39@psu.edu
Enrollment (1995): B: 16 (2) M: 21 (3) D: 12 (0)
Professor:
Douglas B. Beegle, (D), Penn State, 1983, Sc
Jean-Marc Bollag, (D), Basel, 1959, Sb
Edward J. Ciolkosz, (D), Wisconsin, 1967, Sd
Daniel D. Fritton, (D), Iowa State, 1968, Sp
Sridhar Komarneni, (D), Wisconsin, 1973, Sc
Gary W. Petersen, (D), Wisconsin, 1965, Sd
Associate Professor:
Peter J. Landschoot, (D), Rhode Island, 1988, So
Gregory W. Roth, (D), Penn State, 1987, So
Assistant Professor:
Rick L. Day, (D), Penn State, 1991, Ou
Research Associate:
Barry M. Evans, (M), Penn State, 1977, Sd
Douglas A. Miller, (M), Penn State, 1987, Sd
Adjunct Professor:
Andrew S. Rogowski, (D), Iowa State, 1964, Sp
Lawrence A. Schardt, (M), Penn State, 1996, Sd
Andrew N. Sharpley, (D), Massey, 1977, Sc

Cooperating Faculty:
Ann M. Wolf, (D), Penn State, 1988, Sc

Dept of Geosciences (B,M,D) (2007)
503 Deike Building
University Park, PA 16802-2714
p. (814) 865-6711 f. (814) 863-7823
bralower@psu.edu
http://www.geosc.psu.edu/
Administrative Assistant: Lou Klindienst
Enrollment (2007): B: 110 (27) M: 30 (16) D: 66 (14)
Head:
Timothy J. Bralower, (D), California (San Diego), 1986, Pe
Distinguished Professor:
James F. Kasting, (D), Michigan, 1979, Oa
Director, Earth & Env Systems Inst:
Susan L. Brantley, (D), Princeton, 1987, Cg
Director, Astrobiology Research Center:
Hiroshi Ohmoto, (D), Princeton, 1969, Cs
Associate Head, Undergraduate Programs:
David M. Bice, (D), California (Berkeley), 1989, Gg
Associate Head, Graduate Programs:
Katherine H. Freeman, (D), Indiana, 1991, Co
Professor:
Richard B. Alley, (D), Wisconsin, 1987, Gl
Charles J. Ammon, (D), Penn State, 1991, Ys
Michael A. Arthur, (D), Princeton, 1979, Ou
Terry Engelder, (D), Texas A&M, 1973, Nr
Donald M. Fisher, (D), Brown, 1988, Gc
Kevin P. Furlong, (D), Utah, 1981, Gt
Tanya Furman, (D), MIT, 1989, Cg
Peter Heaney, (D), Johns Hopkins, 1989, Gz
Lee R. Kump, (D), S Florida, 1986, Cl
Chris Marone, (D), Columbia, 1988, Yx
Andrew A. Nyblade, (D), Michigan, 1992, Yg
Richard R. Parizek, (D), Illinois, 1961, Hw
Rudy L. Slingerland, (D), Penn State, 1977, Gs
Senior Scientist:
Raymond Fletcher, (D), Brown, 1967, Gc
Todd Sowers, (D), Rhode Island, 1991, Cs
Faculty Fellow:
Peter D. Wilf, (D), Penn, 1998, Py
Director, Earth & Mineral Science Museum:
Russell W. Graham, (D), Texas, 1976, Pv
Associate Professor:
Sridhar Anandakrishnan, (D), Wisconsin, 1990, Ys
Christopher H. House, (D), California (Los Angeles), 1999, Py
James D. Kubicki, (D), Yale, 1990, Cg
Michael E. Mann, (D), Yale, 1998, Oa
Mark E. Patzkowsky, (D), Chicago, 1992, Po
Demian M. Saffer, (D), California (Santa Cruz), 1999, Hw
Assistant Professor:
Matthew S. Fantle, (D), California (Berkeley), 2005, Cs
Maureen D. Feinemann, (D), California (Berkeley), 2004, Cp
Klaus Keller, (D), Princeton, 2000, Og
Eric Kirby, (D), MIT, 2001, Gm
Peter C. LaFemina, (D), Miami, 2005, Yd
Jennifer L. Macalady, (D), California (Davis), 2000, Py
Eliza Richardson, (D), MIT, 2002, Ys
Kamini Singha, (D), Stanford, 2004, Hy
Emeritus:
Shelton S. Alexander, (D), Caltech, 1963, Ys
Hubert L. Barnes, (D), Columbia, 1958, Cp
C. Wayne Burnham, (D), Caltech, 1955, Cp
Roger J. Cuffey, (D), Indiana, 1966, Pg
Alan Davis, (D), Durham (UK), 1965, Ec
Peter Deines, (D), Penn State, 1967, Cs
David H. Eggler, (D), Colorado, 1967, Cp

David P. Gold, (D), McGill, 1963, Gc
Earl K. Graham, Jr., (D), Penn State, 1969, Yx
Roy J. Greenfield, (D), MIT, 1965, Yg
Albert L. Guber, (D), Illinois, 1962, Pe
Benjamin F. Howell, Jr., (D), Caltech, 1949, Ys
Derrill M. Kerrick, (D), California (Berkeley), 1968, Gp
Peter M. Lavin, (D), Penn State, 1962, Ye
Arthur W. Rose, (D), Caltech, 1958, Ce
Rustum Roy, (D), Penn State, 1948, Cg
Robert F. Schmalz, (D), Harvard, 1959, Gu
Robert Scholten, (D), Michigan, 1950, Gt
William Spackman, (D), Harvard, 1949, Pb
Charles P. Thornton, (D), Yale, 1953, Gv
Alfred Traverse, (D), Harvard, 1951, Pl
Barry Voight, (D), Columbia, 1964, Gv
William B. White, (D), Penn State, 1962, Cg
Eugene G. Williams, (D), Penn State, 1957, Gr
Lauren A. Wright, (D), Caltech, 1951, Gg

Earth & Mineral Sciences Museum (2004)
122 Steidle Building
University Park, PA 16802
p. (814) 865-6427 f. (814) 863-7708
sicree@geosc.psu.edu
Enrollment: No data reported since 1995

Director & Curator:
Andrew A. Sicree, (D), Penn State, 1999, Gz

Dept of Meteorology (B,M,D) (2006)
503 Walker Building
University Park, PA 16802-5013
p. (814) 865-0478 f. (814) 865-3663
meteodept@ems.psu.edu
http://www.met.psu.edu

Head:
William H. Brune, (D), Johns Hopkins, 1978, Ow

Professor:
Peter R. Bannon, (D), Colorado, 1979, Ow
Jenni L. Evans, (D), Monash, 1990, Ow
William M. Frank, (D), Colorado State, 1976, Ow
James F. Kasting, (D), Michigan, 1979, Ow
Dennis Lamb, (D), Washington, 1970, Ow
Dennis W. Thomson, (D), Wisconsin, 1968, Ow
John C. Wyngaard, (D), Penn State, 1967, Ow
George S. Young, (D), Colorado State, Ow

Associate Head:
Hampton N. Shirer, (D), Penn State, 1978, Ow

Associate Professor:
John H. E. Clark, (D), Florida State, 1969, Ow
Eugene E. Clothiaux, (D), Brown, Ow
Kenneth J. Davis, (D), Colorado, 1992, Ow
Raymond G. Najjar, (D), Princeton, 1990, Ow
Nelson L. Seaman, (D), Penn State, 1977, Ow
Johannes Verlinde, (D), Colorado State, 1992, Ow

Associate Research Professor:
David R. Stauffer, (D), Penn State, 1990, Ow

Assistant Professor:
Jerry Y. Harrington, (D), Colorado State, 1997, Ow
Sukyoung Lee, (D), Princeton, 1991, Ow

Research Assistant:
Glenn Hunter, (M), Penn State, 1994, Ow
Annette Lario-Gibbs, (M), Penn State, 1986, Ow
Robert Lesher, (M), Penn State, 1990, Ow
Todd Miner, (M), Penn State, 1995, Ow
William F. Ryan, (M), Maryland, 1990, Ow
Jeffrey M. Warner, (M), Penn State, 1999, Ow

Director, Meteorology Computing:
Charles Pavloski, (D), Penn State, 2004, Ow

Research Associate:
Aijun Deng, (D), Penn State, 1999, Ow
Harry W. Henderson, (D), Penn State, 1982, Ow
Arthur Person, (M), Penn State, 1983, Ow
William Syrett, (M), Penn State, 1987, Ow

Instructor:
Frederick J. Gadomski, (M), Penn State, 1983, Ow
Lee M. Grenci, (M), Penn State, 1984, Ow
Paul Knight, (M), Penn State, 1977, Ow

Distinguished Professor:
J. Michael Fritsch, (D), Colorado State, 1978, Ow

Emeritus:
Alfred K. Blackadar, (D), New York, 1950, Ow
Craig F. Bohren, (D), Arizona, 1975, Ow
John J. Cahir, (D), Penn State, 1971, Ow
Toby N. Carlson, (D), Imperial Coll, 1965, Ow
Rosa G. de Pena, (D), Buenos Aires, 1949, Ow
John A. Dutton, (D), Wisconsin, Ow
Alistair B. Fraser, (D), Imperial Coll, 1968, Ow
Charles L. Hosler, (D), Penn State, 1951, Ow

Mining Eng Program, Dept of Energy and Mineral Eng (B,M,D) (2007)
110 Hosler Building
University Park, PA 16802
p. (814) 865-3437 f. (814) 865-3248
mngeng@ems.psu.edu
http://www.eme.psu.edu/mng/
Enrollment (2000): B: 16 (6) M: 6 (2)

Chair:
R. Larry Grayson, (D), West Virginia, 1986, Nm

Associate Professor:
Maochen Ge, (D), Penn State, 1988, Nm
Vladislav J. Kecojevic, (D), Belgrade, 2000, Nm

Assistant Professor:
Jamal Rostami, (D), Colorado Mines, 1997, Nm

Point Park College

Dept of Natural Sciences & Eng Tech (A,B) (2004)
201 Wood Street
Pittsburgh, PA 15222-1994
p. (412) 392-3900 f. (412) 391-1980
Administrative Assistant: Roberta T. Gallick
Enrollment (1996): B: 26 (4)

Head:
Mark O. Farrell, (D), Carnegie Mellon, Oo

Professor:
Sushanta K. Banerjee, (D), Texas A&M, Oo
John J. Kudlac, (D), Pittsburgh, Ng

Associate Professor:
Joseph H. Sunder, (M), Pittsburgh, Oo

Shippensburg University

Geography-Earth Science Dept (B,M) (2008)
1871 Old Main Drive
Shippensburg, PA 17257
p. (717) 477-1685 f. (717) 477-4029
jpment@ship.edu
http:www.ship.edu/~geog/
Administrative Assistant: Judy P. Mentzer

Chair:
William L. Blewett, (D), Michigan State, 1991, Oy

Professor:
Paul G. Marr, (D), Denver, 1996, Oi
Christopher J. Woltemade, (D), Wisconsin, 1993, Hg

Associate Professor:
Michael T. Applegarth, Arizona State, 2001, Or
Alison E. Feeney, (D), Michigan State, 1998, Oo
Thomas P. Feeney, (D), Georgia, 1997, Gm

Kurtis G. Fuellhart, (N)
George M. Pomeroy, (N), Akron, 1999
Janet S. Smith, Georgia, 1999, Oi
Kay R. Williams, (D), Georgia, 1995, Oa
Assistant Professor:
Sean Cornell, (D), Cincinnati
Scott A. Drzyzga, (N), Michigan State, 2003, Oi
Timothy W. Hawkins, (N), Arizona State, 2004, Oa
Claire A. Jantz, (D), Maryland, 2005
Joseph Zume, Oklahoma, 2007

Slippery Rock University

Dept of Geography, Geology, and the Env (B) (2006)
Slippery Rock, PA 16057
p. (724) 738-2048 f. (724) 738-4762
james.hathaway@sru.edu
http://www.sru.edu/pages/668.asp
Administrative Assistant: Bonita L. Vinton
Enrollment (2006): B: 50 (9)
Associate Professor:
Patrick A. Burkhart, (D), Lehigh, 1994, Hg
Assistant Professor:
Patricia A. Campbell, (D), Pittsburgh, 1994, Gc
Xianfeng Chen, (D), West Virginia, 2005, Or
Tamra A. Schiappa, (D), Boise State, 1999, Pi
Julie A. Snow, (D), Rhode Island, 2002, Oa
Michael G. Stapleton, (D), Delaware, 1995, So
Michael J. Zieg, (D), Johns Hopkins, 2001, Gi

Susquehanna University

Dept of Earth & Environmental Science (B) (2006)
514 University Ave
Selinsgrove, PA 17870
p. (570) 372-4216
resslerd@susqu.edu
http://www.susqu.edu/ees
Enrollment (2006): B: 27 (2)
Chair:
Daniel E. Ressler, (D), Iowa State, 1998, Sp
Associate Professor:
Jennifer M. Elick, (D), Tennessee, 1999, Pe
Assistant Professor:
Derek J. Straub, (D), Colo, Oa
Katherine H. Straub, (D), Colorado State, 2002, Ow
Matthew M. Uliana, (D), Texas, 2000, Hw

Temple University

Dept of Geology (B,M) (2008)
1901 N. 13th Street
Beury Hall, Rm. 326
Philadelphia, PA 19122-6081
p. (215) 204-8227 f. (215) 204-3496
scox@temple.edu
http://www.temple.edu/geology
Administrative Assistant: Shelah Cox
Enrollment (2005): M: 11 (0)
Chair:
David E. Grandstaff, (D), Princeton, 1974, Cl
Professor:
George H. Myer, (D), Yale, 1965, Gz
Gene C. Ulmer, (D), Penn State, 1964, Cp
Associate Professor:
Jonathan Nyquist, (D), Wisconsin, 1986, Yg
Laura Toran, (D), Wisconsin, 1986, Hw
Asst. Lab Manager/Bldg. Coordinator:
Donald Deigh-Kai, (M), Temple, 2008

The State Museum of Pennsylvania

Section of Paleontology & Geology (2007)
300 North Street
Harrisburg, PA 17120-0024
p. (717) 783-9897 f. (717) 214-2990
rsullivan@state.pa.us
Senior Curator:
Robert M. Sullivan, (D), Michigan State, 1980, PvgPs

Thiel College

Dept of Environmental Science (B) (2004)
Greenville, PA 16125
p. (412) 589-2821 f. (412) 589-2021
jbarton@thiel.edu
Enrollment (2004): B: 23 (3)
Professor:
James H. Barton, (D), N Colorado, 1977, Oy
Adjunct Professor:
John Moreittini, (M), Hg
Francois Pale, (D), Gd
Dennis Puko, (B), IUP, Ou

University of Pennsylvania

Dept of Earth & Environmental Science (B,M,D) (2005)
240 S. 33rd Street
Philadelphia, PA 19104-6316
p. (215) 898-5724 f. (215) 898-0964
earth@sas.upenn.edu
http://www.sas.upenn.edu/earth/
Enrollment (1998): B: 93 (32) M: 19 (1) D: 16 (2)
Chair:
Frederick N. Scatena, (D), Johns Hopkins, 1987, Gm
Professor:
Peter Dodson, (D), Yale, 1974, Pv
Robert F. Giegengack, Jr., (D), Yale, 1968, Gg
Arthur H. Johnson, (D), Cornell, 1975, So
Hermann W. Pfefferkorn, (D), Muenster, 1968, Pb
Senior Scientist:
Gomaa I. Omar, (D), Pennsylvania, 1985, Cc
David R. Vann, (D), Pennsylvania, 1993, Py
Associate Professor:
Stephen P. Phipps, (D), Princeton, 1984, Gc
Assistant Professor:
Benjamin P. Horton, (D), Durham (UK), 1997, On
Lecturer:
Yvette Bordeaux, (D), Pennsylvania, 2000, Po
Edward L. Doheny, (D), Indiana, 1967, Ng
Susan Gill, (D), Pennsylvania, 1996, Pe
Stanley L. Laskowski, (M), Drexel, 1973, Ge
Willig B. Sarah, (D), Pennsylvania, 1988, Ge
Elaine Wright, (M), Widner, 1976, Oo
Visiting Scholar:
John L. Spiesberger, (D), California (San Diego), 1980, Ob
Emeritus:
R. Ian Harker, (D), Cambridge, 1954, Cp
Reginald Shagam, (D), Princeton, 1956, Gt
W. John Smith, (D), Harvard, 1961, Py
Mary E. Wagner, (D), Bryn Mawr, 1972, Gp
Department Coordinator:
Arlene Mand, (B), 1971, Oo

University of Pittsburgh

Dept of Geology & Planetary Science (B,M,D) (2007)
200 SRCC Building
4107 O'Hara Street
Pittsburgh, PA 15260-3332
p. (412) 624-8780 f. (412) 624-3914

romickm@pitt.edu
http://www.geology.pitt.edu
Enrollment (1998): B: 41 (7)

Chair:
Brian W. Stewart, (D), California (Los Angeles), 1990, Cc
Professor:
Thomas H. Anderson, (D), Texas, 1969, Gt
Associate Professor:
Rosemary C. Capo, (D), California (Los Angeles), 1990, Cl
William P. Harbert, (D), Stanford, 1987, Ym
Michael S. Ramsey, (D), Arizona State, 1996, Or
Assistant Professor:
Mark B. Abbott, (D), Minnesota, 1995, Gs
Daniel J. Bain, (D), Johns Hopkins, 2004, Hg
Emily M. Elliott, (D), Johns Hopkins, 2003, Hg
Michael F. Rosenmeier, (D), Florida, 2003, Gn
Ian Skilling, (D), 1988, Gv
Lecturer:
Mark Collins, (M), Pittsburgh, 1985, Oo
Charles E. Jones, (D), Oxford, 1992, Gg
Adjunct Professor:
Robbie Ali, (D), Columbia, 1988, Oo
Robert Anderson, (D), Pittsburgh, Ge
K. Christopher Beard, (D), North Carolinal, 1984, Pv
David K. Brezinski, (D), Pittsburgh, 1985, Ps
David Crown, (D), Arizona State, 1991
Robert S. Hedin, (D), Rutgers, 1987
Matthew C. Lamanna, (D), Pennsylvania, 2004, Pv
Emeritus:
William Cassidy, (D), 1969
Jack D. Donahue, (D), Columbia, 1967, Ga
Bruce W. Hapke, (D), Cornell, 1962, Xy
Chiao-Min Hsieh, (D), Syracuse, 1953, Oy
Edward G. Lidiak, (D), Rice, 1963, Gi
Harold B. Rollins, (D), Columbia, 1967, Pe
Cooperating Faculty:
George D. Gatewood, (D), Pittsburgh, 1972, Xg
Geology Librarian:
G. Lance Lugar, Oo

University of Pittsburgh at Johnstown

Dept of Geology & Planetary Science (B) (2007)
Johnstown, PA 15904
p. (814) 269-2945 f. (814) 269-2022
beuthin@pitt.edu
www.upj.pitt.edu/geology
Enrollment (2007): B: 28 (6)

Chair:
Jack D. Beuthin, (D), North Carolina, 1989, Gr
Associate Professor:
Uldis Kaktins, (D), Boston, 1975, Hg
Assistant Professor:
Christopher F. Kopf, (D), Massachusetts, 1999, Gp
Instructor:
Michael C. Layton, (M), S Florida, 1982, Yg
Stephen R. Lindberg, (M), SUNY (Stony Brook), 1983, Og
Teresa K. McConnell, (B), Indiana (Penn), 1980, Gg
Emeritus:
William R. Brice, (D), Cornell, 1971, Gg

University of Pittsburgh, Bradford

Dept of Geology & Environmental Science (B) (2007)
300 Campus Drive
Bradford, PA 16701-2898
p. (814) 362-7569 f. (814) 362-5088
aap@pitt.edu
http://www.pitt.edu/~aap/
Department Secretary: Kim Murphy
Enrollment (2000): B: 19 (4)

Program Director:
Assad I. Panah, (D), Oklahoma, 1966, Gc
Associate Professor:
Edgar M. Hopkins, (D), Virginia, 1974, Gs
Adjunct Professor:
Tooran Panah, (D), London, 1979, Pi

West Chester University

Dept of Geology & Astronomy (B,M) (2007)
750 South Church Street
Boucher Hall
West Chester, PA 19383
p. (610) 436-2727 f. (610) 436-3036
jholste@wcupa.edu
Administrative Assistant: Kathryn Wilson
Enrollment (2006): B: 67 (0) M: 26 (0)

Chair:
C. Gil Wiswall, (D), Montana, 1979, Gc
Professor:
Richard M. Busch, (D), Pittsburgh, 1984, Oe
LeeAnn Srogi, (D), Pennsylvania, 1988, Gp
John Stolar, (D), Penn State, 1978, Og
Associate Professor:
Steven C. Good, (D), Colorado, 1993, Pe
Allen H. Johnson, (D), Case Western, 1972, Yg
Timothy M. Lutz, (D), Pennsylvania, 1979, Gq
Arthur Smith, (D), Pennsylvania, 1975, Oe
Assistant Professor:
Cynthia G. Fisher, (D), Colorado, 1991, Ou
Marc R. Gagne, (D), Georgia, 1994, Oo
Rajul E. Pandya, (D), Washington, 1996, Oa

Wilkes University

Dept of Environmental Eng & Earth Sciences (B) (2006)
84 West South Street
Wilkes-Barre, PA 18766
p. 570-408-4610 f. 570-408-7865
marleen.troy@wilkes.edu
http://www.wilkes.edu/pages/388.asp
Enrollment: No data reported since 1995

Chair:
Marleen Troy, (D), Drexel, 1989, Hw
Professor:
Dale A. Bruns, (D), Idaho State, 1981, Ou
James M. Case, (D), Dalhousie, 1979, Ob
Sid P. Halsor, (D), Michigan Tech, 1989, Gi
Kenneth M. Klemow, (D), SUNY (Syracuse), 1982, Py
Brian T. Redmond, (D), RPI, 1982, Gs
Michael A. Steele, (D), Wake Forest, 1988, Py
Associate Professor:
Dan M. Pindzola, (D), Brooklyn Polytech, 1962, Ow
Brian E. Whitman, (D), Michigan Tech, 1998, Hw
Research Associate:
Brian F. Oram, (M), Penn State, 1987, Hg

York College of Pennsylvania

Dept of Physical Science (A) (2004)
York, PA 17405
p. (717) 846-7788 (Ext. 333) f. (717) 849-1619
wkreiger@ycp.edu

Assistant Professor:
William Kreiger, (D), Penn State, 1976, Gi
Adjunct Professor:
Ralph Eisenhart, (M), Penn State, 1994, Gg
Jeri L. Jones, (B), Catawba, 1977, Ga

Puerto Rico

Puerto Rico

University of Puerto Rico

Dept of Agronomy & Soils (B,M) (2004)
College of Agricultural Sciences
P.O. Box 5000
Mayaguez, PR 00681
p. (809) 265-3851 f. (809) 265-0860
Enrollment (1995): B: 26 (4) M: 27 (6)

Acting Head:
Hector M. Lugo, (D), North Carolina State, 1972, Oo
Professor:
Friedrich Beinroth, (D), Stuttgart, 1965, Sd
Nancy D. Cavallaro, (D), Cornell, 1981, Sc
Jose Villarrubia, (D), North Carolina State, 1980, Sf
Senior Scientist:
Victor Snyder, (D), Cornell, 1980, Sp
On Leave:
Gustavo Martinez, (D), Ohio State, 1994, Sc
Miguel A. Munoz, (D), Ohio State, 1988, Sc

Dept of Geology (B,M) (2007)
Box 9017
Mayaguez, PR 00681-9017
p. (787) 265-3845 f. (787) 265-3845
geol@uprm.edu
http://geology.uprm.edu/
Enrollment (2004): B: 19 (19) M: 2 (2)

Professor:
James Joyce, (D), Northwestern, 1985, Ng
Johannes H. Schellekens, (D), Syracuse, 1993, Eg
Assistant Professor:
Eugenio Asencio, (D), South Carolina, 2002, Ys
Fernando Gilbes, (D), S Florida, 1996, Or
Thomas E. Miller, (D), McMaster, 1986, Gm
Wilson R. Ramirez, (D), Tulane, 2000, Gd
Hernan Santos, (D), Colorado, 1999, Pi

Rhode Island

Brown University

Dept of Geological Sciences (B,M,D) (2005)
Box 1846, 324 Brook Street
Providence, RI 02912
p. (401) 863-3338 f. (401) 863-2058
Geological_Sciences@brown.edu
http://www.geo.brown.edu
Administrative Assistant: Nancy Fjeldheim
Enrollment (2005): B: 36 (23) M: 3 (9) D: 49 (7)

Chair:
Warren L. Prell, (D), Columbia, 1974, Ou
Professor:
Reid F. Cooper, (D), Cornell, 1983, Gy
Karen M. Fischer, (D), MIT, 1988, Ys
Donald W. Forsyth, (D), MIT/Woods Hole, 1974, Yr
James W. Head, III, (D), Brown, 1969, Xg
Timothy D. Herbert, (D), Princeton, 1987, Pe
John F. Hermance, (D), Toronto, 1967, Yg
Paul C. Hess, (D), Harvard, 1968, Gi
E. Marc Parmentier, (D), Cornell, 1975, Yg
Carle M. Pieters, (D), MIT, 1977, Or
Malcolm J. Rutherford, (D), Johns Hopkins, 1968, Cp
Peter H. Schultz, (D), Texas, 1972, Xg
Jan A. Tullis, (D), California (Los Angeles), 1971, Gc
Terry E. Tullis, (D), California (Los Angeles), 1971, Yx
Thompson Webb, III, (D), Wisconsin, 1971, Pe
Associate Professor:
Steven C. Clemens, (D), Brown, 1990, Ou
L. Peter Gromet, (D), Caltech, 1979, Cc
Yongsong Huang, (D), Bristol (UK), 1997, Co
Yan Liang, (D), Chicago, 1994, Cp
John F. Mustard, (D), Brown, 1990, Or
Assistant Professor:
James M. Russell, (D), Minnesota, 2004, Gn
Alberto E. Saal, (D), MIT/Woods Hole, 2000, Cg
Emeritus:
Bruno J. Giletti, (D), Columbia, 1957, Cc
John Imbrie, (D), Yale, 1951, Pe
Robley K. Matthews, (D), Rice, 1965, Gr
Richard A. Yund, (D), Illinois, 1960, Gy
Department Coordinator:
Ruth Crane, (B), Emerson Coll, 1985, Oo

Community College of Rhode Island

Dept of Physics (Geology & Ocean Div) (A) (2007)
400 East Avenue
Warwick, RI 02886
p. (401) 825-2175 f. (401) 825-2418
pbanerjee@ccri.edu
Administrative Assistant: Charlene Pelski

Associate Professor:
karen Kortz, (M), Gg
Assistant Professor:
Paul White, (D)
Lab Tech:
Bruce Herrick, (B), Rhode Island, 1978, Oo

Roger Williams University

College of Arts & Sciences (B) (2005)
Bristol, RI 02809
p. (401) 254-3087 f. (401) 254-3286
jborden@rwu.edu
Department Secretary: Valerie Catalano
Enrollment (1997): B: 160 (43)

Head:
Mark D. Gould, (D), Rhode Island, 1973, Ob
Professor:
Thomas Doty, (D), Rhode Island, 1977, Ob
Richard Heavers, (D), Rhode Island, 1977, Op
Thomas J. Holstein, (D), Brown, 1969, Oo
Martine Villalard-Bohnsack, (D), Rhode Island, 1971, Ob
Assistant Professor:
Tim Scott, (D), SUNY (Stony Brook), 1993, Ob

University of Rhode Island

Dept of Geosciences (B,M,D) (2007)
317 Woodward Hall
Kingston, RI 02881
p. (401) 874-2265 f. (401) 874-2190
veeger@uri.edu
http://www.uri.edu/cels/geo/
Administrative Assistant: Cheryl Grasso
Enrollment (2007): B: 31 (5) M: 10 (6) D: 4 (1)

Professor:
Jon C. Boothroyd, (D), South Carolina, 1974, Gs
J. Allan Cain, (D), Northwestern, 1962, Gx
David E. Fastovsky, (D), Wisconsin, 1986, Pg
O Don Hermes, (D), North Carolina, 1967, Gi
Daniel P. Murray, (D), Brown, 1976, Gp
Associate Professor:
Thomas B. Boving, (D), Arizona, 1999, Hw
Anne I. Veeger, (D), Arizona, 1991, Hw
Assistant Professor:
Brian Savage, (D), Cal Tech, 2004, Ys

Research Associate:
Nasir Hamidzada, (M), Rhode Island, 1988, Gx
Lecturer:
Geoffrey W. Cook, (D), Washington State, 2008, Ggi
Adjunct Professor:
Cheryl Hapke, (D), On

Graduate School of Oceanography (M,D) (2005)
215 South Ferry Road
Narragansett, RI 02882
p. (401) 874-6222 f. (401) 874-6889
TheDean@gso.uri.edu
http://www.gso.uri.edu
Enrollment (1997): M: 4 (0) D: 12 (2)
Associate Dean:
Mark Wimbush, (D), California (San Diego), 1969, Op
Research Professor:
Theodore J. Smayda, (D), Oslo, 1967, Ob
Professor:
Robert D. Ballard, (D), Rhode Island, 1974, Ga
Steven N. Carey, (D), Rhode Island, 1982, Ou
Jeremy S. Collie, (D), MIT/Woods Hole, 1985, Ob
Peter Cornillon, (D), Cornell, 1973, Op
Steven L. D'Hondt, (D), Princeton, 1989, Ou
Edward G. Durbin, (D), Rhode Island, 1976, Ob
Isaac Ginis, (D), Inst Experimental Meteorology, 1986, Op
Tetsu Hara, (D), MIT, 1990, Op
Paul E. Hargraves, (D), William & Mary, 1968, Ob
David L. Hebert, (D), Dalhousie, 1988, Op
Christopher Kincaid, (D), Johns Hopkins, 1990, Ou
John King, (D), Minnesota, 1983, Ou
Roger Larson, (D), California (San Diego), 1970, Ou
Margaret Leinen, (D), Rhode Island, 1980, Ou
John T. Merrill, (D), Colorado, 1976, Oc
S. Bradley Moran, (D), Dalhousie, 1991, Oc
Scott W. Nixon, (D), North Carolina, 1970, Ob
Candace Oviatt, (D), Rhode Island, 1967, Ob
Hans Thomas Rossby, (D), MIT, 1966, Op
Lewis Rothstein, (D), Hawaii, 1983, Op
Haraldur Sigurdsson, (D), Durham (UK), 1970, Ou
Jennifer Specker, (D), Oregon State, 1980, Ob
Robert Tyce, (D), California (San Diego), 1976, Ou
D. Randolph Watts, (D), Cornell, 1973, Op
Karen Wishner, (D), California (San Diego), 1979, Ob
Associate Professor:
Brian G. Heikes, (D), Michigan, 1984, Oa
Yang Shen, (D), Brown, 1994, Ou
David C. Smith, (D), California (San Diego), 1994, Ob
Marine Research Scientist:
Percy Donaghay, (D), Oregon State, 1980, Ob
Kathleen Donohue, (D), Rhode Island, 1996, Op
Dian J. Gifford, (D), Dalhousie, 1986, Ob
Alfred K. Hanson, Jr., (D), Rhode Island, 1981, Oc
Robert D. Kenney, (D), Rhode Island, 1984, Ob
Barbara K. Sullivan-Watts, (D), Oregon State, 1977, Og
Associate Dean:
John Farrell, (D), Brown, 1991, Ou
Adjunct Professor:
Lawrence J. Buckley, (D), New Hampshire, 1975, Oc
Richard J. Pruell, (D), Rhode Island, 1984, Co
Charles T. Roman, (D), Delaware, 1981, Ob
Emeritus:
H. Perry Jeffries, (D), Rutgers, 1959, Ob
John A. Knauss, (D), California, 1959, Op
Theodore A. Napora, (D), Yale, 1964, Ob
Michael E. Pilson, (D), California (San Diego), 1964, Oc
James G. Quinn, (D), Connecticut, 1967, Oc
Kenneth A. Rahn, (D), Michigan, 1971, Oc
Saul B. Saila, (D), Cornell, 1952, Ob
Jean-Guy Schilling, (D), MIT, 1966, Ou
John M. Sieburth, (D), Minnesota, 1954, Ob
Elijah V. Swift, (D), Johns Hopkins, 1967, Ob

South Carolina

Clemson University

Bob Campbell Geology Museum (2004)
103 Garden Trail
Clemson, SC 29634-0130
p. 864-656-4602 f. 864-656-6230
http://www.clemson.edu/geomuseum
Director:
Carolyn R. Rebbert, (D), Oregon, 1995, Gx
Curator:
Christian M. Cicimurri, (M), SD Sch of Mines & Tech, 1999, Pv
David J. Cicimurri, (M), SD Sch of Mines & Tech, 1998, Pv

Dept of Geological Sciences (B,M) (2006)
340 Brackett Hall
Clemson, SC 29634-0919
p. (864) 656-3438 f. (864) 656-1041
gravelc@clemson.edu
www.ces.clemson.edu/geology
Administrative Assistant: Cynthia Rae Gravely
Enrollment (2006): B: 28 (7) M: 12 (6)
Director:
Alan W. Elzerman, (D), Wisconsin, 1976, Oo
Professor:
Ronald W. Falta, (D), California (Berkeley), 1990, Hq
John R. Wagner, (D), South Carolina, 1993, Oe
Richard D. Warner, (D), Stanford, 1971, Gz
Associate Professor:
James W. Castle, (D), Illinois, 1978, Gd
Lawrence C. Murdoch, (D), Cincinnati, 1991, Hw
Assistant Professor:
Treavor A. Kendall, (D), Virginia Tech, 2003, Cl
Stephen M.J. Moysey, (D), Stanford, 2005, Hw
Research Associate:
Scott E. Brame, (M), Clemson, 1993, Hw
Lecturer:
W. Gray Dean, (D), Tennessee, 2003, Pg
Lois B. Krause, (D), Clemson, 1996, Oe
Adjunct Professor:
Rolf K. Aadland, (D), Idaho, 1980, Hw
Charles W. Clendenin, (D), Witwatersrand, 1989, Eg
Randolph J. Cumbest, (D), Virginia Tech, 1988, Yg
Vaneaton Price, (D), North Carolina, 1969, Ce
David C. Prowell, (D), California (Santa Cruz), 1974, Gc
Visiting Professor:
Peggy Cain, (D), South Carolina, 1988, Oe
Emeritus:
Raymond A. Christopher, (D), Louisiana State, 1971, Pl
Brian E. Davies, (D), Wales, 1963, Sc
George M. Haselton, (D), Ohio State, 1967, Gl

Coker College

Geology Dept (2007)
300 E. College Avenue
Hartsville, SC 29550
p. (803) 383-8177 f. (803) 383-8048
fedinger@coker.edu
Professor:
Paul F. Edinger, (D), South Carolina, 1991, Oe
Adjunct Professor:
Gerald A. Porter, (M), South Carolina, 1978, Oe

College of Charleston

Dept of Geology & Env Geosciences (B,M) (2007)
66 George Street
Charleston, SC 29424
p. (843) 953-5589 f. (843) 953-5446
colganm@cofc.edu
http://www.cofc.edu/~geology/
Administrative Assistant: Joan E. Newell
Enrollment (2007): B: 105 (25) M: 13 (5)

Chair:
Mitchell W. Colgan, (D), California (Santa Cruz), 1990, PeGe
Professor:
James L. Carew, (D), Texas, 1978, Po
Michael P. Katuna, (D), North Carolina, 1974, Gu
Robert L. Nusbaum, (D), Missouri (Rolla), 1984, Gz
Associate Professor:
Erin K. Beutel, (D), Northwestern, 2000, Gct
Timothy J. Callahan, (D), New Mexico Tech, 2001, Hw
M. Scott Harris, (D), Delaware, 2000, GseGa
Steven C. Jaumé, (D), Columbia, 1994, Ys
Cassandra R. Runyon, (D), Hawaii, 1988, XgOse
Leslie R. Sautter, (D), South Carolina, 1990, Ob
Assistant Professor:
Briget C. Doyle, (D), Missouri (Rolla), 2003, Ng
A. Kem Fronabarger, (D), Tennessee, 1984, Gi
Norman S. Levine, (D), Purdue, 1995, OiGeOu
Vijay M. Vulava, (D), Swiss Fed Inst Tech, 1998, CgHwSc
Emeritus:
Alexander W. Ritchie, (D), Texas, 1975, Gc
Senior Instructor:
Elizabeth F K Rhodes, (M), Charleston, 1999, Gg

Furman University

Dept of Earth & Environmental Sciences (B) (2007)
3300 Poinsett Highway
Greenville, SC 29613
p. (864) 294-2052 f. (864) 294-3585
bill.ranson@furman.edu
http://ees.furman.edu
Administrative Assistant: Nina Anthony
Enrollment (2007): B: 22 (11)

Chair:
William A. Ranson, (D), Massachusetts, 1979, GxzGp
Professor:
John M. Garihan, (D), Penn State, 1973, Gc
Associate Professor:
C. Brannon Andersen, (D), Syracuse, 1994, Cl
Assistant Professor:
Weston R. Dripps, (D), Wisconsin, 2003, HwsGe
Suresh Muthukrishnan, (D), Purdue, 2002, Oi
Emeritus:
Kenneth A. Sargent, (D), Oklahoma, 1973, Hy

University of South Carolina

Dept of Geological Sciences (B,M,D) (2004)
701 Sumter Street
Columbia, SC 29208
p. (803) 777-4535 f. (803) 777-6610
thunell@geol.sc.edu
http://www.geol.sc.edu/
Business Manager: Linda Mabry
Enrollment (2000): B: 14 (3)

Chair:
James N. Kellogg, (D), Princeton, 1981, Yg
Research Professor:
Philip M. Astwood, (D), South Carolina, 1979, Oe
Gwendelyn Geidel, (D), South Carolina, 1982, Hw
Ian Lerche, (D), Manchester, 1965, Yg
Willard S. Moore, (D), SUNY (Stony Brook), 1969, Cc
Director, Graduate Studies:
Robert C. Thunell, (D), Rhode Island, 1978, Pe
Professor:
Arthur D. Cohen, (D), Penn State, 1968, Ec
Christopher G. Kendall, (D), Imperial Coll (UK), 1966, Gs
Thomas J. Owens, (D), Utah, 1984, Ys
Pradeep Talwani, Ys
Douglas F. Williams, (D), Rhode Island, 1976, Cs
Associate Professor:
Miguel Goni, (D), Washington, 1992, Co
James H. Knapp, (D), MIT, 1989, Ye
Venkataraman Lakshmi, (D), Princeton, 1995, Hg
Evangelos K. Paleologos, (D), Arizona, 1994, Hw
Raymond Torres, (D), California (Berkeley), 1997, Hy
Assistant Professor:
David Barbeau, Jr., (D), Arizona, 2003, Gs
Claudia Benitez-Nelson, (D), MIT/Woods Hole, 1999, Oc
Camelia Knapp, (D), Cornell, 2000, Gt
Matthew J. Kohn, (D), RPI, 1991, Cs
Richard Styles, (D), Rutgers, 1998, Op
George Voulgaris, (D), Southampton, 1992, On
Scott M. White, (D), California (Santa Barbara), 2001, Yr
Alicia M. Wilson, (D), Johns Hopkins, 1999, Hw
Gene M. Yogodzinkski, (D), Cornell, 1993, Gi
Adjunct Professor:
Ian F. Clark, (D), Adelaide, 1995, Oe
Brian W. Horn, (D), Colorado Mines, 1997, Gs
Timothy Kana, (D), South Carolina, 1979, On
Steven Mittwede, (D), South Carolina, 1989, Gg
Emeritus:
John R. Carpenter, (D), Florida State, 196, Oe
Leonard R. Gardner, (D), Penn State, 1968, Cl

Marine Science Program (B,M,D) (2005)
Columbia, SC 29208
p. (803) 777-2692 f. (803) 777-3935
thunell@geol.sc.edu
http://www.msci.sc.edu
Enrollment (1996): B: 238 (33) M: 14 (3) D: 14 (1)

Director:
Bjorn Kjerfve, (D), Louisiana State, 1973, Op
Chair:
Robert C. Thunell, (D), Rhode Island, 1978, Pm
Professor:
C. Marjorie Aelion, (D), North Carolina, 1988, Hw
Ronald Benner, (D), Georgia, 1984, Yr
Arthur D. Cohen, (D), Penn State, 1968, Pl
Bruce C. Coull, (D), Lehigh, 1968, Ob
Robert J. Feller, (D), Washington, 1977, Ob
Madilyn Fletcher
Thomas J. Hilbish, (D), SUNY (Stony Brook), 1984, Ob
Christopher G. Kendall, (D), Imperial Coll (UK), 1966, Gs
James T. Morris, (D), Yale, 1979, Ob
Stephen E. Stancyk, (D), Florida, 1974, Ob
Douglas F. Williams, (D), Rhode Island, 1976, Cs
Sarah A. Woodin, (D), Washington, 1972, Ob
Richard G. Zingmark, (D), California (Santa Barbara), 1969, Ob
Associate Professor:
Joan Bernhard
G. Thomas Chandler, (D), Louisiana State, 1986, Ob
Alan W. Decho, (D), Louisiana State, 1987, Ob
Charles R. Lovell, (D), Purdue, 1984, Ob
Evan Paleologos
Joseph M. Quattro, (D), Rutgers, 1991, Ob

Timothy J. Shaw, (D), California (San Diego), 1988, Oc
Richard M. Showman, (D), Washington, 1979, Ob
Garriet Smith
Richard Vogt
David S. Wethey, (D), Michigan, 1979, Ob

Assistant Professor:
Claudia Benitez-Nelson
Renae Brodie
Miguel A. Goni, (D), Washington, 1992, Oc
Brian Helmuth
Venkat Lakshmi
Dwayne E. Porter, (D), South Carolina, 1995, Or
Marcel Reichert, Yr
Richard Styles
Raymond Torres
George Voulgaris

Research Associate:
Dennis M. Allen, (D), Lehigh, 1978, Ob
Alan J. Lewitus, (D), MIT, 1990, Ob
Paul A. Sandifer, (D), Virginia, 1972, Ob

Instructor:
Wendy B. Allen, (M), South Carolina (Coastal), 1980, Ob

Adjunct Professor:
Loren Coen
Carla Curran
Margaret Davidson
Will Heyman
Fred Holland
Roberta Marinelli
Adele Militello
Amy Ringwood
Henk van der Veer, (D), Groningen (Neth), 1986, Ob

Winthrop University

Dept of Chemistry, Physics, & Geology (2006)
Sims Science Building
Winthrop University
Rock Hill, SC 29733
p. (803) 323-4949 f. (803) 323-2246
bolandi@winthrop.edu
http://chem.winthrop.edu

Professor:
Marsha S. Bollinger, (D), South Carolina, 1986, Cm

Associate Professor:
Irene B. Boland, (D), South Carolina, 1996, Gt

Assistant Professor:
Gwen M. Daley, (D), Virginia Tech, 1999, Pe

Wofford College

Geology Department (2006)
Wofford College
429 North Church Street
Spartanburg, SC 29303-3663
p. 864-597-4527
fergusonta@wofford.edu
http://dept.wofford.edu/geology

Director:
Terry A. Ferguson, (D), Tennessee, 1988, Ga

South Dakota

Black Hills State University

Dept of Physical Science (B) (2004)
College of Applied Science & Technology
Spearfish, SD 57799
p. (605) 642-6133 f. (605) 642-6214
Enrollment: No data reported since 1995

Professor:
Everett Follette, (D), Colorado, 1970, Oe
Mark Gabel, (D), Iowa State, 1982, Pb

Assistant Professor:
Steven W. Anderson, (D), Arizona State, 1990, Gv

South Dakota School of Mines & Tech

Dept of Atmospheric Sciences (B,M,D) (2006)
501 E. St. Joseph Street
Rapid City, SD 57701-3995
p. (605) 394-2291 f. (605) 394-6061
Pamela.Cox@sdsmt.edu
http://www.ias.sdsmt.edu/
Enrollment (2004): B: 7 (0) M: 14 (6) D: 6 (3)

Chair:
Mark R. Hjelmfelt, (D), Chicago, 1980, Oa

Professor:
Andrew G. Detwiler, (D), SUNY (Albany), 1980, Oa
John H. Helsdon, (D), SUNY (Albany), 1979, Oa
Patrick R. Zimmerman, (D), Colorado State, 1996, Oa

Associate Professor:
William J. Capehart, (D), Penn State, 1997, Oa

Assistant Professor:
P. V. Sundareshwar, (D), South Carolina, 2002, Oo

Emeritus:
Harold D. Orville, (D), Arizona, 1965, Oa
Paul L. Smith, (D), Carnegie Inst, Oo

Dept of Geology & Geological Eng (B,M,D) (2007)
501 East St. Joseph Street
Rapid City, SD 57701-3995
p. (605) 394-2461 f. (605) 394-6703
geologyinfo@sdsmt.edu
http://geology.sdsmt.edu
Secretary: Pamela G. Fenner
Enrollment (2007): B: 92 (0) M: 35 (0) D: 6 (0)

Chair:
Maribeth H. Price, (D), Princeton, 1995, Oi

Professor:
Arden D. Davis, (D), SD Sch of Mines & Tech, 1983, Hw
Edward F. Duke, (D), Dartmouth, 1984, Gi
James E. Fox, (D), Wyoming, 1972, Gs
James E. Martin, (D), Washington, 1979, Ps
Colin J. Paterson, (D), Otago (NZ), 1978, Eg
William M. Roggenthen, (D), Princeton, 1980, Yg

Field Station Director:
Nuri Uzunlar, (D), SD Sch of Mines & Tech, Gg

Associate Professor:
Larry D. Stetler, (D), Washington State, 1993, Ng

Postdoc:
Darrin Pagnac, (D), California (Riverside), Pv

Museum/Collections Manager:
Search pending, Pv

Assistant Professor:
Gerald Grellet-Tinner, (D), USC, 2005, Pv
Michael P. Terry, (D), Massachusetts, 2000, Gc

Emeritus:
Alvis L. Lisenbee, (D), Penn State, 1972, Gc
Perry H. Rahn, (D), Penn State, 1965, Ng
Jack A. Redden, (D), Harvard, 1956, Gp

South Dakota State University

Plant Science Dept (A,B,M,D) (2007)
Soil Science Program
SAG 219 Box 2207-A
Brookings, SD 57007-1096
p. (605) 688-5123 f. (605) 688-4602
howard.woodard@sdstate.edu

http://plantsci.sdstate.edu/
Enrollment (2007): M: 30 (6) D: 14 (2)

Dean:
Gary D. Lemme, (D), Nebraska, 1979, Sd
Distinguished Professor:
Douglas D. Malo, (D), North Dakota State, 1975, Sd
Professor:
Dwayne L. Beck, (D), South Dakota State, 1983, Ou
Bruce H. Bleakley, (D), Florida, 1986, Sb
Charles G. Carlson, (D), South Dakota State, 1978, Hw
David E. Clay, (D), Minnesota, 1988, Sb
James J. Doolittle, (D), Texas A&M, 1991, Sc
Ronald H. Gelderman, (D), North Dakota State, 1987, Sc
Thomas E. Schumacher, (D), Michigan State, 1982, Sp
Howard J. Woodard, (D), Rutgers, 1985, ScoGg
Associate Professor:
Robert K. Berg, Jr., (D), Iowa State, 1987, Sb
Research Associate:
Anthony G. Bly, (M), South Dakota State, 1992, Sc
Joseph Schumacher, (M), South Dakota State, 1990, Sp
Adjunct Professor:
Paul E. Fixen, (D), Colorado State, 1979, Sc
Dianne H. Rickerl, (D), Auburn, 1986, Sf
Emeritus:
James R. Gerwing, (M), Minnesota, 1978, Sc
Robert Kohl, Utah State, 1967

University of South Dakota

Dept of Earth Sciences & Physics (B) (2007)
414 East Clark Street
Vermillion, SD 57069-2390
p. (605) 677-5649 f. (605) 677-6121
esci@usd.edu
http://www.usd.edu/esci
Administrative Assistant: Jan Strom
Enrollment (2007): B: 23 (4)

Chair:
Timothy H. Heaton, (D), Harvard, 1988, PvOg
Associate Professor:
Kelly A. Dilliard, (D), Washington State, 2006, Gsg
Assistant Professor:
Brennan T. Jordan, (D), Oregon State, 2002, GiOw
Mark R. Sweeney, (D), Washington State, 2004, Gs
Instructor:
Kay L. Kassube, (M), South Dakota, 1997, Gg

Tennessee

Austin Peay State University

Geosciences Dept (B) (2007)
601 College St
Clarksville, TN 37044
p. (931) 221-7454 f. (931) 221-7284
shipleym@apsu.edu
www.apsu.edu/geosciences
Enrollment (2002): B: 75 (16)

Interim Director:
Robert A. Sirk, (D), Kent State, 1991, OywOo
Professor:
Phyllis A. Camilleri, (D), Wyoming, 1994, Gct
Daniel L. Frederick, (D), Tennessee, 1994, Gsr
Phillip R. Kemmerly, (D), Okalahoma State, 1973, GmeHg
Gregory D. Ridenour, (D), OiaOo
Associate Professor:
Jack Deibert, (D), Gsr
Emeritus:
D. M. S. Bhatia, (D), Missouri (Rolla), 1976, Ca
James X. Corgan, (D), Louisiana State, 1967, Pg

Laboratory Director:
Richard Wheeler, (M), Oo

Rhodes College

Geology Dept (2004)
2000 N Parkway
Memphis, TN 38112
p. (901) 843-3089 f. (901) 843-3565
cekstrom@rhodes.edu

Head:
Carol L. Ekstrom, (M), George Washington, 1972, Gg

Tennessee Technological University

Dept of Earth Sciences (B) (2004)
PO Box 5125
Cookeville, TN 38505
p. (931) 372-3121 f. (931) 372-3363
LKnox@tntech.edu
http://www.tntech.edu/earth/Default.htm
Administrative Assistant: Martha Gentry Smith
Enrollment (2004): B: 1 (0)

Chair:
Larry W. Knox, (D), Indiana, 1974, Pm
Professor:
H. Wayne Leimer, (D), Missouri, 1969, Gz
Hugh H. Mills, III, (D), Washington, 1975, Gm
Frank W. Stapor, Jr., (D), Florida State, 1973, Gs
Associate Professor:
Ping-Chi Li, (D), Iowa, 1992, Oi
Assistant Professor:
Michael J. Harrison, (D), Illinois (Urbana), 2002, Gc
Evan A. Hart, (D), Tennessee, 2000, Oy
Adjunct Professor:
Jason E. Duke, (M), Tennessee Tech, 1995, Oi
Geology Librarian:
Regina Lee, Gg

University of Memphis

Center for Earthquake Research & Information (CERI) (M,D) (2006)
3876 Central Avenue, Suite 1
Memphis, TN 38152-3050
p. (901) 678-2007 f. (901) 678-4734
gradadmin@ceri.memphis.edu
http://www.ceri.memphis.edu/
Enrollment (2002): M: 5 (1) D: 4 (0)

Director:
Archibald C. Johnston, (D), Colorado, 1979, Ys
Associate Director:
Christine A. Powell, (D), Princeton, 1976, Ys
Professor:
Jer-Ming Chiu, (D), Cornell, 1982, Ys
Howard Hwang, (D), Columbia, 1976, Ne
Charles A. Langston, (D), Caltech, 1976, Ys
Shahram Pezeshk, (D), Illinois, 1989, Ne
Associate Professor:
Paul A. Bodin, (D), Colorado, 1992, Ys
Michael A. Ellis, (D), Washington State, 1984, Gt
Jose M. Pujol, (D), Wyoming, 1985, Ye
Paul A. Rydelek, (D), California (Los Angeles), 1983, Yg
Robert Smalley, Jr., (D), Cornell, 1988, Yd
Mitchell M. Withers, (D), New Mexico Tech, 1997, Ys
Associate Scientist:
Stephen P. Horton, (D), Nevada (Reno), 1992, Ys
Adjunct Professor:
Chris Cramer, (D), Stanford, 1976, Ys
Eugene S. Schweig, (D), Stanford, 1985, Gt

Emeritus:
H. James Dorman, (D), Columbia, 1961, Ys
Research Associate:
Christy Chiu, (M), Cornell, 1982, Ys
Ann G. Metzger, (M), Memphis, 1983, Ys
Information Services Director:
Gary Patterson, (M), Memphis, 1998, Gg
Cooperating Faculty:
Mervin J. Bartholomew, (D), Virginia Tech, 1971, Gt
Roy B. Van Arsdale, (D), Utah, 1979, Gc

Dept of Earth Sciences (B,M,D) (2006)
001 Johnson Hall
Memphis, TN 38152-3550
p. (901) 678-2177 or 678-2386 f. (901) 678-2178
jbrthlm1@memphis.edu
http://cas.memphis.edu/des/
Enrollment (2002): M: 9 (5)
Chair:
Mervin J. Bartholomew, (D), Virginia Tech, 1971, Gt
Professor:
Jer-Ming Chiu, (D), Cornell, 1982, Ys
Michael A. Ellis, (D), Washington State, 1984, Gt
Archibald C. Johnston, (D), Colorado, 1979, Ys
Hsiang-Te Kung, (D), Tennessee, 1980, Oy
Charles A. Langston, (D), Caltech, 1976, Ys
David N. Lumsden, (D), Illinois (Urbana), 1965, Gd
Christine A. Powell, (D), Princeton, 1976, Ys
George H. Swihart, (D), Chicago, 1987, Gx
Roy B. Van Arsdale, (D), Utah, 1979, Gc
Associate Professor:
Randel T. Cox, (D), Missouri, 1995, Gm
David H. Dye, (D), Washington, 1980, Ga
Daniel Larsen, (D), New Mexico, 1994, Cl
Jose Pujol, (D), Wyoming, 1985, Ye
Thad Wasklewicz, (D), Arizona State, 1996, Oy
Assistant Professor:
Arleen A. Hill, (D), South Carolina, 2002, Oo
Andrew M. Mickelson, (D), Ohio State, 2002, Ga
Esra Ozdenerol, (D), Louisiana State, 2000, Oi
Lensyl Urbano, (D), Minnesota, 2001, Hw
Research Associate:
Chris H. Cramer, (D), Stanford, 1976, Ys
Xiang Li, (D), Chinese (Hong Kong), 2004, Oi
M. Beatrice Magnani, Perugia (Italy), 1999, Ys
Robert, Jr., Smalley, (D), Cornell, 1988, Ys
Mitchell M. Withers, (D), New Mexico Tech, 1997, Ys
Instructor:
M. Adam M. Dosi, (D), Kansas, 2003, Oo
Pamela M. Riddick, (M), Memphis, 1990, Oo
Adjunct Professor:
Paul A. Bodin, (D), Colorado, 1992, Ys
Joan S. Gomberg, (D), California (San Diego), 1986, Ys
Jennifer Houghton, (D), Minnesota, 2003, Co
Katherine R. Mickelson, (D), Ohio State, 2002, Ga
Steven Newkirk, (M), Memphis, 2000, Oe
Eugene S. Schweig, (D), Stanford, 1985, Gt
Emeritus:
Phili B. Deboo, (D), Louisiana State, 1963, Pg
Robert W. Deininger, (D), Rice, 1964, Gx
James Dorman, (D), Columbia, 1961, Ys
W. Theodore Mealor, Jr, (D), Georgia, 1973, Ou
John A. Sobol, (D), Michigan, 1960, Oo

University of Tennessee, Chattanooga

Dept of Physics, Geology, and Astronomy (B) (2007)
100 Bretske Hall
615 McCallie Avenue
Chattanooga, TN 37404
p. (423) 755-4404 f. (423) 755-4683
habte-churnet@utc.edu
Administrative Assistant: Kelly W. Locke
Enrollment (2007): B: 22 (4)
Head:
Habte G. Churnet, (D), Tennessee, 1979, GxgOg
Professor:
Tatiana B. Allen, (D), St. Petersburg, 1988, Oo
Robert Marlowe, (D), Cincinnati, 1983, Oo
Jonathan W. Mies, (D), North Carolina, 1990, GctHg
Ling-Jun Wang, (D), Delaware, 1984, Oo
Assistant Professor:
Ann E. Holmes, (D), Columbia, 1997, Gs
Tracy Jones, (D), Alabama, 1998, Oo
Jin Wang, (D), Nebraska, 2001, Oo
Adjunct Professor:
Gregory Brodie, (M), Purdue, 1979, Ge
Kristin B. Whitson, (D), Vanderbilt, 2003, OooOo
Emeritus:
Richard E. Bergenback, (D), Penn State, 1965, Gd
Eric T. Lane, (D), Rice, 1967, Oo

University of Tennessee, Knoxville

Dept of Earth & Planetary Sciences (B,M,D) (2004)
306 Geological Sciences Building
Knoxville, TN 37996-1410
p. (865) 974-2366 f. (865) 974-2368
cmora@utk.edu
http://geoweb.gg.utk.edu
Enrollment (1998): B: 46 (21) M: 33 (11) D: 18 (5)
Head:
Claudia I. Mora, (D), Wisconsin, 1988, Cs
Distinguished Scientist:
Robert D. Hatcher, Jr., (D), Tennessee, 1965, Gc
Professor:
Thomas W. Broadhead, (D), Iowa, 1978, Pi
Steven G. Driese, (D), Wisconsin, 1982, Gs
William M. Dunne, (D), Bristol (UK), 1980, Gc
Theodore C. Labotka, (D), Caltech, 1978, Gp
Michael L. McKinney, (D), Yale, 1985, Po
Harry Y. McSween, Jr., (D), Harvard, 1977, Xm
Kula C. Misra, (D), W Ontario, 1973, Eg
Lawrence A. Taylor, (D), Lehigh, 1968, Cp
Associate Professor:
G. Michael Clark, (D), Penn State, 1966, Gm
Larry D. McKay, (D), Waterloo, 1991, Hy
Richard T. Williams, II, (D), Virginia Tech, 1979, Yg
Assistant Professor:
Linda Kah, (D), Harvard, 1997, Gs
Edmund Perfect, (D), Cornell, 1986, Sp
Maria Uhle, (D), Virginia, 1997, Cl
Instructor:
Gray Dean, (M), Tennessee, 1998, Py
Lecturer:
Colin Sumrall, (D), Texas, 1997, Pi
Adjunct Professor:
David R. Cole, (D), Penn State, 1980, Ca
William E. Doll, (D), Wisconsin, 1983, Yg
Philip M. Jardine, (D), Virginia Tech, 1985, Sc
Norbert Thonnard, (D), Kentucky, 1971, Cc
Emeritus:
Don W. Byerly, (D), Tennessee, 1966, Ge
Otto C. Kopp, (D), Columbia, 1958, Gz
Kenneth R. Walker, (D), Yale, 1969, Pe
Research Assistant Professor:
Lawrence A. Anovitz, (D), Michigan, 1987, Cp
Mostafa Fayek, (D), Saskatchewan, 1996, Cg

Jeffery E. Moersch, (D), Cornell, 1997, Xg
Lee R. Riciputi, (D), Wisconsin, 1991, Cc
Research Associate Professor:
Daniel Britt, (D), Brown, 1991, Xg

University of Tennessee, Martin

Dept of Geology, Geography & Physics (B) (2004)
215 Johnson EPS Building
Martin, TN 38238
p. (901) 587-7430 f. (901) 587-7434
jrogers@utm.edu
http://www.utm.edu/departments/ggp/home.htm
Principal Secretary: Janice F. Lee
Enrollment (2000): B: 14 (3)

Professor:
Michael A. Gibson, (D), Tennessee, 1988, Pe
Robert P. Self, (D), Rice, 1971, Gs
Helmut C. Wenz, (M), W Michigan, 1968, Oo
Associate Professor:
Aley K. El-Shazly, (D), Stanford, 1991, Gp
Assistant Professor:
Jefferson S. Rogers, (D), Illinois, 1995, Oo
Robert M. Simpson, (D), Indiana State, 2000, Ow
Instructor:
Gloria C. Mansfield, (M), Tennessee, 1981, Gg

University of the South

Dept of Forestry & Geology (B) (2006)
735 University Avenue
Sewanee, TN 37383-1000
p. (931) 598-1271 f. (931) 598-1145
mknoll@sewanee.edu
http://www.sewanee.edu/EnvStudies
Enrollment (1997): B: 66 (0)

Chair:
Scott Torreano, (D), Georgia, 1991, Sf
Professor:
Martin A. Knoll, (D), Texas (El Paso), 1988, Hw
Donald B. Potter, Jr., (D), Massachusetts, 1985, Gc
Stephen A. Shaver, (D), Stanford, 1984, Eg
Associate Professor:
C. Ken Smith, (D), Florida, 1996, Sf
Adjunct Professor:
Glendon W. Smalley, (D), Tennessee, 1975, Sf

Vanderbilt University

Earth and Environmental Sciences (B,M,D) (2006)
Science & Engineering Building
Room 5717, Stevenson Center
Box 35-1805, Station B
Nashville, TN 37235
p. (615) 322-2976 f. (615) 322-2138
jewell.williams@vanderbilt.edu
http://sitemason.vanderbilt.edu/ees/
Administrative Assistant: Jewell B. Williams
Enrollment (2006): M: 8 (8) D: 6 (1)

Professor:
John C. Ayers, (D), RPI, 1991, Cp
David J. Furbish, (D), Colorado, 1985, Hg
Calvin F. Miller, (D), California (Los Angeles), 1977, Gi
Molly F. Miller, (D), California (Los Angeles), 1977, Pe
Associate Professor:
Steven L. Goodbred, Jr., (D), William & Mary, 1999, Gs
Assistant Professor:
Kaye S. Savage, (D), Stanford, 2001, Ge
Lecturer:
Brendan R. Bream, (D), Tennessee, 2003, Gt
Jonathan M. Gilligan, (D), Yale, 1991, Ge
N. Kinzly Moore, (D), Tennessee, 1978, Ge
Emeritus:
Leonard P. Alberstadt, (D), Oklahoma, 1967, Gd
Authur L. Reesman, (D), Missouri, 1966, Gg
William G. Siesser, (D), Cape Town, 1971, Pm
Richard G. Stearns, (D), Northwestern, 1953, Yg
Media Technology Tech:
Alan Wiseman, Oo
Related Staff:
Teri Sparkman, (A), Oo

Texas

Amarillo College

Dept of Physical Science (A) (2005)
P.O. Box 447
Amarillo, TX 79178
p. (806) 371-5326 f. (806) 371-5976
hobbs-rd@actx.edu
Enrollment (2005): A: 5 (0)

Associate Professor:
Richard D. Hobbs, (D), Wyoming, 1998, Or
Instructor:
Darryl W. Maddox, (M), West Texas A&M, 1977, Go
David Pertl, (M), West Texas A&M, 1984, Gg

Angelo State University

Dept of Physics (Geology) (B) (2007)
ASU Station #10904
San Angelo, TX 76909
p. (325) 942-2242 f. (325) 942-2188
joseph.satterfield@angelo.edu
Enrollment: No data reported since 1995

Assistant Professor:
Christian Poppeliers, (D), Indiana, 2001, Ys
Joseph I. Satterfield, (D), Rice, 1995, Gc

Austin Community College

Dept of Geology (A) (2004)
1020 Grove Boulevard
Austin, TX 78741
p. (512) 223-6002 f. (512) 223-6769
rjohns@austincc.edu
http://www.austincc.edu/physci_tf/
Enrollment (2000): A: 6 (0)

Chair:
Ronald A. Johns, (D), Texas, 1993, Pi
Professor:
Robert Blodgett, (D), Texas
Janice L. Smith, (M), Texas, 1991, Gp
Chair, Env Sci & Tech Dept:
George M. Staff, (D), Texas A&M, 1983, Pe
Adjunct Professor:
Heather L. Beatty, (M), Texas Tech, 1992, Ge
Leslie S. Davis, (M), Florida State, 1986, Op
David W. Hill, (M), Texas, 1993, Hw
Frank M. Mikan, (M), Ohio State, 1973, Cg
Salvadore Valastro, (M), Texas, 1975, Cc
Peter J. Wehner, (M), Vanderbilt, 1992, Gv
Cooperating Faculty:
David J. Froehlich, (D), Texas, 1996, Pv

Baylor University

Dept of Geology (B,M,D) (2007)
One Bear Place #97354
101 Bagby Ave.

BSB, 4th Floor, Rm. D409
Waco, TX 76798-7354
p. (254) 710-2361 f. (254) 710-2673
paulette_penney@baylor.edu
http://www.baylor.edu/~Geology/
Administrative Assistant: Janelle Atchley
Administrative Assistant: Jamie J. Ruth
Enrollment (2007): B: 38 (14) M: 16 (8) D: 11 (1)

Chair:
Steven G. Driese, (D), Wisconsin, 1982, SaGe
Professor:
Peter M. Allen, (D), S Methodist, 1977, HgNg
Rena M. Bonem, (D), Oklahoma, 1975, Pei
Vincent S. Cronin, (D), Texas A&M, 1988, Gc
Stephen I. Dworkin, (D), Texas, 1991, ClGd
Don M. Greene, (D), Oklahoma, 1980, Oyw
Lee C. Nordt, (D), Texas A&M, 1996, SdGa
Don F. Parker, (D), Texas, 1976, Giv
Joe C. Yelderman, Jr., (D), Wisconsin, 1983, Hw
Associate Professor:
Stacy C. Atchley, (D), Nebraska, 1990, Gro
John A. Dunbar, (D), Texas, 1989, YgHg
Zhaodong Jordan Feng, (D), Kansas, GmOy
Joseph D. White, (D), Montana, Or
Visiting Professor:
Jeffrey G. Arnold, (D), Purdue, 1992, Hs
Arthur O. Beall, (D), Stanford, 1964, Go
Emeritus:
Harold H. Beaver, (D), Wisconsin, 1954, Go
William G. Brown, (D), Alaska, 1987, Gc
Thomas T. Goforth, (D), S Methodist, 1973, Yg
Oliver T. Hayward, (D), Wisconsin, 1957, Gm
Office Manager:
Paulette Penney, Oo

Brookhaven College

Science/Math Div - Geology Dept (2007)
3939 Valley View Lane
Farmers Branch, TX 75244
p. (214) 620-4758
Smt2460@dcccd.edu

Head:
Susan M. Maxey, (M), Texas (Dallas), 1976, Guo

Del Mar College

Dept of Natural Sciences (A) (2006)
Corpus Christi, TX 78404
p. (512) 886-1240 f. (512) 886-1595
jhalcomb@delmar.edu
Enrollment (1995): A: 0 (2) B: 43 (0)

Instructor:
Walter V. Kramer, (M), Texas (El Paso), Gg
Roger T. Steinberg, (M), Tennessee, 1981, Gr
Emeritus:
Mary S. Thorpe, (M), Baylor, 1966, Ge

Hardin-Simmons University

Dept of Geology & Environmental Science (B,M) (2007)
Box 16164
2200 Hickory Street
Abilene, TX 79698
p. (325) 670-1383 f. (325) 677-8351
ouimette@hsutx.edu
http://www.hsutx.edu/academics/holland/geology/
Enrollment (2007): B: 9 (2) M: 17 (5)

Head:
Mark A. Ouimette, (D), Texas (El Paso), 1994, GeiGt

Houston Community College System

Math/Science Div (A) (2005)
22 Waugh Drive
Houston, TX 77077
p. (713) 868-0758 f. (713) 868-0766
ossie.nye@hccs.edu
Enrollment: No data reported since 1995

Head:
Jane F-J Binford, (M), Houston, 1969, Go
Instructor:
Steven D. Schafersman, (D), Rice, 1983, Gg
Adjunct Professor:
Albert F. Allong, (D), Ohio State, 1960, Oo
Robert F. Brandt, (M), Houston, 1968, Oo
Dwight S. Krantz, (M), Texas A&M, 1974, Oo
Daniel A. Sullivan, (D), DePauw, 1965, Oo

Kilgore College

Dept of Engineering Science (A) (2004)
Kilgore, TX 75662
p. (903) 984-8531 (Ext. 253)
bblack@kilgore.edu
Enrollment (2001): A: 4 (1)

Instructor:
R. Bernard Black, (D), Tulsa, 1986, Gd

Lamar University

Dept of Earth and Space Sciences (B) (2006)
P.O. Box 10031
Beaumont, TX 77710
p. (409) 880-8236 f. (409) 880-8246
geology@hal.lamar.edu
Administrative Assistant: Karen M. Woods
Enrollment (1999): B: 35 (2)

Chair:
Roger W. Cooper, (D), Minnesota, 1978, Gi
Professor:
Jim L. Jordan, (D), Rice, 1975, Cc
Donald E. Owen, (D), Kansas, 1963, Gr
James B. Stevens, (D), Texas, 1969, Gr
Associate Professor:
James W. Westgate, (D), Texas, 1988, Pe
Instructor:
Jeff G. Pittman, (D), Texas, 1992, Pv
Adjunct Professor:
Margaret S. Stevens, (M), Michigan, 1969, Pv

Laredo Community College

Science Dept (B) (2004)
West End Washington Street
Laredo, TX 78040
p. (956) 721-5195 f. (956) 721-5892
Administrative Assistant: Dora San Miguel
Enrollment (1997): B: 20 (0)

Associate Professor:
Janet Earle, (D), Colorado, 1997, Gp
Terry Tilton, (D), Utah, 1991, Gp

Lee College

Dept of Biological & Physical Sciences (A) (2006)
511 South Whiting
Baytown, TX 77522
p. 281-425-6332 f. 281-425-6425
cfoster@lee.edu
Enrollment (2006): A: 10 (0)

Chair:
Carolyn Foster, Oo

Midwestern State University

Dept of Geosciences (B) (2007)
3410 Taft
Wichita Falls, TX 76308
p. (940) 397-4250 f. (940) 397-4893
pamela.stephens@mwsu.edu
www.mwsu.edu/htmldocs/departments/geology/
Enrollment (2007): B: 50 (6)

Chair:
Pamela A. Stephens, (D), Texas (Dallas), 2001, Pm
Associate Professor:
Brent A. Elliott, (D), Helsinki, 2001, Gz
Adjunct Professor:
Brian S. Brister, (D), New Mexico Tech, 1990, Got
Emeritus:
John Kocurko, (D), Texas Tech, 1972, Gs

North Harris College

Geology Dept (A) (2007)
2700 W.W. Thorne Drive
Houston, TX 77073-3499
p. 281-618-5685 f. 281-618-7130
tom.hobbs@nhmccd.edu
Enrollment (2007): A: 20 (0)

Head:
Thomas M. C. Hobbs, (M), Texas (El Paso), 1979, Gg
Professor:
Peter E. Price, (M), Kentucky, 1979, Oi
Adjunct Professor:
Eliane G. Ferreira, (D)
Paula Kirst, (M)
Penni Major, (M)
Michelle Mc Mahon, (D), Aberdeen, 1993, Go
Michael Mulcahey, (M)
Victor S. Resnic, (D), Nat Pet Inst (Russia), 1971, Go
Linda C. Tran, (B), Texas A&M, 1999, Oi

Odessa College

Dept of Geology, Anthropology & Geography (A) (2006)
Odessa, TX 79762
p. (915) 335-6558 f. (915) 335-6860
bmcafee@odessa.edu
Enrollment (2000): A: 6 (0)

Associate Professor:
Gerald B. McAfee, (M), Sul Ross State, 1966, Pi

Rice University

Center for Computational Geophysics (M,D) (2004)
MS-126
PO Box 1892
Houston, TX 77251-1892
p. 713-348-3574 f. 713-348-5214
dmberry@rice.edu
http://terra.rice.edu/department/ccg/

Associate Director:
Alan Levander, (D), Stanford, 1984, Ys
William S. Symes, (D), Harvard, 1975, Oo
Professor:
Richard G. Baraniuk, (D), Illinois, 1992, Oo
C. Sidney Burrus, (D), Stanford, 1965, Oo
Richard G. Gordon, (D), Stanford, 1979, Gt
Dale S. Sawyer, (D), MIT, 1982, Ys
Associate Professor:
Liliana Borcea, (D), Stanford, 1996, Oo
Colin A. Zelt, (D), British Columbia, 1989, Ys
Assistant Professor:
Adrian Lenardic, (D), California (Los Angeles), 1995, Gc

Dept of Earth Science (B,M,D) (2007)
MS 126
PO Box 1892
Houston, TX 77251-1892
p. (713) 348-4880 f. (713) 348-5214
geol@rice.edu
http://earthscience.rice.edu/
Department Administrator: Linda Rassinier
Department Coordinator: Julie Kudelka
Enrollment (2006): B: 22 (0) M: 19 (0) D: 35 (0)

Chair:
Alan R. Levander, (D), Stanford, 1984, Ye
Professor:
John B. Anderson, (D), Florida State, 1972, Gu
Hans G. Ave Lallemant, (D), Leiden (Neth), 1967, Gc
Richard G. Gordon, (D), Stanford, 1979, Ym
William P. Leeman, (D), Oregon, 1974, Gi
Manik Talwani, (D), Columbia, 1959, Yr
Associate Professor:
Gerald R. Dickens, (D), Michigan, 1996, Yr
André W. Droxler, (D), Miami, 1984, Gs
Andreas Luttge, (D), Tubingen, 1990, Cg
Dale S. Sawyer, (D), MIT, 1982, Yr
Colin A. Zelt, (D), British Columbia, 1989, Yg
Assistant Professor:
Adrian Lenardic, (D), California, 1995, Yg
Julia Morgan, (D), Cornell, 1993, Gt
Adjunct Assistant Professor:
Vitor Abreu, (D), Rice, 1998, Gg
Scott A. Morton, (D), Illinois, 1991, Gg
Yitian Xiao, (D), Yale, 1995, Gg
Adjunct Professor:
K. K. Bissada, (D), Washington (St. Louis), 1967, Cg
Carlos A. Cramez, (D), Neuchatel, 1966, Go
Stephen H. Danbom, (D), Connecticut, 1975, Gg
Jeffrey J. Dravis, (D), Rice, 1980, Gs
Rob Dunbar, (D), California, 1981, Gu
Paul M. Harris, (D), Miami, 1977, Yr
Garry D. Jones, (D), Delaware, 1981, Yr
Walter C. Riese, (D), New Mexico, 1980, Eg
Paul D. Spudis, (D), Arizona State, 1982, Xg
Turhan Taner, (M), Tech (Istanbul), 1950, Ye
Gabor Tari, (D), Rice, 1994, Oo
John Van Wagoner, (D), Rice, 1977, Oo
James Lee Wilson, (D), Yale, 1949, Gd
Emeritus:
Donald R. Baker, (D), Princeton, 1955, Co
Albert W. Bally, (D), Zurich, 1953, Gg
H. C. Clark, (D), Stanford, 1966, Ng
Jean-Claude DeBremaecker, (D), California (Berkeley), 1952, Yg
Gerald H. F. Gardner, (D), Princeton, 1953, Yg
Dieter Heymann, (D), Amsterdam, 1958, Xm
John C. Stormer, Jr., (D), California (Berkeley), 1971, Gi
Peter R. Vail, (D), Northwestern, 1959, Gr

Dept of Environmental Science & Eng (M,D) (2004)
P.O. Box 1892
MS 317
Houston, TX 77251-1892
p. (713) 348-4951 f. (713) 348-5203
envi@rice.edu
http://www.ruf.rice.edu/~ceedept/
Enrollment (1997): M: 8 (7) D: 16 (3)

Chair:
Joseph B. Hughes, (D), Iowa, 1992, Oo
Professor:
Philip B. Bedient, (D), Florida, 1975, Hw

Arthur A. Few, Jr., (D), Rice, 1969, Oa
Mason B. Tomson, (D), Oklahoma State, 1972, Cg
Calvin H. Ward, (D), Cornell, 1960, Og
Mark R. Wiesner, (D), Johns Hopkins, 1985, Og
Assistant Professor:
Matthew P. Fraser, (D), Caltech, 1997, Oa
Lecturer:
James B. Blackburn, (D), Texas, 1972, Ou
Adjunct Associate Professor:
Stanley M. Pier, (D), Purdue, 1952, Co
Adjunct Assistant Professor:
Charles J. Newell, (D), Rice, 1989, Hw
Adjunct Professor:
Jean-Yves Bottero, (D), Nancy, 1979, Oo
Geology Librarian:
John Hunter, Oo

Saint Mary's University

Dept of Physics and Earth Sciences (2007)
One Camino Santa Maria
San Antonio, TX 78228-8569
p. (210) 436-3235 f. (210) 431-8005
dfitzgerald@stmarytx.edu
http://www.stmarytx.edu/acad/physicsandearth-science
Enrollment (1998): B: 21 (2)
Chair:
David Fitzgerald, (M), Iowa, 1977, Gd
Associate Professor:
Gene W. Lene, (D), Texas, 1981, Ge

Sam Houston State University

Geology Program (B) (2004)
Huntsville, TX 77341
p. (409) 294-1566 f. (409) 294-3940
bio_bjc@shsu.edu
Enrollment (2000): B: 15 (3)
Coordinator:
Brian J. Cooper, (D), Virginia Tech, 1988, Gz
Professor:
Christopher T. Baldwin, (D), Liverpool, 1976, Gs
Dwight D. Brown, (D), Missouri, 1969, Pl
Associate Professor:
Dennis I. Netoff, (D), Colorado, 1977, Gm
Assistant Professor:
Betsy D. Torrez, (D), Alabama, 1994, Gc

San Antonio College

Dept of Chemistry/Earth Sciences/Astronomy (A) (2007)
1300 San Pedro Avenue
Chemistry-Geology Rm 207
San Antonio, TX 78212
p. (210) 733-2704
gstanley@mail.accd.edu
www.accd.edu/sac/earthsci/default.htm
Enrollment (1999): A: 2 (0)
Professor:
Abdul Qudus, (M), Karachi, 1983, GgOgg
Arthur R. Troell, (D), Rice, 1964, GgOg
Associate Professor:
Dean Lambert, (D), Texas, 1992, Oyi
Assistant Professor:
Anne D. Dietz, (M), Texas A&M Kingsville, 1989, PgOg
David A. Wood, (D), Arizona, 2000, Oo
Instructor:
Alfred N. Alaniz, (M), Colorado, Oo

Adjunct Professor:
Thomas S. Girhard, (M), SW Texas State, 1991, Owy
Rene Hefner, (M), Texas (San Antonio), 1994, Ge
Dwight Jurena, (M), UT-San Antonio, GgOg
Donald E. Pimpler, (M), Tx State, Oi
Ryan E. Rudnicki, (D), Penn State, Oir
Charles K. Smith, (M), Tx State, 2005, Oy
Cooperating Faculty:
Steve Dingman, Oo

South Plains College

Dept of Science (Geology) (A) (2004)
Levelland, TX 79336
p. (806) 894-9611 (Ext. 2322) f. (806) 894-5274
wwheeler@spc.cc.tx.us
Department Secretary: DeeLaine Ruiz
Enrollment (1998): B: 1 (0)
Associate Professor:
William A. Wheeler, (M), Texas Tech, 1983, Gc

Southern Methodist University

Dept of Geological Sciences (B,M,D) (2007)
Box 750395
3225 Daniel Avenue
Dallas, TX 75275-0395
p. (214) 768-2750 f. (214) 768-2701
geol@mail.smu.edu
http://www.smu.edu/geology
Administrative Assistant: Stephanie L. Schwob
Enrollment (2006): B: 22 (0) M: 10 (0) D: 15 (0)
Chair:
Robert T. Gregory, (D), Caltech, 1981, Cs
Professor:
David D. Blackwell, (D), Harvard, 1968, Yh
Eugene T. Herrin, Jr., (D), Harvard, 1957, Ys
Louis L. Jacobs, (D), Arizona, 1977, Pv
A. Lee McAlester, (D), Yale, 1960, Pe
Brian W. Stump, (D), California (Berkeley), 1979, Ys
John V. Walther, (D), California (Berkeley), 1978, Cg
Crayton J. Yapp, (D), Caltech, 1980, Cs
Associate Professor:
Bonnie F. Jacobs, (D), Arizona, 1983, Pl
Associate Scientist:
Kurt M. Ferguson, (D), S Methodist, 1990, Cs
Christopher T. Hayward, (D), S Methodist, 1997, Ys
Ian J. Richards, (D), Tennessee, 1994, Cs
Assistant Professor:
Neil J. Tabor, (D), California (Davis), 2002, Sd
Adjunct Professor:
Anthony R. Fiorillo, (D), Pennsylvania, 1989, Pv
Alisa Winkler, (D), S Methodist, 1990, Pv
Dale A. Winkler, (D), Texas, 1985, Pv
Emeritus:
James E. Brooks, (D), Washington, 1954, Gr
Michael J. Holdaway, (D), California (Berkeley), 1963, Gp
Robert L. Laury, (D), Wisconsin, 1966, Gd

Stephen F. Austin State University

Dept of Geology (B,M) (2006)
PO Box 13011 SFA Station
Nacogdoches, TX 75962
p. (936) 468-3701 f. (936) 468-2437
mielson@sfasu.edu
http://www.geology.sfasu.edu/
Administrative Assistant: Shana R. Scott
Enrollment (2003): B: 36 (4) M: 10 (3)

Acting Chair:
Russell L. Nielson, (D), Utah, 1981, Gr
Professor:
Volker W. Gobel, (D), Colorado Mines, 1972, Gi
Ernest B. Ledger, (D), Texas A&M, 1981, Ca
Assistant Professor:
Chris A. Barker, (D), South Carolina, 1998, Gc
Instructor:
Patricia S. Sharp, (M), Stephen F. Austin, 1978, Gg

Sul Ross State University

Dept of Earth & Physical Sciences (B,M) (2005)
Box C-139
Alpine, TX 79832
p. (915) 837-8259 f. (915) 837-8692
kevinu@sulross.edu
http://www.sulross.edu/~geology
Administrative Assistant: Lori Usher
Enrollment (2005): B: 15 (6) M: 10 (4)

Chair:
Kevin Urbanczyk, (D), Washington State, 1993, Gi
Professor:
David Rohr, (D), Oregon State, 1978, Pi
James L. Whitford-Stark, (D), Brown, 1980, Xg
Adjunct Professor:
Julius Dasch, (D), Yale
David Mattison, (D), Penn State, Gx
Elizabeth A. Measures, (D), Idaho, 1992, Gd

Tarleton State University

Dept of Chemistry, Geosciences, and Env Sci (B,M) (2007)
Box T-540
Stephenville, TX 76402
p. (254) 968-9143 f. (254) 968-9953
cthompson@tarleton.edu
http://www.tarleton.edu/~physci
Administrative Assistant: Kate Caballero
Enrollment (2007): B: 25 (6)

Professor:
Phillip A. Murry, (D), S Methodist, 1982, Pv
Associate Professor:
Stephen W. Field, (D), Massachusetts, 1988, Gz
Carol A. Thompson, (D), Iowa, 1993, Hy
Assistant Professor:
Bethany D. Rinard, (D), S Methodist, 2004, Gv

Tarrant County Junior College

Natural Science Dept (A) (2004)
5301 Campus Drive
Fort Worth, TX 76119
p. (817) 534-4861 (Ext. 267) f. (817) 534-4516
rball@tccd.net
Administrative Assistant: Patsy Hemmings
Enrollment: No data reported since 1995

Chair:
Rayford L. Ball, (D), New Mexico State, 1994, Og
Associate Professor:
Herb J. Hudgens, (M), Oklahoma, 1966, Og

Texas A&M University

Center For Tectonophysics (2004)
College Station, TX 77843-3113
p. (409) 845-3251 f. (409) 845-3002
johnson@geo.tamu.edu
Department Secretary: Amber Jones

Director:
Brann Johnson, (D), Penn State, 1975, Ng
Assistant Head:
Andrew Hajash, (D), Texas A&M, 1975, Cl
Professor:
Andreas K. Kronenberg, (D), Brown, 1983, Gc
James E. Russell, (D), Northwestern, 1966, Nm
John H. Spang, (D), Brown, 1971, Gc
David V. Wiltschko, (D), Brown, 1977, Gc
Associate Professor:
Fred Chester, (D), Texas A&M, 1988, Gc
Lecturer:
Judith Chester, (D), Texas A&M, 1992, Gc
Emeritus:
Neville L. Carter, (D), California (Los Angeles), 1963, Yx
Melvin Friedman, (D), Rice, 1961, Gc
John M. Logan, (D), Oklahoma, 1965, Gc

Dept of Atmospheric Sciences (B,M,D) (2007)
3150 TAMU
College Station, TX 77843-3150
p. (979) 845-7671 f. (979) 862-4466
k-bowman@tamu.edu
http://www.met.tamu.edu/
Enrollment (2007): B: 143 (27) M: 34 (15) D: 28 (9)

Department Head:
Kenneth P. Bowman, (D), Princeton, 1984, Ow
Distinguished Professor:
Gerald North, (D), 1966, Oa
Professor:
Ping Chang, (D), Princeton, 1988, Op
Andrew Dessler, (D), Harvard, 1994, Oo
John Nielsen-Gammon, (D), MIT, 1990, Oa
Richard E. Orville, (D), Arizona, 1966, Oa
R. Saravanan, (D), Princeton, 1990, Oa
Thomas T. Wilheit, (D), MIT, 1970, Ow
Renyi Zhang, (D), MIT, 1993, Oa
Associate Professor:
Donald R. Collins, (D), Caltech, 1999, Oa
Mark Lemmon, (D), Arizona, 1994, Oa
Richard L. Panetta, (D), Wisconsin, 1978, Ow
Ping Yang, (D), Utah, 1996, Oa
Fuqing Zhang, (D), North Carolina, 2000, Oa
Assistant Professor:
Craig Epifanio, (D), Oa
Robert Korty, (D), MIT, 2005, Oa
Shaima Nasiri, (D), Wisconsin, 2004, Oa
Gunnar Schade, (D), 1997, Oa
Courtney Schumacher, (D), Washington, 2002, Oa
Lecturer:
Marion E. Alcorn, (M), Naval Postgrad Sch, 1974, Ow
Adjunct Professor:
Larry Carey, (D), Colorado State, 1999, Oa
Christopher A. Davis, (D), MIT, 1990, Ow
Alex Dessler, (D), Duke, 1956, Oa
Hung-Lung Allen Huang, Wisconsin, 1989, Or
Christian D. Kummerow, (D), Minnesota, 1987, Oa
Steve Lyons, (D), Hawaii, 1981, Ow
Chris Snyder, (D), MIT, 1989, Oa
Wei-Kuo Tao, (D), Illinois, 1982, Oa
Emeritus:
Phanindramohan Das, (D), Chicago, 1963, Ow
George L. Huebner, (D), Texas A&M, 1953, Ow
James P. McGuirk, (D), Colorado State, 1977, Ow
James R. Scoggins, (D), Penn State, 1966, Ow

Dept of Geography (B,M,D) (2008)
810 Eller O&M Building
3147 TAMU
College Station, TX 77843-3147

p. (979) 845-7141 f. (979) 862-4487
cbruton@geog.tamu.edu
http://geog.tamu.edu/
Enrollment (2006): B: 149 (0) M: 30 (0) D: 39 (0)

Professor:
Robert S. Bednarz, (D), Chicago, 1975, Oe
Sarah W. Bednarz, (D), Texas A&M, 1992, Oe
John R. Giardino, (D), Nebraska, 1979, Gm
Michael R. Waters, (D), Arizona, 1983, Ga

Associate Professor:
David M. Cairns, (D), Iowa, 1995, Oy
Andrew Klein, (D), Cornell, 1997, Or
Daniel Z. Sui, (D), Goergia, 1993, Or
Vatche P. Tchakerian, (D), California (Los Angeles), 1989, Gm

Assistant Professor:
Anne Chin, (D), Arizona State, 1994, Gm

Research Associate:
Jean A. Bowman, (M), Rutgers, 1984, Hg

Emeritus:
Clarissa T. Kimber, (D), Wisconsin, 1969, Oy

Dept of Geology & Geophysics (B,M,D) (2007)
3115 TAMU
College Station, TX 77843-3115
p. (979) 845-2451 f. (979) 845-6162
schorm@geo.tamu.edu
http://geoweb.tamu.edu
Enrollment (2007): M: 44 (13) D: 33 (10)

Acting Head:
John H. Spang, (D), Brown, 1970, Gt

Professor:
Wayne M. Ahr, (D), Rice, 1967, Gs
Richard L. Carlson, (D), Washington, 1976, Ou
Frederick M. Chester, (D), Texas A&M, 1988, Gc
Mark Everett, (D), Toronto, 1991, Ym
Jeff Fox, (D), Columbia, 1972, Gu
John R. Giardino, (D), Nebraska, 1979, Gm
Ethan L. Grossman, (D), S California, 1982, Cs
Andrew Hajash, (D), Texas A&M, 1975, Gx
Bruce E. Herbert, (D), California (Riverside), 1992, Ge
Luc T. Ikelle, (D), Paris VII, 1986, Yg
Andreas K. Kronenberg, (D), Brown, 1983, Nr
Christopher C. Mathewson, (D), Arizona, 1971, Ng
Robert K. Popp, (D), Virginia Tech, 1975, Cp
Philip D. Rabinowitz, (D), Columbia, 1973, Yr
Anne L. Raymond, (D), Chicago, 1982, Pb
Mary J. Richardson, (D), MIT, 1980, Gu
William W. Sager, (D), Hawaii, 1979, Ym
David V. Wiltschko, (D), Brown, 1977, Gc
Thomas E. Yancey, (D), California (Berkeley), 1971, Ps

Associate Professor:
Tom Blasingame, (D), Texas A&M, 1989, Np
Judith S. Chester, (D), Texas A&M, 1992, Gc
Richard L. Gibson, Jr., (D), MIT, 1991, Ys
Jerry Jensen, (D), Texas, 1986, Oo
Will Lamb, (D), Wisconsin, 1987, Gp
Franco Marcantonio, (D), Columbia, 1994, Cg
David Sparks, (D), Brown, 1992, Oo
Vatche P. Tchakerian, (D), California (Los Angeles), 1989, Gm
Hongbin Zhan, (D), Nevada (Reno), 1996, Hw

Associate Research Scientist:
Renald N. Guillemette, (D), Stanford, 1983, Gz

Assistant Professor:
Benchun Duan, (D), California (Los Angeles), 2006
John R. Hopper, (D), Columbia, 1994, Yr
Jennifer T. McGuire, (D), Michigan State, 2002, Oo
Brent V. Miller, (D), Dalhousie, 1997, Gg
Julie Newman, (D), Rochester, 1993, Gg
Thomas Olszewski, (D), Penn State, 2000, Pe
Mike Tice, (D), Stanford, 2006
Bridget Wade, (D), Edinburgh, 2001

Emeritus:
Anthony F. Gangi, (D), California (Los Angeles), 1959, Ys
Thomas W. C. Hilde, (D), Tokyo, 1973, Yr
Brann Johnson, (D), Penn State, 1975, Nr
James E. Russell, (D), Northwestern, 1966, Nm
Joel Watkins, (D), Texas, 1961, Gg

Dept of Oceanography (M,D) (2007)
MS 3146
College Station, TX 77843-3146
p. (979) 845-7211 f. (979) 845-6331
lcaldwell@ocean.tamu.edu
http://oceanography.tamu.edu
Enrollment (2007): M: 35 (11) D: 33 (4)

Executive Vice President:
Luis A. Cifuentes, (D), Delaware, 1987, Oc

Dean, College of Geosciences:
Bjorn Kjerfve, (D), Louisiana State, 1973, Op

Distinguished Professor:
Gerald R. North, (D), Wisconsin, 1966, Ow

Director, Texas Sea Grant College Program:
Robert R. Stickney, (D), Florida State, 1971, Ob

Director, Science Operator, IODP:
Paul J. Fox, (D), Columbia, 1972, Ou

Professor:
Jack G. Baldauf, (D), California (Berkeley), 1984, Ou
Thomas Bianchi, (D), Maryland, 1987, Oc
Douglas C. Biggs, (D), MIT/Woods Hole, 1976, Ob
David A. Brooks, (D), Miami, 1975, Op
William R. Bryant, (D), Chicago, 1966, Ou
Lisa Campbell, (D), SUNY (Stony Brook), 1985, Ob
Ping Chang, (D), Princeton, 1988, Op
Piers Chapman, (D), North Wales, 1982, Oc
Wilford D. Gardner, (D), MIT/Woods Hole, 1978, Ou
George A. Jackson, (D), Caltech, 1976, Op
Mahlon C. Kennicutt, III, (D), Texas A&M, 1980, Oc
Mitch Lyle, (D), Oregon State, 1978, Ou
William J. Merrell, (D), Texas A&M, 1971, Op
John W. Morse, (D), Yale, 1973, Oc
Philip D. Rabinowitz, (D), Columbia, 1973, Ou
Mary Jo Richardson, (D), MIT/Woods Hole, 1980, Gu
Gilbert T. Rowe, (D), Duke, 1968, Ob
William W. Sager, (D), Hawaii, 1983, Yr
Peter H. Santschi, (D), Switzerland (Berne), 1975, Oc
Niall C. Slowey, (D), MIT, 1991, Ou
Robert H. Stewart, (D), California (San Diego), 1969, Op
Andrew C. Vastano, (D), Texas A&M, 1967, Op
John H. Wormuth, (D), California (San Diego), 1971, Ob

Senior Scientist:
Norman Guinasso, (D), Texas A&M, 1975, Op
Ann E. Jochens, (D), Texas A&M, 1977, Op
Adam Klaus, (D), Hawaii, 1991, Gu
Terry L. Wade, (D), Rhode Island, 1978, Oc

Associate Research Professor:
Annette Lyle, (D), Michigan, 1989, Ou

Associate Professor:
Rainier Amon, (D), Texas, 1995, Ob
Steven F. DiMarco, (D), Texas (Dallas), 1991, Op
Benjamin S. Giese, (D), Washington, 1989, Op
Robert D. Hetland, (D), Florida State, 1999, Op
Patrick Louchouarn, (D), Quebec (Montreal), 1997, Oc
Alejandro H. Orsi, (D), Texas A&M, 1993, Op
Daniel L. Roelke, (D), Texas A&M, 1997, Ob
Martha L. Scott, (D), Rice, 1966, Oc
Achim Stoessel, (D), Max Planck, 1990, Op

Associate Scientist:
Steven K. Baum, (D), Texas A&M, 1996, Op
Matthew K. Howard, (M), Texas A&M, 1992, Op
Link Ji, (D), Texas A&M, 1991, Op
Jose L. Sericano, (D), Texas A&M, 1993, Cm
Assistant Professor:
Ayal Anis, (D), Oregon State, 1993, Op
Robin Brinkmeyer, (D), Bremen, 2003, Ob
Timothy M. Dellapenna, (D), William & Mary, 1999, Ou
Heath Mills, (D), Georgia Tech, 2004, Ob
Antonietta Quigg, (D), Monash, 2000, Ob
Matthew Schmidt, (D), California (Davis), 2005, Ou
Anja Schulze, (D), Victoria, 2001, Ob
Debbie Thomas, (D), North Carolina, 2002, Ou
Daniel C.O Thornton, (D), Queen Mary (London), 1995, Ob
Shari A. Yvon-Lewis, (D), Miami, 1994, Oc
Research Associate:
Howard F. Seidel, (D), Texas A&M, 1999, Op
Distinguished Professor Emeritus:
Robert A. Duce, (D), MIT, 1964, Oc
Worth D. Nowlin, Jr., (D), Texas A&M, 1966, Op
Robert O. Reid, (M), Scripps, 1948, Op
Emeritus:
Leo D. Berner, Jr., (D), Scripps, 1957, Ob
Rezneat M. Darnell, (D), Minnesota, 1953, Ob
Greta A. Fryxell, (D), Texas A&M, 1975, Ob
Bobby J. Presley, (D), California, 1969, Oc
David R. Schink, (D), California (San Diego), 1962, Oc

Dept of Soil & Crop Sciences (B,M,D) (2007)
Mail Stop 2474
College Station, TX 77843-2474
p. (979) 845-4678 f. (979) 845-0456
hallmark@tamu.edu
http://soilcrop.tamu.edu
Administrative Assistant: Carol J. Rhodes
Enrollment (2007): B: 10 (3) M: 24 (11) D: 10 (5)
Resident Director:
Jaroy Moore, (D), Texas A&M, 1973, Sc
State Soil Env Specialist:
Sam E. Feagley, (D), Missouri, 1979, Sc
Professor & Geology Librarian:
Richard H. Loeppert, Jr., (D), Florida, 1976, Sc
Professor:
Vince A. Haby, (D), Montana State, 1975, Sc
Charles T. Hallmark, (D), Ohio State, 1977, Sd
Frank M. Hons, (D), Texas A&M, 1978, Sc
Robert Lascano, (D), Texas A&M, 1982, Sp
John Matocha, (D), Texas A&M, 1968, Sc
Kevin J. McInnes, (D), Kansas State, 1985, Sp
Edward C. A. Runge, (D), Iowa State, 1963, Sd
David A. Zuberer, (D), S Florida, 1976, Sb
Associate Professor:
Mark L. McFarland, (D), Texas A&M, 1988, Sc
Bill Payne, (D), Texas A&M, Sp
Tony L. Provin, (D), Purdue, 1995, Sc
Robert Wiedenfeld, (D), Texas A&M, 1978, Sc
Assistant Professor:
Terry Gentry, (D), Arizona, Sb
Cristine L. S. Morgan, (D), Wisconsin, 2003, Sp
Emeritus:
Joe B. Dixon, (D), Wisconsin, 1958, Sc
Larry P. Wilding, (D), Illinois, 1962, Sd

Texas A&M University, Commerce

Dept of Biological & Environmental Sci (B,M) (2006)
Commerce, TX 75429
p. (903) 886-5378
Haydn_Fox@tamu-commerce.edu
http://www.tamu-commerce.edu/biology/
Enrollment: No data reported since 1995
Head:
Don Royce Lee, (D), Oregon State, Oy
Assistant Professor:
Hayden A. Fox, (D), South Carolina, 1994, Hg
Jeffery G. Kopachena, (D), Toronto, 1994, Hg

Texas A&M University, Corpus Christi

Environmental Science Program (B,M) (2004)
6300 Ocean Drive
Corpus Christi, TX 78412
p. (361) 825-2754 f. (361) 835-2742
jsengle@falcon.tamucc.edu
Coordinator:
Jennifer M. Smith-Engle, (D), Georgia, 1983, Gs
Professor:
Steven S. Barnes, (D), California (San Diego), 1967, Cm
Robert Benson, (D), Texas A&M, 1985, Yg
C. Alan Berkebile, (D), Boston, 1964, Hw
Grady Price Blount, (D), Arizona State, 1988, Or
Gary A. Jeffress, (D), Maine, 1991, Or
Joe C. Loter, (D), Kansas, 1970, Cg
David A. McKee, (D), Texas A&M, 1986, Ob
Associate Professor:
Katherine Price Blount, (D), Indiana State, 1985, Gm
Joanna Mott, (D), Texas A&M, 1984, Sb
Assistant Professor:
Thomas Naehr, (D), Christian-Albrechts (Germany), Gu
James Silliman, (D), Michigan, Co
Lecturer:
Tania Anders, (D), Gs
Adjunct Professor:
David A. Jensen, (M), Texas A&I, 1974, Hg

Geosciences Program (B,M) (2004)
6300 Ocean Drive
Corpus Christi, TX 78412
p. (361) 825-2754 f. (361) 825-2742
jsengle@falcon.tamucc.edu
Enrollment (1999): B: 21 (5)
Professor:
C. Alan Berkebile, (D), Boston, 1964, Hw
Grady Price Blount, (D), Arizona State, 1988, Or
Jennifer M. Smith-Engle, (D), Georgia, 1983, Gs
Associate Professor:
Katherine Price Blount, (D), Indiana State, Gm
Coordinator, Geology Program:
Thomas H. Naehr, (D), GEOMAR (Kiel), 1996, Gu
Assistant Professor:
James Silliman, (D), Michigan, Co
Lecturer:
Tania Anders, (D), Gs

Texas A&M University, Kingsville

Dept of Geosciences (B) (2006)
Campus Box 164
Kingsville, TX 78363
p. (512) 595-3310 f. (512) 595-3107
kftlm00@tamuk.edu
Enrollment (2006): B: 14 (2)
Professor:
Jim R. Norwine, (D), Indiana State, 1971, Oa
Associate Professor:
Thomas L. McGehee, (D), Texas (Dallas), 1987, Cl
Assistant Professor:
Jaehyung Yu, (D), Texas A&M, 2005, Oi

Lecturer:
John S. Buckley, (D), Texas, Pv
Visiting Professor:
Aiguo Bian, (D), Texas A&M, 2006, Hq

Texas Christian University

Dept of Geology (B,M) (2007)
TCU Box 298830
2950 West Bowie
Fort Worth, TX 76129
p. (817) 257-7270 f. (817) 257-7789
r.hanson@tcu.edu
http://www.geo.tcu.edu/
Administrative Assistant: Terri Mabe
Enrollment (2006): B: 20 (0) M: 12 (0)

Chair:
Richard E. Hanson, (D), Columbia, 1983, GxvGt
Professor:
John A. Breyer, (D), Nebraska, 1974, GdsGo
R. Nowell Donovan, (D), Newcastle upon Tyne, 1972, GdcGt
Ken M. Morgan, (D), Wisconsin, 1978, OrGeOi
Associate Professor:
Arthur B. Busbey, (D), Chicago, 1982, PgvGr
Ranjan S. Muttiah, (D), Purdue, 1992, OirHg
Assistant Professor:
Helge Alsleben, (D), Southern California, 2005, GctGg
Adjunct Professor:
Floyd Henk, Jr., (M), Texas Christian, 1981, Go
Timothy J. McCoy, (D), Hawaii, 1994, Xm
Emeritus:
Arthur J. Ehlmann, (D), Utah, 1953, GzXm

Texas Tech University

Dept of Geosciences (B,M,D) (2007)
Box 41053
Broadway & University
Lubbock, TX 79409-1053
p. (806) 742-3102 f. (806) 742-0100
barbara.graham@ttu.edu
http://www.ttu.edu/~geosc/
Administrative Assistant: Barbara A. Graham
Enrollment (1998): B: 45 (2) M: 44 (7) D: 16 (0)

Chair:
Calvin G. Barnes, (D), Oregon, 1982, Gi
Professor:
James E. Barrick, (D), Iowa, 1978, Ps
Sankar Chatterjee, (D), Calcutta, 1970, Pv
Necip Guven, (D), Gottingen (Ger), 1962, Gz
Colleen A. Leary, (D), Washington, 1979, Oa
Thomas M. Lehman, (D), Texas, 1985, Gs
Hua-Wei Zhou, (D), Caltech, 1990
Associate Professor:
Chia-Bo Chang, (D), Florida State, 1980, Oa
Harold Gurrola, (D), California (San Diego), 1995, Ys
Haraldur R. Karlsson, (D), Chicago, 1988, Cl
Seiichi Nagihara, (D), Texas, 1992, Oi
Moira Ridley, (D), Nebraska, 1997, Cl
John Schroeder, (D), Texas Tech, 1999, Oa
Aaron Yoshinobu, (D), Southern California, 1999
Assistant Professor:
Sukanta Basu, (D), Minnesota, 2004, OaaOa
Peter F. Holterhoff, (D), Cincinnati, 1993
David M. Leverington, (D), Manitoba (Canada), 2001
Christopher C. Weiss, (D), Oklahoma, 2004, OaaOa
Kyle Cameron Wiens, (D), Colorado State, 2005, OaaOa
Research Associate:
Melanie A. Barnes, (D), Texas Tech, 2001, Cg

Emeritus:
George B. Asquith, (D), Wisconsin, 1966, Go
Richard E. Peterson, (D), Missouri, 1971, Oa

The University of Texas at San Antonio

Dept of Earth and Environmental Science (2005)
6900 North Loop 1604 West
San Antonio, TX 78249-0663
p. (210) 458-4455 f. (210) 458-4469
robert.smith1@utsa.edu
http://www.utsa.edu/eps

Chair:
Robert K. Smith, (D), Iowa, 1972, Gz

Trinity University

Dept of Geosciences (B) (2007)
One Trinity Place, #45
San Antonio, TX 78212-7200
p. (210) 999-7606 f. (210) 999-7090
gkroeger@trinity.edu
http://www.trinity.edu/departments/geosciences/
Administrative Assistant: Blanca E. Kirkman
Enrollment (2001): B: 22 (3)

Chair:
Glenn C. Kroeger, (D), Stanford, 1987, Yg
Distinguished Professor:
Thomas W. Gardner, (D), Cincinnati, 1978, Gm
Professor:
Walter Coppinger, (D), Miami (Ohio), 1974, Gc
Diane R. Smith, (D), Rice, 1984, Gi
Emeritus:
Robert L. Freed, (D), Michigan, 1966, Gz
Perry O. Roehl, (D), Wisconsin, 1955, Gs
Edward C. Roy, Jr., (D), Ohio State, 1964, Pg
Technician:
Clancy Donnelly, (B), James Madison, 1995, Gg

University of Houston

Allied Geophysical Lab (2004)
Houston, TX 77204-4231
p. (713) 743-9150 f. (713) 743-9164
http://www.agl.uh.edu
Administrative Secretary: Becky Craft

Director:
Kurt Marfurt, (D), Columbia, 1978, Ye
Associate Director:
Hua-Wei Zhou, (D), Caltech, 1989, Ys
Senior Research Associate:
K. K. Sekharan, (D), Kentucky, 1976, Ye

Dept of Geosciences (B,M,D) (2007)
312 Science and Research Building 1
Houston, TX 77204-5007
p. (713) 743-3399 f. (713) 748-7906
jfcasey@uh.edu
http://www.geosc.uh.edu
Advising Assistant: Sylvia Marshall
Enrollment (2006): B: 133 (8) M: 83 (13) D: 89 (1)

Chair:
John F. Casey, (D), SUNY (Albany), 1980, GtiGu
Distinguished Research Professor:
John F. Dewey, (D), Imperial Coll (London), 1960, GtcGg
Fred Hilterman, (D), Colorado Mines, 1970, Ye
Professor:
Janok P. Bhattacharya, (D), McMaster, 1989, Gs
Kevin Burke, (D), London, 1953, Gt
Daewon Byun, (D), North Carolina State, 1987, Oa

Texas

John P. Castagna, (D), Texas, 1983, Ye
Henry S. Chafetz, (D), Texas, 1970, Gd
Stuart A. Hall, (D), Newcastle, 1976, Ym
Christopher L. Liner, (D), Colorado Mines, 1989, YesYg
Rosalie F. Maddocks, (D), Kansas, 1965, Pm
Kurt Marfurt, (D), Columbia, 1978, Ye
Arch M. Reid, (D), Pittsburgh, 1964, Gi
Arthur Weglein, (D), CUNY, 1980, Ye

Research Professor:
Adry Bissada, (D), Washington, 1967, Co
Gennady Goloshubin, (D), Inst Physics of Earth (Moscow), Ye
De-hua Han, (D), Stanford, Ye

Associate Professor:
Bernhard Rappenglueck, (D), Munich, 1996, Oa
Regina M. Capuano, (D), Arizona, 1988, Hw
William R. Dupre, (D), Stanford, 1975, On
Ian Evans, (D), Texas A&M, 1971, Gr
James Lawrence, (D), Caltech, 1970, Cl
Hua-Wei Zhou, (D), Caltech, 1989, Ys

Research Associate Professor:
Donald Van Niewenhuise, (D), South Carolina, 1978, Ps
Robert Wiley, (D), Colorado Mines, Ys

Assistant Professor:
Thomas Lapen, (D), Wisconsin, 2005, Gz
Barry Lefer, (D), New Hampshire, 1997, Oa
Aibing Li, (D), Brown, 2000, Ys
Michael Murphy, (D), California (Los Angeles), 2000, Gc
Alexander Robinson, (D), California (Los Angeles), 2005, Gc
Jonathon Snow, (D), MIT/Woods Hole, 1992, Ca

Lecturer:
Jennifer N. Lytwyn, (D), Houston, 1993, Gi

Adjunct Professor:
Vitor Abreu, (D), Rice, 1997, Gr
Peter Bartok, (D), Go
Richard T. Beauboeuf, (D), Houston, 1990, Gr
Kurt M. Strack, (D), Cologne, Ye

Emeritus:
Max F. Carman, Jr., (D), California (Los Angeles), 1954, Gi
Robert E. Sheriff, (D), Ohio State, 1950, Ye

IT Staff:
Jay Jay Krishnan, (B), Houston, 2001, Oo

University of Houston Downtown

Dept of Natural Sciences (2004)
1 Main Street
Houston, TX 77002
p. (713) 221-8015 f. (713) 221-8528
merrillg@uhd.edu

Professor:
Glen K. Merrill, (D), Louisiana State, 1968, Pi

Associate Professor:
Penny A. Morris-Smith, (D), California (Berkeley), 1975, Pi

Lecturer:
Roberto Garza, (D), Colorado, 1980, Gm
Kenneth Johnson, (D), Texas Tech, 1995, Gi

University of North Texas

Dept of Geography (B,M) (2004)
P.O. Box 305279
ENV 210
Denton, TX 76203-5279
p. (940) 565-2091 f. (940) 369-7550
geog@unt.edu
http://www.geog.unt.edu
Administrative Assistant: Jo Clark
Enrollment: No data reported since 1995

Chair:
C. Reid Ferring, (D), Texas (Dallas), 1993, Ga

Professor:
Miguel F. Acevedo, (D), California (Berkeley), 1980, Ge
Paul F. Hudak, (D), California (Santa Barbara), 1991, Hw
F. Andrew Schoolmaster, III, (D), Kent State, 1979, Hs

Associate Professor:
Donald Lyons, (D), Colorado, 1991, Ou
Kent M. McGregor, (D), Kansas, 1982, Oa
Joseph R. Oppong, (D), Alberta, 1992, Oe
Harry F. L. Williams, (D), Simon Fraser, 1989, Gm

Assistant Professor:
Minhe Ji, (D), South Carolina, 1996, Oy
Lisa A. Nagaoka, (D), Washington, 03/1, Ga

Adjunct Professor:
Kenneth W. Iles, (M), E Texas State, 1984, Og
George F. Maxey, (M), North Texas, 1991, Gg

University of Texas at Austin

Dept of Marine Science (M,D) (2007)
750 Channel View Drive
Port Aransas, TX 78373-5015
p. (361) 749-6730 f. (361) 749-6777
lee.fuiman@mail.utexas.edu
http://www.utmsi.utexas.edu
Enrollment (2007): M: 14 (3) D: 19 (1)

Chair:
Lee A. Fuiman, (D), Michigan, 1983, Ob

Professor:
Edward J. Buskey, (D), Rhode Island, 1983, Ob
Kenneth H. Dunton, (D), Alaska, 1985, Ob
Wayne S. Gardner, (D), Wisconsin, 1971, Ob
G. J. Holt, (D), Texas A&M, 1976, Ob
Peter Thomas, (D), Leicester (UK), 1977, Ob
Tracy A. Villareal, (D), Rhode Island, 1989, Ob

Associate Professor:
Henrietta N. Edmonds, (D), MIT/Woods Hole, 1991, Ob

Assistant Professor:
Deana L. Erdner, (D), MIT/Woods Hole, 1997, Ob
Izhar A. Khan, (D), Banaras Hindu, 1990, Ob
James W. McClelland, (D), Boston, 1998, Ob
Dong-Ha Min, (D), Scripps, 1999, Ob
B. Scott Nunez, (D), Louisiana State, 1996, Ob
Tamara K. Pease, (D), North Carolina, 2000, Ob
G. Christophe Shank, (D), North Carolina, 2003, Ob

University of Texas at Austin, Jackson School of Geosciences

Dept of Geological Sciences (B,M,D) (2007)
Department of Geological Sciences
1 University Station C1100
Austin, TX 78712-0254
p. (512)471-5172 f. (512)471-9425
jbird@jsg.utexas.edu
http://www.geo.utexas.edu
Enrollment (2006): B: 193 (43) M: 69 (0) D: 86 (0)

Dean:
Eric J. Barron, (D), Miami, 1980, Og

Chair:
Clark R. Wilson, (D), California (San Diego), 1975, Yg

Director, Inst for Geophysics:
Paul L. Stoffa, (D), Columbia, 1974, Ye

Director, Center for Int'l Energy & Env Policy:
Charles G. Groat, (D), Texas, 1970, Eg

Director, Bureau of Economic Geology:
Scott W. Tinker, (D), Colorado, 1996, Go

Professor:
Jay L. Banner, (D), SUNY (Stony Brook), 1986, Cl
Philip Bennett, (D), Syracuse, 1988, Hg

William D. Carlson, (D), California (Los Angeles), 1980, Gp
Mark P. Cloos, (D), California (Los Angeles), 1981, Gc
Ian W. D. Dalziel, (D), Edinburgh, 1963, Gt
William L. Fisher, (D), Kansas, 1961, Gr
Omar Ghattas, (D), Duke, 1988, Gq
Stephen P. Grand, (D), McGill, 1986, Yg
Charles Kerans, (D), Carleton, 1982, Gs
Gary A. Kocurek, (D), Wisconsin, 1980, Gs
J. Richard Kyle, (D), W Ontario, 1977, Em
Leon E. Long, (D), Columbia, 1959, Cc
Sharon Mosher, (D), Illinois, 1978, Gc
Terry Quinn, (D), Brown, 1989, Pe
Timothy B. Rowe, (D), California (Berkeley), 1987, Pv
Mrinal K. Sen, (D), Hawaii, 1987, Ye
John M. Sharp, Jr., (D), Illinois, 1974, Hq
James T. Sprinkle, (D), Harvard, 1971, Pi
Ronald J. Steel, (D), Glasgow (UK), 1970, Gs
Robert H. Tatham, (D), Columbia, 1975, Yg

Research Scientist; Professor:
Peter B. Flemings, (D), Cornell, 1990

Associate Director:
Clifford A. Frohlich, (D), Cornell, 1976

Senior Scientist:
James A. Austin, Jr., (D), MIT, 1979
Nathan L. Bangs, (D), Columbia, 1990
Shirley P. Dutton, (D), Texas, 1986
Craig S. Fulthorpe, (D), Northwestern, 1988
John A. Goff, (D), MIT, 1990
Bob A. Hardage, (D), Oklahoma State, 1967
Michael R. Hudeck, (D), Wyoming, 1990
Martin P. A. Jackson, (D), Cape Town, 1976
Richard A. Ketcham, (D), Texas, 1995, Gz
Stephen E. Laubach, (D), Illinois, 1986
Lawrence A. Lawver, (D), Scripps, 1976
Robert G. Loucks, Texas, 1976
F. Jerry Lucia, (M), Minnesota, 1954
W. Paul Mann, (D), SUNY (Albany), 1983
Kitty L. Milliken, (D), Texas, 1985, Gd
Stephen C. Ruppel, (D), Tennessee, 1979
Bridget R. Scanlon, Kentucky
Thomas H. Shipley, (D), Rice, 1975
Frederick W. Taylor, (D), Cornell, 1978

Associate Professor:
Christopher J. Bell, (D), California (Berkeley), 1997, Pv
James N. Connelly, (D), Memorial, 1991, Cc
Sergey B. Fomel, (D), Stanford, 1991
James E. Gardner, (D), Rhode Island, 1993, Gv
Brian Horton, (D), Arizona, 1998, Gs
Randall A. Marrett, (D), Cornell, 1990, Gc
David Mohrig, (D), Hs
Zong-Liang Yang, (D), Macquarie, 1992, Ow

Research Scientist:
Wulf Achim Gose, (D), S Methodist, 1970, Ym
Todd B. Housh, (D), Washington (St. Louis), 1989, Gi
Fred W. McDowell, (D), Columbia, 1966, Cc

Assistant Professor:
Bayani Cardenas, (D), New Mexico Tech, 2006, Hw
Robert J. Ferguson, (D), Calgary, 2000, Ye
John Lassiter, (D), California (Berkeley), 1995, Cc

Research Associate:
Brian K. Arbic, (D), MIT, 2000
Ginny A. Catania, (D), Washington, 1994
Matthew W. Colbert, (D), Texas, 1999, Gq
Julia F. W. Gale, (D), Exeter, 1987
Ursula Hammes, Colorado, 1992
Matthew J. Hornbach, (D), Wyoming, 1994
Charles S. Jackson, (D), Chicago, 1998
Xavier Janson, (D), Miami, 2002
John Lansdown, (D), Washington, 1992, Cs
Luc L. Lavier, (D), Columbia, 1999
Jessica A. Maisano, (D), Yale, 2000, Pv
Angela McDonnell, (D), Univ Coll (Dublin), 2001
Timothy A. Meckel, (D), Texas, 2003
Jean-Philippe Nicot, (D), Texas, 1998
Diana C. Sava, (D), Stanford, 2004
Harm J. Van Avendonk, (D), Scripps, 1998
Wayne R. Wright, (D), Univ Coll (Dublin), 2001
Christopher K. Zahm, (D), Colorado Mines, 2002

Senior Lecturer:
Mark A. Helper, (D), Texas, 1985, Gc

Emeritus:
Milo M. Backus, (D), MIT, 1956, Ye
Daniel Barker, (D), Princeton, 1961, Gi
Robert E. Boyer, (D), Michigan, 1959, Gt
Richard T. Buffler, (D), California (Berkeley), 1967, Gu
Stephen E. Clabaugh, (D), Harvard, 1950, Gp
Peter T. Flawn, (D), Yale, 1951, Eg
Robert L. Folk, (D), Penn State, 1952, Gd
William E. Galloway, (D), Texas, 1971, Gs
Edward C. Jonas, (D), Illinois, 1954, Gz
Lynton S. Land, (D), Lehigh, 1966, Cl
Wann Langston, Jr., (D), California (Berkeley), 1952, Pv
Ernest L. Lundelius, (D), Chicago, 1954, Pv
John C. Maxwell, (D), Princeton, 1946, Gt
Earle F. McBride, (D), Johns Hopkins, 1960, Gd
William R. Muehlberger, (D), Caltech, 1954, Gc
Yosio Nakamura, (D), Penn State, 1963, Ys
Amos Salvador, (D), Stanford, 1950, Gr
Douglas Smith, (D), Caltech, 1969, Gi
John A. Wilson, (D), Michigan, 1941, Pv

Inst for Geophysics (2007)
JJ Pickle Research Campus
10100 Burnet Road, Bldg. 196 (ROC)
Austin, TX 78758
p. (512) 471-6156 f. (512) 471-8844
utig@ig.utexas.edu
http://www.ig.utexas.edu/

Director:
Paul L. Stoffa, (D), Columbia, 1974, Ye

Associate Director:
Ian W. D. Dalziel, (D), Edinburgh, 1963, Gt
Cliff Frohlich, (D), Cornell, 1976, Ys

Research Professor:
William E. Galloway, (D), Texas, 1971, Gs
Stephen P. Grand, (D), Caltech, 1986, Ys
Yosio Nakamura, (D), Penn State, 1963, Ys
Mrinal K. Sen, (D), Hawaii, 1987, Ye

Senior Scientist:
James A. Austin, Jr., (D), MIT/Woods Hole, 1978, Gu
Nathan L. Bangs, (D), Columbia, 1990, Gu
John A. Goff, (D), MIT/Woods Hole, 1990, Yr
Lawrence A. Lawver, (D), California (San Diego), 1976, Yr
Paul Mann, (D), SUNY (Albany), 1983, Gt
Thomas H. Shipley, (D), Rice, 1975, Yr
Frederick W. Talyor, (D), Cornell, 1979, Pe

Research Scientist:
Donald D. Blankenship, (D), Wisconsin, 1989, Yg
Gail L. Christeson, (D), MIT/Woods Hole, 1994, Yr
Craig S. Fulthorpe, (D), Northwestern, 1988, Gs
Kirk D. McIntosh, (D), California (Santa Cruz), 1992, Yr
Robert J. Pulliam, (D), California (Berkeley), 1991, Ys

Research Associate:
Sean S. Gulick, (D), Lehigh, 1999, Yr
John W. Holt, (D), Caltech, 1997, Ye
Charles Jackson, (D), Chicago, 1998, Pe

Luc L. Lavier, (D), Columbia, 1999, Yg
David L. Morse, (D), Washington, 1997, Yg
Hillary C. Olson, (D), Stanford, 1988, Ps
Robert B. Scott, (D), McGill, 1999, Og
Roustam K. Seifoullaev, (D), Baku State, 1979, Ye
Harm Van Avendonk, (D), Scripps, 1998, Yg

Emeritus:
Milo M. Backus, (D), MIT, 1956, Ye
Arthur E. Maxwell, (D), Scripps, 1959, Og

Postdoc:
Christina Holland, (D), S Florida, 2003, Og
Matthew Hornbach, (D), Wyoming, 2004, Yg
Timothy Whiteaker, (D), Texas, 2004, Gs

Project Coordinator:
Patricia E. Ganey-Curry, (B), Texas A&M, 1978, Oo

Program Manager:
Katherine K. Ellins, (D), Columbia, 1988, Oo

Related Staff:
Mark Wiederspahn, (B), Bucknell, 1975, Oo

University of Texas, Arlington

Dept of Earth & Environmental Sci (B,M,D) (2007)
Box 19049
500 Yates Street
Arlington, TX 76019
p. (817) 272-2987 f. (817) 272-2628
geology@uta.edu
http://www.uta.edu/ees/
Enrollment (2007): B: 100 (13) M: 50 (8) D: 11 (1)

Professor:
John M. Holbrook, (D), Indiana, 1992, Gs
Merlynd K. Nestell, (D), Oregon State, 1966, Pm
Christopher R. Scotese, (D), Chicago, 1985, Gt
John S. Wickham, (D), Johns Hopkins, 1969, Gc

Assistant Professor:
Andrew Hunt, (D), Liverpool, 1988, GeCg
Arne M. Winguth, (D), Hamburg, 1997, Opa

Adjunct Professor:
John E. Damuth, (D), Columbia, 1973, Gu
Gloria Eisenstadt, (D), Johns Hopkins, 1991, Gc
Brooks Ellwood, (D), Rhode Island, 1977, Yg
Gary Lowell, (D), New Mexico Tech, 1970, Gx
Galena P. Nestell, (D), Vsegi (Russia), 1990, Pm
Shan Shanmugan, (D), Gs
Larry Standlee, (D), Rice, Gc
Cornelia Winguth, (D), Hamburg, 1998, GugOm

Emeritus:
William L. Balsam, (D), Brown, 1973, Gu
Burke Burkart, (D), Rice, 1965, Gx
Rex Crick, (D), Kansas, 1971, Py
Donald F. Reaser, (D), Texas, 1960, Gr

University of Texas, Dallas

Dept of Geosciences (B,M,D) (2006)
P.O. Box 830688
Richardson, TX 75083-0688
p. (972) 883-2401 f. (972) 883-2537
ferguson@utdallas.edu
Administrative Assistant: Virginia L. Anderson
Administrative Assistant: Jean Boyer
Enrollment (2006): B: 40 (2) M: 18 (6) D: 30 (7)

Head:
John F. Ferguson, (D), S Methodist, 1981, Yg

Director, Center for Lithospheric Studies:
George A. McMechan, (M), Toronto, 1972, Ys

Professor:
Carlos L. V. Aiken, (D), Arizona, 1976, Yv
William I. Manton, (D), Witwatersrand, 1968, Cc
Robert H. Rutford, (D), Minnesota, 1969, Gl
Robert J. Stern, (D), California (San Diego), 1979, Gt

Research Professor:
Robert B. Finkelman, (D), Maryland, 1980, Gb

Senior Scientist:
Rodger E. Denison, (D), Texas, 1966, Cs
Douglas Kirkland, (D), New Mexico, 1963, Cg

Associate Professor:
Tom H. Brikowski, (D), Arizona, 1987, Hw
James L. Carter, (D), Rice, 1965, Eg

Lecturer:
Ignacio Pujana, (D), Texas (Dallas), 1997, Pm

Retired Professor:
Richard M. Mitterer, (D), Florida State, 1966, Co
Emile A. Pessagno, Jr., (D), Princeton, 1960, Pm

Emeritus:
David E. Dunn, (D), Texas, 1964, Gc
Anton L. Hales, (D), Cape Town, 1936, Ys
Dean C. Presnall, (D), Penn State, 1963, Cp

Science/Mathematics Education Program (M) (2004)
P. O. Box 830688 FN33
Richardson, TX 75080-9688
p. 972 883-2496 f. 972-883-6796
mont@utdallas.edu
http://www.utdallas.edu/dept/SciMathEd

Professor:
Thomas R. Butts, (D), Michigan State, 1973, Oo
Fred L. Fifer, (D), Vanderbilt, 1973, Oo

Associate Professor:
Cynthia E. Ledbetter, (D), Texas A&M, 1987, Oo

Assistant Professor:
Homer A. Montgomery, (D), Pg
Mary Urquhart, (D), Xy

Instructor:
Barbara Curry, (M), Texas (Dallas), Oo

University of Texas, El Paso

Dept of Geological Sciences (B,M,D) (2007)
500 W. University Avenue
El Paso, TX 79968-0555
p. (915) 747-5501 f. (915) 747-5073
ladewig@geo.utep.edu
http://www.geo.utep.edu
Administrative Assistant: Sandra S. Ladewig
Enrollment (2007): B: 44 (2) M: 24 (7) D: 43 (8)

Director:
Diane I. Doser, (D), Utah, 1984, Ys

Professor:
Elizabeth Y. Anthony, (D), Arizona, 1986, Gi
Kate C. Miller, (D), Stanford, 1991, Yg
Terry L. Pavlis, (D), Utah, 1982, Gc
Nicholas E. Pingitore, Jr., (D), Brown, 1973, Cl
Laura F. Serpa, (D), Cornell, 1986, Ye

Senior Scientist:
Steven H. Harder, (D), Texas (El Paso), 1986, Yg

Associate Professor:
Thomas E. Gill, (D), California (Davis), 1995, Ge
Philip C. Goodell, (D), Harvard, 1970, Ce
Jasper G. Konter, (D), Scripps, 2007, GgCl
Richard P. Langford, (D), Utah, 1989, Gs
Aaron A. Velasco, (D), California (Santa Cruz), 1993, Yx

Assistant Professor:
David M. Borrok, (D), Notre Dame, 2005, Cg
Andre S. Ellis, (D), Illinois, 2004, Hg
Jose M. Hurtado, (D), MIT, 2002, Gt
Bridget R. Smith-Konter, (D), Scripps, 2005, YgOr

Lecturer:
Nicholas L. Miller, (M), Marshall, 1994, Ou
Emeritus:
Kenneth F. Clark, (D), New Mexico, 1966, Eg
William C. Cornell, (D), California (Los Angeles), 1972, Pm
G. Randy Keller, (D), Texas Tech, 1973, Ye
David V. Le Mone, (D), Michigan State, 1964, Ps
Robert H. Schmidt, Jr., (D), California (Los Angeles), 1968, Oy
Science/Eng Research Tech:
Galen M. Kaip, (M), Texas (El Paso), 1998, Yx
Beata Maciejewska, (M), Lublin (Poland), 1973, Cg
Minghua Ren, (D), Baylor, 2002, Ca
Network Manager:
Carlos J. Montana, (M), Texas (El Paso), 1992, Yg
Director, GLOBE Program:
Roberta J. Hoffer, (D), Texas (El Paso), 1982, Gg
Cooperating Faculty:
John Walton, (D), Idaho, 1991, Hw

University of Texas, Pan American

Dept of Physics & Geology (B) (2006)
1201 W. University Drive
Edinburg, TX 78539
p. (512) 381-3523 f. (210) 381-3527
bhatti@panam.edu
Enrollment: No data reported since 1995
Assistant Professor:
Ruben A. Mazariegos, (D), Texas A&M, 1993, Ye
Eric R. Rieken, (D), Washington State, 1993, Gt

University of Texas, Permian Basin

Dept of Geology (B,M) (2006)
4901 E. University Boulevard
Odessa, TX 79762-0001
p. (432) 552-2243 f. (432) 552-3230
mutis_e@utpb.edu
Enrollment (2005): B: 18 (4) M: 12 (0)
Chair:
Emilio Mutis-Duplat, (D), Texas, 1972, Gp
Lecturer:
William L. Basham, (D), Oklahoma, 1978, Ye
Emily L. Stoudt, (D), Ohio State, 1975, Gd
Robert C. Trentham, (D), Texas (El Paso), 1981, Go

University of Texas, San Antonio

Dept of Earth & Environmental Science (B,M) (2007)
One UTSA Circle
San Antonio, TX 78249-0663
p. (210) 458-4455 f. (210) 458-4469
Eric.Swanson@utsa.edu
http://www.utsa.edu/eps
Administrative Assistant: Laurie Gay
Enrollment (1998): B: 162 (15) M: 219 (25)
Professor:
Robert K. Smith, (D), Iowa, 1972, Gp
Eric R. Swanson, (D), Texas, 1977, Gv
Associate Professor:
Stuart J. Birnbaum, (D), Cambridge, 1977, Cl
Alan Dutton, (D), Hw
Weldon W. Hammond, Jr., (D), Texas, 1984, Hw
Lance L. Lambert, (D), Pg
Assistant Professor:
Judy Haschenburger, (D), Gm
Kyle Murray, (D), Hw
Debajyoti Paul, (D), Cs
Lecturer:
Rene Hefner, (M), Gg

West Texas A&M University

Dept of Life, Earth & Environmental Sci (B) (2007)
P. O. Box 60808, WT Station
Canyon, TX 79016-0001
p. (806) 656-2570 f. (806) 656-2928
dbingham@mail.wtamu.edu
Department Secretary: Debi Adams
Enrollment (2007): B: 15 (1)
Professor:
Joseph C. Cepeda, (D), Texas, 1977, Gi
Gerald Schultz, (D), Michigan, 1963, Pv
Instructor:
Roberta D. Speer, (M), West Texas State, 1974, Og
Emeritus:
Robert C. Burton, (D), New Mexico, 1965, Ps

Utah

Brigham Young University

Dept of Agronomy & Horticulture (B,M) (2004)
275 Widtsoe Building
P.O. Box 25183
Provo, UT 84602-5183
p. (801) 378-2760 f. (801) 378-7499
ret@email.byu.edu
Department Secretary: Dorine Jesperson
Enrollment: No data reported since 1995
Chair:
Richard E. Terry, (D), Purdue, 1976, Sb
Professor:
Von D. Jolley, (D), Iowa State, 1976, Sc
Sheldon D. Nelson, (D), California (Riverside), 1971, Sp
Associate Scientist:
Bruce L. Webb, (M), Brigham Young, 1978, Sc

Dept of Geography (B,M) (2004)
690 SWKT
Provo, UT 84602
p. (801) 378-3851 f. (801) 378-5978
geog-sec@email.byu.edu
Department Secretary: Karen R. Bryce
Chair:
Matthew J. Shumway, (D), Indiana, 1991, Oo
Professor:
Richard H. Jackson, (D), Clark, 1970, Ou
Associate Professor:
James A. Davis, (D), Arizona State, 1992, Oo
Chad Emmett, (D), Chicago, 1991, Oo
Perry J. Hardin, (D), Utah, 1989, Or
Assistant Professor:
Matthew F. Bekker, (D), Iowa, 2002, Oy
Jeffrey O. Durrant, (D), Hawaii, 2001, Ge
Mark W. Jackson, (D), South Carolina, 2001, Or
Samuel M. Otterstrom, (D), Louisiana State, 1997, Ou
Brandon Plewe, (D), Buffalo, 1997, Oi
Instructor:
Jeffry S. Bird, (M), Brigham Young, 1990, Oy
Adjunct Professor:
Lloyd E. Hudman, (D), Kansas, 1968, Oo

Dept of Geology (B,M) (2005)
S 389 ESC
Provo, UT 84602
p. (801) 422-3918 f. (801) 422-0267
marge_morgan@byu.edu
http://www.geologyindy.byu.edu
Administrative Assistant: Kristine B. Mortenson
Enrollment (2005): B: 74 (23) M: 47 (18)

Professor:
Eric H. Christiansen, (D), Arizona State, 1981, Gi
Michael J. Dorais, (D), Georgia, 1987, Gi
Dana T. Griffen, (D), Virginia Tech, 1976, Gz
Ron Harris, (D), London (UK), 1989, Gc
Jeffrey D. Keith, (D), Wisconsin, 1982, Eg
Bart J. Kowallis, (D), Wisconsin, 1981, Gc
Alan L. Mayo, (D), Idaho, 1982, Hw
John H. McBride, (D), Cornell, 1987, Ye
Thomas H. Morris, (D), Wisconsin, 1986, Gr
Scott M. Ritter, (D), Wisconsin, 1986, Ps
Associate Professor:
Stephen T. Nelson, (D), California (Los Angeles), 1991, Cs
Assistant Professor:
Barry R. Bickmore, (D), Virginia Tech, 2000, Cg
Brooks B. Britt, (D), Calgary, 1993, Pg
Research Associate:
David G. Tingey, (M), Brigham Young, 1989, Oo
Instructor:
Randall Skinner, (M), Brigham Young, 1996, Gg
Adjunct Professor:
R. Paul Nixon, (D), Brigham Young, 1972, Go
Emeritus:
James L. Baer, (D), Brigham Young, 1968, Go
William K. Hamblin, (D), Michigan, 1958, Gr
Lehi F. Hintze, (D), Columbia, 1951, Gr
Wade E. Miller, (D), California (Berkeley), 1968, Pv
Best G. Myron, (D), California (Berkeley), 1961, Go
Morris S. Petersen, (D), Iowa, 1962, Pi
William R. Phillips, (D), Utah, 1954, Gz
J. Keith Rigby, (D), Columbia, 1952, Pe
Other:
Marge Morgan, (A), Brigham Young, 1965

Snow College

Dept of Geology (A) (2007)
Ephraim, UT 84627
p. 435 283-7519 f. 435 283-7501
renee.faatz@snow.edu
Enrollment (2007): A: 7 (2)
Chair:
Renee M. Faatz, (M), Ohio State, 1985, Gg
Assistant Professor:
Ted L. Olson, (M), Utah, 1976, Ys

Southern Utah University

Dept of Physical Science (B) (2007)
351 West University Blvd.
Cedar City, UT 84720
p. (435) 586-7900 f. (435) 865-8051
eves@suu.edu
http://www.suu.edu/sci/physci/geosciences/
Administrative Assistant: Rhonda Riley
Enrollment (2007): B: 18 (6)
Professor:
Robert L. Eves, (D), Washington State, 1991, GrOeCl
C. Frederick Lohrengel, II, (D), Brigham Young, 1968, GsrPs
Assistant Professor:
Mark R. Colberg, (D), Georgia, 2001, GpeGt
Adjunct Professor:
Sue Ann Finstick, (D), Hawaii, 1996, Hw

University of Utah

Dept of Geography (B,M,D) (2004)
260 S. Central Campus Drive
Rm 270
Salt Lake City, UT 84112-9155
p. (801) 581-8218 f. (801) 581-8219
thomas.kontuly@geog.utah.edu
http://www.geog.utah.edu
Administrative Assistant: Susan Van Roosendaal
Enrollment: No data reported since 1995
Chair:
Thomas M. Kontuly, (D), Pennsylvania, 1978, Oo
Professor:
Donald R. Currey, (D), Kansas, 1969, Gm
Philip C. Emmi, (D), North Carolina, 1979, Ou
George F. Hepner, (D), Arizona State, 1979, Oy
Chung M. Lee, (D), Michigan, 1961, Ou
Harvey J. Miller, (D), Ohio State, 1991, Oi
Merrill K. Ridd, (D), Northwestern, 1963, Or
Assistant Professor:
Thomas J. Cova, (D), California (Santa Barbara), 1999, Oi
Richard R. Forster, (D), Cornell, 1997, Or
Katrina A. Moser, (D), McMaster, 1997, Pe
Adjunct Professor:
Jeffrey R. Keaton, (D), Texas A&M, 1988, Oo
Elliott W. Lips, (M), Colorado State, 1990, Gm
Emeritus:
Roger M. Mccoy, (D), Kansas, 1967, Or
Professor-Lecturer:
Arthur Hampson, (D), Hawaii, 1980, Oy
Assistant Professor:
Trevor J. Davis, (D), British Columbia, 1999, Oi
Other:
Fred E. May, (D), Virginia Tech, 1976, Oo

Dept of Geology & Geophysics (B,M,D) (2004)
135 South 1460 East
Room 717
Salt Lake City, UT 84112-0111
p. (801) 581-6553 f. (801) 581-7065
jbollis@mines.utah.edu
http://www.mines.utah.edu/geo
Administrative Assistant: Kim Atwater
Administrative Assistant: Jennifer B. Brown
Enrollment (2004): B: 74 (12) M: 42 (15) D: 31 (8)
Professor:
John M. Bartley, (D), MIT, 1981, Oi
John R. Bowman, (D), Michigan, 1976, Cs
Francis H. Brown, (D), California (Berkeley), 1971, Gg
Ronald L. Bruhn, (D), Columbia, 1976, Gc
Thure E. Cerling, (D), California (Berkeley), 1977, Cl
Marjorie A. Chan, (D), Wisconsin, 1982, Gs
David S. Chapman, (D), Michigan, 1976, Yh
Allan A. Ekdale, (D), Rice, 1974, Pe
Richard D. Jarrard, (D), California (San Diego), 1975, Gu
Barbara P. Nash, (D), California (Berkeley), 1971, Gi
Peter H. Roth, (D), ETH (Switzerland), 1970, Pm
Gerard T. Schuster, (D), Columbia, 1984, Ye
Robert B. Smith, (D), Utah, 1967, Ys
Douglas K. Solomon, (D), Waterloo, 1992, Hw
Michael S. Zhdanov, (D), Moscow State, 1968, Ye
Research Professor:
Walter J. Arabasz, (D), Caltech, 1971, Ys
Associate Professor:
Susan L. Halgedahl, (D), California (Santa Barbara), 1981, Ym
Paul W. Jewell, (D), Princeton, 1989, Hg
William P. Johnson, (D), Colorado, 1993, Ng
Erich U. Petersen, (D), Michigan, 1983, Em
Scott D. Sampson, (D), Toronto, 1993, Pv
Research Associate Professor:
James C. Pechmann, (D), Caltech, 1983, Ys

Research Assistant Professor:
Linda K. Ayliffe, (D), Australian National, 1991, Cl
Nikolay G. Golubev, (D), Moscow (Russia), 1983, Ye
Kristine L. Pankow, (D), California, 1999, Ys
Assistant Professor:
David A. Dinter, (D), MIT, 1994, Gt
Robert N. Harris, (D), Utah, 1996, Gu
Cari Johnson, (D), Stanford, 2003, Gs
Fulvio Tonon, (D), Colorado, 2000, Nr
Research Assistant Professor:
Desmond E. Moser, (D), Queen's, 1993, Cc
Adjunct Associate Professor:
Harley M. Benz, (D), Utah, 1986, Ys
Stephen R. Brown, (D), Columbia, 1984, Nr
Charles M. Meertens, (D), Colorado, 1987, Gt
Virginia B. Sisson, (D), Princeton, 1985, Gp
Adjunct Assistant Professor:
Stephan Husen, (D), GEOMAR (Kiel), 1999
Gerilynn R. Moline, (D), Wisconsin, 1992, Cl
Michael E. Perkins, (D), Utah, 1998, Gt
Adjunct Professor:
Richard Allis, (N)
Lukas Baumgartner, (N)
S. Robert Bereskin, (D), California (Santa Barbara), 1969
John M. Harris, (D), Texas, Pv
James Kirkland, (N)
Bruce J. MacFadden, (D), Columbia, 1976, Pv
Matthew Mikulich, (N)
Michael L. Oristaglio, (D), Oxford, 1978, Ye
William G. Pariseau, (D), Minnesota, 1966, Nr
Phillip E. Wannamaker, (D), Utah, 1983, Ye
Emeritus:
William T. Parry, (D), Utah, 1961, Cg
M. Dane Picard, (D), Princeton, 1963, Gd

Dept of Mining Engineering (B,M,D) (2004)
135 South 1460 East
Room 313
Salt Lake City, UT 84112-0113
p. (801) 581-7198 f. (801) 585-5410
mineeng@mines.utah.edu
http://www.mines.utah.edu/mining
Enrollment (2001): B: 38 (6) M: 3 (2) D: 1 (1)
Chair:
Michael K. McCarter, (D), Utah, 1972, Nm
Professor:
William A. Hustrulid, (D), Minnesota, Nm
William G. Pariseau, (D), Minnesota, 1966, Nr
Associate Professor:
Felipe Calizaya, (D), Colorado Mines, 1985, Nm
Michael G. Nelson, (D), West Virginia, 1989, Nm
Adjunct Associate Professor:
Helmut H. Doelling, (D), Utah, 1964, Gg
Duane L. Whiting, (B), Utah, 1959, Hw
Adjunct Professor:
Robert G. Heers, (B), California, 1940, Nm
Krishna P. Sinha, (D), Minnesota, 1979, Nr
Mining Librarian:
H. Diane Christopherson, Oo

Energy & Geoscience Inst (2005)
423 Wakara Way
Suite 300
Salt Lake City, UT 84108
p. (801) 581-5126 f. (801) 585-3540
rlevey@egi.utah.edu
http://www.egi.utah.edu
Coordinator, Calgary Office: Kenneth L. Shaw
Program Coordinator: Laura A. Garon
Director:
Raymond A. Levey, (D), South Carolina, 1981, Eo
Business Development Manager:
David K. Curtiss, (M), South Carolina, 1995, Eo
Research Professor:
Joseph N. Moore, (D), Penn State, 1975, Gv
Robert M. Ressetar, (D), South Carolina, 1979, Go
Howard P. Ross, (D), Penn State, 1965, Ye
Phillip E. Wannamaker, (D), Utah, 1983, Ye
Research Scientist:
Michael C. Adams, (M), Utah, 1983, Cg
Richardson B. Allen, (D), Columbia, 1983, Go
Jeffrey B. Hulen, (B), Utah, 1969, Eg
Glenn W. Johnson, (D), South Carolina, 1997, Gq
Susan J. Lutz, (M), Utah, 1989, Gg
Daniel D. Schelling, (D), Colorado, 1989, Gc
John A. Stodt, (D), Utah, 1983, Yg
Senior Scientist:
James W. Collister, (D), Indiana, 1992, Cs
Elizabeth A. Dudley-Murphy, (D), Utah, 1996, Or
William N. Krebs, (D), California (Davis), 1977, Pm
Gregory N. Nash, (D), Utah, 1996, Or
Michal Nemcok, (D), Comenius (Czech), 1991, Gc
Paul J. Sikora, (D), Rutgers, 1989, Ps
Associate Professor:
Anthony C. Gary, (D), South Carolina, 1988, Gq
Gregory D. Nash, (D), Utah, 1996, Or
Research Staff:
Corinne E. Bacon, (M), California (Santa Cruz), 1994, Ca
Nicolas F. Dahdah, (M), South Carolina, 1992, Go
Christopher Kesler, (B), Utah, 1997, Or
Phaedra M. Kilbourn, (B), Utah, 1993, Ca
Theresa Lowe, (M), Utah State, 1999, Cm
Gosia Skowron, (B), Utah, 1997, Gc
Louise Spann, (B), Nevada, 1973, Gz
Jocelyn Vedder-Greene, (M), California (Santa Cruz), 1996, Ca
Research Assistant Professor:
Walter A. Jarman, (D), California (Santa Cruz), 1991, Ca
Peter E. Rose, (D), Utah, 1993, Np
Laboratory Director:
Michael A. Abrams, (M), S California, 1979, Ce

Utah State University

Dept of Geology (B,M,D) (2007)
4505 Old Main Hill
Logan, UT 84322-4505
p. (435) 797-1273 f. (435) 797-1588
Geology@usu.edu
http://www.usu.edu/geoldept
Enrollment (2006): B: 47 (5) M: 19 (2)
Department Head:
John W. Shervais, (D), California (Santa Barbara), 1979, Gi
Professor:
James P. Evans, (D), Texas A&M, 1987, Gc
W. David Liddell, (D), Michigan, 1979, Po
Associate Professor:
Donald W. Fiesinger, (D), Calgary, 1975, Gi
Susanne U. Janecke, (D), Utah, 1991, Gt
Thomas E. Lachmar, (D), Idaho, 1989, Hw
Joel L. Pederson, (D), New Mexico, 1999, Gm
Assistant Professor:
Carol M. Dehler, (D), New Mexico, 2001, Gs
Anthony R. Lowry, (D), Utah, 1994, Yg
Lecturer:
Susan K. Morgan, (M), Utah State, 1988, Gd

Weber State University

Dept of Geosciences (B) (2007)
2507 University Circle
Ogden, UT 84408-2507
p. (801) 626-7139 f. (801) 626-7445
ayonkee@weber.edu
http://weber.edu/geosciences/
Administrative Assistant: Susan Swider
Enrollment (2007): B: 62 (11)

Professor:
Jeffrey G. Eaton, (D), Colorado, 1987, Ps
Richard L. Ford, (D), California (Los Angeles), 1997, Gm
Marek Matyjasik, (D), Kent State, 1997, Hg
James R. Wilson, (D), Utah, 1976, Gz
William Adolph Yonkee, (D), Utah, 1990, Gc

Assistant Professor:
Michael W. Hernandez, (D), Utah, 2004, Oi

Emeritus:
Richard W. Moyle, (D), Iowa, 1963, Ps

Vermont

Castleton State College

Dept of Natural Sciences (Geology) (B) (2006)
Castleton State College
233 South Street
Castleton, VT 05735
p. (802) 468-1478 f. (802) 468-1170
grovert@castleton.edu
Enrollment (2006): B: 11 (2)

Professor:
Helen N. Mango, (D), Dartmouth, 1992, Cg

Associate Professor:
Timothy W. Grover, (D), Oregon, 1988, Gp

Johnson State College

Dept of Environmental & Health Sciences (A,B,M) (2007)
337 College Hill
Johnson, VT 05656-9464
p. (802) 635-1329 f. (802) 635-1461
Tania.Bacchus@jsc.edu
Administrative Assistant: Susan Mann
Enrollment (2006): B: 45 (15)

Professor:
Tania S. Bacchus, (D), Maine, 1993, GuOw
Leslie H. Kanat, (D), Cambridge, 1986, Gc

Lyndon State College

Dept of Meteorology (B) (2004)
Lyndonville, VT 05851
p. (802) 626-6254 f. (802) 626-9770
metchair@apollo.lsc.vsc.edu
Department Secretary: Brenda Sweet
Enrollment (1996): B: 73 (13) M: 70 (0)

Chair:
William A. Fingerhut, (D), Colorado State, 1980, Oa

Professor:
Bruce F. Berryman, (D), Wisconsin, 1974, Oa
Patrick T. Gannon, (D), Miami, 1977, Oa

Associate Professor:
Barry A. Richwien, (M), Wisconsin, 1977, Ow

Related Staff:
Peter M. Schmid, (B), Lyndon State, 1992, Ow

Middlebury College

Geology Dept (B) (2007)
Bicentennial Hall
Middlebury, VT 05753
p. (802) 443-5029 f. (802) 443-2072
dwest@middlebury.edu
www.middlebury.edu/academics/ump/majors/geol/
Administrative Assistant: Eileen Fahey
Enrollment (2006): B: 22 (13)

Chair:
David P. West, (D), Maine (Orono), 1993, Gc

Professor:
Raymond A. Coish, (D), W Ontario, 1977, Gi
Patricia L. Manley, (D), Columbia, 1989, Yr

Associate Professor:
Peter C. Ryan, (D), Dartmouth, 1994, Cl

Assistant Professor:
Jeffrey S. Munroe, (D), Wisconsin, 2001, Ge

Visiting Professor:
Thomas O. Manley, (D), Columbia, 1981, Op

Norwich University

Dept of Geology (B) (2005)
158 Harmon Drive
Northfield, VT 05663
p. (802) 485-2310 f. (802) 485-2333
westy@norwich.edu
http://www.norwich.edu
Enrollment (2001): B: 38 (10)

Chair:
David S. Westerman, (D), Lehigh, 1972, Gx

Assistant Professor:
Richard K. Dunn, (D), Delaware, 1998, Gs

Emeritus:
Frederick D. Larsen, (D), Massachusetts, 1972, Gl

Program in Environmental Science (B) (2004)
Northfield, VT 05663
p. (802) 485-2310 f. (802) 485-2580

Director:
David S. Westerman, (D), Lehigh, 1972, Gx

University of Vermont

Dept of Geology (B,M,D) (2004)
Burlington, VT 05405-0122
p. (802) 656-3396 f. (802) 656-0045
bdoolan@zoo.uvm.edu
Administrative Assistant: Patricia Frank
Enrollment (1998): B: 38 (7) M: 13 (4)

Chair:
Barry L. Doolan, (D), SUNY (Binghamton), 1970, Gx

Professor:
Eugene A. Cassell, (D), North Carolina, 1964, Hs
Charlotte J. Mehrtens, (D), Chicago, 1979, Gs
Rolfe S. Stanley, (D), Yale, 1962, Gc

Associate Professor:
Paul R. Bierman, (D), Washington, 1993, Gm
John C. Drake, (D), Harvard, 1967, Cl

Assistant Professor:
Keith A. Klepeis, (D), Texas, 1993, Gc
Andrea Lini, (D), ETH (Switzerland), 1994, Cs
Tracy Rushmer, (D), ETH (Switzerland), 1991, Cp

Staff:
Christine A. Massey, (M), Washington, 1995, Gg

Adjunct Professor:
Howard W. Jaffe, (D), Geneva, 1972, Gz
Stephen F. Wright, (D), Minnesota, 1988, Gc

Emeritus:
David P. Bucke, (D), Oklahoma, 1969, Gg
Allen S. Hunt, (D), Harvard, 1964, Pi

Dept of Plant & Soil Science (B,M,D) (2005)
Hills Building
105 Carrigan Drive
Burlington, VT 05405-0082
p. (802) 656-2630 f. (802) 656-4656
anne-marie.resnik@uvm.edu
Enrollment (1996): B: 65 (0) M: 13 (4) D: 2 (0)

Virginia

College of William & Mary

Dept of Geology (B) (2007)
PO Box 8795
McGlothlin-Street Hall
Williamsburg, VA 23187-8795
p. (757) 221-2440 f. (757) 221-2093
rhmacd@wm.edu
http://www.wm.edu/geology
Administrative Assistant: Carol Roe
Enrollment (2007): B: 53 (21)

Chair:
R. Heather Macdonald, (D), Wisconsin, 1984, GsOe
Professor:
Christopher M. Bailey, (D), Johns Hopkins, 1994, Gc
Associate Professor:
Gregory S. Hancock, (D), California (Santa Cruz), 1998, Gm
Rowan Lockwood, (D), Chicago, 2001, Po
Brent E. Owens, (D), Washington (St. Louis), 1992, Gz
Research Associate:
Carl R. Berquist, Jr., (D), William & Mary, 1986, Ou
David J. Campagna, (D), Purdue, 1990, Gc
Visiting Professor:
Linah N. Ababneh, (D), Arizona, 2006, GeCg
Karen M. Layou, (D), Georgia, 2007, Po
Emeritus:
Kenneth F. Bick, (D), Yale, 1958, Gr
Stephen C. Clement, (D), Cornell, 1964, Gz
Bruce K. Goodwin, (D), Lehigh, 1959, Gc
Gerald H. Johnson, (D), Indiana, 1965, Pg
Cooperating Faculty:
Karen K. Berquist, Oo
Director of Laboratories & Technical Support:
Linda D. Morse, (B), Virginia Tech, 1983, Ge

School of Marine Science (M,D) (2004)
Virginia Institute of Marine Science
P. O. Box 1346
Gloucester Point, VA 23062-1246
p. (804) 684-7103 f. (804) 684-7097
wells@vims.edu
http://www.vims.edu
Enrollment (1998): M: 64 (13) D: 81 (6)

Director/Dean:
L. Donelson Wright, (D), Louisiana State, 1970, Ou
Professor:
Rebecca M. Dickhut, (D), Wisconsin, 1989, Oc
Steven A. Kuehl, (D), North Carolina State, Ou
Jerome P-Y. Maa, (D), Florida, 1986, Op
John D. Milliman, (D), Miami, 1966, Ou
Associate Professor:
James E. Bauer, (D), Maryland, 1989, Oc
John M. Brubaker, (D), Oregon State, 1979, Op
Elizabeth A. Canuel, (D), North Carolina, 1992, Oc
Carl J. Friedrichs, (D), MIT/Woods Hole, 1993, On
Carl H. Hobbs, (D), Mississippi, 1997, Ou
Harry Wang, (D), Johns Hopkins, 1983, Op
Assistant Professor:
Courtney K. Harris, (D), Virginia, On
Jesse E. McNinch, (D), North Carolina, Ou

Emory & Henry College

Neff Center for Teacher Education (A) (2004)
Emory, VA 24327
p. (703) 944-3121
rediss@ehc.edu
Enrollment: No data reported since 1995

Director:
Jack L. Mason, (D), Virginia, 1976, Gg

George Mason University

Dept of Env Sci & Policy - Earth Sci Prg (B,M,D) (2006)
Mail Stop 5F2
4400 University Drive
Fairfax, VA 22030-4444
p. (703) 993-1043 f. (703) 993-1066
rmcbride@gmu.edu
http://mason.gmu.edu/~espp/
Administrative Assistant: Carol Lawrence
Enrollment: No data reported since 1995

Chair:
R. Christian Jones, (D), Wisconsin, 1980, Hs
Professor:
Richard J. Diecchio, (D), North Carolina, 1980, Gr
Robert M. Hazen, (D), Harvard, 1975, Om
Associate Professor:
Donald P. Kelso, (D), Hawaii, 1970, Ob
Randolph A. McBride, (D), Louisiana State, 1997, On
Assistant Professor:
Stephen S. Harlan, (D), New Mexico, 1992, Ym
Instructor:
Giuseppina Kysar Mattietti, (D), George Washington, 2001, Gi
Stacey Verardo, (D), CUNY, 1995, Pe
Adjunct Professor:
Walter W. Hays, (D), Washington, 1961, Yg
Derrick A. Kysar, (M), George Washington, 1994, Pg
Edward C. Parsons, (D), Oxford, 1995, Ob
Graduate Coordinator:
Ronald E. Stewart, (D), Oregon State, 1970, Oo
Other:
Andrea C. Englert, (B), George Mason, 2001, Gg
Cooperating Faculty:
Julia Nord, (D), CUNY (Brooklyn), 1989, Cp

Hampton University

Center for Marine & Coastal Env Studies (B) (2006)
100 Queen Street
Hampton, VA 23668
p. (757) 727-5783 f. (757) 727-5740
george.burbanck@hamptonu.edu
Administrative Assistant: Evelyn Ray
Enrollment (2005): B: 30 (1)

Chair:
George P. Burbanck, (D), Delaware, 1981, On
Professor:
Benjamin E. Cuker, (D), North Carolina State, 1981, Ob
Associate Professor:
Robert A. Jordan, (D), Michigan, 1970, Ob
Assistant Professor:
Deidre M. Gibson, (D), Georgia, 2000, Ob
Research Associate:
Eric Wooden, (M), Virginia Inst of Marine Sci, 2000, Ob
Adjunct Faculty:
Emory Morgan, (M), Oe

J. Sargeant Reynolds Community College

Geology Program (A) (2005)
P.O. Box 85622
Richmond, VA 23285-5622
p. (804) 371-3242 f. (804) 371-3311
rmoore@jsr.vccs.edu
Division Secretary: Mary Dolliver

Associate Professor:
David H. Walz, (M), Texas, 1974, Hw

Adjunct Professor:
Robert H. Stebbins, (M), Columbia, 1955, Oo

James Madison University

Dept of Geology & Environmental Science (B) (2006)
MSC 6903
Memorial Hall
Harrisonburg, VA 22807
p. (703) 568-6130 f. (703) 568-8058
ulansksl@jmu.edu
http://www.jmu.edu/geology/
Department Secretary: Sandra Delawder
Enrollment (1995): B: 65 (21)

Head:
Stanley L. Ulanski, (D), Virginia, 1977, Og

Professor:
Roddy V. Amenta, (D), Bryn Mawr, 1971, Gc
Lynn S. Fichter, (D), Michigan, 1972, Gr
Lance E. Kearns, Delaware, Gz
William C. Sherwood, (D), Lehigh, 1961, So

Associate Professor:
Steven J. Baedke, (D), Indiana, 1998, Hy
Lewis S. Eaton, (D), Virginia, 1999, Gm
Eric J. Pyle, (D), Georgia, 1995, Oe
Kristen E. St. John, (D), Ohio State, 1998, Ou

Assistant Professor:
Steven J. Whitmeyer, (D), Boston, Gc

Instructor:
Michael J. Harris, (D), W Ontario

Lynchburg College

Environmental Science (B) (2007)
1501 Lakeside Dr.
Lynchburg, VA 24501
p. 434-544-8369 f. 434-544-8646
perault@lynchburg.edu
http://www.lynchburg.edu/envsci.xml
Enrollment (2007): B: 32 (11)

Associate Professor:
David R. Perault, (D), Oklahoma, 1998, Ge

Mary Washington College

Dept of Geography (B) (2007)
1301 College Avenue
Fredericksburg, VA 22401-5358
p. (540) 654-1470 f. (540) 654-1074
shanna@umw.edu
http://www.mwc.edu/geog
Enrollment (2006): B: 80 (0)

Chair:
Joseph W. Nicholas, (D), Georgia, 1991, Oy

Associate Professor:
Donald N. Rallis, (D), Penn State, 1992, Oo

Assistant Professor:
Dawn S. Bowen, (D), Queen's, 1998, Oo
Erin H. Fouberg, (D), Nebraska, 1997, Oo
Stephen P. Hanna, (D), Kentucky, 1997, Or
Farhang Rouhani, (D), Arizona, 2001, Oo

Northern Virginia Comm Coll-Loudoun

Natural and Applied Sci Div - Dept of Geology (2005)
1000 Harry Flood Byrd Highway
Sterling, VA 20164
p. (703) 450-2575 f. (703) 450-2665
wbour@nvcc.edu
http://www.nvcc.edu/loudoun/scidiv/GOL/

Professor:
Richard R. Graus, (D), Rochester, 1971, Pe
Kenneth A. Rasmussen, (D), North Carolina, 1989, Pi
Paul D. Vespucci, (D), George Washington, 1988, Ca

Old Dominion University

Dept of Ocean, Earth & Atompheric Sci (B,M,D) (2004)
4600 Elkhorn Avenue
Norfolk, VA 23529-0496
p. (757) 683-4285 f. (757) 683-5303
rzimmerm@odu.edu
http://www.ocean.odu.edu
Enrollment (2000): B: 30 (0) M: 28 (5) D: 28 (6)

Graduate Program Director:
David J. Burdige, (D), California (San Diego), 1983, Oc

Professor:
Larry P. Atkinson, (D), Dalhousie, 1972, Op
Gregory A. Cutter, (D), California (Santa Cruz), 1982, Oc
Dennis A. Darby, (D), Wisconsin, 1971, Ou
Ann Gargett, (D), British Columbia, 1970, Op
Chester E. Grosch, (D), Stevens Inst of Tech, 1967, Op
Eileen E. Hofmann, (D), North Carolina State, 1980, Op
John M. Klinck, (D), Iowa, 1980, Op
George Oertel, (D), Iowa, 1971, Ou
Thomas Royer, (D), Texas A&M, 1969, Op
Joseph H. Rule, (D), Missouri, 1972, Ou
Donald J. P. Swift, (D), North Carolina, 1964, Ou
George T. F. Wong, (D), MIT/Woods Hole, 1976, Oc

Associate Professor:
Frederdick C. Dobbs, (D), Florida State, 1987, Ob
Ronald E. Johnson, (D), Oregon State, 1971, Op
John R. McConaugha, (D), S California, 1977, Ob
Arnoldo Valle-Levinson, (D), SUNY (Stony Brook), 1992, Op
G. Richard Whittecar, Jr., (D), Wisconsin, 1979, Ou

Assistant Professor:
Alexander Bochdansky, (D), Memorial, 1997, Ob
Margaret Mulholland, (D), Maryland, 1998, Ob
Nora Noffke, (D), Oldenburg, 1997, Ge

Radford University

Dept of Geology (B) (2007)
Box 6939
Radford University
Radford, VA 24142
p. (540) 831-5652 f. (540) 831-5732
geology@radford.edu
http://www.radford.edu/~geol-web
Secretary: Sharon Holloway
Enrollment (2007): B: 57 (12)

Chair:
Stephen W. Lenhart, (D), Kentucky, 1985, Pi

Professor:
Kimbell L. Knight, (D), Iowa, 1985, Gd
Parvinder S. Sethi, (D), North Carolina State, 1994, Gz
Chester F. Watts, (D), Purdue, 1983, Ng

Associate Professor:
Jonathan L. Tso, (D), Virginia Tech, 1987, Gp

Research Associate:
Judy Ehlen, (D), Birmingham, 1990, Gm

Adjunct Faculty:
Rhett B. Herman, (D), Montana State, 1996, Yg
Research Faculty:
Robert C. Whisonant, (D), Florida State, 1967, Ga
Laboratory Director:
Amanda H. Patterson, (M), Mississippi, 2005, Gx

Randolph-Macon College

Environmental Studies Program (B) (2007)
P.O. Box 5005
Ashland, VA 23005
p. (804)752-3745 f. (804)752-4724
mfenster@rmc.edu
http://www.rmc.edu/directory/academics/evst
Associate Professor:
Michael S. Fenster, (D), Boston, 1995, On

Tidewater Community College

Geophysical Sciences Dept (A) (2004)
Virginia Beach, VA 23456
p. (804) 427-7189 f. (804) 427-7326
geoinfntry@aol.com
Enrollment: No data reported since 1995
Chair:
Michael E. Lyle, (M), East Carolina, 1984, On
Associate Professor:
James F. Coble, (M), East Carolina, 1977, Gi
Instructor:
Roseann J. Carlson, (M), Washington, 1969, Pg
Gary Lennon, (M), Old Dominion, 1983, Og
Terri Matthews, (M), William & Mary, 1978, Cg
Azam Tabrizi, (M), London, 1978, Pm
Sandra Weiterman, (M), Florida State, 1989, Pm
Adjunct Professor:
Debbie Duffey, (M), Old Dominion, 1989, Gr
Michelle Fults, (M), Old Dominion, 1980, Hy
Miles Pocta, (M), Old Dominion, 1987, Ou
Robert Rose, (M), Old Dominion, 1979, Ow
Technical Staff:
Arthur C. Esenberg, (M), Old Dominion, 1977, Og

University of Mary Washington

Dept of Earth and Environmental Sciences (B) (2008)
1301 College Avenue
Fredericksburg, VA 22401-5358
p. (540) 654-1016 f. (540) 654-1081
gwoodwel@umw.edu
http://www.umw.edu/cas/envisci_geol/
Enrollment (2007): B: 73 (18)
Chair:
Grant R. Woodwell, (D), Yale, 1985, Gc
Professor:
Michael L. Bass, (D), Virginia Tech, 1976, Oo
Associate Professor:
Jodie L. Hayob, (D), Michigan, 1994, Gp
Assistant Professor:
Ben O. Kisila, (D), Arkansas, 2002, Hs
Neil E. Tibert, (D), Massachusetts, 2002, Pm
Charles Whipkey, (D), Pittsburgh, 1999, Cg
Adjunct Professor:
Douglas C. Albach, (M), Louisiana State, Ge
John W. Donley, (M), Virginia Tech, Oo
Emeritus:
Robert L. McConnell, (D), California (Santa Barbara), 1972, Ge

University of Virginia

Blandy Experimental Farm & the Orland E. White Arboretum (2004)
Clark Hall
Charlottesville, VA 22903
p. (540) 837-1758
Director:
Edward F. Connor, (D), Florida, 1979, Oo

Dept of Environmental Sciences (B,M,D) (2006)
Clark Hall
Charlottesville, VA 22903
p. (804) 924-7761 f. (804) 982-2137
jcz@virginia.edu
http://www.evsc.virginia.edu
Administrative Assistant: Paulette Racer
Enrollment: No data reported since 1995
Chair:
Bruce P. Hayden, (D), Chicago, 1968, Oa
Professor:
Robert Dolan, (D), Louisiana State, 1965, Gm
James N. Galloway, (D), California (San Diego), 1972, Oa
Janet S. Herman, (D), Penn State, 1982, Cl
George M. Hornberger, (D), Stanford, 1970, Hq
Alan D. Howard, (D), Johns Hopkins, 1970, Gm
Stephen A. Macko, (D), Texas, 1981, Co
Aaron L. Mills, (D), Cornell, 1975, So
Herman H. Shugart, Jr., (D), Georgia, 1971, Og
Patricia L. Wiberg, (D), Washington, 1987, Og
Henry M. Wilbur, (D), Michigan, 1971, Oo
Joseph C. Zieman, (D), Miami, 1970, Ob
Associate Professor:
Robert Davis, (D), Delaware, 1988, Oa
Wallace E. Reed, (D), Chicago, 1967, Or
David E. Smith, (D), Texas A&M, 1982, Ob
Thomas M. Smith, (D), Tennessee, 1982, Og
Research Associate Professor:
Linda K. Blum, (D), Cornell, 1980, Sb
Assistant Professor:
Jose D. Fuentes, (D), Guelph, 1991, Oa
Karen McGlathery, (D), Cornell, 1992, Ob
Jeff P. Raffensperger, (D), Johns Hopkins, 1993, Hw
Research Associate:
Peter Berg, (D), Tech (Denmark), 1988, So
Jack Cosby, (D), Virginia, 1982, Hs
William Emanuel, (D), Oklahoma State, 1975, Cg
R. Michael Erwin, (D), Maryland, 1975, Oo
William Keene, (M), Virginia, 1981, Oa
Patrick Michaels, (D), Wisconsin, 1979, Oa
Jennie L. Moody, (D), Michigan, 1986, Oa
John H. Porter, (D), Virginia, 1988, Or
G. Carleton Ray, (D), Columbia, 1960, Ob
William Rickards, (D), Miami, 1970, Ob
Professor Emeritus:
Bruce W. Nelson, (D), Illinois, 1955, Gs
Emeritus:
William F. Ruddiman, (D), Columbia, 1969, Gu

Global Systems Analysis Program (2004)
Clark Hall
Charlottesville, VA 22903
p. (804) 914-7642
Director:
Herman H. Shugart, Jr., (D), Georgia, 1971, Og

Program of Interdisciplinary Res in Contaminant Hydrogeology (2004)
Clark Hall
291 McCormick Road
Charlottesville, VA 22904

p. (804) 924-0553 f. (804) 982-2137
http://atlantic.evsc.virginia.edu/~alm7d/pirch/

Director:
Janet S. Herman, (D), Penn State, 1982, Cl

Professor:
George M. Hornberger, (D), Stanford, 1970, Hq
Aaron L. Mills, (D), Cornell, 1975, So

Associate Professor:
Teresa B. Culver, (D), Cornell, 1991, Hy
Erik J. Fernandez, (D), California (Berkeley), 1989, Oo
Roseanne M. Ford, (D), Pennsylvania, 1989, Oo
James A. Smith, (D), Princeton, 1992, Hy

Assistant Professor:
Susan E. Burns, (D), Georgia Tech, 1997, Hy

Shenandoah Watershed Acidification Study (2004)
Clark Hall
Charlottesville, VA 22903
p. (804) 324-1303

Director:
James N. Galloway, (D), California (San Diego), 1972, Oa

Virginia Coast Reserve/Long Term Ecological Research (2004)
Clark Hall
Charlottesville, VA 22903
p. (804) 982-3809

Director:
Bruce P. Hayden, (D), Chicago, 1968, Oa

Virginia State Climatology Office (2004)
Clark Hall
Charlottesville, VA 22903
p. (804) 324-0548

Director:
Patrick J. Michaels, (D), Wisconsin, 1979, Oa

University of Virginia College, Wise

Dept of Natural Science (2004)
1 College Avenue
Wise, VA 24293
p. (540) 328-0200 f. (540) 328-0247
vwd@uvawise.edu
Department Secretary: Brenda Whitaker
Enrollment (1998): B: 23 (7)

Chair:
Philip C. Shelton, (D), Purdue, 1966, Oo

Assistant Professor:
Robert D. VanGundy, (M), North Carolina, 1983, Gs

Virginia Polytechnic Inst & State Univ

Dept of Civil Engineering (M,D) (2004)
Blacksburg, VA 24061
p. (703) 231-6635 f. (703) 231-7532
knocke@vt.edu
Administrative Assistant: Vickie Graham
Enrollment (1995): M: 9 (6) D: 4 (1)

Head:
William R. Knocke, (D), Missouri, 1978, Oo

Associate Professor:
Panos Diplas, (D), Minnesota, 1985, Gs
G. V. Loganathan, (D), Purdue, 1981, Hg
Mark A. Widdowson, (D), Auburn, 1987, Hw

Dept of Crop & Soil Environmental Sci (B,M,D) (2004)
240 Smyth Hall
Blacksburg, VA 24061-0404
p. (540) 231-6305 f. (540) 231-3431
jrhall3@vt.edu
http://sudan.cses.vt.edu
Department Secretary: Nancy Shields
Enrollment (2001): B: 98 (0) M: 14 (0) D: 7 (0)

Head:
John R. Hall, III, (D), Ohio State, 1971, So

Professor:
Marcus M. Alley, (D), Virginia Tech, 1975, Sc
James C. Baker, (D), Virginia Tech, 1978, Sd
Walter L. Daniels, (D), Virginia Tech, 1985, Sd
Stephen J. Donohue, (D), Purdue, 1974, Sc
Gregory K. Evanylo, (D), Georgia, 1982, Sc
Charles Hagedorn, (D), Iowa, 1974, Sb
Gregory L. Mullins, (D), Purdue, 1985, Sc
Raymond B. Reneau, (D), Florida, 1969, Sc
Lucian W. Zelazny, (D), Virginia Tech, 1970, Sc

Associate Professor:
Duane F. Berry, (D), Michigan State, 1984, Sb
Matthew J. Eick, (D), Delaware, 1995, Sc
Naraine Persaud, (D), Florida, 1978, Sp

Assistant Professor:
John M. Galbraith, (D), Cornell, 1997, Sp
Carl E. Zipper, (D), Virginia Tech, 1986, Oe

Adjunct Professor:
Domy C. Adriano, (D), Kansas State, 1970, Sc
V. C. Baligar, (D), Mississippi State, 1975, Sc
Pamela J. Thomas, (D), Virginia Tech, 1998, Sd

Dept of Geography (B,M) (2004)
115 Major Williams
Blacksburg, VA 24061-0115
p. (540) 231-6886 f. (540) 231-6367
geog@vt.edu
Department Secretary: Jane Price

Head:
James B. Campbell, (D), Kansas, 1976, Or

Professor:
Charles M. Good, (D), Chicago, 1970, Oo
Robert W. Morrill, (D), Clark, 1973, Oo
Bonham C. Richardson, (D), Wisconsin, 1970, Oo

Associate Professor:
Susan R. Brooker-Gross, (D), Illinois, 1977, Ou
Laurence W. Carstensen, (D), North Carolina, 1981, Oo
Lawrence S. Grossman, (D), Australian National, 1979, Oo
Gerard O. Toal, (D), Syracuse, 1989, Oo

Dept of Geosciences (B,M,D) (2007)
4044 Derring Hall
Blacksburg, VA 24061
p. (540) 231-6521 f. (540) 231-3386
wilcar@vt.edu
http://www.geos.vt.edu
Administrative Assistant: Carolyn S. Williams
Enrollment (2006): B: 91 (19) M: 14 (10) D: 42 (7)

Chair:
Robert J. Tracy, (D), Massachusetts, 1975, Gp

Research Professor:
Ross J. Angel, (D), Cambridge, 1986, Gz
Robert P. Lowell, (D), Oregon State, 1972, Yr

Professor:
Robert J. Bodnar, (D), Penn State, 1985, Cg
Patricia M. Dove, (D), Princeton, 1991, Cl
Kenneth A. Eriksson, (D), Witwatersrand, 1977, Gs
Michael F. Hochella, Jr., (D), Stanford, 1981, Cl
Scott D. King, (D), Caltech, 1990, Yg
Michal J. Kowalewski, (D), Arizona, 1995, Py
Richard D. Law, (D), London, 1981, Gc

J. Fred Read, (D), W Australia, 1971, Gs
J. Donald Rimstidt, (D), Penn State, 1979, Cl
Nancy L. Ross, (D), Arizona State, 1985, Gz
A. Krishna Sinha, (D), California (Santa Barbara), 1969, Gt

Associate Professor:
Thomas J. Burbey, (D), Nevada (Reno), 1994, Hw
John A. Hole, (D), British Columbia, 1993, Ye
Madeline E. Schreiber, (D), Wisconsin, 1999, Hw
James A. Spotila, (D), Caltech, 1998, Gt
Chester J. Weiss, (D), Texas A&M, 1998, Ye
Shuhai Xiao, (D), Harvard, 1998, Pi

Research Assistant Professor:
Martin C. Chapman, (D), Virginia Tech, 1998, Ys

Assistant Professor:
Barbara M. Bekken, (D), Stanford, 1990, Eg
Jacob O. Sewall, (D), California (Santa Cruz), 2004, Oa
Ying Zhou, (D), Princeton, 2004, Ys

Sr. Research Associate:
Jing Zhao, (D), Chinese Academy of Sci, 1997

Research Scientist:
Erin R. Kraal, (D), California (Santa Cruz), 2006

Research Associate:
Nizhou Han, (D), Iowa State, 1996, Sc

Instructor:
John A. Chermak, (D), Virginia Tech, 1989, Cg

Adjunct Professor:
James S. Beard, (D), California (Davis), 1985, Gi
Nicholas C. Fraser, (D), Aberdeen, 1984, Pv
William S. Henika, (M), Virginia, 1969, Ng
Maureen M. Julian, (D), Cornell, 1966, Gz
Matthew J. Mikulich, (D), Utah, 1971, Ye
Chester F. Watts, (D), Purdue, 1983, Ngg

Emeritus:
Richard K. Bambach, (D), Yale, 1969, Po
F. Donald Bloss, (D), Chicago, 1951, Gz
G. A. Bollinger, (D), St. Louis, 1967, Ys
Cahit Coruh, (D), Istanbul, 1970, Ye
John K. Costain, (D), Utah, 1960, Yg
James R. Craig, (D), Lehigh, 1965, Eg
Gordon C. Grender, (D), Penn State, 1960, Go
David A. Hewitt, (D), Yale, 1970, Cp
Wallace D. Lowry, (D), Rochester, 1943, Gc
Dewey M. McLean, (D), Stanford, 1969, Pe
Paul H. Ribbe, (D), Cambridge, 1963, Gz
Edwin S. Robinson, (D), Wisconsin, 1964, Yg
J. Arthur Snoke, (D), Yale, 1969, Ys

Educational Administrator for Outreach:
S. Llyn Sharp, (M), Virginia Tech, 1990, Oe

Cooperating Faculty:
Lucian W. Zelazny, (D), Virginia Tech, 1970, So

Geosciences Librarian:
Edward Lener, (M), Oo

Virginia Western Community College

Geology Dept (A) (2004)
3095 Colonial Avenue
Roanoke, VA 24038
p. (540) 857-7273 f. (540) 857-7544
abalog-szabo@vw.vccs.edu
Enrollment: No data reported since 1995

Head:
Roy G. Miles, (D), Virginia Tech, 1976, Gg

Associate Professor:
Anna Balog-Szabo, (D), Virginia Tech, 1996, Gs

Washington & Lee University

Dept of Geology (B) (2004)
Lexington, VA 24450
p. (540) 463-8800 f. (540) 463-8142
woodruff.s@wlu.edu
Administrative Assistant: Deborah Hubbard
Enrollment (2000): B: 3 (0)

Head:
Frederic L. Schwab, (D), Harvard, 1968, Gs

Professor:
Edgar W. Spencer, (D), Columbia, 1957, Gc

Associate Professor:
David J. Harbor, (D), Colorado State, 1990, Gm

Assistant Professor:
Christopher D. Connors, (D), Princeton, 1999, Gc
Lisa Greer, (D), Miami, 2001, Pe
Elizabeth P. Knapp, (D), Virginia, 1997, Cl

Technician:
Robert C. Thren, (A), Penn State, 1970, Oo

Geology Librarian:
Hugh Blackmer, Oo

Washington

Central Washington University

Dept of Geological Sciences (B,M) (2007)
400 East University Way
Ellensburg, WA 98926-7418
p. (509) 963-2701 f. (509) 963-2821
chair@geology.cwu.edu
http://www.geology.cwu.edu
Administrative Assistant: Vicki Potts
Enrollment (2006): B: 80 (0) M: 19 (3)

Chair:
Wendy A. Bohrson, (D), California (Los Angeles), 1993, Gvi

Professor:
Lisa L. Ely, (D), Arizona, 1992, Gm
Jeffrey Lee, (D), Stanford, 1990, Gc
M. Meghan Miller, (D), Stanford, 1987, Gt
Charles M. Rubin, (D), Caltech, 1990, Gt

Senior Scientist:
Frank Ramos, (D), California (Los Angeles), 2000, Cc

Associate Professor:
Carey A. Gazis, (D), Caltech, 1994, Cs
Timothy I. Melbourne, (D), Caltech, 1998, Ys

GPS Data Analyst:
Marcelo Santillan, (M), Memphis, 2003, Yd

Assistant Professor:
Beth Pratt-Sitaula, (D), California (Santa Barbara), 2005, Oe

Scientific Instructional Tech Supervisor:
Nick Zentner, (M), Idaho State, 1989, Gg

Adjunct Professor:
Marie A. Ferland, (D), Sydney, 1991, Gu

Emeritus:
James R. Hinthorne, (D), California (Santa Barbara), 1974, Ca

Eastern Washington University

Dept of Geology (B,M) (2006)
130 Science Building
Cheney, WA 99004-2439
p. (509) 359-2286 f. (509) 359-4386
geol@ewu.edu
http://www.ewu.edu/geology
Administrative Assistant: Carolyn L. Harbolt
Enrollment (2004): B: 25 (6) M: 1 (0)

Professor:
John P. Buchanan, (D), Colorado State, 1985, Hw
Ernest H. Gilmour, (D), Montana, 1967, Pi
Linda B. McCollum, (D), SUNY (Binghamton), 1980, Po

Associate Professor:
Richard L. Orndorff, (D), Kent State, 1994, Ge
Jennifer A. Thomson, (D), Massachusetts, 1992, Gp
Assistant Professor:
P. Ted Doughty, (D), Queen's, 1995, Gc
Emeritus:
James I. Hoffman, (D), Michigan State, 1969, Em
Mohammed Ikramuddin, (D), Miami, 1974, Ca
Eugene P. Kiver, (D), Wyoming, 1968, Gl
James R. Snook, (D), Washington, 1962, Gc
William K. Steele, (D), Case Western, 1970, Yg
Scientific Instructional Tech:
Charles R. Strout, (B), Montana Min Sci & Tech, 1978, Gg

Grays Harbor College

Dept of Geology & Oceanography (A) (2004)
Aberdeen, WA 98520
p. (206) 538-4299 f. (206) 532-6716
dsamuels@ghc.ctc.edu
Enrollment (1997): A: 0 (1) B: 2 (0)
Chair:
James B. Phipps, (D), Oregon State, 1974, Gu

Green River Community College

Dept of Geology (A) (2007)
12401 S.E. 320th
Auburn, WA 98092-3699
p. (253) 833-9111 (Ext. 4324) f. (253) 288-3464
bfilson@greenriver.edu
http://www.instruction.greenriver.edu/filson
Instructor:
Robert H. Filson, (D), Washington, 1979, Oe
Kathryn A. Hoppe, (D), Princeton, 1999, Gg

Lower Columbia College

Earth Sciences (Natural Sciences Dept.) (A) (2006)
P.O. Box 3010
Longview, WA 98632
p. (360) 442-2883 f. (360) 442-2659
dcordero@lcc.ctc.edu
http://lowercolumbia.edu/Common/programs---degrees/transfer-degrees/earth-sciences/
Enrollment (2001): A: 8 (4)
Professor:
David I. Cordero, (M), Portland State, 1997, Og

Olympic College

Mathematics, Eng, Sciences, and Health Div (A) (2007)
Bremerton, WA 98310
p. (360) 475-7700 f. (360) 475-7705
sdigby@oc.ctc.edu
Enrollment: No data reported since 1995
Professor:
Susan A. Digby, (D), California (Los Angeles), 2004, Oy
Instructor:
Steve E. Macias, (M), Washington, 1996, Gg

Pacific Lutheran University

Dept of Geosciences (B) (2007)
12180 S. Park
Tacoma, WA 98447
p. (253) 535-7378 f. (253) 536-5055
geos@plu.edu
http://www.plu.edu/~geos/
Enrollment (2007): B: 24 (8)
Professor:
Steven R. Benham, (D), Indiana, 1979, Gd
Duncan Foley, (D), Ohio State, 1978, Hy
Brian E. Lowes, (D), Washington, 1972, Gc
Jill M. Whitman, (D), California (San Diego), 1989, Ou
Associate Professor:
Rosemary McKenney, (D), Penn State, 1997, Gm

Seattle Central Community College

Div of Science & Mathematics (2007)
1701 Broadway
Seattle, WA 98122
p. (206) 587-3858 f. (206) 344-4390
jhull@sccd.ctc.edu
http://seattlecentral.edu/sci-math/
Administrative Assistant: Carole Schmidt
Instructor:
Katie Gagnon, (D), California (San Diego), 2007, Oug
Joseph M. Hull, (D), Rochester, 1988, Gc

Shoreline Community College

Science Div (A) (2007)
16101 Greenwood Avenue North
Seattle, WA 98133
p. (206) 546-4659 f. (206) 546-5865
ebaer@shoreline.edu
http://www.shoreline.edu/science/geology.htm
Enrollment: No data reported since 1995
Professor:
Emma Baer, (D), Gg

Tacoma Community College

Dept of Earth Sciences (A) (2005)
6501 South 19th Street
Tacoma, WA 98466
p. (253) 566-5060 f. (253) 566-5202
rhitz@tcc.ctc.edu
Enrollment: No data reported since 1995
Instructor:
Ralph B. Hitz, (D), California (Santa Barbara), 1997, Og

University of Puget Sound

Dept of Geology (B) (2005)
1500 N. Warner Street
Tacoma, WA 98416
p. (253) 879-3129 f. (253) 879-3352
mvalentine@ups.edu
http://www.ups.edu/geology/home.shtml
Administrative Assistant: Kathy Samms
Enrollment (2005): B: 21 (4)
Chair:
Michael J. Valentine, (D), Massachusetts, 1990, Ym
Professor:
Barry Goldstein, (D), Minnesota, 1985, Gl
Associate Professor:
Tepper Jeffrey, (D), Washington, 1991, Gi
Assistant Professor:
Travis Horton, (D), Stanford, 2004, Ge
Instructor:
Ken Clark, (M), W Washington, 1989, Gg
Emeritus:
Norman R. Anderson, (D), Utah, 1965, Gm
Z. Frank Danes, (D), Charles (Prague), 1949, Yv
Albert A. Eggers, (D), Dartmouth, 1972, Gv
J. Stewart Lowther, (D), Michigan, 1957, Pb
Cooperating Faculty:
Elizabeth Knight, (N), Oo

University of Washington

Dept of Atmospheric Sciences (B,M,D) (2004)
Box 351640
Seattle, WA 98195
p. 206-543-4250 f. 206-543-0308
durrand@atmos.washington.edu
http://www.atmos.washington.edu

Chair:
Dennis L. Hartmann, (D), Princeton, Oa
Research Professor:
Robert A. Brown, (D), Washington, Oa
David S. Covert, (D), Washington, Oa
Thomas C. Grenfell, (D), Washington, Oa
Dean A. Hegg, (D), Washington, Oa
Gary A. Maykut, (D), Washington, Oa
Professor:
Marcia B. Baker, (D), Washington, Oa
David S. Battisti, (D), Washington, Oa
Christopher S. Bretherton, (D), MIT, Oa
Dale R. Durran, (D), MIT, Oa
Peter V. Hobbs, (D), Imperial Coll (UK), Oa
James R. Holton, (D), MIT, Oa
Robert A. Houze, (D), MIT, Oa
Clifford F. Mass, (D), Washington, Oa
Peter B. Rhines, (D), Trinity Coll (Cambridge), Op
Edward S. Sarachik, (D), Brandeis, Oa
John Michael Wallace, (D), MIT, Oa
Stephen G. Warren, (D), Harvard, Yg
Research Associate Professor:
Bradley F. Smull, (D), Washington, Oa
Research Assistant Professor:
Igor V. Kamenkovich, (D), MIT, Oa
Mark T. Stoelinga, (D), Washington, Oo
Sandra E. Yuter, (D), Washington, Oa
Assistant Professor:
David C. Catling, (D), Oxford, Oa
Qiang Fu, (D), Utah, Oa
Gregory J. Hakim, (D), SUNY (Albany), Oa
Lyatt Jaegle, (D), Caltech, Oa
Research Associate:
Bonnie Light, (D), Washington, Oa
Robert Wood, (D), Manchester, Oa
Ping Zhu, (D), Miami, Oa
Affiliate Professor:
D. Edmunds Harrison, (D), Harvard, Oa
Kristina B. Katsaros, (D), Washington, Oa
James E. Overland, (D), New York, Oa
William J. Plant, (D), Purdue, Oa
Lawrence F. Radke, (D), Washington, Oa
Adjunct Associate Professor:
D. Andrew Rothrock, (D), Cambridge, Oa
Adjunct Professor:
Robert E. Breidenthal, (D), Caltech, Oa
Richard H. Gammon, (D), Harvard, Oa
Daniel A. Jaffe, (D), Washington, Oa
Ka-Kit Tung, (D), Harvard, Oa
Emeritus:
Franklin I. Badgley, (D), New York, Oa
Joost A. Businger, (D), State (Utrecht), Oa
Robert J. Charlson, (D), Washington, Oa
Robert G. Fleagle, (D), New York, Oa
Halstead Harrison, (D), Stanford, Oa
Edward R. LaChapelle, (D), Puget Sound, Oa
Conway B. Leovy, (D), MIT, Oa
Richard J. Reed, (D), MIT, Oa
James E. Tillman, (M), MIT, Oa
Norbert Untersteiner, (D), Innsbruck, Oa
Affiliate Assistant Professor:
M. Joan Alexander, (D), Colorado, Oa
Nathan J. Mantua, (D), Washington, Oa
Philip Mote, (D), Washington, Oo
Affiliate Associate Professor:
Timothy S. Bates, (D), Washington, Oa
Nicholas A. Bond, (D), Washington, Oa
Shuyi S. Chen, (D), Penn State, Oa
Bradley R. Colman, (D), MIT, Oa
Steven J. Ghan, (D), MIT, Oa

Dept of Earth & Space Sciences (B,M,D) (2008)
070 Johnson Hall
Box 351310
Seattle, WA 98195-1310
p. (206) 543-1190 f. (206) 543-0489
advising@ess.washington.edu
http://www.ess.washington.edu
Enrollment (2007): B: 91 (38) M: 12 (12) D: 66 (14)

Professor:
George W. Bergantz, (D), Johns Hopkins, 1988, Gi
John R. Booker, (D), California (San Diego), 1968, YmGt
Joanne (Jody) Bourgeois, (D), Wisconsin, 1980, GsrGh
J. Michael Brown, (D), Minnesota, 1980, Gy
Roger Buick, (D), W Australia, 1986, PeoPm
Darrel S. Cowan, (D), Stanford, 1972, Gct
Kenneth C. Creager, (D), California (San Diego), 1984, Ys
Alan R. Gillespie, (D), Caltech, 1982, GmOlr
Bernard Hallet, (D), California (Los Angeles), 1975, GmOl
Robert H. Holzworth, (D), California (Berkeley), 1977, YxXy
Heidi B. Houston, (D), Caltech, 1987, YsGt
I. Stewart (Stu) McCallum, (D), Chicago, 1968, Gi
David R. Montgomery, (D), California (Berkeley), 1991, Gm
Bruce K. Nelson, (D), California (Los Angeles), 1985, CcGi
Charles A. Nittrouer, (D), Washington, 1978, GuYr
John E. Vidale, (D), Caltech, 1986, YsGt
Edwin D. Waddington, (D), British Columbia, 1981, Ol
Peter D. Ward, (D), McMaster, 1976, Po
Stephen G. Warren, (D), Harvard, 1973, Oal
Robert M. Winglee, (D), Sydney, 1984, XyYx
Research Professor:
Patricia M. Anderson, (D), Brown, 1982, Ple
Howard B. Conway, (D), Canterbury (NZ), 1986, Ol
Gonzalo Hernandez, (D), Rochester, 1962, Yx
James A. Mercer, (D), Washington, 1983, GuYr
Dale P. Winebrenner, (D), Washington, 1985, Olr
Research Associate Professor:
Evan H. Abramson, (D), MIT, 1985, Gy
Paul A. Bodin, (D), Colorado, 1992, Ys
Victor C. Kress, (D), California (Berkeley), 1990, Cp
Michael P. McCarthy, (D), Washington, 1988, XyYx
Robert I. Odom, (D), Washington, 1980, YrGu
Jaakko K. Putkonen, (D), Washington, 1997, GmCcl
Ronald S. Sletten, (D), Washington, 1995, OsSb
Adjunct Associate Professor:
Katherine Sian Davies-Vollum, (D), Oxford (UK), 1994, GsrPe
Associate Professor:
Elizabeth (Liz) Nesbitt, (D), California (Berkeley), 1982, Pim
Gerard H. Roe, (D), MIT, 1999, OaGmOl
Eric J. Steig, (D), Washington, 1996, CsOl
John O.H Stone, (D), Cambridge (UK), 1986, Cca
Research Assistant Professor:
Joshua L. Bandfield, (D), Arizona State, 2000, XgOr
Gary B. Hansen, (D), Washington, 1996, XyOr
Erika M. Harnett, (D), Washington, 2003, XyGev
Kenichi Matsuoka, (D), Hokkaido, 2002, Ol
Adjunct Assistant Professor:
Christian A. Sidor, (D), Chicago, 2000, Pv

Assistant Professor:
Olivier Bachmann, (D), Geneva (Switzerland), 2000, Giv
Drew J. Gorman-Lewis, (D), Notre Dame, 2006, PyCl
Senior Lecturer:
Terry W. Swanson, (D), Washington, 1994, CcGe
Adjunct Professor:
David S. Battisti, (D), Washington, 1988, Oa
Donald E. Brownlee, (D), Washington, 1971, Xgy
John R. Delaney, (D), Arizona, 1977, GuYr
H. Paul Johnson, (D), Washington, 1972, Gu
John D. Sahr, (D), Cornell, 1990, YgXy
William S.D Wilcock, (D), MIT, 1992, Gu
Research Professor Emeritus:
Stephen D. Malone, (D), Nevada (Reno), 1972, YsGt
Gary A. Maykut, (D), Washington, 1969, OalYr
Research Associate Professor Emeritus:
John T.A Ely, (D), Washington, 1969, XyYx
Professor Emeritus:
John B. Adams, (D), Washington, 1961, GcXgOr
Marcia B. Baker, (D), Washington, 1971, OaYg
Eric S. Cheney, (D), Yale, 1964, Eg
Robert S. Crosson, (D), Stanford, 1966, YsGct
Bernard W. Evans, (D), Oxford (UK), 1959, Gp
Subrata Ghose, (D), Chicago, 1959, Gy
Conway B. Leovy, (D), MIT, 1963, Oa
Ronald T. Merrill, (D), California (Berkeley), 1967, Ygm
George K. Parks, (D), California (Berkeley), 1966, YxXy
Stephen C. Porter, (D), Yale, 1962, Gl
Charles F. Raymond, (D), Caltech, 1969, Ol
John M. Rensberger, (D), California (Berkeley), 1967, Pv
Stewart W. Smith, (D), Caltech, 1961, Ys
Minze Stuiver, (D), Groningen (Neth), 1958, Cc
Norbert Untersteiner, (D), Innsbruck (Austria), 1950, Oo
Associate Professor Emeritus:
Joseph A. Vance, (D), Washington, 1957, Gi

Quaternary Research Center (B,M,D) (2007)
Box 351360
Seattle, WA 98195-1360
p. (206) 543-1166 f. (206) 543-3836
advising@ess.washington.edu
http://depts.washington.edu/qrc/index.cgi
Department Manager: Marty Schneider
Director:
Bernard Hallet, (D), California (Los Angeles), 1975, Gl
Professor:
Stephen C. Porter, (D), Yale, 1962, Gl
Associate Professor:
Paul D. Quay, (D), Columbia, 1977, Oc
Assistant Professor:
John O. H. Stone, (D), Cambridge, 1986, Cc
Research Associate Professor:
Patricia M. Anderson, (D), Brown, 1982, Pl
Lecturer:
Terry W. Swanson, (D), Washington, 1994, Gm
Research Assistant Professor:
Ronald S. Sletten, (D), Washington, 1995, So
Adjunct Professor:
Linda Brubaker, (D), Michigan, 1973, Pe
Angela E. Close, (D), Cambridge, 1976, Oo
Gerald G. Eck, (D), California (Berkeley), 1977, Pv
Alan R. Gillespie, (D), Caltech, 1982, Or
Donald K. Grayson, (D), Oregon, 1973, Ga
Dennis L. Hartmann, (D), Princeton, 1976, Oa
Alan J. Kohn, (D), Yale, 1957, Pi
Estella B. Leopold, (D), Yale, 1955, Pl
David M. Montgomery, (D), California (Berkeley), 1991, Gm
Charles F. Raymond, (D), Caltech, 1969, Oo
Jeffrey E. Richey, (D), California (Davis), 1973, Cg
Julie K. Stein, (D), Minnesota, 1980, Ga
Matsuo Tsukada, (D), Osaka, 1961, Pl
Stephen G. Warren, (D), Harvard, 1973, Oa
Emeritus:
Minze Stuiver, (D), Groningen (Neth), 1958, Cc
Albert Lincoln Washburn, (D), Yale, 1942, Gm
Affiliate Professor:
Brian F. Atwater, (D), Delaware, 1980, Gr

School of Oceanography (B,M,D) (2004)
Box 357940
Seattle, WA 98195-7940
p. (206) 543-5060 f. (206) 543-6073
admin@ocean.washington.edu
http://www.ocean.washington.edu/
Enrollment (2003): B: 62 (13) M: 45 (14) D: 47 (10)
Director:
Russell E. McDuff, (D), California (San Diego), 1978, Oc
Research Professor:
Mark L. Holmes, (D), Washington, 1975, Ou
Ronald L. Shreve, (D), Caltech, 1959, Yr
Professor:
Knut Aagaard, (D), Washington, 1966, Op
John A. Baross, (D), Washington, 1972, Ob
Roy Carpenter, (D), California (San Diego), 1968, Oo
William O. Criminale, Jr., (D), Johns Hopkins, 1960, Op
Eric A. D'Asaro, (D), MIT/Woods Hole, 1980, Op
John R. Delaney, (D), Arizona, 1977, Yr
Jody W. Deming, (D), Maryland, 1981, Ob
Allan H. Devol, (D), Washington, 1975, Oc
Steven R. Emerson, (D), Columbia, 1974, Oc
Charles C. Eriksen, (D), MIT, 1977, Op
Bruce W. Frost, (D), California (San Diego), 1969, Ob
Richard H. Gammon, (D), Harvard, 1970, Oc
Michael C. Gregg, (D), California (San Diego), 1971, Op
G. Ross Heath, (D), Scripps, 1968, Cm
Barbara M. Hickey, (D), California (San Diego), 1975, Op
H. Paul Johnson, (D), Washington, 1972, Yr
Marvin D. Lilley, (D), Oregon State, 1983, Ob
Seelye Martin, (D), Johns Hopkins, 1967, Op
James W. Murray, (D), MIT/Woods Hole, 1973, Oc
Charles A. Nittrouer, (D), Washington, 1978, Ou
Arthur R. M. Nowell, (D), British Columbia, 1975, Ou
Paul D. Quay, (D), Columbia, 1977, Oc
Peter B. Rhines, (D), Cambridge, 1967, Op
Jeffrey E. Richey, (D), California (Davis), 1973, Oc
Thomas B. Sanford, (D), MIT, 1967, Op
Associate Professor:
Virginia E. Armbrust, (D), MIT/Woods Hole, 1990, Ob
Susan L. Hautala, (D), Washington, 1992, Op
Bruce M. Howe, (D), California (San Diego), 1986, Op
Mitsuhiro Kawase, (D), Princeton, 1986, Op
Richard G. Keil, (D), Delaware, 1991, Oc
Deborah S. Kelley, (D), Dalhousie, 1990, Ou
Evelyn J. Lessard, (D), Rhode Island, 1984, Ob
Parker MacCready, (D), Washington, Op
Stephen C. Riser, (D), Rhode Island, 1980, Op
David A. Rothrock, (D), Cambridge, 1968, Op
LuAnne Thompson, (D), MIT/Woods Hole, 1990, Op
Mark J. Warner, (D), California (San Diego), 1988, Op
William S. D. Wilcock, (D), MIT/Woods Hole, 1992, Ou
Kevin L. Williams, (D), Washington State, 1985, Op
Research Assistant Professor:
Miles G. Logsdon, (D), Washington, 1997, Oi
Andrea S. Ogston, (D), Washington, 1997, Ou
Assistant Professor:
Andrew Barclay, (D), Oregon, 1998, Gu

Daniel Grunbaum, (D), Cornell, 1992, Ob
Anitra E. Ingalls, (D), SUNY (Stony Brook), 2002
Jeffrey D. Parsons, (D), Illinois, 1998, Ou
Gabrielle L. Rocap, (D), MIT/Woods Hole, 2000, Ob

Senior Lecturer:
Christina M. Emerick, (D), Oregon State, 1985, Ou

Lecturer:
Richard M. Strickland, (M), Washington, 1975, Ob

Adjunct Professor:
Rose Ann Cattolico, (D), SUNY (Stony Brook), 1973, Oo
Robert Francis, (D), Washington, 1970, Ob
Barbara B. Krieger-Brockett, (D), Wayne State, 1976, Ob
Ronald T. Merrill, (D), California (Berkeley), 1967, Ym
Bruce K. Nelson, (D), California (Los Angeles), 1985, Cc
Edward Sarachik, (D), Brandeis, 1966, Oa
Robert C. Spindel, (D), Yale, 1971, Op

Emeritus:
George C. Anderson, (D), Washington, 1954, Ob
Karl Banse, (D), Kiel, 1955, Ob
Joe S. Creager, (D), Texas A&M, 1958, Ou
Alyn C. Duxbury, (D), Texas A&M, 1963, On
Terry E. Ewart, (D), Washington, 1965, Op
Joyce C. Lewin, (D), Yale, 1953, Ob
Brian T. R. Lewis, (D), Wisconsin, 1970, Yr
Dean A. McManus, (D), Kansas, 1959, Ou
Maurice Rattray, Jr., (D), Caltech, 1951, Op
Gunnar I. Roden, (M), California (Los Angeles), 1956, Op
Richard W. Sternberg, (D), Washington, 1965, Ou

Cooperating Faculty:
Matthew H. Alford, (D), California (San Diego), 1998, Op
Edward T. Baker, (D), Washington, 1973, Ou
Laurie Balistrieri, (M), Washington, 1977, Oc
John Bullister, (D), California (San Diego), 1984, Oc
David A. Butterfield, (D), Washington, 1990, Yr
Glenn A. Cannon, (D), Johns Hopkins, 1969, Op
Meghan F. Cronin, (D), Rhode Island, 1993, Op
Brian D. Dushaw, (D), California (San Diego), 1992, Op
Richard A. Feely, (D), Texas A&M, 1974, Oc
Don E. Harrison, (D), Harvard, 1977, Op
Albert Hermann, (N), Op
Robin T. Holcomb, (D), Stanford, 1981, Ou
Gregory C. Johnson, (D), MIT/Woods Hole, 1990, Op
Peter A. Jumars, (D), California (San Diego), 1974, Ob
Kathryn A. Kelly, (D), California (San Diego), 1983, Op
William S. Kessler, (D), Washington, 1989, Op
Eric L. Kunze, (D), Washington, 1985, Op
Craig M. Lee, (D), Washington, 1995, Op
Michael J. McPhaden, (D), California (San Diego), 1980, Oo
Curtis D. Mobley, (D), Maryland, 1977, Ow
Harold O. Mofjeld, (D), Washington, 1970, Op
Dennis W. Moore, (D), Harvard, 1968, Op
James H. Morison, (D), Washington, 1980, Op
Jeffrey Napp, (D), California (San Diego), 1986, Ob
Jan Newton, (D), Washington, 1989, Ob
Jeffrey A. Nystuen, (D), California (San Diego), 1985, Op
Joan M. Oltman-Shay, (D), California, 1986, Op
Mary Jane Perry, (D), California (San Diego), 1974, Ob
Thomas Pratt, (N), Gu
Joseph A. Resing, (D), Hawaii, 1997, Oc
Christopher L. Sabine, (D), Hawaii, 1992, Oc
Randy Shuman, (D), Washington, 1978, Ob
Laurenz A. Thomsen, (D), Kiel, 1992, Ob
Cynthia T. Tynan, (D), California (San Diego), 1993, Ob
Rebecca A. Woodgate, (D), Oxford, 1994, Op

Washington State University

Dept of Crop & Soil Sciences (B,M,D) (2004)
201 Johnson Hall
P.O. Box 646420
Pullman, WA 99164-6420
p. (509) 335-3475 f. (509) 335-8674
alexande@mail.wsu.edu
http://css.wsu.edu/
Enrollment (2000): B: 8 (1) M: 13 (4) D: 11 (2)

Chair:
Thomas A. Lumpkin, (D), Hawaii, 1978, Oo

Professor:
David F. Bezdicek, (D), Minnesota, 1967, Sb
James B. Harsh, (D), California (Berkeley), 1983, Sc
Shiou Kuo, (D), Maine, 1973, Sc
William L. Pan, (D), North Carolina, 1983, Sb
John P. Reganold, (D), California (Davis), 1980, Ou

Senior Scientist:
Robert G. Stevens, (D), Colorado State, 1971, Sb

Associate Professor:
Craig G. Cogger, (D), Cornell, 1979, Ou
Joan R. Davenport, (D), Guelph, 1985, Sb
Bruce E. Frazier, (D), Wisconsin, 1975, Or
Frank J. Peryea, (D), California, 1983, Sc

Assistant Professor:
Markus Flury, (D), Swiss Fed Inst of Tech, 1994, Sp

Adjunct Professor:
Alan J. Busacca, (D), California (Davis), 1982, Sa
Ann C. Kennedy, (D), North Carolina State, 1985, Sb
Robert I. Papendick, (D), South Dakota State, 1962, Sp
Jeffery L. Smith, (D), Washington State, 1983, Sb

GeoAnalytical Lab (2006)
1228 Webster Physical Sciences Building
Pullman, WA 99164-2812
p. (509) 335-1626 f. (509) 335-7816
geolab@mail.wsu.edu
http://www.wsu.edu/~geolab

Professor:
Franklin F. Foit, Jr., (D), Michigan, 1968, Gz

Associate Professor:
John A. Wolff, (D), London (UK), 1983, Gi

Assistant Professor:
Garret L. Hart, (D), Wisconsin, 2002, Cc
Jeffrey D. Vervoort, (D), Cornell, 1993, Cc

Research Technologist:
Charles Knaack, (M), Washington State, 1991, Ca

Research Lab Supervisor:
Diane J. Cornelius, (M), Washington State, 1988, Ca

Microprobe Manager:
Scott B. Cornelius, (M), California (Berkeley), 1971, Ca

Program in Env Sci & Regional Planning (B,M,D) (2006)
PO Box 644430
Troy Hall 305
Pullman, WA 99164-4430
p. (509) 335-8538 f. (509) 335-7636
http://esrp.wsu.edu
Administrative Assistant: Elaine O'Fallon
Enrollment (1997): B: 124 (47) M: 119 (42) D: 18 (2)

Chair:
William W. Budd, (D), Pennsylvania, 1983, Oo

Professor:
Gerald L. Young, (D), Indiana, 1968, Oo

Senior Scientist:
Antone L. Brooks, (D), Cornell, 1967, Oo

Associate Professor:
F. Andrew Ford, (D), Dartmouth, 1975, Oo
Eldon H. Franz, (D), Illinois, 1971, Sf
R. Gene Schreckhise, (D), Colorado State, 1974, Oo

Assistant Professor:
Edward J. Brook, (D), MIT/Woods Hole, 1993, Gl

Sally D. Hacker, (D), Brown, 1996, Ob
Theodora A. Tsongas, (D), Colorado, 1976, Oo
Adjunct Professor:
James L. Conca, (D), Caltech, 1985, Oo
John R. Johnson, (D), British Columbia, 1973, Oo
Emmett B. Moore, (D), Minnesota, 1956, Oo
C. David Whiteman, (D), Colorado State, 1980, Oo
James A. Wise, (D), Washington, 1970, Oo
Judith V. Wright, (D), Oregon, 1985, Oo
Emeritus:
George W. Hinman, (D), Carnegie Mellon, 1952, Oo
Geology Librarian:
Cindy Kaag, Oo

School of Earth & Environmental Sci (B,M,D) (2007)
Webster 1228
P.O. Box 642812
Pullman, WA 99164-2812
p. (509) 335-3009 f. (509) 335-3700
libey@wsu.edu
http://www.sees.wsu.edu
Administrative Assistant: Karen S. Libey
Enrollment (2007): B: 41 (8) M: 33 (8) D: 10 (5)
Director:
Stephen Bollens, (D), Washington, 1990, Ob
Associate Director:
C. Kent Keller, (D), Waterloo, 1987, Hw
Professor:
Franklin F. Foit, Jr., (D), Michigan, 1968, Gz
Andrew Ford, (D), Dartmouth Coll, 1975, Ou
David R. Gaylord, (D), Wyoming, 1983, Gs
Peter B. Larson, (D), Caltech, 1983, Cs
A. John Watkinson, (D), Imperial Coll (UK), 1972, Gc
John A. Wolff, (D), Imperial Coll (UK), 1983, Gi
Associate Director:
Michael C. Pope, (D), Virginia Tech, 1995, Gr
Associate Professor:
Richard A. Gill, (D), Colorado State, 1998, Hw
Dirk Schulze-Makuch, (D), Wisconsin (Milwaukee), 1996, Hw
Jeffrey D. Vervoort, (D), Cornell, 1994, Ca
Adjunct Professor:
Richard M. Conrey, (D), Washington State, 1991, Gi
Regan L. Patton, (D), Washington State, 1997, Gc
Stephen P. Reidel, (D), Washington State, 1978, Gi
Emeritus:
Peter R. Hooper, (D), Birmingham, 1959, Gi
Philip E. Rosenberg, (D), Penn State, 1960, Cp
Gary D. Webster, (D), California, 1966, Pi

Washington State University, Vancouver

Science Dept (2004)
1420 NE Salmon Creek Avenue
Vancouver, WA 98686
p. (360) 546-9630 f. (360) 546-9064
hartlieb@vancouver.wsu.edu
Assistant Professor:
Edward J. Brook, (D), MIT, 1993, Ou
Adjunct Professor:
Susan L. Harder, (D), Washington, 1996, Ou

Western Washington University

Dept of Geology (B,M) (2006)
516 High Street
Bellingham, WA 98225-9080
p. (360) 650-3582 f. (360) 650-7302
geology@cc.wwu.edu
http://geology.wwu.edu/
Academic Department Manager: Chris Sutton
Enrollment (2006): B: 69 (28) M: 25 (5)
Chair:
R. Scott Babcock, (D), Washington, 1970, Cg
Professor:
David C. Engebretson, (D), Stanford, 1983, Gt
Thor A. Hansen, (D), Yale, 1978, Pg
Associate Professor:
Douglas H. Clark, (D), Washington, 1995, Gl
Juliet G. Crider, (D), Stanford, 1998, Gt
Susan M. DeBari, (D), Stanford, 1991, Gx
Bernard A. Housen, (D), Michigan, 1994, Ym
Scott R. Linneman, (D), Wyoming, 1990, Oe
Robert J. Mitchell, (D), Michigan Tech, 1996, Hw
Elizabeth R. Schermer, (D), MIT, 1989, Gt
Christopher A. Suczek, (D), Stanford, 1977, Gs
Assistant Professor:
Jackie Caplan-Auerbach, (D), Hawaii, Ys
David M. Hirsch, (D), Texas, 2000, Gp
Research Associate:
M. Clark Blake, (D), Stanford, 1965, Gt
Russell F. Burmester, (D), Princeton, 1974, Ym
Charles A. Ross, (D), Yale, 1964, Pm
Emeritus:
Myrl E. Beck, (D), California (Riverside), 1969, Ym
Edwin H. Brown, (D), California (Berkeley), 1966, Gp
Robert A. Christman, (D), Princeton, 1950, Oe
Don J. Easterbrook, (D), Washington, 1962, Gm
Ross C. Ellis, (D), Washington, 1968, Gc
Maurice L. Schwartz, (D), Columbia, 1966, On
James L. Talbot, (D), Adelaide, 1962, Gc
Technician:
George Mustoe, (M), Oo
Related Staff:
Vicki Critchlow, (B), Oo

Whitman College

Dept of Geology (B) (2005)
345 Boyer
Walla Walla, WA 99362
p. (509) 527-5225 f. (509) 527-5904
mosspm@whitman.edu
http://www.whitman.edu/geology
Administrative Assistant: Patti Moss
Enrollment (2004): B: 39 (15)
Professor:
Robert J. Carson, (D), Washington, 1970, Gm
Kevin R. Pogue, (D), Oregon State, 1993, Gc
Patrick K. Spencer, (D), Washington, 1984, Pg
John D. Winter, (D), Washington, 1978, Gp
Visiting Professor:
Brennan T. Jordan, (D), Oregon State, 2002, Gx
Related Staff:
Tara Gregg, (B), Whitman, 2004, Gg

Whitworth College

Dept of Physics & Geology (A,B) (2004)
Spokane, WA 99251
p. (509) 466-3268 f. (509) 466-3221
rstevens@whitworth.edu
Department Secretary: Sally Brumgartner
Enrollment: No data reported since 1995
Chair:
Delbert Friesen, Oo
Adjunct Professor:
Edwin A. Olson, (D), Columbia, 1963, Gg

Yakima Valley College

Dept of Geology (A) (2006)
P. O. Box 22520
16th Ave. & Nob Hill Blvd.
Yakima, WA 98907
p. (509) 575-2350 (Ext. 2366) f. 509-574-6860
dhuycke@yvcc.edu
http://www.yvcc.edu/programs/Geology/
Enrollment (2005): A: 4 (0)

Instructor:
David Huycke, (M), Wyoming, 1979, Gg

West Virginia

Concord University

Dept of Geology & Physical Sciences (B) (2007)
Athens, WV 24712-1000
p. (304) 384-5327
allenj@concord.edu
http://faculty.concord.edu/allenj/geosci.html
Department Secretary: Janet Heaton
Enrollment (2007): B: 22 (2)

Chair:
Joseph L. Allen, (D), Kentucky, 1994, Gc

Assistant Professor:
David L. Matchen, (D), West Virginia, 2004, Gr

Adjunct Professor:
Alyce Lee, (D), Texas A&M, 2008, Og

Emeritus:
Winton G. Covey, Jr., (D), Texas A&M, 1965, Gg

Cooperating Faculty:
George W. Towers, (D), Arizona, 1991, Oi
Jan Westerik, (M), North Carolina, 1970, Oy

Glenville State College

Div of Science & Mathematics (B) (2006)
Glenville, WV 26351
p. (304) 462-7361 f. (304) 462-4407
Enrollment: No data reported since 1995

Marshall University

Dept of Geology (B,M) (2007)
100 John Marshall Drive
Huntington, WV 25755
p. (304) 696-6720 f. (304) 696-3243
geology@marshall.edu
marshall.edu/geology
Administrative Assistant: Janet M. Clark
Enrollment (2007): B: 26 (5) M: 1 (0)

Chair:
Dewey D. Sanderson, (D), Michigan State, 1972, Hw

Professor:
Ronald L. Martino, (D), Rutgers, 1981, Gs

Associate Professor:
Aley El Shazly, (D), Stanford, 1991, Cg
William Niemann, (D), Missouri (Rolla), 1999, Ng

Potomac State College

Dept of Geology & Geography (A) (2004)
Keyser, WV 26726
p. (304) 788-6800 f. (304) 788-6940
jjninesteel@mail.wvu.edu
Enrollment (1996): B: 1 (0)

Instructor:
Judy J. Ninesteel, (M), West Virginia, 1995, Hg

West Virginia University

Dept of Geology & Geography (B,M,D) (2007)
330 Brooks Hall
P.O. Box 6300
98 Beechurst Ave.
Morgantown, WV 26506-6300
p. (304) 293-5603 f. (304) 293-6522
helen.lang@mail.wvu.edu
http://www.geo.wvu.edu
Enrollment (2007): B: 64 (26) M: 30 (8) D: 9 (1)

Chair:
Trevor Harris, (D), Hull (UK), 1982, Oy

Professor:
Robert E. Behling, (D), Ohio State, 1971, Gm
Timothy R. Carr, (D), Wisconsin, 1981, GoYe
Thomas W. Kammer, (D), Indiana, 1982, Pi
Henry W. Rauch, (D), Penn State, 1972, Hw
John J. Renton, (D), West Virginia, 1965, Ec
Richard A. Smosna, (D), Illinois, 1973, Gr
Timothy A. Warner, (D), Purdue, 1992, Or
Thomas H. Wilson, (D), West Virginia, 1980, Ye

Associate Chair:
Helen M. Lang, (D), Oregon, 1983, Gpz

Associate Professor:
Joseph J. Donovan, (D), Penn State, 1992, Hq
J. Steven Kite, (D), Wisconsin, 1983, Gm
Jaime Toro, (D), Stanford, 1998, Gc

Assistant Professor:
Dorothy Vesper, (D), Penn State, 2002, Hw

Adjunct Professor:
Katherine Lee Avary, (M), North Carolina, 1978, Go
Mary Behling, (M), Miami, Oi
Katherine E. Bruner, (D), West Virginia, 1991, Gd
David Campagna, (D), Purdue, 1990, Gc
Blaine Cecil, (D), West Virginia, 1965, Ec
Nick Fedorko, (M), Ec
William Gillespie, (M), West Virginia, 1954, Pb
William Grady, (M), West Virginia, 1978, Ec
Michael E. Hohn, (D), Indiana, 1976, Gq
Ronald McDowell, (D), Colorado Mines, 1987, Gg
Dave Oldham, (D), West Virginia, 1997, Go
Douglas Patchen, (D), Syracuse, 1971, Go
Thomas Repine, (D), West Virginia, 2001, Gg
Carl Smith, (M), Indiana, 1969, Gg

Emeritus:
Alan C. Donaldson, (D), Penn State, 1959, Gs
Milton T. Heald, (D), Harvard, 1949, Gd
Robert C. Shumaker, (D), Cornell, 1960, Go

West Virginia University, Parkersburg

Div of Natural Sciences/Math (Geology) (A) (2004)
Route 5
Box 167A
Parkersburg, WV 26101-9577
p. (304) 424-8299 f. (304) 424-8315
gary.waggoner@mail.wvu.edu
Enrollment (2001): A: 4 (1)

Associate Professor:
Edward L. Crisp, (D), Indiana, 1975, Go

Wisconsin

Beloit College

R.D. Salisbury Dept of Geology (B) (2005)
700 College Street
Beloit, WI 53511
p. (608) 363-2223 f. (608) 363-2052
mendelsn@beloit.edu

http://geology.beloit.edu
Enrollment (2005): B: 26 (5)

Chair:
Carl V. Mendelson, (D), California (Los Angeles), 1981, Pm
Professor:
Carol Mankiewicz, (D), Wisconsin, 1987, Gs
Assistant Professor:
James R. Rougvie, (D), Texas, 1999, Gp
Susan K. Swanson, (D), Wisconsin, 2001, Hw
Emeritus:
Richard C. Stenstrom, (D), Chicago, 1964, Ge
Tech & Safety Officer:
Stephen M. Ballou, (B), Beloit, 1998, Gg

Lawrence University

Dept of Geology (B) (2006)
115 S. Drew Street
P.O. Box 599
Appleton, WI 54912
p. (920) 832-6733 f. (920) 832-6962
bjornerm@lawrence.edu
http://www.lawrence.edu/dept/geology/
Administrative Assistant: Leslie Trudell
Enrollment (2006): B: 18 (5)

Professor:
Marcia Bjornerud, (D), Wisconsin, 1987, Gc
Associate Professor:
Jeffrey J. Clark, (D), Johns Hopkins, 1997, Gm
Assistant Professor:
Andrew Knudsen, (D), Idaho, 2002, Cg
Emeritus:
John C. Palmquist, (D), Iowa, 1961, Gc
Theodore W. Ross, (D), Washington State, 1969, Gg
Ronald W. Tank, (D), Indiana, 1962, Ge

Milwaukee Area Technical College

Dept of Natural Science (A) (2004)
Milwaukee, WI 53233
p. (414) 297-7427 f. (414) 297-7990
millerr@milwaukee.tec.wi.us
Administrative Assistant: Barb Geddes

Professor:
Richard H. Miller, (D), Wisconsin (Milwaukee), 1987, Gg
Instructor:
Scott R. Johnston, (B), Wisconsin (Milwaukee), 1981, Og
Adjunct Professor:
George Stone, (D), Colorado, 1967, Gl

Milwaukee Public Museum

Dept of Geology (2004)
800 W. Wells Street
Milwaukee, WI 53233
p. (414) 278-2741 f. (414) 278-6100
http://www.mpm.edu/collect/geology/geosec-noframes.html

Chair:
Peter M. Sheehan, (D), California (Berkeley), 1971, Pi

Northland College

Dept of Geoscience (B) (2007)
1411 Ellis Avenue
Ashland, WI 54806
p. (715) 682-1312 f. (715) 682-1849
bgoetz@northland.edu
Enrollment (2007): B: 11 (2)

Professor:
Bruce A. Goetz, (M), S Illinois, 1969, GgmHg
Associate Professor:
Thomas J. Fitz, (D), Delaware, 1999, GgcEg

Saint Norbert College

Geology Dept (B) (2007)
100 Grant Street
De Pere, WI 54115
p. (920) 403-3987 f. (920) 403-4033
tim.flood@snc.edu
http://www.snc.edu
Enrollment (2007): B: 11 (5)

Professor:
Tim P. Flood, (D), Michigan State, 1987, Gv
Associate Professor:
Nelson R. Ham, (D), Wisconsin, 1994, Gl
Assistant Professor:
Elizabeth A. Gordon, (D), SUNY (Binghamton), 1987, Gs

University of Wisconsin Colleges

Dept of Geography & Geology (2007)
cary.komoto@uwc.edu

Associate Professor:
Keith Montgomery, (D), Waterloo, 1986, Oy
Emeritus:
Gene E. Musolf, (D), Wisconsin, 1970, Oy

University of Wisconsin, Eau Claire

Dept of Geology (B) (2006)
157 Phillips Hall
Eau Claire, WI 54702-4004
p. (715) 836-3732 f. (715) 836-5627
steinklm@uwec.edu
http://www.uwec.edu/academic/geology
Administrative Assistant: Lorilie M. Steinke
Enrollment (2006): B: 55 (9)

Chair:
Robert L. Hooper, (D), Washington State, 1983, Gz
Professor:
Karen G. Havholm, (D), Texas, 1991, Oe
Phillip D. Ihinger, (D), Caltech, 1991, Gx
J. Brian Mahoney, (D), British Columbia, 1994, Gs
Kent M. Syverson, (D), Wisconsin, 1992, Gl
Assistant Professor:
Katherine R. Grote, (D), California (Berkeley), 2003, Hw
Senior Lecturer:
Lori D. Snyder, (M), British Columbia, 1994, Gx

University of Wisconsin, Extension

Dept of Environmental Sciences (2007)
3817 Mineral Point Road
Madison, WI 53705-5100
p. (608) 262-1705 f. (608) 262-8086
kmzwettl@wisc.edu

Professor:
John W. Attig, (D), Wisconsin, 1984, Gl
Kenneth R. Bradbury, (D), Wisconsin, 1982, Hw
Thomas J. Evans, (D), Wisconsin, 1994, Gg
James M. Robertson, (D), Michigan, 1972, Eg
Associate Professor:
Madeline B. Gotkowitz, (M), New Mexico, 1993, Hw
David J. Hart, (D), Wisconsin, 2000, Hw
Thomas S. Hooyer, (D), Iowa State, 1999, Gl
Assistant Professor:
Patrick I. McLaughlin, (D), Cincinnati, 2006, Gr
Emeritus:
Lee Clayton, (D), Illinois, 1965, Gl

University of Wisconsin, Green Bay

Dept of Natural and Applied Science - Earth Science Program (A,B,M) (2004)
Green Bay, WI 54302
p. (414) 465-2371 f. (414) 465-2376
dutchs@uwgb.edu
Department Secretary: Fran Baugnet
Program Assistant: Barb Maenpaa
Enrollment (1995): B: 17 (5)

Head:
Steven I. Dutch, (D), Columbia, 1976, Gc

Professor:
Joseph M. Moran, (D), Wisconsin, 1972, Oa
Ronald D. Stieglitz, (D), Illinois, 1972, Gr

Assistant Professor:
Kevin J. Fermanich, (D), Wisconsin, 1995, Sp

Emeritus:
Harold J. Day, (D), Wisconsin, 1963, Hg
Thomas H. McIntosh, (D), Iowa State, 1962, So

University of Wisconsin, Madison

Dept of Atmospheric & Oceanic Sci (B,M,D) (2007)
1225 W. Dayton Street
Atmospheric, Oceanic & Space Science Building
Madison, WI 53706-1695
p. (608)262-2828 f. (608)262-0166
aos@aos.wisc.edu
http://www.aos.wisc.edu
Enrollment (2007): B: 32 (16) M: 32 (10) D: 30 (3)

Joint Executive:
John M. Norman, (D), Wisconsin, 1971, So

Professor:
Steven A. Ackerman, (D), Colorado State, 1987, Or
Jonathan A. Foley, (D), Wisconsin, 1993, Og
Matthew H. Hitchman, (D), Washington, 1985, Oa
Zhengyu Liu, (D), MIT, 1991, Op
Jonathan E. Martin, (D), Washington, 1992, Ow
Michael C. Morgan, (D), MIT, 1994, Ow
Grant W. Petty, (D), Washington, 1990, Or
Gregory J. Tripoli, (D), Colorado State, 1986, Ow
Pao-Kuan Wang, (D), California (Los Angeles), 1978, Oa

Senior Scientist:
Edwin W. Eloranta, (D), Wisconsin, 1972, Yr

Associate Professor:
Ralf Bennartz, (D), Free (Berlin), 1997, Or

Assistant Professor:
Ankur R. Desai, (D), Penn State, 2006, Oa
Eric T. DeWeaver, (D), Maryland, 1999, Oa
Galen Mc Kinley, (D), MIT, 2002, Ob
Daniel J. Vimont, (D), Washington, 2001, Oa

Adjunct Associate Professor:
Jeffrey R. Key, (D), Colorado, 1988, Or

Adjunct Assistant Professor:
Andrew Heidinger, (D), Colorado State, 1998, Or

Adjunct Professor:
John F. Le Marshall, (D), Monash, 1972, Or
W. Paul Menzel, (D), Wisconsin, 1974, Or

Emeritus:
Francis P. Bretherton, (D), Cambridge, 1961, Og
Reid A. Bryson, (D), Chicago, 1948, Og
Stefan L. Hastenrath, (D), Bonn, 1959, Oa
David D. Houghton, (D), Washington, 1963, Oa
Donald R. Johnson, (D), Wisconsin, 1965, Oa
John E. Kutzbach, (D), Wisconsin, 1966, Oa
Robert A. Ragotzkie, (D), Wisconsin, 1953, Ob
William L. Smith, (D), Wisconsin, 1966, Or
Charles R. Stearns, (D), Wisconsin, 1967, Ow
John A. Young, (D), MIT, 1966, Oa

Cooperating Faculty:
Tracey Holloway, (D), Princeton, 2001, Oa

Dept of Geology & Geophysics (B,M,D) (2007)
1215 West Dayton Street
Madison, WI 53706
p. (608) 262-8960 f. (608) 262-0693
geodept@geology.wisc.edu
http://www.geology.wisc.edu
Enrollment (2007): B: 45 (23) M: 25 (19) D: 37 (5)

Chair:
Jean M. Bahr, (D), Stanford, 1986, Hw

Professor:
Mary P. Anderson, (D), Stanford, 1973, Hw
Philip E. Brown, (D), Michigan, 1980, Em
Charles W. Byers, (D), Yale, 1973, Pe
Alan R. Carroll, (D), Stanford, 1991, Gs
D. Charles DeMets, (D), Northwestern, 1988, Ym
Dana H. Geary, (D), Harvard, 1986, Po
Laurel Goodwin, (D), California (Berkeley), 1988, Gc
Clark M. Johnson, (D), Stanford, 1986, Cc
Eric E. Roden, (D), Maryland, 1990, Py
Bradley S. Singer, (D), Wyoming, 1990, Cc
Clifford H. Thurber, (D), MIT, 1981, Ys
Basil Tikoff, (D), Minnesota, 1994, Gc
John W. Valley, (D), Michigan, 1980, Gp
Herbert F. Wang, (D), MIT, 1971, Yg

Senior Scientist:
John Fournelle, (D), Johns Hopkins, 1989, Gi

Associate Professor:
Kurt Feigl, (D), MIT, 1991, Yd
D. Clay Kelly, (D), North Carolina, 1999, Pm
Nita Sahai, (D), Johns Hopkins, 1998, Cl
Harold Tobin, (D), California (Santa Cruz), 1995, Yg

Associate Scientist:
Brian L. Beard, (D), Wisconsin, 1992, Cc

Assistant Professor:
Anders Carlson, (D), Oregon, 2006, Pe
Shanan Peters, (D), Chicago, 2003, Gs
Huifang Xu, (D), Johns Hopkins, 1993, Gz

Adjunct Professor:
Randy J. Hunt, (D), Wisconsin, 1993, Hw

Emeritus:
Charles R. Bentley, (D), Columbia, 1959, Yg
Carl J. Bowser, (D), California (Los Angeles), 1965, Cl
Nikolas I. Christensen, (D), Wisconsin, 1963, Yx
David L. Clark, (D), Iowa, 1957, Pm
Clarence S. Clay, (D), Wisconsin, 1950, Ye
Robert H. Dott, Jr., (D), Columbia, 1955, Gs
Louis J. Maher, Jr., (D), Minnesota, 1961, Pl
Gordon L. Medaris, (D), California, 1966, Gi
David M. Mickelson, (D), Ohio State, 1971, Gl
Lloyd C. Pray, (D), Caltech, 1952, Gs

Museum Director:
Richard Slaughter, (D), Iowa, 2001, Pg

Geology Librarian:
Marie Dvorzak, (M), Oo

Related Staff:
John W. Attig, (D), Wisconsin, 1984, Gl
Kenneth R. Bradbury, (D), Wisconsin, 1982, Hw
Lee Clayton, (D), Illinois, 1965, Gl
Dante Fratta, (D), Georgia Tech, 1999, Ng
Madeline B. Gotkowitz, (M), New Mexico Tech, 1993, Hw
David J. Hart, (D), UW-Madison, 2000, Hg
Thomas S. Hooyer, (D), Iowa State, 1999, Hg
David Krabbenhoft, (D), UW-Madison, 1988, Hg
James M. Robertson, (D), Michigan, 1972, Eg

Dept of Soil Science (B,M,D) (2007)
1525 Observatory Drive
Madison, WI 53706-1299
p. (608) 262-2633 f. (608) 265-2595
tcbusby@wisc.edu
http://www.soils.wisc.edu
Enrollment (2005): B: 29 (16) M: 11 (7) D: 13 (1)

Chair:
Stephen J. Ventura, (D), Wisconsin, 1989, Or

Professor:
William L. Bland, (D), Wisconsin, 1984, Sp
William F. Bleam, (D), Cornell, 1984, Sc
James G. Bockheim, (D), Washington, 1972, Sf
William J. Hickey, (D), California (Riverside), 1990, Sb
King-Jau S. Kung, (D), Cornell, 1984, Sp
Birl Lowery, (D), Oregon, 1980, Sp
Frederick W. Madison, (D), Wisconsin, 1972, Sd
Kevin McSweeney, (D), Illinois, 1984, Sd
John M. Norman, (D), Wisconsin, 1971, Sp

Associate Professor:
Teri C. Balser, (D), California (Berkeley), 2000, Sb
Phillip W. Barak, (D), Hebrew (Israel), 1988, Sc
Joel A. Pedersen, (D), California (Los Angeles), 2001, Sb
J. Mark Powell, (D), Texas A&M, 1989, So

Assistant Professor:
Nick J. Balster, (D), Idaho, 1999, Sp
Carrie A.M Laboski, (D), Minnesota, 2001, So

Emeritus:
Larry G. Bundy, (D), Iowa, 1973, So
Robin F. Harris, (D), Wisconsin, 1972, Sb
Philip A. Helmke, (D), Wisconsin, 1971, Sc
Keith A. Kelling, (D), Wisconsin, 1974, So
Wayne R. Kusssow, (D), Wisconsin, 1966, So
E. Jerry Tyler, (D), North Carolina, 1975, Sd

Geological Engineering Program (B,M,D) (2007)
1415 Engineering Drive
Room 2212 Engineering Hall
Madison, WI 53706-1691
p. (608) 262-3491 f. (608) 262-2453
benson@engr.wisc.edu
http://www.engr.wisc.edu/interd/gep
Enrollment (2000): B: 47 (8) M: 3 (1) D: 1 (0)

Chair:
Craig H. Benson, (D), Texas, 1989, Ng

Professor:
Mary P. Anderson, (D), Stanford, 1973, Hw
Jean M. Bahr, (D), Stanford, 1987, Hw
Tuncer B. Edil, (D), Northwestern, 1973, Ng
Laurel Goodwin, (D)
Michael E. Plesha, (D), Northwestern, 1983, Oo
Kenneth W. Potter, (D), Johns Hopkins, 1976, Hg
Herbert F. Wang, (D), MIT, 1971, Yx

Assistant Professor:
Dante Fratta, (D), Georgia Tech, 1999, Yx
Basil Tikoff, (D), Minnesota, 1997, Gc
Chin Wu, (D), MIT, Hs

Related Staff:
Lynn D. Maertz, (B), Wisconsin (Eau Claire), 1975, Oo

Limnology & Marine Sci Graduate Program (M,D) (2005)
680 North Park Street
Madison, WI 53706
p. (608) 263-3264 f. (608) 265-2340
lkgraham@wisc.edu
http://www.engr.wisc.edu/interd/limnology/http://www.engr.wisc.edu/interd/limnology/
Enrollment (1998): M: 3 (0) D: 9 (2)

Chair:
Kenneth Potter, (D), Johns Hopkins

Professor:
Michael S. Adams, (D), California (Riverside), 1968, Ob
Anders W. Andren, (D), Florida State, 1972, Oc
David E. Armstrong, (D), Wisconsin, 1966, Oc
Steve Carpenter, (D)
Calvin B. DeWitt, (D), Michigan, 1963, Ob
Stanley I. Dodson, (D), Washington, 1970, Ob
John Foley, (D), Wisconsin, 1993
Linda K. Graham, (D), Michigan, 1975, Ob
John A. Hoopes, (D), MIT, 1965, Oo
James F. Kitchell, (D), Colorado, 1970, Ob
John E. Kutzbach, (D), Wisconsin, 1966, Pe
William C. Sonzogni, (D), Wisconsin, 1974, Oc
Joy Zedler, (D), Wisconsin, 1968

Associate Professor:
Emily Stanley, (D)

Assistant Professor:
Sarah Hotchkiss, (D), Minnesota, 1998
Carol Lee, (D), Washington, 1998
Zheng-yu Liu, (D), MIT, 1991, Op
Katherine McMahon, (N), 2002
Jake Vander Zanden, (D), McGill, 1999, On
Chin Wu, (D), MIT

University of Wisconsin, Milwaukee

Dept of Geosciences (B,M,D) (2007)
P.O. Box 413
Milwaukee, WI 53201-0413
p. (414) 229-4561 f. (414) 229-5452
grundl@uwm.edu
http://www.uwm.edu/Dept/Geosciences/
Enrollment (2007): B: 36 (0) M: 20 (1)

Professor:
Douglas S. Cherkauer, (D), Princeton, 1972, Hw
Timothy J. Grundl, (D), Colorado, 1988, HwCl
John L. Isbell, (D), Ohio State, 1990, Gs
William F. Kean, Jr., (D), Pittsburgh, 1972, Ym
Keith A. Sverdrup, (D), Scripps, 1981, Ys

Associate Professor:
Barry I. Cameron, (D), N Illinois, 1998, Gvi
Mark T. Harris, (D), Johns Hopkins, 1988, Gs

Assistant Professor:
Dyanna M. Czeck, (D), Minnesota, 2001, Gc
Stephen Q. Dornbos, (D), Southern California, 2003
Margaret L. Fraiser, (D), Southern California, 2005, Pi
Lindsay J. McHenry, (D), Rutgers, 2004, Gz
Shangping Xu, (D), Princeton, 2005, Hg

Adjunct Professor:
Daniel T. Feinstein, (M), Wisconsin, 1986
Peter M. Sheehan, (D), California (Berkeley), 1971, Pi

University of Wisconsin, Oshkosh

Geography Dept (B) (2004)
800 Algoma Avenue
Oshkosh, WI 54901
p. (414) 424-4105 f. (414) 424-7317

Chair:
John A. Cross, (D), Illinois, 1979, Oy

Associate Professor:
Todd A. Fonstad, (D), Oklahoma, 1975, Oy

Assistant Professor:
Harold E. Gulley, (D), Univ Coll (UK), 1988, Oo
Peter C. Johnson, (D), Oregon, 1985, Oo
Jeffrey S. Torguson, (D), Georgia, 1993, Oo
Kasimierz J. Zaniewski, (D), Wisconsin (Milwaukee), 1987, Oo

Lecturer:
Robert Greuel, (M), Nebraska (Omaha), 1970, Oy
Gebeyehu Mulugeta, (D), Michigan State, 1991, Oo

Geology Dept (B) (2007)
800 Algoma Boulevard
Oshkosh, WI 54901-8649
p. (920) 424-4460 f. (920) 424-0240
brownt@uwosh.edu
http://www.uwosh.edu/departments/geology/
Administrative Assistant: Teresa Brown
Enrollment (2007): B: 55 (4)
Chair:
William N. Mode, (D), Colorado, 1980, Gl
Professor:
Daniel J. Lehrmann, (D), Kansas, 1993, Gr
Associate Professor:
Eric E. Hiatt, (D), Colorado, 1997, Gs
Assistant Professor:
George J. Hudak, (D), Minnesota, 1996, Gv
Maureen A. Muldoon, (D), Wisconsin, 1999, Hw
Timothy S. Paulsen, (D), Illinois, 1997, Gc
Jennifer M. Wenner, (D), Boston, 2001, Gi
Lecturer:
Christie M. Demosthenous, (M), Illinois, 1996, Gz
Emeritus:
Charles W. Fetter, Jr., (D), Indiana, 1971, Hw
Norris W. Jones, (D), Virginia Tech, 1968, Gi
Gene L. La Berge, (D), Wisconsin, 1963, Eg
Thomas S. Laudon, (D), Wisconsin, 1963, Yg
James W. McKee, (D), Louisiana State, 1967, Gr
Brian K. McKnight, (D), Oregon State, 1970, Gd
Instrumentation Specialist:
Thomas J. Suszek, (M), Minnesota (Duluth), 1991, Gg

University of Wisconsin, Parkside
Dept of Geosciences (B) (2007)
Box 2000
900 Wood Road
Kenosha, WI 53141
p. (414) 595-2236 f. (414) 595-2056
christine.evans@uwp.edu
Enrollment (2006): B: 30 (0)

University of Wisconsin, Platteville
Dept of Geosciences (B) (2005)
Platteville, WI 53818
p. (608) 342-1791
nplasca@uwm.edu
Enrollment: No data reported since 1995
Professor:
Charles W. Collins, (D), Nova, 1985, Gm
Associate Professor:
Richard A. Waugh, (D), Wisconsin, 1995, Gc
Assistant Professor:
Todd Stradford, (D), Oklahoma, 1994, Or
Mari A. Vice, (D), S Illinois, 1993, Gd
Emeritus:
William A. Broughton, (B), Wisconsin, 1938, Em
Kenneth A. Shubak, (M), Michigan, 1966, Pi

University of Wisconsin, River Falls
Dept of Plant & Earth Science (B) (2006)
410 S. Third Street
River Falls, WI 54022
p. (715) 425-3345 f. (715) 425-3785
michael.d.middleton@uwrf.edu
http://www.uwrf.edu/pes/geol/
Program Assistant: Sue Freiermuth
Enrollment (2006): B: 42 (19)
Professor:
Robert W. Baker, (D), Minnesota, 1976, Gl
William S. Cordua, (D), Indiana, 1973, Gx
Michael D. Middleton, (D), Colorado, 1983, Pv
Eric M. Sanden, (D), Texas Tech, 1993, Or
Donavon Taylor, (D), Minnesota, 1981, Sp
Ian S. Williams, (D), California (Santa Barbara), 1981, Yg
Associate Professor:
Kerry L. Keen, (D), Minnesota, 1992, Hw
Assistant Professor:
Laine C. Vignona, (D), Rice, 1999, Cl

University of Wisconsin, Stevens Point
Dept of Geography-Geology (B) (2007)
2001 Fourth Avenue
Stevens Point, WI 54481
p. (715) 346-2629 f. (715) 346-3372
krice@uwsp.edu
http://www.uwsp.edu/geo/
Administrative Assistant: Diane Stelzer
Enrollment (2007): B: 60 (17)
Chair:
Keith W. Rice, (D), Kansas, 1989, Oir
Professor:
Karen A. Lemke, (D), Iowa, 1988, Gm
David L. Ozsvath, (D), SUNY (Binghamton), 1985, Hw
Michael E. Ritter, (D), Indiana, 1986, Ow
Associate Professor:
Kevin P. Hefferan, (D), Duke, 1992, GcxGt
Neil C. Heywood, (D), Colorado, 1988, Oy
Eric J. Larson, Oregon State, 2001, Or
Assistant Professor:
Samantha W. Kaplan, (D), Wisconsin, 2003, GnPlGd
Karl E. Ryavec, (D), Minnesota, 2002, Oui
Lisa J. Theo, (M), Wisconsin, Ou

University of Wisconsin, Superior
Dept of Biology & Earth Science (B) (2004)
Superior, WI 54880
p. (715) 394-8311 f. (715) 394-8107
jzaengle@uwsuper.edu
Enrollment: No data reported since 1995
Professor:
John F. Zaengle, (D), Michigan, 1995, Gs
Assistant Professor:
William Bajjali, (D), Ottawa, 1994, Gg

University of Wisconsin, Whitewater
Geography and Geology (2005)
800 W Main Street
Whitewater, WI 53190
p. 262-472-1071 or 5257 f. 262-472-5238
hangerr@uww.edu
Director:
Frank R. Luther, (D), Lehigh, 1976, Gp
Assistant Professor:
Rex Hanger, (D), Yr

Wyoming

Casper College
Geology Dept (A) (2007)
Division of Physical Sciences
125 College Drive

Casper, WY 82601
p. (307) 268-3308 f. (307) 268-2017
ksundell@caspercollege.edu
http://www.caspercollege.edu/physcience/geology/
Administrative Assistant: Rachel Wright
Enrollment (2007): A: 19 (8)

Head:
Kent A. Sundell, (D), California (Santa Barbara), 1985, Gg
Instructor:
Melissa Connely, (M), Utah State, 2000, Gs
Gerald E. Nelson, (D), Kansas, 1983, Gm
Dana P. Van Burgh, (M), N Colorado, 1962, Oe
Adjunct Professor:
Albert E. Allen, (M), Oklahoma, 1958, Gr
Terry Logue, (D), Wyoming, 1975, Gg

University of Wyoming

Dept of Geology and Geophysics (B,M,D) (2007)
Dept. #3006
1000 E. University Ave.
Laramie, WY 82071
p. (307) 766-3386 f. (307) 766-6679
snoke@uwyo.edu
http://home.gg.uwyo.edu
Administrative Assistant: Carol L. Pribyl
Enrollment (2006): B: 120 (27) M: 34 (7) D: 21 (5)

Head:
Arthur W. Snoke, (D), Stanford, 1972, Gc
Associate Head:
Carol D. Frost, (D), Cambridge, 1984, Cc
Professor:
Carrick M. Eggleston, (D), Stanford, 1991, Cl
B. Ronald Frost, (D), Washington, 1973, Gp
Paul L. Heller, (D), Arizona, 1983, Gs
W. Steven Holbrook, (D), Stanford, 1989, Ys
Robert R. Howell, (D), Arizona, 1980, Xg
Neil F. Humphrey, (D), Washington, 1987, Gm
Barbara E. John, (D), California (Santa Barbara), 1987, Gc
James D. Myers, (D), Johns Hopkins, 1979, Gi
Maureen B. Steiner, (D), Texas (Dallas), 1974, Ym
Senior Scientist:
Steven Boese, (D), Wyoming, 1978, Ca
Kevin R. Chamberlain, (D), Washington (St. Louis), 1990, Cc
Susan M. Swapp, (D), Yale, 1982, Gp
Associate Professor:
Michael J. Cheadle, (D), Cambridge, 1989, Yg
Kenneth G. Dueker, (D), Oregon, 1994, Ys
Assistant Professor:
Barbara Carrapa, (D), Vrije (Amsterdam), 2002, Gs
Mark T. Clementz, (D), California (Santa Cruz), 2002, Py
Ye Zhang, (D), Indiana, 2005, Hy
Senior Lecturer:
Randi S. Martinsen, (M), N Arizona, 1975, Go
Lecturer:
Erin A. Campbell-Stone, (D), Wyoming, 1997, Gc
Adjunct Professor:
J. Michael Boyles, (D), Texas, 1983, Go
Warren B. Hamilton, (D), California (Los Angeles), 1951, Gt
Peter H. Hennings, (D), Texas, 1991, Gc
David A. Stephenson, (D), Illinois, 1965, Hy
State Geologist:
Ronald C. Surdam, (D), California (Los Angeles), 1967, Go
Emeritus:
Leon E. Borgman, (D), California (Berkeley), 1963, Gq
Donald W. Boyd, (D), Columbia, 1957, Pi
James I. Drever, (D), Princeton, 1968, Cl
William E. Frerichs, (D), S California, 1967, Pm
Robert S. Houston, (D), Columbia, 1954, Eg
Jason A. Lillegraven, (D), Kansas, 1968, Pv
Ronald W. Marrs, (D), Colorado Mines, 1973, Or
James E. McClurg, (D), Michigan, 1969, Gs
Brainerd Mears, Jr., (D), Columbia, 1950, Gm
Scott B. Smithson, (D), Oslo, 1963, Ys
James R. Steidtmann, (D), Michigan, 1968, Gr
Geology Librarian:
Sally J. Scott, (M), Wisconsin, 1962, Og
Curator:
Brent H. Breithaupt, (M), Wyoming, 1981, Pv
Department Secretary:
Shawn Sheen, Oo
Academic Coordinator:
Sondra S. Cawley, Oo

Dept of Soil Science (M,D) (2004)
P.O. Box 3354
Laramie, WY 82071
p. (307) 766-3103 f. (307) 766-5549
gfv@uwyo.edu
Department Secretary: Ginny Alm
Department Secretary: Kim Malody
Enrollment: No data reported since 1995

Professor:
Larry C. Munn, (D), Montana State, 1977, Sd
Steve E. Williams, (D), North Carolina State, 1977, Sb
Assistant Professor:
Renduo Zhang, (D), Arizona, 1990, Sp
Adjunct Professor:
Gerald E. Schuman, (D), Nebraska, 1974, Sf
Visiting Professor:
Katta J. Reddy, (D), Colorado State, 1986, Sc
Peter D. Stahl, (D), Wyoming, 1988, Sb
Geology Librarian:
George F. Vance, Oo

Australia

Australian National University

Dept of Earth & Marine Sciences (B,M,D) (2005)
47 Daley Road
Canberra, ACT 0200
p. 02 6125 2056 f. 02 6125 5544
ems@anu.edu.au
http://ems.anu.edu.au/

Reader:
D C "Bear" McPhail, (D), Princeton, 1991, Cl
Professor:
Patrick De Deckker, (D), 1981, Gu

Research School of Earth Sciences (2006)
Mills Rd.
Canberra, A.C.T. 0200
p. +61 2 6125 3406
Director.rses@anu.edu.au
http://rses.anu.edu.au/

Curtin University of Technology

School of Applied Geology (2004)
GPO Box 1987
Perth, WA
lindsay@lithos.curtin.edu.au
http://www.curtin.edu.au/curtin/dept/geology/

Dept of Exploration Geophysics (B,M,D) (2007)
G.P.O. Box U1987
Perth, WA 6845
p. +61 8 9266-3565 f. +61 8 9266-3407

D.Hollingsworth@curtin.edu.au
http://www.geophysics.curtin.edu.au/
Enrollment (2007): B: 83 (23) M: 3 (1) D: 21 (3)

Head:
Anton W. Kepic, (D), British Columbia, 1990, Ye

Professor:
Brian J. Evans, (D), Curtin, 1996, Ye
Boris Gurevich, (D), Moscow, 1988, Ye

Senior Research Fellow:
Paul G. Wilkes, (M), London, 1968, Ye

Senior Lecturer:
Andrej Bona, (D), Calgary, 2002, Yx
Milovan Urosevic, (D), Curtin, 1998, YerYx

Research Fellow:
Miroslav Brajanovski, (D), Curtin, 2004, Yex
Aleksandar Dzunic, (M), Belgrade, 1989, Ygs
Brett D. Harris, (D), Curtin, 2002, HwYve
Maxim Lebedev, (D), Mowcow Inst Phys & Tech, 1990, Yx
Andrew Peter Squelch, (D), Nottingham, 1998, Nmr
Michael P. Sykes, (D), Curtin, 2000, Yve

Associate Professor:
Bruce M. Hartley, (D), W Australia, 1969, Yx
Jayson Meyers, (D), Miami, 1993, Yx

Postdoc:
Mark J. Lwin, (D), QUT, 2002, Ye

Adjunct Professor:
John A. McDonald, (D), S Methodist, 1972, Yg
Norman F. Uren, (D), Houston, 1989, Ye

Laboratory Director:
Dominic J. Howman, Curtin, 1994, Ye

Flinders University

School of Geography, Population & Environmental Management (B,M,D) (2007)
GPO Box 2100
Adelaide, SA 5001
clive.forster@flinders.edu.au
http://www.ssn.flinders.edu.au/geog/
Enrollment (2006): M: 50 (0) D: 20 (0)

James Cook University

School of Earth Sciences (B,M,D) (2004)
Townsville, QLD 4811
earthsciences@jcu.edu.au
http://www.es.jcu.edu.au/

Latrobe University

Environmental Geoscience (2007)
Melbourne, VIC 3086
p. +61 3 9479 1273 f. +61 3 9479 1272
john.webb@latrobe.edu.au
http://www.latrobe.edu.au/envsci/

Associate Professor:
John A. Webb, (D), Queensland, 1982, Ge

Macquarie University

Dept of Earth and Planetary Sciences (2007)
North Ryde, NSW 2109
p. 61-2-98506314 f. 61-2-98506904
kdadd@els.mq.edu.au
http://www.es.mq.edu.au/geology/

Div of Environmental & Life Sciences (2004)
del@atlas.es.mq.edu.au
http://www.els.mq.edu.au/

Graduate School of the Environment (2004)
North Ryde
NSW 2109
jmonckto@gse.mq.edu.au
http://www.gse.mq.edu.au/

Monash University

School of Geosciences (2005)
PO Box 28E
Clayton, VIC 3800
p. +61 3 99054884 f. +61 3 99054903
hos@sci.monash.edu.au
http://www.earth.monash.edu.au/

Queensland University of Technology

School of Natural Resource Sciences (2006)
GPO Box 2434
2 George Street
Brisbane, QLD 4001
p. 61 7 3864 2324 f. 61 7 3864 1535
d.gust@qut.edu.au
http://nrs.sci.qut.edu.au/
Administrative Assistant: Janet Nulty

Associate Professor:
David A. Gust, (D), RSES - ANU, 1982, Gi

Professor:
Peter R. Grace, (D), Queensland, 1989, Sb

Associate Professor:
Malcolm E. Cox, (D), Auckland (NZ), 1986, Hw

Research Associate:
Andy Hammond, (D), Massey (NZ), 1998, Gm
Micaela Preda, (D), QUT, Oi

Senior Lecturer:
Stephen McLoughlin, (D), Queensland, 1990, Pg
Gregory E. Webb, (D), Queensland, 1989, Gs

Lecturer:
Gary J. Huftile, (D), Oregon State, 1992, Gc
Robin N. Thwaites, (D), Queensland, 2001, Os

Laboratory Director:
William Kwiecien, (M), QUT, Ca

Other:
Nate Peterson, (B), Oi
John Rigby, Pb

Southern Cross University

Faculty of Resource Science & Management (2004)
P. O. Box 157
Lismore, NSW
dgartsid@scu.edu.au
http://www.scu.edu.au/schools/esm

Professor:
Donald Gartside, (D), Melbourne
David Murray McConchie, Ge
Peter Saenger, (D), Melbourne
Leon Zann, (D), New England

Associate Professor:
Bill Boyd, (D), Glasgow, Pe
Bradley Eyre, (D), Queensland Univ of Tech, Gs
Vicki Harriot, (D), James Cook
Peter Harrison, (D), James Cook
Graham Jones, (D), James Cook, Ca
Stephan Schneirer, (D), Queensland
Alison Specht, (D), Queensland, Sf
Leigh Sullivan, (D), Sydney, Sc

Research Associate:
Malcolm William Clark, (D), Ca

Lecturer:
Danny Bucher, (D), Southern Cross

Murray Cullen, (M), Griffith, Sc
Stephen Duthy, (B), Queensland, Ou
Simon Hartley, (B), New England
Amanda Reichelt-Brushett, (D), Southern Cross

The Flinders University of South Australia

School of Earth Sciences (2005)
Bedford Park, SA
Warren.Lawrance@flinders.edu.au
http://www.es.flinders.edu.au/

University of Adelaide

National Centre for Petroleum (M,D) (2004)
NCPGG
Adelaide, SA 5005
p. 61-8-8303-4299 f. 61-8-8303-4345
admin@ncpgg.adelaide.edu.au
http://www.ncpgg.adelaide.edu.au/

University of Canberra

School of Resource, Environmental and Heritage Sciences (2004)
P. O. Box 1
Belconnen, ACT
rehshos@science.canberra.edu.au
http://scides.canberra.edu.au/rehs/

Freshwater Ecology (2004)
cullen@lake.canberra.edu.au
enterprise.canberra.edu.au/WWW/www-crcfe.nsf

University of Melbourne

School of Earth Sciences (2005)
Grattan Street
Melbourne, VIC 3010
p. +61 3 8344 6520 f. +61 3 8344 7761
head@unimelb.edu.au
http://www.earthsci.unimelb.edu.au/

University of New England

Dept of Geology & Geophysics (2004)
Armidale, NSW
pashley@metz.une.edu.au
http://www.une.edu.au/~geology/home.html

Div of Earth Sciences (2004)
Armidale, NSW 2351
pflood@metz.une.edu.au
http://www.une.edu.au/~geology/home.html

University of New South Wales

School of Biological, Earth & Environmental Sciences (B,M,D) (2007)
Sydney, NSW 2052
p. 61 2 93852961
bees@unsw.edu.au
http://www.bees.unsw.edu.au/index.html
Enrollment (2007): B: 81 (29) M: 6 (0) D: 7 (3)

Professor:
Ian Acworth, (D), Birmingham (UK), 1981, Hw
Professor:
David R. Cohen, (D), New South Wales, 1990, Ce
Ian Graham, (D), UTS Sydney, 1995, GxEg
Paul G. Lennox, (D), Monash, 1985, Gc
Associate Professor:
Geoff R. Taylor, (D), New England, 1977, Oi
Adjunct Professor:
Derecke Palmer, (D), New South Wales, 2001, Ye
Visiting Professor:
Colin R. Ward, (D), New South Wales, 1971, Gs

University of Newcastle

Discipline of Earth Sciences (B,M,D) (2007)
School of Environmental & Life Sciences
University Drive
Callaghan, NSW 2308
p. +61 2 4921 5411 f. +61 2 4921 6925
Phil.Geary@newcastle.edu.au
www.newcastle.edu.au/school/environ-life-science

Head:
Phil Geary, (D), UWS, HwOsHw
Associate Professor:
Ron L. Boyd, (D), Gu
Robin Offler, (D), Gp
Philip K. Seccombe, (D), Manitoba, 1973, Em
Lecturer:
Judy Bailey, (D), Ec
Russell Drysdale, (D), New South Wales, 1995, Gm
Silvia Frisia, (D), ClcCl
Gregory Hancock, (D), OiGmm
Bill Landenberger, (D), Newcastle, Gi
Glen Phillips, (D), Melbourne, GcpGc

University of Queensland

Dept of Geology & Mineralogy (2004)
Brisbane, QLD
j.jell@earthsciences.uq.edu.au
http://www.earthsciences.uq.edu.au/

University of Sydney

School of Geosciences (2006)
Madsen Building F09
Sydney, NSW
geoffc@mail.usyd.edu.au
http://www.usyd.edu.au/su/geosciences/

Head:
Geoffrey L. Clarke, (D), Melbourne, 1987, Gp

University of Tasmania

School of Earth Sciences (2005)
Private Bag 79
Hobart, TAS 7001
p. 61 3 6226 2472 f. 61 3 6226 7662
D.Cooke@utas.edu.au
http://fcms.its.utas.edu.au/scieng/earthsci

Director:
Ross R. Large, (D), New England, Eg
Head:
J Bruce Gemmell, (D), Dartmouth Coll, Eg
Professor:
Anthony J. Crawford, (D), Melbourne, Gi
Jocelyn McPhie, (D), New England, Gv
Associate Professor:
Clive Burrett, (D), Pg
David R. Cooke, (D), Monash, 1992, Eg
Lecturer:
Garry J. Davidson, (D), Tasmania, Cs
Peter J. McGoldrick, (D), Melbourne, Ce
James Reid, (D), Macquarie, Yg
Michael Roach, (N), Tasmania, Yg

University of Western Australia
School of Earth & Geographical Sciences (2007)
Mail Delivery Point M004
35 Stirling Highway
Nedlands, WA
p. 6488 2673
david.haig@uwa.edu.au
http://www.segs.uwa.edu.au/

University of Wollongong
School of Earth & Environmental Sciences (B,M,D) (2004)
Northfields Avenue
Wollongong, NSW 2522
p. 61 2 4221 4419 f. 61 2 4221 4250
colin_murray-wallace@uow.edu.au
http://www.uow.edu.au/science/eesc/

Austria

Univ. Leoben
Dept of Geology (2005)
Montauniversitat Leoben
Kaiser Franz Josef-Strasse 18
Leoben, A-8700
geologie@unileoben.ac.at
www.unileoben.ac.at/~buero62/geologie/geologie.html

Univ. of Vienna
Dept of Geology (2005)
Althanstrasse 14
Vienna, A-1090
p. 00431427753401
Geologie@univie.ac.at
http://www.univie.ac.at/Geologie/

Head:
Thilo Hofmann, (D)

University of Innsbruck
Inst of Geology and Paleontology (2006)
Innrain
A-6020
Innsbruck, 6020
p. +43-512-507-5581 f. +43-512-507-2914
Geologie@uibk.ac.at
http://www.uibk.ac.at/geologie

Azerbaijan

Baku State University
The Faculty of Geology (2005)
23 Z. Khalilov Street
370145 Baku
geology@bsu.az
http://www.ceebd.co.uk/ceeed/un/az/az003.htm

Belgium

K.U. Leuven
Geology Section (2006)
Celestijnenlaan 200E
B-3001 Heverlee- Leuven
Leuven, 3001
p. 00 32 16 326438 f. 00 32 16 326401
noel.vandenberghe@geo.kuleuven.be
http://www.kuleuven.ac.be/facdep/geo/hsg/

Universite de Liege
Dept of Geology (B,M,D) (2007)
Batiment B18 (secretariat)
Sart Tilman
Liege, B-4000
p. 32 4 366 22 51 f. 32 4 366 29 21
V.Miocque@ulg.ac.be
http://www.ulg.ac.be/geolsed/geologie

chargé de cours:
Nathalie Fagel, (D), GsuGe
Hans-Balder Havenith, (D), Liège, Ge

Professor:
Frederic P. Boulvain, (D), Brussels, 1990, Gs
Andre-Mathieu Fransolet, (D), Liege, Gz
Emmanuelle Javaux, (D), Dalhousie, 1999, Pl
Edouard Poty, (D), Liege, Pi
Jacqueline Vander Auwera, (D), Louvain-la-Neuve, Gx

Vrije University Brussel
Geology Dept (2007)
Faculty of Sciences
VUB - DGLG
Pleinlaan 2
Brussels, 1050
p. 0113226293394 f. 0113226293391
phclaeys@vub.ac.be
http://we.vub.ac.be/dglg

Head:
Philippe Claeys, (D), California, 1993, CgPyXc

Professor:
Edward Keppens, (D), Vrije, 1983, CsGeg

Brazil

Federal University of Bahia Geophysics
CPGG (2004)
Salvador, BA
yeda@cpgg.ufba.br
http://www/pppg.ufba.br/

Universidade De Brasilia
Inst de Geosciencias (2004)
Campus Universitario - Asa Norte
CEP 70910-900
Brasilia - DF
gasp@unb.br
http://www.unb.br/ig/indexenglish.html

University of Campinas Geoscience Institute
Inst de Geosciencias (2004)
Caixa Postal: 6152
13083-970
Campinas, SP,
newpe@ige.unicamp.br
http://www.ige.unicamp.br/

Bulgaria

Bulgarian Academy of Sciences (BAS)
Geological Inst (M,D) (2004)
Acad G Bonchev St
Block 24
Sofia, 1113
hrischo@geology.bas.bg
http://www.geology.bas.bg/

Sofia University St. Kliment Ohridski

Climatology and Hydrology (2004)
1504 Sofia
15 Tsar Osvobodtel Blvd.
General Climatology, Climate Change
room 263
topli@gea.uni-sofia.bg
http://www.uni-sofia.bg/newweb/faculties/geo/departments/cli

Dept of Geology and Paleontology (2005)
1504 Sofia
15 Tsar Osvobodtel Blvd.
Geology of Fossil, Fuel, Organic Petrology, Organic Geochemistry
room 276
decan@gea.uni-sofia.bg
http://www.uni-sofia.bg/newweb/faculties/geo/departments/geo

Dept of Cartography & GIS (2006)
1504 Sofia
15 Tsar Osvobodtel Blvd.
GIS, General Physical Geography, Landscape Ecology, Agroecology
Sofia, Sofia 1504
p. 9308261
popov@gea.uni-sofia.bg
http://www.gis.gea.uni-sofia.bg/

Univ of Mining and Geology, St. Ivan Rilski

Faculty of Geology (2004)
Studentski grad
1100 Sofia
webmaster@mgu.bg
http://www.mgu.bg/frame/html

Canada

Acadia University

Dept of Earth and Environmental Science (B,M) (2007)
Huggins Science Hall
12 University Avenue
Wolfville, NS B4P 2R6
p. (902) 585-1208 f. (902) 585-1816
ees@acadiau.ca
http://ace.acadiau.ca/science/ees/
Enrollment (2005): B: 31 (25) M: 8 (4)

Head:
Robert P. Raeside, (D), Calgary, 1982, Gp
Professor:
Linda A. Lusby, (M), Alberta, 1978, Oo
Professor:
Sandra M. Barr, (D), British Columbia, 1973, Gi
Ian S. Spooner, (D), Calgary, 1994, Ge
Associate Professor:
Clifford R. Stanley, (D), British Columbia, 1988, Cg
Assistant Professor:
Peir K. Pufahl, (D), British Columbia, 2001, Gs
Instructor:
David McMullin, (D), British Columbia, 1991, Oe
Geology Tech:
Don Osburn, (N), Gg

Athabasca University

Centre for Natural and Human Science (2005)
Box 10,000
Athabasca, AB T9S 1A1
p. (403) 675-6168 f. (403) 675-6186
jackp@athabascau.ca
Administrative Assistant: Susan Stahl

Associate Professor:
Donald R. Kvill, (D), Alberta, 1984, Gm

Brandon University

Dept of Geology (B) (2005)
270 18th Street
Brandon, MB R7A 6A9
p. (204) 727-9680 f. (204) 728-7346
monita@brandonu.ca
http://www.brandonu.ca/geology
Enrollment (2004): B: 31 (10)

Professor:
Robert K. Springer, (D), California (Davis), 1971, Gi
Associate Professor:
A. Hamid Mumin, (D), W Ontario, 1994, Eg
Simon A. J Pattison, (D), McMaster, 1992, Gs
Assistant Professor:
Rong-Yu Li, (D), Alberta, 2002, Pg
Emeritus:
Harvey R. Young, (D), Queen's, 1973, Gd
Instructional Associate:
Peter J. Adamo, (B), Brandon, 2000, Gg

Brock University

Dept of Earth Sciences (B,M) (2006)
500 Glenridge Avenue
St. Catharines, ON L2S 3A1
p. (965) 688-5550 (Ext. 3526) f. (965) 682-9020
earth@brocku.ca
http://www.brocku.ca/earthsciences/
Enrollment (1999): B: 63 (6) M: 6 (1)

Professor:
Uwe Brand, (D), Ottawa, 1979, Cl
Richard J. Cheel, (D), McMaster, 1984, Gs
Frank Fueten, (D), Toronto, 1989, Gc
Martin J. Head, (D), Aberdeen, 1990, Pl
Wayne T. Jolly, (D), SUNY (Binghamton), 1970, Gx
Francine G. McCarthy, (D), Dalhousie, 1992, Pm
John Menzies, (D), Edinburgh, 1976, Gl
Keith J. Tinkler, (D), Liverpool, 1968, Gm
Associate Professor:
Gregory C. Finn, (D), Memorial, 1989, Gx
Daniel P. McCarthy, (D), Saskatchewan, 1993, Gl
Adjunct Professor:
Stephen J. Cook, (D), Utah, 1992, Hw
Andrew W. Panko, (D), McMaster, 1985, Ge

Cambrian College of Applied Arts & Tech

School of Computer and Engineering Tech (A) (2006)
1400 Barrydowne Road
Sudbury, ON P3A 3V8
p. (705) 566-8101 Ext. 7839 f. (705) 521-0561
gpdalton@cambrianc.on.ca
http://www.cambrianc.on.ca
Enrollment (2004): A: 48 (0)

Professor:
Frank Ingham, (B), Queen's, 1974, Nm
Anthony Insinna, (B), Waterloo, 1984, Ye
Robert H. Johnson, (B), Queen's, 1974, Nm
Timothy A. Jones, (B), McGill, 1975, Eg
Randy Junnila, (M), W Ontario, Eg
Elena Murphy, (M), Laurentian, 1999, Nm
Jana Wicke, (M), Mining Univ of Czech, Nm

Dean:
Geoffrey Dalton, (M), CUNY, 1987

Capilano College

Geology Dept (B) (2004)
Pure and Applied Sciences Division
2055 Purcell Way
North Vancouver, BC V7J 3H5
p. (604) 984-1771 f. (604) 984-1718
dathaide@capcollege.bc.ca
Administrative Assistant: Kathy Chute
Enrollment (1999): B: 33 (0)

Head:
Dileep J. Athaide, (M), British Columbia, 1975, Oe
Instructor:
Arnold G. Thomlinson, (M), British Columbia, 1954, Oe
Geology Librarian:
George Modenesi, Oo

Carleton University

Dept of Earth Sciences (B,M,D) (2004)
1125 Colonel By Drive
2240 Herzberg Laboratories
Ottawa, ON K1S 5B6
p. (613) 520-2600, Ext. 8912 f. (613) 520-2569
earthsciences@carleton.ca
http://www.earthsci.carleton.ca
Enrollment (2002): B: 42 (9) M: 21 (5) D: 18 (4)

Professor:
Gail M. Atkinson, (D), W Ontario, 1993, Ne
Keith Bell, (D), Oxford, 1964, Cc
R. Timothy Patterson, (D), California, 1986, Pm
Giorgio Ranalli, (D), Illinois, 1970, Yg
Claudia Schroder-Adams, (D), Dalhousie, 1986, Pm
Richard Taylor, (D), Leicester, 1980, Cg
David H. Watkinson, (D), Penn State, 1965, Em
Associate Professor:
John Blenkinsop, (D), British Columbia, 1972, Cc
George Dix, (D), Syracuse, 1988, Gs
Frederick A. Michel, (D), Waterloo, 1982, Hw
Assistant Professor:
Sharon D. Carr, (D), Carleton, 1990, Gt
Instructor:
Ildiko Munro, (M), Waterloo, 1975, Pm
Adjunct Research Professor:
Brian Cousens, (D), California, 1990, Cc
J. Allan Donaldson, (D), Johns Hopkins, 1960, Gs
Adjunct Professor:
Robert Berman, (D), British Columbia, 1983, Gp
Harold Gibson, (D), Carleton, 1990, Em
Mark D. Hannington, (D), Toronto, 1989, Em
Jarmila Peck, (D), Charles (Prague), 1962, Pi
Emeritus:
F. Kenneth North, (D), Oxford, 1951, Go
George B. Skippen, (D), Johns Hopkins, 1966, Cg
Distinguished Research Professor:
Richard L. Brown, (D), Edinburgh, 1964, Gt

Dept of Earth Sciences (B,M,D) (2005)
1125 Colonel By Drive
Ottawa, ON K1S 5B6
p. (613) 520-2600, Ext 8912 f. (613) 520-3552
earth_sciences@carleton.ca

Chair:
George R. Dix, (D), Syracuse, 1988, Gs
Professor:
Keith Bell, (D), Oxford, 1964, Cc
R. Timothy Patterson, (D), California (Los Angeles), 1986, Pm
Giorgio Ranalli, (D), Illinois, 1970, Yg
Claudia Schroder-Adams, (D), Dalhousie, 1986, Pm
George B. Skippen, (D), Johns Hopkins, 1966, Cg
Richard P. Taylor, (D), Leicester, 1980, Cg
David H. Watkinson, (D), Penn State, 1965, Em
Associate Professor:
Gail M. Atkinson, (D), W Ontario, 1993, Ne
John Blenkinsop, (D), British Columbia, 1972, Cc
Sharon D. Carr, (D), Carleton, 1990, Gt
Fred A. Michel, (D), Waterloo, 1982, Hw
Lecturer:
Ildi Munro, (M), Waterloo, 1975, Pg
Adjunct Professor:
Robert Berman, (D), British Columbia, 1983, Gp
Brian L. Cousens, (D), California (Santa Barbara), 1990, Cc
Steve L. Cumbaa, (D), Florida, 1975, Pv
J. Allan Donaldson, (D), Johns Hopkins, 1960, Gs
T. Scott Ercit, (D), Manitoba, 1986, Gz
Harold Gibson, (D), Carleton, 1990, Em
Simon Hanmer, (D), Chelsea (UK), 1977, Gc
Mark D. Hannington, (D), Toronto, 1989, Em
Jarmila Kukalova-Peck, (D), Charles (Prague), 1962, Pi
Dale A. Leckie, (D), McMaster, 1983, Gs
R. R. Rainbird, (D), Western, 1991, Gs
Emeritus:
F. K. North, (D), Oxford, 1951, Go

Ottawa-Carleton Geoscience Centre (B,M,D) (2004)
Dept of Earth Sciences
College of Natural Sciences
1125 Colonel By Drive
Ottawa, ON K1S 5B6
p. (613) 520-2600 (Ext. 8769) f. (613) 520-5613
sheila_thayer@carleton.ca
http://www.earthsci.carleton.ca/OCGC/index.html

Director:
John Blenkinsop, (D), British Columbia, 1972, Cc

Dalhousie University

Dept of Earth Sciences (B,M,D) (2004)
Halifax, NS B3H 3J5
p. (902) 494-2358 f. (902) 494-6889
earth.sciences@dal.ca
http://www.dal.ca/~es
Administrative Assistant: Jane Barrett
Enrollment (1999): B: 69 (27) M: 14 (5) D: 20 (6)

Chair:
Martin R. Gibling, (D), Ottawa, 1978, Gs
Professor:
D. Barrie Clarke, (D), Edinburgh, 1969, Gi
Rebecca A. Jamieson, (D), Memorial, 1979, Gx
Peter H. Reynolds, (D), British Columbia, 1967, Cc
David B. Scott, (D), Dalhousie, 1977, Pm
Grant D. Wach, (D), Oxford, 1993, Go
Associate Research Professor:
Gunter K. Muecke, (D), Oxford, 1969, Cg
Associate Professor:
Nicholas Culshaw, (D), Ottawa, 1983, Gc
John C. Gosse, (D), Lehigh, 1994, Cg
Djordje Grujic, (D), ETH (Switzerland), 1992, Gt
Patrick J.C. Ryall, (D), Dalhousie, 1974, Yg
Assistant Professor:
Lawrence Plug, (D), Alaska (Fairbanks), 2000, Gm
Research Associate:
Isabelle Coutand, (D), Rennes, Gg
Lubomir F. Jansa, (D), Charles (Prague), 1967, Gr
Robert Raeside, (D), Calgary, 1982, Gx
Alan Ruffman, (M), Dalhousie, 1966, Ys

John Waldron, (D), Edinburgh, 1981, Gc
Senior Instructor:
Peter I. Wallace, (M), McMaster, 1975, Gg
Instructor:
Charles C. Walls, (M), Dalhousie, 1996, Oi
Honorary Research Associate:
Prasanta Mukhopadhyay, (D), Jadavpur, 1971, Ec
Graham Williams, (D), Sheffield, 1964, Pl
Adjunct Professor:
Juergen Adam, (D), Tech (Berlin)
Alan J. Anderson, (D), Queen's, Cg
Sandra Barr, (D), British Columbia, 1973, Gx
Hugo Beltrami, (D), Quebec (Montreal), Gg
John Calder, (D), Dalhousie, 1991, Ec
Sonya Dehler, (D), British Columbia, Gg
Jarda Dostal, (D), McMaster, 1974, Cg
Don Fox, (D), Dalhousie, Gg
Paul T. Gayes, (D), SUNY (Stony Brook), 1986, Gs
Peter E. Jones, (D), British Columbia, 1963, Op
Lisa M. Kellman, (D), Quebec (Montreal), 1998, Cg
Daniel Kontak, (D), Queen's, Gg
Joel Kronfeld, (D), Rice, 1972, Cg
Michael Melchin, (D), Western, 1987, Pg
Peta J. Mudie, (D), Dalhousie, 1980, Pl
J. Brendan Murphy, (D), McGill, 1982, Gx
Michael Parsons, (D), Stanford, Gg
Georgia Pe-Piper, (D), Cambridge, 1971, Gv
David Piper, (D), Cantab, 1969, Gs
Paul T. Robinson, (D), California (Berkeley), 1964, Gv
Andre Rochon, (D), Quebec, Gg
Matthew Salisbury, (D), Washington, 1974, Yr
Ralph Stea, (D), Dalhousie, 1995, Gu
Hans J. Wielens, (D), Utrecht (Neth), 1979, Go
Emeritus:
H.B. S. Cooke, (D), Witwatersrand, 1947, Pv
Franco Medioli, (D), Parma, 1959, Pm
G. Clinton Milligan, (D), Harvard, 1961, Gc
Marcos Zentilli, (D), Queen's, 1974, Eg
On Leave:
Dorothy I. Godfrey-Smith, (D), Simon Fraser, 1991, Cc
Cooperating Faculty:
Christopher Beaumont, (D), Dalhousie, 1973, Ys

Dept of Oceanography (M,D) (2005)
Life Sciences Centre
1355 Oxford Street
Halifax, NS B3H 4J1
p. (902) 494-3557 f. (902) 494-3877
oceanography@dal.ca
http://www.dal.ca/oceanography
Administrative Officer: Dianne Crouse
Enrollment (2005): M: 24 (5) D: 27 (3)
Chair:
Bernard P. Boudreau, (D), Yale, 1985, Cl
Professor:
Christopher Beaumont, (D), Dalhousie, 1973, Yg
John J. Cullen, (D), California (San Diego), 1980, Ob
Jonathan Grant, (D), South Carolina, 1981, Ob
Richard J. Greatbatch, (D), Cambridge, 1981, Op
Alex E. Hay, (D), British Columbia, 1981, Op
Paul S. Hill, (D), Washington, 1992, Gs
Marlon R. Lewis, (D), Dalhousie, 1984, Ob
Keith E. Louden, (D), MIT, 1976, Yr
Robert M. Moore, (D), Southampton, 1977, Oc
Barry R. Ruddick, (D), MIT, 1977, Op
Keith R. Thompson, (D), Liverpool, 1979, Op
Associate Professor:
Dan Kelley, (D), Dalhousie, 1986, Op
Anna Metaxas, (D), Dalhousie, 1994, Ob
Jinyu Sheng, (D), Memorial, 1991, Op
Christopher T. Taggart, (D), McGill, 1986, Ob
Helmuth Thomas, (D), Rostock, 1997, Oc
Assistant Professor:
Markus Kienast, (D), British Columbia, 2002, Cs
Tetjana Ross, (D), Manitoba, 2003, Op
Adjunct Professor:
Carl M. Boyd, (D), California (San Diego), 1962, Ob
Monica V. Bricelj, (D), SUNY (Stony Brook), 1984, Ob
Peter Cranford, (D), Dalhousie, 1998, Ob
Katja Fennel, (D), Rostock, 1998, Ob
Kenneth Frank, (D), Toledo, 1978, Ob
Barry T. Hargrave, (D), Dalhousie, 1969, Ob
Glen Harrison, (D), North Carolina State, 1974, Ob
Jocelyn Hellou, (D), British Columbia, 1985, Oc
Ritske Huismans, (D), Vrije, 1999, Ou
Bruce D. Johnson, (D), Dalhousie, 1979, Oc
Richard H. Karsten, (D), Alberta, 1998
Maurice M. E. LeVasseur, (D), British Columbia, 1990, Ob
William K. W. Li, (D), Dalhousie, 1978, Ob
David C. Mosher, (D), Dalhousie, 1993, Ou
Neil S. Oakey, (D), McMaster, 1967, Op
David J.W. Piper, (D), Cambridge, 1969, Gu
Harold C. Ritchie, (D), McGill, 1982, Ow
Shubha Sathyendranath, (D), P & M Curie (France), 1981, Ob
Peter C. Smith, (D), MIT/Woods Hole, 1973, Op
Alain F. Vezina, (D), McGill, 1985, Ob
Daniel G. Wright, (D), British Columbia, 1978, Op
Emeritus:
Anthony J. Bowen, (D), California (San Diego), 1967, On
Robert O. Fournier, (D), Rhode Island, 1967, Ob
Eric L. Mills, (D), Yale, 1964, Ob

Douglas College

Dept of Geology, Faculty of Science and Technology (A) (2006)
P.O. Box 2503
New Westminster, BC V3L 5B2
p. (604) 527-5471/5222/5710 f. (604) 527-5095
waddingtond@douglas.bc.ca
Enrollment: No data reported since 1995
Chair:
David Waddington, (M), Queen's (Kingston), 1991, Gg
Chair, Department of Geology:
Michael C. Wilson, (D), Calgary, 1981, Ga
Instructor:
Brett Gilley, (M), Simon Fraser, 2003, Gg
Safdar Kakar, (D), Aberystwyth (Wales), 1971, Gg
Dean, Faculty of Science & Technology:
T. Desmond Wilson, (M), British Columbia, 1983, Eg
Related Staff:
Cindy Hansen, (B), Simon Fraser, Gg

Ecole Polytechnique

Dept de Genie Mineral (B,M,D) (2004)
C.P. 6079 Succ. Centre-ville
Montreal, PQ H3C 3A7
p. (514) 340-4777 f. (514) 340-4191
Administrative Assistant: Gaby Roy
Enrollment (2000): B: 107 (0) M: 11 (0) D: 10 (0)
Chair:
Robert Chapuis, (D), Ecole Polytechnique, 1976, Hw
Professor:
Alex C. Brown, (D), Michigan, 1968, Cl
Michel Chouteau, (D), Ecole Polytechnique, 1982, Ye
Jorgen Elbrond, (M), Copenhague, 1953, Nm
Denis E. Gill, (D), McGill, 1970, Nr

Marc G. Tanguay, (D), Purdue, 1969, Or
Associate Professor:
Michel Aubertin, (D), Ecole Polytechnique, 1989, Nm
Robert Corthesy, (D), Ecole Polytechnique, 1989, Nm
Richard Darling, (D), Stanford, 1968, Ce
Denis Marcotte, (D), Ecole Polytechnique, 1987, Oo
Research Associate:
Andrew Calvert, (D), Cambridge, 1985, Ye
Dean Livelybrooks, (D), Oregon, 1990, Ye
Lecturer:
Louis Bernier, (D), McGill, 1992, Gz
Adjunct Professor:
Jean Berard, (D), Laval, 1964, Ng
Martin Doyon, (D), Ecole Polytechnique, 1992, Hw
Jacques Oullet, (D), Oo
Andre Piche, (D), Ecole Polytechnique, 1984, Nm
Donald J. White, (D), British Columbia, 1989, Ye

Lakehead University

Dept of Geology (B,M) (2007)
955 Oliver Road
Thunder Bay, ON P7B 5E1
p. (807) 343-8461 f. (807) 346-7853
kristine.carey@lakeheadu.ca
http://www.lakeheadu.ca/~geology
Administrative Assistant: Kristine M. Carey
Enrollment (2007): B: 56 (18) M: 11 (3)
Professor:
Graham J. Borradaile, (D), Liverpool, 1971, Gc
Philip W. Fralick, (D), Toronto, 1985, Gs
Mary Louise Hill, (D), Princeton, 1985, Gc
Stephen A. Kissin, (D), Toronto, 1974, Em
Associate Professor:
Peter N. Hollings, (D), Saskatchewan, 1998, Eg
Assistant Professor:
Andrew G. Conly, (D), Toronto, 2003, Eg
Emeritus:
Manfred M. Kehlenbeck, (D), Queen's, 1971, Gc
Edward L. Mercy, (D), Imperial Coll (UK), 1955, Cg
Roger H. Mitchell, (D), McMaster, 1969, Gi
Geology Tech:
Anne Hammond, (B), Lakehead, 1999, Oo
Geology Map Librarian:
S. T. Spivak, (N), Oo

Laurentian University, Sudbury

Dept of Earth Sciences (B,M,D) (2007)
Ramsey Lake Road
Sudbury, ON P3E 2C6
p. (705) 675-1151 (Ext. 6575) f. (705) 675-4898
rmehes@nickel.laurentian.ca
Administrative Assistant: Roxane J. Bourgouin-Mehes
Enrollment (2007): B: 48 (14) M: 32 (4) D: 6 (0)
Chair:
Andrew M. McDonald, (D), Carleton, 1992, Gz
Canada Research Chair, Precambrian Geology:
Balz Kamber, (D), Bern, 1995, Cs
Professor:
Michael Lesher, (D), W Australia, 1984, Em
Darrel Long, (D), W Ontario, 1976, Gs
David A. Pearson, (D), London, 1967, Oe
Associate Professor:
Ann Gallie, (D), British Columbia, 1990, Or
Harold L. Gibson, (D), Carleton, 1990, Em
Daniel J. Kontak, (D), Queens, 1985, Eg
Bruno Lafrance, (D), New Brunswick, 1990, Gc
Stephen Piercey, (D), British Columbia, 2001, Ce
Assistant Professor:
Pedro J. Jugo, (D), Alberta, 2003, Gi
Doug Tinkham, (D), Alabama, 2002, Gp
Elizabeth Turner, (D), Queens, 1999, Pg
Emeritus:
Anthony E. Beswick, (D), London, 1965, Gi
Paul Copper, (D), London, 1965, Pi
James F. Davies, (D), Toronto, 1963, Em
Richard James, (D), Manchester, 1967, Gp
Reid R. Keays, (D), McMaster, 1968, Em
Don H. Rousell, (D), Manitoba, 1965, Gc
Robert E. Whitehead, (D), New Brunswick, 1973, Ce

McGill University

Dept of Atmospheric & Oceanic Sci (B,M,D) (2006)
805 Sherbrooke Street West
Room 945
Montreal, PQ H3A 2K6
p. (514) 398-3764 f. (514) 398-6115
graduateinfo.aos@mcgill.ca
http://www.mcgill.ca/meteo
Administrative Assistant: Ornella N. Cavaliere
Enrollment: No data reported since 1995
Professor:
Jacques F. Derome, (D), Michigan, 1968, Oa
John R. Gyakum, (D), MIT, 1981, Ow
Henry G. Leighton, (D), Alberta, 1968, Oa
Charles A. Lin, (D), MIT, 1979, Oa
Lawrence A. Mysak, (D), Harvard, 1966, Op
Ronald E. Stewart, (D), Oa
Man Kong Yau, (D), MIT, 1977, Oa
Isztar I. Zawadzki, (D), McGill, 1972, Oa
Associate Professor:
Parisa A. Ariya, (D), York, 1996, Oa
Peter Bartello, (D), McGill, 1988, Oa
Frederic Fabry, (D), McGill, 1994, Oa
David Straub, (D), Washington, 1990, Op
Assistant Professor:
Michel Bourqui, (D), ETH, Oa
Bruno Tremblay, (D), McGill, Op
Adjunct Professor:
Gilbert Brunet, (D), McGill, 1989, Oa
Pierre Gauthier, (D), McGill, 1988, Oa
Stephane Laroche, (D), McGill, 1995, Oa
Hai Lin, (D), McGill, 1995, Oa
Richard Menard, (D), McGill, 1994, Oa
Ayrton Zadra, (D), McGill, 2001, Oa

Dept of Earth & Planetary Sciences (B,M,D) (2007)
3450 University Street
Room 238
Montreal, PQ H3A 2A7
p. (514) 398-6767 f. (514) 398-4680
kiki@eps.mcgill.ca
http://www.eps.mcgill.ca
Enrollment (2007): B: 35 (12) M: 17 (5) D: 23 (1)
Chair:
John Stix, (D), Toronto, 1989, Gv
Retired:
Reinhard Hesse, (D), Tech (Munich), 1964, Gs
Robert F. Martin, (D), Stanford, 1969, Gi
Professor:
Don Baker, (D), Penn State, 1985, Cp
Don Francis, (D), MIT, 1974, Gi
Andrew J. Hynes, (D), Cambridge, 1972, Gc
Olivia G. Jensen, (D), British Columbia, 1971, Yg
Alfonso O. Mucci, (D), Miami, 1981, Cm
Anthony E. Williams-Jones, (D), Queen's, 1973, Ce

Diector, Electron Microscopy Res:
H. Vali, (D), Munich, 1983, Om
Associate Professor:
Bruce Hart, (D), W Ontario, 1990, Go
Jeanne Paquette, (D), SUNY (Stony Brook), 1991, Gz
Michael Riedel, (D), Victoria, 2001, Ye
Earth System Science:
Jeffrey M. McKenzie, (D), Syracuse, 2005, Hw
Boswell Wing, (D), Johns Hopkins, 2004, Cs
Director, TEAL; Earth System Science:
William G. Minarik, (D), RPI, 1993, Cp
Adjunct Professor:
Mathieu Duchesne, (D), Laval, 2005, Yr
Hans Hofmann, (D), McGill, Gs
Heather Short, (D), Maine, 2006, Gc
Bjorn Sundby, (D), Bergen (Norway), 1966, Cm
Emeritus:
Jafar Arkani-Hamed, (D), MIT, 1969, Xy
Wallace H. MacLean, (D), McGill, 1968, Em
Eric W. Mountjoy, (D), Toronto, 1960, Gd

Dept of Mining & Metallurgical Eng (B,M,D) (2005)
3450 University Street
Montreal, PQ H3A 2A7
p. (514) 398-4352 f. (514) 398-4492
robin.drew@mcgill.ca
Administrative Assistant: Carol Rousseau
Enrollment (2000): B: 6 (0)

Chair:
Robin A.L. Drew, (D), Newcastle, 1980, Nx
Professor:
George Demopoulos, (D), McGill, 1982, Nx
James A. Finch, (D), McGill, 1973, Nx
John E. Gruzleski, (D), Toronto, 1967, Nx
Roderick I. Guthrie, (D), London, 1967, Nx
Ferri Hassani, (D), Nottingham, 1981, Nm
John J. Jonas, (D), Cambridge, 1960, Nx
Gordon W. Smith, (D), McGill, 1967, Nx
Associate Professor:
Michel L. Bilodeau, (D), McGill, 1975, Nm
Philip A. Distin, (D), London, 1967, Nx
Ralph Harris, (D), McGill, 1980, Oo
Mainul Hasan, (D), McGill, 1987, Nx
Andre R. LaPlante, (D), Toronto, 1980, Nx
Hani Mitri, (D), Nottingham, 1981, Nm
Frank Mucciardi, (D), McGill, 1980, Nx
Jaques Ouellet, (D), Ecole Polytechnique, 1992, Nm
Jerzy Szpunar, (D), Kracow, 1969, Oo
Stephen Yue, (D), Leeds, 1979, Nx
Assistant Professor:
Janusz A. Kozinski, (D), Krakow, 1986, Nx
Lecturer:
Moe Momayezzadeh, (D), McGill, 1993, Nm
John W. Mossop, (B), McGill, 1955, Nm
Adjunct Professor:
Wilfrid Comeau, (M), Penn State, 1977, Nm
Roussos Dimitrakopoulos, (D), Ecole Polytechnique, 1989, Nm
Bryn G. Harris, (D), Birmingham, 1972, Nx
Ahmad Hemani, (D), Salford (UK), Nm
Yves Lizotte, (M), Ecole Polytechnique, 1980, Nm
Bibhu Mohanty, (D), Nm
Malcolm J. Scoble, (D), Nottingham, 1981, Nm
Josef Stachulak, (D), McGill, 1992, Nm
William T. Thompson, (D), Toronto, 1970, Nx
Gus Van Weert, (D), Netherlands, 1956, Nx
Albert E. Wraith, (D), Newcastle, 1968, Nx

Emeritus:
William M. Williams, (D), Toronto, 1960, Nx

McMaster University

School of Geography & Earth Sciences (B,M,D) (2005)
1280 Main Street West
Burke Science Building
Room 235
Hamilton, ON L8S 4M1
p. (905) 525-9140 (Ext. 23535) f. (905) 546-0463
wallann@mcmaster.ca
http://www.science.mcmaster.ca/geo/geomain.html
Enrollment (2005): B: 272 (0) M: 0 (11) D: 0 (2)

Director:
John J. Drake, (D), McMaster, 1973, Yr
Professor:
Vera Chouinard, (D), McMaster, 1986, Oo
Alan P. Dickin, (D), Oxford, 1981, Cc
Susan J. Elliott, (D), McMaster, 1992, Yr
Carolyn H. Eyles, (D), Toronto, 1986, Gs
John D. Eyles, (D), London, 1983, Yr
Fred L. Hall, (D), Chicago, 1972, Yr
Richard S. Harris, (D), Queen's, 1981, Yr
Pavlos S. Kanaroglou, (D), McMaster, 1987, Yr
Kao Lee Liaw, (D), Clark, 1973, Yr
William A. Morris, (D), Open, 1974, Yg
Ming-ko Woo, (D), British Columbia, 1972, Hg
Associate Professor:
Altaf M. Arain, (D), Arizona, 1997, Hg
Joseph I. Boyce, (D), Toronto, 1997, Yr
K. Bruce Newbold, (D), McMaster, 1994, Yr
Edward G. Reinhardt, (D), Carleton, 1996, Pm
William J. Rink, (D), Florida State, 1990, Cc
James E. Smith, (D), Waterloo, 1995, Hw
Michael Waddington, (D), York, 1995, Pe
Lesley A. Warren, (D), Toronto, 1994, Yr
Allison M. Williams, (D), York, 1997, Oo
Robert D. Wilton, (D), S California, 1999, Oo
Assistant Professor:
Paulin Coulibaly, (D), Laval, 2000, Hg
Darren R. Grocke, (D), Oxford, 2001, Cs
H. Antonio Paez, (D), Tohoku, 2000, Oo
Darren M. Scott, (D), McMaster, 2000, Oo
Lecturer:
Gordon R. Hopper, (D), York, 1993
Walter G. Peace, (D), McMaster, 1996, Yr
Susan Vajoczki, (M), McMaster, 1993, Yr
University Professor:
Henry P. Schwarcz, (D), Caltech, 1960, Cs
Emeritus:
Andrew Burghardt, (D), Yr
Paul Clifford, (D), Yr
James H. Crocket, (D), MIT, 1961, Yr
John Davies, (D), Yr
Derek C. Ford, (D), Oxford, 1963, Gm
Doug Grundy, (D), Yr
James R. Kramer, (D), Michigan, 1958, Cl
Gerry V. Middleton, (D), Yr
Yorgos Papageorgiou, Yr
Michael J. Risk, (D), S California, 1971, Po
Wayne R. Rouse, (D), McGill, 1965, Oa
Gerd Westermann, Yr

Memorial University of Newfoundland

Dept of Earth Sciences (B,M,D) (2004)
Centre for Earth Resources Research, Room ER 4063
Alexander Murray Building
St. John's, NF A1B 3X5

p. (709) 737-8142 f. (709) 737-2589
mmoore@mun.ca
http://www.esd.mun.ca
Assistant to the Department Head: Maureen L. Moore
Secretary, CERR: Louise Green
Enrollment: No data reported since 1995

Head:
James A. Wright, (D), Toronto, 1968, Yh

Professor:
Ali E. Aksu, (D), Dalhousie, 1980, Gu
Elliott T. Burden, (D), Calgary, 1982, Pl
Gregory R. Dunning, (D), Memorial, 1984, Cc
John E. Gale, (D), California (Berkeley), 1975, Nr
Jeremy Hall, (D), Glasgow, 1971, Ys
Richard N. Hiscott, (D), McMaster, 1977, Gs
Joseph P. Hodych, (D), Toronto, 1971, Ym
George A. Jenner, (D), Tasmania, 1982, Gi
Hugh G. Miller, (D), British Columbia, 1973, Yv
John S. Myers, (D), Imperial Coll, 1968, Eg
Garry M. Quinlan, (D), Dalhousie, 1981, Gt
Toby C. J. S. Rivers, (D), Ottawa, 1976, Gp
Harold Williams, (D), Toronto, 1961, Gt
Derek H.C. Wilton, (D), Memorial, 1984, Eg

Associate Professor:
Tomas J. Calon, (D), Leiden (Neth), 1978, Gc
Charles A. Hurich, (D), Wyoming, 1988, Ys
Aphrodite D. Indares, (D), Montreal, 1989, Gp
Roger A. Mason, (D), Aberdeen, 1978, Gz
Gummuluru S. Murthy, (D), Alberta, 1969, Ym
Andrew J. Pulham, (D), Wales, 1988, Go
Michael A. Slawinski, (D), Calgary, 1996, Ys
Paul J. Sylvester, (D), Washington (St. Louis), 1984, Ca
Moire A. Wadleigh, (D), McMaster, 1989, Cs
Mark R. Wilson, (D), Saskatchewan, 1987, Cs

Assistant Professor:
Alison Leitch, (D), Australian National, 1986, Yg

Research Associate:
Richard Cox, (D), Memorial, 2000, Gp
James Crowley, (D), Carleton, 1997, Gc
Stephen W. Grimes, (D), Gc

First Year Lab Instructor:
Roberta (Robbie) Hicks, (M), Yr

Adjunct Professor:
Alan J. Anderson, (D), Queen's, 1991, Gi

Emeritus:
Henry Longerich, (D), Indiana, 1967, Cg
Michael G. Rochester, (D), Utah, 1959, Yg

Undergraduate Officer:
Raymund Patzold, (M), Memorial, 1972, Oe

Geology Librarian:
Geraldine Starkes, Oo

Mount Royal College

Dept of Earth Sciences (A) (2006)
4825 Mount Royal Gate SW
Calgary, AB T3E 6K6
p. (403) 240-6165 f. (403) 240-6333
ikubicek@mtroyal.ca
http://www.mtroyal.ab.ca/scitech/earth.shtml
Administrative Assistant: Ines Kubicek
Enrollment: No data reported since 1995

Chair:
Barbara McNicol, (D), Calgary, 1997, Og

Instructor:
Katherine Boggs, (D), Calgary, 2004, Gt
John Cox, (D), Aberdeen, 1994, Eo
Paul Johnston, (D), W Australia, 1986, Pi
Pamela MacQuarrie, (M), Calgary, 1988, Oy

Related Staff:
Michael Clark, (B), Adams, 1976, Og

Northern Alberta Institute of Tech

Dept of Geological & Hydrogeological Technologies (2004)
11762 106 Street
Edmonton, AB T5G 2R1
p. (403) 471-7776 f. (403) 471-8831
brucer@nait.ca
Administrative Assistant: Maria Martinez

Head:
Bruce Reinders, S Alberta Inst of Tech, 1985, Np

Assistant Program Head:
Ikhtyar Omar, (D), Sheffield, 1979, Gp

Asst Program Head:
Valdis Gislason, (M), Calgary, 1986, Ge

Instructor:
Glenn S. Hartley, (B), Alberta, 1977, Em

Queen's University

Dept of Geological Sciences (B,M,D) (2005)
Kingston, ON K7L 3N6
p. (613) 545-2597 f. (613) 545-6592
peterson@geol.queensu.ca
Enrollment (2004): B: 50 (0) M: 30 (0) D: 30 (0)

Head:
Herwart Helmstaedt, (D), New Brunswick, 1968, Gc

Professor:
Dugald M. Carmichael, (D), California (Berkeley), 1967, Gp
Alan H. Clark, (D), Manchester, 1964, Em
John M. Dixon, (D), Connecticut, 1974, Gc
Edward Farrar, (D), Toronto, 1966, Cc
Noel P. James, (D), McGill, 1972, Gd
Guy M. Narbonne, (D), Ottawa, 1981, Pi
Thomas H. Pearce, (D), Queen's, 1967, Gi
Raymond A. Price, (D), Princeton, 1958, Gt
Peter L. Roeder, (D), Pennsylvania, 1960, Cp
Colin J. Thomson, (D), Cambridge, 1981, Ys

Associate Professor:
Robert W. Dalrymple, (D), McMaster, 1977, Gs
John A. Hanes, (D), Toronto, 1979, Cc
Heather E. Jamieson, (D), Queen's, 1982, Cp
Ronald C. Peterson, (D), Virginia Tech, 1980, Gz

Assistant Professor:
Victoria H. Remenda, (D), Waterloo, 1993, Hw

Research Associate:
Doug A. Archibald, (D), Queen's, 1982, Cc

Adjunct Professor:
Lawrence D. Dyke, (D), Texas A&M, 1978, Ng
Sandra L. McBride, (D), Queen's, 1977, Gg

Emeritus:
W. Alan Gorman, (D), McGill, 1956, Gm

Royal Ontario Museum

Dept of Earth Sciences (2004)
100 Queen's Park
Toronto, ON M5S 2C6
p. (416) 856-5811 f. (416) 586-5814
annetteb@rom.on.ca
Administrative Assistant: Annette Boenke-Jarek
Enrollment: No data reported since 1995

Head:
Frederick J. Wicks, (D), Oxford, 1969, Gz

Curator:
T. E. Krogh, (D), MIT, 1964, Cc

Canada

Senior Scientist:
Fernando Corfu, (D), ETH (Switzerland), 1980, Cc
Donald W. Davis, (D), Alberta, 1978, Cc
Curatorial Assistant:
Vincent Vertolli, (M), Toronto, 1974, Gx

Dept of Mineralogy (2004)
100 Queen's Park
Toronto, ON M5S 2C6
p. (416) 586-5816 f. (416) 586-5814
robertg@rom.on.ca
Chair:
Robert I. Gait, (D), Manitoba, 1967, Gz
Curator:
Frederick J. Wicks, (D), Oxford, 1969, Gz
Associate Curator:
Darko B. Sturman, (M), Queen's, 1971, Gz
Emeritus:
Joseph A. Mandarino, (D), Michigan, 1958, Gz

Dept of Palaeobiology (2004)
100 Queen's Park
Toronto, ON M5S 2C6
p. (416) 586-5591 f. (416) 586-5863
Administrative Assistant: Joan Burke
Curator-in-Charge:
Desmond H. Collins, (D), Iowa, 1966, Pi
Senior Curator:
Chris McGowan, (D), London, 1969, Pv
Hans-Dieter Sues, (D), Harvard, 1984, Pv
Peter H. von Bitter, (D), Kansas, 1972, Pm
Assistant Curator:
Kevin L. Seymour, (D), Toronto, 1999, Pv
Assistant Curator:
David M. Rudkin, (B), Toronto, 1975, Pi
Assistant Curator:
Janet B. Waddington, (M), Toronto, 1972, Pi

Royal Tyrrell Museum of Palaeontology

Royal Tyrrell Museum of Palaeontology (2006)
P.O. Box 7500
Drumheller, AB T0J 0Y0
p. (403) 823-7707 f. (403) 823-7131
tyrrell.info@gov.ab.ca
http://www.tyrrellmuseum.com/
Enrollment: No data reported since 1995
Director:
Bruce G. Naylor, (D), Alberta, 1978, Pv
Curator:
Dennis R. Braman, (D), Calgary, 1981, Pl
Donald B. Brinkman, (D), McGill, 1979, Pv
Philip J. Currie, (D), McGill, 1981, Pv
David A. Eberth, (D), Toronto, 1987, Gs
Paul A. Johnston, (D), Australian National, 1986, Pi
Adjunct Research Scientist:
Russell L. Hall, (D), McMaster, 1976, Pi
Leonard V. Hills, (D), Alberta, 1965, Pl
Anthony P. Russell, (D), London, 1972, Pv
Xiao-chun Wu, (D), McGill, Pv
Related Staff:
Andrew G. Neuman, (M), Alberta, 1986, Pv

Saint Francis Xavier University

Dept of Earth Sciences (B,M) (2005)
P.O Box 5000
Antigonish, NS B2G 2W5
p. (902) 867-5109 f. (902) 867-2414
jfarrell@stfx.ca
Enrollment: No data reported since 1995
Professor:
Alan J. Anderson, (D), Queen's, 1990, Gx
Hugo Beltrami, (D), UQAM, Ge
Michael J. Melchin, (D), Western, 1987, Pi
J. Brendan Murphy, (D), McGill, 1982, Gt
Associate Professor:
Lisa M. Kellman, (D), Quebec (Montreal), 1997, Ge
Assistant Professor:
Evelise Bourlon, (D), UQAM
Ozeas Costa, (D), Plymouth (UK)
Grant Ferguson, (D), Manitoba
Instructor:
Cindy Murphy, (M), McGill, 1984, Gg
Colette Rennie, (M), Queen's, 1987, Gg
Sid Taylor, (M), Memorial, 1977, Gg

Simon Fraser University

Dept of Earth Sciences (B,M) (2006)
8888 University Drive
Burnaby, BC V5A 1S6
p. (604) 291-5387 f. (604) 291-4198
wendyb@sfu.ca
http://www.sfu.ca/earth-sciences
Enrollment (2003): B: 546 (0) M: 25 (0)
Chair:
Diana M. Allen, (D), Carleton, 1996, Hy
Professor:
John J. Clague, (D), British Columbia, 1973, Gl
Douglas Stead, (D), Nottingham, Ng
Associate Professor:
Andrew J. Calvert, (D), Cambridge, 1985, Ys
James A. Mac Eachern, (D), Alberta, 1994, Gs
Dan D. Marshall, (D), Lausanne, 1995, Ca
Peter S. Mustard, (D), Carleton, 1990, Gs
Derek J. Thorkelson, (D), Carleton, 1992, Gt
Brent C. Ward, (D), Alberta, 1992, Ge
Assistant Professor:
Gwenn Flowers, (D), British Columbia, 2000, Gl
H. Daniel (Dan) Gibson, (D), Carleton, 2003, Gc
Dirk Kirste, (D), Calgary, 2001, Cg
Glyn Williams-Jones, (D), Open, 2001, Gv
Lecturer:
Kevin Cameron, (M), Memorial, 1986, Gg
Roberta Dunlop, (M), British Columbia, 1984, Gx

Sir Sandford Fleming College

Natural Resources Div - Geology Programs (B) (2004)
Lindsay, ON K9V 5E6
p. (705) 324-9144 (Ext. 3249) f. (705) 878-9312
pkorzeni@flemingc.on.ca
Administrative Assistant: Cathy Alldred
Enrollment: No data reported since 1995
Head:
Paul W. Korzeniowski, C.E.T., Algonquin, 1970, Ng
Professor:
Grant E. Bashford, Sir Sandford Fleming, 1978, Ye
Ronald D. Date, Haileybury, 1958, Eg
Brian S. Gerry, (B), Maine, 1972, Ng
Robert F. Harrington, (M), Waterloo, 1975, Hw
David A. McAuslan, (D), Edinburgh, 1967, Eg
Stephen H. Watts, (D), Sydney, 1975, Gm
Geology Librarian:
Gale Butterill, Oo

Sir Wilfred Grenfell College

Environmental Science Unit (B) (2005)

Division of Science
University Drive
Corner Brook, NF A2H 6P9
p. (709) 637-6289 f. (709) 639-8125
ccampbel@swgc.mun.ca
http://www.swgc.mun.ca
Department Secretary: Phyllis Langdon
Enrollment: No data reported since 1995

Chair:
William J. Iams, (D), Memorial, 1977, Og
Assistant Professor:
Pierre M. Rouleau, (D), Alberta, 1994, Yg
Michael P. Rutherford, (D), Alberta, 1991, Sb
Adjunct Professor:
Antony R. Berger, (D), Liverpool (UK), 1967, Ge
Geology Librarian:
Elizabeth Behrens, Oo

The Manitoba Museum

Geology Dept (2007)
190 Rupert Avenue
Winnipeg, MB R3B 0N2
p. (204) 956-2830 f. (204) 942-3679
gyoung@manitobamuseum.ca
http://www.manitobamuseum.ca
Administrative Assistant: Claire Zimmerman

Museum Librarian:
Cindi Steffan, (N), Oo
Curator:
Graham A. Young, (D), New Brunswick, 1988, Pi

Universite de Montreal

Dept de Geologie (B,M,D) (2004)
Pavillon Ecole Polytechnique
2900 Edouard Montpetit
Montreal, PQ H3T 1J4
p. (514) 343-6820 f. (514) 343-5782
geologie@ere.umontreal.ca
Administrative Assistant: Joanna Frongillo
Enrollment: No data reported since 1995

Acting Chair:
Bernard L. Mamet, (D), Brussels, 1962, Ps
Professor:
Michel Bouchard, (D), McGill, 1980, Gm
Christopher Brooks, (D), Australian National, 1966, Cc
Hans Hoffman, (D), McGill, 1962, Pm
Claude Hubert, (D), McGill, 1965, Gt
Walter E. Trzcienski, Jr., (D), McGill, 1971, Gx
Associate Professor:
John Stix, (D), Toronto, 1989, Gi
Adjunct Professor:
Francois Robert, (D), Ecole Polytechnique, 1984, Gt
William W. Shilts, (D), Syracuse, 1970, Oo
Related Staff:
Jean Beaulieu, (M), Montreal, 1977, Gg
Lan Tu Dang-Nguyen, (D), Faculte des Sci de Saigon, 1969, Pi
Gilles Gauthier, (M), Montreal, 1978, Cc
Mona Kerba-Kachaami, (B), Le Caire, 1968, Cg

Universite du Quebec

INRS Georessources (Quebec Geoscience Center) (2004)
2535 Laurier Boulevard
C.P. 7500
Ste-Foy, PQ G1V 4C7
p. (418) 654-2604 f. (418) 654-2615
cgq@gsc.nrcan.gc.ca
Administrative Assistant: Pascale Cote

Head:
Daniel Lebel, (D), McGill, 1993, Gc
Chair:
Aicha Achab, (D), Alger, 1970, Pl
Professor:
Mario Bergeron, (D), McMaster, 1985, Ca
Normand Bergeron, (D), SUNY (Buffalo), 1994, Gm
Yvon Heroux, (D), Montreal, 1975, Co
Marc R. LaFleche, (D), Montpellier II, 1991, Ct
Rene Lefebvre, (D), Laval, 1994, Hw
Bernard Long, (D), Paul Sabatier, 1975, On
Richard Martel, (D), Laval, 1996, Hw
Normand Tasse, (D), McGill, 1982, Ge
Alain Tremblay, (D), Laval, 1989, Gc
Senior Scientist:
Jean H. Bedard, (D), Montreal, 1985, Gi
Christian Begin, (D), Laval, 1991, Pe
Andree Bolduc, (D), Carleton, 1992, Gl
Guoxiang Chi, (D), Quebec (Chicoutimi), 1991, Eg
Louise Corriveau, (D), McGill, 1989, Gp
Benoit Dube, (D), Quebec (Chicoutimi), 1990, Em
Denis Lavoie, (D), Laval, 1988, Gs
Yves Michaud, (D), Laval, 1991, Gm
Leopold Nadeau, (D), Carleton, 1990, Gt
Michel Parent, (D), W Ontario, 1987, Gl
Didier Perret, (D), Laval, 1995, Ng
Associate Scientist:
Rudolf Bertrand, (D), Neuchatel, 1987, Eo
Andre Chagnon, (D), Neuchatel, 1988, Co
Serge Paradis, (M), Ottawa, 1980, Gl
Research Associate:
Esther Asselin, (M), Laval, 1986, Pl
Eric Boisvert, (M), Quebec (Montreal), 1994, Oo
Edith Bourque, (M), Laval, 1994, Hw
Pierre Brouillette, (B), Laval, 1982, Gg
Christine DeBlonde, (B), Quebec (Montreal), 1988, Oo
Nathalie Fagnan, (M), INRS Georessources, 1998, Hw
Jocelyn Hamel, (B), Laval, 1991, Oo
Kathleen Lauziere, (M), Quebec (Chicoutimi), 1989, Gg
Marc R. Luzincourt, (B), Quebec (Montreal), 1982, Cs
Anna Smirnov, (M), Memorial, 1997, Cs
Azzedine Soufiane, (M), Quebec (Montreal), 1992, Pl
Visiting Professor:
Kees Schrijver, (D), Leiden, 1973, Eg

Universite du Quebec a Chicoutimi

Module des Sciences de la Terre (B,M,D) (2004)
555, Boulevard de l'Universite
Chicoutimi, PQ G7H 2B1
p. (418) 545-5011 (Ext. 5202) f. (418) 545-5012
cerm@uqac.uquebec.ca
Administrative Assistant: Michel Tremblay
Enrollment: No data reported since 1995

Head:
Jacques Carignan, (D), Ecole Polytechnique, 1980, Cg
Professor:
Guy Archambault, (D), Ecole Polytechnique, 1972, Nm
Raynald Duberger, (M), Laval, 1971, Yg
Michael D. Higgins, (D), McGill, 1980, Cc
Wulf U. Mueller, (D), Heidelberg, 1986, Gv
Adam Nagy, (D), Toulouse (France), 1968, Gz
Alain Rouleau, (D), Waterloo, 1984, Hw
Denis W. Roy, (D), Princeton, 1976, Gc
Gerard Woussen, (D), Montreal, 1975, Gi
Associate Professor:
Sarah J. Barnes, (D), Toronto, 1983, Ct

Pierre Cousineau, (D), Laval, 1986, Gv
Real Daigneault, (D), Laval, 1991, Gc
Huan-Zhang Lu, (D), Pennsylvania, 1983, Em
Instructor:
Denis Cote, (M), Quebec (Chicoutimi), 1986, Gi
Adjunct Professor:
Francis Chartrand, (D), Ecole Polytechnique, 1986, Em
Sylvie Gentier, (D), Orleans (France), 1987, Nr
Joelle Riss, (D), Orleans (France), 1988, Gq
Alain Simard, (D), Montreal, 1987, Ce
Ghislain Tourigny, (D), Montreal, 1988, Gc
Pierre Verpaelst, (D), Montreal, 1985, Gg
Emeritus:
Edward H. Chown, (D), Johns Hopkins, 1963, Gx
Jayanta Guha, (D), Jadavpur, 1967, Eg
Related Staff:
Paul Bedard, (D), Quebec (Chicoutimi), 1992, Gi
Tremblay Christian, (M), Quebec (Chicoutimi), 1990, Ct
Geology Librarian:
Francoise Lange, (M), Quebec (Chicoutimi), 1986, Oo

Universite du Quebec a Montreal

Dept des Sciences de la Terre (B,M,D) (2004)
C.P. 8888, Succursale Centre-ville
Montreal, PQ H3C 3P8
p. (514) 987-3000 (Ext. 4620) f. (514) 987-7749
Administrative Assistant: Villagrasa Cosette
Assistant to the Director: Claudette Avard
Enrollment: No data reported since 1995
Professor:
Jean-Pierre Blanchet, Toronto, 1984, Oo
James H. Bourne, (D), Queen's, 1974, Gp
Bernard de Boutray, (D), Clermont (France), 1961, Gz
Anne de Vernal, (D), Montreal, 1986, Pm
Helene Gaonac'h, (D), Montreal, 1994, Gv
Clement Gariepy, (D), Montreal, 1984, Cc
Michel Gauthier, (D), Ecole Polytechnique, 1982, Em
Normand Goulet, (D), Queen's, 1976, Gc
Alfred B. Jaouich, (D), Minnesota, 1975, Sc
Michel Lamothe, (D), W Ontario, 1985, Gl
Rene Laprise, Toronto, 1988, Oa
Marc Lucotte, (D), McGill, 1987, Oc
Jean-Claude Mareschal, (D), Texas A&M, 1975, Yg
Maurice Morency, (D), Kansas, 1972, Ct
Pierre Page, (D), Paris, 1982, Cs
Gilbert Prichonnet, (D), Bordeaux, 1967, Gs
Ross K. Stevenson, (D), Arizona, 1989, Cg
Gilles Tasse, (D), Paris, 1974, Oo
Enrico Torlaschi, Ecole Polytechnique (Milan), 1976, Oo
Peter Zwack, McGill, 1973, Ow
Research Associate:
Marc Bardoux, (D), Carleton, 1989, Gc
Nuno Machado, (D), Montreal, 1985, Cc
Adjunct Dean:
Michel Jebrak, (D), Orleans, 1984, Em

Universite du Quebec a Rimouski

Inst des sciences de la mer de Rimouski (M,D) (2006)
310, Allee des Ursulines
Rimouski, PQ G5L 3A1
p. (418) 723-1986 (Ext. 1742) f. (418) 724-1842
andre_rochon@uqar.qc.ca
www.ismer.ca
Enrollment: No data reported since 1995
Director:
Serge Demers, (D), Laval, 1981, Ob
Professor:
Celine Audet, (D), Laval, 1985, Ob
Jean-Claude Brethes, (D), Aix-Marseille, 1978, Ob
Jean-Francois Dumais, (D), Alberta, 1980, Og
Jean-Pierre Gagne, (D), Montreal, 1993, Oc
Michel Gosselin, (D), Laval, 1990, Ob
Vladimir G. Koutitonsky, (D), SUNY (Stony Brook), 1985, Op
Jocelyne Pellerin, (D), Laval, 1982, Ob
Emilien Pelletier, (D), McGill, 1983, Oc
Suzanne Roy, (D), Dalhousie, 1986, Ob
Bjorn Sundby, (D), Bergen (Norway), 1966, Oc
Bruno Zakardjian, (D), Paris, 1994, Ob

Universite Laval

Dept de geologie et de genie geologique (B,M,D) (2008)
Pavillon Pouliot
Faculte des Sciences et de genie
Ste-Foy, PQ G1K 7P4
p. (418) 656-2193 f. (418) 656-7339
ggl@ggl.ulaval.ca
http://www.ggl.ulaval.ca
Enrollment (2002): B: 127 (27) M: 30 (6) D: 20 (3)
Head:
Josee Duchesne, (D), Laval, 1993, Gz
Professor:
Georges Beaudoin, (D), Ottawa, 1991, Em
Daniel Cassidy, (D), HwGe
Richard Fortier, (D), Montreal, 1994, Yg
Paul W. Glover, (D), East Anglia (UK), 1989, Yx
Rejean J. Hebert, (D), Brest, 1985, Gi
Jacques E. Locat, (D), Sherbrooke, 1982, Ng
Fritz Neuweiler, (D), Berlin, 1995, GsdPo
Rene Therrien, (D), Waterloo, 1993, Hw
Eletron Microprobe Specialist:
Marc Choquette, (D), Laval, 1988, Oo
Associate Scientist:
Pauline Dansereau, (M), Laval, 1988, Gs
Andre Levesque, (B), Laval, 1979, Gz
Pierre Therrien, (M), Laval, 1986, Gq
Research Associate:
Danielle Cloutier, (D), Laval, GeOg
Adjunct Professor:
Marc Constantin, (D), Brest, 1995, Gi
Benoit Fournier, (D), Laval, 1993, NgOmGg
Technician:
Eric David, (B), Laval, Oo
Jean Frenette, (B), Laval, 1985, Gz
Martin Plante, Cg

University College of Cape Breton

Dept of Mathematics & Natural Sciences (B) (2004)
Subdepartment of Earth Sciences
P.O. Box 5300
Sydney, NS B1P 6L2
p. (902) 539-5300 f. (902) 562-0119
Enrollment: No data reported since 1995
Instructor:
Fenton M. Isenor, (M), Acadia, 2000, Gg
Emeritus:
Erwin L. Zodrow, (D)

University College of the Cariboo

Dept of Sciences & Engineering (A) (2004)
Box 3010
Kamloops, BC V2C 5N3
p. (250) 828-5400 f. (250) 828-5450
james@cariboo.bc.ca
Enrollment: No data reported since 1995

Associate Dean:
F. Colin James, (D), Keele, 1966, Gg
Chair:
O. Rajora, (D), Dalhousie, Oo
Instructor:
Kenneth P. Klein, (M), Windsor, 1980, Gg

University of Alberta

Dept of Earth & Atmospheric Sciences (B,M,D) (2005)
126 Earth Sciences Building
Edmonton, AB T6G 2E3
p. (780) 492-3265 f. (780) 492-2030
Brian.Jones@ualberta.ca
http://www.ualberta.ca/EAS/
Department Secretary: Doreen B. Haugan
Enrollment (2000): B: 277 (44) M: 45 (22) D: 42 (6)

Chair:
Brian Jones, (D), Ottawa, 1974, Ps
Associate Chair:
Robert W. Luth, (D), California (Los Angeles), 1985, Cp
Martin J. Sharp, (D), Aberdeen, 1982, Gm
Canada Research Chair:
S. George Pemberton, (D), McMaster, 1979, Gr
Professor:
Octavian Catuneanu, (D), Toronto, 1996, Gs
Brian D. Chatterton, (D), Australian National, 1970, Pi
Robert A. Creaser, (D), La Trobe, 1990, Cc
David M. Cruden, (D), London, 1969, Nr
John England, (D), Colorado, 1974, Gm
Philippe Erdmer, (D), Queen's, 1982, Gc
Larry M. Heaman, (D), McMaster, 1986, Cc
M. John Hodgson, (D), Toronto, 1973, Or
Edgar L. Jackson, (D), Toronto, 1974, Oo
Edward P. Lozowski, (D), Toronto, 1970, Ow
Hans G. Machel, (D), McGill, 1985, Go
Karlis Muehlenbachs, (D), Chicago, 1971, Cs
Gerhard Reuter, (D), McGill, 1985, Ow
Jeremy P. Richards, (D), Australian National, 1990, Em
John Shaw, (D), Reading, 1969, Gm
John W.F Waldron, (D), Edinburgh, 1981, Gc
John D. Wilson, (D), Guelph, 1980, Ow
Associate Professor:
Thomas Chacko, (D), North Carolina, 1987, Gx
G. Peter Kershaw, (D), Alberta, 1983, Oy
Carl A. Mendoza, (D), Waterloo, 1993, Hw
R. Bruce Rains, (D), Alberta, 1969, Gm
Benoit Rivard, (D), Washington (St. Louis), 1990, Or
Benjamin J. Rostron, (D), Alberta, 1995, Hw
G. Arturo Sanchez-Azofeifa, (D), New Hampshire, 1996, Or
Assistant Professor:
Andrew B.G Bush, (D), Toronto, 1995, Oo
Michael W. Caldwell, (D), McGill, 1995, Pv
Theresa D. Garvin, (D), McMaster, 1999, Oo
Karen E. Smoyer, (D), Minnesota, 1997, Oo
Alan B. Woodland, (D), Northwestern, 1989, Gz
Emeritus:
Halfdan Baadsgaard, (D), ETH (Switzerland), 1955, Cc
Ronald A. Burwash, (D), Minnesota, 1955, Gp
Ian A. Campbell, (D), Colorado, 1968, Gm
Henry A. Charlesworth, (D), Glasgow, 1956, Gc
Kenneth J. Fairbairn, (D), Melbourne, 1968, Oo
Robert E. Folinsbee, (D), Minnesota, 1942, Em
Richard C. Fox, (D), Kansas, 1965, Pv
Keith D. Hage, (D), Chicago, 1957, Ow
R. Geoffrey Ironside, (D), Durham, 1965, Oo
Leszek A. Kosinski, (D), Warsaw, 1958, Oo
Jiri Krupicka, (D), Prague, 1947, Gp
Arleigh H. Laycock, (D), Minnesota, 1957, Hg
John F. Lerbekmo, (D), California (Berkeley), 1956, Gs
Harry J. McPherson, (D), McGill, 1967, Gm
Roger D. Morton, (D), Nottingham, 1959, Em
Nathaniel W. Rutter, (D), Alberta, 1966, Ge
O.F. George Sitwell, (D), Toronto, 1968, Oo
Dorian G. W. Smith, (D), Cambridge, 1963, Gz
Peter J. Smith, (D), Edinburgh, 1964, Oo
Charles R. Stelck, (D), Stanford, 1951, Ps
József Tóth, (D), Utrecht (Neth), 1965, Hw
William C. Wonders, (D), Toronto, 1951, Oo

Inst of Geophysical Research (B,M,D) (2007)
Mailstop #615
CEB/Physics
Edmonton, AB T6G 2G7
p. (780) 492-3521 f. (780) 492-0714
Inst-GR@Phys.UAlberta.CA
http://www-geo.phys.ualberta.ca/institute/
Administrative Assistant: Lee Grimard
Enrollment: No data reported since 1995

Director:
Douglas R. Schmitt, (D), Caltech, 1987, Yx
Professor:
Robert A. Creaser, (D), La Trobe, 1992, Cc
T. Bryant Moodie, (D), Toronto, 1972, Oo
Robert Rankin, (D), North Wales, 1984, Xy
Gerhard W. Reuter, (D), McGill, 1985, Oa
Wojciech Rozmus, (D), Inst Nuc Res (Poland), Oo
John C. Samson, (D), Alberta, 1971, Xy
Martin J. Sharp, (D), Aberdeen, Gl
John Shaw, (D), Reading, 1969, Gl
Samuel S. Shen, (D), Wisconsin, Oa
Bruce R. Sutherland, (D), Toronto, 1994, Oo
Gordon E. Swaters, (D), British Columbia, 1985, Op
Richard D. Sydora, (D), Texas, 1985, Xy
Martyn Unsworth, (D), Cambridge, Oo
John D. Wilson, (D), Guelph, 1980, Oa
Associate Professor:
Andrew B. G. Bush, (D), Toronto, Oa
Carl Mendoza, (D), Waterloo, 1993, Hw
Benoit Rivard, (D), Washington, 1990, Or
Ben Rostron, (D), Alberta, Oo
Mauricio D. Sacchi, (D), British Columbia, Ys
Assistant Professor:
Francis Fenrich, (D), Alberta, Xy
Jeff Gu, (D), Harvard, 2001, Ys
Moritz Heimpel, (D), Johns Hopkins, 1995, Oo
Vadim Kravchinsky, (D), Irkutsk, 1996, Ym
Paul Myers, (D), Victoria, Op
Emeritus:
Michael E. Evans, (D), Australian National, 1969, Ym
D. Ian Gough, (D), Witwatersrand, Yg
Keith D. Hage, (D), Chicago, 1957, Oa
F. Walter Jones, (D), McGill, 1968, Ym
Edward P. Lozowski, (D), Toronto, 1970, Oa
Roger D. Morton, (D), Nottingham, 1959, Em
Edo Nyland, (D), California (Los Angeles), 1967, Ys
David Rankin, (D), Alberta, 1960, Ym
Gordon Rostoker, (D), British Columbia, 1966, Xy
J. Timothy Spanos, (D), Alberta, 1977, Ys

University of British Columbia

Dept of Earth & Ocean Sciences (B,M,D) (2007)
6339 Stores Road
Vancouver, BC V6T 1Z4
p. (604) 822-2449 f. (604) 822-6088
head@eos.ubc.ca
http://www.eos.ubc.ca

Canada

Enrollment (2007): B: 271 (64) M: 77 (21) D: 85 (6)

Dean, Science:
Simon Muir Peacock, (D), California (Los Angeles), 1985, Gg
Professor & Head:
Paul L. Smith, (D), McMaster, 1981, Pg
Honorary Professor:
Mati Raudsepp, (D), Manitoba, 1984, Gz
Director, Mineral Deposits Research Unit:
Richard Tosdal, (D), California (Santa Barbara), 1988, Em
Canadian Research Chair:
Dominique Weis, (D), Libre de Bruxelles, 1982, Cc
Canada Research Chair:
Roger Francois, (D), British Columbia, 1987, Cm
Associate Head:
Raymond J. Andersen, (D), California (San Diego), 1975, Oc
Associate Dean, Graduate Studies:
Douw G. Steyn, (D), British Columbia, 1980, Oa
Associate Dean, Faculty of Science:
R. Grant Ingram, (D), MIT, 1971, Op
Professor:
Neil Balmforth, (D), Cambridge, 1990, Oa
R. Marc Bustin, (D), British Columbia, 1979, Ec
Gregory M. Dipple, (D), Johns Hopkins, 1991, Gp
Lee A. Groat, (D), Manitoba, 1988, Gz
Michael Healey, (D), Aberdeen, 1969, Ob
Oldrich Hungr, (D), Alberta, 1981, Gm
James K. Mortensen, (D), California (Santa Barbara), 1983, Cc
Douglas W. Oldenburg, (D), California (Santa Barbara), 1974, Yg
James Kelly Russell, (D), Calgary, 1984, Gi
J. Leslie Smith, (D), British Columbia, 1978, Hw
Roland B. Stull, (D), Washington, 1975, Oa
Associate Professor:
Susan E. Allen, (D), Cambridge, 1988, Op
Philip Austin, (D), Washington, 1987, Oa
Michael G. Bostock, (D), Australian National, 1991, Ys
Erik Eberhardt, (D), Saskatchewan, 1998, Ng
Kurt A. Grimm, (D), California (Santa Cruz), 1992, Gs
Catherine L. Johnson, (D), California (San Diego), 1994, Xg
Lori Kennedy, (D), Texas A&M, 1996, Gc
Maya G. Kopylova, (D), Moscow, 1990, Gi
Ulrich Mayer, (D), Waterloo, 1999, Nm
Kristin J. Orians, (D), California (Santa Cruz), 1992, Cm
Evgeny Pakhomov, (D), Russian Acad of Sci, 1992, Ob
James S. Scoates, (D), Wyoming, 1994, Gi
Canada Research Chair:
Maria Maldonado, (D), McGill, 1999, Ob
Christian Schoof, (D), Oxford, 2002, Gl
Assistant Professor:
Elizabeth H. Hearn, (D), Oregon, 1998, Yg
Felix J. Herrmann, (D), Delft (Neth), 1997, Yx
Kenneth Hickey, (D), James Cook, 1995, Gz
Mark Jellinek, (D), Australian National, 1999, Gg
Philippe Tortell, (D), Princeton, 2001, Ob
Research Associate:
Jane Barling, (D), Monash, 1990, Cc
Farhad Bouzari, (D), Queens, 2003, Eg
Laxmi Chikatamarla, (D), James Cook, 2000, Ec
Jason Cole, (D), Penn State, 2005, Oa
Richard Friedman, (D), British Columbia, 1988, Cc
Philip T. Hammer, (D), California (San Diego), 1991, Ys
Bruno Kieffer, (D), Grenoble, 2002, Cs
Ilkay Kuscu, (D), Middle East Tech, 1997, Eg
Henryk Modzelewski, (D), British Columbia, 2004, Oa
Nigel Phillips, (M), British Columbia, 2001, Yg
Roger Pieters, (D), California (Santa Barbara), Op
Kirstie Simpson, (D), Tasmania, 2001, Gv
Lin-Ping Song, (D), Sichuan (China), 1996, Ng
Dave E. Williams, (D), British Columbia, 1987, Oc
Zhen Zeng, (D), Chinese Academy of Sci, 2003, Oa
Instructor:
Mary Lou Bevier, (D), California (Santa Barbara), 1982, Cc
Sara Harris, (D), Oregon State, 1998, Ou
Stuart Sutherland, (D), Leicester (UK), 1992, Pg
Lecturer:
Brett Gilley, (M), Simon Fraser, 2003, Gg
Francis H. M. Jones, (M), British Columbia, 1987, Ye
Ben Kennedy, (D), McGill, 2006, Gv
Honorary Research Associate:
Terry L. Clark, (D), Toronto, 1971, Oa
Honorary Lecturer:
Dileep Athaide, (M), British Columbia, 1975, Gg
Adjunct Professor:
Robert G. Anderson, (D), Carleton, 1983, Gt
Stephen Billings, (D), Sydney, 1998, Yg
Edward Carmack, (D), Washington, 1972, Op
Claudio DiBacco, (D), California (San Diego), 1999, Ob
Michael G. G. Foreman, (D), British Columbia, 1984, Op
James Haggart, (D), California (Davis), 1984, Pg
Catherine J. Hickson, (D), UBC, 1987, Gv
Mark Holzer, (D), Simon Fraser, 1990, Oa
John L. Jambor, (D), Carleton, 1966, Gz
Murray Journeay, (D), Queen's, 1986, Gc
R. Lynn Kirlin, (D), Utah State, 1968, Oa
Barry Narod, (D), British Columbia, 1979, Yx
Ian Okabe, (D), UBC, 1995, Ow
Thomas F. Pedersen, (D), Edinburgh, 1979, Ou
Stephen Rowins, (D), W Australia, 1994, Em
K. Wayne Savigny, (D), Alberta, 1980, Ng
Barbara H. Scott Smith, (D), Edinburgh, 1977, Gz
John F. H. Thompson, (D), Toronto, 1984, Em
Richard E. Thomson, (D), British Columbia, 1971, Op
Knut von Salzen, (D), Hamburg, 1997, Oa
Chi S. Wong, (D), California (San Diego), 1968, Cm
Director, Lithoprobe:
Ronald M. Clowes, (D), Alberta, 1969, Ys
Emeritus:
Peter M. Bradshaw, (D), Durham (UK), 1965, Ce
Stephen E. Calvert, (D), California (San Diego), 1964, Cm
William E. Cameron, (D), California, 1951, Ob
Richard L. Chase, (D), Princeton, 1963, Gu
Garry K. C. Clarke, (D), Toronto, 1967, Yg
Wilbert R. Danner, (D), Washington, 1957, Gd
Robert M. Ellis, (D), Alberta, 1964, Ys
William K. Fletcher, (D), Imperial Coll (UK), 1968, Ce
Alan G. Lewis, (D), Hawaii, 1961, Ob
Stephen G. Pond, (D), British Columbia, 1965, Op
R. Doncaster Russell, (D), Toronto, 1954, Yg
Alastair J. Sinclair, (D), British Columbia, 1964, Em
Frank J. R. Taylor, (D), Cape Town, 1965, Ob
Tadeusz J. Ulrych, (D), British Columbia, 1962, Yg
Director, Geological Engineering Program:
Roger D. Beckie, (D), Princeton, 1992, Hg
Chair, ATSC Program:
William Hsieh, (D), British Columbia, 1981, Op
Associate Dean, Faculty of Science:
Curtis Suttle, (D), British Columbia, 1987, Ob
On Leave:
Richard A. Pawlowicz, (D), MIT/Woods Hole, 1994, Op
Systems Manager:
John Amor, Oo
Secretary to the Head:
Deborah Varley, (N), Oo
Graduate Secretary:
Acting Graduate Secretary, Oo
Administrator:
Paul de Leon, Oo

Related Staff:
Carol Leven, (N), Oo
Kathy Scott, (N), Oo
Teresa Woodley, (N), Oo
Mandy Wu, Oo

Soil Science (B,M,D) (2004)
2357 Main Mall
Room 139 MacMillan Building
Vancouver, BC V6T 1Z4
p. (604) 822-2783 f. (604) 822-8639
Department Secretary: Giuliana Villegas

Professor:
Timothy M. Ballard, (D), Washington, 1968, Sf
T. Andrew Black, (D), Wisconsin, 1969, Sp
Leslie M. Lavkulich, (D), Wisconsin, 1969, Sd
Peter A. Murtha, (D), Cornell, 1968, Or
Hans D. Schreier, (D), British Columbia, 1976, Og

Associate Professor:
Arthur A. Bomke, (D), Illinois, 1972, Sc
Michael D. Novak, (D), British Columbia, 1981, Sp

Adjunct Professor:
Shannon M. Berch, (D), Laval, 1983, Sb
Bernie J. Zebarth, (D), Saskatchewan, 1988, Sc

University of Calgary

Dept of Geology & Geophysics (B,M,D) (2005)
2500 University Drive NW
Calgary, AB T2N 1N4
p. (403) 220-5841 f. (403) 284-0074
lrlines@ucalgary.ca
http://www.geo.ucalgary.ca/
Enrollment (2003): B: 268 (73) M: 61 (19) D: 27 (4)

Head:
Larry R. Lines, (D), British Columbia, 1976, Ye

Professor:
Laurence R. Bentley, (D), Princeton, 1990, Hw
James R. Brown, (D), Uppsala (Sweden), 1972, Ys
Frederick A. Cook, (D), Cornell, 1980, Ys
Kenneth Duckworth, (D), Leeds, 1964, Ye
Edward D. Ghent, (D), California, 1964, Gp
Terence M. Gordon, (D), Princeton, 1969, Gq
Charles M. Henderson, (D), Calgary, 1988, Pm
Ian E. Hutcheon, (D), Carleton, 1977, Cl
Federico F. Krause, (D), Calgary, 1979, Gs
Edward S. Krebes, (D), Alberta, 1980, Ys
Donald C. Lawton, (D), Auckland, 1979, Ye
James W. Nicholls, (D), California, 1969, Gi
Gerald D. Osborn, (D), California, 1972, Gm
David R. Pattison, (D), Edinburgh, 1985, Gp
Ronald J. Spencer, (D), Johns Hopkins, 1981, Cg
Deborah A. Spratt, (D), Johns Hopkins, 1980, Gc
Robert R. Stewart, (D), MIT, 1983, Ye
Patrick Wu, (D), Toronto, 1982, Ys

Associate Professor:
Russell L. Hall, (D), McMaster, 1976, Pi
Masaki Hayashi, (D), Waterloo, 1996, Hw
Alan R. Hildebrand, (D), Arizona, 1992, Xm
John C. Hopkins, (D), McGill, 1972, Gs
Gary F. Margrave, (D), Alberta, 1981, Ye
Bernhard Mayer, (D), Ludwig-Maximilians, 1993, Cs
Brian J. Moorman, (D), Carlton, 1997, Gm
Cynthia L. Riediger, (D), Waterloo, 1991, Co
Cathy Ryan, (D), Waterloo, 1994, Hw

Assistant Professor:
Jean-Michel Maillol, (D), Alberta, 1992, Ym

Senior Instructor:
Jon W. Jones, (D), Calgary, 1972, Gp

Honorary Professor:
Brian Norford, (D), Yale, 1959, Gr

Adjunct Professor:
John C. Bancroft, (D), Brigham Young, 1975, Ng
Benoit Beauchamp, (D), Calgary, 1987, Pm
Philip J. Currie, (D), McGill, 1981, Pv
Martin G. Fowler, (D), Newcastle upon Tyne, 1984, Co
Susan L. Gordon, (D), Waterloo, 1999, Hw
Stephen E. Grasby, (D), Calgary, 1997, Hw
William D. Gunter, (D), Johns Hopkins, 1974, Cl
J. Bruce Jamieson, (D), Calgary, 1996, Ne
Apostolos Kantzas, (D), Waterloo, 1988, Ng
Thomas F. Moslow, (D), South Carolina, 1980, Co
Peter E. Putnam, (D), Calgary, 1985, Gs
Gerald M. Ross, (D), Carleton, 1983, Gt
Selim Sayegh, (D), McGill, 1980, Gs
John-Paul Zonneveld, (D), Alberta, 1999, Pm

Emeritus:
Peter Bayliss, (D), New South Wales, 1967, Gz
Finley A. Campbell, (D), Princeton, 1958, Em
Peter E. Gretener, (D), ETH (Switzerland), 1953, Ng
Leonard V. Hills, (D), Alberta, 1965, Pl
Alfred A. Levinson, (D), Michigan, 1952, Eg
Alan E. Oldershaw, (D), Liverpool, 1967, Gs
Philip S. Simony, (D), London, 1963, Gc
Norman C. Wardlaw, (D), Glasgow, 1960, Go

Manager, Student Programs:
M. Graham, Oo

University of Guelph

Dept of Land Resource Science (B,M,D) (2006)
Guelph, ON N1G 2W1
p. (519) 824-4120 (Ext. 52456) f. (519) 824-5730
earnaud@uoguelph.ca
http://www.lrs.uoguelph.ca
Enrollment (2005): B: 89 (31) M: 48 (11) D: 20 (3)

Chair:
Stewart G. Hilts, (D), Toronto, 1981, Oo

Director, Kemptville College:
Michael J. Goss, (D), Reading, So

Acting Chair:
Beverley A. Hale, (D), Guelph, 1989, Ct

Professor:
Michael E. Brookfield, (D), Reading (UK), 1973, Pg
Les J. Evans, (D), Wales, 1974, Cl
Raymond G. McBride, (D), Guelph, 1982, Ou
R. Paul Voroney, (D), Saskatchewan, 1983, Sb

Associate Professor:
Gary Parkin, (D), Guelph, 1994, Hw
H. Peter Van Straaten, (D), Goettingen, 1974
Claudia Wagner Riddle, (D), Guelph, 1992, Ow

Assistant Professor:
Emmanuelle Arnaud, (D), McMaster, 2002, Gl
Kari Dunfield, (D), Saskatchewan, Sb
Susan Glasauer, (D), Tech (Munich), 1995, Py
Richard Heck, (D), Saskatchewan, Sd
John Lauzon, (D), Guelph, So
Ivan O'Halloran, (D), Saskatchewan, 1986, So
Jon Warland, (D), Guelph, 1999, Oa

Emeritus:
Eric G. Beauchamp, (D), Cornell, 1965, So
Ward Chesworth, (D), McMaster, 1967, Cl
David E. Elrick, (D), Wisconsin, 1957, Sp
Terry J. Gillespie, (D), Guelph, 1968, Ow
Pieter H. Groenevelt, (D), Wageningen (Neth), 1969, Sp
Beverly D. Kay, (D), Purdue, 1969, Sc
Kenneth M. King, (D), Wisconsin, 1956, Oa
I. Peter Martini, (D), McMaster, 1966, Gs

Murray H. Miller, (D), Purdue, 1957, So
George W. Thurtell, (D), Wisconsin, 1965, Oa

University of Manitoba

Dept of Civil & Geological Engineering (B,M,D) (2005)
342 Engineering Building
Winnipeg, MB R3T 5V6
p. (204) 474-8596 f. (204) 261-9534
Jay_Doering@UManitoba.ca
Administrative Assistant: Ingrid Trestrail
Enrollment (1999): B: 10 (11) M: 6 (3) D: 5 (0)

Associate Dean:
Brian Stimpson, (D), London, 1964, Nr
Professor:
Emery Z. Lajtai, (D), Toronto, 1966, Nr
Allan D. Woodbury, (D), British Columbia, 1973, Hw
Associate Professor:
Daryl M. McCartney, (D), Manitoba, 1991, Oo

Dept of Geological Sciences (B,M,D) (2006)
240 Wallace Building
125 Dysart Road
Winnipeg, MB R3T 2N2
p. (204) 474-9371 f. (204) 474-7623
n_chow@umanitoba.ca
http://www.umanitoba.ca/geoscience
Administrative Assistant: Brenda Miller
Enrollment (2005): B: 41 (5) M: 13 (2) D: 4 (0)

Head:
Nancy Chow, (D), Memorial, 1986, Gs
Professor:
Robert J. Elias, (D), Cincinnati, 1979, Pi
Ian J. Ferguson, (D), Australian National, 1988, Ye
Norman M. Halden, (D), Glasgow, 1983, Cg
Frank C. Hawthorne, (D), McMaster, 1973, Gz
William M. Last, (D), Manitoba, 1980, Gs
Barbara L. Sherriff, (D), McMaster, 1988, Gz
James T. Teller, (D), Cincinnati, 1970, Gs
Senior Scholar:
Lorne D. Ayres, (D), Princeton, 1969, Gv
George S. Clark, (D), Columbia, 1967, Cc
Wooil Moon, (D), British Columbia, 1976, Ys
Allan C. Turnock, (D), Johns Hopkins, 1960, Gp
Associate Professor:
Mostafa Fayek, (D), Saskatchewan, 1996, Cg
Elena Sokolova, (D), Moscow, 1997, Gz
Assistant Professor:
Alfredo Camacho, (D), Australian National, 1998, Gt
Anton Chakhmouridian, (D), St. Petersburg, 1997, Gz
Andrew Frederiksen, (D), British Columbia, 2000, Ys
Research Associate:
Yassir Abdu, (D), Uppsala, 2004, Gz
Rong Liu, (D), National (Singapore), 2005, Om
Michael Schindler, (D), Johann-Wolfgang Goethe, 1997, Gz
Instructor:
William Mandziuk, (M), Manitoba, 1989, Gg
Jeffrey Young, (M), Manitoba, 1992, Gg
Lecturer:
Karen Ferreira, (M), Manitoba, 1984, Gg
Adjunct Professor:
Joel D. Grice, (D), Manitoba, 1973, Gz
David C. Peck, (D), Melbourne, 1990, Eg
R.F. Jon Scoates, (D), Manitoba, 1972, Gc
Nikolay Sidenko, (D), Russian Acad of Sci, 2001, Ge
Graham A. Young, (D), New Brunswick, 1988, Pi
Emeritus:
William C. Brisbin, (D), California (Los Angeles), 1970, Gc
Petr Cerny, (D), Czech Acad of Sci, 1966, Gz
Robert B. Ferguson, (D), Toronto, 1948, Gz
Donald H. Hall, (D), British Columbia, 1959, Ym
Laboratory Director:
Panseok Yang, (D), Memorial, 2002, Ca
Related Staff:
Neil Ball, (B), Manitoba, 1982, Gz
Mark Cooper, (M), Manitoba, 1997, Gz
Sergio Mejia, (M), Manitoba, 1984, Cg
Gregg Morden, (M), Manitoba, 1986, Cg

University of New Brunswick

Dept of Geology (B,M,D) (2004)
Fredericton, NB E3B 5A3
p. (506) 453-4803 f. (506) 453-5055
geology@unb.ca
Administrative Secretary: Merrill Ann Beatty
Enrollment: No data reported since 1995

Chair:
Joseph C. White, (D), W Ontario, 1979, Gc
Professor:
Bruce E. Broster, (D), W Ontario, 1982, Gl
John Todd Dunn, (D), Alberta, 1983, Gi
Ronald K. Pickerill, (D), Liverpool, 1974, Pg
John G. Spray, (D), Cambridge, 1980, Gp
Paul F. Williams, (D), Sydney, 1969, Gc
Associate Professor:
Nicholas J. Susak, (D), Princeton, 1981, Cg
Assistant Professor:
Tom A. Al, (D), Waterloo, 1996, Hw
Karl Butler, (D), British Columbia, 1996, Yg
Cliff S.J. Shaw, (D), W Ontario, 1994
Research Associate:
James Whitehead, (D), New Brunswick, 1998, Gg
Honorary Research Professor:
Henk W. Van De Poll, (D), Swansea, 1970, Gs
Adjunct Professor:
Richard A. F. Grieve, (D), Toronto, 1970, Xg
David R. Lentz, (D), Ottawa, 1992, Cg
Randall F. Miller, (D), Waterloo, 1984, Pg
Emeritus:
Arnold L. McAllister, (D), McGill, 1950, Em
Geology Librarian:
Eszter Schwenke, Oo

University of New Brunswick, Saint John

Dept of Biology (B,M,D) (2007)
PO Box 5050
Saint John, NB E2L 4L5
p. (506) 648-5605 f. (506) 648-5948
lwilson@unbsj.ca
http://www.unbsj.ca/sase/biology/
Administrative Assistant: Kim Banks
Enrollment (2007): B: 2 (0) M: 1 (0)

Associate Professor:
Lucy A. Wilson, (D), Paris VI, 1986, Ga
Emeritus:
Alan Logan, (D), Durham (UK), 1962, Ob

University of Ottawa

Dept of Earth Sciences (B,M,D) (2004)
140 Louis Pasteur
PO Box 450, Station A
Ottawa, ON K1N 6N5
p. (613)562-5800 Ext 6870 f. (613) 562-5192
hdegouf@science.uottawa.ca
http://www.science.uottawa.ca/est/eng/welcome.html

Administrative Assistant: Ronald Labelle
Enrollment (2004): B: 30 (6) M: 31 (4) D: 29 (5)
Chair:
Keiko Hattori, (D), Tokyo, 1977, Cg
Professor:
R. William C. Arnott, (D), Alberta, 1987, Gs
Ian D. Clark, (D), Paris - Sud, 1988, Hw
Anthony D. Fowler, (D), McGill, 1980, Gv
Mark Hannington, (D), Toronto, 1989, Eg
Andre E. Lalonde, (D), McGill, 1985, Gz
Associate Professor:
Keith B. Benn, (D), Montpellier II, 1989, Gc
Andre Desrochers, (D), Memorial, 1986, Gs
Danielle Fortin, (D), Québec, 1992, Py
Michel R. Robin, (D), Waterloo, 1991, Hw
Adjunct Professor:
Frits P. Agterberg, (D), Utrecht (Neth), 1961, Gq
John Ayer, (D), Ottawa, 1999, Cg
Graeme F. Bonham-Carter, (D), Toronto, 1966, Oi
David Andrew Fisher, (D), Copenhagen, 1978, Yg
William K. Fyson, (D), Reading (UK), 1960, Gc
Wayne D. Goodfellow, (D), New Brunswick, 1975, Cg
Jeffrey Wayne Hedenquist, (D), Auckland, 1983, Cg
Donald D. Hogarth, (D), McGill, 1959, Gz
Thomas G. Kotzer, (D), Saskatchewan, 1993, Cg
Michael Carl Lesher, (D), W Australia, 1984, Gv
Dogan Paktunc, (D), Ottawa, 1983, Gz
John A. Percival, (D), Queen's, 1981, Gp
Laurier Poissant, (D), Québec, 1991
Andreas Prokoph, (D), Tubingen, 1994, Gq
Patricia Rasmussen, (D), Waterloo, 1993, Ge
Cees Van Staal, (D), New Brunswick, 1984, Gt
Emeritus:
Jan Veizer, (D), Australian National, 1971, Cl

University of Regina

Dept of Geology (B,M,D) (2004)
Regina, SK S4S 0A2
p. (306) 585-4147 f. (306) 585-5205
geology@max.cc.ureginaa.ca
Administrative Assistant: R. Ruda
Enrollment: No data reported since 1995
Head:
Stephen L. Bend, (D), Newcastle (UK), 1990, Eo
Professor:
Pier L. Binda, (D), Alberta, 1970, Eg
John F. Lewry, (D), Wales, 1964, Gt
Brian R. Watters, (D), Cape Town, 1974, Gi
Associate Professor:
Katherine M. Bergman, (D), McMaster, 1987, Gs
Assistant Professor:
Jisuo Jin, (D), Saskatoon, 1988, Oo
Instructor:
Syed Abbass-Hasanie, (D), Australian National, 1978, Gp
Evanna Simpson, (M), New Brunswick, 1985, Eg
Adjunct Professor:
Kenneth E. Ashton, (D), Queen's, 1986, Gt
Leslie S. Beck, (D), Leeds, 1966, Em
Ralph Chessman, (D), Imperial, 1955, Eg
David Eberth, (D), Toronto, 1987, Gs
Roger Walker, (D), Oxford, 1968, Gs
Emeritus:
Donald M. Kent, (D), Alberta, 1968, Gs
Laurence W. Vigrass, (D), Stanford, 1961, Eo

University of Saskatchewan

Dept of Geological Sciences (B,M,D) (2007)
114 Science Place
Saskatoon, SK S7N 5E2
p. (306) 966-5683 f. (306) 966-8593
kevin.ansdell@usask.ca
http://www.usask.ca/geology
Administrative Assistant: Brenda L. Britton
Enrollment (2000): B: 147 (32) M: 29 (3) D: 13 (0)
Head:
Kevin M. Ansdell, (D), Saskatchewan, 1992, EgCgGt
Professor:
James F. Basinger, (D), Alberta, 1979, Pb
Graham George, (D), Oo
Jim Hendry, (D), Waterloo, 1984, Hw
Chris Holmden, (D), Alberta, 1995, Csc
Robert W. Kerrich, (D), Imperial Coll, 1975, Cg
James B. Merriam, (D), York, 1976, Yg
Yuanming Pan, (D), W Ontario, 1990, Gx
William P. Patterson, (D), Michigan, 1995, CsPe
Brian R. Pratt, (D), Toronto, 1989, Pg
Robin W. Renaut, (D), London, 1982, Gs
Associate Professor:
Sam Butler, (D), Toronto, 2000, Yg
Igor Morozov, (D), Moscow State, 1985, Ys
Ingrid Pickering, (D), Oo
Assistant Professor:
Gabriela Mangano, (D), PsGs
Research Associate:
Bhaskar I. Pandit, (D), Toronto, 1971, Ys
Adjunct Professor:
Irvine Annesley, (D), Gi
Dennis Braman, (D), Pm
Harold Bryant, (D), Toronto, 1989, Pv
Ning Chen, (D), Gz
Tom Kotzer, (D), CgEg
Len Wassenaar, (D), Waterloo, 1990, Hw
Derek A. Wyman, (D), Saskatchewan, 1990, Cg
Emeritus:
Willi K. Braun, (D), Tubingen, 1958, Ps
William G. E. Caldwell, (D), Glasgow, 1957, Ps
Leslie C. Coleman, (D), Princeton, 1955, Gz
Donald J. Gendzwill, (D), Saskatchewan, 1969, Ye
Zoltan Hajnal, (D), Manitoba, 1970, Ys
Mel R. Stauffer, (D), Australian National, 1964, Gc
Cooperating Faculty:
Alec E. Aitken, (D), McMaster, 1987, Gm
Ernest G. Walker, (D), Texas, 1980, Pv
Geology Librarian:
Jane Lamothe, (U), Oo

University of Toronto

Dept of Geology (B,M,D) (2007)
Earth Sciences Centre
22 Russell Street
Toronto, ON M5S 3B1
p. (416) 978-3022 f. (416) 978-3938
welcome@geology.utoronto.ca
http://www.geology.utoronto.ca
Business Officer: Silvanna Papaleo
Enrollment: No data reported since 1995
Chair:
Alexander Cruden, (D), Uppsala, 1989, Gt
Professor:
Richard C. Bailey, (D), Cambridge, 1970, Yg
Nicholas Eyles, (D), Leicester (UK), 1978, Gl
Grant F. Ferris, (D), Guelph, 1985, Gn
Henry C. Halls, (D), Toronto, 1970, Ym
Kenneth W.F. Howard, (D), Birmingham, 1979, Hw
Andrew D. Miall, (D), Ottawa, 1969, Gs
Barbara Sherwood Lollar, (D), Waterloo, 1990, Cs

Edward T. C. Spooner, (D), Manchester (UK), 1976, En
Peter H. von Bitter, (D), Kansas, 1971, Pm
Director, Jack Satterly Geochronology Lab:
Michael Andrew Hamilton, (D), Massachusetts, 1993, Cc
Associate Professor:
James Brenan, (D), RPI, 1990, Gi
Donald W. Davis, (D), Alberta, 1978, Cc
Grant S. Henderson, (D), W Ontario, 1983, Gz
James E. Mungall, (D), McGill, 1993, Ec
Russell N. Pysklywec, (D), Toronto, 1998, Yg
Daniel J. Schulze, (D), Texas, 1982, Gi
Assistant Professor:
Jörg Bollmann, (D), Swiss Fed Inst of Tech, 1995, Pm
Rebecca Ghent, (D), S Methodist, 2002, Or
Jochen Halfar, (D), Stanford, 1999, Pe
Gopalan Srinivasan, (D), Maharaja SayajiRao, 1995, Ow
Ulrich B. Wortmann, Tech (Munich), 96, Gg
Research Associate:
Colin Bray, (D), Oxford, 1980, Cs
Lecturer:
Carl-Georg Bank, (D), British Columbia, 2002, Og
Adjunct Professor:
Jean-Bernard Caron, (D), Toronto, 2005, Pi
Marianne Douglas, (D), Queens, 1993, Gn
Martin J. Head, (D), Aberdeen, 1991, Pi
John H. McAndrews, (D), Minnesota, 1964, Pl
Myrna Joyce S. Simpson, (D), Alberta, 1999
Emeritus:
G M. Anderson, (D), Toronto, 1961, Cp
J. J. Fawcett, (D), Manchester, 1961, Gp
John Gittins, (D), Cambridge, 1959, Gi
Alan M. Goodwin, (D), Wisconsin, 1953, Gg
Thomas E. Krogh, (D), MIT, 1964, Cc
Anthony J. Naldrett, (D), Queen's, 1964, Ec
Geoffrey Norris, (D), Cambridge, 1964, Pl
Pierre-Yves F. Robin, (D), MIT, 1974, Gc
John C. Rucklidge, (D), Manchester, 1962, Ge
W. M. Schwerdtner, (D), Free (Berlin), 1961, Gc
Steven D. Scott, (D), Penn State, 1968, Cp
John A. Westgate, (D), Alberta, 1964, Gl
Frederick J. Wicks, (D), Oxford, 1969, Gz
Chief Librarian:
Bruce Garrod, Oo

Dept of Physics, Geophysics Div (B,M,D) (2004)
60 St. George Street
Toronto, ON M5S 1A7
p. (416) 978-5175 f. (416) 978-7606
mariaw@physics.utoronto.ca
http://www.physics.utoronto.ca
Administrative Assistant: Maria Wong
Enrollment (2004): M: 4 (3) D: 10 (0)
Professor:
Richard C. Bailey, (D), Cambridge, 1970, Yg
David J. Dunlop, (D), Toronto, 1968, Ym
Richard N. Edwards, (D), Cambridge, 1970, Yr
Bernd Milkereit, (D), Kiel, 1984, Ye
Jerry X. Mitrovica, (D), Toronto, 1991, Yd
Gordon F. West, (D), Toronto, 1960, Ye
Derek York, (D), Oxford, 1960, Cc
Research Associate:
Norman M. Evensen, (D), Minnesota, 1978, Cc
Ozden Ozdemir, (D), Newcastle upon Tyne, 1979, Ym
Emeritus:
Ronald M. Farquhar, (D), Toronto, 1954, Cc

University of Victoria

School of Earth & Ocean Sciences (B,M,D) (2006)
P.O. Box 3055
3800 Finnerty Road
Victoria, BC V8W 3P6
p. (250) 721-6120 f. (250) 721-6200
seosuvic@uvic.ca
http://web.uvic.ca/seos/
Enrollment: No data reported since 1995
Director:
Kathryn Gillis, (D), Dalhousie, 1987, Gp
Project Director:
Christopher R. Barnes, (D), Ottawa, 1964, Pm
Professor:
Brian D. Bornhold, (D), MIT, 1973, Gu
Ross N. Chapman, (D), British Columbia, 1975, Op
Christopher J. R. Garrett, (D), Cambridge, 1968, Op
Thomas F. Pedersen, (D), Edinburgh, 1979, Cm
David F. Strong, (D), Edinburgh, 1970, Em
Verena Tunnicliffe, (D), Yale, 1980, Ob
Andrew J. Weaver, (D), British Columbia, 1987, Op
Michael J. Whiticar, (D), Kiel, 1978, Co
Associate Professor:
Dante Canil, (D), Alberta, 1989, Gi
Stanley E. Dosso, (D), British Columbia, 1990, Op
Stephen T. Johnston, (D), Alberta, 1993, Gc
George D. Spence, (D), British Columbia, 1984, Ys
Eileen Van Der Flier-Keller, (D), W Ontario, 1985, Cl
Assistant Professor:
Kevin Telmer, (D), Ottawa, 1997, Cl
Adjunct Professor:
Vaughn Barrie, (D), Wales, 1986, Gu
Melvyn E. Best, (D), MIT, 1970, Ye
George J. Boer, (D), MIT, 1970, Oa
Eddy C. Carmack, (D), Washington, 1972, Op
William R. Crawford, (D), British Columbia, 1976, Op
Kenneth L. Denman, (D), British Columbia, 1972, Og
Kenneth L. Denman, (D), British Columbia, 1972, Op
David M. Farmer, (D), British Columbia, 1972, Op
Gregory M. Flato, (D), Dartmouth, 1991, Oa
Howard J. Freeland, (D), Dalhousie, 1973, Op
John C. Fyfe, (D), McGill, 1987, Oa
John R. Harper, (D), Louisiana State, 1978, Gu
Richard J. Hebda, (D), British Columbia, 1977, Pl
Roy D. Hyndman, (D), Australian National, 1967, Ys
C. Peter Keller, (D), W Ontario, 1985, Or
Dave Lefebure, (D), Carleton, 1986, Eg
Victor M. Levson, (D), Alberta, 1995, Gl
Rolf G. Lueck, (D), British Columbia, 1979, Op
Robie W. Macdonald, (D), Dalhousie, 1972, Oc
David L. Mackas, (D), Dalhousie, 1977, Ob
Norman McFarlane, (D), Michigan, 1974, Oa
Suzanne Paradis, (D), Carleton, 1990, Gv
Garry C. Rogers, (D), British Columbia, 1983, Ys
George J. Simandl, (D), Ecole Polytechnique, 1992, En
Robert W. Stewart, (D), Cantab, 1952, Op
Richard Thomson, (D), British Columbia, 1971, Og
Kelin Wang, (D), W Ontario, 1989, Yg
Michael Wilmut, (D), Queen's, 1971, Oo
C. S. Wong, (D), California (San Diego), 1968, Oc
Hidekatsu Yamazaki, (D), Texas A&M, 1984, Op
Francis Zwiers, (D), Dalhousie, 1980, Oo
Emeritus:
John T. Weaver, (D), Saskatchewan, 1959, Ym
Administrative Officer:
Terry P. Russell, (B), Victoria, 2000, Oo

University of Waterloo

Dept of Earth Sciences (B,M,D) (2005)
Waterloo, ON N2L 3G1

p. (519) 885-1211 (Ext. 3231) f. (519) 746-7484
klalbrec@sciborg.uwaterloo.ca
http://www.science.uwaterloo.ca/earth/
Secretary: Lorraine Albrecht
Enrollment: No data reported since 1995

Professor:
James F. Barker, (D), Waterloo, 1979, Co
David W. Blowes, (D), Waterloo, 1990, Cg
John A. Cherry, (D), Illinois, 1966, Hw
Mario Coniglio, (D), Memorial, 1985, Gr
Maurice B. Dusseault, (D), Alberta, 1977, Ng
Thomas W. D. Edwards, (D), Waterloo, 1987, Cs
Stephen George Evans, (D), Alberta, 1983
Shaun K. Frape, (D), Queen's, 1979, Cl
Robert W. Gillham, (D), Illinois, 1973, Hq
Alan V. Morgan, (D), Birmingham, 1970, Gl
Eric J. Reardon, (D), Penn State, 1974, Cl
David L. Rudolph, (D), Waterloo, 1989, Hq
Sherry L. Schiff, (D), Columbia, 1986, Ng
Edward A. Sudicky, (D), Waterloo, 1983

Associate Professor:
Anthony E. Endres, (D), British Columbia, 1991, Pe
David E. Lawson, (D), Reading (UK), 1971, Gs
Shoufa Lin, (D), New Brunswick, 1992, Oo
Robert Linnen, (D), McGill, 1992, Eg

Research Assistant Professor:
Brent Wolfe, (D), Oo

Assistant Professor:
Andre Unger, (D), Waterloo, 1995

Adjunct Professor:
Tom Al, (N)
Peter Annan, (D), Memorial, 1974, Ym
Edward C. Appleyard, (D), Cambridge, 1963, Gp
Estella Atekwana, (N)
Michael Brookfield, (N)
Robert W. Cleary, (D), Massachusetts, 1971, Gp
Donald W. Davis, (N)
Rick Devlin
Harvey F. Edwin
Michael English
Derek C. Ford, (D), Oxford, 1963, Pe
Neil Foster
John F. Gartner, Oo
John J. Gibson, (D), Waterloo, 1996, Cs
John P. Greenhouse
Daniel Hammerlund
Edwin Harvey, (N)
Masaki Hayashi
Larry Heaman
Francois Holtz
Daniel Hunkeler
John L. Jambor, (D), Carleton, 1966, Ym
Rosemary Knight, (N)
David R. Lee, (D), Virginia Tech, 1976, Hw
Uli Mayer
John Molson
Sorab Panday
Gary Parkin, (N)
Terry D. Prowse, (D), Canterbury, 1981, Cg
R. Gwilym Roberts, (D), McGill, 1966, Em
Nathalie Ross
Houston C. Saunderson, (N)
James Smith
Rene Therrien, (N)
Garth Van der Kamp, (D), Amsterdam, 1973, Hw
Cees van Staal
Leonard Wassenaar
Owen L. White, (D), Illinois, 1970, Ng

Emeritus:
Emil O. Frind, (D), Toronto, 1971, Hq
P.F. Karrow, Gl

Research Professor:
Ramon Aravena, (D), Waterloo, 1993, Co

Research Associate Professor:
Ronald V. Nicholson, (D), Waterloo, 1984, Hw
Carol J. Ptacek, (D), Waterloo, 1992, Cg
Will Robertson, (D), Waterloo, 1992, Hw

Research Assistant Professor:
Gui Lai, (D), Oo
Marek Odziemkowski, (D), Oo
Beth L. Parker, (D), Waterloo, 1996, Hw

Other:
Brewster Conant, (N)
Young-Jin Park, (N)

University of Western Ontario

Dept of Earth Sciences (B,M,D) (2008)
Biology & Geology Building
Room 153/154/155
1151 Richmond Street North
London, ON N6A 5B7
p. (519) 661-3187 f. (519) 661-3198
mrice@uwo.ca
http://www.uwo.ca/earth/
Enrollment (2007): B: 11 (9) M: 24 (5) D: 29 (4)

Provost & Vice-President:
Frederick J. Longstaffe, (D), McMaster, 1978, Cs

Graduate Chair:
Jisuo Jin, (D), Saskatchewan, 1988, Pi

Canada Research Chair:
Gail M. Atkinson, (D), W Ontario, 1993, YssYs
Gordon Southam, (D), Guelph, 1990, Py

Assistant Dean, Graduate & International Research:
Richard A. Secco, (D), W Ontario, 1988, Gy

Professor:
Stephen R. Hicock, (D), W Ontario, 1980, Gl
H Wayne Nesbitt, (D), Johns Hopkins, 1975, Cl
A Guy Plint, (D), Oxford, 1980, Gs

Acting Chair (P.Eng.,P.Geo.):
Robert A. Schincariol, (D), Ohio State, 1993, HwGg

Associate Professor:
Norman A. Duke, (D), Manitoba, 1983, Eg
Roberta L. Flemming, (D), Queen's, 1997, Gz
Dazhi Jiang, (D), New Brunswick, 1996, GcgGg
Sean Shieh, (D), Hawaii, 1998, Yg

NSERC Industrial Research Chair:
Kristy F. Tiampo, (D), Colorado, 2000, Yd

Joint appointment with Geography, U.W.O.:
Desmond Moser, (D), Queens, 1993, GtCcXg

Assistant Professor:
Neil Banerjee, (D), Victoria, 2001, Cs
Patricia Corcoran, (D), Dalhousie, 2001, Gd
Gordon R. Osinski, (D), New Brunswick, 2004, XgcGp
Robert Shcherbakov, (D), Cornell, 2002, YgsYd
Cameron J. Tsujita, (D), McMaster, 1995, Pi
Elizabeth Webb, (D), W Ontario, 2001, Cs

Research Associate:
Brian R. Hart, (D), W Ontario, 1995, Ca
Li Huang, (M), Concordia, 1992, Yr
Po-Yu (Paul) Shen, (D), W Ontario, 1975, Yh

Research Professor:
Lalu Mansinha, (D), British Columbia, 1962, Ys

Professor:
Susan F. Koval, (D), London (UK), Oo

Chair, Dept.Geology/Geophysics, U.Calgary:
David W.S. Eaton, (D), Calgary, 1992, YseYg

Associate Professor of Biology:
  Irena F. Creed, (D), Toronto, 1998, OoSfPy
Adjunct Research Professor:
  Paul Copper, London (UK), 1965, Pg
Adjunct Professor:
  Maria Cioppa, (D), Ym
  Claudia Cochrane, (M), W Ontario, Go
  Savka Dineva, (D), Russian Acad of Sci, 1986, Ys
  Penelope L. King, (D), Arizona State, 1999, Gi
  David T. Lescinsky, (D), Arizona State, 1999, Gv
  Wulf Mueller, (D), Heidelberg, 1985, Gv
  Hamid A. Mumin, (D), W Ontario, 1994, Eg
  Allen Pratt, (D), W Ontario, 1995, Cg
  John H. Sass, (D), Australian National, 1965, Yh
  Cliff Shaw, (D), W Ontario, Gi
  Yves Thibault, (D), W Ontario, 1990, Gi
Research Professor:
  Neil D. MacRae, (D), McMaster, 1966, Gx
Emeritus:
  W. Glen E. Caldwell, (D), Glasgow, 1957, Pm
  William R. Church, (D), Wales, 1961, Gt
  Aleksis Dreimanis, (D), Latvia, 1938, Gl
  Michael E. Fleet, (D), Manchester, 1963, Gz
  William S. Fyfe, (D), Otago (NZ), 1952, Cg
  Robert W. Hodder, (D), California (Berkeley), 1959, Eg
  Alfred C. Lenz, (D), Princeton, 1959, Pi
  Robert F. Mereu, (D), W Ontario, 1962, Ys
  H. Currie Palmer, (D), Princeton, 1963, Ym
  Helmuth H. Schloessin, (D), Cologne, 1955, Yx
  C. Gordon Winder, (D), Cornell, 1953, Gg
  Grant M. Young, (D), Glasgow, 1967, Gc
Research Scientist:
  Menghua Liu, (D), W Ontario, 2000, Oo
  Charles T. Wu, (D), W Ontario, 1984, Oo

University of Windsor

Dept of Earth Sciences (B,M,D) (2004)
401 Sunset Avenue
Windsor, ON N9B 3P4
  p. (519) 253-3000 (Ext. 2486) f. (519) 973-7081
  earth@uwindsor.ca
  http://www.uwindsor.ca/earth
  Administrator: Sharon Horne
  Enrollment (2001): B: 52 (0) M: 8 (0)
Head:
  Ihsan S. Al-Aasm, (D), Ottawa, 1985, Gd
Professor:
  Brian J. Fryer, (D), MIT, 1971, Cc
  V. Chris Lakhan, (D), Toronto, 1984, Or
  Iain M. Samson, (D), Strathclyde, 1983, Eg
  Frank Simpson, (D), Jagiellonian, 1968, Gs
  Alan S. Trenhaile, (D), Wales, 1969, Gm
Associate Professor:
  Cyril G. Rodrigues, (D), Carleton, 1980, Pm
Assistant Professor:
  Maria T. Cioppa, (D), Lehigh, 1997, Ym
  David A. Fowle, (D), Notre Dame, 1999, Cl
  Phil A. Graniero, (D), Toronto, 2001, Oi
  Stanley Reitsma, (D), Queen's, 1998, Hw
Instructor:
  Denis Tetrault, (D), Western, 2002, Gg
Emeritus:
  Peter P. Hudec, (D), RPI, 1965, Ng
  Terence E. Smith, (D), Wales, 1963, Gi
  David T. A. Symons, (D), Toronto, 1965, Ym
Related Staff:
  Melissa Flore, (B), Cs
  Alice Grgicak-Mannnion, (B), Waterloo, 1996, Oi
  Heather Patterson

York University

Dept of Earth & Atmospheric Science (B,M,D) (2005)
4700 Keele Street
Toronto, ON M3J 1P3
  p. (416) 736-5245 f. (416) 736-5817
  eas@yorku.ca
  http://www.yorku.ca/esse
  Administrative Assistant: Paola Panaro
  Enrollment (2003): B: 63 (0)
Professor:
  Keith D. Aldridge, (D), MIT, 1967, Yg
  Qiuming Cheng, (D), Ottawa, 1994, Oi
  Gary T. Jarvis, (D), Cambridge, 1978, Yg
  John (Jack) C. McConnell, (D), Queen's (Belfast), 1969, Oa
  Ian C. McDade, (D), Belfast, 1979, Or
  John Miller, (D), Saskatchewan, 1969, Or
  Peter A. Taylor, (D), Bristol, 1967, Oa
Associate Professor:
  Mary Ann Jenkins, (D), Toronto, 1986, Oa
  Gary P. Klaassen, (D), Toronto, 1983, Oa
  Diane V. Michelangeli, (D), California, 1988, Oa
  Spiros Pagiatakis, (D), New Brunswick, 1988, Yd
  Anthony M. K. Szeto, (D), Australian National, 1982, Yg
  Vincent C. Tao, (D), Calgary, 1997, Oi
  James Whiteway, (D), York, 1994, Oa
Assistant Professor:
  Baoxin Hu, (D), Boston, 1998, Or
Emeritus:
  Gordon G. Shepherd, (D), Toronto, 1956, Oo
  Douglas E. Smylie, (D), Toronto, 1963, Yg

Chile

Universidad de Chile

Dept de Geofisica (2005)
Blanco Encalada 2002
Casilla 2777
Santiago
  p. (56 2)696 6563 f. (56 2)696 8686
  ekausel@dgf.uchile.cl
  http://www.dgf.uchile.cl/index.html

China

China University of Geosciences

Geosciences (2005)
Wuhan
  jgzh@cug.edu.cn
  http://www.cug.edu.cn/cugnew/english/index.html

University of Hong Kong

Dept of Earth Sciences (2006)
Pokfulam Road
Hong Kong
  p. 852 2859 1084 f. 852 2517 6912
  earthsci@hkusub.hku.hk
  http://www.hku.hk/earthsci/rock.html

Czech Republic

Charles University

Inst of Geology and Paleontology (2005)
Faculty of Science
Albertov 6
Prague, 128 43

p. +420221951466 f. +420221951452
kachlik@mail.natur.cuni.cz
http://www.natur.cuni.cz/IGP/

Inst of Geochemistry, Mineralogy & Mineral Res (2004)
Albertov 6
128 43 Praha 2
mihal@natur.cuni.cz
http://www.natur.cuni.cz/~ugmnz/

Inst of Hydrogeology, Eng Geology and App Geophysics (2005)
Albertov 6
128 43 Praha 2
uhigug@natur.cuni.cz
http://prfdec.natur.cuni.cz/~geophys/ustav/katedra.htm

Inst of Petrology and Structural Geology (2007)
Albertov 6
Praha 2, 128 43
petrol@natur.cuni.cz
http://www.natur.cuni.cz/~petrol/

Prof:
Shah Wali Faryad, (A), 2000, GpzGi

Lecturer:
Jiri Konopasek, (D), 2000, Gpg

Dept of Geophysics (2004)
KG MFF UK
V Holesovickach 2
180 00 Prague 8, Czech Republic
Prague, Czech Republic 180 00
p. f. 00420 221912555
jz@karel.troja.mff.cuni.cz
http://geo.mff.cuni.cz

Masaryk University

Dept of Geology and Paleontology (2004)
Kotlarska 2
611 37 Brno
dino@gap.muni.cz
http://www.sci.muni.cz/index_en.htm

Dept of Mineralogy, Petrology and Geochemistry (2004)
Kotlarska 2
611 37 Brno
losos@sci.muni.cz
http://www.muni.cz/sci/structure/315020.html

Dept of Geography (2004)
Kotlarska 2
611 37 Brno
prosek@porthos.geogr.muni.cz
http://www.muni.cz/sci/structure/315030.html

Denmark

Aarhus University

Dept of Earth Sciences (2005)
C. F. Mollers Alle
DK-800 Arhus C
bo@geo.au.dk
http://www.geo.aau.dk.english/index_f.html

Technical Univ. of Denmark

Dept of Geology and Geotechnical Engineering (2004)
Anker Engelunsved
DK-2800 Lyngby
iggav@pop.dtu.dk
http://www.igg.dtu.dk/index_z.htm

University of Copenhagen

Geological Inst (2005)
Oster Voldgage 5-7
DK-1350 Copenhagen K,
p. +4535322400 f. +4535322499
info@geol.ku.dk
http://www.geol.ku.dk

Dept of Geophysics (2005)
Juliane Maries Vej 30
Copenhagen, 2100 Copenhagen
p. +45 353 20605 f. +45 35 36 53 57
awh@gfy.ku.dk
http://www.nbi.ku.dk/theinstitute/page52794.htm

Estonia

University of Tartu

Inst of Geology (2004)
46, Vanemuise Street
Tartu, 51014
p. +372 7 375 836 f. +372 7 375 836
geol@ut.ee
http://www.ut.ee/BGGL/g_ava-e.htm

Fiji

University of the South Pacific

Div of Earth and Environment Science (2007)
Private Mail Bag
GPO Suva
Suva, Fiji Islands
bonato_j@usp.ac.fj
http://www.sidsnet.org/pacific/usp/earth/

Finland

Helsinki University of Technology

Dept of Materials Science and Rock Engineering (2004)
Otakaari 1
SF-02150 Espoo
Kari.Heiskanen@hut.fi
http://www.hut.fi/Units/Departments/MK/

University of Helsinki

Dept of Geology (B,M,D) (2007)
Gustaf Hällströmink 2
P.O. BOX 64
Helsinki, 00014
p. 358-9-19150828 f. 358-9-19150826
Juha.Karhu@helsinki.fi
http://www.helsinki.fi/geology/
Enrollment (2007): B: 0 (6) M: 26 (15) D: 6 (0)

Head:
Juha A. Karhu, (D), Helsinki, 1993, Cg

University of Oulu

Dept of Geology (2005)
Linnanmaa
FIN-90014
Oulu
tuomo.alapieti@oulu.fi
http://cc.oulu.fi/~geolwww/Geology.htm

Dept of Geophysics (2007)
sofianos@metal.ntua.gr
http://www.gh.oulu.fi/

France

Ecole des Mines de Paris

(2004)
60-62, Boulevard Saint Michel
75272 PARIS cedex 06
deveughele@cgi.ensmp.fr
www.ensmp.fr/Eng/ENSMP/aboutENSMP.html

Ecole Nationale Supérieure de Géologie (ENSG)

ENSGéologie (2006)
Rue du Doyen Marcel Roubault
B.P. 40
VANDOEUVRE-LES-NANCY, Lorraine F-54501
p. 33 (0)3 83 59 59 59 f. 33 (0)3 83 59 64 64
amaury.legait@ensg.inpl-nancy.fr
http://www.ensg.inpl-nancy.fr/

Director:
Legait Amaury, (D), Toulouse III, 1985, Eg

Ecole Nat Superieure des Mines de Nancy

Ecole des Mines de Nancy (2004)
Parc de Saurupt
54042 Nancy cedex
ensmn@mines.u-nancy.fr
http://www.mines.u-nancy.fr/

Ecole Normale Superieure de Lyon

Laboratoire Des Sciences De La Terre (2007)
46 allee d'Italie
69364 Lyon Cedex 07
ricard@ens-lyon.fr
http://www.univ-lyon1.fr/ucbl/recherche/SCIENC-ES%20DE%20LA%2

Professor:
Francis Albarede
Pascal Allemand
Isabelle Daniel
Gilles Dromart
Philippe Gillet
Stephane Labrosse
Pierre Thomas

Senior Scientist:
Janne Blichert toft
Bruno Reynard
Yanick Ricard

Associate Professor:
Herve Bertrand
Frederic Chambat
Eva Chamorro Perez
Nicolas Coltice
Gweltaz Maheo

Associate Scientist:
Fabien Dubuffet
Philippe Herve Leloup
Laurence Lemelle
Jan Matas
philippe Oger
Anne Replumaz

Universite de Bourgogne

Centre des Sciences de la Terre (2004)
6 boulevard Gabriel
21100 Dijon
infogeneral@u-bourgogne.fr
http://babel.altavista.com/translate.dyn?urltext=http%3a%2f%

Universite de Pau

Dept of Geology (2004)
I.P.R.A. - Geologie
Avenue de L'Universite
Pau 64230
Pierre.Thore@elf-p.fr
http://www.univ-pau.fr/RECHERCHE/GEOPHY/

Universite Pierre et Marie Curie (Paris VI)

Inst De Physique Du Globe De Paris (2004)
Tour 15 - 4e etage
4 place Jussieu
75252 Paris Cedex 05
webmaster@ipgp.jussieu.fr
http://www.ipgp.jussieu.fr/

University Claude Bernard

Dept des Sciences de la Terre (2004)
Lab. de Geologie, Sedimentologie, Petrologie, Mineralogie
27-43 bd du 11 Novembre 1918
69622 Villeurbanne,
lardeaux@univ-lyon1.fr
http://www.univ-lyon1.fr/ucbl/recherche/SCIENC-ES%20DE%20LA%2

University of Lille

Dept of Geology (2004)
59655, Villeneuve D
Ascq Cedex
Jean.Carpentier@univ-lillel.fr
http://www.univ-lille1.fr/geosciences/

University of Nice-Sopia Antipollis-1

Laboratoire des Sciences de la Terre (2004)
Rue A. Einstein
06560 Valbonne
viri@faille.unice.fr
http://www-geoazur.unice.fr/

University of Pierre & Marie Curie (Paris VI)

Dept de Geol Sed (2004)
Tour 15 - 4e etage
4, place Jussieu
75252 Paris Cedex 05
webmaster@ipgp.jussieu.fr
http://www.ipgp.jussieu.fr/

University of Rennes I

Inst de Geologie (2004)
Campus de Beaulieu
Avenue de General Leclerc
35042 Rennes Cedex
thierry.pilorge@cnrs-dir.fr
http://www.geosciences.univ-rennes1.fr/

Germany

Aachen University of Technology

Inst of Structural Geology, Tect and Geomechanics (2005)
Lochnerstr 4-20

D-52056
J.Urai@ged.rwth-aachen.de
http://www.ged.rwth-aachen.de

Bonn Universitat

Mineralogisch-Petrologisches Inst und Museum (2004)
Popperlsdorfer SchloB
D-53115
Bonn
klapper@uni-bonn.de
http://www.min.uni-bonn.de/english/index.htm

Freien Universitat Berlin

Inst fur Palaontologie (2004)
Maltese-Strasse
74-100, Haus D
D-12249 Berlin
palaeont@zedat.fu-berlin.de
userpage.fu-berlin.de/~palaeont/WELCOME.HTM

Georg-August University of Gottingen

Dept of Geobiology (2007)
Goldschmidstr. 3
Gottingen, Lower Saxony 37077
p. +49 551 397951 f. +49 551 397918
jreitne@gwdg.de
http://www.geobiologie.uni-goettingen.de/

Karlsruhe University

Geophysical Inst (M,D) (2004)
Hertzstr. 16
Karlsruhe, 76187
p. +49-721-6084431 f. +49-721-71173
Friedemann.Wenzel@gpi.uni-karlsruhe.de
http://www-gpi.physik.uni-karlsruhe.de/

Koln University

Inst of Geophysics and Meteorology (2004)
mk@meteo.uni-koeln.de
www-f.rrz.uni-koeln.de/math-nat-fak/geomet/meteo/

LMU - University of Munich

Dept of Earth & Enviromental Sciences (B,M,D) (2007)
Theresienstr. 41/III
Munich, 80333
p. 0049/89/21804250 f. 0049/89/21804176
dingwell@lmu.de
http://min.geo.uni-muenchen.de/
Enrollment (2007): B: 173 (28) M: 12 (0) D: 24 (0)

Director:
Donald Bruce Dingwell, Giv

Professor:
Alexander Altenbach, Pg
Wladyslaw Altermann, Gg
Michael Amler, Pg
Valerian Bachtadse, Yg
Hans-Peter Bunge, Yg
Christina De Campos, Gz
Karl Thomas Fehr, Gz
Friedrich Frey, Gz
Anke Friedrich, (D)
Stuart Gilder, Yg
Wolfgang Heckel, Gz
Ernst Hegner, Gg
Soraya Heuss-Aßbichler, Gz
Heiner Igel, Yg
Harald Immel, Pg
Bernd Lammerer, Gg
Reinhold Leinfelder, Pg
Rocco Malservisi, Yg
Robert Marschik, Gg
Ludwig Masch, Gz
Wolfgang Moritz, Gz
Bettina Reichenbacher, Pg
Wolfgang Schmahl, Gz
Klaus Weber-Diefenbach, Gg
Christian Wolkersdorfer, Gg

Other:
Kirill Aldushin, Gz
Greta Barbi, Yg
Robert Barsch, Yg
Johannes Birner, Gg
Florian Bleibinhaus, Yg
Hans Boysen, Gz
Gilbert Britzke, Yg
Benoit Cordonnier, Gz
Alexander Dorfman, Gz
Thomas Dorfner, Gz
Barbara Emmer, Yg
Werner Ertel-Ingrisch, Gz
Andreas Fichtner, Yg
Michaela Frei, Gg
Frantisek Gallovic, Yg
Helmut Gebrande, Yg
Alexander Gigler, Gz
Peter Gille, Gz
Stefan Grießl, Gz
Hagen Göttlich, Gz
Marc Hennemeyer, Gz
Katja Henßel, Pg
Kai-Uwe Hess, Gz
Maria Linda Iaccheri, Gg
Giampiero Iaffaldano, Yg
Guntram Jordan, Gz
Ines Kaiser-Bischoff, Gz
Thorsten Kowalke, Pg
Thomas Kunzmann, Gz
Martin Käser, Yg
Markus Lackinger, Gz
Yan Lavallee, Yg
Maike Lübbe, Gz
Götz Meisterernst, Gz
Timo Casjen Merkel, Gz
Marcus Mohr, Yg
Lena Müller, Gz
Dieter Müller-Sohnius, Gz
Malik Naumann, Pg
Jens Oser, Yg
Sohyun Park, Gz
Karin Paschert, Gg
Rossitza Pentcheva, Gz
Nikolai Petersen, Yg
Helen Pfuhl, Yg
Antonio Sebastian Piazzoni, Yg
Christina Plattner, Yg
Josep Puente Alvarez de la
Oliver Riedel, Gz
Alexander Rocholl, Gz
Javier Rubio-Sierra, Gz
Gertrud Rößner, Pg
Dieter Schmid, Pg
Julius Schneider, Gz
Bernhard Schuberth, Yg
Oliver Spieler, Gz

Germany

Robert Stark, Gz
Katja Steffens, Gg
Stefan Strasser, Gz
Marco Stupazzini, Yg
Frank Söllner, Gg
Frank Trixler, Gz
Ferdinand Walther, Gz
Joachim Wassermann, Yg
Laura Wehrmann, Pg
Michael Winklhofer, Yg
Ayhan Yurtsever, Gz
Matthias Zeitlhöfler, Gg
Albert Zink, Gz

Section of Palaeontology & Hist. Geol.(2004)
Richard-Wagner-Strasse 10
80333 Munchen,
p. f. +49-89-2180-6601
leinfelder@lmu.de
http://www.palaeo.de/munich

Ludwig-Maximilians-Universitat Munich

Dept for Earth and Enviromental Sciences (2007)
Theresienstr. 41
Muenchen, 80333
p. + 49 89 2180 4250 f. + 49 89 2180 4253
GeoUmwelt@lmu.de
http://www.geo.uni-muenchen.de/depearthenv/index.html

Martin-Luther-Universitaet Halle-Wittenberg

Inst for Geosciences (2006)
Von-Seckendorff-Platz 3
D-06120 Halle
p. ++49-045-55 26010 f. ++49-045-55 27365
direktor@geo.uni-halle.de
http://www.geo.uni-halle.de

Director:
Herbert Pöllmann, (D), 1984, Gz

Technical University Clausthal

Inst of Geology and Paleontology (2007)
TU Clausthal
LeibnizstraBe 10
D-38678 Clausthal-Zellerfeld, 38678
p. 0049532372230 f. 0049532372903
Office@geologie.tu-clausthal.de
www.inggeo.tu-clausthal.de/english/Welcome-e.html

Univ. of Bochum

Ruhr-Universitat, Inst of Geologie (2004)
Post Box 102148
4630 Bochum
Bernhard.Stoeckhert@ruhr-uni-bochum.de
www.ruhr-uni-bochum.de/exogeol/geowiss.html

Univ. of Erlangen-Nuremberg

Applied Geology (2004)
Schlossgarten 5
D-91054 Erlangen
htobscha@geol.uni-erlangen.de
www.rrze.uni-erlangen.de:80/docs/FAU/fakultaet/natIII

Univ. of Freiburg

Geologisches Inst. (2004)
Albertstr. 23-B
W-7800 Freiburg

schaeben@orion.hrz.tu-freiberg.de
http://www.geo.tu-freiberg.de/institut/

Universitaet Muenster

Geologisch-Palaeontologisches Inst (2007)
Correnstr. 24
48149 Muenster
bahlbur@uni-muenster.de
http://www.uni-muenster.de/GeoPalaeontologie/

Universitat Bonn

Geologisches Inst (2007)
Nu ssallee 8
Bonn, D 53115
p. 0049 228 732461 f. 0049 228 739037
jthein@uni-bonn.de
http://granit.geo.uni-bonn.de/

Universitat Hannover

Inst fur Geol. & Palaontologie (2004)
Fachbeleich Erdwissenschaften
Callinstrasse 30
D-30167 Hannover
lueer@mbox.geowi.uni-hannover.de
http://www.unics.uni-hannover.de/igp/Geohome.htm

Universitat Leipzig

Inst fur Geologie (2004)
Talstrasse 35
0-7010 Leipzig
http://www.uni-leipzig.de/~gasse/geoph.html

Universitat Wurzburg

Inst fur Geologie (2005)
Plelcherwall 1
D-97070
sekretariat@geologie.uni-wuerzburg.de
http://www.geologie.uni-wuerzburg.de/index_e.htm

University of Cologne

Dept of Geology and Mineralogy (2006)
Zuelpicherstr. 49a
50674 Koeln
p. +49-221-470-5619 f. +49-221-470-5149
ellen.stefan@uni-koeln.de
http://www.uni-koeln.de/math-nat-fak/geologie

University of Erlangen

Geology Dept (2005)
Schlossgarten 5
91054 Erlangen
geologie@geol.uni-erlangen.de
http://www.geol.uni-erlangen.de/index.html

Head:
Werner Buggisch, (D), Gg

University of Hamburg

Geology - Paleolontology Inst (2004)
Bundesstrasse 55
W-2000 Hamburg 13
dirk.gajewski@zmaw.de
http://www.geowiss.uni-hamburg.de/i-geolo/

University of Mainz

Dept of Geosciences (2007)

P.O. Box 3980
Mainz, 55099
p. (49) 613-1 39 24373 f. (49) 613-1 39 23070
nikoley@mail.uni-mainz.de
http://www.geowiss.uni-mainz.de/

Universität Kiel
Inst fur Geowissenschaften (2007)
Ludewig-Meyn-Strasse 10
D-24118 Kiel
sekretariat@gpi.uni-kiel.de
http://www.ifg.uni-kiel.de/

Universität Tübingen
Geology Dept (2004)
72076 Tübingen
Sigwartstr. 10
e16info@uni-tuebingen.de
http://www.uni-tuebingen.de/geo/ifg

Greece

National Technical University of Athlens
Dept of Mining Engineering and Metallurgy (2004)
Zographou Campus
15 780 Athens
tsoflias@metal.ntua.gr
http://www.metal.ntua.gr/div-geology/index.html

University of Athens
Dept of Geology, Univ Campus (2004)
National & Capodistran Univ. of Athens
Panepistimioupolis, Ilistra
Athens
sideriscos@geol.uoa.gr
http://www.geol.uoa.gr/engindex.htm

Hungary

Eotvos Lorand University
Dept of Mineralogy (2004)
Muzeum KRT 4/A
H-1088 Budapest,
http://iris.elte.hu/geo/

Dept of Paleontology (2004)
Ludovika ter 2
H-1083 Budapest,
monost@ludens.elte.hu
http://iris.elte.hu/geo/

Eotvos University
Dept of Petrology and Geochemistry (2007)
Pazmany P setany 1/C
Budapest H-1117
Budapest, 1117
p. (+36) 1 209 0555 f. (+36) 1 381 2108
cszabo@elte.hu
http://lrg.elte.hu/

Dept of Physical and Historical Geology (2004)
Muzeum krt. 4/A
Budapest H-1088
gtorfo@ludens.elte.hu
http://iris.elte.hu/geo/

Univesity of Szeged
Dept of Physical Geography (2007)
SZTE Termeszeti Foldrajzi Tanszek
Szeged, 6720 Egyetem utca 2. PF:653
Szeged
p. 0036 62544158
mezosi@earth.geo.u-szeged.hu
http://www.geo.u-szeged.hu/ANG/HPfirst.html

India

Presidency College
Dept of Geology (2005)
College Street
Calcutta
bbhaduri@uval.geo.purdue.edu
www.concentric.net/~slahiri/pc/Department/Geology/geo

Roorkee University
Dept of Earth Sciences (2004)
Roorkee
earth@rurkiu.ernet.in
http://members.tripod.com/rurkiu/acd-earth/index

Ireland

Trinity College
Dept of Geology (2006)
Dublin 2
earth@tcd.ie
http://www.tcd.ie/Geology/

University College Cork
Dept of Geology (B,M,D) (2005)
Donovans Road
Cork
p. +353 21 4902657 f. +353 21 4271565
j.gamble@ucc.ie
http://www.ucc.ie/ucc/depts/geology/
Administrative Assistant: Patricia Hegarty

Head:
John Gamble, Gi

Professor:
Ken Higgs, Pl

Research Associate:
Tara Davis, Yg
Jim Smith, Pl

Lecturer:
Alistair Allen, Gp
Bettie Higgs, Yg
Ed Jarvis, Pl
Ivor MacCarthy, (D), 1974, Gs
Pat Meere, (D), NUI (Ireland), 1992, Gc
John Reavy, Gi
Andy J. Wheeler, (D), Cambridge (UK), 1994, Gu

Other:
Mary Lehane, Oo
Mick O'Callaghan, Oo
Dan Rose, Oo

University College Dublin
UCD School of Geological Sciences (2007)
Science Centre West
Belfield
Dublin 4
p. +353 1 716 2331 f. +353 1 283 7733

geology@ucd.ie
www.ucd.ie/geology

Israel

Ben Gurion University of the Negev
Dept of Geological and Env Sciences (2004)
P.O. Box 653
84105 Beer Sheva
edav@bgumail.bgu.ac.il
http://www.bgu.ac.il/geol/

Hebrew University of Jerusalem
Inst of Earth Sciences (2004)
Jerusalem 91904
mali@vms.huji.ac.il
http://earth.es.huji.ac.il/machon/

Tel Aviv University
Dept of Geophysics & Planetary Sciences (2004)
69978 Tel Aviv
akiva@terra.tau.ac.il
http://www.tau.ac.il/geophysics/

Italy

Messina Universita
Dept di Scienze della Terra (2004)
Salita Sperone 31 - 98166 Messina-Sant'Agata
laura@labcart.unime.it
http://ww2.unime.it/scienze-terra/

Universita de Bologna
Dept di Scienze della Terra e Geologico (2004)
Via Zamboni 65
Bologna
direttore@geomin.unibo.it
http://www.geomin.unibo.it/default.htm

Universita de Firenze
Dept Scienze della Terra (2004)
via La Pira 4
50121 Firenze
direttore@geo.unifi.it
http://www.geo.unifi.it/

Universita de Pavia
Dept Scienze della Terra (2007)
Strada Nuova 65
27100 Pavia
dvagnini@unipv.it
http://manhattan.unipv.it/

Universita Degli Studi
Dept di Scienze Geologiche (2005)
pucci@dst.unipi.it
www.mat.uniroma3.it/facolta/GEOLOGIA/DipGeo.html

Universita di Bari
Dept Geomineralogico (2004)
Via E. Orabona
4-70125 Bari
scandale@geomin.uniba.it
http://www.geomin.uniba.it/frame.htm

Universita di Cagliari
Dept di Geoingegneria e Tecnologie Ambientali (2006)
Via Marengo,3
09124 Cagliari
p. +39 70 675 52/29 f. +39 70 675 23
mazzella@unica.it
http://geoing.unica.it/digita.htm
Professor:
Prof. Antonio AM MAZZELLA, Gq

Universita di Calabria
Dept Scienze della Terra (2005)
87036 Arcavacata di Rende
Calabria
derosa@unical.it
http://home.unical.it/dipartimenti/scienze-terra/

Universita di Milano
Dept Scienze della Terra (2004)
via Mangiagalli 34
20133 Milano,
Caterina.Deleonardis@unimi.it
http://www.gp.terra.unimi.it/

Università di Modena e Reggio Emilia
Dept di Scienze della Terra (2007)
Largo S. Eufemia, 19
Modena, 41100
p. 390592055885 f. 390592055887
cecilia.docchio@unimore.it
http://www.terra.unimore.it/

Universita di Palermo
Dept Geologia Geodesia (2004)
Corso Tukory 131
90134 Palermo
leofelix@mbox.unipa.it
http://www.unipa.it/~dipgeopa/

Universita di Siena
Dept di Scienze della Terra (2005)
il nostro indirizzo
Via Laterina 8 53100
lazzarotto@unisi.it
http://www.dst.unisi.it/

Universita di Torino
Dept Scienze della Terra (2005)
via Academia Scienze 5
10123 Torino
direzione.dst@unito.it
hal9000.cisi.unito.it/wf/DIPARTIMEN/Scienze_de2

Universita di Udine
Dept di Georigorse e Territorio (2004)
Istituto Scienze della Terra
Viale Ungheria 43
33100 Udine
beinat@dgt.uniud.it
http://udgtls.dgt.uniud.it/

Universita Pisa
Dept Scienze della Terra (D) (2004)
via S. Maria 53
56126 Pisa

p. +39050847260 f. +39050500932
agostini@dst.unipi.it
http://www.dst.unipi.it/

Universita' Di Napoli

Dept di Scienze della Terra (2004)
Ligo S. Marcellino, 10
80138 Naples
incorona@unina.it
http://www.dst.unina.it/Homeita.htm

Dept of Geophysics and Volcanology (2004)
80138 Napoli
luongius@unina.it
http://www.dgv.unina.it/dgv_eng.html

University of Parma

Dept of Earth Sciences (2005)
Viale delle Scienze No. 78
43100 Parma
giovanni.mezzadri@unipr.it
http://www.unipr.it/arpa/dipgeo/ENGLISH.html

Univsita di Padova

Dept Geologia Paleont. Geof. (2004)
Via Giotto 1
31537 Padova
domenico@geol.unipd.it
http://www.geol.unipd.it/

Japan

Ehime University

Dept of Earth Sciences (2004)
yinouchi@sci.ehime-u.ac.jp
http://www.ehime-u.ac.jp/~cutie/index.html

Hirosaki University

Dept of Earth Science (2004)
wata@cc.hirosaki-u.ac.jp
http://sci.hirosaki-u.ac.jp/~earth/index.html

Hiroshima University

Dept of Earth and Planetary Systems Science (2005)
Kagami-yama 1-3-1
Higashi-Hiroshima
Hiroshima 739
chikako@sci.hiroshima-u.ac.jp
http://www.geol.sci.hiroshima-u.ac.jp/index_e.html

Kagoshima University

Faculty of Science (2004)
koho@sci.kagoshima-u.ac.jp
http://earth.sci.kagoshima-u.ac.jp/index.html

Kanazawa University

Dept of Earth Sciences (2004)
Kakuma-machi
Kanazawa 920-1192
admin@earth.s.kanazawa-u.ac.jp
http://earth.s.kanazawa-u.ac.jp/

Kobe University

Earth & Planetary Sciences (2004)
watanabe@komadori.planet.sci.kobe-u.ac.jp
http://shidahara1.earth.s.kobe-u.ac.jp/index.html

Kumamoto University

Dept of Earth and Environment (2007)
2-39-1, Kurokami
Kumamoto City, 860-8555
p. 81-96-342-3411 f. 81-96-342-3411
tadao@sci.kumamoto-u.ac.jp
http://www.sci.kumamoto-u.ac.jp/~shibuya/

Professor:
Hidetoshi Shibuya, (D), Ymg

Miyazaki University

Earth Science (2005)
nishim@edugeo.miyazaki-u.ac.jp
http://edugeo.miyazaki-u.ac.jp/earth/earth.html

Nagoya University

Dept of Earth and Planetary Sciences (2006)
www-admin@eps.nagoya-u.ac.jp
http://www.eps.nagoya-u.ac.jp/

Nihon University

Dept of Earth Systems Sciences (2005)
inf-geos@chs.nihon-u.ac.jp
http://www.geo.chs.nihon-u.ac.jp/

Shimane University

Dept of Geoscience (B,M,D) (2005)
1060 Nishikawatsu
Matsue, Shimane Prefecture 690-8504
p. +81-852-32-6455 f. +81-852-32-6469
akasaka@riko.shimane-u.ac.jp
http://terra.riko.shimane-u.ac.jp/
Enrollment (2005): B: 200 (0) M: 40 (0) D: 10 (0)

Professor:
Takasu Akira, (D), Gp
Ishiga Hiroaki, (D), Ge
Komuro Hiroaki, (D), Gc
Kimura Junichi, (D), Ge
Akasaka Masahide, (D), 1981, Gz
Yokota Shuichiro, (D), Ng
Sampei Yoshikazu, (D), Eo
Sawada Yoshiyuki, (D), Ng

Associate Professor:
Hayashi Hiroki, (D), Ps
Kamei Takeshi, (D), Tech, Nr
Sakai Tetsuya, (D), Gs
Irizuki Toshiaki, (D), Pg

Assistant Professor:
Ohira Hiroto, (D), En
Shibi Toshihide, (D), Gq

Lecturer:
Masumoto Akira, (D), Hw
Kamei Atsushi, (D), Gi
Roser P. Barry, (D), Cg

Related Staff:
Nakamura Manabu, (M), Ca

Shizuoka University

Inst of Geosciences (2005)
Shizuoka 422-8529
setmasu@ipc.shizuoka.ac.jp
http://www.sci.shizuoka.ac.jp/~geo/Welcome.html

Tohoku University
Dept of Mineralogy, Petrology and Econ Geology (2005)
Sendai 980-8578
www-admin@ganko.tohoku.ac.jp
http://www.ganko.tohoku.ac.jp/

Toyama University
Dept of Earth Sciences (2005)
3190 Gofuku
Toyama 930-8555
Toyama, Toyama Prefecture 930-8555
p. 076-445-6649
hsakai@sci.toyama-u.ac.jp
http://core.sci.toyama-u.ac.jp/index.html

Tsukuba University
Dept of Geoscience (2004)
emato@atm.geo.tsukuba.ac.jp
http://www.geo.tsukuba.ac.jp/

University of Tokyo
Geological Inst (2004)
3-1, Hongo 7chome, Bunkyo-ku
Tokyo
webmaster@eri.u-tokyo.ac.jp
http://www.eri.u-tokyo.ac.jp/

Yokohama National University
Geological Inst (2004)
ishikawa@ed.ynu.ac.jp
http://chigaku.ed.ynu.ac.jp/geology-e.html

Korea, South

Seoul National University
Dept of Earth Science Education (2008)
Shinrim Dong, Kwanak-ku
151-742 Seoul
mkznhs@plaza.snu.ac.kr
http://earth.snu.ac.kr/

Kuwait

Kuwait University
Dept of Geology (2006)
http://139.141.135.14/Frame.html

Latvia

University of Latvia
Faculty of Geographical and Earth Sciences (2004)
10 Alberta Str.
202-203 Room
Riga
zeme@lanet.lv
http://www.lu.lv/e_strukt/fakult/geog/info/

Lithuania

Vilnius University
Dept of Geology and Mineralogy (B,M,D) (2004)
Ciurlionio str. 21/27
2009LT Vilnius
p. +370 5 2398272 f. +370 5 2398272
petras.musteikis@gf.vu.lt
http://www.gf.vu.lt/depts/geology/geominka.html
Enrollment (2003): B: 24 (0) M: 11 (0) D: 1 (0)

Malaysia

Universiti Putra Malaysia
Dept of Environmental Sciences (2004)
43400 UPM Serdang
ramdzani@fsas.upm.edu.my
http://fsas.upm.edu.my/~sas/envpage/Dept.html

Universiti Tecnologi Malaysia
Faculty of Geoinformation Science and Eng (2004)
81310 UTM Skudai
Johor Bahru
Johor
dekan@fksg.utm.my
http://www.fksg.utm.my/

University of Malaya
Dept of Geology (B,M,D) (2007)
Dept. of Geology
University of Malaya
50603 Kuala Lumpur, Malaysia.
Kuala Lumpur, Selangor 50603
p. 03-79674203 f. 03-79675149
sci_geo@um.edu.my
http://www.um.edu.my

Head:
Wan Hasiah Abdullah, (D), Newcastle upon Tyne, 1994, Ec

Professor:
Teh Guan Hoe, (D), Heidelberg, Ca
Lee Chai Peng, (D), Liverpool, Pg
John Kuna Raj, (D), Malaya, 1983, Ng
Denis N.K Tan, (M), Malaya, Go

Associate Professor:
Abd Rashid Ahmad, (D), Oxford, Sc
Azman Abdul Ghani, (D), Liverpool, Gi
Mohamed Ali Hasan, (M), Ulster, Hg
Azhar Hj Hussin, (D), London, Gs
Tajul Anuar Jamaluddin, (D), Wales, 1997, Ng
Mustaffa Kamal Shuib, (M), London, 1986, Gc
Samsudin Hj. Taib, (D), Durham, Yg

Lecturer:
Nuraiteng Tee Abdullah, (D), London, Gr
Ahmad Tajuddin Hj Ibrahim, (D), Newcastle upon Tyne, Ng
Che Noorliza Lat, (M), Nevada (Reno), 1989, Yg
Mat Ruzlin Maulud, (B), Malaya, Oo
Nur Iskandar Taib, (D), Indiana, Gi
Ismail Yusoff, (D), Norwich, Hg

Mexico

Centro de Estudios Sup del Estado Sonora
Escuela Superior de Geociencias (B) (2004)
Col. Las Quintas
Hermosillo, SO 83240
p. 621/5-37-78 f. 621/5-70-47
Department Secretary: Manuel Valenzuela-Renteria

Director:
Marco A. Gonzalez-Juarez, (B), Sonora, 1977, Gg

Associate Professor:
Jesus E. Cruz-Teran, (B), Sonora, 1979, Oo
Leopoldo Diaz-Encinas, (B), Chihuahua, 1988, Em
Gustavo E. Durazo-Tapia, (B), Sonora, 1981, Em
Francisco A. Esparza-Yanez, (B), Sonora, 1983, Gg
Raul A. Gongora-Jurado, (B), Autonoma de Chihuahua, 1977, Nm

Maria B. Hurtado-De La Ree, (B), Sonora, 1987, Nx
Leobardo Lopez-Pineda, (B), Nacional Auton, 1980, Yg
Jesus F. Maytorena-Silva, (B), Sonora, 1981, Gr
Rogelio Monreal-Saavedra, (D), Texas (Dallas), 1989, Gr
Gerardo Monteverde-Gutierrez, (B), Sonora, 1980, Nm
Arnulfo Salazar-Avila, (B), Sonora, 1981, Nx
Angel Slistan-Grijalva, (B), Sonora, 1981, Nm
Instructor:
Luis G. Vite-Picazo, (B), Nacional Auton, 1946, Nm

Centro de Investigación Científica y de Ed Sup de Ensenada

Earth Sciences Div (M,D) (2007)
Km. 107 Carr. Tijuana-Ensenada
Ensenada, Baja California 22860
p. (01152-664)1750500 f. (01152-664)1750559
dir-ct@cicese.mx
http://www.cicese.mx

Associate Professor:
Juan M. Espinosa, (M), CICESE (Ensenada), 1983, Ye
Alejandro Hinojosa, (M), CICESE (Ensenada), 1988, Or
Titular researcher:
Luis A. Delgado-Argote, (D), UNAM, 2000, Git
Assistant Professor:
Jesus M. Brassea, (M), Cinvestav, 1986, Ye
Research Associate:
Jose G. Acosta, (M), CICESE (Ensenada), 1980, Ne
Jose J. Gonzalez, (M), CICESE (Ensenada), 1986, Yd
Mario Gonzalez-Escobar, (D), CICESE, 2002, Gu
Carlos I. Huerta, (D), Texas, 2002, Ne
Luis H. Mendoza, (M), CICESE (Ensenada), 1982, Ne
Alfonso Reyes, (B), California (San Diego), Ne
Titular researcher:
Weber Bodo, (D), Gp
Edgardo Canon, (D), Hawaii, 1991, Ym
Raul Castro, (D), Nevada (Reno), 1991, Ys
Juan Contreras, (D), Columbus, 1999, Gt
Francisco Esparza, (D), CICESE (Ensenada), 1991, Ye
John Fletcher, (D), Utah, 1994, Gpt
Carlos Flores, (D), Toronto, 1986, Ye
Jose Frez, (M), California (Los Angeles), 1980, Ys
Luis A. Gallardo-Delgado, (D), Lancaster (UK), 2004, Ge
Juan Garcia, (D), Oregon State, 1990, Yv
Ewa Glowacka, (D), Polish Acad of Sci, 1991, Ys
Enrique Gomez, (D), Toronto, 1981, Ye
Antonio Gonzalez-Fernandez, (D), Complutense, 1996, Gu
Javier Helenes, (D), Stanford, 1980, Gr
Thomas Kretzschmar, (D), Tubingen, 1995, Ge
Margarita Lopez, (D), Toronto, 1985, Cc
Juan Madrid, (D), Toronto, 1972, Ys
Arturo Martin, (D), Paris XI Orsay, 1988, Gd
Luis Munguia, (D), California (San Diego), 1982, Ys
Alejandro Nava, (D), California (San Diego), 1980, Ys
Marco A. Perez, (D), CICESE, 1995, Ye
Jose M. Romo, (D), CICESE (Ensenada), 2002, Ye
Pratap Sahay, (D), Alberta, 1986, Ys
Stephen V. Smith, (D), Hawaii, 1970, Cm
Francisco Suarez, (M), San Diego State, 1984, Gc
Rogelio Vazquez, (D), CICESE (Ensenada), 2002, Ye
Antonio Vidal, (D), CICESE (Ensenada), 2001, Ys
Victor Wong, (D), CICESE (Ensenada), 2001, Ys

Facultad de Minas, Metalurgia y Minas

Dept of Geology (B) (2004)
Ex-Hacienda San Matias S/N
Guanajuato, GJ 36020
p. 473/23864

Director:
Antonio Nieto Antunez, (M), Stanford, 1970, Nm
Professor:
Angelica Casillas, (B), Inst Politecnico Nacional, 1976, Hw
Esteban Cedillo, (D), Heidelberg, 1988, Em
Juventino Martinez, (D), Paris VI, 1980, Gt
Edgardo Meave, (B), Guanajuato, 1947, Gg
Francisco Medina, (B), Guanajuato, 1982, Og
Ricardo Navarro, (B), Inst Politecnico Nacional, 1973, Nm
Salvador Ulloa, (B), Guanajuato, 1947, Eg
Fernando Vasallo, (D), Lomonosov (USSR), 1982, Em
Carlos Yanez, (B), Inst Politecnico Nat, 1976, Ce

Mexico Universidad Nacional Autonoma

Inst de Geofisica (2004)
webmaster@tlacaelel.igeofcu.unam.mx
http://tlacaelel.igeofcu.unam.mx/index.eng.html

Univ Autonoma de Baja California Sur

Dept of Marine Geology (B) (2004)
Carretera al Sur km 5.5, Box 19-B
La Paz, BS 23000
p. 682/2-47-55/2-01-40/2-45-69 f. 682/2-48-80

Chair:
Alejandro Alvarez-Arellano, (M), Nacional Auton, 1984, Gs
Professor:
Rodolfo Cruz-Orozco, (D), Louisiana State, 1974, Gu
Javier Gaitan-Moran, (M), Intl Inst-Aerospace Sur & Earth Sci, 1986, Or
Carlos A. Galli-Olivier, (D), Utah, 1968, Gd
Jose I. Peredo-Jaime, (B), Auto de Baja California, 1978, Og
Luis R. Segura-Vernis, (D), Nacional Autonoma, 1977, Pi
Associate Professor:
Alejandro J. Carillo-Chavez, (M), Cincinnati, 1981, Gp
Efrain Cornejo-Luna, (B), Nacional Auton, 1969, Gq
Genaro Martinez-Gutierrez, (B), Inst Politecnico Nac, 1982, Gr
Cesar Martinez-Noriega, (B), Nacional Auton, Ou
Jaime I. Monroy-Sanchez, (M), 1987, Ge
Miriam Nuñez-Velazco, (B), Tech (Madero), 1980, Go
Jose A. Perez-Venzor, (B), Autonoma (SLP), 1978, Gi
Ramon Pimentel-Hernandez, (B), Inst Pol Nacional, 1980, Yg
Humberto Rojas-Soriano, (B), Nacional Auton, 1983, Gz
Paulino Rojo-Garcia, (B), Inst Politecnico Nac, 1983, Gc
Assistant Professor:
Luis A. Herrera-Gil, (B), Nacional Auton, 1980, Pg
Lecturer:
Cesar A. Lopez-Ferreira, (B), Nacional Auton, 1979, Hw
Cooperating Faculty:
Oscar Rodriguez-Plasencia, (B), Escuela Militar de Met, 1959, Ow

Universidad Autonoma de Chihuahua

Facultad de Ingenieria (B,M) (2004)
Apartado Postal 1528, Suc. "C"
Chihuahua, CH 31170
p. 14/13-37-11

Director:
Arturo Leal-Bejarano, (B), Autonoma de Chihuahua, 1976, Oo
Head:
Arturo Lujan-Lopez, (M), Essex, 1975, Oo
Chair:
Socorro I. Aguirre-Moriel, (B), ITR (Switzerland), 1979, Nx
Hector M. Mendoza-Aguilar, (B), Auto Chihuahua, 1980, Nx
Professor:
Rafael Chavez-Aguirre, (M), AutoChihuahua, 1993, Hw
Adolfo Chavez-Rodriguez, (D), Arizona, 1987, Hy
Miguel Franco-Rubio, (M), Nac Autonoma, 1978, Eg

Rafael Madrigal-Rubio, (B), Nac Auton, 1969, Ng
Teodulo Mena-Zambrano, (B), Inst Politecnico Nac, 1964, Ng
Hector Minor-Velazquez, (B), Auto Chihuahua, 1978, Nx
Ignacio A. Reyes-Cortes, (B), Texas (El Paso), 1997, Gx
Manuel Reyes-Cortes, (B), Nacional Auton, 1970, Gi
Miguel Royo-Ochoa, (M), Auto Chihuahua, 1997, Hw
David H. Ruiz-Cisneros, (B), Auto Chihuahua, 1974, Nx
Luis M. Trevizo-Cano, (B), Auto Chihuahua, 1978, Nx

Univ Autonoma de San Luis Potosi

Area de Ciencias de la Terra (B,M) (2004)
Facultad de Ingenieri#2.a
Av. Dr. Manuel Nava No. 8
San Luis Potosi#2., SL 78290
p. (48) 13-82-22 f. (48) 13-09-24
Department Secretary: Juan Manuel Torres Aguilera

Director:
Hector David Atisha-Castillo, (M), Texas, 1978, Oo

Head:
Joel Milan-Navarro, (M), Nancy (France), 1979, Or

Professor:
Jose Refugio Acevedo-Arroyo, (B), Nac Auton, 1957, Ng
Luis Garcia-Gutierrez, (M), Stanford, 1951, Eg
Panfilo R. Martinez-Macias, (M), Nacional Auton, 1995, Gc
Francisco Javier Orozco-Villasenor, (M), Colorado State, 1983, Eg
Ramon Ortiz-Aguirre, (M), Madrid, 1979, Hg
Carlos Francisco Puente-Muniz, (B), Auto (SLP), 1979, Hw
Delfino C. Ruvalcaba-Ruiz, (D), Colorado State, 1982, Em
Juan Manuel Torres-Agulera, (B), Auto (SLP), 1980, Gx

Librarian:
Panfilo R. Martinez-Macias, Oo

Univ Nacional Autonoma de Mexico

Inst de Geologia (M,D) (2004)
Apdo. postal 70-296
Ciudad Universitaria
Delegacion Coyoacan, DF 04510
p. (52) 5622 4314 f. (52) 5550 8432
dirigl@geologia.igeolcu.unam.mx
http://132.248.20.1/geol.htm
Administrative Assistant: Ana María Rodriguez Simental

Director:
Dante J. Moran-Zenteno, (D), Nacional Auton, 1992, Cc

Head:
Susana A. Alaniz Alvarez, (D), Nacional Auton, 1995, Gc
Luca Ferrari Pedraglio, (D), Milan, 1992, Gt
Sergio Cevallos Ferriz, (D), Alberta, 1990, Po
Carlos M. Gonzalez Leon, (D), Arizona, 1992, Gr
Klavdia Oleschko Loutkova, (D), Lomosov (Moscow), 1984, So
Maria S. Lozano Garcia, (D), d'Aix (Marseilles), 1979, Pe
Francisco J. Vega Vera, (D), Nacional Auton, 1988, Py

Chair:
Gustavo Tolson Jones, (D), Nacional Auton, 1998, Gc

Senior Scientist:
Shelton Applegate-Pleasants, (D), Chicago, 1961, Pv
Jorge Aranda, (D), Oregon, 1982, Gi
Blanca E. Buitron-Sanchez, (D), Nacional Auton, 1974, Pi
Oscar Carranza-Castaneda, (D), Nacional Auton, 1989, Pv
Gerardo Carrasco-Nunez, (D), Michigan Tech, 1993, Gv
Ana L. Carreno, (D), d'Orsay (Paris), 1979, Pm
Miguel Carrillo Martinez, (D), Paris VI, 1976, Pb
Liberto De Pablo, (D), Ohio State, 1958, Gz
Rodolfo Del Arenal-Capetillo, (B), Nac Auton, 1960, Hw
Ismael Ferrusquia, (D), Houston, 1971, Gr
David Flores-Roman, (D), Nacional Auton, 1981, So
Celestina Gonzalez-Arreola, (D), Nacional Auton, 1989, Pi
Jose C. Guerrero-Garcia, (D), Texas (Dallas), 1975, Ym
John D. Keppie, (D), Glasgow, 1964, Gt
Victor M. Malpica-Cruz, (D), Bordeaux, 1980, Gs
Enrique Martinez Hernandez, (D), Michigan State, 1979, Pl
Juventino Martinez-Reyes, (D), Paris VI, 1980, Gt
Luis M. Mitre-Salazar, (D), Paris IV, 1978, Ge
Adrian Ortega, (D), Waterloo, 1993, Hw
Fernando Ortega-Gutierrez, (D), Leeds (UK), 1975, Gp
Sergio Palacios-Mayorga, (M), Nacional Auton, 1972, So
Jerjes Pantoja-Alor, (M), Arizona, 1963, Gg
Maria del Carmen Perrilliat-Montoya, (D), Nac Auton, 1969, Pi
Angel Nieto Samaniego, (D), Nacional Auton, 1994, Gc
Christina D. Siebe Grabach, (D), Hohenheim (Germany), 1993, Sc
Alicia Silva-Pineda, (D), Nacional Auton, 1980, Po
Max Suter-Cargnelutti, (D), Basel, 1978, Gt
Jordi Tritlla, (D), Barcelona, 1994, Eg
Ana B. Villasenor-Martinez, (D), Nacional Auton, 1991, Pi
Reinhard Weber-Gobel, (D), Tubingen, 1967, Po

Associate Scientist:
Gerardo J. Aguirre Diaz, (D), Texas, 1993, Gv
Thierry Calmus, (D), Paris VI, 1983, Gu
Antoni Camprubi, (D), Barcelona, 1998, Eg
Alejandro J. Carillo Chavez, (D), Wyoming, 1996, Hg
Elena Centeno Garcia, (D), Arizona, 1994, Gt
Rodolfo Corona-Esquivel, (M), Nacional Auton, 1985, Em
Mariano Elias-Herrera, (M), Nacional Auton, 1982, Gp
Maria L. Flores-Delgadillo, (M), Nacional Auton, 1987, So
Gilberto Hernandez-Silva, (D), Nacional Auton, 1983, So
Rafael Huizar-Alvarez, (D), Franche Comte, 1989, Hw
Cesar Jacques Ayala, (D), Cincinnati, 1983, Gr
Marisol Montellano-Ballesteros, (D), California (Berkeley), 1986, Pv
Amabel M. Oretega Rivera, Cc
Odranoel Quintero, (D), Sorbonne, 1995, Gc
Jose L. Rodriguez-Castaneda, (M), Pittsburgh, 1984, Gg
Jaime Roldan Quintana, (M), Iowa, 1976, Gv
Gerardo Sanchez-Rubio, (M), Imperial Coll (UK), 1984, Gv
Jesu Sole, (D), Barcelona, 1996, Cc
Luis F. Vassallo-Morales, (D), Lomonosov, 1981, Em
Maria G. Villasenor-Cabral, (M), Leeds (UK), 1974, Ca

Research Associate:
Irma Aguilera-Ortiz, (B), Iberoameric, 1971, Ca
Victor M. Davila-Alcocer, (M), Texas (Dallas), 1986, Ps
Jose G. Solorio-Munguia, (B), Nacional Auton, 1958, Gz

Adjunct Professor:
J. Duncan Keppie, (D), Glasgow (Scotland), 1967, Gt
Luis Silva-Mora, (D), d'Aix (Marseilles), 1979, Gi

Emeritus:
Gloria Alencaster-Ybarra, (D), Nacional Auton, 1969, Pi
Zoltan De Csema, (D), Columbia, 1955, Gt

Geology Librarian:
Teresa Soledad Medina Malagon, (B), Oo

Div de Estudios de Posgrado-Campus Morelos (2004)
C.P. 62550
Facultad de Ingeneria
Jiutepec, MR
p. (73) 19 39 57 (Ext. 135)

Director:
Alvaro Munoz Mendoza, (D), Strathclyde, 1989, Oo

Acting Head:
Hugo Acosta Borbon, (M), Nacional Auton, 1992, Oo
Esperanza Ramirez Camperos, (M), Nac Auton, 1989, Oo

Netherlands

Universiteit Utrecht

Faculty of Geosciences (2007)

Budapestlaan 4
3584 CD Utrecht
p. +31 30 253 5050
n.meijer@geo.uu.nl
http://www.geo.uu.nl/

VU University Amsterdam
Faculty of Earth and Life Sciences (2007)
De Boelelaan 1085
1081 HV Amsterdam,
webredactie@falw.vu.nl
http://www.falw.vu.nl/

New Zealand

University of Auckland
Dept of Geology (2005)
Private Bag 92019
Auckland
p. 64 9 3737599 f. 64 9 3737435
ie.smith@auckland.ac.nz
http://www.auckland.ac.nz/glg/geology.htm

University of Canterbury
Dept of Geology (B,M,D) (2005)
Private Bag 4800
Christchurch,
p. 64.3.364.2766 f. 64.3.364.2769
geology@canterbury.ac.nz
http://www.geol.canterbury.ac.nz/title.html

Professor:
James Cole, (D), Wellington, Gv
Senior Lecturer:
Kari N. Bassett, (D), Minnesota, 1995, Gs
Associate Professor:
John Bradshaw, (D), London, Gt
Dave MacKinnon, (D), Belfast, Po
David Shelley, (D), Bristol, 1964, Gx
Jamie P. Shulmeister, (D), Australian National, 1991, Pe
Lecturer:
David Bell, (B), New South Wales, Ng
Jocelyn Karen Campbell, (M), Canterbury, 1964, Gc
James McKean, (D), California (Berkeley), Gm
Jarg Pettinga, (D), Auckland, Gt

University of Otago
Dept of Geology (B,M,D) (2007)
P. O. Box 56
Leith Street
Dunedin, 9054
p. +64 3 479-7519 f. +64 3 479-7527
geology@otago.ac.nz
http://www.otago.ac.nz/geology/
Enrollment (2005): M: 20 (0) D: 22 (0)

On leave:
Rick H. Sibson, (D), Imperial Coll (UK), Gc
Professor:
Alan F. Cooper, (D), Otago, 1970, Gg
Richard J. Norris, (D), Oxford, 1969, Gc
Head of Department:
R. Ewan Fordyce, (D), Canterbury (NZ), 1981, Pv
Associate Professor:
Dave Craw, (D), Otago, 1981, Eg
Senior Lecturer:
Daphne E. Lee, (D), Otago, 1980, Pi
J. Michael Palin, (D), Yale, 1992, Gi
Associate Professor:
James D. L. White, (D), California (Santa Barbara), 1989, Gv
Gary S. Wilson, (D), Victoria (NZ), 1993, Gl
Lecturer:
Andrew R. Gorman, (D), British Columbia, 2001, Yg
Candace E. Martin, (D), Yale, 1990, Cg
Emeritus:
Douglas S. Coombs, (D), Gi
Chuck A. Landis, (D), Otago, 1969, Gg
Anthony Reay, (D), Leeds, 1964, Gi

Victoria Univ of Wellington
School of Geography, Env and Earth Sciences (2006)
P. O. Box 600
Wellington,
p. +6444635346 f. +6444635186
enquiries@geo.vuw.ac.nz
http://www.geo.vuw.ac.nz/

Norway

University of Bergen
Dept of Earth Science (2007)
Allegaten 41
N-5007 Bergen
heidi.espedal@geo.uib.no
http://www.geo.uib.no/

University of Tromso
Geology Dept (2004)
N-9037 Tromso
hanken@ibg.uit.no
http://www.ibg.uit.no/geologi/geo_eng_end.html

Poland

Jagiellonian University
Inst of Geological Sciences (2004)
UL Oleandry 2a Str.
30063 Krakow,
slaczka@ing.uj.edu.pl
http://www2.uj.edu.pl/ING/

Silesian University
Faculty of Mining and Geology (2004)
Wydzial Gornictwa I Geologii
ul. Akademicka 2
44-100 Gliwice
dolipski@polsl.gliwice.pl
http://www.polsl.gliwice.pl/alma.mater/wydzialy/gorniczy/ein

Portugal

Oporto University
Dept of Geology (2004)
Faculdade de Ciencias do Porto
Praca de Gomes Teixeira
4099-002 PORTO
sec_geo@fc.up.pt
http://www.fc.up.pt/depts/geo/indexi.html

Universidade de Coimbra
Dept de Ciencias da Terra (2007)
Largo Marques de Pombal
Coimbra, 3000-272 Coimbra

p. +351-239 860 500 f. +351-239 860 501
dct@dct.uc.pt
http://www.dct.uc.pt

Associate Professor:
Alcides Castilho Pereira, (D), Economic Geology, 1992, GebOr

Universidade de Lisboa

Dept de Geologia (2005)
Faculdade de Ciencias
Edificio C6, Piso 4, Campo Grande
Lisboa, 1749-016
p. +351 217500066 f. +351 217500064
geofcul@fc.ul.pt
http://geologia.fc.ul.pt/geoFCUL/index.htm

Universidade de Tras-os-Montes

Dept de Geologia (2004)
Apartado # 202
Vila Real 5001-911
mgomes@utad.pt
http://www.utad.pt/Seccoes/geologia/

Universidade do Minho

Dept de Ciencias da Terra (B,M,D) (2004)
Campus de Gualtar
Braga, 4710-057
p. 351 253 604 300 f. 351 253 678 206
sec@dct.uminho.pt
http://www.dct.uminho.pt/index/index.html

Qatar

University of Qatar

Dept of Geology (2004)
PO Box 2713
Doha
nasir54@qu.edu.qa
http://www.angelfire.com/ms/GeoQU/

Professor:
Sobhi J. Nasir, (D), Germany, 1986, Gx

Associate Professor:
Hamad A. Al-Saad, (D), Egypt, 1996, Oo
Abdulali A. Sadiq, (D), Southampton (UK), 1995, Or

Assistant Professor:
Latifa B. AL-Nouimy, (D), Egypt, 1994, Yg
Sief A. Alhajari, (D), North Carolina, 1992, Oo

Romania

Babes-Bolyai University

Faculty of Biology and Geology, Dept of Geology (2007)
Str. M. Kogalniceanu nr.1
RO - 400084 Cluj-Napoca
Cluj-Napoca, 400084
p. +40-264-405371 f. +40-264-591906
ibucur@bioge.ubbcluj.ro
http://www.ubbcluj.ro/faculties/biolgeol.html

Chair:
Sorin Filipescu, (D), Babes-Bolyai, 1996, GrPm

Professor:
Ioan Bucur, (D), Babes-Bolyai, 1991, PgmGs
Vlad Codrea, (D), Babes-Bolyai, 1995, PvGo

Iasi University A1.I. Cuza

Dept of Mineralogy and Geochemistry (2004)
B-dul Copou nr.20 A
6600 IASI
gpopa@terminus.geo.uaic.ro
http://www.uaic.ro/geo/indexold.html

Russia

Moscow State University

Geological Faculty (B,M,D) (2007)
Vorobjovy Gory
Moscow, 119899
p. (095)9392970 f. (095)9328889
admin@geol.msu.ru
http://www.geol.msu.ru/
Enrollment (2002): B: 65 (44) M: 22 (15)

Perm State University

Faculty of Geology (2004)
15, Bukireva str.
Perm 614000
WebMaster@psu.ru
http://www.psu.ru/general/f/geol/index_e.html

Rostov State University

Faculty of Geology and Geography (2005)
40 Zorge St.
Rostov-on-Don 344090
geofac@rnd.runnet.ru
http://www.rnd.runnet.ru/rsu/geofac.html

Saint-Petersburg State University

Faculty of Geology (2005)
Universitetskaya nab. 7-9
199034 St. Petersburg
olga@dean.geol.pu.ru
www.spbu.ru/e/TEXTS/Departments/geology.html

Slovakia

Comenius University

Dept of Geology and Paleontology (2006)
Faculty of Science
Mlynska dolina G
Bratislava, Slovakia 842 15
p. 00421260296214 f. 00421260296214
kgp@fns.uniba.sk
http://www.fns.uniba.sk/~kgp/

Professor:
Dusan Plasienka, (D), Gt

Slovenia

Ljubljana University

Oddelek za Geologijo (2004)
Askereeva 12
Ljubljana
tadej.dolenec@uni-lj.si
http://www.ntfgeo.uni-lj.si/

South Africa

Rhodes University

Dept of Geology (B,M,D) (2004)
PO Box 94

Grahamstown, Cape Province 6140
p. +27 46 6038309 f. +27 46 6229715
jenny.wright@ru.ac.za
www.rhodes.ac.za/academic/departments/geology/

University of Cape Town
Dept of Geology (2005)
Private Bag
Rondebosch 7700
dept@geology.uct.ac.za
http://www.uct.ac.za/depts/geolsci/

University of Durban-Westville
Dept of Geology (2004)
Faculty of Science
Private Bag X54001
Durban 4000
smccourt@pixie.udw.ac.za
pixie.udw.ac.za/UDW/homepages/geology.html

University of Port Elizabeth
Dept of Geology (B,M,D) (2004)
PO Box 1600
Port Elizabeth, 6000
p. +27 41 5042325 f. +27 41 5042340
glarws@upe.ac.za
http://www.upe.ac.za/geology/geology.html
Enrollment (2002): B: 65 (22) M: 7 (0) D: 1 (0)
Associate Professor:
R. W. Shone, (D), Port Elizabeth, Gd
Lecturer:
Callum Robert Anderson, (M), Port Elizabeth, 1991, Gi
Peter W. K. Booth, (D), Cape Town, Gc
Connan Hempel, (M), Port Elizabeth, Gg

Spain

Coruna University
University Inst of Geology (D) (2007)
Edificio Servicios Centrales de Investigación
Campus de Elviña
La Coruna, 15071
p. 0034 981167000 f. 0034 981167172
xeoloxia@udc.es
http://www.udc.es/iux
Enrollment (2007): D: 6 (3)
Director:
Juan Ramon Vidal Romaní, (D), Comp (Madrid), 1983, Gm
Professor:
Antonio Paz Gonzalez, (D), Santiago de Comp, 1982, Sd
Paleontologist:
Aurora Grandal d'Anglade, (D), Coruña, 1993, Pv
Associate Professor:
Elena Pilar de Uña Alvarez, (D), Santiago de Comp, 1986, Oy
Jorge Delgado Martín, (D), Barcelona, 1993, Gg
María Teresa Taboada Castro, (D), Santiago de Comp, 1990, Sd
Associate Scientist:
Daniel Fernandez-Mosquera, (D), Coruña, 2002, Xc

Univ Complutense de Madrid
Facultad de Ciencias Geologicas (2008)
Cuidad Universitaria
Madrid, E-28040
p. 34 91394 4819 f. 34 91394 5109
geodec@geo.ucm.es
http://www.ucm.es/centros/webs/fgeo
Dean:
Eumenio Ancochea, (D), Fac CC Geol (Madrid), 1972, Gvv
International Relations Vicedean:
Agustin P. Pieren, (D), Fac CC Geologicas, 1982, GrEgg

Universidad de Cadiz
Dept de Cristalografia y Mineralogia, Estratigrafia, Geodinamica, Petrologia, Geoquimica (2004)
Facultad de Ciencias
Pol. Rio San Pedro s/n
11510 Puerto Real (Cadiz),
cristalografia.scanner@uca.es
http://www2.uca.es/dept/geologia/

Universidad de Granada
Dept de Geologia (2004)
Andalusian Institute of Geophysics
Campus Universitario de la Cartuja
18071 Granada,
abril@iag.ugr.es
http://www.ugr.es/iag/iagpds.html

Universidad de Jaen
Departamento de Geologia (B,M,D) (2007)
Edificio B-3
Campus Universitario
Jaen, 23071
p. +34-953-212295 f. +34-953-2112946
jmmolina@ujaen.es
http://geologia.ujaen.es/

Universidad de Oviedo
Dept de Geologia (2005)
Campus de Llamaquique
Jesus Arias de Velasco, s/n
33005 Oviedo, E-33005
dirección@geol.uniovi.es
http://www.geol.uniovi.es/

Universidad de Zaragoza
Facultad de Ciencias (2004)
50.009 Zaragoza,
anperez@posta.unizar.es
http://wzar.unizar.es/acad/fac/geolo/adepo.html

Museo Paleontologico de la Universidad (2004)
salberto@ebro.unizar.es
http://ebro.unizar.es/uz/museo/

Universitat de Barcelona
Dept of Geology (2004)
c/Marti i Franques, S/N
Zona Universitaria Pedralbes
08028 Barcelona,
webmaster@gpre.ub.es
http://www.ub.es/div-fac/div3/facultat/geologia/geo.htm

Dept de Geoquimica, Petrologia I Prospeccio Geologica (2004)
Marti Franques, s/n
Barcelona 8028
dep-gppg@d3.ub.es
http://www.ub.es/geoquimi/index.html

Estratigrafia, Paleontologia i Geociencies marines (2006)

c/Marti Franques, s/n
08028
Barcelona
p. +33 934021384 f. +33 934021384
dep-ep@d3.ub.es
http://www.ub.es/dpep/1welcom3.htm

Universitat de les Illes Balears
Dept de Ciencies de la Terra (2004)
lpomar@uib.es
http://www.uib.es/depart/dctweb/home.htm

Sweden

Chalmers University of Technology
Dept of Geology (2005)
S-412 96 Goteborg
Gothenburg
lars.o.ericsson@geo.chalmers.se
http://geo.chalmers.se/index.htm

Karlstads University
Geoventenskap & Geografi (2005)
Magnus.Johansson@kau.se
http://www.nature.kau.se/geo/

University of Stockholm
Dept of Geology and Geochemistry (2006)
106 91 Stockholm,
webmaster@geo.su.se
http://www.geo.su.se/

Uppsala Universitet
Dept of Earth Sciences (2007)
Villavagen 16
UPPSALA, SE-752 36
PREFEKT@geo.uu.se
http://www.geo.uu.se/default.asp?pageid=1&lan=1

Switzerland

Basel University
Inst of Mineralogy and Petrography (2004)
Bernoullistrasse 30
Basel 4056
Bernadette.Oberlein@unibas.ch
http://therion.minpet.unibas.ch/minpet/index.html

Dept of Earth Sciences (2005)
Bernoullistr.32
CH-4056 Basel
Joelle.Glanzmann@unibas.ch
http://www.unibas.ch/earth/

Bern University
Group of Geomorphology (2004)
kienholz@giub.unibe.ch
http://www.giub.unibe.ch/gfg/

ETH Zurich
Dept of Earth Sciences (2005)
ETH-Zenrum
CH-8092 Zurich,
info@erdw.ethz.ch
http://www.erdw.ethz.ch/

Univ de Lausanne
Earth Sciences Dept (2005)
BFSH-2
Lausanne CH-1015,
Jean.Hernandez@img.unil.ch
http://www-sst.unil.ch/

University of Basel
Geol. & Palaeontological Inst (2005)
Bernouillistrasse 32
CH-4056 Basale,
joelle.glanzmann@unibas.ch
http://www.unibas.ch/earth/GPI/paleo/index.htm

University of Bern
Geologisches Inst (2005)
Baltzerstrasse 1
CH-3012 Bern,
urs@mpi.unibe.ch
http://www.earthsci.unibe.ch//

University of Fribourg
Geol. & Palaeonological Inst (2004)
Chemin du Musee 6
Perolles
CH-1700 Fribourg,
michele.caron@unifr.ch
http://www.unifr.ch/geology/

University of Geneva
Section of Earth Sciences (2005)
13, rue des Maraichers
CH-1205 Geneva,
ursula.eigenmann@terre.unige.ch
http://www.unige.ch/sciences/terre/

University of Neuchatel
Inst de Geologie (2005)
Rue Emile-Argand 11
CH-2007 Neuchatel,
martin.burkhard@unine.ch
http://www-geol.unine.ch/

Turkey

Ankara University
Dept of Geological Engineering (2005)
Muhendislik Fakultesi
Jeoloji Muhendisligi Bolumu
Tandogan
Ankara, TR-06100
p. 90 312 212 67 20 f. 90 312 215 04 87
Jeoloji.Muh.Sekreterligi@eng.ankara.edu.tr
http://www.eng.ankara.edu.tr/geological_eng/

Head:
Nizamettin Kazanci, (D), Ankara, Gs

Professor:
Ergun Gokten, (D), Ankara, Gc
Recep Kilic, (D), Ankara, Ng
Gurol Seyitoglu, (D), Leicester, 1992, Gt
Nurettin Sonel, (D), Go
Baki Varol, (D), Ankara, Gd

Hacettepe Universitesi
Geological Engineering Dept (2005)
Muhendislik Fakultesi

Beytepe Kampusu,
aay@hacettepe.edu.tr
http://www.jeo.hun.edu.tr/800x600.htm

Hacettepe University

Hydrogeological Engineering Dept (B,M,D) (2006)
Beytepe Campus
Ankara, 06532
p. +90 (312) 2977770 f. +90 (312) 2977770
ssimsek@hacettepe.edu.tr
http://www.hidro.hacettepe.edu.tr/

Head:
Alparslan Arikan, (D), Marie et Pierre Curie, 1976, Hg
Professor:
Sakir Simsek, (D), Istanbul, 1982, Hg
Serdar Bayari, (D), Hacettepe, 1991, Hw
Mehmet Ekmekci, (D), Hg
Assistant Professor:
Levent Tezcan, (D), Hacettepe, 1993, Hw

Middle East Technical University

Dept of Geological Engineering (B,M,D) (2004)
Inonu Bulvari
ODTU
Ankara, TR-06531
p. +90-312-2105725 f. +90-312-2101263
erdin@metu.edu.tr
http://www.metu.edu.tr/home/www64/

Ukraine

State Mining University of Ukraine

19 Karl Marx Avenue
Dnepropetrovsk
320600
dfr@nmuu.dp.ua
http://www.apex.dp.ua/english/uageo/ukraine.html

United Kingdom

Bristol University

Dept of Earth Sciences (2004)
Wills Memorial Building
Queens Road
Bristol, BS8B 1RJ
Margaret.Wilkins@bris.ac.uk
http://www.gly.bris.ac.uk/

Brunel University

Geography & Earth Sciences (2004)
Uxbridge UB8 3PH
callum.firth@brunel.ac.uk
http://www.brunel.ac.uk/depts/geo/

Cardiff University

School of Earth, Ocean and Planetary Sciences (B,M,D) (2006)
Main Building
Park Place
Cardiff, Wales CF10 3YE
p. +44 (0)29 2087 4830 f. +44 (0)29 2087 4326
earth-ug@cf.ac.uk
http://www.earth.cf.ac.uk
Enrollment (2002): B: 174 (0) M: 18 (0) D: 28 (0)

Head:
Dianne Edwards, (D), Pb
Professor:
J A. Cartwright, (D), Ys
A Harris
C Harris, Ge
B Leake
R J. Lisle, Gc
R J. Parkes
J A. Pearce, Cg
P N. Pearson, Pe
D Rickard, Cg
V P. Wright, Gs
Commander:
N Rodgers, (D), Ow
Lecturer:
Rhoda Ballinger, (D), Oo
C M. Berry, (D), Pb
P J. Brabham, (D), Ye
L C. Cherns, Pg
J H. Davies, Yg
G Della Porta, Gs
N Fry, Gc
I Fryett, Oo
I Hall, Oo
M Haschke
A Hemsley, Pl
T Jones, Ge
A C. Kerr, Gx
C Lear, Ou
C MacLeod, Gu
N C. Mitchell, Gu
R Perkins, Oo
J Pike, Oo
H M. Prichard, Gx
R Riding, Py
H Sass
H D. Smith, Oo
G M. Steed, Gg
S J. Wakefield, Cm
C F. Wooldridge, Oo
Y Yang, Hg

Durham University

Dept of Geological Sciences (B,M,D) (2004)
Science Laboratories
South Road
Durham, DH1 3LE
p. +44 (0)191 374 2520 f. +44 (0)191 374 2510
durham.geolsci@durham.ac.uk
http://www.dur.ac.uk/geolsci/

Chair:
Jon P. Davidson, (D), Leeds, Cg
Professor:
Neil R. Goulty, (D), Cambridge, Yg
Robert N. Thompson, (D), Oxford, 1965, Gi
Maurice E. Tucker, (D), Reading, Gs
Research Associate:
Bruce L. Charlier, (D), Open, Cg
Dougal A. Jerram, (D), Liverpool, Oo
Lecturer:
Grenville Holland, (D), Oxford, Gp
Colin G. Macpherson, (D), London, 1994, Cg
Kenneth J. McCaffrey, (D), Durham, Gc
Glen A. Milne, (D), Toronto, Yg
Trevor J. Morse, (D), Durham, Pg
Christine Peirce, (D), Cambridge, Yg
Moyra E J Wilson, (D), London, Gs
Fred Worrall, (D), Cambridge, Cg

Emeritus:
- Martin H P Bott, (D), Cambridge, Yg
- Henry Emeleus, (D), Oxford, Oo
- G A J Johnson, (D), Durham, Oo

Senior Research Officer:
- Geoff M. Nowell, (D), Open, 1993, Cg
- Chris J. Ottley, (D), Essex, Cg

Senior Lecturer:
- Howard A. Armstrong, (D), Nottingham, 1983, Pg
- Colin T. Scrutton, (D), Oxford, Ps
- Brian R. Turner, (D), Witwatersrand, Gs

Senior Experimental Officer:
- David L. Stevenson, (B), Durham, Oo

Research Tech:
- Tim J. Bourne, (D), Durham, Pg

Research Assistant:
- Neville A. Hallam, (B), Oxford Poly, Oo
- Jonathan Imber, (D), Durham, Gc
- Anthony J. Mallon, (D), Cork, Go

Reader:
- Gillian R. Foulger, (D), Durham, Yg
- Robert E. Holdsworth, (D), Leeds, Gc
- Graham Pearson, (D), Leeds, Cg
- Richard E. Swarbrick, (D), Cambridge, Go

Edinburgh University

Grant Inst of Earth Science (2006)
Grant Institute
West Mains Road
Edinburgh, EH9 3JW
info@geos.ed.ac.uk
http://www.geos.ed.ac.uk/

Imperial College

T H Huxley School of Env, Earth Sciences and Eng (2004)
RSM Building
Prince Consort Rd
London, SW7 2BP
j.beddington@ic.ac.uk
http://www.huxley.ic.ac.uk/

Keele University

Earth Sciences (2007)
Staffordshire ST5 5BG
p. +44 1782583615 f. +44 1782715261
geo@esci.keele.ac.uk
http://www.esci.keele.ac.uk/

Professor:
- Peter Styles, (D), Newcastle, Ge
- Graham D. Williams, (D), Gc

Research Associate:
- Ranjit Premasiri, (D)

Lecturer:
- Nigel John Cassidy, (D), Keele, Ge
- Stuart Clark, (D), Keele, Gsd
- Stuart S. Egan, (D), Keele, Go
- Ralf Gertisser, (D), Freiburg, GivGz
- Stefan Krause, (D), Potsdam, Ge
- Michael Montenari, (D), Tubingen, Pm
- Brian O'Driscoll, (D), Dublin, Gpi
- Jamie K. Pringle, (D), Liverpool, Ga
- Ian G. Stimpson, (D), Cardiff, Gc

Emeritus:
- Gilbert Kelling, (D), Edinburgh
- Philip D. Lane, (D), Newcastle
- Hugh S. Torrens, (D), Leicester, Gh
- John A. Winchester, (D), Oxford, 1970, Gp

Kingston University

School of Earth Science and Geography (2007)
Penrhyn Road Centre
Penrhyn Road
Kingston upon Thames KT1 2E
p. 0044 (0)20 8547 2000 f. 0044 (0)20 8547 7497
G.Gillmore@kingston.ac.uk
http://www.kingston.ac.uk/geolsci/

Admissions Tutor:
- Andy Swan, (D), Ggb

Leicester University

Geology Dept (2007)
University Road
Leicester, LE1 7RH
p. 0116 252 3933 f. 0116 252 3918
geology@le.ac.uk
http://www.le.ac.uk/geology/gl.html

Open University

Dept of Earth and Environmental Sciences (B,D) (2007)
Walton Hall
Milton Keynes, MK7 6AA
p. +44 (0)1908 653102 f. +44 (0)1908 655151
Earth-Sciences-HoD@open.ac.uk
http://www.open.ac.uk/science/earthsciences/

Oxford Brooks University

Dept of Geology (2004)
Headington
Oxford OX3 OBP
rsd'lemos@brookes.ac.uk
http://www.brookes.ac.uk/geology/

Staffordshire Univeristy

Geology Div (2004)
G.W.Dagger@staffs.ac.uk
www.staffs.ac.uk/sands/scis/geology/geology.html

The University of Liverpool

Dept of Earth and Ocean Sciences (2006)
Jane Herdman Building
4 Brownlow Street
Liverpool, Merseyside L69 3GP
p. +44 151 794 5162 f. +44 151 794 5196
bamberi@liv.ac.uk
http://www.liv.ac.uk/earth_sciences/dept/

University of Aberdeen

Dept of Geology and Petroleum Geology (2006)
Meston Building
King's College
Aberdeen AB24 3UE,
geology@abdn.ac.uk
http://www.abdn.ac.uk/geology/

University of Birmingham

School of Geography, Earth & Environmental Sci (2005)
Edgbaston
Birmingham, B15 2TT
p. +44 121 414 6751 f. +44 121 414 4942
m.p.smith@bham.ac.uk
http://www.gees.bham.ac.uk/

University of Derby
Div of Earth Sciences (2004)
Kedleston Road
Derby, DE22 1GAB
D.Mackenzie@derby.ac.uk
http://www.derby.ac.uk/schools/seas/geology/#

University of Exeter
Camborne School of Mines (2004)
Redruth, Cornwall TR15 3SE
K.Atkinson@csm.ex.ac.uk
http://www.ex.ac.uk/CSM/

University of Glasgow
Dept of Geographical and Earth Sciences (2007)
The Gregory Building
Lilybank Gardens
Glasgow, G12 8QQ
C.Burton@ges.gla.ac.uk
http://www.ges.gla.ac.uk:443/

University of Keele
School of Earth Sciences and Geography (2004)
Keele
Staffordshire, ST5 5BG
i.j.fairchild@esci.keele.ac.uk
http://www.esci.keele.ac.uk/

University of Leeds
School of Earth and Environment (B,M,D) (2007)
Leeds, LS2 9JT
p. +44 (0)113 343 5222 f. +44 (0)113 343 5259
enquiries@see.leeds.ac.uk
http://www.see.leeds.ac.uk

University of Newcastle Upon Tyne
Dept of Civil Engineering (2004)
Drummond Building
Newcastle upon Tyne, NE1 7RU
C.M.Earle-Storey@newcastle.ac.uk
http://www.ncl.ac.uk/~ncivil/index.htm

University of Nottingham
School of Life & Environmental Sciences (2004)
Nottingham NG7 2RD
j.laybourn-parry@nottingham.ac.uk
http://vsb.nott.ac.uk/life-env/

University of Oxford
Dept of Earth Sciences (2007)
Parks Road
Oxford, Oxfordshire OX1 3PR
p. 01865 272000 f. 01865 272072
enquiries@earth.ox.ac.uk
http://www.earth.ox.ac.uk/

University of Portsmouth
School of Earth and Environmental Sciences (2008)
Burnaby Bldg., Burnaby Road
Portsmouth
Hampshire PO1 3QL
rob.strachan@port.ac.uk
http://www.port.ac.uk/sees

University of Reading
Dept of Soil Science (B,M,D) (2006)
PO Box 233
Whiteknights
Reading, RG6 6DW
p. +44 (0)118 378 8911 f. +44 (0)118 378 6660
shes@reading.ac.uk
http://www.soils.rdg.ac.uk/

Head:
Mark E. Hodson, (D), Edinburgh, 1994, Cl

Professor:
Stephen Nortcliff, (D), Sd

Associate Professor:
Timothy Astin, (D), Cambridge, 1982, Yg
Sam Baxter, (D), Sd
Stuart Black, (D), Lancaster, Cc
Chris Collins, (D), Sc
McGoff Hazel, (D), Liverpool, Pg
Andrew Parker, (D), Cl
Steve Robinson, (D), Sf
Liz Shaw, (D), Sb
Anne Verhoef, (D), Sp

University of Sheffield
Evironmental and Geological Sciences (2004)
Beaumont Building
Brookhill,
d.h.lewis@shef.ac.uk
http://www.shef.ac.uk/uni/academic/D-H/es/

University of Wales, Aberystwyth
The Inst of Geography & Earth Sciences (2004)
Llandinam Building
Aberystwyth, SY23 3DB
iml@aber.ac.uk
http://www.aber.ac.uk/~ieswww/

Zimbabwe

University of Zimbabwe
Dept of Geology (2005)
PO Box MP167
Mount Pleasant, Harrare
nematura@yahoo.co.uk
http://www.uz.ac.zw/science/geology/

Alabama

Geological Survey of Alabama

Geological Survey of Alabama (2007)
420 Hackberry Lane
P.O. Box 869999
Tuscaloosa, AL 35486-6999
p. (205) 349-2852 f. (205) 349-2861
info@gsa.state.al.us
http://www.gsa.state.al.us
Administrative Assistant: Irene T. Burgess
Administrative Assistant: Yvonne Massey
Enrollment (1995): B: 2 (0)

State Geologist:
Berry H. Tew, (D), Alabama, 1999, Gg
Deputy Director:
Robert M. Mink, (M), Alabama, 1981, Go

Alaska

Div of Geological & Geophysical Surveys

Alaska Div of Geological & Geophysical Surveys (2007)
3354 College Road
Fairbanks, AK 99709-3707
p. (907) 451-5010 f. (907) 451-5050
dggspubs@dnr.state.ak.us
http://www.dggs.dnr.state.ak.us
Administrative Assistant: Vickie J. Butherus

Director & State Geologist:
Robert F. Swenson, Oo
Deputy Director:
Rodney A. Combellick, Oo
Geophysical Research:
Laurel E. Burns, Oo
Geologist:
Jennifer E. Athey, Oo
Katherine Bull, Oo
Cheryl Cameron, Oo
Larry K. Freeman, Oo
Simone Montayne, Oo
Christopher J. Nye, Oo
Kenneth Papp, Oo
Paige Peapples, Oo
John R. Reeder, Oo
Rocky R. Reifenstuhl, Oo
Janet R. G. Schaefer, Oo
De Anne S. Stevens, Oo
David Szumigala, Oo
Marwan Wartes, Oo
Melanie B. Werdon, Oo
Chief, Geologic Communications:
Paula Davis, Oo
Senior Scientist:
James G. Clough, Oo
Publications Tech:
Joni Robinson, Oo
Cartographer:
Alfred G. Sturmann, Oo
Microcomputer/Network Specialist:
Barbara Kirchner, Oo
Analyst/Programmer:
Susan Seitz, Oo
Seth Snediger, Oo

Arizona

Arizona Geological Survey

Arizona Geological Survey (2004)
416 West Congress Street
Suite 100
Tucson, AZ 85701
p. (520) 770-3500 f. (520) 770-3505
Lee.allisonazgs.az.gov
http://www.azgs.state.az.us
Assistant Director of Administration: R. E. McDonnell

Director & State Geologist:
M. Lee Allison, (D), Wisconsin, 1962, Gg
Senior Geologist:
Jon E. Spencer, (D), MIT, 1981, Gg
Research Geologist:
Philip A. Pearthree, (D), Arizona, 1990, Gm
Stephen M. Richard, (D), California (Santa Barbara), 1988, Gg
Oil & Gas Manager:
Steven L. Rauzi, (M), Utah State, Go
IT Manager:
Richard A. Trapp, (M), Arizona, Gg
Geology Librarian:
Thomas G. McGarvin, (B), California Lutheran, Gg

Arkansas

Arkansas Geological Commission

Arkansas Geological Commission (2005)
Vardelle Parham Geology Center
3815 West Roosevelt Road
Little Rock, AR 72204
p. (501) 296-1877 f. (501) 663-7360
agc@arkansas.gov
http://www.state.ar.us/agc/agc.htm
Administrative Assistant: Sheila Lewis

State Geologist:
Bekki White, (M), Centenary, 1993, Go
Geology Supervisor/Mineralogist:
James M. Howard, (M), Arkansas, 1974, Gg
Geologist Supervisor:
M. Ed Ratchford, (D), Idaho, 1994, Gc
Chief Geologist:
John D. McFarland, III, (M), Arkansas, 1975, Gg
Senior Scientist:
Charles G. Stone, (M), Arkansas, Gc
Geologist:
Scott Ausbrooks, (M), Arkansas (Little Rock), Gg
Associate Scientist:
Doug Hanson, (M), Memphis State, 1991, Gg
William L. Prior, (M), Memphis State, 1979, Hw
William Shelton, (M), Arkansas, 1956, Cg
Assistant Director:
Mac B. Woodward, (B), S State, 1957, Go
Cartographic Supervisor:
Susan L. Young, Oo

California

California Geological Survey

California Geological Survey (2004)
801 K Street
Mail Stop 12-01
Sacramento, CA 95814-3531
p. (916) 445-1825 f. (916) 445-5718
http://www.consrv.ca.gov/dmg/
Department Secretary: JoAnn Guidera

State Geologist:
James F. Davis, (D), Wisconsin, 1965, Oo
Supervising Geologist:
Trinda L. Bedrossian, (M), California (Davis), 1970, Ng
Charles R. Real, (M), S California, 1972, Ys
Michael Reichle, (D), California (San Diego), 1978, Ys
Anthony F. Shakal, (D), MIT, 1980, Ys
Principal Geologist:
David J. Beeby, (B), California (Davis), 1970, En

Senior Seismologist:
Moh J. Huang, (D), Caltech, 1983, Ne
Tousson R. Toppozada, (D), New Mexico Tech, 1974, Ys
Senior Geologist:
William Bryant, (B), Los Angeles State, 1975, Ng
Ron C. Churchill, (D), Minnesota, 1980, Cg
Robert L. Hill, (M), California (Los Angeles), 1972, Eg
Michael P. Kennedy, (D), California (Riverside), 1973, Gu
Russell V. Miller, (B), Oregon State, 1965, Eg
David L. Wagner, (M), San Jose State, 1975, Gg
Senior Engineering Geologist:
Allan G. Barrows, (D), California (Los Angeles), 1969, Gg
Mark J. Delisle, (M), San Diego State, 1963, Ou
Timothy P. McCrink, (M), New Mexico Tech, 1982, Ng
Theodore C. Smith, (B), Humboldt State, 1972, Oo
Thomas E. Spittler, Oo
Robert H. Sydnor, (M), California (Riverside), 1975, Ng
Associate Seismologist:
Tianqing Cao, (D), MIT, 1986, Ys
Chris H. Cramer, (D), Stanford, 1976, Ys
Vladimir Graizer, Oo
Praveen K. Malhotra, (D), Rice, 1991, Ne
Mark D. Petersen, Oo
Associate Geologist:
Stephen P. Bezore, (M), California (Davis), 1972, Eg
Donald Dupras, (B), Cal State (Sacramento), En
Chris T. Higgins, (M), California (Davis), 1977, Gg
Susan L. Kohler-Antablin, (B), Cal State (Fullerton), 1975, Eg
Hasmukhrai H. Majmundar, Oo
Marcia E. Mattison, (B), Sonoma State, 1980, Gg
George J. Saucedo, (B), San Francisco, 1972, Gg
Dinah Shumway, (M), Dartmouth, Eg
Gary Taylor, (B), Cal State (Fresno), Em
Associate Engineering Geologist:
Julie A. Bawcom, Oo
James N. Falls, Oo
Richard B. Greenwood, (M), Cal State (Long Beach), 1981, Gg
Wayne Haydon, Oo
Pamela J. Irvine, (M), California (Berkeley), 1977, Gg
Ralph C. Loyd, (B), Cal State (Sonoma), Em
Michael W. Manson, Oo
Mary Anne McKittrick, Oo
Jack R. McMillan, (B), Humboldt State, 1977, Ng
Florante Perez, Oo
Cindy L. Pridmore, (M), San Diego State, 1983, Gt
Michael Sandecki, (B), Humboldt State, 1979, Ge
John P. Schlosser, Oo
Michael A. Silva, (B), California (Davis), 1978, En
Siang S. Tan, (M), California (Los Angeles), 1972, Gg
Jerome A. Treiman, (B), California (Los Angeles), 1972, Ng
Christopher J. Wills, (M), Wisconsin, Ng
Rick I. Wilson, (B), Fresno State, 1987, Ge
Leslie G. Youngs, (B), Cal State (Sacramento), 1975, Yg
Assistant Geologist:
Dale M. Stickney, (B), California (Berkeley), 1969, Gi
Retired:
Roger W. Sherburne, (D), Penn State, 1974, Ys
On Leave:
Robert B. Darragh, (D), California (Berkeley), 1987, Ys
Geology Librarian:
Rosemary Guerin-Place, Oo

Colorado

Colorado Geological Survey

Colorado Geological Survey (2004)
1313 Sherman Street
Room 715
Denver, CO 80203
p. (303) 866-2611 f. (303) 866-2461
http://geosurvey.state.co.us/
Acting State Geologist:
Ronald W. Cattany
Director, Colorado Avalanche Information Center:
Knox Williams, (M), Colorado State, 1970, Oa
Chief, Mineral Resources:
Jim Cappa, (M), New Mexico Tech, 1975, Gz
Chief, GIS & Technical Services:
Randal C. Phillips, (M), Denver, 1991, Gq
Chief, Env Geology Section:
Matthew A. Sares, (M), Colorado Mines, 1991, Ge
Chief, Engineering Geology:
Dave Noe, (M), Texas, 1984, Ng
Chief, Administration & Business Services:
Patricia Young, (B), Metropolitan, 1994, Oo
Senior Science Advisor:
Vince Matthews, (D), California (Santa Cruz), 1973, Gc

Connecticut

Geological and Natural History Survey

Dept of Environmental Protection (2004)
Environmental & Geographic Info Ctr
79 Elm Street
Hartford, CT 06106-5127
p. (860) 424-3540 f. (860) 424-4058
nancy.mchone@po.state.ct.us
http://dep.state.ct.us
Administrative Assistant: Marilyn Diaz
Geologist:
Nancy W. McHone, (M), Wesleyan, 1998, Gg
Margaret A. Thomas, (M), Connecticut, 1983, Gg
Biologist:
Dawn McKay, (M), Connecticut, 1989, Oo
Kenneth Metzler, (M), Connecticut, 1977, Oo
Nancy Murray, (M), S Connecticut, 1982, Oo
Karen L. Zyko, (B), Dubuque, 1996, Oi
Senior Scientist:
Bruce Hunter, (B), C Connecticut, 1977, Hs
Civil Engineer:
Thomas E. Nosal, (M), C Connecticut, 1992, Oo
Graphic Artist:
Janet Zeh, (B), S Connecticut, 1975, Oo

Delaware

University of Delaware

Delaware Geological Survey (2007)
257 Academy Street, Room 205
Newark, DE 19716-7501
p. (302) 831-2833 f. (302) 831-3579
delgeosurvey@udel.edu
http://www.udel.edu/dgs/
Administrative Assistant: Karen L. D'Amato
Administrative Assistant: Laura K. Wisk
Director:
John H. Talley, (M), Franklin & Marshall, 1974, Hw
Hydrogeologist:
Alan S. Andres, (M), Lehigh, 1984
Senior Scientist:
Peter P. McLaughlin, Jr., (D), Louisiana State, 1989, Ps
Scientist:
Thomas E. McKenna, (D), Texas, 1997, Hw
Kelvin W. Ramsey, (D), Delaware, 1988, Gr
William S. Schenck, (M), Delaware, 1997, Gg

Hydrogeologist:
Changming He, (D), Nevada, 2004
Geodatabase Administrator:
Miriam L. Pomilio, (B), Delaware, 1984
Research Associate:
Stefanie J. Baxter, (M), Ocean/Marine Geology, 1994, Og
Senior Research Technician III:
Paul S. McCreary
Senior Research Tech:
Charles T. Smith, (B), Penn State, 1979, Gg
GIS Specialist/Cartographer:
Lillian T. Wang, (B), Delaware, 1991, Oy
System Administrator:
Steven V. Bertsche, (M), Illinois State, 1975

Florida

Florida Geological Survey

Florida Dept of Environmental Protection (2004)
903 West Tennessee Street
FSU Campus/Gunter Building
Tallahassee, FL 32304-7700
p. (904) 488-4191 f. (904) 488-8086
http://www.dep.state.fl.us/geo/
Administrative Assistant: Sandie Ray
Administrative Secretary: Cindy Collier
Secretary Specialist: Evelyn Jordan
Secretary Specialist: LaMarr A. Mitchell
Secretary Specialist: Carolyn F. Stringer

Head:
Walter Schmidt, (D), Florida State, 1983, Gg
Professional Geologist:
Paul J. Attwood, (M), Kansas, 1977, Go
Richard C. Green, (M), Florida State, 1993, Gr
Senior Scientist:
James H. Balsillie, (B), Portland State, 1971, On
Kenneth M. Campbell, (M), Florida State, 1979, Gs
Bob Caughey, (B), Arizona, 1976, Go
Rodney S. Deltan, (D), Florida State, 1973, Ca
Ed Gambrell, (B), Mississippi State, 1960, Go
Ronald W. Hoenstine, (D), Florida State, 1982, Ge
Burke E. Lane, (M), Penn State, 1968, Hw
Jacqueline M. Lloyd, (M), Chicago, 1979, Ge
Thomas M. Scott, (D), Florida State, 1986, Gr
Associate Scientist:
Jonathan D. Arthur, (D), Florida State, 1994, Gi
Paulette A. Bond, (M), North Carolina, 1972, Cg
Joel G. Duncan, (B), Alabama, 1977, Go
Henry Freedenberg, (M), SUNY (Buffalo), 1976, Go
Edward W. Garrett, (B), Florida State, 1983, Go
Lucien J. Ladner, (B), S Mississippi, 1970, Go
Frank R. Rupert, (M), Florida State, 1980, Pm
Steven M. Spencer, (B), Florida State, 1981, Eg
Research Associate:
Don L. Hargrove, Ye
Systems Project Analyst:
Jesse A. Fairley,Jr., Oo
Professional Engineer:
James D. LeBar, (B), Michigan Tech, 1978, Go
Lab Tech:
Franklin R. Rush.Jr., Oo
Engineer:
Ted B. Kiper, (M), Florida State, 1993, Oo
Custodian:
Jessie L. Hawkins, Sr., Oo
Administrator:
Lawrence D. Curry, (B), S Florida, 1973, Go
Geology Librarian:
Deborah Mekeel, (M), Florida State, 1993, Oo

Georgia

Georgia Dept of Natural Resources

Georgia Geological Survey (2004)
Environmental Protection Division
19 M.L. King, Jr. Drive, Room 400
Atlanta, GA 30334-9004
p. (404) 656-3214 f. (404) 657-8379
http://www.dnr.state.ga.us/dnr/environ/
Administrative Op Assistant: Cindy A. Hynote
Secretary: Landria J. Toogood
Secretary 2: Quander Robertson

State Geologist:
William H. McLemore, (D), Georgia, 1971, Ng
Assistant Branch Chief:
B. Roger Carter, (M), North Carolina State, 1967, Ng
Advanced Geologist:
Mark D. Cocker, (D), Ohio State, 1978, Eg
John O. Costello, (D), South Carolina, 1993, Gp
James S. Guenert, (M), Michigan State, 1988, Hw
Mark E. Hall, (M), Georgia, 1991, Gp
Charles S. Walker, (M), Idaho, 1983, Hw
Senior Geologist:
Charles H. Combs, (B), Georgia State, 1978, Hw
John C. Donahue, (M), Georgia, 1982, Hw
Alan J. Giles, (B), Boston, 1974, Og
David R. Hipple, (B), Indiana, 1972, Oi
Bijan Rahbar, (B), Emporia State, 1986, Hw
Sandra J. Robertson, (M), Tulsa, 1984, Hw
Roy Rountree, (B), Georgia Southwest, 1975, Hw
Victoria P. Trent, (B), West Georgia, 1984, Oi
Geologist:
Tyler J. Boyles, (B), West Georgia, 1999, Gg
Env Specialist:
Michael K. Berry, (B), Georgia Coll, 1995, Oo
Anthony E. McCook, Oo
Lora A. Overacre, (B), Montana, 1999, Oo
Program Manager:
William G. Smith, (B), Emory, 1966, Hw
Vincent P. Williams, (B), Rhode Island, 1969, Oo
Core Drill Crew Chief:
Harold D. Baker, Oo
Michael Perrien, Oo
Core Drill Operator:
Mark A. Choran, Oo
John A. Sides, Oo

Hawaii

Hawai'i Geological Survey

Commission on Water Resources Management (2004)
Dept of Land & Natural Resources
P.O. Box 621
1151 Punchbowl Street
Honolulu, HI 96809
p. (808) 587-0400 f. (808) 587-0219
http://kumu.icsd.hawaii.gov/dlnr/Welcome.html

State Geologist:
Glenn R. Bauer, Gg

Idaho

University of Idaho

Idaho Geological Survey (2008)

Morrill Hall, Third Floor
University of Idaho
Moscow, ID 83844-3014
p. (208) 885-7991 f. (208) 885-5826
igs@uidaho.edu
http://www.idahogeology.org/

Director & State Geologist:
Roy M. Breckenridge, (D), Wyoming, 1975, Gg
Professor:
Kurt L. Othberg, (D), Idaho, 1991, Gg
John A. Welhan, (D), Hw
Associate Professor:
Virginia S. Gillerman, (D), California (Berkeley), 1982, Eg
Reed S. Lewis, (D), Oregon State, 1990, Gg
Assistant Professor:
William M. Phillips, (D), Arizona, Gm
Senior Geologist:
Dean L. Garwood, (M), Idaho, 2001, Gg
Research Associate:
John D. Kauffman, (M), Idaho, 1975, Gg
Victoria E. Mitchell, (M), Idaho
Loudon R. Stanford, (M), Idaho, 1982, Gm
Lecturer:
Michael J. Weaver, (B), Eg
Staff:
Jane S. Freed, (B), Idaho, 1995, Oi
Office Assistant:
Sherry Pixley
Financial Technician:
Tracy Kanikkeberg
Related Staff:
Roger C. Stewart, (M)

Illinois

Illinois State Geological Survey

Office of the Chief (2007)
615 East Peabody Drive
Champaign, IL 61820-6964
p. (217) 333-5111 f. (217) 244-7004
shilts@isgs.uiuc.edu
http://www.isgs.uiuc.edu
Administrative Assistant: Tamra S. Montgomery

Chief:
William W. Shilts, (D), Syracuse, 1970, Gl
Chief Scientist:
E. Donald McKay, (D), Illinois, 1977, Gl
Director, Geologic Mapping Program:
Richard Berg, (D), Illinois, 1979, Ge
Assistant to the Chief for Env Initiatives:
Beverly Herzog, (M), Hw
Senior Scientist:
Donald E. Luman, (D), Or
Other:
Debra A. Griest, (B), Gg
Torie L. Strole, (B), Gg

Energy & Earth Resource Center (2007)
615 E. Peabody Drive.
Champaign, IL 61820-6964
p. 217-244-2430 f. 217-333-2830
finley@isgs.uiuc.edu
Administrative Assistant: Pam Cookus

Center Director:
Robert J. Finley, (D), South Carolina, 1975, Eo
Senior Scientist:
Richard A. Cahill, (D), Illinois, 1980, Ca
Scott M. Frailey, (D), Missouri (Rolla), 1989, Ng
David G. Morse, (D), Johns Hopkins, 1979, Eo
Massoud Rostam-Abadi, (D), Wayne State, 1982, Oo
William R. Roy, (D), Illinois, 1985, Sc
Scientist:
Mei-In (Melissa) Chou, (D), Michigan State, 1977, Co
Sheng-Fu Joseph Chou, (D), Michigan State, 1977, Co
Joseph A. Devera, (M), S Illinois, 1985, Gr
Latif A. Khan, (D), Tech, 1971, Nx
Ivan G. Krapac, (M), Illinois, 1987, Ca
Zakaria Lasemi, (D), Miami, 1990, En
Hannes E. Leetaru, (D), Illinois, 1997, Eo
Donald G. Mikulic, (D), Oregon State, 1979, Ps
Beverly Seyler, (M), SUNY, 1978, Eo
Associate Scientist:
Cheri A. Chenoweth, (B), Illinois, 1979, Ec
Joan E. Crockett, (B), Illinois, 1983, Eo
John P. Grube, (M), Colorado Mines, 1984, Eo
Bryan G. Huff, (M), Illinois, 1984, Eo
Rex A. Knepp, (M), Go
Assistant Scientist:
F. Brett Denny, (B), Missouri (Rolla), 1985, Gr
Scott D. Elrick, (M), California (Riverside), 1998, Ec
Kathleen M. Henry, (B), Illinois State, 1982, Gg
Christopher P. Korose, (B), Illinois, 1995, Ec
Vinodkumar A. Patel, (B), Inst of Tech, 1973, Oo

Geologic Mapping and Hydrogeology Center (2005)
615 East Peabody Drive
Champaign, IL 61820-6964
p. (217) 244-2430 f. (217) 244-7004
keefer@isgs.uiuc.edu
Administrative Assistant: Bonnie Renfrew

Head:
Donald Keefer, (D), Hw
Senior Scientist:
Robert A. Bauer, (M), Illinois, 1983, Nr
Leon R. Follmer, (D), Illinois, 1970, Sd
Keith C. Hackley, (M), Illinois, 1984, Cs
Ardith K. Hansel, (D), Illinois, 1980, Gl
Samuel V. Panno, (M), S Illinois, 1978, Cg
Scientist:
Michael L. Barnhardt, (D), Illinois, 1979, Gm
Michael J. Chrzastowski, (D), Delaware, 1986, Ou
Brandon B. Curry, (D), Illinois, 1995, Pe
David R. Larson, (M), Nebraska, 1976, Hw
Edward Mehnert, (D), Illinois, 1997, Hw
Andre J.M. Pugin, (D), Geneva, 1989, Yg
Christopher J. Stohr, (D), Illinois, 1996, Ng
C. Pius Weibel, (D), Illinois, 1987, Gr
Associate Scientist:
William S. Dey, (M), Illinois, 1983, Hw
David A. Grimley, (D), Illinois, 1996, Ym
Hue-Hwa Hwang, (D), Illinois, 1985, Cs
Richard J. Rice, (B), Illinois State, 1980, Hw
John C. Sieving, (M), Illinois State, 1997, Hw
Wen-June Su, (M), Illinois, 1985, Ng
Robert C. Vaiden, (M), Illinois, 1985, Hw
Hong Wang, (D), Illinois, 1996, Cc
Assistant Scientist:
Sallie E. Greenberg, (M), Illinois, 1997, Cs
Andrew C. Phillips, (D), Illinois (Chicago), 1993, Gm
Edward C. Smith, (B), Illinois State, 1985, Hw
Andrew J. Stumpf, (D), New Brunswick, 2001, Gl

Geoscience Information Center (2005)
615 E. Peabody Dr.
Champaign, IL 61820-6964
p. 217-244-2486 f. 217-333-2830

dufour@isgs.uiuc.edu
http://www.isgs.uiuc.edu
Administrative Assistant: Pamela J. Cookus
Manager, Library & Public Information:
Mary P. Krick, (M), Illinois, 1971, Gg
Geoscience Outreach Coordinator:
Wayne T. Frankie, (M), Kentucky, 1990, Gg
Coordinating Technical Writer/Editor:
Cheryl K. Nimz, (M), Illinois, 1973, Gg
Center Director:
Marie-France Dufour, (B), Ottawa, 1973, Gg
Section Manager:
Robert J. Krumm, (M), S Illinois, 1984, Gg
Associate Scientist:
Curtis C. Abert, (B), Illinois, 1990, Gg
Sheena K. Beaverson, (M), Minnesota, 1993, Oi
Sally L. Denhart, (A), Parkland, 1975, Oo
Jane E. Domier, (M), Kentucky, 1986, Oi
Alison B. Lecouris, (B), Illinois, 1983, Gg
Daniel O. Nelson, (M), Illinois, 1993, Oi
Lisa R. Smith, (B), Illinois State, 1982, Gg
Other:
Damon A. Garner, (B), Gg
Richard R. Hansen, Oo

Transportation & Environment Center (2004)
615 E. Peabody Dr.
Champaign, IL 61820-6964
p. 217-244-0028 f. 217-265-8214
miller@isgs.uiuc.edu
http://www.isgs.uiuc.edu
Administrative Assistant: Barbara A. Frahm
Center Director:
Michael V. Miller, (D), Illinois, 1984, Ge
Associate Scientist:
Daniel J. Adomaitis, (B), Illinois, 1987, Ge
Phyllis L. Bannon-Nilles, (M), New Haven, 1988, Og
Steven S. Benton, (M), W Michigan, 1992, Hw
Mark R. Collier, (B), Lake Forest, 1994, Ge
Scott R. Ellis, (B), Illinois, 1996, Gg
Gregory A. Kientop, (M), Texas A&M, 1988, Ng
James J. Miner, (M), N Illinois, 1979, Ge
Dale R. Schmidt, (M), Wisconsin, 1994, Ge
C. Brian Trask, (D), Syracuse, 1976, Ge
Assistant Scientist:
E. Lee Bray, (M), Brigham Young, 1999, Ge
Nancy I. Caldwell, (B), W Ontario, 1993, Ge
Keith W. Carr, (M), W Ontario, 1994, Ge
Somendra Chakravorty, (M), Illinois (Chicago), 1990, Ge
James W. Geiger, (B), S Illinois, 1984, Ge
Geoffrey E. Pociask, (B), Wisconsin, 1997, Ge
Bonnie J. Robinson, (M), W Ontario, 1998, Ge
Kelli D. Weaver, (M), Nevada, 2001, Ge
Mark A. Yacucci, (M), Illinois State, 1998, Ge

Illinois State Water Survey

Office of the Chief (2004)
2204 Griffith Drive
Champaign, IL 61820
p. (217) 244-5459 f. (217) 333-6540
dwinstan@uiuc.edu

Illinois State Water Survey (2004)
2204 Griffith Drive
Champaign, IL 61820
p. (217) 333-2210 f. (217) 333-6540
Principal Scientist:
Nani G. Bhowmik, (D), Colorado State, 1968, Hs
Senior Professional Scientist:
H. Vernon Knapp, (M), Kansas, 1980, Hs
Associate Hydrogeologist:
Steven D. Wilson, (M), Illinois, 1988, Hw
Professional Scientist:
William C. Bogner, (M), Illinois, 1983, Hs
Associate Professional Scientist:
Walton R. Kelly, (D), Virginia, 1993, Hw
George S. Roadcap, (M), Ohio State, 1990, Hw
Renjie Xia, (D), Illinois, 1991, Hs
Assistant Professional Scientist:
Deva K. Borah, (D), Mississippi, 1979, Hq
Assistant Hydrogeologist:
Randall A. Locke, III, (M), Iowa, 1994, Hw
Scott C. Meyer, (M), North Carolina, 1987, Hw

Analytical Chemistry and Technology (2004)
2204 Griffith Drive
Champaign, IL 61820
p. (217) 333-9321 f. (217) 244-3054
smothers@sws.uiuc.edu
http://www.sws.uiuc.edu
Head, Analytical Chemistry & Tech:
Kent W. Smothers, (B), Blackburn, 1980, Cg
Senior Chemist:
Gary R. Peyton, (M), North Texas, 1968, Oo
Senior Scientist:
Michael E. Caughey, (D), Texas, 1988, Co
Thomas R. Holm, (D), Caltech, 1978, Hw
Shundar Lin, (D), Syracuse, 1967, Hs
Michael L. Machesky, (D), Wisconsin, 1986, Cg
Donald P. Roseboom, (M), Bradley, 1976, Hy
Associate Chemist:
Karen S. Harlin, (M), Illinois, 1991, Ca
Jane E. Rothert, (M), Washington State, 1977, Og
Lab Director:
Daniel Webb, (M), Illinois, Hg
Laboratory Director:
Loretta M. Skowron, (B), Illinois, 1976, Ca
Cooperating Faculty:
Mark E. Peden, (B), Illinois, 1974, Og

Atmospheric Sciences Div (2004)
2204 Griffith Drive
Champaign, IL 61820
p. (217) 333-2210 f. (217) 333-6540

Indiana

Indiana Geological Survey

Indiana Geological Survey (2008)
611 North Walnut Grove
Bloomington, IN 47405
p. (812) 855-5067 f. (812) 855-2862
jsteinm@indiana.edu
http://igs.indiana.edu/
State Geologist & Director:
John C. Steinmetz, (D), Miami, 1978, Pm
Assistant Director, Technical Services:
Richard T. Hill, (B), Oi
Assistant Director, Research:
John A. Rupp, (M), E Washington, 1980, Go
Assistant Director, Business:
Charlotte A. Smith, (B), Indiana, 2005
Head, Env Geology:
Nancy R. Hasenmueller, (M), Ohio State, Ge
Basin Research Coordinator:
Brian D. Keith, (D), RPI, 1974, Go

Assistant Director, Center for Geospatial Data Analysis:
Sally L. Letsinger, (D), Indiana, 2001, Hq
Senior Scientist:
Tracy Branam, (M), Indiana, Cg
Margaret V. Ennis, (B), St. Lawrence, 1964, Cg
Denver Harper, (D), Ec
Walter Hassenmueller, (M), Ohio State, 1970, En
Maria Mastalerz, (D), Ec
Todd Thompson, (D), Indiana, 1987, Gs
Licia A. Weber, Ec
Charles W. Zuppann, (M), Vanderbilt, 1974, Go
Associate Scientist:
Christopher Dintaman, (M), Indiana, Hw
Agnieszka Drobniak, (D), Ec
Rebecca A. Meyer, Ec
Glacial Geologist:
Marni Dickson Karaffa, (M), New Brunswick, 2001, Gl
Research Affiliate:
Carl Rexroad, (D), Pm
Assistant Geologist:
Jennifer Olejnik, (B), Ge

Iowa

Iowa Dept of Natural Resources

Iowa Geological Survey (2007)
109 Trowbridge Hall
Iowa City, IA 52242-1319
p. (319) 335-1575 f. (319) 335-2754
lois.bair@dnr.iowa.gov
http://www.state.ia.us/government/dnr/index.html

Other:
Bill J. Bunker, Oo
Paul E. VanDorpe, Oo

Kansas

University of Kansas

Kansas Geological Survey (2006)
1930 Constant Avenue
Campus West
Lawrence, KS 66047-3726
p. (785) 864-3965 f. (785) 864-5317
douglass@kgs.ku.edu
http://www.kgs.ku.edu

Director:
William E. Harrison, (D), Louisiana State, 1976, Co
Research Assistant:
Kenneth A. Nelson, (B), Kansas, 1993, Oy
Associate Director:
Rex C. Buchanan, (M), Wisconsin, 1982, Oe
Assistant Director:
Lawrence H. Skelton, (M), Wichita State, 1991, Gg
Executive Director:
Rolfe D. Mandel, (D), Kansas, 1991, Ga
Lynn W. Watney, (D), Kansas, 1985, Gr
Senior Scientist:
Lawrence L. Brady, (D), Kansas, 1971, Gg
Robert W. Buddemeier, (D), Washington, 1969, Hg
James J. Butler, Jr., (D), Stanford, 1987, Hw
Timothy R. Carr, (D), Wisconsin, 1981, Go
John H. Doveton, (D), Edinburgh, 1969, Go
Marios A. Sophocleous, (D), Alberta, 1978, Hw
Donald O. Whittemore, (D), Penn State, 1973, Hw
Associate Scientist:
Jianghai Xia, (D), Kansas, 1992, Ye
Evan K. Franseen, (D), Wisconsin, 1989, Gs
David A. Grisafe, (D), Penn State, 1968, Om
Greg A. Ludvigson, (D), Iowa, 1988, Gr
Richard D. Miller, (M), Kansas, 1983, Yx
Gregory C. Ohlmacher, (D), Purdue, 1991, Ng
Assistant Scientist:
Saibal Bhattacharya, (M), Kansas, 1996, Np
Geoffrey C. Bohling, (D), Kansas, 1999, Hw
Alan P. Byrnes, (M), Chicago, 1977, Yg
Peter A. Macfarlane, (D), Kansas, 1993, Hw
James R. McCauley, (D), Kansas, 1977, Or
Kerry D. Newell, (D), Kansas, 1996, Go
Susan E. Nissen, (D), Columbia, 1992, Yg
Choon-byong Park, (D), Kansas, 1995, Yg
Geologist:
Daniel R. Suchy, (D), McGill, 1992, Gg
Research Associate:
Larry M. Magnuson, (B), Kansas, 1967, Ca
Abdelmoneam E. Raef
Robert S. Sawin, (M), Kansas State, 1977, Gg
Margaret A. Townsend, (M), Texas, 1976, Hw
Research Assistant:
Martin K. Dubois, (M), Kansas, 1980, Go
Willard J. Guy, Jr., (B), Colorado, 1955, Go
Blake Wilson, (M), Kansas State, 1993, Oi
Emeritus:
Pieter Berendsen, (D), California (Riverside), 1971, Cg
John C. Davis, (D), Kansas, 1967, Gq
Lee C. Gerhard, (D), Kansas, 1964, Gr
William W. Hambleton, (D), Kansas, 1951, Yg
Daniel F. Merriam, (D), Kansas, 1961, Gg
Ricardo Olea, (D), Kansas, 1982, Gq
Truman Waugh, (B), Washburn, 1963, Ca
Research Tech:
Zachary J. Wenz, (M), Alaska (Fairbanks), 2004, Gg
Research Assistant:
Galen Worthington, (B), Kansas, 2003, Ge
David P. Young, (M), Kansas, 1992, Oo
Geologist:
Brett Wedel, (B), Emporia State, 2002, Oo
Assistant Scientist:
Scott W. White, (D), Wisconsin, 1998, Oo

Kentucky

University of Kentucky

Kentucky Geological Survey (2007)
228 Mining & Minerals Resources Building
Lexington, KY 40506-0107
p. (859) 257-5500 f. (859) 257-1147
cobb@uky.edu
http://www.uky.edu/KGS

State Geologist & Director:
James C. Cobb, (D), Illinois, 1981, Ec
Assistant State Geologist:
John D. Kiefer, (D), Illinois, 1970, Ng
Head, Water Resources:
James S. Dinger, (D), Nevada (Reno), 1977, Hg
Head, Geospatial Analysis:
Gerald A. Weisenfluh, (D), South Carolina, 1982, Ec
Head, Geoscience Information:
Steven J. Cordiviola, (M), Cincinnati, 1974, Eg
Head, Energy & Minerals:
David C. Harris, (M), Stoney Brook, 1982, Go
Head, Administration:
Jackie Silvers, (M), St. Mary's, 1990, Oo
Manager, Well Sample & Core Library:
Patrick J. Gooding, (M), E Kentucky, 1983, Eo
Manager, Lab Services:
Henry E. Francis, (B), W Kentucky, 1966, Ca

Technology Transfer Officer:
Michael J. Lynch, (B), E Kentucky, 1975, Oo
Head, Western Kentucky Office:
David A. Williams, (M), E Kentucky, 1979, Eg
Head, Geologic Hazards Section:
Zhenming Wang, (D), Kentucky, 1998, Ys

Louisiana

Louisiana State University

Louisiana Geological Survey (2007)
3079-Energy, Coast and Environment Bldg.
Baton Rouge, LA 70803
p. (225) 578-5320 f. (225) 578-3662
cjohn@lsu.edu
http://www.lgs.lsu.edu
State Geologist and Professor:
Chacko J. John, (D), Delaware, 1977, GosGe
Assistant Professor:
Douglas A. Carlson, (D), Wisconsin (Milwaukee), 2001, Hw
Marty R. Horn, (D), Texas (Arlington), 1996, EoGrg
Marty R. Horn, (D), Texas (Arlington), 1996, EoGgr
Thomas P. Van Biersel, (D), Wisconsin (Milwaukee), 1999, Hw
GIS Coordinator:
Hampton Peele, (M), Louisiana State, 1997, Oi
Research Associate:
Brian J. Harder, (B), Louisiana State, 1981, Eo
Paul V. Heinrich, (M), Illinois, 1982, Gs
Bobby L. Jones, (B), Louisiana State, 1953, Go
Richard P. McCulloh, (M), Texas, 1977, Gg
Lloyd R. Milner, (B), Louisiana State, 1985, Gg
Patrick M. O'Neill, (B), Louisiana State, 1985, Sf
Robert L. Paulsell, (B), Louisiana State, 1987, Oy
Lisa G. Pond, (B), Louisiana State, 1987, Og
Asheka Rahman, (D), LSU, 1995, HwNg
Warren Schulingkamp, (M), Utah, 1972, EoGgEc
Cartographic Manager:
John I. Snead, (B), Louisiana State, 1978, Gm
Assistant Director:
John E. Johnston, III, (M), Texas, 1977, EoGe
Office Coordinator:
Frances A. Tircuit, (U), Oo
Accountant Technician:
Jeanne Johnson, (U), Oo

Basin Research Energy Section (2004)
Louisiana Geological Survey/LSU
208 Howe Russell Geoscience Complex
Baton Rouge, LA 70803-4101
p. (225) 578-8328 f. (225) 578-3662
http://www.bri.lsu.edu
Office Coordinator: Cherri B. Webre
Associate Professor:
Ronald K. Zimmerman, (D), Louisiana State, 1966, Go
Assistant Professor:
Clayton F. Breland, (D), Tennessee, 1980, Ye
John B. Echols, (D), Louisiana State, 1966, Go
Research Associate:
Brian J. Harder, (B), Louisiana State, 1981, Eo
Bobby L. Jones, (B), Louisiana State, 1953, Go
Phillip W. Lemay, (B), Centenary, 1999, Gg
Michael B. Miller, (M), North Carolina, 1982, Go
Lloyd R. Milner, (B), Louisiana State, 1985, Gg
Patrick M. O'Neill, (B), Louisiana State, 1985, Sf
Computer Analyst:
Reed J. Bourgeois, (B), Louisiana State, 1983, Oo

Accountant Technician:
Carla Domingue, Oo

Maine

Dept of Conservation

Maine Geological Survey (2004)
Natural Resources Information & Mapping Center
State House Station, Number 22
Augusta, ME 04333
p. (207) 287-2801 f. (207) 287-2353
http://www.state.me.us/doc/nrimc/mgs/mgs.htm
Department Secretary: Christine Sansouci
State Geologist:
Robert G. Marvinney, (D), Syracuse, 1986, Gc
Senior Geologist:
Daniel B. Locke, (B), Maine, 1982, Hw
Craig D. Neil, (B), Maine, 1982, Hw
Physical Geologist:
Henry N. Berry IV, (D), Massachusetts, 1989, Gg
Woodrow B. Thompson, (D), Ohio State, 1975, Gl
Marine Geologist:
Stephen M. Dickson, (M), Rhode Island, 1985, On
Joseph T. Kelley, (D), Lehigh, 1980, Gs
Hydrogeologist:
Marc C. Loiselle, (D), MIT, 1979, Hw
Thomas K. Weddle, (D), Boston, 1991, Hw
Geologist:
Michael E. Foley, (B), Maine, 1973, Gg
Research Associate:
Robert A. Johnston, (B), Bridgewater State, 1973, Gl

Maryland

Maryland Dept of Natural Resources

Maryland Geological Survey (2006)
2300 St. Paul Street
Baltimore, MD 21218-5210
p. (410) 554-5500 f. (410) 554-5502
ecleaves@dnr.state.md.us
http://www.mgs.md.gov/
Administrative Assistant: Jeanne Gary
Director:
Emery T. Cleaves, (D), Johns Hopkins, 1973, Gm

Maryland Geological Survey

Coastal & Estuarine Geology Program (2007)
2300 St. Paul Street
Baltimore, MD 21218-5210
p. (410) 554-5537 f. (410) 554-5502
jhalka@dnr.state.md.us
http://www.mgs.md.gov/coastal/index.html
Program Chief:
Jeffrey P. Halka, (M), Rochester, 1972, Gu
Related Staff:
Robert D. Conkwright, (B), Franklin & Marshall, 1973, Gu
E. Lamere Hennessee, (M), George Washington, 1985, On
James M. Hill, (D), Northwestern, 1984, Cl
Katherine A. Offerman, (B), Susquehanna, 2001, Ge
Richard A. Ortt, Jr., (B), Johns Hopkins, 1991, Op
Stephen Van Ryswick, (B), Maryland, 2002, Ge
Darlene V. Wells, (B), Florida State, 1973, Gs

Environmental Geology & Mineral Resources Program (2005)
2300 St. Paul Street
Baltimore, MD 21218-5210

p. (410) 554-5500 f. (410) 554-5502
gbaum@dnr.state.md.us
http://www.mgs.md.gov

Program Chief:
Gerald R. Baum, (D), North Carolina, 1977, Gr
Principal Geologist:
James P. Reger, (D), West Virginia, 1977, Gm
Geologist:
David K. Brezinski, (D), Pittsburgh, 1984, Gr
Sacha J. Lanham, (B), Towson, 2003, Oi
Heather Quinn, (M), Florida, 1988, Ge
Dale W. Shelton, (B), Towson, 1986, Gg
John M. Wilson, (B), Maryland, 1976, Gg
Librarian:
Mary Ruth Buchness, (N), Oo

Hydrogeology & Hydrology Program (2005)
2300 St. Paul Street
Baltimore, MD 21218-5210
p. (410) 554-5500 f. (410) 554-5502
dbolton@mgs.md.gov
http://www.mgs.md.gov
Administrative Aide: Donajean M. Appel

Program Chief:
Harry J. Hansen, (D), Columbia, 1967, Hw
Hydrogeologist:
Grufon Achmad, (D), Missouri, 1973, Hw
David C. Andreasen, (B), Maryland, 1985, Hy
David W. Bolton, (M), W Michigan, 1988, Hw
David D. Drummond, (M), George Washington, 1988, Hy
Mark T. Duigon, (M), Indiana, 1977, Hy

Massachusetts

Office of the State Geologist

Dept of Geosciences (2004)
Univ of Massachusetts (Amherst)
611 North Pleasant Street
Amherst, MA 01002
p. 413-545-2286 f. 413-545-1200

State Geologist:
Stephen B. Mabee, (D), Massachusetts, 1992, Hw

Michigan

Michigan Dept of Environmental Quality

Geological and Land Management Div (2004)
525 W. Allegan St.
P.O. Box 30256
Lansing, MI 48909-7756
p. (517) 241-1515 f. (517) 241-1595
http://www.michigan.gov/deq

State Geologist:
Harold R. Fitch, (U), Oo

Minnesota

University of Minnesota, Twin Cities

Minnesota Geological Survey (2007)
2642 University Ave W
Saint Paul, MN 55114-1057
p. (612) 627-4780 (Ext. 0) f. (612) 627-4778
mgs@umn.edu
http://www.geo.umn.edu/mgs

Director:
Leonard Harvey Thorleifson, (D), Colorado, 1989, Gl
Chief Geologist:
Anthony C. Runkel, (D), Texas, 1988, Gs
Hydrogeologist:
Robert G. Tipping, (M), Minnesota, 1992, Hw
Geophysicists:
Val W. Chandler, (D), Purdue, 1977, Ye
Geologist:
Bruce A. Bloomgren, (B), Minnesota (Duluth), 1970, Gr
Terrence J. Boerboom, (M), Minnesota (Duluth), 1987, Gi
Kenneth L. Harris, (D), North Dakota, 1975, Gl
Howard C. Hobbs, (D), North Dakota, 1975, Gl
Carrie E. Jennings, (D), Minnesota, 1996, Gl
Mark A. Jirsa, (M), Minnesota (Duluth), 1980, Gc
Alan R. Knaeble, (B), Minnesota, 1982, Gl
Barbara A. Lusardi, (M), Maine, 1992, Gl
Gary N. Meyer, (M), North Dakota, 1979, Gl
John H. Mossler, (D), Iowa, 1970, Gr
Technical Editor:
Lori A. Robinson, (M), Alaska (Fairbanks), 1999, Oo
Information Technology Supervisor:
Richard S. Lively, (M), Michigan State, 1977, Cc
Information Technology:
Timothy E. Wahl, (B), Minnesota, 1977, Oo
Associate Director:
Dale R. Setterholm, (M), Carlson Sch Mgmt, 1999, Gg
Associate Administrator:
Linda L. McDonald, Oo
Geologist:
Emily J. Bauer, (B), Wisconsin (River Falls), 1989, Gg

Mississippi

Mississippi Office of Geology

Energy & Coastal Geology Div (2004)
P.O. Box 20307
2380 Highway 80 West
Jackson, MS 39289-1307
p. (601) 961-5500 f. (601) 961-5521

Division Director:
Jack S. Moody, (B), East Carolina, 1970, Go
Geologist:
Steven D. Champlin, (B), Alabama, 1976, Go
Rick L. Ericksen, (M), N Illinois, 1975, Go
Edward J. Henderson, (B), South Carolina, 1948, Go
Peter S. Hutchins, (B), Millsaps, 1990, Oo
Env Scientist:
Jennifer D. Oltman, (M), Mississippi State, 1997, Oy
GIS Specialist:
Barbara E. Yassin, (B), Illinois State, 1989, Oy

Environmental Geology Div (2004)
P.O. Box 20307
2380 Highway 80 West
Jackson, MS 39289-1307
p. (601) 961-5500 f. (601) 354-7151

Division Director:
John C. Marble, (B), Mississippi State, 1974, Ge

Mining & Reclamation Div (2004)
P.O. Box 20307
2380 Highway 80 West
Jackson, MS 39289-1307
p. (601) 961-5500 f. (601) 961-5521
Ken_McCarley@deq.state.ms.us
Secretary: Tamara Duckworth
Secretary: Sandra Saik

Division Director:
J. Kendrick McCarley, (M), Mississippi, 1996, Ng

Assistant Division Director:
Stanley C. Thieling, (M), Iowa, 1973, Go
Geologist:
Michael Akin, (B), Mississippi State, 1995, Ge
James L. Matheny, (B), Delta State, 1994, Oo
Jim F. McMullin, (B), Millsaps, 1959, Gg
Thomas M. Ray, (M), S Mississippi, 1975, Gg
Biologist:
David J. Wickens, (B), Mississippi, 1991, Oo
Related Staff:
Robert J. Millette, (B), SW Louisiana, 1989, Oo
James E. Starnes, (B), Millsaps, 1996, Gg

Mississippi Dept of Environmental Quality (2004)
P.O. Box 20307
2380 Highway 80 West
Jackson, MS 39289-1307
p. (601) 961-5500 f. (601) 354-7151
http://www.deq.state.ms.us/
Administrative Assistant: Jean Inman
Director & State Geologist:
S. Cragin Knox, (M), Mississippi, 1987, Ye
Assistant Director:
Michael B. E. Bograd, (B), Mississippi State, 1971, Gg

Surface Geology Div (2004)
P.O. Box 20307
2380 Highway 80 West
Jackson, MS 39289-1307
p. (601) 961-5500 f. (601) 354-7151
Division Director:
David T. Dockery, (D), Tulane, 1991, Pi
Geologist:
Kenneth D. Davis, (B), Mississippi State, 1968, Gr
Stephen L. Ingram, (B), Millsaps, 1985, Gr
David E. Thompson, (B), Mississippi State, 1986, Gr
GIS Specialist:
Daniel W. Morse, (B), Texas, 1982, Oo

Missouri

Missouri Dept of Natural Resources

Geological Survey & Resource Assessment Div (2004)
PO Box 250
Buehler Building/111 Fairgrounds Rd
Rolla, MO 65402-0250
p. (573) 368-2101 f. (573) 368-2111
gspgeol@mail.dnr.gov
www.dnr.state.mo.us/geology/homegsrad.htm
Designated Principal Assistant: Tami Allison
Division Director & State Geologist:
Mimi R. Garstang, (B), Ge
Head:
James W. Duley, Oo
Deputy Director:
Mike Wells, (M), Oo

Dam & Reservoir Safety (2004)
PO Box 250
Buehler Bldg/111 Fairgrounds Rd
Rolla, MO 65401-0250
p. (573) 368-2175 f. (573) 368-2111
Program Secretary: Tammy Theilen
Program Director:
James L. Alexander, (B), Oo
Professional Staff:
Russell C. Adams, (M), Oo
Robert Clay, (M), Oo

Div of Geology and Land Survey (2007)
PO Box 250
111 Fairgrounds Road
Rolla, MO 65402-0250
p. (573) 368-2100 f. (573) 368-2111
mimi.garstang@dnr.mo.gov
http://www.dnr.state.mo.us/dgls/geosrv
Administrative Assistant: Tami L. Allison
State Geologist & Division Director:
Mimi R. Garstang, Ge
Deputy Director & Assistant State Geologist:
James W. Duley, (B), C Missouri, 1975, Hy

Geological Survey Program (2004)
PO Box 250
Buehler Building/111 Fairgrounds Rd
Rolla, MO 65401-0250
p. (573) 368-2160 f. (573) 368-2111
nrsturs@mail.dnr.state.mo.us
Program Secretary: Deborrah Breuer
Program Director:
Steven W. Sturgess, (M), Missouri, 1986, Oo
Geologist:
William W. Little, (D), Colorado, 1995, Gr
Cathy Smith, (B), Oo
Professional Staff:
Robert Archer, (M), Oo
Peter Bachle, Oo
Hal Baker, (M), Gg
James Brown, (B), Oo
Neil Elfrink, (M), Ge
Joe Gillman, (B), Oo
David Hoffman, (M), Ne
Kurt R. Hollman, (B), Ge
Evan Kifer, (B), Go
Brian Martin, (B), Oo
Mark A. Middendorf, (B), Gg
James Palmer, (B), Gg
Matt Parker, (B), Oo
Larry Pierce, (M), Oo
Jerry Prewett, (B), Oo
Peter Price, (B), Missouri (Rolla), Ge
Myrna Rueff, (B), Ge
Cheryl M. Seeger, (M), Gg
Michael A. Siemens, (B), Ge
David C. Smith, (B), Gg
Gary St Ivany, (B), Ge
Edith Starbuck, (M), Ge
Chris Vierrether, (D), Oo
Geology Librarian:
David Erickson, Ge
Glen Young, Ge
Geological Tech:
Cecil Boswell, Oo
Hairl Dayton, Oo
Billy Reed, Oo
Joe Schlueter, Oo
Fred Shaw, Oo
Secretary:
Loetta Miller, Oo
Sandy Miller, Oo

Land Survey Program (2004)
PO Box 250
Buehler Building/111 Fairgrounds Rd
Rolla, MO 65401-0250
p. (573) 368-2300

Program Director:
J. Michael Flowers, Oo
Professional Staff:
O. Dan Lashley, Oo
John Paulsmyer, Oo
Darrell Pratte, Oo

Water Resources Program (2004)
PO Box 250
Buehler Building/111 Fairgrounds Rd
Rolla, MO 65401-0250
p. (573) 368-2190 f. (573) 368-2111
http://www.dnr.state.mo.us/dgls/
Department Secretary: Sharon Hankins
Program Director:
Steve McIntosh, (M), Hg
Chief:
James E. Vandike, (M), Hw
Geologist:
Scott Kaden, Oo
Bruce Netzler, (M), Go
Beth Warrick, Hy
Associate Scientist:
Robert Bacon, Hs
Scotty D. Baumgartner, (M), SD Sch of Mines & Tech, 1979, Hw
Jeffrey C. Jaquess, (M), Missouri (Rolla), 1989, Hw
Professional Staff:
Jeanette Barnett, (B), Oo
Cynthia Brookshire, (B), Hw
John Drew, (B), Hs
Charles Du Charme, (B), Hs
Richard Gaffney, (M), Hs
Charles R. Hays, (M), Oo
Pat Mulvany, (D), Oo
Geological Tech:
Richard Bates, Oo
Secretary:
Mary Woodland, Oo

Montana

Montana Tech - University of Montana

Montana Bureau of Mines & Geology (2006)
1300 West Park Street
Butte, MT 59701-8997
p. (406) 496-4180 f. (406) 496-4451
edeal@mtech.edu
http://www.mbmg.mtech.edu
Administrative Assistant: Charlotte McKenzie
Director & State Geologist:
Edmond G. Deal, (D), New Mexico, 1973, Gg
Assistant Director:
Marvin R. Miller, (M), Indiana, 1965, Hw
Director, Contracts & Grants:
Lynn M. Job, (B), Montana Tech, Oo
Publications Editor:
Susan A. Barth, (B), California (Davis), 1992, Oo
Manager, Ground-water Assessment Program:
Thomas W. Patton, (M), Montana Tech, 1987, Hw
Director, Earthquake Studies Office:
Michael C. Stickney, (M), Montana, 1980, Ys
Senior Research Hydrogeologist:
John R. Wheaton, (M), Montana, 1987, Hw
Senior Research Geologist, Museum Curator:
Richard B. Berg, (D), Montana, 1964, En
Senior Research Geologist:
David A. Lopez, (D), Colorado Mines, 1981, Gg
Research Division Chief:
John J. Metesh, (D), Montana, 2003, Hw
Associate Research Hydrogeologist:
Robert N. Bergantino, (B), Montana, 1967, Gg
John I. La Fave, (M), Texas, 1987, Hw
Jon C. Reiten, (M), North Dakota, 1983, Hw
Associate Research Geologist:
Larry N. Smith, (D), New Mexico, 1988, Hw
Associate Research Mining Engineer:
Robin B. McCulloch, (M), Montana Tech, 1999, Nm
Associate Research Hydrogeologist:
Terence E. Duaime, (B), Montana Tech, 1978, Hw
Assistant Research Hydrogeologist:
Ginette Abdo, (M), Penn State, 1990, Hw
Camela Carstarphen, (M), Oregon State, 1991, Hw
James P. Madison, (M), Montana, 1993, Hw
Peter M. Norbeck, (M), Idaho, 1975, Hw
John L. Olson, (M), Louisiana (Lafayette), 1990, Hw
Fred A. Schmidt, (B), Montana Tech, 1973, Hw
Assistant Research Geologist:
Phyllis Hargrave, (M), Montana Tech, 1990, Gg
Jeffrey Lonn, (M), Montana, 1985, Gg
Susan M. Vuke, (M), Montana, 1982, Gg
Hydrogeologist:
Michael D. Kerschen, (B), Mesa State, 1983, Hw
Field Bio-Remediation Engineer:
Tom Bowler, (B), Montana Tech, 1988, Nx
Research Associate:
Jaqueline Hunt, (B), Montana Tech, 1990, Cg
GIS Specialist:
Ken Sandau, (A), Montana Tech, Oi
Paul Thale, (M), Montana State, 1994, Oi
Research Hydrogeologist:
Gary Icopini, (D), Michigan State, 2000, Gg
Associate Research Professor:
Steve McGrath, (D), Montana Tech, Ca
Geologic Cartographer:
Susan M. Smith, (B), Montana State, 1970, Oo
Computer Software Eng/Applications:
Luke Buckley, Oo

Nebraska

Unversity of Nebraska - Lincoln

Survey Div, School of Natural Resources (2007)
Conservation & Survey Division
3310 Holdrege Street
616 Hardin Hall
Lincoln, NE 68583-0996
p. (402) 472-3471 f. (402) 472-2946
jvogel2@unl.edu
http://snr5.unl.edu/csd/
Director:
Mark S. Kuzila, (D), Nebraska, 1988, Sd
Professor:
Marvin P. Carlson, (D), Nebraska, 1969, Gt
Anatoly Gitelson, (D), Inst Radio Tech (Russia), 1972, Oo
James W. Goeke, (M), Colorado State, 1970, Hw
David C. Gosselin, (D), SD Sch of Mines & Tech, 1987, CgHw
James W. Merchant, (D), Kansas, 1984, Oru
Donald C. Rundquist, (D), Nebraska, 1977, Ori
James B. Swinehart, (M), Nebraska, 1979, Gs
Senior Scientist:
Susan Olafsen-Lackey, (B), SD Sch of Mines & Tech, 1982, Hw
Steven S. Sibray, (M), New Mexico, 1977, HwGg
Jozsef Szilagyi, (D), California (Davis), 1997, Hq
Associate Professor:
F. Edwin Harvey, (D), Waterloo, 1995, Hw

Matt Joeckel, (D), Iowa, 1993, GsrGg
Assistant Professor:
Xun-Hong Chen, (D), Wyoming, 1994, Hq
Paul Hanson, (D), Nebraska, 2005, Gm
Research Associate:
Steven L. Hartung, (M), Nebraska, 1983, Sd
Leslie M. Howard, (M), Nebraska, 1989, Oyi
Professor:
Duane A. Eversoll, (M), Nebraska, 1977, Ng
Emeritus:
Robert F. Diffendal, (D), Nebraska, 1971, Gr

Nevada

University of Nevada

Nevada Bureau of Mines and Geology (2006)
Mail Stop 178
Reno, NV 89557-0088
p. (775) 784-6691 f. (775) 784-1709
jprice@unr.edu
http://www.nbmg.unr.edu/
Administrative Assistant: Terri Garside

Director:
Jonathan G. Price, (D), California (Berkeley), 1977, Gg
Senior Scientist:
John W. Bell, (M), Arizona State, 1974, Ng
Geoffrey Blewitt, (D), Caltech, 1996, Yd
Stephen B. Castor, (D), Nevada (Reno), 1972, Eg
Craig M. de Polo, (D), Nevada (Reno), 1998, Ng
James E. Faulds, (D), New Mexico, 1989, Gc
Larry J. Garside, (M), Nevada (Reno), 1968, Gg
Christopher D. Henry, (D), Texas, 1975, Gg
P. Kyle House, (D), Arizona, 1996, Gm
Paul J. Lechler, (D), Nevada (Reno), 1995, Cg
John Muntean, (D), Stanford, 1998
Hans-Peter Plag, (D), Free (Berlin), 1988, Yd
Alan R. Ramelli, (M), Nevada (Reno), 1988, Ng
Lisa Shevenell, (D), Nevada (Reno), 1990, Hw
Research Assistant Professor:
William Hammond, (D), Oregon (Eugene), 2000, Yd
Cornelis Kreemer, (D), SUNY (Stony Brook), 2001, Yd
GIS Supervisor:
Ronald H. Hess, (M), Nevada (Reno), 1998, Oy
Geologic Information Specialist:
David A. Davis, (M), Nevada (Reno), 1990, Gg
Assistant Geochemist:
Mario O. Desilets, (M), Nevada (Reno), 1992, Cg
Research Associate:
Daphne D. LaPointe, (M), Montana, 1977, Gg
Publications Manager:
Susan L. Tingley, (B), California (Los Angeles), 1966, Oy
Emeritus:
Harold F. Bonham, Jr., (M), Nevada (Reno), 1963, Em
John W. Erwin, (M), Yg
Liang-Chi Hsu, (D), California (Los Angeles), 1966, Cp
Joseph V. Tingley, (M), Nevada (Reno), 1963, Em
Publications Manager:
Elizabeth Crouse, (B), Kansas, 1996, Gg

New Hampshire

New Hampshire Geological Survey

New Hampshire Dept of Environmental Services (2007)
29 Hazen Drive, P.O.Box 95
Concord, NH 03302-0095
p. (603) 271-3503 f. (603) 271-3305
geology@des.state.nh.us
http://www.des.state.nh.us

State Geologist:
David R. Wunsch, (D), Kentucky, 1992, Hw
Hydrogeologist:
Frederick Chormann, (M), New Hampshire, Hw
Hydrogeologist:
Gregory A. Barker, (B), URI, 1985, Gg
Program Assistant:
Debbie McDonnell, Oo
Outreach Coordinator:
Lee Wilder, (M), New Hampshire, 1964, Gg
Env Tech:
Genevieve Al-Egaily, (B), New Hampshire, 2003, Ge

New Jersey

New Jersey Geological Survey

New Jersey Geological Survey (2007)
PO Box 427
Trenton, NJ 08625
p. (609) 292-1185 f. (609) 633-1004
karl.muessig@dep.state.nj.us
http://www.njgeology.org
Administrative Assistant: Tenika Jacobs

State Geologist:
Karl W. Muessig, (D), Princeton, 1979, Cg
Bureau Chief:
Richard F. Dalton, (B), Rutgers, 1967, Gr
David L. Pasicznyk, (B), Temple, 1979, Yg
Section Chief:
Robert J. Canace, (B), Rutgers, 1975, Hw
William P. Graff, (B), Clark, 1979, Oy
David W. Hall, (M), W Michigan, 1983, Yg
Research Scientist:
Gregory C. Herman, (D), Rutgers, 1997, Gc
Jeffrey L. Hoffman, (M), Princeton, 1981, Hq
Michael E. Serfes, (D), Rutgers, 2005, Hy
Scott D. Stanford, (D), Rutgers, 2001, Gl
Peter J. Sugarman, (D), Rutgers, 1995, Gr
Richard A. Volkert, (M), Montclair State, 1984, Gx
Jeffrey S. Waldner, (M), Drexel, 1994, Yr
Supervising Geologist:
Stephen W. Johnson, (B), Richard Stockton, 1974, Hw
Thomas F. Seckler, (B), Montclair State, 1979, Ge
Steven E. Spayd, (B), Montclair State, 1979, Hw
Jane Uptegrove, (M), Rutgers, 2003, Gr
Supervising Env Specialist:
John Curran, (B), Kean, 1975, Hw
Bill Mennel, Hw
Helen L. Rancan, (M), Stevens, 1995, Hs
Supervising Env Engineer:
Richard Shim-Chim, (B), Toronto, 1973, Hw
Research Scientist:
James T. Boyle, (M), New Mexico Tech, 1984, Hw
John H. Dooley, (M), New Mexico Tech, 1983, Cg
Suhas L. Ghatge, (M), W Michigan, 1984, Yv
Donald H. Monteverde, (M), Lehigh, 1984, Gr
Laura J. Nicholson, (M), Drexel, 1995, Hw
Ronald W. Witte, (M), Lehigh, 1988, Gl
GIS Specialist:
Zehdreh Allen-Lafayette, (M), Syracuse, 1994, Oi
Mark A. French, (B), Rutgers, 1989, Hq
Ted J. Pallis, (M), Montclair State, 1994, Oi
Ronald Pristas, (B), Penn State, 1990, Oi
Principal Geologist:
Andrea L. Friedman, (B), Bryn Mawr, 2001, Ge
Eric W. Roman, (M), Temple, 1999, Hw
Principal Environmental Specialist:
Raymond T. Bousenberry, (B), Fairleigh Dickinson, 1997, Ge

Steven E. Domber, (M), Wisconsin, 2000, Hw
Gregg M. Steidl, (B), Rutgers, 1995, Ge
Senior Geologist:
Emily Godfrey, (M), Montana, 2004, Cl
GIS Specialist:
Michael W. Girard, (B), Bloomsberg, 1996, Oi
Former State Geologist:
Haig F. Kasabach, (M), Michigan, 1959, Hw
Secretary:
Octavia L. Sutphin, Oo
Principal Env Tech:
Doug Rivedal, (B), Coll of New Jersey, 1975, Ge
Investigator:
Walter Marzulli, Oy
Joseph A. Rich, Oy

New Mexico

New Mexico Institute of Mining & Tech

New Mexico Bureau of Geology & Mineral Res (2007)
801 Leroy Place
Socorro, NM 87801-4796
p. (575) 835-5420 f. (575) 835-6333
bureau@gis.nmt.edu
http://geoinfo.nmt.edu/

State Geologist:
Peter A. Scholle, (D), Princeton, 1970, Gs
Senior Volcanologist:
William C. McIntosh, (D), New Mexico Tech, 1990, Cc
Field Geologist:
Bruce Allen, (D), New Mexico, 1993, Gm
Senior Geologist & Associate Director:
Paul W. Bauer, (D), New Mexico Tech, 1988, Gp
Senior Field Geologist:
Steven M. Cather, (D), Texas, 1986, Gs
Adjunct Faculty:
George S. Austin, (D), Iowa, 1971, En
David W. Love, (D), New Mexico, 1980, Ge
Senior Env Geologist Emeritus:
John W. Hawley, (D), Illinois, 1962, Ge
Senior Chemist:
Lynn A. Brandvold, (M), North Dakota State, 1964, Ca
Emeritus:
Charles E. Chapin, (D), Colorado Mines, 1966, Gt
Robert Weber, Oo
Webmaster/Geologist:
Adam S. Read, (M), New Mexico, 1997, Gg
Subsurface Fluids Geologist:
Patrick Walsh, (M), Texas, 2000, Gg
Senior Mining Engineer:
Robert W. Eveleth, (B), New Mexico Tech, 1969, Nm
Senior Lab Associate:
Lisa Peters, (M), Texas (El Paso), 1987, Gg
Senior Geologist & Associate Director:
L Greer Price, (M), Washington, 1974, Gg
Senior Geologic Lab Associate:
Kelly Donahue, (M), New Mexico Tech, 2002, Cg
Stacy Timmons, (M), Oregon State, 2002, Gg
Senior Field Geologist:
Richard M. Chamberlin, (D), Colorado Mines, 1980, Gg
Principal Senior Petroleum Geologist:
Ronald F. Broadhead, (M), Cincinnati, 1979, Go
Principal Senior Geophysicist:
Marshall A. Reiter, (D), Virginia Tech, 1969, Yh
Minerals Outreach Liason:
Virginia McLemore, (D), Texas (El Paso), 1994, Em
Mineralogist/Economic Geologist:
Virgil W. Lueth, (D), Texas, 1988, Gg
Manager, X-Ray Facility:
Christopher G. McKee, (M), New Mexico Tech, 1989, Gz
Manager, Geologic Mapping Program:
Michael Timmons, (D), New Mexico, 2004, Gg
Manager, Digital Cartography Lab:
Glen Jones, (B), New Mexico Tech, 1989, Oo
Hydrogeologist:
Peggy Johnson, (M), New Mexico Tech, 1990, Hw
Lewis A. Land, (D), North Carolina, 1999, Hg
GIS Cartographer:
David J. McCraw, (M), New Mexico, 1985, Oy
Geologist:
Shari Kelly, (D), S Methodist, 1984, Yg
Geological Librarian:
Maureen Wilks, (D), New Mexico Tech, 1991, Gp
Geochronologist:
Matthew T. Heizler, (D), California (Los Angeles), 1993, Cc
Field Geologist:
Sean D. Connell, (M), California (Riverside), 1995, Gm
Geoffrey Rawling, (D), New Mexico Tech, 2002, Gg
Economic Geologist:
Douglas Bland, (M), Wyoming, 1982, Eg
Chemistry Lab Manager:
Bonnie A. Frey, (M), New Mexico Tech, 2002, Cg
Associate Director, Operations:
James M. Barker, (M), California (Santa Barbara), 1972, En
Associate Director:
Nelia W. Dunbar, (D), New Mexico Tech, 1989, Gi
Related Staff:
Ibrahim H. Gundiler, (D), New Mexico Tech, 1975, Nx
Gretchen K. Hoffman, (M), Arizona, 1979, Ec
Cooperating Faculty:
Michael Pullin, (D), Kent State, 1999, Oo

New York

New York State Geological Survey

New York State Geological Survey (2007)
3140 Cultural Education Center
Albany, NY 12230
p. (518) 474-5816 f. (518) 486-2034
djornov@mail.nysed.gov
Administrative Assistant: Donna Jornov

Director:
William M. Kelly, (D), Massachusetts, 1979, Eg
Senior Scientist:
Andrew Kozlowski, (D), W Michigan, 2002, Gl
Marian Lupulescu, (D), Jassy (Romania), 1987, Gz
Langhorne Smith, (D), Virginia Polytechnic, 1996, Go
Charles Ver Straeten, (D), Rochester, 1996, Gr
Associate Scientist:
Ed Landing, (D), Michigan, 1979, Ps
Related Staff:
Michael E. Hawkins, (B), USNY Regents, 1982, Gz
Janet E. Manchester, (M), Texas, 1989, Oo
Richard E. Nyahay, (B), SUNY (Plattsburgh), 1982, Eo
Michael Pascussi, (A), Schenectady Comm Coll, 1999, Go
James Sandor, (A), SUNY (Cobleskill), 2001, Go
Linda A. VanAller-Hernick, (B), Saint Rose, 1974, Pg
Cartographer:
John B. Skiba, (A), Hudson Valley Comm Coll, 1989, Oo

North Carolina

North Carolina Geological Survey

Dept of Environment & Natural Resources (2004)
1612 Mail Service Center
Raleigh, NC 27699-1612

p. (919) 733-2423 f. (919) 733-0900
Rick.Wooten@ncmail.net
http://www.geology.enr.state.nc.us
Administrative Assistant: Derek Buckley
Administrative Assistant: Tonya M. Merritt

Director:
Charles H. Gardner, (M), Emory, 1961, Gg
Chief Geologist:
Timothy W. Clark, (M), Duke, 1998, Gg
Project Geologist:
John G. Nickerson, (B), Campbell, 1983, Og
Mary E. Watson, (M), Arizona, 1980, Gn
Senior Geologist:
Kathleen M. Farrell, (D), Louisiana State, 1989, Gs
Charles W. Hoffman, (M), Portland State, 1981, Gg
Carl E. Merschat, (M), Tennessee, 1967, Gg
Geologist:
Mark W. Carter, (M), Tennessee, 1994, Gt
Geologist:
Jeffrey C. Reid, (D), Georgia, Gg
Engineering Geologist:
Richard M. Wooten, (M), Georgia, 1980, Ng
Cartographic Technician:
Michael A. Medina, (B), California (Berkeley), 1991, Oo
Related Staff:
Paula Maynor, Oo

North Dakota

North Dakota Geological Survey

North Dakota Geological Survey (2005)
600 East Boulevard Avenue
Bismarck, ND 58505-0840
p. (701) 328-8000 f. (701) 328-8010
ndgspubs@state.nd.us
http://www.state.nd.us/ndgs/
Department Secretary: Karen Gutenkunst

Paleontologist:
John Hoganson, (D), North Dakota, 1985, Pg
Geologist:
Ed Murphy, (M), North Dakota, 1983, Oo

Ohio

Ohio Dept of Natural Resources

Div of Geological Survey (2007)
2045 Morse Road Bldg. C-1
Columbus, OH 43229
p. (614) 265-6576 f. (614) 447-1918
geo.survey@dnr.state.oh.us
http://www.ohiodnr.com/geosurvey/
Administrative Assistant: Betty Lewis

Chief:
Lawrence H. Wickstrom, (M), Kent State, 1982, Go
Assistant Chief:
Edward Mac Swinford, (M), E Kentucky, 1983, Gr
Geologist:
Mark T. Baranoski, (M), Toledo, 1982, Go
Glenn E. Larsen, (M), Ohio State, 1984, Gr
Ronald A. Riley, (M), Bowling Green, 1980, Go
Gregory A. Schumacher, (M), Cincinnati, 1984, Gr
Douglas L. Shrake, (M), Wright State, 1989, Gr
Ernie R. Slucher, (M), E Kentucky, 1986, Gr
Erik R. Venteris, (D), Ohio State, 2002, Sc
Mark E. Wolfe, (B), Ohio, 1980, En
Administrator:
Richard R. Pavey, (M), Purdue, 1983, Gl
Geologist:
Christopher Gordon, (B), Ohio State, 2006, Ec
Database Administrator:
Joseph G. Wells, (B), West Virginia, 1994, Oy
GIMS Specialist:
James McDonald, (M), Ohio State, 1990, Yg
Kelli Vogt, (M), Miami, 2007, OiiOi
GIMS Administrator:
Donovan M. Powers, (B), Ohio State, 2000, OiiOi
Geologic Assistant:
Madge R. Fitak, (B), Mt Union, 1972, Gg
Electronic Design Coordinator:
Lisa Van Doren, (B), Wittenberg, 1984, Oo
Natural Resources Administrator:
Edward V. Kuehnle, (B), Cincinnati, 1976, Oo
GIMS Specialist:

Lake Erie Geology Group (2004)
Great Lakes Center
1634 Sycamore Line
Sandusky, OH 44870-4132
p. (419) 626-4296 f. (419) 626-8767
Secretary 1: Karen L. Harold

Geologist Supervisor:
Constance J. Livchak, (M), Bowling Green, 1993, Hy
Researcher:
Dale L. Liebenthal, (U), Oo
Geologist:
Jonathan A. Fuller, (M), W Michigan, 1978, On
Donald E. Guy, Jr., (M), Bowling Green, 1983, On
Bruce E. Gerke, (M), Bowling Green, 2002, Ge
Ryan P. Murphy, (B), Bowling Green, 2002, Gs

Oklahoma

University of Oklahoma

Oklahoma Geological Survey (2006)
100 East Boyd
Energy Center, Room N-131
Norman, OK 73019-0628
p. (405) 325-3031 f. (405) 325-7069
ogs@ou.edu
http://www.ogs.ou.edu/

Director:
Charles J. Mankin, (D), Texas, 1958, Gg
Geologist:
Richard D. Andrews, (M), Wisconsin, 1976, Go
Brian J. Cardott, (M), S Illinois, 1981, Ec
James R. Chaplin, (M), Houston, 1962, Ps
Robert O. Fay, (D), Kansas, 1961, Gg
Kenneth V. Luza, (D), SD Sch of Mines & Tech, 1972, Ng
Assistant Director, OGS Publications:
Stanley T. Krukowski, (D), New Mexico Tech, 1990, En
Assistant Director, Geological Programs:
Neil H. Suneson, (D), California (Santa Barbara), 1980, Gc
Geologist:
Galen W. C. Miller, (M), Oklahoma, 2001, Gc
Chief Geophysicist:
James E. Lawson, (D), Tulsa, 1972, Ys
Geophysicist:
Raymon L. Brown, (D), MIT, 1975, Ye

Oregon

Dept of Geology & Mineral Industries

Oregon Dept of Geology and Mineral Industries (2007)
800 NE Oregon Street
Suite 965

Portland, OR 97232-2162
p. (971) 673-1555 f. (971) 673-1562
james.roddey@dogami.state.or.us
http://www.oregongeology.com

State Geologist:
Vicki S. McConnell, (D), Alaska (Fairbanks), 1996, Gv
Regional Geologist:
Mark L. Ferns, (M), 1978, Gg
Thomas J. Wiley, (M), Stanford, 1983, Gg
Chief Scientist:
Ian P. Madin, (M), Oregon State, Gg
Regional Geologist:
Jonathan C. Allen, (D), Canterbury (NZ), 1998, Gm
Regional Coastal geologist:
Rob Witter, (D), Oregon, 1999
Reclamationist:
Robert Brinkmann, Ge
Nancy Collins, Ge
Randall Moore, Ge
Ben Mundie, Ge
E. Frank Schnitzer, (M), 1982, Ge
Industrial Minerals Geologist:
Clark Neuendorp, (M), Eg
Geotechnical Earthquake Engineer:
Yumei Wang, (M), California (Berkeley), 1988, Ne
Geologist:
Lina Ma, (M)
Field Geologist:
Jason McClaughry, (M), WSU, 2003
Engineering Gelogist:
Bill Burns, (M), Portland State, 1999
Natural Resource Specialist, Oil and Gas Program:
Robert Houston, Ge
Earth Sciences Information Officer:
James Roddey, (B), Wofford, Oe
Assistant Director - MLRR:
Gary Lynch, Ge
Assistant Director - Geologic Survey:
Don Lewis, (M)

Baker City Field Office (2004)
1510 Campbell St
Baker City, OR 97814
p. (541) 523-3133 f. (541) 523-5992
mark.ferns@dogami.state.or.us
Office Specialist: Jan Durflinger

Regional Geologist:
Mark L. Ferns, (M), Oregon, 1979, Oo
Field Geologist:
Vicki S. McConnell, (D), Alaska (Fairbanks), 1995, Oo

Coastal Field Office (2006)
PO Box 1033
Newport, OR 97365
p. 541-574-6658 f. 541-265-5241
Jonathan.Allan@dogami.state.or.us

Regional Geologist:
Rob Witter, (D), Oregon, 1999, Og
Geologist:
George R. Priest, (D), Oregon State, 1980, Gg
Coastal Section Supervisor:
Jonathan C. Allen, (D), Canterbury (NZ), 1998, Gm

Grants Pass Field Office (2004)
5375 Monument Drive
Grants Pass, OR 97526-8513
p. (541) 476-2496 f. (541) 476-2496
tom.wiley@dogami.state.or.us
Department Secretary: Kathleen McGee

Resident Geologist:
Frank Hladky, (M), Idaho State, 1986, Gg
Regional Geologist:
Tom Wiley, (M), Stanford, 1983, Gg

Pennsylvania

Bureau of Topo & Geologic Survey

DCNR-Pennsylvania Geological Survey (2004)
3240 Schoolhouse Road
Middletown, PA 17057-3534
p. (717) 702-2017 f. (717) 702-2065
http://www.dcnr.state.pa.us/topogeo
Administrative Assistant: Elizabeth C. Lyon
Administrative Officer: Lynn M. Goodling

Bureau Director & State Geologist:
Jay B. Parrish, (D), Penn State, 1985, Yg
Assistant Director:
Samuel W. Berkheiser, Jr., (M), E Kentucky, 1974, Eg
Geologist Supervisor:
Rodger T. Faill, (D), Columbia, 1966, Gc
John A. Harper, (D), Pittsburgh, 1977, Pi
Michael E. Moore, (B), Penn State, 1975, Hw
Robert C. Smith, (D), Penn State, 1973, Cg
Geologist Manager:
Christine E. Miles, (M), Franklin & Marshall, 1971, Gg
Senior Geologic Scientist:
John H. Barnes, (M), SUNY (Buffalo), 1972, Gg
Gale C. Blackmer, (D), Penn State, 1992, Gc
Kristin M. Carter, (M), Lehigh, 1993, Go
Clifford H. Dodge, (M), Northwestern, 1976, Gr
Gary M. Fleeger, (M), Illinois, 1980, Hw
William E. Kochanov, (M), West Virginia, 1983, Gg
Christopher D. Laughrey, (B), Pittsburgh (Johnstown), 1977, Co
Antonette K. Markowski, (M), S Illinois, 1990, Ec
Thomas A. McElroy, (M), Massachusetts, 1987, Hw
Caron O'Neil, (M), Pittsburgh, 1986, Gg
James R. Shaulis, (M), Penn State, 1985, Ec
Thomas G. Whitfield, (B), West Virginia, 1973, Gg
Geologist Manager:
Jon D. Inners, (D), Massachusetts, 1975, Gg
IT Generalist Administrator:
Kristin J.H. Warner, (M), Penn State, 1993, Oo
Senior Geologic Scientist:
Viktoras W. Skema, (B), Michigan Tech, 1972, Gr
Geologic Scientist:
Karen L. Andrachick, (M), California (Penn), 1996, Gg
Helen L. Delano, (M), SUNY (Binghamton), 1979, Gg
Jaime Kostelnik, (M), Wright State, 2001, Gg
Anne B. Lutz, (D), Penn State, 1973, Pg
John C. Neubaum, (B), Kutztown, 1983, Ec
Stuart O. Reese, (M), Tennessee, 1986, Oi
Kristen L. Reinertsen, (B), Nicholls State, 1999, Eg
Joseph R. Tedeski, (B), Indiana (Penn), 1992, Eo
Lab Tech:
Leslie T. Chubb, (U), Oo
IT Generalist Administrator:
Cheryl L. Cozart, (A), Point Park, 1967, Oo
IT Generalist:
Janice A. Hayden, (B), Robert Morris, 2001, Oo
Cartographer:
Lajos J. Balogh, (A), Columbia Tech, 1966, Oo
John G. Kuchinski, Oo
Geology Librarian:
Richard Cameron Keen, (M), Villanova, 1982, Oo

Puerto Rico

Puerto Rico Bureau of Geology

Dept of Natural & Environmental Resources (2004)
Apartado 9066600
Puerta de Tierra Station
San Juan, PR 00906-6600
p. (787) 722-2526 f. (787) 723-4255

Director:
Vanessa del S. Rodriguez, Oo

Rhode Island

University of Rhode Island

Rhode Island Geological Survey (2007)
Department of Geosciences
340 Woodward Hall
9 E Alumni Rd
Kingston, RI 02881
p. (401) 874-2265/2191 f. (401) 874-2190
jon_boothroyd@uri.edu, rigsurv@etal.uri.edu
www.uri.edu/cels/geo/ri_geological_survey.htm
Administrative Assistant: Cheryl Grasso

State Geologist:
Jon C. Boothroyd, (D), South Carolina, 1974, GslOn

South Carolina

South Carolina Geological Survey

Dept of Natural Resources (2004)
5 Geology Road
Columbia, SC 29210-4089
p. (803) 896-7700 f. (803) 896-7695
scgs@dnr.state.sc.us
http://www.dnr.sc.gov/geology/
Department Secretary: Agnes K. Streeter

Director:
Charles W. Clendenin, Jr., (D), Witwatersrand, 1989, Eg

Professional Staff:
William R. Doar, III, (M), East Carolina, 1998, Eg
Arthur H. Maybin, III, (M), Georgia, 1976, Eg
Clark A. Niewendorp, (M), W Michigan, 1985, Eg
Paul G. Nystrom, Jr., (M), South Carolina, 1972, Eg
Ralph H. Willoughby, (M), Virginia Tech, 1977, Eg

South Dakota

Dept of Env & Natural Resources

Geological Survey Program (2004)
Akeley-Lawrence Science Center
University of South Dakota
414 East Clark Street
Vermillion, SD 57069-2390
p. (605) 677-5227 f. (605) 677-5895
http://www.sdgs.usd.edu

State Geologist:
Derric L. Iles, (M), Iowa State, 1977, Hw

Natural Resources Administrator:
Tim C. Cowman, (M), South Dakota, 1998, Cg

Geology Specialist:
Dennis W. Tomhave, (B), St. Cloud State, 1976, Gl

Hydrology Specialist:
Stan F. Pence, (M), North Dakota, 1996, Hw
J. Foster Sawyer, (M), SD Sch of Mines & Tech, 1990, Hw

Senior Geologist:
Sarah A. Chadima, (M), Iowa State, 1982, Gg
Mark D. Fahrenbach, (D), SD Sch of Mines & Tech, 1995, Gg
Ann R. Jensen, (B), South Dakota, 1982, Gg
Kelli A. McCormick, (D), Johns Hopkins, 1996, Gg
Layne D. Schulz, (B), South Dakota, 1985, Gl

Senior Hydrologist:
Dragan Filipovic, (M), Belgrade (Yugoslavia), 1985, Hw
Thomas B. Rich, (M), Utah State, 1981, Hw

Tennessee

Tennessee Division of Geology

Dept of Environment & Conservation (2007)
Division of Geology
13th Floor, L & C Tower, 401 Church Street
Nashville, TN 37243
p. (615) 532-1500 f. (615) 532-1517
Ronald.Zurawski@state.tn.us
http://www.state.tn.us/environment/tdg

State Geologist:
Ronald P. Zurawski, (M), Vanderbilt, 1973, GgEog

Geologist 4:
Elaine P. Foust, (B), Austin Peay, 1979, OiEoGg
Albert B. Horton, (M), Vanderbilt, 1981, GgOi

Chief Geologist:
Peter J. Lemiszki, (D), Tennessee, 1992, GcOiGg

Assistant State Geologist:
Michael L. Hoyal, (B), Tennessee, 1979, Gog

Geologist 3:
Vince Antonacci, (M), Ball State, 1987, GgOi
Ronald J. Clendening, (B), Tennessee Tech, 1986, GgOi
Martin S. Kohl, (M), California (Los Angeles), 1978, GgOi
Barry W. Miller, (M), Tennessee, 1989, EcOiGg
Erin F. Sutton, (M), Radford, 2001, GgOi

Secretary:
Juanita C. Griggs
Becky M. Hawkins

Texas

University of Texas at Austin

Bureau of Economic Geology (2005)
University Station, Box X
Austin, TX 78713-8924
p. (512) 471-1534 f. (512) 471-0140
begmail@beg.utexas.edu
http://www.beg.utexas.edu
Administrative Assistant: Lynda A. Miller
Administrative Assistant: Glynis G. Morse
Executive Assistant: Wanda L. LaPlante

Director:
Scott W. Tinker, (D), Colorado, 1996, Go

Associate Director:
Ian Duncan, (D), British Columbia, 1982, Gg
Jay P. Kipper, (B), Trinity (San Antonio), 1983, Oo
Eric C. Potter, (M), Oregon State, 1975, Eo

Senior Research Fellow:
F. Jerry Lucia, (M), Minnesota, 1954, Go

Senior Scientist:
Shirley P. Dutton, (D), Texas, 1986, Gd
Bob A. Hardage, (D), Oklahoma State, 1967, Yg
Martin P. A. Jackson, (D), Cape Town, 1976, Gc
Charles Kerans, (D), Carleton, 1982, Gd
Stephen E. Laubach, (D), Illinois, 1986, Gc
Bob Loucks, (D), Texas, 1976, Go
Stephen C. Ruppel, (D), Tennessee, 1979, Gd
Bridget R. Scanlon, (D), Kentucky, 1985, Hw

Research Scientist:
William A. Ambrose, (M), Texas, 1983, Gs
Alan R. Dutton, (D), Texas, 1982, Hw

Sergey Fomel, (D), Stanford, 2002, Yx
Edgar H. Guevara, (D), Texas, 1974, Go
Susan D. Hovorka, (D), Texas, 1990, Gd
Mike Hudec, (D), Wyoming, 1990, Gg
James W. Jennings, Jr., (D), Texas A&M, 1983, Oo
Jeffrey G. Paine, (M), Washington, 1991, Gr
Jay A. Raney, (D), Texas, 1976, Gd
Daniel D. Schultz-Ela, (D), Minnesota, 1988, Gc
Bruno C. Vendeville, (D), Rennes (France), 1987, Gc
Fred P. Wang, (D), Stanford, Np
William A. White, (M), Texas, 1972, Ge
Lesli J. Wood, (D), Colorado State, 1992, Gr
Hongliu Zeng, (D), Texas, 1994, Gs

Senior Research Fellow:
Robert J. Graebner, (M), Colorado, 1953, Oo
Donald E. Wagner, (D), St. Louis, 1972, Yg

Research Professor:
Milo Backus, (D), 1956, Oo
L. Frank Brown, (D), Wisconsin, 1955, Gg

Database Coordinator:
Lisa E. Remington, (B), Texas, 1992, Gg

Research Associate:
Beverly Blakeney DeJarnett, (M), Penn State, 1986, Gg
Renaud Bouroullec, (D), Imperial (London), 2001, Gc
Alan L. Brown, (D), Louisiana State, 2002, Gg
Sigrid J. Clift, (B), Texas, 1989, Gg
Edward W. Collins, (M), Stephen F. Austin, 1978, Ge
Luciano Correa, (M), Texas, 2002, Eo
Micheal V. DeAngelo, (M), Texas (El Paso), 1988, Ye
Khaled Fouad, (B), Alexandria (Egypt), 1982, Yg
James C. Gibeaut, (D), S Florida, 1991, Gu
Roberto Gutierrez, (D), Texas, 1990, Yd
Tucker F. Hentz, (M), Kansas, 1982, Go
Mark H. Holtz, (B), Wisconsin, 1983, Gg
Xavier Janson, (D), Miami, 2002, Gr
Dave Jennette, (M), Cincinnati, 1986, Gg
Jeffrey A. Kane, (M), Texas (Permian), 1993, Yg
Eugene M. Kim, (D), Texas, 1998, Eg
Paul R. Knox, (M), Cal State (Long Beach), 1989, Gg
Jean-Philippe Nicot, (D), Texas, 1998, Ng
Robert C. Reedy, (M), New Mexico Tech, 1996, Hg
Shinichi Sakurai, (B), Tohoku, 1971, Go
Diana Sava, (D), Stanford, 2004, Yg
Paul Sava, (D), Stanford, 2004, Yg
Julia Stowell Gale, (D), Exeter (UK), 1987, Gc
Mark Tomasso, (D), Birmingham (UK), 2001, Gg
Thomas A. Tremblay, (M), Texas, 1992, Oy
Ramon H. Trevino, (M), Texas (Arlington), 1988, Go
Joseph S. Yeh, (B), Fu-Jen Catholic, 1977, Gm
Laura Zahm, (M), Texas, 1997, Gg

Postdoc:
Tim Dooley, (D), London (UK), 1994, Gc
David Pyles, (D), Colorado, 2004, Gg

Research Scientist Associate:
John R. Andrews, (B), North Carolina, 1990, Gg
Jerome A. Bellian, (M), Miami, 1998, Gs
Florence Bonnaffe, (M), Pierre & Marie Curie, 2002, Yg
Caroline Breton, (B), Texas, 2001, Oi
Dallas B. Dunlap, (B), Texas, 1997, Ye
Ursula Hammes, (D), Colorado, 1992, Gg
Tiffany Hepner, (M), S Florida, 2000, Og
Rebecca H. Jones, (M), Wyoming, 1999, Ye
Paul Murray, (M), Colorado Mines, 1999, Yx
Seay Nance, (M), Texas, 1988, Gg
Robert Reed, (D), Texas, 1999, Gc
Randy L. Remington, (B), Texas, 1992, Ye
Rebecca C. Smyth, (M), Texas, 1995, Hg
John Andrew Tachovsky, (M), Gg
Rachel Waldinger, (M), Texas, 2000, Or

Research Fellow:
Luis Sanchez-Barreda, (D), Texas, 1981, Gg

Media Technology Interim Manager:
Joel L. Lardon, (B), SW Texas State, 1984, Gg

Manager, Computer Group:
Ron Russell, (M), Oklahoma, 1993, Oo

Editor in Charge:
Susann V. Doenges, (M), Texas, 1982, Oo

Bldg. Facilities/Core Research Center:
George T. Bush, (B), Texas Tech, 1971, Gg

Utah

Utah Geological Survey

Utah Dept of Natural Resources (2004)
1594 West North Temple, Suite 3110
P.O. Box 146100
Salt Lake City, UT 84114-6100
p. (801) 537-3300 f. (801) 537-3400
nrugs.rallis@state.ut.us
http://ugs.utah.gov
Administrative Assistant: Nicole C. King
Department Secretary: Cheryl Ostlund

Director:
Richard G. Allis, (D), Toronto, 1977, Eo

Acting Director:
Kimm M. Harty, (M), Alberta, 1984, Gm

Associate Director:
John Kingsley, Oo

Senior Geologist:
Michael O. Hylland, (M), Oregon State, 1990, Ng

Budget & Accounting Officer:
Daniel Kelly, (B), Westminster, 1984, Oo

Accounting Technician:
Linda R. Bennett, Oo

Energy and Mineral Resources Program (2004)
1594 West North Temple, Ste 3110
Box 146100
Salt Lake City, UT 84114-6100
p. (801) 537-3373 f. (801) 537-3400
Administrative Assistant: Cheryl Gustin

Economic Program Manager:
David E. Tabet, (M), Wisconsin, 1974, Ec

Senior Geologist:
Robert W. Gloyn, (M), Princeton, 1967, Em
J. Wallace Gwynn, (D), Utah, 1970, En

Manager, Minerals Section:
Bryce T. Tripp, (B), Utah, 1978, En

Geologist:
Robert E. Blackett, (M), Utah, 1979, Eg
Roger L. Bon, (B), Utah State, 1968, Eg
Craig D. Morgan, (B), Utah, 1975, Eo
Jeffrey Quick, (D), Canterbury, 1992, Ec

Senior Scientist:
Thomas C. Chidsey, (M), Brigham Young, 1977, Eo

Geotechnician:
Kevin P. McGlure, (B), Utah, 1996, Og
Carolyn Olsen, Oo

Environmental Sciences Program (2004)
1594 West North Temple, Ste 3110
PO Box 146100
Salt Lake City, UT 84114-6100
p. (801) 537-3389 f. (801) 537-3400

Manager:
Mike Lowe, (M), Utah State, 1985, Hw

Geologist:
Charles E. Bishop, (B), Utah, 1986, Hw
Hugh A. Hurlow, (D), Washington, 1982, Hw
Janae Wallace, (M), N Arizona, 1993, Hw
Senior Geologist:
James I. Kirkland, (D), Colorado, 1990, Pg
Senior Scientist:
David B. Madsen, (D), Missouri, 1973, Ga
Research Associate:
Martha C. Hayden, (B), Utah, 1978, Pv
Geotechnician:
Alison Corey, Oo

Geologic Hazards Program (2004)
1594 West North Temple
P O Box 146100
Salt Lake City, UT 84114-6100
p. (801) 537-3304 f. (801) 537-3400
Manager:
Gary E. Christenson, (M), Arizona State, 1976, Ng
Geologist:
Francis X. Ashland, (M), Massachusetts, 1991, Ng
Richard E. Giraud, (M), Idaho, 1986, Ng
Barry J. Solomon, (M), California (Santa Barbara), 1974, Ng
Senior Scientist:
William R. Lund, (B), Idaho, 1970, Ng
GIS Analyst:
Neil D. Storey, (B), Weber State, 2001, Oo
Geotechnician:
Gregory N. McDonald, (B), Utah, 1992, Oo

Geologic Information and Outreach Program (2004)
1594 West North Temple Ste 3110
PO Box 146100
Salt Lake City, UT 84114-6100
p. (801) 537-3325 f. (801) 537-3400
Manager:
Sandra N. Eldredge, (B), Skidmore, 1978, Oe
Geologist:
William F. Case, (B), Westminster, 1967, Ng
Mark R. Milligan, (M), Utah, 1995, Gg
Christine M. Wilkerson, (B), Utah, 1985, Gg
Executive Secretary:
Jo Lynn Camplbell, Oo
Bookstore Manager:
Patricia Stokes, Oo
Geology Librarian:
Mage Yonetani, Oo

Geologic Mapping Program (2004)
1594 West North Temple, Ste 3110
PO Box 146100
Salt Lake City, UT 84114-6100
p. (801) 537-3355 f. (801) 537-3400
Manager:
Grant C. Willis, (M), Brigham Young, 1983, Gg
Geologist:
Robert Biek, (M), N Illinois, 1988, Gg
Jonathan K. King, (M), Wyoming, 1984, Gg
Senior Scientist:
Hellmut H. Doelling, (D), Utah, 1964, Gg
GIS Analyst:
Basia Matyjasik, (B), Warsaw, 1988, Gg
Geotechnician:
Kent D. Brown, Oo

Vermont

Dept of Environmental Conservation

Vermont Geological Survey (2005)
103 South Main Street
The Laundry Building
Waterbury, VT 05671-0301
p. (802) 241-3608 f. (802) 241-3273
laurence.becker@anr.state.vt.us
http://www.anr.state.vt.us/geology/vgshmpg.htm
State Geologist:
Laurence Becker, (M), Vermont, 1980, Gg
Geologist-Env Scientist:
Marjorie H. Gale, (M), Vermont, 1980, Gg
Geologist:
Jonathan Kim, (D), Buffalo, 1995, Gg

Virginia

Division of Mineral Resources

Virginia Dept of Mines, Minerals & Energy (2007)
Fontaine Research Park
900 Natural Resources Drive
Suite 500
Charlottesville, VA 22903
p. 434-951-6350 f. (434) 951-6366
ed.erb@dmme.virginia.gov
http://www.dmme.virginia.gov/dmr
State Geologist:
Edward E. Erb, (D), New Mexico, 1979, GvEg
Manager, Geological Info & Support Services:
Michael L. Upchurch, (M), North Carolina, 1972, Gd
Manager, Geologic Mapping:
Matthew J. Heller, (M), North Carolina State, 1996, Gg
Manager, Economic Geology:
William L. Lassetter, (M), Nevada (Reno), 1996, Hw
Professional Staff:
Carl R. Berquist, Jr., (D), William & Mary, 1986, Gr
Elizabeth V. M. Campbell, (M), Virginia Tech, 1985, Gg
Mark W. Carter, (M), Tennessee, 1994, Gr
Cathy B. Enomoto, (B), Virginia Tech, 1981, Eo
Amy K. Gilmer, (M), Texas, 2001, Eg
David A. Hubbard, Jr., (M), Virginia, 1977, Ge
James A. Lovett, (B), Wittenberg, 1978, Eg
Alfred R. Taylor, (B), North Carolina, 1955, Gr
William W. Whitlock, (M), Brigham Young, 1983, Gr
Gary F. Wilder, (B), Ball State, 1972, Gg
Gerald P. Wilkes, (B), Ashland (Ohio), 1974, Gg
Scott T. Williams, (M), Tennessee, 2000, Gc
Publication Sales Specialist:
Delores J. Green, Oo
Geologist Tech Specialist:
E. Wayne Marshall, Oo

Southwestern Minerals & Geology Section (2004)
P.O. Box 144
453 West Main St
Abingdon, VA 24210
p. (540) 676-5577 f. (540) 676-5555

Washington

Division of Geology & Earth Resources

Washington Dept of Natural Resources (2006)
1111 Washington Street, SE, MS 47007
Olympia, WA 98504-7007
p. (360) 902-1450 f. (360) 902-1785
geology@wadnr.gov

http://www.dnr.wa.gov/ geology/
Acting State Geologist:
Ron Teissere, (N), Gg
Geology Librarian:
Lee Walkling, (M), California (Los Angeles), 1970, Gg
Assistant State Geologist:
David K. Norman, (M), Utah, 1980, Gg

West Virginia

Geological & Economic Survey

West Virginia Geological & Economic Survey (2005)
Mont Chateau Research Center
1 Mont Chateau Road
Morgantown, WV 26508-8079
p. (304) 594-2331 f. (304) 594-2575
info@geosrv.wvnet.edu
http://www.wvgs.wvnet.edu
State Geologist:
Michael Ed. Hohn, (D), Indiana, 1946, Gq

Wisconsin

University of Wisconsin - Extension

Wisconsin Geological and Nat History Survey (2007)
3817 Mineral Point Road
Madison, WI 53705-5100
p. (608) 262-1705 f. (608) 262-8086
askageologist@uwex.edu
http://www.uwex.edu/wgnhs/
Department Secretary: Rilla M. Hinkes
Director:
James M. Robertson, (D), Michigan, 1972, Eg
Assistant Director:
Thomas J. Evans, (D), Wisconsin, 1994, Eg
Professor:
John W. Attig, (D), Wisconsin, 1984, Gl
Kenneth R. Bradbury, (D), Wisconsin, 1982, Hw
Frederick W. Madison, (D), Wisconsin, 1972, So
Senior Scientist:
Bruce A. Brown, (D), Manitoba, 1984, Gc
Associate Professor:
Madeline B. Gotkowitz, (M), New Mexico Tech, 1993, Hw
David J. Hart, (D), Wisconsin, 2000, Hw
Thomas S. Hooyer, (D), Iowa State, 1999, Gl
Assistant Professor:
Patrick I. McLaughlin, (D), Cincinnati, 2006, Gr
Emeritus:
Lee Clayton, (D), Illinois, 1965, Gl
Ronald G. Hennings, (M), Wisconsin, 1977, Hw
Stanley A. Nichols, (D), Wisconsin, 1971, Oo
Alexander Zaporozec, (D), Wisconsin, 1975, Hw
Administrative Program Manager:
Kathie M. Zwettler, (B), Edgewood Coll, 1994, Oo
Related Staff:
William G. Batten, (B), Wisconsin, 1973, Hw
Bill C. Bristoll, (M), Wisconsin (Milwaukee), 1995, Oo
Peter M. Chase, (B), Wisconsin (Milwaukee), 1985, Oo
Michael L. Czechanski, (M), Wisconsin, 1977, Oi
Donna M. Duffey, Oo
Susan L. Hunt, (B), Mankato State, 1966, Oo
Mindy C. James, (B), Illinois, 1975, Oo
Marcia J. Jesperson, Oo
Kathy A. Kane, (B), San Francisco State, 1972, Oo
Irene D. Lippelt, (B), Manitoba, 1978, Gg
Deborah L. Patterson, (B), Wisconsin, 1984, Oi
Roger M. Peters, (B), Wisconsin, 1969, Gg
Kathy Campbell Roushar, (B), Iowa, 1979, Oi
Apichart Santipiromkul, (B), Thailand, 1997, Oo
Peter R. Schoephoester, (M), Indiana, 1999, Oi
Virginia L. Trapino

Wyoming

Wyoming State Geological Survey

Wyoming State Geological Survey
PO Box 1347
Laramie, WY 82073
p. (307) 766-2286 f. (307) 766-2605
wsgs-info@uwyo.edu
http://www.wsgs.uwyo.edu/
State Geologist:
Ronald C. Surdam, (D), California (Los Angeles), 1967, Go
Manager, Regional Geology:
Alan J. Ver Ploeg, (M), Iowa State, 1973, Gr
Manager, Energy & Water Resources:
Rodney H. De Bruin, (M), Iowa State, 1975, Eo

Federal Agencies

Argonne National Laboratory

Chemical Technology Div (2004)
9700 South Cass Avenue
Building 205
Argonne, IL 60439
p. (630) 252-4383 f. (630) 252-5528
lewisd@cmt.anl.gov

Senior Scientist:
Milton Blander, (D), Yale, 1953, Xc
Associate Scientist:
Allen J. Bakel, (D), Oklahoma, 1990, Co
Ronald P. Chiarello, (D), Northeastern, 1990, Om
Donald G. Graczyk, (D), Wisconsin, 1975, Cs
Ben D. Holt, (D), Illinois Inst of Tech, 1969, Ca
James J. Mazer, (D), Northwestern, 1987, Cl
David J. Wronkiewicz, (D), New Mexico Tech, 1989, Cg
Adjunct Professor:
Neil C. Sturchio, (D), Washington (St. Louis), 1983, Cg
Related Staff:
Alice M. Essling, (B), St. Xavier, 1956, Ca
Edmund A. Huff, (M), Chicago, 1957, Ca
Francis J. Markun, (B), Lewis, 1961, Cc
Florence P. Smith, (B), Southern, 1968, Ca

Chemistry Div (2005)
9700 South Cass Avenue
Argonne, IL 60439
p. (630) 972-3570 f. (630) 972-5287
gruen@anlchm.chm.anl.gov

Senior Scientist:
Dieter M. Gruen, (D), Chicago, 1951, Om
Michael J. Pellin, (D), Illinois, 1978, Om
Associate Scientist:
Wallis F. Calaway, (D), Indiana, 1975, Om

Energy Systems Div (2005)
9700 South Cass Avenue
Bldg 362 E-340
Argonne, IL 60439-4815
p. (630) 252-3392 f. (630) 252-7288
energy_systems@anl.gov
Administrative Assistant: Barbara Sullivan

Head:
Donald O. Johnson, (D), Illinois, 1972, Gr
Senior Scientist:
Lyle D. McGinnis, (D), Illinois, 1965, Yg
Scientist:
Kenneth L. Brubaker, (D), Wisconsin, 1972, Oa
Dorland E. Edgar, (D), Purdue, 1976, Gm
Steven F. Miller, (B), Knox, 1984, Gg
R. Eric Zimmerman, (D), Northwestern, 1972, Ng
Research Associate:
Paul C. Heigold, (D), Illinois, 1969, Yg
Theresa C. Scholtz, (B), N Illinois, 1985, Ge
Michael D. Thompson, (M), N Illinois, 1989, Yg
Related Staff:
John F. Schneider, (M), N Illinois, 1977, Ca
Linda M. Shem, (M), Northwestern, 1991, Ge
Patrick L. Wilkey, (M), Illinois, 1976, Ng

Energy Technology Div (2004)
9700 South Cass Avenue
Energy Tech Bldg 212
Argonne, IL 60439
p. (630) 252-4930 f. (630) 252-4798
b_baudino@anl.gov
Administrative Assistant: Bonnie K. Baudino

Materials Engineer:
Amy B. Hull, (D), Northwestern, 1987, Cl

Environmental Research Div (2005)
9700 South Cass Avenue
Argonne, IL 60439
p. (630) 252-3879 f. (630) 252-2959
bmlesht@anl.gov
http://www.anl.gov/ER/

Senior Scientist:
Jeffrey S. Gaffney, (D), California (Riverside), 1975, Oa
Raymond M. Miller, (D), Illinois State, 1975, Sb
Marvin L. Wesely, (D), Wisconsin, 1970, Oa
Associate Scientist:
Jacqueline C. Burton, (D), Tennessee, 1978, Cg
Richard L. Coulter, (D), Penn State, 1976, Oa
Paul V. Doskey, (D), Wisconsin, 1982, Oa
Paul A. Fenter, (D), Pennsylvania, 1990, Gy
Julie D. Jastrow, (D), Chicago, 1994, Sb
Lorraine M. LaFreniere, (D), Wisconsin, 1980, Oo
In Young Lee, (D), California (Los Angeles), 1975, Oa
Barry M. Lesht, (D), Chicago, 1977, Op
Nancy A. Marley, (D), Florida State, 1984, Oa
William T. Meyer, (D), Imperial Coll (UK), 1973, Cg
Robert A. Sedivy, (M), Georgia Tech, 1979, Hw
Jack D. Shannon, (D), Oklahoma, 1975, Ow
Douglas L. Sisterson, (M), Wyoming, 1975, Oa
Mohamed Sultan, (D), Washington (St. Louis), 1984, Cg
John L. Walker, (D), Imperial Coll (UK), 1964, Cg
Y Eugene Yan, (D), Ohio State, 1998, Gg
Research Associate:
Richard H. Becker, (M), Washington (St. Louis), Or
David R. Cook, (M), Penn State, 1977, Ow
Clyde B. Dennis, (M), Florida, 1979, Oo
Richard L. Hart, (B), Illinois Inst of Tech, 1970, Oa
Timothy J. Martin, (B), Beloit, 1974, Oa
Barney W. Nashold, (M), Illinois (Chicago Circle), 1976, Oo
Kent A. Orlandini, (B), Illinois, 1957, Cc
Candace M. Rose, (B), Benedictine Coll, 1985, Oo

Geosciences & Information Technology Section (2006)
9700 South Cass Avenue
Argonne, IL 60439
p. (630) 252-6034 f. (630) 252-3611
djmiller@anl.gov
Department Secretary: Sue Baumann

Manager, Geosciences & Information Tech:
Lisa A. Durham, (M), Purdue, 1989, Hw
Assistant System Engineer:
Cheong-yip R. Yuen, (D), Wisconsin (Milwaukee), 1986, Ge
Associate Scientist:
Robert L. Johnson, (D), Cornell, 1991, Oo
Research Associate:
John Ditmars, (D), Caltech, Oo
Jennifer Herbert, (B), N Illinois, 1997, Gg
Zhenhua Jiang, (D), Duke, 1992, Oo
David S. Miller, (D), Johns Hopkins, 1995, Ge
Terri L. Patton, (M), NE Illinois, 1989, Cl
John J. Quinn, (M), Minnesota, 1992, Hq
Gus T. Williams, (D), Northwestern, 1994, Hq
Geology Librarian:
Swati Wagh, Oo

Dept of Agriculture

Office of the Secretary (2004)
1400 Independence Avenue, SW
Washington, DC 20250
p. (202) 720-2791
http://www.usda.gov

Agricultural Research Service (2004)
Beltsville, MD 20705
p. (202) 720-3656
http://www.ars.usda.gov

Forest Service (2004)
P.O. Box 96090
Washington, DC 20090-6090
p. 202-720-USDA
http://www.fs.fed.us

Natural Resources Conservation Service (2004)
14th and Independence Ave., SW
Washington, DC 20250
p. (202) 720-7246
http://www.nrcs.usda.gov

Office of the Chief Economist (2004)
1400 Independence Avenue, SW
Washington, DC 20250
p. (202) 720-4164

Dept of Commerce

Office of the Secretary (2004)
14th Street & Constitution Avenue, NW
Washington, DC 20230
p. (202) 219-3605
http://www.doc.gov

National Oceanic & Atmospheric Administration (2004)
Silver Spring Metro Center 3
1315 East-West Highway
Silver Spring, MD 20910-3282
p. (202) 482-3436
http://www.noaa.gov

National Geophysical Data Center (2004)
David Skaggs Research Center
325 Broadway
Boulder, CO 80303
p. (303) 497-6215
http://www.ngdc.noaa.gov/

National Inst of Standards & Technology (2004)
100 Bureau Drive
Stop 3460
Gaithersburg, MD 20899-3460
p. (301) 975-2758 f. (301) 948-3716
http://www.nist.gov/

Dept of Defense

Office of the Secretary (2004)
The Pentagon
Washington, DC 20301
p. (703) 545-6700
http://www.defenselink.mil

Dept of the Air Force (2004)
The Pentagon
Washington, DC 20330-1000
p. (703) 697-6061 f. (703) 696-7273
http://www.af.mil/

U.S. Special Operations Command (Air Force) (2004)
100 Bartley Street
Hurlburt Field, FL 32544-5273
p. (850) 884-2323
http://www.af.mil/sites/afsoc.html

Air Education and Training Command (2004)
One F Street
Randolph Air Force Base, TX 78150-4324
p. (210) 652-5512
http://www.aetc.af.mil

Headquarters, Air Force Reserve (2004)
1150 Air Force
The Pentagon
Washington, DC 20330-1150
p. (703) 695-9225
http://www.re.hq.af.mil/

Pacific Air Forces (2004)
The Pentagon
Washington, DC 20330-1000
p. (703) 697-6061 f. (703) 696-7273
http://www.af.mil/

Electronic Security Command (2004)
The Pentagon
Washington, DC 20330-1000
p. (703) 697-6061 f. (703) 696-7273
http://www.af.mil/

Air Force Operational Test & Evaluation Center (2004)
1650 Air Force
The Pentagon
Washington, DC 20330-1150
p. (703) 697-4774

Air Combat Command (2004)
205 Dodd Boulevard
Langley Air Force Base, VA 23665-2788
p. (757) 764-3204
http://www.acc.af.mil

Air Mobility Command (2004)
402 Scott Drive, Unit 3EC
Scott Air Force Base, IL 62225-5310
p. (618) 229-3205
http://www.af.mil/sites/amc.html

Air Force Materiel Command (2004)
4375 Chidlaw Road
Wright-Patterson Air Force Base, OH 45433-7129
p. (937) 257-6033

Space Command (2004)
150 Vandenberg Street
Peterson Air Force Base, CO 80914-4020
p. (719) 554-3001
http://www.af.mil/sites/afspc.shtml

Air National Guard Support Center (2004)
Jefferson Plaza I
1411 Jefferson Davis Highway
Arlington, VA 22202-3259
p. (703) 607-2388
http://www.ang.af.mil

Air Weather Service (2004)
106 Peacekeeper Drive
Offutt Air Force Base, NE 68113-4039
p. (402) 294-5749

Air Force Center for Environmental Excellence (2004)
3207 North Road
Brooks Air Force Base, TX 78235-5363
p. (210) 536-2162

Air National Guard Units (Mobilization) (2004)
Jefferson Plaza I

1411 Jefferson Davis Highway
Arlington, VA 22202-3259
p. (703) 607-2388
http://www.ang.af.mil

Dept of the Army (2004)
The Pentagon
Washington, DC 20310
p. (703) 695-0363 f. (703) 614-5520
http://www.army.mil/csa/

U.S. Army Test & Evaluation Command (2004)
4501 Ford Avenue
Alexandria, VA 22302-1458
p. (703) 681-9360

U.S. Army Corps of Engineers (2004)
441 G Street, NW
Washington, DC 20314
p. (202) 761-0001
http://www.usace.army.mil/working.html

Immediate Offiof the Chief of Staff of the Army (2004)
200 Army
The Pentagon
Washington, DC 20310-0200
p. (703) 695-2077
http://www.army.mil/csa/

U.S. Army Forces Command (2004)
The Pentagon
Washington, DC 20310
p. (703) 695-0363 f. (703) 614-5520
http://www.army.mil/csa/

U.S. Army Medical Command (2004)
The Pentagon
Washington, DC 20310
p. (703) 681-8022
http://www.army.mil/csa/

Field Operations Off of Office of Secretary of Army (2004)
The Pentagon
Washington, DC 20310
p. (703) 695-0363 f. (703) 614-5520
http://www.army.mil/csa/

Fld Op AG of Army Staff Resourced OA-22 (2004)
The Pentagon
Washington, DC 20310
p. (703) 695-0363 f. (703) 614-5520
http://www.army.mil/csa/

U.S. Army Training & Doctrine Command (2004)
66 Ingalls Road, Building 27
Ft. Monroe, VA 23651-5000
p. (757) 788-3514
http://www-tradoc.monroe.army.mil

U.S. Army Chem & Bio Defense Command (2004)
The Pentagon
Washington, DC 20310
p. (703) 695-0363 f. (703) 614-5520
http://www.army.mil/csa/

U.S. Army Research Laboratory Command (2004)
The Pentagon
Washington, DC 20310
p. (703) 695-0363 f. (703) 614-5520
http://www.army.mil/csa/

U.S. Army Tank-Automotive & Armament Comm (2004)
The Pentagon
Washington, DC 20310
p. (703) 695-0363 f. (703) 614-5520
http://www.army.mil/csa/

Defense Threat Reduction Agency (2004)
8725 John T. Kingman Road
MS 6201
Ft. Belvoir, VA 22060-6201
p. (703) 767-4883
http://www.dtra.mil

Dept of the Navy (2004)
The Pentagon
Washington, DC 20350
p. (703) 697-9020
http://www.navy.mil

Immediate Off of the Chief of Naval Operations (2004)
The Pentagon
Washington, DC 20350-2000
p. (703) 695-5664 f. (703) 693-9408

Office of Naval Research (2004)
Ballston Towers #1
800 North Quincy Street
Arlington, VA 22203
p. (703) 696-4767
http://www.onr.navy.mil/

Naval Intelligence Command (2004)
4251 Suitland Road
Washington, DC 20395-5720
p. (301) 669-4000
http://www.nmic.navy.mil/

Naval Air Systems Command (2004)
47123 Buse Road
Unit #IPT Building 2272, Suite 540
Patuxent River, MD 20670-1547
p. (301) 757-7825
http://www.navair.navy.mil/

Naval Sea Systems Command (2004)
National Center #3
Zachary Taylor Building
2531 Jefferson Davis Highway
Arlington, VA 20360
p. (703) 602-1580 f. (703) 602-1454
http://www.navsea.navy.mil/homepage.html

Naval Facilities Engineering Command (2004)
Washington Navy Yard
1322 Patterson Avenue, SE
Washington, DC 20374-5065
p. (202) 685-9002
http://www.navfac.navy.mil/

U.S. Marine Corps (2004)
Columbia Pike & Southgate Road Arlington Annex
Arlington, VA 22204
p. (703) 614-2500
http://www.usmc.mil/

Space & Naval Warfare Systems Command (2004)
4301 Pacific Highway
San Diego, CA 92110-3127
p. (619) 524-7053 f. (619) 524-3469
http://www.spawar.navy.mil/

Naval Oceanography Command (2004)
U.S. Naval Observatory
34th Street & Massachusetts Avenue, NW
Washington, DC 20007
p. (202) 762-1020 f. (202) 693-1746
http://www.oceanographer.navy.mil

U.S. Pacific Fleet, Commander in Chief (2004)
The Pentagon
Washington, DC 20350
p. (703) 697-9020
http://www.navy.mil

Dept of Energy

Office of the Secretary (2004)
1000 Independence Ave., SW
Washington, DC 20585
p. 800-dial-DOE
http://www.ma.doe.gov/energy/comments.html

Coal & Power Systems (2004)
Forrestal Building
1000 Independence Avenue, SW
Washington, DC 20585-0320
p. (202) 586-1650 f. (202) 586-7085

Off of Coal, Nuclear, Electric & Alternate Fuels (2004)
COMSAT Building
950 L'Enfant Plaza, SW
Washington, DC 20024
p. (202) 287-7990 f. (202) 287-1933

Energy Information Administration (2004)
Forrestal Building
1000 Independence Avenue, SW
Washington, DC 20585-0601
p. (202) 586-8800 f. (202) 586-0329
http://www.eia.doe.gov/

Asst Secretary for Environment, Safety & Health (2004)
Forrestal Building
1000 Independence Avenue, SW
Washington, DC 20585-0119
p. (202) 586-6151 f. (202) 586-0956
http://www.eh.doe.gov/

Asst Secretary for Environmental Management (2004)
Forrestal Building
1000 Independence Avenue, SW
Washington, DC 20585-0113
p. (202) 586-7709 f. (202) 586-7757
http://www.em.doe.gov/

Assistant Secretary for Fossil Energy (2004)
Forrestal Building
1000 Independence Avenue, SW
Washington, DC 20585-0301
p. (202) 586-6660 f. (202) 586-7847
http://www.fe.doe.gov/

Office of Nuclear Energy, Science & Technology (2004)
Forrestal Building
1000 Independence Avenue, SW
Washington, DC 20585-0117
p. (202) 586-6630 f. (202) 586-8353
http://www.ne.doe.gov/

Office of Oil & Gas (2004)
Forrestal Building
1000 Independence Avenue, SW
Washington, DC 20585-0640
p. (202) 586-6012 f. (202) 586-9739

Asst Secty for Energy Efficiency & Renewable Energy (2004)
Forrestal Building
1000 Independence Avenue, SW
Washington, DC 20585-0121
p. (202) 586-9220 f. (202) 586-9260
http://www.eren.doe.gov/

Office of Resource Management (2004)
Forrestal Building
1000 Independence Avenue, SW
Washington, DC 20585-0620
p. (202) 586-3521 f. (202) 586-0552

Office of Science (2004)
Forrestal Building
1000 Independence Avenue, SW
Washington, DC 20585
p. (202) 586-5430 f. (202) 586-1009
http://www.sc.doe.gov/

Office of Energy Projects (2004)
Federal Energy Regulatory Commission
888 First Street, NE
Washington, DC 20426
p. (202) 219-2700 f. (202) 219-0205

Dept of Health & Human Services

Office of the Secretary (2004)
Hubert H. Humphrey Building
200 Independence Avenue, SW
Washington, DC 20201
p. (202) 690-7000
http://www.dhhs.gov

Agency for Toxic Substances & Disease Registry (2004)
1600 Clifton Road
Atlanta, GA 30333
p. (404) 639-7000
http://www.atsdr.cdc.gov/atsdrhome.html

Centers for Disease Control & Prevention (2004)
1600 Clifton Rd.
Atlanta, GA 30333
p. (404) 639-3535
http://www.cdc.gov

Dept of Justice

Office of the Attorney General (2004)
10th Street & Constitution Avenue, NW
Washington, DC 20530
p. (202) 514-2007
http://www.usdoj.gov

Federal Bureau of Investigation (2004)
J. Edgar Hoover Building
935 Pennsylvania Ave., NW
Washington, DC 20535-0001
p. (202) 324-3000

http://www.fbi.gov

Dept of Labor

Office of the Secretary (2004)
200 Constitution Avenue, NW
Washington, DC 20210
p. (202) 219-7316
http://www.dol.gov

Mine Safety & Health Administration (2004)
Balston Tower #3
4015 Wilson Boulevard
Arlington, VA 22203
p. (703) 235-1385
http://www.msha.gov

Dept of State

Intl Boundary & Water Commission, US & Mexico (2004)
The Commons
4171 North Mesa, Suite C-310
El Paso, TX 79902-1441
p. 800-262-8857

Dept of the Interior

Office of the Secretary (2004)
1849 C. Street N.W.
Washington, DC 20240
p. (202) 208-3171
http://www.doi.gov

Bureau of Land Management (2004)
Office of Public Affairs
1849 C Street, Room 406-LS
Washington, DC 20240
p. (202) 208-6731 f. (202) 208-5242
http://www.blm.gov

Bureau of Indian Affairs (2004)
1849 C. Street N.W.
Washington, DC 20240
p. (202) 208-7163
http://www.doi.gov/bureau-indian-affairs.html

Bureau of Reclamation (2004)
1849 C. Street N.W.
Washington, DC 20240
p. (202) 513-0501 f. (202) 513-0314
http://www.usbr.gov

U.S. Geological Survey (2004)
12201 Sunrise Valley Drive
Reston, VA 20192
p. (703) 648-4000 f. (703) 648-4454
http://www.usgs.gov

USGS Eastern Region (2004)
12201 Sunrise Valley Drive
Reston, VA 20192
p. (703) 648-6660

USGS Central Region (2004)
Denver Federal Center
P.O. Box 25046
Denver, CO 80225
p. (303) 236-5438

USGS Western Region (2004)
345 Middlefield Road
Menlo Park, CA 94025
p. (650) 329-5100

USGS Mid-continent Mapping Center (2004)
1400 Independence Road
Rolla, MO 65401
p. (573) 308-3500
http://www.mcmcweb.er.usgs.gov/

USGS Rocky Mountain Mapping Center (2004)
Denver Federal Center
P.O. Box 25046
Denver, CO 80225
p. (303) 202-4200
http://www.rockyweb.cr.usgs.gov/

USGS Western Mapping Center (2004)
345 Middlefield Road
Menlo Park, CA 94025
p. (650) 329-4251
http://www-wmc.wr.usgs.gov/

USGS Hydrologist of the Northeastern Region (2004)
12201 Sunrise Valley Drive
Reston, VA 20192
p. (703) 648-5828

USGS Hydrologist of the Central Region (2004)
Denver Federal Center
P.O. Box 25046
Denver, CO 80225
p. (303) 236-5021

USGS Hydrologist of the Southern Region (2004)
Spalding Woods Office Park
3850 Holcomb Bridge Road, Suite 160
Norcross, GA 30092
p. (770) 409-7700

USGS Hydrologist of the Western Region (2004)
345 Middlefield Road
Menlo Park, CA 94025
p. (650) 329-4422

National Park Service (2004)
1849 C. Street N.W.
Washington, DC 20240
p. (202) 208-4621 f. (202) 219-0910
http://www.nps.gov

U.S. Fish & Wildlife Service (2004)
1849 C. Street N.W.
Washington, DC 20240
p. (202) 208-4545 f. (202) 208-6965
http://www.fws.gov/

Off of Surface Mining, Reclamation & Enforcement (2004)
1849 C. Street N.W.
Washington, DC 20240
p. (202) 208-4006 f. (202) 219-3106
http://www.osmre.gov/

Minerals Management Service (2004)
1849 C. Street N.W.
Washington, DC 20240
p. (202) 208-3500

http://www.mms.gov/

Dept of the Treasury

Office of the Secretary (2004)
1500 Pennsylvania Avenue, NW
Washington, DC 20220
p. (202) 622-2970
http://www.treas.gov

Internal Revenue Service (2004)
1111 Constitution Avenue, NW
Washington, DC 20224
p. (202) 622-9511 f. (202) 622-8653
http://www.irs.ustreas.gov

Dept of Transportation

Office of the Secretary (2004)
400 Seventh Street, SW
Washington, DC 20590
p. (202) 366-5580
http://www.dot.gov

U.S. Coast Guard (2004)
2100 Second Street, SW
Washington, DC 20593
p. (202) 366-4000
http://www.uscg.mil

Federal Aviation Administration (2004)
800 Independence Avenue, SW
Washington, DC 20591
p. (202) 267-3484 f. (202) 267-3505
http://www.faa.gov

Federal Highway Administration (2004)
400 Seventh Street, SW
Washington, DC 20590
p. (202) 366-0660 f. (202) 366-3244
http://www.fhwa.dot.gov

Environmental Protection Agency

Office of the Administrator (2004)
Ariel Rios Building
1200 Pennsylvania Ave, NW
Washington, DC 20460
p. (202) 564-4700
http://www.epa.gov/

Office of Atmospheric Programs (2004)
501 Third Street, NW
Washington, DC 20001
p. (202) 564-9140 f. (202) 565-2149

Office of Ground Water & Drinking Water (2004)
1200 Pennsylvania Avenue, NW
Washington, DC 20460
p. (202) 260-5400 f. (202) 260-5394

Asst Administrator for Environmental Information (2004)
1200 Pennsylvania Avenue, NW
Washington, DC 20460
p. (202) 564-6665 f. (202) 501-1622

Safety, Health & Env Management Div (2004)
1200 Pennsylvania Avenue, NW
Washington, DC 20460
p. (202) 564-1640 f. (202) 564-0215

National Air & Radiation Environmental Lab (2004)
540 South Morris Avenue
Montgomery, AL 36115-2601
p. (334) 270-3404 f. (334) 270-3454

National Center for Environmental Assessment (2004)
1200 Pennsylvania Avenue, NW
Washington, DC 20460
p. (202) 564-3333 f. (202) 565-0090

National Center for Environmental Research (2004)
Ronald Reagan Building
1300 Pennsylvania Avenue, NW
Washington, DC 20004
p. (202) 564-6825 f. (202) 565-2444

National Enforcement Investigations Center Div (2004)
P.O. Box 25277
Denver Federal Building 53
Denver, CO 80225
p. (303) 236-5100

Office of Science & Technology (2004)
1200 Pennsylvania Avenue, NW
Washington, DC 20460
p. (202) 260-5400 f. (202) 260-5394

Office of Science Policy (2004)
Ronald Reagan Building
1300 Pennsylvania Avenue, NW
Washington, DC 20004
p. (202) 564-6705 f. (202) 565-2911

Office of Solid Waste (2004)
Crystal Station One
2800 Crystal Drive
Arlington, VA 22202
p. (703) 308-8895 f. (703) 308-0513

Office of Emergency & Remedial Response (Superfund/Oil Programs) (2004)
Crystal Gateway One
1235 Jefferson Davis Highway
Arlington, VA 22202
p. (703) 603-8960 f. (703) 603-9146
http://www.epa.gov/superfund

Office of Underground Storage Tanks (2004)
Crystal Gateway One
1235 Jefferson Davis Highway
Arlington, VA 22202
p. (703) 603-9900 f. (703) 603-9163

Office of Wastewater Management (2004)
1200 Pennsylvania Avenue, NW
Washington, DC 20460
p. (202) 564-0748 f. (202) 501-2346

Assistant Administrator for Water Programs (2004)
1200 Pennsylvania Avenue, NW
Washington, DC 20460
p. (202) 260-5700 f. (202) 260-5711

Office of Wetlands, Oceans & Watersheds (2004)
Fairchild Building
499 South Capitol Street, SW
Washington, DC 20003
p. (202) 260-7166 f. (202) 260-6294

Federal Emergency Management Agency

Office of the Director (2004)
500 C Street, SW
Washington, DC 20472
p. (202) 646-4600
http://www.fema.gov/

Jet Propulsion Laboratory

Earth & Space Sciences Div (2004)
Geology & Planetology Section
Pasadena, CA 91109
p. (818) 354-3440 f. (818) 354-0966
Department Administrative Manager: Murray Geller

Lead Scientist:
Bruce E. Banerdt, (D), S California, 1983, Xy
Section Manager:
Ronald G. Blom, (D), California (Santa Barbara), 1987, Or
Section Member:
Michael J. Abrams, (M), Caltech, 1973, Or
Ronald E. Alley, (M), Northwestern, 1972, Or
Diana L. Blaney, (D), Hawaii, 1990, Xg
Bonnie J. Buratti, (D), Cornell, 1983, Or
Robert E. Crippen, (D), California (Santa Barbara), 1989, Or
Joy A. Crisp, (D), Princeton, 1984, Gv
Thomas C. Duxbury, (M), Purdue, 1966, Oo
Diane L. Evans, (D), Washington, 1981, Gm
Tom G. Farr, (D), Washington, 1981, Gm
Matthew P. Golombek, (D), Massachusetts, 1981, Gc
Ken E. Herkenhoff, (D), Caltech, 1989, Xg
Simon J. Hook, (D), Durham (UK), 1989, Or
Erik R. Ivins, (M), California (Los Angeles), 1976, Gt
Anne B. Kahle, (D), California (Los Angeles), 1975, Or
Harold R. Lang, (D), Calgary, 1983, Gr
Kyle C. McDonald, (D), Michigan, 1991, Oo
Robert M. Nelson, (D), Pittsburgh, 1978, Xg
Eni G. Njoku, (D), MIT, 1976, Or
Frank D. Palluconi, (M), Penn State, 1963, Or
David C. Pieri, (D), Cornell, 1979, Gt
Jeffrey J. Plant, (D), Washington (St. Louis), 1991, Oo
Jeffrey Plescia, (D), S California, 1985, Oo
Carol A. Raymond, (D), Columbia, 1989, Yr
Suzanne E. Smrekar, (D), S Methodist, 1990, Xy
Linda J. Spilker, (D), California (Los Angeles), 1992, Xy
Ellen R. Stofan, (D), Brown, 1989, Xg
Glenn J. Veeder, (D), Caltech, 1974, Oo
Senior Scientist:
Ronald S. Saunders, (D), Brown, 1970, Gr

Lawrence Berkeley National Laboratory

Earth Sciences Div (M,D) (2004)
1 Cyclotron Road
MS 90-1116
Berkeley, CA 94720
p. 510-642-4148 f. 510-643-5098
doner@nature.berkeley.edu
http://www-esd.lbl.gov

Group Leader:
Joseph S. Y. Wang, (D), California (Berkeley), 1973, Hw
Senior Scientist:
Chin-Fu Tsang, (D), California (Berkeley), 1969, Hw
Yvonne Y.W. Tsang, (D), California (Berkeley), 1970, Hw
Glenn A. Waychunas, (D), California (Los Angeles), 1979, Gz
Geological Scientist:
Timothy J. Kneafsey, (D), California (Berkeley), 1996, Hw
Associate Scientist:
Jonny Rutqvist, (D), Royal Inst of Tech (Sweden), 1995, Hw
Rohit Salve, (D), California (Berkeley), 1995, Hg
Research Associate:
Robert A. TerBerg, (M), New Mexico Tech, 1994, Hw
Robert C. Trautz, (M), Arizona, 1984, Hw
Faculty Senior Scientist:
Thomas V. McEvilly, (D), St. Louis, 1964, Ys
Staff Scientist:
B. Mack Kennedy, (D), Washington (St. Louis), 1981, Cg
Staff Geological Scientist:
Curtis M. Oldenburg, (D), California (Santa Barbara), 1989, Hw

Lawrence Livermore National Laboratory

Atmospheric, Earth and Energy Div (2007)
7000 East Avenue
PO Box 808
Livermore, CA 94551
p. (925) 423-4412 f. (925) 423-4908
webmaster@cmels.llnl.gov
cmels.llnl.gov

Environmental Science (2004)
7000 East Avenue
L-203
Livermore, CA 94550
p. (925) 423-7184 f. (925) 422-0208
aines1@llnl.gov
Administrative Assistant: Laura Long

Head:
Kenneth J. Jackson, (D), California (Berkeley), 1983, Cl
Senior Scientist:
Roger D. Aines, (D), Caltech, 1984, Cg
Bill L. Bourcier, (D), Oregon State, 1983, Cg
Carol J. Bruton, (D), California (Berkeley), 1981, Cg
Elizabeth A. Burton, (D), Missouri (Washington), 1988, Oe
Thomas A. Buscheck, (D), California (Berkeley), 1984, Ng
Steven Carle, (D), California (Davis), 1996, Hw
Charles R. Carrigan, (D), California (Los Angeles), 1977, Yg
Susan A. Carroll, (D), Northwestern, 1988, Cl
M Lee Davisson, (M), California (Davis), 1992, Cg
Quingyun Duan, (D), Arizona, 1991, Hg
Robert C. Finkel, (D), California (San Diego), 1974, Cg
Samuel (Julio) J. Friedmann, (D), MIT, Gg
William E. Glassley, (D), Washington, 1973, Cg
Joanne Horn, (D), California (Berkeley), 1988, Sb
Jim W. Johnson, (D), Arizona, 1987, Cg
Richard B. Knapp, (D), Arizona, 1978, Ng
Kevin G. Knauss, (D), S California, 1977, Gg
Stefan Kollet, (D), Nebraska, 2003, Gg
Kenrick H. Lee, (M), New Mexico Tech, 1986, Nm
Roald N. Leif, (D), Oregon State, 1993, Cg
Reed M. Maxwell, (D), California (Berkeley), 1998, Ng
Robin L. Newmark, (D), Columbia, 1984, Ys
Gregory Nimz, (D), California (Santa Cruz), 1989, Yx
John J. Nitao, (D), California (Berkeley), 1980, Hq
Gayle A. Pawloski, (B), Cal State (Hayward), 1979, Gg
Abelardo L. Ramirez, (M), Purdue, 1979, Ng
Joseph A. Rard, (D), Iowa State, 1973, Cg
Sarah K. Roberts, (M), Arizona State, 1990, Cl
Andrew F. Tompson, (D), Princeton, 1985, Ng
Brian E. Viani, (D), California (Berkeley), 1981, Cg
Jeffrey L. Wagoner, (M), California (Riverside), 1977, Gs
Ananda M. Wijesinghe, (D), MIT, 1978, Nr
Sun Yunwei, (D), Israel Inst of Tech, 1995, Ng

Lockheed Martin Idaho Technologies

Advanced Field Studies (2004)
P.O. Box 1625
Idaho Falls, ID 83415-2110

p. (208) 526-1970
Group Leader:
Wayne C. Downs, (D), Florida, 1993, Oo
Senior Scientist:
Joel M. Hubbell, (M), Idaho, 1981, Hw
Indrek Porro, (D), New Mexico State, 1990, So
Jerry R. Weidner, (D), Penn State, 1968, Cp
Allan H. Wylie, (M), Montana, 1991, Hw
Scientist:
Thomas K. Honeycutt, (M), S Illinois, 1988, Gg
Scientific Specialist:
James B. Sisson, (D), New Mexico State, 1980, Hq
Robert W. Smith, (D), New Mexico Tech, 1984, Cg

Geologic & Geophysical Studies (2004)
P.O. Box 1625
Idaho Falls, ID 83415-2110
p. (208) 526-1970
Group Leader:
Richard P. Smith, (D), Colorado, 1975, Gt
Senior Scientist:
Nicholas E. Josten, (M), Colorado Mines, 1985, Yg
Mary C. Pfeifer, (D), Colorado Mines, 1991, Oo
Robert C. Starr, (D), Waterloo, 1988, Hg
Scientist:
Carolyn W. Bishop, (M), Arizona, 1991, Hg
Glen S. Carpenter, (B), Colorado Mines, 1983, Ye
Deborah L. McElroy, (M), New Mexico Tech, 1987, Hg
Julie A. Tullis, (B), Idaho State, 1979, Gg

Geotechnology Dept (2004)
P.O. Box 1625
Idaho Falls, ID 83415-2110
p. (208) 526-1970 f. (208) 526-0875
Department Manager:
William E. Harrison, (D), Louisiana State, 1976, Co

Subsurface Transport Studies (2004)
P.O. Box 1625
Idaho Falls, ID 83415-2110
p. (208) 526-1970
Group Leader:
Charles P. Thomas, (D), Georgia Tech, 1968, Np
Senior Scientist:
Ronald C. Arnett, (M), Washington, 1986, Hq
David D. Faulder, (B), Wyoming, 1981, Np
Swen O. Magnuson, (M), New Mexico Tech, 1988, Hq
James M. McCarthy, (D), California (Los Angeles), 1988, Oo
Annette L. Schafer-Perini, (D), New Mexico Tech, 1990, Hq
George M. Shook, (M), Texas, 1988, Np
Scientist:
Eric P. Robertson, (M), Wyoming, 1990, Oo
Andrus J. Sondrup, (M), Utah, 1991, Hq
Associate Scientist:
Peter Martian, (M), Utah, 1991, Oo

Los Alamos National Laboratory

Earth & Environmental Sciences Div (2004)
P.O. Box 1663
Los Alamos, NM 87545
p. (505) 667-3644 f. (505) 667-3494
wallacet@lanl.gov
Enrollment (1996): B: 17 (17) M: 6 (6)
Director:
Terry C. Wallace, Jr., (D), Caltech, 1983, Ys
Technical Staff:
Paul Aamodt, (M), Nevada, 1991, Oo
James N. Albright, (D), Chicago, 1969, Yg
Douglas Alde, (D), Illinois, 1979, Oo
M. James Aldrich, (D), New Mexico, 1972, Gc
William Armstrong, (D), Colorado, Yg
W. Scott Baldridge, (D), Caltech, 1979, Gi
Fairley J. Barnes, (D), New Mexico State, 1986, Oo
Naomi M. Becker, (D), Wisconsin, 1991, Hg
Kay H. Birdsell, (M), Colorado, 1985, Hq
David L. Bish, (D), Penn State, 1977, Gz
James D. Blacic, (D), California (Los Angeles), 1971, Nr
Rainer Bleck, (D), Penn State, 1968, Ow
James E. Bossert, (D), Colorado State, 1990, Oa
Christopher R. Bradley, (D), MIT, 1993, Yg
Thomas L. Brake, (M), New Mexico State, 1985, Nr
David D. Breshears, (D), Colorado State, 1993, Py
David E. Broxton, (M), New Mexico, 1976, Gx
Wendee M. Brunish, (D), Illinois, 1981, Oo
Gilles Y. Bussod, (D), California (Los Angeles), 1988, Yx
Katherine Campbell, (D), New Mexico, 1979, Gq
James W. Carey, (D), Harvard, 1990, Ca
Theodore C. Carney, (M), Colorado State, 1981, Ng
William E. Clements, (D), New Mexico Tech, 1974, Oo
Allen H. Cogbill, (D), Northwestern, 1979, Yg
Gregory L. Cole, (D), Arizona, 1990, Og
Steffanie Coonley, (M), New Mexico, 1984, Oo
Keeley R. Costigan, (D), Colorado State, 1992, Oa
William Cottingame, (D), Texas, 1984, Oa
James L. Craig, (B), Nevada, 1974, Gg
Bruce M. Crowe, (D), California (Santa Barbara), 1974, Gv
Zora V. Dash, (M), New Mexico, 1988, Oo
Deborah J. Daymon, (M), Idaho, 1994, Oo
Micheline Devaurs, (M), Utah State, 1983, Hg
Kalpak Dighe, (D), Clemson, 1994, Oo
John C. Dinsmoor, (B), Colorado Mines, 1989, Nm
Alison M. Dorries, (D), Harvard, 1986, Oo
Donald S. Dreesen, (B), New Mexico, 1968, Np
David V. Duchane, (D), Michigan, 1978, Oo
Michael H. Ebinger, (D), Purdue, 1988, So
C L Edwards, (D), New Mexico Tech, 1975, Yg
C. James Elliott, (D), Yale, Oo
Scott M. Elliott, (D), California, 1983, Oa
Perry D. Farley, (D), Oklahoma, 1981, Oo
George T. Farmer, (D), Cincinnati, 1968, Hw
David N. Fogel, (M), Cal State, 1997, Oo
Carl W. Gable, (D), Harvard, 1989, Gt
Edward S. Gaffney, (D), Caltech, 1973, Yg
Anthony F. Gallegos, (D), Colorado State, 1970, Py
Jamie N. Gardner, (D), California (Davis), 1985, Gg
Fraser Goff, (D), California (Santa Cruz), 1977, Cg
Carl R. Hagelberg, (D), Oregon State, 1992, Oo
Jeffrey C. Hansen, (B), Weber State, 1971, Oo
Charles D. Harrington, (D), Indiana, 1970, Gm
Hans E. Hartse, (D), New Mexico Tech, 1991, Ys
Ward L. Hawkins, (B), Nevada (Reno), 1977, Og
Grant Heiken, (D), California (Santa Barbara), 1972, Gv
Donald D. Hickmott, (D), MIT, 1988, Gp
Steve T. Hildebrand, (D), Texas (Dallas), 1993, Ys
Emil F. Homuth, (D), Washington, 1974, Ye
Leigh S. House, (D), Columbia, 1982, Ys
Lianjie Huang, (D), Paris, 1994, Yg
Paul A. Johnson, (M), Arizona, 1984, Yg
Eric M. Jones, (D), Wisconsin, 1970, Oa
Larry Allan Jones, (B), Nebraska, 1972, Ng
Hemendra N. Kalia, (D), Missouri Sch of Mines, 1970, Nm
Jim C. Kao, (D), Illinois, 1985, Oa
Danny Katzman, (M), New Mexico, 1991, Gg
Elizabeth Keating, (D), Wisconsin, 1995, Gg
Sharad Kelkar, (M), Texas, 1979, Np
C. F. Keller, Jr., (D), Indiana, 1969, Oa

Richard G. Kovach, (M), Naval Postgrad Sch, 1979, Nm
Donathon J. Krier, (M), New Mexico, 1980, Gv
Thomas D. Kunkle, (D), Hawaii, 1978, Xy
Edward M. Kwicklis, (M), Colorado, 1987, Gg
Carol M. LaDelfe, (M), New Hampshire, 1978, Yg
Chung Chieng A. Lai, (D), Texas A&M, 1984, Ow
Schon S. Levy, (M), Texas, 1975, Gi
Peter C. Lichtner, (D), Mainz, 1974, Oo
Rodman Linn, (D), New Mexico State, 1997, Ng
Lynn McDonald, (M), Cal State, Oo
Maureen A. McGraw, (D), California (Berkeley), 1996, Oo
Laurie A. McNair, (D), Carnegie Mellon, 1995, Oo
Wayne R. Meadows, (B), Cal State (Bakersfield), 1974, Yg
Theodore Mockler, (M), Carnegie Mellon, 1986, Oo
Orrin B. Myers, (D), Colorado State, 1992, Py
Balu Nadiga, (D), Caltech, 1992, Oo
Brent D. Newman, (D), New Mexico Tech, 1996, Gg
John W. Nyhan, (D), Colorado State, 1972, So
Ronald D. Oliver, (B), Oregon Inst of Tech, 1972, Yg
Howard J. Patton, (D), MIT, 1978, Yg
Frank V. Perry, (D), California, 1988, Gv
William S. Phillips, (D), MIT, 1985, Yg
Eugene W. Pokorny, (B), Missouri, 1979, Nm
William M. Porch, (D), Washington, 1971, Yg
Allyn R. Pratt, (M), Boise State, 1977, Ge
George Randall, (D), SUNY (Binghamton), Yg
Steen Rasmussen, (D), Tech (Denmark), 1985, Yg
Jon M. Reisner, (D), Iowa State, Oa
Steven L. Reneau, (D), California (Berkeley), 1988, Gm
Douglas O. Revelle, (D), Michigan, 1974, Oa
Peter Roberts, (D), MIT, 1989, Ys
Bruce A. Robinson, (D), MIT, 1985, Oo
R. Roussel-Dupre, (D), Colorado, 1979, Oo
Thomas J. Shankland, (D), Harvard, 1966, Yx
Catherine H. Smith, (D), New Mexico State, 1995, Ca
Wendy E. Soll, (D), MIT, 1991, Hw
Everett P. Springer, (D), Utah State, 1983, Hq
Lee Steck, (D), California, Gg
William J. Stone, (D), North Dakota, 1973, Gg
Brian W. Stump, (D), California (Berkeley), 1979, Ys
Robert P. Swift, (D), Washington, 1969, Nr
E.M.D. Symbalisty, (D), Chicago, 1984, Oo
Steven R. Taylor, (D), MIT, 1980, Ys
James Tencate, (D), Texas, 1992, Oo
Bryan J. Travis, (D), Florida State, 1974, Hq
Lynn Trease, (B), New Mexico, Oo
Peng-Hsiang Tseng, (D), Colorado, 1991, Ng
Gregory A. Valentine, (D), California (Santa Barbara), 1988, Yg
David T. Vaniman, (D), California (Santa Cruz), 1976, Gx
Aaron A. Velasco, (D), California (Santa Cruz), 1993, Yg
Richard G. Warren, (M), New Mexico, 1972, Gi
Douglas J. Weaver, (M), Nevada, 1995, Oo
Thomas A. Weaver, (D), Chicago, 1973, Yg
Rodney W. Whitaker, (D), Indiana, 1976, Oa
Earl M. Whitney, (D), Utah, Oo
Judith L. Winterkamp, (B), Texas, 1977, Oo
Kenneth H. Wohletz, (D), Arizona State, 1980, Gv
Giday WoldeGabriel, (D), Case Western, 1987, Gx
Andrew V. Wolfsberg, (D), Stanford, 1993, Hw
Dongxiao Zhang, (D), Arizona, 1993, Hg
George A. Zyvoloski, (D), California (Santa Barbara), 1975, Ng

Project Leader:
Mark T. Peters, (D), Chicago, 1992, Yg

Technical Staff:
Donald W. Brown, (M), California (Los Angeles), 1961, Nr

Group Leader:
Michael C. Fehler, (D), MIT, 1979, Ys

Deputy Group Leader:
George D. Guthrie, (D), Johns Hopkins, 1989, Gz

Deputy Division Leader (Acting):
David C. Pearson, (D), S Methodist, 1993, Yg

Other:
Christina "Tina" Behr-Andres, (D), Michigan Tech, 1992, Oo

Technical Staff:
James L. Hollins, (M), Phoenix, 2002, Ng

Chemistry Div (B,M,D) (2004)
P.O. Box 1663
Los Alamos, NM 87545
p. (505) 667-4457 f. (505) 665-4355
olivares@lanl.gov
http://pearl1.lanl.gov/external/default.htm

Division Leader:
Alexander J. Gancarz, (D), Caltech, 1976, Cc

Staff:
Kent D. Abney, (D), Colorado State, 1987, Oo
Stephen F. Agnew, (D), Washington State, 1981, Oo
Moses Attrep, (D), Arkansas, 1965, Oe
Timothy M. Benjamin, (D), Caltech, 1979, Ca
Scott M. Bowen, (D), New Mexico, 1983, Ca
James R. Brainard, (D), Indiana, 1979, Oo
Jeffrey C. Bryan, (D), Washington, 1988, Oo
Carol J. Burns, (D), California (Berkeley), 1987, Oo
Timothy P. Burns, (D), Nebraska, 1986, Co
Gilbert W. Butler, (D), California (Berkeley), 1967, Cc
Edwin P. Chamberlain, (D), Texas A&M, 1971, Cc
David L. Clark, (D), Indiana, 1986, Oo
Dean A. Cole, (D), Iowa, 1985, Co
David B. Curtis, (D), Oregon State, 1974, Cc
Paul R. Dixon, (D), Yale, 1989, Oo
Robert J. Donahoe, (D), North Carolina State, 1985, Oo
Stephen K. Doorn, (D), Northwestern, 1989, Oo
Clarence J. Duffy, (D), British Columbia, 1977, Cg
Deward W. Efurd, (D), Arkansas, 1975, Ct
Phillip G. Eller, (D), Ohio State, 1971, Oo
June T. Fabryka-Martin, (D), Arizona, 1988, Hg
Bryan L. Fearey, (D), Iowa State, 1986, Cc
David L. Finnegan, (D), Maryland, 1984, Oa
John R. Fitzpatrick, (M), New Mexico, 1983, Oo
Malcolm M. Fowler, (D), Washington (St. Louis), 1972, Oa
Sammy R. Garcia, (B), New Mexico Highlands, 1972, Ca
Russell E. Gritzo, (B), New Mexico State, 1983, Oo
Richard C. Heaton, (D), Illinois, 1973, Oo
Sara B. Helmick, (B), Southwestern, 1955, Oo
David R. Janecky, (D), Minnesota, 1982, Cl
Kung King-Hsi, (D), Cornell, 1989, Oo
Scott Kinkead, (D), Idaho, 1983, Oo
Gregory J. Kubas, (D), Northwestern, 1970, Oa
Pat J. Langston-Unkefer, (D), Texas A&M, 1978, Co
Patrick A. Longmire, (D), New Mexico, 1991, Oo
Michael MacInnes, (M), Wisconsin, 1969, Oo
Allen S. Mason, (D), Miami, 1974, Oa
Charles M. Miller, (D), Stanford, 1980, Cc
Geoffrey G. Miller, (D), RPI, 1984, Cc
Terrance L. Morgan, (M), Colorado State, 1984, Ng
David Morris, (D), North Carolina State, 1984, Cc
Eugene J. Mroz, (D), Maryland, 1976, Oa
Michael T. Murrell, (D), California (San Diego), 1980, Cc
Allen E. Ogard, (D), Chicago, 1957, Hw
Jose A. Olivares, (D), Iowa State, 1985, Cs
Hain Oona, (D), Arizona, 1979, Oo
Kevin C. Ott, (D), Caltech, 1983, Oo
Edward S. Patera, (D), Arizona State, 1982, Cg
Richard E. Perrin, (B), Denver, 1952, Cc
Eugene J. Peterson, (D), Arizona State, 1976, Ct

Dennis Phillips, (D), Hawaii, 1976, Co
Jane Poths, (D), Chicago, 1982, Cc
Pamela Z. Rogers, (D), California (Berkeley), 1981, Cl
Donald J. Rokop, (D), Lake Forest, 1985, Cc
Robert S. Rundberg, (D), CUNY, 1978, Gq
Nancy N. S. Sauer, (D), Iowa State, 1986, Oo
Norman C. Schroeder, (D), Iowa State, 1985, Cc
Louis A. Silks, (B), Suffolk, 1978, Oo
Paul H. Smith, (D), California (Berkeley), 1987, Oo
Zita V. Svitra, (B), Roosevelt, 1967, Oo
Basil I. Swanson, (D), Northwestern, 1969, Oo
C. Drew Tait, (D), North Carolina State, 1984, Cp
Wayne A. Taylor, (B), New Mexico, 1978, Cl
Kimberly W. Thomas, (D), California (Berkeley), 1978, Ct
Joseph L. Thompson, (D), Penn State, 1963, Ct
Ines Triay, (D), Miami, 1985, Hg
Clifford J. Unkefer, (D), Minnesota, 1981, Oo
David J. Vieira, (D), California (Berkeley), 1978, Oo
Jerry B. Wilhelmy, (D), California (Berkeley), 1969, Oo
Kurt Wolfsberg, (D), Washington (St. Louis), 1959, Cc
William H. Woodruff, (D), Purdue, 1972, Oo
Mary Anne Yates, (D), Carnegie Mellon, 1976, Oo

Emeritus:
Ernest A. Bryant, (D), Washington (St. Louis), 1956, Cg
Merle E. Bunker, (D), Indiana, 1950, Oo
William R. Daniels, (D), New Mexico, 1965, Ct
Donald L. Hull, (M), Iowa State, 1974, Oo

National Aeronautics & Space Administration

NASA Headquarters (2004)
Washington, DC 20546-0001
p. (202) 358-2345 f. (202) 358-2810
http://www.hq.nasa.gov/

Ames Research Center (2004)
Building BN200
Moffett Field, CA 94035-1000
p. (650) 604-5111

Langley Research Center (2004)
11 Langley Boulevard
Hampton, VA 23681-0001
p. (757) 864-4111

Goddard Space Flight Center (2004)
Greenbelt Road
Greenbelt, MD 20771-0001
p. (301) 286-5121

George C. Marshall Space Flight Center (2004)
Marshall Space Flight Center, AL 35812-0001
p. (256) 544-1910

Lyndon B. Johnson Space Center (2004)
Houston, TX 77058-3696
p. (281) 483-5309

National Science Foundation

Office of the Director (2004)
4201 Wilson Boulevard
Arlington, VA 22230
p. (703) 292-5111 f. (703) 292-9232
http://www.nsf.gov

Directorate for Geosciences (2004)
4201 Wilson Boulevard
Arlington, VA 22230
p. (703) 292-8500 f. (703) 292-9042
http://www.geo.nsf.gov/

Atmospheric Sciences Div (2004)
4201 Wilson Boulevard
Arlington, VA 22230
p. (703) 292-8520 f. (703) 292-9022
http://www.geo.nsf.gov/atm/

Earth Sciences Div (2004)
4201 Wilson Boulevard
Arlington, VA 22230
p. (703) 292-8550 f. (703) 292-9025
http://www.geo.nsf.gov/ear/

Ocean Sciences Div (2004)
4201 Wilson Boulevard
Arlington, VA 22230
p. (703) 292-8580 f. (703) 292-9085
http://www.geo.nsf.gov/oce/

Office of Polar Programs (2004)
4201 Wilson Boulevard
Arlington, VA 22230
p. (703) 292-8030 f. (703) 292-9081
http://www.nsf.gov/od/opp/

National Transportation Safety Board

Office of Managing Director (2004)
490 L'Enfant Plaza East, SW
Washington, DC 20594
p. (202) 314-6100
http://www.ntsb.gov

Nuclear Regulatory Commission

NRC Headquarters (2004)
One White Flint North Building
11555 Rockville Pike
Rockville, MD 20852
p. (301) 415-8200
http://www.nrc.gov/

Oak Ridge National Laboratory

Physical Sciences Directorate (2004)
P.O. Box 2008, MS 6230
1 Bethel Valley Road
Oak Ridge, TN 37831-6230
p. (865) 574-1144 f. (865) 574-0323
buchananmv@ornl.gov
Administrative Assistant: Regina G. Violet

Director:
Michelle V. Buchanan, (D), Wisconsin, 1978, Oo

Distinguished Research Scientist:
Donald A. Palmer, (D), W Australia, 1970, Cg

Senior Scientist:
David R. Cole, (D), Penn State, 1980, Cs
Lee R. Riciputi, (D), Wisconsin, 1991, Ca
John M. Simonson, (D), California (Berkeley), 1983, Cg
David J. Wesolowski, (D), Penn State, 1984, Cg

Research Geochemist:
James G. Blencoe, (D), Stanford, 1974, Cp

Associate Scientist:
Pascale Benezeth-Gisquet, (D), Paul-Sabatier, 1994, Cl
Ariel A. Chialvo, (D), Clemson, 1988, Oo
Miroslaw S. Gruszkiewicz, (D), Brigham Young, 1993, Oo
Juske Horita, (D), Tokyo Inst, 1987, Cg

Visiting Professor:
Lawrence M. Anovitz, (D), Michigan, 1987, Cp

Mostafa Fayek, (D), Saskatchewan, 1996, Cg

Chemical Technology Div (2005)
P.O. Box 2008, MS 6228
Oak Ridge, TN 37831
p. (615) 574-6147 f. (615) 574-6866
buchananmv@ornl.gov
Director:
L. E. McNeese, (D), Tennessee, 1967, Oo

Energy Div (2004)
P.O. Box 2008
Oak Ridge, TN 37831-6187
p. (615) 574-5510 f. (615) 574-5788
milorasl@ornl.gov
Administrative Assistant: Teresa D. Ferguson
Director:
Robert B. Shelton, (D), S Illinois, 1970, Oo
Chair:
Donald W. Lee, (D), Michigan, 1977, Hw
Senior Scientist:
Richard H. Ketelle, (M), Tennessee, 1977, Ge
Russell Lee, (D), McMaster, 1978, Oy
William P. Staub, (D), Iowa State, 1969, Ng
Research Associate:
Arthur C. Curtis, (M), Colorado State, 1993, Ge
Robert O. Johnson, (D), Tennessee, 1984, Hg
Richard R. Lee, (M), Temple, 1982, Gr
John D. Tauxe, (D), Texas, 1994, Hw

Environmental Sciences Div (2004)
P.O. Box 2008
Mail Stop 6035
Oak Ridge, TN 37831
p. (865) 574-7374 f. (865) 574-7287
envsci@ornl.gov
http://www.esd.ornl.gov
Director:
Stephen G. Hildebrand, (D), Michigan, 1973, Oo
Senior Scientist:
Marshall Adams, (D), North Carolina, 1974, Ob
Jeff Amthor, (D), Yale, 1987, Sf
Tom Ashwood, (M), Murray State, 1975, Ge
David Bell, Og
Mark Bevelhimer, (D), Tennessee, 1990, Hs
Terence J. Blasing, (D), Wisconsin, 1975, Oa
Thomas Boden, (M), Miami, 1985, Oa
Craig Brandt, (M), Tennessee, 1988, Ge
Scott C. Brooks, (D), Virginia, 1994, Cl
Robert S. Burlage, (D), Tennessee, 1990, Sb
Meng-Dawn Cheng, (D), Illinois, 1986, Oa
Robert B. Cook, (D), Columbia, 1981, Cg
William E. Doll, (D), Wisconsin, 1983, Yg
Thomas O. Early, (D), Washington (St. Louis), 1969, Cl
Baohua Gu, (D), California (Berkeley), 1991, Cl
Philip M. Jardine, (D), Virginia Tech, 1985, Sc
Liyuan Liang, (D), Caltech, 1988, Cl
Steven E. Lindberg, (D), Florida State, 1979, Cl
Gregg H. Marland, (D), Minnesota, 1972, Ge
John F. McCarthy, (D), Rhode Island, 1975, Co
Gerilynn R. Moline, (D), Wisconsin, 1992, Hw
Tony V. Palumbo, (D), North Carolina State, 1980, Sb
Tommy J. Phelps, (D), Wisconsin, 1985, Sb
Ellen D. Smith, (M), Wisconsin, 1979, Hg
Brian P. Spalding, (D), Cornell, 1976, Sc
Robert S. Turner, (D), Pennsylvania, 1983, Cl
David B. Watson, (M), New Mexico Tech, 1983, Hw
Olivia M. West, (D), MIT, 1991, Ng

Associate Scientist:
Tammy Beaty, Oi
Mary Anna Bogle, (M), Miami, 1975, Ge
Norman D. Farrow, (B), Oregon State, 1974, Hg
Deputy Director:
Gary K. Jacobs, (D), Penn State, 1981, Cg

Health Sciences Research Div (2004)
P.O. Box 2008
Oak Ridge, TN 37831-6124
p. (615) 574-5845 f. (615) 574-3036
acobsgk@ornl.gov
Associate Scientist:
Kathryn S. Dickerson, (M), Colorado, 1993, Ge
Frank G. Gardner, (B), E Michigan, 1980, Ge
Richard R. Knott, (B), Mesa, 1981, Gg
Chris A. Muhr, (B), Mesa, 1982, Ge
Doug A. Pickering, (B), Mesa, 1978, Ge
Judith A. Rice, (B), Mesa, 1980, Gg
Robert M. Schlosser, (B), Mesa, 1983, Ge
Gloria H. Stevens, (B), Mesa, 1981, Ge
Mary J. Wilson, (B), Mesa, 1984, Ge
John L. Zutman, (B), Mesa, 1984, Ge
Research Associate:
Marcy L. Espegren, (B), Colorado State, 1977, Ge
Peter M. Kearl, (M), Nevada, 1981, Hw
Alan D. Laase, (M), New Mexico Tech, 1989, Hw
Maureen T. Muck, (M), Tennessee, 1992, Ge

Pacific Northwest National Laboratory

Applied Geology & Geochemistry (2004)
Environmental Technology Div
Wayne J. Martin, Manager
PO Box 999 MSIN K6-81
Richland, WA 99352
p. (509) 376-5952 f. (509) 376-5368
wayne.martin@pnl.gov
http://www.pnl.gov/agg
Administrative Assistant: Charissa J. Chou
Head:
George R. Holdren, (D), Johns Hopkins, 1977, Cl
Staff Scientist:
Christopher J. Murray, (D), Stanford, 1992, Gq
Lab Fellow:
Bernard P. McGrail, (D), Columbia Southern, 1996, Cg
Senior Scientist:
Douglas B. Barnett, (M), E Washington, 1985, Em
Bruce N. Bjornstad, (M), E Washington, 1980, Gs
Kirk J. Cantrell, (D), Georgia Tech, 1989, Cl
Amy P. Gamerdinger, (D), Cornell, 1989, Cl
Tyler J. Gilmore, (M), Idaho, 1987, Hg
Floyd N. Hodges, (D), Texas, 1975, Cg
Duane G. Horton, (D), Illinois, 1983, Gz
Kenneth M. Krupka, (D), Penn State, 1984, Cl
George V. Last, (M), Washington State, 1997, Ge
Jonathan W. Lindberg, (M), Washington State, 1995, Ec
Shas V. Mattigod, (D), Washington State, 1976, Cl
Alan C. Rohay, (D), Washington, 1982, Ys
Herbert T. Schaef, (M), Texas Tech, 1991, Cg
R. Jeffrey Serne, (B), Washington, 1969, Cg
Mark D. Sweeney, (B), C Washington, 1985, Yg
Bruce A. Williams, (B), Colorado Mines, 1980, Ng
Associate Scientist:
Yi-Ju Chien, (M), Stanford, 1998, Gq
Jonathan P. Icenhower, (D), Oklahoma, 1995, Cl
David C. Lanigan, (B), Michigan Tech, 1980, Hg
Virginia L. Legore, (B), Oregon State, 1974, Cg
Clark W. Lindenmeier, (M), E Washington, 1995, Cg

Paul F. Martin, (B), Pacific Lutheran, 1982, Cg
Kent E. Parker, (M), Washington State, 1995, Cg

Research Associate:
Alexandra B. Amonette, (M), CUNY, 1976, Cg
Deborah S. Burke, (B), E Washington, 1992, Sc
Elsa A. Camacho, (B), Heritage, 1996, Cg
Matthew J. O'Hara, (B), Montana, 1996, Cg
Robert D. Orr, (B), N State, 1994, Ca

Hydrology (2004)
Environmental Technology Div
PO Box 999 MSIN K9-36
Richland, WA 99352
p. (509) 372-6091 f. (509) 372-6089

Head:
Mark D. Freshley, (M), Arizona, 1982, Hw

Senior Research Engineer:
Mark D. White, (D), Colorado State, 1986, Hy

Senior Scientist:
Diana H. Bacon, (D), Washington State, 1997, Hq
Marcel P. Bergeron, (M), Indiana, 1979, Hw
Ashokkumar Chilakapati, (D), Rice (India), 1993, Hq
Charles R. Cole, (M), Washington State, 1975, Hw
Michael J. Fayer, (D), Massachusetts, 1984, Hq
Glendon W. Gee, (D), Washington State, 1966, Hq
Charles T. Kincaid, (D), Utah State, 1979, Hw
Randy R. Kirkham, (D), Washington, 1993, Hq
Philip E. Long, (D), Stanford, 1977, Gi
Philip D. Meyer, (D), Illinois, 1992, Hw
Martinus Oostrom, (B), Wageningen Ag, 1984, Hq
Marshall C. Richmond, (D), Iowa, 1987, Hs
Timothy D. Scheibe, (D), Stanford, 1993, Hw
Carver S. Simmons, (D), California, 1975, Hq
Gary P. Streile, (D), California (Riverside), 1984, Hq
Lance W. Vail, (M), Montana State, 1982, Hg
Anderson L. Ward, (D), Guelph, 1993, Hw
Gene Whelan, (D), Utah State, 1992, Hg
Mark S. Wigmosta, (D), Washington, 1991, Hs
Steven B. Yabusaki, (M), Washington, 1986, Hg

Associate Scientist:
William E. Nichols, (M), Oregon State, 1990, Hq
Cynthia L. Rakowski, (M), Utah State, 1996, Oo
Randal Y. Taira, (M), Washington State, 1997, Hg
Signe K. Wurstner, (M), Indiana, 1989, Hw

Research Associate:
Diana Felmy, (D), Washington State, 1997, Oo
William A. Perkins, (B), Oregon State, 1985, Hg
Jason C. Ritter, (M), Oregon State, 1998, Sc

Field Hydrology and Chemistry (2004)
Environmental Technology Div
PO Box 999 MSIN K6-96
Richland, WA 99352
p. (509) 376-4067 f. (509) 372-1704

Head:
Joseph L. Devary, (D), Minnesota, 1975, Gq

Senior Scientist:
James G. Bush, (M), SD Sch of Mines & Tech, 1972, Nx
P. Evan Dresel, (D), Penn State, 1989, Cg
John C. Evans, (D), California (San Diego), 1971, Og
Jon S. Fruchter, (D), California (San Diego), 1971, Cl
Mary J. Hartman, (M), Idaho, 1986, Hw
Vernon G. Johnson, (D), Oregon State, 1979, Oc
Stuart P. Luttrell, (M), Idaho, 1982, Hg
Alex V. mitroshkov, (D), Inst of Phys Chem (Moscow), 1993, Cg
Susan M. Narbutovskih, (D), Penn State, 1983, Yg
Khris B. Olsen, (M), Wyoming, 1976, Cg
Robert E. Peterson, (D), Oregon State, 1977, Cg
Robert G. Riley, (D), SUNY (Syracuse), 1974, Cg
Ronald Schalla, (B), California, 1973, Gg
Debbie S. Sklarew, (D), Arizona, 1978, Og
Ronald M. Smith, (M), Washington State, 1977, Hg
Frank A. Spane, (D), Nevada (Reno), 1977, Hg
James E. Szecsody, (D), Arizona, 1988, So
Christopher J. Thompson, (D), Washington, 1994, Ca
Paul D. Thorne, (M), Arizona, 1983, Hg
Edward C. Thornton, (D), Minnesota, 1983, Cg
Mark D. Williams, (M), Indiana, 1993, Hw

Associate Scientist:
Barbara M. Gillespie, (B), S Colorado, 1979, Oo
John P. McDonald, (B), E Washington, 1993, Hw
Richard B. Mercer, (M), Wright State, 1978, Hw
Chris A. Newbill, (B), Washington, 1971, Gq
Darrell R. Newcomer, (M), Montana Min Sci & Tech, 1988, Hw
Vincent R. Vermeul, (M), Oregon State, 1990, Hg

Research Associate:
Debi L. Morgan, (B), Washington State, 1988, Oo
JoAnne T. Rieger, (B), SD Sch of Mines & Tech, 1979, Oo
Dorothy L. Stewart, (B), Tennessee, 1964, Oo
William D. Webber, (M), Brigham Young, 1992, Gg

Smithsonian Institution

National Museum of Natural History (2004)
1000 Jefferson Drive, SW
Washington, DC 20560
p. (202) 357-2664 f. (202) 786-2328
http://www.mnh.si.edu/

Physical Anthropology Dept (2004)
National Museum of Natural History
Tenth Street and Constitution Avenue, NW
Washington, DC 20560
p. (202) 786-2504 f. (202) 786-2328

Smithsonian Astrophysical Observatory (2004)
60 Garden Street
Cambridge, MA 02138
p. (617) 495-7100 f. (617) 495-7105

Center for Earth & Planetary Studies (2004)
National Air & Space Museum
Sixth Street & Independence Avenue, SW
Washington, DC 20560
p. (202) 357-1426 f. (202) 786-2566
http://www.nasm.edu/ceps/

Smithsonian Center for Ed & Museum Studies (2004)
1000 Jefferson Drive, SW
Washington, DC 20560
p. (202) 357-2111 f. (202) 357-2116

Education, Museum and Scholarly Services (2004)
1000 Jefferson Drive, SW
Washington, DC 20560
p. (202) 357-2627
http://www.si.edu

Smithsonian Environmental Research Center (2004)
P.O. Box 28
Edgewater, MD 21037
p. (443) 482-2205 f. (443) 482-2380
http://www.serc.si.edu

Invertebrate Zoology Dept (2004)
National Museum of Natural History

Tenth Street & Constitution Avenue, NW
Washington, DC 20560
p. (202) 357-2030 f. (202) 357-3043

Mineral Sciences Dept (2004)
National Museum of Natural History
Tenth Street & Constitution Avenue, NW
Washington, DC 20560
p. (202) 357-4809 f. (202) 357-2476

National Science Resources Center (2004)
Arts & Industries Building
900 Jefferson Drive, SW
Washington, DC 20560
p. (202) 357-4894 f. (202) 287-2070

Paleobiology Dept (2004)
National Museum of Natural History
Tenth Street & Constitution Avenue, NW
Washington, DC 20560
p. (202) 357-1893 f. (202) 786-2832

Smithsonian Center for Materials Res & Ed (2004)
Museum Support Center
4210 Silver Hill Road
Suitland, MD 20746
p. ((30) 1) -238- (Ext. 00 ext. 105) f. (301) 238-3709

Vertebrate Zoology Dept (2004)
National Museum of Natural History
Tenth Street & Constitution Avenue, NW
Washington, DC 20560
p. (202) 357-2740

American Association of Petroleum Geologists
1444 South Boulder Avenue
P.O. Box 979
Tulsa, OK 74101-0979
p. (918) 584-2555 f. (918) 560-2626
aapg@aapg.org
http://www.aapg.org
President:
Willard R. Green
President Elect:
Scott W. Tinker
Sections Vice President:
John M. Armentrout
Regions Vice President:
John R. Hogg
Secretary:
Edward A. Beaumont
Treasurer:
Randi Martinsen
Chairman, House of Delegates:
Martin D. Hewitt
Editor:
Richard D. Fritz
Executive Director:
Richard D. Fritz

American Association of Stratigraphic Palynologists
c/o Fredrick J. Rich
Georgia Southern University
Dept. of Geology and Geography, P.O. Box 8149
Statesboro, GA 30460-8149
f. (912) 681-0668
frich@georgiasouthern.edu
http://www.palynology.org
President:
Francine McCarthy
Past President:
Carlos A. Jaramillo
President Elect:
Frederick Rich
Secretary-Treasurer:
Thomas D. Demchuk
Director at Large:
Joyce Lucas-Clark
Sarah de la Rue
Barry Dale
Yow-Yuh Chen
Managing Editor:
James B. Riding

American Geological Institute
4220 King Street
Alexandria, VA 22302-1502
p. (703) 379-2480 f. (703) 379-7563
agi@agiweb.org
http://www.agiweb.org/
President:
Peter J. McCabe
President Elect:
Peter A. Scholle
Secretary:
Mary M. Poulton
Treasurer:
Thomas W. Dignes
Member at Large:
Lucy E. Edwards
Richard M. Powers
Laurie Scheuing
Past President:
Gail M. Ashley
Foundation Chair:
Roger P. Pinkerton
Executive Director:
P. Patrick Leahy

American Geophysical Union
2000 Florida Avenue, NW
Washington, DC 20009-1277
p. (202) 462-6900 f. (202) 328-0566
service@agu.org
http://www.agu.org/
President:
Timothy L. Killeen
President Elect:
Timothy L. Grove
General Secretary:
Carol Finn
International Secretary:
Jaime Urrutia Fucugauchi
Past President:
John A. Orcutt
Executive Director:
A. F. Spilhaus Jr.

American Institute of Hydrology
300 Village Green Circle, Suite 201
Smyrna, GA, 30080
p. (770) 384-1634 f. (770) 438-6172
AIHydro@aol.com
http://www.aihydro.org
President:
Antonius Laenen
Past President:
Miguel A. Marino
President Elect:
Miguel A. Medina Jr.
Vice President for Academic Affairs:
John Nieber
Vice President for Institute Development:
T. Allen J. Gookin
Vice President for International Affairs:
Neven Kresic
General Secretary:
Emitt C. Witt III
Treasurer:
Robert M. Hordon
Board of Registration Chairman:
Rolando Bravo
Board of Registration Secretary:
G. Douglas Glysson
Office Manager:
C. P. Seaburn

American Institute of Professional Geologists
1400 West 122nd Avenue, Suite 250
Westminster, CO 80234
p. (303) 412-6205 f. (303) 253-9220
aipg@aipg.org
http://www.aipg.org

President:
Daniel J. St. Germain
President Elect:
John L. Bognar
Past President:
Kelvin J. Buchanan
Vice President:
Michael D. Lawless
Secretary:
David P. Palmer
Treasurer:
Ronald J. Wallace
Advisory Board Representative:
Charles W. Drake
James F. Howard
Mark W. Rogers
Mark B. Sweatman
Editor:
Gail G. Gibson
Executive Director:
William J. Siok

American Rock Mechanics Association

600 Woodland Terrace
Alexandria, VA, 22302
p. (703) 683-1808 f. (703) 683-1815
smeallie@armarocks.org
http://www.armarocks.org

President:
David P.Yale
Vice President and President-Elect:
Sarah H. Wilson
Treasurer:
John L. Hill III
Secretary:
Anthony Iannacchione
Immediate Past President:
John P. Tinucci
Director:
Ahmed Abou-Sayed
Don Banks
Judy Ehlen
Jami Girard Dwyer
Sidney J. Green
Keith A. Heasley
Paul La Pointe
Azra Nur Tutuncu
Executive Director:
Peter H. Smeallie

American Society of Limnology and Oceanography

5400 Bosque Boulevard, Suite 680
Waco, TX, 76710
p. (254) 399-9635 f. (254) 776-3767
business@aslo.org
http://www.aslo.org/

President:
Sybil Putnam Seitzinger
Past-President:
Jonathan Cole
President-Elect:
Carlos M. Duarte
Secretary:
M. Robin Anderson
Treasurer:
Lynda Shapiro
Member at Large:
Beatrix Beisner
Carla E. Caceres
John A. Downing
Marta Estrada
Wilhelm Graneli
Patricia A. Matrai
James McManus
Wayne Wurtsbaugh
Business Manager:
Helen Schneider Lemay

Association for Women Geoscientists

P.O. Box 30645
Lincoln, NE, 68503-0645
f. (402) 470-3110
office@awg.org
http://www.awg.org

President:
Jessica Moore Ali-Adeeb
President-Elect:
Jean Hsieh
Secretary:
Sarah Welna
Treasurer:
Katie Matthews
Past President:
Nadine Langley
Editor:
Kristina Brody
Business Manager:
Janelle Gerry

Association of American State Geologists

Louisiana Geological Survey
Louisiana State University
3079 Energy, Coast & Environmental Bldg.
Baton Rouge, LA 70803
p. (225) 578-5320 f. (225) 578-3662
cjohn@lsu.edu
http://www.stategeologists.org

President:
Chacko J. John
President Elect:
Berry H. Tew
Vice President:
David R. Wunsch
Past President:
Scott W. Tinker
Secretary:
Vicki S. McConnell
Treasurer:
Vince Mathews
Historian:
James C. Cobb
Statistician:
L. Harvey Thorleifson
Editor:
Karl W. Muessig

Association of Earth Science Editors

c/o Mary Ann Schmidt
554 Chess Street
Pittsburgh, PA 15205-3212

p. (412) 268-4708
maschmidt66@hotmail.com
http://www.aese.org

President:
Rowena Mills
Vice President:
Thomas Overton
Treasurer:
Lowell Lindsay
Secretary:
Margaret Rutka
Director:
Carol Christopher
Karen MacFarlane
Carole Ziegler
Past President:
Fred Spilhaus

Association of Environmental and Engineering Geologists

P.O. Box 460518
Denver, CO, 80246
p. (303) 757-2926 f. (303) 757-2969
aeg@aegweb.org
http://www.aegweb.org

President:
Dorian E. Kuper
Past President:
Terry R. West
Vice President:
Mark P. Molinari
Secretary:
Bruce R. Hilton
Treasurer:
Duane Kreuger
Meeting Manager:
Julie Keaton
Chief Operating Officer:
Becky Roland

The Clay Minerals Society

3635 Concorde Pkwy Suite 500
Chantilly, VA, 20151-1125
p. (703) 652-9960 f. (703) 652-9951
cms@clays.org
http://www.clays.org

President:
Andrew R. Thomas
Vice President:
Derek C. Bain
Secretary:
Warren Huff
Treasurer:
J. Reed Glasmann
Past President:
Ray E. Ferrell Jr.
Editor-in-Chief:
Joseph W. Stucki
Manager:
Michelle Johnson

Council on Undergraduate Research, Geosciences Div.

734 15th Street NW, Suite 550
Washington, DC, 20005
p. (202) 783-4810 f. (202) 783-4811
cur@cur.org
http://www.cur.org

Chair:
Jeffrey G. Ryan
Geoscience Councilor:
C. Brannon Anderson
Lori Bettison-Varga
Patrick Burkhart
William Dinklage
Lydia K. Fox
Karen H. Fryer
Laura A. Guertin
Edward Hansen
Terry Lahm
Patricia Manley
Jeffrey S. Marshall
Bill Miller
Kirsten Nicolaysen
Christopher Oze
Jeanette Pope
Linda Reinen
Kathleen Surpless
Deanna Van Dijk
Stephen Van Horn
Andrew Wulff
Executive Officer:
Nancy Hensel

Environmental and Engineering Geophysical Society

1720 S. Bellaire Street, Suite 110
Denver, CO, 80222
p. (303) 531-7517 f. (303) 820-3844
staff@eegs.org
http://www.eegs.org

President:
Jeffrey G. Paine
President-Elect:
William Brown
Vice-President SAGEEP:
John Stowell
Vice-President Elect SAGEEP:
Susan Pullan
Vice-President Committees:
Dennis Mills
Vice-President Elect Committees:
Jennifer Holt
Past President:
John Clark
Member at Large:
Gregory S. Baker
Fred Boadu
Dwain K. Butler
William E. Doll
Douglas Labrecque
Bruce D Smith
International Board Liaison:
Micki Allen
SAGEEP 2008 General Chair:
Jonathan Nyquist
SAGEEP 2008 Technical Chair:
Ronald Kaufmann
FastTIMES Editor:
Jeffrey G. Paine
JEEG Editor:
Janet Simms

Executive Director:
Kathie Barstnar

Friends of Mineralogy

c/o Michael Kokinos
4620 Doe Street
Shingle Springs, CA 95682
shaffern@indiana.edu
http://www.friendsofmineralogy.org

President:
Virgil W. Lueth
Vice President:
Julian Gray
Secretary:
Gloria Staebler
Treasurer:
Regina Aumente
Past President:
Bill Dameron
Newsletter Editor:
Carol Smith

Geo-Institute of ASCE

1801 Alexander Bell Drive
Reston, VA, 20191-4400
p. (703) 295-6350 f. (703) 295-6351
geo-institute@asce.org
http://www.geoinstitute.org

President:
Arthur G. Hoffmann Jr.
Past President:
John E. Anderson
Vice-President:
Jean-Louis Briaud
Treasurer:
Edward J. Kavazanjian, Jr.
Governor:
Craig H. Benson
Larry P. Jedele
Philip King
Jay A. Padgett
International Secretary to ISSMGE:
Robert D. Holtz
Secretary:
Carol W. Bowers
Director:
Carol W. Bowers

Geological Society of America

3300 Penrose Place
P.O. Box 9140
Boulder, CO, 80301-9140
p. (303) 357-1000 f. (303) 357-1074
jhess@geosociety.org
http://www.geosociety.org

President:
John M. Sharp Jr.
Vice President:
Judith Totman Parrish
Past President:
Stephen G. Wells
Treasurer:
Robbie Rice Gries
Councilor:
J. David Applegate
Elena Centeno-Garcia
Darrel S. Cowan
Jerome V. DeGraff
John W. Geissman
Monica E. Gowan
Jacqueline E. Huntoon
G. Randy Keller
Nancy J. McMillan
Jonathan G. Price
Jill S. Schneiderman
Diane R. Smith
Executive Director:
John W. Hess

Geological Society of London

Burlington House
Piccadilly
London W1J 0BG
UNITED KINGDOM
p. 44 (0)20 7434 9944 f. 44 (0)20 7439 8975
enquiries@geolsoc.org.uk
http://www.geolsoc.org.uk

President:
Richard Fortey
Vice President:
Iain Bartholomew
Martin Culshaw
Paul Henderson
Past President:
Peter Styles
Representative for Foreign & External Affairs:
Edward Derbyshire
Professional Representative:
David Shilston
Publications Representative:
Nick Rogers
Science Representative:
Tony Watts
Treasurer:
Andy Fleet
Executive Secretary:
Edmund Nickless

Geoscience Information Society

c/o American Geological Institute
4220 King Street
Alexandria, VA 22302-1502
suzanne.larsen@colorado.edu
http://www.geoinfo.org/

President:
Suzanne T. Larsen
Vice President/President Elect:
Rusty Kimball
Immediate Past President:
Patricia B. Yocum
Secretary:
Andrea Twiss-Brooks
Treasurer:
Renee Davis
Publications Manager:
Elaine Clement
Newsletter Editor:
Adonna Fleming

History of Earth Sciences Society

P.O. Box 455
Poncha Springs, CO 81242

p. (719) 539-4113 f. (719) 539-4542
erogers@geology-books.com
http://www.historyearthscience.org/

President:
Naomi Oreskes
Past President:
Martin Rudwick
Secretary:
Stephen M. Testa
Treasurer:
Emma C. Rainforth
Program Officer:
Kerry Magruder
Councilor:
Ana Carneiro
Barry J. Cooper

International Association of Hydrogeologists/U.S. National Chapter

c/o Prof. John M. Sharp Jr.
University of Texas
Dept. of Geological Sciences
One University Station, C1100
Austin, TX 78712-0254
p. (512) 471-3317 f. (512) 471-9425
jmsharp@mail.utexas.edu
http://www.iah.org

Chairman:
John M. Sharp, Jr.
Secretary-Treasurer:
Todd Halihan
Past Chairperson:
Leonard F. Konikow

International Basement Tectonics Association

675 South, 400 East
Salt Lake City, UT 84111-3926
p. (801) 328-8541 f. (801) 363-6243
info@tectonics.com
http://www.tectonics.org

Chief Trustee:
Robert D. Hatcher Jr.
Trustee/Treasurer:
John H. McBride
Trustee/Secretary:
John P. Hogan
Trustee:
Karl E. Karstrom
G. Randy Keller
Adjunct Trustee:
Alfred Kröner
Past Chief Trustee:
Marvin P. Carlson

Mineralogical Society of America

3635 Concorde Pkwy Suite 500
Chantilly, VA 20151-1125
p. (703) 652-9950 f. (703) 652-9951
business@minsocam.org
http://www.minsocam.org

President:
Peter J. Heaney
Past President:
Barbara L. Dutrow
Vice-President:
Nancy L. Ross
Secretary:
Mickey E. Gunter
Treasurer:
John M. Hughes
Councilor:
Peter C. Burns
Carol D. Frost
Klaus Mezger
Jean Morrison
Simon A.T. Redfern
Roberta L. Rudnick
MSA Executive Director:
J. Alexander Speer

National Association of Black Geologists and Geophysicists

4212 San Felipe, Suite 420
Houston, Texas 77027-2902
p. (713) 393-4856 f. (713) 393-4802
nabgg-us@hotmail.com
http://www.nabgg.org

President:
Robert J. Johnson
Vice President:
Aisha Ragas
Secretary:
Geraldine Grant
Asst. Secretary:
Teresa Dimmer
Treasurer:
Marian Walters
Asst. Treasurer:
Michael S. Mitchell
Parliamentarian:
Deborah Branch
Past President:
Conrad Allen
Member At Large:
Roxanne Lamb
Jerome Murphy

National Association of Geoscience Teachers

c/o Carleton College, B-SERC
One North College Street
Northfield, MN, 55057
p. (507) 646-7096 f. (507) 646-5175
cmanduca@carleton.edu
http://www.nagt.org

President:
Eric M. Riggs
1st Vice-President:
David N. Steer
2nd Vice-President:
Michael Taber
Past President:
Karen G. Havholm
Secretary/Treasurer:
Eugene F. Pearson
Councilor-at-Large:
Pamela J.W. Gore
Laura A. Guertin
Jeffrey R. Knott
Jeffrey G. Ryan

Linda Selvig
Richard F. Yuretich
Editor, Journal of Geoscience Education:
Carl N. Drummond
Executive Director:
Cathryn A. Manduca

National Association of State Boards of Geology

P.O. Box 11591
Columbia, SC, 29211-1591
p. (803) 739-5676 f. (803) 739-8874
asbog@asbog.org
http://www.asbog.org
President:
Wilson H. Herrod
President Elect:
Lisa E. Hosey
Treasurer:
Michael R. Kunz
Secretary:
Richard K. Spruill
Immediate Past President:
William G. Dixon Jr.
Executive Director:
Sam Christiano

National Earth Science Teachers Association

c/o Roberta Marie Johnson
University Corporation for Atmospheric Research (UCAR)
P.O. Box 3000
Boulder, CO 80307-3000
p. (303) 497-2591 f. (303) 497-2598
rmjohnsn@ucar.edu
http://www.nestanet.org/
President:
Michael J. Passow
President Elect:
Ardis Herrold
Secretary:
Missy Holzer
Treasurer:
Bruce Hall
Retiring President:
Parker Pennington IV
Board of Directors Representative:
Lisa Alter
Executive Director:
Roberta Marie Johnson

National Speleological Society

2813 Cave Avenue
Huntsville, AL, 35810
p. (256) 852-1300 f. (256) 851-9241
nss@caves.org
http://www.caves.org
President:
William Tozer
Executive Vice President:
Gordon Birkhimer
Administrative Vice President:
Ray Keeler
Operations Vice President:
Cheryl Jones
Secretary-Treasurer:
Peri Frantz
Operations Manager:
Stephanie Cothron Searles

North American Commission on Stratigraphic Nomenclature

c/o Jared D. Morrow
Department of Geological Sciences
San Diego State University
5500 Campanile Drive, 237 GMCS
San Diego, CA 92182-1020
p. (619) 594-1395 f. (619) 594-4372
jmorrow@sciences.sdsu.edu
http://www.agiweb.org/nacsn/
Chair:
Jared D. Morrow
Vice-Chair:
Berry H. (Nick) Tew, Jr.

Paleobotanical Section of the Botanical Society of America

c/o Michael Dunn
Deptartment of Biology
Cameron University
Lawton, OK 73505
p. (580) 581-2287 f. (580) 591-8003
michaeld@cameron.edu
http://www.dartmouth.edu/~daghlian/paleo/
President:
Alexandru Tomescu
Secretary-Treasurer:
Michael Dunn
Editor, Bibliography of American Paleobotany:
Walt L. Cressler

Paleontological Research Institution

1259 Trumansburg Road
Ithaca, NY, 14850-1313
p. (607) 273-6623 f. (607) 273-6620
info@museumoftheearth.org
http://www.priweb.org
President:
David H. Taube
Past President:
Patricia H. Kelley
Secretary, Past President:
Philip Bartels
PRI Director:
Warren D. Allmon

Paleontological Society

c/o Russell Wright, Association Manager
Allen Marketing and Management
P.O. Box 7065
Lawrence, KS 66044-7065
p. (785) 843-1235 ext 215 f. (785) 843-1274
paleosoc@allenpress.com
http://www.paleosoc.org
President:
Derek E.G. Briggs
Past President:
David J. Bottjer

President-elect:
Douglas H. Erwin
Secretary:
Roger D.K. Thomas
Treasurer:
Mark E. Patzkowsky
Councilor (Under 40):
Dena M. Smith
Councilor (Unrestricted):
Peg (Margaret M.) Yacobucci
Program Coordinator:
Rowan Lockwood
Education Coordinator:
Michael A. Gibson
Student Representative:
Phoebe A. Cohen
Christy C. Visaggi
Co-Editor, Journal of Paleontology:
Richard Lupia
Stephen Westrop
Co-Editor, Paleobiology:
Matthew Carrano
Peter J. Wagner
Editor, Special Publications:
Susan H. Butts
Editor, Priscum:
Lisa M. Amati
Association Manager:
Russell Wright

Petroleum History Institute

P.O. Box 165
Oil City, PA 16301-0165
wbrice@pitt.edu
http://www.petroleumhistory.org/

President:
Daniel Leech
1st Vice President:
Gerald Friedman
2nd Vice President:
Vacant for 2008
Secretary:
Marilyn Black
Treasurer:
Maureen Leech
Editor:
William Brice

Seismological Society of America

201 Plaza Professional Building
El Cerrito, CA, 94530-4007
p. (510) 525-5474 f. (510) 525-7204
info@seismosoc.org
http://www.seismosoc.org

President:
William L. Ellsworth
Vice-President:
Rachel E. Abercrombie
Secretary:
Keith L. Knudsen
Treasurer:
Mitchell M. Withers
Director:
John G. Anderson
Allison L. Bent
Roland Burgmann
James W. Dewey
William L. Ellsworth
Arthur D. Frankel
Carol Prentice
Paul G. Richards
Frank Scherbaum
David W. Simpson
Brian E. Tucker
Christa von Hillebrandt-Andrade
BSSA Editor:
Andrew J. Michael
SRL Editor:
Luciana Astiz
Executive Director:
Susan B. Newman

SEPM (Society for Sedimentary Geology)

6128 East 38th Street, Suite 308
Tulsa, OK 74135-5814
p. (918) 610-3361 f. (918) 621-1685
hharper@sepm.org
http://www.sepm.org

President:
Mary J. Kraus
President-Elect:
Dale Leckie
Secretary-Treasurer:
John W. Robinson
International Councilor:
Cam Nelson
Councilor for Paleontology:
James A. MacEachern
Councilor for Research Activities:
Christopher R. Fielding
Councilor for Sedimentology:
Gerilyn S. Soreghan
Co-editor, Journal of Sedimentary Research:
Kitty Lou Milliken
Colin P. North
Co-Editor, Palaios:
Stephen T. Hasiotis
Edith L. Taylor
Co-Editor, Special Publications:
Donald F. McNeill
Gary Nichols
Executive Director:
Howard E. Harper Jr.

Society for Mining, Metallurgy, and Exploration

8307 Shaffer Parkway
Littleton, CO, 80127-7002
p. (303) 948-4210 f. (303) 973-3845
sme@smenet.org
http://www.smenet.org

President:
George W. Luxbacher
President Elect:
William H. Wilkinson Jr.
Past President:
James R. Arnold
Executive Director:
David L. Kanagy

The Society for Organic Petrology

c/o American Geological Institute
4220 King Street
Alexandria, VA 22302-1502
pwarwick@usgs.gov
http://www.tsop.org

President:
Leslie F. Ruppert
Vice President:
Hamed Sanei
Secretary:
Paul Hackley
Treasurer:
Michael P. Avery
Councilor:
Isabel Suarez Ruiz
Sharon Swanson
Editor:
Rachel Walker

Society of Economic Geologists

7811 Shaffer Parkway
Littleton, CO, 80127-3732
p. (720) 981-7882 f. (720) 981-7874
seg@segweb.org
http://www.segweb.org

President:
Neil Williams
President-Elect:
John F. H. Thompson
Past President:
Richard J. Goldfarb
Treasurer:
Harold J. Noyes
Vice President, Regional Affairs:
Antonio Arribas R. Jr.
Vice President for Student Affairs:
C. Eric Seedoriff
SEG Foundation President:
Ronald L. Parratt
SEG Publications Board Chair:
Stuart F. Simmons
Executive Director:
Brian G. Hoal

Society of Exploration Geophysicists

P.O. Box 702740
Tulsa, OK, 74170-2740
p. (918) 497-5500 f. (918) 497-5560
mfleming@seg.org
http://www.seg.org

President:
Fred Aminzadeh
President-Elect:
Laurence R. Lines
First Vice President:
Eugenio Loinger
Second Vice President:
Louise Pellerin
Vice-President:
Guillaume Cambois
Secretary-Treasurer:
Daniel A. Ebrom
Editor:
Kees Wapenaar
Executive Director:
Mary L. Fleming

Society of Independent Professional Earth Scientists

4925 Greenville Avenue, Suite 1106
Dallas, TX, 75206-4019
p. (214) 363-1780 f. (214) 363-8195
sipes@sipes.org
http://www.sipes.org

President:
George S. Johnson
Vice President:
Lanny O. Butner
Vice President of National Energy:
H. Jack Naumann, Jr.
Secretary:
Clifford A. Walker
Treasurer:
Lee M. Petersen
Director:
Michael N. Austin
Raymond N. Blackhall
Michael G. Cruson
Joseph M. Finger, Jr.
William R. Finley
Bobby M. Greenwood
Owen R. Hopkins
Kenneth J. Huffman
Peter MacKenzie
Marc D. Maddox
J. Phil Martin, Jr.
Jon B. Selby
Thomas J. Smith
Executive Director:
Diane M. Finstrom

Society of Mineral Museum Professionals

c/o Virgil W. Lueth
New Mexico Bureau of Mines and Mineral Resources
Mineral Museum
801 Leroy Place
Socorro, NM 87801-4796
p. (505) 835-5140 f. (505) 835-6333
vwlueth@nmt.edu
http://www.smmp.net

President:
Virgil W. Lueth
Vice President:
Terry E. Huizing
Past President:
Anthony R. Kampf
Secretary:
Jean F. DeMouthe
Treasurer:
Anna M. Domitrovic
Director:
Jamie Newman
Terri Ottaway
Jeffrey Post
Penny Williamson

Society of Vertebrate Paleontology

111 Deer Lake Road
Deerfield, IL 60015

p. (847) 480-9095 f. (847) 480-9282
svp@vertpaleo.org
http://www.vertpaleo.org

President:
Catherine Badgley
Vice President:
Blair Van Valkenburgh
Past President:
Annalisa Berta
Secretary:
Christopher A. Brochu
Treasurer:
Ted J. Vlamis
Member-at-Large:
Ana M. Baez
Michael D. Gottfried
P. David Polly
Administrative Director:
Kate VanZanten

Soil Science Society of America

677 South Segoe Road
Madison, WI, 53711
p. (608) 273-8080 f. (608) 273-2021
ebergfeld@soils.org
http://www.soils.org

President:
Gary A. Peterson
President-Elect:
Paul M. Bertsch
Past President:
Rattan Lal
Chief Executive Officer:
Ellen G.M. Bergfeld

U.S. Permafrost Association

P.O. Box 750141
Fairbanks, AK 99775-0852
p. (302) 831-0852 f. (302) 831-6654
info@uspermafrost.org
http://www.uspermafrost.org/

President:
Jon E. Zufeldt
President Elect:
Kenneth M. Hinkel
Secretary:
Oliver W. Frauenfeld
Treasurer:
Michael R. Lilly
Past President:
Frederick E. Nelson
Board Member:
Jennifer W. Harden
Board Member:
Yuri Shur
IPA Representative:
Larry D. Hinzman

American Geological Institute
GEOSCIENCE CURRENTS
No. 1
29 November 2007

Student Aid for Geoscience Students to top $2.7 Million in 2008

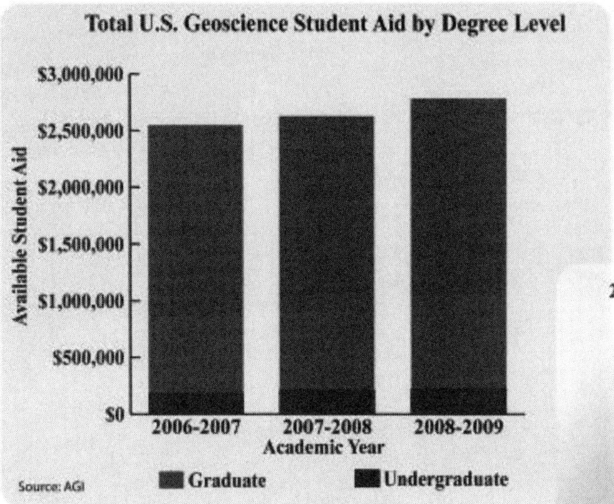

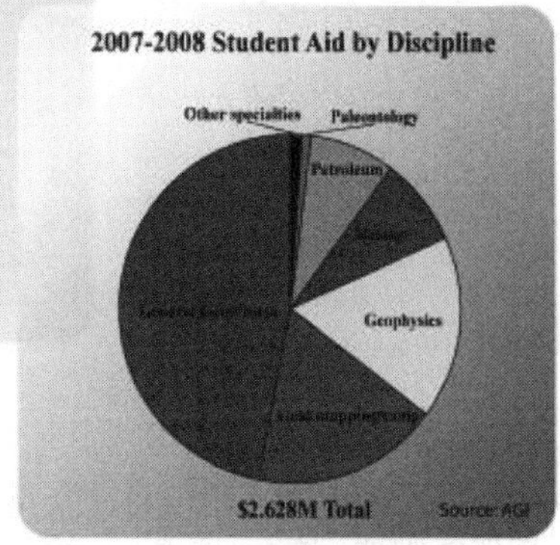

Direct support for geoscience students has been rising for the last two years. The trend looks to continue in 2008-2009 with a projected 6% increase in available funds. These opportunities for student support include funds from government agencies (60%) and non-profit societies (40%, which includes support from private foundations and companies).

Graduate student support comprises 91% of all awards in the 2007-2008 academic year – over $2.4 M spread across 570 individual awards. The largest student support program is the NSF Graduate Student Fellowship program. This program provided more than $1.13 million dollars in support to geoscience graduate students in 2007 from a total program budget of $40.5 million.

Several programs have significantly grown in recent years, including the Society of Exploration Geophysicists scholarship program's 65% increase in awarded support since 2006-2007. Funding for student travel grants has increased by more than $32,000 since that time, though field camp scholarships have remained steady.

- Cindy Martinez and Christopher Keane

American Geological Institute
GEOSCIENCE CURRENTS
No. 2
5 December 2007

Geoscience NSF Graduate Student Fellowships Decline

Total Funding of Geoscience NSF Fellowship

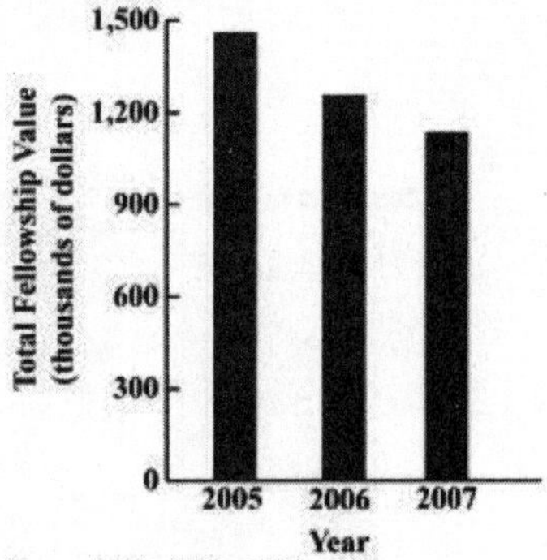

Source: National Science Foundation

Universities with the Most Geoscience NSF Fellows

University	Total Fellowships 2005-2007
Massachusetts Institute of Technology	10
University of Washington	7
California Institute of Technology	7
Columbia University	6
University of California - Berkeley	6
Harvard University	6
Penn State University	5
Stanford University	4
Boston University	3
University of Colorado at Boulder	3
Brown University	3

Source: National Science Foundation

The number of NSF Graduate Fellowships awarded to geoscience students has fallen 22% between 2005 and 2007 while the total budget has remained stable at $40.5 million. However, geoscience students continue to earn over 3% of the total awards, which is higher than the 2.5% of all Ph.D. students studying geoscience. The rate of decline of the number of geoscience fellowships is slightly higher than the total reduction in the number of awards given during the same 3-year period, which was 10%. NSF Graduate Fellowships have been well-distributed across the sub-specialties within in the geoscience community.

Geoscience NSF Graduate Fellows attended 38 individual institutions between 2005 and 2007, including two international programs. MIT tops the list with 10 fellowship winners since 2005, and the top 11 schools listed account for 64% of the awardees.

www.agiweb.org/workforce/ AGI workforce@agiweb.org

American Geological Institute
GEOSCIENCE CURRENTS
No. 3
14 December 2007

Competitive Starting Salaries for Geoscience Graduates

Geoscience starting salaries were competitive with other science and engineering fields in 2007. Bachelors geoscience graduates, generally employed in the environmental and hydrology industry, earned an average of $31,366 p.a. compared to $31,258 for life scientists and $32,500 for chemistry students.

Recent Masters recipients saw the highest starting salaries in the Oil and Gas industry, with an average of $81,300 p.a., according to a new study of recent geoscience graduates by AGI and the American Geophysical Union. This salary level is significantly higher than the average starting salary of all science Masters degree recipients, who earned an average of $46,873 p.a.

New doctorate recipients in all fields of science earned an average of $62,059 p.a. in the private sector, while new geosciences doctorates commanded an average of $72,600.

- Cindy Martinez

Starting Salaries for New Geoscience Masters, 2007

	Average Salary	Median Salary
Oil and Gas	81,300	82,500
Environmental Firm	47,500	45,500
Any Government	46,200	45,000

Starting Salaries for New Geoscience PhDs, 2007

	Average Salary	Median Salary
Postdoc – Academe	43,100	42,000
Postdoc - Government	55,200	53,000
Potentially Perm. Academe	51,900	52,500
Private Sector	72,600	71,000

Source: AGI/AGU study on Recent PhD and Masters Degree Recipients, 2007

SPECIALTY CODES

Specialty codes are used to indicate the research or teaching specialities of faculty members listed in the directory. Bold numbers are totals for each major category. Numbers in parentheses are individual specialty totals.

GEOLOGY **4,436**

Gg	General Geology (668)
Ga	Archaeological Geology (61)
Ge	Environmental Geology (278)
Gm	Geomorphology (350)
Gl	Glacial Geology (168)
Gu	Marine Geology (158)
Gz	Mineralogy & Crystallography (317)
Gn	Paleolimnology (26)
Go	Petroleum Geology (139)
Gx	General Petrology (141)
Gi	Igneous Petrology (329)
Gp	Metamorphic Petrology (193)
Gd	Sedimentary Petrology (117)
Gs	Sedimentology (469)
Gr	Physical Stratigraphy (199)
Gc	Structural Geology (519)
Gt	Tectonics (243)
Gv	Volcanology (152)
Gq	Mathematical Geology (42)
Gy	Mineral Physics (42)
Gh	History of Geology (10)
Gb	Geomedicine (3)
Gf	Forensic Geology (1)

ECONOMIC GEOLOGY **406**

Eg	General Economic Geology (174)
Ec	Coal (54)
Em	Metals (109)
En	Non-Metals (23)
Eo	Oil and Gas (49)

GEOCHEMISTRY **1,379**

Cg	General Geochemistry (366)
Ca	Analytical Geochemistry (111)
Cp	Experimental Petrology/Phase Equilibria (76)
Ce	Exploration Geochemistry (23)
Cc	Geochronology & Radioisotopes (214)
Cl	Low-temperature Geochemistry (224)
Cm	Marine Geochemistry (106)
Co	Organic Geochemistry (98)
Cs	Stable Isotopes (136)
Ct	Trace Element Distribution (36)

GEOPHYSICS **1,402**

Yg	General Geophysics (402)
Yx	Experimental Geophysics (65)
Ye	Exploration Geophysics (165)
Yd	Geodesy (36)
Ym	Geomagnetism & Paleomagnetism (112)
Yv	Gravity (19)
Yh	Heat Flow (19)
Ys	Seismology (388)
Yr	Marine Geophysics (221)

PALEONTOLOGY **1,221**

Pg	General Paleontology (160)
Ps	Paleostratigraphy (91)
Pm	Micropaleontology (126)
Pb	Paleobotany (54)
Pl	Palynology (56)
Pq	Quantitative Paleontology (4)
Pv	Vertebrate Paleontology (224)
Pi	Invertebrate Paleontology (196)
Po	Paleobiology (101)
Pe	Paleoecology & Paleoclimatology (163)
Py	Geobiology (67)

HYDROLOGY **1,498**

Hg	General Hydrology (167)
Hw	Ground Water/Hydrogeology (526)
Hq	Quantitative Hydrology (78)
Hs	Surface Waters (80)
Hy	Geohydrology (66)
Sp	Soil Physics/Hydrology (89)
Sc	Soil Chemistry/Mineralogy (162)
Sd	Pedology/Classification/Morphology (78)
Sf	Forest Soils/Rangelands/Wetlands (45)
Sb	Soil Biology/Biochemistry (85)
Sa	Paleopedology/Archeology (7)
So	Other Soil Science (122)

SOIL SCIENCE **980**

Sp	Soil Physics/Hydrology (89)
Sc	Soil Chemistry/Mineralogy (162)
Sd	Pedology/Classification/Morphology (78)
Sf	Forest Soils/Rangelands/Wetlands (45)
Sb	Soil Biology/Biochemistry (85)
Sa	Paleopedology/Archeology (7)
So	Other Soil Science (122)

ENGINEERING GEOLOGY **392**

Ng	General Engineering Geology (160)
Ne	Earthquake Engineering (20)
Nx	Mining Tech/Extractive Metallurgy (44)
Nm	Mining Engineering (84)
Np	Petroleum Engineering (32)
Nr	Rock Mechanics (53)

OCEANOGRAPHY **1,210**

Og	General Oceanography (78)
Ob	Biological Oceanography (411)
Oc	Chemical Oceanography (174)
Ou	Geological Oceanography (135)
Op	Physical Oceanography (352)
On	Shore and Nearshore Processes (64)

PLANETOLOGY **214**

Xc	Cosmochemistry (42)
Xg	Extraterrestrial Geology (70)
Xy	Extraterrestrial Geophysics (66)
Xm	Meteorites & Tektites (42)

OTHER **3,640**

Og	General Earth Sciences (163)
Oa	Atmospheric Sciences (637)
Oe	Earth Science Education (142)
Oy	Physical Geography (202)
Oo	Ocean Engineering/Mining (32)
Or	Remote Sensing (229)
Os	Soil Science (23)
Ow	Meteorology (217)
Om	Material Science (36)
Ou	Land Use/Urban Geology (68)
Oi	Geographic Information Systems (188)
Ol	Glaciology (15)
Oo	Not Elsewhere Classified (1734)

TOTAL **16,778**

A

Aagaard, Knut, (206) 543-8942 aagaard@apl.washington.edu, Univ of Washington – Op
Aalto, Kenneth R., (707) 826-4978 kral@axe.humboldt.edu, Humboldt State Univ – Gr
Aamodt, Paul, (505) 665-1331 plaamod@lanl.gov, Los Alamos National Lab – Oo
Aay, Henry, (616) 526-7033 aay@calvin.edu, Calvin College – Oo
Ababneh, Linah N., (757) 221-2951 lababneh@wm.edu, College of William & Mary – GeCg
Abbott, Lon, 928-350-2201 Prescott College – Gt
Abbott, Lon, 303-492-6172 lon.abbott@colorado.edu, Univ of Colorado – Gg
Abbott, Mark B., (412) 624-1408 mabbott1@pitt.edu, Univ of Pittsburgh – Gs
Abbott, Mark R., (541) 737-5195 mark@coas.oregonstate.edu, Oregon State Univ – Obr
Abbott, Patrick L., (619) 594-5591 pabbott@geology.sdsu.edu, San Diego State Univ – Gs
Abbott, Richard N., (828) 262-3055 abbottrn@appstate.edu, Appalachian State Univ – Gz
Abdel-Fattah, Ahmadabdel-fa@ohio.edu, Ohio Univ, Lancaster – Hw
Abdelsalam, Mohamed, 573-341-4100 abdelsam@umr.edu, Univ of Missouri, Rolla – Or
Abdo, Ginette, 406-496-4152 gabdo@mtech.edu, Montana Tech of The Univ of Montana – Hw
Abdu, Yassir, (204) 474-7356 abdu@cc.umanitoba.ca, Univ of Manitoba – Gz
Abdullah, Nuraiteng Tee, 03-79674229 naiteng@um.edu.my, Univ of Malaya – Gr
Abdullah, Wan Hasiah, 03-79674232 wanhasia@um.edu.my, Univ of Malaya – Ec
Abell, Benjamin F., (314) 977-3335 abell@eas.slu.edu, Saint Louis Univ – Ow
Abend, Martin, (201) 200-3161 mabend@njcu.edu, New Jersey City Univ – Oy
Aber, James S.aberjame@emporia.edu, Emporia State Univ – Gl
Aber, Susan W., 620-342-5977 abersusa@esumail.emporia.edu, Emporia State Univ – Gz
Abercrombie, Rachelrea@bu.edu, Boston Univ – Gt
Abers, Geoffrey A., (617) 353-2616 abers@bu.edu, Boston Univ – Yg
Abert, Curtis C., 217-244-2188 abert@isgs.uiuc.edu, Illinois State Geological Survey – Gg
Abney, Kent D., (505) 665-3894 Los Alamos National Lab – Oo
Abousleiman, Younane, 405-325-2900 yabousle@ou.edu, Univ of Oklahoma – Nr
Abrajano, Jr., Teofilo A., (518) 276-6036 abrajt@rpi.edu, Rensselaer Polytechnic Institute – Cs
Abrams, Lewis J., 910-962-2350 abramsl@uncw.edu, Univ of North Carolina Wilmington – Gu
Abrams, Michael A., (801) 581-8856 mabrams@egi.utah.edu, Univ of Utah – Ce
Abrams, Michael J., (818) 354-0937 Jet Propulsion Lab – Or
Abramson, Evan H., 206-616-4388 evan@ess.washington.edu, Univ of Washington – Gy
Abreu, Vincent J., (313) 747-3640 Univ of Michigan – Og
Abreu, Vitor, 713-431-7214 vitor.abreu@exxonmobil.com, Univ of Houston – Gr
Acevedo, Miguel F., (940) 565-2381 acevedo@unt.edu, Univ of North Texas – Ge
Achab, Aicha, (418) 654-2603 aachab@inrs.uquebec.ca, Universite du Quebec – Pl
Acharya, Kumud, (702) 862-5371 Kumud.acharya@dri.edu, Univ of Nevada, Reno – Oo
Achmad, Grufon, (410) 554-5556 gachmad@mgs.md.gov, Maryland Geological Survey – Hw
Ackerman, Steven A., (608)263-3647 stevea@ssec.wisc.edu, Univ of Wisconsin, Madison – Or
Acosta, Jose G., jchang@cicese.mx, Centro de Inv Cientifica y de Ed Superior de Ensenada – Ne
Acosta, Patricia E., (413) 597-2221 patricia.e.acosta@williams.edu, Williams College – Oo
Acton, Gary, (530) 752-1861 acton@geology.ucdavis.edu, Univ of California, Davis – Ym
Acworth, Ian, 612 9385 5766 i.acworth@unsw.edu.au, Univ of New South Wales – Hw
Adamo, Peter J., 204-727-9683 adamop@brandonu.ca, Brandon Univ – Gg
Adams, Daryl, (507) 389-1778 daryl.adams@mnsu.edu, Minnesota State Univ – Oe
Adams, Dean, (775) 784-1474 vdadams@unr.nevada.edu, Univ of Nevada, Reno – Oo
Adams, Donald D., (518) 564-4037 adamsdd@plattsburgh.edu, Plattsburgh State Univ (SUNY) – Cl
Adams, Gerald E., 312-344-7540 gadams@colum.edu, Columbia College – Gi
Adams, Herbert G., (818) 677-2575 herb.adams@csun.edu, California State Univ, Northridge – Ng
Adams, John B.john@rad.ess.washington.edu, Univ of Washington – GcXgOr
Adams, Kenneth, (775) 673-7345 ken.adams@dri.edu, Univ of Nevada, Reno – Gam
Adams, Kenneth D., (775) 673-7345 kadams@dri.edu, Desert Research Institute – Gm
Adams, Mark G., (828) 262-3049 Appalachian State Univ – Gp
Adams, Marshall, (865) 574-7335 sma@ornl.gov, Oak Ridge National Lab – Ob
Adams, Michael C., (801) 585-7784 madams@egi.utah.edu, Univ of Utah – Cg
Adams, Michael S., (608) 263-5994 Univ of Wisconsin, Madison – Ob
Adams, Peter N., (352) 846-0825 adamsp@ufl.edu, Univ of Florida – Gm
Adamsen, Floyd, (602) 437-1702 Fadamsen@uswcl.ars.ag.gov, Univ of Arizona – So
Adegoke, Jimmyadegokej@umkc.edu, Univ of Missouri-Kansas City – Oa
Adkins, Jess F., (626) 395-8550 jess@gps.caltech.edu, California Institute of Technology – Cm
Adomaitis, Daniel J., 217-244-8872 adomaitis@isgs.uiuc.edu, Illinois State Geological Survey – Ge
Adovasio, James M., 814-824-2581 adovasio@mercyhurst.edu, Mercyhurst College – Ga
Adrain, Jonathan M., (319) 335-1539 jonathan-adrain@uiowa.edu, Univ of Iowa – Pi
Adrain, Tiffany S., (319) 335-1822 tiffany-adrain@uiowa.edu, Univ of Iowa – Oo
Adriano, Domy C., (803) 725-2752 Univ of Georgia – Sc
Agee, Carl B., (505) 277-1644 agee@unm.edu, Univ of New Mexico – XcGyCp
Agee, Ernest M., (765) 494-3282 eagee@purdue.edu, Purdue Univ – Oa
Agnew, Duncan C., (858) 534-2590 dagnew@ucsd.edu, Univ of California, San Diego – Ys
Agnew, Stephen F., (505) 665-1764 Los Alamos National Lab – Oo
Agrawal, Abinash, 937 775-3455 abinash.agrawal@wright.edu, Wright State Univ – Co
Agterberg, Frits P., (613) 996-2374 Frits.Agterberg@nrcan-rncan.gc.ca, Univ of Ottawa – Gq
Ague, Jay J., (203) 432-3171 jay.ague@yale.edu, Yale Univ – Gx
Aharon, Paul, 205-348-2528 paharon@geo.ua.edu, Univ of Alabama – Ge
Ahearn, Sean C., (212) 772-5327 Graduate School of the City Univ of New York – Or
Ahearn, Sean C., (212) 772-5327 sca@geo.hunter.cuny.edu, Hunter College (CUNY) – Oi
Ahern, Judson L., (405) 325-4480 jahern@ou.edu, Univ of Oklahoma – Yg
Ahlquist, Jon E., (850) 644-1558 ahlquist@met.fsu.edu, Florida State Univ – Ow
Ahmad, Abd Rashid, 03-79674156 abrashid@um.edu.my, Univ of Malaya – Sc
Ahr, Wayne M., (979) 845-2447 ahr@tamu.edu, Texas A&M Univ – Gs
Ahrens, Donald C., (209) 575-6300 Modesto Junior College – Og
Ahrens, Thomas J., (626) 395-6906 tja@gps.caltech.edu, California Institute of Technology – Yx
Aiello, Ivano, (831) 771-4400 iaiello@mlml.calstate.edu, Moss Landing Marine Laboratories – Gu
Aiken, Carlos L., (972) 883-2450 aiken@utdallas.edu, Univ of Texas, Dallas – Yv
Aines, Roger D., (925) 423-7184 aines1@llnl.gov, Lawrence Livermore National Lab – Cg
Ainsley, William F., (901) 962-3493 ainsleyf@uncw.edu, Univ of North Carolina Wilmington – Og
Aitken, Alec E., (306) 966-5672 aaitken@arts.usask.ca, Univ of Saskatchewan – Gm

Aiyyer, Anantha, 919-515-7973 aaiyyer@ncsu.edu, North Carolina State Univ – Oa

Aja, Stephen U., (718) 951-5405 Graduate School of the City Univ of New York – Cl

Aja, Stephen U., (718) 951-5000 (Ext. 2881) suaja@brooklyn.cuny.edu, Brooklyn College (CUNY) – Cl

Akin, Michael, (601) 961-5533 michael_akin@deq.state.ms.us, Mississippi Office of Geology – Ge

Aksu, Ali E., (709) 737-8385 aaksu@sparky2.esd.mun.ca, Memorial Univ of Newfoundland – Gu

Al, Tom A., (506) 447-3189 tal@unb.ca, Univ of New Brunswick – Hw

Al-Aasm, Ihsan S., (519) 253- 2494) alaasm@uwindsor.ca, Univ of Windsor – Gd

Al-Egaily, Genevieve, (603) 271-1973 galegaily@des.state.nh.us, New Hampshire Geological Survey – Ge

Al-Eqabi, Ghassan, (314) 935-5817 ghassan@lithos.wustl.edu, Washington Univ – Ys

Al-Saad, Hamad A.alsaadh@qu.edu.qa, Univ of Qatar – Oo

Alaniz, Alfred N., (210) 733-2831 aalaniz@mail.accd.edu, San Antonio College – Oo

Alavi, Hedy, (410) 516-7091 hedy.alavi@jhu.edu, The Johns Hopkins Univ – Oo

Albach, Douglas C., (540) 654-1392 dalbach@mwc.edu, Univ of Mary Washington – Ge

Albanese, James R., (607) 436-2615 albanejr@oneonta.edu, SUNY, Oneonta – Pg

Alberstadt, Leonard P., 615-322-2160 leonard.p.alberstadt@vanderbilt.edu, Vanderbilt Univ – Gd

Alberts, E E., 573-882-1144 Univ of Missouri, Columbia – Hw

Albin, Edward F.ealbin@gpc.edu, Georgia Perimeter College – Xg

Albrecht, Jochen, 212-772-5221 jochen@hunter.cuny.edu, Hunter College (CUNY) – Oi

Albright, James N., (505) 667-4318 j_albright@lanl.gov, Los Alamos National Lab – Yg

Alcorn, Marion E., (409) 845-4329 alcorn@ariel.tamu.edu, Texas A&M Univ – Ow

Alde, Douglas, (505) 667-0488 dxa@lanl.gov, Los Alamos National Lab – Oo

Aldrich, L. Thomas, 202 478 8820 dept@dtm.ciw.edu, Carnegie Institution of Washington – Yg

Aldrich, M. James, (505) 667-1495 jaldrich@lanl.gov, Los Alamos National Lab – Gc

Aldridge, Keith D., (416) 736-2100 (Ext. 77705) keith@yorku.ca, York Univ – Yg

Aldushin, Kirill, 089/2180 4337 kirill.aldushin@lrz.uni-muenchen.de, LMU - Univ of Munich – Gz

Alexander, Clark R., 912-598-2329 clark@skio.peachnet.edu, Georgia Southern Univ – Ou

Alexander, Conel M., (202) 478-8478 alexander@dtm.ciw.edu, Carnegie Institution of Washington – Xc

Alexander, Shelton S., (814) 863-7246 shel@geosc.psu.edu, Pennsylvania State Univ, Univ Park – Ys

Alexander, Toni A., (334) 844-3417 alexato@auburn.edu, Auburn Univ – Oy

Alexander, Jr., Clark R., 912-598-2329 clark@skio.peachnet.edu, Georgia Southern Univ – Ou

Alexander, Jr., E. Calvin, 612-624-3517 alexa001@umn.edu, Univ of Minnesota, Twin Cities – Cc

Alford, Matthew H., 206-221-3257 malford@apl.washington.edu, Univ of Washington – Op

Algeo, Catherine, (270) 745-5922 katie.algeo@wku.edu, Western Kentucky Univ – Oi

Algeo, Thomas J., (513) 556-4195 thomas.algeo@uc.edu, Univ of Cincinnati – Gs

Algin, Barbara, (212) 854-2905 ba110@columbia.edu, Columbia Univ – Oo

Ali, Robbie, (412) 624-8780 Univ of Pittsburgh – Oo

All, John, (270) 745-5975 john.all@wku.edu, Western Kentucky Univ – Oy

Allan, Craig J., 704-687-5999 cjallan@email.uncc.edu, Univ of North Carolina, Charlotte – Hg

Allan, Deborah L., (612) 625-3158 dallan@soils.umn.edu, Univ of Minnesota, Twin Cities – Sc

Allard, Gilles O., (706) 542-2420 Univ of Georgia – Eg

Allard, Stephen T., 507-457-2739 sallard@winona.edu, Winona State Univ – Gc

Allen, Albert E., (307) 234-4019 Casper College – Gr

Allen, Alistair, +353 21 902769 a.allen@ucc.ie, Univ College Cork – Gp

Allen, Bruce, 505-366-2531 allenb@gis.nmt.edu, New Mexico Institute of Mining & Technology – Gm

Allen, Dennis M., (803) 546-3623 Univ of South Carolina – Ob

Allen, Diana M., (604) 291-3967 dallen@sfu.ca, Simon Fraser Univ – Hy

Allen, Donald B.dballen@colby.edu, Colby College – Eg

Allen, Douglasdallen@salemstate.edu, Salem State College – Cl

Allen, Gary C.gallen@uno.edu, Univ of New Orleans – Gp

Allen, John S., (541) 737-2928 jallen@coas.oregonstate.edu, Oregon State Univ – Op

Allen, Jonathan C., (541) 574-5568 jonathan.allen@dogami.state.or.us, Oregon Dept of Geology & Mineral Industries – Gm

Allen, Joseph L., (304) 384-5238 allenj@concord.edu, Concord Univ – Gc

Allen, Peter M., (254)710-2189 peter_allen@baylor.edu, Baylor Univ – HgNg

Allen, Richard M., (510) 642-1275 rallen@seismo.berkeley.edu, Univ of California, Berkeley – Ys

Allen, Richardson B., (801) 581-7574 pallen@egi.utah.edu, Univ of Utah – Go

Allen, Susan E., (604) 822-2828 sallen@eos.ubc.ca, Univ of British Columbia – Op

Allen, Tatiana B., (423) 425-4520 tatiana-allen @.utc.edu, Univ of Tennessee, Chattanooga – Oo

Allen, Ted T., (604) 291-3967 tallen@ju.edu, Jacksonville Univ – Oo

Allen, Timothy T., (603) 358-2571 tallen@keene.edu, Keene State College – Hw

Allen-King, Richelle, (716) 645-6800 (Ext. 3963) richelle@geology.buffalo.edu, SUNY, Buffalo – Cg

Allen-Lafayette, Zehdreh, (609) 292-2576 zehdreh.allen-lafayette@dep.state.nj.us, New Jersey Geological Survey – Oi

Allensworth, John, 404 413-5772 jallensworth@gsu.edu, Georgia State Univ – Oy

Aller, Josephine Y., (631) 632-8655 jaller@notes.cc.sunysb.edu, SUNY, Stony Brook – Ob

Aller, Robert C., (631) 632-8746 robert.aller@sunysb.edu, SUNY, Stony Brook – Cm

Alley, Marcus M., (540) 231-9777 malley@vt.edu, Virginia Polytechnic Institute & State Univ – Sc

Alley, Richard B., (814) 863-1700 ralley@essc.psu.edu, Pennsylvania State Univ, Univ Park – Gl

Alley, Ronald E., (818) 354-0751 ron@lithos.jpl.nasa.gov, Jet Propulsion Lab – Or

Allis, Richard G., (801) 537-3300 nrugs.rallis@state.ut.us, Utah Geological Survey – Eo

Allison, David T., 251 460-6381 dallison@jaguar1.usouthal.edu, Univ of South Alabama – Gc

Allison, Mead A., (504) 862-3197 malliso@tulane.edu, Tulane Univ – Gs

Allison, Tami L.tami.allison@dnr.mo.gov, Missouri Dept of Natural Resources – Oo

Allmendinger, Richard W., (607) 255-3376 allmendin@geology.cornell.edu, Cornell Univ – Gc

Allmon, Warren D., (607) 273-6623 wda1@cornell.edu, Cornell Univ – Pg

Allmon, Warren D., (607) 273-6623 (Ext. 14) wdal@cornell.edu, Paleontological Research Institution – Pe

Allong, Albert F., (713) 495-2102 Houston Comm College System – Oo

Alm, Ginny, (307) 766-3114 Univ of Wyoming – Oo

Almendinger, James E., 651-433-5953 dinger@smm.org, Univ of Minnesota, Twin Cities – Pe

Almy, Charles C., (336) 316-2263 calmy@guilford.edu, Guilford College – Gg

Alsdorf, Douglas E., 614 247-6908 alsdorf@geology.ohio-state.edu, Ohio State Univ – Hg

Alsleben, Helge, 817-257-5455 h.alsleben@tcu.edu, Texas Christian Univ – GctGg

Alt, Jeffrey C., (734) 764-8380 jalt@umich.edu, Univ of Michigan – Ou

Altaner, Stephen P., (217) 244-1244 altaner@uiuc.edu, Univ of Illinois, Urbana-Champaign – Gz

Altenbach, Alexander, 089/2180 6598 a.altenbach@lrz.uni-muenchen.de, LMU - Univ of Munich – Pg

Altermann, Wladyslaw, 089/2180 6552 wlady.altermann@iaag.geo.uni-muenchen.de, LMU - Univ of Munich – Gg
Altin, Deniz, 678-407-5445 daltin@gpc.edu, Georgia Perimeter College – Gz
Altizer, Sonia, 404 727-4252 saltize@emory.edu, Emory Univ – Sf
Altman, Kathleen A., (775) 784-1813 kaltman@mines.unr.edu, Univ of Nevada, Reno – Nx
Alumbaugh, David L., (510) 215-4227 dalumbaugh@berkeley.edu, Univ of California, Berkeley – Ye
Alvarez, Walter, 510-642-2602 platetec@berkeley.edu, Univ of California, Berkeley – Grh
Alvarez, Walter, (510) 642-2602 platetec@garnet.berkeley.edu, Univ of California, Berkeley – Gr
Alwash, Suzi, (310) 660-3368 El Camino College – Gg
Amati, Lisa M., (315) 267-3482 amatilm@potsdam.edu, SUNY, Potsdam – Pi
Amato, Jeffrey M., 575-646-3017 amato@nmsu.edu, New Mexico State Univ, Las Cruces – Gc
Amaury, Legait, +33 3 83 59 64 00 Ecole Nationale Supérieure de Géologie (ENSG) – Eg
Ambinakudige, Shrinidhi, 662-325-2905 shrinidhi@geosci.msstate.edu, Mississippi State Univ – Oiy
Ambrose, William A., (512) 471-0258 william.ambrose@beg.utexas.edu, Univ of Texas at Austin, Jackson School of Geosciences – Gs
Ambruster, W. Scott, (907) 474-7161 ffwsa@aurora.alaska.edu, Univ of Alaska, Fairbanks – Oo
Amelung, Falk, 305 421-4949 famelung@rsmas.miami.edu, Univ of Miami – Oo
Amend, Jan P., (314) 935-8651 amend@wustl.edu, Washington Univ – Co
Amenta, Roddy V., (540) 568-6674 amentarv@jmu.edu, James Madison Univ – Gc
Ames, Daniel P., 208-282-7851 amesdani@isu.edu, Idaho State Univ – Oi
Amin, Isam E., 330-941-2293 ieamin@ysu.edu, Youngstown State Univ – Hw
Amin, Shahalam M., (570) 389-4790 smamin@bloomu.edu, Bloomsburg Univ – SoOyn
Amini, Hassan, (714) 278-3561 hamini@fullerton.edu, California State Univ, Fullerton – Ge
Amler, Michael, 089/2180 6602 amler@lrz.uni-muenchen.de, LMU - Univ of Munich – Pg
Ammon, Charles J., (814) 865-2310 cammon@geosc.psu.edu, Pennsylvania State Univ, Univ Park – Ys
Amon, Rainier, 409 740 4719 amonr@tamug.edu, Texas A&M Univ – Ob
Amonette, Alexandra B., (509) 376-5019 alexandra.amonette@pnl.gov, Pacific Northwest National Lab – Cg
Amoozegar, Aziz, (919) 515-3967 North Carolina State Univ – Sp
Amor, John, (604) 822-6933 jamor@eos.ubc.ca, Univ of British Columbia – Oo
Amthor, Jeff, (865) 576-2773 Oak Ridge National Lab – Sf
Anandakrishnan, Sridhar, (814) 863-6742 sak@essc.psu.edu, Pennsylvania State Univ, Univ Park – Ys
Anastasio, David J., (610) 758-5117 dja2@lehigh.edu, Lehigh Univ – Gc
Anders, Mark H., (845) 365-8361 manders@ldeo.columbia.edu, Columbia Univ – Gc
Anders, Tania, (361) 825-3755 tanders@falcon.tamucc.edu, Texas A&M Univ, Corpus Christi – Gs
Andersen, C. Brannon, (864) 294-3366 brannon.andersen@furman.edu, Furman Univ – Cl
Andersen, David W., (408) 924-5014 andersen@geosun.sjsu.edu, San Jose State Univ – Gs
Andersen, Raymond J., (604) 822-4511 randersn@eos.ubc.ca, Univ of British Columbia – Oc
Anderson, Alan J., (902) 867-2309 Memorial Univ of Newfoundland – Gi
Anderson, Alan J., (902) 867-2309 aanderso@stfx.ca, Saint Francis Xavier Univ – Gx
Anderson, Alan J., 902-863-5413 aanderso@stfx.ca, Dalhousie Univ – Cg
Anderson, Bruce, (617) 353-4807 brucea@bu.edu, Boston Univ – Oy
Anderson, Callum Robert, +27 41 5042811 ocacra@upe.ac.za, Univ of Port Elizabeth – Gi
Anderson, David M., 303-497-6237 Univ of Colorado – Og
Anderson, Don L., (626) 395-6901 dla@gps.caltech.edu, California Institute of Technology – Ys
Anderson, F. S., 808-956-6887 anderson@higp.hawaii.edu, Univ of Hawai'i, Manoa – Xy
Anderson, G M., (416) 978-2062 Univ of Toronto – Cp
Anderson, J. Lawford, (213) 740-6727 anderson@usc.edu, Univ of Southern California – Gi
Anderson, James G., (617) 495-5922 anderson@huarp harvard.edu, Harvard Univ – Oa
Anderson, James L., 808-974-7640 jamesa@hawaii.edu, Univ of Hawai'i at Hilo – Gc
Anderson, James L., (612) 625-0279 jandersn@soils.umn.edu, Univ of Minnesota, Twin Cities – Sd
Anderson, Jennifer L., 507-457-2457 jlanderson@winona.edu, Winona State Univ – Yg
Anderson, John, (775) 784-4265 jga@seismo.unr.edu, Univ of Nevada, Reno – Ys
Anderson, John B., (713) 348-4652 Rice Univ – Gu
Anderson, John G., (775) 784-4265 jga@seismo.unr.edu, Univ of Nevada, Reno – Ys
Anderson, John L., (502) 852-6844 Univ of Louisville – Oo
Anderson, John R., (770) 551-3125 janderso@gpc.edu, Georgia State Univ – Pg
Anderson, Ken B., 618-453-7389 kanderson@geo.siu.edu, Southern Illinois Univ Carbondale – Co
Anderson, Laurie C., 225-578-2153 laurie@geol.lsu.edu, Louisiana State Univ – Pi
Anderson, Mark R., (402) 472-6656 manderson4@unl.edu, Univ of Nebraska, Lincoln – Oa
Anderson, Mary P., (608) 262-2396 andy@geology.wisc.edu, Univ of Wisconsin, Madison – Hw
Anderson, Neil L., (573) 341-4852 nanders@umr.edu, Univ of Missouri, Rolla – Ye
Anderson, Orson L., (310) 825-2386 Univ of California, Los Angeles – Yx
Anderson, Orson L., (310)825-2386 Univ of California, Los Angeles – Yx
Anderson, Patricia M., 206-685-7682 pata@u.washington.edu, Univ of Washington – Ple
Anderson, Raymond R., (319) 335-1589 randerson@igsb.uiowa.edu, Univ of Iowa – Gr
Anderson, Robert, (412) 624-8780 Univ of Pittsburgh – Ge
Anderson, Robert, 303-735-4684 Robert.S.Anderson@Colorado.EDU, Univ of Colorado – Gm
Anderson, Robert F., (845) 365-8508 boba@ldeo.columbia.edu, Columbia Univ – Oc
Anderson, Robert G., (604) 822-2449 Univ of British Columbia – Gt
Anderson, Robert S., (831) 459-3342 randerson@es.ucsc.edu, Univ of California, Santa Cruz – Gm
Anderson, Roger N., (845) 365-8335 anderson@ldeo.columbia.edu, Columbia Univ – Yh
Anderson, Roger Y., (505) 277-1639 Univ of New Mexico – Gn
Anderson, Sharolyn, 303-871-3378 sander24@du.edu, Univ of Denver – Oi
Anderson, Stephen H., 573-882-6303 andersons@missouri.edu, Univ of Missouri, Columbia – Sp
Anderson, Steven W., (605) 642-6506 Black Hills State Univ – Gv
Anderson, Suzanne, 303-765-0951 suzanne.anderson@colorado.edu, Univ of Colorado – Gm
Anderson, Thomas B., (707) 664-2176 tom.anderson@sonoma.edu, Sonoma State Univ – Gs
Anderson, Thomas F., 217-333-0355 tfanders@uiuc.edu, Univ of Illinois, Urbana-Champaign – Cs
Anderson, Thomas H., (412) 624-8870 taco@pitt.edu, Univ of Pittsburgh – Gt
Anderson, Virginia L., (972) 883-2410 gander@utdallas.edu, Univ of Texas, Dallas – Oo
Anderson, Wayne I., (319) 273-3125 wayne.anderson@uni.edu, Univ of Northern Iowa – Pg
Anderson, William, (617) 358-0208 bander@bu.edu, Boston Univ – Oo
Anderson, William T., (305) 348-2693 andersow@fiu.edu, Florida International Univ – Cs
Anderson, Jr., Alfred T., (773) 702-8138 Univ of Chicago – Gv
Anderson, Jr., John R., 770-274-5077 janderso@gpc.edu, Georgia Perimeter College – Ps
Anderton, John B., (906) 227-2500 janderto@nmu.edu, Northern Michigan Univ – Ga
Andrachick, Karen L., 412.442.5826 kandrachic@state.pa.us, Pennsylvania Bureau of Topographic & Geologic Survey – Gg
Andreasen, David C., (410) 260-8814 Maryland Geological Survey – Hy
Andren, Anders W., (608) 262-2470 Univ of Wisconsin, Madison – Oc

Andres, A. S., 302-831-2833 asandres@udel.edu, Univ of Delaware – Hy
Andres, Robert, 505.662.7973 andres@space.edu, New Mexico Institute of Mining and Technology – Gv
Andrew, Joseph E.jeandrew, Youngstown State Univ – Gc
Andrews, John R., (512) 471-4951 john.andrews@beg.utexas.edu, Univ of Texas at Austin, Jackson School of Geosciences – Gg
Andrews, John T., 303-492-5183 john.t.andrews@colorado.edu, Univ of Colorado – Gl
Andrews, Joy, (510) 885-3492 joy.andrews@csueastbay.edu, California State Univ, East Bay – Cl
Andrews, Richard D., (405) 325-3031 rdandrews@ou.edu, Univ of Oklahoma – Go
Andrus, C. Fred T., 205-348-5177 fandrus@geo.ua.edu, Univ of Alabama – Ge
Andrus, Richard E., (607) 777-2453 Binghamton Univ – Py
Aneja, Viney P., 91951557808 viney_aneja@ncsu.edu, North Carolina State Univ – Oa
Angel, Ross J., (540) 231-7974 rangel@vt.edu, Virginia Polytechnic Institute & State Univ – Gz
Angelo, Joseph A., (407) 768-8000 Florida Institute of Technology – Og
Angelopoulos, Vassilis, 310-794-7090 vassilis@ucla.edu, Univ of California, Los Angeles – XyOa
Angino, Ernest E., (785) 843-7503 Univ of Kansas – Cl
Anglen, Brandy, (559) 442-8039, brandy.anglen@fresnocitycollege.edu, Fresno City College – Gg
Anis, Ayal, (409) 740-4987 anisa@tamug.tamu.edu, Texas A&M Univ – Op
Anooshehpoor, Rasool, (775) 784-1954 rasool@seismo.unr.edu, Univ of Nevada, Reno – Ys
Anovitz, Lawrence M., (865) 574-7041 anovitzlm@ornl.gov, Oak Ridge National Lab – Cp
Ansdell, Kevin M., (306) 966-5698 kevin.ansdell@usask.ca, Univ of Saskatchewan – EgCgGt
Anstey, Robert L., 517-353-9009 anstey@msu.edu, Michigan State Univ – Pi
Antar, Ali A., (860) 832-2931 antar@ccsu.edu, Central Connecticut State Univ – Oo
Anthony, Elizabeth Y., (915) 747-5483 eanthony@geo.utep.edu, Univ of Texas, El Paso – Gi
Anthony, Kovschak, 970-248-1947 akovscha@mesastate.edu, Mesa State College – Gg
Anthony, Leona M., (808) 956-8763 lanthony@soest.hawaii.edu, Univ of Hawai'i, Manoa – Oo
Anthony, Nina, 864-294-2052 nina.anthony@furman.edu, Furman Univ – Oo
Antonacci, Vince, (615) 532-1507 Vince.Antonacci@state.tn.us, Tennessee Division of Geology – GgOi
Antonellini, Marcomarcoa@pangea.stanford.edu, Stanford Univ – Yr
Appel, Christopher (Chip) S., (805) 756-1691 cappel@calpoly.edu, California Polytechnic State Univ – Sc
Appel, Donajean M., (410) 554-5550 dmappel@mgs.md.gov, Maryland Geological Survey – Oo
Appelgate, T. Bruce, (808) 956-9720 bruce@soest.hawaii.edu, Univ of Hawai'i, Manoa – Yr
Applegarth, Michael T., (717) 477-1712 mtappl@ship.edu, Shippensburg Univ – Or
Appold, Martin, 573/882-0701 AppoldM@missouri.edu, Univ of Missouri – Hw
April, Richard, (315) 228-7212 rapril@mail.colgate.edu, Colgate Univ – Cg
Arabas, Karen, 503-370-6666 karabas@willamette.edu, Willamette Univ – Oo
Arabasz, Walter J., (801) 581-7410 arabasz@seis.utah.edu, Univ of Utah – Ys
Arain, Altaf M., (905) 525-9140 (Ext. 27941) arainm@mcmaster.ca, McMaster Univ – Hg
Arakawa, Akio, (310) 825-9874 aar@atmos.ucla.edu, Univ of California, Los Angeles – Oa
Archambault, Guy, (418) 545-5011 Univ du Quebec a Chicoutimi – Nm
Archer, Allen W., (785) 532-2244 aarcher@ksu.edu, Kansas State Univ – Gr
Archer, David, (773) 702-0823 archer@popeye.uchicago.edu, Univ of Chicago – Yr
Archer, Robert, (573) 368-2161 Missouri Dept of Natural Resources – Oo
Archibald, Doug A., (613) 545-6594 Queen's Univ – Cc
Archuleta, Ralph J., (805) 893-4477 Univ of California, Santa Barbara – Ys
Archuleta, Ralph J., (805) 893-8441 ralph@quake.crustal.ucsb.edu, Univ of California, Santa Barbara – Ys
Arehart, Greg B., (775) 784-6470 arehart@unr.edu, Univ of Nevada, Reno – Eg
Arens, Nan, (510) 642-3281 nanarens@violet.berkeley.edu, Univ of California, Berkeley – Pb
Arens, Nan Crystal, 315-781-3930 arens@hws.edu, Hobart & William Smith Colleges – Po
Argast, Anne S., 260-481-6252 argast@ipfw.edu, Indiana Univ / Purdue Univ, Fort Wayne – Gdz
Argenbright, Robert T., (910) 962-3498 argenbrightr@uncw.edu, Univ of North Carolina Wilmington – Oo
Argow, Britt A., 781-283-3165 bargow@wellesley.edu, Wellesley College – Gs
Arikan, Alparslan, +90 (312) 2977720 arikana@hacettepe.edu.tr, Hacettepe Univ – Hg
Ariya, Parisa A., (514) 398-3615 parisa.ariya@mcgill.ca, McGill Univ – Oa
Arkani-Hamed, Jafar, 514-398-6767 jafar@eps.mcgill.ca, McGill Univ – Xy
Arking, Albert, (301) 299-2478 arking@jhu.edu, The Johns Hopkins Univ – Oa
Armbrust, Virginia E., (206) 616-1783 armbrust@ocean.washington.edu, Univ of Washington – Ob
Armi, Laurence, (619) 534-6843 larmi@ucsd.edu, Univ of California, San Diego – Op
Armour, Jake, 704-687-5968 Univ of North Carolina, Charlotte – Gg
Armstrong, David E., (608) 262-2470 Univ of Wisconsin, Madison – Oc
Armstrong, Felicia P., 330-941-1385 fparmstrong@ysu.edu, Youngstown State Univ – Sb
Armstrong, Howard A., 0191 3744780 h.a.armstrong@durham.ac.uk, Durham Univ – Pg
Armstrong, Phillip A., (714) 278-3169 parmstrong@fullerton.edu, California State Univ, Fullerton – Yg
Armstrong, Robert, (631) 632-3088 SUNY, Stony Brook – Ob
Armstrong, Robert A., (609) 258-5260 raa@splash.princeton.edu, SUNY, Stony Brook – Oc
Armstrong, Robert L.roarmstr@nmsu.edu, New Mexico State Univ, Las Cruces – Oa
Armstrong, William, (505) 667-9517 wta@lanl.gov, Los Alamos National Lab – Yg
Arnason, John G., (518) 442-4474 arnason@albany.edu, SUNY, Albany – Gx
Arnaud, Emmanuelle, (519) -82-4-41 (Ext. ext 58087) earnaud@uoguelph.ca, Univ of Guelph – Gl
Arnold, Anthony J., (850) 644-4228 arnold@gly.fsu.edu, Florida State Univ – Pm
Arnold, David L., 301/687-4053 dlarnold@frostburg.edu, Frostburg State Univ – Oa
Arnold, Deborahanold@fiu.edu, Florida International Univ – Gg
Arnold, Jayne H., 765-983-1429 arnolja@earlham.edu, Earlham College – Oo
Arnold, Jeffrey G., (254) 710-2361 Baylor Univ – Hs
Arnold, Zach M., (510) 642-1821 Univ of California, Berkeley – Po
Arnone, John, (775) 673-7445 jarnone@dri.edu, Desert Research Institute – Py
Arnone, John J., (775) 673-7445 jarnone@dri.edu, Univ of Nevada, Reno – Sp
Arnott, R. William C., (613)562-5800 Ext 6854 warnott@uottawa.ca, Univ of Ottawa – Gs
Aronson, James L., 603-646-3338 James.Aronson@dartmouth.edu, Dartmouth College – Cc
Aronson, Richard B., (334) 861-7567 raronson@jaguar1.usouthal.edu, Univ of South Alabama – Ob
Arrhenius, Gustaf, (858) 534-2961 arrhenius@ucsd.edu, Univ of California, San Diego – Cg
Arrigo, Kevin, (650) 723-3599 arrigo@stanford.edu, Stanford Univ – Og
Arritt, Raymond W., (515) 294-9870 rwarritt@iastate.edu, Iowa State Univ of Science & Technology – Ow
Arrowsmith, Ramon, (480) 965-3541 ramon.arrowsmith@asu.edu, Arizona State Univ – Gc
Arthur, Jonathan D., (850) 488-9380 Florida Geological Survey – Gi
Arthur, Michael A., (814) 863-6054 arthur@geosc.psu.edu, Pennsylvania State Univ, Univ Park – Ou
Artiola, Janick F., (520) 621-3516 jartiola@ag.arizona.edu, Univ of Arizona – Sc
Arundale, Wendy H., (907) 474-7039 Univ of Alaska, Fairbanks – Oo

Arundel, Samantha, (928) 523-6651 Sam.Arundel@nau.edu, Northern Arizona Univ – Oiy
Arvidson, Raymond E., (314) 935-5679 arvidson@wunder.wustl.edu, Washington Univ – Or
Arvidson, Raymond E., (314) 935-5679 arvidson@wunder.wustl.edu, Washington Univ – Oo
Arya, Satyapal S., (919) 515-7002 pal_arya@ncsu.edu, North Carolina State Univ – Oa
Asencio, Eugenio, 787 2653845 asencio@uprm.edu, Univ of Puerto Rico – Ys
Ash, Richard, (301) 405-7504 rdash@geol.umd.edu, Univ of Maryland – Ca
Ashby, Michael, (718) 262-2597 York College (CUNY) – Ca
Asher, Pranoti M., (912) 681-0338 PAsher@GeorgiaSouthern.Edu, Georgia Southern Univ – Gi
Ashland, Francis X., (801) 537-3380 nrugs.fashland@state.ut.us, Utah Geological Survey – Ng
Ashley, Gail M., (732) 445-2221 gmashley@rci.rutgers.edu, Rutgers, The State Univ of New Jersey – Gs
Ashley, Jr., Joseph M., (406) 994-6904 jashley@montana.edu, Montana State Univ – Oy
Ashour-Abdalla, Maha, (310) 825-8881 mabdalla@igpp.ucla.edu, Univ of California, Los Angeles – Oo
Ashton, Kenneth E., (306) 787-1909 Univ of Regina – Gt
Ashwood, Tom, (865) 574-7542 Oak Ridge National Lab – Ge
Ashworth, Allan C., 701-231-7919 allan.ashworth@ndsu.edu, North Dakota State Univ – Pe
Asimow, Paul D., 626.395.4133 asimow@gps.caltech.edu, California Institute of Technology – Cp
Askren, Daniel R., 229-931-2329 daskren@canes.gsw.edu, Georgia Southwestern State Univ – Gz
Aslan, Andres, (970) 248-1614 aaslan@mesastate.edu, Mesa State College – Gm
Asmerom, Yemane, (505) 277-4204 asmerom@unm.edu, Univ of New Mexico – Cc
Asper, Vernon L., (228) 688-3178 vernon.asper@usm.edu, Univ of Southern Mississippi – Ou
Asphaug, Erik, (831) 459-2260 easphaug@es.ucsc.edu, Univ of California, Santa Cruz – Xg
Asquith, George B., (806) 742-3154 george.asquith@ttu.edu, Texas Tech Univ – Go
Asselin, Esther, (418) 654-2612 easselin@nrcan.gc.ca, Universite du Quebec – Pl
Aster, Richard C., (505) 835-5924 aster@nmt.edu, New Mexico Institute of Mining and Technology – Ys
Astin, Timothyt.r.astin@rdg.ac.uk, Univ of Reading – Yg
Atchley, Stacy C., (254) 710-2196 stacy_atchley@baylor.edu, Baylor Univ – Gro
Atekwana, Eliot A., (317) 274-7969 eatekwan@iupui.edu, Indiana Univ / Purdue Univ, Indianapolis – Hw
Athaide, Dileep, (604) 984-1771 dathaide@eos.ubc.ca, Univ of British Columbia – Gg
Athaide, Dileep J.dathaide@capcollege.bc.ca, Capilano College – Oe
Atkinson, Gail M., 519-661-4207 gatkins6@uwo.ca, Univ of Western Ontario – YssYs
Atkinson, Larry P., (757) 683-4926 atkinson@ccpo.odu.edu, Old Dominion Univ – Op
Atkinson, Shannon K., (808) 956-8725 shannon@soest.hawaii.edu, Univ of Hawai'i, Manoa – Oo
Atkinson, Jr., William W., 303-492-6103 william.atkinson@colorado.edu, Univ of Colorado – Em
Atreya, Sushil K., (313) 936-0489 Univ of Michigan – Oa
Attig, John W., (608) 262-6131 jwattig@wisc.edu, Univ of Wisconsin, Extension – Gl
Attoh, Kodjopa, (607) 255-1039 attoh@geology.cornell.edu, Cornell Univ – Eg
Attrep, Moses, (505) 667-0088 Los Alamos National Lab – Oe
Attwood, Paul J., (941) 338-2362 Florida Geological Survey – Go
Atudorei, Nieu-Viorelatudorei@unm.edu, Univ of New Mexico – Cs
Atwater, Kim, (801) 581-6553 katwater@mines.utah.edu, Univ of Utah – Oo
Atwater, Tanya M., (805) 893-4483 Univ of California, Santa Barbara – Yr
Atwood, Tracy, (508) 289-3629 Woods Hole Oceanographic Insti – Cm
Au, Whitlow W. L., (808) 291-6432 Univ of Hawai'i, Manoa – Oo
Aubert, John E., (916) 484-8637 aubertj@arc.losrios.edu, American River College – Oy
Aubertin, Michel, (514) 340-4046 Ecole Polytechnique – Nm
Aubry, Marie-Pierre, (732) 445-2044 aubry@isem.univ-montp2.fr, Rutgers, The State Univ of New Jersey – Yr
Audet, Celine, (418) 723-1986 (Ext. 1744) celine_audet@uqar.qc.ca, Universite du Quebec a Rimouski – Ob
Auerbach, Michael, (775) 673-7409 auerbach@dri.edu, Desert Research Institute – Py
Aughenbaugh, Nolan B., (662) 915-5819 naughenb@olemiss.edu, Univ of Mississippi – Nr
Aulbach, Sonja, (301) 405-4088 saulbach@geol.umd.edu, Univ of Maryland – Cg
Aurnou, Jonathan M., (202) 478-8828 aurnou@dtm.ciw.edu, Univ of California, Los Angeles – XyYmg
Ausbrooks, Scott, 501 683-0119 scott.ausbrooks@arkansas.gov, Arkansas Geological Commission – Gg
Ausich, William I., (614) 292-3353 ausich.1@osu.edu, Ohio State Univ – Po
Austin, George S., (505) 835-5230 george@gis.nmt.edu, New Mexico Institute of Mining & Technology – En
Austin, Philip, (604) 822-2175 paustin@eos.ubc.ca, Univ of British Columbia – Oa
Austin, Jr., James A., (512) 471-0450 jamie@ig.utexas.edu, Univ of Texas at Austin, Jackson School of Geosciences – Gu
Autin, Whitney J., 585-395-5738 dirtguy@esc.brockport.edu, SUNY College at Brockport – Gs
Avard, Claudette, (514) 987-3000 avard.claudette@uqam.ca, Universite du Quebec a Montreal – Oo
Avary, Katherine L., 304 594 2331 avary@geosrv.wvnet.edu, West Virginia Univ – Go
Ave Lallemant, Hans G., (713) 348-4889 ave@rice.edu, Rice Univ – Gc
Avent, Jon C.jon_avent@csufresno.edu, California State Univ, Fresno – Gr
Awramik, Stanley M., (805) 893-3830 Univ of California, Santa Barbara – Po
Axen, Gary, 505.835.5178 gaxen@ees.nmt.edu, New Mexico Institute of Mining and Technology – Gc
Ayad, Yasser M.yayad@clarion.edu, Clarion Univ – Oi
Aydin, Atilla, (650) 725-8708 aydin@pangea.stanford.edu, Stanford Univ – Gc
Ayer, John, (705) 670-5996 john.ayer@ndn.gov.on.ca, Univ of Ottawa – Cg
Ayers, John C., 615-322-2158 john.c.ayers@vanderbilt.edu, Vanderbilt Univ – Cp
Ayliffe, Linda K., 801-587-7865 layliffe@mines.utah.edu, Univ of Utah – Cl
Ayres, Lorne D., (204) 474-9371 Univ of Manitoba – Gv
Azam, Farooq, (858) 534-6850 fazam@ucsd.edu, Univ of California, San Diego – Ob
Azmy, Karem, (702) 895-1239 Univ of Nevada, Las Vegas – Gd

B

Baadsgaard, Halfdanbud.baadsgaard@ualberta.ca, Univ of Alberta – Cc
Baarli, Gudveig, (413) 597-2491 gudveig.baarli@williams.edu, Williams College – Ps
Babaei, Abbed, (216) 687-3506 a.babaei@csuohio.edu, Cleveland State Univ – Gc
Babaie, Hassan A., (404) 413-5766 hbabaie@gsu.edu, Georgia State Univ – Gc
Babb, David M., (814) 863-3918 dbabb@essc.psu.edu, The Pennsylvania State Univ, Univ Park – Ow
Babcock, Loren E., (614) 292-0358 babcock.5@osu.edu, Ohio State Univ – Po
Babcock, R. S., (360) 650-3592 babcock@wwu.edu, Western Washington Univ – Cg
Bacchus, Tania S., (802) 635-1329 Tania.Bacchus@jsc.edu, Johnson State College – GuOw
Bachhuber, Frederick W., (702) 895-3120 bachhubf@nevada.edu, Univ of Nevada, Las Vegas – Pl
Bachle, Peter, (573) 368-2161 Missouri Dept of Natural Resources – Oo
Bachmann, Olivier, 206-543-7615 bachmano@u.washington.edu, Univ of Washington – Giv
Bachtadse, Valerian, 089/2180 4237 valerian@geophysik.uni-muenchen.de, LMU - Univ of Munich – Yg
Backhus, Debera A., (812) 855-0563 dbackhus@indiana.edu, Indiana Univ, Bloomington – Cl
Backus, David H., (413) 597-4086 dbackus@williams.edu, Williams College – Ps

Backus, George E., (858) 534-2468 gebackus@ucsd.edu,
Univ of California, San Diego – Yg
Backus, Milo M., (512) 471-6942 backus@mail.utexas.edu,
Univ of Texas at Austin, Jackson School of Geosciences – Ye
Bacon, Corinne E., (801) 581-5126 Univ of Utah – Ca
Bacon, Diana H., (509) 372-6132 diana.bacon@pnl.gov,
Pacific Northwest National Lab – Hq
Bacon, Michael P., (508) 289-2559 mbacon@whoi.edu,
Woods Hole Oceanographic Institution – Oc
Bacon, Robert, (573) 526-0807 nrbaconV,
Missouri Dept of Natural Resources – Hs
Bacon, Steven N., 775-673-7473 Steve.Bacon@dri.edu,
Desert Research Institute – Gm
Bada, Jeffrey L., (858) 534-4258 jbada@ucsd.edu,
Univ of California, San Diego – Co
Badger, Robert L., (315) 267-2624 badgerrl@potsdam.edu,
SUNY, Potsdam – Gi
Badgley, Catherine E., (734) 763-6448 cbadgley@umich.edu,
Univ of Michigan – Pv
Badurek, Chris, (828) 262-7054 badurekca@appstate.edu,
Appalachian State Univ – Oi
Baedke, Steven J., (540) 568-6156 baedkesj@jmu.edu,
James Madison Univ – Hy
Baer, Emma, 206-546-4659 ebaer@shoreline.edu,
Shoreline Community College – Gg
Baer, James L., (801) 374-9368 jim_baer_l@byu.edu,
Brigham Young Univ – Go
Baeseman, Jenny L., (609) 258-6294 baeseman@princeton.edu,
Princeton Univ – Ob
Bagtzoglou, Ross, (212) 854-3154 Columbia Univ – Hw
Baharlou, Alan, 217-581-2626 cfab@ux1.eiu.edu,
Eastern Illinois Univ – Cg
Bahee-Cody, Doris, 505-368-3630 Dine College, Shiprock – Oo
Bahr, Jean M., (608) 262-5513 jmbahr@geology.wisc.edu,
Univ of Wisconsin, Madison – Hw
Bahrmann, Chad, 814-865-9500 cbahrmann@psu.edu,
The Pennsylvania State Univ, Univ Park – Ow
Bai, Baojun, 573-341-4016 baib@umr.edu,
Univ of Missouri, Rolla – Np
Baidya Roy, Somnath, (217) 244-1123 sbroy@atmos.uiuc.edu,
Univ of Illinois, Urbana-Champaign – Oa
Bailey, Christopher M., (757) 221-2445 cmbail@wm.edu,
College of William & Mary – Gc
Bailey, David G., (315) 859-4142 dbailey@hamilton.edu,
Hamilton College – Gi
Bailey, Jack B., (309) 298-1481 Western Illinois Univ – Pg
Bailey, Judy, 61 02 4921 5415 Judy.Bailey@newcastle.edu.au,
Univ of Newcastle – Ec
Bailey, Richard C., (416) 978-3231 bailey@geophy.physics.utoronto.ca,
Univ of Toronto – Yg
Bailey, Richard C., (416) 978-3231 bailey@physics.utoronto.ca,
Univ of Toronto – Yg
Bailey, Richard H., (617) 373-3181 r.bailey@neu.edu,
Northeastern Univ – Pi
Bailey, Sonja S., (618)453-4362 sbailey@geo.siu.edu,
Southern Illinois Univ Carbondale – Oo
Bailey, Thomas G., (407) 465-2400 Univ of South Florida, Saint Petersburg – Ob
Bailie, Catherine M., (719) 389-6621 cbailie@coloradocollege.edu,
Colorado College – Oo
Bain, Daniel J., (412) 624-8780 dbain@pitt.edu,
Univ of Pittsburgh – Hg
Bair, Edwin S., (614) 292-6197 bair.1@osu.edu,
Ohio State Univ – Hw
Baird, Gordon C., (716) 673-3840 baird@fredonia.edu,
SUNY, Fredonia – Ps
Baird, Ronald C., (713) 301-2448 Univ of South Florida, Saint Petersburg – Ob
Bajjali, William, (715) 394 8056 wbajjali@uwsuper.edu,
Univ of Wisconsin, Superior – Gg
Bakel, Allen J., (630) 252-5486 Argonne National Lab – Co
Baker, Cathy, (479) 968-0661 cbaker@atu.edu,
Arkansas Tech Univ – Gg
Baker, Dennis G., (313) 763-6238 Univ of Michigan – Oa
Baker, Don, 514-398-7485 donb@eps.mcgill.ca,
McGill Univ – Cp
Baker, Donald R., (713) 348-4063 geol@rice.edu,
Rice Univ – Co
Baker, Edward T., (206) 526-6251 baker@pmel.noaa.gov,
Univ of Washington – Ou
Baker, Hal, (573) 368-2143 Missouri Dept of Natural Resources – Gg
Baker, Harold D., 404-656-3214 Georgia Dept of Natural Resources – Oo
Baker, James C., (540) 231-9785 suebrown@vt.edu,
Virginia Polytechnic Institute & State Univ – Sd
Baker, Joel E., (410) 326-7205 baker@cbl.umces.edu,
Univ of Maryland – Co
Baker, John M., (612) 625-4249 jbaker@soils.umn.edu,
Univ of Minnesota, Twin Cities – Sp
Baker, Marcia B., 206-685-3799 marcia@ess.washington.edu,
Univ of Washington – OaYg
Baker, Paul A., (919) 684-6450 pbaker@duke.edu,
Duke Univ – Cl
Baker, Richard G., (319) 335-1827 dick-baker@uiowa.edu,
Univ of Iowa – Pl
Baker, Robert W., (715) 425-3345 robert.w.baker@uwrf.edu,
Univ of Wisconsin, River Falls – Gl
Baker, Sophie, 775-673-7434 Sophie.baker@dri.edu,
Desert Research Institute – Gm
Baker, Victor R., (520) 621-7875 baker@hwr.arizona.edu,
Univ of Arizona – Xg
Baker, Victor R., (520) 621-7875 baker@hwr.arizona.edu,
Univ of Arizona – Gm
Baksi, Ajoy K., (225) 388-3422 abaksi@geol.lsu.edu,
Louisiana State Univ – Cc
Baldauf, Jack G., (979) 845-9297 jbaldauf@ocean.tamu.edu,
Texas A&M Univ – Ou
Baldridge, W. Scott, (505) 667-4338 sbaldridge@lanl.gov,
Los Alamos National Lab – Gi
Baldwin, Christopher T., 936-294-1593 baldwin@shsu.edu,
Sam Houston State Univ – Gs
Baldwin, Ewart M., (541) 346-4573 Univ of Oregon – Gr
Baldwin, Michael, (765) 496-9517 baldwin@purdue.edu,
Purdue Univ – Oa
Baldwin, Suzanne L., (315) 443-2672 sbaldwin@syr.edu,
Syracuse Univ – Cc
Baldwin, Jr., A. Dwight, (513) 529-3216 Miami Univ – Hw
Bales, Roger C.rbales@eng.ucmerced.edu,
Univ of Arizona – Cg
Balistrieri, Laurie, (206) 543-8966 balistri@ocean.washington.edu,
Univ of Washington – Oc
Ball, Elizabeth, (775) 682-7351 ball@mines.unr.edu,
Univ of Nevada, Reno – Oo
Ball, Lary A., (508) 289-3356 lball@whoi.edu,
Woods Hole Oceanographic Institution – Oc
Ball, Neil, (204) 474-8075 na_ball@umanitoba.ca,
Univ of Manitoba – Gz
Ball, Rayford L., (817) 531-4537 rball@tccd.net,
Tarrant County Junior College – Og
Ball, William P., (410) 516-5434 bball@jhu.edu,
The Johns Hopkins Univ – Oo
Ballard, Robert D., (401) 874-6115 bballard@gso.uri.edu,
Univ of Rhode Island – Ga
Ballard, Timothy M., (604) 822-2300 Univ of British Columbia – Sf
Ballentine, Robert J., (315) 312-2804 SUNY, Oswego – Ow
Ballester, Gilda E., (313) 747-3670 gilda@sunshine.sprl.umich.edu,
Univ of Michigan – Oa
Balli, Evangelina, (559) 278-3086 vengie_balli@csufresno.edu,
California State Univ, Fresno – Oo
Ballinger, RhodaballingerRC@cf.ac.uk,
Cardiff Univ – Oo
Ballou, Stephen M., (608) 363-2388 ballous@beloit.edu,
Beloit College – Gg
Bally, Albert W., (713) 348-6063 geol@rice.edu,
Rice Univ – Gg
Balmforth, Neil, (604) 822-9835 njb@math.ubc.ca,
Univ of British Columbia – Oa
Balog-Szabo, Anna, (540) 857-7222 Virginia Western Community College – Gs
Balogh, Lajos J., 412.442.4231 lbalogh@state.pa.us,
Pennsylvania Bureau of Topographic & Geologic Survey – Oo
Balsam, William L., (817) 272-2987 Univ of Texas, Arlington – Gu
Balser, Teri C., (608) 262-0132 tcbalser@wisc.edu,
Univ of Wisconsin, Madison – Sb
Balsillie, James H., (850) 488-9380 Florida Geological Survey – On
Balster, Nick J., (608) 263-5719 njbalster@wisc.edu,
Univ of Wisconsin, Madison – Sp
Baltz, Donald M., (225) 388-6512 Louisiana State Univ – Ob
Bambach, Richard K., (617) 484-7457 rbambach@oeb.harvard.edu,
Virginia Polytechnic Institute & State Univ – Po
Bancroft, John C., (403) 220-5026 bancroft@geo.ucalgary.ca,

Univ of Calgary – Ng
Bandfield, Joshua L., 206-543-1190 joshband@u.washington.edu, Univ of Washington – XgOr
Bandopadhyay, Sukumar, (907) 474-6876 ffs0b@aurora.alaska.edu, Univ of Alaska, Fairbanks – Nm
Bane, Jr., John G., (919) 962-0172 bane@marine.unc.edu, Univ of North Carolina, Chapel Hill – Op
Banerdt, Bruce E., (818) 354-5413 Jet Propulsion Lab – Xy
Banerjee, Neilnbanerj3@uwo.ca, Univ of Western Ontario – Cs
Banerjee, Subir K., (612) 624-5722 banerjee@tc.umn.edu, Univ of Minnesota, Twin Cities – Ym
Banerjee, Sushanta K., (412) 392-3895 Point Park College – Oo
Banfield, Jillian, 510-642-9488 jill@seismo.berkeley.edu, Univ of California, Berkeley – CoGz
Bangs, Nathan L., (512) 471-0424 nathan@ig.utexas.edu, Univ of Texas at Austin, Jackson School of Geosciences – Gu
Banjanin, Tom, 850-474-2746 Univ of West Florida – Ou
Bank, Carl-Georg, (416) 978-4381 bank@geology.utoronto.ca, Univ of Toronto – Og
Bank, Tracy L., (716)645-6800 (Ext. 6985) tlbank@buffalo.edu, SUNY, Buffalo – Cg
Banks, Philip O.pob@po.cwru.edu, Case Western Reserve Univ – Cc
Bankston, Lindabankslm@auburn.edu, Auburn Univ – Oo
Banner, Jay L., (512) 471-5016 banner@mail.utexas.edu, Univ of Texas at Austin, Jackson School of Geosciences – Cl
Bannister, Bryant, (520) 621-2414 Univ of Arizona – Oo
Bannon, Peter R., (814) 863-1309 bannon@ems.psu.edu, The Pennsylvania State Univ, Univ Park – Ow
Bannon-Nilles, Phyllis L., 217-244-0081 bannon@isgs.uiuc.edu, Illinois State Geological Survey – Og
Banse, Karl, (206) 543-5079 banse@ocean.washington.edu, Univ of Washington – Ob
Bao, Huiming, (225) 578-3419 bao@lsu.edu, Louisiana State Univ – Cs
Barabas, Arthur H., (559) 278-2912 art_barabas@csufresno.edu, California State Univ, Fresno – Em
Barak, Phillip W., (608) 263-5450 pwbarak@wisc.edu, Univ of Wisconsin, Madison – Sc
Baraniuk, Richard G., (713) 348-5132 richb@rice.edu, Rice Univ – Oo
Baranoski, Mark T., 614-265-6586 mark.baranoski@dnr.state.oh.us, Ohio Dept of Natural Resources – Go
Barazangi, Muawia, (607) 255-6411 barazangi@geology.cornell.edu, Cornell Univ – Ys
Barber, Donald C., (610) 526-5110 dbarber@brynmawr.edu, Bryn Mawr College – Gs
Barber, Richard T., (919) 728-2111 rbarber@duke.edu, Duke Univ – Ob
Barbi, Greta, 089/2180 4234 barbi@gophysik.uni-muenchen.de, LMU - Univ of Munich – Yg
Barcilon, Albert I., (850) 644-4596 barcilon@met.fsu.edu, Florida State Univ – Ow
Barcilon, Victor, (773) 702-8120 vb39@midway.uchicago.edu, Univ of Chicago – Or
Barclay, Andrew, 206-543-8956 andrew@ocean.washington.edu, Univ of Washington – Gu
Barclay, David J., (607) 753-2921 barclayd@cortland.edu, SUNY, Cortland – Gl
Barclay, Julie L.barclayj@cortland.edu, SUNY, Cortland – Gg
Bardach, John E., (808) 944-7285 Univ of Hawai'i, Manoa – Oo
Barger, M. Susan, (505) 277-5925 Univ of New Mexico – Om
Barile, Diane D., (321) 674-8096 dmes@fit.edu, Florida Institute of Technology – Ou
Barker, Charles F., (313) 831-7279 cfbarker@earthlink.net, Wayne State Univ – Gg
Barker, Chris A., (936) 468-2340 Stephen F. Austin State Univ – Gc
Barker, Colin, (918) 631-3014 colin-barker@utulsa.edu, Univ of Tulsa – Cg
Barker, Daniel, (512) 471-5502 danbarker@mail.utexas.edu, Univ of Texas at Austin, Jackson School of Geosciences – Gi
Barker, Gregory A., (603) 271-7332 gbarker@des.state.nh.us, New Hampshire Geological Survey – Gg
Barker, James M., (505) 825-5114 jbark@nmt.edu, New Mexico Institute of Mining & Technology – En
Barker, Jeffrey S., (607) 777-2522 Binghamton Univ – Ys
Barker, John R., (313) 763-6239 Univ of Michigan – Oa
Barker, Nancy, 207-778-7361 barker@maine.edu, Univ of Maine, Farmington – Oo
Barksdale, John D., 850-474-2746 M.S. Geology, Univ of West Florida – Ge
Barling, Jane, (604) 827-3043 jbarling@eos.ubc.ca, Univ of British Columbia – Cc
Barlow, George W., (510) 642-2947 barlowgw@garnet.berkeley.edu, Univ of California, Berkeley – Ob
Barlow, Jay P., (858) 546-7178 Univ of California, San Diego – Ob
Barnard, Walther M., (716) 673-3303 barnard@fredonia.edu, Univ of Hawai'i at Hilo – Gz
Barnard, Walther M., (716) 673-3841 barnard@fredonia.edu, SUNY, Fredonia – Gz
Barnbaum, Cecilia S., (229) 249-2645 cbarnbau@valdosta.edu, Valdosta State Univ – Oo
Barnes, Calvin G., (806) 742-3106 cal.barnes@ttu.edu, Texas Tech Univ – Gi
Barnes, Charles "Chuck" W., (928) 523-8822 chuck.barnes@nau.edu, Northern Arizona Univ – Gc
Barnes, Christopher R., (250) 721-8847 crbarnes@uvic.ca, Univ of Victoria – Pm
Barnes, David, (269) 387-5493 david.barnes@wmich.edu, Western Michigan Univ – Gd
Barnes, Fairley J., (505) 667-4933 fyb@lanl.gov, Los Alamos National Lab – Oo
Barnes, Hubert L., (814) 865-7573 barnes@geosc.psu.edu, Pennsylvania State Univ, Univ Park – Cp
Barnes, Jeffery R., (541) 737-5685 barnes@coas.oregonstate.edu, Oregon State Univ – Oa
Barnes, John H., 717.702.2025 jbarnes@state.pa.us, Pennsylvania Bureau of Topographic & Geologic Survey – Gg
Barnes, Lawrence G., (213) 763-3329 lbarnes@nhm.org, Los Angeles County Museum of Natural History – Pv
Barnes, Melanie A., (806) 742-3204 melanie.barnes@ttu.edu, Texas Tech Univ – Cg
Barnes, Randal J., (612) 625-5828 Univ of Minnesota, Twin Cities – Oo
Barnes, Sarah J., (541) 737-5685 sjbarnes@uqac.uquebec.ca, Universite du Quebec a Chicoutimi – Ct
Barnes, Steven S., (361) 825-2688 sbarnes@falcon.tamucc.edu, Texas A&M Univ, Corpus Christi – Cm
Barnett, Albert P., (919) 560-3324 North Carolina Central Univ – Or
Barnett, Bruce A., (609) 258-1303 bbarnett@princeton.edu, Princeton Univ – Yr
Barnett, Douglas B., (509) 376-3416 brent.barnett@pnl.gov, Pacific Northwest National Lab – Em
Barnett, Roger T.rbarnett@pacific.edu, Univ of the Pacific – Oy
Barnett, Susansbarnett@broward.edu, Broward Community College, Central Campus – Gg
Barnhardt, Michael L., (217) 244-2766 barnhart@isgs.uiuc.edu, Illinois State Geological Survey – Gm
Barnhisel, Richard I., (859) 257-8627 rbarnhis@uky.edu, Univ of Kentucky – Gz
Barnosky, Anthony D., (510) 642-9487 Univ of California, Berkeley – Pv
Baron, Dirk, (661) 654-3044 dbaron@csub.edu, California State Univ, Bakersfield – Cl
Baross, John A., (206) 543-0833 Univ of Washington – Ob
Barr, Sandra M., (902) 585-1340 sandra.barr@acadiau.ca, Acadia Univ – Gi
Barrash, Warren, (208) 426-1229 wb@cgiss.boisestate.edu, Boise State Univ – Hw
Barrera, Enriqueta, (703) 292-8551 ebarrera@nsf.gov, Case Western Reserve Univ – Yr
Barrett, Janejmbarret@is.dal.ca, Dalhousie Univ – Oo
Barrett, Linda R., (330) 972-7620 barrett@uakron.edu, Univ of Akron – Sd
Barrett, Mary L., (318) 869-5232 mbarrett@centenary.edu, Centenary College of Louisiana – Gs
Barrick, James E., (806) 742-3107 jim.barrick@ttu.edu, Texas Tech Univ – Ps
Barrie, Vaughn, (250) 363-6424 Univ of Victoria – Gu
Barron, Eric J., 512-471-6048 ebarron@jsg.utexas.edu, Univ of Texas at Austin, Jackson School of Geosciences – Og
Barron, Robert, (906) 487-2096 rjbarron@mtu.edu, Michigan Technological Univ – Og
Barros, Tony, (305) 237-3754 jbarros@mdcc.edu, Miami-Dade Community College (Wolfson Campus) – Gc
Barrows, Allan G., (213) 620-5032 California Geological Survey – Gg

Barry, Roger G., (303) 492-5488 Univ of Colorado – Oa
Barsch, Robert, 089/2180 4201 barsch@geophysik.uni-muenchen.de, LMU - Univ of Munich – Yg
Bart, Henry A., (215) 951-1268 bart@lasalle.edu, La Salle Univ – Gs
Bart, Philip J., (225) 388-3109 pbart@geol.lsu.edu, Louisiana State Univ – Gr
Bartek, III, Louis R., (919) 962-0687 bartek@email.unc.edu, Univ of North Carolina, Chapel Hill – Gu
Bartello, Peter, (514) 398-8075 peter.bartello@mcgill.ca, McGill Univ – Oa
Bartels, Karen S., (773) 442-6052 k-bartels@neiu.edu, Northeastern Illinois Univ – Gi
Bartels, William S., (517) 629-0313 wbartels@albion.edu, Albion College – Pv
Barth, Andrew P., (317) 274-1243 ibsz100@iupui.edu, Indiana Univ / Purdue Univ, Indianapolis – Gi
Barth, Jack A., (541) 737-1607 barth@coas.oregonstate.edu, Oregon State Univ – Op
Barth, Susan A., 406-496-4687 sbarth@mtech.edu, Montana Tech of The Univ of Montana – Oo
Bartholomew, Alexander J., 845-257-3765 barthola@newpaltz.edu, SUNY New Paltz – GrPi
Bartholomew, Mervin J., (901) 678-1613 jbrthlml@memphis.edu, Univ of Memphis – Gt
Bartlein, Patrick J., (541) 346-4967 Univ of Oregon – Pe
Bartleson, Bruce L., (970) 943-2138 bbartleson@western.edu, Western State College of Colorado – Gs
Bartlett, Douglas H., (858) 534-5233 dbartlett@ucsd.edu, Univ of California, San Diego – Ob
Bartlett, Wendy, 740-376-4775 Marietta College – Ge
Bartley, John M., (801) 585-1670 jbartley@mines.utah.edu, Univ of Utah – Oi
Bartley, John W., (231) 777-0521 john.bartley@muskegoncc.edu, Muskegon Community College – GgPi
Bartley, Julie, (678) 839-4054 jbartley@westga.edu, Univ of West Georgia – CsGtOe
Bartman, Frederick L., (313) 764-7210 Univ of Michigan – Oa
Bartok, Peter, 713-743-3399 Univ of Houston – Go
Barton, Christopher, (727) 893-3100 Univ of South Florida, Saint Petersburg – Gg
Barton, James H., (412) 589-2821 Thiel College – Oy
Barton, Mark D., (520) 621-8529 mdbarton@email.arizona.edu, Univ of Arizona – Eg
Barton, Michael, (614) 292-3132 barton.2@osu.edu, Ohio State Univ – Gx
Bartos, Paul, 303-273-3823 pbartos@mines.edu, Colorado School of Mines – Oo
Bascom, Johnathan, (616) 526-7053 jbascom@calvin.edu, Calvin College – Oo
Basham, William L., 432 552-2057 basham_w@utpb.edu, Univ of Texas, Permian Basin – Ye
Basinger, James F., (306) 966-5684 basinger@usask.ca, Univ of Saskatchewan – Pb
Baskaran, Mark M., (313) 577-3262 Baskaran@wayne.edu, Wayne State Univ – Cc
Baskerville, Charles A., (860) 832-3188 baskerville@ccsu.edu, Central Connecticut State Univ – Ng
Baskin, Perry A., (229) 259-5052 pbaskin@valdosta.edu, Valdosta State Univ – Oo
Bass, Jay D., (217) 333-1018 jaybass@uiuc.edu, Univ of Illinois, Urbana-Champaign – Gy
Bass, Michael L., (540) 654-1424 mbass@mwc.edu, Univ of Mary Washington – Oo
Bassem, Allambassem.allam@stonybrook.edu, SUNY, Stony Brook – Ob
Bassett, Kari N., ((03)) 3-66-7 (Ext. 1 ext 7732) kari.bassett@canterbury.ac.nz, Univ of Canterbury – Gs
Bassett, Scott, (775) 784-1434 sbassett@unr.edu, Univ of Nevada, Reno – Ou
Bassett, William A., (607) 255-7502 bassett@geology.cornell.edu, Cornell Univ – Gz
Basta, Nicholas T., 614-292-6282 basta.4@osu.edu, Ohio State Univ – Sc
Basu, Abhijit, 8128556654/5581 basu@indiana.edu, Indiana Univ, Bloomington – Gd
Basu, Asish R., (585) 275-2413 abasu@earth.rochester.edu, Univ of Rochester – Gx
Basu, Sukanta, 806-742-1139 sukanta.basu@ttu.edu, Texas Tech Univ – OaaOa

Basye, Bruce, (785) 628-1200 Fort Hays State Univ – Ge
Batchelder, Harold P., (541) 737-4500 hbatchelder@coas.oregonstate.edu, Oregon State Univ – Ob
Batchelor, James W., (318) 475-5771 McNeese State Univ – Pm
Bates, Richard, 573-368-2486 nrbater@mail.dnr.state.mo.us, Missouri Dept of Natural Resources – Oo
Batt, Richard J., (716) 878-4736 battrj@buffalostate.edu, Buffalo State College – PiGr
Batteen, Mary L., (408) 656-3265 batteen@oc.nps.navy.mil, Naval Postgraduate School – Op
Batten, William G., (608) 262-9903 wgbatten@wisc.edu, Univ of Wisconsin - Extension – Hw
Battisti, David S., (206) 543-2019 david@atmos.washington.edu, Univ of Washington – Oa
Batzle, Michael L., (303) 384-2067 mbatzle@mines.edu, Colorado School of Mines – Yg
Bauder, James W., (406) 994-5685 jbauder@gemini.oscs.montana.edu, Montana State Univ – Sf
Bauer, Carl J.cjbauer@email.arizona.edu, Univ of Arizona – Hs
Bauer, Emily J., (612)627-4780, Ext. 200 bauer010@umn.edu, Univ of Minnesota, Twin Cities – Gg
Bauer, Glenn R., (808) 587-0263 gbauer@lava.net, Hawai'i Geological Survey – Gg
Bauer, James E., (804) 684-7136 bauer@vims.edu, College of William & Mary – Oc
Bauer, Jeffrey A., (740) 351-3421 jbauer@shawnee.edu, Shawnee State Univ – Ps
Bauer, Paul W., (505) 835-5106 bauer@gis.nmt.edu, New Mexico Institute of Mining & Technology – Gp
Bauer, Paul W., (505) 835-5106 bauer@nmt.edu, New Mexico Institute of Mining and Technology – Gc
Bauer, Robert A., 217-244-2394 bauer@isgs.uiuc.edu, Illinois State Geological Survey – Nr
Bauer, Robert L., (573) 882-3759 BauerR@missouri.edu, Univ of Missouri – Gc
Baum, Gerald R., 410-554-5525 gbaum@dnr.state.md.us, Maryland Geological Survey – Gr
Baum, Steven K., (979) 845-0793 sbaum@ocean.tamu.edu, Texas A&M Univ – Op
Baumann, Karsten, (404) 894-0583 kb@eas.gatech.edu, Georgia Institute of Technology – Oa
Baumgartner, Scotty D., (573) 368-2113 nrbaums@mail.dnr.state.mo.us, Missouri Dept of Natural Resources – Hw
Baumiller, Tomasz K., (734) 764-7543 tomaszb@umich.edu, Univ of Michigan – Pg
Bawcom, Julie A., (707) 459-7426 California Geological Survey – Oo
Baxter, Ethan F.efb@bu.edu, Boston Univ – Cp
Baxter, Sams.j.baxter@reading.ac.uk, Univ of Reading – Sd
Baxter, Stefanie J., 302/831/2833 steff@udel.edu, Univ of Delaware – Og
Bayari, Serdar, +90 (312) 2977740 serdar@hacettepe.edu.tr, Hacettepe Univ – Hw
Bayliss, Peter, (403) 220-5026 Univ of Calgary – Gz
Bayly, M. Brianbaylym@rpi.edu, Rensselaer Polytechnic Institute – Gc
Bazard, David R., (707) 476-4224 dave-bazard@redwoods.edu, College of the Redwoods – Gg
Beall, Arthur O., (254) 710-2361 Baylor Univ – Go
Beane, Rachel J., (207) 725-3160 rbeane@bowdoin.edu, Bowdoin College – Gp
Beard, Brian L., 608-262-1806 beardb@geology.wisc.edu, Univ of Wisconsin, Madison – Cc
Beard, James S., (540) 666-8611 Virginia Polytechnic Inst & State Univ – Gi
Beard, K. Christopher, (412) 622-5782 Carnegie Museum of Nat History – Pv
Beard, K. Christopher, (412) 624-8780 Univ of Pittsburgh – Pv
Beard, Kenneth V., (217) 333-1676 k-beard@uiuc.edu, Univ of Illinois, Urbana-Champaign – Oa
Beatty, Heather L., (512) 339-2929 Austin Community College – Ge
Beatty, Merrill Ann, (506) 453-4803 mbeatty@unb.ca, Univ of New Brunswick – Oo
Beatty, Susan W., (303) 492-8310 susan.beatty@colorado.edu, Univ of Colorado – So
Beaty, Tammy, (865) 574-0119 taw@ornl.gov, Oak Ridge National Lab – Oi
Beauboeuf, Richard T., (713) 743-3399 Univ of Houston – Gr
Beauchamp, Eric G., (519) 824-4120 (Ext. 53239) ebeaucha@lrs.uoguelph.ca, Univ of Guelph – So

Beaudoin, Georges, (418) 656-3141 beaudoin@ggl.ulaval.ca,
Universite Laval – Em
Beaumont, Christopher, (902) 494-3779 chris.beaumont@dal.ca,
Dalhousie Univ – Yg
Beauregard, Allison, 850-729-5469 beaurega@owc.edu,
Okaloosa-Walton College – Ge
Beaver, Frank W., (701) 777-5294 frank_beaver@mail.und.nodak.edu,
Univ of North Dakota – Hw
Beaver, Harold H., (254) 710-2184 harold_beaver@baylor.edu,
Baylor Univ – Go
Beaverson, Sheena K., 217-244-9306 Illinois State Geological Survey – Oi
Bebout, Gray E., (610) 758-5831 geb0@lehigh.edu,
Lehigh Univ – Gp
Beck, Charles, (734) 763-5089 chbeck@umich.edu,
Univ of Michigan – Pg
Beck, Colleen M., (702) 795-8077 colleen@dri.edu,
Desert Research Institute – Sa
Beck, Dwayne L., (605) 224-6357 Dwayne.Beck@sdstate.edu,
South Dakota State Univ – Ou
Beck, John, 617-552-8300 john.beck.1@bc.edu,
Boston College – Pb
Beck, Leslie S., (306) 584-0013 Univ of Regina – Em
Beck, Myrl E., (360) 650-3597 beck@cc.wwu.edu,
Western Washington Univ – Ym
Beck, Susan L., (520) 621-4827 slbeck@email.arizona.edu,
Univ of Arizona – Ys
Beck, Warren, (520) 621-4277 wbeck@physics.arizona.edu,
Univ of Arizona – Oo
Becker, Alex, (510) 643-9181 Univ of California, Berkeley – Yg
Becker, Harryhbecker@geol.umd.edu,
Univ of Maryland – Cc
Becker, Janet M., (808) 956-6514 jbecker@soest.hawaii.edu,
Univ of Hawai'i, Manoa – Oo
Becker, Keir, 305 421-4661 kbecker@rsmas.miami.edu,
Univ of Miami – Yr
Becker, Laurence, (541) 737-9504 beckerla@geo.oregonstate.edu,
Oregon State Univ – Oo
Becker, Laurence, (802) 241-3496 larryb@dec.anr.state.vt.us,
Agency of Natural Res, Dept. of Environmental Conservation – Gg
Becker, Matthew W., (716) 645-6800 (Ext. 3960) mwbecker@geology.buffalo.edu, SUNY, Buffalo – Hw
Becker, Naomi M., (505) 667-2165 nmb@lanl.gov,
Los Alamos National Lab – Hg
Becker, Richard H., (630) 252-7595 Argonne National Lab – Or
Becker, Thorsten W., 213-740-8365 twb@usc.edu,
Univ of Southern California – Ys
Becker, Udo, (734) 615-6894 ubecker@umich.edu,
Univ of Michigan – Oo
Beckie, Roger D., (604) 822-6462 rbeckie@eos.ubc.ca,
Univ of British Columbia – Hg
Bedard, Jean H., (418) 654-2671 jbedard@nrcan.gc.ca,
Universite du Quebec – Gi
Bedard, Paul, (418) 545-5011 Universite du Quebec a Chicoutimi – Gi
Bedford, Keith W., (614) 292-7338 Ohio State Univ – Oo
Bedient, Philip B., (713) 348-4953 bedient@rice.edu,
Rice Univ – Hw
Bednarski, Marsha, (860) 832-2943 bednarski@ccsu.edu,
Central Connecticut State Univ – Oe
Bednarz, Robert S., (979) 845-7187 r-bednarz@tamu.edu,
Texas A&M Univ – Oe
Bednarz, Sarah W., (979) 845-1579 s-bednarz@tamu.edu,
Texas A&M Univ – Oe
Bedrossian, Trinda L., (916) 327-6977 California Geological Survey – Ng
Beeby, David J., (916) 323-8562 California Geological Survey – En
Beegle, Douglas B., (814) 863-1016 dbb@psu.edu,
Pennsylvania State Univ, Univ Park – Sc
Beem, Kenneth A., (301) 279-5231 Montgomery College – Po
Beerbower, James R., (607) 777-2454 Binghamton Univ – Pg
Beeton, Jared M., (719) 587-7357 jmbeeton@adams.edu,
Adams State College – OySd
Beg, Mirza A., 205/247-3624 mbeg@gsa.state.al.us,
Geological Survey of Alabama – Gz
Beget, James E.ffjeb1@uaf.edu,
Univ of Alaska, Fairbanks – Gm
Begin, Christian, (418) 654-2675 cbegin@nrcan.gc.ca,
Universite du Quebec – Pe
Behl, Richard J., (562) 985-5850 behl@csulb.edu,
California State Univ, Long Beach – Gs
Behling, Mary, 304 594 2331 behling@geosrv.wvnet.edu,
West Virginia Univ – Oi
Behling, Robert E., (304) 293-5603 rbehling@wvu.edu,
West Virginia Univ – Gm
Behn, Mark D., 508-289-3637 mbehn@whoi.edu,
Woods Hole Oceanographic Institution – Yr
Behr-Andres, Christina "Tina", (505) 667-3644 behr-andres@lanl.gov,
Los Alamos National Lab – Oo
Behrensmeyer, Anna K., (202) 633-1307 Smithsonian Institution / National Museum of Natural History – Pv
Behrensmeyer, Anna K.behrensmeyer.kay@nmnh.si.edu,
Univ of Arizona – Pe
Beiersdorfer, Raymond E., (330) 941-1753 rebeiersdorfer@ysu.edu,
Youngstown State Univ – Cg
Beilfuss, Meredith, 812-237-2362 mbeilfuss@isugw.indstate.edu,
Indiana State Univ – Oe
Bein, F. L., (317) 274-1100 Indiana Univ, Indianapolis – Oy
Beinroth, Friedrich, (809) 833-2865 Univ of Puerto Rico – Sd
Beitler Bowen, Brenda, (765) 496-2290 bbowen@purdue.edu,
Purdue Univ – Gs
Bekken, Barbara M., (540) 231-4466 bekken@vt.edu,
Virginia Polytechnic Institute & State Univ – Eg
Bekker, Matthew F., (801) 422-1961 matthew_bekker@byu.edu,
Brigham Young Univ – Oy
Belanger, Thomas V., (321) 674-8096 belanger@marine.fit.edu,
Florida Institute of Technology – Hw
Belanger, William A., (508) 929-8158 wbelanger@worcester.edu,
Worcester State College – Og
Belasky, Paul, 510-659-6000-5103 pbelasky@ohlone.edu,
Ohlone College – Pq
Belknap, Daniel F., (207) 581-2159 belknap@maine.edu,
Univ of Maine – Gu
Bell, Christopher J., (512) 471-7301 cjbell@mail.utexas.edu,
Univ of Texas at Austin, Jackson School of Geosciences – Pv
Bell, David, (865) 574-2855 btd@ornl.gov,
Oak Ridge National Lab – Og
Bell, David, (03) 3642-717 d.bell@geol.canterbury.ac.nz,
Univ of Canterbury – Ng
Bell, James C., (612) 625-6703 jbell@soils.umn.edu,
Univ of Minnesota, Twin Cities – Sd
Bell, Jeffrey F., (808) 956-3136 Univ of Hawai'i, Manoa – Xg
Bell, John W., (775) 784-1939 jbell@unr.edu,
Univ of Nevada, Reno – Ng
Bell, Keith, (613) 520-2600 Carleton Univ – Cc
Bell, Keithkeith_bell@carleton.ca,
Carleton Univ – Cc
Bell, Laura, 713-743-3402 lbell4@uh.edu,
Univ of Houston – Oo
Bell, Peter M., (202) 478-8900 Carnegie Institution of Washington – Yx
Bell, Robin E., 845-365-8827 Columbia Univ – Yr
Bellian, Jerome A., (512) 471-8920 jerome.bellian@beg.utexas.edu,
Univ of Texas at Austin, Jackson School of Geosciences – Gs
Bellovich, Steven J., 918-631-2478 steven-bellovich@utusla.edu,
Univ of Tulsa – Oy
Belt, Edward S., (413) 542-2712 esbelt@amherst.edu,
Amherst College – Gs
Beltrami, Hugo, 902-867-2326 hbeltram@stfx.ca,
Saint Francis Xavier Univ – Ge
Beltz, John F., (330) 832-3430 jfbeltz@sssnet.com,
Kent State Univ, Tuscarawas – Ge
Beltz, John F., 330-972-7630 jfb4@uakron.edu,
Univ of Akron – Pm
Ben-Ami, Leah, 617-373-3176 l.ben-ami@neu.edu,
Northeastern Univ – OooOo
Ben-Zion, Yehuda, (213) 740-6734 ybz@earth.usc.edu,
Univ of Southern California – Ys
Benacquista, Matt, (406) 657-2341 Montana State Univ, Billings – Xy
Bend, Stephen L., (306) 585-4021 Univ of Regina – Eo
Bender, Edward E., (714) 432-5681 ebender@mail.occ.cccd.edu,
Orange Coast College – Giz
Bender, John F., 704-687- 5956 jfbender@email.uncc.edu,
Univ of North Carolina, Charlotte – GiCtOu
Bender, Michael L., 609-258-2936 bender@princeton.edu,
Princeton Univ – Cg
Benedetti, Michael M., (910) 962-7650 benedettim@uncw.edu,
Univ of North Carolina Wilmington – Gm
Benezeth-Gisquet, Pascale, (865) 574-4960 benezethp@ornl.gov,
Oak Ridge National Lab – Cl
Benfield, Mark C., (225) 388-6372 mbenfie@lsu.edu,
Louisiana State Univ – Ob
Bengeyfield, Peter, 406-683-5538 apbengey@msn.com,
Univ of Montana, Western – Hg

Benham, Steven R., (253) 535-7378 benhamsr@plu.edu, Pacific Lutheran Univ – Gd
Benison, Kathleen C., (989) 774-3179 benis1kc@cmich.edu, Central Michigan Univ – Gs
Benitez-Nelson, Claudia, 777-0018 cbnelson@geol.sc.edu, Univ of South Carolina – Oc
Benjamin, Patricia A., (508) 929-8606 pbenjamin@worcester.edu, Worcester State College – Oy
Benjamin, Timothy M., (505) 667-5154 Los Alamos National Lab – Ca
Benn, Keith B., (613)562-5800 Ext 6858 kbenn@uottawa.ca, Univ of Ottawa – Gc
Bennartz, Ralf, (608)265-2249 bennartz@aos.wisc.edu, Univ of Wisconsin, Madison – Or
Benner, Jacob, 617-627-2207 jacob.benner@tufts.edu, Tufts Univ – Pge
Bennett, Andrew F., (541) 737-9352 bennett@coas.oregonstate.edu, Oregon State Univ – Op
Bennett, Gerry, (906) 635-2224 Lake Superior State Univ – Eg
Bennett, Gordon E.bennett@platte.unk.edu, Univ of Nebraska, Kearney – Oy
Bennett, Hugh F.hbennett@msu.edu, Michigan State Univ – Ys
Bennett, Linda R., (801) 537-3303 nrugs.lbennett@state.ut.us, Utah Geological Survey – Oo
Bennett, Philip, (512) 471-3587 pbennett@mail.utexas.edu, Univ of Texas at Austin, Jackson School of Geosciences – Hg
Bennett, Steven W., (309) 298-1256 Western Illinois Univ – Hw
Benninger, Larry K., (919) 962-0699 lbenning@email.unc.edu, Univ of North Carolina, Chapel Hill – Cl
Bennington, J Bret, 516 463-5568 geojbb@hofstra.edu, Hofstra Univ – Pi
Benoit-Bird, Kelly, (541) 737-2063 kbenoit@coas.oregonstate.edu, Oregon State Univ – Ob
Benson, Andrew A., (858) 534-4300 abenson@ucsd.edu, Univ of California, San Diego – Ob
Benson, Craig H., (608) 262-7242 benson@engr.wisc.edu, Univ of Wisconsin, Madison – Ng
Benson, David A., 303-273-3800 Colorado School of Mines – Hq
Benson, Donald J., 205-348-7007 Univ of Alabama – Gd
Benson, Donna M., (480) 461-7247 benson_d@mc.maricopa.edu, Mesa Community College – Gg
Benson, Jane L., (270) 809-3106 jane.benson@murraystate.edu, Murray State Univ – Oi
Benson, Robert, (361) 825-5888 benson@cbi.tamucc.edu, Texas A&M Univ, Corpus Christi – Yg
Benson, Robert D., (303) 273-3455 Colorado School of Mines – Ye
Benson, Robert G., (719) 587-7921 rgbenson@adams.edu, Adams State College – Gg
Bentley, Charles R., (608) 238-8873 bentley@geology.wisc.edu, Univ of Wisconsin, Madison – Yg
Bentley, Laurence R., (403) 220-4512 Univ of Calgary – Hw
Bentley, Samuel J., (225) 388-2961 Louisiana State Univ – Ob
Benton, Steven S., 217-244-0082 benton@isgs.uiuc.edu, Illinois State Geological Survey – Hw
Benz, Harley M.benz@gldfs.cr.usgs.gov, Univ of Utah – Ys
Berard, Jean, (514) 340-4731 Ecole Polytechnique – Ng
Berchem, Jean, (310) 206-6484 jberchem@igpp.ucla.edu, Univ of California, Los Angeles – Oo
Berekian, Beverly J., (714)432-5669 bberekian@occ.cccd.edu, Orange Coast College – Gzi
Berelson, William M., (213) 740-5828 berelson@usc.edu, Univ of Southern California – Cm
Berendsen, Pieter, 785-864-2141 pieterb@kgs.ku.edu, Univ of Kansas – Cg
Beresnev, Igor A., (515) 294-7529 beresnev@iastate.edu, Iowa State Univ of Science & Technology – Yg
Berg, Christopher A., (678) 839-4059 cberg@westga.edu, Univ of West Georgia – Gpi
Berg, Jonathan H., (815) 753-0523 jberg@niu.edu, Northern Illinois Univ – Gp
Berg, Peter, (804) 924-1318 pb8n@virginia.edu, Univ of Virginia – So
Berg, Richard, (217) 244-2776 berg@isgs.uiuc.edu, Illinois State Geological Survey – Ge
Berg, Richard B., 406-496-4172 dberg@mtech.edu, Montana Tech of The Univ of Montana – En
Berg, Jr., Robert K., (605) 563-2989 Robert.Berg@sdstate.edu, South Dakota State Univ – Sb
Bergantino, Robert N., 406-496-4150 bbergantino@mtech.edu, Montana Tech of The Univ of Montana – Gg
Bergantz, George W., 206-685-4972 bergantz@u.washington.edu, Univ of Washington – Gi
Bergenback, Richard E., (423) 425-4404 Univ of Tennessee, Chattanooga – Gd
Berger, Antony R., (604) 480-0840 Sir Wilfred Grenfell College – Ge
Berger, Glenn W., (775) 673-7354 glenn.berger@dri.edu, Univ of Nevada, Reno – Gg
Berger, Wolfgang H., (858) 822-2545 wberger@ucsd.edu, Univ of California, San Diego – Gs
Bergeron, Marcel P., (509) 372-6132 marcel.bergeronv, Pacific Northwest National Lab – Hw
Bergeron, Mario, (418) 654-2611 mbergeron@inrs.uquebec.ca, Universite du Quebec – Ca
Bergeron, Normand, (418) 654-3703 nbergeron@inrs.uquebec.ca, Universite du Quebec – Gm
Berggren, William A., 732-445-8523 wberggren@whoi.edu, Rutgers, The State Univ of New Jersey – Pm
Berggren, William A., (508) 289-2593 wberggren@whoi.edu, Woods Hole Oceanographic Institution – Ps
Bergin, Michael H., (404) 894-9723 mhbergin@ce.gatech.edu, Georgia Institute of Technology – Oa
Bergman, Katherine M., (306) 585-4679 Univ of Regina – Gs
Bergslien, Elisa T., 716-878-3793 bergslet@buffalostate.edu, Buffalo State College – ClHw
Bergstrom, Stig M., (614) 292-4473 stig@geology.ohio-state.edu, Ohio State Univ – Ps
Bergwerff, Kenneth A., (616) 526-6371 kbergwer@calvin.edu, Calvin College – Oe
Berkebile, C. Alan, (361) 825-3346 alanb@falcon.tamucc.edu, Texas A&M Univ, Corpus Christi – Hw
Berkheiser, Jr., Samuel W., 717.702.2055 sberkheise@state.pa.us, Pennsylvania Bureau of Topographic & Geologic Survey – Eg
Berkley, John L., (716) 673-3303 berkley@fredonia.edu, SUNY, Fredonia – GxXgGe
Berlin, Graydon L., (928) 523-6566 lenn.berlin@nau.edu, Northern Arizona Univ – Or
Berline, Leo, (541) 737-6397 leo@coas.oregonstate.edu, Oregon State Univ – Ob
Berman, David S., (412) 622-3248 Carnegie Museum of Nat History – Pv
Berman, Robert, (613) 788-4400 Carleton Univ – Gp
Berner, Elizabeth K., (203) 432-3183 Yale Univ – Cl
Berner, Robert A., (203) 432-3183 berner@hess.geology.yale.edu, Yale Univ – Cl
Bernhard, Joan M., 508-289-3480 jbernhard@whoi.edu, Woods Hole Oceanographic Institution – Cg
Bernier, Louis, (514) 340-4557 Ecole Polytechnique – Gz
Bernknopf, Richard, (650) 725-9692 Stanford Univ – Oo
Beroza, Gregory C., (650) 723-4958 beroza@pangea.stanford.edu, Stanford Univ – Ys
Berquist, Karen K.kaberq@wm.edu, College of William & Mary – Oo
Berquist, Jr., Carl R., (757) 221-2448 crberq@wm.edu, College of William & Mary – Ou
Berquist, Jr., Carl R., 757-221-2448 rick.berquist@dmme.virginia.gov, Division of Mineral Resources – Gr
Berquó, Thelma, 612 624-5274 berqu013@umn.edu, Univ of Minnesota, Twin Cities – Ym
Berry, C M.Chris.Berry@earth.cf.ac.uk, Cardiff Univ – Pb
Berry, David R., 909-869-3455 drberry@csupomona.edu, California State Polytechnic Univ, Pomona – Pg
Berry, Duane F., (540) 231-9792 duberry@vt.edu, Virginia Polytechnic Institute & State Univ – Sb
Berry, Henry N., (207) 287-2801 henry.n.berry@state.me.us, Univ of Maine – Gg
Berry, Kate, (775) 784-6344 kberry@unr.nevada.edu, Univ of Nevada, Reno – Oy
Berry, Michael K., 404-656-3214 Georgia Dept of Natural Resources – Oo
Berry, Richard W., (619) 594-6394 San Diego State Univ – Gz
Berry, Ron F., 61 3 6226 2456 Ron.Berry@utas.edu.au, Univ of Tasmania – Gc
Berry, William B. N., 510-642-3925 bberry@berkeley.edu, Univ of California, Berkeley – OuGe
Berryman, Bruce F., (802) 626-6478 Lyndon State College – Oa
Berryman, James G., (650) 723-5911 berryman@sep.stanford.edu, Stanford Univ – Ye
Bersch, Michael G., 205-348-1576 mbersch@geo.ua.edu, Univ of Alabama – Gi
Bert, Theresa M., (727) 896-8626 Univ of South Florida, Saint Peters-

Faculty Index -B

burg – Ob
Berta, Susan M., (812) 237-2444 geberta@isugw.indstate.edu, Indiana State Univ – Gm
Bertine, Kathe K., (619) 594-6369 kbertine@geology.sdsu.edu, San Diego State Univ – Cl
Bertog, Janet, 859-572-1523 bertogj@nku.edu, Northern Kentucky Univ – Gd
Bertrand, Rudolf, (418) 654-2642 ess-inrs-rbertrand@x1.nrcan.gc.ca, Universite du Quebec – Eo
Bertsch, Paul M., (803) 725-2752 Univ of Georgia – Sc
Besancon, James R., (781) 283-3030 jbesancon@wellesley.edu, Wellesley College – Gz
Beske-Diehl, Suzanne, (906) 487-2534 sbeske-d@mtu.edu, Michigan Technological Univ – Ym
Best, James L., 217-244-1839 jimbest@uiuc.edu, Univ of Illinois, Urbana-Champaign – Gs
Best, Melvyn E., (250) 658-0791 best@islandnet.com, Univ of Victoria – Ye
Beste, Edward, (410) 742-8780 Univ of Maryland – Oo
Beswick, Anthony E., (705) 675-1151 Laurentian Univ, Sudbury – Gi
Betancourt, Julio, 520-670-6821 jlbetanc@usgs.gov, Univ of Arizona – Oo
Bethke, Craig M., (217) 333-3369 bethke@uiuc.edu, Univ of Illinois, Urbana-Champaign – Hw
Bettis III, E. A., (319) 335-1831 art-bettis@uiowa.edu, Univ of Iowa – Oo
Betzer, Peter R., (727) 553-1634 pbetzer@marine.usf.edu, Univ of South Florida, Saint Petersburg – Cm
Beus, Stanley S.beuss@myexcel.com, Northern Arizona Univ – Pi
Beutel, Erin K., (843) 953-5591 beutele@cofc.edu, College of Charleston – Gct
Beuthin, Jack D., (814) 269-2945 beuthin@pitt.edu, Univ of Pittsburgh at Johnstown – Gr
Beutner, Edward C., (541) 488-4226 ed.beutner@fandm.edu, Franklin and Marshall College – Gt
Bevelhimer, Mark, (865) 576-0266 mb2@ornl.gov, Oak Ridge National Lab – Hs
Bevier, Mary Lou, (604) 822-4655 mbevier@eos.ubc.ca, Univ of British Columbia – Cc
Bevis, Kenneth A., 812-866-7307 bevis@hanover.edu, Hanover College – Gl
Bevis, Michael G., 614 247 5067 bevis.6@osu.edu, Ohio State Univ – Yx
Beyer, Patricia J., (570) 389-4570 pbeyer@bloomu.edu, Bloomsburg Univ – HsOyGm
Bezdicek, David F., (509) 335-3644 bezdicek@wsu.edu, Washington State Univ – Sb
Bezore, Stephen P., (916) 323-8561 California Geological Survey – Eg
Bhattacharji, Somdev, (718) 951-5500 Graduate School of the City Univ of New York – Gt
Bhattacharya, Janok P., 713-743-3399 jbhattacharya@uh.edu, Univ of Houston – Gs
Bhattacharya, Saibal, (785) 864-2058 saibal@kgs.ku.edu, Univ of Kansas – Np
Bhowmik, Nani G., (217) 333-6775 nbhowmik@uiuc.edu, Illinois State Water Survey – Hs
Bian, Aiguo, 361-593-2261 Mail kfab003@tamuk.edu, Texas A&M Univ, Kingsville – Hq
Bianchi, Thomas, 979 845 5137 tbianchi@tamu.edu, Texas A&M Univ – Oc
Biasi, Glenn P., (775) 784-4576 glenn@seismo.unr.edu, Univ of Nevada, Reno – Ys
Bice, David M., 814-865-4477 dbice@geosc.psu.edu, Pennsylvania State Univ, Univ Park – Gg
Bice, Karen L., (508) 289-3320 kbice@whoi.edu, Woods Hole Oceanographic Institution – Pe
Bickel, Charles E., (415) 338-1963 bickel@sfsu.edu, San Francisco State Univ – Gx
Bickford, M. E., (315) 443-2672 Syracuse Univ – Cc
Bickmore, Barry R., 801-422-4680 brb88@geology.byu.edu, Brigham Young Univ – Cg
Bidigare, Robert R., (808) 956-6567 bidigare@soest.hawaii.edu, Univ of Hawai'i, Manoa – Ob
Biegel, Ronald L., 213-821-5519 biegel@usc.edu, Univ of Southern California – Yg
Biek, Robert, (801) 537-3356 nrugs.bbiek@state.ut.us, Utah Geological Survey – Gg
Bieler, David B., (318) 869-5234 dbieler@centenary.edu, Centenary College of Louisiana – Gt
Bienfang, Paul K., (804) 259-7952 Univ of Hawai'i, Manoa – Oo
Bierly, Gregory, (812) 237-4515 gebierly@isugw.indstate.edu, Indiana State Univ – Oa
Bierman, Don E., (502) 852-6844 Univ of Louisville – Oo
Bierman, Paul R., (802) 656-4411 pbierman@zoo.uvm.edu, Univ of Vermont – Gm
Biggar, James W., (530) 752-0681 jwbiggar@ucdavis.edu, Univ of California, Davis – So
Biggs, Douglas C., (979) 845-3423 dbiggs@ocean.tamu.edu, Texas A&M Univ – Ob
Bigham, Jerry M., (614) 292-9066 bigham.1@osu.edu, Ohio State Univ – Sd
Bilanovic, Dragoljub D., 218-755-2801 dbilanovic@bemidjistate.edu, Bemidji State Univ – GePyCa
Bilek, Susan L., 505.835.6510 sbilek@nmt.edu, New Mexico Institute of Mining and Technology – Ys
Bilham, Roger G., (303) 492-6189 roger.bilham@colorado.edu, Univ of Colorado – Yg
Bill, Steven D., (603) 358-2552 sbill@keene.edu, Keene State College – Pg
Billen, Magali I., 530-754-5696 billen@geology.ucdavis.edu, Univ of California, Davis – Yg
Billings, Stephensbillings@skyri.com, Univ of British Columbia – Yg
Billups, Katharina, 302-645-4249 kbillups@udel.edu, Univ of Delaware – Ou
Billups, Katharina, 302-645-4249 kbillups@udel.edu, Univ of Delaware – Pe
Bilodeau, Michel L., (514) 398-4755 McGill Univ – Nm
Bilodeau, William L., (805) 493-3264 bilodeau@clunet.edu, California Lutheran Univ – GcsGg
Bina, Craig R., (847) 491-5097 craig@earth.northwestern.edu, Northwestern Univ – Yg
Binda, Pier L., (306) 585-4583 Univ of Regina – Eg
Bindeman, Ilya N., 541-346-3817 Ilya@uoregon.edu, Univ of Oregon – Cs
Binford, Jane F-J, (713) 974-0659 Houston Comm College System – Go
Binford, Micheal W., (352) 392-0494 mbinford@geog.ufl.edu, Univ of Florida – Or
Bingham, James M., (270) 745-5980 Western Kentucky Univ – Og
Binkley, Mark S., (662) 325-3473 binkley@geosci.msstate.edu, Mississippi State Univ – Ow
Binzel, Richard P., (617) 253-6486 rpb@mit.edu, Massachusetts Institute of Technology – Xm
Biondi, Biondo L., (650) 723-9831 biondo@sep.stanford.edu, Stanford Univ – Ye
Biondi, Franco, (775) 784-2967 fbiondi@unr.edu, Univ of Nevada, Reno – Sf
Birch, Francis S., (603) 862-1718 fsb@cisunix.unh.edu, Univ of New Hampshire – Yg
Bircher, Harry, 330-549-9051 harrybircher@buckeyecivildesign.com, Youngstown State Univ – Ge
Birchfield, G. Edward, (847) 491-7460 edward@earth.northwestern.edu, Northwestern Univ – Pe
Bird, Dennis K., (650) 723-1664 bird@pangea.stanford.edu, Stanford Univ – Cg
Bird, G. Peter, (310) 825-1126 pbird@ess.ucla.edu, Univ of California, Los Angeles – Yg
Bird, Janelle, 402-472-2663 Univ of Nebraska, Lincoln – Oo
Bird, Jeffry S., (801) 378-2929 jeff_bird@byu.edu, Brigham Young Univ – Oy
Bird, John M., (607) 255-6437 bird@geology.cornell.edu, Cornell Univ – Gt
Birdsell, Kay H., (505) 665-0260 khb@lanl.gov, Los Alamos National Lab – Hq
Birkeland, Peter W., (303) 492-6985 birkelap@stripe.colorado.edu, Univ of Colorado – Sd
Birnbaum, Stuart J., (210) 458-5467 Univ of Texas, San Antonio – Cl
Birner, Johannes, 089/2180 6565 johannes.birner@web.de, LMU - Univ of Munich – Gg
Birnie, Richard W., 603-646-2666 Dartmouth College – Gz
Biscaye, Pierre E., 845-365-8429 Columbia Univ – Cm
Bischoff, William D., (316) 978-6659 Wichita State Univ – Cl
Bish, David L., (812) 855-2039 bish@indiana.edu, Indiana Univ, Bloomington – Gz
Bish, David L., (505) 667-1165 bish@lanl.gov, Los Alamos National Lab – Gz
Bishop, Cathy, 337-482-0678 Univ of Louisiana, Lafayette – Gg
Bishop, Charles E., (801) 537-3361 nrugs.cbishop@state.ut.us, Utah Geological Survey – Hw

Bishop, Gale A., (912) 681-5361 Georgia Southern Univ – Pi
Bishop, Gregory V., (714)432-5669 gbishop@occ.cccd.edu, Orange Coast College – OsHwOy
Bishop, James K., (510) 495-2457 bishop@atmos.berkeley.edu, Univ of California, Berkeley – Oc
Bishop, Kim, 323 343-2409 Kbishop@calstatela.edu, California State Univ, Los Angeles – Gc
Bishop, Michael P., 402-554-4808 mpbishop@mail.unomaha.edu, Univ of Nebraska at Omaha – Or
Bissada, Adry, 713-743-3399 Univ of Houston – Co
Bissett, William P., (202) 767-8278 Univ of South Florida, Saint Petersburg – Ob
Bjornerud, Marcia, (920) 832-7015 marcia.bjornerud@lawrence.edu, Lawrence Univ – Gc
Bjornstad, Bruce N., (509) 373-6948 bruce.bjornstad@pnl.gov, Pacific Northwest National Lab – Gs
Blacic, James D., (505) 667-6815 jblacic@lanl.gov, Los Alamos National Lab – Nr
Black, Jill, 417-836-5300 aab208f@smsu.edu, Southwest Missouri State Univ – Oo
Black, Kathryn, (973) 655-4448 Montclair State Univ – Oy
Black, R. Bernard, (903) 983-8253 Kilgore College – Gd
Black, Robert X., (404) 894-1756 rob.black@eas.gatech.edu, Georgia Institute of Technology – Oa
Black, Ross A., (785) 864-2740 black@ku.edu, Univ of Kansas – Ye
Black, Stuarts.black@rdg.ac.uk, Univ of Reading – Cc
Black, T. Andrew, (604) 822-2730 Univ of British Columbia – Sp
Blackadar, Alfred K., (814) 865-2951 akb1@psu.edu, The Pennsylvania State Univ, Univ Park – Ow
Blackburn, James B., (713) 524-1012 blackbur@rice.edu, Rice Univ – Ou
Blackerby, Bruce A., (559) 278-2955 California State Univ, Fresno – Gi
Blackett, Robert E., (435) 865-8139 Blackett@suu.edu, Utah Geological Survey – Eg
Blackmer, Alfred M., (515) 294-7284 ablackmr@iastate.edu, Iowa State Univ of Science & Technology – Sc
Blackmer, Gale C., 717.702.2032 gblackmer@state.pa.us, Pennsylvania Bureau of Topographic & Geologic Survey – Gc
Blackwelder, Patricia L., (305) 361-4677 pblackwelder@rsmas.miami.edu, Nova Southeastern Univ – Py
Blackwelder, Patricia L., 305 421-4677 pblackwelder@rsmas.miami.edu, Univ of Miami – Po
Blad, Blaine L., (402) 472-9873 Univ of Nebraska, Lincoln – Oa
Bladh, Katherine L., (937) 327-7333 klbladh@wittenberg.edu, Wittenberg Univ – Gi
Bladh, Kenneth W., (937) 327-7334 kbladh@wittenberg.edu, Wittenberg Univ – Gz
Blair, Neal E., 847-491-5349 n-blair@northwestern.edu, Northwestern Univ – Co
Blair, Neal E., (919) 515-7833 neal_blair@ncsu.edu, North Carolina State Univ – Co
Blake, Daniel B., (217) 333-3833 dblake@uiuc.edu, Univ of Illinois, Urbana-Champaign – Pi
Blake, David E., (910) 962-3387 blaked@uncw.edu, Univ of North Carolina Wilmington – Gp
Blake, Don, (949) 824-4195 drblake@uci.edu, Univ of California, Irvine – Yr
Blake, Geoffrey A., (626) 395-6296 gab@gps.caltech.edu, California Institute of Technology – Xc
Blake, M. Clark, (360) 650-3595 Western Washington Univ – Gt
Blake, Norman J., (727) 553-1521 nblake@marine.usf.edu, Univ of South Florida, Saint Petersburg – Ob
Blake, Ruth E., (203) 432-3191 ruth.blake@yale.edu, Yale Univ – Cg
Blakely, Robert F., (812) 855-1339 Indiana Univ, Bloomington – Ys
Blakeney DeJarnett, Beverly, 713-896-6740 bev.dejarnett@beg.utexas.edu, Univ of Texas at Austin, Jackson School of Geosciences – Gg
Blakey, Ronald C., (928) 523-2740 ronald.blakey@nau.edu, Northern Arizona Univ – Gr
Blanchard, Frank N., (352) 392-2296 Univ of Florida – Gz
Blanchet, Jean-Pierre, (514) 987-3000 blanchet@maia.sca.uqam.ca, Universite du Quebec a Montreal – Oo
Bland, Douglas, (505) 466-6696 dmbland@comcast.net, New Mexico Institute of Mining & Technology – Eg
Bland, William L., (608) 262-0221 wlbland@wisc.edu, Univ of Wisconsin, Madison – Sp
Blander, Milton, (630) 252-4548 Argonne National Lab – Xc
Blaney, Diana L., (818) 354-5419 Jet Propulsion Lab – Xg
Blanford, William J., (225) 578-3955 blanford@geol.lsu.edu, Louisiana State Univ – Hw
Blank, Carrine E., 314-935-4456 blank@levee.wustl.edu, Washington Univ – Oo
Blankenship, Donald D., (512) 471-0489 blank@ig.utexas.edu, Univ of Texas at Austin, Jackson School of Geosciences – Yg
Blasing, Terence J., (865) 574-7368 blasingtj@ornl.gov, Oak Ridge National Lab – Oa
Blasingame, Tom, (979) 845-2292 t-blasingame@spindletop.tamu.edu, Texas A&M Univ – Np
Bleakley, Bruce H., (605) 688-5498 Bruce.Bleakley@sdstate.edu, South Dakota State Univ – Sb
Bleam, William F., (608) 262-9956 wfbleam@wisc.edu, Univ of Wisconsin, Madison – Sc
Blechman, Jerome B., (607) 436-3322 SUNY, Oneonta – Oa
Bleck, Rainer, (505) 665-9150 bleck@lanl.gov, Los Alamos National Lab – Ow
Bledsoe, Caroline S., (530) 752-0388 csbledsoe@ucdavis.edu, Univ of California, Davis – So
Bleibinhaus, Florian, 089/2180 4202 bleibi@geophysik.uni-muenchen.de, LMU - Univ of Munich – Yg
Bleistein, Norman, (303) 273-3557 norm3980@aol.com, Colorado School of Mines – Yg
Blencoe, James G., (865) 574-7041 blencoejg@ornl.gov, Oak Ridge National Lab – Cp
Blenkinsop, Johnjblenkin@ccs.carleton.ca, Carleton Univ – Cc
Blenkinsop, John, (613) 520-2600 (Ext. 4398) john_blenkinsop@carleton.ca, Carleton Univ – Cc
Blewett, William L., (717) 477-1513 wlblew@ship.edu, Shippensburg Univ – Oy
Blewitt, Geoffrey, (775) 784-6691 (Ext. 171) gblewitt@unr.edu, Univ of Nevada – Yd
Blewitt, Geoffrey, (775) 682-8778 gblewitt@unr.edu, Univ of Nevada, Reno – Yd
Bloch, Jonathanjbloch@flmnh.ufl.edu, Univ of Florida – Pv
Blom, Ronald G., (818) 354-4681 Jet Propulsion Lab – Or
Bloom, Amy M., (309) 438-8115 ambloom@ilstu.edu, Illinois State Univ – Gn
Bloom, Arthur L., (607) 255-5232 bloom@geology.cornell.edu, Cornell Univ – Gm
Bloom, Paul R., (612) 625-4711 pbloom@soils.umn.edu, Univ of Minnesota, Twin Cities – Sc
Bloomer, Sherman H., (541) 737-4811 bloomers@geo.oregonstate.edu, Oregon State Univ – Gi
Bloomgren, Bruce A., (612)627-4780, Ext. 201 bloom005@umn.edu, Univ of Minnesota, Twin Cities – Gr
Bloss, F. Donald, (540) 231-5151 bloss@vt.edu, Virginia Polytechnic Institute & State Univ – Gz
Blount, Grady P., (361) 825-2358 blount@falcon.tamucc.edu, Texas A&M Univ, Corpus Christi – Or
Blount, Grady P., (361) 825-2358 bunch@falcon.tamucc.edu, Texas A&M Univ, Corpus Christi – Or
Blount, Katherine P., (361) 825-5721 kpblount@falcon.tamucc.edu, Texas A&M Univ, Corpus Christi – Gm
Bloxham, Jeremy, (617) 495-9517 bloxham@eps.harvard.edu, Harvard Univ – Yg
Bluemle, John P., (701) 224-4109 Univ of North Dakota – Gm
Blum, Joel D., (734) 764-1435 jdblum@umich.edu, Univ of Michigan – Cc
Blum, Linda K., (434) 924-0560 lkb2e@virginia.edu, Univ of Virginia – Sb
Blum, Michael, (225) 578-5735 mike@geol.lsu.edu, Louisiana State Univ – Gs
Blusztajn, Jurek, (508) 289-2692 jblusztajn@whoi.edu, Woods Hole Oceanographic Institution – Cc
Bluth, Gregg J. S., (906) 487-3554 gbluth@mtu.edu, Michigan Technological Univ – Or
Bly, Anthony G., (605) 688-4768 Anthony.Bly@sdstate.edu, South Dakota State Univ – Sc
Blythe, Ann, 323 259-2553 ablythe@oxy.edu, Occidental College – Cc
Blythe, Ann E., (213) 821-1094 blythe@usc.edu, Univ of Southern California – Gt
Boadu, Fred K., (919) 660-5432 boadu@akoto.egr.duke.edu, Duke Univ – Yg
Boardman, Mark R., (513) 529-3230 boardman@muohio.edu, Miami Univ – On

Boardman, II, Darwin, 405-744-6358 amm0001@okstate.edu, Oklahoma State Univ – Ps
Boast, Charles W., (217) 333-1278 Univ of Illinois, Urbana-Champaign – Sp
Boateng, Samuel, 859-572-6683 boatengs@nku.edu, Northern Kentucky Univ – Hw
Bobyarchick, Andy R., 704-687-5998 arbobyar@email.uncc.edu, Univ of North Carolina, Charlotte – Gc
Bochdansky, Alexander, 757-6834933 abochdan@odu.edu, Old Dominion Univ – Ob
Bockheim, James G., (608) 263-5903 bockheim@wisc.edu, Univ of Wisconsin, Madison – Sf
Boctor, Nabil, (202) 478-8949 n.boctor@gl.ciw.edu, Carnegie Institution of Washington – Cg
Boden, Thomas, (865) 241-4842 tab@ornl.gov, Oak Ridge National Lab – Oa
Bodenbender, Brian E., (616) 395-7541 bodenbender@hope.edu, Hope College – Pi
Bodenheimer, Peter, (831) 459-2064 peter@ucolick.org, Univ of California, Santa Cruz – Xc
Bodenman, John E., (570) 389-4697 boden@bloomu.edu, Bloomsburg Univ – Oo
Bodin, Paul A., (901) 678-4845 bodin@ceri.memphis.edu, Univ of Memphis – Ys
Bodin, Paul A., 206-616-7315 pbodin@ess.washington.edu, Univ of Washington – Ys
Bodnar, Robert J., (540) 231-7455 bubbles@vt.edu, Virginia Polytechnic Institute & State Univ – Cg
Bodo, Weberbweber@cicese.mx, Centro de Inv Científica y de Educación Superior de Ensenada – Gp
Bodolay, Sylvia, (913) 532-7222 Kansas State Univ – So
Boehm, David A.dirtboy@rochester.rr.com, SUNY College at Brockport – Gg
Boer, George J., (250) 363-8226 george.boer@ec.gc.ca, Univ of Victoria – Oa
Boerboom, Terrence J., (612)627-4780, Ext. 202 boerb001@umn.edu, Univ of Minnesota, Twin Cities – Gi
Boering, Kristie, (510) 642-3472 boering@cchem.berkeley.edu, Univ of California, Berkeley – Oa
Boese, Steven, (307) 766-4859 sboese@uwyo.edu, Univ of Wyoming – Ca
Boesenberg, Joseph, 212-769-5383 bosenbrg@amnh.org, American Museum of Natural History – Xm
Boettcher, Lisa, 989-774-7517 boett1lm@cmich.edu, Central Michigan Univ – Gg
Boggs, Katherine, (403) 440-6645 kboggs@mtroyal.ca, Mount Royal College – Gt
Boggs, Jr., Sam, (541) 346-4553 sboggs@uoregon.edu, Univ of Oregon – Gd
Bogle, Mary Anna, (865) 574-7824 amb@ornl.gov, Oak Ridge National Lab – Ge
Bogner, Jean, (630) 665-0872 Univ of Illinois, Chicago – Hw
Bogner, William C., (217) 333-9546 Illinois State Water Survey – Hs
Bograd, Michael B. E., (601) 961-5528 michael_bogard@deq.state.ms.us, Mississippi Office of Geology – Gg
Bogucki, Donald J., (518) 564-4030 Plattsburgh State Univ (SUNY) – Or
Bogue, Scott W., (323) 259-2563 bogue@oxy.edu, Occidental College – Ym
Bohlen, Walter F., (860) 445-3468 bohlen@uconnvm.uconn.edu, Univ of Connecticut – Op
Bohling, Geoffrey C., (785) 864-2093 geoff@kgs.ku.edu, Univ of Kansas – Hw
Bohlke, Johnjkbohlke@usgs.gov, Univ of Maryland – Cg
Bohn, Bryce A., 406-683-6585 bbohn@fs.fed.us, Univ of Montana, Western – Sf
Bohnenstiehl, Del, 919-515-7449 delwayne_bohnenstiehl@ncsu.edu, North Carolina State Univ – Yr
Bohnenstiehl, DelWayne R., 845-365-8382 del@ldeo.columbia.edu, Columbia Univ – Gu
Bohren, Craig F., (814) 865-2951 bohren@ems.psu.edu, The Pennsylvania State Univ, Univ Park – Ow
Bohrson, Wendy A., (509) 963-2835 bohrson@geology.cwu.edu, Central Washington Univ – Gvi
Boicourt, William, (410) 221-8426 boicourt@hpl.umces.edu, Univ of Maryland – Op
Boisvert, Eric, (418) 654-3705 eboisvert@nrcan.gc.ca, Universite du Quebec – Oo
Bokuniewicz, Henry J., (631) 632-8674 henry.bokuniewicz@sunysb.edu, SUNY, Stony Brook – Yr
Boland, Irene B., 803-323-4949 bolandi@winthrop.edu, Winthrop Univ – Gt
Boland, John J., (410) 516-7103 jboland@jhu.edu, The Johns Hopkins Univ – Oo
Bolduc, Andree, (418) 654-2635 abolduc@nrcan.gc.ca, Universite du Quebec – Gl
Boles, James R., (805) 893-3719 Univ of California, Santa Barbara – Gd
Bolge, Louiselbolge@salemstate.edu, Salem State College – Ca
Bollag, Jean-Marc, (814) 863-0843 jmbollag@psu.edu, Pennsylvania State Univ, Univ Park – Sb
Bollens, Stephen, 509-335-3009 bollens@vancouver.wsu.edu, Washington State Univ – Ob
Bollinger, G. A., (540) 231-6521 Virginia Polytechnic Institute & State Univ – Ys
Bollinger, Marsha S., 803-323-4944 bollingerm@winthrop.edu, Winthrop Univ – Cm
Bollmann, Jörg, (416) 978-2061 bollmann@geology.utoronto.ca, Univ of Toronto – Pm
Bolmer, S. Thompson, (508) 289-2628 tbolmer@whoi.edu, Woods Hole Oceanographic Institution – Yr
Bolt, John R., (312) 922-9410 (Ext. 413) Univ of Illinois, Chicago – Pv
Bolt, John R., (312) 665-7629 jbolt@fmnh.org, Field Museum of Natural History – Pv
Bolton, David W., (410) 554-5561 dwbolton@mgs.md.gov, Maryland Geological Survey – Hw
Bolton, Edward W., (203) 432-3149 edward.bolton@yale.edu, Yale Univ – Gq
Bombolakis, Emanuel G., (617) 552-3648 bombolak@bc.edu, Boston College – Gc
Bomke, Arthur A., (604) 822-6534 Univ of British Columbia – Sc
Bon, Roger L., (801) 537-3363 nrugs.rbon@state.ut.us, Utah Geological Survey – Eg
Bona, Andrej, +61 8 9266 7194 A.Bona@curtin.edu.au, Curtin Univ of Technology – Yx
Bonanos, Alceste Z., (202) 478-8859 bonanos@dtm.ciw.edu, Carnegie Institution of Washington – Oo
Bonatti, Enrico, 845-365-8699 Columbia Univ – Yr
Bond, Paulette A., (850) 488-9380 Florida Geological Survey – Cg
Bonem, Rena M., (254) 710-2187 rena_bonem@baylor.edu, Baylor Univ – Pei
Bonham-Carter, Graeme F., (613) 996-3387 Graeme.Bonham-Carter@nrcan-rncan.gc.ca, Univ of Ottawa – Oi
Bonine, Michael E., (520) 621-5469 bonine@email.arizona.edu, Univ of Arizona – Ou
Bonini, William E., (609) 258-3598 bonini@princeton.edu, Princeton Univ – Yg
Bonnaffe, Florence, 512-232-2336 florence.bonnaffe@beg.utexas.edu, Univ of Texas at Austin, Jackson School of Geosciences – Yg
Bonneau, Laurent, (203) 432-3142 laurent.bonneau@yale.edu, Yale Univ – Oa
Bonse, Marjorie J., (612) 625-3740 mbonse@soils.umn.edu, Univ of Minnesota, Twin Cities – Oo
Bonuso, Nicole, 714 278-8451 nbonuso@fullerton.edu, California State Univ, Fullerton – Pie
Booker, John R., 206-543-9492 booker@ess.washington.edu, Univ of Washington – YmGt
Boorstein, Margaret F., (516) 299-2318 maboorst@liu.edu, Long Island Univ, C.W. Post Campus – Og
Booth, Colin J., (815) 753-7933 cbooth@niu.edu, Northern Illinois Univ – Hw
Booth, Peter W. K., +27 41 5042372 glapwb@upe.ac.za, Univ of Port Elizabeth – Gc
Booth, Robert K.rkb205@lehigh.edu, Lehigh Univ – Oo
Boothroyd, Jon C., (401) 874-2265 jon_boothroyd@uri.edu, Univ of Rhode Island – GslOn
Boothroyd, Jon C., (401) 874-2191 jon_boothroyd@uri.edu, Univ of Rhode Island – Gs
Bopp, Richard F., (518) 276-3075 boppr@rpi.edu, Rensselaer Polytechnic Institute – Co
Borah, Deva K., (217) 244-8856 Illinois State Water Survey – Hq
Borcea, Liliana, (713) 348-5723 borcea@caam.rice.edu, Rice Univ – Oo
Bord, Don, (313) 593-5483 dbord@umich.edu, Univ of Michigan, Dearborn – Xy
Bordeaux, Yvette, (215) 898-9191 bordeaux@sas.upenn.edu, Univ of Pennsylvania – Po

Borella, Peter E., (949) 582-4549 pborella@saddleback.cc.ca.us, Saddleback Community College – Gu
Borgman, Leon E., (307) 766-3386 borgman@uwyo.edu, Univ of Wyoming – Gq
Bork, Kennard B., (740) 587-6486 bork@denison.edu, Denison Univ – Pi
Bornhold, Brian D., (250) 472-4482 bornhold@uvic.ca, Univ of Victoria – Gu
Bornhorst, Theodore J., (906) 487-2721 tjb@mtu.edu, Michigan Technological Univ – Eg
Borns, Jr., Harold W., (207) 581-2196 borns@maine.maine.edu, Univ of Maine – Gl
Boroughs, Terry J., (916) 331-8596 BorougT@arc.losrios.edu, American River College – GgXgCg
Borowski, Walter S., (859) 622-1277 w.borowski, Eastern Kentucky Univ – Gu
Borradaile, Graham J., (807) 343-8594 graham.borradaile@lakeheadu.ca, Lakehead Univ – Gc
Borrok, David M., 915-747-5850` borrok@geo.utep.edu, Univ of Texas, El Paso – Cg
Boryta, Mark D., (916) 278-4683 boryta@csus.edu, California State Univ, Sacramento – Cg
Bosak, Tanjatbosak@mcb.harvard.edu, Massachusetts Institute of Technology – Py
Bosart, Lance F., (518) 442-4564 bosart@atmos.albany.edu, SUNY, Albany – Oa
Boss, Alan P., (202) 478-8858 boss@dtm.ciw.edu, Carnegie Institution of Washington – Oo
Boss, Stephen K., (479) 575-7134 sboss@comp.uark.edu, Univ of Arkansas, Fayetteville – Yr
Bossak, Brian H., (229) 333-5471 bhbossak@valdosta.edu, Valdosta State Univ – Ow
Bossert, James E., (505) 667-6268 bossert@lanl.gov, Los Alamos National Lab – Oa
Bostater, Charles R., (321) 674-8096 bostater@fit.edu, Florida Institute of Technology – Op
Bostater, Charles R., (321) 674-8096 bostater@marine.fit.edu, Florida Institute of Technology – Op
Bostock, Michael G., (604) 822-2082 bostock@eos.ubc.ca, Univ of British Columbia – Ys
Boswell, Cecil, 573-368-2146 nrboswc@mail.dnr.state.mo.us, Missouri Dept of Natural Resources – Oo
Bothner, Wallace A., (603) 862-1718 wally.bothner@unh.edu, Univ of New Hampshire – Gc
Bott, Martin H., 0191 3742511 m.h.p.bott@durham.ac.uk, Durham Univ – Yg
Bottero, Jean-Yvesbottero@arbois.cerege.fr, Rice Univ – Oo
Bottjer, David J., (213) 740-6100 dbottjer@usc.edu, Univ of Southern California – Pe
Bouchard, Michel, (514) 343-7542 bouchami@ere.umontreal.ca, Universite de Montreal – Gm
Boucot, Arthur J., (541) 737-1251 Univ of Oregon – Pi
Boudreau, Alan E., (919) 684-5646 boudreau@duke.edu, Duke Univ – Gi
Boudreau, Bernard P., (902) 494-3557 bernie.boudreau@dal.ca, Dalhousie Univ – Cl
Boudrias, Michel A., (619) 260-4600 Bourn@acusd.edu, Univ of San Diego – Ob
Bouker, Polly A., 678-407-5042 pbouker@gpc.edu, Georgia Perimeter College – Oo
Boulvain, Frederic P., 32 4 366 22 52 fboulvain@ulg.ac.be, Universite de Liege – Gs
Bouma, Arnold H., (225) 388-6186 Louisiana State Univ – Gu
Bourassa, Mark A., (850) 644-6923 bourassa@coaps.fsu.edu, Florida State Univ – Ow
Bourcier, Bill L., (925) 423-3745 bourcier1@llnl.gov, Lawrence Livermore National Lab – Cg
Bourgeois, Joanne (Jody), 206-685-2443 jbourgeo@u.washington.edu, Univ of Washington – GsrGh
Bourgeois, Reed J., (225) 388-8879 reed@lgs.bri.lsu.edu, Louisiana State Univ – Oo
Bourgouin-Mehes, Roxane J., (705) 673-6575 rmehes@laurentian.ca, Laurentian Univ, Sudbury – Oo
Bourke, Robert H., (408) 656-2673 bourke@oc.nps.navy.mil, Naval Postgraduate School – Op
Bourne, James H., (514) 987-3371 bourne.james@uqam.ca, Universite du Quebec a Montreal – Gp
Bourne, Tim J., 0191 3744709 Durham Univ – Pg
Bourque, Edith, (418) 654-2655 ebourque@nrcan.gc.ca, Universite du Quebec – Hw
Bourqui, Michel, (514) 398-5450 michel.bourqui@mail.mcgill.ca, McGill Univ – Oa
Bousenberry, Raymond T., (609) 984-6587 raymond.bousenberry@dep.state.nj.us, New Jersey Geological Survey – Ge
Boutt, David, 413-545-2724 dboutt@geo.umass.edu, Univ of Massachusetts – Hw
Bouwer, Edward John, (410) 516-7437 bouwer@jhu.edu, The Johns Hopkins Univ – Oo
Bouwkamp, John C., (301) 405-4334 jb65@umail.umd.edu, Univ of Maryland – Oo
Bouzari, Farhad, (604) 822-1874 fbouzari@eos.ubc.ca, Univ of British Columbia – Eg
Bovard, Kelly R., (714) 278-3561 kbovard@fullerton.edu, California State Univ, Fullerton – Gm
Boving, Thomas B., (401) 874-2265 Univ of Rhode Island – Hw
Bowen, Anthony J., (902) 494-7082 tony.bowen@dal.ca, Dalhousie Univ – On
Bowen, Brenda B., 989-774-3179 beitl1b@cmich.edu, Central Michigan Univ – Gs
Bowen, David, (406) 994-3331 Montana State Univ – Gd
Bowen, Dawn S., (540) 654-1491 dbowen@mwc.edu, Mary Washington College – Oo
Bowen, Gabriel, (765) 496-9344 gabe@purdue.edu, Purdue Univ – Cs
Bowen, Scott M., (505) 667-4313 Los Alamos National Lab – Ca
Bower, Kathleen M., 217-581-7016 cfkmbl@eiu.edu, Eastern Illinois Univ – Ng
Bowersox, J. Richard, (859) 257-6376 Jr.bowersox@uky.edu, Univ of Kentucky – Oo
Bowersox, Joe, 503-370-6220 jbowerso@willamette.edu, Willamette Univ – Oo
Bowin, Carl O., (508) 289-2572 cbowin@whoi.edu, Woods Hole Oceanographic Institution – Yr
Bowler, Tom, 406-496-4752 tbowler@mtech.edu, Montana Tech of The Univ of Montana – Nx
Bowman, David D., (714) 278-3882 dbowman@fullerton.edo, California State Univ, Fullerton – Ys
Bowman, Jean A., (409) 862-6544 jeanann@csrp.tamu.edu, Texas A&M Univ – Hg
Bowman, John R., (801) 581-7250 jrbowman@mines.utah.edu, Univ of Utah – Cs
Bowman, Kenneth P., (409) 845-4060 bowman@csrp.tamu.edu, Texas A&M Univ – Ow
Bowman, Malcolm J., (631) 632-8669 mbowman@notes.cc.sunysb.edu, SUNY, Stony Brook – Op
Bowman, Robert S., (505) 835-5992 bowman@nmt.edu, New Mexico Institute of Mining and Technology – So
Bowring, Samuel A., (617) 253-3775 sbowring@mit.edu, Massachusetts Institute of Technology – Gg
Bowser, Carl J., (608) 262-8955 bowser@geology.wisc.edu, Univ of Wisconsin, Madison – Cl
Box, Jason E., (614) 688-5758 box.11@osu.edu, Ohio State Univ – Oa
Boyce, Joseph I., (905) 525-9140 (Ext. 24188) boycej@mcmaster.ca, McMaster Univ – Yr
Boyd, Carl M., (902) 494-2413 carl.boyd@dal.ca, Dalhousie Univ – Ob
Boyd, Daniel T., (405) 325-3031 dtboyd@ou.edu, Univ of Oklahoma – Go
Boyd, Donald W., (307) 766-3280 boyd@uwyo.edu, Univ of Wyoming – Pi
Boyd, John P., (313) 764-3338 Univ of Michigan – Op
Boyd, Ron L., 61 02 4921 5744 Ron.Boyd@newcastle.edu.au, Univ of Newcastle – Gu
Boyd, Timtboyd@coas.oregonstate.edu, Oregon State Univ – Op
Boyer, Jean, (972) 883-2406 jean.boyer@utdallas.edu, Univ of Texas, Dallas – Oo
Boyer, Paul S., (973) 443-8790 psboyer@fdu.edu, Fairleigh Dickinson Univ – Pi
Boyer, Robert E., (512) 471-7228 reboyer@mail.utexas.edu, Univ of Texas at Austin, Jackson School of Geosciences – Gt
Boyle, Edward A., (617) 253-3388 eaboyle@mit.edu, Massachusetts Institute of Technology – Oc
Boyle, James T., (609) 984-6587 jim.boyle@dep.state.nj.us, New Jersey Geological Survey – Hw
Boyles, J. M., (307) 766-3386 Univ of Wyoming – Go
Boyles, Tyler J., 404-656-3214 Georgia Dept of Natural Resources – Gg

Boynton, William V., (520) 621-6941 wboynton@lpl.arizona.edu, Univ of Arizona – Xc
Boysen, Hans, 089/2180 4333 boysen@lmu.de, LMU - Univ of Munich – Gz
Bozzato, Edda, (705) 675-1151 (Ext. ext. 2272) ebozzato@laurentian.ca, Laurentian Univ, Sudbury – OooOo
Brabander, Daniel J., 781-283-3056 dbraband@wellesley.edu, Wellesley College – Cl
Brabham, P J.brabham@cf.ac.uk, Cardiff Univ – Ye
Brabson, Elizabeth, (301) 405-4378 lizb@essic.umd.edu, Univ of Maryland – Cg
Brace, Larry H., (313) 764-9460 Univ of Michigan – Gg
Brace, William F., (617) 253-3391 wfbrace@mit.edu, Massachusetts Institute of Technology – Gg
Bradbury, Kenneth R., (608) 263-7921 krbradbu@wisc.edu, Univ of Wisconsin, Extension – Hw
Bradbury, Kenneth R., (608) 263-7921 krbradu@wisc.edu, Univ of Wisconsin - Extension – Hw
Bradley, Christopher R., (505) 665-6713 cbradley@lanl.gov, Los Alamos National Lab – Yg
Bradley, Deborah, (405) 325-2921 dbradley@ou.edu, Univ of Oklahoma – Oo
Bradley, Jeffrey, (660) 562-1818 jbradle@nwmissouri.edu, Northwest Missouri State Univ – Og
Bradley, Michael, (313) 487-0218 Eastern Michigan Univ – Gc
Bradley, Michael D., (520) 621-3865 mdb@hwr.arizona.edu, Univ of Arizona – Oo
Bradley, Raymond S., (413) 545-2120 rbradley@geo.umass.edu, Univ of Massachusetts – Oa
Bradshaw, John, (03) 36 671 7779, j.bradshaw@geol.canterbury.ac.nz, Univ of Canterbury – Gt
Bradshaw, Peter M., (604) 681-8600 pbradshaw@firstpointminerals.com, Univ of British Columbia – Ce
Brady, John B., (413) 585-3953 jbrady@email.smith.edu, Smith College – Gx
Brady, Lawrence L., (785) 864-2159 lbrady@kgs.ku.edu, Univ of Kansas – Gg
Brady, Roland H., (559) 278-2391 roland_brady@csufresno.edu, California State Univ, Fresno – Ng
Braginsky, Stanislav I., (310) 794-1331 sbragins@igpp.ucla.edu, Univ of California, Los Angeles – Ym
Brahana, John V., (479) 575-2570 brahana@uark.edu, Univ of Arkansas, Fayetteville – Hw
Braile, Lawrence W., (765) 494-5979 braile@purdue.edu, Purdue Univ – Ys
Brainard, James R., (505) 667-0150 Los Alamos National Lab – Oo
Brajanovski, Miroslav, +61 8 9266 3519 M.Brajanovski@curtin.edu.au, Curtin Univ of Technology – Yex
Brake, Sandra S., (812) 237-2270 gebrake@isugw.indstate.edu, Indiana State Univ – Ge
Brake, Thomas L., (702) 794-7828 brake_thomas_l@lanl.gov, Los Alamos National Lab – Nr
Bralower, Timothy J., 814-863-1240 bralower@psu.edu, Pennsylvania State Univ, Univ Park – Pe
Braman, Dennis R., (403) 823-7707 Royal Tyrrell Museum of Palaeontology – Pl
Brame, Roderic I., 937 775-3455 Wright State Univ – Oe
Brame, Scott E., (864) 656-7167 scott@clemson.edu, Clemson Univ – Hw
Bramel-Cox, Paula J., (913) 532-7238 Kansas State Univ – So
Bramer, Mary Ann, 616-331-3728 bramerm@gvsu.edu, Grand Valley State Univ – Oo
Branam, Tracy, (812) 855-2687 tbranam@indiana.edu, Indiana Geological Survey – Cg
Branan, Yvonne K.ybranan@iup.edu, Indiana Univ of Pennsylvania – Gv
Brand, Leonard R., (909) 824-4530 Loma Linda Univ – Pv
Brand, Uwe, (905) 688-5550 (Ext. 3529) Uwe.Brand@brocku.ca, Brock Univ – Cl
Brande, Barbara L.brandeb@um.montevallo.edu, Univ of Montevallo – Ys
Brandon, Mark T., (203) 432-3135 mark.brandon@yale.edu, Yale Univ – Gc
Brandriss, Mark E.mbrandi@science.smith.edu, Williams College – Gi
Brandriss, Mark E., (413) 585-3585 mbrandri@email.smith.edu, Smith College – Gx
Brandt, Craig, (865) 574-1921 fcb@ornl.gov, Oak Ridge National Lab – Ge
Brandt, Danita S., 517-355-6595 brandt@msu.edu, Michigan State Univ – Pi
Brandt, Robert F., (713) 868-3723 Houston Community College System – Oo
Brandvold, Lynn A., (505) 835-5517 lynnb@nmt.edu, New Mexico Institute of Mining & Technology – Ca
Brannon, Joyce C., (314) 935-6204 joyce@isochem.wustl.edu, Washington Univ – Cc
Branstrator, Jon W., (765) 983-1339 jonb@earlham.edu, Earlham College – Pg
Brant, Lynn A., (319) 273-6160 lynn.brant@uni.edu, Univ of Northern Iowa – Gn
Brantley, Susan L., (814) 863-1739 brantley@geosc.psu.edu, Pennsylvania State Univ, Univ Park – Cg
Bras, Rafael L., (617) 253-2117 rlbras@mit.edu, Massachusetts Institute of Technology – Hg
Brassea, Jesus M.jbrassea@cicese.mx, Centro de Inv Científica y de Educación Superior de Ensenada – Ye
Brassell, Simon C., (812) 855-3786 simon@indiana.edu, Indiana Univ, Bloomington – Co
Braun, Duane D., (570) 389-4139 dbraun@bloomu.edu, Bloomsburg Univ – GmHwGl
Braun, Willi K., (306) 966-5688 Univ of Saskatchewan – Ps
Bray, Colin, (416) 978-6516 cjbray@quartz.geology.utoronto.ca, Univ of Toronto – Cs
Bray, E. Lee, 708-547-7223 bray@isgs.uiuc.edu, Illinois State Geological Survey – Ge
Bream, Brendan R., 615-343-7831 brendan.r.bream@vanderbilt.edu, Vanderbilt Univ – Gt
Brearley, Adrian J., (505) 277-4163 brearley@unm.edu, Univ of New Mexico – Gz
Breckenridge, Andrew, 814-824-3646 abreckenridge@mercyhurst.edu, Mercyhurst College – Gl
Breckenridge, Roy M., (208) 885-7991 roybreck@uidaho.edu, Univ of Idaho – Gg
Breitenbeck, Gary A., (504) 388-2110 Louisiana State Univ – Sb
Breithaupt, Brent H., (307) 766-2646 uwgeoms@uwyo.edu, Univ of Wyoming – Pv
Breland, Clayton F., (225) 388-8300 clayton@lgs.bri.lsu.edu, Louisiana State Univ – Ye
Brem, Gerald F.gbrem@fufllerton.edu, California State Univ, Fullerton – Cg
Brenan, James, (416) 978-0281 brenan@geology.utoronto.ca, Univ of Toronto – Gi
Brenner, Mark, (352) 392-9617 brenner@ufl.edu, Univ of Florida – Gn
Breshears, David D., (505) 665-2803 daveb@lanl.gov, Los Alamos National Lab – Py
Bretherton, Francis P., (608)262-7497 fbretherton@charter.net, Univ of Wisconsin, Madison – Og
Brethes, Jean-Claude, (418) 724-1779 jean-claude_brethes@uqar.uquebec.ca, Universite du Quebec a Rimouski – Ob
Breton, Caroline, 512-471-0322 cari.breton@beg.utexas.edu, Univ of Texas at Austin, Jackson School of Geosciences – Oi
Brett, Carlton E., (513) 556-4556 carlton.brett@uc.edu, Univ of Cincinnati – Pe
Brett, Carlton E., (513) 556-4556 carlton.brett@uc.edu, Cincinnati Museum Center – Po
Breuer, Deborrah, (573) 368-2160 Missouri Dept of Nat Res – Oo
Brevik, Eric C., (229) 249-2744 ecbrevik@valdosta.edu, Valdosta State Univ – Sd
Brewer, Kevin E., 815-928-5632 kbrewer2@olivet.edu, Olivet Nazarene Univ – HwOiYg
Brewer, Micheal, (202) 994-6185 mbrewer@gwu.edu, George Washington Univ – Or
Brewer, Thomas, (617) 287-5282 Univ of Massachusetts, Boston/Harbor Campus – Hw
Breyer, John A., 817-257-6275 j.breyer@tcu.edu, Texas Christian Univ – GdsGo
Brezinski, David K., (412) 624-8780 Univ of Pittsburgh – Ps
Brezinski, David K., 410-554-5526 dbrezinski@dnr.state.md.us, Maryland Geological Survey – Gr
Brice, William, (814) 269-2942 wbrice@pitt.edu, Cornell Univ – Gg
Brice, William R.wbrice@pitt.edu, Univ of Pittsburgh at Johnstown – Gg
Bricelj, Monica V., (902) 426-8005 monica.bricelj@nrc.ca, Dalhousie Univ – Ob
Bridge, John S., (607) 777-2831 Binghamton Univ – Gs
Bridges, Frank, (831) 459-2893 bridges@cats.ucsc.edu,

Univ of California, Santa Cruz – Yx
Briedis, Anita, 708-209-3128 Anita.Briedis@cuchicago.edu, Concordia Univ – OooOo
Brigham-Grette, Julie, (413) 545-4840 brigham-grette@geo.umass.edu, Univ of Massachusetts – Gl
Brikowski, Tom H., (972) 883-6242 brikowi@utdallas.edu, Univ of Texas, Dallas – Hw
Brimhall, George H., 510.642.5868 brimhall@eps.berkeley.edu, Univ of California, Berkeley – Eg
Briner, Jason P., (645) -68-00 x (Ext. 00 ext 3986) jbriner@buffalo.edu, SUNY, Buffalo – Gl
Brinkman, Donald B., (403) 823-7707 Royal Tyrrell Museum of Palaeontology – Pv
Brinkmeyer, Robinbrinkmer@tamug.edu, Texas A&M Univ – Ob
Brisbin, William C., (204) 474-9454 w_brisbin@umanitoba.ca, Univ of Manitoba – Gc
Briskin, Madeleine, (513) 556-3009 madeleine.briskin@uc.edu, Univ of Cincinnati – Pe
Brister, Brian S., 940 397-4250 bbrister@gunnoil.com, Midwestern State Univ – Got
Bristoll, Bill C., (608) 263-7386 bcbristoll@wisc.edu, Univ of Wisconsin - Extension – Oo
Britt, Brooks B., 801-422-7316 bbb26@geology.byu.edu, Brigham Young Univ – Pg
Britt, Daniel, (865) 974-2366 dbritt@utk.edu, Univ of Tennessee, Knoxville – Xg
Britzke, Gilbert, 089/2180 4209 gilbert.brietzke@geophysik.uni-muenchen.de, LMU - Univ of Munich – Yg
Broadfoot, Lyle A., (520) 621-4303 broadfoot@vega.lpl.arizona.edu, Univ of Arizona – Or
Broadhead, Ronald F., (505) 835-5202 ron@gis.nmt.edu, New Mexico Institute of Mining & Technology – Go
Broadhead, Ronald F., (505) 835-5202 ron@gis.nmt.edu, New Mexico Institute of Mining and Technology – Eo
Broadhead, Thomas W., (865) 974-6002 Univ of Tennessee, Knoxville – Pi
Broadway, Michael J., (906) 227-2500 mbroadwa@nmu.edu, Northern Michigan Univ – Oo
Brochu, Christopher A., (319) 353-1808 christopher-brochu@uiowa.edu, Univ of Iowa – Pv
Brock, Amy L., (309) 298-1516 Western Illinois Univ – GmSa
Brock, Pamela, (718) 997-3300 Queens College (CUNY) – Gg
Brock, Patrick W. G., (718) 997-3328 Graduate School of the City Univ of New York – Gg
Brockhaus, John A., (845) 938-2063 United States Military Acad – Yd
Broda, James E., (508) 289-2466 jbroda@whoi.edu, Woods Hole Oceanographic Institution – Gs
Brodie, Gregory, (423) 751-2784 Univ of Tennessee, Chattanooga – Ge
Brodsky, Emily, 831-459-1854 ebrodsky@pmc.ucsc.edu, Univ of California, Santa Cruz – Ys
Broecker, Wallace S., (845) 365-8413 broecker@ldeo.columbia.edu, Columbia Univ – Oc
Brofferio, Giulia F., (805) 893-8281 giulia@crustal.ucsb.edu, Univ of California, Santa Barbara – Oo
Brogan, George, 949-451-5687 gbrogan@ivc.edu, Irvine Valley College – Gc
Broker, Steven P., (203) 932-7107 Univ of New Haven – Oo
Bromwich, David H., (614) 688-5314 bromwich.1@osu.edu, Ohio State Univ – Oa
Brook, Edward J., (509) 546-9762 Washington State Univ – Gl
Brook, Edward J., (509) 546-9762 brook@vancouver.wsu.edu, Washington State Univ, Vancouver – Ou
Brook, Edward J., (541) 737-8197 brooke@geo.oregonstate.edu, Oregon State Univ – Ou
Brooker-Gross, Susan R., (540) 231-3341 Virginia Polytechnic Institute & State Univ – Ou
Brookfield, Michael E., (519) 824-4120 (Ext. 52654) mbrookfi@lrs.uoguelph.ca, Univ of Guelph – Pg
Brooks, Antone L., (509) 372-7335 tbrooks@tricity.wsu.edu, Washington State Univ – Oo
Brooks, Benjamin, 808-956-7864 bbrooks@soest.hawaii.edu, Univ of Hawai'i, Manoa – Gt
Brooks, Christopher, (514) 343-7035 Universite de Montreal – Cc
Brooks, David A., (979) 845-5527 dbrooks@ocean.tamu.edu, Texas A&M Univ – Op
Brooks, Debra A., (714) 564-4788 Santiago Canyon College – Yg
Brooks, Greg R., (727) 864-8992 Univ of South Florida, Saint Petersburg – Ou
Brooks, Gregg R., (727) 864-8992 brooksgr@eckerd.edu, Eckerd College – Gu
Brooks, Paul D., (520) 621-3424 brooks@hwr.arizona.edu, Univ of Arizona – Cg
Brooks, Scott C., (865) 574-6398 3sb@ornl.gov, Oak Ridge National Lab – Cl
Broome, Stephen W., (919) 515-2643 North Carolina State Univ – Sf
Brophy, James G., (812) 855-6417 brophy@indiana.edu, Indiana Univ, Bloomington – Gi
Broster, Bruce E., (506) 453-4804 broster@unb.ca, Univ of New Brunswick – Gl
Brotak, Edward A., (828) 251-6149 brotak@unca.edu, Univ of North Carolina, Asheville – Oa
Brothers, Timothy S., (317) 274-1101 Indiana Univ / Purdue Univ, Indianapolis – Oy
Brothers, Timothy S., (317) 274-1101 Indiana Univ, Indianapolis – Oy
Brouillette, Pierre, (418) 654-2567 pbrouill@nrcan.gc.ca, Universite du Quebec – Gg
Brower, James C., (315) 443-2672 Syracuse Univ – Po
Brown, Aaron, 323 953-4000 (Ext. 2606) brownad@lacitycollege.edu, Los Angeles City College – Cg
Brown, Alex C., (514) 340-4558 Ecole Polytechnique – Cl
Brown, Bradford D., (208) 722-6701 Univ of Idaho – Oo
Brown, Bruce A., (608) 263-3201 babrown1@wisc.edu, Univ of Wisconsin - Extension – Gc
Brown, Cathe, (202) 633-1788 brownc@si.edu, Smithsonian Institution / National Museum of Natural History – Gz
Brown, Charles Q., (252) 328-1365 brownc@mail.ecu.edu, East Carolina Univ – Gs
Brown, David L.DLBrown@csuchico.edu, California State Univ, Chico – Hg
Brown, David P., 603-862-7052 david.brown@unh.edu, Univ of New Hampshire – Oa
Brown, Donald W., (505) 667-1926 dwb@lanl.gov, Los Alamos National Lab – Nr
Brown, Dorothy, (301) 405-4082 geology@geol.umd.edu, Univ of Maryland – Oo
Brown, Dwight D., (409) 294-1593 Sam Houston State Univ – Pl
Brown, Edwin H., (360) 650-3645 ehbrown@cc.wwu.edu, Western Washington Univ – Gp
Brown, Erik T., (218) 726-8891 etbrown@d.umn.edu, Univ of Minnesota, Duluth – Og
Brown, Francis H., (801) 581-8767 fbrown@mines.utah.edu, Univ of Utah – Gg
Brown, J. Michael, 206-616-6058 brown@ess.washington.edu, Univ of Washington – Gy
Brown, James, (912) 825-6610 Fort Valley State Univ – Oo
Brown, James, (352) 392-1804 Missouri Dept of Natural Res – Oo
Brown, James R., (403) 220-7484 jbrown@geo.ucalgary.ca, Univ of Calgary – Ys
Brown, Jennifer B., (801) 581-7162 jbollis@mines.utah.edu, Univ of Utah – Oo
Brown, Kent D., (352) 392-1804 nrugs.kbrown@state.ut.us, Utah Geological Survey – Oo
Brown, Kevin M., (858) 534-5368 kmbrown@ucsd.edu, Univ of California, San Diego – Gu
Brown, Larry D., (607) 255-7357 brown@geology.cornell.edu, Cornell Univ – Ye
Brown, Laurie, (413) 545-0245 lbrown@geo.umass.edu, Univ of Massachusetts – Ym
Brown, Lewis M., (906) 635-2155 lbrown@lssu.edu, Lake Superior State Univ – Pi
Brown, Michael, (301) 405-4080 mbrown@geol.umd.edu, Univ of Maryland – Gx
Brown, Michael E., (626) 395-8423 mbrown@gps.caltech.edu, California Institute of Technology – Xy
Brown, Michael E., (662) 325-2906 mebrown@ra.msstate.edu, Mississippi State Univ – Ow
Brown, Paul, (507) 389-6613 paul.brown@mnsu.edu, Minnesota State Univ – Ga
Brown, Paul W., (520) 621-1319 pbrown@ag.arizona.edu, Univ of Arizona – Ow
Brown, Philip E., (206) 543-9419 pbrown@geology.wisc.edu, Univ of Wisconsin, Madison – Em
Brown, Randall B., (352) 392-1804 (Ext. 344) rbb@ufl.edu, Univ of Florida – Sd
Brown, Raymon L., (405) 325-3031 raybrown@ou.edu, Univ of Oklahoma – Ye
Brown, Rob, (828) 262-7222 brownrn@appstate.edu, Appalachian State Univ – Oo
Brown, Robert H., (520) 626-9045 rhb@lpl.arizona.edu,

Univ of Arizona – Xc
Brown, Teresa, 920-424-4460 brownt@uwosh.edu,
Univ of Wisconsin, Oshkosh – Oo
Brown, Vernon M., (419) 530-4571 vernon.brown@utoledo.edu,
Univ of Toledo – Gi
Brown, William G., (254) 710-2075 william_brown@baylor.edu,
Baylor Univ – Gc
Brown, Jr., Gordon E., (650) 723-9168 gordon@pangea.stanford.edu,
Stanford Univ – Gz
Brownawell, Bruce J., (631) 632-8658 bbrownawell@notes.cc.sunysb.edu,
SUNY, Stony Brook – Oc
Browne, Brandon, (714) 278-3054 bbrowne@fullerton.edu,
California State Univ, Fullerton – Gv
Browne, Kathleen M., (609) 896-5408 Rider Univ – Oge
Browning, James V., 732-445-3368 jvb@rci.rutgers.edu,
Rutgers, The State Univ of New Jersey – Oo
Brownlee, Donald E., 206-543-8575 brownlee@astro.washington.edu,
Univ of Washington – Xgy
Broxton, David E., 505-667-2492 broxton@lanl.gov,
Los Alamos National Lab – Gx
Brubaker, John M., (804) 684-7222 brubaker@vims.edu,
College of William & Mary – Op
Brubaker, Kenneth L., (630) 252-7630 Argonne National Lab – Oa
Brudzinski, Michael, 513-529-9758 brudzimr@muohio.edu,
Miami Univ – Ys
Brueckner, Hannes K., (718) 997-3322 hannes@ldeo.columbia.edu,
Queens College (CUNY) – Cg
Brueseke, Matthew E., (785) 532-1908 brueseke@ksu.edu,
Kansas State Univ – GivGt
Brugger, Keith A., (320) 589-6310 bruggeka@mrs.umn.edu,
Univ of Minnesota, Morris – Gl
Bruhn, Ronald L., (801) 581-7162 rlbruhn@mines.utah.edu,
Univ of Utah – Gc
Bruland, Kenneth W., (831) 459-4587 bruland@cats.ucsc.edu,
Univ of California, Santa Cruz – Oc
Brumbaugh, David S., (928) 523-7191 david.brumbaugh@nau.edu,
Northern Arizona Univ – Ys
Brun, Lynn J., (701) 231-8903 North Dakota State Univ – Oa
Brune, James N., (775) 784-4974 brune@seismo.unr.edu,
Univ of Nevada, Reno – Ys
Brune, William H., (814) 865-3286 brune@ems.psu.edu,
The Pennsylvania State Univ, Univ Park – Ow
Bruner, Katherine E., (304) 293-5603 bruner@geo.wvu.edu,
West Virginia Univ – Gd
Brunet, Gilbertgilbert.brunet@ec.gc.ca,
McGill Univ – Oa
Brunish, Wendee M., (505) 667-5724 wb@lanl.gov,
Los Alamos National Lab – Oo
Brunkhorst, Bonnie J., (909) 537-5612 bbrunkho@csusb.edu,
California State Univ, San Bernardino – Oe
Brunner, Charlotte A., (228) 688-3402 charlotte.brunner@usm.edu,
Univ of Southern Mississippi – Ou
Bruns, Dale A., (570) 408-4603 Wilkes Univ – Ou
Brunskill, Jeffrey C., 570-389-4355 jbrunski@bloomu.edu,
Bloomsburg Univ – OiwOy
Brunton, George D., (662) 236-4267 gebruntn@olemiss.edu,
Univ of Mississippi – Ng
Brunty, Lois, (513) 761-3004 Ohio State Univ – Oo
Brush, Grace S., (410) 516-7107 gbrush@jhu.edu,
The Johns Hopkins Univ – Pl
Brush, Nigel, (419) 289-5271 nbrush@ashland.edu,
Ashland Univ – Gx
Brusseau, Mark L., (520) 621-3244 brusseau@ag.arizona.edu,
Univ of Arizona – Hw
Bruton, Carol J., (925) 423-1936 bruton1@llnl.gov,
Lawrence Livermore National Lab – Cg
Bryan, Bonnie L., 765-658-4654 bbryan@depauw.edu,
DePauw Univ – Oo
Bryan, Jonathan R., (850) 729-5246 bryanj@owc.edu,
Okaloosa-Walton College – Pi
Bryan, Wilfred B., (508) 289-2582 Woods Hole Oceanographic Inst – Gu
Bryant, Ernest A., (505) 667-2422 Los Alamos National Lab – Cg
Bryant, Harold, (306) 966-5697 hbryant@sk.sympatico.ca,
Univ of Saskatchewan – Pv
Bryant, Marita, (207) 786-6452 mbryant@bates.edu,
Bates College – Gg
Bryant, William, (916) 323-9672 California Geological Survey – Ng
Bryant, William R., (979) 845-2680 wbryant@ocean.tamu.edu,
Texas A&M Univ – Ou
Bryce, Julia G., 603-862-3139 julie.bryce@unh.edu,
Univ of New Hampshire – Cg
Bryce, Karen R., (801) 378-5470 karen_bryce@byu.edu,
Brigham Young Univ – Oo
Bryson, Reid A., (608)262-5814 rabryson@facstaff.wisc.edu,
Univ of Wisconsin, Madison – Og
Buchanan, Donald G., (909) 888-6511 San Bernardino Valley Coll– Og
Buchanan, Georgegbuchana@mc3.edu,
Montgomery County Community College – Ng
Buchanan, John P., (509) 359-7493 jbuchanan@mail.ewu.edu,
Eastern Washington Univ – Hw
Buchanan, Michelle V., (865) 574-4986 buchananmv@ornl.gov,
Oak Ridge National Lab – Oo
Buchanan, Rex C., (785) 864-2106 rex@kgs.ku.edu,
Univ of Kansas – Oe
Buchannon, Robin C., (662) 915-7482 rcb@olemiss.edu,
Univ of Mississippi – Oo
Buchheim, H. Paul, (909) 824-4300 Loma Linda Univ – Gn
Buchness, Mary Ruth, 410-554-5524 mbushness@dnr.state.md.us,
Maryland Geological Survey – Oo
Buck, Brenda J., (702) 895-1694 buckb@unlv.nevada.edu,
Univ of Nevada, Las Vegas – Sd
Buck, Roger W., 845-365-8592 Columbia Univ – Yr
Buck, IV, W. R., (845) 365-8592 buck@ldeo.columbia.edu,
Columbia Univ – Yg
Bucke, David P., (802) 899-3584 david.bucke@uvm.edu,
Univ of Vermont – Gg
Buckingham, Michael J., (858) 534-7977 mbuckingham@ucsd.edu,
Univ of California, San Diego – Op
Buckley, Brendon, 845-365-8782 Columbia Univ – Oo
Buckley, Derek, (919) 733-2423 (Ext. 405) Derek.Buckley@ncmail.net,
North Carolina Geological Survey – Yr
Buckley, Jill S., 505.835.5405 jill@prrc.nmt.edu,
New Mexico Institute of Mining and Technology – Np
Buckley, John S.kfjsb00@tamuk.edu,
Texas A&M Univ, Kingsville – Pv
Buckley, Lawrence J., (401) 874-6671 lbuckley@gso.uri.edu,
Univ of Rhode Island – Oc
Buckley, Luke, 406-496-4677 lbuckley@mtech.edu,
Montana Tech of The Univ of Montana – Oo
Bucur, Ioan, +40-264-405371 ibucur@bioge.ubbcluj.ro,
Babes-Bolyai Univ – PgmGs
Budd, Ann F., (319) 335-1817 ann-budd@uiowa.edu,
Univ of Iowa – Po
Budd, David A., (303) 492-3988 david.budd@colorado.edu,
Univ of Colorado – Gd
Budd, Judith W., (906) 487-2590 jrbudd@mtu.edu,
Michigan Technological Univ – Or
Budd, William W., (509) 335-8536 budd@wsu.edu,
Washington State Univ – Oo
Buddemeier, Robert W., (785) 864-2112 buddrw@kgs.ku.edu,
Univ of Kansas – Hg
Budikova, Dagmar, (309)438-7643 dbudiko@ilstu.edu,
Illinois State Univ – Ge
Buening, Nancy, (916) 278-4683 buening@geology.ucdavis.edu,
California State Univ, Sacramento – Gg
Buesseler, Ken O., (508) 289-2309 kbuesseler@whoi.edu,
Woods Hole Oceanographic Institution – Oc
Buggisch, Werner, +49 9131 8522615 buggisch@geol.uni-erlangen.de,
Univ of Erlangen – Gg
Bugher, Lindsay, (812) 855-5581 lbugher@indiana.edu,
Indiana Univ, Bloomington – Oo
Buick, Roger, 206-543-1913 buick@ess.washington.edu,
Univ of Washington – PeoPm
Bukowinski, Mark S., (510) 642-0977 Univ of California, Berkeley – Gy
Bulkley, Scott, (760) 252-2411 (Ext. 7301) sbulkley@barstow.cc.ca.us,
Barstow Community College – Oo
Bull, William B., (520) 621-6024 Univ of Arizona – Gm
Bullamore, Henry W., (301) 687-4413 hbullamore@frostburg.edu,
Frostburg State Univ – Oo
Bullard, Reuben G., 859-572-6907 bullardr@nku.edu,
Northern Kentucky Univ – Gd
Bullard, Thomas F., (775) 673-7420 tbullard@dri.edu,
Desert Research Institute – Gm
Bullister, John, (206) 526-6741 bullister@pmel.noaa.gov,
Univ of Washington – Oc
Bullock, Emma, (202) 633-1815 bullocke@si.edu,
Smithsonian Insti / National Museum of Natural History – CaXm
Bundy, Larry G., (608) 263-2889 lgbundy@wisc.edu,
Univ of Wisconsin, Madison – So
Bunge, Hans-Peter, 089/2180 4225 bunge@lmu.de,

LMU - Univ of Munich – Yg
Buonaiuto, Frank, 212-650-3092 frank.buonaiuto@hunter.cuny.edu, Hunter College (CUNY) – On
Buras, Nathan, 520-621-5082 buras@hwr.arizona.edu, Univ of Arizona – Hq
Buratti, Bonnie J., (818) 354-7427 Jet Propulsion Lab – Or
Burau, Richard G., (530) 752-0194 rgburau@ucdavis.edu, Univ of California, Davis – So
Burbanck, George P., (757) 727-5783 george.burbanck@hamptonu.edu, Hampton Univ – On
Burbank, Douglas W., (805) 893-7858 burbank@crustal.ucsb.edu, Univ of California, Santa Barbara – Gm
Burbey, Thomas J., (540) 231-6696 tjburbey@vt.edu, Virginia Polytechnic Institute & State Univ – Hw
Burchfiel, B. C., (617) 253-7919 bcburch@mit.edu, Massachusetts Institute of Technology – Gc
Burden, Elliott T., (709) 737-8388 etburden@mun.ca, Memorial Univ of Newfoundland – Pl
Burdige, David J., (757) 683-4930 dburdige@odu.edu, Old Dominion Univ – Oc
Burger, H. Robert, (413) 585-3942 rburger@email.smith.edu, Smith College – Gc
Burger, Paul V., (505) 277-3827 pvburger@unm.edu, Univ of New Mexico – Gz
Burgess, Irene T., 205/247-3677 ithompson@gsa.state.al.us, Geological Survey of Alabama – Oo
Burgess, Tony L., (845) 365-8550 tlb29@columbia.edu, Columbia Univ – Oo
Burghardt, Andrew, (905) 525-9140 (Ext. 23534) burghard@mcmaster.ca, McMaster Univ – Yr
Burgmann, Roland, (510) 643-9545 burgmann@seismo.berkeley.edu, Univ of California, Berkeley – Gt
Burkart, Michael R., (515) 294-5809 mburkart@iastate.edu, Iowa State Univ of Science & Technology – Hw
Burkart, Michael R.mburkart@iastate.edu, Iowa State Univ of Science & Technology – Hy
Burke, Collette D., (316) 978-3140 Wichita State Univ – Pm
Burke, Deborah S., (509) 372-2483 deborah.burke@pnl.gov, Pacific Northwest National Lab – Sc
Burke, Joan, (416) 586-5591 joanb@rom.on.ca, Royal Ontario Museum – Oo
Burke, Kevin, (713) 743-3397 kburke@uh.edu, Univ of Houston – Gt
Burke, Raymond M., (707) 826-4292 rmb2@axe.humboldt.edu, Humboldt State Univ – Gm
Burke, Roger A., (706) 542-2652 Univ of Georgia – Cl
Burkhart, Patrick A., (724) 738-2502 patrick.burkhart@sru.edu, Slippery Rock Univ – Hg
Burkle, Garry L., (814) 865-9141 glb1@psu.edu, Pennsylvania State Univ, Altoona – Gg
Burks, Rachel J., (410) 704-3005 rburks@towson.edu, Towson Univ – Gc
Burlage, Robert S., (865) 574-7321 Oak Ridge National Lab – Sb
Burmeister, Kurtis C., (209) 946-2398 kburmeister@pacific.edu, Univ of the Pacific – GctEg
Burmester, Russell F., (360) 650-3654 russb@cc.wwu.edu, Western Washington Univ – Ym
Burnett, Donald S., (626) 395-6117 burnett@gps.caltech.edu, California Institute of Technology – Xc
Burnett, Earl E., (928) 329-1030 Arizona Western College – HwGc
Burnett, William C., 850-644-6703 wburnett@mailer.fsu.edu, Florida State Univ – Oc
Burney, Curtis, (954) 262-3652 burney@ocean.nova.edu, Nova Southeastern Univ – Oc
Burnham, Charlesburnham_c@fortlewis.edu, Fort Lewis College – Gz
Burnham, Charles W.burnham_c@fortlewis.edu, Harvard Univ – Gz
Burnham, Robyn J., (734) 764-0489 rburnham@umich.edu, Univ of Michigan – Pb
Burnham, Rosemary C., (916) 278-6337 burnhamr@csus.edu, California State Univ, Sacramento – Oo
Burnley, Pamela C.burnley@physics.unlv.edu, Georgia State Univ – Gy
Burns, Alan G., (313) 763-6246 Univ of Michigan – Og
Burns, Carol J., (505) 665-1765 Los Alamos National Lab – Oo
Burns, Peter C., (574) 631-5380 peter.burns.50@nd.edu, Univ of Notre Dame – Gz
Burns, Sandra, (860) 832-2934 burns@ccsu.edu, Central Connecticut State Univ – Oe
Burns, Scott F., (503) 725-3389 burnss@pdx.edu, Portland State Univ – Ng
Burns, Stephen J., (413) 545-0142 sburns@geo.umass.edu, Univ of Massachusetts – Cs
Burns, Susan E.seb8n@virginia.edu, Univ of Virginia – Hy
Burns, Timothy P., (505) 667-4600 Los Alamos National Lab – Co
Burrage, Mark D., (313) 936-0502 Univ of Michigan – Og
Burras, Lee, (318) 482-6163 Univ of Louisiana, Lafayette – Sd
Burras, Lee, (515) 294-0559 lburras@iastate.edu, Iowa State Univ of Science & Technology – Sd
Burrett, Clive, 61 3 6226 2470 Clive.Burrett@utas.edu.au, Univ of Tasmania – Pg
Burrill, Anne, 978-665-3378 aburrill@fsc.edu, Fitchburg State College – Oo
Burrus, C. Sidney, (713) 348-4009 deng@rice.edu, Rice Univ – Oo
Bursik, Marcus I., (716) 645-6800 (Ext. 3992) mib@geology.buffalo.edu, SUNY, Buffalo – Gv
Bursnall, John T., (315) 379-5432 jbursnall@stlawu.edu, St. Lawrence Univ – Gx
Burt, Donald M., (480) 965-6180 donald.burt@asu.edu, Arizona State Univ – Eg
Burton, Elizabeth A., (925) 422-7330 burton14@llnl.gov, Lawrence Livermore National Lab – Oe
Burton, Jacqueline C., (630) 252-8795 jcburton@anl.gov, Argonne National Lab – Cg
Burton, Robert C., (806) 655-3327 West Texas A&M Univ – Ps
Burton, Ronald S., (858) 534-7827 rburton@ucsd.edu, Univ of California, San Diego – Ob
Burton, Jr., Vinston, (919) 560-6269 North Carolina Central Univ – Gm
Burwash, Ronald A., (780) 492-2571 Univ of Alberta – Gp
Burwell, David C., (727) 553-1137 Univ of South Florida, Saint Petersburg – Op
Busacca, Alan J., (509) 335-1859 busacca@wsu.edu, Washington State Univ – Sa
Busalacchi, Antonio, (301) 405-5599 tonyb@essic.umd.edu, Univ of Maryland – Og
Busbey, Arthur B., (817) 257-7301 a.busbey@tcu.edu, Texas Christian Univ – PgvGr
Busby, Cathy J., (805) 893-4068 Univ of California, Santa Barbara – Gs
Busby, Michael R., (270) 809-3370 michael.busby@murraystate.edu, Murray State Univ – Oo
Busch, Richard M., (610) 436-2716 rbusch@wcupa.edu, West Chester Univ – Oe
Busch, William H., 504-280-7230 wbusch@uno.edu, Univ of New Orleans – Ou
Buscheck, Thomas A., (925) 423-9390 buscheck1@llnl.gov, Lawrence Livermore National Lab – Ng
Buseck, Peter R., (480) 965-3945 pbuseck@asu.edu, Arizona State Univ – Cg
Busemann, Henner, (202) 478-8463 busemann@dtm.ciw.edu, Carnegie Institution of Washington – Ca
Bush, Andrew B., (780) 492-0351 andrew.bush@ualberta.ca, Univ of Alberta – Oo
Bush, Andrew M., (860) 486-9358 andrew.bush@uconn.edu, Univ of Connecticut – Pg
Bush, Andrew B. G., (780) 492-0351 andrew.bush@ualberta.ca, Univ of Alberta – Oa
Bush, David M., (678) 839-4057 dbush@westga.edu, Univ of West Georgia – Ou
Bush, George T., (512) 475-9594 george.bush@beg.utexas.edu, Univ of Texas at Austin, Jackson School of Geosciences – Gg
Bush, James G., (509) 376-6555 jg.bush@pnl.gov, Pacific Northwest National Lab – Nx
Buskey, Edward J., (361) 749-6794 ed.buskey@mail.utexas.edu, Univ of Texas at Austin – Ob
Buss, Leo W., (203) 432-3869 Yale Univ – Po
Busse, Friedrich H., (310) 825-7698 Univ of California, Los Angeles – Oo
Bussod, Gilles Y., (505) 667-7220 gbussod@lanl.gov, Los Alamos National Lab – Yx
Bustin, R. Marc, (604) 822-6179 mbustin@eos.ubc.ca, Univ of British Columbia – Ec
Butcher, Elaine, (330) 972-7630 butch@uakron.edu, Univ of Akron – Oo
Butcher, Patricia M., (714) 278-3561 pbutcher@fullerton.edu, California State Univ, Fullerton – Hw
Butler, Gilbert W., (505) 667-6005 Los Alamos National Lab – Cc
Butler, James M., (785) 864-3965 jbutler@ku.edu, Univ of Kansas – Hw

Butler, Karl, (506) 453-4804 Univ of New Brunswick – Yg
Butler, R. Paul, (202) 478-8866 paul@dtm.ciw.edu,
Carnegie Institution of Washington – Oo
Butler, Sam, (306) 966-5702 sam.butler@usask.ca,
Univ of Saskatchewan – Yg
Butler, Jr., James J., (785) 864-2116 jbutler@kgs.ku.edu,
Univ of Kansas – Hw
Butterfield, David A., (206) 526-6722 dab3@u.washington.edu,
Univ of Washington – Yr
Butts, Susanbuttss1@southernct.edu,
Southern Connecticut State Univ – Pi
Buyce, M. Raymondrbuyce@mercyhurst.edu,
Mercyhurst College – Gs
Buynevich, Ilya, 508-289-2841 ibuynevich@whoi.edu,
Woods Hole Oceanographic Institution – Yr
Buys, Cunera, (847) 491-2910 c-buys@northwestern.edu,
Northwestern Univ – Oo
Buzas, Martin A., (202) 633-1313 Smithsonian Institution / National Museum of Natural History – Pm
Byerly, Don W., (865) 974-6007 dbyerly@utk.edu,
Univ of Tennessee, Knoxville – Ge
Byerly, Gary R., (225) 578-5318 glbyer@lsu.edu,
Louisiana State Univ – Gi
Byers, Charles W., (608) 262-2361 cwbyers@geology.wisc.edu,
Univ of Wisconsin, Madison – Pe
Bykerk-Kauffman, Ann, (530) 898-6305 abykerk-kauffman@csuchico.edu,
California State Univ, Chico – Gc
Byrd, Steven, (402) 280-2641 sbyrd@creighton.edu,
Creighton Univ – Yg
Byrne, Robert H., (727) 553-1508 byrne@marine.usf.edu,
Univ of South Florida, Saint Petersburg – Cm
Byrne, Roger, (510) 643-8108 arbyrne@uclink4.berkeley.edu,
Univ of California, Berkeley – Pl
Byrne, Timothy, (860) 486-3142 tim.byrne@uconn.edu,
Univ of Connecticut – Gc
Byrnes, Alan P., (785) 864-2177 abyrnes@kgs.ku.edu,
Univ of Kansas – Yg
Byun, Daewon, 713-743-3399 dbyun@uh.edu,
Univ of Houston – Oa

C

Caballero, Kate, (254) 968-9143 Tarleton State Univ – Oo
Cable, Jaye E., (225) 388-6515 Louisiana State Univ – Ob
Cablk, Mary, (775) 673-7371 mcablk@dri.edu,
Desert Research Institute – Or
Cabrera, Miguel L., (706) 542-1242 Univ of Georgia – So
Cady, Sherry L., (503) 725-3377 cadys@pdx.edu,
Portland State Univ – Py
Cahill, Kevin, (508) 289-2925 kcahill@whoi.edu,
Woods Hole Oceanographic Institution – Cm
Cahill, Richard A., 217-244-2532 cahill@isgs.uiuc.edu,
Illinois State Geological Survey – Ca
Cahir, John J., (814) 863-8358 cahir@ems.psu.edu,
The Pennsylvania State Univ, Univ Park – Ow
Cai, Ming, (850) 645-1551 cai@met.fsu.edu,
Florida State Univ – Ow
Cain, J. Allan, (401) 792-2184 Univ of Rhode Island – Gx
Cain, Peggycain@clemson.edu,
Clemson Univ – Oe
Caine, T. Nelson, (303) 492-5053 Univ of Colorado – Gm
Cairns, David M., (979) 845-2783 cairns@geog.tamu.edu,
Texas A&M Univ – Oy
Calais, Eric, (765) 496-2915 ecalais@purdue.edu,
Purdue Univ – Yg
Calaway, Wallis F., (630) 972-3586 Argonne National Lab – Om
Calcote, Randy, 612-624-8526 calco001@umn.edu,
Univ of Minnesota, Twin Cities – Py
Calder, Eliza, (716) -64-5-68 (Ext. x6100) ecalder@buffalo.edu,
SUNY, Buffalo – Gv
Calder, John, (902) 424-2778 Dalhousie Univ – Ec
Calderone, Gary, (623) 845-3654 Glendale Community College – Ym
Caldwell, Douglas R.dcaldwell26@comcast.net,
Oregon State Univ – Op
Caldwell, Nancy L, 217-265-0268 caldwell@isgs.uiuc.edu,
Illinois State Geological Survey – Ge
Caldwell, Roy, (510) 642-1391 rcaldwell@berkeley.edu,
Univ of California, Berkeley – Po
Caldwell, Roy L., (510) 642-1391 4roy@socrates.berkeley.edu,
Univ of California, Berkeley – Pi
Caldwell, Todd G., 775-673-7368 Todd.Caldwell@dri.edu,
Desert Research Institute – Hw
Caldwell, W. Glen E., (519) 661-3857 gcaldwel@uwo.ca,
Univ of Western Ontario – Pm
Caldwell, William G. E., (519) 661-3857 Univ of Saskatchewan – Ps
Calegari, Pat, (212) 346-1502 Pace Univ, New York Campus – Oo
Calengas, Peter L., (309) 298-1151 Western Illinois Univ – En
Cales, Marcia, (620)441-5297 cales@cowley.edu,
Cowley County Community College – Oo
Calhoun, Frank G., (330) 263-3722 calhoun.2@osu.edu,
Ohio State Univ – Sd
Calhoun, Joseph, 717-872-3289 jcalhoun@hearst.com,
Millersville Univ – Oa
Calizaya, Felipe, (801) 581-5422 fcalizaya@mines.utah.edu,
Univ of Utah – Nm
Callahan, John E., (828) 262-2746 callahnje@appstate.edu,
Appalachian State Univ – Eg
Callahan, Timothy J., (843) 953-8278 callahant@cofc.edu,
College of Charleston – Hw
Calon, Tomas J., (709) 737-8398 tcalon@sparky2.esd.mun.ca,
Memorial Univ of Newfoundland – Gc
Calvert, Andrew, (514) 340-4974 Ecole Polytechnique – Ye
Calvert, Andrew J., (604) 291-5387 acalvert@sfu.ca,
Simon Fraser Univ – Ys
Calvert, David V., (561) 468-3922 (Ext. 117) dvc@gnv.ifas.ufl.edu,
Univ of Florida – Sc
Calvert, Stephen E., (604) 822-5210 calvert@eos.ubc.ca,
Univ of British Columbia – Cm
Calvin, Wendy, (775) 784-1785 wcalvin@unr.edu,
Univ of Nevada, Reno – Or
Calvin, Wendy M., (775) 784-1785 wcalvin@mines.unr.edu,
Univ of Nevada, Reno – Or
Camacho, Alfredo, (204) 474-7413 camacho@cc.umanitoba.ca,
Univ of Manitoba – Gt
Camacho, Elsa A., (509) 376-5473 elsa.camachov,
Pacific Northwest National Lab – Cg
Camann, Eleanor J., 912 681-5943 ECamann@GeorgiaSouthern.edu,
Georgia Southern Univ – On
Cambardella, Cynthia, (515) 294-2921 cindyc@nstl.gov,
Iowa State Univ of Science & Technology – So
Cambiotti, Laura J.ljc0@lehigh.edu,
Lehigh Univ – Oo
Cambray, F William, 517-355-9568 cambray@msu.edu,
Michigan State Univ – Gc
Cameron, Barry I., 414-229-3136 bcameron@uwm.edu,
Univ of Wisconsin, Milwaukee – Gvi
Cameron, Douglas, (406) 496-4247 dcameron@mtech.edu,
Montana Tech of the Univ of Montana – Ca
Cameron, Kevin, (604) 291-4703 kjc@sfu.ca,
Simon Fraser Univ – Gg
Camille, Michael A., (318) 342-1886 Univ of Louisiana, Monroe – Oy
Camilleri, Phyllis A., 931-221-7317 camillerip@apsu.edu,
Austin Peay State Univ – Gct
Cammack, Rex, (417) 836-5173 RexCammack@smsu.edu,
Southwest Missouri State Univ – Or
Camp, Mark J., (419) 530-2398 mark.camp@utoledo.edu,
Univ of Toledo – Pi
Camp, Victor E., (619) 594-7170 vcamp@geology.sdsu.edu,
San Diego State Univ – Gv
Campagna, David, 757 229 1661 david@campagna-associates.com,
West Virginia Univ – Gc
Campagna, David J., (757) 229-1661 College of William & Mary – Gc
Campana, Michael E., (541) 737-2413 michael.campana@oregonstate.edu,
Oregon State Univ – Hg
Campbell, Andrew, (301) 405-4086 ajc@geol.umd.edu,
Univ of Maryland – Gz
Campbell, Andrew R., (505) 835-5327 campbell@nmt.edu,
New Mexico Institute of Mining and Technology – Cs
Campbell, Bruce A., 202-633-2472 Smithsonian Institution / National Air & Space Museum – Xy
Campbell, David L., (319) 335-1314 david-l-campbell@uiowa.edu,
Univ of Iowa – Yg
Campbell, Elizabeth V. M., (434) 951-6343 elizabeth.campbell@dmme.virginia.gov, Division of Mineral Resources – Gg
Campbell, Finley A., (403) 220-6801 Univ of Calgary – Em
Campbell, Glenn A., (859) 622-1423 glenn.campbell@eku.edu,
Eastern Kentucky Univ – Oo
Campbell, Ian A.ian.campbell@ualberta.ca,
Univ of Alberta – Gm
Campbell, J. B., (706) 291-2121 Shorter College – Oo
Campbell, James B., (540) 231-5841 Virginia Polytechnic Institute &

State Univ – Or
Campbell, Janet W., (603) 862-1070 janet.campbell@unh.edu,
Univ of New Hampshire – Oa
Campbell, Jocelyn K., ((03)) 3-667- (Ext. 1 ext 7775) jocelyn.campbell@canterbury.ac.nz, Univ of Canterbury – Gc
Campbell, Katherine, (505) 667-2799 ksc@lanl.gov,
Los Alamos National Lab – Gq
Campbell, Kenneth E., (213) 763-3425 kcampbel@nhm.org,
Los Angeles County Museum of Natural History – Pv
Campbell, Kenneth M., (850) 488-9380 Florida Geological Survey – Gs
Campbell, Lisa, (979) 845-5706 lcampbell@ocean.tamu.edu,
Texas A&M Univ – Ob
Campbell, Patricia A., (724) 738-4405 patricia.campbell@sru.edu,
Slippery Rock Univ – Gc
Campbell-Stone, Erin A., (307) 766-2053 erincs@uwyo.edu,
Univ of Wyoming – Gc
Canace, Robert J., (609) 984-6587 bob.canace@dep.state.nj.us,
New Jersey Geological Survey – Hw
Canales Cisneros, Juan Pablo, (508) 289-2893 jpcanales@whoi.edu,
Woods Hole Oceanographic Institution – Yr
Cande, Steven C., (858) 534-1552 scande@ucsd.edu,
Univ of California, San Diego – Yr
Candela, Philip A., (301) 405-2783 candela@geol.umd.edu,
Univ of Maryland – Cg
Cane, Mark A., (845) 365-8344 mcane@ldeo.columbia.edu,
Columbia Univ – Op
Canil, Dante, (250) 472-4180 dcanil@uvic.ca,
Univ of Victoria – Gi
Canon, Edgardoecanon@cicese.mx,
Centro de Inv Científica y de Educación Superior de Ensenada – Ym
Cantarero, Debra A., (626) 585-7138 dacantarero@paccd.cc.ca.us,
Pasadena City College – Oo
Cantin, Mark, (812) 237-3991 ancantin@isugw.indstate.edu,
Indiana State Univ – Ga
Cantrell, Kirk J., (509) 376-2136 kirk.cantrell@pnl.gov,
Pacific Northwest National Lab – Cl
Canuel, Elizabeth A., (804) 684-7134 ecanuel@vims.edu,
College of William & Mary – Oc
Cao, Hongsheng, 316-978-3140 Wichita State Univ – Cl
Cao, Tianqing, (916) 323-7967 California Geological Survey – Ys
Capehart, William J., (605) 394-2291 wjc@ias.sdsmt.edu,
South Dakota School of Mines & Technology – Oa
Caplan-Auerbach, Jackie, 360-650-4153 jackie@geol.wwu.edu,
Western Washington Univ – Ys
Capo, Rosemary C., (412) 624-8873 rcapo@pitt.edu,
Univ of Pittsburgh – Cl
Capoccia, Mary, (614) 292-8522 capoccia.6@osu.edu,
Ohio State Univ – Oo
Caponigri, Winifred F., (219) 239-8417 wcaponigri@hcc-nd.edu,
Holy Cross College – Gg
Cappa, Jim, (303) 866-3293 jim.cappa@state.co.us,
Colorado Geological Survey – Gz
Capriulo, Gerard M., (914) 251-6640 SUNY, Purchase – Ob
Capuano, Regina M., (713) 743-3426 capuano@uh.edu,
Univ of Houston – Hw
Caputo, Mario V., (909) 594-5611 (Ext. 4439) mvcaputo@earthlink.net,
Mount San Antonio College – Gg
Carbotte, Suzanne, 845-365-8895 Columbia Univ – Yr
Cardelino, Carlos, (404) 894-1751 carlos@sosona.eas.gatech.edu,
Georgia Institute of Technology – Oa
Cardenas, Andres, 727-553-1198 acardenas@marine.usf.edu,
Univ of South Florida, Saint Petersburg – Oc
Cardenas, Bayani, 512-471-9425 cardenas@mail.utexas.edu,
Univ of Texas at Austin, Jackson School of Geosciences – Hw
Cardimona, Steve, (707) 468-3219 scardimo@mendocino.edu,
Mendocino College – Ys
Cardon, Zoe, (860) 486-3868 zoe.cardon@uconn.edu,
Univ of Connecticut – Py
Cardott, Brian J., (405) 325-3031 bcardott@ou.edu,
Univ of Oklahoma – Ec
Carew, James L., (843) 953-5592 carewj@cofc.edu,
College of Charleston – Po
Carey, Anne E., (614) 292-2375 carey@geology.ohio-state.edu,
Ohio State Univ – Hq
Carey, James W., (505) 667-5540 bcarey@lanl.gov,
Los Alamos National Lab – Ca
Carey, Kristine M., (807) 343-8461 kristine.carey@lakeheadu.ca,
Lakehead Univ – Oo
Carey, Larry, 256-961-7909 larry.carey@nsstc.uah.edu,
Texas A&M Univ – Oa
Carey, Steven N., (401) 874-6209 scarey@gso.uri.edu,
Univ of Rhode Island – Ou
Carey, Tara, 309-796-5274 careyt@bhc.edu,
Black Hawk College – Oo
Carey, Jr., Andrew G.careya@onid.orst.edu,
Oregon State Univ – Ob
Carignan, George R., (313) 763-2174 Univ of Michigan – Oo
Carignan, Jacques, (418) 545-5011 jcarigna@uqac.uquebec.ca,
Universite du Quebec a Chicoutimi – Cg
Carina, Dorothy, (740) 368-3615 ddcarina@owu.edu,
Ohio Wesleyan Univ – Oo
Carl, James D.carljd@slic.com,
SUNY, Potsdam – Gz
Carle, Steven, (925) 423-5039 carle1@llnl.gov,
Lawrence Livermore National Lab – Hw
Carlisle, Donald, (310) 825-1934 carlisle@ess.ucla.edu,
Univ of California, Los Angeles – Eg
Carlson, Anders, 608-262-1921 acarlson@geology.wisc.edu,
Univ of Wisconsin, Madison – Pe
Carlson, Barry A.bacarlso@delta.edu,
Delta College – Yg
Carlson, Catherine A., (860) 465-5218 carlsonc@easternct.edu,
Eastern Connecticut State Univ – Hw
Carlson, Charles G., (605) 688-4761 Gregg.Carlson@sdstate.edu,
South Dakota State Univ – Hw
Carlson, Diane H., (916) 278-6382 carlsondh@csus.edu,
California State Univ, Sacramento – Gc
Carlson, Douglas A., 225/578-3671 dcarlson@lsu.edu,
Louisiana State Univ – Hw
Carlson, Ernest H., (330) 672-3778 ecarlson@kent.edu,
Kent State Univ – Ce
Carlson, Gaylen R., (714) 278-3942 gcarlson@fullerton.edu,
California State Univ, Fullerton – Oe
Carlson, Keith J., (507) 933-7307 jcarlson@gac.edu,
Gustavus Adolphus College – Pv
Carlson, Marvin P., (402) 472-7549 mcarlson@unl.edu,
Unversity of Nebraska - Lincoln – Gt
Carlson, Richard E., (515) 294-9868 richard@iastate.edu,
Iowa State Univ of Science & Technology – Ow
Carlson, Richard L., (979) 845-1371 carlson@geo.tamu.edu,
Texas A&M Univ – Ou
Carlson, Richard W., (202) 478-8474 carlson@dtm.ciw.edu,
Carnegie Institution of Washington – Cc
Carlson, Roseann J., (804) 427-7189 Tidewater Comm College – Pg
Carlson, Sandra J., (530) 752-2834 carlson@geology.ucdavis.edu,
Univ of California, Davis – Po
Carlson, Toby N., (814) 863-1582 tnc@psu.edu,
The Pennsylvania State Univ, Univ Park – Ow
Carlson, William D., (512) 471-4770 wcarlson@mail.utexas.edu,
Univ of Texas at Austin, Jackson School of Geosciences – Gp
Carmack, Eddy C., (250) 363-6585 carmack@ios.bc.ca,
Univ of Victoria – Op
Carmack, Edward, (250) 363-6585 carmack@dfo-mpo.gc.ca,
Univ of British Columbia – Op
Carman, Jessie C., 410-293-6551 carman@usna.edu,
United States Naval Academy – Op
Carman, Mary R., (508) 289-2987 mcarman@whoi.edu,
Woods Hole Oceanographic Institution – Gu
Carmichael, Dugald M., (319) 335-1824 Queen's Univ – Gp
Carmichael, Ian S., (510) 642-2577 ian@eps.berkeley.edu,
Univ of California, Berkeley – Gi
Carmichael, Robert S., (319) 335-1824 robert-carmichael@uiowa.edu,
Univ of Iowa – Ye
Carney, Cindy K., (513) 873-3455 Miami Univ – Oo
Carney, Cindy K., 937 775-3455 cindy.carney@wright.edu,
Wright State Univ – Gd
Carney, Robert S., (225) 388-6511 Louisiana State Univ – Ob
Carney, Theodore C., (505) 667-3415 tedc@lanl.gov,
Los Alamos National Lab – Ng
Caron, Jean-Bernard, (416) 586-5753 jcaron@rom.on.ca,
Univ of Toronto – Pi
Carp, Jana, (828) 262-7091 carpje@appstate.edu,
Appalachian State Univ – Oo
Carpenter, Ken, (303) 370-6392 KCarpenter@dmns.org,
Denver Museum of Nature & Science – Pv
Carpenter, Phillip J., (815) 753-1523 phil@geol.niu.edu,
Northern Illinois Univ – Yg
Carpenter, Roy, (206) 543-8535 rcarp@u.washington.edu,
Univ of Washington – Oo
Carpenter, Scott J., (319) 335-1818 scott-j-carpenter@uiowa.edu,

Univ of Iowa – Cs
Carr, James R., (775) 784-4244 carr@unr.edu,
Univ of Nevada, Reno – Ng
Carr, Keith W., 217-265-0267 carr@isgs.uiuc.edu,
Illinois State Geological Survey – Ge
Carr, Mark, (831) 459-3958 carr@biology.ucsc.edu,
Univ of California, Santa Cruz – Ob
Carr, Michael J., (732) 932-7494 carr@rutgers.edu,
Rutgers, The State Univ of New Jersey – Gv
Carr, Sharon D.scarr@ccs.carleton.ca,
Carleton Univ – Gt
Carr, Timothy R., (785) 864-2135 tcarr@kgs.ku.edu,
Univ of Kansas – Go
Carr, Timothy R., (785) 864-2135 tcarr@ku.edu,
Univ of Kansas – Go
Carr, Timothy R., 304-293-9660 timothy.carr@mail.wvu.edu,
West Virginia Univ – GoYe
Carrano, Matthew T., (202) 633-1506 Smithsonian Institution / National Museum of Natural History – Pv
Carrapa, Barbara, (307) 766-3386 Univ of Wyoming – Gs
Carrigan, Charles R., (925) 422-3941 carrigan1@llnl.gov,
Lawrence Livermore National Lab – Yg
Carrigan, Charles W., (815) 939-5346 CCarriga@olivet.edu,
Olivet Nazarene Univ – CcGxc
Carrington, Thomas J., (334) 844-4282 carritj@auburn.edu,
Auburn Univ – Gr
Carroll, Alan R., (608) 262-2368 acarroll@geology.wisc.edu,
Univ of Wisconsin, Madison – Gs
Carroll, Mary Anne, (313) 763-0271 Univ of Michigan – Oa
Carroll, Monica, (815) 753-7945 carroll@niu.edu,
Northern Illinois Univ – Cl
Carroll, Richard E., 205/247-3551 rcarroll@gsa.state.al.us,
Geological Survey of Alabama – Pl
Carroll, Susan A., (925) 423-7552 Lawrence Livermore National Lab – Cl
Carson, Bobbbc00@lehigh.edu,
Lehigh Univ – Gu
Carson, Robert J., 509-527-5224 carsonrj@whitman.edu,
Whitman College – Gm
Carstarphen, Camela, 406-496-4633 ccarstarphen@mtech.edu,
Montana Tech of The Univ of Montana – Hw
Carstens, Ken, (270) 809-6755 Murray State Univ – Oo
Carstensen, Laurence W., (540) 231-2600 Virginia Polytechnic Institute & State Univ – Oo
Carter, B. Roger, (404) 656-3214 Georgia Dept of Nat Resources – Ng
Carter, Burchard D., 229-931-2325 noid@canes.gsw.edu,
Georgia Southwestern State Univ – Ps
Carter, James L., (972) 883-2455 jcarter@utdallas.edu,
Univ of Texas, Dallas – Eg
Carter, Joseph G., (919) 962-0685 clams@email.unc.edu,
Univ of North Carolina, Chapel Hill – Po
Carter, Kristin M., 412.442.4233 krcarter@state.pa.us,
Pennsylvania Bureau of Topographic & Geologic Survey – Go
Carter, Mark W., (704) 251-6208 North Carolina Geological Survey – Gt
Carter, Mark W., 434-951-6357 mark.carter@dmme.virginia.gov,
Division of Mineral Resources – Gr
Carter, Neville L., (409) 845-1371 Texas A&M Univ – Yx
Carton, James, (301) 405-5365 Univ of Maryland – Op
Cartwright, J A.cartwrightJA@cf.ac.uk,
Cardiff Univ – Ys
Carver, Gary A., (907) 487-4551 Humboldt State Univ – Gmt
Cary, Jeffrey, 970-247-7620 cary_j@fortlewis.edu,
Fort Lewis College – Eg
Cary, Kevin, (270) 745-2981 kevin.cary@wku.edu,
Western Kentucky Univ – Oi
Casavant, Robert R., (520)626-3785 casavant@geo.arizona.edu,
Univ of Arizona – Gm
Casciotti, Karen L., (508) 289-3738 kcasciotti@whoi.edu,
Woods Hole Oceanographic Institution – Oc
Case, James M., (570) 408-4618 Wilkes Univ – Ob
Case, William F., (801) 537-3340 nrugs.bcase@state.ut.us,
Utah Geological Survey – Ng
Casey, John F., (713) 743-3390 jfcasey@uh.edu,
Univ of Houston – GtiGu
Casey, William H., (530) 752-1511 whcasey@ucdavis.edu,
Univ of California, Davis – Cl
Cashman, Katharine V., (541) 346-4323 cashman@oregon.uoregon.edu,
Univ of Oregon – Gv
Cashman, Patricia H., (775) 784-6924 pcashman@mines.unr.edu,
Univ of Nevada, Reno – Gc
Cashman, Susan M., (707) 826-3114 smc1@humboldt.edu,
Humboldt State Univ – Gc
Caskey, John, (415) 405-0353 caskey@sfsu.edu,
San Francisco State Univ – Gc
Cass, Glen R., (404) 894-3948 Georgia Institute of Technology – Oa
Cassar, Nicolas, (609) 258-7435 ncassar@princeton.edu,
Princeton Univ – Ob
Cassel, Donald K., (919) 515-1457 North Carolina State Univ – Sp
Cassell, Eugene A., (802) 656-2620 ecassell@zoo.uvm.edu,
Univ of Vermont – Hs
Cassidy, Daniel, 418-656-8720 daniel.cassidy@ggl.ulaval.ca,
Universite Laval – HwGe
Cassidy, Daniel P., (269) 387-5324 daniel.cassidy@wmich.edu,
Western Michigan Univ – Hw
Cassidy, Nigel J., +44 01782 583180 n.j.cassidy@esci.keele.ac.uk,
Keele Univ – Ge
Cassie, Robert M., (585) 395-5716 rcassie@weather.brockport.edu,
SUNY College at Brockport – Gc
Casso, Susan A., (508) 289-2248 scasso@whoi.edu,
Woods Hole Oceanographic Institution – Oo
Castagna, John P., (405) 325-6697 castagna@ou.edu,
Univ of Oklahoma – Ye
Castagna, John P., 713-743-3399 jcastagna@uh.edu,
Univ of Houston – Ye
Castillo, Paterno R., (858) 534-0383 pcastillo@ucsd.edu,
Univ of California, San Diego – Gi
Castillon, David A., (417) 836-5607 Southwest Missouri State Univ – Gm
Castle, James W., (864) 656-5015 jcastle@clemson.edu,
Clemson Univ – Gd
Castor, Stephen B., (775) 784-6691 (Ext. 146) scastor@unr.edu,
Univ of Nevada – Eg
Castro, Jonathan M., 202-633-1810 castroj@si.edu,
Smithsonian Institution / National Museum of Natural History – Gvi
Castro, Jonathan M., (440) 775-6701 jonathan.castro@oberlin.edu,
Oberlin College – Gv
Castro, Maria Clara, 734-615-3812 mccastro@umich.edu,
Univ of Michigan – Hy
Castro, Mark S., (301) 689-3115 castro@al.umces.edu,
Univ of Maryland – Oa
Castro, Raulraul@cicese.mx,
Centro de Inv Científica y de Educación Superior de Ensenada – Ys
Cataneo, Robert, 217-581-2626 cfrc2@eiu.edu,
Eastern Illinois Univ – Oa
Catau, John C., (417) 836-5173 Southwest Missouri State Univ – Oo
Cather, Steven M., (505) 835-5153 New Mexico Institute of Mining and Technology – Gc
Cather, Steven M., (505) 835-5153 steve@gis.nmt.edu,
New Mexico Institute of Mining & Technology – Gs
Catherine, Forster A., (631) 444-8203 catherine.forster@sunysb.edu,
SUNY, Stony Brook – Pv
Cathles, Lawrence M., (607) 255-7135 cathles@geology.cornell.edu,
Cornell Univ – Eg
Catlos, Elizabeth J., 405-744-9246 catlos@okstate.edu,
Oklahoma State Univ – Gz
Cattolico, Rose Ann, (206) 543-9363 racat@u.washington.edu,
Univ of Washington – Oo
Catuneanu, Octavian, (780) 492-6569 octavian.catuneanu@ualberta.ca,
Univ of Alberta – Gs
Caughey, Bob, (941) 338-2362 Florida Geological Survey – Go
Caughey, Michael E., (217) 333-1611 mcaughey@ux1.cso.uiuc.edu,
Illinois State Water Survey – Co
Caupp, Craig L., (301) 687-4755 ccaupp@frostburg.edu,
Frostburg State Univ – OoHs
Cavaliere, Ornella N., (514) 398-3758 ornella.cavaliere@mcgill.ca,
McGill Univ – Oo
Cavallaro, Nancy D., (809) 265-3851 Univ of Puerto Rico – Sc
Cawley, Sondra S., (307) 766-3389 acadcoord.geol@uwyo.edu,
Univ of Wyoming – Oo
Cawlfield, Jeff, 573-341-4557 jdc@umr.edu,
Univ of Missouri, Rolla – Hy
Cecil, Blaine, 703 648 6415 bcecil@usgs.gov,
West Virginia Univ – Ec
Cecil, Cammy, (480) 955-5768 cam@asu.edu,
Arizona State Univ – Oo
Ceh, Brian, (812) 237-2251 geceh@isugw.indstate.edu,
Indiana State Univ – Oi
Celestian, Aaron, (270) 745-5977 aaron.celestian@wku.edu,
Western Kentucky Univ – GzYgGg
Celestian, Stan, (623) 845-3681 stan.celestian@gcmail.maricopa.edu,
Glendale Community College – Oe

Celia, Michael A., (609) 258-5425 Princeton Univ – Hw
Cemen, Ibrahim, 405-744-6358 icemen@okstate.edu, Oklahoma State Univ – Gc
Center, Teresa J., (479) 575-3159 tcenter@comp.uark.edu, Univ of Arkansas, Fayetteville – Oo
Centorbi, Tracey, (301) 405-6965 tlcento@geol.umd.edu, Univ of Maryland – Cg
Cepeda, Joseph C., (806) 651-2584 jcepeda@mail.wtamu.edu, West Texas A&M Univ – Gi
Cercone, Karen Rose, (724) 357-5623 kcercone@iup.edu, Indiana Univ of Pennsylvania – Cl
Cerling, Thure E., (801) 581-5558 tcerling@mines.utah.edu, Univ of Utah – Cl
Cerny, Petr, (204) 474-8765 p_cerny@umanitoba.ca, Univ of Manitoba – Gz
Cerrato, Robert M., (631) 632-8666 rcerrato@notes.cc.sunysb.edu, SUNY, Stony Brook – Ob
Cervantes, Brandy, (541) 737-4414 bkuebel@coas.oregonstate.edu, Oregon State Univ – Op
Cervantes, Enedina, (520) 621-2027 enedina@email.arizona.edu, Univ of Arizona – Oo
Cervato, Cinzia C., (515) 294-7583 cinzia@iastate.edu, Iowa State Univ of Science & Technology – Gg
Cess, Robert D., (631) 632-8321 rcess@notes.cc.sunysb.edu, SUNY, Stony Brook – Oa
Cessi, Paola, (858) 534-0622 pcessi@ucsd.edu, Univ of California, San Diego – Op
Cetin, Haluk, (270) 809-2085 haluk.cetin@murraystate.edu, Murray State Univ – Oi
Chacko, Elizabeth, (202) 994-6185 echack@gwu.edu, George Washington Univ – Oy
Chacko, Thomas, (780) 492-5395 tom.chacko@ualberta.ca, Univ of Alberta – Gx
Chadima, Sarah A., (605) 677-5227 schadima@usd.edu, South Dakota Dept of Env & Natural Resources – Gg
Chadwick, John, 704-687-5947 Univ of North Carolina, Charlotte – Or
Chafetz, Henry S., (713) 743-3427 hchafetz@uh.edu, Univ of Houston – Gd
Chagnon, Andre, (418) 654-2610 ess-inrs-achagnon@x1.nrcan.gc.ca, Universite du Quebec – Co
Chakhmouridian, Anton, (204) 474-7278 chakhmou@ms.umanitoba.ca, Univ of Manitoba – Gz
Chakravorty, Somendra, 217-333-6470 samchak@isgs.uiuc.edu, Illinois State Geological Survey – Ge
Chaldu, Ronald S., (315) 312-2802 chaldu@oswego.edu, SUNY, Oswego – Oo
Challender, Stuart, 406-994-7566 schallender@montana.edu, Montana State Univ – Oi
Chamberlain, Edwin P., (718) 951-5416 Los Alamos National Lab – Cc
Chamberlain, John A., (718) 951-5926 Graduate School of the City Univ of New York – Po
Chamberlain, John A., (718) 951-5000 (Ext. 2885) johnc@brooklyn.cuny.edu, Brooklyn College (CUNY) – Po
Chamberlain, Kevin R., (307) 766-2914 kchamber@uwyo.edu, Univ of Wyoming – Cc
Chamberlin, Richard M., (505) 835-5310 richard@gis.nmt.edu, New Mexico Institute of Mining & Technology – Gg
Chambers, Jeanne, (775) 784-5329 chambers@equinox.unr.edu, Univ of Nevada, Reno – Py
Chambers, John E., (202) 478-8855 chambers@dtm.ciw.edu, Carnegie Institution of Washington – Oo
Chameides, William L., (404) 894-1749 Georgia Institute of Tech – Oa
Chameides, William L., 613-8004 bill.chameides@duke.edu, Duke Univ – OaGe
Champlin, Steven D., (601) 961-5506 stephen_champlin@deq.state.ms.us, Mississippi Office of Geology – Go
Chan, Kwan M., (562) 985-4817 kmchan@csulb.edu, California State Univ, Long Beach – Oc
Chan, Lui-Heung, (225) 388-6439 lchan@geol.lsu.edu, Louisiana State Univ – Oc
Chan, Marjorie A., (801) 581-6551 machan@mines.utah.edu, Univ of Utah – Gs
Chandler, Cyndy, (508) 289-2765 cchandler@whoi.edu, Woods Hole Oceanographic Institution – Og
Chandler, G. Thomas, (803) 777-9481 Univ of South Carolina – Ob
Chandler, Val W., 612-627-4780 (Ext. 203) chand004@umn.edu, Univ of Minnesota, Twin Cities – Ye
Chaney, Phil L., (334) 844-3420 chanepl@auburn.edu, Auburn Univ – Oy
Chang, Chia-Bo, (806) 742-3143 chia-bo.chang@ttu.edu, Texas Tech Univ – Oa
Chang, Edmund, kchang@notes.cc.sunysb.edu, SUNY, Stony Brook – Oa
Chang, Michael E., (404) 894-0838 chang@sosona.eas.gatech.edu, Georgia Institute of Technology – Oa
Chang, Ping, (979) 845-8196 chang@shark.tamu.edu, Texas A&M Univ – Op
Chang, Ping, (979) 845-8196 ping@ocean.tamu.edu, Texas A&M Univ – Op
Channell, James T., 352-392-3658 jetc@geology.ufl.edu, Univ of Florida – YmGt
Chanton, Jeffery P.chanton@ocean.fsu.edu, Florida State Univ – Cm
Chao, Shenn-Yu, (410) 221-8427 chao@hpl.umces.edu, Univ of Maryland – Op
Chapin, Charles E., (505) 835-5613 chapin@gis.nmt.edu, New Mexico Institute of Mining & Technology – Gt
Chapin, F. Stuart, (510) 642-1003 Univ of California, Berkeley – Pe
Chaplin, James R., (405) 325-3031 Univ of Oklahoma – Ps
Chapman, David S., (801) 581-7642 dchapman@mines.utah.edu, Univ of Utah – Yh
Chapman, Demian D., (954)262-3657 chapmand@nova.edu, Nova Southeastern Univ – Ob
Chapman, Marshall, (606) 783-5397 m.chapman@moreheadstate.edu, Morehead State Univ – Gv
Chapman, Martin C., (540) 231-5036 mcc@vt.edu, Virginia Polytechnic Institute & State Univ – Ys
Chapman, Piers, 979 845 7211 piers.chapman@tamu.edu, Texas A&M Univ – Oc
Chapman, Ross N., (250) 472-4340 chapman@uvic.ca, Univ of Victoria – Op
Chapman, Thomas E., 912 681-0174 TChapman@GeorgiaSouthern.edu, Georgia Southern Univ – OioOo
Chappell, James R., (970) 491-5147 Jim_Chappell@partner.nps.gov, Colorado State Univ – Ge
Chapuis, Robert, (514) 340-4777 Ecole Polytechnique – Hw
Charette, Matthew A., (508) 289-3205 mcharette@whoi.edu, Woods Hole Oceanographic Institution – Cm
Charles, Christopher D., (858) 534-5911 ccharles@ucsd.edu, Univ of California, San Diego – Pe
Charlesworth, Henry A., (780) 492-3265 Univ of Alberta – Gc
Charlevoix, Donna J., (217) 244-9575 charlevo@atmos.uiuc.edu, Univ of Illinois, Urbana-Champaign – Oa
Charlier, Bruce L., 0191 3741143 b.l.charlier, Durham Univ – Cg
Chase, Clement G., (520) 621-2417 cgchase@geo.arizona.edu, Univ of Arizona – Yg
Chase, Peter M., (608) 265.6003 pmchase@wisc.edu, Univ of Wisconsin - Extension – Oo
Chase, Richard L., (604) 822-3086 rchase@eos.ubc.ca, Univ of British Columbia – Gu
Chase, Robert W., 740-376-4776 chaser@marietta.edu, Marietta College – Np
Chase, Ronald B., (269) 387-5500 ronald.chase@wmich.edu, Western Michigan Univ – Gp
Chase, Zanna, (541) 737-5192 zanna@coas.oregonstate.edu, Oregon State Univ – OcCm
Chatelain, Edward E., (229) 333-5758 echatela@valdosta.edu, Valdosta State Univ – Pi
Chatterjee, Lata R., (617) 353-8345 lata@bu.edu, Boston Univ – Ou
Chatterjee, Sankar, (806) 742-3108 sankar.chatterjee@ttu.edu, Texas Tech Univ – Pv
Chatterton, Brian D., (780) 492-3983 brian.chatterton@ualberta.ca, Univ of Alberta – Pi
Chaubey, Indrajeet, (765) 494-5013 ichaubey@purdue, Purdue Univ – Hg
Chaudhuri, Sambhudas, (785) 532-2246 ksuncsc@ksu.edu, Kansas State Univ – Cc
Cheadle, Michael J., (307) 766-3206 cheadle@uwyo.edu, Univ of Wyoming – Yg
Checkley, David M., (858) 534-4228 dcheckley@ucsd.edu, Univ of California, San Diego – Ob
Cheek, William H., (417) 836-5249 Southwest Missouri State Univ – Oo
Cheel, Richard J., (905) 688-5550 (Ext. 3512) RCheel@brocku.ca, Brock Univ – Gs
Chelton, Dudley B., (541) 737-4017 dchelton@coas.oregonstate.edu, Oregon State Univ – Opr
Chen, Chu-Yung, (217) 333-0942 cychen7@uiuc.edu, Univ of Illinois, Urbana-Champaign – Cg

Chen, Dake, (845) 365-8496 Columbia Univ – Op
Chen, Gang, (907) 474-6875 ffgc@aurora.alaska.edu,
Univ of Alaska, Fairbanks – Nm
Chen, Gao, (404) 894-9565 Georgia Institute of Technology – Oa
Chen, Jiquan, (419) 530-2664 jiquan.chen@utoledo.edu,
Univ of Toledo – Oo
Chen, Jiuhua, (631) 632-8058 juihua.chen@sunysb.edu,
SUNY, Stony Brook – Gy
Chen, Lifo, 205/247-3622 lchen@gsa.state.al.us,
Geological Survey of Alabama – Ca
Chen, Robert, (727) 553-1590 chen@marine.usf.edu,
Univ of South Florida, Saint Petersburg – Oo
Chen, Tsing-Chang, (515) 294-9874 tmchen@iastate.edu,
Iowa State Univ of Science & Technology – Oa
Chen, Wang-Ping, (217) 333-2744 wpchen@uiuc.edu,
Univ of Illinois, Urbana-Champaign – Ys
Chen, Xianfeng, (724) 738-2385 xianfeng.chen@sru.edu,
Slippery Rock Univ – Or
Chen, Xun-Hong, (402) 472-0772 x-chen2@unl.edu,
Unversity of Nebraska - Lincoln – Hq
Cheney, Eric S., 206-543-1163 vaalbara@u.washington.edu,
Univ of Washington – Eg
Cheney, John T., (413) 542-2311 Amherst College – Gi
Cheng, H. H., (612) 625-1793 hcheng@soils.umn.edu,
Univ of Minnesota, Twin Cities – Sb
Cheng, Hai, 612-624-9598 cheng021@umn.edu,
Univ of Minnesota, Twin Cities – Cc
Cheng, Meng-Dawn, (423) 241-5918 ucn@ornl.gov,
Oak Ridge National Lab – Oa
Cheng, Qiuming, (416) 736-2100 (Ext. 22842) qiuming@yorku.ca,
York Univ – Oi
Cheng, Songlin, 937 775-3455 songlin.cheng@wright.edu,
Wright State Univ – Hw
Cheng, Zhongqi, (718) 951-5000 (Ext. 2647) zcheng@brooklyn.cuny.edu,
Brooklyn College (CUNY) – Ca
Chenoweth, Cheri A., 217-244-4610 chenoweth@isgs.uiuc.edu,
Illinois State Geological Survey – Ec
Cherkauer, Douglas S., 414-229-4563 aquadoc@uwm.edu,
Univ of Wisconsin, Milwaukee – Hw
Chermak, John A., 540-231-1785 jchermak@vt.edu,
Virginia Polytechnic Institute & State Univ – Cg
Cherniak, Daniele J., (518) 276-3358 chernd@rpi.edu,
Rensselaer Polytechnic Institute – Gx
Chernoff, Barry, 860 6852452 bchernoff@wesleyan.edu,
Wesleyan Univ – Ge
Chernosky, Jr., Joseph V., (207) 581-2151 josephc@maine.edu,
Univ of Maine – Gx
Cherns, L C.cherns@cf.ac.uk,
Cardiff Univ – Pg
Cherukupalli, Nehru, 516 463-6545 geonec@hofstra.edu,
Hofstra Univ – Gg
Cherukupalli, Nehru E., 718-951-4365 nehru@brooklyn.cuny.edu,
American Museum of Natural History – Gi
Cherukupalli, Nehru E., (718) 951-5000 (Ext. 4365) nehru@brooklyn.cuny.edu, Brooklyn College (CUNY) – Gi
Chesley, John J., (520) 621-9639 jchesley@email.arizona.edu,
Univ of Arizona – Cg
Chesner, Craig A., 217-581-6323 cfcac@eiu.edu,
Eastern Illinois Univ – Gx
Chester, Frederick M., (979) 845-3296 chesterf@geo.tamu.edu,
Texas A&M Univ – Gc
Chester, Judith S., (979) 845-1380 chesterj@geo.tamu.edu,
Texas A&M Univ – Gc
Chestnut, Donald R., (859) 257-5500 chesnut@uky.edu,
Univ of Kentucky – Ec
Chesworth, Ward, (519) 824-4120 (Ext. 52457) wcheswor@lrs.uoguelph.ca, Univ of Guelph – Cl
Chi, Guoxiang, (418) 654-3710 guchi@nrcan.gc.ca,
Universite du Quebec – Eg
Chialvo, Ariel A., (865) 574-1252 chialvoaa@ornl.gov,
Oak Ridge National Lab – Oo
Chiappe, Luis M., (213) 863-3323 lchiappe@nhm.org,
Los Angeles County Museum of Natural History – Pv
Chiarello, Ronald P., (630) 252-9327 Argonne National Lab – Om
Chidsey, Thomas C., (801) 537-3364 nrugs.tchidsey@state.ut.us,
Utah Geological Survey – Eo
Chien, Yi-Ju, (509) 373-4822 Yi-Hu.chien@pnl.gov,
Pacific Northwest National Lab – Gq
Chikatamarla, Laxmi, (604) 822-0745 laxmi@eos.ubc.ca,
Univ of British Columbia – Ec
Chilakapati, Ashokkumar, (509) 372-6077 chilakapati@pnl.gov,
Pacific Northwest National Lab – Hq
Childers, Susan E., 208-885-4289 childers@uidaho.edu,
Univ of Idaho – Gg
Chillrud, Steven, (845) 365-8893 Columbia Univ – Cg
Chilton, Phyllis A., (702) 895-3971 Central State Univ – Oo
Chimonas, George, (404) 894-3985 Georgia Institute of Tech – Oa
Chin, Anne, (409) 845-7141 Texas A&M Univ – Gm
Chin, Karen, 303-735-3074 karen.chin@colorado.edu,
Univ of Colorado – Po
Chin, Mian, (301) 286-5488 chin@rondo.gsfc.nasa.gov,
Georgia Institute of Technology – Oa
Chin, Yu-Ping, (614) 292-6953 yo@geology.ohio-state.edu,
Ohio State Univ – Hg
Chipping, David H., (805) 756-1695 dchippin@calpoly.edu,
California Polytechnic State Univ – Ng
Chirenje, Tait, 609-652-4588 Richard Stockton Coll of New Jersey – Sc
Chiu, Ching-Sang, (831) 656-3239 chiu@nps.navy.mil,
Naval Postgraduate School – Op
Chiu, Christy, (901) 678-2007 chi@ceri.memphis.edu,
Univ of Memphis – Ys
Chiu, Jer-Ming, (901) 678-2007 chiu@ceri.memphis.edu,
Univ of Memphis – Ys
Chiu, Jer-Ming, (901) 678-4839 chiu@ceri.memphis.edu,
Univ of Memphis – Ys
Choi, Christine S., (805) 893-4831 choi@geology.ucsb.edu,
Univ of California, Santa Barbara – Gg
Choi, Jinmu, (662) 325-5803 jc778@msstate.edu,
Mississippi State Univ – Oi
Chopping, Mark J., 973-655-4448 choppingm@mail.montclair.edu,
Montclair State Univ – Or
Choquette, Marcchoquette@ggl.ulaval.ca,
Universite Laval – Oo
Choran, Mark A., 404-656-3214 Georgia Dept of Nat Resources – Oo
Chormann, Frederick, (603) 271-1975 fchormann@des.state.nh.us,
New Hampshire Geological Survey – Hw
Chorover, Jonathan D., (520) 626-5635 chorover@cals.arizona.edu,
Univ of Arizona – Cg
Chou, Charissa J., (509) 372-3804 Charissa.chou@pnl.gov,
Pacific Northwest National Lab – Gq
Chou, Mei-In (Melissa), 217-244-0312 chou@isgs.uiuc.edu,
Illinois State Geological Survey – Co
Chou, Sheng-Fu Joseph, 217-244-2744 jchou@isgs.uiuc.edu,
Illinois State Geological Survey – Co
Chouinard, Vera, (905) 525-9140 (Ext. 23518) chouinar@mcmaster.ca,
McMaster Univ – Oo
Chouteau, Michel, (514) 340-4703 Ecole Polytechnique – Ye
Chow, Nancy, (204) 474-6451 n_chow@umanitoba.ca,
Univ of Manitoba – Gs
Chowdhury, Dipak K., (260) 481-6249 chowdhur@ipfw.edu,
Indiana Univ / Purdue Univ, Fort Wayne – Ys
Chowdhury, Shafiul H.chowdhus@newpaltz.edu,
SUNY New Paltz – Hw
Chown, Edward H., (418) 545-5011 Univ du Quebec a Chicoutimi – Gx
Chowns, Timothy M., (678) 839-4052 tchowns@westga.edu,
Univ of West Georgia – Gs
Christensen, Beth A., 516-877-4174 christensen@adelphi.edu,
Adelphi Univ – PmeGs
Christensen, Beth A., (516) 877-4174 christensen@adelphi.edu,
Georgia State Univ – Gs
Christensen, Douglas, doug@giseis.alaska.edu,
Univ of Alaska, Fairbanks – Ys
Christensen, Nikolas I., (608) 265-4469 chris@geology.wisc.edu,
Univ of Wisconsin, Madison – Yx
Christensen, Philip R., (480) 965-7105 phil.christensen@asu.edu,
Arizona State Univ – Xg
Christenson, Gary E., (801) 537-3304 nrugs.gchriste@state.ut.us,
Utah Geological Survey – Ng
Christeson, Gail L., (512) 471-6156 gail@ig.utexas.edu,
Univ of Texas at Austin, Jackson School of Geosciences – Yr
Christian, Tremblay, (418) 545-5011 fminier@uqac.uquebec.ca,
Universite du Quebec a Chicoutimi – Ct
Christiansen, Eric H., 801422 2113 eric_christiansen@byu.edu,
Brigham Young Univ – Gi
Christie-Blick, Nicholas, (845) 365-8821 ncb@ldeo.columbia.edu,
Columbia Univ – Gs
Christman, Robert A., (360) 650-3587 xman@cc.wwu.edu,
Western Washington Univ – Oe
Christopher, Raymond A.christ7@clemson.edu,
Clemson Univ – Pl

Christopherson, Gary, 520-621-6267 garych@casa.arizona.edu,
Univ of Arizona – Oi
Chrzan, Daryl, (510) 643-1624 dcchrzan@garnet.berkeley.edu,
Univ of California, Berkeley – Om
Chrzastowski, Michael J., (217) 284-2194 chrzasto@isgs.uiuc.edu,
Illinois State Geological Survey – Ou
Chu, Peter C., (831) 656-3688 chu@nps.navy.mil,
Naval Postgraduate School – Op
Chuang, Patrick Y., (831) 459-1501 pchuang@es.ucsc.edu,
Univ of California, Santa Cruz – Oa
Chuang, Patrick Y., (831) 459-1501 pchuang@pmc.ucsc.edu,
Univ of California, Santa Cruz – Oa
Chubb, Leslie T., 717.702.2022 lchubb@state.pa.us,
Pennsylvania Bureau of Topographic & Geologic Survey – Oo
Church, Thomas M., (302) 831-2558 tchurch@udel.edu,
Univ of Delaware – Cm
Church, William R., (519) 661-3192 wrchurch@uwo.ca,
Univ of Western Ontario – Gt
Churchill, Ron C., (916) 327-0745 California Geological Survey – Cg
Churnet, Habte G., (423) 425-4404 habte-churnet@utc.edu,
Univ of Tennessee, Chattanooga – GxgOg
Chyba, Christopher F., (650) 725-6468 chyba@seti.org,
Stanford Univ – Gh
Chyi, Lindgren L., (330) 972-7635 lchyi@uakron.edu,
Univ of Akron – Eg
Ciannelli, Lorenzo, (541) 737-3142 lciannelli@coas.oregonstate.edu,
Oregon State Univ – Ob
Cicerone, Robertrcicerone@bridgew.edu,
Bridgewater State College – Ys
Cichanski, Marek A., (408) 864-8664 mac7151@mercury.fhda.edu,
De Anza College – Gg
Ciciarelli, John A., (412) 773-3867 jac7@psu.edu,
Pennsylvania State Univ, Monaca – Gg
Cicimurri, Christian M., (864) 650-8456 cmcici@clemson.edu,
Clemson Univ – Pv
Cicimurri, David J., (864) 656-4601 dcheech@clemson.edu,
Clemson Univ – Pv
Ciesielski, Paul F., (352) 392-2231 pfcinfl@nervm.nerdc.ufl.edu,
Univ of Florida – Pm
Ciesla, Fred, (202) 478-8812 ciesla@dtm.ciw.edu,
Carnegie Institution of Washington – Xm
Cifelli, Richard L., (405) 325-4712 Univ of Oklahoma – Pv
Cifuentes, Luis A., (979) 845-3380 lcifuentes@ocean.tamu.edu,
Texas A&M Univ – Oc
Cihacek, Larry J., (701) 231-8572 North Dakota State Univ – Og
Ciolkosz, Edward J., (814) 865-1530 f8i@psu.edu,
Pennsylvania State Univ, Univ Park – Sd
Cioppa, Mariamcioppa@uwindsor.ca,
Univ of Western Ontario – Ym
Cioppa, Maria T., (519) -25-3-30 (Ext. ext. 2502) mcioppa@uwindsor.ca,
Univ of Windsor – Ym
Cirmo, Christopher P., (607) 753-2924 cirmoc@cortland.edu,
SUNY, Cortland – Hg
Cisne, John L., (607) 255-3698 cisne@geology.cornell.edu,
Cornell Univ – Pg
Civan, Faruk, (405) 325-6778 fcivan@ou.edu,
Univ of Oklahoma – Np
Claerbout, Jon F., (650) 723-3717 Stanford Univ – Ye
Claeys, Philippe, 322 629-3391 phclaeys@vub.ac.be,
Vrije Univ Brussel – CgPyXc
Clague, David A., (808) 967-8819 Univ of Hawai'i, Manoa – Gv
Clague, David A.clague@mbari.org,
Stanford Univ – Ob
Clague, John J., (604) 291-5387 Simon Fraser Univ – Gl
Clapham, Matthew E.mclapham@pmc.ucsc.edu,
Univ of California, Santa Cruz – Pe
Clapham, Jr., Wentworth B., (216) 687-4820 Cleveland State Univ – Ge
Clapp, Charles E., (612) 625-2767 eclapp@soils.umn.edu,
Univ of Minnesota, Twin Cities – Sb
Clarey, Timothy L., (989) 686-9252 tlclarey@delta.edu,
Delta College – HwGc
Clark, Alan H., (613) 545-6187 Queen's Univ – Em
Clark, Allen L., (204) 474-8857 Univ of Hawai'i, Manoa – Eg
Clark, David L., (505) 665-0005 Los Alamos National Lab – Oo
Clark, David L., (608) 262-4972 dlc@geology.wisc.edu,
Univ of Wisconsin, Madison – Pm
Clark, Douglas H., (360) 650-7939 doug.clark@wwu.edu,
Western Washington Univ – Gl
Clark, G. Michael, (865) 974-6006 Univ of Tennessee, Knoxville – Gm
Clark, George S., (204) 474-7343 gs_clark@umanitoba.ca,
Univ of Manitoba – Cc
Clark, H. C., (713) 527-4887 hcclark@owlnet.rice.edu,
Rice Univ – Ng
Clark, Ian D., (613) 562-5800 Ext 6137 idclark@uottawa.ca,
Univ of Ottawa – Hw
Clark, James A., (630) 752-5163 james.a.clark@wheaton.edu,
Wheaton College – Gm
Clark, James S., (919) 660-7402 jimclark@duke.edu,
Duke Univ – Oo
Clark, Janet M., (304) 696-6720 dudley@marshall.edu,
Marshall Univ – Oo
Clark, Jeffrey J., (920) 832-6733 Lawrence Univ – Gm
Clark, Jo, (940) 565-2091 jclark@unt.edu,
Univ of North Texas – Yr
Clark, John H., 814-863-1581 clark@ems.psu.edu,
The Pennsylvania State Univ, Univ Park – Ow
Clark, Jordan, (805) 897-7838 Univ of California, Santa Barbara – Hg
Clark, Joseph C., (724) 357 5622 Indiana Univ of Pennsylvania – Gr
Clark, Karen, (613) 562-5838 Napa Valley College – Oo
Clark, Ken, (253) 879-3138 kclark@ups.edu,
Univ of Puget Sound – Gg
Clark, Kenneth F., (915) 581-8371 Kclark1@rr.elp.com,
Univ of Texas, El Paso – Eg
Clark, Lori, 909-941-2385 lori.clark@chaffey.edu,
Chaffey College – Cm
Clark, Marin, 734-615-0484 marinkc@umich.edu,
Univ of Michigan – Gm
Clark, Mark W., (352) 392-1803 (Ext. 316) clarkmw@gnv.ifas.ufl.edu,
Univ of Florida – Sf
Clark, Michael, (403) 440-8944 mdclark@mtroyal.ca,
Mount Royal College – Og
Clark, Murlene W., (251) 460-6381 mclark@jaguar1.usouthal.edu,
Univ of South Alabama – Pi
Clark, Peter U., (541) 737-1247 clarkp@geo.oregonstate.edu,
Oregon State Univ – Gl
Clark, Richard D., 717-872-3930 Richard.clark@millersville.edu,
Millersville Univ – Oa
Clark, Richard E., (406) 683-7151 r_clark@umwestern.edu,
Univ of Montana, Western – Oo
Clark, Robert A., (520) 621-3842 clark@hwr.arizona.edu,
Univ of Arizona – Hs
Clark, Robert O., (928) 523-1321 robert.clark@nau.edu,
Northern Arizona Univ – Oo
Clark, Russell G., (517) 629-0312 rgclark@albion.edu,
Albion College – Gi
Clark, Shannonskclark@unm.edu,
Univ of New Mexico – Oo
Clark, Stuart, +44 01782 583171 s.clark@esci.keele.ac.uk,
Keele Univ – Gsd
Clark, Terry L., (604) 822-0278 tclark@eos.ubc.ca,
Univ of British Columbia – Oa
Clark, Timothy W., (919) 733-2423 (Ext. 401) Tyler.Clark@ncmail.net,
North Carolina Geological Survey – Gg
Clark, II, George R., (785) 532-2242 grc@ksu.edu,
Kansas State Univ – Poy
Clarke, Amanda, 480-965-5081 geology@asu.edu,
Arizona State Univ – Gv
Clarke, Anthony O., (502) 852-6844 Univ of Louisville – Gm
Clarke, Antony D., (808) 956-6215 tclarke@soest.hawaii.edu,
Univ of Hawai'i, Manoa – Oa
Clarke, D. Barrie, (902) 494-3438 Dalhousie Univ – Gi
Clarke, Garry K. C., (604) 822-3602 gclarke@eos.ubc.ca,
Univ of British Columbia – Yg
Clarke, Geoffrey L., +61293513244 geoffc@mail.usyd.edu.au,
Univ of Sydney – Gp
Clarke, John T., (313) 747-3540 Univ of Michigan – Oa
Clarke, Julia A., 919-515-3711 North Carolina State Univ – Pv
Clarke, Thomas A., (808) 236-7425 tomcat@soest.hawaii.edu,
Univ of Hawai'i, Manoa – Ob
Clarke, Jr., Roy S., (202) 633-1789 clarker@si.edu,
Smithsonian Institution / National Museum of Natural History – Xm
Clary, Renee M., (662) 325-5924 rclary@geosci.msstate.edu,
Mississippi State Univ – Oe
Class, Connie, (845) 365-8712 Columbia Univ – Cg
Clauer, C. Robert, (313) 763-6248 Univ of Michigan – Og
Clausen, Ben, (909) 824-4548 Loma Linda Univ – Yg
Clay, Clarence S., (608) 262-9473 clay@geology.wisc.edu,
Univ of Wisconsin, Madison – Ye
Clay, David E., (605) 688-5081 David.Clay@sdstate.edu,
South Dakota State Univ – Sb

Clayson, Carol Ann, (850) 644-0922 clayson@met.fsu.edu, Florida State Univ – Ow
Clayton, Jordan A., 40404135791 jclayton@gsu.edu, Georgia State Univ – Oo
Clayton, Lee, (608) 263-6839 lclayton@wisc.edu, Univ of Wisconsin - Extension – Gl
Clayton, Robert N., (773) 702-7777 Univ of Chicago – Cs
Clayton, Robert W., (626) 395-6909 clay@gps.caltech.edu, California Institute of Technology – Ye
Clayton, Robert W., 208-496-1906 claytonr@byui.edu, Brigham Young Univ, Idaho – GcYgNg
Cleary, W. A., (407) 674-8096 Florida Institute of Technology – Oo
Cleary, William J., 910-962-2320 clearyw@uncw.edu, Univ of North Carolina Wilmington – Gu
Cleaveland, Malcolm K., (479) 575-6741 mcleavel@uark.edu, Univ of Arkansas, Fayetteville – Oi
Cleaves, Emery T., (410) 554-5503 ecleaves@mgs.md.gov, Maryland Dept of Natural Resources – Gm
Clebnik, Sherman M., (860) 465-4323 clebniks@easternct.edu, Eastern Connecticut State Univ – Gl
Clemens, Steven C., (401) 863-1964 Steven_Clemens@Brown.edu, Brown Univ – Ou
Clemens, William A., (510) 642-6675 bclemens@berkeley.edu, Univ of California, Berkeley – Pv
Clemens-Knott, Diane, (714) 278-2369 dclemensknott@fullerton.edu, California State Univ, Fullerton – Ca
Clement, Bradford M., (305) 348-3085 clementb@fiu.edu, Florida International Univ – Ym
Clement, William, (208)426-4307 billc@cgiss.boisestate.edu, Boise State Univ – Yg
Clements, Lee Ann J., (904) 745-7300 lclemen@ju.edu, Jacksonville Univ – Ob
Clements, William E., (505) 667-2637 clements@lanl.gov, Los Alamos National Lab – Oo
Clementz, Mark T., (307) 766-6048 mclement1@uwyo.edu, Univ of Wyoming – Py
Clemo, Thomas M., (208) 426-1416 tclemo@boisestate.edu, Boise State Univ – Yg
Clendenin, Charles W., (803) 896-7702 clendenin@dnr.state.sc.us, Clemson Univ – Eg
Clendenin, Jr., Charles W., (803) 896-7708 clendenin@water.dnr.state.sc.us, South Carolina Geological Survey – Eg
Clendening, Ronald J., (615) 532-1508 Ron.Clendening@state.tn.us, Tennessee Division of Geology – GgOi
Cleveland, Cutler J., (617) 353-3083 cutler@bu.edu, Boston Univ – Oo
Clifford, Paul, (905) 525-9140 (Ext. 27694) McMaster Univ – Yr
Clifford, Sheila A., (508) 289-2328 sclifford@whoi.edu, Woods Hole Oceanographic Institution – Oo
Clift, Sigrid J., (512) 471-0320 sigrid.clift@beg.utexas.edu, Univ of Texas at Austin, Jackson School of Geosciences – Gg
Cline, Anita M., (678) 839-4051 acline@westga.edu, Univ of West Georgia – Oo
Cline, Jean S., (702) 895-1091 jcline@unlv.nevada.edu, Univ of Nevada, Las Vegas – Em
Cloos, Mark P., (512) 471-4170 cloos@mail.utexas.edu, Univ of Texas at Austin, Jackson School of Geosciences – Gc
Close, Angela E., (206) 543-2078 Univ of Washington – Oo
Closs, L. Graham, (303) 273-3856 lcloss@mines.edu, Colorado School of Mines – Ce
Clothiaux, Eugene E., 814-865-2915 cloth@meteo.psu.edu, The Pennsylvania State Univ, Univ Park – Ow
Clough, Gene A., (207) 786-6396 gclough@bates.edu, Bates College – Ym
Clough, Linda, (785) 532-6724 lindasd@ksu.edu, Kansas State Univ – Oo
Clougherty, Leoleo-clougherty@uiowa.edu, Univ of Iowa – Oo
Cloutier, Danielle, (418) 656-7679 danielle.cloutier@ggl.ulaval.ca, Universite Laval – GeOg
Clowes, Ronald M., (604) 822-4138 clowes@eos.ubc.ca, Univ of British Columbia – Ys
Clyde, William C., (603) 862-3148 will.clyde@unh.edu, Univ of New Hampshire – Pg
Clynch, James R., (408) 656-3268 jclynch@oc.nps.navy.mil, Naval Postgraduate School – Yd
Coakley, Bernard, 907-474-5385 bernard.coakley@gi.alaska.edu, Univ of Alaska, Fairbanks – Yg
Coakley, James A., (541) 737-5686 coakley@coas.oregonstate.edu, Oregon State Univ – Oar
Coash, John R., (805) 664-3054 California State Univ, Bakersfield – Gg
Cobb, James C., (859) 257-5500 cobb@kgs.mm.uky.edu, Univ of Kentucky – Ec
Cobb, James C., (859) 257-5500 (Ext. 130) cobb@uky.edu, Univ of Kentucky – Ec
Coble, James F., (804) 427-7189 Tidewater Community College – Gi
Coble, Paula G., (727) 553-1631 pcoble@marine.usf.edu, Univ of South Florida, Saint Petersburg – Cm
Coch, Nicholas K., (718) 997-3326 Graduate School of the City Univ of New York – Gs
Coch, Nicholas K., (718) 997-3326 ncoch.geol.qc, Queens College (CUNY) – Gs
Cochran, Elizabeth, elizabeth.cochran@ucr.edu, Univ of California, Riverside – Ys
Cochran, J. Kirk, (631) 632-8733 kcochran@notes.cc.sunysb.edu, SUNY, Stony Brook – Oc
Cochran, James R., (845) 365-8396 jrc@ldeo.columbia.edu, Columbia Univ – Yv
Cockburn, Jaclyn, 518-388-8015 cockburj@union.edu, Union College – HgGeHs
Cocker, Mark D., (404) 656-3214 Georgia Dept of Nat Resources – Eg
Codrea, Vlad, (+40) -26-4-40 (Ext. 00 ext 5205) vcodrea@bioge.ub-bcluj.ro, Babes-Bolyai Univ – PvGo
Cody, Anita M., (515) 294-1714 amcody@iastate.edu, Iowa State Univ of Science & Technology – Cl
Cody, George D., (202) 478-8980 g.cody@gl.ciw.edu, Carnegie Institution of Washington – Co
Cody, Robert D., (515) 294-1714 rdcody@iastate.edu, Iowa State Univ of Science & Technology – Gs
Coe, Douglas A., (406) 496-4207 dcoe@mtech.edu, Montana Tech of the Univ of Montana – Cg
Coe, Robert S., (831) 459-2393 rcoe@pmc.ucsc.edu, Univ of California, Santa Cruz – Ym
Coffroth, Mary Alice, 716-645-2363 (Ext. 187) coffroth@bufalo.edu, SUNY, Buffalo – Gu
Cogbill, Allen H., (505) 667-1049 ahc@lanl.gov, Los Alamos National Lab – Yg
Cogger, Craig G., (206) 840-4512 cogger@coopext.cahe.wsu.edu, Washington State Univ – Ou
Cohen, Andrew S., (520) 621-4691 cohen@email.arizona.edu, Univ of Arizona – Ps
Cohen, Anne L., (508) 289-2958 acohen@whoi.edu, Woods Hole Oceanographic Institution – Pe
Cohen, Arthur D., (803) 777-4502 Univ of South Carolina – Pl
Cohen, Barbara A., (505) 277-2747 bcohen@unm.edu, Univ of New Mexico – Xcg
Cohen, David R., 612 9385 8084 d.cohen@unsw.edu.au, Univ of New South Wales – Ce
Cohen, Joel E., (212) 327-8883 cohen@rockvax.rockefeller.edu, Columbia Univ – Oo
Cohen, Michael R., (803) 777-4502 Indiana Univ / Purdue Univ, Indianapolis – Oe
Cohen, Ronald E., (202) 478-8937 r.cohen@gl.ciw.edu, Carnegie Institution of Washington – Gy
Cohen, Saul B., (212) 772-5412 Graduate School of the City Univ of New York – Oy
Cohen, Shaul E., (541) 346-4500 scohen@uoregon.edu, Univ of Oregon – Oo
Cohoon, Richard R., (479) 964-0816 rcohoon@atu.edu, Arkansas Tech Univ – Ge
Coish, Raymond A., (802) 443-5423 coish@middlebury.edu, Middlebury College – Gi
Cok, Anthony E., 516 877 4171 cok@adelphi.edu, Adelphi Univ – OuGeu
Colberg, Mark R., (435)865-8331 colberg@suu.edu, Southern Utah Univ – GpeGt
Colbert, Matthew W., (512) 471-5257 colbert@mail.utexas.edu, Univ of Texas at Austin, Jackson School of Geosciences – Gq
Colburn, Ivan P., (323) 343-2413 California State Univ, Los Angeles – Gs
Colby, Bonnie C., (520) 621-4775 bcolby@email.arizona.edu, Univ of Arizona – Oo
Colby, Jeff, (828) 262-7126 colbyj@appstate.edu, Appalachian State Univ – Oi
Cole, Charles R., (509) 372-6068 charles.cole@pnl.gov, Pacific Northwest National Lab – Hw
Cole, David R., (865) 574-5473 coledr@ornl.gov, Oak Ridge National Lab – Cs
Cole, Dean A., (505) 665-0832 Los Alamos National Lab – Co
Cole, Gregory L., (505) 667-1858 gcole@lanl.gov, Los Alamos National Lab – Og

Cole, James, (03) 3642-766 j.cole@geol.canterbury.ac.nz, Univ of Canterbury – Gv
Cole, Jasonjasonc@atmosp.physics.utoronto.ca, Univ of British Columbia – Oa
Cole, Jennifer S., (617) 373-3039 j.cole@neu.edu, Northeastern Univ – GeHw
Cole, Julia E., (520) 626-2341 jecole@email.arizona.edu, Univ of Arizona – Pe
Cole, Kevin C., 616-331-3791 colek@gvsu.edu, Grand Valley State Univ – Gz
Cole, Rex D., (970) 248-1599 rcole@mesastate.edu, Mesa State College – Gs
Cole, Rick, (727) 553-1522 rcole@marine.usf.edu, Univ of South Florida, Saint Petersburg – Op
Cole, Ron, (814) 332-2876 rcole@allegheny.edu, Allegheny College – Gt
Cole, S., (207) 859-5790 swcole@colby.edu, Colby College – Oo
Coleman, Drew S., 919-962-0705 dcoleman@email.unc.edu, Univ of North Carolina, Chapel Hill – Cg
Coleman, James M., (310) 825-1776 chanjc@lsu.edu, Louisiana State Univ – Gs
Coleman, Leslie C., (306) 966-5690 les.coleman@sk.sympatico.ca, Univ of Saskatchewan – Gz
Coleman, Paul J., (310) 825-1776 pcoleman@igpp.ucla.edu, Univ of California, Los Angeles – Xy
Coleman, Robert G., (650) 723-9205 coleman@pangea.stanford.edu, Stanford Univ – Gi
Coleman, Tommy L., (205) 851-5462 tcoleman@aamu.edu, Alabama A&M Univ – So
Coleman, Jr., Paul J., (310) 825-1776 Univ of California, Los Angeles – Gc
Coles, Kenneth S., 724-357-5626 kcoles@iup.edu, Indiana Univ of Pennsylvania – Oe
Colgan, Mitchell W., (843) 953-7171 colganm@cofc.edu, College of Charleston – PeGe
Colgan, Patrick, 616-313-3201 colganp@gvsu.edu, Grand Valley State Univ – Gl
Colinvaux, Paul A., (352) 392-2231 pcolinva@mbl.edu, Univ of Florida – Yr
Colle, Brian, (631) 632-3174 colle@cyclone.msrc.sunysb.edu, SUNY, Stony Brook – Oa
Collett, Jr., Jeffrey L., (970) 491-8697 collett@lamar.colostate.edu, Colorado State Univ – Oa
Collie, Jeremy S., (401) 874-6859 jcollie@gso.uri.edu, Univ of Rhode Island – Ob
Collier, James D., (970) 247-7129 collier_j@fortlewis.edu, Fort Lewis College – Cg
Collier, Mark R., 312-793-7384 collier@isgs.uiuc.edu, Illinois State Geological Survey – Ge
Collier, Robert W., (541) 737-4367 rcollier@coas.oregonstate.edu, Oregon State Univ – Cm
Collins, Charles D., (417) 836-5144 Southwest Missouri State Univ – Oy
Collins, Chrisc.d.collins@reading.ac.uk, Univ of Reading – Sc
Collins, Curtis A., (408) 656-3271 collins@oc.nps.navy.mil, Naval Postgraduate School – Op
Collins, Desmond H., (416) 586-5593 Royal Ontario Museum – Pi
Collins, Edward W., (512) 471-6247 eddie.collins@beg.utexas.edu, Univ of Texas at Austin, Jackson School of Geosciences – Ge
Collins, John A., (508) 289-2733 jcollins@whoi.edu, Woods Hole Oceanographic Institution – Ys
Collins, Laurel S., (305) 348-1732 collinsl@fiu.edu, Florida International Univ – Po
Collins, Lorence G.lorencec@sysmatrix.net, California State Univ, Northridge – Gz
Collins, Mark, (412) 624-6615 mookie@pitt.edu, Univ of Pittsburgh – Oo
Collins, Mary E., (352) 392-1951 mec@gnv.ifas.ufl.edu, Univ of Florida – Sd
Collins, Nicholas, (301) 405-6630 collinsn@geol.umd.edu, Univ of Maryland – Cg
Collins, William D., (510) 495-2407 wdcollins@berkeley.edu, Univ of California, Berkeley – Oa
Collinson, James W., (614) 292-2721 Ohio State Univ – Ps
Collister, James W., (801) 585-7852 Univ of Utah – Cs
Colman, Steven M., (218) 726-6723 scolman@d.umn.edu, Univ of Minnesota, Duluth – Gn
Colucci, Steven, (607) 255-1752 sjc25@cornell.edu, Cornell Univ – Oa
Colville, Alan A., (323) 343-2407 acolvil@calstatela.edu, California State Univ, Los Angeles – Gz
Colwell, Frederick, (541) 737-5220 rcolwell@coas.oregonstate.edu, Oregon State Univ – GuPy
Combi, Michael R., (313) 764-7226 mcombi@engin.umich.edu, Univ of Michigan – Oa
Combs, Charles H., 404-656-3214 Georgia Dept of Nat Resources – Hw
Comeau, Wilfrid, (514) 289-6322 McGill Univ – Nm
Comer, John B., (812) 855-2687 jcomer@indiana.edu, Indiana Geological Survey – Cg
Comerford, Nicholas B., (352) 392-1951 nbc@mail.ifas.ufl.edu, Univ of Florida – Sf
Comrie, Andrew, (520) 621-1585 comrie@email.arizona.edu, Univ of Arizona – Oy
Conca, James L., (509) 375-3268 Washington State Univ – Oo
Conder, James, (314) 935-7965 conder@mantle.wustl.edu, Washington Univ – Ys
Condie, Kent C., (505) 835-5531 kcondie@nmt.edu, New Mexico Institute of Mining and Technology – Ct
Condit, Christopher D., (413) 545-0272 ccondit@geo.umass.edu, Univ of Massachusetts – Gi
Condrey, Richard E., (225) 388-6456 Louisiana State Univ – Ob
Congleton, John D., (678) 839-4066 jconglet@westga.edu, Univ of West Georgia – Pv
Conkin, James E., (502) 852-6821 Univ of Louisville – Pg
Conkwright, Robert D., 410-554-5545 rconkwright@dnr.state.md.us, Maryland Geological Survey – Gu
Conly, Andrew G., 807-343-8463 andrew.conly@lakeheadu.ca, Lakehead Univ – Eg
Connell, Sean D., 505.366.2534 connell@gis.nmt.edu, New Mexico Institute of Mining & Technology – Gm
Connell, Sean D., 505.366.2534 connell@gis.nmt.edu, New Mexico Institute of Mining and Technology – Gm
Connelly, Elsie E., (402) 472-6706 econnelly1@unl.edu, Univ of Nebraska, Lincoln – Oo
Connelly, James N., (512) 471-6166 connelly@mail.utexas.edu, Univ of Texas at Austin, Jackson School of Geosciences – Cc
Connelly, Jeffrey B., (501) 569-3543 jbconnelly@ualr.edu, Univ of Arkansas, Little Rock – Gc
Connely, Melissa, 307-268-2017 mconnely@caspercollege.edu, Casper College – Gs
Connolly, Robert P., (318) 926-3314 povpoint@bayou.com, Univ of Louisiana, Monroe – Ga
Connolly, Jr., Harold, 212-769-5081 American Museum of Natural History – Xm
Connor, Charles B., (813) 974-0325 cconnor@chuma1.cas.usf.edu, Univ of South Florida, Tampa – Gv
Connor, Edward F., (804) 924-0554 Univ of Virginia – Oo
Connors, Christopher D., (540) 458-8170 connors@wlu.edu, Washington & Lee Univ – Gc
Conover, David O., (631) 632-8667 dconover@notes.cc.sunysb.edu, SUNY, Stony Brook – Ob
Conrey, Richard M., (509) 332-6610 conrey@mail.wsu.edu, Washington State Univ – Gi
Constable, Catherine G., (858) 534-3183 cconstable@ucsd.edu, Univ of California, San Diego – Ym
Constable, Steven C., (858) 534-2409 sconstable@ucsd.edu, Univ of California, San Diego – Yr
Constantin, Marc, 418-656-3139 mconstan@ggl.ulaval.ca, Universite Laval – Gi
Constantopoulos, James T., (505) 562-2651 Eastern New Mexico Univ – Em
Conti, Marla, (323) 953-4000 (Ext. 2611) contimr@lacitycollege.edu, Los Angeles City College – Oy
Contreras, Juanjuanc@cicese.mx, Centro de Inv Científica y de Educación Superior de Ensenada – Gt
Conway, F M.michael.conway@azwestern.edu, Arizona Western College – Gv
Conway, Howard B., 206-685-8085 conway@ess.washington.edu, Univ of Washington – Ol
Coogan, James, (970) 943-3425 jcoogan@western.edu, Western State College of Colorado – Gc
Cook, David R., (630) 252-5840 drcook@anl.gov, Argonne National Lab – Ow
Cook, Edward R., (845) 365-8618 Columbia Univ – Hw
Cook, Frederick A., (403) 220-6594 cook@litho.ucalgary.ca, Univ of Calgary – Ys
Cook, Geoffrey W., 401-874-9384 gwcook@mail.uri.edu, Univ of Rhode Island – Ggi
Cook, Kerry H., (607) 255-9716 khc6@cornell.edu, Cornell Univ – Oa

Cook, Marlon R., 205/247-3692 mcook@gsa.state.al.us,
Geological Survey of Alabama – Hw
Cook, Robert B.cookrb@ornl.gov,
Oak Ridge National Lab – Cg
Cook, Robert B., (334) 844-4282 cookrob@auburn.edu,
Auburn Univ – Eg
Cook, Stanton A.no email,
Univ of Oregon – Oo
Cook, Stephen J., ((90) 5) -688- (Ext. 50 ext. 5469) sjcook@brocku.ca,
Brock Univ – Hw
Cook, Steve, (541) 737-0962 cooks@geo.oregonstate.edu,
Oregon State Univ – Oo
Cook, William B., (313) 763-5344 Univ of Michigan – Oa
Cooke, David R., 61 3 6226 7605 d.cooke@utas.edu.au,
Univ of Tasmania – Eg
Cooke, III, William H., 662-325-9479 whc5@geosci.msstate.edu,
Mississippi State Univ – Or
Cooke-Andresen, Michele L., (413) 547-3142 cooke@geo.umass.edu,
Univ of Massachusetts – Gc
Cookus, Pamela J., 217-244-2486 cookus@isgs.uiuc.edu,
Illinois State Geological Survey – Oo
Coolbaugh, Mark, (775) 784-1415 mfc@unr.nevada.edu,
Univ of Nevada, Reno – Gg
Coolen, Marco J., (508) 289-2931 mcoolen@whoi.edu,
Woods Hole Oceanographic Institution – Cm
Coombs, Douglas S., +64 3 479-7505 doug.coombs@otago.ac.nz,
Univ of Otago – Gi
Coombs, Margery C., (413) 545-0514 Univ of Massachusetts – Oo
Cooney, Timothy M., (319) 273-2918 timothy.cooney@uni.edu,
Univ of Northern Iowa – Oe
Coonley, Steffanie, (505) 665-2330 scoonley@lanl.gov,
Los Alamos National Lab – Oo
Cooper, Alan F., +64 3 479-7515 Alan.Cooper@stonebow.otago.ac.nz,
Univ of Otago – Gg
Cooper, Barbara C., (765) 494-3048 bccooper@purdue.edu,
Purdue Univ – Hw
Cooper, Brian J., (409) 294-1566 Sam Houston State Univ – Gz
Cooper, Garry V., (828) 262-7051 coopergv@appstate.edu,
Appalachian State Univ – Oo
Cooper, Jon C., (914) 251-6647 SUNY, Purchase – Oo
Cooper, Kari M., (530) 754-8826 kmcooper@geology.ucdavis.edu,
Univ of California, Davis – Gi
Cooper, Mark, (204) 474-8075 Univ of Manitoba – Gz
Cooper, Patricia A., (808) 956-9513 cooper@soest.hawaii.edu,
Univ of Hawai'i, Manoa – Ys
Cooper, Reid F., (401) 863-2160 Reid_Cooper@Brown.edu,
Brown Univ – Gy
Cooper, Richard A., (860) 445-3437 msiadm01@uconnvm.uconn.edu,
Univ of Connecticut – Ob
Cooper, Roger W., (409) 880-8239 Lamar Univ – Gi
Cooper, Terence H., (612) 625-7747 tcooper@soils.umn.edu,
Univ of Minnesota, Twin Cities – Sd
Coorough, Patricia J., (414) 278-6156 coorough@mpm.edu,
Milwaukee Public Museum – Pg
Cope, Tim D., (765)-658-6443 tcope@depauw.edu,
DePauw Univ – GsOi
Copeland, Jonathan, (912) 598-2462 copeland@GaSoU.edu,
Georgia Southern Univ – Ob
Copper, Paul, (705) -67-5-11 (Ext. ext. 6575) Univ of Western Ontario – Pg
Copper, Paulpcopper@laurentian.ca,
Laurentian Univ, Sudbury – Pi
Coppinger, Walter, (210) 999-7654 wcopping@trinity.edu,
Trinity Univ – Gc
Corbato, Charles E., 614 247-4246 corbato.1@osu.edu,
Ohio State Univ – Ye
Corbett, David R., (252) 328-1367 East Carolina Univ – Oc
Corcoran, Deborah, 417-836-6889 dbc135f@smsu.edu,
Southwest Missouri State Univ – Oo
Corcoran, Patricia, (519) -66-1-21 (Ext. , ext. 86836) pcorcor@uwo.ca,
Univ of Western Ontario – Gd
Corcoran, Thomas, (561) 297-3310 Florida Atlantic Univ – Oo
Corcoran, William, (417) 836-5781 wtc382f@smsu.edu,
Southwest Missouri State Univ – Oa
Cordell, Ann S., (352) 392-1721 (Ext. 491) cordell@flmnh.ufl.edu,
Univ of Florida – Gx
Cordero, David I., 360-442-2883 dcordero@lcc.ctc.edu,
Lower Columbia College – Og
Cordiviola, Steven J., (859) 257-5500 scordiviola@kgs.mm.uky.edu,
Univ of Kentucky – Eg
Cordonnier, Benoit, 089/2180 4271 cordonnier@min.uni-muenchen.de,
LMU - Univ of Munich – Gz
Cordua, William S., (715) 425-3139 william.s.cordua@uwrf.edu,
Univ of Wisconsin, River Falls – Gx
Corey, Alison, (801) 537-3122 nrugs.acorey@state.ut.us,
Utah Geological Survey – Oo
Corfu, Fernando, (416) 586-5811 Royal Ontario Museum – Cc
Corliss, Bruce H., (919) 684-2951 bcorliss@geo.duke.edu,
Duke Univ – Pm
Cormier, Marie-Helene, (573) 882-8608 cormierm@missouri.edu,
Univ of Missouri – Yr
Cormier, Vernon F., (860) 486-3547 vernon.cormier@uconn.edu,
Univ of Connecticut – Ys
Cornebise, Michael W., 217-581-5400 cfmwc@eiu.edu,
Eastern Illinois Univ – Og
Cornelius, Diane J., (509) 335-4486 geolab@mail.wsu.edu,
Washington State Univ – Ca
Cornelius, Michael, 410-293-6554 corneliu@usna.edu,
United States Naval Academy – Og
Cornelius, Scott B., (509) 335-6770 cornelius@wsu.edu,
Washington State Univ – Ca
Cornell, William C., (915) 747-5218 cornell@geo.utep.edu,
Univ of Texas, El Paso – Pm
Cornell, Winton, 918-631-3248 wcornell@utulsa.edu,
Univ of Tulsa – Gi
Cornillon, Peter, (401) 874-6283 pcornillon@gso.uri.edu,
Univ of Rhode Island – Op
Cornwell, Jeffery C., (410) 221-8445 cornwell@hpl.umces.edu,
Univ of Maryland – Oc
Cornwell, Kevin J., (916) 278-6667 cornwell@csus.edu,
California State Univ, Sacramento – Gm
Coron, Cynthia R., (203) 392-5840 coroncl@southernct.edu,
Southern Connecticut State Univ – Em
Corona, Thomas J., (303) 556-8525 Metropolitan State College of Denver – Ow
Correa, Luciano, 512-471-6737 luciano.correa@beg.utexas.edu,
Univ of Texas at Austin, Jackson School of Geosciences – Eo
Corriveau, Louise, (418) 654-2672 lcorrive@nrcan.gc.ca,
Universite du Quebec – Gp
Corsetti, Frank A., (213) 740-6123 fcorsett@earth.usc.edu,
Univ of Southern California – Gs
Corson, Mark W., (660) 562-1672 mcorson@nwmissouri.edu,
Northwest Missouri State Univ – Oy
Corthesy, Robert, (514) 340-4555 Ecole Polytechnique – Nm
Coruh, Cahit, (540) 231-6894 coruh@vt.edu,
Virginia Polytechnic Institute & State Univ – Ye
Cosby, Jack, (804) 324-7787 bjc4a@virginia.edu,
Univ of Virginia – Hs
Cosette, Villagrasavillagrasa.cosette@uqam.ca,
Universite du Quebec a Montreal – Oo
Costa, Daniel P., (831) 459-2789 costa@biology.ucsc.edu,
Univ of California, Santa Cruz – Ob
Costain, John K., (540) 231-8912 costain@vt.edu,
Virginia Polytechnic Institute & State Univ – Yg
Costello, David K., (727) 553-3953 dcostello@marine.usf.edu,
Univ of South Florida, Saint Petersburg – Op
Costello, John, (770) 928-2713 john_costello@dnr.state.ga.us,
Georgia State Univ – Gc
Costello, John O., 404-656-3214 Georgia Dept of Nat Resources – Gp
Costello, Margaret, (562) 985-4809 mcostel2@csulb.edu,
California State Univ, Long Beach – Oo
Costigan, Keeley R., (505) 665-4788 krc@lanl.gov,
Los Alamos National Lab – Oa
Cote, Denis, (418) 545-5011 Universite du Quebec a Chicoutimi – Gi
Cote, Pascale, (418) 654-2601 pacote@nrcan.gc.ca,
Universite du Quebec – Oo
Cotter, James F., (320) 589-6312 cotterjf@mrs.umn.edu,
Univ of Minnesota, Morris – Gl
Cottingame, William, (505) 667-8339 wcottingame@lanl.gov,
Los Alamos National Lab – Oa
Cotton, William R., (970) 491-8593 cotton@atmos.colostate.edu,
Colorado State Univ – Oa
Cottrell, Elizabeth, (202) 633-1859 cottrelle@si.edu,
Smithsonian Institution / National Museum of Natural History – Cp
Cottrell, Rory D.rory@earth.rochester.edu,
Univ of Rochester – Ym
Coulibaly, Paulin, (905) 525-9140 (Ext. 27903) couliba@mcmaster.ca,
McMaster Univ – Hg
Coull, Bruce C., (803) 777-2692 Univ of South Carolina – Ob
Coulter, Richard L., (630) 252-5833 rl_coulter@anl.gov,
Argonne National Lab – Oa

Cousens, Brian L.brian_cousens@carleton.ca,
Carleton Univ – Cc
Cousineau, Pierre, (418) 545-5011 pcousine@uqac.uquebec.ca,
Universite du Quebec a Chicoutimi – Gv
Couture, Rex A., (314) 935-4194 rex@levee.wustl.edu,
Washington Univ – Ca
Cova, Thomas J., (801) 581-7930 tom.cova@geog.utah.edu,
Univ of Utah – Oi
Coveney, Raymond M., (816) 235-2980 coveneyr@umkc.edu,
Univ of Missouri-Kansas City – Em
Cowan, Clinton A., (507) 646-7021 ccowan@carleton.edu,
Carleton College – Gs
Cowan, Darrel S., 206-543-4033 darrel@u.washington.edu,
Univ of Washington – Gct
Cowan, Ellen A., (828) 262-2260 cowanea@appstate.edu,
Appalachian State Univ – Gm
Cowart, James B., (850) 644-5784 jcowart@mailer.fsu.edu,
Florida State Univ – Cc
Cowen, James P., (808) 956-7124 jcowen@soest.hawaii.edu,
Univ of Hawai'i, Manoa – Cm
Cowen, Richard, (530) 752-1859 cowen@geology.ucdavis.edu,
Univ of California, Davis – Po
Cowgill, Eric S., 530-754-6574 cowgill@geology.ucdavis.edu,
Univ of California, Davis – Gc
Cowles, Deborah, 937 775-3455 deborah.cowles@wright.edu,
Wright State Univ – Oo
Cowles, Timothy J., (541) 737-3966 tjc@coas.oregonstate.edu,
Oregon State Univ – Ob
Cowley, Scott W., 303-273-3638 scowley@mines.edu,
Colorado School of Mines – Oo
Cowman, Tim C., (605) 677-5227 tcowman@usd.edu,
South Dakota Dept of Env & Natural Resources – Cg
Cox, Charles S., (858) 534-3235 cscox@ucsd.edu,
Univ of California, San Diego – Op
Cox, Helen M., 818-677-3532 helen.m.cox@csun.edu,
California State Univ, Northridge – Oa
Cox, John, (403) 440-6160 jcox@mtroyal.ca,
Mount Royal College – Eo
Cox, Leif, (406) 496-4216 lcox@mtech.edu,
Montana Tech of the Univ of Montana – Ye
Cox, Malcolm E., 61 7 3138 1649 m.cox@qut.edu.au,
Queensland Univ of Technology – Hw
Cox, Randel T., (901) 678-4361 randycox@memphis.edu,
Univ of Memphis – Gm
Cox, Richard, (709) 737-8014 rcox@mun.ca,
Memorial Univ of Newfoundland – Gp
Cox, Ronadh, (413) 597-2297 ronadh.cox@williams.edu,
Williams College – Gs
Cox, Shelah, 215 204-8227 scox@temple.edu,
Temple Univ – Oo
Cox, Shelley M., (323) 857-6318 scox@bcf.usc.edu,
Los Angeles County Museum of Natural History – Pv
Cozart, Cheryl L., 412.442.4234 ccozart@state.pa.us,
Pennsylvania Bureau of Topographic & Geologic Survey – Oo
Crabtree, Roy E., (727) 896-8626 Univ of South Florida, Saint Petersburg – Ob
Craddock, John P., (651) 696-6620 craddock@macalester.edu,
Macalester College – Gc
Craddock, Robert A., 202-633-2474 Smithsonian Institution / National Air & Space Museum – Xg
Craig, Harmon, (858) 534-3260 Univ of California, San Diego – Cs
Craig, James L., (505) 665-7996 jlcraig@lanl.gov,
Los Alamos National Lab – Gg
Craig, James R., (540) 231-5222 jrcraig@vt.edu,
Virginia Polytechnic Institute & State Univ – Eg
Craig, Mitchell, (510) 885-3425 mitchell.craig@csueastbay.edu,
California State Univ, East Bay – Yg
Cramer, Chris, (901) 678-2007 cramer@ceri.memphis.edu,
Univ of Memphis – Ys
Cramer, Chris H., (916) 322-9312 California Geological Survey – Ys
Cramer, Chris H., 6784992 cramer@ceri.memphis.edu,
Univ of Memphis – Ys
Crampton, Jeremy W., (404) 413-5771 jcrampton@gsu.edu,
Georgia State Univ – Oi
Crane, Kathleen, (212) 772-5265 Graduate School of the City Univ of New York – Yh
Crane, Peter R., (773) 922-9410 Univ of Chicago – Po
Crane, Ruth, (401) 863-2449 ruth_Crane@brown.edu,
Brown Univ – Oo
Cranford, Peter, (902) 426-3277 cranfordp@mar.dfo-mpo.gc.ca,
Dalhousie Univ – Ob
Cranganu, Constantin, (718) 951-5000 (Ext. 2878) cranganu@brooklyn.cuny.edu,
Brooklyn College (CUNY) – Go
Craw, Dave, +64 3 479-7529 dave.craw@otago.ac.nz,
Univ of Otago – Eg
Crawford, Anthony J., 61 3 6226 2490 Tony.Crawford@utas.edu.au,
Univ of Tasmania – Gi
Crawford, Maria Luisa B., (610) 526-5111 mcrawfor@brynmawr.edu,
Bryn Mawr College – Gx
Crawford, Nicholas, (270) 745-5889 Western Kentucky Univ – Hg
Crawford, Thomas J., (678) 839-4062 Univ of West Georgia – Eg
Crawford, Vernon J., (541) 552-6479 crawford@sou.edu,
Southern Oregon Univ – Gg
Crawford, Waynecrawford@ipgp.jussieu.fr,
Woods Hole Oceanographic Institution – Ou
Crawford, William A., (610) 526-5112 Bryn Mawr College – Cg
Crawford, William R., (250) 363-6369 craw@ios.bc.ca,
Univ of Victoria – Op
Creager, Kenneth C., 206-685-2803 kcc@ess.washington.edu,
Univ of Washington – Ys
Crease, James, (302) 645-4240 Univ of Delaware – Op
Creaser, Robert A., (780) 492-2942 robert.creaser@ualberta.ca,
Univ of Alberta – Cc
Creasy, John W., (207) 786-6153 jcreasy@bates.edu,
Bates College – Gx
Creed, Irena F., (519) -66-1-21 (Ext. ext.84265) icreed@uwo.ca,
Univ of Western Ontario – OoSfPy
Crelling, John C., (618) 453-7361 jcrelling@geo.siu.edu,
Southern Illinois Univ Carbondale – Ec
Crepeau, Richard J., (828) 262-7052 crepeaurj@appstate.edu,
Appalachian State Univ – Oo
Crespi, Jean M., (860) 486-0601 crespi@geol.uconn.edu,
Univ of Connecticut – Gc
Crews, Tim, 928-350-2215 tcrews@prescott.edu,
Prescott College – Sb
Crider, Juliet G., (360) 650-3589 criderj@cc.wwu.edu,
Western Washington Univ – Gt
Criminale, Jr., William O., (206) 543-9506 lascala@amath.washington.edu,
Univ of Washington – Op
Crippen, Robert E., (818) 354-2475 robert.e.crippen@jpl.nasa.gov,
Jet Propulsion Lab – Or
Crisp, Edward L., (304) 424-8255 West Virginia Univ, Parkersburg – Go
Crisp, Joy A., (818) 354-9036 Jet Propulsion Lab – Gv
Criss, Robert E., (314) 935-7441 criss@levee.wustl.edu,
Washington Univ – Cs
Critchlow, Vicki, (360) 650-3582 critch@geol.wwu.edu,
Western Washington Univ – Oo
Crocket, James H., (905) 525-9140 (Ext. 27694) crocketj@mcmaster.ca,
McMaster Univ – Yr
Crockett, Joan E., 217-244-2388 crockett@isgs.uiuc.edu,
Illinois State Geological Survey – Eo
Croft, Paul J., (908) 737-3720 pcroft@kean.edu,
Kean Univ – Ow
Croll, Don, (831) 459-3610 dcroll@cats.ucsc.edu,
Univ of California, Santa Cruz – Ob
Cromartie, William J., (609) 652-4413 iaprod193@pollux.stockton.edu,
Richard Stockton College of New Jersey – Oo
Cronin, Meghan F., (206) 526-6449 Meghan.F.Cronin@noaa.gov,
Univ of Washington – Op
Cronin, Vincent S., (254) 710-2174 Vince_Cronin@baylor.edu,
Baylor Univ – Gc
Cronoble, James M., (303) 556-3070 Metropolitan State College of Denver – Gg
Crook, Keith A. W., (808) 956-9429 Univ of Hawai'i, Manoa – Gt
Crook, Keith A. W., (808) 956-9429 Univ of Hawai'i, Manoa – Gs
Crosby, Benjamin T., 208-282-2949 crosbenj@isu.edu,
Idaho State Univ – GmtHs
Cross, Aureal T., 517-355-4630 cross1@msu.edu,
Michigan State Univ – Pb
Cross, John A., (414) 424-7112 Univ of Wisconsin, Oshkosh – Oy
Crossen, Kristine J., (907) 786-6838 Univ of Alaska, Anchorage – Gl
Crossey, Laura J., (505) 277-5349 lcrossey@unm.edu,
Univ of New Mexico – Cl
Crossley, David J., (314) 977-3153 crossley@eas.slu.edu,
Saint Louis Univ – Yg
Crosson, Robert S., 206-543-6505 crosson@ess.washington.edu,
Univ of Washington – YsGct
Crouse, David A., (919) 515-7302 North Carolina State Univ – Sc
Crouse, Dianne, (902) 494-1434 dianne.crouse@dal.ca,

Dalhousie Univ – Oo
Crouse, Elizabeth, (775) 78-4-66 (Ext. ext. 122) crouse@unr.edu, Univ of Nevada – Gg
Crow, Clay, (702) 895-3116 ccrow@nevada.edu, Univ of Nevada, Las Vegas – Ca
Crowder, Margaret, (270) 745-5973 margaret.crowder@wku.edu, Western Kentucky Univ – Oe
Crowe, Bruce M., (702) 794-7206 bmc@lanl.gov, Los Alamos National Lab – Gv
Crowe, Douglas E., 706-542-2382 crowe@gly.uga.edu, Univ of Georgia – Em
Crowell, John C., (805) 893-8231 Univ of California, Santa Barbara – Gt
Crowell, John C., (805) 893-8231 crowell@quake.crustal.ucsb.edu, Univ of California, Santa Barbara – Gc
Crowley, James, (709) 737-3046 jcrowley@mun.ca, Memorial Univ of Newfoundland – Gc
Crowley, Peter D., (413) 542-2715 pdcrowley@amherst.edu, Amherst College – Gc
Croxen III, Fred W., (928) 344-7586 fred.croxen@azwestern.edu, Arizona Western College – HwPvOi
Crozaz, Ghislaine, (314) 935-6257 gcw@wuphys.wustl.edu, Washington Univ – Xm
Crozier, Carl, (919) 793-4428 North Carolina State Univ – Sb
Crozier, George F., (334) 861-7557 gcrozier@jaguar1.usouthal.edu, Univ of Alabama, Dauphin Island – Oo
Cruden, Alexander, (416) 978-3021 chair@geology.utoronto.ca, Univ of Toronto – Gt
Cruden, David M., (780) 492-5798 dave.cruden@ualberta.ca, Univ of Alberta – Nr
Cruikshank, Kenneth M., (503) 725-3383 cruikshankk@pdx.edu, Portland State Univ – Gc
Crumbly, Isaac J., (912) 825-6454 Fort Valley State Univ – Oo
Cruse, Anna, 405-744-6358 anna.cruse@okstate.edu, Oklahoma State Univ – Co
Cruse, Richard M., (515) 294-7850 Iowa State Univ of Science & Technology – So
Csatho, Beata, (716) 645-6800 (Ext. 3921) bcsatho@buffalo.edu, SUNY, Buffalo – Yg
Cudaback, Cynthia N., 919-515-3711 North Carolina State Univ – Gu
Cuffey, Roger J., (814) 865-1293 cuffey@ems.psu.edu, Pennsylvania State Univ, Univ Park – Pg
Cuker, Benjamin E., (757) 727-5783 benjamin.cuker@hamptonu.edu, Hampton Univ – Ob
Cullen, James L.jcullen@salemstate.edu, Salem State College – Gs
Cullen, John J., (902) 494-6667 john.cullen@dal.ca, Dalhousie Univ – Ob
Cullen, Stephen J., (805) 893-3758 sjc@geog.geog.ucsb.edu, Univ of California, Santa Barbara – Hw
Cullers, Robert L., (785) 532-2240 rcullers@ksu.edu, Kansas State Univ – Ct
Culshaw, Nicholas, (902) 494-3501 Dalhousie Univ – Gc
Culver, Stephen J., (252) 328-6360 culvers@mail.ecu.edu, East Carolina Univ – Pm
Culver, Teresa B., (804) 924-6375 tbc4e@virginia.edu, Univ of Virginia – Hy
Culver, William C., (727) 341-4381 culverw@email.spjc.cc.fl.us, Saint Petersburg Junior College – Gg
Cumbest, Randolph J., (803) 725-5284 Clemson Univ – Yg
Cummings, Michael L., (503) 725-3395 cummingsm@pdx.edu, Portland State Univ – Gg
Cummins, R. Hays, (513) 529-1338 Miami Univ – Oo
Cundall, Peter A., (612) 626-0369 71212.1255@compuserve.com, Univ of Minnesota, Twin Cities – Nr
Cunningham, Mary A., 845-437-5547 macunningham@vassar.edu, Vassar College – Oi
Cunningham, Philip, (850) 644-4334 cunningham@met.fsu.edu, Florida State Univ – Ow
Cunnold, Derek M., (404) 894-3814 cunnold@eas.gatech.edu, Georgia Institute of Technology – Oa
Curran, H. Allen, (413) 585-3943 acurran@email.smith.edu, Smith College – Pg
Curran, John, (609) 984-6587 john.curran@dep.state.nj.us, New Jersey Geological Survey – Hw
Curray, Joseph R., (858) 534-3299 jcurray@ucsd.edu, Univ of California, San Diego – Gu
Currey, Donald R., (801) 581-8218 Univ of Utah – Gm
Currie, Brian S., (513) 529-7578 curriebs@muohio.edu, Miami Univ – Gs
Currie, Philip J., (403) 823-7707 Royal Tyrrell Mus of Palaeo – Pv
Currie, Philip J., (403) 823-7707 Univ of Calgary – Pv
Currie, William, (301) 689-3115 currie@al.umces.edu, Univ of Maryland – Cg
Curry, Brandon B., (217) 244-5787 Illinois State Geological Survey – Pe
Curry, Janel M., (616) 526-6869 Calvin College – Oo
Curry, Joan E., (520) 626-5081 curry@ag.arizona.edu, Univ of Arizona – Sc
Curry, Lawrence D., (850) 487-2219 Florida Geological Survey – Go
Curry, William B., (508) 289-2591 wcurry@whoi.edu, Woods Hole Oceanographic Institution – Pm
Curry Rogers, Kristina A., 651-696-6288 krogers@macalester.edu, Macalester College – Pv
Curtice, Joshua M., (508) 289-2618 jcurtice@whoi.edu, Woods Hole Oceanographic Institution – Cc
Curtin, Tara M., 315-781-3928 curtin@hws.edu, Hobart & William Smith Colleges – Pe
Curtis, Arthur C., (615) 576-7642 Oak Ridge National Lab – Ge
Curtis, Charles M., (562) 938-4643 ccurtis@lbcc.cc.ca.us, Long Beach City College – Gs
Curtis, David B., (505) 667-4845 Los Alamos National Lab – Cc
Curtis, Garniss H., (510) 845-0333 Univ of California, Berkeley – Cc
Curtis, Jason H., (352) 392-2296 curtis@ufl.edu, Univ of Florida – Pe
Curtis, John B., (303) 273-3887 jbcurtis@mines.edu, Colorado School of Mines – Go
Curtis, Lynn A., (954) 201-6942 lcurtis@broward.edu, Broward Community College, Central Campus – Gg
Curtiss, David K., (801) 585-9248 dcurtiss@egi.utah.edu, Univ of Utah – Eo
Cushman, John, (765) 494-8040 jcushman@purdue.edu, Purdue Univ – Sp
Cushman, Robert A., (909) 824-4530 Loma Linda Univ – Ps
Custer, Stephan G., (406) 994-6906 uessc@montana.edu, Montana State Univ – Hg
Cutter, Gregory A., (757) 683-4929 gcutter@odu.edu, Old Dominion Univ – Oc
Cvancara, Alan M., (307) 266-6733 Univ of North Dakota – Pi
Czarnetzki, Alan C., (319) 273-2152 alan.czarnetzki@uni.edu, Univ of Northern Iowa – Ow
Czartoski, Heather M., (734) 995-4624 Concordia Univ – Ge
Czechanski, Michael L., (608) 263-7393 mlczecha@wisc.edu, Univ of Wisconsin - Extension – Oi
Czeck, Dyanna M., (440) 775-5957 Oberlin College – Gc
Czeck, Dyanna M., 414-229-3948 dyanna@uwm.edu, Univ of Wisconsin, Milwaukee – Gc

D

D'Alberto, Lucio, (828) 262-3049 Appalachian State Univ – Gc
D'Allura, Jad A., (541) 552-6480 dallura@sou.edu, Southern Oregon Univ – Gc
D'Andrea, Anthony, (541) 737-8079 dandrea@coas.oregonstate.edu, Oregon State Univ – Ob
D'Antonio, Carla, (510) 643-6341 Univ of California, Berkeley – Pe
D'Asaro, Eric A., (206) 545-2982 dasaro@apl.washington.edu, Univ of Washington – Op
D'Hondt, Steven L., (401) 792-6808 dhondt@gso.uri.edu, Univ of Rhode Island – Ou
Da Bai, George S., (205) 851-5462 gdabai@aamu.edu, Alabama A&M Univ – Nm
Daemen, Jaak, (775) 7844309 daemen@mines.unr.edu, Univ of Nevada, Reno – Nr
Dagdelen, Kadri, (303) 273-3711 kdagdele@mines.edu, Colorado School of Mines – Nm
Dagg, Michael J., (225) 851-2800 Louisiana State Univ – Ob
Dahdah, Nicolas F., (801) 585-7853 ndahdah@egi.utah.edu, Univ of Utah – Go
Dahl, Jason M., 218-755-4106 jdahl@bemidjistate.edu, Bemidji State Univ – Xgm
Dahl, Peter S., (330) 672-2218 pdahl@kent.edu, Kent State Univ – Gp
Dahlen, F. A., (609) 258-4130 fad@princeton.edu, Princeton Univ – Ys
Dahlen, F. Anthony, (609) 258-4130 Princeton Univ – Ys
Dahlgren, Randy A., (530) 752-2814 radahlgren@ucdavis.edu, Univ of California, Davis – So
Daida, Jason M., (313) 747-4581 daida@eecs.umich.edu, Univ of Michigan – Oa
Daigneault, Real, (418) 545-5011 rdaignea@uqac.uquebec.ca, Universite du Quebec a Chicoutimi – Gc
Dakan, Arthur W., (502) 852-6844 Univ of Louisville – Oo

Daley, Gwen M., 803-323-4973 daleyg@winthrop.edu,
Winthrop Univ – Pe
Dalldorf, Graham, 775-673-7616 Graham.dalldorf@dri.edu,
Desert Research Institute – Gma
Dallmeyer, R. David, (706) 542-7448 dallmeyr@uga.edu,
Univ of Georgia – Gt
Dalrymple, G. Brentbrentandsharon@comcast.net,
Oregon State Univ – Ou
Dalrymple, Robert W., (613) 545-6186 Queen's Univ – Gs
Dalton, Richard F., (609) 292-2576 richard.dalton@dep.state.nj.us,
New Jersey Geological Survey – Gr
Daly, Christopher, (541) 737-2531 daly@oce.oregonstate.edu,
Oregon State Univ – Ow
Daly, Julia F., 207-778-7403 dalyj@maine.edu,
Univ of Maine, Farmington – GmlGu
Daly, Raymond J.daly@hawk.ucc.edu,
Union County College – Pv
Dalziel, Ian W. D., (512) 471-0431 ian@ig.utexas.edu,
Univ of Texas at Austin, Jackson School of Geosciences – Gt
Dalziel, Ian W. D., (512) 471-0431 ian@utig.ig.utexas.edu,
Univ of Texas at Austin, Jackson School of Geosciences – Gt
Dam, Hans G., (860) 405-9098 hans.dam@uconn.edu,
Univ of Connecticut – Ob
Damuth, John E., (817) 272-2987 damuth@uta.edu,
Univ of Texas, Arlington – Gu
Dana, Gayle L., (775) 674-7538 gdana@dri.edu,
Univ of Nevada, Reno – Or
Dando, William A., (812) 237-7874 gedando@isugw.indstate.edu,
Indiana State Univ – Oo
Dane, Jake H.jdane@acesag.auburn.edu,
Auburn Univ – Sp
Daniel, Christopher G., (570) 577-1133 cdaniel@bucknell.edu,
Bucknell Univ – Gp
Daniel, Stephen R.sdaniel@mines.edu,
Colorado School of Mines – Oo
Daniels, Jeffrey J., (614) 292-1039 jeff@geology.ohio-state.edu,
Ohio State Univ – Yg
Daniels, Walter L., (540) 231-7175 wdaniels@vt.edu,
Virginia Polytechnic Institute & State Univ – Sd
Daniels, William R., (505) 667-4546 Los Alamos National Lab – Ct
Danko, George, (775) 784-4284 danko@mines.unr.edu,
Univ of Nevada, Reno – Hq
Danko, George, (775) 784-4284 danko@unr.edu,
Univ of Nevada, Reno – Nm
Danner, Wilbert R., (604) 822-6892 Univ of British Columbia – Gd
Dannevik, William P., 314 977 3115 dannevik@eas.slu.edu,
Saint Louis Univ – Oa
Dansereau, Paulinedansero@ggl.ulaval.ca,
Universite Laval – Gs
Daoust, Mario, 417-836-5301 Southwest Missouri State Univ – Oa
Darby, Brian, (225) 578-5810 bdarby@geol.lsu.edu,
Louisiana State Univ – Gt
Darby, Dennis A., (757) 683-4701 ddarby@odu.edu,
Old Dominion Univ – Ou
Darby, Jeannie, (916) 752-5670 Univ of California, Davis – Hg
Darby, Steve B., 205-247-3611 sdarby@gsa.state.al.us,
Geological Survey of Alabama – Oi
Dark, Shawna J., 818-677-3532 shawna.dark@csun.edu,
California State Univ, Northridge – Oy
Darling, Richard, (514) 340-4743 Ecole Polytechnique – Ce
Darling, Robert S., (607) 753-2923 darlingr@cortland.edu,
SUNY, Cortland – Gx
Darmody, Robert G., (217) 333-9489 Univ of Illinois, Urbana-Champaign – Sd
Darnell, Rezneat M., (979) 845-5756 rdarnell@ocean.tamu.edu,
Texas A&M Univ – Ob
Daroub, Samira H., (561) 993-1500 sdaroub@mail.ifas.ufl.edu,
Univ of Florida – So
Darragh, Robert B., (916) 327-1059 California Geological Survey – Ys
Darrell, James H., (912) 681-5757 JDarrell@GeorgiaSouthern.edu,
Georgia Southern Univ – Pl
Das, Phanindramohan, (409) 845-0633 das@ariel.tamu.edu,
Texas A&M Univ – Ow
Das, Sarah B., 508-289-2464 sdas@whoi.edu,
Woods Hole Oceanographic Institution – Pm
Dash, Zora V., (505) 667-1923 zvd@lanl.gov,
Los Alamos National Lab – Oo
Dattilo, Benjamin F., 260-481-6253 dattilob@ipfw.edu,
Indiana Univ / Purdue Univ, Fort Wayne – GrPiGs
Daugherty, Carolyn M., (928) 523-0984 carolyn.daugherty@nau.edu,
Northern Arizona Univ – Oo
Daugherty, LeRoy A., (505) 646-3406 New Mexico State Univ, Las Cruces – Sd
Davenport, Joan R., (509) 786-9384 jdavenp@tricity.wsu.edu,
Washington State Univ – Sb
Davi, Jodie, 603-646-1898 Jodie.Davi@dartmouth.edu,
Dartmouth College – Oo
David, Eric, (418) 656-8116 eric.david@ggl.ulaval.ca,
Universite Laval – Oo
David, Mark B., (217) 333-4308 m-david@uiuc.edu,
Univ of Illinois, Urbana-Champaign – Sf
Davidson, Cameron, 507-646-7144 cdavidso@carleton.edu,
Carleton College – Gx
Davidson, Fiona M., (479) 575-3879 fdavidso@comp.uark.edu,
Univ of Arkansas, Fayetteville – Oo
Davidson, Garry J., 61 3 6226 2815 Garry.Davidson@utas.edu.au,
Univ of Tasmania – Cs
Davidson, Gregg R., (662) 915-5824 davidson@olemiss.edu,
Univ of Mississippi – Hw
Davidson, Jon P., 0191 3742528 j.p.davidson@durham.ac.uk,
Durham Univ – Cg
Davie, Jr., Robert N., (813) 341-4306 davier@email.spjc.cc.fl.us,
Saint Petersburg Junior College – Yg
Davies, Brian E., (864) 646-7876 bdavies@clemson.edu,
Clemson Univ – Sc
Davies, Bruce, (415) 338-1755 San Francisco State Univ – Oo
Davies, Caroline P.daviesc@umkc.edu,
Univ of Missouri-Kansas City – Pl
Davies, J H.huw@earth.cf.ac.uk,
Cardiff Univ – Yg
Davies, James F., (705) 675-1151 (Ext. ext. 2298) Laurentian Univ, Sudbury – Em
Davies, John, (905) 525-9140 (Ext. 23533) davies.j@sympatico.ca,
McMaster Univ – Yr
Davies, Rondi M., 212-769-5314 rdavies@amnh.org,
American Museum of Natural History – Gz
Davies-Vollum, Katherine Sian, 206-692-4626 ksdavies@u.washington.edu, Univ of Washington – GsrPe
Davis, Alan, 910-270-0186 ddu@psu.edu,
Pennsylvania State Univ, Univ Park – Ec
Davis, Arden D., (605) 394-2473 Arden.Davis@sdsmt.edu,
South Dakota School of Mines & Technology – Hw
Davis, Bruce, (859) 622-1418 bruce.davis@eku.edu,
Eastern Kentucky Univ – Oi
Davis, Carolyn Y., (251) 460-6381 Univ of South Alabama – Oo
Davis, Curtiss O., (541) 737-4432 cdavis@coas.oregonstate.edu,
Oregon State Univ – Ob
Davis, Daniel M., 631-632-8217 daniel.davis@sunysb.edu,
SUNY, Stony Brook – Yg
Davis, David A., (775) 784-6691 (Ext. 133) ddavis@unr.edu,
Univ of Nevada – Gg
Davis, Donald R., (520) 621-3801 drdavis@hwr.arizona.edu,
Univ of Arizona – Hq
Davis, Donald W., (416) 586-5811 dond@rom.on.ca,
Royal Ontario Museum – Cc
Davis, Donald W., (416) 946-0365 dond@geology.utoronto.ca,
Univ of Toronto – Cc
Davis, Douglas D., (404) 894-4008 douglas.davis@eas.gatech.edu,
Georgia Institute of Technology – Oa
Davis, George H., (520) 621-1856 gdavis@u.arizona.edu,
Univ of Arizona – Gc
Davis, J. Matthew, (603) 862-1718 matt.davis@unh.edu,
Univ of New Hampshire – Hw
Davis, James A., 217-581-5528 cfjad2@eiu.edu,
Eastern Illinois Univ – Og
Davis, James A., (801) 378-3852 james_davis@byu.edu,
Brigham Young Univ – Oo
Davis, James F., (814) 865-6544 California Geological Survey – Oo
Davis, Jim, 417-836-4897 jad068f@smsu.edu,
Southwest Missouri State Univ – Oo
Davis, John C.john.davis5@mchsi.com,
Univ of Kansas – Gq
Davis, Kenneth D., (601) 354-6328 Mississippi Office of Geology – Gr
Davis, Kenneth J., (814) 863-8601 davis@met.psu.edu,
The Pennsylvania State Univ, Univ Park – Ow
Davis, Leslie S., (512) 223-1795 (Ext. 22763) ldavis@austin.cc.tx.us,
Austin Community College – Op
Davis, Owen K., (520) 621-7953 odavis@email.arizona.edu,
Univ of Arizona – Pl
Davis, P. Thompson, (781) 891-3479 pdavis@bentley.edu,

Bentley College – Gl
Davis, Paul M., (310) 825-1343 pdavis@ess.ucla.edu, Univ of California, Los Angeles – YsmYg
Davis, R. Laurence, (203) 932-7108 rldavis@charger.newhaven.edu, Univ of New Haven – Ge
Davis, Ralph K., (479) 575-4515 ralphd@comp.uark.edu, Univ of Arkansas, Fayetteville – Hw
Davis, Robert, (804) 924-0579 red3u@virginia.edu, Univ of Virginia – Oa
Davis, Stanley N., (520) 621-5082 sndavis@email.arizona.edu, Univ of Arizona – Hw
Davis, Tara, +353 21 4903696 t.davis@ucc.ie, Univ College Cork – Yg
Davis, Terry E., (323) 343-2411 California State Univ, Los Angeles – Cc
Davis, Thomas L., (303) 273-3938 tdavis@mines.edu, Colorado School of Mines – Ye
Davis, Trevor J., (801) 587-9019 Univ of Utah – Oi
Davis, Jr., Richard A., (727) 974-2773 Univ of South Florida, Saint Petersburg – Gs
Davis, Jr., Richard A., (813) 974-2773 rdavis@chuma.cas.usf.edu, Univ of South Florida, Tampa – Gs
Davison, Ian R., (207) 581-2983 Univ of Maine – Ob
Davisson, M L., (925) 423-5993 davisson2@llnl.gov, Lawrence Livermore National Lab – Cg
Davydov, Vladimir I., (208) 426-1119 vdavydov@boisestate.edu, Boise State Univ – Pg
Dawers, Nancye H., (504) 865-5198 ndawers@tulane.edu, Tulane Univ – Gc
Dawsey, Cyrus B., (334) 844-3418 dawsecb@auburn.edu, Auburn Univ – Oy
Dawson, James C., (518) 564-4035 james.dawson@plattsburgh.edu, Plattsburgh State Univ (SUNY) – Ge
Dawson, Jane P., (515) 294-6302 jpdawson@iastate.edu, Iowa State Univ of Science & Technology – Gp
Dawson, Mary R., (412) 622-3246 Carnegie Museum of Nat History – Pv
Day, Harold J., (414) 465-2369 Univ of Wisconsin, Green Bay – Hg
Day, Howard W., (530) 752-2882 day@geology.ucdavis.edu, Univ of California, Davis – Gp
Day, James E., (309) 438-8678 jeday@ilstu.edu, Illinois State Univ – Pi
Day, Rick L., (814) 863-1615 r4d@psu.edu, Pennsylvania State Univ, Univ Park – Ou
Day, Steven M., (619) 594-2663 day@moho.sdsu.edu, San Diego State Univ – Ys
Day, Jr., John W., (504) 388-6508 Louisiana State Univ – Ob
Daymon, Deborah J., (505) 667-9021 deba@lanl.gov, Los Alamos National Lab – Oo
Dayton, Hairl, 573-368-2138 nrdayth@mail.dnr.state.mo.us, Missouri Dept of Natural Resources – Oo
Dayton, Paul K., (858) 534-6740 pdayton@ucsd.edu, Univ of California, San Diego – Ob
De Beus, Barbara, (973) 655-4448 debeusb@mail.montclair.edu, Montclair State Univ – Oo
De Boer, Jelle Z., (860) 685-2254 jdeboer@wesleyan.edu, Wesleyan Univ – YmGta
de Boutray, Bernard, (514) 987-8523 de_boutray@uqam.ca, Universite du Quebec a Montreal – Gz
De Bruin, Rodney H., (307)766-2286e226 rdebruin@uwyo.edu, Wyoming State Geological Survey – Eo
De Campos, Christina, 089/2180 4264 campos@min.uni-muenchen.de, LMU - Univ of Munich – Gz
de Caprariis, Pascal P., (317) 274-7732 pdecaprr@iupui.edu, Indiana Univ / Purdue Univ, Indianapolis – Hq
De Carlo, Eric H., (808) 956-6473 edecarlo@soest.hawaii.edu, Univ of Hawai'i, Manoa – Ca
De Carlo, Eric H., (808) 956-6473 edecarlo@soest.hawaii.edu, Univ of Hawai'i, Manoa – Cm
De Deckker, Patrick, +61-2-6125 2070 patrick.dedeckker@anu.edu.au, Australian National Univ – Gu
de Foy, Benjamin, 314-977-3122 foy@eas.slu.edu, Saint Louis Univ – Oa
De Hon, Rene A., (318) 342-1894 Univ of Louisiana, Monroe – Gz
de Hoop, Maarten, (765) 496-6439 mdehoop@math.purdue.edu, Purdue Univ – Gq
De Jong, Kees A.kees.dejong@uc.edu, Univ of Cincinnati – Gt
De Kock, Walter E., (319) 266-6577 Univ of Northern Iowa – Oe
de Leon, Paul, (604) 822-2789 pdeleon@eos.ubc.ca, Univ of British Columbia – Oo
de Menocal, Peter B., (845) 365-8483 peter@ldeo.columbia.edu, Columbia Univ – Pe
De Nault, Kenneth J., (319) 273-2033 kenneth.denault@uni.edu, Univ of Northern Iowa – Gz
de Polo, Craig M., (775) 784-6691 (Ext. 149) cdepolo@unr.edu, Univ of Nevada – Ng
de Silva, Shanika, (541) 737-1212 desilvsh@geo.oregonstate.edu, Oregon State Univ – Gv
de Szoeke, Roland, (541) 737-3160 szoeke@coas.oregonstate.edu, Oregon State Univ – Op
de Uña Alvarez, Elena P., 00 34 988387137 edeuna@uvigo.es, Coruna Univ – Oy
de Vargas, Colomban, (732) -93-2-65 (Ext. ext. 236) vargas@imcs.rutgers.edu, Rutgers, The State Univ of New Jersey – Pm
de Vernal, Anne, (514) 98-7 30 (Ext. ext 8599#) devernal.anne@uqam.ca, Universite du Quebec a Montreal – Pm
de Wet, Carol B., (717) 291-4388 Carol.deWet@FandM.edu, Franklin and Marshall College – Gs
Deakin, Ann K., (716) 673-3303 deakin@fredonia.edu, SUNY, Fredonia – OyiGg
Deal, Edmond G., 406-496-4180 edeal@mtech.edu, Montana Tech of The Univ of Montana – Gg
Deal, Richard, (814) 732-1733 rdeal@edinboro.edu, Edinboro Univ of Pennsylvania – Oi
Dean, Claude S., (859) 622-2317 Eastern Kentucky Univ – Gc
Dean, Gray, (865) 974-2366 wdean1@utk.edu, Univ of Tennessee, Knoxville – Py
Dean, Jeffrey S., (520) 621-2320 jdean@ltrr.arizona.edu, Univ of Arizona – Oo
Dean, Lewis S., 205/247-3548 ldean@gsa.state.al.us, Geological Survey of Alabama – Eg
Dean, Nicholeimcs@nova.edu, Nova Southeastern Univ – Oo
Dean, Robert, 717-245-1109 deanr@dickinson.edu, Dickinson College – Gi
Dean, W. G., (864) 656-7655 wgdean@clemson.edu, Clemson Univ – Pg
deAngelis, Marie, 718-409-7380 mdeangelis, SUNY Maritime College – Oc
DeAngelo, Micheal V., (512) 232-3373 michael.deangelo@beg.utexas.edu, Univ of Texas at Austin, Jackson School of Geosciences – Ye
DeBari, Susan M., (360) 650-3588 debari@geol.wwu.edu, Western Washington Univ – Gx
Debes, John H., (202) 478-8862 debes@dtm.ciw.edu, Carnegie Institution of Washington – Oo
DeBlonde, Christine, (418) 654-3718 chdeblon@nrcan.gc.ca, Universite du Quebec – Oo
Deboo, Phili B., (901) 678-4424 pdeboo@memphis.edu, Univ of Memphis – Pg
DeBremaecker, Jean-Claude, (713) 348-4886 deb@rice.edu, Rice Univ – Yg
Debusk, William F., (352) 392-1803 (Ext. 319) wdebusk@ufl.edu, Univ of Florida – Sf
DeCaria, Alex J., 717-871-4739 Alex.Decaria@millersville.edu, Millersville Univ – Oa
DeCelles, Peter G., (520) 621-4910 decelles@email.arizona.edu, Univ of Arizona – Gd
Decker, Edward R., (207)-581-2152 dianne.perro@umit.maine.edu, Univ of Maine – Gc
Decker, Robert W., (808) 882-7569 lavadecker@aol.com, Univ of Hawai'i at Hilo – Oo
DeConto, Robert, (413) 545-3426 deconto@geo.umass.edu, Univ of Massachusetts – Oa
Deen, Patricia A., (760) 744-1150 (Ext. xt. 2512) pdeen@palomar.edu, Palomar College – Og
Deffeyes, Kenneth S., (609) 258-3597 deffeyes@geo.princeton.edu, Princeton Univ – Eg
Deffeyes, Kenneth S., 609-258-3597 deffeyes@princeton.edu, Princeton Univ – Eg
DeGaetano, Arthur, (607) 255-0385 atd2@cornell.edu, Cornell Univ – Oa
DeGoosh, Stephen R., (906) 227-2904 sdegoosh@nmu.edu, Northern Michigan Univ – Oo
Dehler, Carol M., (435) 797-0764 carol.dehler@usu.edu, Utah State Univ – Gs
Dehne, Kevin T., (989) 686-9326 ktdehne@delta.edu, Delta College – Og
Deibert, Edward J., (701) 231-8578 North Dakota State Univ – Og
Deibert, Jack, (931) 221-6318 deibertj@apsu.edu,

Austin Peay State Univ – Gsr
Deines, Peter, (814) 865-7152 p7d@psu.edu, Pennsylvania State Univ, Univ Park – Cs
Deininger, Robert W., (901) 678-2177 Univ of Memphis – Gx
Deitzer, Gerald F., (301) 405-4335 gd3@umail.umd.edu, Univ of Maryland – Oo
Del Genio, Anthony D., (212) 678-5588 adelgenio@giss.nasa.gov, Columbia Univ – Oa
Delaney, Jeremy S., (732) 445-3616 jsd@rci.rutgers.edu, Rutgers, The State Univ of New Jersey – Gx
Delaney, John, 206-534-4830 jdelaney@u.washington.edu, American Museum of Natural History – Gg
Delaney, John R., (206) 543-4830 jdelaney@u.washington.edu, Univ of Washington – Yr
Delaney, Margaret L., (831) 459-4736 delaney@cats.ucsc.edu, Univ of California, Santa Cruz – Oc
Delano, Helen L., 717.702.2031 hdelano@state.pa.us, Pennsylvania Bureau of Topographic & Geologic Survey – Gg
Delano, John W., (518) 442-4479 jdelano@atmos.albany.edu, SUNY, Albany – Cp
Delawder, Sandradelawdsa@jmu.edu, James Madison Univ – Oo
Delgado Martín, Jorge, 00 34 981 167000 delgado@iccp.udc.es, Coruna Univ – Gg
Delgado-Argote, Luis A.ldelgado@cicese.mx, Centro de Inv Científica y de Educación Superior de Ensenada – Git
Delisle, Mark J., (916) 322-9320 California Geological Survey – Ou
Della Porta, GportaGD@cf.ac.uk, Cardiff Univ – Gs
Della Valle, Richard S., (707) 253-3161 Napa Valley College – Cg
Dellapenna, Timothy M., (409) 740-4952 dellapet@tamug.tamu.edu, Texas A&M Univ – Ou
Dellwig, Louis F., 785-864-4974 Univ of Kansas – Or
Delson, Eric, (718) 960-8405 eedlc@cunyvm.cuny.edu, Graduate School of the City Univ of New York – Pv
Deltan, Rodney S., (850) 488-9380 Florida Geological Survey – Ca
Delusina, Irina, (530) 752-1861 delusina@geology.ucdavis.edu, Univ of California, Davis – Pl
DeMaster, David J., (919) 515-7026 david_demaster@ncsu.edu, North Carolina State Univ – Oc
Demchak, Jennifer, 570-662-4613 jdemchak@mansfield.edu, Mansfield Univ – Hs
deMenocal, Peter, (845) 365-8483 Columbia Univ – Pe
Demerjian, Kenneth L., (518) 442-4557 kld@asrc.cestm.albany.edu, SUNY, Albany – Oa
Demers, Serge, (418) 723-1986 (Ext. 1483) Universite du Quebec a Rimouski – Ob
DeMets, D. Charles, (608) 262-8598 chuck@geology.wisc.edu, Univ of Wisconsin, Madison – Ym
Demicco, Robert V., (607) 777-2604 Binghamton Univ – Gs
Deming, David, (405) 325-6304 ddeming@ou.edu, Univ of Oklahoma – Yh
Deming, Jody W., (206) 543-0845 jdeming@u.washington.edu, Univ of Washington – Ob
Demopoulos, George, (514) 398-4755 McGill Univ – Nx
Demosthenous, Christie M., (920) 424-7167 demosthe@uwosh.edu, Univ of Wisconsin, Oshkosh – Gz
Dempsey, David P., (415) 338-7716 ddempsey@norte.sfsu.edu, San Francisco State Univ – Oa
Deng, Aijun, (814) 863-8253 deng@essc.psu.edu, The Pennsylvania State Univ, Univ Park – Ow
Deng, Baolin, (505) 835-5505 New Mexico Institute of Mining and Technology – Cg
Dengler, Lorinda, (707) 826-3115 lad1@axe.humboldt.edu, Humboldt State Univ – Yg
Denhart, Sally L., 217-333-5102 denhart@isgs.uiuc.edu, Illinois State Geological Survey – Oo
Denicourt, Raymond, (813) 974-2236 Univ of South Florida, Tampa – Gz
DeNiro, Michael J., (805) 893-8032 Univ of California, Santa Barbara – Cs
Denison, Rodger E., (972) 883-2453 denison@utdallas.edu, Univ of Texas, Dallas – Cs
Denizman, Can, (229) 333-5752 Valdosta State Univ – Hw
Denman, Kenneth L., (250) 363-6346 denman@ios.bc.ca, Univ of Victoria – Og
Denman, Kenneth L., (250) 363-8230 denmank@ec.gc.ca, Univ of Victoria – Op
Dennett, Keith E., (775) 784-4056 dennett@scs.unr.edu, Univ of Nevada, Reno – Hs
Denning, A.Scott, (970) 491-8359 denning@atmos.colostate.edu, Colorado State Univ – Oa
Dennis, Clyde B., (630) 252-5999 cbdennis@anl.gov, Argonne National Lab – Oo
Dennison, John M., (919) 962-0686 dennison@email.unc.edu, Univ of North Carolina, Chapel Hill – Gr
Denniston, Rhawn F., (319) 895-4306 RDenniston@cornellcollege.edu, Univ of Iowa – Cs
Denniston, Rhawn F., (319) 895-4306 rdenniston@cornellcollege.edu, Cornell College – Ge
Denny, F. Brett, 618-985-3394 denny@isgs.uiuc.edu, Illinois State Geological Survey – Gr
Denton, Gary R. W., (671) 735-2690 Univ of Guam – Ob
Denton, George H., (207) 581-2193 debbies@maine.edu, Univ of Maine – Gl
DePaolo, Donald J., (510) 642-7686 depaolo@socrates.berkeley.edu, Univ of California, Berkeley – Cc
Dependahl, Jan, (805) 965-0581 (Ext. 2316) dependj@sbcc.net, Santa Barbara City College – Oe
DePodesta, Daniel, (203) 932-7107 Univ of New Haven – Oy
dePolo, Diane, (775) 784-4976 diane@seismo.unr.edu, Univ of Nevada, Reno – Ys
Dera, Przemyslaw, (202) 478-8940 p.dera@gl.ciw.edu, Carnegie Institution of Washington – Gy
Deriso, Richard B., (858) 546-7020 rderiso@ucsd.edu, Univ of California, San Diego – Ob
Derome, Jacques F., (514) 398-5350 jacques.derome@mcgill.ca, McGill Univ – Oa
Derry, Louis A., (607) 255-4726 derry@geology.cornell.edu, Cornell Univ – Cl
Derstler, Kraig L., (504) 280-6799 kdersle@uno.edu, Univ of New Orleans – Pv
Desai, Ankur R., (608) 265-9201 desai@aos.wisc.edu, Univ of Wisconsin, Madison – Oa
Desiderio, Russell H., (541) 737-2415 desi@coas.oregonstate.edu, Oregon State Univ – Op
Desilets, Mario O., (775) 784-6691 (Ext. 125) desilets@unr.edu, Univ of Nevada – Cg
deSouza, Anthony R., 202-334-2744 adesouza@nas.edu, National Academy of Sciences/National Research Council – Og
Desrochers, Andre, (613)562-5800 Ext 6861 adesro@uottawa.ca, Univ of Ottawa – Gs
Dessene, Lynn J., (318) 482-6163 Univ of Louisiana, Lafayette – Sf
Dessler, Alexander, (520) 621-4589 dessler@vega.lpl.arizona.edu, Univ of Arizona – Oo
Dessler, Andrew, 979-862-1427 adessler@tamu.edu, Texas A&M Univ – Oo
Dessureault, Sean, (520) 621-2359 sdessure@email.arizona.edu, Univ of Arizona – Nx
Dethier, David P., (413) 597-2078 david.p.dethier@williams.edu, Williams College – Gm
Detournay, Emmanuel M., (612) 625-5522 detou001@tc.umn.edu, Univ of Minnesota, Twin Cities – Nr
Detrick, Robert S., (508) 289-3335 rdetrick@whoi.edu, Woods Hole Oceanographic Institution – Yr
Dettman, David, (520) 621-4618 dettman@email.arizona.edu, Univ of Arizona – Cs
Detwiler, Andrew G., (605) 394-1995 Andrew.Detwiler@sdsmt.edu, South Dakota School of Mines & Technology – Oa
Detwiler, Daniel P., (805) 664-3129 California State Univ, Bakersfield – Oo
Deuser, Werner G., (508) 289-2551 wdeuser@whoi.edu, Woods Hole Oceanographic Institution – Cs
Devane, S.J., John F., (617) 552-8134 devane@bc.edu, Boston College – Ym
Devary, Joseph L., (509) 376-8345 joe.devary@pnl.gov, Pacific Northwest National Lab – Gq
Devaurs, Micheline, (505) 667-1519 devaurs_micheline@lanl.gov, Los Alamos National Lab – Hg
Dever, Edward, (541) 737-2749 edever@coas.oregonstate.edu, Oregon State Univ – Op
Devera, Joseph A., 618-985-3394 devera@isgs.uiuc.edu, Illinois State Geological Survey – Gr
Devlin, J. R., 785-864-4994 devlin@ku.edu, Univ of Kansas – Hw
Devol, Allan H., (206) 543-1292 devol@u.washington.edu, Univ of Washington – Oc
Dewar, William K.dewar@ocean.fsu.edu, Florida State Univ – Op
DeWeaver, Eric T., (608)265-5438 deweaver@aos.wisc.edu, Univ of Wisconsin, Madison – Oa
DeWeese, Georgina G., (678) 839-4065 gdeweese@westga.edu,

Univ of West Georgia – Oy
Dewers, Thomas A., (405) 325-4430 tdewers@ou.edu,
Univ of Oklahoma – Cl
deWet, Andrew P., (717) 291-3815 andy.dewet@fandm.edu,
Franklin and Marshall College – Ge
Dewey, Christopher P., (662) 325-2909 dewey@geosci.msstate.edu,
Mississippi State Univ – Pm
Dewey, John, (530) 754-7472 dewey@geology.ucdavis.edu,
Univ of California, Davis – Gt
Dewey, John F., 713-743-3399 dewey@geology.ucdavis.edu,
Univ of Houston – GtcGg
DeWitt, Calvin B., (608) 263-7771 Univ of Wisconsin, Madison – Ob
Dexter, Barbara L., (914) 251-6641 SUNY, Purchase – Ob
Dexter, Leland R., (928) 523-6535 lee.dexter@nau.edu,
Northern Arizona Univ – Oy
Dey, William S., 217-244-2779 dey@isgs.uiuc.edu,
Illinois State Geological Survey – Hw
DeYoung, Damon P., 619-594-6443 ddeyoung@geology.sdsu.edu,
San Diego State Univ – Gg
Di Girolamo, Larry, (217) 333-3080 larry@atmos.uiuc.edu,
Univ of Illinois, Urbana-Champaign – Or
Dias, Nelson, 812-237-2444 ndias@isugw.indstate.edu,
Indiana State Univ – Or
DiBacco, Claudio, (902) 426-3258 DiBaccoC@MAR.dfo-mpo.gc.ca,
Univ of British Columbia – Ob
Dibb, Jack E., 603-862-3063 jack.dibb@unh.edu,
Univ of New Hampshire – Oa
Dick, Henry J B., (508) 289-2590 hdick@whoi.edu,
Woods Hole Oceanographic Institution – Gi
Dick, Jeffrey C.jcdick@ysu.edu,
Youngstown State Univ – Ng
Dick, Richard P., 614-247-7605 dick.78@osu.edu,
Ohio State Univ – Sb
Dick, Warren A., (330) 263-3877 dick.5@osu.edu,
Ohio State Univ – Sb
Dickerhoof, Dean W.ddickerh@mines.edu,
Colorado School of Mines – Oo
Dickerson, Richard E., (310) 825-5864 Univ of California, Los Angeles – Oo
Dickey, Jr., D. Neil, (815) 753-7932 neil@geol.niu.edu,
Northern Illinois Univ – Gi
Dickhut, Rebecca M., (804) 684-7253 rdickhut@vims.edu,
College of William & Mary – Oc
Dickin, Alan P., (905) 525-9140 (Ext. 24365) dickin@mcmaster.ca,
McMaster Univ – Cc
Dickinson, Robert E., (404) 385-1509 robted@eas.gatech.edu,
Georgia Institute of Technology – Oa
Dickinson, William R., (520) 299-5220 wrdickin@dakotacom.net,
Univ of Arizona – Gt
Dickman, Steven R., (607) 777-2857 Binghamton Univ – Yg
Dickson, Andrew G., +1 (858) 822 2990 adickson@ucsd.edu,
Univ of California, San Diego – Cm
Dickson, Frank, (775) 784-4002 fdickson@mines.unr.edu,
Univ of Nevada, Reno – Gg
Dickson, Loretta D., (570) 484-2068 ldickson@lhup.edu,
Lock Haven Univ – GicGe
Dickson, Stephen M., (207) 287-7174 stephen.m.dickson@maine.gov,
Univ of Maine – Gu
Diebold, Frank E., (406) 496-4215 Montana Tech of the Univ of
Montana – Cl
Diebold, John, (845) 365-8524 Columbia Univ – Yr
Diecchio, Richard J., (703) 993-1208 rdiecchi@gmu.edu,
George Mason Univ – Gr
Dieckmann, Melissa S., (859) 622-1276 Melissa.Dieckmann@eku.edu,
Eastern Kentucky Univ – Hs
Diehl, Jimmy F., (906) 487-2665 jdiehl@mtu.edu,
Michigan Technological Univ – Ym
Diem, Jeremy E., (404) 413-5770 jdiem@gsu.edu,
Georgia State Univ – Oa
Diemer, John A., 704-687-5994 jadiemer@email.uncc.edu,
Univ of North Carolina, Charlotte – Gs
Dieterich, James H., (951) 827-2976 james.dieterich@ucr.edu,
Univ of California, Riverside – Yg
Dieterle, Dwight A., (727) 553-1114 ddieterle@marine.usf.edu,
Univ of South Florida, Saint Petersburg – Ob
Dietrich, Richard V., (989) 772-5303 dietr1rv@cmich.edu,
Central Michigan Univ – Gx
Dietrich, William E., (510) 642-2633 bill@geomorph.berkeley.edu,
Univ of California, Berkeley – Gm
Dietsch, Craigcraig.dietsch@uc.edu,
Univ of Cincinnati – Gp
Dietz, Anne D., (210) 785-6090 adietz@mail.accd.edu,
San Antonio College – PgOg
Dietz, Richard D., (970) 351-2950 rdietz@unco.edu,
Univ of Northern Colorado – Xg
Diffenbaugh, Noah S., (765) 494-0754 diffenbaugh@purdue.edu,
Purdue Univ – Pe
Diffendal, Robert F., (402) 472-7546 rfd@unl.edu,
University of Nebraska - Lincoln – Gr
Digby, Susan A., 360-475-7840 sdigby@oc.ctc.edu,
Olympic College – Oy
Diggins, Thomas P., 330-941-3605 tpdiggins@ysu.edu,
Youngstown State Univ – Py
Dighe, Kalpak, (505) 665-1701 kdighe@lanl.gov,
Los Alamos National Lab – Oo
DiGiacomo-Cohen, Mary, 860 405 9015 lisrc@uconnvm.uconn.edu,
Univ of Connecticut – Gl
Dilcher, David L., (352) 392-1721 dilcher@flmnh.ufl.edu,
Univ of Florida – Pb
Dilek, Yildirm, (513) 529-2212 dileky@muohio.edu,
Miami Univ – Gt
DiLeonardo, Christopher G., (415) 949-7491 Foothill College – Gc
Dilles, John H., (541) 737-1245 dillesj@geo.oregonstate.edu,
Oregon State Univ – Em
Dilliard, Kelly A., (605) 624-1984 Kelly.Dilliard@usd.edu,
Univ of South Dakota – Gsg
DiMarco, Steven F., (979) 862-4168 dimarco@skeefum.tamu.edu,
Texas A&M Univ – Op
DiMichele, William A., (202) 633-1319 Smithsonian Institution /
National Museum of Natural History – Pb
Dimmick, Charles W., (860) 832-2936 dimmick@ccsu.edu,
Central Connecticut State Univ – Ge
DiNaso, Steven, 217-581-2626 cfsd2@eiu.edu,
Eastern Illinois Univ – Oi
Ding, Kang, 612-626-1860 mlcd@umn.edu,
Univ of Minnesota, Twin Cities – Cg
Dinger, James S., (859) 257-5500 dinger@kgs.mm.uky.edu,
Univ of Kentucky – Hg
Dinger, James S., (859) 257-5500 (Ext. 163) dinger@uky.edu,
Univ of Kentucky – Hw
Dingle, A. Nelson, (313) 764-3335 Univ of Michigan – Oa
Dingman, S. Lawrence, (603) 862-1718 ldingman@cisunix.unh.edu,
Univ of New Hampshire – Hq
Dingus, Delmar D., (805) 756-2753 ddingus@calpoly.edu,
California Polytechnic State Univ – So
Dingwell, Donald Bruce, 089/21804136 dingwell@min.uni-muenchen.de,
LMU - Univ of Munich – Giv
Dinsmoor, John C., (702) 295-6189 john_dinsmoor@lanl.gov,
Los Alamos National Lab – Nm
Dintaman, Christopher, (812) 855-7428 cdintama@indiana.edu,
Indiana Geological Survey – Hw
Dinter, David A., (801) 581-7937 dadinter@mines.utah.edu,
Univ of Utah – Gt
Dinterman, Philip A., 205-247-3559 pdinterman@gsa.state.al.us,
Geological Survey of Alabama – Gg
Diouf, Arona, 336-334-7979 North Carolina Agricultural & Tech State
Univ – Gg
DiPietro, Angelina, (617) 552-3641 dipietro@bc.edu,
Boston College – Gg
DiPietro, Joseph A., (812) 465-7041 dipietro@usi.edu,
Univ of Southern Indiana – Gc
Diplas, Panos, (703) 231-6069 Virginia Polytechnic Institute & State
Univ – Gs
Dipple, Gregory M., (604) 822-2624 gdipple@eos.ubc.ca,
Univ of British Columbia – Gp
Distin, Philip A., (514) 398-4755 McGill Univ – Nx
Ditmars, John, (630) 252-5953 Argonne National Lab – Oo
Dittmer, Eric, (541) 552-6496 dittmer@sou.edu,
Southern Oregon Univ – Gg
DiVenere, Victor, (516) 299-2318 divenere@liu.edu,
Long Island Univ, C.W. Post Campus – Ym
Dix, George R.georgedix@pigeon.carleton.ca,
Carleton Univ – Gs
Dixon, Grady, (662) 325-5801 grady.dixon@msstate.edu,
Mississippi State Univ – Ow
Dixon, Jackie E., 305-284-6792 j.dixon@miami.edu,
Univ of Miami – GipGv
Dixon, Jacqueline E., 305 421-4150 jdixon@rsmas.miami.edu,
Univ of Miami – Cp
Dixon, Joe B., (979) 845-8323 j-dixon@tamu.edu,
Texas A&M Univ – Sc

Dixon, John C., (479) 575-5808 jcdixon@comp.uark.edu, Univ of Arkansas, Fayetteville – Gm
Dixon, John M., (613) 545-6172 Queen's Univ – Gc
Dixon, Timothy, 305 421-4660 tdixon@rsmas.miami.edu, Univ of Miami – Or
Doar, III, William R., (803) 896-7708 South Carolina Geological Survey – Eg
Dobbs, Frederdick C., (757) 683-4301 fdobbs@odu.edu, Old Dominion Univ – Ob
Dobler, Scott, (270) 745-7078 scott.dobler@wku.edu, Western Kentucky Univ – Oa
Dobrzhinetskaya, Larissa F., (951) 827-2028 larissa@ucrac1.ucr.edu, Univ of California, Riverside – Gx
Dobson, David M., (336) 316-2278 ddobson@guilford.edu, Guilford College – Gu
Dobson, Patrick F., (510) 486-5373 pfdobson@lbl.gov, Lawrence Berkeley National Lab – Cg
Dockal, James A., (910) 962-3494 dockal@uncw.edu, Univ of North Carolina Wilmington – Gd
Dockery, David T., (601) 354-6328 Mississippi Office of Geology – Pi
Dodd, David W., 228-688-7127 David.Dodd@usm.edu, Univ of Southern Mississippi – Oi
Dodd, J. Robert, (516) 632-8204 dodd@indiana.edu, Indiana Univ, Bloomington – Po
Dodd, Stephen R., (928) 523-8242 steve.dodd@nau.edu, Northern Arizona Univ – Oo
Dodge, Clifford H., 717.702.2036 cdodge@state.pa.us, Pennsylvania Bureau of Topographic & Geologic Survey – Gr
Dodge, Rebecca L., (678) 839-4067 rdodge@westga.edu, Univ of West Georgia – Oy
Dodge, Richard E., 954-262-3651 dodge@nsu.nova.edu, Nova Southeastern Univ – Py
Dodge, Sheridan L., (201) 595-3409 William Paterson College – Oo
Dodson, Peter, (215) 898-8784 dodsonp@vet.upenn.edu, Univ of Pennsylvania – Pv
Dodson, Russell L., (570) 662-4614 rdodson@mnsfld.edu, Mansfield Univ – Gm
Dodson, Stanley I., (608) 262-6395 Univ of Wisconsin, Madison – Ob
Doehring, Donald O., 970-491-5661 dond@cnr.colostate.edu, Colorado State Univ – Ge
Doel, Ronald E., (541) 737-1243 doelr@geo.oregonstate.edu, Oregon State Univ – Gh
Doelling, Hellmut H., (801) 537-3306 mnet.nrregion.hdoellin@state.ut.us, Utah Geological Survey – Gg
Doenges, Susann V., (512) 471-0217 susie.doenges@beg.utexas.edu, Univ of Texas at Austin, Jackson School of Geosciences – Oo
Dogwiler, Toby, 507-457-5267 tdogwiler@winona.edu, Winona State Univ – GmHqOw
Doheny, Edward L., (215) 898-6085 Univ of Pennsylvania – Ng
Doherty, Cheryl, (302) 831-2569 sparrow@udel.edu, Univ of Delaware – Oo
Dolan, James F., (213) 740-8599 dolan@usc.edu, Univ of Southern California – Gt
Dolan, Robert, (540) 924-0544 rd5q@virginia.edu, Univ of Virginia – Gm
Dolbier, Rachel, (775) 784-4528 rdolbier@unr.edu, Univ of Nevada, Reno – Gg
Dolgoff, Anatole, (212) 346-1502 Pace Univ, New York Campus – Og
Doll, William E., d8e@ornl.gov, Oak Ridge National Lab – Yg
Dollase, Wayne A., (310) 825-3823 dollase@ess.ucla.edu, Univ of California, Los Angeles – Gz
Dollhopf, Douglas J., (406) 944-5594 dollhopf@montana.edu, Montana State Univ – Oo
Domack, Cynthia R., (315) 859-4710 cdomack@hamilton.edu, Hamilton College – Pg
Domack, Eugene W., (315) 859-4711 edomack@hamilton.edu, Hamilton College – Gs
Domber, Steven E., (609) 984-6587 steven.domber@dep.state.nj.us, New Jersey Geological Survey – Hw
Domier, Jane E., 217-244-2513 domier@isgs.uiuc.edu, Illinois State Geological Survey – Oi
Domingue, Carla, (225) 388-8407 carla@lgs.bri.lsu.edu, Louisiana State Univ – Oo
Dominic, David F., 937 775-3455 david.dominic@wright.edu, Wright State Univ – Gs
Donaghay, Percy, (401) 874-6944 donaghay@gso.uri.edu, Univ of Rhode Island – Ob
Donahoe, Robert J., (505) 667-7603 Los Alamos National Lab – Oo
Donahoe, Rona J., 205-348-1879 rdonahoe@geo.ua.edu, Univ of Alabama – Cl
Donahue, Jack D., (412) 624-8780 jdonahue@pitt.edu, Univ of Pittsburgh – Ga
Donahue, John C., (404) 656-3214 Georgia Dept of Nat Res – Hw
Donahue, Kelly, (505) 835-5249 kdonahue@gis.nmt.edu, New Mexico Institute of Mining & Technology – Cg
Donahue, Thomas, (313) 763-2390 Univ of Michigan – Oa
Donaldson, Alan C., (304) 293-5603 adonalds@wvu.edu, West Virginia Univ – Gs
Donaldson, J. Allan, (613) 520-2600 Carleton Univ – Gs
Donaldson, Paul R., (208) 426-3639 pdonalds@boisestate.edu, Boise State Univ – Ye
Donaldson, Sue, (775) 784-4848 Univ of Nevada, Reno – Cg
Doney, Scott C., 508-289-3776 sdoney@whoi.edu, Woods Hole Oceanographic Institution – Cm
Dong, Hailiang, (513) 529-2517 dongh@muohio.edu, Miami Univ – Gz
Donley, John W., (540) 654-1392 jdonley@mwc.edu, Univ of Mary Washington – Oo
Donnellan, Andrea, 818-354-4737 Andrea.Donnellan@jpl.nasa.gov, Univ of Southern California – Yx
Donnelly, Clancy, (210) 999-7093 clancy.donnelly@trinity.edu, Trinity Univ – Gg
Donnelly, Jeffrey, 508-289-2994 jdonnelly@whoi.edu, Woods Hole Oceanographic Institution – Gu
Donnelly, Joseph, (727) 553-1648 jdonnelly@marine.usf.edu, Univ of South Florida, Saint Petersburg – Ob
Donnelly, Thomas W., (607) 777-2323 Binghamton Univ – Gg
Donoghue, Joseph F., (850) 644-2703 jdonoghu@mailer.fsu.edu, Florida State Univ – Gu
Donohue, Kathleen, (401) 874-6615 kdonohue@gso.uri.edu, Univ of Rhode Island – Op
Donohue, Stephen J., (540) 231-9740 donohue@vt.edu, Virginia Polytechnic Institute & State Univ – Sc
Donovan, Jeff C., (727) 553-1116 jdonovan@marine.usf.edu, Univ of South Florida, Saint Petersburg – Op
Donovan, John F., 507-457-5292 Winona State Univ – Eg
Donovan, Joseph J., (304) 293-5603 donovan@geo.wvu.edu, West Virginia Univ – Hq
Donovan, R. Nowell, (817) 257-7214 r.donovan@tcu.edu, Texas Christian Univ – GdcGt
Doolan, Barry L., (802) 656-0248 bdoolan@zoo.uvm.edu, Univ of Vermont – Gx
Dooley, John H., (609) 292-2576 john.dooley@dep.state.nj.us, New Jersey Geological Survey – Cg
Doolittle, James J., (605) 688-4750 James.Doolittle@sdstate.edu, South Dakota State Univ – Sc
Dorais, Michael J., (801) 422-1347 dorais@byu.edu, Brigham Young Univ – Gi
Dorale, Jeffrey A., (319) 335-0822 jeffrey-dorale@uiowa.edu, Univ of Iowa – Cs
Doran, Peter T., (312) 413-7275 pdoran@uic.edu, Univ of Illinois, Chicago – Hw
Dore, Melissa L., (954)262-3610 missy@nova.edu, Nova Southeastern Univ – Oo
Dorfman, Alexander, 089/2180 4275 LMU - Univ of Munich – Gz
Dorfner, Thomas, 089/2180 4278 dorfner@min.uni-muenchen.de, LMU - Univ of Munich – Gz
Dorman, Clive E., (619) 594-5707 cdorman@geology.sdsu.edu, San Diego State Univ – Op
Dorman, H. James, (901) 678-2007 dorman@ceri.memphis.edu, Univ of Memphis – Ys
Dorman, Leroy M., (858) 534-2406 ldorman@ucsd.edu, Univ of California, San Diego – Yg
Dorries, Alison M., (505) 665-6952 adorries@lanl.gov, Los Alamos National Lab – Oo
Dorsey, Rebecca J., (541) 346-4431 Univ of Oregon – Gr
Dort, Jr., Wakefield, (785) 864-4974 Univ of Kansas – Gm
Dortch, Quay, (225) 851-2800 Louisiana State Univ – Ob
Doser, Diane I., (915) 747-5851 doser@geo.utep.edu, Univ of Texas, El Paso – Ys
Doskey, Paul V., (630) 252-7662 pvdoskey@anl.gov, Argonne National Lab – Oa
Doss, Paul K., (812) 465-7132 pdoss@usi.edu, Univ of Southern Indiana – Hw
Dosso, Stanley E., (250) 472-4341 sdosso@uvic.ca, Univ of Victoria – Op
Dostal, Jarda, (902) 420-5747 Dalhousie Univ – Cg
Dott, Jr., Robert H., (608) 262-1856 rdott@geology.wisc.edu, Univ of Wisconsin, Madison – Gs

Faculty Index -D

Doty, Thomas, (401) 254-3066 Roger Williams Univ – Ob
Dougan, Jane, (954)262-3621 douganj@nova.edu,
Nova Southeastern Univ – Oo
Doughty, Christine, 15104866453 cadoughty@lbl.gov,
Lawrence Berkeley National Lab – Hw
Doughty, P. T., (509) 359-4257 pdoughty@mail.ewu.edu,
Eastern Washington Univ – Gc
Douglas, Arthur V., (402) 280-2464 sonora@creighton.edu,
Creighton Univ – Og
Douglas, Bruce, 812-855-3848 douglasb@indiana.edu,
Indiana Univ, Bloomington – Gc
Douglas, Marianne, (780) 492-0055 marianne.douglas@ualberta.ca,
Univ of Toronto – Gn
Douglas, Robert G., (213) 740-8275 rdouglas@usc.edu,
Univ of Southern California – Pe
Douglass, Daniel, 617 373-4381 d.douglass@neu.edu,
Northeastern Univ – Gl
Douglass, David N., (626) 585-7036 dndouglass@paccd.cc.ca.us,
Pasadena City College – Ym
Dove, Patricia M., (540) 231-2444 pdove@vt.edu,
Virginia Polytechnic Institute & State Univ – Cl
Doveton, John H., (785) 864-2100 doveton@kgs.ku.edu,
Univ of Kansas – Go
Doveton, John H., (785) 864-3965 doveton@kgs.ku.edu,
Univ of Kansas – Gq
Dowd, John F., (706) 542-2383 jdowd@gly.uga.edu,
Univ of Georgia – Hy
Dowdy, Robert H., (612) 625-7058 bdowdy@soils.umn.edu,
Univ of Minnesota, Twin Cities – Sc
Downs, James W.jdowns@geology.ohio-state.edu,
Ohio State Univ – Gz
Downs, Robert T., (520) 626-8092 rdowns@email.arizona.edu,
Univ of Arizona – Gi
Dowse, Mary E., (505) 538-6352 dowsem@wnmu.edu,
Western New Mexico Univ – Og
Doyle, Briget C., (843) 953- 0877 Doyleb@cofc.edu,
College of Charleston – Ng
Doyle, James A., (530) 752-7591 Univ of California, Davis – Pb
Doyle, Josephj6doyle@bridgew.edu,
Bridgewater State College – Gg
Doyle, Larry J., (727) 553-1512 ldoyle@marine.usf.edu,
Univ of South Florida, Saint Petersburg – Gs
Drahovzal, James A., (859) 257-5500 (Ext. 155) drahovzal@uky.edu,
Univ of Kentucky – Eo
Drake, John C., (802) 656-0244 jdrake@zoo.uvm.edu,
Univ of Vermont – Cl
Drake, John J., (905) 525-9140 (Ext. 24018) drake@mcmaster.ca,
McMaster Univ – Yr
Drake, Lon D., (319) 335-1826 Univ of Iowa – Hy
Drake, Michael J., (520) 621-6962 drake@lpl.arizona.edu,
Univ of Arizona – Xc
Draper, David S., (505) 277-1607 draper@unm.edu,
Univ of New Mexico – GiCpXc
Draper, Grenville, (305) 348-3087 draper@fiu.edu,
Florida International Univ – Gc
Drayson, S. Roland, (313) 936-0484 Univ of Michigan – Oa
Dreesen, Donald S., (505) 667-1913 dreesen@lanl.gov,
Los Alamos National Lab – Np
Dreger, Douglas S., (510) 643-1719 dreger@seismo.berkeley.edu,
Univ of California, Berkeley – Ys
Dreimanis, Aleksis, (519) 661-2253 mschell@uwo.ca,
Univ of Western Ontario – Gl
Drescher, Andrew, (612) 625-2374 Univ of Minnesota, Twin Cities – So
Dreschoff, Gisela, (785) 864-4517 Univ of Kansas – Ce
Dresel, P. Evan, (509) 376-8341 evan.dresel@pnl.gov,
Pacific Northwest National Lab – Cg
Drever, James I., (307) 766-6434 drever@uwyo.edu,
Univ of Wyoming – Cl
Drew, Douglas A., (404) 496-4202 ddrew@mtech.edu,
Montana Tech of the Univ of Montana – Cg
Drew, Robin A., (514) 398-4755 McGill Univ – Nx
Drews, Patricia L., (660) 562-1273 drews@nwmissouri.edu,
Northwest Missouri State Univ – Oi
Drexler, John W., (303) 492-5251 john.drexler@colorado.edu,
Univ of Colorado – Ca
Driese, Steven G., (865) 974-6002 sdriese@utk.edu,
Univ of Tennessee, Knoxville – Gs
Driese, Steven G., (254) 710-2177 Steven_Driese@baylor.edu,
Baylor Univ – SaGe
Driever, Steven L., (816) 235-2971 drievers@umkc.edu,
Univ of Missouri-Kansas City – Oo
Dripps, Weston R., (864) 294-3392 wes.dripps@furman.edu,
Furman Univ – HwsGe
Driscoll, John R., (607) 753-2926 driscollj@cortland.edu,
SUNY, Cortland – Oo
Drobniak, Agnieszka, (812) 855-2687 agdrobni@indiana.edu,
Indiana Geological Survey – Ec
Drobny, Gerry, (916) 691-7204 Cosumnes River College – Gg
Droser, Mary L., 951-827-3797 mary.droser@ucr.edu,
Univ of California, Riverside – Po
Droxler, André W., (713) 348-4885 andre@rice.edu,
Rice Univ – Gs
Druckenmiller, Patrick S., 907-474-6954 ffpsd@uaf.edu,
Univ of Alaska, Fairbanks – Pv
Druffel, Ellen R. M., (949) 824-2116 edruffel@uci.edu,
Univ of California, Irvine – Oc
Drummond, Carl N., (260) 481-5750 drummond@ipfw.edu,
Indiana Univ / Purdue Univ, Fort Wayne – Gs
Drummond, David D., (410) 554-5551 drummond@mgs.md.gov,
Maryland Geological Survey – Hy
Dryden, Nancy, 405-744-6358 ndryden@okstate.edu,
Oklahoma State Univ – Oo
Drysdale, Russell, 61 02 4921 5749 Russell.Drysdale@newcastle.edu.au,
Univ of Newcastle – Gm
Drzewiecki, Peter A., 860-465-4322 DrzewieckiP@easternct.edu,
Eastern Connecticut State Univ – Gs
Drzyzga, Scott A., (717) 477-1307 sadrzy@ship.edu,
Shippensburg Univ – Oi
Duaime, Terence E., 406-496-4157 tduaime@mtech.edu,
Montana Tech of The Univ of Montana – Hw
Duan, Jennifer, (702) 895-0452 gduan@dri.edu,
Univ of Nevada, Reno – Hs
Duan, Quingyun, (925) 422-7704 duan2@llnl.gov,
Lawrence Livermore National Lab – Hg
Duba, Alfred, 212-769-5398 duba@amnh.org,
American Museum of Natural History – Gg
Dube, Benoit, (418) 654-2669 bdube@nrcan.gc.ca,
Universite du Quebec – Em
Duberger, Raynald, (418) 545-5011 rduberge@uqac.uquebec.ca,
Universite du Quebec a Chicoutimi – Yg
Dubois, Lisa, (760) 744-1150 (Ext. xt. 2369) ldubois@palomar.edu,
Palomar College – Og
Dubois, Martin K., (785) 864-2178 mdubois@kgs.ku.edu,
Univ of Kansas – Go
Dubois, Robert L., (405) 325-3253 Univ of Oklahoma – Ym
Duby, Paul F., (212) 854-2928 pfd1@columbia.edu,
Columbia Univ – Nx
Duce, Robert A., (979) 845-5756 rduce@ocean.tamu.edu,
Texas A&M Univ – Oc
Ducea, Mihai N., (520) 621-5171 ducea@email.arizona.edu,
Univ of Arizona – Gt
Duchane, David V., (505) 667-9893 duchane@lanl.gov,
Los Alamos National Lab – Oo
Duchesne, Josee, (418) 656-2177 duchesne@ggl.ulaval.ca,
Universite Laval – Gz
Duchesne, Mathieumathieu.duchesne@mcgill.ca,
McGill Univ – Yr
Duckstein, Lucien, (520) 621-2274 duckstein@engref.fr,
Univ of Arizona – Hq
Duckworth, Kenneth, (403) 220-7375 Univ of Calgary – Ye
Dudash, Leigh, 909-941-2385 Chaffey College – Yg
Dudley, Mark A.dudley@ucmo.edu,
Univ of Central Missouri – Gg
Dudley, Walter, (808) 933-3311 Univ of Hawai'i, Manoa – Ou
Dudley-Murphy, Elizabeth A., (801) 585-5982 bmurphy@egi.utah.edu,
Univ of Utah – Or
Duebendorfer, Ernest M., (928) 523-7510 ernie.d@nau.edu,
Northern Arizona Univ – Gc
Duedall, Iver W., (321) 674-8096 duedall@fit.edu,
Florida Institute of Technology – Oc
Dueker, Kenneth G., (307) 766-2657 dueker@uwyo.edu,
Univ of Wyoming – Ys
Duennebier, Frederick K., (808) 956-4779 fred@soest.hawaii.edu,
Univ of Hawai'i, Manoa – Yr
Duer, Douglasdeur@unr.edu,
Univ of Nevada, Reno – Oo
Duex, Timothy W., (337) 482-6222 twd7693@louisiana.edu,
Univ of Louisiana, Lafayette – Hg
Duffey, Debbie, (804) 427-7189 Tidewater Community College – Gr
Duffey, Donna M., (608) 265-2129 dmduffey@wisc.edu,

Univ of Wisconsin - Extension – Oo
Duffey, Patricia, 785-628-5389 pduffey@fhsu.edu,
Fort Hays State Univ – Oo
Duffield, Wendell A., (928) 523-4852 Northern Arizona Univ – Gv
Duffy, Clarence J., (505) 667-5154 Los Alamos National Lab – Cg
Duffy, Thomas, (609) 258-6769 duffy@princeton.edu,
Princeton Univ – Gy
Duffy, Thomas, 609-258-6769 duffy@princeton.edu,
Princeton Univ – Yx
Dufour, Marie-France, 217-333-5115 dufour@isgs.uiuc.edu,
Illinois State Geological Survey – Gg
Dugolinsky, Brent K., (607) 436-2741 dugolibk@oneonta.edu,
SUNY, Oneonta – Ou
Duigon, Mark T., (410) 554-5552 mtduigon@mgs.md.gov,
Maryland Geological Survey – Hy
Duke, Edward F., (605) 394-2388 Edward.Duke@sdsmt.edu,
South Dakota School of Mines & Technology – Gi
Duke, Jason E., (931) 372-3121 jduke@tntech.edu,
Tennessee Technological Univ – Oi
Duke, Norman A., (519) 661-3199 nduke@uwo.ca,
Univ of Western Ontario – Eg
Duley, James W., (573) 368-2101 nrduleb@mail.dnr.state.mo.us,
Missouri Dept of Natural Resources – Oo
Duley, James W., 573-368-2105 bill.duley@dnr.mo.state,
Missouri Dept of Natural Resources – Hy
Dumais, Jean-Francois, (418) 724-1753 jean-francois_dumais@uqar.uquebec.ca,
Universite du Quebec a Rimouski – Og
Dumitru, Trevor, (650) 723-1328 trevor@pangea.stanford.edu,
Stanford Univ – Gc
Dunbar, John A., (254) 710-2191 john_dunbar@baylor.edu,
Baylor Univ – YgHg
Dunbar, Nelia W., (505) 835-5783 nelia@nmt.edu,
New Mexico Institute of Mining & Technology – Gi
Dunbar, Nelia W., (505) 835-5783 nelia@nmt.edu,
New Mexico Institute of Mining and Technology – Cg
Dunbar, Rob, (713) 527-4880 Rice Univ – Gu
Dunbar, Robert B., (650) 725-6830 dunbar@pangea.stanford.edu,
Stanford Univ – On
Duncan, Janet E., (617) 287-5281 Univ of Massachusetts, Boston/
Harbor Campus – Ou
Duncan, Joel G., (850) 488-9380 Florida Geological Survey – Go
Duncan, Robert A., (541) 737-5206 rduncan@coas.oregonstate.edu,
Oregon State Univ – Ou
Duncan, Thomas, (541) 737-5206 Univ of California, Berkeley – Pb
Dundas, Robert G., (559) 278-6984 rdundas@csufresno.edu,
California State Univ, Fresno – Pv
Dunfield, Kari, ((51) 9) -824- (Ext. 20 xt 58088) kdunfield@lrs.uoguelph.ca,
Univ of Guelph – Sb
Dunlap, Dallas B., (512) 471-4858 dallas.dunlap@beg.utexas.edu,
Univ of Texas at Austin, Jackson School of Geosciences – Ye
Dunlop, David J., (905) 828-3968 dunlop@physics.utoronto.ca,
Univ of Toronto – Ym
Dunlop, Roberta, (604) 291-4995 rdunlop@sfu.ca,
Simon Fraser Univ – Gx
Dunn, David E., (972) 883-4044 ddunn@utdallas.edu,
Univ of Texas, Dallas – Gc
Dunn, Douglas, (216) -23-1-46 (Ext. ext 3233) ddunn@cmnh.org,
Cleveland Museum of Natural History – Pg
Dunn, John Todd, (506) 453-4804 Univ of New Brunswick – Gi
Dunn, Richard K., (802) 485-2304 rdunn@norwich.edu,
Norwich Univ – Gs
Dunn, Robert A., 808-956-3728 dunnr@hawaii.edu,
Univ of Hawai'i, Manoa – Yr
Dunn, Steven R., 413-538-2531 sdunn@mtholyoke.edu,
Mount Holyoke College – Gp
Dunn-Norman, Shari, 573-341-4840 caolila@umr.edu,
Univ of Missouri, Rolla – Np
Dunne, George C., (818) 677-2511 george.dunne@csun.edu,
California State Univ, Northridge – Gc
Dunne, William M., (423) 974-4161 wdunne@utk.edu,
Univ of Tennessee, Knoxville – Gc
Dunning, Gregory R., (709) 737-8481 dunning@esd.mun.ca,
Memorial Univ of Newfoundland – Cc
Dunning, Jeremy D., (812) 856-4448 dunning@indiana.edu,
Indiana Univ, Bloomington – Gc
Dunton, Kenneth H., (361) 749-6744 ken.dunton@mail.utexas.edu,
Univ of Texas at Austin – Ob
Dupras, Donald, (916) 323-0111 California Geological Survey – En
Dupre, William R., (713) 743-3425 wdupre@uh.edu,
Univ of Houston – On
Durako, Michael J., (910) 256-3721 Univ of South Florida, Saint
Petersburg – Og
Durbin, Edward G., (401) 874-6850 edurbin@gso.uri.edu,
Univ of Rhode Island – Ob
Durbin, James, 812-465-1208 jdurbin@usi.edu,
Univ of Southern Indiana – Gm
Durflinger, Jan, (541) 523-3133 jan.durflinger@state.or.us,
Oregon Dept of Geology & Mineral Industries – Oo
Durham, Lisa A., (630) 252-3170 ladurham@anl.gov,
Argonne National Lab – Hw
Durrant, Jeffrey O., (801) 422-4116 jod2@email.byu.edu,
Brigham Young Univ – Ge
Durski, Scott M., (541) 737-8857 sdurski@coas.oregonstate.edu,
Oregon State Univ – Op
Dushaw, Brian D., (206) 685-4198 dushaw@apl.washington.edu,
Univ of Washington – Op
Dutch, Steven I., (920) 465-2246 dutchs@uwgb.edu,
Univ of Wisconsin, Green Bay – Gc
Dutcher, Russell R., (618) 453-7362 Southern Illinois Univ Carbondale – Ec
Dutro, Thomas, (202) 633-1322 Smithsonian Institution / National
Museum of Natural History – Pi
Dutrow, Barbara L., (225) 388-2525 dutrow@geol.lsu.edu,
Louisiana State Univ – Gz
Dutta, Prodip K., (812) 237-2268 gedutta@isugw.indstate.edu,
Indiana State Univ – Gs
Dutton, Alan, 210-458-5746 Alan.Dutton@utsa.edu,
Univ of Texas, San Antonio – Hw
Dutton, Alan R., (512) 471-5739 alan.dutton@beg.utexas.edu,
Univ of Texas at Austin, Jackson School of Geosciences – Hw
Dutton, John A., 814-865-1534 dutton@ems.psu.edu,
The Pennsylvania State Univ, Univ Park – Ow
Dutton, Shirley P., (512) 471-0329 shirley.dutton@beg.utexas.edu,
Univ of Texas at Austin, Jackson School of Geosciences – Gd
Duxbury, Thomas C., (818) 354-4301 Jet Propulsion Lab – Oo
Dvorkin, Jack, (650) 725-9296 Stanford Univ – Yg
Dvorzak, Marie, (608) 262-8956 mdvorzak@geology.wisc.edu,
Univ of Wisconsin, Madison – Oo
Dworkin, Stephen I., (254) 710-2186 steve_dworkin@baylor.edu,
Baylor Univ – ClGd
Dwyer, Daryl F., 419-5302661 daryl.dwyer@utoledo.edu,
Univ of Toledo – Oo
Dwyer, Gary S., (919) 681-8164 gsd3@duke.edu,
Duke Univ – Gs
Dwyer, Joseph R., 321-674-7208 dwyer@pss.fit.edu,
Florida Institute of Technology – Oa
Dyar, M. Darby, (413) 538-3073 mdyar@mtholyoke.edu,
Mount Holyoke College – Gz
Dye, David H., 901-678-3330 daviddye@memphis.edu,
Univ of Memphis – Ga
Dyer, Jamie, (662) 325-5802 jamie.dyer@msstate.edu,
Mississippi State Univ – Ow
Dyke, Lawrence D., (613) 545-6173 Queen's Univ – Ng
Dyl, II, Stanley J., 906-487-2572 sjdyl@mtu.edu,
Michigan Technological Univ – Gz
Dymek, Robert F., (314) 935-5344 bob_d@levee.wustl.edu,
Washington Univ – Gp
Dymek, Robert F., (314) 935-5344 bob_d@levee.wustl.edu,
Washington Univ – Gi
Dyreson, Eric G., (406) 683-7275 e_dyreson@umwestern.edu,
Univ of Montana, Western – Gq
Dyroff, Terry L., (301) 279-5234 Montgomery College – Gm
Dziewonski, Adam M., (617) 495-2510 dziewons@eps.harvard.edu,
Harvard Univ – Ys
Dzunic, Aleksandar, +61 8 9266 3521 A.Dzunic@curtin.edu.au,
Curtin Univ of Technology – Ygs

E

Earle, Janet, (956) 721-5195 Laredo Community College – Gp
Earls, Sandy, 405-744-6358 sandrae@okstate.edu,
Oklahoma State Univ – Oo
Early, Thomas O.eot@ornl.gov,
Oak Ridge National Lab – Cl
Earnest, Daniel, (301) 405-4090 earnest@geol.umd.edu,
Univ of Maryland – Gz
Easson, Gregory L., (662) 915-5995 geasson@olemiss.edu,
Univ of Mississippi – Or
Easter-Pilcher, Andrea L., (406) 683-7910 Univ of Montana, Western – Py
Easterbrook, Don J., (360) 650-3583 dbunny@cc.wwu.edu,

Western Washington Univ – Gm
Eastin, Matt, 704-687-5914 Univ of North Carolina, Charlotte – Ow
Eastler, Thomas E., 207-778-7401 eastler@maine.edu, Univ of Maine, Farmington – GeOrGg
Eastman, Gina, (701) 777-2248 Univ of North Dakota – Yr
Eastoe, Christopher J., (520) 621-1638 eastoe@email.arizona.edu, Univ of Arizona – Cs
Eaton, David W., 403-220-8863 eatond@ucalgary.ca, Univ of Western Ontario – YseYg
Eaton, Jeffrey G., (801) 626-6225 jeaton@weber.edu, Weber State Univ – Ps
Eaton, Lewis S., (540) 568-3339 eatonls@jmu.edu, James Madison Univ – Gm
Ebel, Denton, 212-769-5381 debel@amnh.org, American Museum of Natural History – Xm
Ebel, John E., (617) 552-3399 ebel@bc.edu, Boston College – Ys
Ebel, John E., 617-552-8319 john.ebel.1@bc.edu, Boston College – Yg
Eberhardt, Erik, (604) 827-5573 eeberhardt@eos.ubc.ca, Univ of British Columbia – Ng
Eberhart, Mark E., (303) 273-3726 meberhar@mines.edu, Colorado School of Mines – Om
Eberle, Jaelyn J., 303-492-8069 jaelyn.eberle@colorado.edu, Univ of Colorado – Pv
Eberle, Jaelyn J.Jaelyn.Eberle@colorado.edu, Univ of Colorado – Pv
Eberli, Gregor P., 305 421-4678 geberli@rsmas.miami.edu, Univ of Miami – Gr
Ebersole, Brian, (909) 621-8924 bebersole@pomona.edu, Pomona College – Oo
Ebersole, Sandy M., 205-247-3613 sebersole@gsa.state.al.us, Geological Survey of Alabama – Og
Ebert, James R., (607) 436-3065 ebertjr@oneonta.edu, SUNY, Oneonta – Gr
Eberth, David A., (403) 823-7707 Royal Tyrrell Mus of Palaeo – Gs
Ebinger, Cynthia J., 585-276-3364 ebinger@earth.rochester.edu, Univ of Rochester – Gt
Ebinger, Michael H., (505) 667-3147 mhe@lanl.gov, Los Alamos National Lab – So
Eble, Cortland F., (859) 257-5500 (Ext. 149) eble@uky.edu, Univ of Kentucky – Pl
Echelmeyer, Keithkechel@gi.alaska.edu, Univ of Alaska, Fairbanks – Gl
Echols, John B., (225) 388-8573 john@lgs.bri.lsu.edu, Louisiana State Univ – Go
Eckert, Donald J., (614) 292-9048 eckert.1@osu.edu, Ohio State Univ – Sc
Eckert, James O., (203) 432-3181 xwn6w3i02@sneakemail.com, Yale Univ – Gp
Eckstein, Yoram, (330) 672-2364 yeckste1@kent.edu, Kent State Univ – Hw
Econ, George D., (517) 796-8526 Jackson Community College – Ge
Eddy-Dilek, Carol A., (513) 529-3218 Miami Univ – Hw
Edgar, Dorland E., (630) 252-7596 Argonne National Lab – Gm
Edgcomb, Virginia, 508-289-3734 vedgcomb@whoi.edu, Woods Hole Oceanographic Institution – Cg
Edgell, Dennis J., (910) 521-6479 dennis.edgell@uncp.edu, Univ of North Carolina, Pembroke – Oa
Edil, Tuncer B., (608) 262-3225 edil@engr.wisc.edu, Univ of Wisconsin, Madison – Ng
Edinger, Paul F., (843) 383-8177 Coker College – Oe
Edlund, Mark B.mbedlund@smm.org, Univ of Minnesota, Twin Cities – Pe
Edmonds, Henrietta N., (361) 749-6772 hnedmonds@mail.utexas.edu, Univ of Texas at Austin – Ob
Edwards, Barbara, (610) 526-5115 bedwards@brynmawr.edu, Bryn Mawr College – Oo
Edwards, Benjamin R., 717-254-8934 edwardsb@dickinson.edu, Dickinson College – Gi
Edwards, C L, (505) 667-8464 cledwards@lanl.gov, Los Alamos National Lab – Yg
Edwards, Cathy, (513) 529-3216 edwardca@muohio.edu, Miami Univ – Oo
Edwards, Dianneedwardsd2@cf.ac.uk, Cardiff Univ – Pb
Edwards, Katrina, 213-821-4390 kje@usc.edu, Univ of Southern California – Py
Edwards, Margo H., (808) 956-5232 margo@soest.hawaii.edu, Univ of Hawai'i, Manoa – Gu
Edwards, Mary E., (907) 474-5014 ffmee@aurora.alaska.edu, Univ of Alaska, Fairbanks – Pe
Edwards, R. Lawrence, 612-626-0207 edwar001@umn.edu, Univ of Minnesota, Twin Cities – Cc
Edwards, Richard N., (416) 978-2267 edwards@core.physics.utoronto.ca, Univ of Toronto – Yr
Efurd, Deward W., (505) 667-2437 Los Alamos National Lab – Ct
Egan, Stuart S., +44 01782 583174 s.s.egan@esci.keele.ac.uk, Keele Univ – Go
Egbert, Gary D., (541) 737-2947 egbert@coas.oregonstate.edu, Oregon State Univ – Yrm
Egenhoff, Sven O., (970) 491-5661 Colorado State Univ – Gs
Eggers, Albert A., (253) 879-3130 eggers@ups.edu, Univ of Puget Sound – Gv
Eggers, Delores M., (828) 251-6654 Univ of North Carolina, Asheville – Oo
Eggler, David H., (814) 863-0014 eggler@geosc.psu.edu, Pennsylvania State Univ, Univ Park – Cp
Eggleston, Carrick M., (307) 766-6769 carrick@uwyo.edu, Univ of Wyoming – Cl
Eggleston, David B., (919) 515-7840 eggleston@ncsu.edu, North Carolina State Univ – Ob
Eglinton, Timothy I., (508) 289-2328 teglinton@whoi.edu, Woods Hole Oceanographic Institution – Co
Ehlen, Judyjudyehlen@hotmail.com, Radford Univ – Gm
Ehler, Stanley W., (913) 532-7205 Kansas State Univ – Sd
Ehlers, Todd A., (734) 763-5112 tehlers@umich.edu, Univ of Michigan – Yg
Ehlmann, Arthur J., (817) 257-6278 a.ehlmann@tcu.edu, Texas Christian Univ – GzXm
Ehlschlaeger, Charles R., (212) 772-5321 Graduate School of the City Univ of New York – Oo
Eichenbaum, Jack, 212-772-5265 jaconet@aol.com, Hunter College (CUNY) – Oy
Eicher, Don L., (303) 492-6187 Univ of Colorado – Pm
Eick, Matthew J., (540) 231-8943 eick@vt.edu, Virginia Polytechnic Institute & State Univ – Sc
Eicken, Hajo, (907) 474-7280 hajo.eicken@gi.alaska.edu, Univ of Alaska, Fairbanks – Og
Eiler, John M., (626) 395-6239 eiler@gps.caltech.edu, California Institute of Technology – Cg
Einaudi, Marco T., (650) 723-0575 marco@pangea.stanford.edu, Stanford Univ – Em
Eisele, Fred L., (303) 497-1483 eisele@ncar.ucar.edu, Georgia Institute of Technology – Oa
Eisenberger, Peter M., (917) 568-6915 petere@ldeo.columbia.edu, Columbia Univ – Oo
Eisenhart, Karen, (814) 732-2398 keisenhart@edinboro.edu, Edinboro Univ of Pennsylvania – Oy
Eisenhart, Ralph, (717) 846-7788 York College of Pennsylvania – Gg
Eisenstadt, Gloria, (817) 272-2987 gloria_eisenstadt@yahoo.com, Univ of Texas, Arlington – Gc
Eivazi, Frieda, 573-681-5459 Univ of Missouri, Columbia – Sc
Ekdale, Allan A., (801) 581-7266 ekdale@mines.utah.edu, Univ of Utah – Pe
Ekeberg, Barb, (651) 696-6105 Macalester College – Oo
Ekman, Kate, (773) 442-6050 K-Ekman@neiu.edu, Northeastern Illinois Univ – Oo
Ekmekci, Mehmet, +90 312 2977730 ekmekci@hacettepe.edu.tr, Hacettepe Univ – Hg
Ekstrom, Carol L., (901) 843-3089 cekstrom@rhodes.edu, Rhodes College – Gg
Ekstrom, Goran A., (617) 496-8276 ekstrom@eps.harvard.edu, Harvard Univ – Ys
El Shazly, Aley, (304) 696-6756 elshazly@marshall.edu, Marshall Univ – Cg
El Tabakah, Mohamadeltabakh@hotmail.com, Queens College (CUNY) – Cl
El-Baz, Farouk, (617) 353-5081 farouk@bu.edu, Boston Univ – Or
El-Kadi, Aly I., (808) 956-6331 elkadi@hawaii.edu, Univ of Hawai'i, Manoa – Hw
El-Shazly, Aley K., (731) 587-7437 aelshazly@utm.edu, Univ of Tennessee, Martin – Gp
Elbert, David C., 410-516-5049 elbert@jhu.edu, The Johns Hopkins Univ – Gz
Elbrond, Jorgen, (514) 340-4923 Ecole Polytechnique – Nm
Elder, Kathryn L., (508) 289-2513 kelder@whoi.edu, Woods Hole Oceanographic Institution – Cs
Elders, Wilfred A.wilfred.elders@ucr.edu,

Univ of California, Riverside – Gi
Eldredge, Niles, (212) 769-5723 American Museum of Natural History – Pi
Eldredge, Niles, (212) 769-5723 Graduate School of the City Univ of New York – Pg
Eldredge, Sandra N., (801) 537-3328 nrugs.selredg@state.ut.us, Utah Geological Survey – Oe
Elias, Robert J., (204) 474-8862 eliasrj@ms.umanitoba.ca, Univ of Manitoba – Pi
Elick, Jennifer M., 570-372-4214 elick@susqu.edu, Susquehanna Univ – Pe
Eller, Phillip G., (505) 667-5830 Los Alamos National Lab – Oo
Ellingson, Robert E., (850) 644-8583 bobe@met.fsu.edu, Florida State Univ – Ow
Ellins, Katherine K., (512) 232-3251 kellins@ig.utexas.edu, Univ of Texas at Austin, Jackson School of Geosciences – Oo
Elliot, David H., (614) 292-5076 elliot.1@osu.edu, Ohio State Univ – Gi
Elliot, James L., (617) 253-6308 jle@mit.edu, Massachusetts Institute of Technology – Oa
Elliott, Brent A., 940-397-4250 brent.elliott@mwsu.edu, Midwestern State Univ – Gz
Elliott, C. James, (505) 667-9517 cje@lanl.gov, Los Alamos National Lab – Oo
Elliott, Colleen G., (406) 496-4763 celliott@mtech.edu, Montana Tech of the Univ of Montana – Gc
Elliott, David K., (928) 523-7188 david.elliott@nau.edu, Northern Arizona Univ – Pv
Elliott, Emily M., (412) 624-8780 eelliott@pitt.edu, Univ of Pittsburgh – Hg
Elliott, Monty A., (541) 552-6482 Southern Oregon Univ – Gr
Elliott, Scott M., (505) 667-0949 sme@lanl.gov, Los Alamos National Lab – Oa
Elliott, Susan J., (905) 525-9140 (Ext. 23139) elliotts@mcmaster.ca, McMaster Univ – Yr
Elliott, W. Crawford, (404) 413-5756 wcelliott@gsu.edu, Georgia State Univ – Cl
Elliott, William S., 541-552-8185 elliottw@sou.edu, Southern Oregon Univ – Gs
Ellis, Andre S., (915) 747-5503 ellis@geo.utep.edu, Univ of Texas, El Paso – Hg
Ellis, Hugh, (901) 678-2007 hugh.ellis@jhu.edu, The Johns Hopkins Univ – Oo
Ellis, Hugh I.ellis@acusd.edu, Univ of San Diego – Ob
Ellis, Michael A., (901) 678-4980 ellis@ceri.memphis.edu, Univ of Memphis – Gt
Ellis, Robert M., (604) 822-6574 rellis@eos.ubc.ca, Univ of British Columbia – Ys
Ellis, Scott R., 217-265-5105 ellis@isgs.uiuc.edu, Illinois State Geological Survey – Gg
Ellsworth, Timothy R., (217) 333-2055 Univ of Illinois, Urbana-Champaign – Sp
Ellsworth, William, (650) 329-4784 Stanford Univ – Yg
Ellwood, Brooks B., (225) 578-3416 ellwood@lsu.edu, Louisiana State Univ – Ym
Elmore, A C., 573-341-6784 elmorea@umr.edu, Univ of Missouri, Rolla – Ge
Elmore, R. Douglas, (405) 325-4493 delmore@ou.edu, Univ of Oklahoma – Ym
Eloranta, Edwin W., (608)262-7327 eloranta@lidar.ssec.wisc.edu, Univ of Wisconsin, Madison – Yr
Elrick, David E., (519) 824-4120 delrick@lrs.uoguelph.ca, Univ of Guelph – Sp
Elrick, Maya, (505) 277-5077 dolomite@unm.edu, Univ of New Mexico – Gs
Elrick, Scott D., 217-333-3222 elrick@isgs.uiuc.edu, Illinois State Geological Survey – Ec
Elston, Wolfgang E., (505) 277-5339 Univ of New Mexico – Eg
Elswick, Erika R., (812) 855-2493 eelswick@indiana.edu, Indiana Univ, Bloomington – Cl
Ely, John T.ely@u.washington.edu, Univ of Washington – XyYx
Ely, Lisa L., (509) 963-2177 ely@cwu.edu, Central Washington Univ – Gm
Elzerman, Alan W., (864) 656-5568 awlzrmn@clemson.edu, Clemson Univ – Oo
Emanuel, Kerry A., (617) 253-2462 emanuel@texmex.mit.edu, Massachusetts Institute of Technology – Ow
Emanuel, William, (804) 924-3762 emanuel@virginia.edu, Univ of Virginia – Cg
Embree, Glenn F., (208) 496-1905 geology@byui.edu, Brigham Young Univ, Idaho – Gx
Emeleus, Henry, 0191 3742502 c.h.emeleus@durham.ac.uk, Durham Univ – Oo
Emerick, Christina M., (206) 543-2491 tina@ocean.washington.edu, Univ of Washington – Ou
Emerson, Steven R., (206) 543-0428 emerson@u.washington.edu, Univ of Washington – Oc
Emmer, Barbara, 089/2180 4231 LMU - Univ of Munich – Yg
Emmett, Chad, (801) 422-7886 chad_emmett@byu.edu, Brigham Young Univ – Oo
Emmi, Philip C., (801) 581-5562 pcemmi@geog.utah.edu, Univ of Utah – Ou
Emry, Robert, (202) 633-1323 Smithsonian Institution / National Museum of Natural History – Pv
Enberg, Dennis P., (919) 560-5172 North Carolina Central Univ – Oo
Encarnacion, John, (314) 977-3119 jpe@slu.edu, Saint Louis Univ – Gx
Endres, Anthony E., (519) 888-4567 (Ext. 3552) Univ of Waterloo – Pe
Engebretson, David C., (360) 650-3595 engebret@cc.wwu.edu, Western Washington Univ – Gt
Engel, Annette S., (225) 578-2469 aengel@geol.lsu.edu, Louisiana State Univ – Py
Engel, Manfred, (906) 635-2224 mengel@lssu.edu, Lake Superior State Univ – Eg
Engel, Michael H., (405) 325-4435 ab1635@ou.edu, Univ of Oklahoma – Co
Engel, Richard E., (406) 994-5295 engel@gemini.oscs.montana.edu, Montana State Univ – Sc
Engelder, Terry, (814) 865-3620 engelder@geosc.psu.edu, Pennsylvania State Univ, Univ Park – Nr
Engelhardt, Hermann, (626) 395-3720 hermann@gps.caltech.edu, California Institute of Technology – Gl
Engelmann, George F., 402-554-4804 gengelmann@mail.unomaha.edu, Univ of Nebraska at Omaha – Pv
England, Anthony W., (313) 763-5534 england@eecs.umich.edu, Univ of Michigan – Oo
England, John, (780) 492-5673 john.england@ualberta.ca, Univ of Alberta – Gm
Englebright, Stephen C., (631) 632-8230 SUNY, Stony Brook – Oe
Englert, Andrea C.aenglert@autometric.com, George Mason Univ – Gg
English, David C., (727) 553-1503 denglish@marine.usf.edu, Univ of South Florida, Saint Petersburg – Op
Engstrom, Daniel R., (612) 433-5953 (Ext. 18) dre@umn.edu, Univ of Minnesota, Twin Cities – Pe
Ennis, Margaret V., (812) 855-2687 mennis@indiana.edu, Indiana Geological Survey – Cg
Enomoto, Cathy B., 434 951 6362 cathy.enomoto@dmme.virginia.gov, Division of Mineral Resources – Eo
Enos, Paul, (785) 864-2744 enos@KU.edu, Univ of Kansas – Gs
Enright, James T., (858) 534-3784 jenright@ucsd.edu, Univ of California, San Diego – Ob
Enright, Richard L., (508) 531-1390 enright@bridgew.edu, Bridgewater State College – Hg
Entekhabi, Dara, (617) 253-9698 darae@mit.edu, Massachusetts Institute of Technology – Hg
Enz, John W., (701) 231-8576 North Dakota State Univ – Oa
Epifanio, Craigcepi@tamu.edu, Texas A&M Univ – Oa
Eppes, Martha C., 704-687-5993 meppes@uncc.edu, Univ of North Carolina, Charlotte – Os
Epstein, Claude M., (609) 652-4611 Richard Stockton Coll of New Jersey – Hw
Erb, Edward E., 434-951-6350 ed.erb@dmme.virginia.gov, Division of Mineral Resources – GvEg
Erdmer, Philippe, (403) 492-2676 philippe.erdmer@ualberta.ca, Univ of Alberta – Gc
Erdner, Deana L., (361) 749-6719 derdner@mail.utexas.edu, Univ of Texas at Austin – Ob
Ericksen, Rick L., (601) 961-5507 rick_ericksen@deq.state.ms.us, Mississippi Office of Geology – Go
Erickson, J. Mark, (315) 379-5198 St. Lawrence Univ – Po
Erickson, Rolfe C., (707) 664-2296 rolfe.erickson@sonoma.edu, Sonoma State Univ – Gx
Eriksen, Charles C., (206) 543-6528 eriksen@u.washington.edu, Univ of Washington – Op
Eriksson, Kenneth A., (540) 231-4680 kaeson@vt.edu,

Virginia Polytechnic Institute & State Univ – Gs
Eriksson, Nancy, (503) 725-3022 erikssonn@pdx.edu, Portland State Univ – Oo
Ernissee, John J.ernissee@clarion.edu, Clarion Univ – Pi
Ernst, W. Gary, (650) 723-2750 ernst@pangea.stanford.edu, Stanford Univ – Cp
Erofeev, Anatoli, (541) 737-4656 aerofeev@coas.oregonstate.edu, Oregon State Univ – Op
Erofeeva, Svetlana, (541) 737-3964 serofeev@coas.oregonstate.edu, Oregon State Univ – Yg
Erslev, Eric A., (970) 491-6375 erslev@cnr.colostate.edu, Colorado State Univ – Gc
Ertel-Ingrisch, Werner, 089/2180 4275 ertel@min.uni-muenchen.de, LMU - Univ of Munich – Gz
Ervin, C. Patrick, (815) 753-1942 pervin@niu.edu, Northern Illinois Univ – Yv
Erwin, Diane, (510) 642-3921 Univ of California, Berkeley – Pb
Erwin, Douglas H., (202) 633-1324 Smithsonian Institution / National Museum of Natural History – Pi
Erwin, John W., (775) 784-6691 Univ of Nevada – Yg
Erwin, R. Michael, (804) 924-3207 rme5g@virginia.edu, Univ of Virginia – Oo
Esbensen, Steven K., (541) 737-5687 esbensen@coas.oregonstate.edu, Oregon State Univ – Oa
Escutia, Carlota, (510) 885-3486 escutia@odp.tamu.edu, California State Univ, East Bay – Ou
Esenberg, Arthur C., (804) 427-7189 Tidewater Comm College – Og
Eshleman, Keith N., (301) 689-3155 eshleman@al.umces.edu, Univ of Maryland – Hg
Esling, Steven P., (618) 453-7363 esling@geo.siu.edu, Southern Illinois Univ Carbondale – Hw
Eslinger, Eric, 518-337-4922 eslingee@strose.edu, The College of Saint Rose – EoGzHy
Esparza, Franciscofesparz@cicese.mx, Centro de Inv Científica y de Educación Superior de Ensenada – Ye
Espinosa, Juan M.jespinos@cicese.mx, Centro de Inv Científica y de Educación Superior de Ensenada – Ye
Essene, Eric J., (734) 936-1555 essene@umich.edu, Univ of Michigan – Gp
Essling, Alice M., (630) 252-3493 Argonne National Lab – Ca
Etherton, Brian J., 704-687-5984 betherto@uncc.edu, Univ of North Carolina, Charlotte – Ow
Ethington, Raymond L., (573) 882-6470 EthingtonR@missouri.edu, Univ of Missouri – Pm
Ethridge, Frank G., (970) 491-6195 fredpet@cnr.colostate.edu, Colorado State Univ – Gs
Ettensohn, Dorothy, (213) 763-3327 dettenso@nhm.org, Los Angeles County Museum of Natural History – Gz
Ettensohn, Frank R., (859) 257-1401 fettensohn@uky.edu, Univ of Kentucky – Ps
Eusden, Jr., J. Dykstra, (207) 786-6152 deusden@bates.edu, Bates College – Gc
Evangelou, V. P., (515) 294-9237 Iowa State Univ of Science & Technology – Sc
Evanoff, Emmett, 970-351-2647 Emmett.Evanoff@unco.edu, Univ of Northern Colorado – Grs
Evans, Barry M., (814) 863-3531 bmel@psu.edu, Pennsylvania State Univ, Univ Park – Sd
Evans, Bernard W., 206-543-1163 bwevans@u.washington.edu, Univ of Washington – Gp
Evans, Brian J., +61 8 9266-7092 B.Evans@curtin.edu.au, Curtin Univ of Technology – Ye
Evans, David G., (916) 278-4331 dave_evans@csus.edu, California State Univ, Sacramento – Hw
Evans, Diane L., (818) 354-2418 Jet Propulsion Lab – Gm
Evans, Ian, (713) 743-3428 yddraig@uh.edu, Univ of Houston – Gr
Evans, J. B., (617) 253-2856 brievans@mit.edu, Massachusetts Institute of Technology – Yx
Evans, James E., (419) 372-2414 evansje@bgnet.bgsu.edu, Bowling Green State Univ – Gs
Evans, James P., (435) 797-1267 james.evans@usu.edu, Utah State Univ – Gc
Evans, Jenni L., (814) 865-3240 evans@meteo.psu.edu, The Pennsylvania State Univ, Univ Park – Ow
Evans, John C., (509) 376-0934 john.evans@pnl.gov, Pacific Northwest National Lab – Og
Evans, Les J., (519) 824-4120 (Ext. 53017) levans@lrs.uoguelph.ca, Univ of Guelph – Cl
Evans, Michael E., (780) 492-5517 evans@phys.ualberta.ca, Univ of Alberta – Ym
Evans, Owen C., (631) 632-8061 owen.evans@sunysb.edu, SUNY, Stony Brook – Cg
Evans, Robert L., (508) 289-2673 revans@whoi.edu, Woods Hole Oceanographic Institution – Yr
Evans, Thomas J., (608) 263-4125 tevans@wisc.edu, Univ of Wisconsin, Extension – Gg
Evans, Thomas J., (608) 263-4125 tevans@wisc.edu, Univ of Wisconsin - Extension – Eg
Evanylo, Gregory K., (540) 231-9739 gevanylo@vt.edu, Virginia Polytechnic Institute & State Univ – Sc
Eveleth, Robert W., (505) 835-5325 beveleth@gis.nmt.edu, New Mexico Institute of Mining & Technology – Nm
Evensen, Norman M., (416) 978-3159 evensen@physics.utoronto.ca, Univ of Toronto – Cc
Evenson, Edward B., (610) 758-3659 ebe0@lehigh.edu, Lehigh Univ – Gl
Everett, Lorne G., (805) 893-2887 Univ of California, Santa Barbara – Hw
Everett, Mark, (979) 862-2129 everett@geo.tamu.edu, Texas A&M Univ – Ym
Eversoll, Duane A., (402) 472-7524 deversoll2@unl.edu, University of Nebraska - Lincoln – Ng
Eves, Robert L., (435)586-1934 eves@suu.edu, Southern Utah Univ – GrOeCl
Ewart, Terry E., (206) 543-1327 ewart@apl.washington.edu, Univ of Washington – Op
Ewers, Ralph O., (859) 622-1278 Ralph.Ewers@eku.edu, Eastern Kentucky Univ – Hw
Ewing, Rodney C., 734763-9295 rodewing@umich.edu, Univ of Michigan – Gz
Ewing, Rodney E., (734) 763-9295 Univ of New Mexico – Gz
Exum, Kenith G., (904) 484-1155 Pensacola Junior College – Gg
Eylander, John, (402) 280-2641 john.eylander@afwa.af.mil, Creighton Univ – Ow
Eyles, Carolyn H., (905) 525-9140 (Ext. 23524) eylesc@mcmail.cis.mcmaster.ca, McMaster Univ – Gs
Eyles, John D., (905) 525-9140 (Ext. 23152) eyles@mcmaster.ca, McMaster Univ – Yr
Eyles, Nicholas, 416-287-7195 eyles@utsc.utoronto.ca, Univ of Toronto – Gl
Eyrich, Terry L., (209) 384-6331 Merced College – Oo

F

Faatz, Renee M., 435 283-7519 renee.faatz@snow.edu, Snow College – Gg
Fabry, Frederic, (514) 398-7733 frederic.fabry@staff.mcgill.ca, McGill Univ – Oa
Fabryka-Martin, June T., (505) 665-2300 Los Alamos National Lab – Hg
Fagel, Nathalie, +32 4 3662209 Nathalie.Fagel@ulg.ac.be, Universite de Liege – GsuGe
Fagents, Sarah, 808-956-3163 gagents@higp.hawaii.edu, Univ of Hawai'i, Manoa – Gv
Fagerlin, Stanley C., (417) 836-5977 scf621f@smsu.edu, Southwest Missouri State Univ – Pm
Fagherazzi, Sergio, 617-353-2092 sergio@bu.edu, Boston Univ – Gm
Fagnan, Nathalie, (418) 654-3713 nfagnan@nrcan.gc.ca, Universite du Quebec – Hw
Fahey, Eileen, (802) 443-5970 efahey@middlebury.edu, Middlebury College – Oo
Fahey, John V., 609-258-4123 jvf@princeton.edu, Princeton Univ – Gg
Fahrenbach, Mark D., (605) 394-2229 mark.fahrenbach@state.sd.us, South Dakota Dept of Env & Natural Resources – Gg
Fahy, Danfahyd@nova.edu, Nova Southeastern Univ – Oo
Faill, Rodger T., 717.702.2041 rfaill@state.pa.us, Pennsylvania Bureau of Topographic & Geologic Survey – Gc
Fairbanks, Richard, (845) 365-8499 Columbia Univ – Cg
Fairbanks, Richard G., (845) 365-8499 fairbanks@LDEO.columbia.edu, Columbia Univ – Pe
Fairbridge, Rhodes W., (516) 267-8329 Columbia Univ – Gm
Fairley, Jerry P., 208-885-9259 jfairley@uidaho.edu, Univ of Idaho – Hw
Fairley,Jr., Jesse A., (850) 488-9380 Florida Geological Survey – Oo
Falkner, Kelly K., (541) 737-3625 kfalkner@coas.oregonstate.edu, Oregon State Univ – Oc
Falkowski, Paul G., (732) 932-6555 (Ext. 370) falko@rci.rutgers.edu,

Rutgers, The State Univ of New Jersey – Ob
Falls, James N., (707) 725-4413 California Geological Survey – Oo
Falls, William W., (813) 253-7000 Univ of South Florida, Saint Petersburg – Ob
Falta, Ronald W., (864) 656-0125 ron@clemson.edu, Clemson Univ – Hq
Fan, Pow-Foong, (808) 956-7748 pffan@soest.hawaii.edu, Univ of Hawai'i, Manoa – Gs
Fan, Weihongfanw@pollux.stockton.edu, Richard Stockton College of New Jersey – Oi
Fang, Jiasong, (515) 294-6583 jsfang@iastate.edu, Iowa State Univ of Science & Technology – Ge
Fantle, Matthew S., 814-863-9968 mfantle@geosc.psu.edu, Pennsylvania State Univ, Univ Park – Cs
Fares, Mary A., (229) 333-5755 mfares@valdosta.edu, Valdosta State Univ – Oo
Fariduddin, Mohammad, 773/442-6059 M-Fariduddin@neiu.edu, Northeastern Illinois Univ – Pe
Farley, Kenneth A., (626) 395-6005 farley@gps.caltech.edu, California Institute of Technology – Cg
Farley, Martin B., 910-521-6478 martin.farley@uncp.edu, Univ of North Carolina, Pembroke – PlOe
Farley, Perry D., (505) 667-2415 dfarley@lanl.gov, Los Alamos National Lab – Oo
Farlow, James O., 260-481-6251 farlow@ipfw.edu, Indiana Univ / Purdue Univ, Fort Wayne – PvoPe
Farmer, David M., (250) 363-6591 dmf@ios.bc.ca, Univ of Victoria – Op
Farmer, Emma C., 516 463-5568 geoecf@hofstra.edu, Hofstra Univ – Ou
Farmer, G. Lang, 303-492-6534 farmer@cires.colorado.edu, Univ of Colorado – Cg
Farmer, George T., (505) 665-0225 gfarmer@lanl.gov, Los Alamos National Lab – Hw
Farmer, Jack D., (480) 965-6748 jfarmer@asu.edu, Arizona State Univ – Py
Farnsworth, Roy L., (207) 783-9936 Bates College – Gl
Farquhar, James, (301) 405-1434 jfarquha@essic.umd.edu, Univ of Maryland – Cs
Farquhar, Ronald M., (416) 978-5175 Univ of Toronto – Cc
Farr, Tom G., (818) 354-9057 tom.farr@jpl.nasa.gov, Jet Propulsion Lab – Gm
Farrar, Edward, (613) 545-6501 Queen's Univ – Cc
Farrar, Stewart S., (859) 622-1279 Stewart.Farrar@eku.edu, Eastern Kentucky Univ – Gc
Farrell, Brian F., (617) 495-2998 Harvard Univ – Ow
Farrell, John, (401) 874-6561 jfarrell@gso.uri.edu, Univ of Rhode Island – Ou
Farrell, Kathleen M., (919) 733-2423 North Carolina Geological Survey – Gs
Farrell, Mark O., (412) 392-3879 Point Park College – Oo
Farrell, Stewart, (609) 652-4245 farrells@stockton.edu, Richard Stockton College of New Jersey – On
Farrell, Stewart C., (609) 652-4245 Stockton State College – On
Farrington, John W., 508-289-3911 jfarrington@whoi.edu, Woods Hole Oceanographic Institution – Oc
Farrow, Norman D.ndf@ornl.gov, Oak Ridge National Lab – Hg
Farthing, Dori J., 585-245-5298 farthing@geneseo.edu, SUNY, Geneseo – Gz
Farver, John R., (419) 372-7203 jfarver@bgnet.bgsu.edu, Bowling Green State Univ – Gy
Faryad, Shah Wali, +420221951521 faryad@natur.cuni.cz, Charles Univ – GpzGi
Fastovsky, David E., (401)874-2185 defastov@uri.edu, Univ of Rhode Island – Pg
Faul, Ulrich, 617-358-4341 ufaul@bu.edu, Boston Univ – Gc
Faulds, James, (775) 682-8751 jfaulds@unr.edu, Univ of Nevada, Reno – Gc
Faulds, James E., (775) 682-8751 jfaulds@unr.edu, Univ of Nevada, Reno – Gc
Faulds, James E., (775) 784-6691 (Ext. 159) jfaulds@unr.edu, Univ of Nevada – Gc
Faulkner, D. John, (858) 534-4259 Univ of California, San Diego – Oc
Faulkner, Stephen P., (225) 388-8792 Louisiana State Univ – Cg
Faure, Gunter, (614) 292-3454 faure.1@osu.edu, Ohio State Univ – Cc
Favor, Michael, (313) 593-5235 Univ of Michigan, Dearborn – Oo
Fawcett, J. J., (416) 978-3027 fawcett@quartz.geology.utoronto.ca, Univ of Toronto – Gp
Fawcett, Peter J., (505) 277-3867 fawcett@unm.edu, Univ of New Mexico – Pe
Fay, Robert O., (405) 325-3031 Univ of Oklahoma – Gg
Fayek, Mostafa, (204) 474-7982 fayek@cc.umanitoba.ca, Univ of Manitoba – Cg
Fayek, Mostafa, (865) 576-4839 fayekm@ornl.gov, Oak Ridge National Lab – Cg
Fayek, Mostafamfayek@utk.edu, Univ of Tennessee, Knoxville – Cg
Fayer, Michael J., (509) 372-6045 mike.fayer@pnl.gov, Pacific Northwest National Lab – Hq
Fayon, Annia K., 612-626-9805 fayon001@umn.edu, Univ of Minnesota, Twin Cities – Gc
Feagley, Sam E., (979) 845-1460 s-feagley@tamu.edu, Texas A&M Univ – Sc
Fearey, Bryan L., (505) 665-2423 Los Alamos National Lab – Cc
Fearn, Miriam L., (251) 460-6381 Univ of South Alabama – Pe
Feather, Russell, (202) 633-1793 featherr@si.edu, Smithsonian Institution / National Museum of Natural History – Gz
Fedorko, Nick, 304 594 2331 fedorko@geosrv.wvnet.edu, West Virginia Univ – Ec
Feeley, Todd C., (406) 994-6917 tfeely@montana.edu, Montana State Univ – Gi
Feely, Richard A., (206) 526-6214 feely@pmel.noaa.gov, Univ of Washington – Oc
Feeney, Alison E., (717) 477-1319 aefeen@ark.ship.edu, Shippensburg Univ – Oo
Feeney, Thomas P., (717) 477-1297 tpfeen@ark.ship.edu, Shippensburg Univ – Gm
Fegel, Larry, 616-331-2760 fegell@gvsu.edu, Grand Valley State Univ – Gg
Fegley, M. Bruce, (314) 935-4852 bfegley@levee.wustl.edu, Washington Univ – Xc
Fehler, Michael C., (505) 662-6561 fehler@lanl.gov, Los Alamos National Lab – Ys
Fehn, Udo, (585) 275-7884 fehn@earth.rochester.edu, Univ of Rochester – Eg
Fehr, Karl Thomas, 089/2180 4256 fehr@min.uni-muenchen.de, LMU - Univ of Munich – Gz
Fei, Yingwei, (202) 478-8936 y.fel@gl.ciw.edu, Carnegie Institution of Washington – Cp
Feibel, Craig S., (732) 932-8853 feibel@rci.rutgers.edu, Rutgers, The State Univ of New Jersey – Ga
Feigenson, Mark D., (732) 445-3149 feigy@rci.rutgers.edu, Rutgers, The State Univ of New Jersey – Cg
Feigl, Kurt, 608-262-0176 feigl@wisc.edu, Univ of Wisconsin, Madison – Yd
Fein, Jeremy B., (574) 631-6101 fein.1@nd.edu, Univ of Notre Dame – Cl
Feinemann, Maureen D., 814) 863-6649 mdf12@psu.edu, Pennsylvania State Univ, Univ Park – Cp
Feinglos, Mark N., (919) 684-4005 feing002@mc.duke.edu, Duke Univ – Gz
Feingold, Joshua, (954) 262-8307 joshua@nova.edu, Nova Southeastern Univ – Ob
Felbeck, Horst, (858) 534-6647 hfelbeck@ucsd.edu, Univ of California, San Diego – Ob
Feldmann, Rodney M., (330) 672-2506 rfeldman@kent.edu, Kent State Univ – Pi
Felipe, Mihalifelipem1@southernct.edu, Southern Connecticut State Univ – Yg
Feller, Robert J., (803) 777-3937 Univ of South Carolina – Ob
Fellows, Larry D., (520) 770-3500 Fellows_Larry@pop.state.az.us, Arizona Geological Survey – Gg
Felmy, Diana, (509) 373-7775 diana.felmy@pnl.gov, Pacific Northwest National Lab – Oo
Felton, Richard M., (660) 562-1569 rfelton@nwmissouri.edu, Northwest Missouri State Univ – Pg
Fendorf, Scott E., (650) 723-5238 fendorf@pangea.stanford.edu, Stanford Univ – Sc
Feng, Huan E., (973) 655-7549 fengh@mail.montclair.edu, Montclair State Univ – Cm
Feng, Xiahong, 603-646-1712 xiahong.feng@dartmouth.edu, Dartmouth College – Cs
Feng, Yang, (617) 287-5287 Univ of Massachusetts, Boston/Harbor Campus – Or
Feng, Yucheng, (334) 844-3967 yfeng@acesag.auburn.edu, Auburn Univ – Sb
Feng, Zhaodong, (973) 655-7811 fengz@mail.montclair.edu,

Montclair State Univ – Oi

Feng, Zhaodong J., (254) 710-2194 Zhaodong_Feng@baylor.edu, Baylor Univ – GmOy

Fenical, William H., (858) 534-2133 wfenical@ucsd.edu, Univ of California, San Diego – Oc

Fennel, Katja, (732) 932-9709 kfennel@marine.rutgers.edu, Dalhousie Univ – Ob

Fenner, Pamela G., (605) 394-2641 pamela.fenner@sdsmt.edu, South Dakota School of Mines & Technology – Oo

Fenrich, Francis, (780) 492-2149 frances@space.ualberta.ca, Univ of Alberta – Xy

Fenster, Michael S., (804)752-3745 mfenster@rmc.edu, Randolph-Macon College – On

Fenstermaker, Lynn, (702) 862-5412 lynn@dri.edu, Desert Research Institute – Or

Fenter, Paul A., (630) 252-7053 fenter@anl.gov, Argonne National Lab – Gy

Ferger, Marisa, (814) 863-4229 mferger@psu.edu, The Pennsylvania State Univ, Univ Park – Ow

Ferguson, Ian J., (204) 474-9154 ij_ferguson@umanitoba.ca, Univ of Manitoba – Ye

Ferguson, John F., (972) 883-2410 ferguson@utdallas.edu, Univ of Texas, Dallas – Yg

Ferguson, Robert J., (512) 471-6405 fergusonr@mail.utexas.edu, Univ of Texas at Austin, Jackson School of Geosciences – Ye

Ferguson, Terry A., (864) 597-4527 fergusonta@wofford.edu, Wofford College – Ga

Ferland, Marie A., 509-963-2829 ferlandm@cwu.edu, Central Washington Univ – Gu

Fermanich, Kevin J., (920) 465-2240 fermanik@uwgb.edu, Univ of Wisconsin, Green Bay – Sp

Fernandez, Erik J., (804) 924-1351 ejf3c@virginia.edu, Univ of Virginia – Oo

Fernandez, George C. J., (775) 784-4206 gcjf@scs.unr.edu, Univ of Nevada, Reno – Hg

Fernandez, Louis A., (909) 537-5024 California State Univ, San Bernardino – Gi

Fernandez-Mosquera, Daniel, 34981167000 xemos@udc.es, Coruna Univ – Xc

Ferns, Mark, 541-523-3133 mark.ferns@dogami.state.or.us, Eastern Oregon Univ – GvcEg

Ferns, Mark L., (541) 523-3133 mark.l.ferns@state.or.us, Oregon Dept of Geology & Mineral Industries – Oo

Ferrand, Lin A., (212) 650-8017 Graduate School of the City Univ of New York – Hq

Ferrari, Raffaele, (617) 253-1291 raffaele@mit.edu, Massachusetts Institute of Technology – Op

Ferre, Eric C., 618-453-7368 eferre@geo.siu.edu, Southern Illinois Univ Carbondale – Gc

Ferre, Paul A., (520) 621-2952 ty@hwr.arizona.edu, Univ of Arizona – Hw

Ferre, Paul (Ty), (520) 621-6082 ty@hwr.arizona.edu, Univ of Arizona – So

Ferreira, Karen, (204) 474-9374 ferreira@ms.umanitoba.ca, Univ of Manitoba – Gg

Ferreira, Maryanne F., (508) 289-2266 mferreira@whoi.edu, Woods Hole Oceanographic Institution – Oo

Ferrell, Jr., Ray E., (225) 388-5306 rferrell@lsu.edu, Louisiana State Univ – Cl

Ferrill, Martha J., (906) 227-1156 jferrill@nmu.edu, Northern Michigan Univ – Oo

Ferring, C. Reid, (940) 565-2993 ferring@unt.edu, Univ of North Texas – Ga

Ferris, Dawn, 419-755-3909 ferris.69@osu.edu, Ohio State Univ – Sf

Ferris, Grant F., 416-978-0526 ferris@quartz.geology.utoronto.ca, Univ of Toronto – Gn

Ferriz, Horacio, (209) 667-3874 hferriz@geology.csustan.edu, California State Univ, Stanislaus – Ng

Ferruzza, David, (717) 361-1348 Elizabethtown College – Oa

Ferry, John M., (410) 516-8121 jferry@jhu.edu, The Johns Hopkins Univ – Gp

Fetter, Jr., Charles W., (920) 424-4460 fetter@uwosh.edu, Univ of Wisconsin, Oshkosh – Hw

Few, Jr., Arthur A., (713) 527-4003 (Ext. 3601) few@rice.edu, Rice Univ – Oa

Fichter, Lynn S., (540) 568-6531 fichtels@jmu.edu, James Madison Univ – Gr

Fichtner, Andreas, 089/2180 4230 andreas.fichtner@geophysik.uni-muenchen.de, LMU - Univ of Munich – Yg

Fick, Walter H., (913) 532-7223 Kansas State Univ – Sf

Field, Cyrus W., (541) 574-6239 fieldc@geo.oregonstate.edu, Oregon State Univ – Em

Field, Stephen W., (254) 968-9887 field@tarleton.edu, Tarleton State Univ – Gz

Fielding, Christopher R., 402-472-9801 cfielding2@unl.edu, Univ of Nebraska, Lincoln – Gs

Fiesinger, Donald W., (435) 797-3072 don.fiesinger@usu.edu, Utah State Univ – Gi

Fifarek, Richard H., (618) 453-7364 fifarek@geo.siu.edu, Southern Illinois Univ Carbondale – Em

Figueroa, Otto, (714) 278-4370 ofigueroa@fullerton.edu, California State Univ, Fullerton – Hw

Filby, Sandra, (530) 752-2502 Univ of California, Davis – Oo

Filer, Jonathan K., (410) 704-2116 jfiler@towson.edu, Towson Univ – Gr

Filipescu, Sorin, (+40) -26-4-40 (Ext. 00 ext 5206) sorin@bioge.ub-bcluj.ro, Babes-Bolyai Univ – GrPm

Filipovic, Dragan, (605) 677-5227 dfilipovic@usd.edu, South Dakota Dept of Env & Natural Resources – Hw

Filippelli, Gabriel M., (317) 274-3795 gfilippe@iupui.edu, Indiana Univ / Purdue Univ, Indianapolis – Cm

Filley, Timothy R., (765) 494-6581 filley@purdue.edu, Purdue Univ – Co

Fillmore, Robert P., (970) 943-2650 rfillmore@western.edu, Western State College of Colorado – Gs

Filson, Robert H., (253) -839-4324 bfilson@greenriver.edu, Green River Community College – Oe

Finch, James A., (514) 398-4755 McGill Univ – Nx

Fingerhut, William A., (802) 626-6257 Lyndon State College – Oa

Fink, Jonathan H., (480) 965-3195 jon.fink@asu.edu, Arizona State Univ – Gv

Fink, Uwe, ((52) 0) -2736 uwefink@lpl.arizona.edu, Univ of Arizona – Oa

Fink, Jr., L. Kenneth, (207) 563-3146 Univ of Maine – Ou

Finkel, Robert C., (925) 422-2044 finkel1@llnl.gov, Lawrence Livermore National Lab – Cg

Finkelman, Robert B., (972) 883-2459 bobf@utdallas.edu, Univ of Texas, Dallas – Gb

Finks, Robertfinksr@union.edu, Union College – Pg

Finks, Robert M., (718) 997-3305 Graduate School of the City Univ of New York – Pi

Finley, Robert J., 217-244-8389 finley@isgs.uiuc.edu, Illinois State Geological Survey – Eo

Finley, Robert J., 217-244-8389 finley@isgs.uiuc.edu, Univ of Illinois, Urbana-Champaign – Go

Finn, Gregory C., (905) 688-5550 (Ext. 3528) Greg.Finn@brocku.ca, Brock Univ – Gx

Finnegan, David L., (505) 667-6548 Los Alamos National Lab – Oa

Finney, Bruce P., (907) 474-7724 Univ of Alaska, Fairbanks – Ou

Finney, Stanley C., (562) 985-8637 scfinney@csulb.edu, California State Univ, Long Beach – Ps

Finstick, Sue A., 435-586-7900 finstick@suu.edu, Southern Utah Univ – Hw

Fiorillo, Anthony R., (214) 421-3466 Southern Methodist Univ – Pv

Firing, Eric, (808) 956-7894 efiring@hawaii.edu, Univ of Hawai'i, Manoa – Op

Fischer, Alfred G., (213) 740-5821 fisher@usc.edu, Univ of Southern California – Gs

Fischer, Howard J., (850) 488-0474 Tallahassee Comm College – Gd

Fischer, Karen M., (401) 863-1360 karen_fischer@brown.edu, Brown Univ – Ys

Fischer, Kathleen M., 252-335-3425 kmfischer@mail.ecsu.edu, Elizabeth City State Univ – Cm

Fischer, Mark P., (815) 753-7939 mfischer@niu.edu, Northern Illinois Univ – Gc

Fischer, Peter J., (818) 677-3574 peter.fischer@csun.edu, California State Univ, Northridge – Gu

Fischer, Tobias, (505) 277-0284 fischer@unm.edu, Univ of New Mexico – Gv

Fischer, William A., (719) 632-4223 bfischer@coloradocollege.edu, Colorado College – Pg

Fish, Jennifer M., 831-459-1235 jfish@es.ucsc.edu, Univ of California, Santa Cruz – Oo

Fish, Lisa, (845) 365-8650 lisaf@ldeo.columbia.edu, Columbia Univ – Oo

Fisher, Andrew T., (831) 459-5598 afisher@es.ucsc.edu, Univ of California, Santa Cruz – Hw

Fisher, Berton, 918-631-3248 jbfisher@utulsa.edu,

Univ of Tulsa – Cg
Fisher, Cynthia G., (610) 436-2108 cfisher@wcupa.edu, West Chester Univ – Ou
Fisher, Daniel C., (734) 764-0488 dcfisher@umich.edu, Univ of Michigan – Pi
Fisher, David A., (613 996-7623 fisher@nrcan-rncan.gc.ca, Univ of Ottawa – Yg
Fisher, David E., (305) 284-3254 dfisher@miami.edu, Univ of Miami – Cs
Fisher, Donald M., (814) 865-3206 fisher@geosc.psu.edu, Pennsylvania State Univ, Univ Park – Gc
Fisher, George W., (410) 516-7237 gfisher@jhu.edu, The Johns Hopkins Univ – Gp
Fisher, Nicholas S., (631) 632-8649 nfisher@notes.cc.sunysb.edu, SUNY, Stony Brook – Oc
Fisher, Richard V., (805) 893-3946 Univ of California, Santa Barbara – Gr
Fisher, Stephen R., (859) 257-5500 (Ext. 158) sfisher@uky.edu, Univ of Kentucky – Cl
Fisher, Timothy G., (219) 980-7122 Indiana Univ Northwest – Gl
Fisher, Timothy G., (219) 980-7122 tfisher@indiana.edu, Indiana Univ / Purdue Univ, Indianapolis – Gl
Fisher, Timothy G., (419) 530-2883 timothy.fisher@utoledo.edu, Univ of Toledo – Gl
Fisher, William L., (512) 471-5600 wfisher@mail.utexas.edu, Univ of Texas at Austin, Jackson School of Geosciences – Gr
Fisher, Jr., Thomas R., (410) 221-8432 fisher@hpl.umces.edu, Univ of Maryland – Ob
Fisk, Lennard A., (313) 747-3660 Univ of Michigan – Oa
Fisk, Martin R., (541) 737-5208 mfisk@coas.oregonstate.edu, Oregon State Univ – Ou
Fiske, Richard S., (202) 633-1794 fisker@si.edu, Smithsonian Institution / National Museum of Natural History – Gv
Fitak, Madge R., 614-265-6585 madge.fitak@dnr.state.oh.us, Ohio Dept of Natural Resources – Gg
Fitch, Harold R., 517-241-1548 fitchh@michigan.gov, Michigan Dept of Environmental Quality – Oo
Fitts, Charles R., (207) 780-5351 cfitts@usm.maine.edu, Univ of Southern Maine – Hw
Fitz, Robert A., (501) 543-8879 Univ of Arkansas, Pine Bluff – Oo
Fitz, Thomas J., 715.682.1852 tfitz@northland.edu, Northland College – GgcEg
Fitzgerald, David, (210) 436-3235 Saint Mary's Univ – Gd
Fitzgerald, Duncan M., (617) 353-2530 dunc@bu.edu, Boston Univ – On
Fitzgerald, Paul G., 315-443-2672 pgfitzge@syr.edu, Syracuse Univ – Gt
Fitzgerald, William F., (860) 405-9158 william.fitzgerald@uconn.edu, Univ of Connecticut – Oc
Fitzjarrald, David R., (518) 437-8735 fitz@asrc.cestm.albany.edu, SUNY, Albany – Oa
Fitzpatrick, John R., (505) 667-4761 Los Alamos National Lab – Oo
Fitzpatrick, Patrick J., (228) 688-1157 fitz@gri.msstate.edu, Mississippi State Univ – Ow
Fitzsimmons, Kevin, (520) 626-3324 kevfitz@ag.arizona.edu, Univ of Arizona – Oo
Fixen, Paul E., (605) 692-6280 pfixen@ppi-far.org, South Dakota State Univ – Sc
Flaherty, Frank A., (229) 333-5665 flaherty@valdosta.edu, Valdosta State Univ – Oo
Flament, Pierre J., (808) 956-6663 Univ of Hawai'i, Manoa – Op
Flato, Gregory M., (250) 363-8223 greg.flato@ec.gc.ca, Univ of Victoria – Oa
Flawn, Peter T., (512) 471-1825 pflawn@po.utexas.edu, Univ of Texas at Austin, Jackson School of Geosciences – Eg
Fleeger, Gary M., 717.702.2045 gfleeger@state.pa.us, Pennsylvania Bureau of Topographic & Geologic Survey – Hw
Fleet, Michael E., (519) 661-3184 mfleet@uwo.ca, Univ of Western Ontario – Gz
Flegal, Russell, (831) 459-2093 rflegal@es.ucsc.edu, Univ of California, Santa Cruz – Cg
Fleischer, Robert L., (518) 388-6985 fleischr@union.edu, Union College – Cc
Fleisher, P. Jay, (607) 436-3375 fleishpj@oneonta.edu, SUNY, Oneonta – Gl
Fleming, Thomas H., 203-392-5837 flemingt1@southernct.edu, Southern Connecticut State Univ – Gi
Flemming, Roberta L., (519) 661-2111 (Ext. 86690) rflemmin@uwo.ca, Univ of Western Ontario – Gz
Flesch, Lucy M., (765) 494-0263 lmflesch@purdue.edu, Purdue Univ – Yg
Fleshin, John, (216) 687-3527 j.fleshin@csuohio.edu, Cleveland State Univ – Gg
Flessa, Karl W., (520) 621-7336 kflessa@email.arizona.edu, Univ of Arizona – Po
Fletcher, Charles H., (808) 956-2582 fletcher@soest.hawaii.edu, Univ of Hawai'i, Manoa – Gs
Fletcher, Johnjfletche@cicese.mx, Centro de Inv Científica y de Educación Superior de Ensenada – Gpt
Fletcher, Peter, (508) 697-1390 Bridgewater State College – Sd
Fletcher, Raymond, (814) 863-1769 Pennsylvania State Univ, Univ Park – Gc
Fletcher, William K., (604) 822-2392 wfletcher@eos.ubc.ca, Univ of British Columbia – Ce
Flierl, Glenn R., (617) 253-4692 glenn@pimms.mit.edu, Massachusetts Institute of Technology – Op
Flood, Roger D., (631) 632-6971 rflood@notes.cc.sunysb.edu, SUNY, Stony Brook – Ou
Flood, Roger D., (631) 632-6971 roger.flood@sunysb.edu, SUNY, Stony Brook – Gu
Flood, Tim P., (920) 403-1356 tim.flood@snc.edu, Saint Norbert College – Gv
Flore, Melissa, ((51) 9) -253- (Ext. 00 ext. 2500) mflore@uwindsor.ca, Univ of Windsor – Cs
Florence, Frank P., (315) 786-2242 fflorence@sunyjefferson.edu, Jefferson Community College – Gg
Flores, Carloscflores@cicese.mx, Centro de Inv Científica y de Educación Superior de Ensenada – Ye
Flores, Mari C., (818) 677-3541 mari.flores@csun.edu, California State Univ, Northridge – Oo
Florsheim, Joan L.florsheim@geology.ucdavis.edu, Univ of California, Davis – Gm
Flory, David M., 515-294-0264 flory@iastate.edu, Iowa State Univ of Science & Technology – Ow
Flory, Richard A., 530-898-5262 rflory@csuchico.edu, California State Univ, Chico – Ps
Flower, Martin F. J., (312) 996-9662 flower@uic.edu, Univ of Illinois, Chicago – Gi
Flowers, George C., (504) 865-5198 flowers@tulane.edu, Tulane Univ – Cg
Flowers, Gwenn, (604) 291-6638 gflowers@sfu.ca, Simon Fraser Univ – Gl
Flowers, Rebecca, 303-492-5135 flowersr@colorado.edu, Univ of Colorado – CgcGt
Fluegeman, Richard H., (765) 285-8267 rfluegem@bsu.edu, Ball State Univ – Pm
Flynn, John J., (212) 769-5806 American Museum of Nat History – Pv
Flynn, Luke P., (808) 956-3154 Univ of Hawai'i, Manoa – Or
Flynn, Natalie, 215.951.1269 La Salle Univ – Gz
Fodor, Ronald V., (919) 515-7177 ron_fodor@ncsu.edu, North Carolina State Univ – Gi
Fogel, David N., (505) 665-3305 fogel@lanl.gov, Los Alamos National Lab – Oo
Fogel, Marilyn L., (202) 686-2410 m.fogel@gl.ciw.edu, Carnegie Institution of Washington – Co
Fogel, Robert A., 212-769-5399 bobby@amnh.org, American Museum of Natural History – Cp
Fogg, Graham E., (530) 752-6810 gefogg@ucdavis.edu, Univ of California, Davis – Hy
Fogleman, Wavell W., (603) 535-2289 wwf@mail.plymouth.edu, Plymouth State College – Oo
Foit, Jr., Franklin F., (509) 335-3093 foit@mail.wsu.edu, Washington State Univ – Gz
Foland, Kenneth A., (614) 292-1350 kfoland@geology.ohio-state.edu, Ohio State Univ – Cc
Foley, Duncan, (253) 535-7568 foleyd@plu.edu, Pacific Lutheran Univ – Hy
Foley, Jonathan A., (608)265-9119 jfoley@wisc.edu, Univ of Wisconsin, Madison – Og
Foley, Michael E., (207) 287-2801 Dept of Conservation – Gg
Folinsbee, Robert E., (780) 492-4205 Univ of Alberta – Em
Folk, Robert L., (512) 471-5294 rlfolk@mail.utexas.edu, Univ of Texas at Austin, Jackson School of Geosciences – Gd
Follette, Everett, (605) 642-6235 Black Hills State Univ – Oe
Follmer, Leon R., (217) 244-6945 follmer@isgs.uiuc.edu, Illinois State Geological Survey – Sd
Follmer, Leon R., (217) 244-6945 follmer@isgs.uiuc.edu, Univ of Illinois, Urbana-Champaign – Sa
Follows, Michael, (617) 253-5939 mick@plume.mit.edu, Massachusetts Institute of Technology – Op
Foltltin, Thomas, (202) 994-7156 George Washington Univ – Oy

Fomel, Sergey, 512-475-9573 sergey.fomel@beg.utexas.edu, Univ of Texas at Austin, Jackson School of Geosciences – Yx
Fonstad, Todd A., (414) 424-4103 Univ of Wisconsin, Oshkosh – Oy
Fontheim, Ernest G., (313) 764-6590 Univ of Michigan – Oo
Foos, Annabelle M., (330) 972-7631 afoos@uakron.edu, Univ of Akron – Cg
Foote, Michael J., (773) 702-4320 mfoote@midway.uchicago.edu, Univ of Chicago – Pg
Forbes, Robert W., (502) 852-6844 Univ of Louisville – Oo
Force, Eric R., 520 320 1002 ejforce@aol.com, Univ of Arizona – Eg
Ford, Andrew, 509-335-7846 fords@wsu.edu, Washington State Univ – Ou
Ford, Derek C., (905) 525-9140 (Ext. 23534) dford@mcmail.cis.mcmaster.ca, McMaster Univ – Gm
Ford, F. Andrew, (509) 335-7846 forda@mail.wsu.edu, Washington State Univ – Oo
Ford, Richard L., (801) 626-6942 rford@weber.edu, Weber State Univ – Gm
Ford, Roseanne M., (804) 924-6283 rmf3f@virginia.edu, Univ of Virginia – Oo
Fordham, William D., (973) 443-8779 fordham@fdu.edu, Fairleigh Dickinson Univ – Cg
Fordyce, R. Ewan, +64 3 479-7510/7535 ewan.fordyce@otago.ac.nz, Univ of Otago – Pv
Foreman, Michael G. G., (250) 363-6306 ForemanM@pac.dfo-mpo.gc.ca, Univ of British Columbia – Op
Forgotson, Jr., James M., (405) 325-4451 jforgot@ou.edu, Univ of Oklahoma – Go
Forman, Steven L., (312) 413-9404 slf@uic.edu, Univ of Illinois, Chicago – Gl
Fornari, Daniel J., (508) 289-2857 dfornari@whoi.edu, Woods Hole Oceanographic Institution – Gu
Forrest, Michael R., (310) 693-0921 Mforrest@rhiohondo.edu, Rio Hondo College – Ys
Forsman, Nels F., (701) 777-4349 nels_forsman@und.edu, Univ of North Dakota – Eg
Forster, Richard R., (801) 581-3611 rick.forster@geog.utah.edu, Univ of Utah – Or
Forsyth, Donald W., (401) 863-1699 donald_forsyth@brown.edu, Brown Univ – Yr
Fortier, Richard, (418) 656-2746 richard.fortier@ggl.ulaval.ca, Universite Laval – Yg
Fortin, Danielle, (613)562-5800 Ext 6423 dfortin@uottawa.ca, Univ of Ottawa – Py
Fortsas, Peter, 773/442-6065 P-Fortsas@neiu.edu, Northeastern Illinois Univ – Oo
Fortsch, David E., (208) 282-2604 fortdavi@isu.edu, Idaho State Univ – Oy
Foss, Donald J., (415) 485-9523 jcprice@metro.net, Marin Community College – Og
Foster, Carolyn, 281-425-6333 cfoster@lee.edu, Lee College – Oo
Foster, David A., (352) 392-7316 dfoster@geology.ufl.edu, Univ of Florida – GtCc
Foster, Gregfjohn@nova.edu, Nova Southeastern Univ – Or
Foster, John H., (714) 278-7066 jfoster@fullerton.edu, California State Univ, Fullerton – Ng
Foster, Kristikfoster@broward.edu, Broward Community College, Central Campus – Ob
Foster, Kristikfoster@nova.edu, Nova Southeastern Univ – Or
Foster, Merrill W., (309) 677-2352 fossil@bradley.bradley.edu, Bradley Univ – Pg
Foster, Stuart, (270) 745-5976 Western Kentucky Univ – Oa
Foster Jr., Charles T., (319) 335-1801 tom-foster@uiowa.edu, Univ of Iowa – Gp
Foti, Pamela, (928) 523-6196 Pam.foti@nau.edu, Northern Arizona Univ – Oo
Fouad, Khaled, (512) 232-1527 khaled.fouad@beg.utexas.edu, Univ of Texas at Austin, Jackson School of Geosciences – Yg
Fouberg, Erin H., (540) 654-1494 efouberg@mwc.edu, Mary Washington College – Oo
Fouch, Matthew J., (480) 965-5081 fouch@asu.edu, Arizona State Univ – Yg
Foufoula-Georgiou, Efi, (612) 626-0369 Univ of Minnesota, Twin Cities – Hg
Fouke, Bruce W., (217) 244-5431 fouke@uiuc.edu, Univ of Illinois, Urbana-Champaign – Gs
Foulger, Gillian R., 0191 3742514 g.r.foulger@durham.ac.uk, Durham Univ – Yg
Fountain, Andrew G., (503) 725-3386 andrew@pdx.edu, Portland State Univ – Gl
Fountain, John C., 919-515-3717 fountain@ncsu.edu, North Carolina State Univ – Hw
Fournelle, John, (608) 262-7964 johnf@geology.wisc.edu, Univ of Wisconsin, Madison – Gi
Fournier, Benoit, (418) 656-3930 benoit.fournier@ggl.ulaval.ca, Universite Laval – NgOmGg
Fournier, Robert O., (902) 494-3666 robert.fournier@dal.ca, Dalhousie Univ – Ob
Foust, Elaine P., (615) 532-1505 Elaine.Foust@state.tn.us, Tennessee Division of Geology – OiEoGg
Fovell, Robert, (310) 206-9956 fovell@atmos.ucla.edu, Univ of California, Los Angeles – Oa
Fowell, Sarah, (907) 474-7810 ffsjf@uaf.edu, Univ of Alaska, Fairbanks – Pl
Fowle, David A., (519) -25-3-30 (Ext. ext. 3763) fowle@uwindsor.ca, Univ of Windsor – Cl
Fowler, Anthony D., (613)562-5800 Ext 6273 afowler@uottawa.ca, Univ of Ottawa – Gv
Fowler, Malcolm M., (505) 667-5439 Los Alamos National Lab – Oa
Fowler, Martin G., (403) 292-7038 fowler@gsc.emr.ca, Univ of Calgary – Co
Fox, David L., 612-624-6361 dlfox@umn.edu, Univ of Minnesota, Twin Cities – Pg
Fox, James E., (605) 394-2468 James.fox@sdsmt.edu, South Dakota School of Mines & Technology – Gs
Fox, Jeff, 979-845-2673 fox@iodp.tamu.edu, Texas A&M Univ – Gu
Fox, Laurel R., (831) 459-2533 fox@biology.ucsc.edu, Univ of California, Santa Cruz – Oo
Fox, Lewislfox@broward.edu, Broward Community College, Central Campus – Oc
Fox, Lydia K., (209) 946-2481 lkfox@pacific.edu, Univ of the Pacific – GizOe
Fox, Neil I., 573-882-2144 foxn@missouri.edu, Univ of Missouri, Columbia – Oa
Fox, Paul J., (979) 845-8480 fox@odpemail.tamu.edu, Texas A&M Univ – Ou
Fox, Richard C., (780) 492-5491 richard.fox@ualberta.ca, Univ of Alberta – Pv
Fox, William T., (413) 597-3927 william.t.fox@williams.edu, Williams College – Gs
Foyle, Anthony M., (912) 598-3310 Georgia Southern Univ – Gu
Frahm, Barbara A., 217-244-0028 frahm, Illinois State Geological Survey – Oo
Frahm, Ellery, 612-624-7370 frah0010@umn.edu, Univ of Minnesota, Twin Cities – Oo
Frailey, Scott M., 217-244-2430 frailey@isgs.uiuc.edu, Illinois State Geological Survey – Ng
Fraiser, Margaret L., 414-229-6463 mfraiser@uwm.edu, Univ of Wisconsin, Milwaukee – Pi
Fralick, Philip W., (807) 343-8288 philip.fralick@lakeheadu.ca, Lakehead Univ – Gs
Francis, Carl A., (617) 495-4758 museum@eps.harvard.edu, Harvard Univ – Gz
Francis, Don, 514-398-4885 donf@eps.mcgill.ca, McGill Univ – Gi
Francis, Henry E., (859) 257-5500 francis@kgs.mm.uky.edu, Univ of Kentucky – Ca
Francis, Robert, (206) 543-7345 Univ of Washington – Ob
Francis, Robert D., (562) 985-4929 jfrancis@csulb.edu, California State Univ, Long Beach – Co
Francisco, Joseph S., (765) 494-3284 francisc@purdue.edu, Purdue Univ – Oa
Francois, Rogerrfrancois@eos.ubc.ca, Univ of British Columbia – Cm
Frandsen, Susansfrandse@broward.edu, Broward Community College, Central Campus – Og
Frank, Charles O., (618) 453-7365 frank@geo.siu.edu, Southern Illinois Univ Carbondale – Gx
Frank, Kenneth, (902) 426-3498 frankk@mar.dfo-mpo.gc.ca, Dalhousie Univ – Ob
Frank, Mark R., (815) 753-8395 mfrank@niu.edu, Northern Illinois Univ – Cp
Frank, Tracy D., 402-472-9799 tfrank2@unl.edu, Univ of Nebraska, Lincoln – CsGdCm
Frank, William M., (814) 865-0470 frank@ems.psu.edu,

The Pennsylvania State Univ, Univ Park – Ow
Franke, Otto L., (212) 650-6984 Graduate School of the City Univ of New York – Gc
Frankie, Wayne T., 217-244-2427 frankie@isgs.uiuc.edu, Illinois State Geological Survey – Gg
Franks, Peter J. S., (858) 534-7528 pfranks.ucsd.edu, Univ of California, San Diego – Ob
Franseen, Evan K., (785) 864-2072 evanf@kgs.ku.edu, Univ of Kansas – Gs
Franseen, Evan K., (785) 864-2072 evanf@kgs.ku.edu, Univ of Kansas – Gd
Fransolet, Andre-Mathieu, 32 4 366 22 06 amfransolet@ulg.ac.be, Universite de Liege – Gz
Franz, Eldon H., (509) 335-7474 franz@wsu.edu, Washington State Univ – Sf
Franz, Kristie, (515) 294-7454 kfranz@iastate.edu, Iowa State Univ of Science & Technology – Hs
Franzen, David W., (701) 231-8882 dfranzen@ndsuext.nodak.edu, North Dakota State Univ – Sc
Franzi, David A., (518) 564-4033 franzida@plattsburgh.edu, Plattsburgh State Univ (SUNY) – Gl
Fraser, Alistair B., (814) 863-0791 abf1@psu.edu, The Pennsylvania State Univ, Univ Park – Ow
Fraser, Matthew P., (713) 348-5883 mpf@rice.edu, Rice Univ – Oa
Fraser, Nicholas C., (540) 666-8611 Virginia Polytechnic Institute & State Univ – Pv
Fraser-Smith, Antony C., (650) 723-3684 Stanford Univ – Ym
Fratta, Dante, 608-265-5644 fratta@wisc.edu, Univ of Wisconsin, Madison – Yx
Fratta, Dante, 608-265-5644 fratta@wisc.edu, Univ of Wisconsin, Madison – Ng
Frazer, L. N., (808) 956-3724 neil@soest.hawaii.edu, Univ of Hawai'i, Manoa – Ys
Frazier, Bruce E., (509) 335-2381 bfrazier@wsu.edu, Washington State Univ – Or
Frazier, William J., (706) 569-3026 frazier_bill@colstate.edu, Columbus State Univ – Gs
Frederick, Daniel L., (931) 221-7455 frederickd@apsu.edu, Austin Peay State Univ – Gsr
Frederick, John E., (773) 702-3237 Univ of Chicago – Oa
Frederiksen, Andrew, (204) 474-9460 andrew_frederiksen@umanitoba.ca, Univ of Manitoba – Ys
Fredricks, Helen, 508-289-3678 hfredricks@whoi.edu, Woods Hole Oceanographic Institution – Oc
Freed, Andrew M., (765) 496-3738 freed@purdue.edu, Purdue Univ – Gg
Freed, Jane S., (208) 885-7479 jfreed@uidaho.edu, Univ of Idaho – Oi
Freed, Robert L., (210) 999-7092 bfreed@trinity.edu, Trinity Univ – Gz
Freedenberg, Henry, (850) 488-9380 Florida Geological Survey – Go
Freedman, Terri, (610) 526-5118 tfreedma@brynmawr.edu, Bryn Mawr College – Oo
Freeland, Howard J., (250) 363-6590 freelandhj@dfo-mpo.gc.ca, Univ of Victoria – Op
Freeman, Katherine H., (814) 863-8177 kate@essc.psu.edu, Pennsylvania State Univ, Univ Park – Co
Freeman, Thomas J., (573) 882-6673 FreemanT@missouri.edu, Univ of Missouri – Gd
Freeman, Veronica, 740-376-4775 Marietta College – Pg
Freeman-Lynde, Raymond, (706) 542-2391 rfreeman@gly.uga.edu, Univ of Georgia – Gu
Frei, Allan, (212) 772-5322 afrei@hunter.cuny.edu, Hunter College (CUNY) – Hs
Frei, Michaela, 089/2180 6590 michaela.frei@iaag.geo.uni-muenchen.de, LMU - Univ of Munich – Gg
Freiermuth, Sue, (715) 425-3345 susan.m.freiermuth@uwrf.edu, Univ of Wisconsin, River Falls – Oo
Freile, Debra, (201) 200-3161 dfreile@njcu.edu, New Jersey City Univ – Ge
Freilich, Michaelmhf@coas.oregonstate.edu, Oregon State Univ – Op
French, Bevan, (202) 633-1326 Smithsonian Institution / National Museum of Natural History – Xm
French, Mark A., (609) 984-6587 mark.french@dep.state.nj.us, New Jersey Geological Survey – Hq
Frenette, Jean, 418-656-8123 frenette@ggl.ulaval.ca, Universite Laval – Gz
Frenkel, Robert E., (541) 737-1207 frenkelr@geo.oregonstate.edu, Oregon State Univ – Oy
Frerichs, William E., (307) 766-3386 Univ of Wyoming – Pm
Freshley, Mark D., (509) 372-6094 mark.freshley@pnl.gov, Pacific Northwest National Lab – Hw
Frew, Nelson M., (508) 289-2489 nfrew@whoi.edu, Woods Hole Oceanographic Institution – Ca
Frey, Bonnie A., (505) 835-5160 bfrey@nmt.edu, New Mexico Institute of Mining & Technology – Cg
Frey, Frederick A., (617) 253-2818 fafrey@mit.edu, Massachusetts Institute of Technology – Ct
Frey, Friedrich, 089/21804332 f.frey@lmu.de, LMU - Univ of Munich – Gz
Frey, Holli M., 518-388-6418 freyh@union.edu, Union College – GvCag
Freymueller, Jeffrey T., (907) 474-7286 jeff@giseis.alaska.edu, Univ of Alaska, Fairbanks – Yd
Frez, Josejofrez@cicese.mx, Centro de Inv Científica y de Educación Superior de Ensenada – Ys
Friberg, La Verne M., (330) 972-8046 lfribe1@uakron.edu, Univ of Akron – Gp
Fricke, Henry C., (719) 389-6514 hfricke@coloradocollege.edu, Colorado College – Cs
Friedl, Mark, (617) 353-5745 friedl@bu.edu, Boston Univ – Or
Friedl, Mark A., (617) 353-4807 Boston Univ – Or
Friedman, Andrea L., 609-292-2576 andrea.friedman@dep.state.nj.us, New Jersey Geological Survey – Ge
Friedman, Gerald, (518) 276-8520 gmfriedman@juno.com, Rensselaer Polytechnic Institute – Gs
Friedman, Gerald M., (518) 276-8520 gmfriedman@juno.com, Brooklyn College (CUNY) – Gs
Friedman, Richard, (604) 822-6564 Univ of British Columbia – Cc
Friedmann, Samuel (Julio) J., (925) 423-0585 friedmann2@llnl.gov, Lawrence Livermore National Lab – Gg
Friedrichs, Carl J., (804) 684-7303 cfried@vims.edu, College of William & Mary – On
Friehauf, Kurt, (610) 683-4446 friehauf@kutztown.edu, Kutztown Univ of Pennsylvania – Em
Friehe, Carl, (949) 824-6159 cfriehe@uci.edu, Univ of California, Irvine – Yr
Frieman, Edward A., (858) 822-1510 efrieman@ucsd.edu, Univ of California, San Diego – Oo
Friend, Donald A., (507) 389-2617 donald.friend@mnsu.edu, Minnesota State Univ – Gm
Fries, David P., (727) 553-3961 dfries@marine.usf.edu, Univ of South Florida, Saint Petersburg – Oo
Frind, Emil O., (519) 888-4567 (Ext. 3959) Univ of Waterloo – Hq
Frisbie, Malcolm P., (859) 622-1273 malcolm.frisbie@eku.edu, Eastern Kentucky Univ – Oo
Frisia, Silvia, +61 2 4921 5402 Silvia.Frisia@newcastle.edu.au, Univ of Newcastle – ClcCl
Frisk, DeAnn, (515) 294-4477 dfrisk@iastate.edu, Iowa State Univ of Science & Technology – Oo
Fritsch, J. Michael, (814) 863-1842 fritsch@ems.psu.edu, The Pennsylvania State Univ, Univ Park – Ow
Fritsche, A. E.a.eugene.fritsche@csun.edu, California State Univ, Northridge – Gd
Fritsen, Christian H., (775) 673-7487 cfritsen@dri.edu, Univ of Nevada, Reno – Og
Fritsen, Christian H., (775) 673-7487 cfritsen@dri.edu, Desert Research Institute – Py
Fritton, Daniel D., (814) 865-1143 ddf@psu.edu, Pennsylvania State Univ, Univ Park – Sp
Fritts, Harold C., (520) 621-1608 hfritts@ltrr.arizona.edu, Univ of Arizona – Pe
Fritz, John O., (913) 532-5539 Kansas State Univ – Sf
Fritz, Sherilyn C., (402) 472-6431 sfritz2@unl.edu, Univ of Nebraska, Lincoln – Gn
Fritz, Steven J., (765) 494-3276 sfrito@purdue.edu, Purdue Univ – Hw
Fritz, William J., (404) 651-1156 wfritz@gsu.edu, Georgia State Univ – Gs
Frizado, Joseph P., (419) 372-7202 frizado@bgnet.bgsu.edu, Bowling Green State Univ – Oi
Froehlich, David J., (512) 223-4894 eohippus@austin.cc.tx.us, Austin Community College – Pv
Froelich, Philip N., (404) 894-3944 Georgia Inst of Technology – Cm
Frohlich, Cliff, (512) 471-0460 cliff@ig.utexas.edu, Univ of Texas at Austin, Jackson School of Geosciences – Ys
Frolking, Stephen E., (603) 862-0244 steve.frolking@unh.edu,

Univ of New Hampshire – Oa
Frolking, Tod A., (740) 587-6222 frolking@denison.edu,
Denison Univ – Oy
Fronabarger, A. Kem, (843) 953-5509 fronabarger@cofc.edu,
College of Charleston – Gi
Frost, B. Ronald, (307) 766-4290 rfrost@uwyo.edu,
Univ of Wyoming – Gp
Frost, Bruce W., 206-543-7186 frost@ocean.washington.edu,
Univ of Washington – Ob
Frost, Carol D., (307) 766-6254 frost@uwyo.edu,
Univ of Wyoming – Cc
Frost, Eric G., (619) 594-5003 eric.frost@sdsu.edu,
San Diego State Univ – Gc
Fruchter, Jon S., (509) 376-3937 john.fruchter@pnl.gov,
Pacific Northwest National Lab – Cl
Fry, Nfryn@cf.ac.uk,
Cardiff Univ – Gc
Fryar, Alan E., (859) 257-4392 alan.fryar@uky.edu,
Univ of Kentucky – Hw
Frye, Robert, (520) 626-7027 Univ of Arizona – Oo
Fryer, Brian J., (519) -25-3-30 (Ext. ext, 3750) bfryer@uwindsor.ca,
Univ of Windsor – Cc
Fryer, Gerard J., (808) 956-7875 gerard@hawaii.edu,
Univ of Hawai'i, Manoa – Ys
Fryer, Karen H., (740) 368-3618 khfryer@owu.edu,
Ohio Wesleyan Univ – Gc
Fryer, Patricia, (808) 956-3146 pfryer@soest.hawaii.edu,
Univ of Hawai'i, Manoa – Gu
Fryer, Patricia B., (808) 956-3146 pfryer@soest.hawaii.edu,
Univ of Hawai'i, Manoa – Gu
Fryett, IfryettI@cf.ac.uk,
Cardiff Univ – Oo
Fryxell, Greta A., (979) 845-4543 Texas A&M Univ – Ob
Fryxell, Joan E., (909) 537-5311 jfryxell@csusb.edu,
California State Univ, San Bernardino – Gc
Fthenakis, Vasilis M., (516) 282-2830 Columbia Univ – Oa
Fu, Rong, (404) 385-0670 fu@eas.gatech.edu,
Georgia Institute of Technology – Oa
Fuelberg, Henry E., (850) 644-6466 fuelberg@met.fsu.edu,
Florida State Univ – Ow
Fuentes, Jose D., (804) 982-2654 jf6s@virginia.edu,
Univ of Virginia – Oa
Fueten, Frank, (905) 688-5550 (Ext. 3856) FFueten@brocku.ca,
Brock Univ – Gc
Fuiman, Lee A., (361) 749-6775 lee.fuiman@mail.utexas.edu,
Univ of Texas at Austin – Ob
Fujita, Kazuya, 517-355-0142 fujita@msu.edu,
Michigan State Univ – Ys
Full, Robert J., (510) 642-9896 rjfull@garnet.berkeley.edu,
Univ of California, Berkeley – Pi
Fullagar, Paul D., (919) 962-0677 fullagar@unc.edu,
Univ of North Carolina, Chapel Hill – Cc
Fuller, Doult O., (202) 994-8073 dfuller@gwu.edu,
George Washington Univ – Oy
Fuller, Jonathan A., 419-626-4296 Ohio Dept of Nat Resources – On
Fuller, Michael, (808) 956-4038 mfuller@soest.hawaii.edu,
Univ of Hawai'i, Manoa – Ym
Fuller, Michael D., (805) 893-3508 Univ of California, Santa Barbara – Ym
Fullmer, James W., (203) 392-5841 fullmerj1@southernct.edu,
Southern Connecticut State Univ – Ow
Fulthorpe, Craig S., (512) 471-0459 craig@ig.utexas.edu,
Univ of Texas at Austin, Jackson School of Geosciences – Gs
Fults, Michelle, (804) 427-7189 Tidewater Community College – Hy
Fung, Inez, (510) 643-9367 inez@atmos.berkeley.edu,
Univ of California, Berkeley – Oa
Furbish, David J., 615-322-2976 david.j.furbish@vanderbilt.edu,
Vanderbilt Univ – Hg
Furlong, Kevin P., (814) 863-0567 kevin@geodyn.psu.edu,
Pennsylvania State Univ, Univ Park – Gt
Furman, Tanya, (814) 865-5782 furman@geosc.psu.edu,
Pennsylvania State Univ, Univ Park – Cg
Fusch, Richard D., (740) 368-3616 rdfusch@owu.edu,
Ohio Wesleyan Univ – Og
Fyfe, John C., (250) 363-8236 john.fyfe@ec.gc.ca,
Univ of Victoria – Oa
Fyfe, William S., (519) 661-3180 pjfyfe@uwo.ca,
Univ of Western Ontario – Cg
Fyson, William K., (613) 745-6645 wfyson@uottawa.ca,
Univ of Ottawa – Gc

G

Gabel, Mark, (605) 642-9035 Black Hills State Univ – Pb
Gabet, Emmanuel, 408 924-5035 manny.gabet@sjsu.edu,
San Jose State Univ – Gm
Gable, Carl W., (505) 665-3533 gable@lanl.gov,
Los Alamos National Lab – Gt
Gabrielov, Andrei, (765) 496-2868 gabriea@purdue.edu,
Purdue Univ – Ys
Gade, Ole, (828) 262-3000 gadeo@appstate.edu,
Appalachian State Univ – Oy
Gadomski, Frederick J., (814) 863-4229 gadomski@mail.meteo.psu.edu,
The Pennsylvania State Univ, Univ Park – Ow
Gaetani, Glenn A., (508) 289-3724 ggaetani@whoi.edu,
Woods Hole Oceanographic Institution – Gi
Gaffney, Edward S., (505) 665-6387 gaffney@lanl.gov,
Los Alamos National Lab – Yg
Gaffney, Eugene S., (212) 769-5801 American Museum of Natural History – Pv
Gaffney, Jeffrey S., (630) 252-5178 gaffney@anl.gov,
Argonne National Lab – Oa
Gagne, Jean-Pierre, (418) 724-1870 jean-pierre_gagne@uqar.uquebec.ca,
Universite du Quebec a Rimouski – Oc
Gagne, Marc R., (610) 436-3014 mgagne@wcupa.edu,
West Chester Univ – Oo
Gagnon, Alan R., (508) 289-2961 agagnon@whoi.edu,
Woods Hole Oceanographic Institution – Cs
Gagnon, Katie, 206-516-3161 KGagnon@sccd.ctc.edu,
Seattle Central Community College – Oug
Gaherty, James B., (404) 894-1992 james.gaherty@eas.gatech.edu,
Georgia Institute of Technology – Yv
Gahn, Forest J., (208) 496-1900 gahnf@byui.edu,
Brigham Young Univ, Idaho – PogGs
Gaidos, Eric J., (808) 956-7897 gaidos@soest.hawaii.edu,
Univ of Hawai'i, Manoa – Yr
Gaillard, Jean-Francois, (847) 467-1376 jf_gaillard@northwestern.edu,
Northwestern Univ – Cl
Gaillou, Eloise, (202) 633-1797 gailloue@si.edu,
Smithsonian Institution / National Museum of Natural History – Gz
Gaines, Robert R., 909-621-8674 Robert.Gaines@pomona.edu,
Pomona College – Gs
Gait, Robert I., (416) 586-5818 Royal Ontario Museum – Gz
Galbraith, John M., (540) 231-9784 ttcf@vt.edu,
Virginia Polytechnic Institute & State Univ – Sp
Gale, John E., (709) 737-8397 jgale@plato.ucs.mun.ca,
Memorial Univ of Newfoundland – Nr
Gale, Marjorie H., (802) 241-3608 marjorie.gale@anr.state.vt.us,
Agency of Natural Resources, Dept. of Env Conservation – Gg
Galewsky, Joseph, 505-277-4204 galewsky@unm.edu,
Univ of New Mexico – Ow
Galicki, Stan, 601-974-1340 galics@millsaps.edu,
Millsaps College – Ge
Gallacher, Heather, (216) 523-7201 h.gallacher@csuohio.edu,
Cleveland State Univ – Gg
Gallegos, Anthony F., (505) 665-0862 agallegos@lanl.gov,
Los Alamos National Lab – Py
Galli, Kenneth G., (617) 552-4504 kenneth.galli@bc.edu,
Boston College – Gg
Gallie, Ann, (705) 675-1151 (Ext. ext, 2349) agallie@laurentian.ca,
Laurentian Univ, Sudbury – Or
Gallovic, Frantisek, 089/2180 4209 frantisek.gallovic@geophysik.uni-muenchen.de, LMU - Univ of Munich – Yg
Galloway, James N., (804) 924-1303 jng@virginia.edu,
Univ of Virginia – Oa
Galloway, William E., (512) 471-0494 galloway@mail.utexas.edu,
Univ of Texas at Austin, Jackson School of Geosciences – Gs
Gallup, Christina D., (218) 726-8984 cgallup@d.umn.edu,
Univ of Minnesota, Duluth – Cc
Gallus, William A., (515) 294-2270 wgallus@iastate.edu,
Iowa State Univ of Science & Technology – Oa
Galperin, Boris, (727) 553-1101 bgalperin@marine.usf.edu,
Univ of South Florida, Saint Petersburg – Op
Gamache, Robert R., (508) 934-3904 Univ of Massachusetts/Lowell – Oa
Gamble, Douglas W., (910) 962-3778 gambled@uncw.edu,
Univ of North Carolina Wilmington – Oy
Gamble, Fredrick L., (904) 484-1152 Pensacola Junior College – Og
Gamble, John, +353 21 4903955 j.gamble@ucc.ie,
Univ College Cork – Gi
Gambrell, Ed, (850) 675-6558 Florida Geological Survey – Go
Gambrell, Robert P., (225) 388-6426 Louisiana State Univ – Cg

Gamerdinger, Amy P., (509) 373-3077 amy.gamerdinger@pnl.gov, Pacific Northwest National Lab – Cl
Gammon, Richard H., (206) 543-1609 gammon@macmail.chem.washington.edu, Univ of Washington – Oc
Gammons, Christopher H., (406) 496-4763 cgammons@mtech.edu, Montana Tech of the Univ of Montana – Cg
Gancarz, Alexander J., (505) 667-2606 Los Alamos National Lab – Cc
Gander, Melodie, (775) 784-6987 Univ of Nevada, Reno – Oo
Ganey-Curry, Patricia E., (512) 471-0408 patty@ig.utexas.edu, Univ of Texas at Austin, Jackson School of Geosciences – Oo
Gangadharan, Anil K., (404) 413-5166 anil@gsu.edu, Georgia State Univ – Oa
Gangi, Anthony F., (979) 845-1381 gangi@geo.tamu.edu, Texas A&M Univ – Ys
Ganguli, Rajive, (907) 474-7631 ffrg@uaf.edu, Univ of Alaska, Fairbanks – Nm
Gann, Delbert E., 601-974-1341 gannde@.millsaps.edu, Millsaps College – Gz
Gannon, Patrick T., (802) 626-6225 Lyndon State College – Oa
Gans, Phillip B., (805) 893-2642 Univ of California, Santa Barbara – Gt
Gansecki, Cheryl A., (808) 974-7302 gansecki@hawaii.edu, Univ of Hawai'i at Hilo – Cc
Gantzer, Clark J., 573-882-0611 Univ of Missouri, Columbia – Os
Gao, Feng, (617) 353-8033 fgao@bu.edu, Boston Univ – Or
Gao, Stephen S., 573-341-6676 sgao@umr.edu, Univ of Missouri, Rolla – Ys
Gao, Yuan, 973-353-1139 yuangao@andromeda.rutgers.edu, Rutgers, The State Univ of New Jersey, Newark – Oa
Gaonac'h, Helene, (514) 987-3000 helene@volcan.geotop.uqam.ca, Universite du Quebec a Montreal – Gv
Garbesi, Karina, 510-885-3172 karina.garbesi.csueastbay.edu, California State Univ, East Bay – Cg
Garces, Milton A., (808) 325-1558 Univ of Hawai'i, Manoa – Ys
Garcia, Antonio, (805) 756-2430 afgarcia@calpoly.edu, California Polytechnic State Univ – Gg
Garcia, Antonio F., 805-756-2430 afgarcia@calpoly.edu, California Polytechnic State Univ – Gm
Garcia, Jan, (323) 259-2823 jangarcia@oxy.edu, Occidental College – Oo
Garcia, Juanjgarcia@cicese.mx, Centro de Inv Científica y de Educación Superior de Ensenada – Yv
Garcia, Marcelo, 217-244-4484 mhgarcia@uiuc.edu, Univ of Illinois, Urbana-Champaign – Gm
Garcia, Michael O., (808) 956-6641 garcia@soest.hawaii.edu, Univ of Hawai'i, Manoa – Gi
Garcia, Oswaldo, (415) 338-2061 ogarcia@sfsu.edu, San Francisco State Univ – Oa
Garcia, Sammy R., (505) 667-4151 Los Alamos National Lab – Ca
Garcia, William, 704-687-5982 wjgarcia@uncc.edu, Univ of North Carolina, Charlotte – Pv
Gardner, Charles H., (919) 733-3833 North Carolina Geological Survey – Gg
Gardner, Gerald H. F., (412) 421-5514 Rice Univ – Yg
Gardner, James, 603-862-3473 jim.gardner@unh.edu, Univ of New Hampshire – Gu
Gardner, James E., (512) 471-0953 gardner@mail.utexas.edu, Univ of Texas at Austin, Jackson School of Geosciences – Gv
Gardner, Jamie N., (505) 667-1799 jgardner@lanl.gov, Los Alamos National Lab – Gg
Gardner, Leonard R., (803) 777-2424 Univ of South Carolina – Cl
Gardner, Robert H., (301) 689-3115 gardner@al.umces.edu, Univ of Maryland – Ou
Gardner, Thomas W., (210) 999-7655 tgardner@trinity.edu, Trinity Univ – Gm
Gardner, Wayne S., (361) 749-6823 wayne.gardner@mail.utexas.edu, Univ of Texas at Austin – Ob
Gardner, Wilford D., (979) 845-7211 wgardner@ocean.tamu.edu, Texas A&M Univ – Ou
Gardulski, Anne F., 617-627-2891 anne.gardulski@tufts.edu, Tufts Univ – Grs
Garfield, Newell, (408) 656-3226 Naval Postgraduate School – Op
Garfield, Newell (Toby), (415) 338-3713 garfield@sfsu.edu, San Francisco State Univ – Op
Gargett, Ann, (757) 683-6009 gargett@ccpo.odu.edu, Old Dominion Univ – Op
Gariepy, Clement, (514) 987-3592 geotop@er.uqam.ca, Universite du Quebec a Montreal – Cc
Garihan, John M., (864) 294-3363 jack.garihan@furman.edu, Furman Univ – Gc
Garner, Damon A., 217-244-2768 garner@isgs.uiuc.edu, Illinois State Geological Survey – Gg
Garnero, Edward, (480) 965-7653 garnero@asu.edu, Arizona State Univ – Ys
Garnett, Julie, 530-898-5262 jgarnett@csuchico.edu, California State Univ, Chico – Oo
Garon, Laura A., (801) 581-5126 lgaron@egi.utah.edu, Univ of Utah – Yr
Garrett, Christopher J. R., (250) 721-7702 garrett@uvphys.phys.uvic.ca, Univ of Victoria – Op
Garrett, Edward W., (850) 487-2219 Florida Geological Survey – Go
Garrison, Ervan G., (706) 542-1097 egarriso@uga.edu, Univ of Georgia – Ga
Garrison, Robert E., (831) 459-5563 Univ of California, Santa Cruz – Gs
Garrod, Bruce, (416) 978-3538 bruce.garrod@utoronto.ca, Univ of Toronto – Oo
Garside, Larry J., (775) 784-6691 (Ext. 137) lgarside@unr.edu, Univ of Nevada – Gg
Garside, Terri, (775) 784-6691 (Ext. 126) tgarside@unr.edu, Univ of Nevada – Oo
Garstang, Mimi R., 573-368-2101 mimi.garstang@dnr.mo.gov, Missouri Dept of Natural Resources – Ge
Gartland, Eugene, ggartland@rochester.rr.com, SUNY College at Brockport – Gg
Gartner, Janette, (781) 891-2901 jgartner@bentley.edu, Bentley College – Hy
Garven, Grant, (410) 516-8689 garven@jhu.edu, The Johns Hopkins Univ – Hw
Garven, Grant, 617-627-3795 Grant.Garven@tufts.edu, Tufts Univ – Hwy
Garver, John I., (518) 388-6770 garverj@union.edu, Union College – GtCcGr
Garvin, Theresa D., (780) 492-4593 theresa.garvin@ualberta.ca, Univ of Alberta – Oo
Garwood, Dean L., (208) 885-5203 dgarwood@uidaho.edu, Univ of Idaho – Gg
Garwood, Roland W.garwood@nps.navy.mil, Naval Postgraduate School – Op
Gary, Anthony C., (801) 585-9768 tgary@egi.utah.edu, Univ of Utah – Gq
Garzione, Carmala N., (585) 273-4572 garzione@earth.rochester.edu, Univ of Rochester – Gs
Gascho, Gary J., (912) 386-3329 Univ of Georgia – Sc
Gaspari-Bridges, Patricia, (609) 258-5483 pattygb@princeton.edu, Princeton Univ – Oo
Gastaldo, Robert A., (207) 859-5807 ragastal@colby.edu, Colby College – PbGs
Gastil, R. Gordon, (619) 594-6211 ggastil@geology.sdsu.edu, San Diego State Univ – Gt
Gaston, Lewis A., (504) 388-1323 lgaston@agctr.lsu.edu, Louisiana State Univ – Sc
Gates, Alexander E., (973) 353-5034 agates@andromeda.rutgers.edu, Rutgers, The State Univ of New Jersey, Newark – Gc
Gatewood, George D., (412) 321-2400 Univ of Pittsburgh – Xg
Gatrell, Jay, 812-237-2256 gejdg@isugw.indstate.edu, Indiana State Univ – Oi
Gaubatz, Piper, (413) 545-0768 gaubatz@geo.umass.edu, Univ of Massachusetts – Oo
Gaudette, Henri E., (603) 862-1718 Univ of New Hampshire – Cc
Gauri, K. L., (502) 852-6821 Univ of Louisville – Ge
Gaurin, Steve, (413) 585-3690 sgaurin@email.smith.edu, Smith College – Ou
Gauthier, Donald J., (818) 778-5514 gauthidj@lavc.edu, Los Angeles Valley College – Oi
Gauthier, Jacques, (203) 432-3150 jacques.gauthier@yale.edu, Yale Univ – Pv
Gauthier, Michel, (514) 987-4560 gauthier.michel@uqam.ca, Universite du Quebec a Montreal – Em
Gauthier, Pierrepierre.gauthier@ec.gc.ca, McGill Univ – Oa
Gavin, Dan, (541) 346-5787 dgavin@uoregon.edu, Univ of Oregon – Oy
Gay, Laurie, 210-458-4455 Laurie.Gay@utsa.edu, Univ of Texas, San Antonio – Oo
Gayes, Paul T., (803) 349-2213 Dalhousie Univ – Gs
Gaylord, David R., (509) 335-8127 gaylordd@wsu.edu, Washington State Univ – Gs
Gazis, Carey A., 509.963-2820 cgazis@geology.cwu.edu, Central Washington Univ – Cs
Ge, Maochen, 814-865-5861 mug10@psu.edu, Pennsylvania State Univ, Univ Park – Nm

Ge, Shemin, 303-492-8323 ges@colorado.edu,
Univ of Colorado – Hw
Geaghan, James P., (225) 851-2800 Louisiana State Univ – Ob
Geary, Dana H., (608) 263-7754 dana@geology.wisc.edu,
Univ of Wisconsin, Madison – Po
Geary, Phil, 61 02 4921 6726 Phil.Geary@newcastle.edu.au,
Univ of Newcastle – HwOsHw
Gebrande, Helmut, 089/2180 4235 gebrande@geophysik.uni-muenchen.de,
LMU - Univ of Munich – Yg
Gedzelman, Stanley, (212) 650-6470 City College (CUNY) – Oa
Gedzelman, Stanley D., (212) 650-6470 Graduate School of the City Univ of New York – Ow
Gee, Glendon W., (509) 372-6096 glendon.gee@pnl.gov,
Pacific Northwest National Lab – Hq
Geerken, Roland, (203) 432-3421 Yale Univ – Or
Gehrels, George E., (520) 621-6026 ggehrels@geo.arizona.edu,
Univ of Arizona – Gt
Gehrels, Tom, (520) 621-6970 tgehrels@lpl.arizona.edu,
Univ of Arizona – Oo
Geidel, Gwendelyn, (803) 777-4511 geidel@geol.sc.edu,
Univ of South Carolina – Hw
Geiger, James W., 618-985-8371 geiger@isgs.uiuc.edu,
Illinois State Geological Survey – Ge
Geisler, Jonathan H., (912) 681-5398 GeislerJ@GeorgiaSouthern.edu,
Georgia Southern Univ – Pv
Geissman, John W., (505) 277-3433 jgeiss@unm.edu,
Univ of Michigan – Ym
Geissman, John W., (505) 277-3433 jgeiss@unm.edu,
Univ of New Mexico – Ym
Geist, Dennis J., (208) 885-6491 dgeist@uidaho.edu,
Univ of Idaho – Gi
Gelbaum, Carol, 770-274-5050 dunsci@gpc.edu,
Georgia Perimeter College – Gg
Gelderman, Ronald H., (605) 688-4770 Ronald.Gelderman@sdstate.edu,
South Dakota State Univ – Sc
Gellasch, Christopher A., (845) 938-4135 bc7537@usma.edu,
United States Military Academy – Gr
Geller, Marvin A., (631) 632-8701 mgeller@notes.cc.sunysb.edu,
SUNY, Stony Brook – Oa
Geller, Michael D., (609) 652-4620 mgeller@stockton.edu,
Richard Stockton College of New Jersey – Oo
Gemmell, J B., 61 3 6226 2893 Bruce.Gemmell@utas.edu.au,
Univ of Tasmania – Eg
Gendzwill, Donald J., (306) 966-5693 don.gendzwill@usask.ca,
Univ of Saskatchewan – Ye
Genereux, David P., (919) 515-7015 genereux@ncsu.edu,
North Carolina State Univ – Hw
Genes, Andrew N., (617) 287-5290 Univ of Massachusetts, Boston/
Harbor Campus – Gl
Gentile, Richard J., (816) 235-2974 gentiler@umkc.edu,
Univ of Missouri-Kansas City – Gr
Gentry, Terry, 979-845-3041 Gentry@tamu.edu,
Texas A&M Univ – Sb
George, Graham, 306-966-5722 g.george@usask.ca,
Univ of Saskatchewan – Oo
George, Milon F., 573-882-2710 Univ of Missouri, Columbia – Os
Gephart, John, (607) 255-5267 gephart@geology.cornell.edu,
Cornell Univ – Eg
Gephart, John W., (607) 255-3538 gephart@geology.cornell.edu,
Cornell Univ – Gc
Geraghty, Emily M., 406-683-7137 e_geraghty@umwestern.edu,
Univ of Montana, Western – Gc
Gerba, Charles P., (520) 621-6906 gerba@ag.arizona.edu,
Univ of Arizona – Sb
Gerbi, Christopher C., 207 581-2153 Univ of Maine – Gz
Gerencher, Jr., Joseph J., (610) 861-1440 gerencher@moravian.edu,
Moravian College – Gg
Gerhard, Lee C.leeg@sunflower.com,
Univ of Kansas – Gr
Gerhard, Lee C., 78538643965 lgerhard@kgs.ku.edu,
Univ of Kansas – Gs
Gerhardt, Hannes, (678) 839-4064 hgerhard@westga.edu,
Univ of West Georgia – Oo
Gerke, Bruce E., 419-626-4296 bruce.gerke@dnr.state.oh.us,
Ohio Dept of Natural Resources – Ge
Gerla, Philip J., (701) 777-3305 phil.gerla@mail.und.nodak.edu,
Univ of North Dakota – Hw
Gerlach, Russel L., (417) 836-5604 Southwest Missouri State Univ – Oo
Gerlak, Andrea K., (845) 365-8550 akg2004@columbia.edu,
Columbia Univ – Oo
German, Chris, 508-289-2853 cgerman@whoi.edu,
Woods Hole Oceanographic Institution – Gt
Germanoski, Dru, (610) 330-5196 germanod@lafayette.edu,
Lafayette College – Gm
Gersmehl, Carol, 212-772-3534 carol.gersmehl@hunter.cuny.edu,
Hunter College (CUNY) – Oe
Gersmehl, Philip, 212-6503534 gersmehl@geo.hunter.cuny.edu,
Hunter College (CUNY) – Oy
Gertisser, Ralf, +44 01782 583181 r.gertisser@esci.keele.ac.uk,
Keele Univ – GivGz
Gertsch, Leslie, 573-341-7278 lgertsch@mtu.edu,
Univ of Missouri, Rolla – Nm
Gerwing, James R., (605) 688-4772 James.Gerwing@sdstate.edu,
South Dakota State Univ – Sc
Ghani, Azman Abdul, 03-79674234 azmangeo@um.edu.my,
Univ of Malaya – Gi
Ghassemi, Ahmad, (701) 777-3213 ahmad_ghassemi@mail.und.nodak.edu,
Univ of North Dakota – Nr
Ghatge, Suhas L., (609) 984-6587 suhas.ghatge@dep.state.nj.us,
New Jersey Geological Survey – Yv
Ghattas, Omar, (512) 232-4304 omar@ices.utexas.edu,
Univ of Texas at Austin, Jackson School of Geosciences – Gq
Ghelerter, Jill, 770-274-5050 jghelert@gpc.edu,
Georgia Perimeter College – Hw
Ghent, Edward D., (403) 220-5847 ghent@geo.ucalgary.ca,
Univ of Calgary – Gp
Ghent, Rebecca, (416) 978-0597 ghentr@geology.utoronto.ca,
Univ of Toronto – Or
Ghil, Michael, (310) 206-2285 ghil@atmos.ucla.edu,
Univ of California, Los Angeles – Oa
Ghoneim, Emansalma@bu.edu,
Boston Univ – Gm
Ghose, Subrata, 206-543-7378 ghose@u.washington.edu,
Univ of Washington – Gy
Ghosheh, Baher A., (814) 732-2207 ghosheh@edinboro.edu,
Edinboro Univ of Pennsylvania – Oo
Giacalone, Joe, 520-626-8365 giacalone@lpl.arizona.edu,
Univ of Arizona – Xy
Gianniny, Gary, (970) 247-7254 gianniny_g@fortlewis.edu,
Fort Lewis College – Gs
Giaramita, Mario J., (209) 667-3558 marty@geology.csustan.edu,
California State Univ, Stanislaus – Gp
Giardino, John R., (979) 845-3628 giardino@geo.tamu.edu,
Texas A&M Univ – Gm
Giardino, John R., (409) 867-9067 rickg@tamu.edu,
Texas A&M Univ – Gm
Gibbs, Gaynellocean@lsu.edu,
Louisiana State Univ – Oo
Gibeaut, James C., (512) 471-0344 jim.gibeaut@beg.utexas.edu,
Univ of Texas at Austin, Jackson School of Geosciences – Gu
Gibling, Martin R., (902) 494-2355 mgibling@is.dal.ca,
Dalhousie Univ – Gs
Gibson, Carl H., (858) 534-3184 cgibson@ucsd.edu,
Univ of California, San Diego – Oo
Gibson, David, 207-778-7402 dgibson@maine.edu,
Univ of Maine, Farmington – GigGz
Gibson, Deidre M., 757.727.5883 deidre.gibson@hamptonu.edu,
Hampton Univ – Ob
Gibson, H. Daniel (Dan), (604) 291-7057 hdgibson@sfu.ca,
Simon Fraser Univ – Gc
Gibson, Harold, (613) 788-4400 Carleton Univ – Em
Gibson, Harold L., (705) 675-1151 (Ext. ext2371) hgibson@laurentian.ca,
Laurentian Univ, Sudbury – Em
Gibson, Lance R., (913) 532-7224 Kansas State Univ – Sd
Gibson, Lay J., (520) 621-7486 Univ of Arizona – Ou
Gibson, Michael A., (731) 587-7435 mgibson@utm.edu,
Univ of Tennessee, Martin – Pe
Gibson, Ronald C., (714) 895-8194 Golden West College – Gc
Gibson, Jr., Richard L., (979) 862-8653 gibson@geo.tamu.edu,
Texas A&M Univ – Ys
Giegengack, Jr., Robert F., (215) 898-5191 gieg@sas.upenn.edu,
Univ of Pennsylvania – Gg
Gielow, Arthur W., (716) 878-4911 gielowaw@buffalostate.edu,
Buffalo State College – Oo
Gierasch, Peter, (607) 255-8544 pjg10@cornell.edu,
Cornell Univ – Oo
Gierke, John S., (906) 487-2535 jsgierke@mtu.edu,
Michigan Technological Univ – Hy

Gierlowski-Kordesch, Elizabeth H., (740) 593-1989 gierlows@ohio.edu, Ohio Univ, Lancaster – Gn
Giese, Benjamin S., (979) 845-2306 bgiese@ocean.tamu.edu, Texas A&M Univ – Op
Giese, Graham S., (508) 289-2297 ggiese@whoi.edu, Woods Hole Oceanographic Institution – On
Giese, Rossman F., (716)645-6800 (Ext. 3008) glgclay@acsu.buffalo.edu, SUNY, Buffalo – Gz
Gieskes, Joris M., (858) 534-4257 jgieskes@ucsd.edu, Univ of California, San Diego – Oc
Gifford, Dian J., (401) 874-6690 dgifford@gso.uri.edu, Univ of Rhode Island – Ob
Gifford-Gonzalez, Dianne, (831) 459-2633 Univ of California, Santa Cruz – Oo
Gigler, Alexander, 089/2180 4185 LMU - Univ of Munich – Gz
Gilbert, Co'Quesie, (619) 534-2470 cag65@columbia.edu, Columbia Univ – Oo
Gilbert, Hersh, (765) 496-9518 hersh@purdue.edu, Purdue Univ – Ys
Gilbert, J. Freeman, (858) 534-2470 fgilbert@ucsd.edu, Univ of California, San Diego – Ys
Gilbert, Kathleen W., 781-283-3086 kgilbert@wellesley.edu, Wellesley College – Cs
Gilbert, M. Charles, (405) 325-3253 mcgilbert@ou.edu, Univ of Oklahoma – Cp
Gilbert, Patricia M., (410) 221-8422 Univ of Maryland – Ob
Gilbes, Fernando, 787 2653845 gilbes@cacique.uprm.edu, Univ of Puerto Rico – Or
Gilchrist, Brian E., (313) 763-6230 gilchrist@engin.umich.edu, Univ of Michigan – Oo
Gilder, Stuart, 089/2180 4239 stuart.gilder@geophysik.uni-muenchen.de, LMU - Univ of Munich – Yg
Gildner, Raymond F., 260-481-0249 gildnerr@ipfw.edu, Indiana Univ / Purdue Univ, Fort Wayne – Pq
Giles, Alan J., (404) 656-3214 Georgia Dept of Natural Resources – Og
Giles, Joseph F., (701) 231-7858 North Dakota State Univ – Sc
Giles, Katherine A., 575-646-2033 kgiles@nmsu.edu, New Mexico State Univ, Las Cruces – Gs
Giletti, Bruno J., (401) 863-2242 Brown Univ – Cc
Gill, Denis E., (514) 340-4220 Ecole Polytechnique – Nr
Gill, James B., 831 459-3842 jgill@pmc.ucsc.edu, Univ of California, Santa Cruz – Gi
Gill, Richard A., 509-335-6422 rgill@wsu.edu, Washington State Univ – Hw
Gill, Shaungshaun@nova.edu, Nova Southeastern Univ – Oo
Gill, Susan, (215) 898-6517 sgill@sas.upenn.edu, Univ of Pennsylvania – Pe
Gill, Swarndeep S., (724) 938-1677 gill@cup.edu, California Univ of Pennsylvania – Oa
Gill, Thomas E., (915) 747-5168 gill@geo.utep.edu, Univ of Texas, El Paso – Ge
Gillam, Mary L.gillam@rmi.net, Fort Lewis College – Gm
Gille, Peter, 089/2180 4355 peter.gille@lrz.uni-muenchen.de, LMU - Univ of Munich – Gz
Gillerman, Virginia S., (208) 332-4420 vgillerm@uidaho.edu, Univ of Idaho – Eg
Gillerman, Virginia S., (208) 426-4002 vgillerm@boisestate.edu, Boise State Univ – Eg
Gillespie, Alan R., 206-685-8265 arg3@.u.washington.edu, Univ of Washington – GmOlr
Gillespie, Barbara M., (509) 376-5802 barbara.gillespie@pnl.gov, Pacific Northwest National Lab – Oo
Gillespie, Terry J., (519) 824-4120 (Ext. 54276) tgillesp@lrs.uoguelph.ca, Univ of Guelph – Ow
Gillespie, Thomas, (609) 771-2569 Trenton State College – Gc
Gillett, Blakeney D., 205/247-3693 bgillett@gsa.state.al.us, Geological Survey of Alabama – Hw
Gillette, David D., (928) 523-1432 david.gillette@nau.edu, Northern Arizona Univ – Pv
Gilley, Brettbgilley@eos.ubc.ca, Univ of British Columbia – Gg
Gilley, Brett, (604) 777-6117 gilleyb@douglas.bc.ca, Douglas College – Gg
Gillham, Douglas M., 662-325-9646 dgillham@geosci.msstate.edu, Mississippi State Univ – Ow
Gilliam, David, (954)262-3634 gilliam@nova.edu, Nova Southeastern Univ – Ob
Gilliam, James W., (919) 515-2040 North Carolina State Univ – Sb
Gilligan, Jonathan M., 615-322-2420 jonathan.gilligan@vanderbilt.edu, Vanderbilt Univ – Ge
Gillis, Kathryn, 250-721-6120 seosdir@uvic.ca, Univ of Victoria – Gp
Gillman, Joe, (573) 368-2161 Missouri Dept of Natural Resources – Oo
Gilmer, Amy K., (434) 951-6368 amy.gilmer@dmme.virginia.gov, Division of Mineral Resources – Eg
Gilmore, Martha S., (860) 685-3129 mgilmore@wesleyan.edu, Wesleyan Univ – XgGmOr
Gilmore, Tyler J., (509) 376-2370 tyler.gilmore@pnl.gov, Pacific Northwest National Lab – Hg
Gilmour, Ernest H., (509) 359-7480 egilmour@mail.ewu.edu, Eastern Washington Univ – Pi
Gilotti, Jane A., (319) 335-1097 jane-gilotti@uiowa.edu, Univ of Iowa – Gc
Gilpin, Bernard J., (714) 895-8233 Golden West College – Ys
Gingerich, Philip D., (734) 764-0490 gingeric@umich.edu, Univ of Michigan – Pv
Ginis, Isaac, (401) 874-6484 iginis@gso.uri.edu, Univ of Rhode Island – Op
Ginsberg, David, (989) 774-1293 Central Michigan Univ – Gp
Ginsburg, Robert N., 305 421-4875 rginsburg@rsmas.miami.edu, Univ of Miami – Gs
Giordano, Alberto, (617) 287-5280 Univ of Massachusetts, Boston/ Harbor Campus – Or
Giordano, Thomas H., 575-646-2708 tgiordan@nmsu.edu, New Mexico State Univ, Las Cruces – Cl
Giorgis, Scott D., 585-245-5293 giorgis@geneseo.edu, SUNY, Geneseo – Gc
Giosan, Liviu, 508-289-2257 lgiosan@whoi.edu, Woods Hole Oceanographic Institution – Gu
Girard, Michael W., (609) 292-2576 mike.girard@dep.state.nj.us, New Jersey Geological Survey – Oi
Giraud, Richard E., (801) 537-3351 nrugs.rgiraud@state.ut.us, Utah Geological Survey – Ng
Girhard, Thomas S., (210) 785-6078 San Antonio College – Owy
Girty, Gary H., (619) 594-2552 ggirty@geology.sdsu.edu, San Diego State Univ – Gc
Gitelson, Anatoly, 402-472-8386 gitelson@calmit.unl.edu, University of Nebraska - Lincoln – Oo
Gittins, John, 416 445 6096 j.gittins@utoronto.ca, Univ of Toronto – Gi
Glasauer, Susan, (519) 824-2453 glasauer@uoguelph.ca, Univ of Guelph – Py
Glaser, Paul H., 612-624-8395 glase001@umn.edu, Univ of Minnesota, Twin Cities – Gn
Glasmann, Joseph R., (541) 737-1237 Oregon State Univ – Gs
Glass, Alex, 919 684 5847 alex.glass@duke.edu, Duke Univ – Ggu
Glass, Billy P., (302) 831-8229 bglass@udel.edu, Univ of Delaware – Xm
Glass, Charles E., (520) 621-6063 Univ of Arizona – Ne
Glassley, William E., (925) 422-6499 glassley@llnl.gov, Lawrence Livermore National Lab – Cg
Glatzmaier, Gary A., (831) 459-5504 glatz@pmc.ucsc.edu, Univ of California, Santa Cruz – Xy
Glawe, Lloyd N., (318) 342-1883 Univ of Louisiana, Monroe – Pm
Glazner, Allen F., (919) 962-0689 afg@unc.edu, Univ of North Carolina, Chapel Hill – Gi
Gleason, Gayle C., (607) 753-2816 gleasong@cortland.edu, SUNY, Cortland – Gc
Glenister, Brian F., (319) 335-1828 brian-glenister@uiowa.edu, Univ of Iowa – Pi
Glenn, Craig R., (808) 956-2200 glenn@soest.hawaii.edu or cglenn@hawaii.edu, Univ of Hawai'i, Manoa – Gs
Glenn, Ed, (520) 626-2664 eglenn@ag.arizona.edu, Univ of Arizona – Oo
Glenn, Nancy F., 208-345-1994 glennanc@isu.edu, Idaho State Univ – Or
Glessner, Justin, (217) 333-2695 jglessne@uiuc.edu, Univ of Illinois, Urbana-Champaign – Cs
Glidden, Karen L, (207) 780-5350 glidden@usm.maine.edu, Univ of Southern Maine – Oo
Glover, David M., (508) 289-2656 dglover@whoi.edu, Woods Hole Oceanographic Institution – Oc
Glover, Leslie J., (501) 543-8536 Univ of Arkansas, Pine Bluff – Sc
Glover, Paul W., 418-656-5180 paul.glover@ggl.ulaval.ca, Universite Laval – Yx
Glowacka, Ewaglowacka@cicese.mx,

Centro de Inv Científica y de Educación Superior de Ensenada – Ys
Gloyn, Robert W., (801) 537-3308 nrugs.rgloyn@state.ut.us, Utah Geological Survey – Em
Gluhovsky, Alexander, (765) 494-0670 aglu@purdue.edu, Purdue Univ – Oa
Glumac, Chair, Bosiljka, (413) 585-3680 bglumac@email.smith.edu, Smith College – Gs
Glynn, William G.wgglynn@gw.dec.ny.us, SUNY College at Brockport – Gg
Gobel, Volker W., (936) 468-2493 vgobel@sfasu.edu, Stephen F. Austin State Univ – Gi
Goble, Ronald J., 402-472-2647 rgoble@unl.edu, Univ of Nebraska, Lincoln – GzCc
Gobler, ChristopherChristopher.Gobler@stonybrook.edu, SUNY, Stony Brook – Ob
Godchaux, Martha M., (208) 882-9062 Mount Holyoke College – Gv
Godfrey, Emily, (609) 984-6587 emily.godfrey@dep.state.nj.us, New Jersey Geological Survey – Cl
Godfrey-Smith, Dorothy I., (902) 494-1451 Dalhousie Univ – Cc
Goedecke, George H.ggoedeck@nmsu.edu, New Mexico State Univ, Las Cruces – Oa
Goeke, James W., (308) 696-6704 jgoeke@unl.edu, Unversity of Nebraska - Lincoln – Hw
Goetz, Alexander, 303-492-5086 goetz@cses.colorado.edu, Univ of Colorado – Or
Goetz, Andrew R., 303-871-2674 agoetz@du.edu, Univ of Denver – Og
Goetz, Bruce A., 715.682.1312 bgoetz@northland.edu, Northland College – GgmHg
Goff, Fraser, (505) 667-8060 fraser@lanl.gov, Los Alamos National Lab – Cg
Goff, John A., (512) 471-0476 goff@ig.utexas.edu, Univ of Texas at Austin, Jackson School of Geosciences – Yr
Goforth, Thomas T., (254) 710-2183 tom_goforth@baylor.edu, Baylor Univ – Yg
Gokten, Ergungokten@eng.ankara.edu.tr, Ankara Univ – Gc
Gold, Aliceagold@calpoly.edu, California Polytechnic State Univ – Oo
Gold, David P., (814) 865-7261 gold@ems.psu.edu, Pennsylvania State Univ, Univ Park – Gc
Goldberg, David S., (845) 365-8674 Columbia Univ – Yr
Goldberg, Edward D., (858) 534-2407 egoldberg@ucsd.edu, Univ of California, San Diego – Oc
Goldfarb, Richard J., 303-492-7025 goldfarb@usgs.gov, Univ of Colorado – Oo
Goldfarb, Theodore D., (631) 632-7908 tgoldfarb@notes.cc.sunysb.edu, SUNY, Stony Brook – Oo
Goldfinger, Chris, (541) 737-5214 gold@coas.oregonstate.edu, Oregon State Univ – Ou
Goldhamer, David A., (559) 646-6500 dagoldhamer@ucdavis.edu, Univ of California, Davis – Oo
Goldman, Daniel, (937) 229-4820 dan.goldman@notes.udayton.edu, Univ of Dayton – Po
Goldman, Stephen L., (419) 530-1540 Univ of Toledo – Oo
Goldreich, Peter M., (626) 395-6193 pmg@gps.caltech.edu, California Institute of Technology – Xy
Goldsmith, Victor, (212) 772-5450 Graduate School of the City Univ of New York – On
Goldstein, Barry, (253) 879-3822 goldstein@ups.edu, Univ of Puget Sound – Gl
Goldstein, Fredric R., (609) 771-2569 Trenton State College – Oe
Goldstein, Joseph I., (413) 545-0300 jigo@ecs.umass.edu, Univ of Massachusetts – Oo
Goldstein, Robert H., (785) 864-2738 gold@ku.edu, Univ of Kansas – Gs
Goldstein, Steven, (845) 365-8787 Columbia Univ – Cg
Goldstein, Steven L., (845) 365-8787 steveg@ldeo.columbia.edu, Columbia Univ – Cs
Goldstein, Susan T., (706) 542-2652 sgoldst@gly.uga.edu, Univ of Georgia – Pm
Golomb, Dan S., (508) 934-2274 Univ of Massachusetts/Lowell – Oa
Golombek, Matthew P., (818) 393-7948 mgolombek@jpl.nasa.gov, Jet Propulsion Lab – Gc
Goloshubin, Gennady, 713-743-3399 Univ of Houston – Ye
Golubev, Nikolay G., (801) 585-3505 ngolubev@mines.utah.edu, Univ of Utah – Ye
Goman, Michelle F., 607255-4726 goman@geology.cornell.edu, Cornell Univ – Oo
Gomberg, Joan S., (901) 678-4858 gomberg@ceri.memphis.edu, Univ of Memphis – Ys
Gombosi, Tamas I., (313) 764-7222 Univ of Michigan – Oa
Gomes, Nicole, (925) 685-1230 (Ext. 2423) Diablo Valley College – Oo
Gomez, Basil, (812) 237-2249 gegomez@isugw.indstate.edu, Indiana State Univ – Gm
Gomez, Enriqueegomez@cicese.mx, Centro de Inv Científica y de Educación Superior de Ensenada – Ye
Gomez, Francisco, (607) 255-6889 fgomez@geology.cornell.edu, Cornell Univ – Gt
Gomez, Francisco, 573/882-9744 fgomez@missouri.edu, Univ of Missouri – Gt
Gomezdelcampo, Enrique, 419 372 2886 egomezd@bgsu.edu, Bowling Green State Univ – Ge
Goncharov, Alexander, (202) 478-8947 agoncharov@gl.ciw.edu, Carnegie Institution of Washington – Gy
Gong, Hongmian, (212) 772-4658 hgong@geo.hunter.cuny.edu, Hunter College (CUNY) – Oo
Goni, Miguel, (541) 737-0578 mgoni@coas.oregonstate.edu, Oregon State Univ – Oc
Gonneea, Meagan, (508) 289-3236 mgonneea@whoi.edu, Woods Hole Oceanographic Institution – Cm
Gonzales, David A., (970) 247-7378 gonzales_d@fortlewis.edu, Fort Lewis College – Gx
Gonzalez, Jose J.javier@cicese.mx, Centro de Inv Científica y de Educación Superior de Ensenada – Yd
Gonzalez, Luis A., 785-864-2743 lgonzalez@ku.edu, Univ of Kansas – Cs
Gonzalez, Maria M., 989-774-3179 gonza1mm@cmich.edu, Central Michigan Univ – Em
Gonzalez, Richard, (619) 260-4600 gonzalez@acusd.edu, Univ of San Diego – Oo
Gonzalez Leon, Carlos M.cmgleon@servidor.unam.mx, Universidad Nacional Autonoma de Mexico – Gr
Gonzalez-Fernandez, Antoniomindundi@cicese.mx, Centro de Inv Científica y de Educación Superior de Ensenada – Gu
Good, Charles M., (540) 231-6467 Virginia Polytechnic Institute & State Univ – Oo
Good, Daniel B., (912) 681-5361 DanGood@GeorgiaSouthern.edu, Georgia Southern Univ – Oo
Good, James W., (541) 737-2525 good@coas.oregonstate.edu, Oregon State Univ – Og
Good, Steven C.sgood@wcupa.edu, West Chester Univ – Pe
Goodbred, Jr., Steven L., 615-343-6424 steven.goodbred@vanderbilt.edu, Vanderbilt Univ – Gs
Goodell, Laurel P.goodell@princeton.edu, Princeton Univ – Yr
Goodell, Laurel P., (609) 258-1043 laurel@princeton.edu, Princeton Univ – Oe
Goodell, Philip C., (915) 747-5593 goodell@geo.utep.edu, Univ of Texas, El Paso – Ce
Goodfellow, Wayne D., (613) 996-8163 wayne.goodfellow@nrcan-rncan.gc.ca, Univ of Ottawa – Cg
Goodge, John W., (218) 726-8486 jgoodge@d.umn.edu, Univ of Minnesota, Duluth – Gp
Goodin, Ruth, 305-284-4253 ruthgoodin@miami.edu, Univ of Miami – Oo
Gooding, Patrick J., (859) 389-8810 gooding@uky.edu, Univ of Kentucky – Eo
Gooding, Timothy, (202) 633-1795 goodingt@si.edu, Smithsonian Institution / National Museum of Natural History – Gg
Goodison, Marjorie, 585-275-5713 margie@earth.rochester.edu, Univ of Rochester – Oo
Goodliffe, Andrewamg@ua.edu, Univ of Alabama – Yg
Goodling, Lynn M., 717.702.2054 lygoodling@state.pa.us, Pennsylvania Bureau of Topographic & Geologic Survey – Oo
Goodman, Karen J., 607 778-5417 goodman_k@sunybroome.edu, Broome Community College – Og
Goodman, Lowell, (701) 777-4590 Univ of North Dakota – Ou
Goodrich, David C., (520) 670-6380 dgoodrich@tucson.ars.ag.gov, Univ of Arizona – Hs
Goodwillie, Andrew, (845)-365-8717 Columbia Univ – Yr
Goodwin, Alan M., (416) 978-5613 Univ of Toronto – Gg
Goodwin, David H., (740) 587-5621 goodwind@denison.edu, Denison Univ – Pi
Goodwin, Laurel, 608-262-8960 laurel@geology.wisc.edu, Univ of Wisconsin, Madison – Gc
Goodwin, Laurel B., (608) 265-4232 laurel@geology.wisc.edu,

New Mexico Institute of Mining and Technology – Gc
Goodwin, Mark B., (505) 835-5178 Univ of California, Berkeley – Pv
Goos, Robert J., (701) 231-8581 North Dakota State Univ – Sc
Gooseff, Michael N., 303-273-3800 Colorado School of Mines – Hs
Gopal, Sucharitasuchi@bu.edu,
Boston Univ – Oi
Gopal, Sucharita, (617) 353-5744 suchi@crsa.bu.edu,
Boston Univ – Oo
Gordon, Arnold L., (845) 365-8325 agordon@ldeo.columbia.edu,
Columbia Univ – Op
Gordon, Christopher, 614 265 6594 chris.gordon@dnr.state.oh.us,
Ohio Dept of Natural Resources – Ec
Gordon, Elizabeth A., (920) 403-3227 elizabeth.gordon@snc.edu,
Saint Norbert College – Gs
Gordon, Richard G., (713) 348-5279 rgg@rice.edu,
Rice Univ – Ym
Gordon, Richard G., (713) 348-5279 rgg@rice.edu,
Rice Univ – Gt
Gordon, Robert B., (203) 432-3125 gordon@hess.geology.yale.edu,
Yale Univ – Nr
Gordon, Steven J., (719) 333-3067 steve.gordon@usafa.af.mil,
United States Air Force Academy – Gm
Gordon, Terence M., (403) 220-8301 tmg@geo.ucalgary.ca,
Univ of Calgary – Gq
Gore, Pamela J. W., 678-891-3754 pgore@gpc.edu,
Georgia Perimeter College – GsOeGn
Gorelick, Steven, (650) 725-2950 gorelick@geo.stanford.edu,
Stanford Univ – Yr
Gorelick, Steven M., (650) 725-2950 gorelick@pangea.stanford.edu,
Stanford Univ – Hw
Gorman, Andrew R., +64 3 479-7516 andrew.gorman@otago.ac.nz,
Univ of Otago – Yg
Gorman, Donald R., (309) 677-2347 rock@bradley.bradley.edu,
Bradley Univ – Gs
Gorman, W. Alan, (613) 545-6176 Queen's Univ – Gm
Gorman-Lewis, Drew J., 206-543-1190 dgormanl@u.washington.edu,
Univ of Washington – PyCl
Gorring, Matthew L., (973) 655-5409 gorringm@mail.montclair.edu,
Montclair State Univ – Gi
Gorsevski, Peter, (419) 372-7201 peterg@bgsu.edu,
Bowling Green State Univ – OirGm
Gorsline, Donn S., (213) 740-5125 gorsline@usc.edu,
Univ of Southern California – Gu
Gose, Wulf A., (512) 471-3596 wulf@mail.utexas.edu,
Univ of Texas at Austin, Jackson School of Geosciences – Ym
Gosnell, Hannah, (541) 737-1222 gosnell@colorado.edu,
Oregon State Univ – Ou
Gosnold, William D., (701) 777-2631 william_gosnold@und.edu,
Univ of North Dakota – Yh
Goss, Michael J., (519) 824-4120 (Ext. 2491) mgoss@uoguelph.ca,
Univ of Guelph – So
Gosse, John, 902-494-6632 jcgosse@is.dal.CA,
Univ of Kansas – Gm
Gosse, John C., 902-494-6632 John.Gosse@Dal.Ca,
Dalhousie Univ – Cg
Gosselin, David C., (402) 472-8919 gosselin@unl.edu,
Unversity of Nebraska - Lincoln – CgHw
Gosselin, Michel, (418) 724-1761 michel_gosselin@uqar.uquebec.ca,
Universite du Quebec a Rimouski – Ob
Gotkowitz, Madeline B., 608-262-1580 mbgotkow@facstaff.wisc.edu,
Univ of Wisconsin, Madison – Hw
Gottfried, Michael D., 517-432-5480 gottfrie@msu.edu,
Michigan State Univ – Pv
Gottgens, Johan F., (419) 530-8451 johan.gottgens@utoledo.edu,
Univ of Toledo – Oo
Goudge, Theodore L., (660) 562-1798 tgoudge@nwmissouri.edu,
Northwest Missouri State Univ – Oy
Goudreau, Joanne, (508) 289-2560 jgoudreau@whoi.edu,
Woods Hole Oceanographic Institution – Cm
Gough, D. Ian, (780) 439-0124 Univ of Alberta – Yg
Gouin, Francis R., (301) 405-4374 fgouin@alsc.umd.edu,
Univ of Maryland – Oo
Gould, Mark D., (401) 254-3087 Roger Williams Univ – Ob
Goulden, Michael L., (949) 824-1983 Univ of California, Irvine – Oo
Goulet, Normand, (514) 987-3375 r27254@er.uqam.ca,
Universite du Quebec a Montreal – Gc
Goulty, Neil R., 0191 3742513 n.r.goulty,
Durham Univ – Yg
Goyne, Keith W., 573-882-0090 Univ of Missouri, Columbia – Sc
Grable, Judy, (229) 333-5752 Valdosta State Univ – Hs
Grace, Cathy A., (662)915-7498 cag@olemiss.edu,
Univ of Mississippi – Gg
Grace, Peter R., 61 7 3138 2610 pr.grace@qut.edu.au,
Queensland Univ of Technology – Sb
Graczyk, Donald G., (630) 252-3489 Argonne National Lab – Cs
Gradie, Jonathan C., (808) 625-5262 Univ of Hawai'i, Manoa – Xg
Grady, William, 304 594 2331 grady@geosrv.wvnet.edu,
West Virginia Univ – Ec
Graebner, Robert J., (512) 232-3422 bob.graebner@beg.utexas.edu,
Univ of Texas at Austin, Jackson School of Geosciences – Oo
Graedel, Thomas E., (203) 432-9733 thomas.graedel@yale.edu,
Yale Univ – Cl
Graetz, Donald A., (352) 392-1803 (Ext. 318) dag@ifas.ufl.edu,
Univ of Florida – Sc
Graf, Jr., Joseph L., (541) 552-6861 graf@sou.edu,
Southern Oregon Univ – Em
Graff, Thomas O., (479) 575-3878 tgraff@comp.uark.edu,
Univ of Arkansas, Fayetteville – Ou
Graff, William P., (609) 292-2576 bill.graff@dep.state.nj.us,
New Jersey Geological Survey – Oy
Graham, Barbara A., (806) 742-3103 barbara.graham@ttu.edu,
Texas Tech Univ – Oo
Graham, David W., (541) 737-4140 dgraham@coas.oregonstate.edu,
Oregon State Univ – Ou
Graham, Elizabeth Y., 205-348-6427 bgraham@geo.ua.edu,
Univ of Alabama – Ca
Graham, Iani.graham@unsw.edu.au,
Univ of New South Wales – GxEg
Graham, John A., (202) 478-8867 Carnegie Insti of Washington – Oo
Graham, Julie, (413) 545-5935 graham@geo.umass.edu,
Univ of Massachusetts – Oo
Graham, Linda K., (608) 262-2640 Univ of Wisconsin, Madison – Ob
Graham, M., (403) 220-5850 margg@geo.ucalgary.ca,
Univ of Calgary – Oo
Graham, Michael, (831) 771-4400 mgraham@mlml.calstate.edu,
Moss Landing Marine Laboratories – Ob
Graham, Peter H., (612) 625-8268 pgraham@soils.umn.edu,
Univ of Minnesota, Twin Cities – Sb
Graham, Russell, (303) 370-8363 RGraham@dmns.org,
Denver Museum of Nature & Science – Pv
Graham, Russell W., (814) 865-6336 graham@ems.psu.edu,
Pennsylvania State Univ, Univ Park – Pv
Graham, Stephan A., (650) 723-0507 graham@pangea.stanford.edu,
Stanford Univ – Go
Graham, Vickie, (207) 872-3244 Virginia Polytechnic Institute & State Univ – Oo
Graham, Jr., Earl K., (814) 865-2273 graham@ems.psu.edu,
Pennsylvania State Univ, Univ Park – Yx
Graham, Jr., James H., (863) 956-1151 jhg@gnv.ifas.ufl.edu,
Univ of Florida – Sb
Grainger, Richard, (270) 809-6301 richard.grainger@murraystate.edu,
Murray State Univ – Oy
Graizer, Vladimir, (916) 445-4764 California Geological Survey – Oo
Grala, Katarzynz, (662) 325-9942 kg160@msstate.edu,
Mississippi State Univ – Oi
Grammer, G M., (269) 387-3667 michael.grammer@wmich.edu,
Western Michigan Univ – Gs
Gramsch, Stephen A., (202) 478-8939 s.gramsch@gl.ciw.edu,
Carnegie Institution of Washington – Gy
Gran, Karen, 218-726-7406 kgran@d.umn.edu,
Univ of Minnesota, Duluth – Gms
Grand, Stephen P., (512) 471-3005 steveg@maestro.geo.utexas.edu,
Univ of Texas at Austin, Jackson School of Geosciences – Ys
Grandal d'Anglade, Aurora, 00 34 981 167000 xeaurora@udc.es,
Coruna Univ – Pv
Grande, Anthony, 212-772-5265 tony.grande@hunter.cuny.edu,
Hunter College (CUNY) – Oo
Grande, Lance, (312) 665-7632 grande@fmnh.org,
Field Museum of Natural History – Pv
Grandstaff, David E., (215) 204-8228 grand@temple.edu,
Temple Univ – Cl
Graney, Joseph R., (607) 777-6347 Binghamton Univ – Cl
Granger, Darryl E., (765) 494-0043 dgranger@purdue.edu,
Purdue Univ – Oa
Graniero, Phil A., (519) 253-3000 (Ext. 2485) graniero@uwindsor.ca,
Univ of Windsor – Oi
Grannell, Roswitha B., (562) 985-4927 grannell@csulb.edu,
California State Univ, Long Beach – Yv
Grant, Gordon E., (541) 750-7328 gordon.grant@oregonstate.edu,
Oregon State Univ – Gm

Grant, James A., (218) 726-7237 jgrant@d.umn.edu,
Univ of Minnesota, Duluth – Gp
Grant, John, (213) 265-8838 East Los Angeles College – Oo
Grant, John A., 202-633-2474 Smithsonian Institution / National Air & Space Museum – Xg
Grant, Jonathan, (902) 494-2021 jon.grant@dal.ca,
Dalhousie Univ – Ob
Grant, Raymond W., (480) 461-7008 rwgrant@mail.mc.maricopa.edu,
Mesa Community College – Gg
Grasmueck, Mark, 305 421-4858 mgrasmueck@rsmas.miami.edu,
Univ of Miami – Yr
Grasso, Cheryl, 401-874-2265 cgrasso@uri.edu,
Univ of Rhode Island – Oo
Grathoff, Georg H., (503) 725-3391 grathoffg@pdx.edu,
Portland State Univ – Gz
Grattan, Stephen R., (530) 752-1130 srgrattan@ucdavis.edu,
Univ of California, Davis – Oo
Graumlich, Lisa J., (520) 621-6465 Univ of Arizona – Pe
Graumlich, Lisa J., (406) 994-5320 lisa@montana.edu,
Montana State Univ – Ou
Graustein, William C., (203) 287-2853 william.graustein@yale.edu,
Yale Univ – Cl
Gravely, Cynthia Rae, (864) 656-3438 gravelc@clemson.edu,
Clemson Univ – Oo
Graves, Alexandria, 919-513-0635 alexandria_graves@ncsu.edu,
North Carolina State Univ – Sb
Graves, Charles E., (314) 977-3121 Saint Louis Univ – Oa
Gray, Lee M., (330) 823-3605 graylm@muc.edu,
Mount Union College – Po
Gray, M. Joan, (860) 486-1386 Ball State Univ – Oo
Gray, Mary Beth, (570) 577-1146 mbgray@bucknell.edu,
Bucknell Univ – Gc
Gray, Norman H.normal.gray@uconn.edu,
Univ of Connecticut – Gx
Gray, Robert S., (805) 965-0581 gray@sbcc.net,
Santa Barbara City College – Gs
Gray, Sarah, (860) 486-1386 sgray@acusd.edu,
Univ of San Diego – Oo
Gray, William G., (219) 631-5942 william.g.gray.3@nd.edu,
Univ of Notre Dame – Hw
Grayson, R. Larry, 814-863-1644 rlg19@psu.edu,
Pennsylvania State Univ, Univ Park – Nm
Greaney, Peter H., (850) 488-0462 Tallahassee Comm College – Gs
Greatbatch, Richard J., (902) 494-6674 richard.greatbatch@dal.ca,
Dalhousie Univ – Op
Greb, Stephen F., (859) 257-5500 (Ext. 136) greb@uky.edu,
Univ of Kentucky – Ec
Greeley, Ronald, (480) 965-7045 greeley@asu.edu,
Arizona State Univ – Xg
Green, Delores J., (434) 951-6346 delores.green@dmme.virginia.gov,
Division of Mineral Resources – Oo
Green, Douglas H., (740) 593-1843 green@ohio.edu,
Ohio Univ, Lancaster – Yg
Green, Harry W., (510) 642-3059 Univ of California, Berkeley – Pv
Green, Jack, (562) 985-4198 jgreen3@csulb.edu,
California State Univ, Long Beach – Gv
Green, John C., (218) 726-7208 jgreen@d.umn.edu,
Univ of Minnesota, Duluth – Gi
Green, Louise, (709) 737-4519 louise@sparky2.esd.mun.ca,
Memorial Univ of Newfoundland – Oo
Green, Richard C., (850) 488-9380 Florida Geological Survey – Gr
Green, II, Harry W., (951) 827-4505 harry.green@ucr.edu,
Univ of California, Riverside – Yx
Greenberg, Jeffrey K., (630) 752-5866 jeffrey.k.greenberg@wheaton.edu,
Wheaton College – Gt
Greenberg, Richard J., (520) 621-6940 greenberg@lpl.arizona.edu,
Univ of Arizona – Xy
Greenberg, Sallie E., 217-244-4068 greenberg@isgs.uiuc.edu,
Illinois State Geological Survey – Cs
Greene, Barbara, (570) 484-2048 bgreene@lhup.edu,
Lock Haven Univ – Oo
Greene, Brian M., 412-395-7323 Brian.Greene@usace.army.mil,
Youngstown State Univ – Ng
Greene, Charles H., (607) 255-5449 chg2@cornell.edu,
Cornell Univ – Ob
Greene, David C., (740) 587-6476 greened@denison.edu,
Denison Univ – Gc
Greene, Don M., (254) 710-2193 don_greene@baylor.edu,
Baylor Univ – Oyw
Greene, Roberta, 315-267-2286 greenera@potsdam.edu,
SUNY, Potsdam – Oo
Greenfield, Roy J., (814) 237-1810 roy@geosc.psu.edu,
Pennsylvania State Univ, Univ Park – Yg
Greenstein, Benjamin J., (319) 895-4307 bgreenstein@cornellcollege.edu,
Cornell College – Gu
Greenwood, James P., 860 685-2545 jgreenwood@wesleyan.edu,
Wesleyan Univ – Xc
Greenwood, Richard B., (213) 620-5023 California Geological Survey – Gg
Greer, Lisa, (814) 865-6723 lgreer@geosc.psu.edu,
Washington & Lee Univ – Pe
Gregg, Jay M., 405-744-6358 jay.gregg@okstate.edu,
Oklahoma State Univ – Gs
Gregg, Michael C., (206) 543-1353 gregg@apl.washington.edu,
Univ of Washington – Op
Gregg, Tara, 509-527-5696 greggtb@whitman.edu,
Whitman College – Gg
Gregg, Tracy K. P., (716) 645-6800 (Ext. 2463) tgregg@geology.buffalo.edu, SUNY, Buffalo – Gv
Gregg, William J., (906) 497-2795 wjgregg@mtu.edu,
Michigan Technological Univ – Gc
Gregor, C. B., 937 775-3455 Wright State Univ – Gs
Gregory, Joseph T., (510) 643-2108 Univ of California, Berkeley – Pv
Gregory, Robert T., (214) 768-3075 Southern Methodist Univ – Cs
Greiner, Georgie, (319) 335-1819 georgiana-greiner@uiowa.edu,
Univ of Iowa – Oo
Grellet-Tinner, Gerald, 605-394-5326 Gerald.Grellet-Tinner@sdsmt.edu,
South Dakota School of Mines & Technology – Pv
Grenci, Lee M., (814) 863-4227 grenci@ems.psu.edu,
The Pennsylvania State Univ, Univ Park – Ow
Grender, Gordon C., (540) 231-6521 Virginia Polytechnic Institute & State Univ – Go
Gretener, Peter E., (403) 220-5849 Univ of Calgary – Ng
Greuel, Robert, (414) 424-4105 Univ of Wisconsin, Oshkosh – Oy
Grew, Edward S., (207) 581-2169 esgrew@maine.edu,
Univ of Maine – Gp
Grew, Priscilla C., (402) 472-7854 pgrew2@unl.edu,
Univ of Nebraska, Lincoln – Gg
Grgicak-Mannnion, Alice, (519) -25-3-30 (Ext. ext. 2170) grgica3@uwindsor.ca,
Univ of Windsor – Oi
Grguric, Gordan, (609) 652-4492 iaprod73@stockton.edu,
Stockton State College – Oc
Grice, Joel D., (613) 952-3510 Univ of Manitoba – Gz
Gries, John C., (316) 978-3140 Wichita State Univ – Gt
Griest, Debra A., 217-244-2390 griest@isgs.uiuc.edu,
Illinois State Geological Survey – Gg
Grieve, Richard A. F., (613) 995-5372 impact@gsc.emr.ca,
Univ of New Brunswick – Xg
Grießl, Stefan, 089/2180 4188 stefan.griessl@uni.muenchen.de,
LMU - Univ of Munich – Gz
Griffen, Dana T., 801-422-2305 dana_griffen@byu.edu,
Brigham Young Univ – Gz
Griffin, John R., (402) 472-4632 jgriffin2@unl.edu,
Univ of Nebraska, Lincoln – Eo
Griffin, Kevin L., (845) 365-8371 griff@ldeo.columbia.edu,
Columbia Univ – Oo
Griffin, Robert J., 603-862-2021 rob.griffin@unh.edu,
Univ of New Hampshire – Oa
Griffin, William F., (906) 482-4810 wgriffin@mtu.edu,
Michigan Technological Univ – Hw
Griffing, David H., 607-431-4629 GriffingD@Hartwick.edu,
Hartwick College – Gd
Griffith, Caitlin, (520) 626-3806 griffith@lpl.arizona.edu,
Univ of Arizona – Oo
Grigg, Richard W., (808) 956-7186 rgrigg@soest.hawaii.edu,
Univ of Hawai'i, Manoa – Ob
Griggs, Gary B., (831) 459-5006 ggriggs@pmc.ucsc.edu,
Univ of California, Santa Cruz – On
Grigsby, Jeffry D., (765) 285-8264 00jdgrigsby@bsu.edu,
Ball State Univ – Gs
Grimaldi, Richard, (607) 436-3069 grimalr@oneonta.edu,
SUNY, Oneonta – Oa
Grimes, Stephen W., (709) 737-4333 sgrimes@mun.ca,
Memorial Univ of Newfoundland – Gc
Grimley, David A., 217-244-7324 grimley@isgs.uiuc.edu,
Illinois State Geological Survey – Ym
Grimm, Kurt A., (604) 822-9258 kgrimm@eos.ubc.ca,
Univ of British Columbia – Gs
Grindlay, Nancy R., (910) 962-2352 grindlayn@uncw.edu,
Univ of North Carolina Wilmington – Yg

Grisafe, David A., (785) 864-2137 dgrisafe@kgs.ku.edu,
Univ of Kansas – Om
Grismer, Mark E., (530) 752-3245 megrismer@ucdavis.edu,
Univ of California, Davis – Hw
Grismer, Mark E., (530) 752-3243 megrismer@ucdavis.edu,
Univ of California, Davis – Sp
Griswold, George B., (505) 299-6192 New Mexico Institute of Mining and Technology – Nx
Gritzo, Russell E., (505) 667-0481 Los Alamos National Lab – Oo
Groat, Charles G., (512) 471-1772 cgroat@mail.utexas.edu,
Univ of Texas at Austin, Jackson School of Geosciences – Eg
Groat, Lee A., (604) 822-4525 lgroat@eos.ubc.ca,
Univ of British Columbia – Gz
Grocke, Darren R., 905-525-9140 (Ext. 26334) grocke@mcmaster.ca,
McMaster Univ – Cs
Groenevelt, Pieter H., (519) 824-4120 (Ext. 53585) pgroenev@lrs.uoguelph.ca, Univ of Guelph – Sp
Groenewold, Gerald H., (701) 777-5000 Univ of North Dakota – Hw
Gromet, L. Peter, (401) 863-1920 peter_gromet@brown.edu,
Brown Univ – Cc
Groppo, John G., (606) 257-0254 Univ of Kentucky – Nm
Grosch, Chester E., (757) 683-4931 enright@ccpo.odu.edu,
Old Dominion Univ – Op
Grose, L. Trowbridge, (303) 273-3806 tgrose@mines.edu,
Colorado School of Mines – Gt
Grosfils, Eric B., (909) 621-8673 egrosfils@pomona.edu,
Pomona College – Xg
Groshong, Jr., Richard H., 205-348-1882 rgroshon@geo.ua.edu,
Univ of Alabama – Gc
Gross, Gerardo W., (505) 835-5132 New Mexico Institute of Mining and Technology – Yx
Gross, Michael, (305) 348-3932 grossm@fiu.edu,
Florida International Univ – Gc
Grossman, Ethan L., (979) 845-0637 grossman@geo.tamu.edu,
Texas A&M Univ – Cs
Grossman, Lawrence, (773) 702-8153 chair@geosci.uchicago.edu,
Univ of Chicago – Xc
Grossman, Lawrence S., (540) 231-5116 Virginia Polytechnic Institute & State Univ – Oo
Groszos, Mark S., (229) 333-5664 msgroszo@valdosta.edu,
Valdosta State Univ – Gc
Grote, Katherine R., 715/836-5485 grotekr@uwec.edu,
Univ of Wisconsin, Eau Claire – Hw
Grove, Karen, (415) 338-2617 kgrove@sfsu.edu,
San Francisco State Univ – Gs
Grove, Timothy L., (617) 253-2878 tlgrove@mit.edu,
Massachusetts Institute of Technology – Gi
Grover, John E., (513) 556-6674 john.grover@uc.edu,
Univ of Cincinnati – Gz
Grover, Timothy W., (802) 468-1289 tim.grover@castleton.edu,
Castleton State College – Gp
Groves, Christopher, 270-745-5974 chris.groves@wku.edu,
Univ of Kentucky – Hg
Groves, Christopher, (270) 745-5974 chris.groves@wku.edu,
Western Kentucky Univ – Hg
Groves, John R., (319) 273-3072 john.groves@uni.edu,
Univ of Northern Iowa – Pg
Grube, John P., 217-244-1716 grube@isgs.uiuc.edu,
Illinois State Geological Survey – Eo
Gruen, Dieter M., (630) 972-3513 Argonne National Lab – Om
Grujic, Djordje, (902) 494-2208 dgrujic@is.dal.ca,
Dalhousie Univ – Gt
Grunbaum, Daniel, (206) 221-6594 grunbaum@ocean.washington.edu,
Univ of Washington – Ob
Grunder, Anita L., (541) 737-1249 grundera@geo.oregonstate.edu,
Oregon State Univ – Gi
Grundl, Timothy J., 414-229-4765 grundl@uwm.edu,
Univ of Wisconsin, Milwaukee – HwCl
Grundy, Doug, (905) 525-9140 (Ext. 27694) grundy@mcmaster.ca,
McMaster Univ – Yr
Grunwald, Sabine, (352) 392-1951 sgrunwald@mail.ifas.ufl.edu,
Univ of Florida – So
Gruszkiewicz, Miroslaw S., (865) 574-4965 gruszkiewicz@ornl.gov,
Oak Ridge National Lab – Oo
Gruzleski, John E., (514) 398-4755 McGill Univ – Nx
Gryta, Jeffrey J., (814) 732-2803 grytaj@edinboro.edu,
Edinboro Univ of Pennsylvania – Gm
Gu, Baohua, 865-574-7286 gub1@ornl.gov,
Oak Ridge National Lab – Cl
Gu, Jeff, (780) 492-2292 jgu@phys.ualberta.ca,
Univ of Alberta – Ys
Guan, Hann-wen, (404) 894-3895 hwg@minitower.gtri.gatech.edu,
Georgia Institute of Technology – Oa
Guard, Charles, (671) 735-2685 Univ of Guam – Ow
Guber, Albert L.guber@ems.psu.edu,
Pennsylvania State Univ, Univ Park – Pe
Guccione, Margaret J., (479) 575-3354 guccione@comp.uark.edu,
Univ of Arkansas, Fayetteville – Gm
Guenert, James S., (404) 656-3214 Georgia Dept of Nat Resources – Hw
Guerin, Gilles, (845)-365-8671 Columbia Univ – Gu
Guerin-Place, Rosemary, (916) 327-1851 California Geological Survey – Oo
Guerrero, Victoria, (202) 282-7374 Univ of the District of Columbia – Oo
Guertal, Elizabeth A.eguertal@acesag.auburn.edu,
Auburn Univ – Sc
Guevara, Edgar H., (512) 471-3159 edgar.guevara@beg.utexas.edu,
Univ of Texas at Austin, Jackson School of Geosciences – Go
Guggenheim, Stephen J., (312) 996-3263 xtal@uic.edu,
Univ of Illinois, Chicago – Gz
Guha, Jayanta, 418 545 5222 jguha@uqac.uquebec.ca,
Universite du Quebec a Chicoutimi – Eg
Guiberson, Patrick F., (775) 784-1434 patg@unr.edu,
Univ of Nevada, Reno – Oy
Guidera, JoAnn, (916) 445-1923 California Geological Survey – Oo
Guilbert, John M., (520) 621-6024 guilbert@geo.arizona.edu,
Univ of Arizona – Em
Guillemette, Renald N., (979) 845-6301 guillemette@geo.tamu.edu,
Texas A&M Univ – Gz
Guinasso, Norman, (979) 862-2323 Texas A&M Univ – Op
Guinness, Edward, (314) 935-5493 guinness@wunder.wustl.edu,
Washington Univ – Or
Guitierrez-Alonso, Gabrielgaby@gugu.usal.es,
Florida International Univ – Gc
Guitjens, John C., (775) 784-4071 guitjens@ers.unr.edu,
Univ of Nevada, Reno – Hq
Gulick, Sean S., (512) 471-3262 sean@ig.utexas.edu,
Univ of Texas at Austin, Jackson School of Geosciences – Yr
Gulley, Harold E., (414) 424-7115 Univ of Wisconsin, Oshkosh – Oo
Gulliver, John S., (612) 625-4080 gulli003@tc.umn.edu,
Univ of Minnesota, Twin Cities – Hs
Gundersen, James N., (316) 978-3140 Wichita State Univ – Ga
Gunderson, Lance, 404 727 8108 lgunder@emory.edu,
Emory Univ – Sf
Gundiler, Ibrahim H., (505) 835-5730 gundiler@gis.nmt.edu,
New Mexico Institute of Mining & Technology – Nx
Gunter, Mickey E., (208) 885-6015 mgunter@uidaho.edu,
Univ of Idaho – Gz
Gunter, William D., (403) 472-4406 Univ of Calgary – Cl
Guo, Laodong, 228-688-1176 Laodong.Guo@usm.edu,
Univ of Southern Mississippi – Oc
Gupta, Anupma, (845) 365-8550 ag2001@columbia.edu,
Columbia Univ – Gt
Gupta, Hoshin V., (520) 626-6974 hoshin@hwr.arizona.edu,
Univ of Arizona – Hs
Gupta, Satish C., (612) 625-1241 sgupta@soils.umn.edu,
Univ of Minnesota, Twin Cities – Sp
Gurevich, Boris, +61 8 9266-7359 B.Gurevich@curtin.edu.au,
Curtin Univ of Technology – Ye
Gurlea, Lawrence P.lisobar@aol.com,
Youngstown State Univ – Cg
Gurney, Kevin, (765) 494-5982 keving@purdue.edu,
Purdue Univ – Oa
Gurnis, Michael C., (626) 395-6979 gurnis@gps.caltech.edu,
California Institute of Technology – Ys
Gurrola, Harold, (806) 742-3299 harold.gurrola@ttu.edu,
Texas Tech Univ – Ys
Gust, David A., 61 7 3138 2217 d.gust@qut.edu.au,
Queensland Univ of Technology – Gi
Gustin, Cheryl, (801) 537-3360 nrugs.cgustin@state.ut.us,
Utah Geological Survey – Oo
Guth, Lawrence R., 978-665-3082 lguth@fsc.edu,
Fitchburg State College – Gc
Guth, Peter L., (410) 293-6560 pguth@usna.edu,
United States Naval Academy – Gc
Guthrie, George D., (505) 665-6340 gguthrie@lanl.gov,
Los Alamos National Lab – Gz
Guthrie, Roderick I., (514) 398-4755 McGill Univ – Nx
Gutierrez, Melida, (417) 836-5967 meg434f@smsu.edu,
Southwest Missouri State Univ – Hy
Gutierrez, Roberto, (512) 471-0342 oskar@mail.utexas.edu,
Univ of Texas at Austin, Jackson School of Geosciences – Yd

Gutmann, James T., (860) 685-2258 jgutmann@wesleyan.edu, Wesleyan Univ – Gv
Gutowski, Vincent P., (217) 581-3825 cfvpg@ux1.eiu.edu, Eastern Illinois Univ – Gm
Gutowski, William J., (515) 294-5632 gutowski@iastate.edu, Iowa State Univ of Science & Technology – Ow
Gutowski, William J., (515) 294-5632 gutowski@iastate.edu, Iowa State Univ of Science & Technology – Oa
Gutzler, David J., (505) 277-3328 gutlzer@unm.edu, Univ of New Mexico – Oa
Guven, Necip, (806) 742-3110 necip.guven@ttu.edu, Texas Tech Univ – Gz
Guy, Jr., Donald E., 419-626-4296 don.guy@dnr.state.oh.us, Ohio Dept of Natural Resources – On
Guy, Jr., Willard J., (785) 864-2076 bguy@kgs.ku.edu, Univ of Kansas – Go
Guza, Robert T., (858) 534-0585 rguza@ucsd.edu, Univ of California, San Diego – On
Guzina, Bojan B., (612) 626-0789 guzina@wave.ce.umn.edu, Univ of Minnesota, Twin Cities – Nr
Gwynn, J. Wallace, (801) 537-3366 nrugs.wgwynn@state.ut.us, Utah Geological Survey – En
Gyakum, John R., (514) 398-6076 john.gyakum@mcgill.ca, McGill Univ – Ow
Göttlich, Hagen, 089/2180 5615 goettlich@ennab.de, LMU - Univ of Munich – Gz

H

Haack, Sheridan K., 517-887-8909 skhaack@srvrdmilns.er.usgs.gov, Michigan State Univ – Sb
Haas, Johnson R., 269-387-2878 johnson.haas@wmich.edu, Western Michigan Univ – Cl
Haase, Jennifer S., (765) 494-8677 jhaase@purdue.edu, Purdue Univ – Yd
Habib, Ayman F., (614) 292-6918 Ohio State Univ – Oo
Habib, Daniel, (718) 997-3333 Graduate School of the City Univ of New York – Pl
Habib, Daniel, (718) 997-3333 daniel_habib@qc.edu, Queens College (CUNY) – Pl
Haby, Vince A., 903-834-6191 VHaby@ag.tamu.edu, Texas A&M Univ – Sc
Hacker, Bradley R., (805) 893-7952 Univ of California, Santa Barbara – Gp
Hacker, David B., 330-672-8831 dhacker@kent.edu, Kent State Univ – Gc
Hacker, Sally D., (360) 546-9634 hacker@vancouver.wsu.edu, Washington State Univ – Ob
Hackley, Keith C., 217-244-2396 hackley@isgs.uiuc.edu, Illinois State Geological Survey – Cs
Haddad, Brent, (831) 459-4149 bhaddad@cats.ucsc.edu, Univ of California, Santa Cruz – Oo
Haddadin, Munther J.munther_haddadin@hotmail.com, Oregon State Univ – Hw
Haddock, Gerald H., (630) 752-5063 Wheaton College – Gi
Haddock, Gregory D., (660) 562-1719 haddock@nwmissouri.edu, Northwest Missouri State Univ – Oi
Haderlie, Eugene C., (408) 656-2918 Naval Postgraduate School – Ob
Hadizadeh, Jafar, (502) 852-6821 Univ of Louisville – Gc
Haff, Peter K., (919) 684-5902 haff@duke.edu, Duke Univ – Oo
Hafner, James A., (413) 545-0778 hafner@geo.umass.edu, Univ of Massachusetts – Oo
Hagan, Robert M., (530) 752-0453 Univ of California, Davis – Hg
Hagedorn, Charles, (540) 231-4895 chagedor@vt.edu, Virginia Polytechnic Institute & State Univ – Sb
Hagelberg, Carl R., (505) 667-3596 drch@lanl.gov, Los Alamos National Lab – Oo
Hageman, Steven J., (828) 262-6609 hagemansj@appstate.edu, Appalachian State Univ – Pi
Hager, Bradford H., (617) 253-0126 bhhager@mit.edu, Massachusetts Institute of Technology – Ys
Hager, Carl A., 410-293-6561 hager@usna.edu, United States Naval Academy – Op
Hager, Richard P., (609) 652-4405 Stockton State College – Oo
Hagerty, Michael, 617-552-8300 michael.hagerty.1@bc.edu, Boston College – Ys
Haggart, JamesJHaggart@nrcan.gc.ca, Univ of British Columbia – Pg
Haggerty, Janet, 918-631-2304 janet-haggerty@utulsa.edu, Univ of Tulsa – Gu
Haggerty, Roy D., (541) 737-1210 haggertr@geo.oregonstate.edu, Oregon State Univ – Hw
Haggerty, Stephen E., 305-348-2617 haggerty@fiu.edu, Florida International Univ – Gz
Hagni, Richard D., 573-341-4657 Univ of Missouri, Rolla – Eg
Haileab, Bereket, (507) 646-4402 bhaileab@carleton.edu, Carleton College – Gx
Haine, Thomas W., (410) 516-7048 thomas.haine@jhu.edu, The Johns Hopkins Univ – Op
Haines, Forest E.fhaines@adrian.edu, Adrian College – Gd
Haines, Meredith A., (727) 553-1625 Univ of South Florida, Saint Petersburg – Og
Haitjema, Hendrick M., (812) 855-5971 haitjema@indiana.edu, Indiana Univ, Bloomington – Hw
Haitjema, Hendrik M., (812) 855-0563 Indiana Univ, Bloomington – Hw
Haj, Adel E., 660-543-8095 haj@ucmo.edu, Univ of Central Missouri – Gm
Hajash, Andrew, (979) 845-2400 hajash@geo.tamu.edu, Texas A&M Univ – Gx
Hajnal, Zoltan, (306) 966-5694 zoltan.hajnal@usask.ca, Univ of Saskatchewan – Ys
Halama, Ralf, (301) 405-4088 rhalama@geol.umd.edu, Univ of Maryland – Cg
Halbach, Karl F., (805) 965-0581 (Ext. 2470) halbach@sbcc.net, Santa Barbara City College – Gm
Halbach, Thomas R., (612) 625-3135 thalbach@soils.umn.edu, Univ of Minnesota, Twin Cities – Sb
Halbig, Joseph B.halbig@wazoo.com, Univ of Hawai'i at Hilo – Ca
Halden, Norman M., (204) 474-6910 nm_halden@umanitoba.ca, Univ of Manitoba – Cg
Hale, Beverley A., (519) 824-4120 (Ext. 53434) bhale@lrs.uoguelph.ca, Univ of Guelph – Ct
Hale, Dave, 303-273-3461 dhale@mines.edu, Colorado School of Mines – Ye
Hale, Leslie, (202) 633-1796 halel@si.edu, Smithsonian Institution / National Museum of Natural History – Gg
Hales, Burke, (541) 737-8121 bhales@coas.oregonstate.edu, Oregon State Univ – Oc
Halfar, Jochen, (905) 828-5419 jochen.halfar@utoronto.ca, Univ of Toronto – Pe
Halfman, John D., (315) 781-3918 halfman@hws.edu, Hobart & William Smith Colleges – Gn
Halgedahl, Susan L., (801) 585-3963 shalg@mines.utah.edu, Univ of Utah – Ym
Halihan, Todd, 405-744-6358 halihan@okstate.edu, Oklahoma State Univ – Hw
Halka, Jeffrey P., 410-554-5543 jhalka@dnr.state.md.us, Maryland Geological Survey – Gu
Hall, Anne M., (404) 727-2863 ahall04@emory.edu, Emory Univ – Gz
Hall, Brenda L., (207) 581-2191 brendah@maine.edu, Univ of Maine – Gl
Hall, Chris, (734) 764-6391 cmhall@umich.edu, Univ of Michigan – Cc
Hall, Clarence A., (310) 825-1010 hall@ess.ucla.edu, Univ of California, Los Angeles – Gt
Hall, David W., (609) 984-6587 dave.hall@dep.state.nj.us, New Jersey Geological Survey – Yg
Hall, Donald H., (204) 474-9371 Univ of Manitoba – Ym
Hall, Frank R., (504) 280-1105 frhall@uno.edu, Univ of New Orleans – Ym
Hall, Fred L., (905) 525-9140 (Ext. 24334) hallfl@mcmaster.ca, McMaster Univ – Yr
Hall, Ihall@cf.ac.uk, Cardiff Univ – Oo
Hall, Jeremy, (709) 737-7569 jhall@waves.esd.mun.ca, Memorial Univ of Newfoundland – Ys
Hall, John C., 205 652 3725 jhall@uwa.edu, Univ of West Alabama – Og
Hall, Jude, (740) 587-6217 hall@denison.edu, Denison Univ – Oo
Hall, Luke D., (805) 642-3211 lhall@vcccd.net, Ventura College – Oy
Hall, Mark E., (404) 656-3214 Georgia Dept of Natural Resources – Gp
Hall, Michael, (727) 553-3916 mhall@seas.marine.usf.edu, Univ of South Florida, Saint Petersburg – Oo
Hall, Paulphall@bu.edu, Boston Univ – Yg
Hall, Robert D., (317) 274-0225 rhall@iupui.edu,

Indiana Univ / Purdue Univ, Indianapolis – Gl
Hall, Russell L., (403) 220-6678 Royal Tyrrell Mus of Palaeo – Pi
Hall, Russell L., (403) 220-6678 Univ of Calgary – Pi
Hall, Stuart A., (713) 743-3416 sahgeo@uh.edu,
Univ of Houston – Ym
Hall, III, John R., (504) 231-6305 jrhall3@vt.edu,
Virginia Polytechnic Institute & State Univ – So
Hallam, Neville A., 0191 3742522 n.a.hallam@durham.ac.uk,
Durham Univ – Oo
Haller, Mick, (541) 737-9141 hallerm@coas.oregonstate.edu,
Oregon State Univ – Oo
Hallet, Bernard, 206-685-2409 hallet@u.washington.edu,
Univ of Washington – GmOl
Halley, Robert B., (727) 893-3100 Univ of South Florida, Saint Petersburg – Gu
Hallmark, Charles T., (979) 845-4678 hallmark@tamu.edu,
Texas A&M Univ – Sd
Hallock, Brent G., (805)756-2436 bhallock@calpoly.edu,
California Polytechnic State Univ – Sf
Hallock-Muller, Pamela, (727) 553-1567 pmuller@marine.usf.edu,
Univ of South Florida, Saint Petersburg – Pm
Halloran, Paula, (231) 777-0289 paula.halloran@muskegoncc.edu,
Muskegon Community College – Oo
Halls, Henry C., 905-828-5363 hlhalls@utm.utoronto.ca,
Univ of Toronto – Ym
Halls, Joanne N., (910) 962-7614 hallsj@uncw.edu,
Univ of North Carolina Wilmington – Oi
Halsor, Sid P., (570) 408-4611 Wilkes Univ – Gi
Halverson, Larry, (515) 294-0495 larryh@iastate.edu,
Iowa State Univ of Science & Technology – So
Ham, Jay M., (913) 532-6119 Kansas State Univ – Sp
Ham, Nelson R., (920) 403-3977 nelson.ham@snc.edu,
Saint Norbert College – Gl
Hambleton, William W., 785-864-4974 Univ of Kansas – Oo
Hambleton, William W.wwhamble@ku.edu,
Univ of Kansas – Yg
Hamblin, William K., 801-422-2469 Brigham Young Univ – Gr
Hamburger, Michael W., (812) 855-2934 hamburg@indiana.edu,
Indiana Univ, Bloomington – Ys
Hameed, Sultan, (631) 632-8319 hameed@notes.cc.sunysb.edu,
SUNY, Stony Brook – Oa
Hamel, Jocelyn, (418) 654-2680 jhamel@nrcan.gc.ca,
Universite du Quebec – Oo
Hames, Willis E., (334) 844-4881 hameswe@auburn.edu,
Auburn Univ – Gp
Hamidzada, Nasir, (401) 874-4071 Univ of Rhode Island – Gx
Hamilton, Gordon S., (207) 581-3446 gordon.hamilton@maine.edu,
Univ of Maine – Gl
Hamilton, Michael Andrew, (416) 946-7424 mahamilton@geology.utoronto.ca,
Univ of Toronto – Cc
Hamilton, Stephen K., 616-671-2231 hamilton@kbs.msu.edu,
Michigan State Univ – Cg
Hamilton, Stuart, 850-474-3207 shamilton@uwf.edu,
Univ of West Florida – Oi
Hamilton, Victoria E., 808-956-3152 hamilton@higp.hawaii.edu,
Univ of Hawai'i, Manoa – Xy
Hamilton, Warren, 303-384-2047 whamilto@mines.edu,
Colorado School of Mines – Ys
Hammen, John L., (701) 777-4589 Univ of North Dakota – Oy
Hammer, Julia E., 808-956-5996 jhammer@hawaii.edu,
Univ of Hawai'i, Manoa – Gv
Hammer, Philip T., (604) 822-5703 phammer@eos.ubc.ca,
Univ of British Columbia – Ys
Hammer, William R., (309) 794-7487 glhammer@augustana.edu,
Augustana College – Pv
Hammes, Ursula, 512-471-1891 ursula.hammes@beg.utexas.edu,
Univ of Texas at Austin, Jackson School of Geosciences – Gg
Hammond, Andy, 61 7 31384185 a.hammond@qut.edu.au,
Queensland Univ of Technology – Gm
Hammond, Anne, (807) 343-8677 anne.hammond@lakeheadu.ca,
Lakehead Univ – Oo
Hammond, Douglas E., (213) 740-5837 dhammond@usc.edu,
Univ of Southern California – Cm
Hammond, Paul E., (503) 725-3387 hammondp@pdx.edu,
Portland State Univ – Gi
Hammond, William, 775 784-6691 Ext 153 whammond@unr.edu,
Univ of Nevada – Yd
Hammond, Jr., Weldon W., 210-458-5440 Univ of Texas, San Antonio – Hw
Hampson, Arthur, (801) 585-5698 spike.hampson@geog.utah.edu,
Univ of Utah – Oy
Hampton, Brian A., 517-432-3595 bhampton@msu.edu,
Michigan State Univ – Gs
Hampton, Duane R., (269) 387-5496 duane.hampton@wmich.edu,
Western Michigan Univ – Hw
Hams, Jacqueline, (818) 778-5566 hamsje@lavc.edu,
Los Angeles Valley College – Og
Han, De-hua, 713749150 Univ of Houston – Ye
Han, Nizhou, (540) 231-2403 nhan@vt.edu,
Virginia Polytechnic Institute & State Univ – Sc
Hanan, Barry B., (619) 594-6710 bhanan@mail.sdsu.edu,
San Diego State Univ – Cc
Hancock, Gregory, 61 02 4921 5090 Greg.Hancock@newcastle.edu.au,
Univ of Newcastle – OiGmm
Hancock, Gregory S., (757) 221-2446 gshanc@wm.edu,
College of William & Mary – Gm
Hand, Bryce M., (315) 443-2672 Syracuse Univ – Gs
Haneberg, William, (505) 255-8005 New Mexico Institute of Mining and Technology – Ng
Hanes, John A., (613) 545-6188 Queen's Univ – Cc
Haney, Christa M., 662-325-9645 meloche@geosci.msstate.edu,
Mississippi State Univ – Ow
Hanger, Rex, (262) 472-5258 hangerr@mail.uww.edu,
Univ of Wisconsin, Whitewater – Yr
Hanke, Brenda, (513) 455-7160 bhanke@cincymuseum.org,
Cincinnati Museum Center – Pi
Hanke, Brenda R., 513-556-3732 Univ of Cincinnati – Pi
Hanke, Steve H., (410) 516-7183 hanke@jhuvms.hcf.jhu.edu,
The Johns Hopkins Univ – Oo
Hanks, Sharon, (201) 595-3457 William Paterson College – Oo
Hanley, Thomas B., (706) 568-2075 hanley_tom@colstate.edu,
Columbus State Univ – Gc
Hanlon, Jr., Edward A., (941) 658-3400 hanlon@gnv.ifas.ufl.edu,
Univ of Florida – So
Hanna, Ruth L., (510) 373-4919 Las Positas College – Gg
Hanna, Stephen P., (540) 654-1490 shanna@mwc.edu,
Mary Washington College – Or
Hannah, Judith L., (970) 491-1329 jhannah@cnr.colostate.edu,
Colorado State Univ – Gi
Hannibal, Joseph T., (216) -23-1-46 (Ext ext 3233) hannibal@cmnh.org,
Cleveland Museum of Natural History – Pi
Hannibal, Joseph T., (216) 231-4600 jhanniba@cmnh.org,
Case Western Reserve Univ – Pi
Hannington, Mark, (613) 562-5292 Mark.Hannington@science.uottawa.ca,
Univ of Ottawa – Eg
Hannington, Mark D., (613) 788-4400 Carleton Univ – Em
Hannula, Kimberly, (970) 247-7463 hannula_k@fortlewis.edu,
Fort Lewis College – Gc
Hanor, Jeffrey S., (225) 388-3418 hanor@geol.lsu.edu,
Louisiana State Univ – Cl
Hansel, Ardith K., (217) 333-5852 hansel@isgs.uiuc.edu,
Illinois State Geological Survey – Gl
Hansen, Cindy, (604) 527-5217 hansenc@douglas.bc.ca,
Douglas College – Gg
Hansen, Edward C.hansen@hope.edu,
Hope College – Gm
Hansen, Gary B., 206-543-6221 ghansen@rad.ess.washington.edu,
Univ of Washington – XyOr
Hansen, Harry J., (410) 554-5554 hhansen@mgs.md.gov,
Maryland Geological Survey – Hw
Hansen, James E., (212) 678-5500 jhansen@giss.nasa.gov,
Columbia Univ – Oa
Hansen, Jeffrey C., (505) 667-5043 jchansen@lanl.gov,
Los Alamos National Lab – Oo
Hansen, Katherine J., (406) 994-6912 khansenmontana.edu,
Montana State Univ – Oa
Hansen, Michael C., (614) 265-6580 Ohio State Univ – Pg
Hansen, Neil C., (320) 589-1711 hansennc@caa.mrs.umn.edu,
Univ of Minnesota, Twin Cities – Sp
Hansen, Richard R., 217-333-5101 hansen@isgs.uiuc.edu,
Illinois State Geological Survey – Oo
Hansen, Thor A., (360) 650-3648 thorenet@cc.wwu.edu,
Western Washington Univ – Pg
Hansen, Vicki L., (218) 726-6211 vhansen@d.umn.edu,
Univ of Minnesota, Duluth – Gc
Hansen, William, (212) 346-1502 Pace Univ, New York Campus – Oa
Hansen, William J., (508) 929-8608 whansen@worcester.edu,
Worcester State College – Oi
Hansen-Speer, Karla, 812-237-2444 khansenspee@isugw.indstate.edu,

Indiana State Univ – Ga
Hanson, Andrew D., (702) 895-1092 Univ of Nevada, Las Vegas – Co
Hanson, B. B., (303) 492-2330 hansonb@spot.colorado.edu, Univ of Colorado – Oo
Hanson, Blaine R., (530) 752-1130 brhanson@ucdavis.edu, Univ of California, Davis – Hg
Hanson, Doug, 501 683-0115 doug.hanson@Arkansas.gov, Arkansas Geological Commission – Gg
Hanson, Gilbert N., 631-632-8210 gilbert.hanson@sunysb.edu, SUNY, Stony Brook – Cc
Hanson, Lindley S.lhanson@salemstate.edu, Salem State College – Gm
Hanson, Paul, (402) 472-7762 phanson2@unl.edu, Unversity of Nebraska - Lincoln – Gm
Hanson, Richard E., (817) 257-7996 r.hanson@tcu.edu, Texas Christian Univ – GxvGt
Hanson, Sarah L., 517-264-3944 slhanson@adrian.edu, Adrian College – Gi
Hanson, Jr., Alfred K., (401) 874-6899 akhanson@gso.uri.edu, Univ of Rhode Island – Oc
Hapke, Bruce W., (412) 624-8876 hapke@pitt.edu, Univ of Pittsburgh – Xy
Hapke, Cheryl, (401)874-2265 chapke@usgs.gov, Univ of Rhode Island – On
Haq, Saad, (765) 496-7206 haq@purdue.edu, Purdue Univ – Gt
Hara, Tetsu, (401) 874-6509 thara@uri.edu, Univ of Rhode Island – Op
Harbaugh, John W., (650) 723-3365 harbaugh@pangea.stanford.edu, Stanford Univ – Gq
Harbert, William P., (412) 624-8874 harbert@pitt.edu, Univ of Pittsburgh – Ym
Harbolt, Carolyn L., (509) 359-2286 charbolt@mail.ewu.edu, Eastern Washington Univ – Oo
Harbor, David J., 540 458 8871 harbord@wlu.edu, Washington & Lee Univ – Gm
Harbottle, Garmangarman@bnl.gov, SUNY, Stony Brook – Cc
Hardage, Bob A., (512) 471-0300 bob.hardage@beg.utexas.edu, Univ of Texas at Austin, Jackson School of Geosciences – Yg
Harder, Brian J., (225) 578-8533 bhardel@lsu.edu, Louisiana State Univ – Eo
Harder, Brian J., (225) 388-8533 brian@lgs.bri.lsu.edu, Louisiana State Univ – Eo
Harder, Steven H., (915) 747-5746 harder@geo.utep.edu, Univ of Texas, El Paso – Yg
Harder, Susan L., (360) 546-9512 Washington State Univ, Vancouver – Ou
Hardie, Lawrence A., (410) 516-7050 hardie@ekman.eps.jhu.edu, The Johns Hopkins Univ – Gs
Hardin, Perry J., (801) 378-6062 perry_hardin@byu.edu, Brigham Young Univ – Or
Harding, Chris, (515) 294-7521 charding@iastate.edu, Iowa State Univ of Science & Technology – Gq
Harding, Lawrence W., (301) 405-6372 larry@kestrel.umd.edu, Univ of Maryland – Ob
Hardwick, Susan W., (541) 346-4557 susanh@uoregon.edu, Univ of Oregon – Oo
Hardy, Douglas R., (802) 649-1829 dhardy@geo.umass.edu, Univ of Massachusetts – Hy
Hardy, Shaun J., (202) 478-7960 hardy@dtm.ciw.edu, Carnegie Institution of Washington – Oo
Hardy, Shaun J., (202) 478-7960 s.hardy@dtm.ciw.edu, Carnegie Institution of Washington – Oo
Hargrave, Barry T., (902) 426-3188 hargraveb@mar.dfo-mpo.gc.ca, Dalhousie Univ – Ob
Hargrave, Phyllis, 406-496-4606 phargrave@mtech.edu, Montana Tech of The Univ of Montana – Gg
Hargraves, Paul E., (401) 874-6241 pharg@gso.uri.edu, Univ of Rhode Island – Ob
Hargreaves, Bruce R., (610) 758-3683 brh0@lehigh.edu, Lehigh Univ – Ob
Hargrove, Don L., (850) 487-2219 Florida Geological Survey – Ye
Harkay, J. Russell, (603) 358-2588 Keene State College – Yg
Harker, R. Ian, (215) 898-5724 rharker@hotmail.com, Univ of Pennsylvania – Cp
Harkrider, David G., (626) 395-6910 California Institute of Tech – Ys
Harlan, Jeffrey, (303) 273-3610 jharlan@mines.edu, Colorado School of Mines – Om
Harlan, Stephen S., (703) 993-1210 George Mason Univ – Ym
Harlan, Thomas P., (520) 621-7680 Univ of Arizona – Oo
Harlin, Karen S., (217) 244-6413 k-harlin@uiuc.edu, Illinois State Water Survey – Ca
Harlow, George E., (212) 769-5378 Graduate School of the City Univ of New York – Gz
Harlow, George E., 212-769-5378 gharlow@amnh.org, American Museum of Natural History – Ga
Harms, Tekla A., (413) 542-2711 taharms@amherst.edu, Amherst College – Gc
Harmsen, Fraka J. M., (559) 278-2395 fraka_harmsen@csufresno.edu, California State Univ, Fresno – Gs
Harnett, Erika M., 206-543-0212 eharnett@ess.washington.edu, Univ of Washington – XyGev
Harold, Karen L., 419-626-4296 Ohio Dept of Natural Resources – Oo
Harper, Denver, (812) 855-7428 dharper@indiana.edu, Indiana Geological Survey – Ec
Harper, John A., 412.442.4230 jharper@state.pa.us, Pennsylvania Bureau of Topographic & Geologic Survey – Pi
Harper, John R., (250) 655-4035 cori@islandnet.com, Univ of Victoria – Gu
Harper, Stephen B., (252) 328-6773 harpers@mail.ecu.edu, East Carolina Univ – Gm
Harper, Jr., Charles W., (405) 325-7725 charper@ou.edu, Univ of Oklahoma – Pi
Harpp, Karen, (315) 228-7211 kharpp@mail.colgate.edu, Colgate Univ – Cg
Harrell, James A., (419) 530-2193 james.harrell@utoledo.edu, Univ of Toledo – Ga
Harries, Peter J., (813) 974-4974 harries@chuma.cas.usf.edu, Univ of South Florida, Tampa – Pg
Harrington, Calvin T., (229) 333-7410 tharring@valdosta.edu, Valdosta State Univ – Oo
Harrington, Charles D., (505) 667-0078 charrington@lanl.gov, Los Alamos National Lab – Gm
Harris, Andrew J., (808) 956-3157 Univ of Hawai'i, Manoa – Gv
Harris, Ann G., 330-941-3613 agharris@cc.ysu.edu, Youngstown State Univ – Ge
Harris, Brett D., +618 9266-3089 B.Harris@curtin.edu.au, Curtin Univ of Technology – HwYve
Harris, Bryn G., (613) 233-9835 McGill Univ – Nx
Harris, CharrisC@cf.ac.uk, Cardiff Univ – Ge
Harris, Courtney K., (804) 684-7194 ckharris@vims.edu, College of William & Mary – On
Harris, David C., (859) 257-5500 dcharris@uky.edu, Univ of Kentucky – Go
Harris, DeVerle P., (520) 621-6024 dharris@geo.arizona.edu, Univ of Arizona – Eg
Harris, Gene, (312) 996-4499 doggone@uic.edu, Univ of Illinois, Chicago – Oo
Harris, Glendon H., (706) 542-2968 Univ of Georgia – Sb
Harris, James B., 601-974-1343 harrijb@millsaps.edu, Millsaps College – Ye
Harris, Jasper L., (919) 530-6394 jharris@wpo.nccu.edu, North Carolina Central Univ – Oy
Harris, Jerry M., (650) 723-0496 Stanford Univ – Ys
Harris, John M., (213) 763-3321 jharris@nhm.org, Los Angeles County Museum of Natural History – Pv
Harris, Jonathan G., 662-325-7947 Mississippi State Univ – Gg
Harris, Kenneth L., (612)627-4780 Ext. 205 harri015@umn.edu, Univ of Minnesota, Twin Cities – Gl
Harris, Lee E., (321) 674-8096 lharris@fit.edu, Florida Institute of Technology – On
Harris, M. Scott, (843) 953-5589 harrisms@cofc.edu, College of Charleston – GseGa
Harris, Mark T., 414-229-5126 or 414-229-5777 mtharris@uwm.edu, Univ of Wisconsin, Milwaukee – Gs
Harris, Ralph, (415) 723-0496 McGill Univ – Oo
Harris, Randa R., (678) 839-4056 rharris@westga.edu, Univ of West Georgia – Hs
Harris, Richard S., (905) 525-9140 (Ext. 27216) harrisr@mcmaster.ca, McMaster Univ – Yr
Harris, Robert N., (801) 587-9366 rnharris@mines.utah.edu, Univ of Utah – Gu
Harris, Robert N., (541) 737-4370 rharris@coas.oregonstate.edu, Oregon State Univ – Yr
Harris, Robin F., (608) 263-5691 rfharris@wisc.edu, Univ of Wisconsin, Madison – Sb
Harris, Ron, (801) 422-9264 rharris@byu.edu, Brigham Young Univ – Gc
Harris, Sara, (604) 822-5674 sara@eos.ubc.ca,

Univ of British Columbia – Ou
Harris, Stewart M., (631) 632-8322 stewart.harris@notes.cc.sunysb.edu, SUNY, Stony Brook – Om
Harris, Trevor, (304) 293-5603 tharris2@wvu.edu, West Virginia Univ – Oy
Harris, Virginia M., (860) 685-2244 vharris@wesleyan.edu, Wesleyan Univ – Oo
Harris, W. Burleigh, (910) 962-3492 harrisw@uncw.edu, Univ of North Carolina Wilmington – Gs
Harris, Walter M., (313) 936-0511 wharris@sunshine.sprl.umich.edu, Univ of Michigan – Oa
Harris, William H., (718) 951-5416 Graduate School of the City Univ of New York – Hw
Harris, Jr., Stanley E., (618) 453-3351 Southern Illinois Univ Carbondale – Ge
Harris, Jr., Willie G., (352) 392-1951 apatite@ufl.edu, Univ of Florida – Sc
Harrison, Bruce I., (505) 835-5864 bruce@nmt.edu, New Mexico Institute of Mining and Technology – Gm
Harrison, Christopher G., 305 421-4610 charrison@rsmas.miami.edu, Univ of Miami – Yr
Harrison, Don E., (206) 526-6225 harrison@pmel.noaa.gov, Univ of Washington – Op
Harrison, Glen, (902) 426-7451 harrisong@mar.dfo-mpo.gc.ca, Dalhousie Univ – Ob
Harrison, Kevin G., (617) 552-4653 harriskg@bc.edu, Boston College – Cg
Harrison, Lee C., (518) 437-8741 lee@asrc.cestm.albany.edu, SUNY, Albany – Oa
Harrison, Michael J., (931) 372-3751 MHarrison@tntech.edu, Tennessee Technological Univ – Gc
Harrison, Robert S., (516) 299-2318 Long Island Univ, C.W. Post Campus – Oy
Harrison, T. Mark, (310) 825-7970 tmh@argon.ess.ucla.edu, Univ of California, Los Angeles – Cc
Harrison, Wendy J., (303) 273-3821 wharriso@mines.edu, Colorado School of Mines – Cl
Harrison, William E., (785) 864-2070 harrison@kgs.ku.edu, Univ of Kansas – Co
Harry, Dennis L., (970) 491-2714 dharry@cnr.colostate.edu, Colorado State Univ – Yg
Harsh, James B., (509) 335-3650 harsh@wsu.edu, Washington State Univ – Sc
Harshvardhan, (765) 494-0693 harsh@purdue.edu, Purdue Univ – Oa
Hart, Brian R.bhart@uwo.ca, Univ of Western Ontario – Ca
Hart, Bruce, 514-398-3677 hart@eps.mcgill.ca, McGill Univ – Go
Hart, David J., 608-262-1580 djhart@wisc.edu, Univ of Wisconsin, Madison – Hg
Hart, Evan A., (931) 372-3121 EHart@TnTech.edu, Tennessee Technological Univ – Oy
Hart, Garret L., 509-335-3009 ghart@wsu.edu, Washington State Univ – Cc
Hart, Richard L., (630) 252-5839 rlhart@anl.gov, Argonne National Lab – Oa
Hart, Robert, (850) 645-1552 rhart@met.fsu.edu, Florida State Univ – Ow
Hart, Stanley R., (508) 289-2837 shart@whoi.edu, Woods Hole Oceanographic Institution – Cg
Hart, William K., (513) 529-3216 hartwk@muohio.edu, Miami Univ – Gi
Harte, Michael, (541) 737-1339 mharte@coas.oregonstate.edu, Oregon State Univ – Og
Hartel, Peter G., (706) 542-0898 Univ of Georgia – Sb
Harter, Robert D., (603) 862-1020 Univ of New Hampshire – Sc
Harter, Thomas L., (530) 752-2709 thharter@ucdavis.edu, Univ of California, Davis – Hw
Hartley, Bruce M., +61 8 9266-2297 B.Hartley@curtin.edu.au, Curtin Univ of Technology – Yx
Hartman, Joseph H., (701) 777-5055 joseph_hartman@und.edu, Univ of North Dakota – Pi
Hartman, Mary J., (509) 373-0028 mary.hartman@pnl.gov, Pacific Northwest National Lab – Hw
Hartse, Hans E., (505) 664-8495 Los Alamos National Lab – Ys
Hartshorn, Truman A., 404413-5750 truman@gsu.edu, Georgia State Univ – Oo
Hartung, Steven L., (402) 472-3236 shartung1@unl.edu, University of Nebraska - Lincoln – Sd
Harty, Kimm M., (801) 537-3313 nrugs.kharty@state.ut.us, Utah Geological Survey – Gm
Harvey, Cyril H., (336) 316-2238 charvey@guilford.edu, Guilford College – Gr
Harvey, F. Edwin, (402) 472-8237 feharvey@unl.edu, University of Nebraska - Lincoln – Hw
Harvey, Ralph P., (216) 368-0198 rph@case.edu, Case Western Reserve Univ – Xm
Harvey, Rodger, (410) 326-7206 harvey@cbl.umces.edu, Univ of Maryland – Co
Harwood, David M., (402) 472-2648 dharwood1@unl.edu, Univ of Nebraska, Lincoln – Pm
Harwood, Richard D., 309-796-5271 harwoodr@bhc.edu, Black Hawk College – Gv
Harz, William H., (805) 965-0581 (Ext. 2316) harz@sbcc.edu, Santa Barbara City College – Hg
Hasan, Mainul, (514) 398-4755 McGill Univ – Nx
Hasan, Mohamed Ali, 03-79674203/4141 alihasan@um.edu.my, Univ of Malaya – Hg
Hasan, Syed E., (816) 235-2976 hasans@umkc.edu, Univ of Missouri-Kansas City – Ng
Haschenburger, Judy, 210-458-4956 Judy.Haschenburger@utsa.edu, Univ of Texas, San Antonio – Gm
Hase, Harold W., (970) 248-1161 shase@mesastate.edu, Mesa State College – Em
Hasenmueller, Nancy R., (812) 855-7428 hasenmue@indiana.edu, Indiana Geological Survey – Ge
Hasiotis, Stephen T., (785) 864-4941 hasiotis@ku.edu, Univ of Kansas – Pe
Hasnath, Syed, (617) 353-4352 has@bu.edu, Boston Univ – Oo
Hassan, Ahmed, (702) 862-5465 hassan@dri.edu, Univ of Nevada, Reno – Oo
Hassani, Ferri, (514) 398-4755 McGill Univ – Nm
Hassenmueller, Walter, (812) 855-2687 whasenmu@indiana.edu, Indiana Geological Survey – En
Hassett, John J., (217) 333-9472 Univ of Illinois, Urbana-Champaign – Sc
Hastenrath, Stefan L., (608)262-3659 slhasten@wisc.edu, Univ of Wisconsin, Madison – Oa
Hatch, Christine, (775) 784-4509 Univ of Nevada, Reno – Gg
Hatcher, Jr., Robert D., (865) 974-6565 bobmap@utk.edu, Univ of Tennessee, Knoxville – Gc
Hatheway, Richard B., 585-245-5291 hatheway@geneseo.edu, SUNY, Geneseo – Gx
Hatten, Jeff, (541) 737-8273 jhatten@coas.oregonstate.edu, Oregon State Univ – Oc
Hattin, Donald E., (812) 855-8232 hattin@indiana.edu, Indiana Univ, Bloomington – Gr
Hattori, Keiko, (613)562-5800 Ext 6866 khattori@uottawa.ca, Univ of Ottawa – Cg
Hatzenbuhler, Elaine C., 701-231-7951 e.hatzenbuhler@ndsu.edu, North Dakota State Univ – Oe
Hauck, II, Steven A., (216) 368-3675 hauck@case.edu, Case Western Reserve Univ – Yg
Haug, Eric J., (314) 977-3136 ejh@eas.slu.edu, Saint Louis Univ – Yr
Haugan, Doreen B., (780) 492-4825 doreen.haugan@ualberta.ca, Univ of Alberta – Oo
Hauksson, Egill, (626) 395-6954 hauksson@gps.caltech.edu, California Institute of Technology – Ys
Hauri, Erik H., (202) 478-8471 hauri@dtm.ciw.edu, Carnegie Institution of Washington – Cc
Hausback, Brian, (916) 278-6521 hausback@csus.edu, California State Univ, Sacramento – Gv
Hauser, Ernest C., 937 775-3455 ernest.hauser@wright.edu, Wright State Univ – Ye
Hausladen, Gary J., (775) 784-6345 hausl@unr.edu, Univ of Nevada, Reno – Oy
Hausmann, Laurie, 314-977-3131 hausmann@eas.slu.edu, Saint Louis Univ – Oo
Hautala, Susan L., (206) 543-0596 susanh@ocean.washington.edu, Univ of Washington – Op
Havenith, Hans-Balder, 32 4 3662035 HB.Havenith@ulg.ac.be, Universite de Liege – Ge
Haverluk, Terry W., (719) 333-8746 United States Air Force Acad – Og
Havholm, Karen G., (715) 836-2945 havholkg@uwec.edu, Univ of Wisconsin, Eau Claire – Oe
Havlin, John L., (913) 532-7211 Kansas State Univ – Sc
Havlin, John L., (919) 515-2655 havlin@ncsu.edu, North Carolina State Univ – Sc

Faculty Index -H

Hawke, B. Ray, (808) 956-3133 Univ of Hawai'i, Manoa – Xg
Hawkins, David P., (740) 587-5788 hawkins@denison.edu, Denison Univ – GxCc
Hawkins, James W., (858) 534-2161 jhawkins@ucsd.edu, Univ of California, San Diego – Gp
Hawkins, Michael E., (518) 486-2011 mhawkins@mail.nysed.gov, New York State Geological Survey – Gz
Hawkins, R. B., (520) 621-7273 rhawkins@ag.arizona.edu, Univ of Arizona – Hs
Hawkins, Timothy W., (717) 477-1662 twhawk@ship.edu, Shippensburg Univ – Oa
Hawkins, Ward L., (505) 667-5835 whawkins@lanl.gov, Los Alamos National Lab – Og
Hawkins, Sr., Jessie L., (850) 488-4191 Florida Geological Survey – Oo
Hawley, John W., (505) 255-4847 hgeomatters@qwest.net, New Mexico Institute of Mining & Technology – Ge
Hawley, Rebecca D., (928) 523-1251 d.hawley@nau.edu, Northern Arizona Univ – Oo
Hawley, Robert L., 603-646-2373 Robert.Hawley@dartmouth.edu, Dartmouth College – Gl
Hawman, Robert B., (706) 542-2398 rob@seismo.gly.uga.edu, Univ of Georgia – Ys
Hawthorne, Frank C., (204) 474-8861 frank_hawthorne@umanitoba.ca, Univ of Manitoba – Gz
Hay, Alex E., (902) 494-6657 alex.hay@dal.ca, Dalhousie Univ – Op
Hay, Richard, 520-615-7575 rhay@geo.arizona.edu, Univ of Arizona – Gd
Hay, Rodrick, (310) 243-3385 rhay@csudh.edu, California State Univ, Dominguez Hills – Or
Hayashi, Masaki, (403) 220-2794 hayashi@ucalgary.ca, Univ of Calgary – Hw
Hayden, Bruce P., (804) 924-0545 bph@virginia.edu, Univ of Virginia – Oa
Hayden, Geoffrey W., (912) 752-2597 Mercer Univ – Om
Hayden, Janice A., 412.442.4287 jhayden@state.pa.us, Pennsylvania Bureau of Topographic & Geologic Survey – Oo
Hayden, Martha C., (801) 537-3311 nrugs.mhayden@state.ut.us, Utah Geological Survey – Pv
Haydon, Wayne, (415) 904-7718 California Geological Survey – Oo
Hayes, Charlotte M., (847) 467-1002 chayes@earth.northwestern.edu, Northwestern Univ – Oo
Hayes, Dennis E., (845) 365-8470 deph@ldeo.columbia.edu, Columbia Univ – Yr
Hayes, Garry F., (209) 667-3090 California State Univ, Stanislaus – Gg
Hayes, Garry F., (209) 575-6294 Modesto Junior College – Gg
Hayes, John M., (914) 365-8470 jhayes@whoi.edu, Indiana Univ, Bloomington – Co
Hayes, John M., (914) 365-8470 jhayes@whoi.edu, Woods Hole Oceanographic Institution – Ct
Hayes, Michael J., (402) 472-4271 mhayes@enso.unl.edu, Univ of Nebraska, Lincoln – Oa
Hayes, Miles O., 504-280-6325 mhayes@uno.edu, Univ of New Orleans – Gs
Haygood, Margo G., (858) 534-5987 mhaygood@ucsd.edu, Univ of California, San Diego – Ob
Haymon, Rachel M., (805) 893-3718 Univ of California, Santa Barbara – Cm
Haynes, Jr., C. Vance, (520) 621-6307 Univ of Arizona – Ga
Hayob, Jodie L., (540) 654-1425 jhayob@mwc.edu, Univ of Mary Washington – Gp
Hays, James D., (845) 365-8403 jimhays@ldeo.columbia.edu, Columbia Univ – Pe
Hays, James F., 520-903-1634 jhays@post.harvard.edu, Univ of Arizona – Cp
Hays, Paul B., (313) 764-7220 Univ of Michigan – Oa
Hays, Phillip D., (479) 575-3355 Univ of Arkansas, Fayetteville – Cs
Hays, Starr, (607) 273-6623 (Ext. 10) srh15@cornell.edu, Paleontological Research Institution – Oo
Hays, Walter W., 703 255-2458 whays10211@aol.com, George Mason Univ – Yg
Hayward, Oliver T., (254) 710-2182 O_T_hayward@baylor.edu, Baylor Univ – Gm
Haywick, Douglas W., (334) 460-6381 dhaywick@jaguar1.usouthal.edu, Univ of South Alabama – Gs
Hazel, Kate V., 212-769-5390 kateh@amnh.org, American Museum of Natural History – Oo
Hazel, McGoffH.J.McGoff@rdg.ac.uk, Univ of Reading – Pg
Hazel, Jr., Joseph E., (928) 523-9145 joseph.hazel@nau.edu, Northern Arizona Univ – Gs
Hazelton, Glen, (508) 356-5372 Univ of Massachusetts, Boston/Harbor Campus – Oo
Hazen, Robert M., (202) 478-8962 r.hazen@gl.ciw.edu, Carnegie Institution of Washington – Gz
Hazen, Robert M., (703) 993-2163 rhazen@gmu.edu, George Mason Univ – Om
Hazlett, Richard W., (909) 621-8676 rhazlett@pomona.edu, Pomona College – Gv
He, Ruoying, 919-513-0943 ruoying_he@ncsu.edu, North Carolina State Univ – Op
Head, Martin J., (905) 688-5550 (Ext. 5216) mjhead@brocku.ca, Brock Univ – Pl
Head, Martin J., (905) 68-8 55 (Ext. x5216) mjhead@brocku.ca, Univ of Toronto – Pi
Head, III, James W., (401) 863-2526 james_head_iii@brown.edu, Brown Univ – Xg
Heald, Milton T., (304) 293-5603 West Virginia Univ – Gd
Healey, Michael, (604) 822-4705 mhealey@eos.ubc.ca, Univ of British Columbia – Ob
Healy, Richard J., (508) 289-3514 rhealy@whoi.edu, Woods Hole Oceanographic Institution – Oo
Heaman, Larry M., (780) 492-2778 larry.heaman@ualberta.ca, Univ of Alberta – Cc
Heaney, Peter, (814) 865-6821 heaney@geosc.psu.edu, Pennsylvania State Univ, Univ Park – Gz
Hearn, Carter, (202) 633-1756 hearnc@si.edu, Smithsonian Institution / National Museum of Natural History – Gi
Hearn, Elizabeth H.ehearn@eos.ubc.ca, Univ of British Columbia – Yg
Hearn, Thomas M., (505) 646-5076 thearn@nmsu.edu, New Mexico State Univ, Las Cruces – Ys
Heath, G. Ross, (206) 543-3153 rheath@u.washington.edu, Univ of Washington – Cm
Heath, Kathleen M., (812) 237-3004 anheath@isugw.indstate.edu, Indiana State Univ – Ga
Heatherington, Ann L., (352) 392-6220 aheath@geology.ufl.edu, Univ of Florida – Cc
Heaton, Janet, (304) 384-5327 Concord Univ – Oo
Heaton, Jill, (775) 784-8056 jheaton@unr.edu, Univ of Nevada, Reno – Oi
Heaton, Richard C., (505) 667-1141 Los Alamos National Lab – Oo
Heaton, Thomas H., (626) 395-6897 heaton@gps.caltech.edu, California Institute of Technology – Ne
Heaton, Timothy H., (605) 677-6122 Timothy.Heaton@usd.edu, Univ of South Dakota – PvOg
Heatwole, Charles A., (212) 772-5323 Graduate School of the City Univ of New York – Oo
Heatwole, Charles A., (212) 772-5323 cah@geo.hunter.cuny.edu, Hunter College (CUNY) – Oo
Heavers, Richard, (401) 254-3095 Roger Williams Univ – Op
Hebda, Richard J., (250) 387-5493 Univ of Victoria – Pl
Hebert, David L., (401) 874-6610 herbert@gso.uri.edu, Univ of Rhode Island – Op
Hebert, Rejean J., 418-656-3137 herbert@ggl.ulaval.ca, Universite Laval – Gi
Heck, Frederick R., (616) 592-2588 heckf@ferris.edu, Ferris State Univ – Gg
Heck, Richard, (519) 824-4120 (Ext. 52450) rheck@lrs.uoguelph.ca, Univ of Guelph – Sd
Heckathorn, Scott A., (419) 530-2925 scott.heckathorn@utoledo.edu, Univ of Toledo – Oo
Heckel, Philip H., (319) 335-1804 philip-heckel@uiowa.edu, Univ of Iowa – Gd
Heckel, Wolfgang, 089/2180 4331 heckl@lmu.de, LMU - Univ of Munich – Gz
Hedenquist, Jeffrey W., (613)562-5800 (Ext. 6534) hedenquist@aol.com, Univ of Ottawa – Cg
Heerschap, Lauren K., 970-247-7278 heerschap_l@fortlewis.edu, Fort Lewis College – Gt
Hefferan, Kevin P., (715) 346-4453 kheffera@uwsp.edu, Univ of Wisconsin, Stevens Point – GcxGt
Hefner, Rene, 210-458-2584 Rene.Hefner@utsa.edu, Univ of Texas, San Antonio – Gg
Hegarty, Patricia, +353 21 4902533 p.hegarty@ucc.ie, Univ College Cork – Oo
Hegner, Ernst, 089/2180 4274 hegner@lmu.de, LMU - Univ of Munich – Gg
Hehr, John G., (479) 575-3684 Univ of Arkansas, Fayetteville – Ow
Heidinger, Andrew, 608-263-6757 heidinger@ssec.wisc.edu, Univ of Wisconsin, Madison – Or

Heien, Gene W., (740) 593-1104 heien@ohio.edu,
Ohio Univ, Lancaster – Gz
Heigold, Paul C., (630) 252-7861 Argonne National Lab – Yg
Heiken, Grant, (505) 667-8477 heiken@lanl.gov,
Los Alamos National Lab – Gv
Heikes, Brian G., (401) 874-6810 zagar@notos.gso.uri.edu,
Univ of Rhode Island – Oa
Heil, Dean, (505) 646-3405 New Mexico State Univ, Las Cruces – Sc
Heimann, William H., (785) 625-5663 Fort Hays State Univ – Ge
Heimlich, Richard A., (330) 672-4124 rheimlic@kent.edu,
Kent State Univ – Gi
Heimpel, Moritz, (780) 492-3519 mheimpel@phys.ualberta.ca,
Univ of Alberta – Oo
Heinrich, Paul V., (225) 578-4398 heinric@lsu.edu,
Louisiana State Univ – Gs
Heinrichs, John F., (785) 628-4536 jheinric@fhsu.edu,
Fort Hays State Univ – OyrOl
Heinz, Dion L., (773) 702-3046 heinz@uchicago.edu,
Univ of Chicago – Yx
Heinzel, Chad E., 319-273-6168 chad.heinzel@uni.edu,
Univ of Northern Iowa – Gag
Heiser, Patricia A., 907 786-6854 afpah1@uaa.alaska.edu,
Univ of Alaska, Anchorage – Pe
Heitz, Leroy F., (671) 735-2691 lheitz@uog.edu,
Univ of Guam – Hs
Heiz, George, (301) 405-1797 gh17@umail.umd.edu,
Univ of Maryland – Cm
Heizler, Matthew T., (505) 835-5244 matt@nmt.edu,
New Mexico Institute of Mining and Technology – Cc
Helenek, Henry L., (309) 677-2350 helenek@bradley.bradley.edu,
Bradley Univ – Gt
Helenes, Javierjhelenes@cicese.mx,
Centro de Inv Científica y de Educación Superior de Ensenada – Gr
Helgers, Karen, (845) 687-5224 helgersk@sunyulster.edu,
SUNY Ulster County Community College – Og
Heliker, Christina C., (808) 967-8807 cheliker@usgs.gov,
Univ of Hawai'i at Hilo – Gv
Heller, Matthew J., (434) 951-6351 matt.heller@dmme.virginia.gov,
Division of Mineral Resources – Gg
Heller, Paul L., (307) 766-2245 heller@uwyo.edu,
Univ of Wyoming – Gs
Hellou, Jocelyn, (902) 426-7451 hellouj@mar.dfo-mpo.gc.ca,
Dalhousie Univ – Oc
Helmberger, Donald V., (626) 395-6998 helm@gps.caltech.edu,
California Institute of Technology – Ys
Helmick, Sara B., (505) 667-9583 Los Alamos National Lab – Oo
Helmke, Philip A., (608) 263-4947 pahelmke@wisc.edu,
Univ of Wisconsin, Madison – Sc
Helmle, Kevin P., (954)262-3656 kevinh@nsu.nova.edu,
Nova Southeastern Univ – Ob
Helmstaedt, Herwart, (613) 545-6175 Queen's Univ – Gc
Helper, Mark A., (512) 471-1009 helper@mail.utexas.edu,
Univ of Texas at Austin, Jackson School of Geosciences – Gc
Helsdon, John H., (605) 394-2291 jhelsdon@ias.sdsmt.edu,
South Dakota School of Mines & Technology – Oa
Hemani, Ahmad, (514) 398-4755 (Ext. 0517) McGill Univ – Nm
Hembree, Daniel, (740) 593-1101 hembree@ohio.edu,
Ohio Univ, Lancaster – Pi
Hemley, Russell J., (202) 478-8951 r.hemley@gl.ciw.edu,
Carnegie Institution of Washington – Yx
Hemmasi, Mohammad, (701) 777-4587 Univ of North Dakota – Oo
Hemming, Sidney R., (845) 365-8417 sidney@ldeo.columbia.edu,
Columbia Univ – Cs
Hemmings, Patsy, (817) 515-4537 Tarrant County Junior College – Oo
Hemmings, Sidney, (845) 365-8417 Columbia Univ – Cg
Hempel, Connan, +27 41 5042243 glacch@upe.ac.za,
Univ of Port Elizabeth – Gg
Hemphill-Haley, Eileenehh@EHHMusic.com,
Humboldt State Univ – Pm
Hemphill-Haley, Mark, (707) 826-3933 mah54@axe.humboldt.edu,
Humboldt State Univ – Gt
Hemsley, AhemsleyAR@cf.ac.uk,
Cardiff Univ – Pl
Hemzacek Laukant, Jean M., 773/442-6056 J-Hemzacek@neiu.edu,
Northeastern Illinois Univ – Sc
Hendershott, Myrl C., (858) 534-3643 mhendershott@ucsd.edu,
Univ of California, San Diego – Op
Henderson, Charles M., (403) 220-6170 henderson@geo.ucalgary.ca,
Univ of Calgary – Pm
Henderson, Edward J., (601) 354-6328 Mississippi Offi of Geology – Go
Henderson, Grant S., (416) 978-6041 henders@geology.utoronto.ca,
Univ of Toronto – Gz
Henderson, Harry W., (814) 863-2473 hwh@psu.edu,
The Pennsylvania State Univ, Univ Park – Ow
Henderson, Paul, (508) 289-3466 phenderson@whoi.edu,
Woods Hole Oceanographic Institution – Cm
Henderson, Stephen W., (770) 784-8345 shender@emory.edu,
Emory Univ-Oxford College – Pe
Henderson, Wayne G., (714) 278-3561 whenderson@fullerton.edu,
California State Univ, Fullerton – Pg
Hendricks, David M., (520) 621-7223 dhendric@ag.arizona.edu,
Univ of Arizona – Sc
Hendricks, Stephanie, 218-755-2920 shendricks@bemidjistate.edu,
Bemidji State Univ – Oo
Hendrickx, Jan M., (505) 835-5892 hendrick@nmt.edu,
New Mexico Institute of Mining and Technology – Hw
Hendrix, Marc S., (406) 243-5278 marc@selway.umt.edu,
Univ of Montana – Gs
Hendrix, Thomas E., 616-331-3728 Grand Valley State Univ – Gc
Hendry, Jim, (306) 966-5720 jim.hendry@usask.ca,
Univ of Saskatchewan – Hw
Hendy, Ingrid, 734615-6892 ihendy@umich.edu,
Univ of Michigan – Py
Henika, William S., (540) 231-4298 bhenika@vt.edu,
Virginia Polytechnic Institute & State Univ – Ng
Hennemeyer, Marc, 089/2180 4340 marc@hennemeyer.de,
LMU - Univ of Munich – Gz
Hennessee, E. L., 410-554-5519 lhennessee@dnr.state.md.us,
Maryland Geological Survey – On
Henning, Stanley J., (515) 294-7846 sjhennin@iastate.edu,
Iowa State Univ of Science & Technology – So
Hennings, Peter H., (307) 766-3386 Univ of Wyoming – Gc
Hennings, Ronald G., (608) 263-7395 rghennin@wisc.edu,
Univ of Wisconsin - Extension – Hw
Henrici, Amy C., (412) 622-3265 Carnegie Mus of Natural History – Pv
Henrickson, Eiler L., (507) 663-1655 Colorado College – Eg
Henry, Christopher D., (775) 784-6691 (Ext. 128) chenry@unr.edu,
Univ of Nevada – Gg
Henry, Darrell J., (225) 388-2693 dhenry@geol.lsu.edu,
Louisiana State Univ – Gp
Henry, Eric J., 910-962-7622 henrye@uncw.edu,
Univ of North Carolina Wilmington – Hw
Henry, Kathleen M., 217-244-4990 henry@isgs.uiuc.edu,
Illinois State Geological Survey – Gg
Henry, Vernon J., (912) 598-2463 Henry@SKIO.Peachnet.edu,
Georgia Southern Univ – Ou
Henry, Vernon J., (912) 598-2463 henry@skio.peachnet.edu,
Georgia Southern Univ – Ou
Henshel, Diane S., (812) 855-4556 dhenshel@indiana.edu,
Indiana Univ, Bloomington – Oo
Henson, Harvey, 618-453-7349 henson@geo.siu.edu,
Southern Illinois Univ Carbondale – Yg
Hentz, Paula A., (303) 370-6473 Phentz@dmns.org,
Denver Museum of Nature & Science – Oo
Hentz, Tucker F., (512) 471-7281 tucker.hentz@beg.utexas.edu,
Univ of Texas at Austin, Jackson School of Geosciences – Go
Henyey, Thomas L., (213) 740-5832 henyey@usc.edu,
Univ of Southern California – Yg
Henßel, Katja, 089/2180 6624 k.henssel@lrz.uni-muenchen.de,
LMU - Univ of Munich – Pg
Hepburn, J. Christopher, (617) 552-3642 hepburn@bc.edu,
Boston College – Gp
Hepburn, Judith, (617) 552-0843 hepburnj@bc.edu,
Boston College – Ge
Hepner, George F., (801) 581-6021 george.hepner@geog.utah.edu,
Univ of Utah – Oy
Hepner, Tiffany, (512) 475-9572 tiffany.hepner@beg.utexas.edu,
Univ of Texas at Austin, Jackson School of Geosciences – Og
Herbers, Thomas H., (408) 656-2917 herbers@oc.nps.navy.mil,
Naval Postgraduate School – Op
Herbert, Bruce E., (979) 845-2405 herbert@geo.tamu.edu,
Texas A&M Univ – Ge
Herbert, Jennifer, (630) 252-0493 Argonne National Lab – Gg
Herbert, Timothy D., (401) 863-1207 timothy_herbert@brown.edu,
Brown Univ – Pe
Herkenhoff, Ken E., (818) 354-3539 ken.e.herkenhoff@jpl.nasa.gov,
Jet Propulsion Lab – Xg
Herman, Ellen K., 570-577-1382 ekh008@bucknell.edu,
Bucknell Univ – Hw
Herman, Gregory C., (609) 984-6587 greg.herman@dep.state.nj.us,

New Jersey Geological Survey – Gc
Herman, Herbert, (631) 632-8480 herbert.herman@notes.cc.sunysb.edu, SUNY, Stony Brook – Oo
Herman, Janet S., (804) 924-0553 jsh5w@virginia.edu, Univ of Virginia – Cl
Herman, Louis M., (808) 538-0067 lherman@hawaii.edu, Univ of Hawai'i, Manoa – Oo
Herman, Rhett B., 540 831-5441 rherman@radford.edu, Radford Univ – Yg
Hermance, John F., (401) 863-3830 john_hermance@brown.edu, Brown Univ – Yg
Hermann, Albert, 206-526-6495 Albert.J.Hermann@noaa.gov, Univ of Washington – Op
Hermes, O D., (401) 874-2192 dhermes@uri.edu, Univ of Rhode Island – Gi
Hernandez, Gonzalo, 206-543-9055 hernandez@u.washington.edu, Univ of Washington – Yx
Heron, Duncan, (919) 684-5321 duncan.heron@duke.edu, Duke Univ – Gr
Heroux, Yvon, (418) 654-2637 yheroux@inrs.uquebec.ca, Universite du Quebec – Co
Heroy, David, 281-425-6332 dheroy@lee.edu, Lee College – Gg
Herrero-Bervera, Emilio, (808) 956-6192 herrero@soest.hawaii.edu, Univ of Hawai'i, Manoa – Ym
Herring, John, 928-350-2227 jherring@prescott.edu, Prescott College – Oa
Herring, Thomas A., (617) 253-5941 tah@mit.edu, Massachusetts Institute of Technology – Yd
Herrmann, Felix J., (604) 822-8628 fherrmann@eos.ubc.ca, Univ of British Columbia – Yx
Herrmann, Robert B., (314) 977-3120 rbh@eas.slu.edu, Saint Louis Univ – Ys
Herrstrom, Eileen A., (217) 333-7732 herrstro@uiuc.edu, Univ of Illinois, Urbana-Champaign – Gi
Herz, Norman, (706) 542-2652 Univ of Georgia – Gx
Herzberg, Claude T., (732) 445-3154 herzberg@rci.rutgers.edu, Rutgers, The State Univ of New Jersey – Cp
Herzog, Beverly, 217 244-2788 herzog@isgs.uiuc.edu, Illinois State Geological Survey – Hw
Hesp, Patrick A., (225) 578-6244 pahesp@lsu.edu, Louisiana State Univ – Gm
Hess, Darrel E., (415) 239-3104 dhess@ccsf.cc.ca.us, City College of San Francisco – Oy
Hess, Donna, (937) 229-3432 donna.hess@notes.udayton.edu, Univ of Dayton – Oo
Hess, Kai-Uwe, 089/2180 4275 hess@min.uni-muenchen.de, LMU - Univ of Munich – Gz
Hess, Paul C., (401) 863-1929 paul_hess@brown.edu, Brown Univ – Gi
Hess, Ronald H., (775) 784-6691 (Ext. 121) rhess@unr.edu, Univ of Nevada – Oy
Hess Tanguay, Lillian, 516 463-6545 geolht@hofstra.edu, Hofstra Univ – Gg
Hess-Tanguay, Lillian, (516) 299-2318 lhess@liu.edu, Long Island Univ, C.W. Post Campus – Gs
Hesse, Reinhard, 514-398-3627 hesse@eps.mcgill.ca, McGill Univ – Gs
Hessler, Angela M., 616-331-3669 hesslera@gvsu.edu, Grand Valley State Univ – Oe
Hessler, Robert R., (858) 534-2665 rhessler@ucsd.edu, Univ of California, San Diego – Ob
Hesterberg, Dean L., (919) 515-2636 North Carolina State Univ – Sc
Hetherington, Jean, (925) 685-1230 (Ext. 462) jhetheri@dvc.edu, Diablo Valley College – Gg
Hetland, Robert D., (979) 458-0096 rhetland@ocean.tamu.edu, Texas A&M Univ – Op
Heuermann, Richard, (314) 935-4614 heuer@levee.wustl.edu, Washington Univ – Oo
Heuss-Aßbichler, Soraya, 089/21804252 soraya@min.uni-muenchen.de, LMU - Univ of Munich – Gz
Hewins, Roger H., (732) 445-3232 hewins@rci.rutgers.edu, Rutgers, The State Univ of New Jersey – Xm
Hewitt, David A., (540) 231-6521 dhewitt@vt.edu, Virginia Polytechnic Institute & State Univ – Cp
Hey, Richard N., (808) 956-8972 hey@soest.hawaii.edu, Univ of Hawai'i, Manoa – Yr
Hey, Richard N., (808) 956-8972 hey@soest.hawaii.edu, Univ of Hawai'i, Manoa – Gu
Heymann, Dieter, (713) 348-4890 dieter@owlnet.rice.edu, Rice Univ – Xm
Heyniger, William C., (908) 737-3689 wheynige@kean.edu, Kean Univ – Oo
Heywood, Neil C., (715) 346-4452 nheywood@uwsp.edu, Univ of Wisconsin, Stevens Point – Oy
Hiatt, Eric E., (920) 424-7001 hiatt@uwosh.edu, Univ of Wisconsin, Oshkosh – Gs
Hibbard, James P., (919) 515-7242 jim_hibbard@ncsu.edu, North Carolina State Univ – Gc
Hibbard, Malcolm J.mhibbard@mines.unr.edu, Univ of Nevada, Reno – Gi
Hibbs, Barry, (323) 343-2414 bhibbs@calstatela.edu, California State Univ, Los Angeles – Hw
Hickcox, C W., (404) 727-0118 geocwh@emory.edu, Emory Univ – Gg
Hickcox, David H., (740) 368-3619 dhhickco@owu.edu, Ohio Wesleyan Univ – Oy
Hickey, Barbara M., (206) 543-4737 bhickey@u.washington.edu, Univ of Washington – Op
Hickey, Craig J., (662) 915-5963 chickey@olemiss.edu, Univ of Mississippi – Ys
Hickey, James, (660) 562-1817 jhickey@nwmissouri.edu, Northwest Missouri State Univ – Ge
Hickey, Kenneth, (604) 822-3765 khickey@eos.ubc.ca, Univ of British Columbia – Gz
Hickey, Leo J., (203) 432-3163 leo.hickey@yale.edu, Yale Univ – Pb
Hickey, William J., (608) 262-9018 wjhickey@wisc.edu, Univ of Wisconsin, Madison – Sb
Hickey-Vargas, Rosemary, (305) 348-3471 hickey@fiu.edu, Florida International Univ – Cg
Hickman, Carole S., (510) 642-3429 caroleh@socrates.berkeley.edu, Univ of California, Berkeley – Po
Hickmott, Donald D., (505) 667-8753 dhickmott@lanl.gov, Los Alamos National Lab – Gp
Hicks, Roberta (Robbie), (709) 737-8349 rhicks@mun.ca, Memorial Univ of Newfoundland – Yr
Hickson, Catherine J., (604) 666-9772 Chickson@nrcan.gc.ca, Univ of British Columbia – Gv
Hickson, Thomas A., (651) 962-5241 tahickson@stthomas.edu, Univ of Saint Thomas – Yr
Hicock, Stephen R., (519) 661-3189 shicock@uwo.ca, Univ of Western Ontario – Gl
Higgins, Charles G.higgins@redshift.com, Univ of California, Davis – Gm
Higgins, Chris T., (916) 322-9997 California Geological Survey – Gg
Higgins, Jerry D., (303) 273-3817 jhiggins@mines.edu, Colorado School of Mines – Ng
Higgins, Michael D., (418) 545-5011 (Ext. 5052) mhiggins@uqac.ca, Universite du Quebec a Chicoutimi – Cc
Higgins, Pennilynloligo@earth.rochester.edu, Univ of Rochester – Cs
Higgs, Bettie, +353 21 4902117 b.higgs@ucc.ie, Univ College Cork – Yg
Higgs, Ken, +353 21 4902290 k.higgs@ucc.ie, Univ College Cork – Pl
Higinbotham, Pamela, 724 938-4180 higinbotham@cup.edu, California Univ of Pennsylvania – Oo
Hilbish, Thomas J., (803) 777-6629 Univ of South Carolina – Ob
Hilde, Thomas W. C., (979) 845-8477 hilde@geo.tamu.edu, Texas A&M Univ – Yr
Hildebrand, Alan R., 403-220-2291 ahildebr@ucalgary.ca, Univ of Calgary – Xm
Hildebrand, John A., (858) 534-4069 jhildebrand@ucsd.edu, Univ of California, San Diego – Yr
Hildebrand, Stephen G.hildebrandsg@ornl.gov, Oak Ridge National Lab – Oo
Hildebrand, Steve T., (505) 667-4318 hildebrand@lanl.gov, Los Alamos National Lab – Ys
Hill, Arleen A., 901-678-2589 aahill@memphis.edu, Univ of Memphis – Oo
Hill, Chris, 619 644 7342 chris.hill@gcccd.net, Grossmont College – Gg
Hill, David W., (512) 239-0115 Austin Community College – Hw
Hill, Douglas, 63164215544 dhill@notes.cc.sunysb.edu, SUNY, Stony Brook – Oo
Hill, Elizabeth A., (412) 622-3247 Carnegie Museum of Natural History – Pv
Hill, James M., (410) 554-5542 jhill@dnr.state.md.us, Maryland Geological Survey – Cl

Hill, John R., (520) 621-6063 jrhill@email.arizona.edu,
Univ of Arizona – Nm
Hill, Joseph C., (570)389-4567 jhill@bloomu.edu,
Bloomsburg Univ – GcpGg
Hill, Lon-Claylhill@broward.edu,
Broward Community College, Central Campus – Xm
Hill, Malcolm, 617 373-4377 m.hill@neu.edu,
Northeastern Univ – OiGz
Hill, Margarita, (301) 405-4341 mh1_60@umail.umd.edu,
Univ of Maryland – Ou
Hill, Mary Louise, (807) 343-8319 mary.louise.hill@lakeheadu.ca,
Lakehead Univ – Gc
Hill, Paul S., (902) 494-2266 paul.hill@dal.ca,
Dalhousie Univ – Gs
Hill, Richard T., (812) 855-9583 hill2@indiana.edu,
Indiana Geological Survey – Oi
Hill, Robert L., (213) 620-4804 California Geological Survey – Eg
Hill, Robert L., (301) 405-1347 rh30@umail.umd.edu,
Univ of Maryland – Sb
Hill, Ronald J., 970-943-2145 rhill@western.edu,
Western State College of Colorado – GoCos
Hill, Ryan, 617 373-4684 r.hill@neu.edu,
Northeastern Univ – Gg
Hill, Tessa M., (530) 752-0179 tmhill@geology.ucdavis.edu,
Univ of California, Davis – Pe
Hiller, Lena, 516 463-5564 geolzh@hofstra.edu,
Hofstra Univ – Oo
Hills, Denise J., 205-247-3694 dhills@gsa.state.al.us,
Geological Survey of Alabama – Gg
Hills, Leonard V., (403) 220-5848 Univ of Calgary – Pl
Hills, Leonard V., (403) 220-5848 Royal Tyrrell Mus of Palaeo – Pl
Hilsenrath, Elmina J., (301) 405-4369 eh68@umail.umd.edu,
Univ of Maryland – Ou
Hilterman, Fred, (713) 743-5802 fhilterman@uh.edu,
Univ of Houston – Ye
Hilton, David R., (858) 822-0639 drhilton@ucsd.edu,
Univ of California, San Diego – Cs
Hilts, Stewart G., (519) 824-4120 (Ext. 52448) shilts@lrs.uoguelph.ca,
Univ of Guelph – Oo
Himmelberg, Glen R.himmelbergg@missouri.edu,
Univ of Missouri – Gp
Hindery, Derrick, (541) 346-6106 dhindery@uoregon.edu,
Univ of Oregon – Oo
Hindle, Tobin, (561) 297-2846 thindle@fau.edu,
Florida Atlantic Univ – Hw
Hindman, Edward E., (212) 650-6469 Graduate School of the City Univ of New York – Ow
Hindman, Edward E., (212) 650-6469 City College (CUNY) – Oa
Hine, Albert C., (727) 553-1161 ahine@marine.usf.edu,
Univ of South Florida, Saint Petersburg – Gs
Hines, Jennifer K., 217-244-2410 hines@isgs.uiuc.edu,
Illinois State Geological Survey – Gg
Hines, Mary E., (910) 962-3012 hinese@uncw.edu,
Univ of North Carolina Wilmington – Ou
Hinkes, Rilla M., (608) 262-1705 rmhinkes@wisc.edu,
Univ of Wisconsin - Extension – Oo
Hinman, George W., (509) 335-8689 ghinman@wsu.edu,
Washington State Univ – Oo
Hinman, Nancy W., (406) 243-5277 nhinman@selway.umt.edu,
Univ of Montana – Co
Hinnov, Linda A., (410) 516-7135 hinnov@jhu.edu,
The Johns Hopkins Univ – Gs
Hinojosa, Alejandroalhinc@cicese.mx,
Centro de Inv Científica y de Educación Superior de Ensenada – Or
Hintz, John G., (570)389-4140 jhintz@bloomu.edu,
Bloomsburg Univ – Oio
Hintze, Lehi F., (801) 378-3904 Brigham Young Univ – Gr
Hinze, William J., (765) 494-5982 Purdue Univ – Ye
Hippensteel, Scott P., 704-687-5992 shippens@email.uncc.edu,
Univ of North Carolina, Charlotte – Gr
Hipple, David R., (404) 656-3214 Georgia Dept of Natl Resources – Oi
Hirsch, David M., (360) 650-2166 hirschd@cc.wwu.edu,
Western Washington Univ – Gp
Hirschboeck, Katherine, 520-621-6466 katie@ltrr.arizona.edu,
Univ of Arizona – Oa
Hirschboeck, Katherine K., (520) 621-6466 hirschbo@ltrr.arizona.edu,
Univ of Arizona – Oa
Hirschboeck, Katherine K., (520) 621-6466 katie@ltrr.arizona.edu,
Univ of Arizona – Pe
Hirschmann, Marc M., 612-625-6698 hirsc022@umn.edu,
Univ of Minnesota, Twin Cities – Gi
Hirt, William H., (530) 938-5255 hirt@siskiyous.edu,
College of the Siskiyous – Gi
Hiscott, Richard N., (709) 737-8394 rhiscott@sparky2.esd.mun.ca,
Memorial Univ of Newfoundland – Gs
Hiskey, J. Brent, (520) 621-6185 Univ of Arizona – Nx
Hitchman, Matthew H., (608)262-4653 matt@aos.wisc.edu,
Univ of Wisconsin, Madison – Oa
Hites, Ronald A., (812) 855-0193 Indiana Univ, Bloomington – Co
Hitz, Ralph B., (253) 566-5299 rhitz@tcc.ctc.edu,
Tacoma Community College – Og
Hitzman, Murray W., (303) 384-2127 mhitzman@mines.edu,
Colorado School of Mines – Em
Hjelmfelt, Mark R., (605) 394-2291 hjelmfel@ias.sdsmt.edu,
South Dakota School of Mines & Technology – Oa
Hladky, Frankfrank.hladky@state.or.us,
Oregon Dept of Geology & Mineral Industries – Gg
Hluchy, Michele M., (607) 871-2838 Alfred Univ – Cl
Ho, David T., (845) 365-8706 david@ldeo.columbia.edu,
Columbia Univ – Hg
Hoare, Richard D., (419) 372-7203 Bowling Green State Univ – Pi
Hobbs, Benjamin F., (410) 516-4681 bhobbs@jhu.edu,
The Johns Hopkins Univ – Oo
Hobbs, Carl H., (804) 684-7271 hobbs@vims.edu,
College of William & Mary – Ou
Hobbs, Howard C., (612)627-4780, Ext. 207 hobbs001@umn.edu,
Univ of Minnesota, Twin Cities – Gl
Hobbs, John D.dhobbs@mtech.edu,
Montana Tech of the Univ of Montana – Oo
Hobbs, Richard D., (806) 371-5333 hobbs-rd@actx.edu,
Amarillo College – Or
Hobbs, Thomas M., 281-618-5796 Tom.Hobbs@nhmccd.edu,
North Harris College – Gg
Hochella, Jr., Michael F., (540) 231-6227 hochella@vt.edu,
Virginia Polytechnic Institute & State Univ – Cl
Hockey, Thomas A., (319) 273-2065 thomas.hockey@uni.edu,
Univ of Northern Iowa – Oo
Hodder, Donald R., (914) 257-3757 SUNY New Paltz – Gg
Hodder, Robert W., (519) 433-9550 rhodder@uwo.ca,
Univ of Western Ontario – Eg
Hode, Tomas, (503) 725-3024 Portland State Univ – Xg
Hodel, Erinhodel@nova.edu,
Nova Southeastern Univ – Oo
Hodell, David A., (352) 392-6137 hodell@geology.ufl.edu,
Univ of Florida – GuPe
Hodges, Floyd N., (509) 376-4627 floyd.hodges@pnl.gov,
Pacific Northwest National Lab – Cg
Hodges, John, (617) 353-9374 jcfh@bu.edu,
Boston Univ – Or
Hodgkiss, Jr., William S., (858) 534-1798 whodgkiss@ucsd.edu,
Univ of California, San Diego – Oo
Hodgson, M. John, (780) 492-0347 john.hodgson@ualberta.ca,
Univ of Alberta – Or
Hodson, Mark E.m.e.hodson@reading.ac.uk,
Univ of Reading – Cl
Hodych, Joseph P., (709) 737-7567 jhodych@mun.ca,
Memorial Univ of Newfoundland – Ym
Hoe, Teh Guan, 03-79674231 tehgh@um.edu.my,
Univ of Malaya – Ca
Hoenstine, Ronald W., (850) 488-9380 Florida Geological Survey – Ge
Hoersch, Alice L., (215) 951-1269 hoersch@lasalle.edu,
La Salle Univ – Gp
Hoff, Jean L., (612) 308-5914 jhoff@stcloudstate.edu,
Saint Cloud State Univ – Gs
Hoffer, Roberta J., (915) 747-7497 robin@geo.utep.edu,
Univ of Texas, El Paso – Gg
Hoffman, Charles W., (919) 733-7353 North Carolina Geological Survey – Gg
Hoffman, Gretchen K., (505) 835-5640 gretchen@gis.nmt.edu,
New Mexico Institute of Mining & Technology – Ec
Hoffman, Hans, (514) 343-7699 hofmann@ere.umontreal.ca,
Universite de Montreal – Pm
Hoffman, James I., (509) 359-4255 james.hoffman@mailserver.ewu.edu,
Eastern Washington Univ – Em
Hoffman, Jeffrey L., 609-984-6587 jeffrey.l.hoffman@dep.state.nj.us,
New Jersey Geological Survey – Hq
Hoffman, Kenneth A., (805) 756-1357 kehoffma@calpoly.edu,
California Polytechnic State Univ – Ym
Hoffman, Paul F., (617) 496-6380 hoffman@eps.harvard.edu,
Harvard Univ – Gc

Hofmann, Eileen E., (757) 683-5334 Old Dominion Univ – Op
Hofmann, Hans, 514-398-2769 hofmann@eps.mcgill.ca, McGill Univ – Gs
Hofmeister, Anne M., (314) 935-7440 hofmeist@levee.wustl.edu, Washington Univ – Gz
Hogan, James F., (520) 626-2910 jhogan@sahra.arizona.edu, Univ of Arizona – Cg
Hogan, John P., (573) 341-4618 jhogan@umr.edu, Univ of Missouri, Rolla – Gi
Hoganson, John, (701) 328-8000 jhoganso@state.nd.us, North Dakota Geological Survey – Pg
Hogarth, Donald D., (613)562-5800 Ext 6207 Univ of Ottawa – Gz
Hoggan, Roger D., (208) 496-1905 geology@byui.edu, Brigham Young Univ, Idaho – PeGg
Hohn, Michael E.hohn@geosrv.wvnet.edu, West Virginia Geological & Economic Survey – Gq
Hoisch, Thomas D., (928) 523-1904 thomas.hoisch@nau.edu, Northern Arizona Univ – Gp
Hojjatie, Barry, (229) 333-5753 bhojjati@valdosta.edu, Valdosta State Univ – Om
Holberg, Jay B., (520) 621-4571 holberg@vega.lpl.arizona.edu, Univ of Arizona – Oa
Holbrook, John M., (573) 651-2338 jholbrook@semovm.semo.edu, Univ of Texas, Arlington – Gs
Holbrook, W. S., (307) 766-2427 steveh@uwyo.edu, Univ of Wyoming – Ys
Holcomb, Robin T., (206) 543-5274 rholcomb@ocean.washington.edu, Univ of Washington – Ou
Holdaway, Michael J., (214) 692-2750 Southern Methodist Univ – Gp
Holden, Gregory S., (303) 273-3855 gholden@mines.edu, Colorado School of Mines – Gx
Holdren, George R., (509) 376-2242 rich.holdren@pnl.gov, Pacific Northwest National Lab – Cl
Holdsworth, Robert E., 0191 3742529 r.e.holdsworth@durham.ac.uk, Durham Univ – Gc
Hole, John A., (540) 231-3858 hole@vt.edu, Virginia Polytechnic Institute & State Univ – Ye
Holgood, Jay S., (614) 292-3999 hobgood.1@osu.edu, Ohio State Univ – Oa
Holk, Gregory J., (562) 985-5006 gholk@csulb.edu, California State Univ, Long Beach – Cs
Holl, Karen D., (831) 459-3668 kholl@cats.ucsc.edu, Univ of California, Santa Cruz – Ou
Hollabaugh, Curtis L., (678) 839-4050 chollaba@westga.edu, Univ of West Georgia – CgGz
Holland, Grenville, 0191 3742526 j.g.holland@durham.ac.uk, Durham Univ – Gp
Holland, Heinrich D., (617) 495-5892 holland@eps.harvard.edu, Harvard Univ – Cg
Holland, Nicholas D., (858) 534-2085 nholland@ucsd.edu, Univ of California, San Diego – Ob
Holland, Steven M., (706) 542-0424 stratum@gly.uga.edu, Univ of Georgia – Gr
Holland, Jr., F. D., (701) 772-1622 budholland@aol.com, Univ of North Dakota – Pi
Hollenbaugh, Kenneth M., (208) 426-3700 khollenb@boisestate.edu, Boise State Univ – Eg
Holliday, Joseph W., (310) 660-3371 El Camino College – Gg
Holliday, Vance, (520) 621-4734 vthollid@email.arizona.edu, Univ of Arizona – Oo
Holliday, Vance, 520-621-4734 vthollid@email.arizona.edu, Univ of Arizona – Ga
Hollings, Peter N., (807) 343-8329 peter.hollings@lakeheadu.ca, Lakehead Univ – Eg
Hollins, James L., (702) 794-7522 james_hollins@lanl.gov, Los Alamos National Lab – Ng
Hollister, Lincoln S., (609) 258-4106 linc@princeton.edu, Princeton Univ – Gp
Hollocher, Kurt T., (518) 388-6518 hollochk@union.edu, Union College – GxCga
Holloway, John R., (480) 965-6907 john.holloway@asu.edu, Arizona State Univ – Cp
Holloway, Sharon, (540) 831-5652 shollawa@radford.edu, Radford Univ – Oo
Holloway, Tracey, 608-262-5356 taholloway@wisc.edu, Univ of Wisconsin, Madison – Oa
Holm, Daniel K., (330) 672-4094 dholm@kent.edu, Kent State Univ – Gc
Holm, Paul E., (513) 529-1935 holmp@muohio.edu, Miami Univ – Gc
Holm, Richard F., (928) 523-7200 richard.holm@nau.edu, Northern Arizona Univ – Gv
Holm, Thomas R., (217) 333-2604 Illinois State Water Survey – Hw
Holman, Robert A., (541) 737-2914 holman@coas.oregonstate.edu, Oregon State Univ – On
Holmden, Chris, (306) 966-5697 Univ of Saskatchewan – Csc
Holmes, Ann E., (423) 425-1704 ann-holmes@utc.edu, Univ of Tennessee, Chattanooga – Gs
Holmes, Mark L., (206) 543-7313 mholmes@ocean.washington.edu, Univ of Washington – Ou
Holmes, Mary Anne, 402-472-5211 mholmes2@unl.edu, Univ of Nebraska, Lincoln – Gs
Holmes, Michael J., (617) 353-2526 mjmh@bu.edu, Boston Univ – Oo
Holroyd, Pat, (510) 642-3733 pholroyd, Univ of California, Berkeley – Pv
Holstein, Thomas J., (401) 254-3097 Roger Williams Univ – Oo
Holt, Ben D., (630) 252-4347 Argonne National Lab – Ca
Holt, G. J., (361) 749-6716 joanholt@mail.utexas.edu, Univ of Texas at Austin – Ob
Holt, John W., 512-471-0487 jack@ig.utexas.edu, Univ of Texas at Austin, Jackson School of Geosciences – Ye
Holt, Robert M., (662) 915-6687 rmholt@olemiss.edu, Univ of Mississippi – Hq
Holt, William E., 631-632-8215 william.holt@sunysb.edu, SUNY, Stony Brook – Ys
Holtz, Barbara J., (970) 491-5662 barbh@cnr.colostate.edu, Colorado State Univ – Oo
Holtz, Mark H., (512) 475-9525 mark.holtz@beg.utexas.edu, Univ of Texas at Austin, Jackson School of Geosciences – Gg
Holtz, Thomas R., (301) 405-4084 tholtz@geol.umd.edu, Univ of Maryland – Pv
Holycross, David, (812) 237-3471 Indiana State Univ – Ga
Holzer, Mark, (604) 822-0531 mholzer@langara.bc.ca, Univ of British Columbia – Oa
Holzworth, Robert H., 206-685-7410 bobholz@ess.washington.edu, Univ of Washington – YxXy
Homsey, Lara, (270) 809-2408 lara.homsey@murraystate.edu, Murray State Univ – Ga
Homuth, Emil F., (702) 794-7351 fred_homuth@lanl.gov, Los Alamos National Lab – Ye
Hon, Ken, (808) 974-7302 kenhon@hawaii.edu, Univ of Hawai'i at Hilo – Gv
Hon, Rudolph, (617) 552-3656 hon@bc.edu, Boston College – Gi
Honjo, Susumu, (508) 289-2589 shonjo@whoi.edu, Woods Hole Oceanographic Institution – Ou
Hons, Frank M., (979) 845-3477 f-hons@tamu.edu, Texas A&M Univ – Sc
Hood, Lonnie L., (520) 621-6936 lon@lpl.arizona.edu, Univ of Arizona – Xy
Hood, Raleigh, (305) 361-4668 raleigh@hpl.umces.edu, Univ of Maryland – Ob
Hood, Teresa A., 305-284-8647 t.hood@miami.edu, Univ of Miami – Gg
Hood, William C., (970) 241-8020 WHood@compuserve.com, Mesa State College – Gg
Hooft, Emilie E., (541) 346-4762 emilie@newberry.uoregon.edu, Univ of Oregon – Yr
Hook, James E., (912) 386-3182 Univ of Georgia – Sp
Hook, Paul B., (406) 944-3724 paulhook@montana.edu, Montana State Univ – Sf
Hook, Simon J., (818) 354-0974 simon@lithos.jpl.nasa.gov, Jet Propulsion Lab – Or
Hooke, Roger L., (207) 581-2203 rhooke@acadia.net, Univ of Maine – Gm
Hooks, W. Gary, (205) 348-1877 Univ of Alabama – Gm
Hooper, Peter R.prhooper@mail.wsu.edu, Washington State Univ – Gi
Hooper, Robert L., (715) 836-4932 hooperrl@uwec.edu, Univ of Wisconsin, Eau Claire – Gz
Hoopes, John A., (608) 262-2977 Univ of Wisconsin, Madison – Oo
Hoover, Karin A., (530) 898-5618 khoover@csuchico.edu, California State Univ, Chico – Hw
Hoover, Kenneth D., (904) 745-7300 khoover@ju.edu, Jacksonville Univ – Oo
Hoover, Michael T., (919) 515-7305 North Carolina State Univ – Sd
Hooyer, Thomas S., 608-263-4175 tshooyer@wisc.edu, Univ of Wisconsin, Madison – Hg
Hopkins, Daniel P., (816) 235-2973 hopkinsd@umkc.edu,

Univ of Missouri-Kansas City – Oo
Hopkins, David G., (701) 231-8948 North Dakota State Univ – Sd
Hopkins, David M., (907) 474-7565 Univ of Alaska, Fairbanks – Gg
Hopkins, Edgar M., (814) 362-7568 emhl@exchange.upb.pitt.edu, Univ of Pittsburgh, Bradford – Gs
Hopkins, John C., (403) 220-5842 hopkins@geo.ucalgary.ca, Univ of Calgary – Gs
Hopkins, Kenneth D., (970) 351-2853 kenneth.hopkins@unco.edu, Univ of Northern Colorado – Gm
Hopkins, Thomas L., (727) 553-1501 thopkins@marine.usf.edu, Univ of South Florida, Saint Petersburg – Ob
Hopkins, Thomas S., (205) 348-1791 Univ of Alabama, Dauphin Island – Ob
Hopmans, Jan W., (530) 752-3060 jwhopmans@ucdavis.edu, Univ of California, Davis – Sp
Hoppe, Kathryn A., (253) -83-3-91 (Ext. ext. 4323) khoppe@green-river.edu, Green River Community College – Gg
Hopper, John R., 979-845-0119 hopper@geo.tamu.edu, Texas A&M Univ – Yr
Hoppie, Bryce W., (507) 389-2315 bryce.hoppie@mnsu.edu, Minnesota State Univ – Ge
Hoppin, Richard A., (319) 335-1805 richard-hoppin@uiowa.edu, Univ of Iowa – Gc
Hopson, Clifford A., (805) 893-3149 Univ of California, Santa Barbara – Gi
Horak, Michael J., (913) 532-7240 Kansas State Univ – Sb
Horita, Juske, (865) 576-2750 horitaj@ornl.gov, Oak Ridge National Lab – Cg
Horn, Joanne, (925) 422-5653 Lawrence Livermore National Lab – Sb
Horn, Marty R., 225-578-5320 mhorn@lsu.edu, Louisiana State Univ – EoGrg
Horn, Marty R., 225-578-5320 mhorn@lsu.edu, Louisiana State Univ – EoGgr
Hornberger, George M., (804) 924-6762 gmh3k@virginia.edu, Univ of Virginia – Hq
Hornberger, George M., (434) 924-6762 gmh3k@virginia.edu, Univ of Virginia – Hq
Horne, Jerry D., (909) 888-6511 San Bernardino Valley College – Ge
Horne, Sharon, ((51) 9) -253- (Ext. 00 ext, 2486) shorne@uwindsor.ca, Univ of Windsor – Oo
Horner, John R., (406) 994-3982 jhorner@montana.edu, Montana State Univ – Pv
Horner, Tim C., (916) 278-5635 hornertc@csus.edu, California State Univ, Sacramento – Hw
Hornsby, Arthur G., (352) 392-1951 agh@mail.ifas.ufl.edu, Univ of Florida – Sp
Horton, Albert B., (615) 532-1509 Albert.Horton@state.tn.us, Tennessee Division of Geology – GgOi
Horton, Benjamin P., (215) 573-5388 bphorton@sas.upenn.edu, Univ of Pennsylvania – On
Horton, Brian, 512-471-5172 horton@mail.utexas.edu, Univ of Texas at Austin, Jackson School of Geosciences – Gs
Horton, Duane G., (509) 376-6868 duane.horton@pnl.gov, Pacific Northwest National Lab – Gz
Horton, Robert, (515) 294-7843 rhorton@iastate.edu, Iowa State Univ of Science & Technology – Sp
Horton, Stephen P., (901) 678-2007 horton@ceri.memphis.edu, Univ of Memphis – Ys
Horton, Travis, 253-879-3936 thorton@ups.edu, Univ of Puget Sound – Ge
Horton, Jr., Robert A., (805) 664-3059 California State Univ, Bakersfield – Gd
Horwath, William R., (530) 754-6029 wrhorwath@ucdavis.edu, Univ of California, Davis – So
Hose, Louise D., 505.234.5561 lhose@cemrc.org, New Mexico Institute of Mining and Technology – Gs
Hosler, Charles L., (814) 863-8358 hosler@ems.psu.edu, The Pennsylvania State Univ, Univ Park – Ow
Hostetler, Steven W., (541) 737-8928 steve@coas.oregonstate.edu, Oregon State Univ – Oa
Hou, Weilin, (727) 553-3988 whou@marine.usf.edu, Univ of South Florida, Saint Petersburg – Op
Hou, Zhangshuan, 716-645-6800 (Ext. 3984) hou2@buffalo.edu, SUNY, Buffalo – Yg
Houghton, Bruce F., (808) 956-2561 bhought@soest.hawaii.edu, Univ of Hawai'i, Manoa – Gv
Houghton, David D., (608)262-0776 ddhought@wisc.edu, Univ of Wisconsin, Madison – Oa
Houghton, Jenniferjhoughtn@memphis.edu, Univ of Memphis – Co
Hounslow, Arthur, 405-372-2328 Oklahoma State Univ – Cl
Hourigan, Jeremy, 831-459-2873 jhourigan@pmc.ucsc.edu, Univ of California, Santa Cruz – Gc
Hourigan, Jeremy K., 831 459-2873 jhourigan@pmc.ucsc.edu, Univ of California, Santa Cruz – Gt
House, Christopher H., (814) 865-8802 chouse@geosc.psu.edu, Pennsylvania State Univ, Univ Park – Py
House, Kyle, (775) 784-6691 khouse@nbmg.unr.edu, Univ of Nevada, Reno – Gm
House, Leigh S., (505) 667-1912 house@lanl.gov, Los Alamos National Lab – Ys
House, P. Kyle, (775) 784-6691 (Ext. 176) khouse@unr.edu, Univ of Nevada – Gm
Housen, Bernard A., (360) 650-3588 bernieh@cc.wwu.edu, Western Washington Univ – Ym
Houser, Christopher, 850-857-6121 chouser@uwf.edu, Univ of West Florida – On
Housh, Todd B., (512) 471-5750 housh@mail.utexas.edu, Univ of Texas at Austin, Jackson School of Geosciences – Gi
Houston, Adam L.ahouston@unl.edu, Univ of Nebraska, Lincoln – Oa
Houston, Heidi B., 206-616-7092 hhouston@u.washington.edu, Univ of Washington – YsGt
Houston, Robert S., (307) 766-3386 Univ of Wyoming – Eg
Hovan, Steven A., (724) 357-7662 hovan@iup.edu, Indiana Univ of Pennsylvania – Ou
Hover, Victoria C., 337-482-1166 vhover@louisiana.edu, Univ of Louisiana, Lafayette – Cg
Hovis, Guy L., (610) 330-5192 hovisguy@lafayette.edu, Lafayette College – Gy
Hovorka, Susan D., (512) 471-4863 susan.hovorka@beg.utexas.edu, Univ of Texas at Austin, Jackson School of Geosciences – Gd
Howard, Alan D., (804) 924-0563 ah6p@virginia.edu, Univ of Virginia – Gm
Howard, Hugh H., 916 484-8805 howardh@arc.losrios.edu, American River College – Oi
Howard, James M., 501 683-0125 mike.howard@Arkansas.gov, Arkansas Geological Commission – Gg
Howard, Jeffrey L., (313) 577-3258 aa2675@wayne.edu, Wayne State Univ – Gs
Howard, Kenneth W., 416-287-7233 Univ of Toronto – Hw
Howard, Leslie M., (402) 472-9192 lhoward@unl.edu, University of Nebraska - Lincoln – Oyi
Howard, Matthew K., (979) 862-4169 mhoward@ocean.tamu.edu, Texas A&M Univ – Op
Howard, Theodore E., (603) 862-1020 Univ of New Hampshire – Oo
Howarth, David A., (502) 852-6844 Univ of Louisville – Oa
Howd, Peter, (727) 553-1158 phowd@marine.usf.edu, Univ of South Florida, Saint Petersburg – Ou
Howden, Stephan, 228-688-5284 Stephan.Howden@usm.edu, Univ of Southern Mississippi – Op
Howe, Abigail W., (860) 486-4432 abigail.howe@uconn.edu, Univ of Connecticut – Gg
Howe, Bruce M., (206) 543-9141 howe@apl.washington.edu, Univ of Washington – Op
Howe, John A., (419) 372-7194 Bowling Green State Univ – Pv
Howe, Julie, (334) 844-3972 jah0020@auburn.edu, Auburn Univ – Sc
Howe, Stephen S., (518) 442-5053 showe@albany.edu, SUNY, Albany – Cs
Howell, Michael W., (803) 777-6943 howell@geol.sc.edu, Univ of South Florida, Saint Petersburg – Ou
Howell, Robert R., (307) 766-6296 rhowell@uwyo.edu, Univ of Wyoming – Xg
Howell, Jr., Benjamin F., (814) 863-0886 howellbf@aol.com, Pennsylvania State Univ, Univ Park – Ys
Hower, James C., (859) 257-0261 hower@caer.uky.edu, Univ of Kentucky – Ec
Howes, David W.dhowes@clarion.edu, Clarion Univ – Or
Howman, Dominic J., +61 8 9266-2329 D.Howman@curtin.edu.au, Curtin Univ of Technology – Ye
Hoyal, Michael L., (615) 532-1504 Mike.Hoyal@state.tn.us, Tennessee Division of Geology – Gog
Hoyt, Greg D., (919) 684-3562 North Carolina State Univ – Sb
Hoyt, William H., (970) 351-2487 william.hoyt@unco.edu, Univ of Northern Colorado – Gu
Hozik, Michael J., (609) 652-4277 hozikm@stockton.edu, Richard Stockton College of New Jersey – Ym
Hozik, Michael J., 609 652-4277 HozikM@Stockton.Edu, Stockton State College – Gm

Hsiao, Theodore C., (530) 752-0691 tchsiao@ucdavis.edu,
Univ of California, Davis – Sb
Hsieh, Chiao-Min, (412) 624-9323 chh57@pitt.edu,
Univ of Pittsburgh – Oy
Hsieh, William, (604) 822-2821 whsieh@eos.ubc.ca,
Univ of British Columbia – Op
Hsu, Liang-Chilihsu@nbmg.unr.edu,
Univ of Nevada, Reno – Cp
Hsu, Liang-Chi, (775) 784-6691 (Ext. 122) lihsu@unr.edu,
Univ of Nevada – Cp
Hsu, Shih A., (225) 578-2962 sahsu@antares.esl.lsu.edu,
Louisiana State Univ – Oa
Hsueh, Yahsueh@ocean.fsu.edu,
Florida State Univ – Op
Hsui, Albert T.hsui@uiuc.edu,
Univ of Illinois, Urbana-Champaign – Yg
Hu, (Max) Qinhong, (510) 495-2338 q_hu@lbl.gov,
Lawrence Berkeley National Lab – Hw
Hu, Baoxin, (416)736-2100 (Ext. 77729) baoxin@yorku.ca,
York Univ – Or
Hu, Feng-Sheng, (217) 244-2982 fshu@life.uiuc.edu,
Univ of Illinois, Urbana-Champaign – Pe
Hu, Jingzhu, (202) 686-2410 hu@1sx17c.nsls.bnl.gov,
Carnegie Institution of Washington – Yx
Hu, Qi S., 402-472-6642 QHU2@unl.edu,
Univ of Nebraska, Lincoln – Oa
Hu, Zhiyong, 850-474-3494 zhu@uwf.edu,
Univ of West Florida – Or
Huang, Alex, (828) 232-5157 Univ of North Carolina, Asheville – Oa
Huang, Kainian, (352) 392-2231 knhuang@geology.ufl.edu,
Univ of Florida – Ym
Huang, Li, (519) 661-3188 LHuang3@uwo.ca,
Univ of Western Ontario – Yr
Huang, Lianjie, (505) 665-1108 ljh@lanl.gov,
Los Alamos National Lab – Yg
Huang, Moh J., (916) 322-9304 California Geological Survey – Ne
Huang, Norden E., (301) 286-8879 Univ of Delaware – Oo
Huang, Scott L., (907) 474-6880 ffslh@aurora.alaska.edu,
Univ of Alaska, Fairbanks – Nr
Huang, Shaopeng, (734) 763-3169 shaopeng@umich.edu,
Univ of Michigan – Yh
Huang, Yongsongyongsong_huang@brown.edu,
Brown Univ – Co
Hubbard, Deborah, (540) 463-8800 hubbardd@wlu.edu,
Washington & Lee Univ – Oo
Hubbard, Dennis K., (440) 775-8346 dennis.hubbard@oberlin.edu,
Oberlin College – On
Hubbard, Kenneth G., (402) 472-8294 Univ of Nebraska, Lincoln – Oa
Hubbard, Susan S., (510) 486-5266 sshubbard@lbl.gov,
Lawrence Berkeley National Lab – Hy
Hubbard, William B., (520) 621-6942 hubbard@lpl.arizona.edu,
Univ of Arizona – Xy
Hubbard, Jr., David A., (434) 951-6353 david.hubbard@dmme.virginia.gov,
Division of Mineral Resources – Ge
Hubenthal, Michael, 607-777-4612 hubenth@iris.edu,
Binghamton Univ – Oe
Huber, Brian T., (202) 633-1328 Smithsonian Institution / National
Museum of Natural History – Pm
Huber, James K., (319) 472-2565 jhuber@fmtcs.com,
Univ of Iowa – Pl
Huber, Matthew, (765) 494-0044 huberm@purdue.edu,
Purdue Univ – Pe
Hubert, Claude, (514) 343-7541 Universite de Montreal – Gt
Hubert, Johnjhubert@geo.umass.edu,
Univ of Massachusetts – Gs
Huckabey, Marsha, (573) 882-2040 HuckabeyM@missouri.edu,
Univ of Missouri – Oo
Hudak, George J., (920) 424-4463 hudak@uwosh.edu,
Univ of Wisconsin, Oshkosh – Gv
Hudak, Paul F., (940) 565-4312 hudak@unt.edu,
Univ of North Texas – Hw
Hudec, Mike, (512) 471-1428 mike.hudec@beg.utexas.edu,
Univ of Texas at Austin, Jackson School of Geosciences – Gg
Hudec, Peter P., ((51) 9) -253- (Ext. 00 ext. 2491) hudec@uwindsor.ca,
Univ of Windsor – Ng
Hudgens, Herb J., (817) 531-4752 herb@tcjc.cc.tx.us,
Tarrant County Junior College – Og
Hudleston, Peter J., 612-625-0046 hudle001@umn.edu,
Univ of Minnesota, Twin Cities – Gc
Hudman, Lloyd E., (801) 378-4346 lloyd_hudman@byu.edu,
Brigham Young Univ – Oo
Hudson, Michael R., (419) 289-5270 mhudson@ashland.edu,
Ashland Univ – Gp
Hudson, Robert J. M., (217) 333-7641 rjhudson@uiuc.edu,
Univ of Illinois, Urbana-Champaign – Cg
Huebert, Barry J., (808) 956-6896 huebert@soest.hawaii.edu,
Univ of Hawai'i, Manoa – Oc
Huerta, Carlos I.huerta@cicese.mx,
Centro de Inv Científica y de Educación Superior de Ensenada – Ne
Huestis, Stephen P., (505) 277-3636 crustal@unm.edu,
Univ of New Mexico – Yg
Huete, Alfredo R., (520) 621-3228 ahuete@email.arizona.edu,
Univ of Arizona – Or
Huey, Gregory L., (404) 894-5541 greg.huey@eas.gatech.edu,
Georgia Institute of Technology – Oa
Huff, Bryan G., 217-244-2509 huff@isgs.uiuc.edu,
Illinois State Geological Survey – Eo
Huff, Edmund A., (630) 252-3633 Argonne National Lab – Ca
Huff, Glenn F., 575-646-2708 huff@nmsu.edu,
New Mexico State Univ, Las Cruces – Hw
Huff, Marilyn C., 575-646-4714 huff@nmsu.edu,
New Mexico State Univ, Las Cruces – Gg
Huff, Warren D., (513) 556-3731 warren.huff@uc.edu,
Univ of Cincinnati – Gz
Huffines, Gary R., 970-351-1071 Gary.Huffines@unco.edu,
Univ of Northern Colorado – Oa
Huffman, Debra E., (727) 553-3930 dhuffman@marine.usf.edu,
Univ of South Florida, Saint Petersburg – Oo
Huffman, Robert L., (912) 752-2704 Mercer Univ – Oo
Huftile, Gary J., 61 7 3138 4470 g.huftile@qut.edu.au,
Queensland Univ of Technology – Gc
Huggett, William, 618-453-7392 huggett@geo.siu.edu,
Southern Illinois Univ Carbondale – Ec
Hughen, Konrad A., (508) 289-3353 khughen@whoi.edu,
Woods Hole Oceanographic Institution – Cc
Hughes, John M., 513-529-3216 hughesjm@muohio.edu,
Miami Univ – Gz
Hughes, Joseph B., (713)348-5603 hughes@rice.edu,
Rice Univ – Oo
Hughes, Malcolm K., (520) 621-6470 mhughes@ltrr.arizona.edu,
Univ of Arizona – Pe
Hughes, Michele, 208-282-3235 hughmich@isu.edu,
Idaho State Univ – Oo
Hughes, Nigel C.nigel.hughes@ucr.edu,
Cincinnati Museum Center – Pi
Hughes, Nigel C., (951) 827-3098 nigel.hughes@ucr.edu,
Univ of California, Riverside – Po
Hughes, Philip, (714) 564-6610 Santa Ana College – Gs
Hughes, Richard G., (405) 325-8141 rghughes@ou.edu,
Univ of Oklahoma – Np
Hughes, Scott S., 208-282-3565 hughscot@isu.edu,
Idaho State Univ – Gx
Hughes, Terence J., (207) 581-2198 debbies@maine.edu,
Univ of Maine – Gl
Hughes III, Richard O., 909-389-3237 rihughes@craftonhills.edu,
Crafton Hills College – Gl
Hugli, Wilbur G., 850-474-3470 whugli@uwf.edu,
Univ of West Florida – Oa
Hugo, Richard, 503-725-3356 hugo@pdx.edu,
Portland State Univ – Gy
Huh, Oscar K., (225) 388-2952 oscar@antares.csi.lsu.edu,
Louisiana State Univ – Or
Huismans, Ritske, (902) 494-6392 ritske.huismans,
Dalhousie Univ – Ou
Hulbe, Christina L., (503) 725-3022 Portland State Univ – Gl
Hulen, Jeffrey B., (801) 581-8497 jhulen@egi.utah.edu,
Univ of Utah – Eg
Hull, Amy B., (630) 252-6631 amyhull@anl.gov,
Argonne National Lab – Cl
Hull, Donald L., (505) 667-4151 Los Alamos National Lab – Oo
Hull, Joseph M.jhull@sccd.ctc.edu,
Seattle Central Community College – Gc
Humayun, Munir, (773) 834-1337 hum8@midway.uchicago.edu,
Univ of Chicago – Cg
Humphrey, John D., (303) 273-3819 jhumphre@mines.edu,
Colorado School of Mines – Gd
Humphrey, Neil F., (307) 766-2728 neil@uwyo.edu,
Univ of Wyoming – Gm
Humphrey, Peggy, (406) 994-5718 peggyh@montana.edu,
Montana State Univ – Oo

Humphreys, Eugene D., (541) 346-5575 gene@newberry.uoregon.edu, Univ of Oregon – Yg
Humphris, Susan E., (508) 289-3451 shumphris@whoi.edu, Woods Hole Oceanographic Institution – Gu
Hunckler, Robert, 810-989-5750 rhunckler@sc4.edu, Saint Clair County Community College – Og
Hundley, Stacey, (937) 229-2261 Stacey.Hundley@notes.udayton.edu, Univ of Dayton – Gg
Hung, Ming-Chih, (660) 562-1797 mhung@nwmissouri.edu, Northwest Missouri State Univ – Or
Hungr, Oldrich, (604) 822-8471 ohungr@eos.ubc.ca, Univ of British Columbia – Gm
Hunsaker, Jr., Donald, (559) 278-3086 California State Univ, Fresno – Ge
Hunt, Allen S., (802) 656-0241 ahunt@zoo.uvm.edu, Univ of Vermont – Pi
Hunt, Andrew, 817 272 2987 andrewhunthome@yahoo.com, Univ of Texas, Arlington – GeCg
Hunt, Gene, (202) 633-1331 Smithsonian Institution / National Museum of Natural History – Pm
Hunt, Graham H., (502) 852-6821 Univ of Louisville – Gg
Hunt, Jaqueline, 406-496-4842 jhunt@mtech.edu, Montana Tech of The Univ of Montana – Cg
Hunt, Kathykh26@umail.umd.edu, Univ of Maryland – Oo
Hunt, Randy J.rjhunt@usgs.gov, Univ of Wisconsin, Madison – Hw
Hunt, Richard G., (508) 929-8641 rhunt@worcester.edu, Worcester State College – Oy
Hunt, Robert M., (402) 472-4604 rhunt2@unl.edu, Univ of Nebraska, Lincoln – Pv
Hunt, Susan L., (608) 263-3418 slhunt@wisc.edu, Univ of Wisconsin - Extension – Oo
Hunten, Donald M., (520) 621-4002 dhunten@lpl.arizona.edu, Univ of Arizona – Oa
Hunter, Bruce, (860) 424-3588 bruce.hunter@po.state.ct.us, Geological and Natural History Survey of Connecticut – Hs
Hunter, Glenn, (814) 863-4636 hunter@mail.meteo.psu.edu, The Pennsylvania State Univ, Univ Park – Ow
Hunter, John R., (858) 546-7108 Univ of California, San Diego – Ob
Huntley, David, (619) 594-5483 dhuntley@geology.sdsu.edu, San Diego State Univ – Hw
Huntoon, Jacqueline E., (906) 487-2531 jeh@mtu.edu, Michigan Technological Univ – Gs
Huntoon, Laura, 520-626-1151 huntoon@email.arizona.edu, Univ of Arizona – Oo
Huntress, Jr., Wesley T., (202) 478-8910 w.huntress@gl.ciw.edu, Carnegie Institution of Washington – Xc
Huntsman, John R., (910) 962-3499 huntsmanj@uncw.edu, Univ of North Carolina Wilmington – Gc
Hurd, David, (814) 732-2493 dhurd@edinboro.edu, Edinboro Univ of Pennsylvania – Og
Hurich, Charles A., (709) 737-2384 chuck@waves.esd.mun.ca, Memorial Univ of Newfoundland – Ys
Hurlbut, Jr., Cornelius S., (617) 495-2351 Harvard Univ – Gz
Hurley, Neil F., (303) 384-2128 nfhurley@mines.edu, Colorado School of Mines – Go
Hurlow, Hugh A., (801) 537-3385 nrugs.hharlow@state.ut.us, Utah Geological Survey – Hw
Hurst, Richard W., (323) 343-2410 rhurst@calstatela.edu, California State Univ, Los Angeles – Cc
Hurst, Stephen D., (217) 333-0205 shurst@uiuc.edu, Univ of Illinois, Urbana-Champaign – Gc
Hurtado, Jose M., (915) 747-5669 hurtado@geo.utep.edu, Univ of Texas, El Paso – Gt
Hurtgen, Matthew T., (847) 491-7539 matt@earth.northwestern.edu, Northwestern Univ – Gs
Husch, Jonathan M., (609) 896-5330 husch@rider.edu, Rider Univ – GieGz
Huss, Gary, (480) 965-9508 gary.huss@asu.edu, Arizona State Univ – Xc
Hussey, II, Arthur M., (207) 666-5526 ahussey@bowdoin.edu, Bowdoin College – Gc
Hussin, Azhar Hj, 03-79674203 azharhh@um.edu.my, Univ of Malaya – Gs
Hustrulid, William A., (801) 585-3029 hustrulid@mines.utah.edu, Univ of Utah – Nm
Hutcheon, Ian E., (403) 220-6744 ian@earth.geo.ucalgary.ca, Univ of Calgary – Cl
Hutchins, David A., (302) 645-4079 dahutch@udel.edu, Univ of Delaware – Ob
Hutchins, Eleanor, (518) 580-5190 ehutchin@skidmore.edu, Skidmore College – Oo
Hutchins, Peter S., (601) 961-5505 peter_hutchins@deq.state.ms.us, Mississippi Office of Geology – Oo
Hutchinson, Charles F., (520) 621-7896 Univ of Arizona – Or
Hutchison, David, 607-431-4730 HutchisonD@hartwick.edu, Hartwick College – GxgGi
Hutchison, Howard J., (510) 642-3733 Univ of California, Berkeley – Pv
Huycke, David, 509-574-4817 dhuycke@yvcc.edu, Yakima Valley College – Gg
Huyer, Adriana J., (541) 737-2108 ahuyer@coas.oregonstate.edu, Oregon State Univ – Op
Huysken, Kristin, (219) 980-6739 Indiana Univ Northwest – Gi
Hwang, Howard, (901) 678-4830 hwang@ceri.memphis.edu, Univ of Memphis – Ne
Hwang, Hue-Hwa, 217-244-9876 hwang@uiuc.edu, Illinois State Geological Survey – Cs
Hyatt, James A., (860) 465-5202 hyattj@easternct.edu, Eastern Connecticut State Univ – Gm
Hyers, Donald A., (413) 664-4511 North Adams State College – Oy
Hylland, Michael O., (801) 537-3300 nrugs.mhylland@state.ut.us, Utah Geological Survey – Ng
Hyndman, David W., 517-353-4442 hyndman@msu.edu, Michigan State Univ – Hy
Hyndman, Roy D., (250) 363-6428 Univ of Victoria – Ys
Hynek, Brian M., 303-735-4312 brian.hynek@colorado.edu, Univ of Colorado – Xg
Hynes, Andrew J., 514-398-5884 andrew@eps.mcgill.ca, McGill Univ – Gc
Hynote, Cindy A., 404-656-3214 Georgia Dept of Natural Resources – Oo
Hysell, David, (607) 255-0630 daveh@geology.cornell.edu, Cornell Univ – Yg

I

Iaccheri, Maria Linda, 089/2180 4274 LMU - Univ of Munich – Gg
Iaffaldano, Giampiero, 089/2180 4220 giampiero@geophysik.uni-muenchen.de, LMU - Univ of Munich – Yg
Ibaraki, Motomu, (614) 292-7528 ibaraki@geology.ohio-state.edu, Ohio State Univ – Hg
Ibrahim, Ahmad Tajuddin Hj, 03-79674230 ahmadt@um.edu.my, Univ of Malaya – Ng
Ibrahim, Mohamed, 212-772-5267 mibrahim@geo.hunter.cuny.edu, Hunter College (CUNY) – Oo
Icenhower, Jonathan P., (509) 372-0078 jon.icenhower@pnl.gov, Pacific Northwest National Lab – Cl
Icopini, Gary, 406-496-4841 gicopini@mtech.edu, Montana Tech of The Univ of Montana – Gg
Ide, Kayo, (310) 206-6484 Univ of California, Los Angeles – Og
Idleman, Bruce D.bdi2@lehigh.edu, Lehigh Univ – Cc
Idone, Vincent P., (518) 442-4577 vpi@atmos.albany.edu, SUNY, Albany – Oa
Ierley, Glenn R., (858) 534-5917 Univ of California, San Diego – Op
Igel, Heiner, 089/2180 4204 heiner.igel@lmu.de, LMU - Univ of Munich – Yg
Ihinger, Phillip D., (715) 836-2158 ihinger@uwec.edu, Univ of Wisconsin, Eau Claire – Gx
Ikelle, Luc T., (979) 862-4296 ikelle@geo.tamu.edu, Texas A&M Univ – Yg
Ikramuddin, Mohammed, (509) 359-7866 mohammed.ikramuddin@mailserver.ewu.edu, Eastern Washington Univ – Ca
Iles, Derric L., (605) 677-5227 diles@usd.edu, South Dakota Dept of Env & Natural Resources – Hw
Iles, Kenneth W., (940) 565-2040 kiles@unt.edu, Univ of North Texas – Og
Illari, Lodovica, (617) 253-2286 illari@squall.mit.edu, Massachusetts Institute of Technology – Ow
Illman, Walter A., (319) 335-1827 walter-illman@uiowa.edu, Univ of Iowa – Hg
Imber, Jonathan, 0191 3742508 jonathan.imber@durham.ac.uk, Durham Univ – Gc
Imbrie, John, (401) 863-3916 john_imbrie@brown.edu, Brown Univ – Pe
Immel, Harald, 089/2180 6615 h.immel@lrz.uni-muenchen.de, LMU - Univ of Munich – Pg
Immonen, Wilma, (406) 496-4182 wimmonen@mtech.edu, Montana Tech of the Univ of Montana – Oo
Inamdar, Shreeram, 302-831-8877 inamdar@udel.edu,

Univ of Delaware – Hg
Ince, Simon, (520) 621-5082 Univ of Arizona – Hs
Indares, Aphrodite D., (709) 737-2456 afin@sparky2.esd.mun.ca, Memorial Univ of Newfoundland – Gp
Ingall, Ellery D., (404) 894-3883 ellery.ingall@eas.gatech.edu, Georgia Institute of Technology – Cm
Ingersoll, Andrew P., 626 395-6167 api@gps.caltech.edu, California Institute of Technology – Oa
Ingersoll, Raymond V., (310) 825-8634 ringer@ess.ucla.edu, Univ of California, Los Angeles – Gt
Ingham, Frank, (705) 566-8101 (Ext. 7806) Cambrian College of Applied Arts & Technology – Nm
Ingle, Jr., James C., (650) 723-3366 ingle@pangea.stanford.edu, Stanford Univ – Pm
Ingram, B. Lynn, (510) 643-1474 ingram@eps.berkeley.edu, Univ of California, Berkeley – Cs
Ingram, Lynningram@socrates.berkeley.edu, Univ of California, Berkeley – Po
Ingram, R. Grant, (604) 822-6436 gingram@eos.ubc.ca, Univ of British Columbia – Op
Ingram, Stephen L., (601) 354-6328 Mississippi Office of Geology – Gr
Inman, Douglas L., (858) 534-4334 dinman@ucsd.edu, Univ of California, San Diego – On
Inman, Jeanjean_inman@deq.state.ms.us, Mississippi Office of Geology – Oo
Inners, Jon D., 717.702.2034 jinners@state.pa.us, Pennsylvania Bureau of Topographic & Geologic Survey – Gg
Inoue, Masamachi, (225) 388-2521 Louisiana State Univ – Op
Insinna, Anthony, (705) 566-8101 (Ext. 7555) toinsinna@cambrianc.on.ca, Cambrian College of Applied Arts & Technology – Ye
Inskeep, William P., (406) 994-5077 binskeep@gemini.oscs.montana.edu, Montana State Univ – Sc
Ioannides, Dimitri, (417) 836-5318 dii608f@smsu.edu, Southwest Missouri State Univ – Ou
Iqbal, Mohammad Z., (319) 273-2998 m.iqbal@uni.edu, Univ of Northern Iowa – Hw
Irlandi, Elizabeth A., (321) 674-8096 irlandi@fit.edu, Florida Institute of Technology – Ob
Ironside, R. Geoffreygeoff.ironside@ualberta.ca, Univ of Alberta – Oo
Irvin, Gene D., 205/247-3542 dirvin@gsa.state.al.us, Geological Survey of Alabama – Gg
Irvine, Pamela J., (213) 620-4784 California Geological Survey – Gg
Irvine, T. Neil, (202) 478-8950 n.irvine@gl.ciw.edu, Carnegie Institution of Washington – Gi
Isaacson, Peter E., (208) 885-7969 isaacson@uidaho.edu, Univ of Idaho – Pi
Isacks, Bryan L., (607) 255-2307 blil@cornell.edu, Cornell Univ – Ys
Isbell, John L., 414-229-2877 jisbell@uwm.edu, Univ of Wisconsin, Milwaukee – Gs
Isenor, Fenton M., (902) 563-1184 Fenton_Isenor@uccb.ca, Univ College of Cape Breton – Gg
Ishman, Scott E., (618) 453-7377 Southern Illinois Univ Carbondale – Pe
Isiorho, Solomon A., 260-481-6254 isiorho@ipfw.edu, Indiana Univ / Purdue Univ, Fort Wayne – HwGeOy
Isley, Douglas C., (505) 368-3581 Dine College, Shiprock – Oi
Ismat, Zeshan, 717-358-4485 zeshan.ismat@fandm.edu, Franklin and Marshall College – Gc
Isphording, Wayne C., (251) 460-7568 wisphord@jaguar1.usouthal.edu, Univ of South Alabama – Ca
Israel, Daniel W., (919) 515-2388 North Carolina State Univ – Sb
Istanbulluoglu, Erkan, 402-472-2661 eistanbulluoglu2@unl.edu, Univ of Nebraska, Lincoln – Hq
Ito, Emi, 612-624-7881 eito@umn.edu, Univ of Minnesota, Twin Cities – Cs
Ito, Garrett T., 808-956-9717 gito@hawaii.edu, Univ of Hawai'i, Manoa – Yr
Iturrino, Geraldo, (845) 365-8656 Columbia Univ – Yr
Ivany, Linda C., (734) 763-9253 ivany@umich.edu, Syracuse Univ – Pi
Ivany, Linda C., 315-443-2672 lcivany@syr.edu, Syracuse Univ – Pe
Iverson, Neal R., (515) 294-8048 niverson@iastate.edu, Iowa State Univ of Science & Technology – Gl
Iverson, Richard L.iverson@ocean.fsu.edu, Florida State Univ – Ob
Ivins, Erik R., (818) 354-4785 Jet Propulsion Lab – Gt
Ivy, Logan D., (303) 370-6474 LIvy@dmns.org, Denver Museum of Nature & Science – Pv
Ivy, Russell L., (561) 297-3295 ivy@fau.edu, Florida Atlantic Univ – Oo
Izold, Mark D., (719) 579-3029 Mark.Izold@ppcc.edu, Pikes Peak Community College – Pm

J

Jablonski, David, (773) 702-8163 djablons@midway.uchicago.edu, Univ of Chicago – Pi
Jackson, Brian P., 603-646-1272 Brian.Jackson@dartmouth.edu, Dartmouth College – Ca
Jackson, Charles, (512) 471-0401 charles@ig.utexas.edu, Univ of Texas at Austin, Jackson School of Geosciences – Pe
Jackson, David D., (310) 825-0421 djackson@ucla.edu, Univ of California, Los Angeles – YsdYg
Jackson, Edgar L., (780) 492-0342 ed.jackson@ualberta.ca, Univ of Alberta – Oo
Jackson, Frankie, 406-994-6642 frankiej@montana.edu, Montana State Univ – Pv
Jackson, George A., (979) 845-0405 jackson@benthos.tamu.edu, Texas A&M Univ – Op
Jackson, Hiram, (916) 691-7605 Cosumnes River College – Gg
Jackson, Jeremy B. C., (858) 822-2432 jbjackson@ucsd.edu, Univ of California, San Diego – Po
Jackson, Justyn D., 662-325-1114 jdj8@msstate.edu, Mississippi State Univ – Ow
Jackson, Karen E., (904) 745-7300 Jacksonville Univ – Oo
Jackson, Kenneth J., (925) 422-6053 jackson8@llnl.gov, Lawrence Livermore National Lab – Cl
Jackson, Mark W., (801) 378-9753 mark_jackson@byu.edu, Brigham Young Univ – Or
Jackson, Martin P. A., (512) 475-9548 martin.jackson@beg.utexas.edu, Univ of Texas at Austin, Jackson School of Geosciences – Gc
Jackson, Michael, 612-624-5274 jacks057@umn.edu, Univ of Minnesota, Twin Cities – Ym
Jackson, Philip L., (541) 737-1203 jacksonp@geo.oregonstate.edu, Oregon State Univ – Oy
Jackson, Richard A., (718) 460-1476 Long Island Univ, Brooklyn Campus – Gc
Jackson, Richard H., (801) 378-6063 richard_jackson@byu.edu, Brigham Young Univ – Ou
Jackson, Robert B., 919 660 7408 jackson@duke.edu, Duke Univ – Oo
Jacob, Daniel J., (617) 495-1794 djj@io.harvard.edu, Harvard Univ – Oa
Jacob, Klaus H., (845) 365-8440 Columbia Univ – Ys
Jacobi, Robert D., (716) 645-6800 (Ext. 2468) rdjacobi@acsu.buffalo.edu, SUNY, Buffalo – Gr
Jacobs, Alan M., 330-941-2933 amjacobs@ysu.edu, Youngstown State Univ – Ge
Jacobs, Gary K.jacobsGK@ornl.gov, Oak Ridge National Lab – Cg
Jacobs, Katharine L., (520) 626-3054 kjacobs@hwr.arizona.edu, Univ of Arizona – Hg
Jacobs, Louis L., (214) 768-2773 Southern Methodist Univ – Pv
Jacobs, Stanley, (845) 365-8326 Columbia Univ – Op
Jacobs, Stanley J., (313) 936-1255 Univ of Michigan – Op
Jacobs, Tenika, (609) 984-6587 tenika.jacobs@dep.state.nj.us, New Jersey Geological Survey – Oo
Jacobsen, Jeffrey S., (406) 994-4605 jefj@montana.edu, Montana State Univ – So
Jacobsen, Stein B., (617) 495-5233 jacobsen@neodymium.harvard.edu, Harvard Univ – Cc
Jacobsen, Steven D., 847-467-1825 steven@earth.northwestern.edu, Northwestern Univ – Gy
Jacobson, Andrew D., (847) 491-3132 adj@earth.northwestern.edu, Northwestern Univ – Cl
Jacobson, Carl E., (515) 294-1837 cejac@iastate.edu, Iowa State Univ of Science & Technology – Gc
Jacobson, Elizabeth, (702) 895-1753 eaj@nevada.edu, Univ of Nevada, Las Vegas – Hg
Jacobson, Roger, (775) 673-7364 roger@dri.edu, Univ of Nevada, Reno – Cg
Jacoby, Gordon, (845) 365-8616 Columbia Univ – Hw
Jaecks, Glennjaecksg@arc.losrios.edu, American River College – GgOgPg
Jaeger, John M., (352) 846-1381 jaeger@geology.ufl.edu, Univ of Florida – Gs
Jaffe, Howard W., (802) 899-4339 Univ of Vermont – Gz
Jaffe, Peter R., (609) 258-4653 jaffe@princeton.edu, Princeton Univ – Hw

Jaguszewski, Janice M., 612-626-0557 j-jagu@umn.edu,
Univ of Minnesota, Twin Cities – Oo
Jahren, A. H., (410) 516-7134 jahren@jhu.edu,
The Johns Hopkins Univ – Pe
Jain, Atul K., (217) 333-2128 jain@atmos.uiuc.edu,
Univ of Illinois, Urbana-Champaign – Oa
Jakosky, Bruce M., 303-492-8004 jakosky@argyre.colorado.edu,
Univ of Colorado – Xg
Jaksha, Lawrence H., (505) 835-5501 New Mexico Institute of Mining and Technology – Ys
Jamaluddin, Tajul Anuar, 03-79674152 taj@um.edu.my,
Univ of Malaya – Ng
Jambor, John L., 604-948-1368 JLJambor@aol.com,
Univ of British Columbia – Gz
James, Corinne, (541) 737-2270 corinne@coas.oregonstate.edu,
Oregon State Univ – Op
James, David E., (202) 478-8838 james@dtm.ciw.edu,
Carnegie Institution of Washington – Ys
James, F. Colin, (250) 828-5402 Univ College of the Cariboo – Gg
James, Matthew J., (707) 664-2301 james@sonoma.edu,
Sonoma State Univ – Pi
James, Mindy C., (608) 263-7394 mcjames@wisc.edu,
Univ of Wisconsin - Extension – Oo
James, Noel P., (613) 545-6170 Queen's Univ – Gd
James, Richard, (705) 675-1151 (Ext. ext. 2271) rjames@laurentian.ca,
Laurentian Univ, Sudbury – Gp
Jamieson, Heather E., (613) 545-6181 Queen's Univ – Cp
Jamieson, J. Bruce, (403) 288-0803 Univ of Calgary – Ne
Jamieson, Rebecca A., (902) 494-3771 Dalhousie Univ – Gx
Janecke, Susanne U., (435) 797-3877 susanne.janecke@cc.usu.edu,
Utah State Univ – Gt
Janecky, David R., (505) 667-7603 Los Alamos National Lab – Cl
Jang-Condell, Hannah, (202) 478-8813 hannah@dtm.ciw.edu,
Carnegie Institution of Washington – Oo
Jannik, Nancy O.njannik@winona.edu,
Winona State Univ – Hw
Janos, Nancy A.nanjan@geo.princeton.edu,
Princeton Univ – Oo
Janowitz, Gerald S., (919) 515-7837 jerry_janowitz@ncsu.edu,
North Carolina State Univ – Op
Jansma, Pamela E., (479) 575-4748 pjansma@comp.uark.edu,
Univ of Arkansas, Fayetteville – Gc
Janson, Xavier, 512-475-9524 xavier.janson@beg.utexas.edu,
Univ of Texas at Austin, Jackson School of Geosciences – Gr
Jaouich, Alfred B., (514) 987-3378 jaouich.alfred@uqam.ca,
Universite du Quebec a Montreal – Sc
Jaquess, Jeffrey C., (573) 368-2148 nrjaquj@mail.dnr.state.mo.us,
Missouri Dept of Natural Resources – Hw
Jardine, Philip M.ipj@ornl.gov,
Oak Ridge National Lab – Sc
Jardine, Philip M., (865) 574-8058 ipj@ornl.gov,
Univ of Tennessee, Knoxville – Sc
Jarman, Walter A., (801) 585-3082 Univ of Utah – Ca
Jarrard, Richard D., (801) 585-3964 jarrard@mines.utah.edu,
Univ of Utah – Gu
Jarvis, Ed, +353 21 4902698 e.jarvis@ucc.ie,
Univ College Cork – Pl
Jarvis, Gary T., (416) 736-2100 (Ext. 77710) jarvis@yorku.ca,
York Univ – Yg
Jarzen, David M., (352) 392-1721 (Ext. 245) dmj@flmnh.ufl.edu,
Univ of Florida – Pb
Jasinski, Jerry P., (603) 358-2563 jjasinsk@keene.edu,
Keene State College – Gz
Jastrow, Julie D., (630) 252-3226 jdjastrow@anl.gov,
Argonne National Lab – Sb
Jaumé, Steven C., (843) 953-1802 jaumes@cofc.edu,
College of Charleston – Ys
Javaux, Emmanuelle, 32 4 366 54 22 EJ.Javaux@ulg.ac.be,
Universite de Liege – Pl
Jawitz, James W., (352) 392-1951 (Ext. 03) jawitz@ufl.edu,
Univ of Florida – Sp
Jay Krishnan, Jay, (713) 743-1697 krishnan@Bayou.UH.EDU,
Univ of Houston – Oo
Jayakumar, Amal, (609) 258-6294 ajayakum@princeton.edu,
Princeton Univ – Cm
Jayko, Angela S.ajayko@usgs.gov,
Humboldt State Univ – Gt
Jaynes, Dan B., (515) 294-8243 jaynes@nstl.gov,
Iowa State Univ of Science & Technology – Sp
Jeanloz, Raymond, (510) 642-2639 jeanloz@uclink.berkeley.edu,
Univ of California, Berkeley – Gy
Jebrak, Michel, (514) 987-3000 (Ext. 3986) jebrak.michel@uqam.ca,
Universite du Quebec a Montreal – Em
Jefferson, Anne, 704-687-5977 ajefferson@uncc.edu,
Univ of North Carolina, Charlotte – HgGm
Jeffery, David L., 740-376-4844 jefferyd@marietta.edu,
Marietta College – Go
Jeffress, Gary A., (361) 825-2720 jeffress@tamucc.edu,
Texas A&M Univ, Corpus Christi – Or
Jeffrey, Tepper, 253-879-3829 jtepper@ups.edu,
Univ of Puget Sound – Gi
Jeffries, H. Perry, (401) 874-6222 jeffries@uri.edu,
Univ of Rhode Island – Ob
Jekeli, Christopher, (614) 292-7117 Ohio State Univ – Yd
Jekeli, Christopher, 614 292 7117 jekeli.1@osu.edu,
Ohio State Univ – Yd
Jellinek, Markmjellinek@eos.ubc.ca,
Univ of British Columbia – Gg
Jenkins, B M., 252-335-3375 bmjenkins@mail.ecsu.edu,
Elizabeth City State Univ – Ge
Jenkins, David M., (607) 777-2736 Binghamton Univ – Gx
Jenkins, Mary Ann, (416) 7362100 (Ext. 22992) maj@yorku.ca,
York Univ – Oa
Jenkins, William J., (508) 289-2554 wjenkins@whoi.edu,
Woods Hole Oceanographic Institution – Oc
Jenner, George A., (709) 737-8387 gjenner@sparky2.esd.mun.ca,
Memorial Univ of Newfoundland – Gi
Jennette, Dave, (512) 471-0359 dave.jennette@beg.utexas.edu,
Univ of Texas at Austin, Jackson School of Geosciences – Gg
Jennings, Anne E., 303-492-7621 anne.jennings@colorado.edu,
Univ of Colorado – Pe
Jennings, Carrie E., (612) 627-4780, Ext. 220 carrie@umn.edu,
Univ of Minnesota, Twin Cities – Gl
Jennings, Eleanor, 918-631-3028 eleanor-jennings@utulsa.edu,
Univ of Tulsa – Sb
Jennings, Stephen P., 205-247-3543 sjennings@gsa.state.al.us,
Geological Survey of Alabama – Hg
Jennings, Jr., James W., (512) 471-4364 jim.jennings@beg.utexas.edu,
Univ of Texas at Austin, Jackson School of Geosciences – Oo
Jensen, Ann R., (605) 677-5227 ajensen@usd.edu,
South Dakota Dept of Env & Natural Resources – Gg
Jensen, David A., (361) 825-3333 Texas A&M Univ, Corpus Christi – Hg
Jensen, Debbie K., (208) 885-6192 debbiej@uidaho.edu,
Univ of Idaho – Oo
Jensen, Jerry, (979) 845-2206 jerry.jensen@pe.tamu.edu,
Texas A&M Univ – Oo
Jensen, Olivia G., 514-398-3587 olivia@eps.mcgill.ca,
McGill Univ – Yg
Jensen, Ryan, (812) 237-2258 gejensen@isugw.indstate.edu,
Indiana State Univ – Or
Jenson, John W., (671) 735-2689 jjenson@uog.edu,
Univ of Guam – Hy
Jercinovic, Michael J., (413) 545-2431 mjj@geo.umass.edu,
Univ of Massachusetts – Gx
Jerde, Eric, (606) 783-5406 e.jerde@moreheadstate.edu,
Morehead State Univ – Cg
Jerram, Dougal A., 0191 3342281 d.a.jerram@durham.ac.uk,
Durham Univ – Oo
Jesperson, Marcia J., (608) 263-7389 mjjesperson@wisc.edu,
Univ of Wisconsin - Extension – Oo
Jessey, David R., 909-869-3457 drjessey@csupomona.edu,
California State Polytechnic Univ, Pomona – Eg
Jeu, Amy, 212-772-4019 amy.jeu@hunter.cuny.edu,
Hunter College (CUNY) – Or
Jewell, Paul W., (801) 581-6636 pwjewell@mines.utah.edu,
Univ of Utah – Hg
Jezek, Kenneth C., (614) 292-6531 jezek.1@osu.edu,
Ohio State Univ – Or
Ji, Link, (979) 862-3134 lji@ocean.tamu.edu,
Texas A&M Univ – Op
Ji, Minhe, (940) 565-2377 jminhe@unt.edu,
Univ of North Texas – Oy
Ji, Wei, (816) 235-1334 jiwei@umkc.edu,
Univ of Missouri-Kansas City – Oy
Jiang, Dazhi, 519-661-3192 djiang3@uwo.ca,
Univ of Western Ontario – GcgGg
Jiang, Fuming, (609) 258-3261 fumingj@princeton.edu,
Princeton Univ – Gy
Jiang, Zhenhua, (410) 436-6864 Argonne National Lab – Oo
Jiang, Zhenying, (760) 252-2411 (Ext. 7202) Barstow Community

Johnston, Robert A., (207) 287-2801 Dept of Conservation – Gl
Johnston, Stephen T., 250 472 4481 stj@uvic.ca, Univ of Victoria – Gc
Johnston, III, John E., (225) 578-8657 hammer@lsu.edu, Louisiana State Univ – EoGe
Jokipii, Jack R., (520) 621-4256 jokipii@lpl.arizona.edu, Univ of Arizona – Xy
Jolley, Robert, (727) 553-1633 bjolley@marine.usf.edu, Univ of South Florida, Saint Petersburg – Oo
Jolley, Von D., (801) 378-2491 Brigham Young Univ – Sc
Jolliff, Bradley L., (314) 935-5622 blj@levee.wustl.edu, Washington Univ – Gx
Jolly, Wayne T., (905) 688-5550 (Ext. 3527) Wayne.Jolly@brocku.ca, Brock Univ – Gx
Jonas, John J., (514) 398-4755 McGill Univ – Nx
Jones, Alan, (607) 777-2518 Binghamton Univ – Ys
Jones, Bobby L., (225) 578-8567 bjone21@lsu.edu, Louisiana State Univ – Go
Jones, Brian, (780) 492-5249 brian.jones@ualberta.ca, Univ of Alberta – Ps
Jones, Charles E., (412) 624-6347 cejones@pitt.edu, Univ of Pittsburgh – Gg
Jones, Craig H., 303-492-6994 cjones@cires.colorado.edu, Univ of Colorado – Ys
Jones, Douglas S., (352) 392-1721 dsjones@flmnh.ufl.edu, Univ of Florida – Pg
Jones, Douglas S., (352) 392-1721 (Ext. 485) dsjones@flmnh.ufl.edu, Univ of Florida – Pg
Jones, Eric M., (505) 667-6386 honais@lanl.gov, Los Alamos National Lab – Oa
Jones, F. Walter, (780) 492-0667 wjones@phys.ualberta.ca, Univ of Alberta – Ym
Jones, Francis H., (604) 822-2138 fjones@eos.ubc.ca, Univ of British Columbia – Ye
Jones, Glen, (505) 835-5627 glen@gis.nmt.edu, New Mexico Institute of Mining & Technology – Oo
Jones, Henry, (408) 656-3270 Naval Postgraduate School – Op
Jones, Jon W., (403) 220-5024 Univ of Calgary – Gp
Jones, Julia A., (541) 737-1224 jonesj@geo.oregonstate.edu, Oregon State Univ – So
Jones, Larry Allan, (505) 667-0142 ljones@lanl.gov, Los Alamos National Lab – Ng
Jones, Lawrence, 970-2481787 lajones@mesastate.edu, Mesa State College – Gs
Jones, Lawrence S., 406-657-1108 jonesl@rocky.edu, Rocky Mountain College – Gs
Jones, Minnie O., (312) 996-3154 mojones@uic.edu, Univ of Illinois, Chicago – Oo
Jones, Norris W., (920) 424-4460 jonesnw@uwosh.edu, Univ of Wisconsin, Oshkosh – Gi
Jones, Peter E., 902-426-3869 Dalhousie Univ – Op
Jones, R. Christian, (703) 993-1127 rcjones@gmu.edu, George Mason Univ – Hs
Jones, Randy M., (918) 542-8441 Northeastern Oklahoma A&M College – Oo
Jones, Rebecca H., 512-471-0257 rebecca.jones@beg.utexas.edu, Univ of Texas at Austin, Jackson School of Geosciences – Ye
Jones, Rhian, 505-277-4204 rjones@unm.edu, Univ of New Mexico – Gz
Jones, Robert L., (217) 333-9490 Univ of Illinois, Urbana-Champaign – Sc
Jones, Stephen C., 205/247-3601 sjones@gsa.state.al.us, Geological Survey of Alabama – Hg
Jones, TjonesTP@cf.ac.uk, Cardiff Univ – Ge
Jones, Tim L., (505) 646-3405 New Mexico State Univ, Las Cruces – Sp
Jones, Timothy A., (705) 566-8101 (Ext. 7797) tajones@cambrianc.on.ca, Cambrian College of Applied Arts & Technology – Eg
Jones, Tracy, (423) 425-4409 tracy-jines@utc.edu, Univ of Tennessee, Chattanooga – Oo
Jones, William W., (812) 855-4556 Indiana Univ, Bloomington – Hs
Jordan, Benjamin R., (208) 496-1853 jordanb@byui.edu, Brigham Young Univ, Idaho – GvOuGx
Jordan, Brad C., (570) 577-3024 jordan@bucknell.edu, Bucknell Univ – Gi
Jordan, Brennan T., (605) 677-6143 Brennan.Jordan@usd.edu, Univ of South Dakota – GiOw
Jordan, Brennan T., 509-527-4934 jordanbt@whitman.edu, Whitman College – Gx
Jordan, Evelyn, (850) 675-6558 Florida Geological Survey – Oo
Jordan, Guntram, 089/2180 4353 guntram.jodan@lrz.uni-muenchen.de, LMU - Univ of Munich – Gz
Jordan, Jim L., (409) 880-8240 Lamar Univ – Cc
Jordan, Lance, (954)262-3619 jordanl@nova.edu, Nova Southeastern Univ – Ob
Jordan, Robert A., (757) 727-5783 robert.jordan@hamptonu.edu, Hampton Univ – Ob
Jordan, Robert R., (302) 831-6415 rrjordan@udel.edu, Univ of Delaware – Gr
Jordan, Teresa A., (607) 255-3596 jordan@geology.cornell.edu, Cornell Univ – Gr
Jordan, Thomas H., (213) 821-1237 tjordan@usc.edu, Univ of Southern California – Ys
Jornov, Donnadjornov@mail.nysed.gov, New York State Geological Survey – Oo
Jorstad, Robert B., (217) 581-6244 cfrbj@ux1.eiu.edu, Eastern Illinois Univ – Gr
Joseph, Lura E., (217) 333-1266 Univ of Illinois, Urbana-Champaign – Go
Joseph, Miranda, 520-621-5839 mirandaj@email.ariozna.edu, Univ of Arizona – Oo
Josephs, Richard L., (701) 777-2131 richard.josephs@und.nodak.edu, Univ of North Dakota – Ga
Journeay, Murray, (604) 666-1130 mjourneay@gsc.nrcan.gc.ca, Univ of British Columbia – Gc
Journel, Andre G., (650) 723-1594 journel@pangea.stanford.edu, Stanford Univ – Gq
Jugo, Pedro J., (705) 675-1151 (Ext. ext. 2106) pjugo@laurentian.ca, Laurentian Univ, Sudbury – Gi
Julian, Maureen M., (540) 231-7123 erie@vt.edu, Virginia Polytechnic Institute & State Univ – Gz
Jull, A. J. Timothy, (520) 621-6816 jull@u.arizona.edu, Univ of Arizona – Cg
Jung, Glenn H., (408) 656-2552 Naval Postgraduate School – Op
Junnila, Randy, (705) 566-8101 (Ext. 7796) Cambrian College of Applied Arts & Technology – Eg
Jurdy, Donna M., (847) 491-7163 donna@earth.northwestern.edu, Northwestern Univ – Yg
Jurena, Dwight, (210) 733-2840 djurena@mail.accd.edu, San Antonio College – GgOg
Jurmanovich, Barb, 989-686-9445 Delta College – Oo
Juster, Thomas C., (813) 974-9691 Univ of South Florida, Tampa – Hw
Jutla, Rajinder S., (417) 836-5298 rsj922f@smsu.edu, Southwest Missouri State Univ – Ou

K

Kadel, Steve, (623) 845-3618 s.kadel@gcmail.maricopa.edu, Glendale Community College – Xg
Kaden, Scott, 573-368-2356 nrkades@mail.dnr.state.mo.us, Missouri Dept of Natural Resources – Oo
Kaesler, Roger L., (785) 864-2751 kaesler@ku.edu, Univ of Kansas – Pm
Kafka, Alan, 617-552-8300 alan.kafka.1@bc.edu, Boston College – Ys
Kah, Linda, (865) 974-2366 Univ of Tennessee, Knoxville – Gs
Kahle, Anne B., (818) 354-7265 anne@aster.jpl.nasa.gov, Jet Propulsion Lab – Or
Kahle, Charles F., (419) 372-2888 ckahle@bgnet.bgsu.edu, Bowling Green State Univ – Gs
Kaip, Galen M., (915) 747-6817 kaip@geo.utep.edu, Univ of Texas, El Paso – Yx
Kaiser-Bischoff, Ines, 089/2180 4314 kaiser-bischoff@lmu.de, LMU - Univ of Munich – Gz
Kakar, Safdar, (604) 527-5230 kakars@douglas.bc.ca, Douglas College – Gg
Kaktins, Uldis, 814-269-2944 kaktins@pitt.edu, Univ of Pittsburgh at Johnstown – Hg
Kalakay, Thomas J., 406-587-4605 Univ of Montana, Western – Gc
Kalakay, Thomas J., 406-657-1101 kalakayt@rocky.edu, Rocky Mountain College – Gc
Kaldor, Michael, (305) 237-3025 mkaldor@mdcc.edu, Miami-Dade Community College (Wolfson Campus) – Gg
Kalia, Hemendra N., (702) 295-5767 hemendra_kalia@lanl.gov, Los Alamos National Lab – Nm
Kallemeyn, Gregory, (310) 825-3202 Univ of California, Los Angeles – Cg
Kaltenbacher, Eric A., (727) 553-3959 eak@marine.usf.edu, Univ of South Florida, Saint Petersburg – Oo
Kamb, Barclay, (626) 395-6456 barclay@gps.caltech.edu, California Institute of Technology – Gl
Kamber, Balz, (705) 675-1151 (Ext. ext. 2249) bkamber@laurentian.ca, Laurentian Univ, Sudbury – Cs
Kamenkovich, Vladimir D., (228) 688-3091 vladimir.kamenkovich@

usm.edu, Univ of Southern Mississippi – Op
Kamerling, Marc J., (805) 893-8435 marc@quake.crustal.ucsb.edu, Univ of California, Santa Barbara – Yr
Kamhi, Samuel R., (718) 460-1476 Long Island Univ, Brooklyn Campus – Gz
Kamilli, Robert J., (520) 670-5576 bkamilli@strider.swfo.arizona.edu, Univ of Arizona – Eg
Kammer, Thomas W., (304) 293-5603 tkammer@wvu.edu, West Virginia Univ – Pi
Kamola, Diane, (785) 864-2724 dlkamola@ku.edu, Univ of Kansas – Gs
Kampf, Anthony R., (213) 763-3328 akampf@nhm.org, Los Angeles County Museum of Natural History – Gz
Kamykowski, Daniel, (919) 515-7894 dan_kamykowski@ncsu.edu, North Carolina State Univ – Ob
Kana, Todd M., (410) 221-8481 kana@hpl.umces.edu, Univ of Maryland – Ob
Kanamori, Hiroo, (626) 395-6914 hiroo@gps.caltech.edu, California Institute of Technology – Ys
Kanaroglou, Pavlos S., 905-525-9140 (Ext. 23525) pavlos@mcmaster.ca, McMaster Univ – Yr
Kanat, Leslie H., (802) 635-1327 kanatl@jsc.vsc.edu, Johnson State College – Gc
Kane, Jeffrey A., (512) 471-6136 jeff.kane@beg.utexas.edu, Univ of Texas at Austin, Jackson School of Geosciences – Yg
Kane, Kathy A., (608) 263-7385 kakane@wisc.edu, Univ of Wisconsin - Extension – Oo
Kanemasu, Edward T., (706) 542-2151 Univ of Georgia – Sp
Kanfoush, Sharon L., (315) 792-3134 skanfoush@utica.edu, Utica College – GgsOg
Kantzas, Apostolos, (403) 220-8907 akantzas@ucalgary.ca, Univ of Calgary – Ng
Kao, Jim C., (505) 667-9226 Los Alamos National Lab – Oa
Kaplan, Alexey, (845) 365-8689 Columbia Univ – Op
Kaplan, Isaac R., (310) 825-5076 Univ of California, Los Angeles – Cg
Kaplan, Samantha W., 715-346-4149 skaplan@uwsp.edu, Univ of Wisconsin, Stevens Point – GnPlGd
Kaplinski, Matthew A., (928) 523-9145 Northern Arizona Univ – Yg
Kaplun, Nicolainkaplun@njcu.edu, New Jersey City Univ – Oo
Kapp, Paul, (520) 626-8763 pkapp@geo.arizona.edu, Univ of Arizona – Oo
Kar, Aditya, (478) 825-6844 cara@mail.fvsu.edu, Georgia State Univ – Gi
Karabinos, Paul, (413) 597-2079 paul.m.karabinos@williams.edu, Williams College – Gc
Karaffa, Marni D., (812) 855-7428 mldickso@indiana.edu, Indiana Geological Survey – Gl
Karhu, Juha A., 358-9-19150834 Juha.Karhu@helsinki.fi, Univ of Helsinki – Cg
Karig, Daniel E.dek9@cornell.edu, Cornell Univ – Yr
Karl, David M., (808) 956-8964 dkarl@soest.hawaii.edu, Univ of Hawai'i, Manoa – Ob
Karl, Tami S., (850) 644-5861 karl@gly.fsu.edu, Florida State Univ – Oo
Karlen, Douglas L., (515) 294-3336 dkarlen@nstl.gov, Iowa State Univ of Science & Technology – So
Karlin, Robert, (775) 784-1770 karlin@mines.unr.edu, Univ of Nevada, Reno – Ou
Karlin, Robert, (775) 784-1770 karlin@unr.edu, Univ of Nevada, Reno – Hg
Karlsson, Haraldur R., (806) 742-3130 hal.karlsson@ttu.edu, Texas Tech Univ – Cl
Karlstrom, Karl E., (505) 277-4346 kek1@unm.edu, Univ of New Mexico – Gt
Karner, Daniel B., (707) 664-2854 karner@sonoma.edu, Sonoma State Univ – Cc
Karner, Frank R., (701) 777-4570 frank_karner@mail.und.nodak.edu, Univ of North Dakota – Gx
Karner, James M., (505) 277-8327 Univ of New Mexico – Gz
Karpilo, Jr, Ronald J., (970) 225-3500 ron_karpilo@partner.nps.gov, Colorado State Univ – Oy
Karwoski, Todd, (301) 405-0084 karwoski@geol.umd.edu, Univ of Maryland – Cg
Kasabach, Haig F., (609) 292-1185 New Jersey Geological Survey – Hw
Kaspar, Thomas A., (515) 294-8873 Iowa State Univ of Science & Technology – So
Kassube, Kay L., 605-677-6403 Kay.Kassube@usd.edu, Univ of South Dakota – Gg
Kastens, Kim A., (845) 365-8836 kastens@ldeo.columbia.edu, Columbia Univ – Gu
Kasting, James F., (814) 865-3207 kasting@essc.psu.edu, Pennsylvania State Univ, Univ Park – Oa
Kasting, James F., (814) 865-3207 kasting@essc.psu.edu, The Pennsylvania State Univ, Univ Park – Ow
Kastner, Miriam, (858) 534-2065 mkastner@ucsd.edu, Univ of California, San Diego – Cl
Kath, Randal L., (678) 839-4063 rkath@westga.edu, Univ of West Georgia – Gc
Kato, Terence T., (916) 898-5262 tkato@csuchico.edu, California State Univ, Chico – Gt
Kattenhorn, Simon A., (208) 885-5063 simkat@uidaho.edu, Univ of Idaho – Gc
Katuna, Michael P., (843) 953-5588 katunam@cofc.edu, College of Charleston – Gu
Katz, Gabrielle, (828) 262-3000 katzgl@appstate.edu, Appalachian State Univ – Hg
Katz, Miriam E., 732-445-3445 mimikatz@rci.rutgers.edu, Rutgers, The State Univ of New Jersey – Pm
Katzman, Danny, (505) 667-0599 katzman@lanl.gov, Los Alamos National Lab – Gg
Kauahikaua, James P., (808) 967-7320 Univ of Hawai'i, Manoa – Gv
Kauffman, Chad, 724 938-5760 kauffman@cup.edu, California Univ of Pennsylvania – Oa
Kauffman, Erle G., (812) 855-5154 kauffman@indiana.edu, Indiana Univ, Bloomington – Po
Kauffman, John D., (208) 885-7540 johndk@uidaho.edu, Univ of Idaho – Gg
Kauffman, Marvin E., (406) 683-7615 marvsuekauffman@hotmail.com, Univ of Montana, Western – Gr
Kaufman, Alan J., (301) 4050395 kaufman@geol.umd.edu, Univ of Maryland – Cg
Kaufman, Darrell S.darrell.kaufman@nau.edu, Northern Arizona Univ – Gm
Kaufman, Darrell S., (928) 523-7192 darrell.kaufman@nau.edu, Northern Arizona Univ – Gm
Kaufmann, Robert K., (617) 353-3940 kaufmann@bu.edu, Boston Univ – Oo
Kaufmann, Ronald S., (619) 260-5904 kaufmann@sandiego.edu, Univ of San Diego – Ob
Kavner, Abby, 310-206-3675 akavner@ucla.edu, Univ of California, Los Angeles – Yg
Kawase, Mitsuhiro, (206) 543-0766 kawase@ocean.washington.edu, Univ of Washington – Op
Kay, Beverly D., (519) 824-4120 (Ext. 52484) bkay@lrs.uoguelph.ca, Univ of Guelph – Sc
Kay, Richard F., (919) 684-2143 rich_kay@baa.mc.duke.edu, Duke Univ – Oo
Kay, Robert W., (607) 255-3461 kay@geology.cornell.edu, Cornell Univ – Gi
Kay, Suzanne M., (607) 255-4701 skay@geology.cornell.edu, Cornell Univ – Gi
Kaye, John M., (662) 325-3915 Mississippi State Univ – Gg
Kays, M. Allan, (541) 346-4578 makays@oregon.uoregon.edu, Univ of Oregon – Gp
Kazanci, Nizamettinkazanci@eng.ankara.edu.tr, Ankara Univ – Gs
Keaffaber, J. J., (321) 674-8096 dmes@fit.edu, Florida Institute of Technology – Oc
Keala, Lori, (909) 621-8675 lkeala@pomona.edu, Pomona College – Oo
Kean, Jr., William F., 414-229-5231 wkean@uwm.edu, Univ of Wisconsin, Milwaukee – Ym
Kearl, Peter M., (303) 248-6211 Oak Ridge National Lab – Hw
Kearney, Micheal S., (301) 405-4057 mk11@umail.umd.edu, Univ of Maryland – Gu
Kearns, Lance E., 540-568-6421 kearnsle@jmu.edu, James Madison Univ – Gz
Keating, Barbara H., (808) 956-8143 keating@soest.hawaii.edu, Univ of Hawai'i, Manoa – Ym
Keating, Elizabeth, (505) 665-6714 ekeating@lanl.gov, Los Alamos National Lab – Gg
Keating, Martha E., (781) 891-2980 mkeating@bentley.edu, Bentley College – Oo
Keaton, Jeffrey R., (801) 581-8218 Univ of Utah – Oo
Keatts, Merida, 330-672-2897 mkeatts@kent.edu, Kent State Univ – Oo
Keays, Reid R., (705) 675-1151 Laurentian Univ, Sudbury – Em

Kebbede, Girma, 413-538-2004 gkebbede@mtholyoke.edu,
Mount Holyoke College – Oy
Kecojevic, Vladislav J., 814-865-4288 vuk2@psu.edu,
Pennsylvania State Univ, Univ Park – Nm
Keefer, Donald, 217 244-2786 keefer@isgs.uiuc.edu,
Illinois State Geological Survey – Hw
Keeler, Gerald, (734) 764-1435 jkeeler@umich.edu,
Univ of Michigan – Oo
Keeler, Gerald J., (313) 936-1836 jkeeler@engin.umich.edu,
Univ of Michigan – Og
Keeling, Charles D., (858) 534-4230 cdkeeling@ucsd.edu,
Univ of California, San Diego – Co
Keeling, David, (270) 745-4555 david.keeling@wku.edu,
Western Kentucky Univ – Og
Keeling, Ralph F., (858) 534-7582 rkeeling@ucsd.edu,
Univ of California, San Diego – Oa
Keen, Cecil S., (507) 389-5169 cecil.keen@mnsu.edu,
Minnesota State Univ – Oa
Keen, Kerry L., (715) 425-3729 kerry.l.keen@uwrf.edu,
Univ of Wisconsin, River Falls – Hw
Keen, Richard C., 717.702.2020 rkeen@state.pa.us,
Pennsylvania Bureau of Topographic & Geologic Survey – Oo
Keene, William, (804) 924-0586 wck@virginia.edu,
Univ of Virginia – Oa
Keeney, Dennis R., (515) 294-8066 drkeeney@iastate.edu,
Iowa State Univ of Science & Technology – So
Keesee, Robert G., (518) 442-4566 rgk@atmos.albany.edu,
SUNY, Albany – Oa
Kehew, Alan E., (269) 387-5495 alan.kehew@wmich.edu,
Western Michigan Univ – Hw
Kehler, Philip L., (501) 569-3545 plkehler@ualr.edu,
Univ of Arkansas, Little Rock – Gr
Kehoe-Forutan, Sandra J., (570) 389-4106 kehoe@bloomu.edu,
Bloomsburg Univ – Oo
Keigwin, Lloyd D., (508) 289-2784 lkeigwin@whoi.edu,
Woods Hole Oceanographic Institution – Cs
Keil, Klaus, (808) 956-3898 Univ of Hawai'i, Manoa – Xm
Keil, Richard G., (206) 616-1947 rickkeil@ocean.washington.edu,
Univ of Washington – Oc
Keilis-Borok, Vladimir, (310) 825-8945 vkb@ess.ucla.edu,
Univ of California, Los Angeles – Ys
Keimig, Francis T., (413) 545-0659 frank@geo.umass.edu,
Univ of Massachusetts – Ow
Keith, Brian D., 812-855-4213 keithb@indiana.edu,
Indiana Univ, Bloomington – Gs
Keith, Brian D., (812) 855-5412 keithb@indiana.edu,
Indiana Geological Survey – Go
Keith, Edward, (954)262-3674 edwardok@nova.edu,
Nova Southeastern Univ – Ob
Keith, Jeffrey D., 801-422-2189 jdk@geology.byu.edu,
Brigham Young Univ – Eg
Kelbert, Anna, (541) 737-4113 anya@coas.oregonstate.edu,
Oregon State Univ – Yr
Kelemen, Peter B., 845-365-8728 peterk@ldeo.columbia.edu,
Columbia Univ – Gi
Kelkar, Sharad, (505) 667-4639 kelkar@vega.lanl.gov,
Los Alamos National Lab – Np
Keller, C. Kent, (509) 335-3040 ckkeller@wsu.edu,
Washington State Univ – Hw
Keller, C. Peter, (250) 721-7333 seamus@uvvm.uvic.ca,
Univ of Victoria – Or
Keller, Dianne M., (315) 228-7893 dkeller@mail.colgate.edu,
Colgate Univ – Cg
Keller, Edward A., (805) 893-4207 Univ of California, Santa Barbara – Gm
Keller, G. R., (915) 747-5850 keller@geo.utep.edu,
Univ of Texas, El Paso – Ye
Keller, George H., (541) 737-7994 george.keller@orst.edu,
Oregon State Univ – Ou
Keller, Gerta, 609-258-4117 gkeller@princeton.edu,
Princeton Univ – Pm
Keller, Jean, (845) 938-4185 United States Military Academy – Oo
Keller, Joanne, (417) 836-5800 cjk766t@smsu.edu,
Southwest Missouri State Univ – Oo
Keller, Klaus, (814) 865-6718 kkeller@geosc.psu.edu,
Pennsylvania State Univ, Univ Park – Og
Keller, Randall, (541) 737-7648 kellerr@geo.oregonstate.edu,
Oregon State Univ – Gg
Keller, Jr., C. F., (505) 667-0920 cfk@lanl.gov,
Los Alamos National Lab – Oa
Kelley, Alice R., (207) 581-2056 arkelley@maine.edu,
Univ of Maine – Ga
Kelley, Cheryl A., (573) 882-8813 KelleyC@missouri.edu,
Univ of Missouri – Co
Kelley, Dan, (902) 494-1694 dan.kelley@dal.ca,
Dalhousie Univ – Op
Kelley, Deborah S., (206) 543-9279 kelley@ocean.washington.edu,
Univ of Washington – Ou
Kelley, Joseph T., (207) 581-2162 jtkelley@maine.edu,
Univ of Maine – Gu
Kelley, Michael, (607) 254-5331 mikek@ee.cornell.edu,
Cornell Univ – Oa
Kelley, Michael B., (845) 938-3938 United States Military Acad– Hs
Kelley, Patricia H., (910) 962-7406 kelleyp@uncw.edu,
Univ of North Carolina Wilmington – Pg
Kelley, Shari A., (505) 661-6171 sakelley@ix.netcom.com,
New Mexico Institute of Mining and Technology – Gt
Kelling, Keith A., (608) 263-2795 kkelling@wisc.edu,
Univ of Wisconsin, Madison – So
Kellman, Lisa M., (902) 867-5086 lkellman@stfx.ca,
Saint Francis Xavier Univ – Ge
Kellman, Lisa M., 902-867-5086 lkellman@stfx.ca,
Dalhousie Univ – Cg
Kellogg, James N., (803) 777-4501 kellogg@geol.sc.edu,
Univ of South Carolina – Yg
Kellogg, Louise H., (530) 752-3690 kellogg@geology.ucdavis.edu,
Univ of California, Davis – Yg
Kellogg, Thomas B., (207) 581-2194 tomk@iceage.umeqs.maine.edu,
Univ of Maine – Pm
Kelly, D. Clay, (608) 262-1698 ckelly@geology.wisc.edu,
Univ of Wisconsin, Madison – Pm
Kelly, Daniel, (801) 537-3310 nrugs.dkelly@state.ut.us,
Utah Geological Survey – Oo
Kelly, Kathryn A., (206) 543-9810 kkelly@apl.washington.edu,
Univ of Washington – Op
Kelly, Shari, (505) 661-6171 sakelley@ix.netcom.com,
New Mexico Institute of Mining & Technology – Yg
Kelly, Walton R., (217) 333-3729 Illinois State Water Survey – Hw
Kelly, William C., (734) 764-1435 billkell@umich.edu,
Univ of Michigan – Eg
Kelly, William M., (518) 474-7559 wkelly@mail.nysed.gov,
New York State Geological Survey – Eg
Kelsey, Harvey M., (707) 826-3991 hmk1@axe.humboldt.edu,
Humboldt State Univ – Gm
Kelso, Donald P., (703) 993-1061 dkelso@gmu.edu,
George Mason Univ – Ob
Kelso, Paul R., (906) 635-2158 pkelso@lssu.edu,
Lake Superior State Univ – Ym
Kelson, Christopher R., (315) 267-2285 kelsoncr@potsdam.edu,
SUNY, Potsdam – Gz
Kelty, Thomas, (562) 985-4589 kelty@csulb.edu,
California State Univ, Long Beach – Gc
Kemeny, John M., (520) 621-4448 kemeny@u.arizona.edu,
Univ of Arizona – Nr
Kemmerly, Phillip R., (931) 221-7471 kemmerlyp@apsu.edu,
Austin Peay State Univ – GmeHg
Kemp, Paul, (631) 632-8796 Kemp@ocean2.msrc.sunysb.edu,
SUNY, Stony Brook – Ob
Kemp, Paul F., (631) 632-8796 paul.kemp@stonybrook.edu,
SUNY, Stony Brook – Ob
Kendall, Christopher G., (803) 777-2410 Univ of South Carolina – Gs
Kendall, Treavor A., (864) 656-1897 treavok@clemson.edu,
Clemson Univ – Cl
Kendrick, David C., 315-781-3929 kendrick@hws.edu,
Hobart & William Smith Colleges – Po
Kendrick, Katherine J., (951) 276-4418 kendrick@gps.caltech.edu,
Univ of California, Riverside – Gm
Kenig, Fabien, (312) 996-3020 fkenig@uic.edu,
Univ of Illinois, Chicago – Ge
Kenna, Timothy, (845)-365-8513 Columbia Univ – Cg
Kennan, Sean, (954)262-3637 skeenan@nova.edu,
Nova Southeastern Univ – Og
Kennedy, Ann C., (509) 335-1554 akennedy@wsu.edu,
Washington State Univ – Sb
Kennedy, B. Mack, (510) 486-6451 bmkennedy@lbl.gov,
Lawrence Berkeley National Lab – Cg
Kennedy, Benbkennedy@eos.ubc.ca,
Univ of British Columbia – Gv
Kennedy, Christina B., (928) 523-0983 tina.kennedy@nau.edu,
Northern Arizona Univ – Oo
Kennedy, Lori, (604) 822-1811 lkennedy@eos.ubc.ca,

Univ of British Columbia – Gc
Kennedy, Martin J., (951) 827-2025 martin.kennedy@ucr.edu, Univ of California, Riverside – Gr
Kennedy, Michael P., (213) 620-5024 California Geological Survey – Gu
Kennel, Charles F., (310) 825-4018 Univ of California, Los Angeles – Oo
Kennel, Charles F., (858) 534-2826 ckennel@ucsd.edu, Univ of California, San Diego – Oo
Kennett, James P., (805) 893-3103 Univ of California, Santa Barbara – Gu
Kenney, Robert D., (401) 874-6664 rkenney@gso.uri.edu, Univ of Rhode Island – Ob
Kennicutt, III, Mahlon C., 979 458 0115 mck2@gerg.tamu.edu, Texas A&M Univ – Oc
Kenny, Ray, (970) 247-7462 kenny_r@fortlewis.edu, Fort Lewis College – Gm
Kent, Adam J., (541) 737-1205 adam.kent@geo.oregonstate.edu, Oregon State Univ – Ca
Kent, Dennis V., (732) 445-7049 dvk@rci.rutgers.edu, Rutgers, The State Univ of New Jersey – Ym
Kent, Donald M., (306) 585-4677 Univ of Regina – Gs
Kent, Douglas, 405-377-0166 Oklahoma State Univ – Hw
Kenworthy, Stephen, (270) 745-8777 stephen.kenworthy@wku.edu, Western Kentucky Univ – Gm
Kenyon, Patricia M., (212) 650-6472 Graduate School of the City Univ of New York – Yg
Kepic, Anton W., +61 8 9266-7503 A.Kepic@curtin.edu.au, Curtin Univ of Technology – Ye
Keppens, Edward, 32-2-6293395 ekeppens@vub.ac.be, Vrije Univ Brussel – CsGeg
Keppie, J. Duncan, 56224303 duncan@servidor.unam.mx, Universidad Nacional Autonoma de Mexico – Gt
Kerans, Charles, (512) 471-1368 charles.kerans@beg.utexas.edu, Univ of Texas at Austin, Jackson School of Geosciences – Gd
Kerans, Charles, (512) 471-4282 ckerans@mail.utexas.edu, Univ of Texas at Austin, Jackson School of Geosciences – Gs
Kerr, A C.KerrA@cf.ac.uk, Cardiff Univ – Gx
Kerr, Dennis R., 918 631 3020 dennis-kerr@utulsa.edu, Univ of Tulsa – Gs
Kerrich, Robert W., (306) 966-5719 robert.kerrich@usask.ca, Univ of Saskatchewan – Cg
Kerrick, Derrill M., (814) 865-7574 kerrick@geosc.psu.edu, Pennsylvania State Univ, Univ Park – Gp
Kerschen, Michael D., 406-496-4142 mkerschen@mtech.edu, Montana Tech of The Univ of Montana – Hw
Kershaw, G. Peter, (780) 492-0346 peter.kershaw@ualberta.ca, Univ of Alberta – Oy
Kerwin, Charles M., 603.358.2405 ckerwin@keene.edu, Keene State College – Gg
Kerwin, Michael W., 303-871-3998 mkerwin@du.edu, Univ of Denver – Gg
Kesel, Richard H., (225) 578-5880 gakesel@lsu.edu, Louisiana State Univ – Hg
Keskinen, Mary J., 907-474-7769 ffmjk@uaf.edu, Univ of Alaska, Fairbanks – Gx
Kesler, Christopher, (801) 585-9771 ckesler@egi.utah.edu, Univ of Utah – Or
Kesler, Stephen E., (734) 763-5057 skesler@umich.edu, Univ of Michigan – Eg
Kessinger, Walter P., (337) 984-3554 Univ of Louisiana, Lafayette – Pm
Kessler, Fritz, (301) 687-4266 fkessler@frostburg.edu, Frostburg State Univ – Or
Kessler, John D., (609) 258-6899 jdkessle@princeton.edu, Princeton Univ – Cm
Kessler, William S., (206) 526-6221 kessler@pmel.noaa.gov, Univ of Washington – Op
Ketcham, Richard A., (512) 471-6942 ketcham@mail.utexas.edu, Univ of Texas at Austin, Jackson School of Geosciences – Gz
Ketterer, Michael, (520) 523-7055 Case Western Reserve Univ – Ca
Kettler, Richard M., (402) 472-0882 rkettler1@unl.edu, Univ of Nebraska, Lincoln – Cl
Key, Jeffrey R., (608)263-2605 jkey@ssec.wisc.edu, Univ of Wisconsin, Madison – Or
Key, Jr., Marcus M., 717-245-1448 key@dickinson.edu, Dickinson College – Pi
Keyantash, John, (310) 243-2363 jkeyantash@csudh.edu, California State Univ, Dominguez Hills – Hg
Keyser, Daniel, (518) 442-4559 keyser@atmos.albany.edu, SUNY, Albany – Oa
Khalequzzaman, Md., (570) 484-2075 mkhalequ@lhup.edu, Lock Haven Univ – HwOni
Khan, Belayet H., (217) 581-6246 cfbhk@ux1.eiu.edu, Eastern Illinois Univ – Oy
Khan, Izhar A., (361) 749-6810 ikhan@mail.utexas.edu, Univ of Texas at Austin – Ob
Khan, Latif A., 217-244-2383 khan@isgs.uiuc.edu, Illinois State Geological Survey – Nx
Khan, Mohamed, (701) 231-8596 mkhan@ndsuext.nodak.edu, North Dakota State Univ – Oa
Khanbilvardi, Reza M., (215) 650-8009 Graduate School of the City Univ of New York – Hy
Khawaja, Ikram U.ikhawaja@cis.ysu.edu, Youngstown State Univ – Ec
Khazanov, George V., (313) 763-6229 Univ of Michigan – Og
Khosrowpanah, Shahram, (671) 735-2694 khosrow@uog.edu, Univ of Guam – Hs
Khurana, Krishan, (310) 825-8240 Univ of California, Los Angeles – Oo
Kiage, Lawrence M., (404) 413-5777 geolkk@langate.gsu.edu, Georgia State Univ – Oo
Kiang, Chia S., (404) 894-1748 chia.kiang@eas.gatech.edu, Georgia Institute of Technology – Oa
Kidane-Mariam, Tadesse, (814) 732-1576 tadesse@edinboro.edu, Edinboro Univ of Pennsylvania – Oo
Kidd, Claren, (405) 325-6217 ckidd@ou.edu, Univ of Oklahoma – Oo
Kidd, William S., (518) 442-4477 wkidd@atmos.albany.edu, SUNY, Albany – Gt
Kidder, David L., (740) 593-1108 kidder@ohio.edu, Ohio Univ, Lancaster – Gs
Kidder, Gerald, (352) 392-1951 kidder@ufl.edu, Univ of Florida – So
Kidwell, Susan M., (773) 702-3008 skidwell@midway.uchicago.edu, Univ of Chicago – Gs
Kiefer, John D., (859) 257-5500 (Ext. 145) kiefer@uky.edu, Univ of Kentucky – Ng
Kieffer, Brunobkieffer@eos.ubc.ca, Univ of British Columbia – Cs
Kieffer, Scott, 303-273-3714 skieffer@mines.edu, Colorado School of Mines – Nr
Kieffer, Susan W., (217) 244-6206 skieffer@uiuc.edu, Univ of Illinois, Urbana-Champaign – Gv
Kienast, Markus, (902) 494-8338 markus.kienast@dal.ca, Dalhousie Univ – Cs
Kiene, Ronald P., (251) 861-7526 rkiene@jaguar1.usouthal.edu, Univ of South Alabama – Oc
Kieniewicz, Johanna M., (740) 587-6487 kieniewiczj@denison.edu, Denison Univ – Pe
Kientop, Gregory A., 708/547-7092 kientop@isgs.uiuc.edu, Illinois State Geological Survey – Ng
Kihm, Allen J., (701) 858-3864 kihm@minotstateu.edu, Minot State Univ – Pv
Kilbourn, Phaedra M., (801) 581-5461 Univ of Utah – Ca
Kilcoyne, John R., (303) 556-4258 Metropolitan State College of Denver – Oy
Kile, Susan, (217) 581-2626 csskk1@ux1.eiu.edu, Eastern Illinois Univ – Oo
Kilic, Recepkilic@eng.ankara.edu.tr, Ankara Univ – Ng
Kilinc, Attila I., (513) 556-5967 attila.kilinc@uc.edu, Univ of Cincinnati – Cp
Killeen, Timothy L., (313) 747-3435 Univ of Michigan – Oa
Killorn, Randy J., (515) 294-1923 rkillorn@iastate.edu, Iowa State Univ of Science & Technology – So
Kilmer, Frank H., (671) 734-9305 Univ of Guam – Ps
Kim, Changjoo, 50738922617 changjoo.kim@mnsu.edu, Minnesota State Univ – Oi
Kim, Eugene M., (512) 471-1806 eugene.kim@beg.utexas.edu, Univ of Texas at Austin, Jackson School of Geosciences – Eg
Kim, Jonathan, (802) 241-3469 jon.kim@anr.state.vt.us, Agency of Natural Resources, Dept. of Env Conservation – Gg
Kim, Kwang, (775) 784-7522 kwangkim@unr.edu, Univ of Nevada, Reno – Oo
Kim, Sangil, (541) 737-4035 skim@coas.oregonstate.edu, Oregon State Univ – Op
Kim, Won-Young, (845) 365-8387 Columbia Univ – Ys
Kimber, Clarissa T., (409) 845-7141 Texas A&M Univ – Oy
Kimberley, Michael M., (919) 515-7831 kimberley@ncsu.edu, North Carolina State Univ – Ce
Kimberly, Paul, (202) 633-1854 kimberlyp@si.edu, Smithsonian Inst / National Museum of Natural History – GvOi
Kimbrough, David L., (619) 594-1385 dkimbrough@sunstroke.sdsu.edu,

San Diego State Univ – Cc
Kimerling, A. Jon, (541) 737-1225 kimerlia@geo.oregonstate.edu, Oregon State Univ – Or
Kimyai, Abbas, (209) 667-3466 geology@science.csustan.edu, California State Univ, Stanislaus – Pl
Kincaid, Charles T., (509) 372-6102 ct_kincaid@pnl.gov, Pacific Northwest National Lab – Hw
Kincaid, Christopher, (401) 874-6571 kincaid@lighthouse.gso.uri.edu, Univ of Rhode Island – Ou
Kind, Thomas C., (270) 809-3110 or (270) 809-2493 tom.kind@murraystate.edu, Murray State Univ – Or
Kineke, Gail C., (617) 552-3655 kinekeg@bc.edu, Boston College – On
King, Elizabeth M., (309) 438-8922 emking@ilstu.edu, Illinois State Univ – Gx
King, Gary M., (207) 563-3146 Univ of Maine – Oc
King, Hobart M., (570) 662-4615 hking@mnsfld.edu, Mansfield Univ – Ge
King, John, (401) 874-6594 jking@gso.uri.edu, Univ of Rhode Island – Ou
King, John S., (716) 645-6800 (Ext. 6100) SUNY, Buffalo – Gc
King, Jonathan K., (801) 537-3354 nrugs.jking@state.ut.us, Utah Geological Survey – Gg
King, Kenneth M., (519) 824-4120 (Ext. 52453) kking@lrs.uoguelph.ca, Univ of Guelph – Oa
King, Nicole C., (801) 537-3300 nrugs.nking@state.ut.us, Utah Geological Survey – Oo
King, Norman R., (812) 464-1794 nking@usi.edu, Univ of Southern Indiana – Gr
King, Penelope L., (505) 277-1643 penking@unm.edu, Univ of New Mexico – GiXcCp
King, Penelope L., 505-277-1643 penking@unm.edu, Univ of Western Ontario – Gi
King, Robert W., (617) 253-7064 rwk@chandler.mit.edu, Massachusetts Institute of Technology – Yg
King, Scott D., 540-231-6521 Virginia Polytechnic Institute & State Univ – Yg
King, Wendell C., (845) 938-4658 United States Military Academy – Hs
King, Jr., David T., (334) 844-4882 kingdat@auburn.edu, Auburn Univ – Gr
King-Rundel, Judith A., 310 243-3377 dendrochick@aol.com, California State Univ, Dominguez Hills – Oe
Kingsley, John, (801) 537-3309 nrugs.cbird@state.ut.us, Utah Geological Survey – Oo
Kinkead, Scott, (505) 665-1760 Los Alamos National Lab – Oo
Kinnicutt, Patrick, 989-774-2294 kinni1p@cmich.edu, Central Michigan Univ – Gq
Kinsland, Gary L., (337) 482-6824 glkinsland@louisiana.edu, Univ of Louisiana, Lafayette – Yg
Kinzler, Rosamond, 212-496-3637 kinzler@amnhl.org, American Museum of Natural History – Gi
Kiper, Ted B., (850) 488-9380 Florida Geological Survey – Oo
Kipper, Jay P., (512) 475-9505 jay.kipper@beg.utexas.edu, Univ of Texas at Austin, Jackson School of Geosciences – Oo
Kipphut, George W., (270) 809-2847 Murray State Univ – Op
Kirby, Carl S., (570) 577-1385 kirby@bucknell.edu, Bucknell Univ – Cl
Kirby, Eric, (814) 865-0732 ekirby@geosc.psu.edu, Pennsylvania State Univ, Univ Park – Gm
Kirby, Matthew E., (714) 278-2158 mkirby@fullerton.edu, California State Univ, Fullerton – Gs
Kirchgasser, William T., (315) 267-2296 kirchgwt@potsdam.edu, SUNY, Potsdam – Ps
Kirchner, James W., (510) 643-8559 kirchner@seismo.berkeley.edu, Univ of California, Berkeley – Ge
Kirkby, Kent C., (612) 624-1392 kirkby@tc.umn.edu, Univ of Minnesota, Twin Cities – Gg
Kirkham, MaryBeth, (913) 532-0422 Kansas State Univ – Sp
Kirkham, Randy R., (509) 372-6038 rr_kirkham@pnl.gov, Pacific Northwest National Lab – Hq
Kirkland, Brenda L., 662-325-2911 kirkland@geosci.msstate.edu, Mississippi State Univ – Gd
Kirkland, Douglas, (972) 883-2453 dougkirk@utdallas.edu, Univ of Texas, Dallas – Cg
Kirkland, James I., (801) 537-3307 nrugs.jkirklan@state.ut.us, Utah Geological Survey – Pg
Kirkman, Blanca E., (210) 999-7092 bkirkman@trinity.edu, Trinity Univ – Oo
Kirkpatrick, Mary R., (918) 542-8441 Northeastern Oklahoma A&M College – Oo
Kirlin, R. Lynn, (250) 721-8681 lkirlin@comcast.net, Univ of British Columbia – Oa
Kirschenfeld, Taylor, 850-474-2746 Univ of West Florida – Ob
Kirschner, David, (314) 977-3128 dkirschn@eas.slu.edu, Saint Louis Univ – Cg
Kirschvink, Joseph L., (626) 395-6136 kirschvink@caltech.edu, California Institute of Technology – Pg
Kirste, Dirk, 604-291-5365 dkirste@sfu.ca, Simon Fraser Univ – Cg
Kish, Stephen A., (850) 644-2064 kish@gly.fsu.edu, Florida State Univ – Eg
Kisila, Ben O., (540) 654-1107 bkisila@umw.edu, Univ of Mary Washington – Hs
Kissel, David E., (706) 542-0900 Univ of Georgia – Sc
Kissick, Reginar.kissick@moreheadstate.edu, Morehead State Univ – Oo
Kissin, Stephen A., (807) 343-8220 stephen.kissin@lakeheadu.ca, Lakehead Univ – Em
Kisslinger, Carl, 303-492-6089 Univ of Colorado – Ys
Kitchell, James F., (608) 262-9512 Univ of Wisconsin, Madison – Ob
Kitchen, Newell R., 573-882-1138 kitchenn@missouri.edu, Univ of Missouri, Columbia – Sp
Kite, J. Steven, (304) 293-5603 kite@geo.wvu.edu, West Virginia Univ – Gm
Kitrell, Carruth, (270) 809-2591 carruth.kitrell@murraystate.edu, Murray State Univ – Oo
Kitts, B. Kathy, (815) 753-5990 kkitts@niu.edu, Northern Illinois Univ – Xc
Kivelson, Margaret G., (310) 825-3435 mkivelson@igpp.ucla.edu, Univ of California, Los Angeles – Xy
Kiver, Eugene P., (509) 359-7959 eugene.kiver@mailserver.ewu.edu, Eastern Washington Univ – Gl
Kjerfve, Bjorn, (803) 777-4529 Univ of South Carolina – Op
Kjerfve, Bjorn, 979 845 3651 bkjerfve@ocean.tamu.edu, Texas A&M Univ – Op
Klaassen, Gary P., (416) 736-2100 (Ext. 77727) gklaass@yorku.ca, York Univ – Oa
Klapper, Gilbertg-klapper@northwestern.edu, Univ of Iowa – Pm
Klasik, John A., 909-869-3453 jaklasik@csupomona.edu, California State Polytechnic Univ, Pomona – Gu
Klaus, Adam, (979) 845-3055 aklaus@odpemail.tamu.edu, Texas A&M Univ – Gu
Klaus, James S., 305-284-3426 Univ of Miami – Gg
Kleffner, Mark A., (419) 995-8208 kleffner.1@osu.edu, Ohio State Univ, Lima – Ps
Klein, Andrew, (979) 845-7179 klein@geog.tamu.edu, Texas A&M Univ – Or
Klein, Cornelis, (505) 277-2023 cklein@unm.edu, Univ of New Mexico – Gz
Klein, Emily M., (919) 684-5965 ek4@duke.edu, Duke Univ – Gi
Klein, Kenneth P., (250) 828-5453 Univ College of the Cariboo – Gg
Kleinhans, Frederick W., (317) 274-6904 fkleinha@iupui.edu, Indiana Univ / Purdue Univ, Indianapolis – Xg
Kleinspehn, Karen L., 612-624-0537 klein004@umn.edu, Univ of Minnesota, Twin Cities – Gt
Kleiss, Harold J., (919) 515-2643 North Carolina State Univ – Sd
Klemas, Victor, (302) 831-8256 klemas@udel.edu, Univ of Delaware – Or
Klemow, Kenneth M., (570) 408-4758 Wilkes Univ – Py
Klemperer, Simon L., (650) 723-8214 sklemp@stanford.edu, Stanford Univ – Yr
Klepeis, Keith A., (802) 656-0247 kklepeis@zoo.uvm.edu, Univ of Vermont – Gc
Kleppa, Ole J., (773) 702-7190 Univ of Chicago – Cp
Kligfield, Roy M., 303-443-2920 Univ of Colorado – Gc
Klinck, John M., (757) 683-6005 Klinck@ccpo.odu.edu, Old Dominion Univ – Op
Klindienst, Lou, (814) 865-4203 blk1@psu.edu, Pennsylvania State Univ, Univ Park – Oo
Kline, Stephen W., (479) 968-0202 skline@atu.edu, Arkansas Tech Univ – Gx
Klinger, Daniel J., (718) 997-3452 daniel_klinger@qc.edu, Queens College (CUNY) – Gm
Klink, Lindseyhabakuk@nova.edu, Nova Southeastern Univ – Oo
Klinkhammer, Gary P., (541) 737-5209 gklinkhammer@coas.oregonstate.edu, Oregon State Univ – OuCm
Kloosterziel, Rudolf C., (808) 956-6895 rudolf@soest.hawaii.edu,

Univ of Hawai'i, Manoa – Op
Klosterman, Sue, (937) 229-2661 Sue.Klosterman@notes.udayton.edu, Univ of Dayton – Gg
Kluitenberg, Gerard J., (913) 532-7215 Kansas State Univ – Sp
Klusman, Ronald W.rklusman@mines.edu, Colorado School of Mines – Ca
Knaack, Charles, (509) 335-6742 knaack@wsu.edu, Washington State Univ – Ca
Knaeble, Alan R., (612)627-4780, Ext. 210 knaeb001@umn.edu, Univ of Minnesota, Twin Cities – Gl
Knapp, Camelia, 777-8491 camelia@geol.sc.edu, Univ of South Carolina – Gt
Knapp, Elizabeth P., (540) 463-8867 knappe@wlu.edu, Washington & Lee Univ – Cl
Knapp, H. Vernon, (217) 333-4423 vknapp@uiuc.edu, Illinois State Water Survey – Hs
Knapp, James, (803) 777-6886 knapp@geol.sc.edu, Cornell Univ – Gc
Knapp, James H., (803) 777-6886 knapp@geol.sc.edu, Cornell Univ – Oo
Knapp, James H., (803) 777-6886 knapp@geol.sc.edu, Univ of South Carolina – Ye
Knapp, Richard B., (925) 423-3328 knapp4@llnl.gov, Lawrence Livermore National Lab – Ng
Knapp, Roy M., 405-325-6829 knapp@ou.edu, Univ of Oklahoma – Np
Knapp, Warren, (607) 255-3034 knapp@nrcc.cornell.edu, Cornell Univ – Oa
Knauss, Daniel M., (303) 273-3625 dknauss@mines.edu, Colorado School of Mines – Oo
Knauss, John A., (401) 874-6141 jknauss@gso.uri.edu, Univ of Rhode Island – Op
Knauss, Kevin G., (925)423-1372 knauss1@llnl.gov, Lawrence Livermore National Lab – Gg
Knauss, Virginia L., (310) 243-3377 vknauss@csudh.edu, California State Univ, Dominguez Hills – Oo
Knauth, L. Paul, (480) 965-2867 knauth@asu.edu, Arizona State Univ – Cs
Kneafsey, Timothy J., (510) 486-4414 tjkneafsey@lbl.gov, Lawrence Berkeley National Lab – Hw
Knepp, Rex A., 217-244-2422 knepp@isgs.uiuc.edu, Illinois State Geological Survey – Go
Knight, Allen W., (530) 752-0453 aknight557@aol.com, Univ of California, Davis – Hs
Knight, David, (518) 442-4204 knight@atmos.albany.edu, SUNY, Albany – Oa
Knight, Kimbell L., (540) 831-5430 kknight@radford.edu, Radford Univ – Gd
Knight, Larry, (847) 925-6589 William Rainey Harper College – Gg
Knight, Paul, (814) 863-4229 knight@mail.meteo.psu.edu, The Pennsylvania State Univ, Univ Park – Ow
Knight, Rosemary J., (650) 723-4746 Stanford Univ – Hy
Knighton, Raymond E., (701) 231-8577 North Dakota State Univ – Sp
Knittle, Elise, (831) 459-4949 eknittle@pmc.ucsc.edu, Univ of California, Santa Cruz – Gy
Knocke, William R., (703) 231-6635 Virginia Polytechnic Institute & State Univ – Oo
Knoll, Andrew H., (617) 495-9306 aknoll@harvard.edu, Harvard Univ – Pb
Knoll, Martin A., (931) 598-1713 mknoll@sewanee.edu, Univ of the South – Hw
Knoll, Michael D., (208) 426-4003 mknoll@cgiss.boisestate.edu, Boise State Univ – Yg
Knopoff, Leon, (310) 825-1885 knopoff@physics.ucla.edu, Univ of California, Los Angeles – Ys
Knott, Jeffrey R., (714) 278-5547 jknott@fullerton.edu, California State Univ, Fullerton – Ge
Knowles, Charles E., (919) 515-7943 ernie_knowles@ncsu.edu, North Carolina State Univ – Op
Knowlton, Nancy, (858) 822-2486 nknowlton@ucsd.edu, Univ of California, San Diego – Ob
Knox, Larry W., (931) 372-3523 lknox@tntech.edu, Tennessee Technological Univ – Pm
Knox, Paul R., (512) 471-7313 paul.knox@beg.utexas.edu, Univ of Texas at Austin, Jackson School of Geosciences – Gg
Knox, S. Cragin, (601) 961-5500 cragin_knox@deq.state.ms.us, Mississippi Office of Geology – Ye
Knudsen, Andrew, (920) 832-6731 knudsena@lawrence.edu, Lawrence Univ – Cg
Knudsen, Guygknudsen@uidaho.edu, Univ of Idaho – Sb
Knuepfer, Peter L. K., (607) 777-2389 Binghamton Univ – Gt
Knyazikhin, Yuri, (617) 353-8843 jknjazi@bu.edu, Boston Univ – Or
Koch, Andrey, (541) 737-2991 akoch@coas.oregonstate.edu, Oregon State Univ – Op
Koch, Dorothy M., (203) 432-6925 dorothy.koch@yale.edu, Yale Univ – Yg
Koch, Evamaria W., (410) 221-8418 Univ of South Florida, Saint Petersburg – Og
Koch, Johannes, 330-263-2368 jkoch@wooster.edu, The College of Wooster – Gl
Koch, Magalymkoch@bu.edu, Boston Univ – Ga
Koch, Paul L., (831) 459-5861 pkoch@es.ucsc.edu, Univ of California, Santa Cruz – Py
Koch, Paul L., (831) 459-5861 pkoch@pmc.ucsc.edu, Univ of California, Santa Cruz – Pv
Kochanov, William E., 717.702.2033 wkochanov@state.pa.us, Pennsylvania Bureau of Topographic & Geologic Survey – Gg
Kochel, R. C., (570) 577-3032 kochel@bucknell.edu, Bucknell Univ – Gm
Kocurek, Gary A., (512) 471-5855 garyk@mail.utexas.edu, Univ of Texas at Austin, Jackson School of Geosciences – Gs
Kocurko, John, 940-397-4250 none, Midwestern State Univ – Gs
Kodama, Kenneth P., (610) 758-3663 kpk0@lehigh.edu, Lehigh Univ – Ym
Koehl, Mimi A. R., (510) 642-8103 Univ of California, Berkeley – Po
Koermer, James P.jamesk@psc.plymouth.edu, Plymouth State College – Ow
Kogan, Mikhail, (845) 365-8882 Columbia Univ – Yd
Kohl, Martin S., (865) 594-5597 Martin.Kohl@state.tn.us, Tennessee Division of Geology – GgOi
Kohler, Kevin, (954)262-3641 kevin@nova.edu, Nova Southeastern Univ – Oo
Kohler, Nicholas, (541) 346-4160 nicholas@uoregon.edu, Univ of Oregon – Oi
Kohler-Antablin, Susan L., (916) 322-2719 California Geological Survey – Eg
Kohlstedt, David L., 612-626-1544 dlkohl@umn.edu, Univ of Minnesota, Twin Cities – Yx
Kohlstedt, Sally G., 612-624-9368 sgk@umn.edu, Univ of Minnesota, Twin Cities – Oo
Kohn, Matthew J., (803) 777-5565 kohn@geol.sc.edu, Univ of South Carolina – Cs
Kolenbrander, Lawrence G., (828) 227-7367 Western Carolina Univ – Oo
Kollar, Albert, (412) 622-3263 Carnegie Mus of Natural History – Pi
Kollet, Stefan, (925)422-9297 kollet2@llnl.gov, Lawrence Livermore National Lab – Gg
Komar, Paul D., (541) 737-5210 pkomar@coas.oregonstate.edu, Oregon State Univ – On
Komarneni, Sridhar, (814) 865-1542 komarneni@psu.edu, Pennsylvania State Univ, Univ Park – Sc
Kominz, Michelle A., (269) 387-5340 michelle.kominz@wmich.edu, Western Michigan Univ – Yr
Konig, Ronald H., (479) 575-3411 rkonig@comp.uark.edu, Univ of Arkansas, Fayetteville – Eg
Konigsberg, Alvin S., (914) 257-3758 SUNY New Paltz – Oa
Konopasek, Jiri, +420221951525 kony@natur.cuni.cz, Charles Univ – Gpg
Kontak, Daniel J., (705) 675-1151 (Ext. ext. 2352) dkontak@laurentian.ca, Laurentian Univ, Sudbury – Eg
Konter, Jasper G., (915) 747-5507 jkonter@geo.utpe.edu, Univ of Texas, El Paso – GgCl
Kontrovitz, Mervin, (318) 342-1750 Univ of Louisiana, Monroe – Pm
Kontuly, Thomas M., (801) 581-8218 thomas.kontuly@geog.utah.edu, Univ of Utah – Oo
Koons, Donaldsond_koons@colby.edu, Colby College – Gm
Koons, Peter O., (207)-581-2158 peter.koons@maine.edu, Univ of Maine – Gt
Kooser, Marilyn A., (951) 827-3440 marilyn.kooser@ucr.edu, Univ of California, Riverside – Gr
Kooyman, Gerald L., (858) 534-2091 gkooyman@ucsd.edu, Univ of California, San Diego – Ob
Kopaska-Merkel, David C., 205/247-3695 davidkm@gsa.state.al.us, Geological Survey of Alabama – Gs
Koper, Keith D., (314) 977-3131 Saint Louis Univ – Ys
Kopf, Christopher F., (814) 269-2942 kopf@pitt.edu,

Univ of Pittsburgh at Johnstown – Gp
Kopp, Otto C., (865) 974-6399 okopp@utk.edu, Univ of Tennessee, Knoxville – Gz
Koppers, Anthony, (541) 737-5425 akoppers@coas.oregonstate.edu, Oregon State Univ – Cmc
Kopylova, Maya G., (604) 822-0865 mkopylova@eos.ubc.ca, Univ of British Columbia – Gi
Koralek, Susankoralek@sou.edu, Southern Oregon Univ – Oo
Koretsky, Carla M., (269) 387-5337 carla.koretsky@wmich.edu, Western Michigan Univ – Cl
Kornreich Wolf, Susan, (309) 794-7369 glskwolf@augustana.edu, Augustana College – Oe
Korom, Scott F., (701) 777-6156 scott_korom@mail.und.nodak.edu, Univ of North Dakota – Hw
Korose, Christopher P., 217-333-7256 korose@staff.cso.uiuc.edu, Illinois State Geological Survey – Ec
Korotev, Randy L., (314) 935-5637 korotev@wustl.edu, Washington Univ – Ct
Korty, Robert, 979-847-9090 korty@tamu.edu, Texas A&M Univ – Oa
Korycansky, Don, (831) 459-5843 Univ of California, Santa Cruz – Xc
Kosinski, Leszek A., 33142883215 Univ of Alberta – Oo
Koskinen, William C., (612) 625-4276 koskinen@soils.umn.edu, Univ of Minnesota, Twin Cities – Sc
Kosro, P. Michael, (541) 737-3079 kosro@coas.oregonstate.edu, Oregon State Univ – Op
Kostelnik, Jaime, 412.442.5828 jkostelnik@state.pa.us, Pennsylvania Bureau of Topographic & Geologic Survey – Gg
Koster Van Groos, August F., (312) 996-8678 kvg@uic.edu, Univ of Illinois, Chicago – Cp
Kosterov, Andrey, 612-625-9985 kosterov@umn.edu, Univ of Minnesota, Twin Cities – Ym
Kostka, Joel, (850) 645-3334 jkostka@ocean.fsu.edu, Florida State Univ – Ob
Kostov, Svilen, 229-931-2321 skostov@canes.gsw.edu, Georgia Southwestern State Univ – Oo
Kotzer, Thomas G., (306) 657-3597 Tom.kotzer@lightsource.ca, Univ of Ottawa – Cg
Koutitonsky, Vladimir G., (418) 724-1986 (Ext. 1763) vgk@uqar.qc.ca, Universite du Quebec a Rimouski – Op
Kovach, Richard G., (702) 295-6180 rkovach@lanl.gov, Los Alamos National Lab – Nm
Kovach, Robert L., (650) 723-4827 Stanford Univ – Ys
Koval, Susan F., (519) 661-3439 skoval@uwo.ca, Univ of Western Ontario – Oo
Kowalewski, Michal J., (540) 231-5951 michalk@vt.edu, Virginia Polytechnic Institute & State Univ – Py
Kowalke, Thorsten, 089/2180 6733 t.kowalke@lrz.uni-muenchen.de, LMU - Univ of Munich – Pg
Kowallis, Bart J., 801-422-2467 bjk@geology.byu.edu, Brigham Young Univ – Gc
Kozinski, Janusz A., (514) 398-4755 McGill Univ – Nx
Koziol, Andrea M., (937) 229-2954 Andrea.Koziol@notes.udayton.edu, Univ of Dayton – Cp
Kozlowski, Andrew, 518 486-2012 akoslows@mail.nysed.gov, New York State Geological Survey – Gl
Kozyra, Janet U., (313) 747-3550 jukozyra@engin.umich.edu, Univ of Michigan – Oa
Krabbenhoft, David, 608-821-3843 dpkrabbe@usgs.gov, Univ of Wisconsin, Madison – Hg
Krall, Donald B., (908) 737-3694 dkrall@kean.edu, Kean Univ – Gl
Kramer, J. Curtis, (209) 946-2482 ckramer@pacific.edu, Univ of the Pacific – Gg
Kramer, James R., (905) 525-9140 (Ext. 24514) kramer@mcmail.cis.mcmaster.ca, McMaster Univ – Cl
Kramer, Walter V., 361.698.1385 wkramer@delmar.edu, Del Mar College – Gg
Krantz, David E., (419) 530-2662 david.krantz@utoledo.edu, Univ of Toledo – Gs
Krantz, Dwight S., (713) 465-8658 Houston Community College System – Oo
Krapac, Ivan G., 217-333-6442 krapac@isgs.uiuc.edu, Illinois State Geological Survey – Ca
Kratt, Christopher, 775-673-7483 Chris.Kratt@dri.edu, Desert Research Institute – Or
Kraus, Mary J., 303-492-7251 mary.kraus@colorado.edu, Univ of Colorado – GsSa
Krause, David W., (631) 444-3117 david.krause@sunysb.edu, SUNY, Stony Brook – Pv
Krause, Federico F., (403) 220-5845 fkrause@ucalgary.ca, Univ of Calgary – Gs
Krause, Lois B., (864) 656-7653 krause@clemson.edu, Clemson Univ – Oe
Krause, Stefan, +44 01782 583186 s.krause@esci.keele.ac.uk, Keele Univ – Ge
Krauskopf, Konrad B., (650) 723-3325 konrad@pangea.stanford.edu, Stanford Univ – Yr
Krauss, David, (617) 552-0804 kraussd@bc.edu, Boston College – Po
Kravchinsky, Vadim, (780) 492-5591 vkrav@phys.ualberta.ca, Univ of Alberta – Ym
Kreamer, David K., (702) 895-3553 kreamerd@nevada.edu, Univ of Nevada, Las Vegas – Hw
Krebes, Edward S., (403) 220-5028 Univ of Calgary – Ys
Krebs, William N., (281) 759-7755 bkrebs@egi.utah.edu, Univ of Utah – Pm
Kreemer, Cornelis, (775) 682-8780 kreemer@unr.edu, Univ of Nevada, Reno – Yd
Kreemer, Cornelis, (775) 78-4-66 (Ext. ext. 154) kreemer@unr.edu, Univ of Nevada – Yd
Kreger, Robert D., (562) 860-2451 (Ext. 668) rkreger@cerritos.edu, Cerritos College – Oy
Kreidenweis, Sonia M., (970) 491-8350 sonia@atmos.colostate.edu, Colorado State Univ – Oa
Kreiger, Williamwkreiger@ycp.edu, York College of Pennsylvania – Gi
Kreitzer, Debra D., (270) 745-5984 debbie.kreitzer@wku.edu, Western Kentucky Univ – Oi
Kremer, James N., (860) 405-9027 jkremer@uconn.edu, Univ of Connecticut – Ob
Kremer, Robert J., 573-882-6408 Univ of Missouri, Columbia – Os
Kresan, Peter L., (520) 621-6000 pkresan@geo.arizona.edu, Univ of Arizona – Gg
Kress, Victor C., 206-616-8512 kress@u.washington.edu, Univ of Washington – Cp
Kressler, Sharon J., 612-625-5068 kress004@umn.edu, Univ of Minnesota, Twin Cities – Oo
Kreutz, Karl J., (207)-581-3011 karl.kreutz@maine.edu, Univ of Maine – Cs
Krick, Mary P., 217-333-5110 krick@isgs.uiuc.edu, Illinois State Geological Survey – Gg
Krieger-Brockett, Barbara B., (206) 543-2216 krieger@cheme.washington.edu, Univ of Washington – Ob
Krier, Donathon J., (505) 665-7834 krier@lanl.gov, Los Alamos National Lab – Gv
Kring, David, (520) 621-2024 kring@lpl.arizona.edu, Univ of Arizona – Xc
Krinsley, David H., (480) 965-5081 Arizona State Univ – Gs
Krinsley, David H., (541) 302-3314 krinsley@oregon.uoregon.edu, Univ of Oregon – Gd
Krishnamurthy, R. V., (269) 387-5501 r.v.krishnamurthy@wmich.edu, Western Michigan Univ – Cs
Krishnamurti, Rubykrishnamurti@ocean.fsu.edu, Florida State Univ – Op
Krishnamurti, T. N., (850) 644-2210 tnk@io.met.fsu.edu, Florida State Univ – Ow
Krishtalka, Leonard, (785) 864-4540 krishtalka@ku.edu, Univ of Kansas – Pv
Krissek, Lawrence A., (614) 292-1924 krissek.1@osu.edu, Ohio State Univ – Gs
Krockover, Gerald H., (765) 494-5795 hawk1@purdue.edu, Purdue Univ – Oe
Kroeger, Edwin E., (907) 474-5150 Univ of Alaska, Fairbanks – Ng
Kroeger, Glenn C., (210) 999-7607 gkroeger@trinity.edu, Trinity Univ – Yg
Kroeger, Timothy J., (218) 755-2783 tkroeger@bemidjistate.edu, Bemidji State Univ – PlGsHw
Kroenke, Loren W., (808) 956-7845 kroenke@soest.hawaii.edu, Univ of Hawai'i, Manoa – Gu
Krogh, Thomas E., (416) 586-5811 tomk@rom.on.ca, Univ of Toronto – Cc
Krogstad, Eirik J., 404413-5750 ejkrogstad@gsu.edu, Georgia State Univ – Cc
Krol, Michael, mkrol@bridgew.edu, Bridgewater State College – Gt
Kroll, Richard L., (908) 737-3698 rkroll@kean.edu, Kean Univ – Gx

Kronenberg, Andreas K., (979) 845-1371 kronenberg@geo.tamu.edu, Texas A&M Univ – Nr
Krot, Alexander N., 808-956-3900 sasha@higp.hawaii.edu, Univ of Hawai'i, Manoa – Xm
Krothe, Noel C., (812) 855-8197 krothen@indiana.edu, Indiana Univ, Bloomington – Hw
Kruge, Michael A., 973-655-7668 krugem@mail.montclair.edu, Montclair State Univ – Co
Kruger, Darrell P., (318) 342-1887 Univ of Louisiana, Monroe – Oy
Krukowski, Stanley T., 405/325-3031 skrukowski@ou.edu, Univ of Oklahoma – En
Krumm, Robert J., 217-333-4085 krumm@isgs.uiuc.edu, Illinois State Geological Survey – Gg
Krupka, Kenneth M., (509) 376-4412 ken.krupka@pnl.gov, Pacific Northwest National Lab – Cl
Kruse, Sarah E., (813) 974-7341 Univ of South Florida, Tampa – Yg
Krygier, John B., (740) 368-3622 jbkrygie@owu.edu, Ohio Wesleyan Univ – Oi
Ku, Teh-Lung, (213) 740-5826 rku@usc.edu, Univ of Southern California – Cg
Ku, Timothy C., (860)685-2265 tcku@wesleyan.edu, Wesleyan Univ – Cl
Kubas, Gregory J., (505) 667-5846 Los Alamos National Lab – Oa
Kubatko, Karrie-Ann H., 305-284-4254 kkubatko@miami.edu, Univ of Miami – Ge
Kubicek, Ines, (403) 440-6165 ikubicek@mtroyal.ca, Mount Royal College – Oo
Kubicki, James D., (814) 865-3951 kubicki@geosc.psu.edu, Pennsylvania State Univ, Univ Park – Cg
Kubo, Atsushi, (609) 258-3261 akubo@princeton.edu, Princeton Univ – Gy
Kuchinski, John G., 717.702.2027 jokuchinsk@state.pa.us, Pennsylvania Bureau of Topographic & Geologic Survey – Oo
Kuchta, Mark, (303) 273-3306 mkuchta@mines.edu, Colorado School of Mines – Nm
Kudela, Raphael M., (831) 459-3290 kudela@cats.ucsc.edu, Univ of California, Santa Cruz – Or
Kudlac, John J., (412) 392-3423 Point Park College – Ng
Kuehl, Steven A., (804) 684-7118 kuehl@vims.edu, College of William & Mary – Ou
Kuehn, Kenneth W., (270) 745-3082 Western Kentucky Univ – Ec
Kuehnle, Edward V., 614-265-6593 ed.kuehnle@dnr.state.oh.us, Ohio Dept of Natural Resources – Oo
Kuentz, David C., (513) 529-5992 kuentzdc@muohio.edu, Miami Univ – Ca
Kues, Barry S., (505) 277-3626 bkues@unm.edu, Univ of New Mexico – Pi
Kuhlman, Robertrkuhlman@mc3.edu, Montgomery County Community College – Gg
Kuhn, William R., (313) 936-0493 Univ of Michigan – Oa
Kuhnhenn, Gary L., (859) 622-8140 gary.kuhnhenn@eku.edu, Eastern Kentucky Univ – Gd
Kujawa, Frank B., 407-823-2800 kujawa@pegasus.cc.ucf.edu, Univ of Central Florida – Cp
Kujawinski, Elizabeth B., (508) 289-3696 ekujawinski@whoi.edu, Woods Hole Oceanographic Institution – Oc
Kulander, Byronbyron.kulander@wright.edu, Wright State Univ – Gc
Kulatilake, Pinnaduwa H. S. W., (520) 621-6064 kulatila@u.arizona.edu, Univ of Arizona – Nr
Kulhawy, Fred H., 607-255-4820 fhk1@cornell.edu, Cornell Univ – Ng
Kulkarni, Shrinivas R., (626) 395-4010 srk@astro.caltech.edu, California Institute of Technology – Xy
Kulp, Mark A., (504) 280-1170 mkulp@uno.edu, Univ of New Orleans – Gr
Kumar, Ajoy, 717-871-2432 Ajoy.Kumar@millersville.edu, Millersville Univ – Op
Kummerow, Christian D., (970) 491-7473 kummerow@atmos.colostate.edu, Colorado State Univ – Yr
Kump, Lee R., (814) 863-1274 lkump@psu.edu, Pennsylvania State Univ, Univ Park – Cl
Kung, Ernest C., 573-882-5909 Univ of Missouri, Columbia – Oa
Kung, Hsiang-Te, (901) 678-4538 hkung@memphis.edu, Univ of Memphis – Oy
Kung, King-Jau S., (608) 262-6530 kskung@wisc.edu, Univ of Wisconsin, Madison – Sp
Kunkle, Thomas D., (505) 667-1259 Los Alamos National Lab – Xy
Kuntz, Karakkuntz@fhsu.edu, Fort Hays State Univ – Oo
Kunze, A. W. Gerhard, (330) 972-7630 akunze@uakron.edu, Univ of Akron – Yg
Kunze, Eric L., (206) 543-8467 kunze@uvic.ca, Univ of Washington – Op
Kunzmann, Thomas, 089/2180 4292 kunzmann@min.uni-muenchen.de, LMU - Univ of Munich – Gz
Kuo, Hsiao-Lan, (773) 702-8107 Univ of Chicago – Oa
Kuo, Shiou, (206) 840-4573 skuo@wsu.edu, Washington State Univ – Sc
Kuperman, William A., (858) 534-1803 wak@mpl.ucsd.edu, Univ of California, San Diego – Op
Kurapov, Alexandre, (541) 737-2865 kurapov@coas.oregonstate.edu, Oregon State Univ – Op
Kursinski, Robert, (520) 626-3338 kursinsk@atmo.arizona.edu, Univ of Arizona – Oa
Kurtz, Andrewkurtz@bu.edu, Boston Univ – Clg
Kurz, Mark D., (508) 289-2328 mkurz@whoi.edu, Woods Hole Oceanographic Institution – Cc
Kuscu, Ilkayikuscu@eos.ubc.ca, Univ of British Columbia – Eg
Kushnir, Yochanan, (845) 365-8669 Columbia Univ – Oa
Kusky, Timothy M., (314) 977-3131 kusky@eas.slu.edu, Saint Louis Univ – Gg
Kusnick, Judith E., (916) 278-4692 California State Univ, Sacramento – Oe
Kusssow, Wayne R., (608) 263-3631 wrkussow@wisc.edu, Univ of Wisconsin, Madison – So
Kustka, Adam B., (609) 258-2612 kustka@princeton.edu, Princeton Univ – Ct
Kuszmaul, Joel S., (662) 915-7499 kuszmaul@olemiss.edu, Univ of Mississippi – Nr
Kutis, Michael, (765) 285-8265 Ball State Univ – Gg
Kutzbach, John E., (608)262-0392 jek@facstaff.wisc.edu, Univ of Wisconsin, Madison – Oa
Kuwabara, James, 415-452-7776 kuwabara@usgs.gov, City College of San Francisco – Og
Kuzila, Mark S., (402) 472-7537 mkuzila@unl.edu, University of Nebraska - Lincoln – Sd
Kvamme, Kenneth L., (617) 353-3415 Boston Univ – Ga
Kvill, Donald R.donald@cs.athabascau.ca, Athabasca Univ – Gm
Kwicklis, Edward M., (505) 665-7408 kwicklis@lanl.gov, Los Alamos National Lab – Gg
Kwiecien, William, 61 7 3138 2323 w.kwiecien@qut.edu.au, Queensland Univ of Technology – Ca
Kyle, Garykyle@nmsu.edu, New Mexico State Univ, Las Cruces – Yg
Kyle, J. Richard, (512) 471-4351 rkyle@mail.utexas.edu, Univ of Texas at Austin, Jackson School of Geosciences – Em
Kyle, Philip R., (505) 835-5995 kyle@nmt.edu, New Mexico Institute of Mining and Technology – Gi
Kysar Mattietti, Giuseppina, (703) 993-1045 gkysar@gmu.edu, George Mason Univ – Gi
Kyte, Frank T., (310) 825-2015 kyte@igpp.ucla.edu, Univ of California, Los Angeles – Ct
Käser, Martin, 089/2180 4138 martin.kaeser@geophysik.uni-muenchen.de, LMU - Univ of Munich – Yg

L

La Berge, Gene L., (920) 424-4460 laberge@uwosh.edu, Univ of Wisconsin, Oshkosh – Eg
La Fave, John I., 406-496-4306 jlafave@mtech.edu, Montana Tech of The Univ of Montana – Hw
La Force, Matthew J., (415) 338-6869 San Francisco State Univ – Hg
La Russa, Carol J., (530) 752-0519 cjlarussa@ucdavis.edu, Univ of California, Davis – Oo
La Tour, Timothy E., 404413-5767 tlatour@gsu.edu, Georgia State Univ – Gp
Laabs, Benjamin J., (507) 933-7442 blaabs@gac.edu, Gustavus Adolphus College – Gl
Laabs, Benjamin J., 585-245-5305 laabs@geneseo.edu, SUNY, Geneseo – GmlGe
Laase, Alan D., (303) 248-6250 Oak Ridge National Lab – Hw
Labandeira, Conrad C., (202) 633-1336 Smithsonian Institution / National Museum of Natural History – Pi
LaBarbera, Michael C., (773) 702-8092 mlabarbe@uchicago.edu, Univ of Chicago – Po
LaBella, Joel, (860)685-2242 jlabella@wesleyan.edu, Wesleyan Univ – Oo
Labelle, Ronald, (613)562-5800 Ext 6872 rlabelle@science.uottawa.ca,

Univ of Ottawa – Oo

Laboski, Carrie A., (608) 263-2795 laboski@wisc.edu, Univ of Wisconsin, Madison – So

Labotka, Theodore C., (865) 974-4805 tlabotka@utk.edu, Univ of Tennessee, Knoxville – Gp

Labuz, Joseph F., (612) 625-9060 jlabuz@umn.edu, Univ of Minnesota, Twin Cities – Nr

Lachmar, Thomas E., (435) 797-1247 tom.lachmar@usu.edu, Utah State Univ – Hw

Lacis, Andrew A., (212) 678-5595 alacis@giss.nasa.gov, Columbia Univ – Oa

Lackinger, Markus, 089/2180 LMU - Univ of Munich – Gz

Lackmann, Gary M., 919-515-1439 gary@ncsu.edu, North Carolina State Univ – Oa

Lacy, Tortlacy@cerritos.edu, Cerritos College – Gg

LaDelfe, Carol M., (505) 667-8474 cladelfe@lanl.gov, Los Alamos National Lab – Yg

Ladewig, Sandra S., (915) 747-5501 ladewig@geo.utep.edu, Univ of Texas, El Paso – Oo

Ladner, Lucien J., (850) 488-9380 Florida Geological Survey – Go

LaFemina, Peter C., 814-865-7326 pfemina@geosc.psu.edu, Pennsylvania State Univ, Univ Park – Yd

Laffoon, Carolyn J., (765) 494-0201 carolyn@purdue.edu, Purdue Univ – Oo

LaFleche, Marc R., (418) 654-2670 mlafleche@inrs.uquebec.ca, Universite du Quebec – Ct

Lafrance, Bruno, (705) 675-1151 (Ext. 2264) blafrance@nickel.laurentian.ca, Laurentian Univ, Sudbury – Gc

LaFreniere, Lorraine M., (630) 252-7969 lafreniere@anl.gov, Argonne National Lab – Oo

Lager, George A., (502) 852-6821 Univ of Louisville – Gz

Lageson, David R., (406) 994-6913 lageson@montana.edu, Montana State Univ – Gc

Lagowski, Alison A., (716) 645-6800 (Ext. 6101) aal@acsu.buffalo.edu, SUNY, Buffalo – Oo

Lai, Chung Chieng A., (505) 665-6635 cal@lanl.gov, Los Alamos National Lab – Ow

Laine, Edward P., (207) 725-3621 edlaine@bowdoin.edu, Bowdoin College – Gu

Laird, David A., (515) 294-1581 dalaird@iastate.edu, Iowa State Univ of Science & Technology – Sc

Laird, Jo, (603) 862-1718 jl@cisunix.unh.edu, Univ of New Hampshire – Gp

Laird, Neil, 315-781-3603 laird@hws.edu, Hobart & William Smith Colleges – Ow

Laity, Julie E., (818) 677-3532 julie.laity@csun.edu, California State Univ, Northridge – Oy

Lajtai, Emery Z., (204) 474-8701 lajtai@cc.umanitoba.ca, Univ of Manitoba – Nr

Lakatos, Stephen, (718) 262-2589 York College (CUNY) – Cc

Lakhan, V. Chris, (519) -25-3-30 (Ext. ext. 2183) lakan@uwindsor.ca, Univ of Windsor – Or

Laki, Sam, (937) 376-6272 Central State Univ – Hg

Lakshmanan, T. R., (617) 353-3084 laksh@bu.edu, Boston Univ – Oo

Lakshmi, Venkataraman, (803) 777-3552 lakshmi@geol.sc.edu, Univ of South Carolina – Hg

Lal, Devendra x., (858) 534-2134 dlal@ucsd.edu, Univ of California, San Diego – Xc

Lal, Rattan, 614-292-9069 lal.1@osu.edu, Ohio State Univ – Sp

Lala, Garland, (518) 437-8703 lala@asrc.cestm.albany.edu, SUNY, Albany – Oa

Lalonde, Andre E., (613)562-5800 Ext 6859 aelsc@uottawa.ca, Univ of Ottawa – Gz

Lam, Phoebe, 508-289-3792 pjlam@whoi.edu, Woods Hole Oceanographic Institution – Cm

Lamanna, Matthew C., (412) 624-8780 Univ of Pittsburgh – Pv

Lamb, Dennis, (814) 865-0174 lno@psu.edu, The Pennsylvania State Univ, Univ Park – Ow

Lamb, John A., (612) 625-1772 jlamb@soils.umn.edu, Univ of Minnesota, Twin Cities – Sc

Lamb, Melissa A., (651) 962-5242 malamb@stthomas.edu, Univ of Saint Thomas – Yr

Lamb, Will, (979) 845-3075 lamb@geo.tamu.edu, Texas A&M Univ – Gp

Lambert, Dean, (210) 733-2828 dlambert@mail.accd.edu, San Antonio College – Oyi

Lambert, Lance L., 210-458-5447 Lance.Lambert@utsa.edu, Univ of Texas, San Antonio – Pg

Lamborg, Carl, 508-289-2556 clamborg@whoi.edu, Woods Hole Oceanographic Institution – Oc

Lamelas, Francisco, 508-929-8553 flamelas@worcester.edu, Worcester State College – Yg

Lammerer, Bernd, 089/2180 6517 lammerer@iaag.geo.uni-muenchen.de, LMU - Univ of Munich – Gg

Lamothe, Jane, (306)966-6049 jane.lamothe@usask.ca, Univ of Saskatchewan – Oo

Lamothe, Michel, (514) 987-3361 lamothe.michel@uqam.ca, Universite du Quebec a Montreal – Gl

Lancaster, Nicholas, (775) 673-7304 nick@dri.edu, Desert Research Institute – Gm

Lancaster, Nick, (775) 673-7304 Nick.Lancaster@dri.edu, Univ of Nevada, Reno – Gm

Lancaster, Nick, (775) 673-7304 nick@dri.edu, Univ of Nevada, Reno – Gm

Lancaster, Stephen, 541-737-9258 lancasts@geo.oregonstate.edu, Oregon State Univ – Hs

Land, Lewis, 505.234.9234 lland@gis.nmt.edu, New Mexico Institute of Mining and Technology – Hy

Land, Lewis A., 505-887-5505 lland@gis.nmt.edu, New Mexico Institute of Mining & Technology – Hg

Land, Lynton S., (804) 453-6605 JandL@rivnet.net, Univ of Texas at Austin, Jackson School of Geosciences – Cl

Landau, Matthew, (609) 652-4578 Stockton State College – Og

Landenberger, Bill, 61 02 4921 6366 Bill.Landenberger@newcastle.edu.au, Univ of Newcastle – Gi

Lander, Mark A., (671) 735-2685 Univ of Guam – Ow

Landin, Michael G., (518) 442-4572 landin@atmos.albany.edu, SUNY, Albany – Ow

Landing, Ed, 518-473-8071 elanding@mail.nysed.gov, New York State Geological Survey – Ps

Landing, William M.landing@ocean.fsu.edu, Florida State Univ – Oc

Landis, Chuck A., +64 3 479-7520 chuck.landis@otago.ac.nz, Univ of Otago – Gg

Landman, Neil H., (212) 769-5712 American Museum of Natural History – Pi

Landry, Michael R., (808) 956-7776 landry@soest.hawaii.edu, Univ of Hawai'i, Manoa – Ob

Landry, Peter B., (508) 289-3443 plandry@whoi.edu, Woods Hole Oceanographic Institution – Ca

Landschoot, Peter J., (814) 863-1017 pjll@psu.edu, Pennsylvania State Univ, Univ Park – So

Lane, Burke E., (850) 488-9380 Florida Geological Survey – Hw

Lane, Charles L., (541) 552-6114 lane@sou.edu, Southern Oregon Univ – Hg

Lane, Eric T., (423) 425-4523 elane@utc.edu, Univ of Tennessee, Chattanooga – Oo

Lang, Harold R., (304) 293-5603 harold@lithos.jpl.nasa.gov, Jet Propulsion Lab – Gr

Lang, Helen M., 304-293-5469 hlang@wvu.edu, West Virginia Univ – Gpz

Lange, Francoise, (418) 545-5011 Univ du Quebec a Chicoutimi – Oo

Lange, Rebecca A., (734) 764-7421 becky@umich.edu, Univ of Michigan – Gi

Langebrake, Larry C., (727) 553-1008 llange@marine.usf.edu, Univ of South Florida, Saint Petersburg – Oo

Langenheim, Jr., Ralph L., (217) 333-1338 rlangenh@staff.uiuc.edu, Univ of Illinois, Urbana-Champaign – Gr

Langer, Arthur, 718-951-4793 alanger@brooklyn.cuny.edu, American Museum of Natural History – Gg

Langford, Richard P., (915) 747-5968 langford@geo.utep.edu, Univ of Texas, El Paso – Gs

Langmuir, Charles H., (617) 384-9948 langmuir@eps.harvard.edu, Harvard Univ – Cg

Langston, Charles A., (901) 678-4869 langston@ceri.memphis.edu, Univ of Memphis – Ys

Langston, Jr., Wann, (512) 471-7736 wannl@mail.utexas.edu, Univ of Texas at Austin, Jackson School of Geosciences – Pv

Langston-Unkefer, Pat J., (505) 665-2556 Los Alamos National Lab – Co

Lanham, Sacha J., 410-554-5562 slanham@dnr.state.md.us, Maryland Geological Survey – Oi

Lanigan, David C., (509) 376-9308 david.lanigan@pnl.gov, Pacific Northwest National Lab – Hg

Lansdown, John, (512) 471-4810 johnl@mail.utexas.edu, Univ of Texas at Austin, Jackson School of Geosciences – Cs

Lansey, Kevin E., (520) 621-2512 lansey@engr.arizona.edu, Univ of Arizona – Hs

Lapen, Thomas, 713-743-3399 tlapen@uh.edu, Univ of Houston – Gz

Lapenis, Andrei, (518) 442-4191 SUNY, Albany – Cg

LaPlante, Andre R., (514) 398-4755 McGill Univ – Nx

LaPlante, Wanda L., (512) 471-7113 wanda.laplante@beg.utexas.edu, Univ of Texas at Austin, Jackson School of Geosciences – Oo

LaPointe, Daphne D., (775) 784-6691 (Ext. 134) dlapoint@unr.edu, Univ of Nevada – Gg

Laprise, Rene, (514) 987-3000 laprise@maia.sca.uqam.ca, Universite du Quebec a Montreal – Oa

Lardon, Joel L., (512) 475-9519 joel.lardon@beg.utexas.edu, Univ of Texas at Austin, Jackson School of Geosciences – Gg

Large, Ross R., 61 3 6226 2819 Ross.Large@utas.edu.au, Univ of Tasmania – Eg

Largen, Stacia, (202) 633-1830 largens@si.edu, Smithsonian Institution / National Museum of Natural History – Oo

Larimer, John, (480) 965-5081 Arizona State Univ – Gg

Larner, Kenneth L.klarner@mines.edu, Colorado School of Mines – Ye

Laroche, Stephanestephane.laroche@ec.gc.ca, McGill Univ – Oa

LaRock, Paul A., (225) 388-6307 Louisiana State Univ – Ob

Larsen, Daniel, (901) 678-4358 dlarsen@memphis.edu, Univ of Memphis – Cl

Larsen, Frederick D., (802) 485-2336 Norwich Univ – Gl

Larsen, Glenn E., (614) 265-6474 glenn.larsen@dnr.state.oh.us, Ohio Dept of Natural Resources – Gr

Larsen, Kristine, (860) 832-2938 larsen@ccsu.edu, Central Connecticut State Univ – Xy

Larsen, Suzanne T., 303 492 4611 suzanne.larsen@colorado.edu, Univ of Colorado – Oo

Larson, David J., 510-885-3132 david.larson@csueastbay.edu, California State Univ, East Bay – Ou

Larson, David R., 217-244-2770 dlarson@isgs.uiuc.edu, Illinois State Geological Survey – Hw

Larson, Edwin E., 303-492-6172 Univ of Colorado – Ym

Larson, Eric J., 715-346-4098 elarson@uwsp.edu, Univ of Wisconsin, Stevens Point – Or

Larson, Grahame J., 517-353-9485 larsong@msu.edu, Michigan State Univ – Gl

Larson, Harold P., (520) 621-6943 hplarson@u.arizona.edu, Univ of Arizona – Oo

Larson, Lawrence T., (775) 784-4002 lawrencelarson@hotmail.com, Univ of Nevada, Reno – Eg

Larson, Peter B., (509) 335-3095 plarson@wsu.edu, Washington State Univ – Cs

Larson, Robert C., (812) 237-2290 gelarsn@isugw.indstate.edu, Indiana State Univ – Oo

Larson, Roger, (401) 874-6165 rlar@gso.uri.edu, Univ of Rhode Island – Ou

LaSage, Danita M., 859-622-1506 Danita.LaSage@eku.edu, Eastern Kentucky Univ – Hw

Lascano, Robert, 806-723-5238 RLascano@ag.tamu.edu, Texas A&M Univ – Sp

Lasemi, Zakaria, 217-244-6944 lasemi@isgs.uiuc.edu, Illinois State Geological Survey – En

LaSeur, Noel E., (850) 644-2814 laseur@met.fsu.edu, Florida State Univ – Ow

Lash, Gary G., (716) 673-3842 lash@fredonia.edu, SUNY, Fredonia – Gr

Lash, Gary S.lash@fredonia.edu, SUNY, Buffalo – Gc

Lasher-Trapp, Sonia, (765) 496-2866 slasher@purdue.edu, Purdue Univ – Oa

Lasker, Howard R., 716-645-2363 (Ext. 184) hlasker@buffalo.edu, SUNY, Buffalo – Gu

Laskowski, Edward A., (805) 664-3040 California State Univ, Bakersfield – So

Laskowski, Stanley L., (215) 573-3164 slaskows@sas.upenn.edu, Univ of Pennsylvania – Ge

Lassetter, William L., 434-9516361 william.lassetter@dmme.virginia.gov, Division of Mineral Resources – Hw

Lassiter, John, (512) 471-4002 lassiter1@mail.utexas.edu, Univ of Texas at Austin, Jackson School of Geosciences – Cc

Last, George V., (509) 376-3961 george.last@pnl.gov, Pacific Northwest National Lab – Ge

Last, William M., (204) 474-8361 wm_last@umanitoba.ca, Univ of Manitoba – Gs

Laswell, Troy J., (662) 325-3915 Mississippi State Univ – Gm

Lat, Che Noorliza, 03-79674157 noorliza@um.edu.my, Univ of Malaya – Yg

Latimer, Jennifer, 812-237-2254 jlatimer@indstate.edu, Indiana State Univ – Cm

Laton, W. R., (714) 278-7514 wlaton@fullerton.edu, California State Univ, Fullerton – Hw

Lattman, Lawrence H., (505) 266-9606 New Mexico Institute of Mining and Technology – Gm

Laubach, Stephen E., (512) 471-6303 steve.laubach@beg.utexas.edu, Univ of Texas at Austin, Jackson School of Geosciences – Gc

Laudon, Robert C., (573) 341-4466 rlaudon@umr.edu, Univ of Missouri, Rolla – Go

Laudon, Thomas S., (414) 424-4464 laudon@uwosh.edu, Univ of Wisconsin, Oshkosh – Yg

Laughrey, Christopher D., 412.442.4232 claughrey@state.pa.us, Pennsylvania Bureau of Topographic & Geologic Survey – Co

Lauretta, Dante, (520) 626-1138 lauretta@lpl.arizona.edu, Univ of Arizona – Oo

Laustsen, Henrietta, 541-962-3132 hlaustsen@eou.edu, Eastern Oregon Univ – OiGfz

Lauziere, Kathleen, (418) 654-2658 klauzier@nrcan.gc.ca, Universite du Quebec – Gg

Lauzon, John, (519) 824-4120 (Ext. 52459) jlauzon@lrs.uoguelph.ca, Univ of Guelph – So

Lavallee, Daniel, (805) 893-8446 daniel@crustal.ucsb.edu, Univ of California, Santa Barbara – Oo

Lavallee, Yan, 089/2180 4221 yanlavallee@hotmail.com, LMU - Univ of Munich – Yg

Lavier, Luc L., 512-471-0455 luc@ig.utexas.edu, Univ of Texas at Austin, Jackson School of Geosciences – Yg

Lavkulich, Leslie M., (604) 822-3477 Univ of British Columbia – Sd

Lavoie, Denis, (418) 654-2571 delavoie@nrcan.gc.ca, Universite du Quebec – Gs

Law, Eric W., (740) 826-8242 ericlaw@muskingum.edu, Muskingum College – Gp

Law, Richard D., (540) 231-6685 rdlaw@vt.edu, Virginia Polytechnic Institute & State Univ – Gc

Lawrence, Carol, (703) 993-1616 clawren2@gmu.edu, George Mason Univ – Oo

Lawrence, David P., (252) 328-4393 lawrenced@mail.ecu.edu, East Carolina Univ – Gc

Lawrence, Henry, (814) 732-1572 hlawrence@edinboro.edu, Edinboro Univ of Pennsylvania – Ou

Lawrence, James, (713) 743-3410 jlawrence@uh.edu, Univ of Houston – Cl

Lawrence, Rick L., (406) 994-5409 Montana State Univ – Or

Lawrence, Robert D., (541) 737-1250 Oregon State Univ – Gc

Laws, Edward A., (808) 956-2912 laws@soest.hawaii.edu, Univ of Hawai'i, Manoa – Cm

Laws, Richard A., (910) 962-4125 laws@uncw.edu, Univ of North Carolina Wilmington – Pm

Lawson, James E., (918) 366-4152 Univ of Oklahoma – Ys

Lawson, Leigh A., 812-237-2444 Indiana State Univ – Gg

Lawson, Merlin P., (402) 472-5358 mlawson1@unl.edu, Univ of Nebraska, Lincoln – Oy

Lawton, Donald C., (403) 220-5718 Univ of Calgary – Ye

Lawton, Timothy F., 575-646-4910 tlawton@nmsu.edu, New Mexico State Univ, Las Cruces – Gr

Lawver, Lawrence A., (512) 471-0433 lawver@ig.utexas.edu, Univ of Texas at Austin, Jackson School of Geosciences – Yr

Lay, Thorne, (831) 459-3164 tlay@pmc.ucsc.edu, Univ of California, Santa Cruz – Ys

Layer, Paul, 907-474-5514 player@gi.alaska.edu, Univ of Alaska, Fairbanks – Ym

Layne, Graham D., (508) 289-2405 glayne@whoi.edu, Woods Hole Oceanographic Institution – Ca

Layou, Karen M.kmlayou@wm.edu, College of William & Mary – Po

Layton, Michael C., (301) 845-2458 Univ of Pittsburgh at Johnstown – Yg

Lazarewicz, Andrew R., (617) 552-8326 lazarewi@bc.edu, Boston College – Oa

Lazarus, Steven M., 321-674-2160 slazarus@fit.edu, Florida Institute of Technology – Oa

Le, Ly N., (408) 656-1003 lely@nps.navy.mil, Naval Postgraduate School – Op

Le, Yanfen, 660-562-1525 le@nwmissouri.edu, Northwest Missouri State Univ – Oi

Le Marshall, John F., (301)763-8000 Univ of Wisconsin, Madison – Or

Le Mone, David V., (915) 747-5275 lemone@geo.utep.edu, Univ of Texas, El Paso – Ps

Lea, David W., (805) 893-8665 Univ of California, Santa Barbara – Oc

Lea, Peter D., (207) 725-3439 plea@bowdoin.edu,
Bowdoin College – Gl
Leafgren, Rita F., (970) 351-1239 rita.leafgren@unco.edu,
Univ of Northern Colorado – Gg
Leake, Martha A., (229) 333-5756 mleake@valdosta.edu,
Valdosta State Univ – Xg
Leakey, Carol, (907) 474-7388 fncll1@aurora.alaska.edu,
Univ of Alaska, Fairbanks – Oo
Leap, Darrell I., (765) 494-3699 terraaqua@iquest.net,
Purdue Univ – Hy
Lear, Ccarrie@earth.cf.ac.uk,
Cardiff Univ – Ou
Leary, Colleen A., (806) 742-3417 colleen.leary@ttu.edu,
Texas Tech Univ – Oa
Leasure, V. Lynn, (270) 809-6761 lynn.shelby@murraystate.edu,
Murray State Univ – Gg
Leatham, W. Britt, (909) 537-5322 bleatham@csusb.edu,
California State Univ, San Bernardino – Ps
Leavens, Peter B., 302-831-8106 pbl@udel.edu,
Univ of Delaware – Gz
Leavitt, Steven W., (520) 621-6468 sleavitt@ltrr.arizona.edu,
Univ of Arizona – Cs
LeBar, James D., (850) 487-2219 Florida Geological Survey – Go
Lebedev, Maxim, +61 8 9266 3519 M.Lebedev@curtin.edu.au,
Curtin Univ of Technology – Yx
Lebel, Daniel, (418) 654-2675 dlebel@nrcan.gc.ca,
Universite du Quebec – Gc
Lebit, Hermann, (337) 264-7360 hlebit@uno.edu,
Georgia State Univ – Gc
LeBlanc, Lynn L., (318) 342-1892 Univ of Louisiana, Monroe – Oa
Lebofsky, Larry A., (520) 621-6947 lebofsky@lpl.arizona.edu,
Univ of Arizona – Og
Lechler, Paul, (775) 682-8773 plechler@unr.edu,
Univ of Nevada, Reno – Cg
Lechler, Paul J., (775) 784-6691 (Ext. 123) plechlor@unr.edu,
Univ of Nevada – Cg
Leckie, James O.leckie@cive.stanford.edu,
Stanford Univ – Cl
Leckie, R. Mark, (413) 545-1948 mleckie@geo.umass.edu,
Univ of Massachusetts – Pm
Leclair, Suzanne, 504-862-3168 leclair@tulane.edu,
Tulane Univ – Gs
Lecouris, Alison B., 217-244-2387 lecouris@isgs.uiuc.edu,
Illinois State Geological Survey – Gg
Ledbetter, Michael T., 501-569-3257 mtledbetter@ualr.edu,
Univ of Arkansas, Little Rock – Og
Ledger, Ernest B., (936) 468-2228 eledger@sfasu.edu,
Stephen F. Austin State Univ – Ca
Ledgerwood, Amie S., (906) 487-2531 asledger@mtu.edu,
Michigan Technological Univ – Oo
Lee, Chung M., (801) 581-8218 chunglee@geog.utah.edu,
Univ of Utah – Ou
Lee, Cindy, (632) 632-8741 cindylee@notes.cc.sunysb.edu,
SUNY, Stony Brook – Co
Lee, Craig M., (206) 685-7656 craig@apl.washington.edu,
Univ of Washington – Op
Lee, Daphne E., +64 3 479-7525 daphne.lee@otago.ac.nz,
Univ of Otago – Pi
Lee, Don Royce, (303) 273-3808 Texas A&M Univ, Commerce – Oy
Lee, In Young, (630) 252-8724 Argonne National Lab – Oa
Lee, Janice F., (731) 587-7430 jlee@utm.edu,
Univ of Tennessee, Martin – Oe
Lee, Jeffrey, (509) 963-2801 jeff@geology.cwu.edu,
Central Washington Univ – Gc
Lee, Jeffrey, (805) 893-7858 jeff@magic.geol.ucsb.edu,
Univ of California, Santa Barbara – Gc
Lee, Jejungleej@umkc.edu,
Univ of Missouri-Kansas City – Gq
Lee, Keenan, (303) 273-3808 klee@mines.edu,
Colorado School of Mines – Or
Lee, Kenrick H., (925) 422-8492 lee23@llnl.gov,
Lawrence Livermore National Lab – Nm
Lee, Michael D., 510-885-3155 michael.lee@csueastbay.edu,
California State Univ, East Bay – Hg
Lee, Ming-Kuo, (334) 844-4898 leeming@auburn.edu,
Auburn Univ – Hw
Lee, Patricia D., (808) 956-8489 plee@soest.hawaii.edu,
Univ of Hawai'i, Manoa – Gg
Lee, Regina, (303) 273-3808 Tennessee Technological Univ – Gg
Lee, Seung Ryeol, (301) 405-4054 leesr@geol.umd.edu,
Univ of Maryland – Cg
Lee, Sukyoung, (814) 863-1587 sl@psu.edu,
The Pennsylvania State Univ, Univ Park – Ow
Lee, Tien-Chang, (951) 827-4506 tien.lee@ucr.edu,
Univ of California, Riverside – Yh
Lee, Zhongping, (727) 553-3952 Univ of South Florida, St. Petersburg – Op
Leeman, William P., (713) 348-4892 leeman@rice.edu,
Rice Univ – Gi
Lees, Jonathan M., (919) 962-0695 Univ of North Carolina, Chapel Hill – Yg
Leetaru, Hannes E., (217) 333-5058 leetaru@isgs.uiuc.edu,
Univ of Illinois, Urbana-Champaign – Go
Leetaru, Hannes E., 217-333-5058 leetaru@isgs.uiuc.edu,
Illinois State Geological Survey – Eo
Leete, Jerome, 252-335-3375 jaybird2@hotmail.com,
Elizabeth City State Univ – Ge
Lefebvre, Rene, (418) 654-2651 Universite du Quebec – Hw
Lefebvre, Richard H., 616-331-3728 Grand Valley State Univ – Gx
Lefer, Barry, 713-743-3399 lefer@ucar.edu,
Univ of Houston – Oa
LeFever, Richard D., (701) 777-3014 richard_lefever@und.edu,
Univ of North Dakota – Gs
Leger, Carol, 603.358.2570 cleger@keene.edu,
Keene State College – Og
Legore, Virginia L., (509) 376-5019 virginia.legore@pnl.gov,
Pacific Northwest National Lab – Cg
Lehane, Mary, +353 21 4902764 m.lehane@ucc.ie,
Univ College Cork – Oo
Lehman, Scott J., (303) 492-8980 lehmans@spot.colorado.edu,
Univ of Colorado – Og
Lehman, Thomas M., (806) 742-3148 tom.lehman@ttu.edu,
Texas Tech Univ – Gs
Lehmann, Charlotte, clehmann@bates.edu,
Bates College – Cl
Lehmann, David, (814) 641-3602 lehmann@juniata.edu,
Juniata College – Gr
Lehre, Andre K., (707) 826-3165 akl1@humboldt.edu,
Humboldt State Univ – Gm
Lehrmann, Daniel J., (920) 424-7001 lehrmann@uwosh.edu,
Univ of Wisconsin, Oshkosh – Gr
Leier, Andrew, (609) 258-7024 aleier@princeton.edu,
Princeton Univ – Gs
Leif, Roald N., (925) 422-2469 Lawrence Livermore National Lab – Cg
Leighton, Henry G., (514) 398-3766 henry.leighton@mcgill.ca,
McGill Univ – Oa
Leighton, Lindsey, (619) 594-6978 lleighton@geology.sdsu.edu,
San Diego State Univ – Pe
Leighton, Morris W., (217) 333-3542 Univ of Illinois, Urbana-Champaign – Eo
Leighty, Robert S., (480) 461-7021 rleighty@mail.mc.maricopa.edu,
Mesa Community College – Og
Leimer, H. Wayne, (931) 372-3522 hwleimer@tntech.edu,
Tennessee Technological Univ – Gz
Leinen, Margaret, (401) 874-6222 mleinen@nsf.gov,
Univ of Rhode Island – Ou
Leinfelder, Reinhold, 089/2180 6629 r.leinfelder@lrz.uni-muenchen.de,
LMU - Univ of Munich – Pg
Leitch, Alison, (709) 737-3306 aleitch@mun.ca,
Memorial Univ of Newfoundland – Yg
Leite, Michael B., (308) 432-6377 mleite@csc.edu,
Chadron State College – Gg
Leithold, Elana L., (919) 515-7282 lonnie_leithold@ncsu.edu,
North Carolina State Univ – Gs
Leitz, Robert E., (303) 556-3072 Metropolitan State Coll of Denver – Gz
Leland, John F., (818) 240-1000 (Ext. 5517) leland@glendale.cc.ca.us,
Glendale Community College – Cc
LeMasurier, Wesley E., (303) 556-2681 Univ of Colorado, Denver – Gv
Lemay, Phillip W., (225) 388-6922 philip@lgs.bri.lsu.edu,
Louisiana State Univ – Gg
Lembke, Chad E., (727) 553-3976 clembke@marine.usf.edu,
Univ of South Florida, Saint Petersburg – Oo
Lemiszki, Peter J., (865) 594-5596 Peter.Lemiszki@state.tn.us,
Tennessee Division of Geology – GcOiGg
Lemke, Karen A., (715) 346-2709 klemke@uwsp.edu,
Univ of Wisconsin, Stevens Point – Gm
Lemke, Larry D., 1-313-5776412 ldlemke@sun.science.wayne.edu,
Wayne State Univ – Ge
Lemme, Gary D., (612) 589-1711 Gary.Lemme@sdstate.edu,
South Dakota State Univ – Sd
Lemmon, Mark, 979-458-8098 lemmon@tamu.edu,

Texas A&M Univ – Oa

Lemmond, Peter C., (508) 289-2457 plemmond@whoi.edu, Woods Hole Oceanographic Institution – Oo

Lenardic, Adrian, (713) 348-4883 adrian@rice.edu, Rice Univ – Gc

Lenczewski, Melissa E., (815) 753-7937 lenczewski@niu.edu, Northern Illinois Univ – Co

Lene, Gene W., (210) 436-3011 glene@stmarytx.edu, Saint Mary's Univ – Ge

Lener, Edward, (540) 231-9249 lener@vt.edu, Virginia Polytechnic Institute & State Univ – Oo

Lenhart, Stephen W., (540) 831-5652 slenhart@radford.edu, Radford Univ – Pi

Lennon, Gary, (804) 427-7189 Tidewater Community College – Og

Lennox, Paul G., + 61 2 9385 8096 p.lennox@unsw.edu.au, Univ of New South Wales – Gc

Lenters, John D.jlenters2@unl.edu, Univ of Nebraska, Lincoln – Oa

Lentz, David R., (506) 547-2070 Univ of New Brunswick – Cg

Lenz, Alfred C., (519) 661-3195 aclenz@uwo.ca, Univ of Western Ontario – Pi

Leo, J. Mia, (845) 356-8633 mia@ldeo.columbia.edu, Columbia Univ – Oo

Leonard, Eric M., (719) 389-6513 eleonard@coloradocollege.edu, Colorado College – Gl

Leonard, Joseph W., (606) 257-3818 Univ of Kentucky – Nm

Leonard, Lynn A., (910) 962-3490 Univ of South Florida, Saint Petersburg – Oo

Leonard, Lynn A., 910-962-2338 lynnl@uncw.edu, Univ of North Carolina Wilmington – Gu

Leonard, Meredith L., 818.778.5595 leonarml@lavc.edu, Los Angeles Valley College – Ow

Leonhart, Leo S., (520) 881-7300 Univ of Arizona – Hw

Leovy, Conway B., 206-543-4952 leovy@atmos.washington.edu, Univ of Washington – Oa

Lepper, Ken, 701-231-6746 ken.lepper@ndsu.edu, North Dakota State Univ – Cc

Lerbekmo, John F., (780) 492-3042 Univ of Alberta – Gs

Lerch, Robert N., 573-882-1402 Univ of Missouri, Columbia – Os

Lerche, Ian, (803) 777-4507 Univ of South Carolina – Yg

Lerczak, James, (541) 737-6128 jlerczak@coas.oregonstate.edu, Oregon State Univ – Op

Lerdau, Manuel, (631) 632-6633 mlerdau@notes.cc.sunysb.edu, SUNY, Stony Brook – Ob

Lerman, Abraham, (847) 491-7385 abe@earth.northwestern.edu, Northwestern Univ – Cg

Lerner-Lam, Arthur L., (845) 365-8356 lerner@ldeo.columbia.edu, Columbia Univ – Ys

Lescinsky, David T., (505) 277-3842 dlescins@unm.edu, Univ of New Mexico – Gvi

Lescinsky, David T., 505-277-3842 dlescins@unm.edu, Univ of Western Ontario – Gv

Lesher, Charles E., (530) 752-9779 lesher@geology.ucdavis.edu, Univ of California, Davis – Gi

Lesher, Michael, (705) 675-1151 (Ext. ext. 2276) lesher@sympatico.ca, Laurentian Univ, Sudbury – Em

Lesher, Michael C., (705)675-1151 Ext 2276 lesher@sympatico.ca, Univ of Ottawa – Gv

Lesher, Robert, (814) 865-3241 lesher@essc.psu.edu, The Pennsylvania State Univ, Univ Park – Ow

Leshin, Laurie, (480) 965-0796 laurie.leshin@asu.edu, Arizona State Univ – Xc

Lesht, Barry M., (630) 252-4208 bmlesht@anl.gov, Argonne National Lab – Op

Leslie, Stephen A., (501) 569-8061 saleslie@ualr.edu, Univ of Arkansas, Little Rock – Po

Lesmes, David P., (617) 552-0839 lesmes@bc.edu, Boston College – Yg

Lessard, Evelyn J., (206) 543-8795 elessard@u.washington.edu, Univ of Washington – Ob

Letelier, Ricardo M., (541) 737-3890 letelier@coas.oregonstate.edu, Oregon State Univ – Ob

Letsinger, Sally L., 812-855-1356 sletsing@indiana.edu, Indiana Geological Survey – Hq

Leung, Irene S., (718) 960-8572 Graduate School of the City Univ of New York – Gz

Leung, Irene S., (718) 960-8572 Lehman College (CUNY) – Gz

Leuthart, Clara A., (502) 852-6844 Univ of Louisville – Hs

Lev, Steven M., (410) 704-2744 slev@towson.edu, Towson Univ – Cl

Levander, Alan R., (713) 348-6064 alan@rice.edu, Rice Univ – Ye

LeVasseur, Maurice M., (418) 656-3207 maruice.levasseur@bio.ulaval.ca, Dalhousie Univ – Ob

Leven, Carol, (604) 822-3278 cleven@eos.ubc.ca, Univ of British Columbia – Oo

Leventer, Amy, (315) 228-7213 aleventer@mail.colgate.edu, Colgate Univ – Ou

Leveson, David J., (718) 951-5330 Graduate School of the City Univ of New York – Oo

Leveson, David J., (718) 951-5000 (Ext. 2879) dleveson@brooklyn.cuny.edu, Brooklyn College (CUNY) – Gh

Levesque, Andrelevesque.andre@ggl.ulaval.ca, Universite Laval – Gz

Levey, Raymond A., (801) 585-3826 rlevey@egi.utah.edu, Univ of Utah – Eo

Levin, Harold L., (314) 935-5681 levin@levee.wustl.edu, Washington Univ – Pi

Levin, Lisa A., (858) 534-3579 llevin@ucsd.edu, Univ of California, San Diego – Ob

Levin, Stewart A., (303) 265-8649 salevin@lgc.com, Stanford Univ – Yg

Levin, Vadim, 732-445-5415 vadim@ldeo.columbia.edu, Rutgers, The State Univ of New Jersey – Ys

Levine, Murray D., (541) 737-3047 mlevine@coas.oregonstate.edu, Oregon State Univ – Op

Levine, Norman S., (843) 953-5589 levinen@cofc.edu, College of Charleston – OiGeOu

Levinson, Alfred A., (403) 220-5846 Univ of Calgary – Eg

Levinton, Jeffrey, (631) 632-8602 levinton@notes.cc.sunysb.edu, SUNY, Stony Brook – Ob

Levson, Victor M., (250) 952-0391 vic.levson@gems9.gov.bc.ca, Univ of Victoria – Gl

Levy, Jonathan, (513) 529-1947 levyj@muohio.edu, Miami Univ – Hw

Levy, Melissa H., (916) 484-8684 levym@arc.losrios.edu, American River College – GgOgy

Levy, Schon S., (505) 667-9504 sslevy@lanl.gov, Los Alamos National Lab – Gi

Lew, Jeffrey, (310) 825-3023 lew@atmos.ucla.edu, Univ of California, Los Angeles – Oa

Lewin, Ralph A., (858) 534-4836 rlewin@ucsd.edu, Univ of California, San Diego – Ob

Lewis, Alan G., (604) 822-3626 alewis@eos.ubc.ca, Univ of British Columbia – Ob

Lewis, Anthony J., (225) 578-6199 galewi@lsu.edu, Louisiana State Univ – Or

Lewis, Betty, (614) 265-6578 betty.lewis@dnr.state.oh.us, Ohio Dept of Natural Resources – Oo

Lewis, Brian T. R., (206) 543-7419 blewis@u.washington.edu, Univ of Washington – Yr

Lewis, Gerald L., (626) 585-7137 gllewis@paccd.cc.ca.us, Pasadena City College – Ps

Lewis, John S., (520) 621-4972 jsl@u.arizona.edu, Univ of Arizona – Xc

Lewis, Jon C., (724) 357-5624 jclewis@iup.edu, Indiana Univ of Pennsylvania – Gc

Lewis, Marlon R., (902) 494-3513 marlon.lewis@dal.ca, Dalhousie Univ – Ob

Lewis, Reed S., (208) 885-7472 reedl@uidaho.edu, Univ of Idaho – Gg

Lewis, Ronald D., (334) 844-4886 lewisrd@auburn.edu, Auburn Univ – Pi

Lewis, Stephen D., (559) 278-6956 slewis@csufresno.edu, California State Univ, Fresno – Yg

Lewry, John F., (306) 585-4270 Univ of Regina – Gt

Leyden, Barbara W., (813) 974-0324 Univ of South Florida, Tampa – Pl

Li, Aibing, 713-743-3399 Univ of Houston – Ys

Li, Baosheng, 631 632-9642 baosheng.li@sunysb.edu, SUNY, Stony Brook – Gy

Li, Chusi, (812) 855-1558 cli@indiana.edu, Indiana Univ, Bloomington – Ce

Li, Gary, 510-885-3165 gary.li@csueastbay.edu, California State Univ, East Bay – Oi

Li, Jie (Jackie), 217-333-7008 jackieli@uiuc.edu, Univ of Illinois, Urbana-Champaign – Gy

Li, Ping-Chi, (931) 372-3752 PLI@TnTech.edu, Tennessee Technological Univ – Oi

Li, Rong-Yu, 204-727-9684 lir@brandonu.ca, Brandon Univ – Pg

Li, Rongxing, (614) 292-6946 li.282@osu.edu,
Ohio State Univ – Oo
Li, William K., (902) 426-6349 lib@mar.dfo-mpo.gc.ca,
Dalhousie Univ – Ob
Li, Xiaowenlix@bu.edu,
Boston Univ – Or
Li, Yanmei, (270) 745-5982 yanmei.li@wku.edu,
Western Kentucky Univ – Ouo
Li, Yaoguo, 303-273-3510 ygli@mines.edu,
Colorado School of Mines – Yv
Li, Yong-Gang, (213) 740-3556 yli@usc.edu,
Univ of Southern California – Ys
Li, Yuan-Hui, (808) 956-6297 yhli@soest.hawaii.edu,
Univ of Hawai'i, Manoa – Cm
Li, Yuncong, (305) 246-7000 yunli@mail.ifas.ufl.edu,
Univ of Florida – Sc
Lian, Jie, 734-647-5704 jlian@umich.edu,
Univ of Michigan – Gz
Liang, George H., (913) 532-7225 Kansas State Univ – So
Liang, Liyuan, 865-241-3933 2ll@ornl.gov,
Oak Ridge National Lab – Cl
Liang, Yan, (401) 863-9477 yan_liang@brown.edu,
Brown Univ – Cp
Liaw, Kao L., (905) 525-9140 (Ext. 23528) rliaw@mcmaster.ca,
McMaster Univ – Yr
Libarkin, Julie C., 517-355-8369 libarkin@msu.edu,
Michigan State Univ – Oe
Liberty, Lee M., (208) 426-1166 lml@cgiss.boisestate.edu,
Boise State Univ – Ys
Libey, Karen S., (509) 335-6227 libey@wsu.edu,
Washington State Univ – Oo
Licari, Gerald R., (213) 265-8838 East Los Angeles College – Pm
Licciardi, Joseph M., 603-862-1718 joe.licciardi@unh.edu,
Univ of New Hampshire – Gl
Licht, Kathy J., (317) 278-1343 Indiana Univ / Purdue Univ, Indianapolis – Gl
Lichtner, Peter C., (505) 667-3420 lichtner@lanl.gov,
Los Alamos National Lab – Oo
Liddell, W. David, (435) 797-1261 dave.liddell@usu.edu,
Utah State Univ – Po
Lidgard, Scott H., (312) 665-7625 lidgard@fmnh.org,
Field Museum of Natural History – Pi
Lidiak, Edward G., (412) 624-8871 egl@pitt.edu,
Univ of Pittsburgh – Gi
Lidicker, Jr., William Z., (510) 642-3059 Univ of California, Berkeley – Pv
Liebe, Richard M., (585) 395-5100 (Ext. 7524) rliebe@weather.brockport.edu, SUNY College at Brockport – Ps
Liebens, Johan, 850-474-2065 liebens@uwf.edu,
Univ of West Florida – So
Liebenthal, Dale L., 419-626-4296 dale.liebenthal@dnr.state.oh.us,
Ohio Dept of Natural Resources – Oo
Lieberman, Bruce S., (785) 864-2741 blieber@ku.edu,
Univ of Kansas – Po
Lieberman, Robert C., (631) 632-8214 robert.liebermann@sunysb.edu,
SUNY, Stony Brook – Gy
Lieberman, Ruth S., (313) 763-7412 Univ of Michigan – Oa
Liebling, Richard, 516 463-6545 georsl@hofstra.edu,
Hofstra Univ – Gg
Liebling, Richard S., (212) 772-5412 Graduate School of the City Univ of New York – Gz
Lienert, Barry R., (808) 956-7815 lienert@soest.hawaii.edu,
Univ of Hawai'i, Manoa – Yg
Liepert, Beate, (845) 365-8870 Columbia Univ – Ow
Lierman, Tom, (859) 622-1280 tom.lierman@eku.edu,
Eastern Kentucky Univ – Gd
Lifton, Nathaniel A., (520) 626-3251 lifton@geo.arizona.edu,
Univ of Arizona – Cg
Lillegraven, Jason A., (307) 766-3386 bagpipe@uwyo.edu,
Univ of Wyoming – Pv
Lilley, Marvin D., (206) 543-0859 lilley@ocean.washington.edu,
Univ of Washington – Ob
Lillie, Robert J., (541) 737-1242 lillier@geo.oregonstate.edu,
Oregon State Univ – Ye
Lilly, Eva, (334) 844-4282 lillyet@auburn.edu,
Auburn Univ – Oo
Lima, Ivan D.ilima@whoi.edu,
Woods Hole Oceanographic Institution – Oc
Lin, Charles A., (514) 398-3760 charles.lin@mcgill.ca,
McGill Univ – Oa
Lin, Douglas, (831) 459-2732 lin@lick.ucsc.edu,
Univ of California, Santa Cruz – Xc
Lin, Gong-yuh, (818) 677-3532 gongyuh.lin@csun.edu,
California State Univ, Northridge – Oa
Lin, Haihai.lin@ec.gc.ca,
McGill Univ – Oa
Lin, Hsing K., (907) 474-6347 ffhkl@aurora.alaska.edu,
Univ of Alaska, Fairbanks – Oo
Lin, Jian, (508) 289-2576 jlin@whoi.edu,
Woods Hole Oceanographic Institution – Gt
Lin, Jing, 919-515-7912 jing_lin@ncsu.edu,
North Carolina State Univ – Hs
Lin, Shoufa, (519) 888-4567 (Ext. 6557) shoufa@sciborg.uwaterloo.ca,
Univ of Waterloo – Oo
Lin, Shundar, (309) 671-3196 Illinois State Water Survey – Hs
Lincoln, Beth Z., (517) 629-0331 Albion College – Gc
Lincoln, Jonathan M., (973) 655-7273 lincolnj@mail.montclair.edu,
Montclair State Univ – Gr
Lincoln, Timothy N., (517) 629-0486 Albion College – Em
Lindberg, David R., (510) 642-3926 drl@uclink4.berkeley.edu,
Univ of California, Berkeley – Po
Lindberg, Jonathan W., (509) 376-5005 jon.lindberg@pnl.gov,
Pacific Northwest National Lab – Ec
Lindberg, Stephen R., (814) 539-7723 slindber@pitt.edu,
Univ of Pittsburgh at Johnstown – Og
Lindberg, Steven E.sll@ornl.gov,
Oak Ridge National Lab – Cl
Lindbo, David, (919) 793-4428 North Carolina State Univ – Sd
Linde, Alan T., (202) 478-8835 linde@dtm.ciw.edu,
Carnegie Institution of Washington – Ys
Lindemann, Richard H., (518) 580-5196 rlindemann@skidmore.edu,
Skidmore College – Pi
Lindemann, William C., (505) 646-3405 New Mexico State Univ, Las Cruces – Sb
Lindenmeier, Clark W., (509) 376-8419 clark.lindenmeier@pnl.gov,
Pacific Northwest National Lab – Cg
Lindline, Jennifer, (505) 454-2046 lindlinej@nmhu.edu,
New Mexico Highlands Univ – GipGe
Lindsay, Everett H., (520) 621-6024 ehlind@geo.arizona.edu,
Univ of Arizona – Pv
Lindsay, Thomas C., (503) 725-3374 tcl@pdx.edu,
Portland State Univ – Pe
Lindsey, Gregory H., (317) 274-8704 glindsey@iupui.edu,
Indiana Univ / Purdue Univ, Indianapolis – Oo
Lindsey, Michal, (504) 388-2110 mlindsey@agctr.lsu.edu,
Louisiana State Univ – So
Lindsley, Donald H., (631) 632-8195 SUNY, Stony Brook – Cp
Lindsley-Griffin, Nancy, (402) 472-2629 nlg@unl.edu,
Univ of Nebraska, Lincoln – Gt
Lindzen, Richard S., (617) 253-2432 rlindzen@mit.edu,
Massachusetts Institute of Technology – Ow
Lineback, Neal G., (828) 262-3000 linebackng@appstate.edu,
Appalachian State Univ – Oy
Lineberry, Gene T., (606) 257-2833 Univ of Kentucky – Nm
Liner, Christopher L., 713-743-3399 cliner@uh.edu,
Univ of Houston – YesYg
Lines, Larry R., (403) 220-5841 Univ of Calgary – Ye
Lini, Andrea, (802) 656-0245 andrea.lini@uvm.edu,
Univ of Vermont – Cs
Link, Curtis A., (406) 496-4165 clink@mtech.edu,
Montana Tech of the Univ of Montana – Ye
Link, Paul K., (208) 282-3846 linkpaul@isu.edu,
Idaho State Univ – Gs
Linky, Edward, 212-772-5265 linky.edward@epamail.epa.gov,
Hunter College (CUNY) – Oo
Linn, Anne M.alinn@nas.edu,
National Academy of Sciences/National Research Council – Gs
Linn, Rodman, (505) 665-6254 rrl@lanl.gov,
Los Alamos National Lab – Ng
Linneman, Scott R., (360) 650-7207 Scott.Linneman@wwu.edu,
Western Washington Univ – Oe
Linsley, Braddock K., (518) 442-4478 blinsley@albany.edu,
SUNY, Albany – Cs
Liou, Juhn G., (650) 723-2716 liou@pangea.stanford.edu,
Stanford Univ – Gp
Liou, Kuo Nan, (415) 723-2716 knliou@atmos.ucla.edu,
Univ of California, Los Angeles – Oa
Lipella, Marc, (202) 633-1820 lipellam@si.edu,
Smithsonian Institution / National Museum of Natural History – Cg
Lippelt, Irene D., (608) 262-7430 ilippelt@wisc.edu,
Univ of Wisconsin - Extension – Gg

Lipps, Jere H., (510) 642-9006 Univ of California, Berkeley – Po
Lips, Elliott W., (801) 581-8218 elliott.lips@geog.utah.edu, Univ of Utah – Gm
Lisenbee, Alvis L., (605) 394-2463 Alvis.Lisenbee@sdsmt.edu, South Dakota School of Mines & Technology – Gc
Lisichenko, Richard, 785-628-4159 rlisiche@fhsu.edu, Fort Hays State Univ – Oi
Lisle, R J.lisle@cf.ac.uk, Cardiff Univ – Gc
Lisle, Thomas, (707) 825-2930 tel7001@humboldt.edu, Humboldt State Univ – Hs
Lithgow-Bertelloni, Carolina R., (734) 647-9938 crlb@umich.edu, Univ of Michigan – Yg
Litman, Gary W., (727) 892-4226 Univ of South Florida, St. Petersburg – Oo
Little, Richard D., (413) 527-8536 rdlittle2000@aol.com, Greenfield Community College – Gg
Little, William W., (208) 496-2427 littlew@byui.edu, Brigham Young Univ, Idaho – GsdGm
Little, William W., 573-368-2480 bill.little@dnr.mo.gov, Missouri Dept of Natural Resources – Gr
Liu, Gousheng, (850) 644-6298 liug@met.fsu.edu, Florida State Univ – Ow
Liu, Jian-yi, (406) 994-6905 uesjl@montana.edu, Montana State Univ – Oo
Liu, Jingpu P., 919-515-3711 jpliu@ncsu.edu, North Carolina State Univ – Gu
Liu, Kelly H., 573-341-6724 liukh@umr.edu, Univ of Missouri, Rolla – Ys
Liu, Lanbo, (860) 486-1388 lanbo.liu@uconn.edu, Univ of Connecticut – Yg
Liu, Menghua, (519) 661-3186 mliu@uwo.ca, Univ of Western Ontario – Oo
Liu, Mian, (573) 882-3784 LiuM@missouri.edu, Univ of Missouri – Yg
Liu, Peng Cheng, (805) 893-4905 pcliu@crustal.ucsb.edu, Univ of California, Santa Barbara – Ys
Liu, Rong, (204) 474-9371 Univ of Manitoba – Om
Liu, Shaw C., (404) 894-1758 Georgia Institute of Technology – Oa
Liu, Zhen-xian, (314) 882-3784 zxliu@bnl.gov, Carnegie Institution of Washington – Gy
Liu, Zhengyu, (608)262-0777 zliu3@wisc.edu, Univ of Wisconsin, Madison – Op
Livaccari, Richard F., (970) 248-1081 rlivacca@mesastate.edu, Mesa State College – Gc
Livchak, Constance J., 419-626-4296 constance.livchak@dnr.state.oh.us, Ohio Dept of Natural Resources – Hy
Lively, Richard S., (612)627-4780, Ext. 211 lively@umn.edu, Univ of Minnesota, Twin Cities – Cc
Livelybrooks, Dean, (514) 340-3710 Ecole Polytechnique – Ye
Livingstone, Daniel A., (919) 684-3264 livingst@duke.edu, Duke Univ – Pe
Lizarralde, Daniel, (404) 894-3893 Georgia Institute of Tech – Yv
Lizarralde, Daniel, 508-289-2942 dlizarralde@whoi.edu, Woods Hole Oceanographic Institution – Ys
Lizotte, Yves, (819) 736-4331 McGill Univ – Nm
Lloyd, Jacqueline M., (850) 488-9380 Florida Geological Survey – Ge
Lo, Howard H., (216) 687-4821 h.lo@csuohio.edu, Cleveland State Univ – Gi
Loague, Keith, (650) 723-0847 keith@pangea.stanford.edu, Stanford Univ – Hq
Loaiciga, Hugo A., (805) 893-8053 Univ of California, Santa Barbara – Hg
Lobben, Amy, (541) 346-4566 lobben@uoregon.edu, Univ of Oregon – OioOo
Locat, Jacques E., (418) 656-2179 locat@ggl.ulaval.ca, Universite Laval – Ng
Loch, James D., (660) 543-8804 loch@ucmo.edu, Univ of Central Missouri – Ps
Lock, Brian E., (337) 482-6823 belock@louisiana.edu, Univ of Louisiana, Lafayette – Gs
Locke, Daniel B., (207) 287-2801 Dept of Conservation – Hw
Locke, David C., (718) 997-3271 Graduate School of the City Univ of New York – Ca
Locke, James L., (415) 485-9526 jim@marin.cc.ca.us, Marin Community College – Og
Locke, Kelly W., (423) 425-4404 kelley-locke@utc.edu, Univ of Tennessee, Chattanooga – Oo
Locke, William W., (406) 994-6918 wlocke@montana.edu, Montana State Univ – Gm
Locke, III, Randall A., (217) 333-3866 Illinois State Water Survey – Hw
Locker, Stanley D., (727) 553-1502 slocker@marine.usf.edu, Univ of South Florida, Saint Petersburg – Ou
Lockley, Martin G., (303) 556-4884 Univ of Colorado, Denver – Pi
Lockwood, John P., (808) 967-7357 geohaz@aloha.net, Univ of Hawai'i at Hilo – Gv
Lockwood, Rowan, (757) 221-2878 rxlock@wm.edu, College of William & Mary – Po
Lodders-Fegley, Katharina, (314) 935-4851 lodders@levee.wustl.edu, Washington Univ – Xc
Loeppert, Jr., Richard H., (979) 845-3663 r-loeppert@tamu.edu, Texas A&M Univ – Sc
Loesch, Arthur Z., (518) 442-4560 azl@atmos.albany.edu, SUNY, Albany – Oa
Loewy, Staci, 661-654-2550 sloewy@csub.edu, California State Univ, Bakersfield – Gc
Lofthouse, Stephen T., (212) 346-1760 slofthouse@fsmail.pace.edu, Pace Univ, New York Campus – Og
Logan, Alan, (506) 648-5715 logan@unbsj.ca, Univ of New Brunswick, Saint John – Ob
Logan, Amelia, (202) 633-1801 logana@si.edu, Smithsonian Institution / National Museum of Natural History – Ca
Logan, John M., (409) 845-1371 Texas A&M Univ – Gc
Logan, John M., (541) 346-4587 jmllogan@aol.com, Univ of Oregon – Nr
Logan, John M., (541) 347-7403 jmllogan@aol.com, Oregon State Univ – Gt
Loganathan, G. V., (703) 231-6211 Virginia Polytechnic Institute & State Univ – Hg
Logsdon, Miles G., (206) 543-5334 mlog@u.washington.edu, Univ of Washington – Oi
Logsdon, Sally D., (515) 294-8265 logsdon@nstl.gov, Iowa State Univ of Science & Technology – Sp
Logue, Terry, 307-265-8887 tlogue@caspercollege.edu, Casper College – Gg
Lohmann, George P., (508) 289-2840 glohmann@whoi.edu, Woods Hole Oceanographic Institution – Pe
Lohmann, Kyger C., (734) 763-2298 kacey@umich.edu, Univ of Michigan – Cs
Lohrengel, II, C. Frederick, 435-586-7941 lohrengel@suu.edu, Southern Utah Univ – GsrPs
Lohrenz, Steven E., (228) 688-3177 steven.lohrenz@usm.edu, Univ of Southern Mississippi – Ob
Loiselle, Marc C., (207) 287-2801 marc.c.loiselle@state.me.us, Univ of Maine – Hy
Lombard, Armand J., (480) 461-7021 lombard@mc.maricopa.edu, Mesa Community College – Gg
London, David, (405) 325-7626 dlondon@ou.edu, Univ of Oklahoma – Cp
Long, Ann D., (217) 244-6172 annlong@uiuc.edu, Univ of Illinois, Urbana-Champaign – Gm
Long, Austin, (520) 621-8888 along@geo.arizona.edu, Univ of Arizona – Cl
Long, Austin, (520) 621-8888 along@geo.arizona.edu, Univ of Arizona – Cs
Long, Bernard, (418) 654-2590 blong@inrs.uquebec.ca, Universite du Quebec – On
Long, Darrel, (705) 675-1151 (Ext. ext. 2268) dlong@laurentian.ca, Laurentian Univ, Sudbury – Gs
Long, David T., 517-353-9618 long@msu.edu, Michigan State Univ – Cl
Long, Leland T., (404) 894-2860 tim.long@eas.gatech.edu, Georgia Institute of Technology – Ys
Long, Leon E., (512) 471-7562 leonlong@mail.utexas.edu, Univ of Texas at Austin, Jackson School of Geosciences – Cc
Long, Philip E., (509) 372-6090 philip.long@pnl.gov, Pacific Northwest National Lab – Gi
Longerich, Henry, (709) 737-8380 henry@sparky2.esd.mun.ca, Memorial Univ of Newfoundland – Cg
Longhi, Johnlonghi@ldeo.columbia.edu, American Museum of Natural History – Gi
Longoria, Jose F., (305) 348-3614 longoria@fiu.edu, Florida International Univ – Ps
Longshore, John D., (707) 826-4182 jdl1@axe.humboldt.edu, Humboldt State Univ – Gi
Longstaffe, Frederick J., (519) 661-3177 flongsta@uwo.ca, Univ of Western Ontario – Cs
Longuet-Higgins, Michael S., (858) 534-3936 mlonguet@ucsd.edu, Univ of California, San Diego – Op
Longworth, Bret, 508-289-3559 blongworth@whoi.edu, Woods Hole Oceanographic Institution – Ou

Lonn, Jeffrey, 406-496-4177 jlonn@mtech.edu,
Montana Tech of The Univ of Montana – Gg
Lonsdale, Darcy J., (631) 632-8712 dlonsdale@notes.cc.sunysb.edu,
SUNY, Stony Brook – Ob
Lonsdale, Peter F., (858) 534-2855 plonsdale@ucsd.edu,
Univ of California, San Diego – Gu
Loope, David B., (402) 472-2647 dloope1@unl.edu,
Univ of Nebraska, Lincoln – Gs
Loose, Pamloose@badlands.nodak.edu,
North Dakota State Univ – Oo
Lopez, David A., 406-657-2632 dlopez@mtech.edu,
Montana Tech of The Univ of Montana – Gg
Lopez, Dina L., (740) 593-9435 lopezd@ohio.edu,
Ohio Univ, Lancaster – Cl
Lopez, Glenn R., (631) 632-8660 glopez@notes.cc.sunysb.edu,
SUNY, Stony Brook – Ob
Lopez, Margarita, marlopez@cicese.mx,
Centro de Inv Científica y de Educación Superior de Ensenada – Cc
Lopez, Robert, (408) 741-2437 blopez@es.ucsc.edu,
West Valley College – Gg
López-Morales, Mercedes, (202) 478-8480 mercedes@dtm.ciw.edu,
Carnegie Institution of Washington – Oo
Lord, Mark L.mlord@wcu.edu,
Western Carolina Univ – Hw
Lorenz, Brenna, (671) 734-9357 blorenz@uog.edu,
Univ of Guam – Gi
Lorenz, Edward N., 617253-4089 Massachusetts Institute of Technology – Ow
Lorenzo, Juan M., (225) 388-2497 juan@geol.lsu.edu,
Louisiana State Univ – Ys
Loring, Arthur P., (718) 262-2079 York College (CUNY) – Gm
Losh, Steven, (607) 255-9518 losh@geology.cornell.edu,
Cornell Univ – Eg
Loter, Joe C., (361) 825-2469 jloter@tamucc.edu,
Texas A&M Univ, Corpus Christi – Cg
Lott, Dempsey E., (508) 289-2929 dlott@whoi.edu,
Woods Hole Oceanographic Institution – Oc
Loubere, Paul, (815) 753-7949 paul@geol.niu.edu,
Northern Illinois Univ – Pe
Louchouarn, Patrick, 409 740 4710 loup@tamug.edu,
Texas A&M Univ – Oc
Louchouarn, Patrick, (212) 854-0479 pl2065@columbia.edu,
Columbia Univ – Co
Loucks, Bob, (512) 471-0366 bob.loucks@beg.utexas.edu,
Univ of Texas at Austin, Jackson School of Geosciences – Go
Louden, Keith E., (902) 494-3452 keith.louden@dal.ca,
Dalhousie Univ – Yr
Louie, John, (775) 784-4219 louie@unr.edu,
Univ of Nevada, Reno – Ye
Louie, John, (775) 784-4219 louie@unr.edu,
Univ of Nevada, Reno – Yg
Louie, John N., (775) 784-4219 louie@seismo.unr.edu,
Univ of Nevada, Reno – Ye
Love, David W., (505) 835-5146 dave@gis.nmt.edu,
New Mexico Institute of Mining and Technology – Ge
Love, Gordon, gordon.love@ucr.edu,
Univ of California, Riverside – Co
Love, Pamela, (775) 784-6469 plove@unr.edu,
Univ of Nevada, Reno – Oo
Lovell, Charles R., (803) 777-7036 Univ of South Carolina – Ob
Lovell, Mark D., (208) 496-1903 lovellm@byui.edu,
Brigham Young Univ, Idaho – HyOie
LoVetere, Crystalclovetere@cerritos.edu,
Cerritos College – Oy
Lovett, James A., (276) 676-5830 james.lovett@dmme.virginia.gov,
Division of Mineral Resources – Eg
Lowe, Donald R., (650) 725-3040 lowe@pangea.stanford.edu,
Stanford Univ – Gd
Lowe, John C., (202) 994-6188 George Washington Univ – Og
Lowe, Mike, (801) 537-3389 nrugs.mlowe@state.ut.us,
Utah Geological Survey – Hw
Lowe, Theresa, (801) 585-1276 Univ of Utah – Cm
Lowell, Gary, 817 272 2987 lowell@uta.edu,
Univ of Texas, Arlington – Gx
Lowell, Robert P., (404) 894-9472 bob.lowell@eas.gatech.edu,
Georgia Institute of Technology – Yr
Lowell, Robert P., 540-231-6004 rlowell@vt.edu,
Virginia Polytechnic Institute & State Univ – Yr
Lowell, Thomas V., (513) 556-4165 thomas.lowell@uc.edu,
Univ of Cincinnati – Gl

Lowenstein, Tim K., (607) 777-4254 Binghamton Univ – Cl
Lower, Steven K., 614 292-1571 lower.9@osu.edu,
Ohio State Univ – Py
Lowery, Birl, (608) 262-2752 blowery@wisc.edu,
Univ of Wisconsin, Madison – Sp
Lowery, Mary Sue, (619) 260-4600 slowery@acusd.edu,
Univ of San Diego – Ob
Lowes, Brian E., (253) 535-7377 lowesbe@plu.edu,
Pacific Lutheran Univ – Gc
Lowman, Judy A., (909) 941-2367 Chaffey College – Gv
Lowry, Anthony R., 435-797-7096 tony.lowry@usu.edu,
Utah State Univ – Yg
Lowry, Wallace D., (540) 231-6521 Virginia Polytechnic Institute & State Univ – Gc
Lowther, J. Stewart, (253) 879-3826 slowther@ups.edu,
Univ of Puget Sound – Pb
Loyd, Ralph C., (916) 322-9207 California Geological Survey – Em
Loynachan, Thomas E., (515) 294-3064 teloynac@iastate.edu,
Iowa State Univ of Science & Technology – So
Lozier, M. Susan, (919) 681-8199 s.lozier@duke.edu,
Duke Univ – Op
Lozinsky, Richard P.rick_lozinsky@fullcoll.edu,
Fullerton College – Gr
Lozowski, Edward P., (780) 492-0348 edward.lozowski@ualberta.ca,
Univ of Alberta – Ow
Lu, Huan-Zhang, (418) 545-5011 hzlu@uqac.uquebec.ca,
Universite du Quebec a Chicoutimi – Em
Lucas, Susan, (814) 732-1573 slucas@edinboro.edu,
Edinboro Univ of Pennsylvania – Oo
Lucas-Clark, Joyce, 415-452-7426 jluclark@comcast.net,
City College of San Francisco – Pg
Lucey, Paul G., (808) 956-3137 Univ of Hawai'i, Manoa – Xy
Lucia, F. Jerry, (512) 471-7367 jerry.lucia@beg.utexas.edu,
Univ of Texas at Austin, Jackson School of Geosciences – Go
Lucotte, Marc, (514) 987-3767 lucotte.marc_michel@uqam.ca,
Universite du Quebec a Montreal – Oc
Ludman, Allan, (718) 997-3271 Graduate School of the City Univ of New York – Gg
Ludman, Allan, (718) 997-3324 allan_ludman@qc.edu,
Queens College (CUNY) – Gg
Ludvigson, Greg A., 785-864-2734 gludvigson@ku.edu,
Univ of Kansas – Gr
Ludvigson, Gregory A.gludvigson@ku.edu,
Univ of Iowa – Gd
Ludwikoski, David J., 410-455-4216 dludwikoski@ccbcmd.edu,
Community College of Baltimore County, Catonsville campus – Oe
Lueck, Rolf G., (250) 721-8918 Univ of Victoria – Op
Lueth, Virgil W., (505) 835-5140 vwlueth@nmt.edu,
New Mexico Institute of Mining & Technology – Gg
Lugo, Hector M., (809) 265-3851 Univ of Puerto Rico – Oo
Lukas, Roger, (808) 956-7896 rlukas@soest.hawaii.edu,
Univ of Hawai'i, Manoa – Op
Lulla, Kamleshkamlesh.p.lulla@jsc.nasa.gov,
Univ of Delaware – Or
Luman, Donald E., 217-244-2179 luman@isgs.uiuc.edu,
Illinois State Geological Survey – Or
Lumpkin, Thomas A., (509) 335-2726 lumpkin@wsu.edu,
Washington State Univ – Oo
Lumsden, David N., (901) 678-4359 dlumsden@memphis.edu,
Univ of Memphis – Gd
Lund, Jay R., (916) 752-5671 Univ of California, Davis – Hs
Lund, Steven P., (213) 740-5835 slund@usc.edu,
Univ of Southern California – Ym
Lund, William R., (801) 865-8126 mnet.nrregion.blund@state.ut.us,
Utah Geological Survey – Ng
Lundblad, Steven P., 808-974-7641 slundbla@hawaii.edu,
Univ of Hawai'i at Hilo – Gr
Lundelius, Ernest L., (512) 471-6556 erniel@mail.utexas.edu,
Univ of Texas at Austin, Jackson School of Geosciences – Pv
Lundin, Robert F., (480) 965-3514 robert.lundin@asu.edu,
Arizona State Univ – Pm
Lundstrom, Craig C., (217) 244-6293 lundstro@uiuc.edu,
Univ of Illinois, Urbana-Champaign – Cp
Lundy, Don A.Don Lundy [dlundy@esnt.com],
Georgia Perimeter College – Hy
Luneburg, Catalina M.hlebit@uno.edu,
Georgia State Univ – Gc
Lunine, Jonathan I., (520) 621-2789 jlunine@lpl.arizona.edu,
Univ of Arizona – Xc
Luo, Chao, (404) 894-0838 Georgia Institute of Technology – Oa

Luo, Yu-Ran, (727) 553-3922 yluo@marine.usf.edu,
Univ of South Florida, Saint Petersburg – Oc
Luo, Zhexi, (412) 622-6578 luoz@clpgh.org,
Carnegie Museum of Natural History – Pv
Lupia, Richard, (405) 325-7229 rlupia@ou.edu,
Univ of Oklahoma – Pm
Lupiani, Michaelalupiani@ocean.fsu.edu,
Florida State Univ – Yr
Lupo, Anthony R., 573-884-1638 lupoa@missouri.edu,
Univ of Missouri, Columbia – Oa
Lupulescu, Marian, 518-474-1432 mlupules@mail.nysed.gov,
New York State Geological Survey – Gz
Lusardi, Barbara A., (612)627-4780, Ext. 212 lusar001@umn.edu,
Univ of Minnesota, Twin Cities – Gl
Lusby, Linda A., 902-585-1323 linda.lusby@acadiau.ca,
Acadia Univ – Oo
Luth, Robert W., (780) 492-2740 robert.luth@ualberta.ca,
Univ of Alberta – Cp
Luther, Douglas S., (808) 956-5875 dluther@soest.hawaii.edu,
Univ of Hawai'i, Manoa – Op
Luther, Frank R., (262) 472-5257 lutherf@mail.uww.edu,
Univ of Wisconsin, Whitewater – Gp
Luther, Mark E., (727) 553-1528 mluther@marine.usf.edu,
Univ of South Florida, Saint Petersburg – Op
Luther, III, George W., (302) 645-4208 luther@udel.edu,
Univ of Delaware – Cm
Luttge, Andreas, (713) 348-6304 aluttge@rice.edu,
Rice Univ – Cg
Luttrell, Stuart P., (509) 376-6023 stuart.luttrell@pnl.gov,
Pacific Northwest National Lab – Hg
Lutz, Anne B., 717.702.2043 anlutz@state.pa.us,
Pennsylvania Bureau of Topographic & Geologic Survey – Pg
Lutz, Susan J., (801) 585-9688 sjlutz@egi.utah.edu,
Univ of Utah – Gg
Lutz, Timothy M., (610) 436-3498 tlutz@wcupa.edu,
West Chester Univ – Gq
Lux, Daniel R., (207) 581-4494 dlux@maine.maine.edu,
Univ of Maine – Gi
Luyendyk, Bruce P., (805) 893-2827 Univ of California, Santa Barbara – Yr
Luza, Kenneth V., (405) 325-3031 kluza@ou.edu,
Univ of Oklahoma – Ng
Luzincourt, Marc R., (418) 654-3715 mluzinco@nrcan.gc.ca,
Universite du Quebec – Cs
Lwin, Mark J., 61 8 9266-2324 M.Lwin@curtin.edu.au,
Curtin Univ of Technology – Ye
Lwiza, Kamazima M., (631) 632-7309 klwiza@notes.cc.sunysb.edu,
SUNY, Stony Brook – Oc
Lübbe, Maike, 089/2180 4337 Meike.Luebbe@lrz.uni-muenchen.de,
LMU - Univ of Munich – Gz
Lyle, Annette, 979 845 7191 alyle@ocean.tamu.edu,
Texas A&M Univ – Ou
Lyle, Michael E., (804) 427-7189 Tidewater Community College – On
Lyle, Mitch, 979 845 3380 mlyle@ocean.tamu.edu,
Texas A&M Univ – Ou
Lyle, Mitchell W., (208) 426-1167 mlyle@cgiss.boisestate.edu,
Boise State Univ – Ou
Lynch, F. L., (662) 325-2910 claynac@ra.msstate.edu,
Mississippi State Univ – Gz
Lynch, Michael J., (859) 257-5500 Univ of Kentucky – Oo
Lyon, Elizabeth C., 717.702.2063 elyon@state.pa.us,
Pennsylvania Bureau of Topographic & Geologic Survey – Oo
Lyon, Ronald J. P., (650) 725-8077 lyon@pangea.stanford.edu,
Stanford Univ – Or
Lyons, Donald, (940) 565-2721 dlyons@unt.edu,
Univ of North Texas – Ou
Lyons, James V., (508) 929-8719 jlyons@worcester.edu,
Worcester State College – Og
Lyons, Lawrence, (310) 206-7876 larry@atmos.ucla.edu,
Univ of California, Los Angeles – Oa
Lyons, Walter A., 970-351-2647 walyons@frii.com,
Univ of Northern Colorado – Oa
Lyons, William B., (614) 292-6531 lyons.142@osu.edu,
Ohio State Univ – Cl
Lytwyn, Jennifer N., (713) 743-3397 jlytwyn@uh.edu,
Univ of Houston – Gi

M

M. Dosi, M. Adam, 901-678-2979 mdosi@memphis.edu,
Univ of Memphis – Oo
Ma, Lena Q., (352) 392-9063 lqma@ufl.edu,
Univ of Florida – Sc
Maa, Jerome P-Y., (804) 684-7270 maa@vims.edu,
College of William & Mary – Op
Maantay, Juliana, 718-960-8574 maantay@aol.com,
Lehman College (CUNY) – Oi
Maas, John, 909-941-2385 Chaffey College – Yg
Maas, Richard P., (828) 251-6366 maas@unca.edu,
Univ of North Carolina, Asheville – Oo
Maasch, Kirk A., (207) 581-2197 kirk@iceage.umeqs.maine.edu,
Univ of Maine – Oa
Mabe, Terri, (817) 257-7270 t.mabe@tcu.edu,
Texas Christian Univ – Oo
Mabee, Stephen B.sbmabee@geo.umass.edu,
Office of the State Geologist – Hw
Mac Eachern, James A., (604) 291-5388 jmaceach@sfu.ca,
Simon Fraser Univ – Gs
Macalady, Donald L.dmacalad@mines.edu,
Colorado School of Mines – Cl
Macalady, Jennifer L., 814-865-6330 jlm80@psu.edu,
Pennsylvania State Univ, Univ Park – Py
MacAyeal, Douglas R., (773) 702-8027 drm7@midway.uchicago.edu,
Univ of Chicago – Yr
MacCarthy, Ivor, +353 21 4902152 i.maccarthy@ucc.ie,
Univ College Cork – Gs
MacCarthy, Patrickpmaccart@mines.edu,
Colorado School of Mines – Co
MacCready, Parker, (206) 685-9588 parker@ocean.washington.edu,
Univ of Washington – Op
MacDonald, Harold C., (479) 575-3355 Univ of Arkansas, Fayetteville – Or
Macdonald, Ken C., (805) 893-4005 Univ of California, Santa Barbara – Yr
Macdonald, R. Heather, (757) 221-2443 rhmacd@wm.edu,
College of William & Mary – GsOe
Macdonald, Robie W., (250) 363-6409 robie@ios.bc.ca,
Univ of Victoria – Oc
MacDonald, William D., (607) 777-2863 Binghamton Univ – Gc
Macdougall, J. Douglas, (619) 534-3294 Univ of California, San
Diego – Cm
MacFadden, Bruce J., (352) 392-1721 (Ext. 496) bmacfadd@flmnh.ufl.
edu, Univ of Florida – Pv
Macfarlane, Andrew W., (305) 348-3980 macfarla@fiu.edu,
Florida International Univ – Em
Macfarlane, Anne, (775) 327-5078 arcplann@att.net,
Univ of Nevada, Reno – Ou
Macfarlane, Peter A., (785) 864-2068 dowser@kgs.ku.edu,
Univ of Kansas – Hw
MacGregor, Kelly, (651) 696-6441 macgregor@macalester.edu,
Macalester College – Gm
Machado, Nunor30404@er.uqam.ca,
Universite du Quebec a Montreal – Cc
Machel, Hans G., (780) 492-5659 hans.machel@ualberta.ca,
Univ of Alberta – Go
Machesky, Michael L., (217) 333-9322 machesky@uiuc.edu,
Illinois State Water Survey – Cg
Macias, Edward, (314) 935-6820 macias@wustl.edu,
Washington Univ – Oa
Macias, Steve E., (360) 475-7711 smacias@oc.ctc.edu,
Olympic College – Gg
Maciejewska, Beata, (915) 747-5263 beatam@utep.edu,
Univ of Texas, El Paso – Cg
MacInnes, Michael, (505) 665-2154 Los Alamos National Lab – Oo
Macintyre, Ian G., (202) 633-1339 Smithsonian Institution / National
Museum of Natural History – Gs
Mack, Gregory H., 575-646-1343 gmack@nmsu.edu,
New Mexico State Univ, Las Cruces – Gs
Mack, John E., (716) 878-5006 mackje@buffalostate.edu,
Buffalo State College – Xg
Mack, Seymour, (559) 278-2948 California State Univ, Fresno – Gi
Mackas, David L., (250) 363-6442 mackas@ios.bc.ca,
Univ of Victoria – Ob
Mackenzie, Fred T., (808) 956-6344 fredm@soest.hawaii.edu,
Univ of Hawai'i, Manoa – Cm
Mackey, Kevin G., 517-355-3436 mackeyke@msu.edu,
Michigan State Univ – Ys
MacKinnon, Dave, (033-667- 17764 dave.mackinnon@canterbury.
ac.nz, Univ of Canterbury – Po
Macko, Stephen A., 434-924-6849 sam8f@virginia.edu,
Univ of Virginia – Co
MacLaren, Bruce R., (859) 622-1503 bruce.maclaren@eku.edu,
Eastern Kentucky Univ – Oo
MacLaughlin, Mary M., (406) 496-4655 marymac@mtech.edu,

Montana Tech of the Univ of Montana – Nr
MacLean, Wallace H., 514-398-4888 whm@eps.mcgill.ca, McGill Univ – Em
MacLeod, Cmacleod@cf.ac.uk, Cardiff Univ – Gu
MacLeod, Kenneth A., (573) 884-3118 macleodk@missouri.edu, Univ of Missouri – Pg
MacNish, Robert, (520) 621-5082 macnish@hwr.arizona.edu, Univ of Arizona – Hw
Macomber, Richard, (718) 460-1476 Long Island Univ, Brooklyn Campus – Pg
Macpherson, Colin G., 0191 3342283 colin.macpherson@durham.ac.uk, Durham Univ – Cg
MacPherson, Glenn J., (202) 633-1803 macphers@si.edu, Smithsonian Inst / National Museum of Natural History – XcGi
Macpherson, Gwen L., (785) 864-2742 glmac@ku.edu, Univ of Kansas – Hw
MacQuarrie, Pamela, (403) 440-6176 pmacquarrie@mtroyal.ca, Mount Royal College – Oy
MacRae, Neil D., (519) 661-3178 nmacrae@uwo.ca, Univ of Western Ontario – Gx
Madden, Theodore R., (617) 253-6384 madden@erl.mit.edu, Massachusetts Institute of Technology – Yg
Maddock, III, Thomas, (520) 621-7120 maddock@hwr.arizona.edu, Univ of Arizona – Hq
Maddocks, Rosalie F., (713) 743-3429 maddocks@uh.edu, Univ of Houston – Pm
Maddox, Darryl W., (806) 371-5330 maddox-dw@actx.edu, Amarillo College – Go
Madej, Mary Annmary_ann_madej@usgs.gov, Humboldt State Univ – Gm
Madison, Frederick W., (608) 263-4004 fredmad@wisc.edu, Univ of Wisconsin, Madison – Sd
Madison, James P., 406-496-4619 jmadison@mtech.edu, Montana Tech of The Univ of Montana – Hw
Madrid, Juanjuaneu@cicese.mx, Centro de Inv Científica y de Educación Superior de Ensenada – Ys
Madsen, David B., (801) 537-3314 nrugs.dmadsen@state.ut.us, Utah Geological Survey – Ga
Madsen, John A., 302-831-1608 jmadsen@udel.edu, Univ of Delaware – Yr
Maertz, Lynn D., (608) 262-3491 maertz@engr.wisc.edu, Univ of Wisconsin, Madison – Oo
Maerz, Norbert H., (573) 341-6714 norbert@umr.edu, Univ of Missouri, Rolla – Nr
Maes, Stephanie, 518-454-2910 maess@strose.edu, The College of Saint Rose – GicEg
Magaard, Lorenz, (808) 956-7509 lorenz@hawaii.edu, Univ of Hawai'i, Manoa – Op
Magistrale, Harold W., (619) 594-6741 harold@hal.sdsu.edu, San Diego State Univ – Ys
Magloughlin, Jerry F., (970) 491-1812 jerrym@cnr.colostate.edu, Colorado State Univ – Gp
Magnani, M. Beatrice, 901-678-4830 mmagnani@ceri.memphis.edu, Univ of Memphis – Ys
Magnusdottir, Gudrun, (949) 824-3520 gudrun@uci.edu, Univ of California, Irvine – Oa
Magnuson, Larry M., (785) 864-2150 magnuson@kgs.ku.edu, Univ of Kansas – Ca
Maguire, David, (313) 232-3617 Charles Stewart Mott Community College – Xg
Maguire, Rory O., 919-513-3033 rory_maguire@ncsu.edu, North Carolina State Univ – Sb
Mahadevan, Amalaamala@bu.edu, Boston Univ – Op
Maher, Jr., Harmon D., 402-554-4807 hmaher@mail.unomaha.edu, Univ of Nebraska at Omaha – Gc
Maher, Jr., Louis J., (608) 262-9595 maher@geology.wisc.edu, Univ of Wisconsin, Madison – Pl
Mahfoud, Robert F., (318) 475-5755 McNeese State Univ – Em
Mahler, Robert L., (208) 885-7025 Univ of Idaho – Oo
Mahmood, Rezaul, (270) 745-5979 rezaul.mahmood@wku.edu, Western Kentucky Univ – Oa
Mahoney, J. Brian, (715) 836-4952 mahonej@uwec.edu, Univ of Wisconsin, Eau Claire – Gs
Mahoney, John, (808) 956-8705 jmahoney@soest.hawaii.edu, Univ of Hawai'i, Manoa – Ca
Mahoney, William, (617) 287-5291 Univ of Massachusetts, Boston/ Harbor Campus – Oo
Mahood, Gail A., (650) 723-1429 gail@pangea.stanford.edu, Stanford Univ – Gi
Mahrt, Larry, (541) 737-5691 mahrt@coas.oregonstate.edu, Oregon State Univ – Oa
Maier, Raina M., (520) 621-7231 rmaier@ag.arizona.edu, Univ of Arizona – Oo
Maillol, Jean-Michel, (403) 220-8393 maillol@ucalgary.ca, Univ of Calgary – Ym
Maisano, Jessica A., (512) 471-0260 maisano@mail.utexas.edu, Univ of Texas at Austin, Jackson School of Geosciences – Pv
Maisey, John G., (212) 769-5811 American Mus of Nat History – Pv
Majmundar, Hasmukhrai H., (415) 904-7717 California Geological Survey – Oo
Major, R. P., (662) 915-5440 rpm@cedar.olemiss.edu, Univ of Mississippi – Gd
Major, Ruth H., 518-629-7131 Hudson Valley Comm College – Ggh
Mak, John E., (631) 632-8673 jemak@notes.cc.sunysb.edu, SUNY, Stony Brook – Oa
Mak, Mankin, (217) 333-8071 mak@atmos.uiuc.edu, Univ of Illinois, Urbana-Champaign – Oa
Maki, Jon R., (859) 622-1507 jon.maki@eku.edu, Eastern Kentucky Univ – Oo
Makovicky, Peter J., (312) 665-7633 pmakovicky@fieldmuseum.org, Field Museum of Natural History – Pv
Malachowski, Judith M., (404) 358-5085 Gordon College – Oo
Malahoff, Alexander, (808) 956-6802 malahoff@hawaii.edu, Univ of Hawai'i, Manoa – Ou
Malanotte-Rizzoli, Paola M., (617) 253-2451 rizzoli@ocean.mit.edu, Massachusetts Institute of Technology – Op
Malcolm, Alisonmalcolm@ima.umn.edu, Massachusetts Institute of Technology – Yx
Malcuit, Robert J.malcuit@denison.edu, Denison Univ – Gx
Maletz, Jorg, (716) 645-6800 (Ext. 3996) jorgm@acsu.buffalo.edu, SUNY, Buffalo – Pg
Malhotra, Praveen K., (916) 323-8556 California Geological Survey – Ne
Malhotra, Renu, (520) 626-5899 renu@lpl.arizona.edu, Univ of Arizona – Oo
Maliekal, Jose A., (585) 395-2582 jmalieka@weather.brockport.edu, SUNY College at Brockport – Oa
Malin, Peter E., (919) 681-8889 p.malin@auckland.ac.nz, Duke Univ – Ys
Malinconico, Lawrence L., (610) 330-5195 malincol@lafayette.edu, Lafayette College – Yg
Malinconico, Mary Ann, (610) 330-5195 Lafayette College – Eo
Malinowski, Jon C., (845) 938-4673 United States Military Acad – Oo
Malinverno, Alberto, (845)-365-8577 Columbia Univ – Gu
Mallarino, Antonio W., (515) 294-6200 apmallar@iastate.edu, Iowa State Univ of Science & Technology – So
Mallinson, David, (252) 328-1344 mallinsond@mail.ecu.edu, East Carolina Univ – Gu
Mallinson, David J., (727) 553-3927 Univ of South Florida, Saint Petersburg – Gu
Mallon, Anthony J., 0191 3744785 a.j.mallon@durham.ac.uk, Durham Univ – Go
Mallon, Connie, (334) 861-7517 Univ of Alabama, Dauphin Island – Oo
Malo, Douglas D., (605) 688-4586 Douglas.Malo@sdstate.edu, South Dakota State Univ – Sd
Malody, Kim, (307) 766-2263 Univ of Wyoming – Oo
Malone, David H., (309) 4387649 dhmalon@ilstu.edu, Illinois State Univ – Gc
Malone, Stephen D., 206-685-3811 steve@ess.washington.edu, Univ of Washington – YsGt
Malone, Thomas F.tom_malone@ncsu.edu, North Carolina State Univ – Oa
Maloof, Adam C., (609) 258-4101 maloof@princeton.edu, Princeton Univ – Ym
Malservisi, Rocco, 089/2180 4202 malservisi@rsmas.miami.edu, LMU - Univ of Munich – Yg
Malzer, Gary L., (612) 625-6728 gmalzer@soils.umn.edu, Univ of Minnesota, Twin Cities – Sc
Mamet, Bernard L., (514) 343-6833 Universite de Montreal – Ps
MANCA, Prof. Pierpaolo P., 070-675-5529 ppmanca#unica.it, Universita di Cagliari – Nm
Manchester, Janet E., (518) 474-5816 New York State Geological Survey – Oo
Manchester, Steven R., (352) 392-1721 steven@flmnh.ufl.edu, Univ of Florida – Pb
Manchester, Steven R., (352) 392-1721 (Ext. 495) steven@flmnh.ufl.edu, Univ of Florida – Pb
Mancia, Laural, xm$geol@qc1.qc.edu,

Faculty Index -M

Queens College (CUNY) – Oo
Mancini, Ernest A., 205-348-4319 emancini@geo.ua.edu, Univ of Alabama – Pe
Mancuso, Joseph J., (419) 372-7196 jmancus@bgnet.bgsu.edu, Bowling Green State Univ – Em
Mand, Arlene, (215) 573-3164 amand@sas.upenn.edu, Univ of Pennsylvania – Oo
Mandarino, Joseph A., (416) 586-5819 Royal Ontario Museum – Gz
Mandel, Rolfe, 785-864-2171 mandel@ku.edu, Univ of Kansas – Ga
Mandel, Rolfe D., (785) 864-2171 mandel@kgs.ku.edu, Univ of Kansas – Ga
Mandernack, Kevin W., (303) 273-2224 kmandern@mines.edu, Colorado School of Mines – Py
Mandeville, Charles W., 212-769-5380 cmandy@amnh.org, American Museum of Natural History – Gv
Mandia, Scott, 631-451-4104 mandias@sunysuffolk.edu, Suffolk County Community College – Ow
Mandra, York T., (415) 338-2061 ytmandra@sfsu.edu, San Francisco State Univ – Pg
Manduca, Cathryn A., (507) 646-7096 cmanduca@carleton.edu, Carleton College – Gi
Mandziuk, William, (204) 474-7826 mandziu0@cc.umanitoba.ca, Univ of Manitoba – Gg
Manecan, Teodosia, (212) 772-5265 tmanecan@geo.hunter.cuny.edu, Hunter College (CUNY) – Gp
Manfrino, Carrie M., (908) 737-3697 cmanfrin@kean.edu, Kean Univ – Gu
Manga, Michael, (510) 643-8532 manga@seismo.berkeley.edu, Univ of California, Berkeley – GvXy
Manganini, Steven J., (508) 289-2778 smanganini@whoi.edu, Woods Hole Oceanographic Institution – Ou
Mangano, Gabriela, 306-966-5730 gabriela.mangano@usask.ca, Univ of Saskatchewan – PsGs
Mangel, Mark S., (831) 459-5785 msmangel@cats.ucsc.edu, Univ of California, Santa Cruz – Oo
Manger, Walter L., (479) 575-3370 wmanger@comp.uark.edu, Univ of Arkansas, Fayetteville – Ps
Manghnani, Murli H., (808) 956-7825 murli@soest.hawaii.edu, Univ of Hawai'i, Manoa – Yx
Mango, Helen N., (802) 468-1478 helen.mango@castleton.edu, Castleton State College – Cg
Mankiewicz, Carol, (608) 363-2371 mankiewicz@beloit.edu, Beloit College – Gs
Mankin, Charles J., (405) 325-3031 cjmankin@ou.edu, Univ of Oklahoma – Gg
Manley, Patricia L., (802) 443-5430 patmanley@middlebury.edu, Middlebury College – Yr
Manley, Thomas O., (802) 443-3114 tmanley@middlebury.edu, Middlebury College – Op
Mann, C. John, (217) 333-1166 mann@hercules.geology.uiuc.edu, Univ of Illinois, Urbana-Champaign – Gq
Mann, Daniel H., (907) 474-5872 Univ of Alaska, Fairbanks – Pe
Mann, Keith O., (740) 368-3620 komann@owu.edu, Ohio Wesleyan Univ – Pi
Mann, Michael E., (814) 863-4075 mann@meteo.psu.edu, Pennsylvania State Univ, Univ Park – Oa
Mann, Paul, (512) 471-0452 paulm@ig.utexas.edu, Univ of Texas at Austin, Jackson School of Geosciences – Gt
Mann, Shannon, (303) 273-3701 smann@mines.edu, Colorado School of Mines – Oo
Mann, Susan, (802) 635-1325 manns@badger.jsc.vsc.edu, Johnson State College – Oo
Manning, Craig E., (310) 206-3290 manning@ess.ucla.edu, Univ of California, Los Angeles – Gp
Manos, Leah D., (660) 562-1385 lmanos@nwmissouri.edu, Northwest Missouri State Univ – Oy
Manous, Joe D., (845) 938-2930 bj0995@usma.edu, United States Military Academy – Hs
Mansell, Robert S., (352) 392-1951 rsm@mail.ifas.ufl.edu, Univ of Florida – Sp
Mansfield, Charles, (812) 885-4233 Vincennes Univ Junior Coll – So
Mansfield, Gloria C., (901) 587-7441 glomans@utm.edu, Univ of Tennessee, Martin – Gg
Mansinha, Lalu, (519) 661-3145 mansinha@uwo.ca, Univ of Western Ontario – Ys
Manson, Michael W., (707) 576-2939 California Geological Survey – Oo
Manspeizer, Warren, (973) 353-5509 mansp@andromeda.rutgers.edu, Rutgers, The State Univ of New Jersey, Newark – Gr
Mantei, Erwin J., (417) 836-5446 ejm893f@smsu.edu, Southwest Missouri State Univ – Ct
Mantha, Phanikumar S., 517-432-0851 phani@msu.edu, Michigan State Univ – Hw
Manton, William I., (972) 883-2441 manton@utdallas.edu, Univ of Texas, Dallas – Cc
Manu, Andrew, (205) 851-5462 amanu@aamu.edu, Alabama A&M Univ – Sd
Manz, Lorraine, (701) 328-8005 lmanz@state.nd.us, North Dakota Geological Survey – Oo
Manzi, Anthony, 718-409-7371 amanzi@sunymaritime.edu, SUNY Maritime College – Ow
Mao, Ho-kwang, (202) 478-8960 h.mao@gl.ciw.edu, Carnegie Institution of Washington – Yx
Mapes, Royal H., (740) 593-1844 mapes@ohiou.edu, Ohio Univ, Lancaster – Pi
Maples, Christopher G.cmaples@indiana.edu, Indiana Univ, Bloomington – Pg
Marble, John C., (601) 354-6328 Mississippi Office of Geology – Ge
Marcano, Eugenio J., 413-538-3237 emarcano@mtholyoke.edu, Mount Holyoke College – Oi
Marcantonio, Franco, 979-845-9240 marcantonio@geo.tamu.edu, Texas A&M Univ – Cg
Marchal, Olivier, (508) 289-3374 omarchal@whoi.edu, Woods Hole Oceanographic Institution – Pe
Marchant, David R., (617) 353-3236 marchant@bu.edu, Boston Univ – Gm
Marchefti, David W., 970 943-2367 dmarchetti@western.edu, Western State College of Colorado – GmCl
Marchisin, John, (212) 346-1502 Pace Univ, New York Campus – Og
Marchisin, John, (201) 200-3161 jmarchisin@njcu.edu, New Jersey City Univ – Gg
Marchitto, Thomas M., 303-492-7739 tom.marchitto@colorado.edu, Univ of Colorado – Oc
Marcotte, Denis, (514) 340-4620 Ecole Polytechnique – Oo
Marcus, Nancy H., 850 644-6700 marcus@ocean.fsu.edu, Florida State Univ – Ob
Marelli, Dan, (727) 896-8626 Univ of South Florida, St. Petersburg – Oo
Mareschal, Jean-Claude, (514) 987-6864 jmc@volcan.geotop.uqam.ca, Universite du Quebec a Montreal – Yg
Marfurt, Kurt, (713) 743-9119 kmarfurt@uh.edu, Univ of Houston – Ye
Margrave, Gary F., (403) 220-4606 Univ of Calgary – Ye
Margulis, Lynn, (413) 545-3244 Univ of Massachusetts – Py
Maria, Tony, 812-461-5326 ahmaria@usi.edu, Univ of Southern Indiana – Gv
Mariant, Judy, (408) 288-3716 jmariant@sjeccd.cc.ca.us, San Jose City College – Gg
Marino, Miguel A., (530) 752-0684 mamarino@ucdavis.edu, Univ of California, Davis – Hw
Marion, Giles, (775) 673-7349 gmarion@dri.edu, Desert Research Institute – Sc
Mark, Bryan G., (614) 247-6180 mark.9@osu.edu, Ohio State Univ – Oa
Markell, Clark, (701) 858-3069 markell@minotstateu.edu, Minot State Univ – Oe
Market, Patrick S., 573-882-1496 markets@missouri.edu, Univ of Missouri, Columbia – Oa
Markey, Richard J., (970) 491-3789 rmarkey@cnr.colostate.edu, Colorado State Univ – Cc
Markley, Michelle J., (413) 538-2814 mmarkley@mtholyoke.edu, Mount Holyoke College – GcgGt
Markowski, Antonette K., 717.702.2038 amarkowski@state.pa.us, Pennsylvania Bureau of Topographic & Geologic Survey – Ec
Marks, Dennis W., 229-244-5159 dmarks@valdosta.edu, Valdosta State Univ – Oo
Markun, Francis J., (630) 252-7521 Argonne National Lab – Cc
Marland, Gregg H.gum@ornl.gov, Oak Ridge National Lab – Ge
Marley, Nancy A., (630) 252-5014 namarley@anl.gov, Argonne National Lab – Oa
Marlowe, Robert, (423) 425-4513 Univ of Tennessee, Chattanooga – Oo
Marone, Chris, (814) 865-7964 cjm38@psu.edu, Pennsylvania State Univ, Univ Park – Yx
Marquez, L. Lynn, 717-872-3288 Lynn.Marquez@millersville.edu, Millersville Univ – Cg
Marr, Paul G., (717) 477-1656 pgmarr@ship.edu, Shippensburg Univ – Oi
Marrett, Randall A., (512) 471-2113 marrett@mail.utexas.edu, Univ of Texas at Austin, Jackson School of Geosciences – Gc
Marrs, Ronald W., (307) 766-2330 rwmarrs@uwyo.edu,

Univ of Wyoming – Or
Marsaglia, Kathleen M., (818) 677-3541 kathleen.marsaglia@csun.edu, California State Univ, Northridge – Gs
Marsan, Yvonne, 910-962-7288 marsany@uncw.edu, Univ of North Carolina Wilmington – Oi
Marschak, Frederic, (805) 965-0581 (Ext. 2316) marschak@sbcc.net, Santa Barbara City College – Xg
Marschik, Robert, 089/2180 6527 robert.marschik@iaag.geo.uni-muenchen.de, LMU - Univ of Munich – Gg
Marsh, Bruce D., (410) 516-4652 bmarsh@jhu.edu, The Johns Hopkins Univ – Gi
Marsh, Stuart E., 520-621-8574 Univ of Arizona – Or
Marshak, Stephen, (217) 333-7705 smarshak@uiuc.edu, Univ of Illinois, Urbana-Champaign – Gc
Marshall, Charles R., (617) 495-2572 marshall@eps.harvard.edu, Harvard Univ – Pg
Marshall, Claude Monte, (619) 594-1395 mmarshall@geology.sdsu.edu, San Diego State Univ – Ym
Marshall, Dan D., (604) 291-5387 Simon Fraser Univ – Ca
Marshall, E. Wayne, (434) 951-6355 wayne.marshall@dmme.virginia.gov, Division of Mineral Resources – Oo
Marshall, Hal G., (313) 936-2310 Univ of Michigan – Oo
Marshall, Jeffrey S., 909-869-3461 marshall@csupomona.edu, California State Polytechnic Univ, Pomona – Gm
Marshall, John C., (617) 253-9615 marshall@gulf.mit.edu, Massachusetts Institute of Technology – Op
Marshall, Sylvia, (713) 743-3401 smarshall@uh.edu, Univ of Houston – Oo
Marston, Sallie A., 520-621-3903 marston@3mail.arizona.edu, Univ of Arizona – Ou
Marta, Patino Douce, (706) 542-2399 mapatino@uga.edu, Univ of Georgia – Gi
Martel, Richard, (418) 654-2683 rmartel@inrs.uquebec.ca, Universite du Quebec – Hw
Martel, Stephen J., (808) 956-7797 martel@soest.hawaii.edu, Univ of Hawai'i, Manoa – Ng
Martens, Christopher S., (919) 962-0152 Univ of North Carolina, Chapel Hill – Oc
Martens, Dean, (520) 670-6380 (Ext. 156) martens@tucson.ars.ag.gov, Univ of Arizona – So
Martin, Aaron, (301) 405-8763 martinaj@geol.umd.edu, Univ of Maryland – Gt
Martin, Anthony J., (740) 368-3621 geoam@learnlink.emory.edu, Emory Univ – Pe
Martin, Arturoamartin@cicese.mx, Centro de Inv Científica y de Educación Superior de Ensenada – Gd
Martin, Barton S., (740) 368-3621 bsmartin@owu.edu, Ohio Wesleyan Univ – Gx
Martin, Brian, (573) 368-2143 Missouri Dept of Natural Resources – Oo
Martin, Candace E., +64 3 479-7526 candace.martin@otago.ac.nz, Univ of Otago – Cg
Martin, Charles W., (765) 983-1672 martich@earlham.edu, Earlham College – Gg
Martin, Deborah, (928) 523-2650 deborah.martin@nau.edu, Northern Arizona Univ – Oo
Martin, Ellen E., (352) 392-2141 emartin@geology.ufl.edu, Univ of Florida – Cl
Martin, Freddie A., (225) 578-2110 fmartin@agctr.lsu.edu, Louisiana State Univ – Oo
Martin, Gerardgerard.martin@state.ma.us, Bridgewater State College – Hw
Martin, James E., (605) 394-2427 James.Martin@sdsmt.edu, South Dakota School of Mines & Technology – Ps
Martin, John W., (408) 298-2181 San Jose City College – Og
Martin, Jonathan B., (352) 392-6219 jbmartin@.ufl.edu, Univ of Florida – ClHg
Martin, Jonathan E., (608)262-9845 jon@aos.wisc.edu, Univ of Wisconsin, Madison – Ow
Martin, Larry D., (785) 864-5639 ldmartin@ku.edu, Univ of Kansas – Pv
Martin, Paul F., (509) 376-2519 paul.martin@pnl.gov, Pacific Northwest National Lab – Cg
Martin, Paul S., (520) 629-9459 pmartin@geo.arizona.edu, Univ of Arizona – Pe
Martin, Randal S., (505) 835-5469 New Mexico Institute of Mining and Technology – Oa
Martin, Robert F., 514-398-7370 bobm@eps.mcgill.ca, McGill Univ – Gi
Martin, Ronald E., (302) 831-2569 daddy@udel.edu, Univ of Delaware – Pm
Martin, Scott C., 330-941-3026 scmartin@ysu.edu, Youngstown State Univ – Ge
Martin, Seelye, (206) 543-6438 seelye@ocean.washington.edu, Univ of Washington – Op
Martin, Susan P., (740) 376-4775 Marietta College – Oo
Martin, Timothy J., (630) 252-8708 tjmartin@anl.gov, Argonne National Lab – Oa
Martin, Walter E., 704-687-5954 wemartin@email.uncc.edu, Univ of North Carolina, Charlotte – Oga
Martin, Wayne D., (513) 529-1361 Miami Univ – Gs
Martin, William R., (508) 289-2836 wmartin@whoi.edu, Woods Hole Oceanographic Institution – Oc
Martin, III, Randolph J., (617) 552-2583 martinrk@bc.edu, Boston College – Gc
Martin-Hayden, James, (419) 530-2634 jhayden@geology.utoledo.edu, Univ of Toledo – Hq
Martinelli, Jason T., (402) 280-3523 jay@creighton.edu, Creighton Univ – Ow
Martinez, Fernando, (808) 956-6882 martinez@soest.hawaii.edu, Univ of Hawai'i, Manoa – Yr
Martinez, Gustavo, (809) 767-9705 Univ of Puerto Rico – Sc
Martinez-Hackert, Bettina, 716-878-6731 martinb@buffalostate.edu, Buffalo State College – Gvm
Martini, Anna M., (413) 542-2067 ammartini@amherst.edu, Amherst College – Cl
Martini, I. Peter, (519) 824-4120 (Ext. 52488) pmartini@lrs.uoguelph.ca, Univ of Guelph – Gs
Martino, Ronald L., (304) 696-2715 martinor@marshall.edu, Marshall Univ – Gs
Martinsen, Randi S., (307) 766-4858 randmart@uwyo.edu, Univ of Wyoming – Go
Martinson, Douglas G., 845-365-8830 dgm@ldeo.columbia.edu, Columbia Univ – Op
Martinson, Tom L., (334) 844-3412 martitl@auburn.edu, Auburn Univ – Oy
Marvinney, Robert G., (207) 287-2801 robert.v.marvinney@state.me.us, Univ of Maine – Gc
Marzen, Luke J., (334) 844-3462 marzelj@auburn.edu, Auburn Univ – Oy
Marzolf, John E., (618) 453-7372 marzolf@geo.siu.edu, Southern Illinois Univ Carbondale – Gs
Marzulli, Walter, (609) 292-2576 walter.marzulli@dep.state.nj.us, New Jersey Geological Survey – Oy
Masahide, Akasaka, +81-852-32-6455 akasaka@riko.shimane-u.ac.jp, Shimane Univ – Gz
Masch, Ludwig, 089/2180 4273 masch@min.uni-muenchen.de, LMU - Univ of Munich – Gz
Maslowski, Wieslaw, (831) 656-3162 maslowsk@nps.navy.mil, Naval Postgraduate School – Op
Mason, Allen S., (505) 667-4140 Los Alamos National Lab – Oa
Mason, Brian H., (202) 633-1804 masonb@si.edu, Smithsonian Institution / National Museum of Natural History – Xm
Mason, Charles E., 606-783-2166 c.mason@morehead-st.edu, Morehead State Univ – Pi
Mason, Jane, (510) 643-9745 Univ of California, Berkeley – Oo
Mason, Owen, (907) 474-6293 Univ of Alaska, Fairbanks – Ga
Mason, Robert P., (410) 326-7387 mason@cbl.umces.edu, Univ of Maryland – Ct
Mason, Roger A., (709) 737-4385 rmason@sparky2.esd.mun.ca, Memorial Univ of Newfoundland – Gz
Massare, Judy A., (585) 395-2419 jmassare@weather.brockport.edu, SUNY College at Brockport – Pv
Massey, Christine A., (802) 656-1344 cmassey@zoo.uvm.edu, Univ of Vermont – Gg
Massey, Julia E., 205 652 3414 jem@uwa.edu, Univ of West Alabama – Oo
Massey, Yvonne, 205/247-3687 ymassey@gsa.state.al.us, Geological Survey of Alabama – Oo
Mastalerz, Maria, (812) 855-5412 mmastale@indiana.edu, Indiana Geological Survey – Ec
Mastalerz, Maria D., 812-855-7416 mmastale@indiana.edu, Indiana Univ, Bloomington – Ec
Masterllark, Timothy L., 205-348-6289 masterlark@geo.ua.edu, Univ of Alabama – GqvYd
Masters, T. Guy, (619) 534-4122 tmasters@ucsd.edu, Univ of California, San Diego – Ys
Matano, Ricardo P., (541) 737-2212 rmatano@coas.oregonstate.edu, Oregon State Univ – Op
Matchen, David L., (304) 384-5315 dmatchen@concord.edu, Concord Univ – Gr

Mateer, Richard S., (606) 257-8032 Univ of Kentucky – Nm
Matheney, Ronald K., (701) 777-4569 ronald_matheney@und.edu, Univ of North Dakota – Cs
Matheny, James L., (601) 605-1507 Mississippi Office of Geology – Oo
Matheson, Richard E., (727) 896-8626 Univ of South Florida, Saint Petersburg – Ob
Mathewson, Christopher C., (979) 845-2488 mathewson@geo.tamu.edu, Texas A&M Univ – Ng
Mathez, Edmond A., (212) 769-5379 Graduate School of the City Univ of New York – Gi
Mathez, Edmond A., 212-769-5379 mathez@amnh.org, American Museum of Natural History – Gx
Mathison, Mark E., (515) 294-9686 mathison@iastate.edu, Iowa State Univ of Science & Technology – Oe
Mathur, Ryan, 814-641-3725 mathur@juniata.edu, Juniata College – Gc
Matisoff, Gerald, (216) 368-3677 gerald.matisoff@case.edu, Case Western Reserve Univ – Cl
Matney, Timothy, 330-972-6892 matney@uakron.edu, Univ of Akron – Ga
Matocha, JohnJMatocha@ag.tamu.edu, Texas A&M Univ – Sc
Matsch, Charles L., (218) 726-7586 cmatsch@d.umn.edu, Univ of Minnesota, Duluth – Gl
Matson, Pamela A., (650) 725-6812 matson@pangea.stanford.edu, Stanford Univ – Sb
Matsumoto, Katsumi, 612-624-0275 katsumi@umn.edu, Univ of Minnesota, Twin Cities – Oc
Matsuoka, Kenichi, 206-685-2887 matsuoka@ess.washington.edu, Univ of Washington – Ol
Matsuyama, Isamu, (202) 478-8863 matsuyama@dtm.ciw.edu, Carnegie Institution of Washington – Oo
Mattheus, Paul E., (907) 474-7758 Univ of Alaska, Fairbanks – Po
Matthews, Debra, (208)426-1561 debramatthews@boisestate.edu, Boise State Univ – Gg
Matthews, Robert A., (916) 752-0179 Univ of California, Davis – Ge
Matthews, Robley K., (401) 863-2757 rkm@brown.edu, Brown Univ – Gr
Matthews, Terri, (804) 427-7189 Tidewater Community College – Cg
Matthews, Vince, (303) 866-3028 vince.matthews@state.co.us, Colorado Geological Survey – Gc
Matthias, Allan D., (520) 621-7226 matthias@ag.arizona.edu, Univ of Arizona – Oa
Mattice, Jack S., (516) 632-6905 SUNY, Stony Brook – Ob
Mattigod, Shas V., (509) 376-4311 shas.mattigod@pnl.gov, Pacific Northwest National Lab – Cl
Mattinson, James M., (805) 893-3219 Univ of California, Santa Barbara – Cc
Mattioli, Glen S., (479) 575-7295 mattioli@uark.edu, Univ of Arkansas, Fayetteville – Gv
Mattison, Marcia E., (916) 322-9579 California Geological Survey – Gg
Mattison, Rebecca, (781) 283-3188 rmattison@wellesley.edu, Wellesley College – Pv
Mattox, Stephen R., 616-331-3734 mattoxs@gvsu.edu, Grand Valley State Univ – Oe
Mattox, Tari N., 616-331-3728 Grand Valley State Univ – Gv
Mattson, Peter H., (718) 997-3300 Graduate School of the City Univ of New York – Gc
Mattson, Peter H., (718) 997-3335 mattson@qcvaxa.acc.qc.edu, Queens College (CUNY) – Gc
Matty, David J., (989) 774-3179 matty1dj@cmich.edu, Central Michigan Univ – Gi
Matty, Jane M., (989) 774-3179 matty1jm@cmich.edu, Central Michigan Univ – Cl
Matyjasik, Basia, (801) 537-3122 nrugs.bmatyjas@state.ut.us, Utah Geological Survey – Gg
Matyjasik, Marek, (801) 626-7726 mmatyjasik@weber.edu, Weber State Univ – Hg
Matzke, David, (313) 593-5036 dmatzke@umich.edu, Univ of Michigan, Dearborn – Xy
Matzke, Gordon E., (541) 737-1217 matzkeg@geo.oregonstate.edu, Oregon State Univ – Og
Mauger, Richard L., (252) 328-6016 maugerr@mail.ecu.edu, East Carolina Univ – Gi
Maul, George A., 321-674-8096 gmaul@fit.edu, Florida Institute of Technology – Op
Maulud, Mat Ruzlin, 03-79674144 matruzlin@um.edu.my, Univ of Malaya – Oo
Maurice, Patricia A., (219) 631-5380 Patricia.A.Maurice.3@nd.edu, Univ of Notre Dame – Ge
Maurrasse, Florentin J-M. R., (305) 348-2350 maurrass@fiu.edu, Florida International Univ – Ps
Mausel, Paul W.gemause@isugw.indstate.edu, Indiana State Univ – Or
Mavko, Gerald M., (650) 723-9438 Stanford Univ – Yg
Maxey, George F., (940) 565-2372 maxey@unt.edu, Univ of North Texas – Gg
Maxey, Susan M., (214) 620-4750 Brookhaven College – Guo
Maxson, Julie A.jmaxson@gac.edu, Gustavus Adolphus College – Gs
Maxson, Kathleenmaxson@nova.edu, Nova Southeastern Univ – Oo
Maxwell, Arthur E., 512-471-0411 art@ig.utexas.edu, Univ of Texas at Austin, Jackson School of Geosciences – Og
Maxwell, John C., (512) 471-5355 maribemax@aol.com, Univ of Texas at Austin, Jackson School of Geosciences – Gt
Maxwell, Keely B., 717 358-4549 keely.maxwell@fandm.edu, Franklin and Marshall College – Oy
Maxwell, Reed M., (925)422-7436 maxwell5@llnl.gov, Lawrence Livermore National Lab – Ng
Maxwell, Ted A., 202-633-2356 Smithsonian Institution / National Air & Space Museum – Xg
May, Bridgett, 228-688-3175 Bridgett.May@usm.edu, Univ of Southern Mississippi – Oo
May, Diane M., (417) 836-6900 DianeMay@smsu.edu, Southwest Missouri State Univ – Oo
May, Fred E., (801) 581-8218 pscem.fmay@state.ut.us, Univ of Utah – Oo
May, James, (662) 325-5806 Mississippi State Univ – Hw
May, Michael, (502) 745-4555 Western Kentucky Univ – Ge
Maybin, III, Arthur H., (803) 896-7708 South Carolina Geological Survey – Eg
Mayborn, Kyle R., (309) 298-1577 Western Illinois Univ – Gp
Mayer, Alex S., (906) 487-3372 asmayer@mtu.edu, Michigan Technological Univ – Hy
Mayer, Bernhard, (403) 220-5389 bmayer@ucalgary.ca, Univ of Calgary – Cs
Mayer, Christine M., (419) 530-8377 christine.mayer@utoledo.edu, Univ of Toledo – Oo
Mayer, James R., (678) 839-4055 jmayer@westga.edu, Univ of West Georgia – Hw
Mayer, Larrylmayer@geo.arizona.edu, Univ of Arizona – Gm
Mayer, Larry A., 603-862-2615 larry.mayer@unh.edu, Univ of New Hampshire – Gu
Mayer, Lawrence M., (207) 563-3146 Univ of Maine – Oc
Mayer, Paul S., (414) 278-6156 mayer@mpm.edu, Milwaukee Public Museum – Pi
Mayer, Ulrich, (604) 822-1539 umayer@eos.ubc.ca, Univ of British Columbia – Nm
Mayer, Victor J., 970-351-2647 mayer.4@osu.edu, Univ of Northern Colorado – Oe
Mayewski, Paul A., (207) 581-3019 paul.mayewski@maine.edu, Univ of Maine – Pe
Mayfield, Matthew G., (918) 540-6266 Northeastern Oklahoma A&M College – Oo
Mayfield, Michael W., (828) 262-7058 mayfldmw@appstate.edu, Appalachian State Univ – Hs
Maykut, Gary A.maykut@atmos.washington.edu, Univ of Washington – OalYr
Maynard, J. Barry, (513) 556-5034 maynarjb@uc.edu, Univ of Cincinnati – Cl
Maynor, Paula, (919) 733-2423 North Carolina Geological Survey – Oo
Mayo, Alan L., 801-422-2338 alan_mayo@byu.edu, Brigham Young Univ – Hw
Mayo, David P., (323) 343-2420 dmayo@calstatela.edu, California State Univ, Los Angeles – Oe
Mazariegos, Ruben A., (210) 381-3523 Univ of Texas, Pan American – Ye
Mazer, James J., (630) 252-7362 Argonne National Lab – Cl
Mazur, April, (724) 357-2379 amazur@iup.edu, Indiana Univ of Pennsylvania – Oo
Mazzella, Prof. Antonio A., 070-675-5542 mazzella@unica.it, Universita di Cagliari – Gq
Mazzullo, Salvatore J., (316) 978-3140 Wichita State Univ – Go
Mbila, Monday O., 256-372-4185 monday.mbila@aamu.edu, Alabama A&M Univ – Sd
Mc Kinley, Galen, 608-262-4817 galen@aos.wisc.edu, Univ of Wisconsin, Madison – Ob
Mc Mahon, Michelle, 281-765-7865 mmckubota@aol.com, North Harris College – Go
McAdoo, Brian G., (845) 437-7703 brmcadoo@vassar.edu,

Vassar College – Yr
McAfee, Gerald B., (432) 335-6558 bmcafee@odessa.edu, Odessa College – Pi
McAlester, A. Lee, (214) 768-2623 Southern Methodist Univ – Pe
McAlice, Bernard J., (207) 563-3146 Univ of Maine – Ob
McAllister, Arnold L., (506) 453-4804 Univ of New Brunswick – Em
McAndrews, John H., (416) 586-5609 Univ of Toronto – Pl
McBirney, Alexander R., (541) 344-2539 Univ of Oregon – Gv
McBride, Earle F., (512) 471-1905 efmcbride@mail.utexas.edu, Univ of Texas at Austin, Jackson School of Geosciences – Gd
McBride, John H., 801 422 5219 john_mcbride@byu.edu, Brigham Young Univ – Ye
McBride, Randolph A., (703) 993-1642 rmcbride@gmu.edu, George Mason Univ – On
McBride, Raymond G., (519) 824-4120 (Ext. 52492) rmcbride@lrs.uoguelph.ca, Univ of Guelph – Ou
McBride, Sandra L., (613) 545-6664 Queen's Univ – Gg
McCaffrey, Kenneth J., 0191 3742523 k.j.mccaffrey, Durham Univ – Gc
McCaffrey, Robert, (518) 276-8521 mccafr@rpi.edu, Rensselaer Polytechnic Institute – Yg
McCall, Earla L.emccall@clarion.edu, Clarion Univ – Oo
McCall, Peter L., (216) 368-3676 plm4@case.edu, Case Western Reserve Univ – Po
McCallum, I. Stewart (Stu), 206-543-9494 mccallum@u.washington.edu, Univ of Washington – Gi
McCallum, Malcolm E., (970) 491-6250 Colorado State Univ – Gz
McCallum, Robert J., (201) 595-3462 William Paterson College – Oo
McCarley, J.Kendrick, (601) 961-5515 ken_mccarley@deq.state.ms.us, Mississippi Office of Geology – Ng
McCarter, Michael K., (801) 581-8603 Univ of Utah – Nm
McCarthy, Daniel P., (905) 688-5550 (Ext. 3864) dmccarth@brocku.ca, Brock Univ – Gl
McCarthy, Francine G., (905) 688-5550 (Ext. 4286) Francine@brocku.ca, Brock Univ – Pm
McCarthy, Gregory J., (701) 231-7193 Greg.McCarthy@ndsu.edu, North Dakota State Univ – Gz
McCarthy, James J., (617) 495-2330 jmccarthy@oeb.harvard.edu, Harvard Univ – Ob
McCarthy, James M., (206) 685-2543 Lockheed Martin Idaho Technologies – Oo
McCarthy, Michael P., 206-685-2543 mccarthy@ess.washington.edu, Univ of Washington – XyYx
McCarthy, Paul, (907) 474-6894 ffpjm@uaf.edu, Univ of Alaska, Fairbanks – Gs
McCartney, Daryl M., (204) 474-6558 mccartn@cc.umanitoba.ca, Univ of Manitoba – Oo
McCartney, Kevin, (207) 768-9482 mccartney@polaris.umpi.maine.edu, Univ of Maine, Presque Isle – Pm
McCaslin, Bobby D., (505) 646-3405 New Mexico State Univ, Las Cruces – Sc
McCauley, James R., (785) 864-2192 jim_mccauley@kgs.ku.edu, Univ of Kansas – Or
McCauley, Marlene, (336) 316-2236 mmcauley@guilford.edu, Guilford College – Cp
McChesney, Ronald J., 740-368- rjmcches@owu.edu, Ohio Wesleyan Univ – Og
McClain, James S., (530) 752-7093 mcclain@geology.ucdavis.edu, Univ of California, Davis – Yr
McClean, Julie L., (408) 656-2437 mcclean@oc.nps.navy.mil, Naval Postgraduate School – Op
McCleave, James D., (207) 581-4392 Univ of Maine – Ob
McClellan, Beth, (785) 864-2723 bethmc@ku.edu, Univ of Kansas – Gp
McClellan, Guerry H., (352) 392-3626 guerry@ufl.edu, Univ of Florida – En
McClelland, James W., (361) 749-6756 jimm@mail.utexas.edu, Univ of Texas at Austin – Ob
McClelland, William C., (208) 885-4704 wmcclell@uidaho.edu, Univ of Idaho – Gt
McClurg, Charles A., (301) 405-4342 cml9@umail.umd.edu, Univ of Maryland – Oo
McClurg, James E., (307) 766-3386 mcclurg@uwyo.edu, Univ of Wyoming – Gs
McCollum, Linda B., (509) 359-7473 lmccollum@mail.ewu.edu, Eastern Washington Univ – Po
McConaugha, John R., (757) 683-4698 jmcconau@odu.edu, Old Dominion Univ – Ob
McConchie, David M., 61 2 6620 3632 Southern Cross Univ – Ge
McConnell, David A., (330) 972-8047 dam6@uakron.edu, Univ of Akron – Gc
McConnell, John (Jack) C., (416) 736-2100 (Ext. 77709) jack@nimbus.yorku.ca, York Univ – Oa
McConnell, Joseph, (775) 673-7348 jmcconn@dri.edu, Univ of Nevada, Reno – Oo
McConnell, Robert L.rmconnel@mwc.edu, Univ of Mary Washington – Ge
McConnell, Teresa K., (814) 269-2944 tkm2@pitt.edu, Univ of Pittsburgh at Johnstown – Gg
McConnell, Vicki S., (541) 523-3133 Vicki.McConnell@state.or.us, Oregon Dept of Geology & Mineral Industries – Oo
McCook, Anthony E., 404-656-3214 Georgia Dept of Natural Resources – Oo
McCord, James T., (505) 844-5157 jtmccor@sandia.gov, New Mexico Institute of Mining and Technology – Hq
McCorkle, Daniel C., (508) 289-2949 dmccorkle@whoi.edu, Woods Hole Oceanographic Institution – Cm
McCormack, John K., (775) 784-4518 mccormac@mines.unr.edu, Univ of Nevada, Reno – Gz
McCormick, George R.erie-@msn.com, Univ of Iowa – Gz
McCormick, Kelli A., (605) 677-5227 kmccormi@usd.edu, South Dakota Dept of Env & Natural Resources – Gg
McCormick, Michael L., 315-859-4832 mmccormi@hamilton.edu, Hamilton College – Py
McCorquodale, Donald, (954) 920-1909 Nova Southeastern Univ – Oc
McCoy, Edward L., (330) 263-3884 mccoy.13@osu.edu, Ohio State Univ – Sp
Mccoy, Roger M., (801) 581-8218 Univ of Utah – Or
McCoy, Timothy J., (202) 633-2206 mccoyt@si.edu, Smithsonian Inst / National Museum of Natural History – Xmg
McCoy, William D., (413) 545-1535 wdmccoy@geo.umass.edu, Univ of Massachusetts – Oy
McCoy, Jr., Floyd W., (808) 235-7318 Univ of Hawai'i, Manoa – Gu
McCraw, David J., (505) 835-5594 djmc@mailhost.nmt.edu, New Mexico Institute of Mining & Technology – Oy
McCray, John E., (303) 384-2181 jmccray@mines.edu, Colorado School of Mines – Hw
McCreary, Julian P., (808) 956-2216 jay@soest.hawaii.edu, Univ of Hawai'i, Manoa – Op
McCrink, Timothy P., (916) 324-2549 California Geological Survey – Ng
McCulley, Dawn, (209) 667-3466 dawn@physics.csustan.edu, California State Univ, Stanislaus – Oo
McCulloch, Debia F.debiamcculloch@aol.com, Georgia Perimeter College – Gg
McCulloch, Robin B., 406-496-4171 rmcculloch@mtech.edu, Montana Tech of The Univ of Montana – Nm
McCulloh, Richard P., (225) 578-5327 mccullo@lsu.edu, Louisiana State Univ – Gg
McCullough, Jr., Edgar J., (520) 621-6024 Univ of Arizona – Ge
McCurdy, Maureen, (318) 257-3165 mm@coes.latech.edu, Louisiana Tech Univ – Hw
McCurry, Michael O., (208) 282-3960 mccumich@isu.edu, Idaho State Univ – Gz
McDade, Ian C., (416) 736-2100 (Ext. 40124) mcdade@yorku.ca, York Univ – Or
McDaniel, Paul A., (208) 885-7012 Univ of Idaho – Sd
McDermott, Thomas M., (585) 395-5718 SUNY College at Brockport – Oo
McDonald, Andrew M., (705) 675-1151 (Ext. ext. 2266) amcdonald@laurentian.ca, Laurentian Univ, Sudbury – Gz
McDonald, Eric, (775) 673-7302 emcdonald@dri.edu, Univ of Nevada, Reno – Sp
McDonald, Eric, (775) 673-7302 emcdonald@dri.edu, Desert Research Institute – Sd
McDonald, Gregory N., (801) 537-3383 nrugs.gmcdonal@state.ut.us, Utah Geological Survey – Oo
McDonald, James, (614) 265-6601 jim.mcdonald@dnr.state.oh.us, Ohio Dept of Natural Resources – Yg
McDonald, John A.mcdonald@geophy.curtin.edu.au, Curtin Univ of Technology – Yg
McDonald, John P., (509) 373-0362 john.p.mcdonald@pnl.gov, Pacific Northwest National Lab – Hw
McDonald, Kyle C., (818) 354-3440 kyle.mcdonald@jpl.nasa.gov, Jet Propulsion Lab – Oo
McDonald, Linda L., (612)627-4780 Ext. 213 l-mcdo@umn.edu, Univ of Minnesota, Twin Cities – Oo
McDonald, Lynn, (505) 667-1582 lmcdonald@lanl.gov, Los Alamos National Lab – Oo
McDonnell, Debbie, (603) 271-4086 dmcdonnell@des.state.nh.us,

New Hampshire Geological Survey – Oo
McDonnell, R. E., (520) 770-3500 Arizona Geological Survey – Oo
McDonough, William F., (301) 405-5561 mcdonoug@geol.umd.edu, Univ of Maryland – Cg
McDowell, Fred W., (512) 471-1672 mcdowell@mail.utexas.edu, Univ of Texas at Austin, Jackson School of Geosciences – Cc
McDowell, Patricia F., (541) 346-4567 Univ of Oregon – Gm
McDowell, Ronald, 304 594 2331 mcdowell@geosrv.wvnet.edu, West Virginia Univ – Gg
McDowell, S. Douglas, (906) 487-2531 sdmcdowe@mtu.edu, Michigan Technological Univ – Gx
McDuff, Russell E., (206) 545-1947 mcduff@ocean.washington.edu, Univ of Washington – Oc
McElroy, Anne, (631) 632-8488 amcelroy@notes.cc.sunysb.edu, SUNY, Stony Brook – Oc
McElroy, Michael B., (617) 495-9261 mbm@io.harvard.edu, Harvard Univ – Oa
McElroy, Thomas A., 717.702.2046 tmcelroy@state.pa.us, Pennsylvania Bureau of Topographic & Geologic Survey – Hw
McElwain, Jenny, (312) 665-7635 jmcelwain@fieldmuseum.org, Field Museum of Natural History – Pb
McElwee, Carl D., (785) 864-3965 cmcelwee@ku.edu, Univ of Kansas – Hg
McEnroe, Maryann, (914) 251-6676 SUNY, Purchase – Oc
McEvilly, Thomas V., (510) 486-7347 Lawrence Berkeley Nat Lab – Ys
McEwen, Alfred, (520) 621-4573 mcewen@lpl.arizona.edu, Univ of Arizona – Xg
McFadden, Bruce J., (352) 846-2000 bmacfadd@flmnh.ufl.edu, Univ of Florida – Pv
McFadden, Leslie M., (505) 277-6121 ldmcfadnm.unm.edu, Univ of New Mexico – Sd
McFarland, Mark L., (979) 845-5366 ml-mcfarland@tamu.edu, Texas A&M Univ – Sc
McFarland, III, John D., 501 683-0111 john.mcfarland@Arkansas.gov, Arkansas Geological Commission – Gg
McFarlane, Norman, (250) 363-8227 norm.mcfarlane@ec.gc.ca, Univ of Victoria – Oa
McFarquhar, Greg M., (217) 265-5458 mcfarq@atmos.uiuc.edu, Univ of Illinois, Urbana-Champaign – Or
McGarvin, Thomas G., (520) 770-3500 Arizona Geological Survey – Gg
McGeary, Susan, (302) 831-8174 smcgeary@udel.edu, Univ of Delaware – Yg
McGee, Kathleen, (908) 527-2518 kathleen.mcgee@state.or.us, Oregon Dept of Geology & Mineral Industries – Oo
McGee, Thomas M., (662) 915-7320 tmm@mmri.olemiss.edu, Univ of Mississippi – Ys
McGehee, Thomas L., (512) 595-3590 kftlm00@tamuk.edu, Texas A&M Univ, Kingsville – Cl
McGhee, Jr., George R., (732) 445-3832 mcghee@rci.rutgers.edu, Rutgers, The State Univ of New Jersey – Po
McGill, George E., (413) 545-0140 Univ of Massachusetts – Gc
McGill, Sally F., (909) 537-5347 smcgill@csusb.edu, California State Univ, San Bernardino – Gt
McGinn, Patrick J., (609) 258-7438 pmcginn@princeton.edu, Princeton Univ – Cm
McGinnis, Lyle D., (630) 252-8722 Argonne National Lab – Yg
McGinnis, Lyle D., (708) 972-8722 Northern Illinois Univ – Yg
McGlathery, Karen, (804) 924-0558 kjm4k@virginia.edu, Univ of Virginia – Ob
McGlure, Kevin P., (801) 537-3374 nrugs.kmcclure@state.ut.us, Utah Geological Survey – Og
McGoldrick, Peter J., 61 3 6226 7209 Univ of Tasmania – Ce
McGowan, Chris, (416) 586-5753 chrism@rom.on.ca, Royal Ontario Museum – Pv
McGowan, John A., 858-534-2074 jmcgowan@ucsd.edu, Univ of California, San Diego – Ob
McGrail, Bernard P., (509) 376-9193 pete.mcgrail@pnl.gov, Pacific Northwest National Lab – Cg
McGrath, Steve, 406-496-4767 smcgrath@mtech.edu, Montana Tech of The Univ of Montana – Ca
McGrath, Jr., Dorn C., (202) 994-6185 dornmcg@gwu.edu, George Washington Univ – Oy
McGraw, Maureen A., (505) 665-8128 mcgraw@lanl.gov, Los Alamos National Lab – Oo
McGreevy, Patrick V.pmcgreev@clarion.edu, Clarion Univ – Oo
McGregor, Kent M., (940) 565-2380 mcgregor@unt.edu, Univ of North Texas – Oa
McGregor, Stuart W., 205/247-3629 smcgregor@gsa.state.al.us, Geological Survey of Alabama – Oo
McGrew, Allen J., (937) 229-3455 Univ of Dayton – Gc
McGrew, Heidi, (937) 229-2261 Heidi.McGrew@notes.udayton.edu, Univ of Dayton – Yg
McGuire, Jeffrey J., (508) 289-3290 jmcguire@whoi.edu, Woods Hole Oceanographic Institution – Ys
McGuire, Jennifer T., 979-845-4520 mcguire@geo.tamu.edu, Texas A&M Univ – Oo
McGuirk, James P., (979) 845-4431 mcguirk@ariel.tamu.edu, Texas A&M Univ – Ow
McGwire, Kenneth C., (775) 673-7324 kenm@dri.edu, Desert Research Institute – Oy
McHargue, Lanny R., (520) 621-6825 mchargue@physics.arizona.edu, Univ of Arizona – Cg
McHenry, Lindsay J., 414-229-3951 lmchenry@uwm.edu, Univ of Wisconsin, Milwaukee – Gz
McHone, J. Gregory, (860) 486-1391 gregory.mchone@uconn.edu, Univ of Connecticut – Gi
McHone, Nancy W., (860) 424-3591 nancy.mchone@po.state.ct.us, Geological and Natural History Survey of Connecticut – Gg
McHugh, Cecilia M., (718) 997-3322 Graduate School of the City Univ of New York – Ou
McIlvin, Matt, 508-289-2884 mmcilvin@whoi.edu, Woods Hole Oceanographic Institution – Oc
McInnes, Kevin J., (979) 845-5986 k-mcinnes@tamu.edu, Texas A&M Univ – Sp
McIntosh, Jennifer C., (520) 626-2282 mcintosh@hwr.arizona.edu, Univ of Arizona – Cg
McIntosh, Kirk D., (512) 471-0480 kirk@ig.utexas.edu, Univ of Texas at Austin, Jackson School of Geosciences – Yr
McIntosh, Thomas H., (414) 465-2747 Univ of Wisconsin, Green Bay – So
McIntosh, William C., (505) 835-5324 mcintosh@nmt.edu, New Mexico Institute of Mining & Technology – Cc
McIntosh, William C., (505) 835-5324 mcintosh@nmt.edu, New Mexico Institute of Mining and Technology – Cc
McIntyre, Andrew, (718) 997-3329 Graduate School of the City Univ of New York – Pe
McIntyre, Marcella R., 205-247-3654 mmcintyre@gsa.state.al.us, Geological Survey of Alabama – Gg
McIsaac, Gregory F., (217) 333-9411 Univ of Illinois, Urbana-Champaign – Hs
McKay, Dawn, (860) 424-3592 dawn.mckay@po.state.ct.us, Geological and Natural History Survey of Connecticut – Oo
McKay, E. Donald, (217) 333-0044 mckay@isgs.uiuc.edu, Illinois State Geological Survey – Gl
McKay, Larry D., (865) 974-0821 lmckay@utk.edu, Univ of Tennessee, Knoxville – Hy
McKay, Robert R., 814-393-1895 mckay@clarion.edu, Clarion Univ – Oo
McKean, James, ((03)) 3-667- (Ext. 1 ext 7073) j.mckean@geol.canterbury.ac.nz, Univ of Canterbury – Gm
McKee, Brent A., (225) 851-2800 Louisiana State Univ – Cg
McKee, Christopher G., (505) 835-5647 xrf@gis.nmt.edu, New Mexico Institute of Mining & Technology – Gz
McKee, David A., (361) 825-2676 dmckee@tamucc.edu, Texas A&M Univ, Corpus Christi – Ob
McKee, James W., (414) 424-4460 mckee@athenet.net, Univ of Wisconsin, Oshkosh – Gr
McKee, Karen L., (225) 388-6429 Louisiana State Univ – Ob
McKeegan, Kevin D., (310) 825-3580 mckeegan@ess.ucla.edu, Univ of California, Los Angeles – Xc
McKenna, Malcolm C., (303) 499-3566 m4pmck@indra.com, Columbia Univ – Pv
McKenna, Thomas E., 302-831-2833 mckennat@udel.edu, Univ of Delaware – Hy
McKenney, Rosemary, 253-535-8726 mckennra@plu.edu, Pacific Lutheran Univ – Gm
McKenzie, Charlotte, 406-496-4180 cmckenzie@mtech.edu, Montana Tech of The Univ of Montana – Oo
McKenzie, Conniemckenzie@coes.latech.edu, Louisiana Tech Univ – Oo
McKenzie, Garry D., (614) 292-0655 mckenzie.4@osu.edu, Ohio State Univ – Gl
McKenzie, Jeffrey M., 514-398-3833 mckenzie@eps.mcgill.ca, McGill Univ – Hw
McKenzie, Phyllis, (202) 633-1860 mckenzie@si.edu, Smithsonian Institution / National Museum of Natural History – Oo
McKibben, Michael A., (951) 827-3444 michael.mckibben@ucr.edu, Univ of California, Riverside – Cg
McKinney, D. Brooks, (315) 781-3304 dbmck@hws.edu,

Hobart & William Smith Colleges – Gx
McKinney, Frank K., (828) 262-2748 mckinneyfk@appstate.edu, Appalachian State Univ – Pi
McKinney, Marg J., (828) 262-2747 mckinnymj@appstate.edu, Appalachian State Univ – Pg
McKinney, Michael L., (865) 974-6359 Univ of Tennessee, Knoxville – Po
McKinnon, William B., (314) 935-5604 Washington Univ – Xg
McKittrick, Mary Anne, (916) 323-8553 California Geological Survey – Oo
McKnight, Brian K., (414) 424-4460 mcknight@uwosh.edu, Univ of Wisconsin, Oshkosh – Gd
McLafferty, Sara L., (212) 772-5224 slm@everest.hunter.cuny.edu, Graduate School of the City Univ of New York – Oo
McLaughlin, Patrick I., (608) 262-8658 pimclaughlin@wisc.edu, Univ of Wisconsin, Extension – Gr
McLaughlin, Richard A., (919) 515-7306 North Carolina State Univ – Sc
McLaughlin, Jr., Peter P., (302) 831-8263 ppmclau@udel.edu, Univ of Delaware – Ps
McLaughlin, Jr., Peter P., 302-831-2833 ppmclau@udel.edu, Univ of Delaware – Gr
McLaurin, Brett T., 570-389-4142 bmclauri@bloomu.edu, Bloomsburg Univ – GrsGd
McLean, Dewey M., (540) 231-6521 dmclean@vt.edu, Virginia Polytechnic Institute & State Univ – Pe
McLelland, Jamesjmclelland@mail.colgate.edu, Colgate Univ – Gp
McLemore, Virginia, (505) 835-5521 ginger@gis.nmt.edu, New Mexico Institute of Mining & Technology – Em
McLemore, Virginia T., (505) 835-5521 ginger@gis.nmt.edu, New Mexico Institute of Mining and Technology – Eg
McLemore, William H., (404) 656-3214 Georgia Dept of Natural Resources – Ng
McLennan, Scott M., (631) 632-8194 scott.mclennan@sunysb.edu, SUNY, Stony Brook – Cg
McLeod, Clara P., (314) 935-5406 cpmcleod@library.wustl.edu, Washington Univ – Oo
McLeod, Samuel A., (213) 763-3325 smcleod@ref.usc.edu, Los Angeles County Museum of Natural History – Pv
McLoughlin, Stephen, 61 7 3138 2329 s.mcloughlin@qut.edu.au, Queensland Univ of Technology – Pg
McManus, Dean A., (206) 543-0587 mcmanus@u.washington.edu, Univ of Washington – Ou
McManus, George B., (860) 445-3474 mcmanus@uconnvm.uconn.edu, Univ of Connecticut – Ob
McManus, Jerry F., (508) 289-3328 jmcmanus@whoi.edu, Woods Hole Oceanographic Institution – Pe
McManus, Jim, (541) 737-8285 mcmanus@coas.oregonstate.edu, Oregon State Univ – GuCm
McManus, Margaret A., (831) 459-4736 Univ of California, Santa Cruz – On
McMechan, George A., (972) 883-2419 mcmec@utdallas.edu, Univ of Texas, Dallas – Ys
McMenamin, Mark, 413-538-2280 mmcmenam@mtholyoke.edu, Mount Holyoke College – PgGst
McMillan, Brock, (507) 389-3202 brock.mcmillan@mnsu.edu, Minnesota State Univ – So
McMillan, Jack R., (916) 322-2718 California Geological Survey – Ng
McMillan, Margaret E., (501) 569-3024 memcmillan@ualr.edu, Univ of Arkansas, Little Rock – Gm
McMillan, Nancy J., 575-646-5000 nmcmilla@nmsu.edu, New Mexico State Univ, Las Cruces – Gi
McMillan, Robert S., (520) 621-6968 bob@lpl.arizona.edu, Univ of Arizona – Oo
McMullin, David, (902) 585-1276 david.mcmullin@acadiau.ca, Acadia Univ – Oe
McMullin, Jim F., (601) 961-5529 jim_mcmullin@deq.state.ms.us, Mississippi Office of Geology – Gg
McMurtry, Gary M., (808) 956-6858 garym@soest.hawaii.edu, Univ of Hawai'i, Manoa – Cm
McNair, Laurie A., (505) 665-3328 mcnair@lanl.gov, Los Alamos National Lab – Oo
McNally, Karen C., (831) 459-4833 kmcnally@pmc.ucsc.edu, Univ of California, Santa Cruz – Ys
McNamara, James P., (208) 426-1354 jmcnamar@boisestate.edu, Boise State Univ – Hq
McNamara, Jodi, (315) 228-7201 jmcnamara@mail.colgate.edu, Colgate Univ – Oo
McNassor, Cathy, (213) 763-3389 mcnassor@bcf.usc.edu, Los Angeles County Museum of Natural History – Pv
McNaught, Mark A., (610) 250-5197 Mount Union College – Gc
McNeal, Brian L., (352) 392-1803 (Ext. 336) blm@gnv.ifas.ufl.edu, Univ of Florida – Sc
McNeal, Karen S., 662-325-9644 ksm163@msstate.edu, Mississippi State Univ – Co
McNeill, Donald F., 305-284-3360 d.mcneill@miami.edu, Univ of Miami – Gg
McNichol, Ann P., (508) 289-3394 amcnichol@whoi.edu, Woods Hole Oceanographic Institution – Oc
McNicol, Barbara, (403) 440-6175 bmcnicol@mtroyal.ca, Mount Royal College – Og
McNinch, Jesse E., (804) 684-7191 mcninch@vims.edu, College of William & Mary – Ou
McNulty, Brendan A., (310) 243-3412 bmcnulty@csudh.edu, California State Univ, Dominguez Hills – Gc
McNutt, Marciamcnutt@mbari.org, Univ of California, Santa Cruz – Yr
McNutt, Marcia K., 831-775-1814 mcnutt@mbari.org, Stanford Univ – Yr
McPhaden, Michael J., (206) 526-6783 mcphaden@pmel.noaa.gov, Univ of Washington – Oo
McPhail, D C "Bear", +61 2 6125 2776 bear@ems.anu.edu.au, Australian National Univ – Cl
McPhee-Shaw, Erika, (831) 771-4400 eshaw@mlml.calstate.edu, Moss Landing Marine Laboratories – Op
McPherron, Robert L., (310) 825-1882 rmcpherr@igpp.ucla.edu, Univ of California, Los Angeles – Xy
McPherron, Robert L., (310) 825-1882 rmcpherron@igpp.ucla.edu, Univ of California, Los Angeles – Xy
McPherson, Bob, (707) 826-5828 bob-mcpherson@redwoods.edu, College of the Redwoods – Ys
McPherson, Robert C., (707) 826-5828 rm4@humboldt.edu, Humboldt State Univ – Gt
McPhie, Jocelyn, 61 3 6226 2892 Jocelyn.McPhie@utas.edu.au, Univ of Tasmania – Gv
McQuarrie, Megan, mmcquarrie@cerritos.edu, Cerritos College – Gu
McQuarrie, Nadine, 609-258-7024 nmcq@princeton.edu, Princeton Univ – Gg
McQuarrie, Nadine S., (609) 258-7024 nmcq@princeton.edu, Princeton Univ – Gt
McRoberts, Christopher A., (607) 753-2925 mcroberts@cortland.edu, SUNY, Cortland – Pg
McSween, Jr., Harry Y., (865) 974-2366 mcsween@utk.edu, Univ of Tennessee, Knoxville – Xm
McSweeney, Kevin, (608) 262-0331 kmcsween@wisc.edu, Univ of Wisconsin, Madison – Sd
McWilliams, James C., (310) 206-2829 jcm@atmos.ucla.edu, Univ of California, Los Angeles – Oa
McWilliams, Michael O., (650) 723-3718 mcwilliams@stanford.edu, Stanford Univ – Ym
McWilliams, Robert G., (513) 529-2595 Miami Univ – Ps
Mead, James I., (928) 523-7184 james.mead@nau.edu, Northern Arizona Univ – Po
Mead, Judson, (812) 855-1008 Indiana Univ, Bloomington – Yg
Meador, Charles, 228-688-3520 Charles.Meador@usm.edu, Univ of Southern Mississippi – Oo
Meadows, Guy A., (313) 764-5235 Univ of Michigan – Op
Meadows, Wayne R., (505) 665-0291 wmeadows@lanl.gov, Los Alamos National Lab – Yg
Mealor, Jr, W. Theodore, 901-678-5393 wmealor@cc.memphis.edu, Univ of Memphis – Ou
Means, Winthrop D., (518) 442-4472 SUNY, Albany – Gc
Mears, Jr., Brainerd, (307) 766-3386 Univ of Wyoming – Gm
Measures, Christopher, (808) 956-5924 chrism@soest.hawaii.edu, Univ of Hawai'i, Manoa – Oc
Measures, Elizabeth A., (915) 837-8259 measures@sulross.edu, Sul Ross State Univ – Gd
Mechoso, Carlos R., (310) 825-3057 mechoso@atmos.ucla.edu, Univ of California, Los Angeles – Oa
Medaris, Gordon L., (608) 262-2708 medaris@geology.wisc.edu, Univ of Wisconsin, Madison – Gi
Medina, Michael A., (919) 733-2423 North Carolina Geological Survey – Oo
Meentemeyer, Ross, 704-687-5944 rkmeente@email.uncc.edu, Univ of North Carolina, Charlotte – Oi
Meere, Pat, +353 21 4903056 p.meere@ucc.ie, Univ College Cork – Gc
Meert, Joseph G., (352) 392-2231 Univ of Florida – YgGt
Meertens, Charles M., (303) 497-8011 Univ of Utah – Gt
Megdal, Sharon B., (520) 792-9591 smegdal@ag.arizona.edu, Univ of Arizona – Hg
Mehegan, James M., (909) 222-8350 james.mehegan@rcc.edu,

Riverside Community College – Gi
Mehnert, Edward, (217) 244-2765 Illinois State Geological Survey – Hw
Mehrtens, Charlotte J., (802) 656-0243 cmehrten@zoo.uvm.edu, Univ of Vermont – Gs
Meigs, Andrew J., (541) 737-1214 meigsa@geo.oregonstate.edu, Oregon State Univ – Gc
Meinert, Lawrence D., (413) 585-2657 lmeinert@email.smith.edu, Smith College – Em
Meintzer, Robert E., 205/247-3594 rmeintzer@gsa.state.al.us, Geological Survey of Alabama – Cg
Meisterernst, Götz, 089/2180 4356 goetz.meisterernst@lrz.uni-muenchen.de, LMU - Univ of Munich – Gz
Meixner, Thomas, 520-626-1532 tmeixner@hwr.arizona.edu, Univ of Arizona – Cg
Mejia, Sergio, (204) 474-9452 mejia@cc.umanitoba.ca, Univ of Manitoba – Cg
Mekeel, Deborah, (850) 488-9380 Florida Geological Survey – Oo
Mekik, Figen A., 616-331-3020 mekikf@gvsu.edu, Grand Valley State Univ – Pe
Meko, David M., (520) 621-3457 dmeko@ltrr.arizona.edu, Univ of Arizona – Hq
Melas, Faye, (212) 772-5265 ffmelas@yahoo.com, Hunter College (CUNY) – Gs
Melbourne, Timothy I., (509) 963-2799 tim@geology.cwu.edu, Central Washington Univ – Ys
Melchin, Michael, (902) 867-5177 Dalhousie Univ – Pg
Melchin, Michael J., (902) 867-5177 mmelchin@stfx.ca, Saint Francis Xavier Univ – Pi
Melchiorre, Erik, (909) 537-7754 emelch@csusb.edu, California State Univ, San Bernardino – Cs
Meldahl, Keith H., (760) 757-2121 (Ext. ext. 6412) kmeldahl@miracosta.edu, MiraCosta College – Gg
Melim, Leslie A., (309) 298-1377 Western Illinois Univ – Gd
Mellem, Karen J., (612) 625-2208 kmellem@soils.umn.edu, Univ of Minnesota, Twin Cities – Oo
Mellors, Robert J., (619) 594-3455 rmellors@geology.sdsu.edu, San Diego State Univ – Ye
Melosh, Henry J., (520) 621-2806 jmelosh@lpl.arizona.edu, Univ of Arizona – Gt
Melson, William G., (202) 633-1809 melsonw@si.edu, Smithsonian Inst / National Museum of Natural History – GvtGa
Meltzer, Anne S., (610) 758-3673 asm3@lehigh.edu, Lehigh Univ – Ys
Melville, W. Kendall, (858) 534-3208 gradchair@sio.ucsd.edu, Univ of California, San Diego – Op
Menard, Richardrichard.menard@ec.gc.ca, McGill Univ – Oa
Mendelson, Carl V., (608) 363-2223 mendelsn@beloit.edu, Beloit College – Pm
Mendelssohn, Irving A., (225) 388-6425 Louisiana State Univ – Ob
Mendoza, Carl A., (780) 492-2664 carl.mendoza@ualberta.ca, Univ of Alberta – Hw
Mendoza, Luis H.lmendoza@cicese.mx, Centro de Inv Científica y de Educación Superior de Ensenada – Ne
Meng, Jin, (212) 496-3337 jmeng@amnh.org, Graduate School of the City Univ of New York – Pv
Menke, William H., (845) 365-8438 menke@ldeo.columbia.edu, Columbia Univ – Ys
Menking, Kirsten M., (845) 437-5545 kimenking@vassar.edu, Vassar College – Gm
Mennel, Bill, 609-984-6587 bill.mennel@dep.state.nj.us, New Jersey Geological Survey – Hw
Menninga, Clarence, (616) 526-7093 menn@calvin.edu, Calvin College – CcGg
Menon, Manoj, (775) 784-4509 mmenon@unr.edu, Univ of Nevada, Reno – Gg
Mensing, Scott A., (775) 784-6346 smensing@unr.edu, Univ of Nevada, Reno – Oy
Mensoian, Michael G., (617) 287-5284 Univ of Massachusetts, Boston/ Harbor Campus – Oy
Mentzer, Judy P., (717) 477-1685 jpment@wharf.ship.edu, Shippensburg Univ – Oo
Menzel, W. Paul, (608)263-4930 paulm@ssec.wisc.edu, Univ of Wisconsin, Madison – Or
Menzies, John, (905) 688-5550 (Ext. 3865) jmenzies@brocku.ca, Brock Univ – Gl
Mercer, James A., 206-543-1361 mercer@apl.washington.edu, Univ of Washington – GuYr
Mercer, Richard B., (509) 376-8945 rb.mercer@pnl.gov, Pacific Northwest National Lab – Hw
Merchant, James W., (402) 472-7531 jmerchant@unl.edu, Unversity of Nebraska - Lincoln – Oru
Merck, Jr., John W., (301) 405-2808 jmerck@wam.umd.edu, Univ of Maryland – Pv
Meretsky, Vicky J., (812) 855-5971 meretsky@indiana.edu, Indiana Univ, Bloomington – Sf
Mereu, Robert F., (519) 661-3605 a424@uwo.ca, Univ of Western Ontario – Ys
Merguerian, Charles M., 516 463-5567 geocmm@hofstra.edu, Hofstra Univ – Gc
Merifield, Paul M., (310) 794-5019 pmerifie@ucla.edu, Univ of California, Los Angeles – Ng
Merino, Enrique, (812) 855-5088 merino@indiana.edu, Indiana Univ, Bloomington – Cl
Merk, George P.merk@msu.edu, Michigan State Univ – Gs
Merkel, Timo Casjen, 089/2180 4337 casjen.merkel@lrz.uni-muenchen.de, LMU - Univ of Munich – Gz
Merrell, William J., (409) 740-4400 Texas A&M Univ – Op
Merriam, Daniel F., (785) 864-2127 merriam@kgs.ku.edu, Univ of Kansas – Gg
Merriam, James B., (306) 966-5716 jim.merriam@usask.ca, Univ of Saskatchewan – Yg
Merrifield, Mark A., (808) 956-6161 markm@soest.hawaii.edu, Univ of Hawai'i, Manoa – Oo
Merrill, Glen K., (713) 221-8168 merrillg@uhd.edu, Univ of Houston Downtown – Pi
Merrill, John T., (401) 874-6715 jmerrill@boreas.gso.uri.edu, Univ of Rhode Island – Oc
Merrill, Robert D., (559) 278-4557 robert_merrill@csufresno.edu, California State Univ, Fresno – Gm
Merrill, Ronald T., 206-685-3829 ron@ess.washington.edu, Univ of Washington – Ygm
Merritt, Tonya M., (919) 733-2423 (Ext. 406) Tonya.Merritt@ncmail.net, North Carolina Geological Survey – Yr
Merritts, Dorothy J., (717) 291-4398 dorothy.merritts@fandm.edu, Franklin and Marshall College – Gm
Merschat, Carl E., (704) 251-6208 North Carolina Geological Survey – Gg
Mertz, Karl A., 916-278-6337 mertz@csus.edu, California State Univ, Sacramento – Gr
Mertzman, Stanley A., (717) 291-3818 stan.mertzman@fandm.edu, Franklin and Marshall College – Gi
Meskhidze, Nicholas, 919-515-7243 nicholas_meskhidze@ncsu.edu, North Carolina State Univ – Oa
Messer, Sharon, 740-351-3421 smesser@shawnee.edu, Shawnee State Univ – Oo
Messina, Paula, (408) 924-5027 pmessina@geosun.sjsu.edu, San Jose State Univ – Oe
Messing, Charles, (954) 262-3658 messingc@ocean.nova.edu, Nova Southeastern Univ – Ob
Metaxas, Anna, (902) 494-3021 Dalhousie Univ – Ob
Metcalf, Rodney V., (702) 895-4442 metcalfr@nevada.edu, Univ of Nevada, Las Vegas – Gp
Metesh, John J., 406-496-4159 jmetesh@mtech.edu, Montana Tech of The Univ of Montana – Hw
Mettee, Maurice F., 205/247-3627 smettee@gsa.state.al.us, Geological Survey of Alabama – Oo
Metz, Paul A., (907) 474-6749 Univ of Alaska, Fairbanks – Em
Metz, Robert, (908) 737-3687 rmetz@kean.edu, Kean Univ – Gr
Metzger, Ann G., (901) 678-2007 metzger@ceri.memphis.edu, Univ of Memphis – Ys
Metzger, Ellen P., (408) 924-5048 metzger@geosun.sjsu.edu, San Jose State Univ – Gp
Metzger, Ronald A., (541) 888-7216 rmetzger@socc.edu, Southwestern Oregon Community College – Pms
Metzler, Christopher V., (619) 757-2121 cmetzler@mcc.miracosta.cc.ca.us, MiraCosta College – Gg
Metzler, Kenneth, (860) 424-3585 kenneth.metzler@po.state.ct.us, Geological and Natural History Survey of Connecticut – Oo
Meyer, David L., (513) 556-4530 david.meyer@uc.edu, Cincinnati Museum Center – Pi
Meyer, David L.david.meyer@uc.edu, Univ of Cincinnati – Pi
Meyer, Gary N., (612)627-4780, Ext. 215 meyer015@umn.edu, Univ of Minnesota, Twin Cities – Gl
Meyer, George L., (760) 776-7272 College of the Desert – Gg
Meyer, Grant A., (505) 277-4204 gmeyer@unm.edu, Univ of New Mexico – Gm

Meyer, Jeff, (805) 965-0531 (Ext. 2313) Santa Barbara City College – Em
Meyer, Judith, (417) 836-5604 jlm545f@smsu.edu, Southwest Missouri State Univ – Oo
Meyer, Michael T., 785-864-3965 mmeyer@usus.gov, Univ of Kansas – Hg
Meyer, Philip D., (503) 417-7552 philip.meyer@pnl.gov, Pacific Northwest National Lab – Hw
Meyer, Rebecca A., (812) 855-2687 reameyer@indiana.edu, Indiana Geological Survey – Ec
Meyer, Scott C., (217) 333-5382 Illinois State Water Survey – Hw
Meyer, Steven J., (402) 472-8768 agme007@unlvm.unl.edu, Univ of Nebraska, Lincoln – Oa
Meyer, Thomas, (805) 664-2104 California State Univ, Bakersfield – Oo
Meyer, William T., (630) 969-6586 wtmeyer@anl.gov, Argonne National Lab – Cg
Meyer-Arendt, Klaus J., 850-474-2792 kjma@uwf.edu, Univ of West Florida – On
Meyers, James H., 507-457-5266 jmeyers@winona.edu, Winona State Univ – Gs
Meyers, Jayson, +61 8 9266-4976 J.Meyers@curtin.edu.au, Curtin Univ of Technology – Yx
Meyers, Philip A., (734) 764-0597 pameyers@umich.edu, Univ of Michigan – Co
Meyers, Samuel P., (225) 388-6457 Louisiana State Univ – Ob
Meyers, Stephen R., 919-962-0675 smeyers@email.unc.edu, Univ of North Carolina, Chapel Hill – Pe
Meylan, Anne, (727) 896-8626 Univ of South Florida, St. Petersburg – Ob
Meylan, Maurice A., (601) 266-4527 mmeylan@otr.usm.edu, Univ of Southern Mississippi – Gu
Miall, Andrew D., (416) 978-8841 miall@quartz.geology.utoronto.ca, Univ of Toronto – Gs
Michael, Peter J., 918 631 3017 pjm@utulsa.edu, Univ of Tulsa – Gi
Michaels, Patrick J., (804) 924-0549 pjm8x@virginia.edu, Univ of Virginia – Oa
Michaels, Paul, (208) 426-1929 pm@cgiss.boisestate.edu, Boise State Univ – Ne
Michalski, Greg, (765) 494-3704 gmichals@purdue.edu, Purdue Univ – Cs
Michaud, Jene D., (808) 933-7411 jene@hawaii.edu, Univ of Hawai'i at Hilo – Hq
Michaud, Yves, (418) 654-2647 Universite du Quebec – Gm
Michel, Fred A.fmichel@ccs.carleton.ca, Carleton Univ – Hw
Michel, Frederick A., (613) 520-2600 Carleton Univ – Hw
Michel, Jacqueline, 504-280-6325 jmichel@uno.edu, Univ of New Orleans – Cg
Michelangeli, Diane V., (416) 736-2100 (Ext. 77713) dvm@yorku.ca, York Univ – Oa
Mickelson, Andrew M., 901-678-4505 amicklsn@memphis.edu, Univ of Memphis – Ga
Mickelson, David M., (608) 262-7863 mickelson@geology.wisc.edu, Univ of Wisconsin, Madison – Gl
Mickelson, Katherine R.kmicklsn@memphis.edu, Univ of Memphis – Ga
Mickler, Patrick, 661-654-3281 pmickler@csub.edu, California State Univ, Bakersfield – Gg
Mickus, Kevin L., (417) 836-6375 klm983f@smsu.edu, Southwest Missouri State Univ – Yg
Middleton, Gerry V., (905) 525-9140 (Ext. 27694) middleto@mcmaster.ca, McMaster Univ – Yr
Middleton, Larry T., (928) 523-2429 larry.middleton@nau.edu, Northern Arizona Univ – Gs
Middleton, Michael D., (715) 425-3139 michael.d.middleton@uwrf.edu, Univ of Wisconsin, River Falls – Pv
Mies, Jonathan W., (423) 425-4404 Jonathan-Mies@utc.edu, Univ of Tennessee, Chattanooga – GctHg
Mika, Robert J., (609) 258-8423 rmika@princeton.edu, Princeton Univ – Cg
Mikan, Frank M., (512) 327-1213 Austin Community College – Cg
Mikols, Stephany, (419) 530-2009 stephany.mikols@utoledo.edu, Univ of Toledo – Oo
Mikulic, Donald G., 217-244-2518 mikulic@isgs.uiuc.edu, Illinois State Geological Survey – Ps
Mikulich, Matthew J., 719-395-6794 mjmikulich@msn.com, Virginia Polytechnic Institute & State Univ – Ye
Milam, Keith A., 740-593-1106 milamk@ohio.edu, Ohio Univ, Lancaster – Xg
Miles, Christine E., 717.702.2044 cmiles@state.pa.us, Pennsylvania Bureau of Topographic & Geologic Survey – Gg
Miles, Randall J., 573-882-6607 Univ of Missouri, Columbia – Sd
Miles, Roy G., (540) 857-7222 Virginia Western Comm College – Gg
Militzer, Burkhardmilitzer@seismo.berkeley.edu, Univ of California, Berkeley – Gy
Milkereit, Bernd, (416) 978-2466 bernd@core.physics.utoronto.ca, Univ of Toronto – Ye
Miller, Arnold I., (513) 556-4022 Miami Univ – Pi
Miller, Arnold I., (513) 556-4022 arnold.miller@uc.edu, Univ of Cincinnati – Pi
Miller, Barry W., (865) 594-5599 Barry.Miller@state.tn.us, Tennessee Division of Geology – EcOiGg
Miller, Brent V., (979) 458-3671 bvmiller@geo.tamu.edu, Texas A&M Univ – Gg
Miller, Bryon C., (402) 280-2461 byron.miller@noaa.gov, Creighton Univ – Oa
Miller, Calvin F., 615-322-2232 calvin.miller@vanderbilt.edu, Vanderbilt Univ – Gi
Miller, Charles B., (541) 737-4524 cmiller@coas.oregonstate.edu, Oregon State Univ – Ob
Miller, Charles M., (505) 667-8415 Los Alamos National Lab – Cc
Miller, Chris, (765) 496-9474 ccmiller@purdue.edu, Purdue Univ – Oo
Miller, Dalton W., (662) 325-2912 dmiller@ra.msstate.edu, Mississippi State Univ – Og
Miller, David S., (630) 252-7191 Argonne National Lab – Ge
Miller, Donald S., 518 478 0758 milled2@rpi.edu, Rensselaer Polytechnic Institute – Cc
Miller, Douglas A., (814) 863-7207 dam8@psu.edu, Pennsylvania State Univ, Univ Park – Sd
Miller, Douglas C., (302) 645-4277 dmiller@udel.edu, Univ of Delaware – Ob
Miller, Elizabeth L., (650) 723-1149 miller@pangea.stanford.edu, Stanford Univ – Gc
Miller, Galen W. C., (405) 325-3031 gwmiller@ou.edu, Univ of Oklahoma – Gc
Miller, Gary L., (828) 232-5184 Univ of North Carolina, Asheville – Oo
Miller, Geoffrey G., (506) 643-2361 Los Alamos National Lab – Cc
Miller, Gerald A., (515) 294-1923 soil@iastate.edu, Iowa State Univ of Science & Technology – So
Miller, Gifford H., 303-492-6962 gmiller@colorado.edu, Univ of Colorado – Cc
Miller, Glenn C., (775) 784-4108 gcmiller@scs.unr.edu, Univ of Nevada, Reno – Hg
Miller, Harvey J., (801) 585-3972 harvey.miller@geog.utah.edu, Univ of Utah – Oi
Miller, Hugh, 303-273-3558 hbmiller@mines.edu, Colorado School of Mines – Nm
Miller, Hugh G., (709) 737-3001 hmiller@waves.esd.mun.ca, Memorial Univ of Newfoundland – Yv
Miller, J W., (828) 232-5182 jwmiller@unca.edu, Univ of North Carolina, Asheville – Ge
Miller, James D., 218-726-8385 mille066@d.umn.edu, Univ of Minnesota, Twin Cities – Gx
Miller, James F., (417) 836-5447 jfm845f@smsu.edu, Southwest Missouri State Univ – Ps
Miller, Jerry R., (828) 227-7367 jmiller@wcu.edu, Western Carolina Univ – Gm
Miller, John, (416)736-2100 (Ext. 77729) jrmiller@yorku.ca, York Univ – Or
Miller, Jonathan S., (408) 924-5015 jsmiller@email.sjsu.edu, San Jose State Univ – Gi
Miller, Kate C., (915) 747-5536 miller@geo.utep.edu, Univ of Texas, El Paso – Yg
Miller, Keith B., (785) 532-2250 kbmill@ksu.edu, Kansas State Univ – Pe
Miller, Kenneth G., (732) 445-3622 kgm@rci.rutgers.edu, Rutgers, The State Univ of New Jersey – Gu
Miller, Loetta, 573-368-2161 nrmilll@mail.dnr.state.mo.us, Missouri Dept of Natural Resources – Oo
Miller, Lynda A., (512) 471-0116 lynda.miller@beg.utexas.edu, Univ of Texas at Austin, Jackson School of Geosciences – Yr
Miller, M. Meghan, (509) 963-2825 meghan@cwu.edu, Central Washington Univ – Gt
Miller, Marli G., (541) 346-4410 Univ of Oregon – Gc
Miller, Marvin R., 406-496-4155 mmiller@mtech.edu, Montana Tech of The Univ of Montana – Hw
Miller, Michael B., (225) 388-3412 byron@lgs.bri.lsu.edu, Louisiana State Univ – Go
Miller, Michael V., 217-333-7093 miller@isgs.uiuc.edu,

Illinois State Geological Survey – Ge
Miller, Molly F., 615-322-3528 molly.miller@vanderbilt.edu, Vanderbilt Univ – Pe
Miller, Murray H., (519) 824-4120 (Ext. 53758) mmiler@lrs.uoguelph.ca, Univ of Guelph – So
Miller, Nicholas L., (915) 747-5501 nmiller_tv@yahoo.com, Univ of Texas, El Paso – Ou
Miller, Randall F., (506) 643-2361 Univ of New Brunswick – Pg
Miller, Raymond M., (630) 252-3395 rmmiller@anl.gov, Argonne National Lab – Sb
Miller, Reuben F., (303) 871-2513 Univ of Denver – So
Miller, Richard D., (785) 864-2091 rmiller@kgs.ku.edu, Univ of Kansas – Yx
Miller, Richard D., (785) 864-3965 rmiller@kgs.ku.edu, Univ of Kansas – Yx
Miller, Richard H., (619) 594-5118 rmiller@geology.sdsu.edu, San Diego State Univ – Ps
Miller, Robert B., (408) 924-5025 rmiller@geosun.sjsu.edu, San Jose State Univ – Gc
Miller, Robert N., (541) 737-4555 miller@coas.oregonstate.edu, Oregon State Univ – Op
Miller, Ronald H., (313) 763-5727 Univ of Michigan – Og
Miller, Russell V., (213) 620-5025 California Geological Survey – Eg
Miller, Sandy, 573-368-2143 nrmills@mail.dnr.state.mo.us, Missouri Dept of Natural Resources – Oo
Miller, Steven F., (506) 643-2361 Argonne National Lab – Gg
Miller, Wade E., 801-422-2321 wade_miller@byu.edu, Brigham Young Univ – Pv
Miller, Wally, (775) 784-4072 wilymalr@ers.unr.edu, Univ of Nevada, Reno – Hg
Miller, William C., (707) 826-3110 wm1@axe.humboldt.edu, Humboldt State Univ – Pi
Miller, William P., (601) 325-2912 Univ of Georgia – Sc
Millette, Robert J., (601) 961-5518 Mississippi Office of Geology – Oo
Millette, Thomas L., (413) 538-2813 tmillett@mtholyoke.edu, Mount Holyoke College – OriGm
Milligan, Mark R., (801) 537-3326 nrugs.milliga@state.ut.us, Utah Geological Survey – Gg
Milliken, Kitty L., (512) 471-6082 kittym@mail.utexas.edu, Univ of Texas at Austin, Jackson School of Geosciences – Gd
Milliman, John D., (804) 684-7112 milliman@vims.edu, College of William & Mary – Ou
Mills, Aaron L., (804) 924-0564 alm7d@virginia.edu, Univ of Virginia – So
Mills, Eric L., (902) 494-3437 e.mills@dal.ca, Dalhousie Univ – Ob
Mills, Heath, 979 845 5105 hmills@ocean.tamu.edu, Texas A&M Univ – Ob
Mills, James G., 765-658-4669 jmills@depauw.edu, DePauw Univ – Gi
Mills, Kathy, (907) 786-6840 Univ of Alaska, Anchorage – Oo
Mills, Patricia, (505) 835-5635 geos@nmt.edu, New Mexico Institute of Mining and Technology – Oo
Mills, Patricia, 970-351-1229 patricia.mills@unco.edu, Univ of Northern Colorado – OooOo
Mills, III, Hugh H., (931) 372-3521 hmills@tntech.edu, Tennessee Technological Univ – Gm
Milne, Glen A., 0191 3742512 g.a.milne@durham.ac.uk, Durham Univ – Yg
Milner, Lloyd R., (225) 578-3410 lmilne1@lsu.edu, Louisiana State Univ – Gg
Milner, Lloyd R., (225) 388-3410 riley@lgs.bri.lsu.edu, Louisiana State Univ – Gg
Milroy, Scott, 228-688-2325 Scott.Milroy@usm.edu, Univ of Southern Mississippi – Ob
Min, Dong-Ha, (361) 749-6743 dongha@mail.utexas.edu, Univ of Texas at Austin – Ob
Min, Kyoungwon, (352) 392-2720 kmin@ufl.edu, Univ of Florida – Cc
Minarik, William G., 514-398-2596 minarik@eps.mcgill.ca, McGill Univ – Cp
Miner, James J., 217-244-5786 miner@isgs.uiuc.edu, Illinois State Geological Survey – Ge
Mink, Robert M., 205/247-3589 bmink@gsa.state.al.us, Geological Survey of Alabama – Go
Minkel, Donald H., 518-743-2325 minkeld@acc.sunyacc.edu, Adirondack Community College – Gc
Minnich, Richard A., (951) 827-5515 richard.minnich@ucr.edu, Univ of California, Riverside – Oy
Minor, Timothy B., (775) 673-7477 tminor@dri.edu, Desert Research Institute – Or
Minster, J. Bernard H., 858-534-5650 jbminster@ucsd.edu, Univ of California, San Diego – Ys
Miranda, Elena A., 818-677-4671 elena.miranda@csun.edu, California State Univ, Northridge – Gc
Miretti, Domenick, (213) 265-8837 East Los Angeles College – Oy
Mirsky, Arthur, 317278-0229 amirsky@iupui.edu, Indiana Univ / Purdue Univ, Indianapolis – Ge
Mishler, Brent, (510) 642-6810 bmishler@garnet.berkeley.edu, Univ of California, Berkeley – Pe
Misra, Kula C., (865) 974-6020 kmisra@utk.edu, Univ of Tennessee, Knoxville – Eg
Mitcham, Samuel W., (318) 342-1885 Univ of Louisiana, Monroe – Oo
Mitchell, Brian J., (314) 977-3123 Saint Louis Univ – Ys
Mitchell, Charles C., (334) 844-5489 mitchc1@auburn.edu, Auburn Univ – Sc
Mitchell, Charles E., (716) 645-6800 (Ext. 3991) cem@geology.buffalo.edu, SUNY, Buffalo – Po
Mitchell, LaMarr A., (360) 650-3591 Florida Geological Survey – Oo
Mitchell, Martin D., (507) 389-1610 martin.mitchell@mnsu.edu, Minnesota State Univ – Ou
Mitchell, N C.neil@ocean.cf.ac.uk, Cardiff Univ – Gu
Mitchell, Robert J., (360) 650-3591 robert.mitchell@geol.wwu.edu, Western Washington Univ – Hw
Mitchell, Roger H., (807) 343-8287 roger.mitchell@lakeheadu.ca, Lakehead Univ – Gi
Mitchell, Steven W., (805) 664-2270 California State Univ, Bakersfield – Ps
Mitchem, Jamie, 724 938-5969 mitchem@cup.edu, California Univ of Pennsylvania – Oa
Mitchneck, Beth A., 520-626-9820 bethm@email.arizona.edu, Univ of Arizona – Oo
Mitchum, Gary T., (727) 553-3941 gmitchum@marine.usf.edu, Univ of South Florida, Saint Petersburg – Op
Mitra, Gautam, (585) 275-5816 mitr@troi.cc.rochester.edu, Univ of Rochester – Gc
Mitra, Shankar, (405) 325-4462 smitra@ou.edu, Univ of Oklahoma – Gc
Mitri, Hani, (514) 398-4755 McGill Univ – Nm
Mitroshkov, Alex V., (509) 373-0026 alex.mitroshkov@pnl.gov, Pacific Northwest National Lab – Cg
Mitrovica, Jerry X., (416) 978-4946 jxm@physics.utoronto.ca, Univ of Toronto – Yd
Mitterer, Richard M., (972) 883-2462 mitterer@utdallas.edu, Univ of Texas, Dallas – Co
Mityga, Henry G., (301) 405-4344 hm12@umail.umd.edu, Univ of Maryland – Oo
Mix, Alan C., (541) 737-5212 amix@coas.oregonstate.edu, Oregon State Univ – Ou
Miyaki, Lawrence, 323 953-4000 (Ext. 2694) miyakiln@lacitycollege.edu, Los Angeles City College – Oy
Miyares, Ines, (212) 772-5443 imiyares@geo.hunter.cuny.edu, Hunter College (CUNY) – Oo
Miyares, Inez, (212) 772-5443 Graduate School of the City Univ of New York – Oo
Miyashiro, Akiho, (518) 442-4466 SUNY, Albany – Gp
Moberly, Ralph, (808) 956-8765 ralph@soest.hawaii.edu, Univ of Hawai'i, Manoa – Gu
Mobley, Curtis D., (206) 230-8166 Univ of Washington – Ow
Mock, R. Stephen, (406) 683-7261 s_mock@umwestern.edu, Univ of Montana, Western – Ca
Mockler, Theodore, (505) 667-4318 mockler@lanl.gov, Los Alamos National Lab – Oo
Mode, William N., (920) 424-7004 mode@uwosh.edu, Univ of Wisconsin, Oshkosh – Gl
Modzelewski, Henryk, (604) 822-3591 hmodzelewski@eos.ubc.ca, Univ of British Columbia – Oa
Moe-Hoffman, Amy P., 303-492-0018 amy.moe@colorado.edu, Univ of Colorado – Pg
Moecher, David P., (859) 257-6939 moker@uky.edu, Univ of Kentucky – Gx
Moeglin, Thomas D., (417) 836-5612 tdm375f@smsu.edu, Southwest Missouri State Univ – Ng
Moersch, Jeffery E., (865) 974-2366 Univ of Tennessee, Knoxville – Xg
Mofjeld, Harold O., (206) 526-6819 mofjeld@pmel.noaa.gov, Univ of Washington – Op
Mogk, David W., (406) 994-6916 mogk@montana.edu, Montana State Univ – Gp
Moglen, Glenn E.moglen@eng.umd.edu, Univ of Maryland – Hw

Mohanty, Bibhu, (514) 398-4755 McGill Univ – Nm
Mohnen, Volker A., (518) 437-8704 vam@atmos.albany.edu, SUNY, Albany – Oa
Mohr, Karen I., (518) 442-4561 mohr@atmos.albany.edu, SUNY, Albany – Hy
Mohr, Marcus, 089/2180 4230 marcus.mohr@geophysik.uni-muenchen.de, LMU - Univ of Munich – Yg
Mohrig, David, 512-471-2282 mohrig@mail.utexas.edu, Univ of Texas at Austin, Jackson School of Geosciences – Hs
Moiseyev, Alexis N., (510) 885-3486 California State Univ, East Bay – Cg
Mojzsis, Stephen J., 303-492-5014 mojzsis@colorado.edu, Univ of Colorado – Xc
Moldowan, J. Michael, (650) 725-0913 moldowan@pangea.stanford.edu, Stanford Univ – Co
Moldwin, Mark, (310) 206-4176 mmoldwin@igpp.ucla.edu, Univ of California, Los Angeles – Xy
Molina, Jean A., (612) 625-6259 jmolina@soils.umn.edu, Univ of Minnesota, Twin Cities – Sb
Molina, Luisa T., (617) 253-1603 ltmolina@mit.edu, Massachusetts Institute of Technology – Oa
Molinari, John E., (518) 442-4562 molinari@atmos.albany.edu, SUNY, Albany – Oa
Mollner, Daniel, (507) 933-7569 dmollner, Gustavus Adolphus College – Oo
Molnar, Peter, 303-492-4936 molnar@terra.colorado.edu, Univ of Colorado – Gt
Molnia, Bruce F., (703) 648-4120 bmolnia@usgs.gov, Duke Univ – Or
Momayez, Moe, (520) 626-5977 mmomayez@email.arizona.edu, Univ of Arizona – Nm
Momayezzadeh, Moe, (514) 398-4755 (Ext. 0530) McGill Univ – Nm
Momen, Nasim, (617) 353-5679 Boston Univ – Oo
Monaghan, Robert J., (314) 966-7719 Meramec Comm College – Oe
Monahan, Edward C., (860) 445-3457 sgoadm01@uconnvm.uconn.edu, Univ of Connecticut – Oa
Moncrief, John F., (612) 625-2771 john.moncrief@soils.umn.edu, Univ of Minnesota, Twin Cities – Sp
Monger, Bruce, (607) 255-4176 monger@geology.cornell.edu, Cornell Univ – Ob
Monger, H. C., (505) 646-3405 New Mexico State Univ, Las Cruces – Sa
Monk, Janice J., 520-621-3836 jmonk@email.arizona.edu, Univ of Arizona – Oo
Monroe, James S., (989) 774-3179 Central Michigan Univ – Gs
Montagne, Cliff, (406) 994-5079 montagne@gemini.oscs.montana.edu, Montana State Univ – Sd
Montagne, John, (406) 585-8307 Montana State Univ – Gl
Montana, Art, (310)825-3880 montana@ess.ucla.edu, Univ of California, Los Angeles – Cp
Montana, Carlos J., (915) 747-5498 montana@geo.utep.edu, Univ of Texas, El Paso – Yg
Montañez, Isabel P., (530) 754-7823 montanez@geology.ucdavis.edu, Univ of California, Davis – Gd
Montenari, Michael, +44 01782 583162 m.montenari@esci.keele.ac.uk, Keele Univ – Pm
Monteverde, Donald H., (609) 292-2576 don.monteverde@dep.state.nj.us, New Jersey Geological Survey – Gr
Monteverdi, John P., (415) 338-7728 montever@sfsu.edu, San Francisco State Univ – Oa
Montgomery, Carla W., (815) 753-1943 datarock@niu.edu, Northern Illinois Univ – Cc
Montgomery, David R., 206-685-2560 dave@ess.washington.edu, Univ of Washington – Gm
Montgomery, Homer A., 972-883-2496 mont@utdallas.edu, Univ of Texas, Dallas – Pg
Montgomery, Keith, (715) 845-9602 Univ of Wisconsin Colleges – Oy
Montgomery, Tamra S., 217-333-5105 montgomery@isgs.uiuc.edu, Illinois State Geological Survey – Oo
Montgomery, William W., (201) 200-3161 wmontgomery@njcu.edu, New Jersey City Univ – Hw
Montlucon, Daniel, (508) 289-2821 dmontlucon@whoi.edu, Woods Hole Oceanographic Institution – Co
Montwill, Gail F., (909) 946-1796 Santiago Canyon College – Gg
Moodie, T. Bryant, (780) 492-5742 bryant.moodie@ualberta.ca, Univ of Alberta – Oo
Moody, Jack S., (601) 961-5522 jack_moody@deq.state.ms.us, Mississippi Office of Geology – Go
Moody, Jennie L., (804) 924-0592 jlm8h@virginia.edu, Univ of Virginia – Oa
Moody, Lynn E., (805)756-2807 lmoody@calpoly.edu, California Polytechnic State Univ – Sd
Mooers, Howard, 218-726-7239 hmooers@d.umn.edu, Univ of Minnesota, Duluth – Gl
Mookerjee, Matty, (707) 664-2002 matty.mookerjee@sonoma.edu, Sonoma State Univ – Gc
Moon, Wooil, (204) 474-9833 wmoon@cc.umanitoba.ca, Univ of Manitoba – Ys
Mooney, Walter, (650) 329-4764 mooney@andreas.wr.usgs.gov, Stanford Univ – Ys
Moore, Alexandra, (607) 255-4644 am113@cornell.edu, Cornell Univ – Oe
Moore, Carleton B., (480) 965-3576 cmoore@asu.edu, Arizona State Univ – Cg
Moore, Daniel K., (208) 496-1902 moored@byui.edu, Brigham Young Univ, Idaho – GxiCp
Moore, Dennis W., (734) 763-0202 dennis.w.moore@noaa.gov, Univ of Washington – Op
Moore, Ellen J., (541) 758-0314 ellen.moore@cmug.com, Oregon State Univ – Pi
Moore, Emmett B., (509) 376-1089 Washington State Univ – Oo
Moore, George W., (541) 737-1244 mooreg@geo.oregonstate.edu, Oregon State Univ – Gt
Moore, Gregory F., (808) 956-6854 gmoore@hawaii.edu, Univ of Hawai'i, Manoa – Gt
Moore, J. Casey, (831) 459-2574 cmoore@pmc.ucsc.edu, Univ of California, Santa Cruz – Gc
Moore, Jaroy, 806-746-4011 JMoore@ag.tamu.edu, Texas A&M Univ – Sc
Moore, Johnnie N., (406) 243-6807 gl_jnm@selway.umt.edu, Univ of Montana – Gs
Moore, Joseph N., (801) 585-6931 jmoore@egi.utah.edu, Univ of Utah – Gv
Moore, Maureen L., (709) 737-2334 mmoore@esd.mun.ca, Memorial Univ of Newfoundland – Oo
Moore, Michael E., 717.702.2024 michmoore@state.pa.us, Pennsylvania Bureau of Topographic & Geologic Survey – Hw
Moore, N. K., 615-322-2976 kinzly.moore@vanderbilt.edu, Vanderbilt Univ – Ge
Moore, Robert M., (902) 494-3871 robert.moore@dal.ca, Dalhousie Univ – Oc
Moore, Theodore C., (734) 763-0202 tedmoore@umich.edu, Univ of Michigan – Ou
Moore, Wayne E., (517) 774-3179 Central Michigan Univ – Eo
Moore, Willard S., (803) 777-2262 moore@geol.sc.edu, Univ of South Carolina – Cc
Mooreman, Thomas B., (515) 294-2308 Iowa State Univ of Science & Technology – So
Moores, Eldridge M., (530) 752-0352 moores@geology.ucdavis.edu, Univ of California, Davis – Gt
Moorhead, Daryl L., (419) 530-2017 daryl.moorhead@utoledo.edu, Univ of Toledo – Oo
Moorhead, Kevin K., (828) 232-5183 moorhead@unca.edu, Univ of North Carolina, Asheville – Sb
Moorman, Brian J., (403) 220-4835 moorman@ucalgary.ca, Univ of Calgary – Gm
Moosavi, Dean, (507) 389-6323 sadredin.moosavi@mnsu.edu, Minnesota State Univ – Oe
Mora, Claudia I., (865) 974-5498 cmora@utk.edu, Univ of Tennessee, Knoxville – Cs
Mora, German, (515) 294-7586 gmora@iastate.edu, Iowa State Univ of Science & Technology – Cs
Moraghan, John T., (701) 231-8821 North Dakota State Univ – Sc
Morales, Michaelmoralesm@esumail.emporia.edu, Emporia State Univ – Pv
Moran, Joseph M., (414) 465-2432 moranj@uwgb.edu, Univ of Wisconsin, Green Bay – Oa
Moran, S. Bradley, (401) 874-6530 moran@gso.uri.edu, Univ of Rhode Island – Oc
Moran, Susan, (520) 670-6380 (Ext. 171) moran@tucson.ars.ag.gov, Univ of Arizona – So
Moran-Zenteno, Dante J.dante@tonatiuh.igeofcu.unam.mx, Universidad Nacional Autonoma de Mexico – Cc
Morden, Gregg, (204) 474-8870 gmorden@ms.umanitoba.ca, Univ of Manitoba – Cg
Morehouse, Barbara, 520-622-9018 morehoub@email.arizona.edu, Univ of Arizona – Oo
Morel, Francois M., 609-258-2416 morel@princeton.edu, Princeton Univ – Cl
Morel, Francois M M., (609) 258-2416 morel@princeton.edu, Princeton Univ – Cl
Morency, Maurice, (514) 987-3381 morency.maurice@uqam.ca,

Universite du Quebec a Montreal – Ct
Moreno, Rafael, (303) 556-8477 Metropolitan State Coll of Denver – Og
Morgan, Craig D., (801) 537-3370 nrugs.cmorgan@state.ut.us,
Utah Geological Survey – Eo
Morgan, Cristine L., (979) 845-3603 cmorgan@ag.tamu.edu,
Texas A&M Univ – Sp
Morgan, Debi L., (509) 376-9116 debi.morgan@pnl.gov,
Pacific Northwest National Lab – Oo
Morgan, Emory, (757) 727-5783 emory.morgan@hamptonu.edu,
Hampton Univ – Oe
Morgan, F D., (617) 253-7857 morgan@erl.mit.edu,
Massachusetts Institute of Technology – Yg
Morgan, George B., (405) 325-2642 gmorgan@ou.edu,
Univ of Oklahoma – Gi
Morgan, Julia, (713) 348-6330 Rice Univ – Gt
Morgan, Ken M., (817) 257-7721 k.morgan@tcu.edu,
Texas Christian Univ – OrGeOi
Morgan, Michael C., (608)265-8159 morgan@aurora.aos.wisc.edu,
Univ of Wisconsin, Madison – Ow
Morgan, Paul, (928) 523-7175 paul.morgan@nau.edu,
Northern Arizona Univ – Yh
Morgan, Siobahn M., (319) 273-2389 siobahn.morgan@uni.edu,
Univ of Northern Iowa – Oo
Morgan, Susan K., (435) 797-2176 sue.morgan@usu.edu,
Utah State Univ – Gd
Morgan, Sven S., (989) 774-1082 morga1ss@cmich.edu,
Central Michigan Univ – Gc
Morgan, Terrance L., (505) 667-0837 Los Alamos National Lab – Ng
Morgan, W. Jasonwjmorgan@princeton.edu,
Princeton Univ – Yg
Morison, James H., (206) 543-1394 morison@apl.washington.edu,
Univ of Washington – Op
Moritz, Wolfgang, 089/2180 4336 w.moritz@lrz.uni-muenchen.de,
LMU - Univ of Munich – Gz
Morozov, Igor, (306) 966-2761 igor.morozov@usask.ca,
Univ of Saskatchewan – Ys
Morra, Matthew J., (208) 885-6315 Univ of Idaho – Sb
Morrill, Robert W., (540) 231-5790 Virginia Polytechnic Institute &
State Univ – Oo
Morrin, M. Elizabeth, 973-353-5100 morrin@andromeda.rutgers.edu,
Rutgers, The State Univ of New Jersey, Newark – Oo
Morris, Brenda, ((76) 0) -744- (Ext. 50 ext. 2734) bmorris@palomar.edu,
Palomar College – Oo
Morris, Catherine, (910) 962-3736 morris@uncw.edu,
Univ of North Carolina Wilmington – Oo
Morris, David, (314) 935-6926 Los Alamos National Lab – Cc
Morris, Donald P., (610) 758-5175 dpm2@lehigh.edu,
Lehigh Univ – Oo
Morris, James T., (210) 458-5450 Univ of South Carolina – Ob
Morris, Julie, (314) 935-6926 jmorris@levee.wustl.edu,
Washington Univ – Cc
Morris, Robert W., (937) 327-7330 rmorris@wittenberg.edu,
Wittenberg Univ – Pi
Morris, Thomas H., (801) 422-3761 tom_morris@byu.edu,
Brigham Young Univ – Gr
Morris, William A., (905) 525-9140 (Ext. 24195) morriswa@mcmaster.ca,
McMaster Univ – Yg
Morris-Smith, Penny A., (713) 221-8178 smithp@uhd.edu,
Univ of Houston Downtown – Pi
Morrison, Ernie, (316)978-6647 Cowley County Comm College – Oo
Morrison, Jean, (213) 740-3551 morrison@usc.edu,
Univ of Southern California – Gp
Morrison, Joel L., (614) 292-3396 Ohio State Univ – Oo
Morrow, Jared R., 594-1395 jmorrow@geology.sdsu.edu,
San Diego State Univ – Gs
Morse, Daniel W., (601) 354-6328 Mississippi Office of Geology – Oo
Morse, David G., 217-244-5527 morse@isgs.uiuc.edu,
Illinois State Geological Survey – Eo
Morse, David L., (512) 232-3241 Univ of Texas at Austin, Jackson
School of Geosciences – Yg
Morse, Glynis G., (512) 471-1254 glynis.morse@beg.utexas.edu,
Univ of Texas at Austin, Jackson School of Geosciences – Yr
Morse, John W., (979) 845-9630 morse@ocean.tamu.edu,
Texas A&M Univ – Oc
Morse, Linda D., (757) 221-2444 ldmors@wm.edu,
College of William & Mary – Ge
Morse, Stearns A., (413) 545-0175 Univ of Massachusetts – Gi
Morse, Trevor J., 0191 3744784 t.j.morse@durham.ac.uk,
Durham Univ – Pg
Morse, Walter, (318) 478-0684 McNeese State Univ – Ge
Mortensen, James K., (604) 822-6208 jmortensen@eos.ubc.ca,
Univ of British Columbia – Cc
Mortenson, Kristine B., 801-422-3919 kris_mortenson@byu.edu,
Brigham Young Univ – Oo
Morton, Allan E., (602) 426-4351 Central Arizona College – Oe
Morton, Douglas M., (951) 276-6397 scamp@ucrac1.ucr.edu,
Univ of California, Riverside – Gp
Morton, Penelope, (218) 726-7962 pmorton@d.umn.edu,
Univ of Minnesota, Duluth – Eg
Morton, Robert, (813) 974-2773 Univ of South Florida, Tampa – Gs
Morton, Roger D., (780) 492-3265 Univ of Alberta – Em
Morton, Ronald L., (218) 726-7218 rmorton@d.umn.edu,
Univ of Minnesota, Duluth – Eg
Mortyn, P. Graham, (559) 278-2585 gmortyn@csufresno.edu,
California State Univ, Fresno – Ou
Moser, Desmond, 519-661-4214 dmoser22@uwo.ca,
Univ of Western Ontario – GtCcXg
Moser, Desmond E., (801) 585-3782 demoser@mines.utah.edu,
Univ of Utah – Cc
Moser, Katrina A., (801) 581-5800 katrina.moser@geog.utah.edu,
Univ of Utah – Pe
Moses, Carl O., (610) 758-5891 carl.moses@lehigh.edu,
Lehigh Univ – Cl
Moses, Lawrence L., 724 938-5864 moses@cup.edu,
California Univ of Pennsylvania – Oy
Mosher, David C., (902) 426-3149 mosher@agc.bio.ns.ca,
Dalhousie Univ – Ou
Mosher, Sharon, (512) 471-4135 mosher@mail.utexas.edu,
Univ of Texas at Austin, Jackson School of Geosciences – Gc
Moshier, Stephen O., (630) 752-5856 stephen.o.moshier@wheaton.edu,
Wheaton College – Gd
Moskowitz, Bruce M., 612-624-1547 bmosk@umn.edu,
Univ of Minnesota, Twin Cities – Ym
Mosley-Thompson, Ellen E., (614) 292-2580 thompson.4@osu.edu,
Ohio State Univ – Oa
Moslow, Thomas F., (403) 269-6911 Univ of Calgary – Co
Moss, Neil E., 205/247-3557 nmoss@gsa.state.al.us,
Geological Survey of Alabama – Hw
Moss, Patti, (509) 527-5225 moss@whitman.edu,
Whitman College – Oo
Mossa, Joann, (352) 392-0494 mossa@geog.ufl.edu,
Univ of Florida – Gm
Mossler, John H., (612)627-4780 Ext. 218 mossl001@umn.edu,
Univ of Minnesota, Twin Cities – Gr
Mossop, John W., (514) 398-4755 McGill Univ – Nm
Mostovoi, Gueorgui V., (662) 325-7191 mostovoi@erc.msstate.edu,
Mississippi State Univ – Ow
Motani, Ryosuke, 530-754-6284 Motani@geology.ucdavis.edu,
Univ of California, Davis – Pv
Motaralli, Peter P., (671) 735-2134 Univ of Guam – So
Motavalli, Peter P., 573-884-3212 motvallip@missouri.edu,
Univ of Missouri, Columbia – Os
Mott, Joanna, (361) 825-6024 Texas A&M Univ, Corpus Christi – Sb
Mouat, David A., (775) 673-7402 dmouat@dri.edu,
Desert Research Institute – Ge
Mouginis-Mark, Peter J., (808) 956-3147 Univ of Hawai'i, Manoa – Xg
Moulding, Alisonmoulding@nova.edu,
Nova Southeastern Univ – Oo
Moulis, Anastasia, 617-552-8300 anastasia.macherides.1@bc.edu,
Boston College – Ys
Moum, James N., (541) 737-2553 jmoum@coas.oregonstate.edu,
Oregon State Univ – Op
Mount, Jeffrey F., (916) 752-7092 Univ of California, Davis – Gs
Mount, Jeffrey F., (530) 752-7092 mount@geology.ucdavis.edu,
Univ of California, Davis – Gm
Mountain, Gregory S., 732-445-0817 gmtn@rci.rutgers.edu,
Rutgers, The State Univ of New Jersey – Gu
Mountain, Keith R., (502) 852-6844 Univ of Louisville – Oa
Mountjoy, Eric W., 514-398-4894 ericm@eps.mcgill.ca,
McGill Univ – Gd
Mousset-Jones, Pierre, (775) 784-6959 mousset@mines.unr.edu,
Univ of Nevada, Reno – Nm
Moyer, Kerry A., (814) 732-2454 kmoyer@edinboro.edu,
Edinboro Univ of Pennsylvania – Oa
Moysey, Stephen M., (864) 656-5019 smoysey@clemson.edu,
Clemson Univ – Hw
Mozley, Peter S., (505) 835-5311 mozley@nmt.edu,
New Mexico Institute of Mining and Technology – Gs
Mroz, Eugene J., (505) 667-7758 Los Alamos National Lab – Oa
Mroz, Paul J.pjm57@earthlink.net,

SUNY College at Brockport – Ow
Mucci, Alfonso O., 514-398-4892 alm@eps.mcgill.ca, McGill Univ – Cm
Mucciardi, Frank, (514) 398-4755 McGill Univ – Nx
Muck, Maureen T., (303) 248-7796 Oak Ridge National Lab – Ge
Muckleston, Keith W., (541) 737-1204 mucklesk@geo.oregonstate.edu, Oregon State Univ – Oo
Mudrey, Jr., Michael G., (608) 263-5495 Northern Illinois Univ – Eg
Mudrick, Stephen E., (573) 882-6721 Univ of Missouri, Columbia – Oa
Muehlberger, William R., (512) 471-4885 wmuehl@mail.utexas.edu, Univ of Texas at Austin, Jackson School of Geosciences – Gc
Muehlenbachs, Karlis, (780) 492-2827 karlis.muehlenbachs@ualberta.ca, Univ of Alberta – Cs
Mueller, Erich M., (334) 460-7136 fkmrc@aol.com, Univ of South Alabama – Ob
Mueller, Karl J., 303-492-7336 karl.mueller@colorado.edu, Univ of Colorado – Gc
Mueller, Paul A., (352) 392-2231 mueller@geology.ufl.edu, Univ of Florida – Cc
Mueller, Raymond G., (609) 652-4209 ray.mueller@stockton.edu, Richard Stockton College of New Jersey – So
Mueller, Thomas, (724) 938-4255 mueller@cup.edu, California Univ of Pennsylvania – Oi
Mueller, Wulf, (418) 545-5013 Wulf_mueller@uqac.ca, Univ of Western Ontario – Gv
Mueller, Wulf U., (418) 545-5013 wmueller@uqac.uquebec.ca, Universite du Quebec a Chicoutimi – Gv
Muessig, Karl W., (609) 292-1185 karl.muessig@dep.state.nj.us, New Jersey Geological Survey – Cg
Muhleman, Duane O., (626) 395-6186 dom@gps.caltech.edu, California Institute of Technology – Xy
Mukasa, Samuel B., (734) 936-3227 mukasa@umich.edu, Univ of Michigan – Cc
Mukhopadhyay, Sujoy, (617) 496-6441 sujoy@eps.harvard.edu, Harvard Univ – Cg
Muldoon, Maureen A., (920) 424-4461 muldoon@uwosh.edu, Univ of Wisconsin, Oshkosh – Hw
Mulholland, Margaret, (757) 683-3972 mmulholl@odu.edu, Old Dominion Univ – Ob
Mulla, David J., (612) 625-6721 mulla003 at umn.edu, Univ of Minnesota, Twin Cities – Sp
Mullen, Robert, 330-263-3785 mullen.91@osu.edu, Ohio State Univ – So
Mullen, Steven L., (520) 621-6842 mullen@atmo.arizona.edu, Univ of Arizona – Oa
Muller, Otto H., (607) 871-2208 Alfred Univ – Gc
Muller, Peter, (808) 956-8081 pmuller@soest.hawaii.edu, Univ of Hawai'i, Manoa – Op
Muller, Peter D., (607) 436-3067 mullerpd@oneonta.edu, SUNY, Oneonta – Gc
Muller, Robert G., (727) 896-8626 Univ of South Florida, Saint Petersburg – Ob
Muller-Karger, Frank E., (727) 553-3335 Univ of South Florida, Saint Petersburg – Or
Mulley, Lauri A., (585) 395-2636 lmulley@brockport.edu, SUNY College at Brockport – Oo
Mullins, Gregory L., (540) 231-4383 gmullins@vt.edu, Virginia Polytechnic Institute & State Univ – Sc
Mullins, Henry T., 315-443-2672 htmullin@syr.edu, Syracuse Univ – Gu
Mulugeta, Gebeyehu, (414) 424-4249 Univ of Wisconsin, Oshkosh – Oo
Mulvaney, Richard L., (217) 333-9467 Univ of Illinois, Urbana-Champaign – Sc
Mulvany, Pat, 573-368-2158 nrmulvp@mail.dnr.state.mo.us, Missouri Dept of Natural Resources – Oo
Mumin, A. Hamid, 204-727-9685 mumin@brandonu.ca, Brandon Univ – Eg
Mumin, Hamid A., (204) 727-9685 mumin@brandonu.ca, Univ of Western Ontario – Eg
Mungall, James E., 416-978-2975 mungall@geology.utoronto.ca, Univ of Toronto – Ec
Munguia, Luislmunguia@cicese.mx, Centro de Inv Científica y de Educación Superior de Ensenada – Ys
Munk, LeeAnn, 907 786-6895 aflm@uaa.alaska.edu, Univ of Alaska, Anchorage – Ge
Munk, Walter H., (619) 534-2877 wmunk@ucsd.edu, Univ of California, San Diego – Op
Munn, Barbara J., (916) 278-6811 California State Univ, Sacramento – Em
Munn, Larry C., (307) 766-2127 Univ of Wyoming – Sd
Munoz, Miguel A., (809) 767-9705 Univ of Puerto Rico – Sc
Munro, Ildiimunro@ccs.carleton.ca, Carleton Univ – Pg
Munroe, Jeffrey S., 802-443-3446 jmunroe@middlebury.edu, Middlebury College – Ge
Munski, Douglas C., (701) 777-4591 Univ of North Dakota – Oe
Munski, Douglas C., 701-777-4591 douglas_munski@und.nodak.edu, North Dakota State Univ – Oo
Murdoch, Lawrence C., (864) 656-2597 lmurdoc@clemson.edu, Clemson Univ – Hw
Murowchick, James B., (816) 235-2979 murowchickj@umkc.edu, Univ of Missouri-Kansas City – Cg
Murphy, Alexander B., (541) 346-4571 abmurphy@uoregon.edu, Univ of Oregon – Oo
Murphy, Charles H., (908) 737-3422 cmurphy@kean.edu, Kean Univ – Ow
Murphy, Cindy, (902) 867-2299 cmurphy@stfx.ca, Saint Francis Xavier Univ – Gg
Murphy, Ed, (701) 328-8002 emurphy@state.nd.us, North Dakota Geological Survey – Oo
Murphy, J. Brendan, (902) 867-2481 Dalhousie Univ – Gx
Murphy, J. Brendan, (902) 867-2481 bmurphy@stfx.ca, Saint Francis Xavier Univ – Gt
Murphy, Jack, (303) 370-6445 JMurphy@dmns.org, Denver Museum of Nature & Science – Gz
Murphy, Kim, (908) 527-2518 Univ of Pittsburgh, Bradford – Oo
Murphy, Michael, (713) 743-3423 mmurphy@bayou.uh.edu, Univ of Houston – Gc
Murphy, Michael A.michael.murphy@ucr.edu, Univ of California, Riverside – Ps
Murphy, Ryan P., 419-626-4296 ryan.murphy@dnr.state.oh.us, Ohio Dept of Natural Resources – Gs
Murphy, Vincent, (617) 552-8300 Boston College – Yg
Murphy, William M., (530) 898-5163 wmurphy@csuchico.edu, California State Univ, Chico – Cl
Murray, A. Bradshaw, (919) 684-5847 abmurray@duke.edu, Duke Univ – Ou
Murray, Alison E., (775) 673-7361 alison@dri.edu, Desert Research Institute – Py
Murray, Bruce C., (626) 395-3780 bcm@gps.caltech.edu, California Institute of Technology – Xg
Murray, Christopher J., (509) 376-5848 chris.murray@pnl.gov, Pacific Northwest National Lab – Gq
Murray, Daniel P., (401) 792-2197 dpmurray@uri.edu, Univ of Rhode Island – Gp
Murray, Haydn H., (812) 855-5583 murrayh@indiana.edu, Indiana Univ, Bloomington – En
Murray, James W., (206) 543-4730 jmurray@u.washington.edu, Univ of Washington – Oc
Murray, Kent S., (313) 436-9129 kmurray@umich.edu, Univ of Michigan, Dearborn – Hw
Murray, Kyle, 210-458-5779 Kyle.Murray@utsa.edu, Univ of Texas, San Antonio – Hw
Murray, Mark, 505.835.6930 murray@ees,nnt,edy, New Mexico Institute of Mining and Technology – Gc
Murray, Nancy, (860) 424-3589 nancy.murray@po.state.ct.us, Geological and Natural History Survey of Connecticut – Oo
Murray, Paul, 512-471-9552 paul.murray@beg.utexas.edu, Univ of Texas at Austin, Jackson School of Geosciences – Yx
Murray, Richard, (617) 353-6532 rickm@bu.edu, Boston Univ – Pg
Murrell, Michael T., (505) 667-0967 Los Alamos National Lab – Cc
Murry, Phillip A., (817) 968-9889 murry@tarleton.edu, Tarleton State Univ – Pv
Murtha, Peter A., (604) 822-6452 Univ of British Columbia – Or
Murthy, Gummuluru S., (709) 737-7573 gmurthy@morgan.ucs.mun.ca, Memorial Univ of Newfoundland – Ym
Murthy, V. Rama, (505) 277-3842 vrmurthy@unm.edu, Univ of New Mexico – XcCc
Musolf, Gene E., (715) 845-9602 Univ of Wisconsin Colleges – Oy
Musselman, Zachary A., 601-974-1344 musseza@millsaps.edu, Millsaps College – Gm
Musser, Linda R., (814) 863-7073 Pennsylvania State Univ, Altoona – Oo
Mustard, John F., (401) 863-1264 john_mustard@brown.edu, Brown Univ – Or
Mustard, Peter S., (604) 291-5389 pmustard@sfu.ca, Simon Fraser Univ – Gs
Mustart, David A., (415) 338-7729 mustart@sfsu.edu, San Francisco State Univ – Gg
Mustoe, George, (360) 650-3585 mustoeg@cc.wwu.edu, Western Washington Univ – Oo

Muthukrishnan, Suresh, 864 294 3361 suresh.muthukrishnan@furman.edu, Furman Univ – Oi
Mutis-Duplat, Emilio, 432 552-2243 mutis_e@utpb.edu, Univ of Texas, Permian Basin – Gp
Mutter, John C., (845) 365-8348 jcm@ldeo.columbia.edu, Columbia Univ – Yr
Mutti, Laurence J., (814) 641-3601 mutti@juniata.edu, Juniata College – Gx
Muttiah, Ranjan S., 817-257-6272 r.muttiah@tcu.edu, Texas Christian Univ – OirHg
Muza, Jay P., (954) 201-6771 jmuza@broward.edu, Broward Community College, Central Campus – Ou
Myer, George H., (215) 204-7173 gmyer@temple.edu, Temple Univ – Gz
Myers, James D., (307) 766-2203 magma@uwyo.edu, Univ of Wyoming – Gi
Myers, Jeffrey A., 503-838-8365 myersj@wou.edu, Western Oregon Univ – Gs
Myers, John S., (709) 737-8417 jmyers@sparky2.esd.mun.ca, Memorial Univ of Newfoundland – Eg
Myers, Orrin B., (505) 665-3742 obm@lanl.gov, Los Alamos National Lab – Py
Myers, Paul, (780) 492-6706 pmyers@ualberta.ca, Univ of Alberta – Op
Myers, Paul B., (610) 758-3665 pbm1@lehigh.edu, Lehigh Univ – Hw
Myers, W. Darwin, (707) 253-3111 Napa Valley College – Gc
Mylavarapu, S R., (352) 846-3833 raom@ufl.edu, Univ of Florida – Sc
Müller, Lena, 089/2180 4340 lena-mueller@lrz.uni-muenchen.de, LMU - Univ of Munich – Gz
Müller-Sohnius, Dieter, 089/2180 4259 mueso@min.uni-muenchen.de, LMU - Univ of Munich – Gz
Mylroie, Joan R., (662) 325-1954 jmylroie@deanas.msstate.edu, Mississippi State Univ – Oe
Mylroie, John E., (662) 325-8774 mylroie@geosci.msstate.edu, Mississippi State Univ – Gm
Myneni, Ranga, 617-353-5742 rmyneni@bu.edu, Boston Univ – Or
Myneni, Satish C B., (609) 258-5848 smyneni@princeton.edu, Princeton Univ – Sc
Myron, Best G., (801) 374-0818 myron_best_g@byu.edu, Brigham Young Univ – Go
Myrow, Paul M., (719) 389-6790 pmyrow@coloradocollege.edu, Colorado College – Gs
Mysak, Lawrence A., (514) 398-3768 lawrence.mysak@mcgill.ca, McGill Univ – Op
Mysen, Bjorn O., (202) 478-8975 b.mysen@gl.ciw.edu, Carnegie Institution of Washington – Cp

N

Nabelek, John L., (541) 737-2757 nabelek@coas.oregonstate.edu, Oregon State Univ – Ys
Nabelek, Peter I., (573) 884-6463 NabelekP@missouri.edu, Univ of Missouri – Ct
Nabighian, Misac, 303-273-3933 mnabighi@mines.edu, Colorado School of Mines – Nm
Nadeau, Leopold, (418) 654-2645 lnadeau@nrcan.gc.ca, Universite du Quebec – Gt
Nadiga, Balu, (505) 667-9466 balu@lanl.gov, Los Alamos National Lab – Oo
Nadon, Gregory C., (740) 593-4212 nadon@ohio.edu, Ohio Univ, Lancaster – Gs
Naehr, Thomas H., (361) 825-2470 tnaehr@falcon.tamucc.edu, Texas A&M Univ, Corpus Christi – Gu
Nagaoka, Lisa A., (940) 565-2510 lnagaoka@unt.edu, Univ of North Texas – Ga
Nagihara, Seiichi, (806) 742-3149 seiichi.nagihara@ttu.edu, Texas Tech Univ – Oi
Nagy, Adam, (418) 545-5011 Universite du Quebec a Chicoutimi – Gz
Nagy, Andrew F., (313) 764-6592 Univ of Michigan – Oa
Nagy, Kathryn L., 312-355-5276 klnagy@uic.edu, Univ of Illinois, Chicago – Cg
Nagy-Shadman, Elizabeth A., 818-677-6151 elizabeth.nagy-shadman@csun.edu, California State Univ, Northridge – Cc
Nahir, Tal M.tnahir@csuchico.edu, California State Univ, Chico – Oa
Nair, Vimala D., (352) 392-1803 (Ext. 324) vdna@mail.ifas.ufl.edu, Univ of Florida – Sc
Najjar, Raymond G., (814) 863-1586 najjar@essc.psu.edu, The Pennsylvania State Univ, Univ Park – Ow
Nakagawa, Masami, 303-384-2132 mnakagaw@mines.edu, Colorado School of Mines – Or
Nakamura, Noboru, (773) 702-3802 nnn@bethel.uchicago.edu, Univ of Chicago – Oa
Nakamura, Yosio, (512) 471-0428 yosio@ig.utexas.edu, Univ of Texas at Austin, Jackson School of Geosciences – Ys
Naldrett, Anthony J., 44 (0)1483 776 944 ajnaldrett@yahoo.com, Univ of Toronto – Ec
Namikas, Steven, (225) 578-6142 snamik1@lsu.edu, Louisiana State Univ – Gm
Nance, Marlon A., (916) 278-4683 nance@csus.edu, California State Univ, Sacramento – Gg
Nance, R. Damian, (740) 593-1107 nance@ohio.edu, Ohio Univ, Lancaster – Gt
Nance, Seay, 512-471-6285 seay.nance@beg.utexas.edu, Univ of Texas at Austin, Jackson School of Geosciences – Gg
Napora, Theodore A., (401) 874-6222 Univ of Rhode Island – Ob
Napp, Jeffrey, (206) 526-4148 jnapp@afsc.noaa.gov, Univ of Washington – Ob
Narayanan, Rangesan, (775) 784-6701 rang@scs.unr.edu, Univ of Nevada, Reno – Hg
Narbonne, Guy M., (613) 545-6168 Queen's Univ – Pi
Narbutovskih, Susan M., (509) 376-9235 susan.narbutovskih@pnl.gov, Pacific Northwest National Lab – Yg
Narod, Barry, (604) 822-2267 narod@eos.ubc.ca, Univ of British Columbia – Yx
Nash, Barbara P., (801) 581-8587 bpnash@mines.utah.edu, Univ of Utah – Gi
Nash, David, david.nash@uc.edu, Univ of Cincinnati – Gm
Nash, Gregory N., (801) 585-3889 gnash@egi.utah.edu, Univ of Utah – Or
Nash, Jonathan, (541) 737-4573 nash@coas.oregonstate.edu, Oregon State Univ – Op
Nashold, Barney W., (630) 252-7698 bwnashold@anl.gov, Argonne National Lab – Oo
Nasir, Sobhi J., 00974-4852207 nasir54@qu.edu.qa, Univ of Qatar – Gx
Nasiri, Shaima, 979-845-8076 snasiri@ariel.met.tamu.edu, Texas A&M Univ – Oa
Naslund, H. Richard, (607) 777-4313 Binghamton Univ – Gi
Nastrom, Gregory D., 320-3083260 nastrom@stcloudstate.edu, Saint Cloud State Univ – Oa
Nater, Edward A., (612) 625-1725 enater@soils.umn.edu, Univ of Minnesota, Twin Cities – Sd
Nations, J. DaleDale.Nations@nau.edu, Univ of Arizona – Pm
Nations, Jack D.dale.nations@nau.edu, Northern Arizona Univ – Pm
Natland, James, 305 421-4123 jnatland@rsmas.miami.edu, Univ of Miami – Gi
Naugle, Burl I., (270) 809-3105 burl.naugle@murraystate.edu, Murray State Univ – Hq
Naumann, Malik, 089/2180 6714 LMU - Univ of Munich – Pg
Naumann, Terry R., (907) 786-6846 aftrn@uaa.alaska.edu, Univ of Alaska, Anchorage – Gi
Nava, Alejandrofnava@cicese.mx, Centro de Inv Cientifica y de Educación Superior de Ensenada – Ys
Navrotsky, Alexandra, (530) 752-3292 anavrotsky@ucdavis.edu, Univ of California, Davis – Om
Naymik, Thomas G., (614) 424-6344 naymikt@battelle.org, Ohio State Univ – Hw
Neal, Clive R., (219) 631-8328 neal.1@nd.edu, Univ of Notre Dame – Gi
Neal, Donald W., (252) 328-4392 neald@mail.ecu.edu, East Carolina Univ – Gr
Neal, William J., 616-331-3381 nealw@gvsu.edu, Grand Valley State Univ – Gs
Nealson, Kenneth, (213) 821-2271 knealson@usc.edu, Univ of Southern California – Py
Nechaev, Dmitri, (228) 688-2573 dmitri.nechaev@usm.edu, Univ of Southern Mississippi – Op
Neelin, J. David, (310) 206-3734 neelin@atmos.ucla.edu, Univ of California, Los Angeles – Oa
Neff, Jason, 303-492-6187 Univ of Colorado – Ge
Negrini, Robert M., (805) 664-2185 California State Univ, Bakersfield – Ym
Neher, Deborah A., (419) 530-2585 deborah.neher@utoledo.edu, Univ of Toledo – Sb
Nehru, Cherukupalli E., (718) 951-5417 Graduate School of the City

Univ of New York – Gi
Neiers, Melissa, (208) 282-3365 neiemeli@isu.edu, Idaho State Univ – Oo
Neil, Craig D., (207) 287-2801 Dept of Conservation – Hw
Nekvasil, Hanna, 631-632-8201 hanna.nekvasil@sunysb.edu, SUNY, Stony Brook – Cp
Nelson, Bruce K., 206-543-4434 bnelson@u.washington.edu, Univ of Washington – CcGi
Nelson, Bruce W., (434) 296-3851 bwn@virginia.edu, Univ of Virginia – Gs
Nelson, Charles L.x00013@stcloudstate.edu, Saint Cloud State Univ – On
Nelson, Daniel O., 217-244-2425 Illinois State Geological Survey – Oi
Nelson, David M., nelsonda@coas.oregonstate.edu, Oregon State Univ – Ob
Nelson, Eric P., (303) 273-3811 enelson@mines.edu, Colorado School of Mines – Gc
Nelson, Gerald E., 307-268-2233 gnelson@caspercollege.edu, Casper College – Gm
Nelson, K. Douglas, (315) 443-3626 kdnelson@mailbox.syr.edu, Cornell Univ – Gt
Nelson, Kathy, (309) 794-7318 ggnelson@augustana.edu, Augustana College – Oo
Nelson, Kenneth A., (785) 864-2164 nelson@kgs.ku.edu, Univ of Kansas – Oy
Nelson, Lauren, (650) 723-3715 lmnelson@stanford.edu, Stanford Univ – Oo
Nelson, Lise, (541) 346-4856 lise@darkwing.uoregon.edu, Univ of Oregon – Oo
Nelson, Marc A., (501) 575-3964 manelson@comp.uark.edu, Univ of Arkansas, Fayetteville – Cl
Nelson, Michael G., (801) 585-3064 mgnelson@mines.utah.edu, Univ of Utah – Nm
Nelson, Pamela J., (623) 845-3680 pam.nelson@gcmail.maricopa.edu, Glendale Community College – Gv
Nelson, Paul D., (314) 966-7717 Meramec Community College – Oe
Nelson, Robert E., (207) 859-5800 Colby College – Pe
Nelson, Robert K., (508) 289-2656 rnelson@whoi.edu, Woods Hole Oceanographic Institution – Co
Nelson, Robert M., (818) 354-1797 Jet Propulsion Lab – Xg
Nelson, Robert S., (309)438-7808 rsnelso@ilstu.edu, Illinois State Univ – Gm
Nelson, Sheldon D., (801) 378-2151 Brigham Young Univ – Sp
Nelson, Stephen, (520) 626-3322 nsteve@ag.arizona.edu, Univ of Arizona – So
Nelson, Stephen A., (504) 862-3194 snelson@tulane.edu, Tulane Univ – Gi
Nelson, Stephen T., 801-422-8688 steve_nelson@byu.edu, Brigham Young Univ – Cs
Nemcok, Michal, (801) 585-9767 mnemcok@egi.utah.edu, Univ of Utah – Gc
Nesbitt, Elizabeth (Liz), (206) 543-5949 lnesbitt@u.washington.edu, Univ of Washington – Pim
Nesbitt, H W., (519) 661-3194 hwn@uwo.ca, Univ of Western Ontario – Cl
Nesbitt, Stephen W., (217) 244-3740 snesbitt@atmos.uiuc.edu, Univ of Illinois, Urbana-Champaign – Oa
Nesse, William D., (970) 351-2830 wdnesse@unco.edu, Univ of Northern Colorado – Gx
Nestell, Galena P., (817) 272-2987 gnestell@exchange.uta.edu, Univ of Texas, Arlington – Pm
Nestell, Merlynd K., (817) 272-2987 nestell@uta.edu, Univ of Texas, Arlington – Pm
Netoff, Dennis I., (409) 294-1454 Sam Houston State Univ – Gm
Netzler, Bruce, 573-368-2485 nrnetzb@mail.dnr.state.mo.us, Missouri Dept of Natural Resources – Go
Neubaum, John C., 717.702.2039 jneubaum@state.pa.us, Pennsylvania Bureau of Topographic & Geologic Survey – Ec
Neubeck, William S., (518) 388-6519 neubeckw@union.edu, Union College – Gm
Neuhauser, Kenneth R., (785) 628-5349 kneuhaus@fhsu.edu, Fort Hays State Univ – Gc
Neuhoff, Philip S., (352) 846-2413 neuhoff@ufl.edu, Univ of Florida – Cl
Neuman, Andrew G., (403) 823-7707 Royal Tyrrell Museum of Palaeontology – Pv
Neuman, Dennis R., (406) 994-4822 dneuman@montana.edu, Montana State Univ – Oo
Neuman, Shlomo P., (520) 621-7114 neuman@hwr.arizona.edu, Univ of Arizona – Hq
Neumann, A. Conrad, (919) 962-0190 neumann@marine.unc.edu, Univ of North Carolina, Chapel Hill – Gu
Neumann, Klaus, (765) 285-8262 kneumann@gw.bsu.edu, Ball State Univ – Cl
Neuweiler, Fritz, (418) 656-7479 fritz.neuweiler@ggl.ulaval.ca, Universite Laval – GsdPo
Nevins, Susan K., (607) 753-2815 GeologyDept@cortland.edu, SUNY, Cortland – Oo
Nevkasil, Hanna, 212-769-5390 hanna.nevkasil@stonybrook.edu, American Museum of Natural History – Gi
Newberry, Rainer J., ffrn@uaf.edu, Univ of Alaska, Fairbanks – Em
Newbill, Chris A., (509) 376-4214 chris.newbill@pnl.gov, Pacific Northwest National Lab – Gq
Newbold, K. B., (905) 525-9140 (Ext. 27948) newbold@mcmaster.ca, McMaster Univ – Yr
Newcomer, Darrell R., (509) 376-1054 darrell.newcomer@pnl.gov, Pacific Northwest National Lab – Hw
Newell, Charles J., (713) 663-6600 Rice Univ – Hw
Newell, Joan E., newellje@cofc.edu, College of Charleston – Oo
Newell, Kerry D., (785) 864-2183 dnewell@kgs.ku.edu, Univ of Kansas – Go
Newell, Richard G., 919 681-8865 richard-newell@duke.edu, Duke Univ – Eg
Newkirk, Steven, (901) 678-2177 Univ of Memphis – Oe
Newman, Brent, 505.667.3201 bnewman@lanl.gov, New Mexico Institute of Mining and Technology – Cs
Newman, Brent D., (505) 667-3021 bnewman@lanl.gov, Los Alamos National Lab – Gg
Newman, Dianne K., (626) 395-6790 dkn@gps.caltech.edu, California Institute of Technology – Py
Newman, Jamie, 212-769-5386 jamien@amnh.org, American Museum of Natural History – Gg
Newman, Julie, 979-845-9283 newman@geo.tamu.edu, Texas A&M Univ – Gg
Newman, Steven B., (860) 832-2940 newman@ccsu.edu, Central Connecticut State Univ – Oa
Newman, William A., (619) 534-2150 wnewman@ucsd.edu, Univ of California, San Diego – Ob
Newman, William I., (310) 825-3912 win@ucla.edu, Univ of California, Los Angeles – Xy
Newmark, Robin L., (925) 423-3644 newmark@llnl.gov, Lawrence Livermore National Lab – Ys
Newsom, Horton E., (505) 277-0375 newsom@unm.edu, Univ of New Mexico – Xc
Newton, Cathryn R., 315-443-2672 crnewton@syr.edu, Syracuse Univ – Po
Newton, Jan, (206) 616-3641 newton@ocean.washington.edu, Univ of Washington – Ob
Newton, Robert M., (413) 585-3946 rnewton@email.smith.edu, Smith College – Gm
Ng, Timothy J., (301) 405-4345 tng@grad.umd.edu, Univ of Maryland – Oo
Ngoy, Kikombo, (908) 737-3718 kngoy@kean.edu, Kean Univ – Oy
Ni, James F., (505) 646-1920 jni@nmsu.edu, New Mexico State Univ, Las Cruces – Ys
Ni-Meister, Wenge, 212-772-5321 wenge.ni-meister@hunter.cuny.edu, Hunter College (CUNY) – Or
Nicholas, Joseph W., (540) 654-1470 jnicola@mwc.edu, Mary Washington College – Oy
Nicholls, James W., (403) 220-7127 Univ of Calgary – Gi
Nichols, Krissa D., (617) 353-9709 krissa@bu.edu, Boston Univ – Oo
Nichols, Kyle K., 518-580-5194 knichols@skidmore.edu, Skidmore College – Gm
Nichols, Liz, 949-451-5561 lnichols@ivc.edu, Irvine Valley College – Oo
Nichols, Stanley A., (608) 262-6556 sanichol@wisc.edu, Univ of Wisconsin - Extension – Oo
Nichols, Terry E., (501) 575-7317 tenichol@comp.uark.edu, Univ of Arkansas, Fayetteville – Oo
Nichols, William E., (509) 372-6040 will.nicholsv, Pacific Northwest National Lab – Hq
Nichols, Jr., Woodrow W., (919) 560-5171 North Carolina Central Univ – Oy
Nicholson, Kirsten N., (765) 285-8268 knichols@gw.bsu.edu, Ball State Univ – Gi
Nicholson, Laura J., (609) 984-6587 laura.nicholson@dep.state.nj.us,

New Jersey Geological Survey – Hw
Nicholson, Sharon E., (850) 644-4022 nicholson@met.fsu.edu, Florida State Univ – Ow
Niciejewski, Richard J., (313) 747-3445 niciejew@engin.umich.edu, Univ of Michigan – Og
Nickerson, John G., (919) 733-7353 North Carolina Geological Survey – Og
Nicolaysen, Kirsten P., 785-532-1908 knic@k-state.edu, Univ of Kansas – Gt
Nicoletti, Mary Annmanicole@princeton.edu, Princeton Univ – Oo
Nicot, Jean-Philippe, 512-471-6246 jp.nicot@beg.utexas.edu, Univ of Texas at Austin, Jackson School of Geosciences – Ng
Nielsen, Donald R., (530) 753-5760 drnielsen@ucdavis.edu, Univ of California, Davis – Sp
Nielsen, Gerald A., (406) 994-5075 nielsen@gemini.oscs.montana.edu, Montana State Univ – Sd
Nielsen, Peter A., (603) 358-2553 pnielsen@keene.edu, Keene State College – Gx
Nielsen, Roger L., (541) 737-1235 nielsenr@geo.oregonstate.edu, Oregon State Univ – Gi
Nielsen-Gammon, Johnn-g@tamu.edu, Texas A&M Univ – Oa
Nielson, Russell L., (936) 468-2248 rnielson@sfasu.edu, Stephen F. Austin State Univ – Gr
Niem, Alan R., (541) 737-1233 niema@geo.oregonstate.edu, Oregon State Univ – Gd
Niemann, Williamniemann@marshall.edu, Marshall Univ – Ng
Niemi, Nathan, 734-764-6377 naniemi@umich.edu, Univ of Michigan – Gc
Niemi, Tina M., (816) 235-5342 niemit@umkc.edu, Univ of Missouri-Kansas City – Gt
Niemitz, Jeffery W., 717-245-1285 niemitz@dickinson.edu, Dickinson College – ClOuGn
Nieto, Alberto S., (217) 333-1068 anieto@hercules.geology.uiuc.edu, Univ of Illinois, Urbana-Champaign – Ng
Niewendorp, Clark A., (803) 896-7708 South Carolina Geological Survey – Eg
Niiler, Pearn P., (619) 534-4100 pniiler@ucsd.edu, Univ of California, San Diego – Op
Nikitina, Daria L., 724 938-4020 nikitina@cup.edu, California Univ of Pennsylvania – Ge
Nimmo, Francis, 831-459-1783 fnimmo@pmc.ucsc.edu, Univ of California, Santa Cruz – Xy
Nimz, Cheryl K., 217-265-5194 nimz@isgs.uiuc.edu, Illinois State Geological Survey – Gg
Nimz, Gregory, (925) 423-2766 nimz1@llnl.gov, Lawrence Livermore National Lab – Yx
Nissen, Barbara K., (315) 469-2432 nissenb@sunyocc.edu, Onondaga Community College – Ge
Nissen, Susan E., (785) 864-2179 snissen@kgs.ku.edu, Univ of Kansas – Yg
Nitao, John J., (925) 423-0297 nitao1@llnl.gov, Lawrence Livermore National Lab – Hq
Nitecki, Matthew H., (312) 665-7093 mnitecki@fmnh.org, Field Museum of Natural History – Pi
Nittler, Larry R., (202) 478-8460 Carnegie Inst of Washington – Xc
Nittrouer, Charles A., 206-543-5099 nittroue@ocean.washington.edu, Univ of Washington – GuYr
Nixon, R. Paul, 801-422-4657 paul_nixon@byu.edu, Brigham Young Univ – Go
Nixon, Scott W., (401) 874-6803 swn@gso.uri.edu, Univ of Rhode Island – Ob
Nixon III, Roy A.RickN41784@aol.com, Georgia Perimeter College – Gm
Niyogi, Dev, (765) 494-9531 climate@purdue.edu, Purdue Univ – Oa
Njoku, Eni G., (818) 354-3693 eni.g.njoku@jpl.nasa.gov, Jet Propulsion Lab – Or
Nkedi-Kizza, Peter, (352) 392-1951 kizza@gnv.ifas.ufl.edu, Univ of Florida – Sp
Noakes, John E., (706) 542-1395 Univ of Georgia – Cc
Noble, Elizabeth B., 252-335-3595 ebnoble@mail.ecsu.edu, Elizabeth City State Univ – Op
Noble, Paula J., (775) 784-6211 noblepj@unr.edu, Univ of Nevada, Reno – Pm
Noblett, Jeffrey B., (719) 389-6681 jnoblett@coloradocollege.edu, Colorado College – Gx
Noe, Dave, (303) 866-2432 dave.noe@state.co.us, Colorado Geological Survey – Ng
Nof, Doronnof@ocean.fsu.edu, Florida State Univ – Op
Noffke, Norannoffke@odu.edu, Old Dominion Univ – Ge
Nolan, Robert P., (718) 951-4242 Graduate School of the City Univ of New York – Co
Nold, John L., (660) 543-8853 nold@ucmo.edu, Univ of Central Missouri – Em
Nolet, Guust, (609) 258-4128 Princeton Univ – Om
Nolet, Guust, 609-258-4128 nolet@princeton.edu, Princeton Univ – Ys
Nolet, Guust M., (609) 258-4128 nolet@princeton.edu, Princeton Univ – Ys
Nolin, Anne, (541) 737-8051 nolina@geo.oregonstate.edu, Oregon State Univ – Or
Noll, Karen, 415-452-7775 knoll@ccsf.edu, City College of San Francisco – Oo
Noll, Mark R., (585) 395-5717 mnoll@esc.brockport.edu, SUNY College at Brockport – Cl
Noll, Michael G., (229) 333-7143 mgnoll@valdosta.edu, Valdosta State Univ – Oo
Noltimier, Hallan C., (614) 292-9796 noltimier.2@osu.edu, Ohio State Univ – Ym
Norbeck, Peter M., 406-496-4121 pnorbeck@mtech.edu, Montana Tech of The Univ of Montana – Hw
Nord, Julia, (703) 993-3395 jnord@gmu.edu, George Mason Univ – Cp
Nordstrom, Darrell K., 303-541-3037 dkn@usgs.gov, Univ of Colorado – Cl
Nordt, Lee C., (254) 710-4288 lee_nordt@baylor.edu, Baylor Univ – SdGa
Norell, Mark A., (212) 769-5804 American Mus of Natural History – Pv
Noren, Anders, 612-626-7889 noren021@umn.edu, Univ of Minnesota, Twin Cities – Gn
Norford, Brian, (403) 292-7000 Univ of Calgary – Gr
Norman, David I., (505) 835-5404 dnorman@nmt.edu, New Mexico Institute of Mining and Technology – Em
Norman, David K., 360-902-1439 dave.norman@wadnr.gov, Washington Division of Geology & Earth Resources – Gg
Norman, John M., (608) 262-4576 jmnorman@wisc.edu, Univ of Wisconsin, Madison – Sp
Norman, John M., (608)262-4576 jmnorman@wisc.edu, Univ of Wisconsin, Madison – So
Norris, Dean R., (321) 674-8096 norris@fit.edu, Florida Institute of Technology – Ob
Norris, Gary, (775) 784-6835 gnorris@scs.unr.edu, Univ of Nevada, Reno – Hs
Norris, Geoffrey, (416) 978-4851 norris@quartz.geology.utoronto.ca, Univ of Toronto – Pl
Norris, Richard J., +64 3 479-7520 richard.norris@otago.ac.nz, Univ of Otago – Gc
Norris, Robert M., (805) 893-3339 Univ of California, Santa Barbara – Gm
Norris-Tull, Delena, 406-683-7043 d_norris@umwestern.edu, Univ of Montana, Western – Oe
Nortcliff, Stephens.nortcliff@reading.ac.uk, Univ of Reading – Sd
North, Geraldg-north@tamu.edu, Texas A&M Univ – Oa
North, Gerald R., (979) 845-8083 Texas A&M Univ – Ow
Northrup, Clyde J., (208) 426-1581 cjnorth@boisestate.edu, Boise State Univ – Gc
Norton, Stephen A., (207) 581-2156 norton@maine.maine.edu, Univ of Maine – Cl
Norwine, Jim R., (512) 595-3589 kfjrn00@tamuk.edu, Texas A&M Univ, Kingsville – Oa
Nosal, Thomas E., 86004243590 thomas.nosal@po.state.ct.us, Geological and Natural History Survey of Connecticut – Oo
Nosker, Sue, (707) 664-2328 nosker@sonoma.edu, Sonoma State Univ – Oo
Nosse, Mary, 831-459-4089 mnosse@es.ucsc.edu, Univ of California, Santa Cruz – Oo
Nourse, Jonathan A., 909-869-3460 janourse@csupomona.edu, California State Polytechnic Univ, Pomona – Gt
Novacek, Michael J., (212) 769-5805 American Mus of Nat History – Pv
Novack-Gottshall, Philip M., (678) 839-4061 pnovackg@westga.edu, Univ of West Georgia – Pg
Novak, Gary A., (323) 343-2406 gnovak@calstatela.edu, California State Univ, Los Angeles – Gz
Novak, Irwin D., (207) 780-5025 novak@usm.maine.edu, Univ of Southern Maine – Gm

Novak, Michael D., (604) 822-2875 Univ of British Columbia – Sp
Nowack, Robert L., (765) 494-5978 nowack@purdue.edu,
Purdue Univ – Ys
Nowell, Arthur R. M., (206) 543-6605 nowell@cofs.washington.edu,
Univ of Washington – Ou
Nowell, Geoff M., +44 (0)191 3342339 g.m.nowell@durham.ac.uk,
Durham Univ – Cg
Nowlin, Jr., Worth D., (979) 845-1443 wnowlin@tamu.edu,
Texas A&M Univ – Op
Nuhfer, Edward, (303) 556-9429 Univ of Colorado, Denver – Ng
Null, E. Jan, (415) 338-2061 San Francisco State Univ – Oa
Numbere, Daopu, 573-341-4758 numbere@umr.edu,
Univ of Missouri, Rolla – Np
Nunez, B. S., (361) 749-6705 scott.nunez@mail.utexas.edu,
Univ of Texas at Austin – Ob
Nunn, Jeffrey A., (225) 388-6657 jeff@geol.lsu.edu,
Louisiana State Univ – Yg
Nur, Amos M., (650) 723-9526 Stanford Univ – Yg
Nusbaum, Robert L., (843) 953-5596 nusbaumr@cofc.edu,
College of Charleston – Gz
Nutter, Paul, 970-351-2398 Paul.Nutter@unco.edu,
Univ of Northern Colorado – Oaw
Nyahay, Richard E., (518) 486-2161 rnyahay@mail.nysed.gov,
New York State Geological Survey – Eo
Nyblade, Andrew A., (814) 863-8341 andy@geosc.psu.edu,
Pennsylvania State Univ, Univ Park – Yg
Nyhan, John W., (505) 667-3163 Los Alamos National Lab – So
Nyland, Edo, (780) 492-5502 edo@phys.ualberta.ca,
Univ of Alberta – Ys
Nyman, Matthewmwnyman@unm.edu,
Univ of New Mexico – Gp
Nyquist, Jonathan, (215) 204-7484 nyq@temple.edu,
Temple Univ – Yg
Nystrom, Jr., Paul G., (803) 896-7708 nystrom@water.dnr.state.sc.us,
South Carolina Geological Survey – Eg
Nystuen, Jeffrey A., (206) 543-1343 nystuen@apl.washington.edu,
Univ of Washington – Op
Nzengung, Valentine A., (706) 542-2699 vnzengun@gly.uga.edu,
Univ of Georgia – Cl

O

O'Boyle, Christina, 323 953-4000 (Ext. 2691) oboylecl@lacitycollege.edu,
Los Angeles City College – Gg
O'Brien, Arnold L., (201) 200-3161 Univ of Massachusetts/Lowell – Hw
O'Brien, James J., (850) 644-4581 obrien@coaps.fsu.edu,
Florida State Univ – Ow
O'Brien, John M., (201) 200-3161 Trenton State College – Gs
O'Brien, Lawrence E., (914) 341-4570 lobrien@mail.sunyorange.edu,
Orange County Community College – Gg
O'Brien, Neal R., (315) 267-2287 obriennr@potsdam.edu,
SUNY, Potsdam – Gs
O'Brien, Rachel, (814) 332-2875 robrien@allegheny.edu,
Allegheny College – Hw
O'Callaghan, Mick, +353 21 4902662 mick.ocallaghan@ucc.ie,
Univ College Cork – Oo
O'Connell, Richard J., (617) 495-2532 oconnell@eps.harvard.edu,
Harvard Univ – Yg
O'Connell, Suzanne B., (860) 685-2262 soconnell@wesleyan.edu,
Wesleyan Univ – GsuGe
O'Connor, George A., (352) 392-1803 (Ext. 329) gao@ufl.edu,
Univ of Florida – Sc
O'Connor, Joel S., (631) 584-5088 oconnor.joel@notes.cc.sunysb.edu,
SUNY, Stony Brook – Og
O'Connor, Yuet-Ling, (562) 938-4794 yoconnor@lbcc.cc.ca.us,
Long Beach City College – Gi
O'Donnell, James, (860) 445-3471 odonnell@uconnvm.uconn.edu,
Univ of Connecticut – Op
O'Driscoll, Brian, 0044-1782-583184 b.odriscoll@esci.keele.ac.uk,
Keele Univ – Gpi
O'Driscoll, Michael A., 252 328 5578 odriscollm@mail.ecu.edu,
East Carolina Univ – Hg
O'Fallon, Elaineofallon@mail.wsu.edu,
Washington State Univ – Cc
O'Halloran, Ivan, (519) 674-1635 iohallo@ridgetownc.uoguelph.ca,
Univ of Guelph – So
O'Hara, Charles G., (662) 325-2067 cgohara@gri.msstate.edu,
Mississippi State Univ – Oir
O'Hara, Kieran D., (859) 257-6931 geokoh@uky.edu,
Univ of Kentucky – Gc
O'Hara, Matthew J., (509) 373-1671 matt.ohara@pnl.gov,
Pacific Northwest National Lab – Cg
O'Hearn, Timothy, (202) 633-1811 ohearnt@si.edu,
Smithsonian Inst / National Museum of Natural History – GvCa
O'Keefe, Jennifer, (606) 783-2349 j.okeefe@moreheadstate.edu,
Morehead State Univ – Gn
O'Leary, Maureen, 631-444-3730 maureen.oleary@sunysb.edu,
SUNY, Stony Brook – Pv
O'Meara, Danielomeara@ou.edu,
Univ of Oklahoma – Np
O'Meara, Stephanie, (970) 225-3584 Stephanie_O'Meara@partner.nps.gov,
Colorado State Univ – Gc
O'Melia, Charles R., (410) 516-7102 omelia@jhu.edu,
The Johns Hopkins Univ – Oo
O'Mullan, Gregory D.gomullan@princeton.edu,
Princeton Univ – Ob
O'Neil, Caron, 717.702.2042 coneil@state.pa.us,
Pennsylvania Bureau of Topographic & Geologic Survey – Gg
O'Neil, James, (734) 764-1435 jro@umich.edu,
Univ of Michigan – Cs
O'Neil, Patrick E., 205/247-3586 poneil@gsa.state.al.us,
Geological Survey of Alabama – Hs
O'Neill, Patrick M., (504) 388-2681 pat@lgs.bri.lsu.edu,
Louisiana State Univ – Sf
O'Rourke, Thomas, (607) 255-6470 tdo1@cornell.edu,
Cornell Univ – Ng
Oakey, Neil S., (902) 426-3147 oakeyn@mar.dfo-mpo.gc.ca,
Dalhousie Univ – Op
Oberdorfer, June A., (408) 924-5026 june@geosun.sjsu.edu,
San Jose State Univ – Hw
Obermeyer, Nancy J., 812-237-4351 nancyo@kiva.net,
Indiana State Univ – Oi
Obia, Godson C., 217-581-5528 cfgo@eiu.edu,
Eastern Illinois Univ – Oo
Oboh-Ikuenobe, Francisca E., (573) 341-6946 ikuenobe@umr.edu,
Univ of Missouri, Rolla – Pl
Obreza, Thomas A., (941) 658-3400 taob@gnv.ifas.ufl.edu,
Univ of Florida – So
Oches, Eric A., (813) 974-7531 oches@cas.usf.edu,
Univ of South Florida, Tampa – Gm
Odom, John W., (334) 844-3966 jodom@acesag.auburn.edu,
Auburn Univ – Sc
Odom, LeRoy A., (850) 644-6706 odom@magnet.fsu.edu,
Florida State Univ – Cg
Odom, Robert I., 206-685-3788 odom@apl.washington.edu,
Univ of Washington – YrGu
Oduor, Peter, (701) 231-7145 Peter.Oduor@ndsu.edu,
North Dakota State Univ – Oi
Oellrich, Peggy, (954)262-3605 hess@nova.edu,
Nova Southeastern Univ – Oo
Oertel, George, (757) 683-4935 goertel@odu.edu,
Old Dominion Univ – Ou
Oertel, Gerhard, (310) 825-1780 oertel@ess.ucla.edu,
Univ of California, Los Angeles – Gc
Offerman, Katherine A., 410-554-5544 kofferman@dnr.state.md.us,
Maryland Geological Survey – Ge
Offler, Robin, 61 02 4921 7044 Robin.Offler@newcastle.edu.au,
Univ of Newcastle – Gp
Ogard, Allen E., (505) 667-6344 Los Alamos National Lab – Hw
Ogg, James G., (765) 494-8681 jogg@purdue.edu,
Purdue Univ – Gs
Oglesby, David D., (951) 827-2036 david.oglesby@ucr.edu,
Univ of California, Riverside – Ys
Oglesby, Elizabeth, 520-616-7242 eoglesby@email.arizona.edu,
Univ of Arizona – Oo
Oglesby, Robert J.roglesby2@unl.edu,
Univ of Nebraska, Lincoln – Oa
Ogram, Andrew V., (352) 392-5790 avo@mail.ifas.ufl.edu,
Univ of Florida – Sb
Ogston, Andrea S., (206) 543-0768 ogston@ocean.washington.edu,
Univ of Washington – Ou
Ohler, Kathy, 612-624-9031 k-ohler@umn.edu,
Univ of Minnesota, Twin Cities – Oo
Ohlmacher, Gregory C., (785) 864-2194 ohlmac@kgs.ku.edu,
Univ of Kansas – Ng
Ohman, Mark D., (619) 534-2754 mohman@ucsd.edu,
Univ of California, San Diego – Ob
Ohmoto, Hiroshi, (814) 865-4074 ohmoto@geosc.psu.edu,
Pennsylvania State Univ, Univ Park – Cs
Ojakangas, Richard W., (218) 726-7923 rojakang@d.umn.edu,
Univ of Minnesota, Duluth – Gd

Ojala, Carl F., (313) 487-0218 Eastern Michigan Univ – Oa
Okabe, Ian, (604) 664-9272 Ian.Okabe@ec.gc.ca, Univ of British Columbia – Ow
Okal, Emile A., (847) 491-3238 emile@earth.northwestern.edu, Northwestern Univ – Ys
Okaya, David A., (213) 740-7452 okaya@earth.usc.edu, Univ of Southern California – Ys
Okubo, Paul G., (808) 967-7328 Univ of Hawai'i, Manoa – Ys
Okulewicz, Steven C., 516 463-6545 geosco@hofstra.edu, Hofstra Univ – Gg
Okunade, Samuel, (937) 376-6455 Central State Univ – Gm
Olafsen-Lackey, Susan, (402) 371-6512 slackey@unl.edu, Unversity of Nebraska - Lincoln – Hw
Olanrewaju, Johnson, 814-871-7453 olanrewa001@gannon.edu, Gannon Univ – Cg
Oldenburg, Curtis M., (510) 486-7419 cmoldenburg@lbl.gov, Lawrence Berkeley National Lab – Hw
Oldenburg, Douglas W., (604) 822-5406 doldenburg@eos.ubc.ca, Univ of British Columbia – Yg
Oldershaw, Alan E., (403) 220-3258 oldersha@geo.ucalgary.ca, Univ of Calgary – Gs
Oldham, Dave, 304 599 0740 dwoldham@aol.com, West Virginia Univ – Go
Oldham, Richard L., 568-3100 (12147) OldhamR@arc.losrios.edu, American River College – Gg
Oldow, John S., (208) 885-7327 oldow@uidaho.edu, Univ of Idaho – Gc
Olea, Ricardo ricardo@oleageostats.com, Univ of Kansas – Gq
Oleinik, Anton, (561) 297-3297 aoleinik@fau.edu, Florida Atlantic Univ – Pe
Olejnik, Jennifer, (812) 855-7428 jolejnik@indiana.edu, Indiana Geological Survey – Ge
Olesik, John W., (614) 292-6954 olesik.2@osu.edu, Ohio State Univ – Ca
Olhoeft, Gary R., (303) 273-3458 golhoeft@mines.edu, Colorado School of Mines – Yg
Olivares, Jose A., (505) 665-2643 Los Alamos National Lab – Cs
Oliver, Dean S., (405) 325-2921 dsoliver@ou.edu, Univ of Oklahoma – Np
Oliver, John E., (812) 237-2260 geoliv@isugw.indstate.edu, Indiana State Univ – Oa
Oliver, Ronald D., (702) 794-7095 ron_oliver@lanl.gov, Los Alamos National Lab – Yg
Olmstead, Wayne, 406-496-4753 wolmstead@mtech.edu, Montana Tech of The Univ of Montana – Ca
Olmsted, Wayne, (406) 496-4151 Montana Tech of the Univ of Montana – Ca
Olness, Alan E., (320) 589-3411 Univ of Minnesota, Twin Cities – So
Olsen, Carolyn, (801) 537-3357 nrugs.colsen@State.ut.us, Utah Geological Survey – Oo
Olsen, Khris B., (509) 376-4114 kb.olsenv, Pacific Northwest National Lab – Cg
Olsen, Kim B., (805) 893-7394 kbolsen@crustal.ucsb.edu, Univ of California, Santa Barbara – Ys
Olsen, Kim B., 619-594-2649 kbolsen@geology.sdsu.edu, San Diego State Univ – Ye
Olsen, Paul E., (845) 365-8491 polsen@ldeo.columbia.edu, Columbia Univ – Pv
Olsen, Phil G., (805) 965-0581 Santa Barbara City College – Ow
Olsen, Sakiko N., (410) 516-7040 olsen@jhu.edu, The Johns Hopkins Univ – Gp
Olson, Charles, (313) 593-5235 caolson@umich.edu, Univ of Michigan, Dearborn – Or
Olson, Edwin A., (509) 466-3268 Whitworth College – Gg
Olson, Hillary C., (512) 471-0455 olson@utig.ig.utexas.edu, Univ of Texas at Austin, Jackson School of Geosciences – Ps
Olson, John L., 406-657-2938 jolson@mtech.edu, Montana Tech of The Univ of Montana – Hw
Olson, Kenneth R., (217) 333-9639 Univ of Illinois, Urbana-Champaign – Sc
Olson, Peter L., (410) 516-7707 olson@jhu.edu, The Johns Hopkins Univ – Yg
Olson, Ted L., 435 283-7533 ted.olson@snow.edu, Snow College – Ys
Olsson, Richard K., (732) 445-3043 olsson@rci.rutgers.edu, Rutgers, The State Univ of New Jersey – Pm
Olszewski, Thomas, 979-845-2465 olszewski@geo.tamu.edu, Texas A&M Univ – Pe
Oltman, Jennifer D., (601) 961-5518 jennifer-oltman@deq.state.ms.us, Mississippi Office of Geology – Oy
Oltman-Shay, Joan M., (425) 644-9660 joan@nwra.com, Univ of Washington – Op
Olyphant, Greg A., (812) 855-5154 olyphant@indiana.edu, Indiana Univ, Bloomington – Hw
Omar, Gomaa I., (215) 898-6908 gomar@sas.upenn.edu, Univ of Pennsylvania – Cc
Omar, Ikhtyar, (403) 471-7775 Northern Alberta Institute of Tech – Gp
Onasch, Charles M., (419) 372-7197 conasch@bgnet.bgsu.edu, Bowling Green State Univ – Gc
Onderdonk, Nate, (562) 985-2654 nonderdo@csulb.edu, California State Univ, Long Beach – Gtm
Onderdonk, Nathan, 323 343 2412 nonderd@exchange.calstatela.edu, California State Univ, Los Angeles – Gt
Ondrovic, Claire A., (631) 632-8535 claire.ondrovic@sunysb.edu, SUNY, Stony Brook – Oo
ONeal, Michael, 302-831-8273 michael@udel.edu, Univ of Delaware – Gm
Onstott, Tullis, 609-258-1622 tullis@princeton.edu, Princeton Univ – Yg
Onstott, Tullis C., (609) 258-6898 tullis@princeton.edu, Princeton Univ – Py
Oona, Hain, (505) 667-5685 Los Alamos National Lab – Oo
Oostdam, Bernard L., 717-871-9928 boostdam@hotmail.com, Millersville Univ – Ou
Oostrom, Martinus, (509) 372-6044 mart.oostrom@pnl.gov, Pacific Northwest National Lab – Hq
Opdyke, Neil D., (352) 392-6127 Univ of Florida – Ym
Ophori, Duke U., (973) 655-7558 ophorid@mail.montclair.edu, Montclair State Univ – Hw
Opp, Susan, (510) 885-3475 susan.opp@csueastbay.edu, California State Univ, East Bay – Oo
Oppenheimer, Michael, (609) 258-2338 omichael@princeton.edu, Princeton Univ – Oa
Opper, Carl, (727) 791-2536 opper.carl@spcollege.edu, Saint Petersburg College, Clearwater – Hw
Oppliger, Gary, (775) 784-7056 oppliger@mines.unr.edu, Univ of Nevada, Reno – Ye
Oppo, Delia W., 508-289-2681 doppo@whoi.edu, Woods Hole Oceanographic Institution – Pm
Oppong, Joseph R., (940) 565-2181 oppong@unt.edu, Univ of North Texas – Oe
Oram, Brian F., (570) 408-4619 Wilkes Univ – Hg
Orcutt, John A., (619) 534-2887 jorcutt@ucsd.edu, Univ of California, San Diego – Yr
Orcutt, Karen, 228-688-3154 Univ of Southern Mississippi – Ob
Orf, Leigh, (828) 232-5159 Univ of North Carolina, Asheville – Oa
Orians, Kristin J., (604) 822-6571 korians@eos.ubc.ca, Univ of British Columbia – Cm
Orlandini, Kent A., (630) 252-4236 Argonne National Lab – Cc
Orme, Amalie, (818) 677-3532 amalie.orme@csun.edu, California State Univ, Northridge – Gm
Orndorff, Richard L., (702) 895-4388 orndorfr@nevada.edu, Univ of Nevada, Las Vegas – Gq
Orndorff, Richard L., (509) 359-2855 rorndorff@mail.ewu.edu, Eastern Washington Univ – Ge
Ornduff, Robert, (510) 642-2089 Univ of California, Berkeley – Pb
Orphan, Victoria, (626) 395-1786 vorphan@gps.caltech.edu, California Institute of Technology – Py
Orr, Robert D., (209) 376-2481 robert.orr@pnl.gov, Pacific Northwest National Lab – Ca
Orr, William N., (541) 346-4577 worr@darkwing.uoregon.edu, Univ of Oregon – Pm
Orsi, Alejandro H., (979) 845-4014 aorsi@ocean.tamu.edu, Texas A&M Univ – Op
Orsi, Thomas H., (504) 645-8778 Univ of Southern Mississippi – Yr
Ort, Michael H., (928) 523-9363 michael.ort@nau.edu, Northern Arizona Univ – Gv
Ortiz, Joseph D., 330-672-2225 jortiz@kent.edu, Kent State Univ – Gs
Ortoleva, Peter J., (812) 855-2717 ortoleva@indiana.edu, Indiana Univ, Bloomington – Cl
Ortt, Jr., Richard A., (410) 554-5541 rortt@dnr.state.md.us, Maryland Geological Survey – Op
Orville, Harold D., (605) 394-2291 horville@ias.sdsmt.edu, South Dakota School of Mines & Technology – Oa
Orville, Richard E., (979) 845-7671 rorville@tamu.edu, Texas A&M Univ – Oa
Osborn, Gerald D., (403) 220-6448 Univ of Calgary – Gm
Osborn, Jeffrey, (816) 785-4017 josborn@truman.edu, Truman State Univ – Pb

Osborn, Thomas R., (410) 516-7039 osborn@jhu.edu,
The Johns Hopkins Univ – Op
Osborne, W. Edward, 205/247-3540 eosborne,
Geological Survey of Alabama – Gg
Osburn, Don, (902) 585-1513 Acadia Univ – Gg
Oser, Jens, 089/2180 4208 oeser@geophysik.uni-muenchen.de,
LMU - Univ of Munich – Yg
Osgood, David T., (203) 931-6056 Univ of New Haven – Cl
Osiensky, James L., (208) 885-4681 osiensky@uidaho.edu,
Univ of Idaho – Hw
Osinski, Gordon R., (519) -66-1-21 (Ext. ext.84208) gosinbski@uwo.ca,
Univ of Western Ontario – XgcGp
Osisanya, Samuel O., (405) 325-6872 sosisanya@ou.edu,
Univ of Oklahoma – Np
Oskin, Michael E., 919-962-4239 oskin@email.unc.edu,
Univ of North Carolina, Chapel Hill – Gm
Osleeb, Jeffrey P., (212) 817-8240 josleeb@gc.cuny.edu,
Graduate School of the City Univ of New York – Oo
Osleger, David A., (530) 754-7824 osleger@geology.ucdavis.edu,
Univ of California, Davis – Gr
Osmond, Deanna L., (919) 515-7303 deanna_osmond@ncsu.edu,
North Carolina State Univ – Sb
Osmond, John K., (850) 644-5860 osmond@gly.fsu.edu,
Florida State Univ – Cc
Ostermann, Dorinda R., (508) 289-2592 dostermann@whoi.edu,
Woods Hole Oceanographic Institution – Pe
Ostlund, Cheryl, (801) 537-3305 nrugs.costlund@state.ut.us,
Utah Geological Survey – Oo
Othberg, Kurt L., (208) 885-7560 othberg@uidaho.edu,
Univ of Idaho – Gg
Ott, Kevin C., (505) 667-4600 Los Alamos National Lab – Oo
Otterstrom, Samuel M., (801) 422-7751 samuel_otterstrom@byu.edu,
Brigham Young Univ – Ou
Ottley, Chris J., 0191 3742525 c.j.ottley@durham.ac.uk,
Durham Univ – Cg
Otto, LeeAnn, (619) 260-4600 lotto@acusd.edu,
Univ of San Diego – Oo
Ottu, Emanuel D., (914) 251-6652 SUNY, Purchase – Oo
Otvos, Ervin G., 228-872-4235 ervin.otvos@usm.edu,
Univ of Southern Mississippi – On
Ou, Hsien Wang, (845) 365-8338 dou@ldeo.columbia.edu,
Columbia Univ – Op
Ou, Li-Tse, (352) 392-1951 lto@gnv.ifas.ufl.edu,
Univ of Florida – Sb
Ouellet, Jaques, (514) 398-4755 (Ext. 0536) McGill Univ – Nm
Ouellette, Stefanie, (954)328-0580 ouellett@nova.edu,
Nova Southeastern Univ – Oo
Ouellette, V., (970) 351-2647 vlouell@unco.edu,
Univ of Northern Colorado – Oo
Ouimette, Mark A., (325) 670-1383 ouimette@hsutx.edu,
Hardin-Simmons Univ – GeiGt
Ould-Mey, Mohameden, (812) 237-2253 Indiana State Univ – Oo
Over, D. Jeffrey, (585) 245-5294 over@geneseo.edu,
SUNY, Geneseo – Ps
Overacre, Lora A., 404-656-3214 Georgia Dept of Nat Resources – Oo
Overpeck, Jonathan T., (520) 622-9065 jto@u.arizona.edu,
Univ of Arizona – Oa
Oviatt, Candace, (401) 874-6132 coviatt@gso.uri.edu,
Univ of Rhode Island – Ob
Oviatt, Charles G., (785) 532-2245 joviatt@ksu.edu,
Kansas State Univ – Gm
Owen, Claudia, (909) 787-3106 owenc@fiu.edu,
Univ of Oregon – Gp
Owen, Donald E., (409) 880-8234 Lamar Univ – Gr
Owen, Lewis, 513-556-42-3 Lewis.Owen@uc.edu,
Univ of Cincinnati – Gl
Owen, Robert M., (734) 763-4593 rowen@umich.edu,
Univ of Michigan – Oc
Owens, Brent E., (757) 221-1813 beowen@wm.edu,
College of William & Mary – Gz
Owens, Don R., (501) 569-3544 drowens@ualr.edu,
Univ of Arkansas, Little Rock – Em
Owens, Thomas J., (803) 777-4530 Univ of South Carolina – Ys
Owensby, Clenton E., (913) 532-7232 Kansas State Univ – Sf
Oza, Rupal, 212-650-3035 rupal.oza@hunter.cuny.edu,
Hunter College (CUNY) – Oo
Ozbay, M. Ugor, (303) 273-3710 Colorado School of Mines – Nr
Ozdemir, Levent, (303) 273-3419 lozdemir@mines.edu,
Colorado School of Mines – Nr
Ozdemir, Ozden, (905) 828-3829 ozdemir@physics.utoronto.ca,
Univ of Toronto – Ym
Ozdenerol, Esra, 901-678-2787 eozdenrl@memphis.edu,
Univ of Memphis – Oi
Ozkan-Haller, H. Tuba, (541) 737-9170 ozkan@coas.oregonstate.edu,
Oregon State Univ – On
Ozsvath, David L., (715) 346-2287 dozsvath@uwsp.edu,
Univ of Wisconsin, Stevens Point – Hw

P

Pachut, Jr., Joseph F., (317) 274-7785 jpachut@iupui.edu,
Indiana Univ / Purdue Univ, Indianapolis – Po
Pacia, Christine, (908) 737-3691 cpacia@kean.edu,
Kean Univ – Oo
Padian, Kevin, (510) 642-7434 kpadian@violet.berkeley.edu,
Univ of California, Berkeley – Pv
Paduan, Jeffrey D., (408) 656-3350 paduan@oc.nps.navy.mil,
Naval Postgraduate School – Op
Paez, H. A., 905-525-9140 (Ext26099) paezha@mcmaster.ca,
McMaster Univ – Oo
Page, Pierre, (514) 987-7786 page.pierre@uqam.ca,
Universite du Quebec a Montreal – Cs
Pagiatakis, Spiros, (416) 736-2100 (Ext. 20644) spiros@yorku.ca,
York Univ – Yd
Pagnac, Darrin, 605-394-2469 Darrin.Pagnac@sdsmt.edu,
South Dakota School of Mines & Technology – Pv
Paige, David A., (310) 825-4268 dap@ess.ucla.edu,
Univ of California, Los Angeles – Xy
Paine, Jeffrey G., (512) 475-9524 jeff.paine@beg.utexas.edu,
Univ of Texas at Austin, Jackson School of Geosciences – Gr
Pair, Donald, (937) 229-2936 Univ of Dayton – Gl
Pakhomov, Evgeny, (604) 827-5564 epakhomov@eos.ubc.ca,
Univ of British Columbia – Ob
Palenik, Brianbpalenik@ucsd.edu,
Univ of California, San Diego – Ob
Paleologos, Evangelos K., 803 777-8125 epal@geol.sc.edu,
Univ of South Carolina – Hw
Palin, J. Michael, +64 3 479-9083 michael.palin@stonebow.otago.ac.nz,
Univ of Otago – Gi
Palladino, Steve D., (805) 654-6400 (Ext. 1365) Ventura College – Oi
Pallis, Ted J., (609) 984-6587 ted.pallis@dep.state.nj.us,
New Jersey Geological Survey – Oi
Pallmann, Albert J., (314) 977-3125 Saint Louis Univ – Oa
Palluconi, Frank D., (818) 354-8362 frank.d.palluconi@jpl.nasa.gov,
Jet Propulsion Lab – Or
Palmer, Arthur N., (607) 436-3064 palmeran@oneonta.edu,
SUNY, Oneonta – Hg
Palmer, Christina, (909) 537-5336 cpalmer@csusb.edu,
California State Univ, San Bernardino – Oo
Palmer, Derecke, 612 9385 8719 d.palmer@unsw.edu.au,
Univ of New South Wales – Ye
Palmer, Donald A., (865) 576-5109 palmerda@ornl.gov,
Oak Ridge National Lab – Cg
Palmer, Donald F., (330) 672-2680 dpalmer@kent.edu,
Kent State Univ – Yg
Palmer, H. C., (519) 661-2111 (Ext. 86749) cpalmer@uwo.ca,
Univ of Western Ontario – Ym
Palmquist, John C., (414) 832-6732 palmquij@lawrence.edu,
Lawrence Univ – Gc
Palumbo, Tony V., (423) 576-8002 avp@ornl.gov,
Oak Ridge National Lab – Sb
Pan, William L., (509) 335-3611 wlpan@wsu.edu,
Washington State Univ – Sb
Pan, Yuanming, (706) 966-5699 yuanming.pan@usask.ca,
Univ of Saskatchewan – Gx
Pan, Zaitao, 314-977-3114 panz@eas.slu.edu,
Saint Louis Univ – Oa
Panah, Assad I., (814) 362-7569 aap@pitt.edu,
Univ of Pittsburgh, Bradford – Gc
Panah, Tooran, (814) 362-7560 Univ of Pittsburgh, Bradford – Pi
Pancha, Aasha, (775) 784-4251 pancha@seismo.unr.edu,
Univ of Nevada, Reno – Ys
Panchang, Vijay, (207) 581-1218 Univ of Maine – Oo
Pandit, Bhaskar I., (306) 966-5702 bhaskar.pandit@usask.ca,
Univ of Saskatchewan – Ys
Pandya, Rajul E., (610) 436-2213 West Chester Univ – Oa
Panero, Wendy R., 614 292 6290 panero.1@osu.edu,
Ohio State Univ – Yg
Panetta, Richard L., (979) 845-1386 panetta@ariel.tamu.edu,
Texas A&M Univ – Ow
Panhorst, Terry L., (662) 915-5825 panhorst@olemiss.edu,

Univ of Mississippi – Em
Pani, Eric A., (318) 342-1878 Univ of Louisiana, Monroe – Oa
Panish, Peter T., (413) 545-2593 panish@acad.umass.edu, Univ of Massachusetts – Gp
Pankow, Kristine L., 801-585-6484 pankow@seis.utah.edu, Univ of Utah – Ys
Panneerselvam, Kuppusamy, (660) 543-4936 panneerselvam@ucmo.edu, Univ of Central Missouri – Cg
Panno, Samuel V., (217) 244-2456 panno@isgs.uiuc.edu, Illinois State Geological Survey – Cg
Pant, Hari, 718-960-5859 hari.pant@lehman.cuny.edu, Lehman College (CUNY) – Cg
Panter, Kurt S., (419) 372-7337 kpanter@bgnet.bgsu.edu, Bowling Green State Univ – Gv
Paola, Christopher, 612-624-8025 cpaola@umn.edu, Univ of Minnesota, Twin Cities – Gs
Papageorgiou, Yorgos, (905) 525-9140 (Ext 23516) yorgos@mcmaster.ca, McMaster Univ – Yr
Papaleo, Silvannaspapaleo@zircon.geology.utoronto.ca, Univ of Toronto – Oo
Papcun, George, 12012993161 gpapcun@njcu.edu, New Jersey City Univ – Oe
Papelis, Charalambos, (702) 895-0453 lambis@dri.edu, Univ of Nevada, Reno – Cl
Papendick, Robert I., (509) 335-1552 papendick@wsu.edu, Washington State Univ – Sp
Papike, James J., (505) 277-1633 jpapike@unm.edu, Univ of New Mexico – Ca
Papike, James J., (505) 277-1646 jpapike@unm.edu, Univ of New Mexico – Gz
Paquette, Jeanne, 514-398-4402 jeannep@eps.mcgill.ca, McGill Univ – Gz
Paquette, Robert G., (408) 656-3253 paquette@oc.nps.navy.mil, Naval Postgraduate School – Op
Paradis, Serge, (418) 654-2556 sparadis@nrcan.gc.ca, Universite du Quebec – Gl
Paradis, Suzanne, (250) 363-6732 suparadi@nrcan.gc.ca, Univ of Victoria – Gv
Paradise, Thomas R., (479) 575-4359 paradise@uark.edu, Univ of Arkansas, Fayetteville – Oi
Parcell, William C., (316)978- 3140 Wichita State Univ – Gr
Pardi, Richard R., (201) 595-2695 William Paterson College – Cc
Parekh, Bhupendra K., (606) 257-0239 Univ of Kentucky – Nm
Parendes, Laurie A., (814) 732-2454 lparendes@edinboro.edu, Edinboro Univ of Pennsylvania – Oo
Parent, Michel, (418) 654-2657 Universite du Quebec – Gl
Pares, Josep M., (734) 615-0472 jmpares@umich.edu, Univ of Michigan – Ym
Parise, John B., (631) 632-8196 john.parise@sunysb.edu, SUNY, Stony Brook – Gz
Pariseau, William G., (801) 581-5164 wgparis@mines.utah.edu, Univ of Utah – Nr
Parizek, Richard R., (814) 865-3012 parizek@ems.psu.edu, Pennsylvania State Univ, Univ Park – Hw
Park, Choon-byong, (785) 864-2162 park@kgs.ku.edu, Univ of Kansas – Yg
Park, Jeffrey J., (203) 432-3172 jeffrey.park@yale.edu, Yale Univ – Ys
Park, Lisa E., (330) 972-7633 lepark@uakron.edu, Univ of Akron – Pi
Park, Sohyun, 089/2180 4333 soohyun.park@physik.uni-muenchen.de, LMU - Univ of Munich – Gz
Park, Stephen K., (951) 827-4501 magneto@ucrmt.ucr.edu, Univ of California, Riverside – Ym
Parker, Andrewa.parker@reading.ac.uk, Univ of Reading – Cl
Parker, Don F., (254) 710-2192 don_parker@baylor.edu, Baylor Univ – Giv
Parker, Gary, 217-244-6161 parkerg@uiuc.edu, Univ of Illinois, Urbana-Champaign – Gm
Parker, Gordon, (313) 593-5488 gparker@umich.edu, Univ of Michigan, Dearborn – Cg
Parker, Kent E., (509) 373-6337 kent.parker@pnl.gov, Pacific Northwest National Lab – Cg
Parker, Marjorie, (207) 725-3628 mparker@bowdoin.edu, Bowdoin College – Oo
Parker, Matt, (619) 534-2475 Missouri Dept of Natural Resources – Oo
Parker, Matthew, 919-513-4367 mdparker@ncsu.edu, North Carolina State Univ – Ow
Parker, Robert L., (619) 534-2475 rlparker@ucsd.edu, Univ of California, San Diego – Ym
Parker, Ronald L., (765) 983-1231 parkero@earlham.edu, Earlham College – Hy
Parker, Stephen R., (406) 496-4185 sparker@mtech.edu, Montana Tech of the Univ of Montana – Oo
Parker, William C., (850) 644-1568 parker@gly.fsu.edu, Florida State Univ – Pi
Parkhurst, David F., (812) 855-4556 Indiana Univ, Bloomington – Oo
Parkin, Gary, (519) 824-4120 (Ext 52452) gparkin@lrs.uoguelph.ca, Univ of Guelph – Hw
Parkinson, Christopher D., 504-280-6792 cparkins@uno.edu, Univ of New Orleans – Gp
Parks, Carlton R., (407) 868-0508 Florida Institute of Technology – Ow
Parks, George A., (650) 725-0578 parks@pangea.stanford.edu, Stanford Univ – Cl
Parks, George K., 510-643-5512 parks@ess.washington.edu, Univ of Washington – YxXy
Parlange, Jean-Yves, (607) 255-2476 jp58@cornell.edu, Cornell Univ – Hg
Parmentier, E. Marc, (401) 863-1700 em_parmentier@brown.edu, Brown Univ – Yg
Parnell, Roderic A., (928) 523-3329 roderic.parnell@nau.edu, Northern Arizona Univ – Cl
Parrish, Jay B., 717.702.2053 jayparrish@state.pa.us, Pennsylvania Bureau of Topographic & Geologic Survey – Yg
Parrish, Judith T., 208-885-5006 jparrish@uidaho.edu, Univ of Idaho – Pe
Parrish, Pia, (619) 594-5586 pparrish@geology.sdsu.edu, San Diego State Univ – Oo
Parry, William T., (801) 581-6217 bparry@mines.utah.edu, Univ of Utah – Cg
Parsley, Ronald L., (504) 862-3191 parsley@tulane.edu, Tulane Univ – Po
Parsons, Edward C.narose@ix.netcom.com, George Mason Univ – Ob
Parsons, Jeffrey D., (206) 221-6627 parsons@ocean.washington.edu, Univ of Washington – Ou
Parsons-Hubbard, Karla M., (440) 775-8353 karla.hubbard@oberlin.edu, Oberlin College – Po
Pasch, Anne D., (907) 786-6896 ahadp@uaa.alaska.edu, Univ of Alaska, Anchorage – Pg
Paschert, Karin, 089/2180 6573 karin.paschert@lmu.de, LMU - Univ of Munich – Gg
Pascussi, Michael, 518-486-2056 mpascucc@mail.nysed.gov, New York State Geological Survey – Go
Pashin, Jack C., 205/247-3558 jpashin@gsa.state.al.us, Geological Survey of Alabama – Gs
Pashin, Jack C., (205) 349-2852 jpashin@gsa.state.al.us, Mississippi State Univ – GoEcGs
Pasicznyk, David L., (609) 984-6587 dave.pasicznyk@dep.state.nj.us, New Jersey Geological Survey – Yg
Pasken, Robert W., (314) 977-3125 Saint Louis Univ – Ow
Pasteris, Jill D., (314) 935-5434 pasteris@levee.wustl.edu, Washington Univ – Em
Pasternack, Gregory B., (530) 754-9243 gpast@ucdavis.edu, Univ of California, Davis – Sp
Patchen, Douglas, 304 594 2331 doug.patchen@mail.wvu.edu, West Virginia Univ – Go
Patchett, P. Jonathan, (520) 621-2070 patchett@geo.arizona.edu, Univ of Arizona – Cc
Patel, Vinodkumar A., 217-244-0639 patel@isgs.uiuc.edu, Illinois State Geological Survey – Oo
Patera, Edward S., (505) 667-4457 Los Alamos National Lab – Cg
Paterson, Colin J., (605) 394-5114 Colin.Paterson@sdsmt.edu, South Dakota School of Mines & Technology – Eg
Paterson, Scott R., (213) 740-6103 paterson@usc.edu, Univ of Southern California – Gc
Patino, Lina C., 517-355-4626 Michigan State Univ – Ca
Patino-Douce, Alberto E., (706) 542-2394 alpatino@uga.edu, Univ of Georgia – Gp
Patrick, David M., (601) 266-4530 Univ of Southern Mississippi – Ng
Patro, Prasanta, (541) 737-4453 patro@coas.oregonstate.edu, Oregon State Univ – Yr
Patt, Anthony, (617) 353-7555 apatt@bu.edu, Boston Univ – Oo
Patten, James T., (727) 553-3957 jpatten@marine.usf.edu, Univ of South Florida, Saint Petersburg – Oo
Patterson, Amanda H., 540 831-6029 ahpatters@radford.edu, Radford Univ – Gx
Patterson, Deborah L., (608) 265-3048 dlpatter@wisc.edu,

Univ of Wisconsin - Extension – Oi
Patterson, Gary, (901) 678-2007 patterson@ceri.memphis.edu, Univ of Memphis – Gg
Patterson, R. Timothy, tpatters@ccs.carleton.ca, Carleton Univ – Pm
Patterson, William P., (306) 966-5691 bill.patterson@usask.ca, Univ of Saskatchewan – CsPe
Pattison, David R., (403) 220-3263 pattison@geo.ucalgary.ca, Univ of Calgary – Gp
Pattison, Simon A., 204-727-7468 pattison@brandonu.ca, Brandon Univ – Gs
Patton, Clyde P., Univ of Oregon – Oo
Patton, Howard J., (505) 667-1003 patton@lanl.gov, Los Alamos National Lab – Yg
Patton, James L., (510) 642-3059 Univ of California, Berkeley – Pv
Patton, Peter C., (860) 685-2268 ppatton@wesleyan.edu, Wesleyan Univ – Gm
Patton, Regan L., (509) 534-3670 Washington State Univ – Gc
Patton, Terri L., (630) 252-3294 Argonne National Lab – Cl
Patton, Thomas W., 406-496-4153 tpatton@mtech.edu, Montana Tech of The Univ of Montana – Hw
Patzkowsky, Mark E., (814) 863-1959 brachio@geosc.psu.edu, Pennsylvania State Univ, Univ Park – Po
Patzold, Raymund, (709) 737-4464 rpatzold@mun.ca, Memorial Univ of Newfoundland – Oe
Paul, Debajyoti, 210-458-5751 Debajyoti.Paul@utsa.edu, Univ of Texas, San Antonio – Cs
Paul, John H., (727) 553-1168 jpaul@marine.usf.edu, Univ of South Florida, Saint Petersburg – Ob
Paulsell, Robert L., (225) 578-8655 rpaulsell@lsu.edu, Louisiana State Univ – Oy
Paulsen, Gary M., (913) 532-7234 Kansas State Univ – Sp
Paulsen, Timothy S., (920) 424-7002 paulsen@uwosh.edu, Univ of Wisconsin, Oshkosh – Gc
Paulson, Clayton A., (541) 737-2528 cpaulson@coas.oregonstate.edu, Oregon State Univ – Op
Paulson, Suzanne, (310) 206-4442 paulson.atmos.ucla.edu, Univ of California, Los Angeles – Oa
Pavelich, Michael J., 303-273-3612 mpavelic@mines.edu, Colorado School of Mines – Oo
Pavey, Richard R., (614) 265-6599 rick.pavey@dnr.state.oh.us, Ohio Dept of Natural Resources – Gl
Pavlakovich, Vera K., 520-626-6969 vkp@email.arizona.edu, Univ of Arizona – Oo
Pavlis, Gary L., (812) 855-5141 pavlis@indiana.edu, Indiana Univ, Bloomington – Ys
Pavlis, Terry L., (504) 280-6797 tpavlis@uno.edu, Univ of New Orleans – Gt
Pavlis, Terry L., 915-747-5570 pavlis@geo.utep.edu, Univ of Texas, El Paso – Gc
Pavloski, Charles, (814) 865-0478 pavloski@essc.psu.edu, The Pennsylvania State Univ, Univ Park – Ow
Pavlovskaya, Marianna, (212) 772-5320 mpavlov@geo.hunter.cuny.edu, Hunter College (CUNY) – Oi
Pavlowsky, Robert T., (417) 836-8473 rtp138f@smsu.edu, Southwest Missouri State Univ – Gm
Pawloski, Gayle A., (925) 423-0437 pawloski1@llnl.gov, Lawrence Livermore National Lab – Gg
Pawlowicz, Richard A., (604) 822-1356 rpawlowicz@eos.ubc.ca, Univ of British Columbia – Op
Paxton, Stanley T., 405-744-6358 pstanle@okstate.edu, Oklahoma State Univ – Go
Payne, Bill, 806-354-5803 WPayne@ag.tamu.edu, Texas A&M Univ – Sp
Paynter, Kennedy T., (410) 405-6893 meesgrad@deans.umd.edu, Univ of Maryland – Oo
Paytan, Adina, (650) 724-4073 apaytan@pangea.stanford.edu, Stanford Univ – Oc
Paz Gonzalez, Antonio, 34 981167000 tucho@udc.es, Coruna Univ – Sd
Pazzaglia, Frank J., (610) 758-3667 fjp3@lehigh.edu, Lehigh Univ – Gm
Pe-Piper, Georgia, (902) 420-5744 Dalhousie Univ – Gv
Peace, Walter G., (905) 525-9140 (Ext. 23517) peacew@mcmaster.ca, McMaster Univ – Yr
Peacock, Mary, (775) 784-1958 mpeacock@scs.unr.edu, Univ of Nevada, Reno – Oo
Peacock, Simon M., (480) 965-1733 simon.peacock.@asu.edu, Arizona State Univ – Gp
Peacock, Simon M., scidean@science.ubc.ca, Univ of British Columbia – Gg
Peacor, Donald R., (734) 764-1452 drpeacor@umich.edu, Univ of Michigan – Gz
Peake, Jeffrey S., 402-554-2726 jpeake@mail.unomaha.edu, Univ of Nebraska at Omaha – Oy
Pearce, Bryan R., (207) 581-1216 Univ of Maine – Oo
Pearce, J A.PearceJA@cf.ac.uk, Cardiff Univ – Cg
Pearce, Thomas H., (613) 545-6174 Queen's Univ – Gi
Pearcy, William G., (541) 737-2601 wgpearcy@coas.oregonstate.edu, Oregon State Univ – Ob
Pearson, Ann, (617) 384-8392 pearson@eps.harvard.edu, Harvard Univ – Cm
Pearson, David A., (705) 675-1151 (Ext. ext. 2336) dpearson@laurentian.ca, Laurentian Univ, Sudbury – Oe
Pearson, David C., (505) 667-3942 cpearson@lanl.gov, Los Alamos National Lab – Yg
Pearson, Eugene F., (209) 946-2926 epearson@pacific.edu, Univ of the Pacific – GsPgOg
Pearson, Graham, 0191 3744701 d.g.pearson@durham.ac.uk, Durham Univ – Cg
Pearson, P N.PearsonP@cf.ac.uk, Cardiff Univ – Pe
Pearthree, Philip A., (520) 770-3500 Pearthree_Philip@pop.state.az.us, Arizona Geological Survey – Gm
Pease, Tamara K., (361) 749-6746 peasetk@mail.utexas.edu, Univ of Texas at Austin – Ob
Peaslee, Graham F., (616) 395-7117 peaslee@hope.edu, Hope College – Oo
Peate, David W., (319) 335-0567 david-peate@uiowa.edu, Univ of Iowa – Cg
Peate, Ingrid A., (319) 335-1824 ingrid-peate@uiowa.edu, Univ of Iowa – Gv
Peavy, Samuel T., 229-931-2330 speavy@canes.gsw.edu, Georgia Southwestern State Univ – Yg
Pechmann, James C., (801) 581-3858 pechmanN@seis.utah.edu, Univ of Utah – Ys
Peck, David C., (204) 677-6885 Univ of Manitoba – Eg
Peck, John A., (330) 972-7630 jpeck@uakronj.edu, Univ of Akron – Gs
Peck, Theodore R., (217) 333-9486 Univ of Illinois, Urbana-Champaign – Sc
Peck, William H., (315) 228-7717 wpeck@mail.colgate.edu, Colgate Univ – Gp
Peden, Mark E., (217) 333-8325 m-peden@uiuc.edu, Illinois State Water Survey – Og
Pedersen, Joel A., (608) 263-4971 joelpedersen@wisc.edu, Univ of Wisconsin, Madison – Sb
Pedersen, Thomas F.tfp@uvic.ca, Univ of British Columbia – Ou
Pedersen, Thomas F., (604) 822-5984 tfp@unixg.ubc.ca, Univ of Victoria – Cm
Pederson, Darryll T., (402) 472-7563 dpederson2@unl.edu, Univ of Nebraska, Lincoln – Hw
Pederson, Joel L., (435) 797-7097 joel.pederson@usu.edu, Utah State Univ – Gm
Pedone, Vicki A., (818) 677-2046 vicki.pedone@csun.edu, California State Univ, Northridge – Gd
Peele, Hampton, (225) 578-3455 hampton@lsu.edu, Louisiana State Univ – Oi
Peery, Larry J., (816) 248-3391 Central Methodist Univ – Yg
Peirce, Christine, 0191 3742515 christine.peirce@durham.ac.uk, Durham Univ – Yg
Pekarek, Alfred H., (320) 308-4127 apekarek@stcloudstate.edu, Saint Cloud State Univ – Gg
Pellerin, Jocelyne, (418) 724-1704 jocelyne_pellerin@uqar.uquebec.ca, Universite du Quebec a Rimouski – Ob
Pelletier, Emilien, (418) 723-1986 (Ext. 1764) emilien_pelletier@uqar.uquebec.ca, Universite du Quebec a Rimouski – Oc
Pelletier, Jon D., (520) 626-2126 jon@geo.arizona.edu, Univ of Arizona – Gm
Pellin, Michael J., (630) 972-3510 Argonne National Lab – Om
Pelto, Mauri S., 508-213-2168 peltoms@nichols.edu, Nichols College – Gl
Pelton, John R., (208) 426-3647 jpelton@boisestate.edu, Boise State Univ – Ys
Peltzer, Gilles, (310) 206-2156 peltzer@ess.ucla.edu, Univ of California, Los Angeles – Gt
Pemberton, S. George, (780) 492-2044 george.pemberton@ualberta.ca, Univ of Alberta – Gr
Pence, Stan F., (605) 677-5227 spence@usd.edu,

South Dakota Dept of Env & Natural Resources – Hw
Peng, Lee Chai, 03-79674233 leecp@um.edu.my, Univ of Malaya – Pg
Peng, Zhan, (314) 935-4517 zpeng@levee.wustl.edu, Washington Univ – Cc
Penland, Shea, 504-280-3119 spenland@uno.edu, Univ of New Orleans – Gm
Penney, Paulette, (254) 710-2361 paulette_penney@baylor.edu, Baylor Univ – Oo
Pennington, Wayne D., (906) 487-3573 wayne@mtu.edu, Michigan Technological Univ – Ys
Penniston-Dorland, Sarah, (301) 405-4087 sarahpd@geol.umd.edu, Univ of Maryland – Gp
Pennock, Jonathan R., (334) 861-7531 jpennock@jaguar1.usouthal.edu, Univ of Alabama, Dauphin Island – Oc
Penry, Deborah L., (510) 643-5915 Univ of California, Berkeley – Ob
Pentcheva, Rossitza, 089/2180 4352 pentcheva@lrz.uni-muenchen.de, LMU - Univ of Munich – Gz
Penton, Toni, (402)472-2655 tpenton1@unl.edu, Univ of Nebraska, Lincoln – Oo
Pepino, Richard V., 717 358-4555 richard.pepino@fandm.edu, Franklin and Marshall College – Ou
Pepper, Ian L., (520) 621-7234 ipepper@ag.arizona.edu, Univ of Arizona – Oo
Perault, David R., 434-544-8370 perault@lynchburg.edu, Lynchburg College – Ge
Percival, John A., (613)995-4723 John.Percival@nrcan-rncan.gc.ca, Univ of Ottawa – Gp
Percy, David, (503) 725-3373 Portland State Univ – Oi
Perdue, Edward Michael, (404) 894-3942 michael.perdue@eas.gatech.edu, Georgia Institute of Technology – Co
Pereira, Alcides C., +351-239 860 563 apereira@dct.uc.pt, Universidade de Coimbra – GebOr
Perez, Florante, (213) 620-5026 California Geological Survey – Oo
Perez, Marco A.mperez@cicese.mx, Centro de Inv Científica y de Educación Superior de Ensenada – Ye
Perez, Richard R., (518) 437-8751 perez@asrc.cestm.albany.edu, SUNY, Albany – Oa
Perfect, Edmund, (865) 974-2366 Univ of Tennessee, Knoxville – Sp
Perfit, Michael R., (352) 392-2128 perfit@geology.ufl.edu, Univ of Florida – Giu
Perg, Lesley A., 612-624-3664 lperg@umn.edu, Univ of Minnesota, Twin Cities – Gm
Perkins, Michael E., (801) 581-6552 mperk@mines.utah.edu, Univ of Utah – Gt
Perkins, RPerkinsR@cf.ac.uk, Cardiff Univ – Oo
Perkins, Robert B.rperkins @pdx.edu, Portland State Univ – Hq
Perkins, Ronald D., (919) 684-3376 rperkins@geo.duke.edu, Duke Univ – Gd
Perkins, William A., (509) 372-6131 wa_perkins@pnl.gov, Pacific Northwest National Lab – Hg
Perkins, III, Dexter, (701) 777-2991 Dexter_Perkins@und.nodak.edu, Univ of North Dakota – Gp
Perlin, Alexander, (541) 737-2990 aperlin@coas.oregonstate.edu, Oregon State Univ – Op
Perlin, Natalie, (541) 737-8191 nperlin@coas.oregonstate.edu, Oregon State Univ – Oa
Perlmutter, Martin, (708) 252-5856 Univ of Illinois, Chicago – Gs
Perret, Didier, (418) 654-2686 dperret@nrcan.gc.ca, Universite du Quebec – Ng
Perrien, Michael, 404-656-3214 Georgia Dept of Nat Resources – Oo
Perrin, Nancy A., (850) 488-0463 Tallahassee Community College – Gs
Perrin, Richard E., (505) 667-4755 Los Alamos National Lab – Cc
Perrin, William F., (619) 546-7096 wperrin@ucsd.edu, Univ of California, San Diego – Ob
Perro, Dianne, (207)-581-2152 dianne.perro@umit.maine.edu, Univ of Maine – Oo
Perry, Baker, (828) 262-7597 perrylb@appstate.edu, Appalachian State Univ – Og
Perry, Eugene C., (815) 753-7935 eperry@niu.edu, Northern Illinois Univ – Cs
Perry, Frank V., (505) 667-1033 fperry@lanl.gov, Los Alamos National Lab – Gv
Perry, Shinisha, 205-247-3541 sperry@gsa.stat.al.us, Geological Survey of Alabama – Ng
Persaud, Naraine, (540) 231-3817 npers@vt.edu, Virginia Polytechnic Institute & State Univ – Sp
Person, Arthur, (814) 863-8568 The Pennsylvania State Univ, Univ Park – Ow
Person, Mark, (812) 855-4404 maperson@indiana.edu, Indiana Univ, Bloomington – Hw
Pertl, Davidpertl-dj@actx.edu, Amarillo College – Gg
Peryea, Frank J., (509) 663-8181 Washington State Univ – Sc
Pessagno, Jr., Emile A., (972) 883-2444 pessagno@utdallas.edu, Univ of Texas, Dallas – Pm
Pestrong, Raymond, (415) 338-2080 rayp@sfsu.edu, San Francisco State Univ – Ng
Petcovic, Heather L., 269-387-5488 heather.petcovic@wmich.edu, Western Michigan Univ – Gi
Peteet, Dorothy M., (845) 365-8420 peteet@ldeo.columbia.edu, Columbia Univ – Pl
Peters, Catherine A., (609) 258-5645 cap@phoenix.princeton.edu, Princeton Univ – Hw
Peters, Esther C., (703) 385-6000 Univ of South Florida, St Petersburg – Ob
Peters, Lisa, (505) 835-5217 lisa@nmt.edu, New Mexico Institute of Mining & Technology – Gg
Peters, Mark T., (202) 586-9279 Mark_Peters@notes.ymp.gov, Los Alamos National Lab – Yg
Peters, Roger M., (608) 263-7387 rpeters@wisc.edu, Univ of Wisconsin - Extension – Gg
Peters, Shanan, 608-262-5987 peters@geology.wisc.edu, Univ of Wisconsin, Madison – Gs
Peters, Shanan, 734-764-6542 shananp@umich.edu, Univ of Michigan – Pi
Peters, Stephen C., 610/758-3660 scp2@lehigh.edu, Lehigh Univ – Cl
Petersen, Erich U., (801) 581-7238 eupeters@mines.utah.edu, Univ of Utah – Em
Petersen, Gary W., (814) 865-1540 gwp2@psu.edu, Pennsylvania State Univ, Univ Park – Sd
Petersen, Mark D., (916) 322-9317 California Geological Survey – Oo
Petersen, Morris S., (801) 373-5494 msp85@comcast.net, Brigham Young Univ – Pi
Petersen, Nikolai, 089/2180 4233 petersen@geophysik.uni-muenchen.de, LMU - Univ of Munich – Yg
Petersen, Ulrich, (617) 495-2353 petersen@eps.harvard.edu, Harvard Univ – Eg
Peterson, BradleyBradley.Peterson@stonybrook.edu, SUNY, Stony Brook – Ob
Peterson, Curt D., (503) 725-3375 petersonc@pdx.edu, Portland State Univ – Gs
Peterson, Eric W., (309)438-7865 ewpeter@ilstu.edu, Illinois State Univ – Hg
Peterson, Eugene J., (505) 667-5182 Los Alamos National Lab – Ct
Peterson, Gary L., (619) 594-5594 gpeterson@geology.sdsu.edu, San Diego State Univ – Gr
Peterson, Jon W.peterson@hope.edu, Hope College – Ge
Peterson, Larry C., 305-284-6821 l.peterson@miami.edu, Univ of Miami – Gr
Peterson, Larry C., 305 421-4692 lpeterson@rsmas.miami.edu, Univ of Miami – Gu
Peterson, Melinda, (541) 737-1238 petersom@geo.oregonstate.edu, Oregon State Univ – Oo
Peterson, Nate, 61 7 3138 2318 n.peterson@qut.edu.au, Queensland Univ of Technology – Oi
Peterson, Peter A., (515) 294-9652 pap@iastate.edu, Iowa State Univ of Science & Technology – Oo
Peterson, Richard E., (806) 742-3101 richard.peterson@ttu.edu, Texas Tech Univ – Oa
Peterson, Robert E., (509) 373-9020 robert.peterson@pnl.gov, Pacific Northwest National Lab – Cg
Peterson, Ronald C., (613) 545-6180 Queen's Univ – Gz
Peterson, Virginia L., (828) 227-7367 peterson@wcu.edu, Grand Valley State Univ – Gc
Petr, Vilem, 303-273-3222 vpetr@mines.edu, Colorado School of Mines – Nm
Petronis, Michael S., (505) 454-3513 mspetro@nmhu.edu, New Mexico Highlands Univ – YmGcv
Petsch, Steven, 413-545-4413 spetsch@geo.umass.edu, Univ of Massachusetts – Co
Pettengill, Gordon H., (617) 253-6384 hgp@mit.edu, Massachusetts Institute of Technology – Og
Pettigrew, Neal R., (207) 581-4384 Univ of Maine – Op
Pettinga, Jarg, ((03)) 3-667- (Ext. 1 ext 7716) j.pettinga@geol.canterbury.ac.nz, Univ of Canterbury – Gt
Petty, Grant W., (608)263-3265 gpetty@aos.wisc.edu,

Univ of Wisconsin, Madison – Or
Pettyjohn, Wayne, 405-372-1981 Oklahoma State Univ – Hw
Petuch, Edward J., (561) 297-2398 epetuch@fau.edu,
Florida Atlantic Univ – Pg
Peucker-Ehrenbrink, Bernhard, (508) 289-2518 Woods Hole Oceanographic Institution – Cm
Peyton, Gary R., (217) 333-5905 Illinois State Water Survey – Oo
Pezeshk, Shahram, 901-678-4727 spezeshk@memphis.edu,
Univ of Memphis – Ne
Pezzopane, Silvio K.earthdogman@hotmailcom,
Humboldt State Univ – Gt
Pfannkuch, Hans O., 612-624-1620 h2olafpf@umn.edu,
Univ of Minnesota, Twin Cities – Hw
Pfeffer, Richard L., (850) 644-5594 pfeffer@met.fsu.edu,
Florida State Univ – Ow
Pfefferkorn, Hermann W., (215) 898-5156 hpfeffer@sas.upenn.edu,
Univ of Pennsylvania – Pb
Pfirman, Stephanie L., (212) 854-5120 spfirman@barnard.columbia.edu,
Columbia Univ – Gv
Pfuhl, Helen, 089/2180 4230 helen.pfuhl@geophysik.uni-muenchen.de,
LMU - Univ of Munich – Yg
Pheifer, Raymond N., 217-581-6248 cfrnp@eiu.edu,
Eastern Illinois Univ – Ec
Phelps, Tommy J., (423) 574-7290 tkp@ornl.gov,
Oak Ridge National Lab – Sb
Phelps, William, 714-564-6690 phelps_william@sac.edu,
Santa Ana College – Pg
Philander, S. George H., (609) 258-5683 gphlder@princeton.edu,
Princeton Univ – Op
Philbrick, Tom, (203) 837-8773 Western Connecticut State Univ – Oo
Phillippi, Nathan E., 910-521-6588 nathan.phillippi@uncp.edu,
Univ of North Carolina, Pembroke – Oy
Phillips, Andrew C., 217-333-2513 phillips@isgs.uiuc.edu,
Illinois State Geological Survey – Gm
Phillips, Brian L., (631) 632-6853 brian.phillips@sunysb.edu,
SUNY, Stony Brook – Gz
Phillips, Dennis, (505) 667-4253 Los Alamos National Lab – Co
Phillips, Fred M., (505) 835-5540 phillips@nmt.edu,
New Mexico Institute of Mining and Technology – Hw
Phillips, Glen, +61 2 4921 5410 Glen.Phillips@newcastle.edu.au,
Univ of Newcastle – GcpGc
Phillips, Melissameliphil@nova.edu,
Nova Southeastern Univ – Oo
Phillips, Michael, 815-224-0394 mike_phillips@ivcc.edu,
Illinois Valley Community College – GgeGm
Phillips, Michael W., (419) 530-4572 michael.phillips@utoledo.edu,
Univ of Toledo – Gz
Phillips, Nathan, (617) 353-2841 nathan@bu.edu,
Boston Univ – Oo
Phillips, Nigel, (604) 822-3765 nphillips@eos.ubc.ca,
Univ of British Columbia – Yg
Phillips, Owen M., (410) 516-4658 omphil@aol.com,
The Johns Hopkins Univ – Op
Phillips, P. L., (910) 522-5787 lee.phillips@uncp.edu,
Univ of North Carolina, Pembroke – Gd
Phillips, Paul, (785) 628-5969 pphillip@fhsu.edu,
Fort Hays State Univ – Oe
Phillips, Randal C., (303) 866-3413 randal.phillips@state.co.us,
Colorado Geological Survey – Gq
Phillips, Roger J., 314-935-6356 phillips@wustite.wustl.edu,
Washington Univ – Xy
Phillips, Shawn, (812) 237-2070 anphilli@isugw.indstate.edu,
Indiana State Univ – Oo
Phillips, William M., (208) 885-8928 phillips@uidaho.edu,
Univ of Idaho – Gm
Phillips, William R., (801) 378-4545 Brigham Young Univ – Gz
Phillips, William S., (505) 667-8106 wsp@lanl.gov,
Los Alamos National Lab – Yg
Philp, R. Paul, (405) 325-4469 pphilp@ou.edu,
Univ of Oklahoma – Co
Philpotts, Anthony R., (860) 486-1394 anthony.philpotts@uconn.edu,
Univ of Connecticut – Gi
Phinney, Robert A., (609) 258-4118 rphinney@princeton.edu,
Princeton Univ – Ys
Phipps, Stephen P., (215) 898-4602 sphipps@sas.upenn.edu,
Univ of Pennsylvania – Gc
Piazzoni, Antonio Sebastian, 089/2180 4230 antonio.piazzoni@geophysik.uni-muenchen.de, LMU - Univ of Munich – Yg
Picard, M. Dane, (801) 581-7991 mdane@mines.utah.edu,
Univ of Utah – Gd
Picardal, Flynn W., (812) 855-0732 Indiana Univ, Bloomington – So
Piccoli, Philip M., (301) 405-6966 piccoli@geol.umd.edu,
Univ of Maryland – Cg
Piche, Andre, (514) 630-9300 Ecole Polytechnique – Nm
Pichler, Thomas, (813) 974-0321 pichler@chuma.cas.usf.edu,
Univ of South Florida, Tampa – Cl
Pickerill, Ronald K., (506) 453-4804 rpickeri@unb.ca,
Univ of New Brunswick – Pg
Pickering, Ingrid, 306-966-5706 ingrid.pickering@usask.ca,
Univ of Saskatchewan – Oo
Pickett, Nicki, 954-201-6677 npickett@broward.edu,
Broward Community College, Central Campus – Oo
Pier, Stanley M., (713) 522-0633 Rice Univ – Co
Pierce, James, (507) 389-1114 james.pierce@mnsu.edu,
Minnesota State Univ – Xc
Pierce, Larry, (573) 368-2161 Missouri Dept of Natural Resources – Oo
Pierce, Robert W., (505) 562-2419 robert.pierce@enmu.edu,
Eastern New Mexico Univ – Gr
Pierce, Stephen D., (541) 737-2425 spierce@coas.oregonstate.edu,
Oregon State Univ – Op
Piercey, Stephen, (705) 675-1151 (Ext. 2364) spiercey@nickel.laurentian.ca, Laurentian Univ, Sudbury – Ce
Pieren, Agustin P., 34 913944819 vdposgrado@geo.ucm.es,
Univ. Complutense de Madrid – GrEgg
Pieri, David C., (818) 354-6299 Jet Propulsion Lab – Gt
Pierrehumbert, Raymond T., (773) 702-8811 Univ of Chicago – Oa
Pierzynski, Gary M., (913) 532-7209 Kansas State Univ – Sc
Pieters, Carle M., (401) 863-2417 pieters@mare.geo.brown.edu,
Brown Univ – Or
Pieters, Roger, (604) 822-4297 rpieters@eos.ubc.ca,
Univ of British Columbia – Op
Pietrafesa, Leonard J., (919) 515-3717 leonard_pietrafesa@ncsu.edu,
North Carolina State Univ – Op
Pietruszka, Aaron, (619) 594-2648 apietruszka@geology.sdsu.edu,
San Diego State Univ – Ct
Pigott, John D., (405) 325-4498 jpigott@ou.edu,
Univ of Oklahoma – Ye
Pihlak, Madis, (301) 405-4350 mp65@umail.umd.edu,
Univ of Maryland – Ou
Pike, J, PikeJ@cf.ac.uk, Cardiff Univ – Oo
Pike, Scott, 503-370-6587 spike@willamette.edu,
Willamette Univ – Gg
Pike, Steven M., (508) 289-2350 spike@whoi.edu,
Woods Hole Oceanographic Institution – Ca
Pilcher, Brian K., (406) 683-7327 b_pilcher@umwestern.edu,
Univ of Montana, Western – Sf
Pilkey, Jr., Orrin H., (919) 684-4238 opilkey@geo.duke.edu,
Duke Univ – Ou
Pilskaln, Cynthia H., (207) 581-4381 Univ of Maine – Oc
Pilson, Michael E., (401) 874-6104 pilson@gso.uri.edu,
Univ of Rhode Island – Oc
Pimpler, Donald E., (210) 733-2704 dpimpler@mail.accd.edu,
San Antonio College – Oi
Pindzola, Dan M., (570) 408-4612 Wilkes Univ – Ow
Pinet, Paul, (315) 228-7656 ppinet@mail.colgate.edu,
Colgate Univ – Ou
Pingitore, Jr., Nicholas E., (915) 747-5754 nick@geo.utep.edu,
Univ of Texas, El Paso – Cl
Pinkel, Robert, (619) 534-2056 rpinkel@ucsd.edu,
Univ of California, San Diego – Op
Pinter, Nicholas, (618) 453-7375 npinter at geo.siu.edu,
Southern Illinois Univ Carbondale – Gm
Pinter, Susann B., 618-453-7375 spinter@geo.siu.edu,
Southern Illinois Univ Carbondale – Gg
Piontkovski, Sergey, (631) 632-8694 spiontkovski@notes.cc.sunysb.edu,
SUNY, Stony Brook – Ob
Piper, David J., (902) 426-6580 dpiper@nrcan.gc.ca,
Dalhousie Univ – Gu
Pipkin, Bernard W., (213) 740-6106 pipkin@usc.edu,
Univ of Southern California – Ng
Pires, E M., (516) 299-2318 mark.pires@liu.edu,
Long Island Univ, C.W. Post Campus – Og
Pirie, Diane H., (305) 348-2876 Florida International Univ – Gg
Pirkle, Earl C., (352) 392-2231 Univ of Florida – Eg
Pisias, Nicklas G., (541) 737-5213 npisias@coas.oregonstate.edu,
Oregon State Univ – Ou
Pitlick, John, (303) 492-5906 pitlick@colorado.edu,
Univ of Colorado – Gm
Pitman, III, Walter C., (845) 365-8397 pitman@ldeo.columbia.edu,
Columbia Univ – Yr

Pittman, Jeff G., (409) 880-8346 pittmanjg@hal.lamar.edu, Lamar Univ – Pv
Pittman, Jeffrey G., (409) 880-8346 Univ of Colorado, Denver – Pv
Pizzuto, James E., (302) 831-2710 pizzuto@udel.edu, Univ of Delaware – Gm
Plag, Hans-Peter, (775) 78-4-66 (Ext. ext 172) hpp@unr.edu, Univ of Nevada – Yd
Plane, David A., (520) 621-1738 Univ of Arizona – Oo
Plank, Owen C., (706) 542-9072 Univ of Georgia – Sc
Plank, Terry A., (617) 353-4213 tplank@bu.edu, Boston Univ – Cg
Plant, Jeffrey J., (818) 393-3799 plant@jpl.nasa.gov, Jet Propulsion Lab – Oo
Plante, Martin, (418) 656-8121 martin.plante@ggl.ulaval.ca, Universite Laval – Cg
Plasienka, Dusan, 0042160296529 plasienka@fns.uniba.sk, Comenius Univ – Gt
Platt, John, 213-821-1194 jplatt@usc.edu, Univ of Southern California – Gc
Platt, Rutherford H., (413) 545-2499 rplatt@geo.umass.edu, Univ of Massachusetts – Ou
Plattner, Christina, 089/21804220 christina.plattner@geophysik.uni-muenchen.de, LMU - Univ of Munich – Yg
Platzman, George W., (private} Univ of Chicago – Oa
Plescia, Jeffrey, (202) 358-0295 Jet Propulsion Lab – Oo
Plesha, Michael E., (608) 262-5741 plesha@engr.wisc.edu, Univ of Wisconsin, Madison – Oo
Plewe, Brandon, (801) 378-4161 brandon_plewe@byu.edu, Brigham Young Univ – Oi
Plint, A G., (519) 661-3179 gplint@uwo.ca, Univ of Western Ontario – Gs
Plotnick, Roy E., (312) 996-2111 plotnick@uic.edu, Univ of Illinois, Chicago – Pi
Plug, Lawrence, 902-494-1200 lplug@is.dal.ca, Dalhousie Univ – Gm
Plumb, Raymond A., (617) 253-6281 rap@rossby.mit.edu, Massachusetts Institute of Technology – Ow
Plummer, Charles C., plummercc@csus.edu, California State Univ, Sacramento – Gp
Plymate, Thomas G., (417) 836-4419 tgp438f@smsu.edu, Southwest Missouri State Univ – Gx
Poage, Michael A., (724) 357-5627 mpoage@iup.edu, Indiana Univ of Pennsylvania – Cs
Pociask, Geoffrey E., 217-265-8212 pociask@isgs.uiuc.edu, Illinois State Geological Survey – Ge
Pocta, Miles, (804) 427-7189 Tidewater Community College – Ou
Podosek, Frank A., (314) 935-7367 fap@levee.wustl.edu, Washington Univ – Cc
Poeter, Eileen P., (303) 273-3829 epoeter@mines.edu, Colorado School of Mines – Ng
Pogue, Kevin R., (509) 527-5955 pogue@whitman.edu, Whitman College – Gc
Pohkarel, Shankar S., (404) 413-5778 geossp@langate.gsu.edu, Georgia State Univ – Oo
Pohll, Greg, (775) 674-7523 greg.pohll@dri.edu, Univ of Nevada, Reno – Ca
Pohll, Greg, (775) 674-7523 pohll@dri.edu, Univ of Nevada, Reno – Hg
Pojeta, John, (202) 633-1347 Smithsonian Institution / National Museum of Natural History – Pi
Pokorny, Eugene W., (702) 295-7496 eugene_pokorny@lanl.gov, Los Alamos National Lab – Nm
Polet, Jascha, 909-869-3459 jpolet@csupomona.edu, California State Polytechnic Univ, Pomona – Ysg
Poliakon, Erin E., 910-521-6588 erin.poliakon@uncp.edu, Univ of North Carolina, Pembroke – Og
Polito, Thomas A., (515) 294-0513 tpolito@iastate.edu, Iowa State Univ of Science & Technology – So
Pollack, Henry N., (734) 763-0084 hpollack@umich.edu, Univ of Michigan – Yh
Pollard, David D., (650) 723-4679 dpollard@pangea.stanford.edu, Stanford Univ – Gc
Pollard, William B., (229) 333-5757 wpollard@valdosta.edu, Valdosta State Univ – Om
Pollock, Meagen, 717-245-1813 pollockm@dickinson.edu, Dickinson College – Gi
Pollock, Stephen G., (207) 780-5353 pollock@usm.maine.edu, Univ of Southern Maine – Gx
Polovina, Jeffrey J., (808) 983-5390 Univ of Hawai'i, Manoa – Oo
Polvani, Lorenzo M., (212) 854-7331 Columbia Univ – Ow
Polyak, Victor J., 505-277-4204 polyak@unm.edu, Univ of New Mexico – Cg
Polzin, Amy, (815) 753-1943 amy@geol.niu.edu, Northern Illinois Univ – Oo
Pond, Lisa G., (225) 578-0401 lgpond@lsu.edu, Louisiana State Univ – Og
Pond, Stephen G., (604) 822-2205 spond@eos.ubc.ca, Univ of British Columbia – Op
Pontius, Steven, (812) 237-2262 spontius@isugw.indstate.edu, Indiana State Univ – Oo
Ponto, Edward S., (315) 469-2432 Onondaga Community College – Og
Poole, Denise, 828-884-8164 pooledm@brevard.edu, Brevard College – Oo
Poole, T. Craig, (559) 442-8427, craig.poole@fresnocitycollege.edu, Fresno City College – Gg
Pope, Gregory A., (973) 655-7385 popeg@mail.montclair.edu, Montclair State Univ – Gm
Pope, Jeanette K., 765-658-4105 jpope@depauw.edu, DePauw Univ – Ge
Pope, John K., (513) 529-3216 Miami Univ – Pi
Pope, John P., (660) 562-1211 jppope@nwmissouri.edu, Northwest Missouri State Univ – Ps
Pope, Lin F., (601) 266-6054 lpope@ocean.st.usm.edu, Univ of Southern Mississippi – Gz
Pope, Michael C., (509) 335-5989 mcpope@wsu.edu, Washington State Univ – Gr
Popenoe, Hugh L., (352) 392-2643 hlp@ufl.edu, Univ of Florida – So
Popp, Brian N., (808) 956-6206 popp@soest.hawaii.edu, Univ of Hawai'i, Manoa – Cm
Popp, Brian N., (808) 956-6206 popp@soest.hawaii.edu, Univ of Hawai'i, Manoa – Cs
Popp, Robert K., (979) 845-2451 popp@geo.tamu.edu, Texas A&M Univ – Cp
Poppeliers, Christian, (325) 942-2230 christian.poppeliers@angelo.edu, Angelo State Univ – Ys
Porch, William M., (505) 667-0971 wporch@lanl.gov, Los Alamos National Lab – Yg
Poreda, Robert J., (585) 275-0051 poreda@earth.rochester.edu, Univ of Rochester – Cg
Porinchu, David F., (614) 247-2614 porinchu.1@osu.edu, Ohio State Univ – Oa
Portell, Roger W., (352) 392-1721 (Ext. 258) portell@flmnh.ufl.edu, Univ of Florida – Pi
Porter, Gerald A., (843) 383-5079 Coker College – Oe
Porter, John H., (804) 924-8999 jhp7e@virginia.edu, Univ of Virginia – Or
Porter, Owen A., (501) 543-8527 Univ of Arkansas, Pine Bluff – Sc
Porter, Stephen C., 206-543-1904 scporter@u.washington.edu, Univ of Washington – Gl
Porter, William A., (252) 335-3378 waporter@mail.ecsu.edu, Elizabeth City State Univ – Oy
Portier, Ralph J., (225) 388-4287 Louisiana State Univ – Ob
Portillo, Danny, (719) 333-3080 United States Air Force Academy – Oi
Portman, Donald J., (313) 936-0538 Univ of Michigan – Oa
Posler, Gerry L., (913) 532-6101 Kansas State Univ – Sf
Post, Donald F., (520) 621-1262 postdf@ag.arizona.edu, Univ of Arizona – Sd
Post, Jeffrey E., (202) 633-1814 postj@si.edu, Smithsonian Institution / National Museum of Natural History – Gz
Poths, Jane, (505) 667-1506 Los Alamos National Lab – Cc
Potter, Eric C., (512) 471-7090 eric.potter@beg.utexas.edu, Univ of Texas at Austin, Jackson School of Geosciences – Eo
Potter, Kenneth W., (608) 262-0400 potter@engr.wisc.edu, Univ of Wisconsin, Madison – Hg
Potter, Lee S., 319-273-2759 lspotter@earthlink.net, Univ of Northern Iowa – Gz
Potter, Paul E., (513) 556-3732 Univ of Cincinnati – Gs
Potter, Jr., Donald B. bpotter@sewanee.edu, Univ of the South – Gc
Potter, Jr., Noel, 717-245-1340 pottern@dickinson.edu, Dickinson College – Gm
Potts, Donald C., (831) 459-4417 potts@biology.ucsc.edu, Univ of California, Santa Cruz – Ob
Poty, Edouard, 32 4 366 52 83 E.Poty@ulg.ac.be, Universite de Liege – Pi
Poulain, Pierre-Marie R. G., (408) 656-3318 poulain@oc.nps.navy.mil, Naval Postgraduate School – Op
Poulsen, Christopher J., (734) 615-2236 poulsen@umich.edu, Univ of Michigan – Pe

Poulson, Simon R., (775)784-1104 poulson@mines.unr.edu, Univ of Nevada, Reno – Cs
Poulton, Mary M., 520 621-8391 mpoulton@email.arizona.edu, Univ of Arizona – Ng
Poussart, Pascale F., (609) 258-4101 poussart@princeton.edu, Princeton Univ – Co
Powell, Christine A., (901) 678-2007 powell@ceri.memphis.edu, Univ of Memphis – Ys
Powell, Christine A., (901) 678-8455 powell@ceri.memphis.edu, Univ of Memphis – Ys
Powell, J. Mark, (608) 890-0070 jmpowel2@wisc.edu, Univ of Wisconsin, Madison – So
Powell, Ross D., (815) 753-7952 ross@geol.niu.edu, Northern Illinois Univ – Gs
Powell, Thomas M., (510) 642-7455 zackp@socrates.berkeley.edu, Univ of California, Berkeley – Op
Powell, Wayne G., (718) 951-5761 wpowell@brooklyn.cuny.edu, Graduate School of the City Univ of New York – Gp
Powell, Wayne G., (718) 951-5416 wpowell@brooklyn.cuny.edu, Brooklyn College (CUNY) – Gp
Power, Mary E., (510) 643-7776 Univ of California, Berkeley – Pe
Powers, Donovan M., (614) 265-6591 donovan.powers@dnr.state.oh.us, Ohio Dept of Natural Resources – OiiOi
Powers, Roger W., (907) 474-6188 Univ of Alaska, Fairbanks – Oo
Prahl, Frederick G., (541) 737-3969 fprahl@coas.oregonstate.edu, Oregon State Univ – Co
Prakash, Anupma, 907-474-1897 prakash@gi.alaska.edu, Univ of Alaska, Fairbanks – Ye
Pranter, Matthew J., 303-492-1461 matthew.pranter@colorado.edu, Univ of Colorado – Go
Prather, Julia, (530) 752-3668 prather@geology.ucdavis.edu, Univ of California, Davis – Oo
Prather, Michael J., (949) 824-5838 mprather@uci.edu, Univ of California, Irvine – Oa
Prather, Thomas L., (970) 943-2092 tprather@western.edu, Western State College of Colorado – Gc
Pratson, Lincoln F., (919) 681-8077 lincoln.pratson@duke.edu, Duke Univ – Gs
Pratt, Allen, 613-992-1376 apratt@NRCan.gc.ca, Univ of Western Ontario – Cg
Pratt, Allyn R., (505) 667-4308 pratt_allyn_r@lanl.gov, Los Alamos National Lab – Ge
Pratt, Brian R., (306) 966-5725 brian.pratt@usask.ca, Univ of Saskatchewan – Pg
Pratt, Lisa M., (812) 855-9203 prattl@indiana.edu, Indiana Univ, Bloomington – Co
Pratt-Sitaula, Beth, (509) 963-2870 psitaula@geology.cwu.edu, Central Washington Univ – Oe
Prave, Anthony R., (212) 650-6471 City College (CUNY) – Gs
Pray, Lloyd C., (608) 262-5366 Univ of Wisconsin, Madison – Gs
Precedo, Laura, 954-201 lprecedo@broward.edu, Broward Community College, Central Campus – Oc
Precht, Francis L., (301) 687-4440 fprecht@frostburg.edu, Frostburg State Univ – Oy
Preda, Micaela, 61 7 3138 1910 m.preda@qut.edu.au, Queensland Univ of Technology – Oi
Preer, James, (202) 282-7371 Univ of the District of Columbia – Hg
Prell, Warren L., (401) 863-3221 warren_prell@brown.edu, Brown Univ – Ou
Prenger, Joseph P., (352) 392-1803 (Ext. 332) jprenger@ufl.edu, Univ of Florida – So
Prentice, Michael L., (603) 862-4783 Univ of Maine – Gu
Prentice, Michael L., 603-535-3249 mlprentice@plymouth.edu, Univ of New Hampshire – Pe
Presley, Bobby J., (979) 845-5136 bpresley@ocean.tamu.edu, Texas A&M Univ – Oc
Presnall, Dean C., (972) 883-2444 presnall@utdallas.edu, Univ of Texas, Dallas – Cp
Prestegaard, Karen L., (301) 405-6982 kpresto@geol.umd.edu, Univ of Maryland – Hg
Prestegaard, Karen L., (301) 405-6982 kpresto@glue.umd.edu, Univ of Maryland – Hw
Preston, Leiph, (775)784-1684 preston@seismo.unr.edu, Univ of Nevada, Reno – Ys
Preston, William L., (805) 756-2210 wpreston@calpoly.edu, California Polytechnic State Univ – Oe
Prevost, Jean-Herve, (609) 258-5424 Princeton Univ – Ng
Prewett, Jerry, (573) 368-2161 Missouri Dept of Nat Resources – Oo
Prezant, Robert S., (718) 997-4105 rprezant@qc.edu, Graduate School of the City Univ of New York – Py
Prezzano, Susan C., (814) 226-2396 sprezzano@clarion.edu, Clarion Univ – Oo
Pribyl, Carol L., (307) 766-2443 Cpribyl@uwyo.edu, Univ of Wyoming – Oo
Price, Douglas M., 330-941-3019 dmprice@ysu.edu, Youngstown State Univ – Cg
Price, Jane, (702) 784-1779 Virginia Polytechnic Institute & State Univ – Oo
Price, Jason R., 717-872-3005 Jason.Price@millersville.edu, Millersville Univ – Gs
Price, Jeff T.jprice@csuchico.edu, California State Univ, Chico – Oa
Price, Jonathan D., (518) 276-2372 pricej@rpi.edu, Rensselaer Polytechnic Institute – Cp
Price, Jonathan G., (775) 784-6691 jprice@nbmg.unr.edu, Univ of Nevada, Reno – Cg
Price, L G., (505) 835-5752 gprice@gis.nmt.edu, New Mexico Institute of Mining & Technology – Gg
Price, Maribeth H., (605) 394-1290 Maribeth.Price@sdsmt.edu, South Dakota School of Mines & Technology – Oi
Price, Marie D., (202) 994-6187 George Washington Univ – Oy
Price, Peter, 573-368-2131 Missouri Dept of Natural Resources – Ge
Price, Peter E., 281-765-7764 Peter.E.Price@nhmccd.edu, North Harris College – Oi
Price, Raymond A., (613) 545-6542 Queen's Univ – Gt
Price, Rene, (305) 348-3119 pricer@fiu.edu, Florida International Univ – Hw
Prichard, H M.prichard@cf.ac.uk, Cardiff Univ – Gx
Prichard, Mary, (323) 953-4000 (Ext. 2611) prichame@lacitycollege.edu, Los Angeles City College – Oy
Prichard, Terry L., (559) 468-2086 tlprichard@ucdavis.edu, Univ of California, Davis – Oo
Prichonnet, Gilbert, (514) 987-3383 prichonnet.gilbert@uqam.ca, Universite du Quebec a Montreal – Gs
Pride, Douglas E., (614) 292-9523 pride.1@osu.edu, Ohio State Univ – Em
Pride, Steven, (510) 495-2823 SRPride@lbl.gov, Univ of California, Berkeley – Ys
Pridmore, Cindy L., (916) 925-6902 California Geological Survey – Gt
Priest, George R., (541) 574-6642 george.priest@dogami.state.or.us, Oregon Dept of Geology & Mineral Industries – Gg
Pringle, James M., (603) 862-5000 jpringle@cisunix.unh.edu, Univ of New Hampshire – Op
Pringle, Jamie K., +44 01782 583163 j.k.pringle@esci.keele.ac.uk, Keele Univ – Ga
Prinn, Ronald G., 617-253-2452 rprinn@mit.edu, Massachusetts Institute of Technology – Oa
Prinz, Martin, (212) 769-5381 Graduate School of the City Univ of New York – Gi
Prior, William L., 501 683-0117 bill.prior@Arkansas.gov, Arkansas Geological Commission – Hw
Pristas, Ronald, (609) 984-6587 ron.pristas@dep.state.nj.us, New Jersey Geological Survey – Oi
Pritchard, Lowell, 404 727 7926 lpritc2@emory.edu, Emory Univ – Sf
Proctor, Beth, (507) 389-5492 bertha.proctor@mnsu.edu, Minnesota State Univ – Hs
Proctor, Elizabeth, 415-452-7150 City College of San Francisco – Oi
Proctor, Lita M., (850) 644-4951 Florida State Univ – Ob
Prokopenko, Maria G., (609) 258-7544 mprokope@princeton.edu, Princeton Univ – Cg
Prokoph, Andreas, (613) 253-3488 aprokocon@aol.com, Univ of Ottawa – Gq
Prothero, Donald R., (323) 259-2557 prothero@oxy.edu, Occidental College – Pv
Prothero, William A., (805) 893-3308 Univ of California, Santa Barbara – Ys
Provin, Tony L., (979) 862-4955 t-provin@tamu.edu, Texas A&M Univ – Sc
Prueher, Libby, (970) 351-2470 Libby.Prueher@unco.edu, Univ of Northern Colorado – Gg
Pruell, Richard J., (401) 782-3000 Univ of Rhode Island – Co
Pruitt, William O., (530) 752-0453 Univ of California, Davis – Ow
Prunty, Lyle D., (701) 231-8580 North Dakota State Univ – Sp
Pruss, Sara B., (413) 585-3948 spruss@email.smith.edu, Smith College – Pg
Puchtel, Igor, 301-405-4054 ipuchtel@geol.umd.edu, Univ of Maryland – Cg
Puckette, James, 405-744-6358 jpucket@okstate.edu, Oklahoma State Univ – Go
Puente, Carlos E., (916) 752-0689 Univ of California, Davis – Hg

Puente, Carlos E., (530) 752-0689 cepuente@ucdavis.edu,
Univ of California, Davis – Hq
Pufahl, Peir K., 902-585-1858 peir.pufahl@acadiau.ca,
Acadia Univ – Gs
Puffer, John H., 973-353-5238 jpuffer@andromeda.rutgers.edu,
Rutgers, The State Univ of New Jersey, Newark – Gi
Pugin, Andre J., 217-244-2183 pugin@isgs.uiuc.edu,
Illinois State Geological Survey – Yg
Pujana, Ignacio, (972) 883-2461 pujana@utdallas.edu,
Univ of Texas, Dallas – Pm
Pujol, Jose M., (901) 678-2007 pujol@ceri.memphis.edu,
Univ of Memphis – Ye
Pulham, Andrew J., (709) 737-8395 apulham@mun.ca,
Memorial Univ of Newfoundland – Go
Pulliam, Robert J., (512) 471-6156 jay@ig.utexas.edu,
Univ of Texas at Austin, Jackson School of Geosciences – Ys
Pullin, Michael, (505) 835-6185 mpullin@nmt.edu,
New Mexico Institute of Mining & Technology – Oo
Pun, Aurora, (505) 277-5629 apun@unm.edu,
Univ of New Mexico – Ca
Purcell, Jennifer E., (410) 221-8431 purcell@hpl.umces.edu,
Univ of Maryland – Ob
Purdom, William B., (541) 552-6494 purdom@sou.edu,
Southern Oregon Univ – Gx
Purdy, Ann, (248) 628-1562 Wayne State Univ – Gg
Purdy, G. Michael, (845) 365-8348 mpurdy@ldeo.columbia.edu,
Columbia Univ – Yr
Purkis, Sampurkis@nova.edu,
Nova Southeastern Univ – Or
Purtle, Jennifer M., (501) 575-7317 jms14@comp.uark.edu,
Univ of Arkansas, Fayetteville – Oo
Purvance, Matthew, (775) 784-4692 mdp@seismo.unr.edu,
Univ of Nevada, Reno – Ys
Putkonen, Jaakko K., 206-543-0689 putkonen@u.washington.edu,
Univ of Washington – GmCcl
Putman, George W., (518) 442-4466 SUNY, Albany – Ge
Putnam, Peter E., (403) 215-7850 Univ of Calgary – Gs
Pyle, Douglas, 808-956-9607 pyled@hawaii.edu,
Univ of Hawai'i, Manoa – Cg
Pyle, Eric J., 540-568-7115 pyleej@jmu.edu,
James Madison Univ – Oe
Pyle, Joseph M., (518) 276-4899 pylej@rpi.edu,
Rensselaer Polytechnic Institute – Gp
Pysklywec, Russell N., (416) 978-4852 russ@geology.utoronto.ca,
Univ of Toronto – Yg
Pöllmann, Herbert, +345 5526110 herbert.poellman@geo.uni-halle.de,
Martin-Luther-Universitaet Halle-Wittenberg – Gz

Q

Qiu, Bo, (808) 956-9505 bo@soest.hawaii.edu,
Univ of Hawai'i, Manoa – Op
Quade, Jay, (520) 792-0454 jquade@geo.arizona.edu,
Univ of Arizona – Sc
Qualls, Robert G., (775) 327-5014 qualls@ers.unr.edu,
Univ of Nevada, Reno – Oo
Quane, Steve, 719-389-6516 steve.quane@coloradocollege.edu,
Colorado College – Gv
Quay, Paul D., (206) 545-8061 pdquay@u.washington.edu,
Univ of Washington – Oc
Qudus, Abdul, (210) 733-2830 aqudus@mail.accd.edu,
San Antonio College – GgOgg
Quebedeaux, Bruno, (301) 405-4336 bq1@umail.umd.edu,
Univ of Maryland – Oo
Quick, Jeffrey, (801) 537-3372 nrugs.jquick@state.ut.us,
Utah Geological Survey – Ec
Quick, Thomas J., (330) 972-6935 tquick@uakron.edu,
Univ of Akron – Ca
Quigg, Antonietta, 409 740 4990 quigga@tamug.tamu.edu,
Texas A&M Univ – Ob
Quinlan, Frank, (828) 232-5160 fquinlan@unca.edu,
Univ of North Carolina, Asheville – Oa
Quinlan, Garry M., (709) 737-3030 gquinlan@sparky2.esd.mun.ca,
Memorial Univ of Newfoundland – Gt
Quinn, Harry M., (619) 349-8703 College of the Desert – Ps
Quinn, Heather, 410-554-5522 hquinn@dnr.state.md.us,
Maryland Geological Survey – Ge
Quinn, James G., (401) 874-6219 jgquinn@gso.uri.edu,
Univ of Rhode Island – Oc
Quinn, John J., (630) 252-5357 quinnj@anl.gov,
Argonne National Lab – Hq
Quinn, Pat, (954)262-3642 quinn@nova.edu,
Nova Southeastern Univ – Ob
Quinn, Terrence M., (727) 553-1130 tquinn@marine.usf.edu,
Univ of South Florida, Saint Petersburg – Cm
Quinn, Terry, 512-471-0377 quinn@ig.utexas.edu,
Univ of Texas at Austin, Jackson School of Geosciences – Pe

R

Rabalais, Nancy N., (225) 851-2800 Louisiana State Univ – Ob
Rabinowitz, Philip D., (979) 862-2994 rabinowitz@geo.tamu.edu,
Texas A&M Univ – Yr
Rack, Franktrack2@unl.edu, Univ of Nebraska, Lincoln – Ou
Raczkowski, Charles, (919) 334-7543 North Carolina Agricultural & Tech State Univ – So
Radcliffe, David E., (706) 542-0897 Univ of Georgia – Sp
Radcliffe, Dennis, 516 463-5565 geodzr@hofstra.edu,
Hofstra Univ – Gz
Rademacher, Laura K., (209) 946-7351 lrademacher@pacific.edu,
Univ of the Pacific – ClHyOs
Radke, Jerry K., (515) 382-4102 jkradke@iastate.edu,
Iowa State Univ of Science & Technology – Sp
Raeside, Robert, (902) 585-1583 Dalhousie Univ – Gx
Raeside, Robert P., (902) 585-1583 rob.raeside@acadiau.ca,
Acadia Univ – Gp
Raffensperger, Jeff P., (804) 924-0581 Univ of Virginia – Hw
Rafferty, Milton D., (417) 836-5800 Southwest Missouri State Univ – Oy
Ragan, Donal M., (480) 965-6939 dragan@asu.edu,
Arizona State Univ – Gc
Ragland, Paul C., (850) 644-5018 ragland@magnet.fsu.edu,
Florida State Univ – Ca
Ragotzkie, Robert A.ragotzkie@wisc.edu,
Univ of Wisconsin, Madison – Ob
Rahbar, Bijan, 404-656-3214 Georgia Dept of Natural Resources – Hw
Rahman, Asheka, 225-578-5320 arahma1@lsu.edu,
Louisiana State Univ – HwNg
Rahn, Kenneth A., (401) 874-6713 krahn@uri.edu,
Univ of Rhode Island – Oc
Rahn, Perry H., (605) 394-2494 Perry.Rahn@sdsmt.edu,
South Dakota School of Mines & Technology – Ng
Rai, Chandra S., (405) 325-6866 crai@ou.edu,
Univ of Oklahoma – Np
Raia, Federicaraia@scisun.sci.ccny.edu,
American Museum of Natural History – Gv
Railsback, L. Bruce, (706) 542-3453 rlsbk@gly.uga.edu,
Univ of Georgia – Gd
Rains, Mark C., 974-0323 Univ of South Florida, Tampa – Hw
Rains, R. Bruce, (780) 492-0350 bruce.rains@ualberta.ca,
Univ of Alberta – Gm
Raj, John K., 03-79674225 jkraj@um.edu.my,
Univ of Malaya – Ng
Rajora, O., (250) 828-5453 Univ College of the Cariboo – Oo
Rakovan, John F., (513) 529-3245 rakovajf@muohio.edu,
Miami Univ – Cl
Rakowski, Cynthia L., (509) 375-6927 cindy.rakowski@pnl.gov,
Pacific Northwest National Lab – Oo
Raleigh, C. Barry, (808) 956-6182 soest@soest.hawaii.edu,
Univ of Hawai'i, Manoa – Nr
Rallis, Donald N., (540) 654-1492 drallis@mwc.edu,
Mary Washington College – Oo
Ramachandran, Kumar, 918-631-2517 kumar-ramachandran@utulsa.edu,
Univ of Tulsa – Ye
Ramage, Joan, 610-758-6410 jmr204@lehigh.e.du,
Lehigh Univ – Or
Raman, Sethu S., (919) 515-7144 sethu_raman@ncsu.edu,
North Carolina State Univ – Oa
Ramananantoandro, R., 717-871-2086 rramana@millersville.edu,
Millersville Univ – Ng
Ramanathan, V., (619) 534-8815 vramanathan@ucsd.edu,
Univ of California, San Diego – Oa
Ramelli, Alan R., (775) 784-6691 (Ext. 148) ramelli@unr.edu,
Univ of Nevada – Ng
Ramirez, Abelardo L., (925) 422-6919 ramirez3@llnl.gov,
Lawrence Livermore National Lab – Ng
Ramirez, Pedro C., (323) 343-2417 pramire@calstatela.edu,
California State Univ, Los Angeles – Gs
Ramirez, Wilson R., (787) 265-3845 ramirezwilson@aol.com,
Univ of Puerto Rico – Gd
Ramos, Frank, (626) 688-4457 ramos@geology.cwu.edu,
Central Washington Univ – Cc
Ramp, Steven R., (408) 656-2201 sramp@nps.navy.mil,

Naval Postgraduate School – Op
Ramsey, Kelvin W., (302) 831-3586 kwramsey@udel.edu, Univ of Delaware – Gr
Ramsey, Kelvin W., 302-831-2833 kwramsey@udel.edu, Univ of Delaware – Gs
Ramsey, Michael S., (412) 624-8772 ramsey@ivis.eps.pitt.edu, Univ of Pittsburgh – Or
Ramspott, Matthew E., 301/687-4412 meramspott@frostburg.edu, Frostburg State Univ – Or
Ranalli, Giorgio, granalli@ccs.carleton.ca, Carleton Univ – Yg
Rancan, Helen L., 609-984-6587 helen.rancan@dep.state.nj.us, New Jersey Geological Survey – Hs
Randall, David A., (970) 491-8407 randall@atmos.colostate.edu, Colorado State Univ – Oa
Randall, George, (505) 667-9483 Los Alamos National Lab – Yg
Randall, Gyles W., (507) 835-3620 Univ of Minnesota, Twin Cities – So
Randazzo, Anthony F., (352) 392-2634 randazzo@geology.ufl.edu, Univ of Florida – Gd
Randerson, James T., (626) 395-2683 jimr@gps.caltech.edu, Univ of California, Irvine – Oo
Randolph, James C., (812) 855-2457 Indiana Univ, Bloomington – Sf
Raney, Jay A., (512) 471-5357 jay.raney@beg.utexas.edu, Univ of Texas at Austin, Jackson School of Geosciences – Gd
Rankey, Eugene, 305 421-4690 Univ of Miami – Oo
Rankin, Robert, (780) 492-5082 rankin@phys.ualberta.ca, Univ of Alberta – Xy
Ransom, Mickey D., (913) 532-7203 Kansas State Univ – Sd
Ranson, William A., (864) 294-3364 bill.ranson@furman.edu, Furman Univ – GxzGp
Rao, Pemmasani D., (907) 474-6748 Univ of Alaska, Fairbanks – Oo
Rao, S. T., (518) 437-3200 strao@dec.state.ny.us, North Carolina State Univ – Oa
Rapp, George R., 218-726-7629 grapp@d.umn.edu, Univ of Minnesota, Duluth – Ga
Rapp, Robert, 631-632-8192 robert.rapp@sunysb.edu, SUNY, Stony Brook – Gy
Rard, Joseph A., (925) 422-6872 Lawrence Livermore National Lab – Cg
Rasbury, E. Troy, 631-632-1488 troy.rasbury@sunysb.edu, SUNY, Stony Brook – Cc
Rasmussen, Craig E., (313) 763-5192 Univ of Michigan – Og
Rasmussen, Kenneth A., (703) 323-2139 krasmussen@nvcc.edu, Northern Virginia Community College - Loudoun Campus – Pi
Rasmussen, Patricia, (613) 941-9868 Pat_Rasmussen@hc-sc.gc.ca, Univ of Ottawa – Ge
Rasmussen, Steen, (505) 665-0052 steen@lanl.gov, Los Alamos National Lab – Yg
Ratajeski, Kent, (859) 257-4444 krata2@email.uky.edu, Univ of Kentucky – Oo
Ratchford, M. E., 683-0118 ed.ratchford@arkansas.gov, Arkansas Geological Commission – Gc
Rathburn, Anthony, 812-237-2269 gerathbu@isugw.indstate.edu, Indiana State Univ – Ob
Rathburn, Sara L., (970) 491-6956 rathburn@cnr.colostate.edu, Colorado State Univ – Gg
Rattray, Jr., Maurice, (206) 543-5189 rattray@ocean.washington.edu, Univ of Washington – Op
Rauber, Robert M., (217) 333-2835 r-rauber@uiuc.edu, Univ of Illinois, Urbana-Champaign – Oa
Rauch, Henry W., (304) 293-5603 hrauch@wvu.edu, West Virginia Univ – Hw
Raudsepp, Mati, (604) 822-6396 mraudsepp@eos.ubc.ca, Univ of British Columbia – Gz
Rauzi, Steven L., (520) 770-3500 Arizona Geological Survey – Go
Ravat, Dhananjay, (859) 257-4726 dhananjay.ravat@uky.edu, Univ of Kentucky
Ravat, Dhananjay, (618) 453-7352 ravat@geo.siu.edu, Southern Illinois Univ Carbondale – Ye
Ravelo, Christina, (831) 459-3722 acr@cats.ucsc.edu, Univ of California, Santa Cruz – Cs
Ravizza, Gregory, 808-956-2916 gravizza@soest.hawaii.edu, Univ of Hawai'i, Manoa – Cm
Rawling, Geoffrey, (505) 366-2535 geoff@gis.nmt.edu, New Mexico Institute of Mining & Technology – Gg
Ray, Evelyn, 757.727.5752 evelyn.ray@hamptonu.edu, Hampton Univ – Oo
Ray, G. Carleton, (804) 924-0551 cr@virginia.edu, Univ of Virginia – Ob
Ray, Peter S., (850) 644-1894 pray@met.fsu.edu, Florida State Univ – Ow
Ray, Thomas M., (601) 961-5531 tom_ray@deq.state.ms.us, Mississippi Office of Geology – Gg
Rayburn, John A.rayburnj@newpaltz.edu, SUNY New Paltz – Gml
Raymo, Maureen E., (617) 253-0474 raymo@mit.edu, Boston Univ – Pe
Raymond, Anne L., (979) 845-0644 raymond@geo.tamu.edu, Texas A&M Univ – Pb
Raymond, Carol A., (818) 354-8690 Jet Propulsion Lab – Yr
Raymond, Charles F., 206-685-3829 charlie@ess.washington.edu, Univ of Washington – Ol
Raymond, Dorothy E., 205/247-3547 draymond@gsa.state.al.us, Geological Survey of Alabama – Gg
Raymond, Loren A., (828) 262-2749 raymondla@appstate.edu, Appalachian State Univ – Gx
Rayne, Todd W., (315) 859-4698 trayne@hamilton.edu, Hamilton College – Hy
Rayner, John N., (614) 292-2514 Ohio State Univ – Oa
Rea, David K., (734) 936-0521 davidrea@umich.edu, Univ of Michigan – Gu
Read, Adam S., (505) 366-2533 adamread@gis.nmt.edu, New Mexico Institute of Mining & Technology – Gg
Read, J. Fred, (540) 231-5124 jread@vt.edu, Virginia Polytechnic Institute & State Univ – Gs
Reagan, Mark K., (319) 335-1802 mark-reagan@uiowa.edu, Univ of Iowa – Gi
Reaka-Kudla, Marjorie, (301) 405-6944 mr9@umail.umd.edu, Univ of Maryland – Ob
Real, Charles R., (916) 323-8550 California Geological Survey – Ys
Reams, Max W., (815) 939-5394 mreams@olivet.edu, Olivet Nazarene Univ – Gsm
Reaser, Donald F., (817) 272-2984 Reaser@uta.edu, Univ of Texas, Arlington – Gr
Reasor, Paul, (850) 644-4056 reasor@met.fsu.edu, Florida State Univ – Ow
Reaven, Sheldon, (631) 632-8765 sheldon.reaven@notes.cc.sunysb.edu du, SUNY, Stony Brook – Oo
Reavy, John, +353 21 4902886 j.reavy@ucc.ie, Univ College Cork – Gi
Reay, Anthony, +64 3 479-7520 tony.reay@otago.ac.nz, Univ of Otago – Gi
Rebbert, Carolyn R., (864) 656-4602 crebber@clemson.edu, Clemson Univ – Gx
Rech, Jason, 513-529-1935 rechja@muohio.edu, Miami Univ – Gm
Rechcigl, John E., (863) 735-1314 rechcigl@gnv.ifas.ufl.edu, Univ of Florida – So
Rector, James W., (510) 643-7820 Univ of California, Berkeley – Ye
Redalje, Donald G., (228) 688-1174 donald.redalje@usm.edu, Univ of Southern Mississippi – Ob
Redden, Jack A., (605) 394-5113 South Dakota School of Mines & Technology – Gp
Reddy, Christopher M., (508) 289-2316 creddy@whoi.edu, Woods Hole Oceanographic Institution – Co
Reddy, Gudigopuram B., 336-334-7543 reddyg@ncat.edu, North Carolina Agricultural & Tech State Univ – So
Reddy, K R., (352) 392-1804 (Ext. 317) krr@ufl.edu, Univ of Florida – Sf
Reddy, Katta J., (307) 766-6658 Univ of Wyoming – Sc
Redmon, Marshamredmon@truman.edu, Truman State Univ – Oo
Redmond, Brian T., (570) 408-4803 Wilkes Univ – Gs
Reeburgh, William S., (949) 824-2986 reeburgh@uci.edu, Univ of California, Irvine – Oo
Reed, Billy, (573) 368-2165 Missouri Dept of Natural Resources – Oo
Reed, Denise J., (504) 280-7395 djreed@uno.edu, Univ of New Orleans – Gm
Reed, Donald L., (408) 924-5036 dreed@sjsu.edu, San Jose State Univ – Gu
Reed, Juliet C., (610) 526-7487 Bryn Mawr College – Gz
Reed, Mark H., (541) 346-5587 mhreed@oregon.uoregon.edu, Univ of Oregon – Em
Reed, Randy, 530-225-7328 rreed@shastacollege.edu, Shasta Community College – Ge
Reed, Robert, (512) 475-8786 rob@alumni.utexas.edu, Univ of Texas at Austin, Jackson School of Geosciences – Gc
Reed, Wallace E., (804) 924-0567 wer@virginia.edu, Univ of Virginia – Or
Reed, Walter E., 661-245-1443 wreed@frazmtn.com,

Univ of California, Los Angeles – Gs
Reeder, Patrice, (330) 263-2380 preeder@wooster.edu,
The College of Wooster – Oo
Reeder, Richard J., 631-632-8208 rjreeder@stonybrook.edu,
SUNY, Stony Brook – Cl
Reedy, Robert C., (512) 471-7244 bob.reedy@beg.utexas.edu,
Univ of Texas at Austin, Jackson School of Geosciences – Hg
Reedy, Robert C., (505) 277-0300 rreedy@unm.edu,
Univ of New Mexico – Xcy
Rees, Margaret N., (702) 895-3890 rees@nevada.edu,
Univ of Nevada, Las Vegas – Gs
Reese, Betsy, (610) 526-7462 breese@brynmawr.edu,
Bryn Mawr College – Yr
Reese, Joseph F., (814) 732-2529 jreese@edinboro.edu,
Edinboro Univ of Pennsylvania – Gc
Reese, Stuart O., 717.702.2028 sreese@state.pa.us,
Pennsylvania Bureau of Topographic & Geologic Survey – Oi
Reesman, Authur L., 615-322-2976 None,
Vanderbilt Univ – Gg
Reeve, Andrew S., (207) 581-2353 asreeve@maine.maine.edu,
Univ of Maine – Hw
Regage, Jenniferrehage@nova.edu,
Nova Southeastern Univ – Oo
Reganold, John P., (509) 335-8856 reganold@wsu.edu,
Washington State Univ – Ou
Reger, James P., 410-554-5523 jreger@dnr.state.md.us,
Maryland Geological Survey – Gm
Regis, Robert S., (906) 227-1176 rregis@nmu.edu,
Michigan Technological Univ – Gl
Rehm, George W., (612) 625-6210 grehm@soils.umn.edu,
Univ of Minnesota, Twin Cities – Sc
Reichard, James S., (912) 681-1153 JReich@GeorgiaSouthern.edu,
Georgia Southern Univ – Hw
Reichenbacher, Bettina, 089/2180 6603 b.reichenbacher@lrz.uni-muenchen.de, LMU - Univ of Munich – Pg
Reichle, Michael, (916) 323-8553 California Geological Survey – Ys
Reicosky, Donald C., (320) 589-3411 reicosky@morris.ars.usda.gov,
Univ of Minnesota, Twin Cities – So
Reid, Arch M., (713) 743-3404 areid@uh.edu,
Univ of Houston – Gi
Reid, James, 61 3 6226 2477 James.Redi@utas.edu.au,
Univ of Tasmania – Yg
Reid, Jeffrey C., (919) 733-2423 (Ext. 403) Jeff.Reid@ncmail.net,
North Carolina Geological Survey – Gg
Reid, John R., (701) 777-2131 john_reidjr@yahoo.com,
Univ of North Dakota – Gm
Reid, Joseph L., (619) 534-2055 jreid@ucsd.edu,
Univ of California, San Diego – Op
Reid, Leslie, (707) 825-2933 lreid@fs.fed.us,
Humboldt State Univ – Gm
Reid, Mary R., (928) 523-7200 mary.reid@nau.edu,
Northern Arizona Univ – Gi
Reid, Robert O., (979) 845-4089 rreid@ocean.tamu.edu,
Texas A&M Univ – Op
Reid, Ruth P., 305 421-4606 preid@rsmas.miami.edu,
Univ of Miami – Gs
Reid, Steven K., 606-783-5293 s.reid@morehead-st.edu,
Morehead State Univ – Gs
Reidel, Stephen P., (509) 376-9932 sp_reidel@pnl.gov,
Washington State Univ – Gi
Reilinger, Robert E., (617) 253-7860 reilinge@erl.mit.edu,
Massachusetts Institute of Technology – Yg
Reimers, Clare E., (541) 737-2426 creimers@coas.oregonstate.edu,
Oregon State Univ – Cm
Reinders, Bruce, (403) 471-7776 Northern Alberta Inst of Tech – Np
Reinen, Linda A., (909) 621-8672 lreinen@pomona.edu,
Pomona College – Gc
Reinersman, Phillip N., (727) 553-3952 reinersman@marine.usf.edu,
Univ of South Florida, Saint Petersburg – Op
Reinertsen, Kristen L., 717.702.2047 kreinertse@state.pa.us,
Pennsylvania Bureau of Topographic & Geologic Survey – Eg
Reinfelder, John R., 732-932-8013 reinfelder@envsci.rutgers.edu,
Rutgers, The State Univ of New Jersey – Cl
Reinfelder, Ying Fan, (732) 445-2044 YFanRein@aol.com,
Rutgers, The State Univ of New Jersey – Hw
Reinhardt, Edward G., (905) 525-9140 (Ext. 27594) ereinhar@mcmail.cis.mcmaster.ca, McMaster Univ – Pm
Reinisch, Bodo W., (508) 934-4983 Univ of Massachusetts/Lowell – Oa
Reisner, Jon M., (505) 665-1889 reisner@lanl.gov,
Los Alamos National Lab – Oa
Reiss, Nathan, (212) 346-1502 Pace Univ, New York Campus – Oa
Reitan, Paul H., (716) 645-6800 (Ext. 3988) preitan@acsu.buffalo.edu,
SUNY, Buffalo – Gp
Reiten, Jon C., 406-657-2630 jreiten@mtech.edu,
Montana Tech of The Univ of Montana – Hw
Reiter, Marshall A., (505) 835-5306 mreiter@nmt.edu,
New Mexico Institute of Mining & Technology – Yh
Reitsma, Stanley, (519) -25-3-30 (Ext. ext. 2507) reitsma@uwindsor.ca,
Univ of Windsor – Hw
Reitz, Elizabeth J., (706) 542-1464 ereitz@arches.uga.edu,
Univ of Georgia – Ga
Remenda, Victoria H., (613) 545-6594 Queen's Univ – Hw
Remington, Lisa E., 512-475-7585 lisa.remington@beg.utexas.edu,
Univ of Texas at Austin, Jackson School of Geosciences – Gg
Remington, Randy L., (512) 475-9599 randy.remington@beg.utexas.edu,
Univ of Texas at Austin, Jackson School of Geosciences – Ye
Remson, Irwin, (650) 723-9191 Stanford Univ – Hw
Ren, Minghua, (915) 747-5843 ren@geo.utep.edu,
Univ of Texas, El Paso – Ca
Renaut, Robin W., (306) 966-5705 robin.renaut@usask.ca,
Univ of Saskatchewan – Gs
Reneau, Raymond B., (540) 231-9779 reneau@vt.edu,
Virginia Polytechnic Institute & State Univ – Sc
Reneau, Steven L., (505) 665-3151 sreneau@lanl.gov,
Los Alamos National Lab – Gm
Renegar, Abigaildrenegar@nova.edu,
Nova Southeastern Univ – Oo
Renfrew, Bonnie, 217-244-2430 renfrew@isgs.uiuc.edu,
Illinois State Geological Survey – Oo
Renne, Paul, (510) 644-1350 prenne@bgc.org,
Univ of California, Berkeley – Cc
Rennie, Colette, (902) 867-2299 crennie@stfx.ca,
Saint Francis Xavier Univ – Gg
Rensberger, John M., 206-685-1735 rensb@u.washington.edu,
Univ of Washington – Pv
Renshaw, Carl E., 603-646-3365 carl.renshaw@dartmouth.edu,
Dartmouth College – Hg
Rensing, Christopher C., (520) 626-8482 rensingc@ag.arizona.edu,
Univ of Arizona – Sb
Renton, John J., (304) 293-5603 jrenton@wvu.edu,
West Virginia Univ – Ec
Repeta, Daniel J., (508) 289-2635 drepeta@whoi.edu,
Woods Hole Oceanographic Institution – Co
Repine, Thomas, 304 594 2331 repine@geosrv.wvnet.edu,
West Virginia Univ – Gg
Repka, James L., (949) 582-4694 jrepka@saddleback.cc.ca.us,
Saddleback Community College – Gm
Resig, Johanna M., (808) 956-7757 jresig@soest.hawaii.edu,
Univ of Hawai'i, Manoa – Pm
Resing, Joseph A., (206) 526-6184 resing@pmel.noaa.gov,
Univ of Washington – Oc
Resio, D. R., (407) 768-8000 Florida Institute of Technology – Op
Resnic, Victor S., 281-618-5800 rvictor@mcleodusa.net,
North Harris College – Go
Resor, Phillip G., 860 6853139 presor@wesleyan.edu,
Wesleyan Univ – Gct
Ressetar, Robert M., (801) 585-3544 Univ of Utah – Go
Ressler, Daniel E., 570-372-4216 resslerd@susqu.edu,
Susquehanna Univ – Sp
Restrepo, Jorge I., (561) 297-2795 restrepo@fau.edu,
Florida Atlantic Univ – Hq
Retallack, Gregory J., (541) 346-4558 gregr@darkwing.uoregon.edu,
Univ of Oregon – Oo
Retelle, Michael J., (207) 786-6155 mretelle@bates.edu,
Bates College – Gl
Rettke, Robert C., (318) 475-5783 McNeese State Univ – Gd
Reusch, Douglas N., 207-778-7463 reusch@main.edu,
Univ of Maine, Farmington – GtCgOu
Reuss, Robert L., 617-627-3494 bert.reuss@tufts.edu,
Tufts Univ – Gzo
Reuter, Gerhard, (780) 492-0358 gerhard.reuter@ualberta.ca,
Univ of Alberta – Ow
Reuter, Gerhard W., (780) 492-0358 gerhard.reuter@ualberta.ca,
Univ of Alberta – Oa
Revelle, Douglas O., (505) 667-1256 revelle@lanl.gov,
Los Alamos National Lab – Oa
Revenaugh, Justin, 612-624-7553 justinr@umn.edu,
Univ of Minnesota, Twin Cities – Ys
Reves-Sohn, Robert A., (508) 289-3616 rsohn@whoi.edu,
Woods Hole Oceanographic Institution – Yr

Revetta, Frank A., (315) 267-2289 revettfa@potsdam.edu, SUNY, Potsdam – Yg
Rex, Arthur B., (828) 262-6911 rexab@appstate.edu, Appalachian State Univ – Oi
Rexroad, Carl, (812) 855-7428 crexroad@indiana.edu, Indiana Geological Survey – Pm
Rexroad, Carl B., (812) 855-7428 Indiana Univ, Bloomington – Pi
Rey, Patrice, 612-624-6801 prey@geosci.usyd.edu.au, Univ of Minnesota, Twin Cities – Gt
Reyes, Alfonsoreyeszca@cicese.mx, Centro de Inv Científica y de Educación Superior de Ensenada – Ne
Reyes, Manuel, (919) 334-7543 North Carolina Agricultural & Tech State Univ – Hw
Reynolds, Barbara C., (828) 232-5186 kreynolds@unca.edu, Univ of North Carolina, Asheville – Oo
Reynolds, James H., (828) 884-8377 reynoljh@brevard.edu, Brevard College – Gr
Reynolds, Peter H., (902) 494-2325 preynold@is.dal.ca, Dalhousie Univ – Cc
Reynolds, Robert W., (541) 383-7557 Central Oregon Community College – Gv
Reynolds, Stephen J., (480) 965-9049 sreynolds@asu.edu, Arizona State Univ – Gt
Reynolds, William R., (662) 234-6449 Univ of Mississippi – Gs
Reynolds, III, John E., (727) 864-8431 Univ of South Florida, Saint Petersburg – Ob
Rhines, Peter B., (206) 543-0593 rhines@ocean.washington.edu, Univ of Washington – Op
Rhode, David E., (775) 673-7310 dave@dri.edu, Desert Research Institute – Pe
Rhodes, Amy L., (413) 585-3947 arhodes@email.smith.edu, Smith College – Cg
Rhodes, Brady P., (714) 278-2942 brhodes@fullerton.edu, California State Univ, Fullerton – Gc
Rhodes, Carol J., (979) 845-3001 cj-rhodes@tamu.edu, Texas A&M Univ – Oo
Rhodes, Dallas D., (912) 681-5361 DRhodes@GeorgiaSouthern.edu, Georgia Southern Univ – Gm
Rhodes, Elizabeth F., (843) 953-8139 Kinge@cofc.edu, College of Charleston – Gg
Rhodes, Frank H. T., (607) 255-6233 Cornell Univ – Pi
Rhodes, J. Michael, (413) 538-2841 jmrhodes@geo.umass.edu, Univ of Massachusetts – Ca
Rhue, R D., (352) 392-1951 rdr@gnv.ifas.ufl.edu, Univ of Florida – Sc
Rial, Jose A., (919) 966-4553 jose_rial@unc.edu, Univ of North Carolina, Chapel Hill – Ys
Ribbe, Paul H., (540) 231-6880 ribbe@vt.edu, Virginia Polytechnic Institute & State Univ – Gz
Rice, Alanarice@amnh.org, American Museum of Natural History – Yg
Rice, Chuck W., (541) 346-3005 Kansas State Univ – Sb
Rice, Jack M., (541) 346-3005 jrice@darkwing.uoregon.edu, Univ of Oregon – Gp
Rice, James R., (617) 495-3445 rice@husm.harvard.edu, Harvard Univ – Yg
Rice, Keith W., 715-346-4454 krice@uwsp.edu, Univ of Wisconsin, Stevens Point – Oir
Rice, Richard J., 217-244-2458 rice@isgs.uiuc.edu, Illinois State Geological Survey – Hw
Rice, Thomas J., (805) 756-2420 trice@calpoly.edu, California Polytechnic State Univ – Sd
Rice-Snow, R. Scott, (765) 285-8269 00rssnow@bsu.edu, Ball State Univ – Gm
Rich, Charles C., (419) 372-7199 Bowling Green State Univ – Gm
Rich, Fredrick J., (912) 681-0849 frich@GeorgiaSouthern.edu, Georgia Southern Univ – Pl
Rich, Jeremy J., (609) 258-2612 jjrich@princeton.edu, Princeton Univ – Sb
Rich, Joseph A., (609) 292-2576 joseph.rich@dep.state.nj.us, New Jersey Geological Survey – Oy
Rich, Mark, (706) 542-2652 Univ of Georgia – Ps
Rich, Thomas B., (605) 677-5227 trich@usd.edu, South Dakota Dept of Env & Natural Resources – Hw
Richard, Benjamin H.benjamin.richard@wright.edu, Wright State Univ – Ye
Richard, Gigi, 970-248-1689 grichard@mesastate.edu, Mesa State College – Hs
Richard, Robert, (310) 825-6663 rrichard@igpp.ucla.edu, Univ of California, Los Angeles – Yg
Richard, Stephen M., (520) 770-3500 Arizona Geological Survey – Gg
Richards, Ian J., (214) 768-2750 Southern Methodist Univ – Cs
Richards, James H., (530) 752-0170 jhrichards@ucdavis.edu, Univ of California, Davis – So
Richards, Jeremy P., (780) 492-3430 jeremy.richards@ualberta.ca, Univ of Alberta – Em
Richards, Kelvin J.rkelvin@hawaii.edu, Univ of Hawai'i, Manoa – Og
Richards, Mark A.markr@seismo.berkeley.edu, Univ of California, Berkeley – Yg
Richards, Norman L., (203) 932-7107 Univ of New Haven – Ou
Richards, Paul G., (845) 365-8389 richards@ldeo.columbia.edu, Columbia Univ – Ys
Richards, Paul L., prichard@brockport.edu, SUNY College at Brockport – Hg
Richards, Vince, (954)262-3657 ardsrich@nova.edu, Nova Southeastern Univ – Ob
Richardson, Bonham C., (540) 231-5514 Virginia Polytechnic Institute & State Univ – Oo
Richardson, Eliza, (814) 865-7326 eliza@geosc.psu.edu, Pennsylvania State Univ, Univ Park – Ys
Richardson, Jimmie L., (412) 357-5625 North Dakota State Univ – Sf
Richardson, Mary J., (979) 845-3651 mrichardson@ocean.tamu.edu, Texas A&M Univ – Gu
Richardson, Mary Jo, (979) 845-7966 mrichardson@ocean.tamu.edu, Texas A&M Univ – Gu
Richardson, Randall M., (520) 621-4950 rmr@u.arizona.edu, Univ of Arizona – Yg
Richey, Jeffrey E., (206) 543-7339 jrichey@u.washington.edu, Univ of Washington – Oc
Richman, James G., (541) 737-3328 jrichman@coas.oregonstate.edu, Oregon State Univ – Op
Richmond, Marshall C., (509) 372-6241 mc_richmond@pnl.gov, Pacific Northwest National Lab – Hs
Richter, Carl, (337) 482-5353 richter@louisiana.edu, Univ of Louisiana, Lafayette – Ym
Richter, Daniel D., 919-613-8031 drichter@duke.edu, Duke Univ – Os
Richter, Frank M., (773) 702-8118 richter@geosci.uchicago.edu, Univ of Chicago – Gt
Richwien, Barry A., (802) 626-6205 Lyndon State College – Ow
Riciputi, Lee R., (865) 576-4839 i79@ornl.gov, Univ of Tennessee, Knoxville – Cc
Rickard, Drickard@cf.ac.uk, Cardiff Univ – Cg
Rickards, William, (804) 924-5965 rickards@virginia.edu, Univ of Virginia – Ob
Rickerl, Dianne H., (605) 688-5541 Diane.Rickerl@sdstate.edu, South Dakota State Univ – Sf
Rickwell, Sheryl, 609-258-6144 srickwel@princeton.edu, Princeton Univ – Gg
Rickwell, Sheryl A., (609) 258-6144 srickwel@princeton.edu, Princeton Univ – Oo
Ridd, Merrill K., (801) 581-7939 merrill.ridd@geog.utah.edu, Univ of Utah – Or
Riddick, Pamela M.pamriddick@memphis.edu, Univ of Memphis – Oo
Ridenour, Gregory D., (931) 221-7454 ridenourg@apsu.edu, Austin Peay State Univ – OiaOo
Ridge, John C., 617-627-3494 jack.ridge@tufts.edu, Tufts Univ – Gln
Ridgway, Kenneth D., (765) 494-3269 ridge@purdue.edu, Purdue Univ – Gs
Riding, Rriding@cf.ac.uk, Cardiff Univ – Py
Ridky, Alice M., (207) 859-5800 amridky@colby.edu, Colby College – Oo
Ridley, John R., (970) 491-5943 jridley@cnr.colostate.edu, Colorado State Univ – Eg
Ridley, Moira, (806) 742-0627 moira.ridley@ttu.edu, Texas Tech Univ – Cl
Riebesell, John, (313) 593-5132 jriebese@umich.edu, Univ of Michigan, Dearborn – Ou
Riedel, Michael, 514-398-2722 mriedel@eps.mcgill.ca, McGill Univ – Ye
Riedel, Oliver, 089/2180 4335 oliver.riedl@lrz.uni-muenchen.de, LMU - Univ of Munich – Gz
Riediger, Cynthia L., (403) 220-8783 riediger@geo.ucalgary.ca, Univ of Calgary – Co
Rieger, JoAnne T., (509) 376-8344 joanne.rieger@pnl.gov, Pacific Northwest National Lab – Oo

Riegl, Bernhard M., (954)262-3671 rieglb@nova.edu, Nova Southeastern Univ – Ob
Rieke, George H., (520) 621-2832 grieke@as.arizona.edu, Univ of Arizona – Oo
Rieke, Herman H., 337-482-6556 hhr8668@louisiana.edu, Univ of Louisiana, Lafayette – Np
Rieken, Eric R., (210) 381-3526 Univ of Texas, Pan American – Gt
Riemersma, Peter E., 616-331-3553 riemersp@gvsu.edu, Grand Valley State Univ – Hw
Rieppel, Olivier C., (312) 665-7630 rieppel@fmnh.org, Field Museum of Natural History – Pv
Rietmeijer, Frans J., (505) 277-5733 fransjmr@unm.edu, Univ of New Mexico – Gp
Rigano, Lance S., (516) 572-7044 riganol@ncc.edu, Nassau Community College – Hw
Rigby, J. Keith, 801-377-5653 Brigham Young Univ – Pe
Rigby, John, 61 7 3138 1638 j.rigby@qut.edu.au, Queensland Univ of Technology – Pb
Rigby, Jr., J. Keith, (219) 631-6245 rigby.1@nd.edu, Univ of Notre Dame – Pv
Riggs, Eric, (765) 496-1974 emriggs@purdue.edu, Purdue Univ – Oe
Riggs, Nancy, (928) 523-9362 nancy.riggs@nau.edu, Northern Arizona Univ – Gv
Riggs, Stanley R., (252) 328-6015 riggss@mail.ecu.edu, East Carolina Univ – Gu
Rigotti, Peter A., (718) 982-2827 College of Staten Island – Yg
Rigsby, Catherine A., (252) 328-4297 rigsbyc@mail.ecu.edu, East Carolina Univ – Gs
Riha, Susan, (607) 255-1729 sjr4@cornell.edu, Cornell Univ – Sf
Riley, James, (520) 626-6681 jjriley@ag.arizona.edu, Univ of Arizona – Oo
Riley, Robert G., (509) 376-1935 robert.riley@pnl.gov, Pacific Northwest National Lab – Cg
Riley, Ronald A., (614) 265-6573 ron.riley@dnr.state.oh.us, Ohio Dept of Natural Resources – Go
Rimmer, Susan M., (859) 257-4607 srimmer@uky.edu, Univ of Kentucky – Ec
Rimstidt, J. Donald, (540) 231-6894 jdr02@vt.edu, Virginia Polytechnic Institute & State Univ – Cl
Rinard, Bethany D., (254) 968-9894 rinard@vms.tarleton.edu, Tarleton State Univ – Gv
Rind, David H., (212) 678-5593 drind@giss.nasa.gov, Columbia Univ – Oa
Rindsberg, Andrew K., 205 652 3416 arindsberg@uwa.edu, Univ of West Alabama – Gg
Ringle, Dean, (614) 462-3900 Ohio State Univ – Oo
Rink, William J., (905) 525-9140 (Ext. 24178) rinkwj@mcmail.cis.mcmaster.ca, McMaster Univ – Cc
Riordan, Allen J., (919) 515-7973 al_riordan@ncsu.edu, North Carolina State Univ – Oa
Ripley, Edward M., (812) 855-1196 ripley@indiana.edu, Indiana Univ, Bloomington – Em
Riser, Stephen C., (206) 543-1187 riser@ocean.washington.edu, Univ of Washington – Op
Risk, Michael J., (905) 525-9140 (Ext. 24183) riskmj@mcmail.cis.mcmaster.ca, McMaster Univ – Po
Risley, Lance, (201) 595-3438 William Paterson College – Oo
Ritchie, Alexander W., (843) 953-5591 ritchiea@cofc.edu, College of Charleston – Gc
Ritchie, Harold C., (902) 494-5192 hritchie@phys.ocean.dal.ca, Dalhousie Univ – Ow
Ritsema, Jeroen, 734-615-6405 jritsema@umich.edu, Univ of Michigan – Ys
Ritter, Charles J., (937) 229-2953 Univ of Dayton – Ct
Ritter, Jason C., (509) 372-6046 rc_ritter@pnl.gov, Pacific Northwest National Lab – Sc
Ritter, John B., (937) 327-7332 jritter@wittenberg.edu, Wittenberg Univ – Gm
Ritter, Michael E., 715-346-4449 mritter@uwsp.edu, Univ of Wisconsin, Stevens Point – Ow
Ritter, Scott M., 801-4224239 scott_ritter@byu.edu, Brigham Young Univ – Ps
Ritterbush, Linda A., (805) 493-3265 ritterbu@clunet.edu, California Lutheran Univ – Pi
Ritz, Richard, (402) 280-2461 richard.ritz@afwa.af.mil, Creighton Univ – Ow
Ritzi, Jr., Robert W., 937 775-3455 robert.ritzi@wright.edu, Wright State Univ – Hw
Rivard, Benoit, (780) 492-0345 benoit.rivard@ualberta.ca, Univ of Alberta – Or
Rivas Camargo, David, (541) 737-9092 drivas@coas.oregonstate.edu, Oregon State Univ – Op
Rivedal, Doug, (609) 984-6587 doug.rivedal@dep.state.nj.us, New Jersey Geological Survey – Ge
Rivers, Toby C. J. S., (709) 737-8392 trivers@sparky2.esd.mun.ca, Memorial Univ of Newfoundland – Gp
Roach, Michael, 61 3 6226 2474 Univ of Tasmania – Yg
Roadcap, George S., (217) 333-7951 Illinois State Water Survey – Hw
Robarge, Wayne P., (919) 515-1454 North Carolina State Univ – Sc
Robbins, Debra C., (828) 251-6441 Univ of North Carolina, Asheville – Oo
Robbins, Gary A., (860) 486-2448 gary.robbins@uconn.edu, Univ of Connecticut – Hw
Roberson, Herman E., (607) 777-2835 Binghamton Univ – Gs
Robert, Neuman, (202) 633-1343 Smithsonian Institution / National Museum of Natural History – Pi
Robert, Pierre C., (612) 625-3125 probert@soils.umn.edu, Univ of Minnesota, Twin Cities – Sd
Roberts, A. Lynn, (410) 516-4387 lroberts@jhu.edu, The Johns Hopkins Univ – Oo
Roberts, Charles E., (561) 297-3254 Florida Atlantic Univ – Or
Roberts, Frank, (504) 388-2964 Montgomery County Community College – Gp
Roberts, Harry H., (225) 388-2964 harry@antares.csi.lsu.edu, Louisiana State Univ – Gs
Roberts, Jennifer A., (785) 864-4997 jenrob@ku.edu, Univ of Kansas – Co
Roberts, Mark L., (508) 289-3654 mroberts@whoi.edu, Woods Hole Oceanographic Institution – Yg
Roberts, Paul H., (310) 206-2707 roberts@math.ucla.edu, Univ of California, Los Angeles – Ym
Roberts, Peter, (505) 667-1199 proberts@lanl.gov, Los Alamos National Lab – Ys
Roberts, R. Gwilym, (604) 291-4657 Univ of Waterloo – Em
Roberts, Sarah K., (925) 423-4112 roberts28@llnl.gov, Lawrence Livermore National Lab – Cl
Roberts, Sheila J., (419) 372-0354 sjrober@bgnet.bgsu.edu, Bowling Green State Univ – CgHw
Roberts, Sheila M., (406) 683-7017 s_roberts@umwestern.edu, Univ of Montana, Western – Ge
Roberts, Steve, (937) 229-2261 Steve.Roberts@notes.udayton.edu, Univ of Dayton – Gg
Robertson, Douglas S., 303-492-3694 Univ of Colorado – Oo
Robertson, James M., (608) 263-7384 jmrober1@facstaff.wisc.edu, Univ of Wisconsin, Madison – Eg
Robertson, James M., (608) 263-7384 jmrober1@wisc.edu, Univ of Wisconsin - Extension – Eg
Robertson, Quander, 404-656-3214 Georgia Dept of Nat Resources – Oo
Robertson, Sandra J., (404) 656-3214 Georgia Dept of Natural Resources – Hw
Robin, Michel R., (613)562-5800 Ext 6852 mrobin@uottawa.ca, Univ of Ottawa – Hw
Robin, Pierre-Yves F., 905 828-5419 Univ of Toronto – Gc
Robinson, Alexander, 713-743-3399 arobinson@uh.edu, Univ of Houston – Gc
Robinson, Allan R., (617) 495-2819 robinson@pacific.harvard.edu, Harvard Univ – Op
Robinson, Bonnie J., 217-265-8213 robinson@isgs.uiuc.edu, Illinois State Geological Survey – Ge
Robinson, Bruce A., (505) 667-1910 robinson@lanl.gov, Los Alamos National Lab – Oo
Robinson, Cordulacordula@bu.edu, Boston Univ – Gm
Robinson, Deloresdmr@geo.ua.edu, Univ of Alabama – Gc
Robinson, Edward, (305) 348-3572 c/o draper@fiu.edu, Florida International Univ – Pm
Robinson, Edwin S., (540) 231-6521 esrobinson@vt.edu, Virginia Polytechnic Institute & State Univ – Yg
Robinson, George W., (906) 487-2572 robinson@mtu.edu, Michigan Technological Univ – Gz
Robinson, Kevin, (619) 594-1386 krobinson@geology.sdsu.edu, San Diego State Univ – Gc
Robinson, Lance, (954)262-3662 lancerob@nova.edu, Nova Southeastern Univ – Oo
Robinson, Laural, robinson@whoi.edu, Woods Hole Oceanographic Institution – Cm
Robinson, Lori A., (612)627-4780 Ext. 221 robin126@umn.edu, Univ of Minnesota, Twin Cities – Oo

Robinson, Paul D., (618) 453-7373 robinson@geo.siu.edu,
Southern Illinois Univ Carbondale – Gz
Robinson, Paul T., (902) 494-2361 Dalhousie Univ – Gv
Robinson, Peter, (413) 545-2286 Univ of Massachusetts – Gc
Robinson, Peter, 303-492-5211 peter.robinson@colorado.edu,
Univ of Colorado – Pv
Robinson, R. Craig, (860) 832-2950 Central Connecticut State Univ – Xy
Robinson, Stephen D., (315) 229-6239 St. Lawrence Univ – Hg
Robinson, Stevej.s.robinson@reading.ac.uk,
Univ of Reading – Sf
Robinson, Walter A., (217) 333-2292 robinson@atmos.uiuc.edu,
Univ of Illinois, Urbana-Champaign – Oa
Robison, Linda, (702) 895-3262 lrobison@ccmail.nevada.edu,
Univ of Nevada, Las Vegas – Oo
Robison, Richard A., (785) 864-2739 robison@ku.edu,
Univ of Kansas – Pg
Robl, Thomas L., 859-257-0272 robl@caer.uky.edu,
Univ of Kentucky – Cg
Roby, Scott W., (315) 312-2790 roby@oswego.edu,
SUNY, Oswego – Oo
Rocha, Guillermo, (718) 951-5000 (Ext. 2887) grocha@brooklyn.cuny.edu, Brooklyn College (CUNY) – Gg
Roche, James E., (225) 388-2707 jroche@geol.lsu.edu,
Louisiana State Univ – Gg
Rochester, Michael G., (709) 737-7565 mrochest@morgan.ucs.mun.ca,
Memorial Univ of Newfoundland – Yg
Rochette, Elizabeth A., (603) 862-0713 Univ of New Hampshire – Sc
Rochette, Scott M., (585) 395-2603 rochette@weather.brockport.edu,
SUNY College at Brockport – Ow
Rocholl, Alexander, 089/2180 4293 rocholl@min.uni-muenchen.de,
LMU - Univ of Munich – Gz
Rockaway, John D., 859-572-5412 rockawayj@nku.edu,
Northern Kentucky Univ – Ng
Rockwell, Thomas K., (619) 594-4441 trockwell@sunstroke.sdsu.edu,
San Diego State Univ – Gm
Rockwood, Anthony A., (303) 556-8399 Metropolitan State College of Denver – Ow
Rodbell, Donald T., (518) 388-6034 rodbelld@union.edu,
Union College – GlmGe
Roden, Eric E., (608) 260-0724 eroden@geology.wisc.edu,
Univ of Wisconsin, Madison – Py
Roden, Gunnar I., (206) 543-5627 giroden@u.washington.edu,
Univ of Washington – Op
Roden, Michael F., (706) 542-2416 mroden@uga.edu,
Univ of Georgia – Gi
Rodgers, David W., 208-282-4634 rodgdavi@isu.edu,
Idaho State Univ – Gc
Rodgers, George P., (510) 659-6155 Ohlone College – Gg
Rodgers, John C., (662) 325-0732 jcr100@msstate.edu,
Mississippi State Univ – Oy
Rodgers, NRodgersN@cf.ac.uk,
Cardiff Univ – Ow
Rodgers, Peter W., (805) 893-4893 Univ of California, Santa Barbara – Ys
Rodi, William, (617) 253-7855 rodi@mit.edu,
Massachusetts Institute of Technology – Yg
Rodland, David L., 315 312 2787 rodland@oswego.edu,
SUNY, Oswego – Pi
Rodolfo, Kelvin S., (312) 733-0617 krodolfo@uic.edu,
Univ of Illinois, Chicago – Gu
Rodrick, Gary E., (352) 392-8001 Univ of South Florida, Saint Petersburg – Ob
Rodrigues, Cyril G., (519) -25-3-30 (Ext. ext. 2499) cgr@uwindsor.ca,
Univ of Windsor – Pm
Rodriguez, Joaquin, (212) 772-5321 Graduate School of the City Univ of New York – Pi
Rodriguez, Vanessa del S., (787) 722-2526 Puerto Rico Bureau of Geology – Oo
Rodriguez-Iturbe, Ignacio, (609) 258-2287 irodrigu@princeton.edu,
Princeton Univ – Hg
Roe, Carol, 757-221-2440 crroex@wm.edu,
College of William & Mary – Oo
Roe, Gerard H., 206-543-4980 gerard@ess.washington.edu,
Univ of Washington – OaGmOl
Roecker, Steven W., (518) 276-6773 roecks@rpi.edu,
Rensselaer Polytechnic Institute – Yg
Roeder, Peter L., (613) 545-6185 Queen's Univ – Cp
Roegiers, Jean-Claude, (405) 325-6787 jroegiers@ou.edu,
Univ of Oklahoma – Nr
Roehl, Perry O., (210) 999-7092 Trinity Univ – Gs
Roelke, Daniel L., (979) 845-0169 droelke@tamu.edu,
Texas A&M Univ – Ob
Roemer, Elizabeth, (520) 621-2897 eroemer@pirlmail.lpl.arizona.edu,
Univ of Arizona – Oo
Roemmich, Dean H., (619) 534-2307 droemmich@ucsd.edu,
Univ of California, San Diego – Op
Roepke, Harlan H., (765) 285-8270 00hhroepke@bsu.edu,
Ball State Univ – Gs
Roering, Joshua J., (541) 346-5574 Univ of Oregon – Gm
Roeske, Sarah M., (530) 752-4933 roeske@geology.ucdavis.edu,
Univ of California, Davis – Gc
Roethel, Frank J., (631) 632-8732 froethel@notes.cc.sunysb.edu,
SUNY, Stony Brook – Oc
Rogers, Garry C., (250) 363-6450 Univ of Victoria – Ys
Rogers, J D., 573-341-6198 rogersda@umr.edu,
Univ of Missouri, Rolla – Ne
Rogers, Jefferson S., (731) 587-7442 jrogers@utm.edu,
Univ of Tennessee, Martin – Oo
Rogers, Jeffrey C., (614) 292-0148 rogers.21@osu.edu,
Ohio State Univ – Oa
Rogers, John J., (919) 962-0695 jrogers@email.unc.edu,
Univ of North Carolina, Chapel Hill – Gg
Rogers, Lewis F., (770) 718-3805 lrogers@hermes.gc.peachnet.edu,
Gainesville College – Ge
Rogers, Pamela Z., (505) 667-1765 Los Alamos National Lab – Cl
Rogers, Raymond R., (651) 696-6434 rogers@macalester.edu,
Macalester College – Gs
Rogers, Robert D., (209) 667-3466 rrogers@geology.csustan.edu,
California State Univ, Stanislaus – GcmGt
Roggenthen, William M., (605) 394-2460 William.Roggenthen@sdsmt.edu,
South Dakota School of Mines & Technology – Yg
Rogowski, Andrew S., (814) 863-8758 asr@psu.edu,
Pennsylvania State Univ, Univ Park – Sp
Rohay, Alan C., (509) 376-6925 alan.rohay@pnl.gov,
Pacific Northwest National Lab – Ys
Rohr, David, (432) 837-8167 drohr@sulross.edu,
Sul Ross State Univ – Pi
Rohs, C. Renee, (660) 562-1201 rrohs@nwmissouri.edu,
Northwest Missouri State Univ – Cg
Rokop, Donald J., (505) 667-4299 Los Alamos National Lab – Cc
Rollins, Harold B., (412) 624-8780 snail@pitt.edu,
Univ of Pittsburgh – Pe
Rollinson, Paul A., (417) 836-5688 paulrollinson@smsu.edu,
Southwest Missouri State Univ – Oo
Rolston, Dennis E., (530) 752-2113 derolston@ucdavis.edu,
Univ of California, Davis – So
Roman, Charles T., (401) 874-6885 croman@gso.uri.edu,
Univ of Rhode Island – Ob
Roman, Eric W., (609) 984-6587 eric.roman@dep.state.nj.us,
New Jersey Geological Survey – Hw
Roman, Micheal R., (410) 221-8425 roman@hpl.umces.edu,
Univ of Maryland – Ob
Romanek, Christopher S., (706) 542-2652 Univ of Georgia – Cs
Romanovsky, Vladimir, (907) 474-7459 ffver@uaf.edu,
Univ of Alaska, Fairbanks – Yg
Romanowicz, Barbara A., 510-643-5690 barbara@seismo.berkeley.edu,
Univ of California, Berkeley – Ys
Romanowicz, Edwin A., (518) 564-2152 romanoea@plattsburgh.edu,
Plattsburgh State Univ (SUNY) – Hw
Romberger, Samuel B., (303) 273-3828 sromberg@mines.edu,
Colorado School of Mines – Em
Romer, Brett, (760) 773-2573 College of the Desert – Oo
Romo, Jose M.jromo@cicese.mx,
Centro de Inv Científica y de Educación Superior de Ensenada – Ye
Rona, Peter A., (732) 932-6555 (Ext. 241) rona@ahab.rutgers.edu,
Rutgers, The State Univ of New Jersey – Yr
Rondenay, Stephane, 617-253-6299 rondenay@mit.edu,
Massachusetts Institute of Technology – Ys
Rondot, Beth, (516) 299-2318 beth.rondot@liu.edu,
Long Island Univ, C.W. Post Campus – Oo
Rooney, Tyrone, 517-432-5522 rooneyt@msu.edu,
Michigan State Univ – Gi
Roper, Robert G., (404) 252-3730 roper@bellsouth.net,
Georgia Institute of Technology – Oa
Rose, Arthur W., (814) 865-7261 rose@ems.psu.edu,
Pennsylvania State Univ, Univ Park – Ce
Rose, Candace M., (630) 252-3499 cmrose@anl.gov,
Argonne National Lab – Oo
Rose, Dan, +353 21 4902189 Univ College Cork – Oo
Rose, Joan B., (727) 553-3928 Univ of South Florida, St Petersburg – Oo
Rose, Patricia, 708-209-3089 Patricia.Rose@cuchicago.edu,

Concordia Univ – Oy
Rose, Peter E., (801) 585-7785 prose@egi.utah.edu, Univ of Utah – Np
Rose, Robert, (804) 427-7189 Tidewater Community College – Ow
Rose, Seth E., 404413-5750 geoser@langate.gsu.edu, Georgia State Univ – Hw
Rose, Timothy, (202) 633-1398 roset@si.edu, Smithsonian Inst / National Museum of Natural History – CaGv
Rose, William I., (906) 487-2367 raman@mtu.edu, Michigan Technological Univ – Gv
Roseboom, Donald P., (309) 671-3196 Illinois State Water Survey – Hy
Rosen, Carl J., (612) 625-8114 crosen@umn.edu, Univ of Minnesota, Twin Cities – Sc
Rosen, Peter S., 617 373-4380 p.rosen@neu.edu, Northeastern Univ – Onu
Rosenberg, Gary D., (317) 274-7468 grosenbe@iupui.edu, Indiana Univ / Purdue Univ, Indianapolis – Po
Rosenberg, Philip E., (509) 335-4368 rosenberg@wsu.edu, Washington State Univ – Cp
Rosenblatt, Richard H., (858) 534-2199 rrosenblatt@ucsd.edu, Univ of California, San Diego – Ob
Rosendahl, Jenny, 214-768-2750 jennyr@smu.edu, Southern Methodist Univ – Oo
Rosenfeld, Charles L., (541) 322-3131 chuck.rosenfeld@osucascades.edu, Oregon State Univ – Gm
Rosenfeld, John L., (310) 825-1505 rosenfel@ucla.edu, Univ of California, Los Angeles – Gp
Rosenfeld, Leslie K., (831) 656-3253 lkrosenf@nps.navy.mil, Naval Postgraduate School – Op
Rosenmeier, Michael F., (412) 624-8775 mrosenme@pitt.edu, Univ of Pittsburgh – Gn
Ross, Daniel K., (808) 956-6193 kross@soest.hawaii.edu, Univ of Hawai'i, Manoa – Gi
Ross, David A., (508) 289-2578 dross@whoi.edu, Woods Hole Oceanographic Institution – Gu
Ross, Gerald M., (403) 292-7000 Univ of Calgary – Gt
Ross, Howard P., (801) 581-5184 Univ of Utah – Ye
Ross, Martin E., (617) 373-3263 m.ross@neu.edu, Northeastern Univ – Gi
Ross, Nancy L., (540) 231-6356 nross@vt.edu, Virginia Polytechnic Institute & State Univ – Gz
Ross, Robert M., (607) 273-6623 (Ext. 18) rmr16@cornell.edu, Paleontological Research Institution – Po
Ross, Tetjana, (902) 494-1327 tetjana.ross@dal.ca, Dalhousie Univ – Op
Ross, Thomas E., (910) 521-6218 tom.ross@uncp.edu, Univ of North Carolina, Pembroke – Oy
Rossbach, Thomas J., (252) 335-3376 tjrossbach@mail.ecsu.edu, Elizabeth City State Univ – Ps
Rossby, Hans T., (401) 874-6521 trossby@gso.uri.edu, Univ of Rhode Island – Op
Rossell, Irene M., (828) 232-5185 irossell@unca.edu, Univ of North Carolina, Asheville – Oo
Rossman, Georgegrr@gps.caltech.edu, American Museum of Natural History – Gz
Rossman, George R., (626) 395-6471 grr@gps.caltech.edu, California Institute of Technology – Gz
Rostam-Abadi, Massoud, 217-244-4977 massoud@isgs.uiuc.edu, Illinois State Geological Survey – Oo
Rostami, Jamal, 814-863-7606 jur17@psu.edu, Pennsylvania State Univ, Univ Park – Nm
Rostoker, Gordon, (780) 492-5286 rostoker@space.ualberta.ca, Univ of Alberta – Xy
Rostron, Ben, (780) 492-2178 ben.rostron@ualberta.ca, Univ of Alberta – Oo
Rostron, Benjamin J., (780) 492-2178 ben.rostron@ualberta.ca, Univ of Alberta – Hw
Roth, Gregory W., (814) 863-1018 gwr@psu.edu, Pennsylvania State Univ, Univ Park – So
Roth, Peter H., (801) 581-6704 phroth@mines.utah.edu, Univ of Utah – Pm
Rothberg, Maryann, (609) 258-4655 mar@cedr.princeton.edu, Princeton Univ – Oo
Rothert, Jane E., (217) 333-7942 Illinois State Water Survey – Og
Rothman, Daniel H., (617) 253-7861 dhr@mit.edu, Massachusetts Institute of Technology – Yg
Rothrock, David A., (206) 545-2262 rothrock@apl.washington.edu, Univ of Washington – Op
Rothstein, Lewis, (401) 874-6517 lrothstein@gso.uri.edu, Univ of Rhode Island – Op
Rothstein, Lewis M., (401) 874-6517 Univ of Hawai'i, Manoa – Oo
Roughgarden, Joan, (650) 723-3648 Stanford Univ – Ob
Rougvie, James R., 608-363-2268 rougviej@beloit.edu, Beloit College – Gp
Rouhani, Farhang, (540) 654-1895 frouhani@mwc.edu, Mary Washington College – Oo
Rouleau, Alain, (418) -54-5-50 (Ext. x5213) arouleau@uqac.ca, Universite du Quebec a Chicoutimi – Hw
Rouleau, Pierre M., (604) 637-6294 Sir Wilfred Grenfell College – Yg
Roumani-Denn, Vivienne, (510) 642-7366 vivienne@garnet.berkeley.edu, Univ of California, Berkeley – Oo
Rounds, Steven W., (916) 278-7828 rounds@csus.edu, California State Univ, Sacramento – Gg
Rountree, Roy, 404-656-3214 Georgia Dept of Natural Resources – Hw
Rouse, Roland C., (734) 763-0952 rouserc@umich.edu, Univ of Michigan – Gz
Rouse, Wayne R., (905) 525-9140 (Ext. 24538) rouse@mcmail.cis.mcmaster.ca, McMaster Univ – Oa
Rouse, Jr., Lawrence J., (225) 388-2953 Louisiana State Univ – Op
Rousell, Don H., (705) 675-1151 (Ext. ext. 2265) drousell@laurentian.ca, Laurentian Univ, Sudbury – Gc
Roush, Ted, (415) 338-2061 San Francisco State Univ – Yg
Roushar, Kathy Campbell, (608) 265-4683 kroushar@wisc.edu, Univ of Wisconsin - Extension – Oi
Rousseau, Carol, (514) 398-4755 carol@minmet.lan.mcgill.ca, McGill Univ – Oo
Roussel-Dupre, R., (505) 667-9228 rroussel-dupre@lanl.gov, Los Alamos National Lab – Oo
Routh, Partha, (208)426-2757 prouth@boisestate.edul, Boise State Univ – Cg
Rouxel, Olivier, (508) 289-3655 orouxel@whoi.edu, Woods Hole Oceanographic Institution – Cm
Rovey, Charles W., (417) 836-6869 charlesrovey@smsu.edu, Southwest Missouri State Univ – Hy
Rowe, Clinton M., (402) 472-1946 crowe1@unl.edu, Univ of Nebraska, Lincoln – Oa
Rowe, Gilbert T., (409) 740-4458 growe@ocean.tamu.edu, Texas A&M Univ – Ob
Rowe, Harold D., 859-257-2770 hrowe@uky.edu, Univ of Kentucky – Cs
Rowe, Michael, (319) 384-3474 michael-rowe@uiowa.edu, Univ of Iowa – Gi
Rowe, Timothy B., (512) 471-1725 rowe@mail.utexas.edu, Univ of Texas at Austin, Jackson School of Geosciences – Pv
Rowell, Albert J., (785) 864-2747 arowell@ku.edu, Univ of Kansas – Pi
Rowell, Bruce F., (610) 683-4437 rowell@kutztown.edu, Kutztown Univ of Pennsylvania – Po
Rowell, Thelma, (510) 642-0268 Univ of California, Berkeley – Pv
Rowins, Stephenrowins@src.sk.ca, Univ of British Columbia – Em
Rowland, F.S., (949) 824-6016 rowland@uci.edu, Univ of California, Irvine – Yr
Rowland, Scott K., (808) 956-3150 Univ of Hawai'i, Manoa – Oo
Rowland, Stephen, (702) 895-3625 srowland@nevada.edu, Univ of Nevada, Las Vegas – Pi
Rowley, David B., (773) 702-8102 rowley@geosci.uchicago.edu, Univ of Chicago – Gc
Roy, David C., (617) 552-3647 roy@bc.edu, Boston College – Gr
Roy, Denis W., (418) 545-5011 dwroy@uqac.uquebec.ca, Universite du Quebec a Chicoutimi – Gc
Roy, Mousumi, (505) 277-2580 mroy@unm.edu, Univ of New Mexico – Yg
Roy, Rustum, (814) 865-3421 Pennsylvania State Univ, Univ Park – Cg
Roy, Suzanne, (418) 723-1986 (Ext. 1748) suzanne_roy@uqar.uquebec.ca, Universite du Quebec a Rimouski – Ob
Roy, William R., 217-333-1197 roy@isgs.uiuc.edu, Illinois State Geological Survey – Sc
Royden, Leigh H., (617) 253-1292 lhroyden@mit.edu, Massachusetts Institute of Technology – Gt
Royer, Dana, (860) 685-2836 droyer@wesleyan.edu, Wesleyan Univ – PeoPb
Royer, Thomas, (757) 683-5547 royer@ccpo.odu.edu, Old Dominion Univ – Op
Rozgonyi, Tibor G., (303) 273-3700 trozgony@mines.edu, Colorado School of Mines – Nm
Rozmus, Wojciech, (780) 492-8486 rozmus@phys.ualberta.ca, Univ of Alberta – Oo
Rubin, Alan E., (310) 825-3202 rubin@igpp.ucla.edu,

Univ of California, Los Angeles – Xm
Rubin, Allan M., (609) 258-1506 arubin@princeton.edu, Princeton Univ – Yg
Rubin, Allan M., 609-258-1506 arubin@princeton.edu, Princeton Univ – Gc
Rubin, Charles M., (509) 963-2827 charlier@cwu.edu, Central Washington Univ – Gt
Rubin, Kenneth H., (808) 956-8973 krubin@hawaii.edu, Univ of Hawai'i, Manoa – Cc
Rubin, Vera C., (202) 478-8861 rubin@gal.ciw.edu, Carnegie Institution of Washington – Oo
Rubio-Sierra, Javier, 089/2180 4317 rubio@lrz.uni-muenchen.de, LMU - Univ of Munich – Gz
Rucklidge, John C., (416) 978-2061 jcr@quartz.geology.utoronto.ca, Univ of Toronto – Ge
Ruddick, Barry R., (902) 494-2505 barry.ruddick@dal.ca, Dalhousie Univ – Op
Ruddiman, William F., 540-348-1963 wfr5c@virginia.edu, Univ of Virginia – Gu
Rudkin, David M., (416) 586-5592 davidru@rom.on.ca, Royal Ontario Museum – Pi
Rudman, Albert J., (812) 855-7310 rudman@indiana.edu, Indiana Univ, Bloomington – Ys
Rudnick, Daniel L., (858) 534-7669 drudnick@ucsd.edu, Univ of California, San Diego – Op
Rudnick, Roberta L., (301) 405-1311 rudnick@geol.umd.edu, Univ of Maryland – Gx
Rudnicki, Ryan E., (210) 733-2704 rrudnicki@mail.accd.edu, San Antonio College – Oir
Rueger, Bruce F., (207) 859-5806 bfrueger@colby.edu, Colby College – Pl
Ruehr, Thomas A., (805) 756-2552 truehr@calpoly.edu, California Polytechnic State Univ – Sb
Ruez, Jr., Dennis R., (334) 844-4315 ruezden@auburn.edu, Auburn Univ – Pe
Ruff, Larry J., (734) 763-9301 ruff@umich.edu, Univ of Michigan – Ys
Ruffin, Nel, 662-325-7459 ruffin@geosci.msstate.edu, Mississippi State Univ – Oy
Ruffman, Alan, (902) 422-6482 Dalhousie Univ – Ys
Ruggiero, Peter, (541) 737-1239 ruggierp@science.oregonstate.edu, Oregon State Univ – On
Ruhle, James L., (714) 278-3561 jruhle@fullerton.edu, California State Univ, Fullerton – Pg
Ruina, Andy L., (607) 255-7108 alr3@cornell.edu, Cornell Univ – Yx
Ruiz, Joaquin, (520) 621-4090 jruiz@u.arizona.edu, Univ of Arizona – Cg
Rule, Joseph H., (757) 683-3274 jrule@odu.edu, Old Dominion Univ – Ou
Rumble, III, Douglas, (202) 478-8990 d.rumble@gl.ciw.edu, Carnegie Institution of Washington – Gp
Rumph, Leslie K., (775) 327-2056 rumph@mines.unr.edu, Univ of Nevada, Reno – Oo
Rumstay, Kenneth S., (229) 333-5754 krumstay@valdosta.edu, Valdosta State Univ – Oo
Rundberg, Robert S., (505) 667-4559 Los Alamos National Lab – Gq
Rundle, Johnjbrundle@ucdavis.edu, Univ of California, Davis – Yd
Rundquist, Donald C., (402) 472-7536 drundquist@unl.edu, Unversity of Nebraska - Lincoln – Ori
Runge, Edward C. A., (979) 845-3041 e-runge@tamu.edu, Texas A&M Univ – Sd
Runkel, Anthony C., (612)627-4780 Ext. 222 runke001@umn.edu, Univ of Minnesota, Twin Cities – Gs
Runnegar, Bruce N., (310) 206-1738 runnegar@ucla.edu, Univ of California, Los Angeles – Po
Runyon, Cassandra R., (843) 953-8279 runyonc@cofc.edu, College of Charleston – XgOse
Ruoff, A. L., (607) 255-4161 alr3@cornell.edu, Cornell Univ – Cp
Rupert, Frank R., (850) 488-9380 Florida Geological Survey – Pm
Rupert, Gerald B., 573-341-4616 Univ of Missouri, Rolla – Yg
Rupp, John A., (812) 855-1323 rupp@indiana.edu, Indiana Geological Survey – Go
Ruppel, Carolyn D., (404) 894-0231 cdr@piedmont.eas.gatech.edu, Georgia Institute of Technology – Yv
Ruppel, Stephen C., (512) 471-2965 stephen.ruppel@beg.utexas.edu, Univ of Texas at Austin, Jackson School of Geosciences – Gd
Ruscher, Paul H., (850) 644-2752 ruscher@met.fsu.edu, Florida State Univ – Ow
Rush.Jr., Franklin R., (850) 488-9380 Florida Geological Survey – Oo
Rushmer, Tracy, (802) 656-8136 trushmer@zoo.uvm.edu, Univ of Vermont – Cp
Rusmore, Margaret E., 323 259 2565 rusmore@oxy.edu, Occidental College – Gc
Rusnak, Barbra J., (719) 333-8278 United States Air Force Acad – Og
Russell, Ann D., (530) 752-3311 russell@geology.ucdavis.edu, Univ of California, Davis – Cm
Russell, Christopher T., (310) 825-3188 ctrussell@igpp.ucla.edu, Univ of California, Los Angeles – Yg
Russell, D. R., (727) 553-3962 rrussell@marine.usf.edu, Univ of South Florida, Saint Petersburg – Oo
Russell, Dale A., (919) 515-1339 dale_russell@ncsu.edu, North Carolina State Univ – Pv
Russell, Ernest E., (662) 325-3915 Mississippi State Univ – Gr
Russell, Gail S., (601) 266-4077 gail.russell@usm.edu, Univ of Southern Mississippi – Cc
Russell, James E., (979) 845-6184 jim.russell@pe.tamu.edu, Texas A&M Univ – Nm
Russell, James E., (409) 845-2241 russell@spindletop.tamu.edu, Texas A&M Univ – Nm
Russell, James K., (604) 822-2703 krussell@eos.ubc.ca, Univ of British Columbia – Gi
Russell, James M., (401) 863-6330 James_Russell@Brown.edu, Brown Univ – Gn
Russell, R. Doncaster, (604) 822-2551 drussell@eos.ubc.ca, Univ of British Columbia – Yg
Russell, Ron, (512) 471-8831 ron.russell@beg.utexas.edu, Univ of Texas at Austin, Jackson School of Geosciences – Oo
Russell, Terry P., (250) 721-6184 trussell@uvic.ca, Univ of Victoria – Oo
Russell, Theresa J., (501) 575-4403 trussell@comp.uark.edu, Univ of Arkansas, Fayetteville – Oo
Russelle, Michael P., (612) 625-8145 russelle@soils.umn.edu, Univ of Minnesota, Twin Cities – Sc
Russo, Raymond, 352-392-6766 russo@geology.ufl.edu, Univ of Florida – Yg
Rustad, James R., 530-754-6021 Rustad@geology.ucdavis.edu, Univ of California, Davis – Cl
Rutberg, Randye L., (212) 772-5326 rrutberg@geo.hunter.cuny.edu, Hunter College (CUNY) – Cc
Rutford, Robert H., (972) 883-6470 rutford@utdallas.edu, Univ of Texas, Dallas – Gl
Rutherford, Malcolm J., (401) 863-1927 Malcolm_Rutherford@Brown.edu, Brown Univ – Cp
Rutherford, Michael P., (604) 637-6328 Sir Wilfred Grenfell Coll – Sb
Rutledge, Steven A., (970) 491-8680 rutledge@atmos.colostate.edu, Colorado State Univ – Oa
Rutqvist, Jonny, (510) 486-5432 JRUTQVIST@LBL.GOV, Lawrence Berkeley National Lab – Hw
Rutstein, Martin S., (914) 257-3763 SUNY New Paltz – Gz
Ruttan, Lore, 404 7274217 lruttan@emory.edu, Emory Univ – Ob
Ruttenberg, Kathleen, 808-956-9371 kcr@soest.hawaii.edu, Univ of Hawai'i, Manoa – Cg
Ruttenberg, Kathleen C.kcr@soest.hawaii.edu, Univ of Hawai'i, Manoa – Cg
Rutter, Nathaniel W., (780) 492-3085 nat.rutter@ualberta.ca, Univ of Alberta – Ge
Ruzicka, Alexander (Alex) M., (503) 725-3372 ruzickaa@pdx.edu, Portland State Univ – Xm
Ryall, Patrick J., (902) 494-3465 pryall@is.dal.ca, Dalhousie Univ – Yg
Ryan, C. Patrick, (541) 346-5875 cpat@newberry.uoregon.edu, Univ of Oregon – Yg
Ryan, Cathy, (403) 220-2793 ryan@geo.ucalgary.ca, Univ of Calgary – Hw
Ryan, Jeffrey G., (813) 974-1598 ryan@chuma.cas.usf.edu, Univ of South Florida, Tampa – Ct
Ryan, Peter C., (802) 443-2557 pryan@middlebury.edu, Middlebury College – Cl
Ryan, Susan, (724) 938-4531 ryan@cup.edu, California Univ of Pennsylvania – Og
Ryan, William B., (845) 365-8312 billr@ldeo.columbia.edu, Columbia Univ – Gu
Ryan, William F., (814) 865-0478 The Pennsylvania State Univ, Univ Park – Ow
Ryan, William B. F., (845) 365-8312 Columbia Univ – Yr
Ryavec, Karl E., 715-346-4450 kryavec@uwsp.edu,

Univ of Wisconsin, Stevens Point – Oui
Ryberg, Paul T.pryberg@clarion.edu, Clarion Univ – Gs
Rydelek, Paul A., (901) 678-4883 par@ceri.memphis.edu, Univ of Memphis – Yg
Rye, Danny M., (203) 432-3174 danny.rye@yale.edu, Yale Univ – Cs
Rye, Lois E., (805) 965-0581 (Ext. 2315) Santa Barbara City Coll – Oe
Rye, Rob, (213) 821-2267 rye@usc.edu, Univ of Southern California – Py
Rygel, Michael C., (315) 267-3401 rygelmc@potsdam.edu, SUNY, Potsdam – Gs
Rößner, Gertrud, 089/2180 6612 g.roessner@lrz.uni-muenchen.de, LMU - Univ of Munich – Pg

S

Saal, Alberto E., (401) 863-7238 Alberto_Saal@Brown.edu, Brown Univ – Cg
Saalfeld, Alan J., 614 292 6665 saalfeld.1@osu.edu, Ohio State Univ – Yd
Saar, Martin O., 612-625-7332 saar@umn.edu, Univ of Minnesota, Twin Cities – Hg
Sabala-Foreman, Susan M., (928) 523-4561 Northern Arizona Univ – Oo
Sabine, Christopher L., (206) 526-4809 chris.sabine@noaa.gov, Univ of Washington – Oc
Sablock, Jeanette M.jsablock@salemstate.edu, Salem State College – Gx
Sablock, Peter, (978) 542-6493 psablock@salemstate.edu, Salem State College – Gc
Sacchi, Mauricio D., (780) 492-1060 sacchi@phys.ualberta.ca, Univ of Alberta – Ys
Saccocia, Peter J.psaccocia@bridgew.edu, Bridgewater State College – Cm
Sacks, I. Selwyn, (202) 478-8839 sacks@dtm.ciw.edu, Carnegie Institution of Washington – Ys
Sacramentogrilo, Isabelle, 619-594-5607 isacramentogrilo@geology.sdsu.edu, San Diego State Univ – Gg
Sadd, James L., 323 259 2518 jsadd@oxy.edu, Occidental College – Oi
Sadiq, Abdulali A.sadiqa@qu.edu.qa, Univ of Qatar – Or
Sadler, E. John, 573-884-1971 sadlerj@missouri.edu, Univ of Missouri, Columbia – Sp
Sadler, Peter M., (951) 827-5616 peter.sadler@ucr.edu, Univ of California, Riverside – Gd
Sadowsky, Michael J., (612) 624-2706 sadowsky@soils.umn.edu, Univ of Minnesota, Twin Cities – Sb
Saffer, Demian M., 814-865-7965 dsaffer@geosc.psu.edu, Pennsylvania State Univ, Univ Park – Hw
Sagebiel, J. C., 909 307-2669 237 jsagebiel@sbcm.sbcounty.gov, San Bernardino County Museum – Pv
Sageman, Bradley B., (847) 467-2257 brad@earth.northwestern.edu, Northwestern Univ – Ps
Sager, William W., (979) 845-9828 wsager@ocean.tamu.edu, Texas A&M Univ – Yr
Saghi Szabo, Gotthard, (202) 478-8970 g.saghi-szabo@gl.ciw.edu, Carnegie Institution of Washington – Gy
Sahagian, Dork L., (603) 862-3875 Univ of New Hampshire – Gm
Sahai, Nita, (608) 262-4972 sahai@geology.wisc.edu, Univ of Wisconsin, Madison – Cl
Sahai, Surinder K., 405-744-6358 sksahai@okstate.edu, Oklahoma State Univ – Yg
Sahay, Pratappratap@cicese.mx, Centro de Inv Científica y de Educación Superior de Ensenada – Ys
Sahr, John D., 206-685-4816 jdsahr@ee.washington.edu, Univ of Washington – YgXy
Saila, Saul B., (401) 874-6485 saila@gso.uri.edu, Univ of Rhode Island – Ob
Saini-Eidukat, Bernhardt, (701) 231-8785 bernhardt.saini-eidukat@ndsu.edu, North Dakota State Univ – Gi
Saint, Prem K.psaint@fullerton.edu, California State Univ, Fullerton – Hw
Saito, Mak A., 508-289-3696 msaito@whoi.edu, Woods Hole Oceanographic Institution – Oc
Saja, David B., (216) 231-4600 (Ext. 3229) dsaja@cmnh.org, Cleveland Museum of Natural History – Gdc
Sak, Peter B., 717-245-1423 sakp@dickinson.edu, Dickinson College – Gc
Sakshaug, Egil, (477) 359-1583 Univ of South Florida, St Petersburg – Oo
Saku, James C., (301) 687-4724 jsaku@frostburg.edu, Frostburg State Univ – Oo
Sakurai, Shinichi, (512) 471-4306 shinichi.sakurai@beg.utexas.edu, Univ of Texas at Austin, Jackson School of Geosciences – Go
Salamon, Miklos D. G., (303) 273-3712 msalamon@mines.edu, Colorado School of Mines – Nr
Saleeby, Jason B., (626) 395-6141 jason@gps.caltech.edu, California Institute of Technology – Gc
Sallenger, Abby H., (727) 893-3100 Univ of South Florida, Saint Petersburg – On
Salmon, Richard L., (858) 534-2090 rsalmon@ucsd.edu, Univ of California, San Diego – Op
Salmun, Haydee, (212) 772-5224 hsalmun@geo.hunter.cuny.edu, Hunter College (CUNY) – Yg
Salters, Vincent J., (850) 644-1934 salters@magnet.fsu.edu, Florida State Univ – Cg
Saltzman, Eric S., 949-824-3936 esaltzma@uci.edu, Univ of California, Irvine – Oa
Saltzman, Matthew R., 614 292-0481 saltzman.11@osu.edu, Ohio State Univ – Gs
Salvador, Amos, (512) 471-6757 amos.salvador@mail.utexas.edu, Univ of Texas at Austin, Jackson School of Geosciences – Gr
Salve, Rohit, (510) 486-6416 R_Salve@lbl.gov, Lawrence Berkeley National Lab – Hg
Salvucci, Guido D., (617) 353-8344 gdsalvuc@crsa.bu.edu, Boston Univ – Hq
Sambrook, Richard A., 859-622-1426 richard.sambrook@eku.edu, Eastern Kentucky Univ – Oy
Sambrotto, Raymond N., (845) 365-8402 Columbia Univ – Ob
Samelson, Roger, (541) 737-4752 rsamelson@coas.oregonstate.edu, Oregon State Univ – Op
Sammarco, Paul W., (225) 851-2800 Louisiana State Univ – Oo
Sammis, Charles G., (213) 740-5836 sammis@usc.edu, Univ of Southern California – Yg
Sammis, Theodore W., (505) 646-3405 New Mexico State Univ, Las Cruces – Sp
Samms, Kathy, (253) 879-3814 ksamms@ups.edu, Univ of Puget Sound – Oo
Sample, James C., (928) 523-7198 Northern Arizona Univ – Cl
Sampson, Ruthie, (318) 342-1878 gesampson@ulm.edu, Univ of Louisiana, Monroe – Oo
Sampson, Scott A., (662) 325-9491 scotts@ext.msstate.edu, Mississippi State Univ – OirOs
Sampson, Scott D., (801) 585-0561 ssampson@umnh.utah.edu, Univ of Utah – Pv
Samson, Iain M., (519) -25-3-30 (Ext. ext. 2489) ims@uwindsor.ca, Univ of Windsor – Eg
Samson, John C., (780) 492-3616 samson@space.ualberta.ca, Univ of Alberta – Xy
Samson, Perry J., (313) 763-6213 Univ of Michigan – Oa
Samson, Scott A., (727) 553-3915 samson@marine.usf.edu, Univ of South Florida, Saint Petersburg – Oo
Samson, Scott D., 315-443-2672 sdsamson@syr.edu, Syracuse Univ – Cc
Samuelson, Alan C., (765) 285-8270 00acsamuelso@bsu.edu, Ball State Univ – Hw
Samuelson, D. James, (408) 288-3716 San Jose City College – Oo
San Juan, Francisco C., (252) 335-3451 fcsanjuan@mail.ecsu.edu, Elizabeth City State Univ – Eg
Sanchez, Charles, (520) 782-3836 sanchez@ag.arizona.edu, Univ of Arizona – Oo
Sanchez-Azofeifa, G. Arturo, (780) 492-1822 arturo.sanchez@ualberta.ca, Univ of Alberta – Or
Sanchez-Barreda, Luis, 512-471-0155 luis.barreda@beg.utexas.edu, Univ of Texas at Austin, Jackson School of Geosciences – Gg
Sandau, Ken, 406-496-4151 ksandau@mtech.edu, Montana Tech of The Univ of Montana – Oi
Sandecki, Michael, (916) 324-4026 California Geological Survey – Ge
Sandel, Bill R., (520) 621-4305 sandel@vega.lpl.arizona.edu, Univ of Arizona – Oa
Sanden, Eric M., (715) 425-3729 eric.m.sanden@uwrf.edu, Univ of Wisconsin, River Falls – Or
Sanders, Laura L., (773) 442-6051 L-Sanders@neiu.edu, Northeastern Illinois Univ – HwGe
Sanders, Richard P., (678) 839-4060 rsanders@westga.edu, Univ of West Georgia – Gp
Sanderson, Dewey D., (304) 696-5435 sanderso@marshall.edu, Marshall Univ – Hw
Sandholm, Scott T., (404) 894-3895 scott.sandholm@eas.gatech.edu, Georgia Institute of Technology – Oa
Sandifer, Paul A., (803) 795-6350 Univ of South Carolina – Ob

Sandor, James, 518 473-9988 New York State Geological Survey – Go
Sandor, Jonathan A., (515) 294-2209 jasandor@iastate.edu, Iowa State Univ of Science & Technology – Sd
Sandvol, Eric A., (573) 884-9616 sandvole@missouri.edu, Univ of Missouri – Ys
Sandwell, David T., (619) 534-7109 dsandwell@ucsd.edu, Univ of California, San Diego – Yr
Sandy, Michael R., (937) 229-3436 Univ of Dayton – Pi
Sanford, Allan R., (505) 835-5212 sanford@nmt.edu, New Mexico Institute of Mining and Technology – Ys
Sanford, Lawrence P., (410) 221-8429 lsanford@hpl.umces.edu, Univ of Maryland – Op
Sanford, Robert A., 217-244-7250 rsanford@uiuc.edu, Univ of Illinois, Urbana-Champaign – Py
Sanford, Thomas B., (206) 543-1365 sanford@apl.washington.edu, Univ of Washington – Op
Sanford, William E., (970) 491-5929 bills@cnr.colostate.edu, Colorado State Univ – Hw
Sankey, Julia, (209) 667-3466 California State Univ, Stanislaus – Pe
Sansone, Francis J., (808) 956-8370 sansone@hawaii.edu, Univ of Hawai'i, Manoa – Co
Santander, Erma, (520) 621-7120 erma@hwr.arizona.edu, Univ of Arizona – Oo
Santelmann, Mary V., (541) 737-1215 santelmm@geo.oregonstate.edu, Oregon State Univ – Oy
Santillan, Marcelo, (509) 963-1107 marcelo@geology.cwu.edu, Central Washington Univ – Yd
Santipiromkul, Apichart, (608) 262-9468 asantipiromk@wisc.edu, Univ of Wisconsin - Extension – Oo
Santos, Hernanhsantos@uprm.edu, Univ of Puerto Rico – Pi
Santschi, Peter H., (409) 740-4476 Texas A&M Univ – Oc
Sanudo, Sergio, 213-821-1302 sanudo@usc,edu, Univ of Southern California – Py
Sanudo-Wilhelmy, Sergio A., (631) 632-8615 ssanudo@notes.cc.sunysb.edu, SUNY, Stony Brook – Oc
Sarachik, Edward, 206-543-6720 sarachik@atmos.washington.edu, Univ of Washington – Oa
Sarah, Willig B., (215) 898-5724 sawillig@earthlink.net, Univ of Pennsylvania – Ge
Saravanan, R., 979-845-0175 r.saravanan@tamu.edu, Texas A&M Univ – Oa
Sares, Matthew A., (303) 866-2073 matt.sares@state.co.us, Colorado Geological Survey – Ge
Sargent, Kenneth A., (864) 294-3362 ken.sargent@furman.edu, Furman Univ – Hy
Sarmiento, Jorge L., (609) 258-6585 jls@princeton.edu, Princeton Univ – Oc
Sartain, Jerry B., (352) 392-1803 (Ext. 330) jbs@gnv.ifas.ufl.edu, Univ of Florida – So
Sarton, E. J., 607-777-2264 jsarton@servtech.com, Binghamton Univ – Oe
Sarwar, A. K. Mostofa, (504) 280-6717 asarwar@uno.edu, Univ of New Orleans – Ye
Sasowsky, Ira D., (330) 972-5389 ids@uakron.edu, Univ of Akron – Hw
Sass, John H., (928) 556-7226 Northern Arizona Univ – Yg
Sass, John H., (519) 661-3605 jhsass@uwo.ca, Univ of Western Ontario – Yh
Sathyendranath, Shubha, (902) 426-8044 shubha.sathyendranath@dal.ca, Dalhousie Univ – Ob
Sato, Yoko, (973) 655-4448 satoy@mail.montclair.edu, Montclair State Univ – Gg
Satterfield, Joseph I., 325-942-2577, Ext. 226 joseph.satterfield@angelo.edu, Angelo State Univ – Gc
Satterly, Thomas G., (406) 683-7869 t_satterly@umwestern.edu, Univ of Montana, Western – Xg
Saucedo, George J., (415) 557-1529 California Geological Survey – Gg
Sauck, William A., (269) 387-4991 bill.sauck@wmich.edu, Western Michigan Univ – Yg
Sauer, Nancy N. S., (505) 665-3759 Los Alamos National Lab – Oo
Sauer, Peter, 812-855-6591 pesauer@indiana.edu, Indiana Univ, Bloomington – Pe
Saul, Lou Ella R., (213) 744-4366 lousaul@earthlink.net, Los Angeles County Museum of Natural History – Pi
Saunders, James A., (334) 844-4884 saundja@auburn.edu, Auburn Univ – Cl
Saunders, Joe, (318) 342-1899 Univ of Louisiana, Monroe – Ga
Saunders, Ralph, 310 243-3284 rsaunders@csudh.edu, California State Univ, Dominguez Hills – Oo
Saunders, Ronald S., (818) 354-2867 saunders@jpl.nasa.gov, Jet Propulsion Lab – Gr
Saunders, W. Bruce, (610) 526-5114 wsaunder@brynmawr.edu, Bryn Mawr College – Pi
Sautter, Leslie R., (843) 953-5586 sautterl@cofc.edu, College of Charleston – Ob
Sava, Paul, 303-384-2362 psava@mines.edu, Colorado School of Mines – Ye
Savage, Brian, (202) 478-8846 savage13@dtm.ciw.edu, Carnegie Institution of Washington – Ys
Savage, Brian, 401-874-5392 savage@uri.edu, Univ of Rhode Island – Ys
Savage, Donald, (510) 642-3499 Univ of California, Berkeley – Pv
Savage, E. Lynn, (718) 951-5000 (Ext. 2874) lsavage@brooklyn.cuny.edu, Brooklyn College (CUNY) – Gs
Savage, Kaye S., 615-322-2976 kaye.s.savage@vanderbilt.edu, Vanderbilt Univ – Ge
Savage, Norman M., (541) 346-4585 nmsavage@oregon.uoregon.edu, Univ of Oregon – Pi
Savigny, K. Wayne, (604) 684-5900 Univ of British Columbia – Ng
Savin, Samuel M., (216) 368-6592 sms7@po.cwru.edu, Case Western Reserve Univ – Cs
Savina, Mary E., (507) 646-4404 msavina@carleton.edu, Carleton College – Gm
Savov, Ivan P., (202) 478-8475 savov@dtm.ciw.edu, Carnegie Institution of Washington – Cg
Savrda, Charles E., (334) 844-4887 savrdce@auburn.edu, Auburn Univ – Gd
Sawin, Robert S., (785) 864-2099 bsawin@kgs.ku.edu, Univ of Kansas – Gg
Sawyer, Dale S., (713) 348-5106 dale@rice.edu, Rice Univ – Yr
Sawyer, J. Foster, (605) 394-2229 South Dakota Dept of Env & Natural Resources – Hw
Saxena, Surenda K., (718) 951-5416 Graduate School of the City Univ of New York – Cg
Saxena, Surendra K., (305) 348-3030 saxenas@fiu.edu, Florida International Univ – Gy
Sayles, Frederick L., (508) 289-2561 fsayles@whoi.edu, Woods Hole Oceanographic Institution – Oc
Saylor, Beverly Z., 216 368 3763 bzs@case.edu, Case Western Reserve Univ – Gs
Sayre, William O., (505) 473-6305 sayre@csf.edu, College of Santa Fe – Hg
Scanlin, Michael A., (717) 361-1323 Elizabethtown College – Ye
Scanlon, Bridget R., (512) 471-8241 bridget.scanlon@beg.utexas.edu, Univ of Texas at Austin, Jackson School of Geosciences – Hw
Scatena, Frederick N., (215) 898-6907 fns@sas.upenn.edu, Univ of Pennsylvania – Gm
Schaaf, Crystal, (617) 358-0503 schaaf@bu.edu, Boston Univ – Or
Schade, Gunnar, (979) 845-7671 schade@ariel.met.tamu.edu, Texas A&M Univ – Oa
Schaef, Herbert T., (509) 373-9949 todd.schaef@pnl.gov, Pacific Northwest National Lab – Cg
Schaefer, Jeffra K., (609) 258-2612 jschaefe@princeton.edu, Princeton Univ – Ct
Schaefer, Joerg, 845-365-8703 Columbia Univ – Pe
Schaetzl, Randall, 517-353-7726 schaetzl@msu.edu, Michigan State Univ – Gm
Schafer, Carl M., (586) 286-2154 schaferc@macomb.edu, Macomb Community College, Center Campus – Gg
Schafer, Tom, 785-628-5969 tschafer@fhsu.edu, Fort Hays State Univ – Oi
Schafer, Jr., John W., (515) 294-3063 jschafer@iastate.edu, Iowa State Univ of Science & Technology – So
Schafersman, Steven D., (713) 666-8765 Houston Community College System – Gg
Schaffer, Jeffery, (707) 253-3161 Napa Valley College – Oy
Schaffer, Linda J.schafm4@aol.com, SUNY College at Brockport – Oe
Schaffrin, Burkhard A., 614 292-0502 schaffrin.1@osu.edu, Ohio State Univ – Yd
Schales, Franklin D., (410) 742-1179 Univ of Maryland – Oo
Schalla, Ronald, (509) 376-5064 ron.schalla@pnl.gov, Pacific Northwest National Lab – Gg
Schaller, Mirjam, 734-615-4286 mirjam@umich.edu, Univ of Michigan – Gm
Schapaugh, Bill T., (913) 532-7242 Kansas State Univ – So
Schardt, Lawrence A., (814) 863-7655 lschardt@psu.edu,

Pennsylvania State Univ, Univ Park – Sd
Scharnberger, Charles K. Charles.scharnberger@millersville.edu, Millersville Univ – Gc
Schauble, Edwin A., (310) 825-3880 Univ of California, Los Angeles – Cg
Schauss, Kim E., 812 464-1701 Univ of Southern Indiana – Oo
Scheibe, Timothy D., (509) 372-6065 tim.scheibe@pnl.gov, Pacific Northwest National Lab – Hw
Schell, Marie, (519) 661-3191 mschell@uwo.ca, Univ of Western Ontario – Oo
Schellekens, Johannes H., (787) 265-3845 schellek@uprm.edu, Univ of Puerto Rico – Eg
Schelling, Daniel D., (801) 585-3539 Univ of Utah – Gc
Schelske, Claire L., (352) 392-9617 Univ of Florida – Pe
Schenck, William S., (302) 831-8262 rockman@udel.edu, Univ of Delaware – Gg
Schenck, William S., 302-831-2833 rockman@udel.edu, Univ of Delaware – Gg
Scherer, Reed P., (815) 753-7951 reed@geol.niu.edu, Northern Illinois Univ – Pm
Schermer, Elizabeth R., (360) 650-3658 schermer@geol.wwu.edu, Western Washington Univ – Gt
Schiappa, Tamra A., (724) 738-2829 tamra.schiappa@sru.edu, Slippery Rock Univ – Pi
Schick, Mary Louise, (312) 996-6123 mlschick@uic.edu, Univ of Illinois, Chicago – Oo
Schieber, Juergen, (812) 856-4740 jschiebe@indiana.edu, Indiana Univ, Bloomington – Gs
Schiebout, Judith A., (225) 578-2717 schiebout@geol.lsu.edu, Louisiana State Univ – Pv
Schiffman, Peter, (530) 752-3669 schiffman@geology.ucdavis.edu, Univ of California, Davis – Gp
Schilling, Jean-Guy, (401) 874-6628 jgs@gso.uri.edu, Univ of Rhode Island – Ou
Schimmelmann, Arndt, (812) 855-7645 aschimme@indiana.edu, Indiana Univ, Bloomington – Cs
Schimmrich, Steven, (845) 687-7683 schimmrs@sunyulster.edu, SUNY Ulster County Community College – Og
Schincariol, Robert A., (519) 661-3732 schincar@uwo.ca, Univ of Western Ontario – HwGg
Schindler, Michael, (204) 474-7178 schindl0@cc.umanitoba.ca, Univ of Manitoba – Gz
Schleifer, Stanley, (718) 262-2079 York College (CUNY) – Ge
Schlesinger, Michael E., (217) 333-2192 schlesin@atmos.uiuc.edu, Univ of Illinois, Urbana-Champaign – Oa
Schlesinger, William H., (919) 613-8004 schlesin@duke.edu, Duke Univ – Oo
Schlische, Roy W., (732) 445-3142 schlisch@rci.rutgers.edu, Rutgers, The State Univ of New Jersey – Gt
Schloessin, Helmuth H., (519) 661-2253 hhsuwo@uwo.ca, Univ of Western Ontario – Yx
Schlosser, John P., (916) 224-4748 California Geological Survey – Oo
Schlosser, Peter, (845) 365-8707 schlosser@ldeo.columbia.edu, Columbia Univ – Hw
Schlue, John W., (505) 835-5426 New Mexico Institute of Mining and Technology – Ys
Schlueter, Joe, (573) 368-2143 Missouri Dept of Nat Resources – Oo
Schmahl, Wolfgang, 089/2180 4311 Wolfgang.Schmahl@lrz.uni-muenchen.de, LMU - Univ of Munich – Gz
Schmalz, Robert F., (814) 865-3836 rfs3@psu.edu, Pennsylvania State Univ, Univ Park – Gu
Schmid, Dieter, 089/2180 6635 d.schmid@lrz.uni-muenchen.de, LMU - Univ of Munich – Pg
Schmid, Ginger, (507) 389-2617 ginger.schmid@mnsu.edu, Minnesota State Univ – Sd
Schmid, Peter M., (802) 626-6328 Lyndon State College – Ow
Schmidt, Christopher J., (269) 387-5490 christopher.schmidt@wmich.edu, Western Michigan Univ – Gc
Schmidt, Dale R., 312-793-7382 schmidt@isgs.uiuc.edu, Illinois State Geological Survey – Ge
Schmidt, David A., 541-346-3005 das@darkwing.uoregon.edu, Univ of Oregon – Yd
Schmidt, Fred A., 406-496-4149 fschmidt@mtech.edu, Montana Tech of The Univ of Montana – Hw
Schmidt, Keegan L., 208 790-2283 klschmidt@lcsc.edu, Lewis-Clark State College – Gc
Schmidt, Mariek, (202) 633-1799 schmidtm@si.edu, Smithsonian Inst / National Museum of Natural History – XgGi
Schmidt, Matthew, 979 862 8342 mschmidt@ocean.tamu.edu, Texas A&M Univ – Ou
Schmidt, Richard C., (216) 464-2656 Case Western Reserve Univ – Em
Schmidt, Walter, (850) 488-4191 Florida Geological Survey – Gg
Schmidt, Jr., Robert H., (915) 544-9736 schmidt@geo.utep.edu, Univ of Texas, El Paso – Oy
Schmitt, Danielle M., (609) 258-7015 dschmitt@princeton.edu, Princeton Univ – Oe
Schmitt, Douglas R., (780) 492-3985 doug@phys.ualberta.ca, Univ of Alberta – Yx
Schmitt, James G., (406) 994-6903 jschmitt@montana.edu, Montana State Univ – Gr
Schmitt, Michael A., (612) 625-7017 mschmitt@soils.umn.edu, Univ of Minnesota, Twin Cities – Sc
Schmittner, Andreas, (541) 737-9952 aschmittner@coas.oregonstate.edu, Oregon State Univ – Yr
Schmitz, Darrel W., (662) 325-2904 schmitz@geosci.msstate.edu, Mississippi State Univ – Hw
Schmitz, Mark D., (202) 478-8458 schmitz@dtm.ciw.edu, Boise State Univ – Cc
Schneider, David P., (508) 289-2873 dschneider@whoi.edu, Woods Hole Oceanographic Institution – Ct
Schneider, John F., (630) 252-8923 Argonne National Lab – Ca
Schneider, Julius, 089/2180 4354 julius.schneider@lrz.uni-muenchen.de, LMU - Univ of Munich – Gz
Schneider, Niklas, (808) 9568383 nschneid@hawaii.edu, Univ of Hawai'i, Manoa – Og
Schneider, Robert J., (508) 289-2756 rschneider@whoi.edu, Woods Hole Oceanographic Institution – Ct
Schneiderman, Jill S., (845) 437-5542 schneiderman@vassar.edu, Vassar College – Gs
Schnitker, Detmar F., (207) 563-3146 Univ of Maine – Ou
Schnoebelen, Douglas J., (319) 358-3617 djschnoe@usgs.gov, Univ of Iowa – Hg
Schoenberger, Erica J., (410) 516-6158 erica@jhu.edu, The Johns Hopkins Univ – Oo
Schoenbohm, Lindsay M., 614 247-4835 schoenbohm.1@osu.edu, Ohio State Univ – Gt
Schoenenberger, Katherine, (937) 229-3432 Katherine.Schoenenberger@notes.udayton.edu, Univ of Dayton – Gl
Schoephoester, Peter R., (608) 262-2320 schoephoeste@wisc.edu, Univ of Wisconsin - Extension – Oi
Scholl, David, (650) 725-4439 dscholl@pangea.stanford.edu, Stanford Univ – Yr
Scholl, Kathy, (507) 933-7333 kscholl@gac.edu, Gustavus Adolphus College – Oo
Scholle, Peter A., (505) 835-5294 scholle1@nmt.edu, New Mexico Institute of Mining & Technology – Gs
Scholtz, Theresa C., (630) 252-6499 Argonne National Lab – Ge
Scholz, Christopher A., 315-443-2672 cascholz@syr.edu, Syracuse Univ – Gs
Scholz, Christopher H., (845) 365-8360 scholz@ldeo.columbia.edu, Columbia Univ – Yx
Schoof, Christian, (604) 822-3063 cschoof@eos.ubc.ca, Univ of British Columbia – Gl
Schoolmaster, III, F. Andrew, (940) 565-2452 andrew@unt.edu, Univ of North Texas – Hs
Schoon, Michelle, (620)441-5204 schoon@cowley.edu, Cowley County Community College – Oo
Schoonen, Martin A., (631) 632-8007 martin.schoonen@sunysb.edu, SUNY, Stony Brook – Cl
Schoonmaker, Jane E., (808) 956-9935 jane@soest.hawaii.edu, Univ of Hawai'i, Manoa – Cl
Schoonmaker, Jane E., (808) 956-9935 jane@soest.hawaii.edu, Univ of Hawai'i, Manoa – Cm
Schopf, J. William, (310) 825-1170 schopf@ess.ucla.edu, Univ of California, Los Angeles – Po
Schorlemmer, Danijel, 213-740-4141 ds@usc.edu, Univ of Southern California – Ys
Schorn, Howard E., (510) 643-5205 Univ of California, Berkeley – Pb
Schouten, Hans, (508) 289-2574 hschouten@whoi.edu, Woods Hole Oceanographic Institution – Yr
Schrag, Daniel P., (617) 495-7676 schrag@eps.harvard.edu, Harvard Univ – Cg
Schrage, Jon M., (402) 280-5759 Creighton Univ – Ow
Schreckhise, R. Gene, (509) 375-9323 Washington State Univ – Oo
Schreiber, B. C., (828) 675-4537 geologo@aol.com, Appalachian State Univ – Gs
Schreiber, B. Charlotte, (718) 997-3300 Queens College (CUNY) – Gd
Schreiber, Madeline E., (540) 231-3377 Virginia Polytechnic Institute & State Univ – Hw
Schreiber, Jr., Joseph F., (520) 621-6024 Univ of Arizona – Gs

Schreier, Hans D., (604) 822-4401 Univ of British Columbia – Og
Schrieber, B. Charlotte, (718) 997-3300 Graduate School of the City Univ of New York – Gd
Schrijver, Kees, (418) 654-2638 kschrijver@inrs.uquebec.ca, Universite du Quebec – Eg
Schriver, David, (310) 825-6663 dave@igpp.ucla.edu, Univ of California, Los Angeles – Yg
Schroder-Adams, Claudia, (613) 520-2600 Carleton Univ – Pm
Schroder-Adams, Claudiacsadams@ccs.carleton.ca, Carleton Univ – Pm
Schroeder, Brock C., (815) 939-5361 bschroed@olivet.edu, Olivet Nazarene Univ – Xg
Schroeder, John, (806) 742-2813 john.schroeder@ttu.edu, Texas Tech Univ – Oa
Schroeder, Kathleen, (828) 262-7055 schroederk@appstate.edu, Appalachian State Univ – Oo
Schroeder, Norman C., (505) 667-0967 Los Alamos National Lab – Cc
Schroeder, Paul A., (706) 542-2384 schroe@gly.uga.edu, Univ of Georgia – Gz
Schroeder, William W., (334) 861-7528 wschroed@jaguar1.usouthal.edu, Univ of Alabama, Dauphin Island – Og
Schubert, Gerald, (310) 825-4577 schubert@ucla.edu, Univ of California, Los Angeles – Yg
Schubert, Wayne H., (970) 491-8521 waynes@atmos.colostate.edu, Colorado State Univ – Oa
Schuberth, Bernhard, 089/2180 4220 bernhard@geophysik.uni-muenchen.de, LMU - Univ of Munich – Yg
Schulingkamp, Warren, 225-578-5320 warrenii@lsu.edu, Louisiana State Univ – EoGgEc
Schulmeister, Marcia K., 620-341-5983 Emporia State Univ – Ge
Schulte, Kimberly, 770-274-5083 kschulte@gpc.edu, Georgia Perimeter College – Ge
Schulte, Mitchell D., 573-884-5712 SchulteMD@missouri.edu, Univ of Missouri – Cg
Schultz, Adam, (541) 737-9832 adam@coas.oregonstate.edu, Oregon State Univ – Yr
Schultz, Gerald, 806-651-2580 gschultz@mail.wtamu.edu, West Texas A&M Univ – Pv
Schultz, Jan, (805) 965-0581 (Ext. 2313) schultz@sbcc.net, Santa Barbara City College – Cl
Schultz, Peter H., (401) 863-2417 peter_schultz@brown.edu, Brown Univ – Xg
Schultz, Richard A., (775) 784-4318 schultz@mines.unr.edu, Univ of Nevada, Reno – Nr
Schultz-Ela, Daniel D., (512) 471-4335 dan.schultzela@beg.utexas.edu, Univ of Texas at Austin, Jackson School of Geosciences – Gc
Schulz, Layne D., (605) 677-5227 lschulz@usd.edu, South Dakota Dept of Env & Natural Resources – Gl
Schulze, Anja, 409 740 4540 aschulzea@tamug.edu, Texas A&M Univ – Ob
Schulze, Daniel J., 905-8283970 dschulze@utm.utoronto.ca, Univ of Toronto – Gi
Schulze-Makuch, Dirk, 509-335-1180 dirksm@wsu.edu, Washington State Univ – Hw
Schumacher, Courtney, (979) 845-7671 courtney@ariel.met.tamu.edu, Texas A&M Univ – Oa
Schumacher, Gregory A., (614) 265-6597 greg.schumacher@dnr.state.oh.us, Ohio Dept of Natural Resources – Gr
Schumacher, Joseph, (605) 688-6309 Joseph.Schumacher@sdstate.edu, South Dakota State Univ – Sp
Schumacher, Thomas E., (605) 688-4762 Thomas.Schumacher@sdstate.edu, South Dakota State Univ – Sp
Schuman, Gerald E., (307) 772-2433 Univ of Wyoming – Sf
Schumann, Arnold W., (863) 956-1151 Univ of Florida – So
Schumm, Stanley A., (970) 482-0680 Colorado State Univ – Gm
Schuster, Gerard T., (801) 581-4373 schuster@mines.utah.edu, Univ of Utah – Ye
Schwab, Brandon E., 707-826-3950 schwab@humboldt.edu, Humboldt State Univ – Gi
Schwab, Fred, (310) 825-3123 schwab@cyclop.ess.ucla.edu, Univ of California, Los Angeles – Ys
Schwab, Frederic L., (540) 463-8870 schwabf@wlu.edu, Washington & Lee Univ – Gs
Schwab, James J., (518) 437-8754 schwab@asrc.cestm.albany.edu, SUNY, Albany – Oa
Schwab, Paul A., (913) 532-7213 Kansas State Univ – Sc
Schwankl, Larry J., (530) 752-1130 ljschwankl@ucdavis.edu, Univ of California, Davis – Oo
Schwarcz, Henry P., (905) 525-9140 (Ext. 24186) schwarcz@mcmaster.ca, McMaster Univ – Cs
Schwartz, David, (408) 479-6495 Cabrillo College – Gu
Schwartz, Frank W., (614) 292-6196 frank@geology.ohio-state.edu, Ohio State Univ – Hg
Schwartz, Hilde, (831) 459-5429 hschwartz@pmc.ucsc.edu, Univ of California, Santa Cruz – Pv
Schwartz, Joshua J., 205-348-1878 jschwartz@geo.ua.edu, Univ of Alabama – GiCc
Schwartz, Matthew, 850-474-3469 mschwartz@uwf.edu, Univ of West Florida – Hy
Schwartz, Maurice L., (360) 650-3586 coastal@cc.wwu.edu, Western Washington Univ – On
Schwartz, Robert K., (614) 292-6196 Allegheny College – Gs
Schwartz, Susan Y., (831) 459-3133 sschwartz@pmc.ucsc.edu, Univ of California, Santa Cruz – Ys
Schweickert, Richard A., (775) 784-6901 richschw@mines.unr.edu, Univ of Nevada, Reno – Gt
Schweig, Eugene S., (901) 678-4974 schweig@usgs.gov, Univ of Memphis – Gt
Schweitzer, Carrie E., 330-672-2505 cschweit@kent.edu, Kent State Univ – Pi
Schweitzer, Carrie E., 330-244-3303 cschweit@kent.edu, Kent State Univ, Stark Campus – Pi
Schweitzer, Mary H., 919-515-7838 mary_schweitzer@ncsu.edu, North Carolina State Univ – Po
Schwerdtner, W. M., 416-978-2062 fried@quartz.geology.utoronto.ca, Univ of Toronto – Gc
Schwert, Donald P., (701) 231-7496 donald.schwert@ndsu.edu, North Dakota State Univ – Pe
Schwimmer, David R., (706) 569-3028 schwimmer_david@colstate.edu, Columbus State Univ – Pv
Schwimmer, Reed A., 609-896-5346 rschwimmer@rider.edu, Rider Univ – GsOgn
Schwinger, Marilyn, (303) 273-3800 mschwing@mines.edu, Colorado School of Mines – Oo
Schwob, Stephanie L., 214-768-2770 sschwob@smu.edu, Southern Methodist Univ – Oo
Sclater, John G., (619) 534-8653 jsclater@ucsd.edu, Univ of California, San Diego – Yh
Scoates, James S., (604) 822-3667 jscoates@eos.ubc.ca, Univ of British Columbia – Gi
Scoble, Malcolm J.malcoms@unixg.ubc.ca, McGill Univ – Nm
Scoggins, James R., (409) 845-7671 Texas A&M Univ – Ow
Scotchmoor, Judith, (510) 642-4877 Univ of California, Berkeley – Pg
Scotese, Christopher R., (817) 272-2987 cscotese@exchange.uta.edu, Univ of Texas, Arlington – Gt
Scotford, David M., (513) 529-3216 Miami Univ – Gp
Scott, Christopher A.cascott@email.arizona.edu, Univ of Arizona – Hw
Scott, Darren M., (905) 525-9140 (Ext. 23534) scottdm@mcmaster.ca, McMaster Univ – Oo
Scott, David B., (902) 494-3604 Dalhousie Univ – Pm
Scott, Durelle T., 402-472-2660 dscott4@unl.edu, Univ of Nebraska, Lincoln – Cl
Scott, Edward R., (808) 956-3955 Univ of Hawai'i, Manoa – Xm
Scott, James H., (202) 478-8908 j.scott@gl.ciw.edu, Carnegie Institution of Washington – Co
Scott, James H., 603-646-1096 james.scott@dartmouth.edu, Dartmouth College – Py
Scott, Jon T., (518) 442-4558 jscott@atmos.albany.edu, SUNY, Albany – Oa
Scott, Kathy, (604) 822-5606 kscott@eos.ubc.ca, Univ of British Columbia – Oo
Scott, Martha L., (979) 845-5179 mscott@astra.tamu.edu, Texas A&M Univ – Oc
Scott, Mary W., (614) 292-6152 scott.36@osu.edu, Ohio State Univ – Oo
Scott, Robert B., 512-471-0375 rscott@ig.utexas.edu, Univ of Texas at Austin, Jackson School of Geosciences – Og
Scott, Robert W., 918-631-3103 robert-scott@utulsa.edu, Univ of Tulsa – Gs
Scott, Sally J., (307) 766-6538 sscott@uwyo.edu, Univ of Wyoming – Og
Scott, Shana R., (936) 468-3701 sscott@sfasu.edu, Stephen F. Austin State Univ – Oo
Scott, Steven D., 416-978-5424 scottsd@geology.utoronto.ca, Univ of Toronto – Cp
Scott, Thomas A., (951) 827-5115 thomas.scott@ucr.edu, Univ of California, Riverside – Oy
Scott, Thomas M., (850) 488-9380 Florida Geological Survey – Gr

Scott, Tim, (401) 254-3108 Roger Williams Univ – Ob
Scott, Verne H., (530) 752-0453 vhscott@ucdavis.edu, Univ of California, Davis – Hs
Scott, Vernon, 405-744-6358 vscott@okstate.edu, Oklahoma State Univ – Oe
Scott Smith, Barbara H., (604) 984-9609 bhssmith@allstream.net, Univ of British Columbia – Gz
Scott, Jr., Thurman E., (405) 325-2900 tscott@ou.edu, Univ of Oklahoma – Gc
Scow, Kate, (530) 752-4632 kmscow@ucdavis.edu, Univ of California, Davis – So
Scranton, Mary I., (631) 632-8735 mscranton@notes.cc.sunysb.edu, SUNY, Stony Brook – Oc
Screaton, Elizabeth, (352) 3924612 screaton@geology.ufl.edu, Univ of Florida – Hg
Scrutton, Colin T., 0191 3742538 Durham Univ – Ps
Scudder, Sylvia J., (352) 392-1721 (Ext. 246) scudder@flmnh.ufl.edu, Univ of Florida – Pg
Scuderi, Louis, 505-277-4204 tree@unm.edu, Univ of New Mexico – Pe
Seager, Richard, (845) 365-8743 Columbia Univ – Ow
Seager, Sara, (202) 478-8868 seager@dtm.ciw.edu, Carnegie Institution of Washington – Oo
Seager, Saraseager@mit.edu, Massachusetts Institute of Technology – Xy
Seager, William R., 575-646-2708 wseager@hotmail.com, New Mexico State Univ, Las Cruces – Gc
Seaman, Nelson L., (814) 863-1583 The Pennsylvania State Univ, Univ Park – Ow
Seaman, Sheila J., (413) 545-2822 sjs@geo.umass.edu, Univ of Massachusetts – Gi
Searl, Gary H., (541) 346-4559 Univ of Oregon – Oo
Sears, James W., (406) 243-5251 jwsears@selway.umt.edu, Univ of Montana – Gt
Sears, Rollie, (913) 532-7245 Kansas State Univ – So
Sebastian, Glenn R., (251) 460-6382 gsebasti@jaguar1.usouthal.edu, Univ of South Alabama – Oy
Sebens, Kenneth P., (301) 405-7978 ks95@umail.umd.edu, Univ of Maryland – Ob
Seber, Dogan, (607) 255-4383 seber@geology.cornell.edu, Cornell Univ – Ys
Sebetich, Michael, (201) 595-2792 William Paterson College – Oo
Secco, Richard A., (519) 661-4079 secco@uwo.ca, Univ of Western Ontario – Gy
Seccombe, Philip K., 61 02 4921 5414 Phil.Seccombe@newcastle.edu.au, Univ of Newcastle – Em
Seckler, Thomas F., (609) 292-2576 tom.seckler@dep.state.nj.us, New Jersey Geological Survey – Ge
Sedivy, Robert A., (630) 252-1897 rasedivy@anl.gov, Argonne National Lab – Hw
Sedlock, Richard L., (408) 924-5020 richard.sedlock@sjsu.edu, San Jose State Univ – GtOe
See, Paul D., (503) 738-5869 Clatsop Community College – Gg
Seeber, Leonardo, (845) 365-8385 Columbia Univ – Ys
Seedorff, Eric, (520) 626-3921 seedorff@geo.arizona.edu, Univ of Arizona – Eg
Seeley, Mark W., (612) 625-4724 mseeley@umn.edu, Univ of Minnesota, Twin Cities – Oa
Seewald, Jeffrey S., (508) 289-2966 jseewald@whoi.edu, Woods Hole Oceanographic Institution – Cp
Segall, Paul, (650) 725-7241 Stanford Univ – Yg
Segars, William P., (706) 542-9072 Univ of Georgia – Sc
Seibel, Erwin, (415) 338-2061 San Francisco State Univ – On
Seidel, Howard F., (979) 458-2729 hseidel@ocean.tamu.edu, Texas A&M Univ – Op
Seidel, Robert, (701) 777-4588 Univ of North Dakota – Oy
Seidemann, David E., (718) 951-5761 Yale Univ – Cg
Seidemann, David E., (718) 951-5000 (Ext. 2882) dsebc@cunyvm.cuny.edu, Brooklyn College (CUNY) – Cc
Seifert, Karl E., (515) 294-5265 kseifert@iastate.edu, Iowa State Univ of Science & Technology – Ct
Seifoullaev, Roustam K., (512) 232-3223 roustam@utig.ig.utexas.edu, Univ of Texas at Austin, Jackson School of Geosciences – Ye
Seilacher, Adolf, (203) 432-3173 Yale Univ – Pi
Seitz, Jeffery C., (510) 885-3438 jeff.seitz@csueastbay.edu, California State Univ, East Bay – Cp
Sekharan, K. K.kksekharan@uh.edu, Univ of Houston – Ye
Selenskas, Sandra L., (203) 932-7107 Univ of New Haven – Oo
Self, Burl E., (417) 836-4849 bes723f@smsu.edu, Southwest Missouri State Univ – Oo
Self, Robert P., (731) 587-7475 rself@utm.edu, Univ of Tennessee, Martin – Gs
Selim, Hussein M., (504) 388-2110 Louisiana State Univ – Sp
Selleck, Bruce, (315) 228-7949 bselleck@mail.colgate.edu, Colgate Univ – Gs
Seltzer, Geoffrey O., 315-443-4980 goseltze@syr.edu, Syracuse Univ – Gl
Selverstone, Jane E., (505) 277-6528 selver@unm.edu, Univ of New Mexico – Gp
Semazzi, Fred H. M., (919) 515-1434 fred_semazzi@ncsu.edu, North Carolina State Univ – Oa
Semerjian, Christopher J., (770) 718-3862 csemerj@hermes.gc.peachnet.edu, Gainesville College – Ge
Semken, Steven C., (505) 368-3630 Dine College, Shiprock – Ge
Semken, Jr., Holmes A., (319) 335-1830 holmes-semken@uiowa.edu, Univ of Iowa – Pv
Semtner, Albert J., (408) 656-3267 sbert@oc.nps.navy.mil, Naval Postgraduate School – Op
Sen, Gautam, (305) 348-2299 seng@fiu.edu, Florida International Univ – Gi
Sen, Mrinal K., (512) 471-0466 Univ of Texas at Austin, Jackson School of Geosciences – Ye
Sen, Mrinal K., (512) 471-0466 mrinal@utig.ig.utexas.edu, Univ of Texas at Austin, Jackson School of Geosciences – Ye
Sen Gupta, Barun K., (225) 388-5984 barun@geol.lsu.edu, Louisiana State Univ – Pm
Sennert, Sally K., (202) 633-1805 kuhns@si.edu, Smithsonian Inst / National Museum of Natural History – GvOr
Senock, Randy S.rsenock@csuchico.edu, California State Univ, Chico – Sf
Senwo, Zachary, 256-372-4216 zsenwo@aamu.edu, Alabama A&M Univ – Sc
Seong, Jeong C., 678839069 jseong@westga.edu, Univ of West Georgia – Oi
Seramur, Keith C., (828) 262-3049 Appalachian State Univ – Hw
Serfes, Michael E., (609) 984-6587 mike.serfes@dep.state.nj.us, New Jersey Geological Survey – Hy
Sericano, Jose L., (979) 862-2323 jose@gerg.tamu.edu, Texas A&M Univ – Cm
Serne, R. Jeffrey, (509) 376-8429 jeff.serne@pnl.gov, Pacific Northwest National Lab – Cg
Serpa, Laura F., 915-747-6058 serpa@geo.utep.edu, Univ of Texas, El Paso – Ye
Serrano, Carmen, (207) 872-3244 serrano@fit.edu, Florida Institute of Technology – Oo
Serrano, Carmen, (321) 674-8096 serrano@fit.edu, Florida Institute of Technology – Oo
Sethi, Parvinder S., (540) 831-5619 psethi@radford.edu, Radford Univ – Gz
Setterholm, Dale R., (612)627-4780 Ext. 223 sette001@umn.edu, Univ of Minnesota, Twin Cities – Gg
Severin, Kenneth P.fnkps@uaf.edu, Univ of Alaska, Fairbanks – Oo
Severinghaus, Jeffrey P.jseveringhaus@ucsd.edu, Univ of California, San Diego – Pe
Severs, Matthew, 330-263-2297 msevers@wooster.edu, The College of Wooster – Gx
Sewall, Jacob O., 540-231-6521 Virginia Polytechnic Institute & State Univ – Oa
Sexton, John L., (618) 453-7374 sexton@geo.siu.edu, Southern Illinois Univ Carbondale – Ye
Seyfried, Jr., William E., 612-624-1333 wes@umn.edu, Univ of Minnesota, Twin Cities – Cl
Seyitoglu, GurolGurol.Seyitoglu@eng.ankara.edu.tr, Ankara Univ – Gt
Seyler, Beverly, 217-244-2389 seyler@isgs.uiuc.edu, Illinois State Geological Survey – Eo
Seymour, Kevin L., (416) 586-5754 kevins@rom.on.ca, Royal Ontario Museum – Pv
Shaak, Graig D., (352) 392-1721 (Ext. 257) gdshaak@flmnh.ufl.edu, Univ of Florida – Pe
Shadwick, Robert E., (619) 534-7973 rshadwick@ucsd.edu, Univ of California, San Diego – Ob
Shaffer, Nelson R., (812) 855-2687 shaffern@indiana.edu, Indiana Geological Survey – En
Shagam, Reginald, (215) 898-6085 rshagam@bgumail.ggu.ac.il, Univ of Pennsylvania – Gt
Shah, Subhash N., (405) 325-6871 subhash@ou.edu, Univ of Oklahoma – Np

Shakal, Anthony F., (916) 322-7481 California Geological Survey – Ys
Shakoor, Abdul, (330) 672-2968 ashakoor@kent.edu, Kent State Univ – Ng
Shalev, Eylon, (919) 684-4780 shalev@duke.edu, Duke Univ – Ys
Shaman, Jeffrey, (541) 737-4915 jshaman@coas.oregonstate.edu, Oregon State Univ – Oa
Shank, G. C., (361) 749-6776 chris.shank@mail.utexas.edu, Univ of Texas at Austin – Ob
Shankland, Thomas J., (505) 667-4907 shanklan@lanl.gov, Los Alamos National Lab – Yx
Shanmugan, Shan, (817) 272-2987 Geology@uta.edu, Univ of Texas, Arlington – Gs
Shannon, Jack D., (630) 252-5807 jack_shannon@anl.gov, Argonne National Lab – Ow
Shannon, Jeremy, (906) 487-2531 jmshanno@mtu.edu, Michigan Technological Univ – Gv
Shapiro, Russell, (530) 898-4300 rshapiro@csuchico.edu, California State Univ, Chico – Ps
Sharkey, Debra, (916) 691-7210 Cosumnes River College – Oy
Sharma, Govind, (205) 851-5462 gsharma@aamu.edu, Alabama A&M Univ – Oo
Sharma, Mukul, 603-646-0024 Mukul.Sharma@dartmouth.edu, Dartmouth College – Ge
Sharma, Shiv K., (808) 956-8476 Univ of Hawai'i, Manoa – Gx
Sharon, Yitzhak Y., (609) 652-4500 Stockton State College – Yr
Sharp, Jonathan H., (302) 645-4259 jsharp@udel.edu, Univ of Delaware – Oc
Sharp, Martin J., (780) 492-4156 martin.sharp@ualberta.ca, Univ of Alberta – Gl
Sharp, Martin J., (780) 492-4156 martin.sharp@ualberta.ca, Univ of Alberta – Gm
Sharp, Patricia S., (936) 468-2095 psharp@sfasu.edu, Stephen F. Austin State Univ – Gg
Sharp, S. L., 540-231-4080 llyn@vt.edu, Virginia Polytechnic Institute & State Univ – Oe
Sharp, Thomas G., (480) 965-5081 tom.sharp@asu.edu, Arizona State Univ – Gz
Sharp, William E., (313) 763-6200 Univ of Michigan – Oa
Sharp, Zachary D., (505) 277-4204 zsharp@unm.edu, Univ of New Mexico – Cs
Sharp, Jr., John M., (512) 471-3317 jsharp@mail.utexas.edu, Univ of Texas at Austin, Jackson School of Geosciences – Hq
Sharpley, Andrew N., (814) 863-0948 ans3@psu.edu, Pennsylvania State Univ, Univ Park – Sc
Sharpton, Virgil, (907) 474-6663 buck.sharpton@gi.alaska.edu, Univ of Alaska, Fairbanks – Xg
Shaulis, James R., 717.702.2037 jshaulis@state.pa.us, Pennsylvania Bureau of Topographic & Geologic Survey – Ec
Shaver, Stephen A., (931) 598-1116 sshaver@sewanee.edu, Univ of the South – Eg
Shaw, Alison, 508-289-3775 ashaw@whoi.edu, Woods Hole Oceanographic Institution – Ou
Shaw, Bruce, (845) 365-8380 Columbia Univ – Ys
Shaw, Christopher A., (323) 857-6317 chrissha@bcf.usc.edu, Los Angeles County Museum of Natural History – Pv
Shaw, Cliff, 506-447-3195 cshaw@unb.ca, Univ of Western Ontario – Gi
Shaw, Colin, 406 994 6760 colin.shaw1@montana.edu, Montana State Univ – GcEmGp
Shaw, Fred, 573-368-2479 nrshawf.mail.dnr.state.mo.us, Missouri Dept of Natural Resources – Oo
Shaw, Frederick C., (212) 642-2202 Lehman College (CUNY) – Ps
Shaw, Frederick C., (718) 960-8565 fcslc@cunyvm.cuny.edu, Graduate School of the City Univ of New York – Pi
Shaw, George, (518) 388-6310 shawg@union.edu, Union College – Yx
Shaw, Joey N., (334) 844-3957 jnshaw@acesag.auburn.edu, Auburn Univ – Sd
Shaw, John, (780) 492-3573 john.shaw@ualberta.ca, Univ of Alberta – Gm
Shaw, John, (617) 495-8008 shaw@eps.harvard.edu, Harvard Univ – Gt
Shaw, Kenneth L., (212) 642-2202 kshaw@egi.utah.edu, Univ of Utah – Ye
Shaw, Liz, e.j.shaw@reading.ac.uk, Univ of Reading – Sb
Shaw, Ping-Tung, (919) 515-7276 ping-tung_shaw@ncsu.edu, North Carolina State Univ – Op
Shaw, Richard F., (225) 388-6734 Louisiana State Univ – Ob
Shcherbakov, Robert, (519) -66-1-21 (Ext. ext.84212) rshcherb@uwo.ca, Univ of Western Ontario – YgsYd
Shchukin, Eugene D., (410) 516-5079 shchukin@jhuvms.hcf.jhu.edu, The Johns Hopkins Univ – Oo
Shearer, Peter M., 858-534-2260 pshearer@ucsd.edu, Univ of California, San Diego – Ys
Shearer, Jr., Charles K., (505) 277-9159 cshearer@unm.edu, Univ of New Mexico – Gi
Shearman, R. K., (541) 737-1866 shearman@coas.oregonstate.edu, Oregon State Univ – Op
Sheehan, Anne F., 303-492-4597 anne.sheehan@colorado.edu, Univ of Colorado – Ys
Sheehan, Peter M., (414) 278-2741 sheehan@mpm.edu, Milwaukee Public Museum – Pi
Sheehan, Peter M., 414-278-2741 sheehan@uwm.edu, Univ of Wisconsin, Milwaukee – Pi
Sheen, Shawn, (307) 766-3386 shawns@uwyo.edu, Univ of Wyoming – Oo
Sheets, H. David, (716) 645-6800 (Ext. 6100) SUNY, Buffalo – Oo
Sheldon, Amy L., (585) 245-5988 sheldon@geneseo.edu, SUNY, Geneseo – Hw
Shell, Karen, (541) 737-0980 kshell@coas.oregonstate.edu, Oregon State Univ – Oa
Shelley, David, ((03)) 3-667- (Ext. 1 ext 7723) david.shelley@canterbury.ac.nz, Univ of Canterbury – Gx
Shellito, Lucinda, (970) 351-2491 Lucinda.Shellito@unco.edu, Univ of Northern Colorado – Oa
Shelton, Dale W., 410-554-5505 dshelton@dnr.state.md.us, Maryland Geological Survey – Gg
Shelton, Kevin L., (573) 882-6568 SheltonKL@missouri.edu, Univ of Missouri – Em
Shelton, Philip C., (540) 328-0200 Univ of Virginia College, Wise – Oo
Shelton, William, 501 683-0112 Arkansas Geological Commission – Cg
Shem, Linda M., (630) 252-3857 Argonne National Lab – Ge
Shen, Po-Yu (Paul), (519) 661-3142 pys@uwo.ca, Univ of Western Ontario – Yh
Shen, Samuel S., (780) 492-0216 shen@ualberta.ca, Univ of Alberta – Oa
Shen, Yang, (401) 874-6848 yshen@gso.uri.edu, Univ of Rhode Island – Ou
Sheng, Jinyu, (902) 494-2718 jinyu.sheng@dal.ca, Dalhousie Univ – Op
Shepard, Michael K., (570) 389-4568 mshepard@bloomu.edu, Bloomsburg Univ – XgYgOr
Shepard, Thomas E., 205/247-3587 tshepard@gsa.state.al.us, Geological Survey of Alabama – Oo
Shepardson, Daniel, (765) 494-8181 dshep@purdue.edu, Purdue Univ – Oe
Shepherd, Gordon G., (416) 736-2100 gordon@yorku.ca, York Univ – Oo
Sheppard, Scott S., (202) 478-8854 sheppard@dtm.ciw.edu, Carnegie Institution of Washington – Oo
Shepson, Paul B., (765) 494-0278 pshepson@purdue.edu, Purdue Univ – Oa
Sherburne, Roger W., (916) 322-9308 California Geological Survey – Ys
Sheridan, Michael F., (716) 645-6800 (Ext. 3984) mfs@geology.buffalo.edu, SUNY, Buffalo – Gv
Sheridan, Robert E., (732) 445-2015 rsheridan@rci.rutgers.edu, Rutgers, The State Univ of New Jersey – Yr
Sheriff, Robert E., (713) 743-3414 rsheriff@bayou.uh.edu, Univ of Houston – Ye
Sheriff, Steven D., (406) 243-6560 gl_sds@selway.umt.edu, Univ of Montana – Ym
Sherman-Morris, Kathleen M., 662-325-4243 kms5@geosci.msstate.edu, Mississippi State Univ – Ow
Sherr, Barry F., (541) 737-4369 sherrb@coas.oregonstate.edu, Oregon State Univ – Ob
Sherr, Evelyn B., (541) 737-4369 sherre@coas.oregonstate.edu, Oregon State Univ – Ob
Sherrell, Robert M., (732) 932-6555 (Ext. 252) sherrell@ahab.rutgers.edu, Rutgers, The State Univ of New Jersey – Cm
Sherriff, Barbara L., (204) 474-9786 bl_sherriff@umanitoba.ca, Univ of Manitoba – Gz
Shervais, John W., (435) 797-1274 john.shervais@usu.edu, Utah State Univ – Gi
Sherwood, William C., (540) 568-6473 sherwowc@jmu.edu, James Madison Univ – So
Sherwood Lollar, Barbara, 416-978-0770 bslollar@chem.utoronto.ca, Univ of Toronto – Cs

Shevenell, Lisa A., (775) 784-1779 lshevenell@charter.net, Univ of Nevada, Reno – Hw
Shew, Roger D., 910-962-7676 shewr@uncw.edu, Univ of North Carolina Wilmington – Eo
Shi, Wei, 919-513-4641 wei_shi@ncsu.edu, North Carolina State Univ – Sb
Shick, J. Malcolm, (207) 581-2562 Univ of Maine – Ob
Shieh, Sean, (519) 661-2111 (Ext. ext. 82467) sshieh@uwo.ca, Univ of Western Ontario – Yg
Shieh, Yuch-Ning, (765) 494-3272 ynshieh@purdue.edu, Purdue Univ – Cs
Shields, William, (309) 438-7479 weshiel@ilstu.edu, Illinois State Univ – Gg
Shiller, Alan M., (228) 688-1178 alan.shiller@usm.edu, Univ of Southern Mississippi – Oc
Shilts, William W., (217) 333-5111 shilts@isgs.uiuc.edu, Illinois State Geological Survey – Gl
Shim, Sang-Heon (Dan), (617) 324-0249 sangshim@mit.edu, Massachusetts Institute of Technology – Yx
Shim-Chim, Richard, (609) 984-6587 rich.shim-chim@dep.state.nj.us, New Jersey Geological Survey – Hw
Shimizu, Nobumichi, (508) 289-2963 Woods Hole Oceanographic Institution – Ca
Shindell, Drew T., (212) 678-5561 dshindell@giss.nasa.gov, Columbia Univ – Oa
Shinn, Eugene, (727) 893-3100 Univ of South Florida, St Petersburg – Gg
Shipley, Susie L., 614 292-7230 shipley.6@osu.edu, Ohio State Univ – Oo
Shipley, Thomas H., (512) 471-0430 tom@ig.utexas.edu, Univ of Texas at Austin, Jackson School of Geosciences – Yr
Shipp, Robert L., (334) 460-6351 rshipp@jaguar1.usouthal.edu, Univ of South Alabama – Ob
Shirer, Hampton N., (814) 863-1992 hns@psu.edu, The Pennsylvania State Univ, Univ Park – Ow
Shirey, Steven B., (202) 478-8473 shirey@dtm.ciw.edu, Carnegie Institution of Washington – Cc
Shirley, Terry, 704-687-5925 trshirle@uncc.edu, Univ of North Carolina, Charlotte – Oaw
Shivji, Mahmood S., (954) 262-3653 mahmood@ocean.nova.edu, Nova Southeastern Univ – Ob
Shock, Everett, 480-727-6404 shock@asu.edu, Arizona State Univ – Oo
Shoemaker, Kurt A., 740-351-3395 kshoemaker@shawnee.edu, Shawnee State Univ – Gi
Sholes, Mark A., (406) 496-4272 msholes@mtech.edu, Montana Tech of the Univ of Montana – Ec
Sholkovitz, Edward R., (508) 289-2346 esholkovitz@whoi.edu, Woods Hole Oceanographic Institution – Oc
Shone, R. W., +27 41 5042277 glarws@upe.ac.za, Univ of Port Elizabeth – Gd
Shor, George G., (619) 534-2853 gshor@ucsd.edu, Univ of California, San Diego – Yr
Shore, Patrick J., (314) 935-7357 patrick@wups.wustl.edu, Washington Univ – Gc
Short, Heatherhashort@eps.mcgill.ca, McGill Univ – Gc
Short, R. T., (727) 553-3990 tshort@marine.usf.edu, Univ of South Florida, Saint Petersburg – Oo
Showers, William J., (919) 515-7143 w_showers@ncsu.edu, North Carolina State Univ – Cs
Showman, Adam, (520) 621-4021 showman@lpl.arizona.edu, Univ of Arizona – Oo
Showman, Richard M., (803) 777-2457 Univ of South Carolina – Ob
Shrady, Catherine H., (315) 229-5264 St. Lawrence Univ – Gc
Shrake, Douglas L., (614) 265-6579 doug.shrake@dnr.state.oh.us, Ohio Dept of Natural Resources – Gr
Shreve, Ronald L., (310) 825-5273 shreve@ess.ucla.edu, Univ of California, Los Angeles – Gm
Shreve, Ronald L.shreve@ess.ucla.edu, Univ of Washington – Yr
Shroder, Jr., John F., 402-554-2770 jshroder@mail.unomaha.edu, Univ of Nebraska at Omaha – Gm
Shropshire, K. Lee, (970) 351-2285 leeshrop@att.net, Univ of Northern Colorado – Pg
Shu, Jinfu, (202) 478-8963 j.shu@gl.ciw.edu, Carnegie Institution of Washington – Yx
Shuford, James W., (205) 851-5462 jshufard@aamu.edu, Alabama A&M Univ – So
Shugart, Jr., Herman H., (804) 924-7642 hhs@virginia.edu, Univ of Virginia – Og
Shuib, Mustaffa Kamal, 03-79674227 mustaffk@um.edu.my, Univ of Malaya – Gc
Shulik, Stephen J.shulik@clarion.edu, Clarion Univ – Ym
Shulmeister, Jamie P., (03) 3642-762 james.shulmeister@canterbury.ac.nz, Univ of Canterbury – Pe
Shum, C. K., 614 292 7118 ckshum@osu.edu, Ohio State Univ – Yd
Shumaker, Robert C., (304) 293-5603 rshumaker@wvu.edu, West Virginia Univ – Go
Shuman, Larry M., (404) 228-7276 Univ of Georgia – Sc
Shuman, Randy, (206) 296-8243 randy.shuman@metrokc.gov, Univ of Washington – Ob
Shumway, Dinah, (714) 275-0434 California Geological Survey – Eg
Shumway, Matthew J., (801) 422-2707 jms7@email.byu.edu, Brigham Young Univ – Oo
Shuster, Robert D., 402-554-2457 rshuster@mail.unomaha.edu, Univ of Nebraska at Omaha – Gi
Shuttleworth, W. James, (520) 621-8787 shuttle@hwr.arizona.edu, Univ of Arizona – Hs
Shyn, Tong-Wha, (313) 763-9940 Univ of Michigan – Oo
Sibert, John R., (808) 956-4109 jsibert@soest.hawaii.edu, Univ of Hawai'i, Manoa – Oo
Sibley, Duncan F., 517-355-4600 sibley@msu.edu, Michigan State Univ – Gs
Sibray, Steven S., (308) 632-1382 ssibray@unl.edu, Unversity of Nebraska - Lincoln – HwGg
Sibson, Rick H., +64 3 479-7506 rick.sibson@otago.ac.nz, Univ of Otago – Gc
Sicree, Andrew A., (814) 865-6427 sicree@geosc.psu.edu, Pennsylvania State Univ, Univ Park – Gz
Siddarthan, Raj, (775) 784-1411 siddhart@scs.unr.edu, Univ of Nevada, Reno – Hs
Siddoway, Christine S., (719) 389-6717 csiddoway@coloradocollege.edu, Colorado College – Gc
Sides, John A., 404-656-3214 Georgia Dept of Natural Resources – Oo
Sidor, Christian A., 206-221-3285 casidor@u.washington.edu, Univ of Washington – Pv
Siebert, Lee, (202) 633-1818 siebertl@si.edu, Smithsonian Institution / National Museum of Natural History – Gv
Siefert, Ronald L.siefert@cbl.umces.edu, Univ of Maryland – Ca
Siegel, Donald I., 315-443-2672 disiegel@syr.edu, Syracuse Univ – Hw
Siegrist, Jr., Henry G., (671) 735-2687 hsiegris@uog.edu, Univ of Guam – Ge
Sieh, Kerry E., (626) 395-6115 sieh@gps.caltech.edu, California Institute of Technology – Gt
Siesser, William G., 615-322-2984 william.g.siesser@vanderbilt.edu, Vanderbilt Univ – Pm
Siever, Raymond, (617) 495-2354 siever@eps.harvard.edu, Harvard Univ – Gd
Sieving, John C., 217-244-2406 sieving@isgs.uiuc.edu, Illinois State Geological Survey – Hw
Siewers, Fredrick D., (270) 745-5988 fred.siewers@wku.edu, Western Kentucky Univ – Pg
Sigler, William V., (419) 530-2897 william.sigler@utoledo.edu, Univ of Toledo – Oo
Sigman, Daniel M., 609-258-2194 sigman@princeton.edu, Princeton Univ – Cg
Signer, Marcia, 303-492-8312 signer@colorado.edu, Univ of Colorado – Oo
Sigurdsson, Haraldur, (401) 874-6596 haraldur@gso.uri.edu, Univ of Rhode Island – Ou
Sikora, Paul J., (801) 581-4122 psikora@egi.utah.edu, Univ of Utah – Ps
Sikora, Todd D., 717-872-3292 Todd.Sikora@millersville.edu, Millersville Univ – Oa
Sikorski, Janelle, 513-529-9662 sikorsj@muohio.edu, Miami Univ – Gm
Silliman, James, (361) 825-3718 silliman@falcon.tamucc.edu, Texas A&M Univ, Corpus Christi – Co
Silliman, Stephen E., (219) 631-5846 Univ of Notre Dame – Hw
Sillman, Sanford, (734) 763-4217 sillman@umich.edu, Univ of Michigan – Ow
Silva, Michael A., (916) 324-0768 California Geological Survey – En
Silva, Tiago, (541) 737-5283 tsilva@coas.oregonstate.edu, Oregon State Univ – Gu
Silvano, Janet, (617) 627-3494 Tufts Univ – Oo
Silver, Eli A., (831) 459-2266 esilver@es.ucsc.edu,

Faculty Index -S

Univ of California, Santa Cruz – Yr
Silver, Eli A., (831) 459-2266 esilver@pmc.ucsc.edu, Univ of California, Santa Cruz – Yr
Silver, Leon T., (626) 395-6490 lsilver@gps.caltech.edu, California Institute of Technology – Cc
Silver, Mary W., (831) 459-2908 msilver@cats.ucsc.edu, Univ of California, Santa Cruz – Ob
Silver, Paul G., (202) 478-8834 silver@dtm.ciw.edu, Carnegie Institution of Washington – Ys
Silvers, Jackie, (859) 257-5500 jsilvers@kgs.mm.uky.edu, Univ of Kentucky – Oo
Silvertooth, Jeffrey C., (520) 621-7228 silver@ag.arizona.edu, Univ of Arizona – So
Simandl, George J., (250) 952-0413 gsimandle@galaxy.gov.bc.ca, Univ of Victoria – En
Simila, Gerald W., (818) 677-3543 gerry.simila@csun.edu, California State Univ, Northridge – Yg
Simkin, Tom, (202) 633-1819 simkint@si.edu, Smithsonian Institution / National Museum of Natural History – Gv
Simmonds, Robert T.simmonds@olympus.net, SUNY, Oneonta – Pg
Simmons, Carver S., (509) 372-6047 carver.simmons@pnl.gov, Pacific Northwest National Lab – Hq
Simmons, Craig, csimmons@mines.edu, Colorado School of Mines – Cp
Simmons, F. William, (217) 333-4424 Univ of Illinois, Urbana-Champaign – Sp
Simmons, Lizanne V., (949) 589-0562 Santiago Canyon College – Gr
Simmons, M. G., (617) 253-6393 gene@hager-richter.com, Massachusetts Institute of Technology – Yg
Simmons, William B., 7342866325 , 5042806325 wsimmons@uno.edu, Univ of Michigan – Gz
Simmons, William B., (504) 280-6791 wsimmons@uno.edu, Univ of New Orleans – Gz
Simms, Alex, 405-744-6358 alex.simms@okstate.edu, Oklahoma State Univ – Gs
Simms, Dorothy F., (918) 456-5511 simms@nsuok.edu, Northeastern State Univ – Pg
Simms, Janet E., (601) 634-3493 Mississippi State Univ – Yg
Simms, John J., (918) 456-5511 simmsj@nsuok.edu, Northeastern State Univ – Gg
Simoneit, Bernd R., (541) 737-2155 Oregon State Univ – Oc
Simons, Mark, (626) 395-6984 simons@caltech.edu, California Institute of Technology – Ys
Simons, Sharon J., (618)453-3352 simons@geo.siu.edu, Southern Illinois Univ Carbondale – Oo
Simonson, Bruce M., (440) 775-8347 bruce.simonson@oberlin.edu, Oberlin College – Gs
Simonson, John M., (865) 574-4962 simonsonjm@ornl.gov, Oak Ridge National Lab – Cg
Simony, Philip S., (403) 220-6679 Univ of Calgary – Gc
Simovich, Marie, (619) 260-4600 simo@acusd.edu, Univ of San Diego – Oo
Simpkins, William W., (515) 294-7814 bsimp@iastate.edu, Iowa State Univ of Science & Technology – Hw
Simpson, Dale R., (610) 758-3664 Lehigh Univ – Gz
Simpson, Edward L., (610) 683-4447 simpson@kutztown.edu, Kutztown Univ of Pennsylvania – Gs
Simpson, Frank, (519) -25-3-30 (Ext. ext. 2487) franks@uwindsor.ca, Univ of Windsor – Gs
Simpson, Kirstie, (604) 822-2392 ksimpson@eos.ubc.ca, Univ of British Columbia – Gv
Simpson, Robert M., (731) 587-7439 msimpson@utm.edu, Univ of Tennessee, Martin – Ow
Simpson, Jr., H. James, (845) 365-8570 simpsonj@ldeo.columbia.edu, Columbia Univ – Cm
Simpson-Baczek, Wanda, (415) 239-3104 415-452-7427, City College of San Francisco – Gz
Sims, Albert L., (218) 281-8619 simsx008@tc.umn.edu, Univ of Minnesota, Twin Cities – Sc
Sims, Hallie J., (319) 335-0995 hallie-sims@uiowa.edu, Univ of Iowa – Pe
Sims, Kenneth W., (508) 289-2634 ksims@whoi.edu, Woods Hole Oceanographic Institution – Gi
Sims, William J., (501) 371-7613 wjsims@ualr.edu, Univ of Arkansas, Little Rock – Gs
Sinclair, Alastair J., (604) 822-3086 asinclair@eos.ubc.ca, Univ of British Columbia – Em
Singer, Bradley S., (608) 262-6366 bsinger@geology.wisc.edu, Univ of Wisconsin, Madison – Cc
Singer, Jill K., (716) 878-4724 singerjk@buffalostate.edu, Buffalo State College – GsOp
Singer, Michael J., (530) 752-7499 mjsinger@ucdavis.edu, Univ of California, Davis – So
Singh, Harbans, (973) 655-7383 singhh@mail.montclair.edu, Montclair State Univ – Oo
Singha, Kamini, 814) 863-6649 ksingha@geosc.psu.edu, Pennsylvania State Univ, Univ Park – Hy
Singler, Charles R., 330-941-3611 crsingler@ysu.edu, Youngstown State Univ – Gs
Sinha, A. Krishna, (540) 231-5580 pitlab@vt.edu, Virginia Polytechnic Institute & State Univ – Gt
Sinha, Aisish, 310 243-3166 asinha@csudh.edu, California State Univ, Dominguez Hills – Gg
Sinha, Binita, (925) -68-5-12 (Ext. X2461) bsinha@dvc.edu, Diablo Valley College – Oi
Sinton, John M., (808) 956-7751 sinton@hawaii.edu, Univ of Hawai'i, Manoa – Gi
Sipiera, Paul P., (847) 925-6726 William Rainey Harper College – Xm
Sirbescu, Mona-Liza C., (989) 774-3179 sirbe1mc@cmich.edu, Central Michigan Univ – Gz
Sirk, Robert A., (931) 221-7473 sirkr@apsu.edu, Austin Peay State Univ – OywOo
Sisterson, Douglas L., (630) 252-5836 dsisterson@anl.gov, Argonne National Lab – Oa
Sitwell, O.F. George, (780) 492-4593 o.f.g.sitwell@ualberta.ca, Univ of Alberta – Oo
Size, William B., (404) 727-0203 wsize@emory.edu, Emory Univ – Gi
Skalac, Priscilla, (815) 928-5477 pskalac, Olivet Nazarene Univ – Oe
Skehan, James W., 617-552-8312 james.skehan.1@bc.edu, Boston College – Gc
Skehan, S.J., James W., (617) 552-8312 skehan@bc.edu, Boston College – Gc
Skelton, Lawrence H., (316) 943-2343 lskelton@kgs.ku.edu, Univ of Kansas – Gg
Skema, Viktoras W., 717.702.2035 vskema@state.pa.us, Pennsylvania Bureau of Topographic & Geologic Survey – Gr
Skiba, John B., (518) 486-2013 jskiba@mail.nysed.gov, New York State Geological Survey – Oo
Skidmore, Mark L., 406-994-7251 skidmore@montana.edu, Montana State Univ – ClPyOy
Skilling, Ian, (412) 624-5873 skilling@pitt.edu, Univ of Pittsburgh – Gv
Skinner, Brian J., (203) 432-3175 brian.skinner@yale.edu, Yale Univ – Cg
Skinner, H. Catherine W., (203) 432-3787 Yale Univ – Gz
Skinner, James, 417-836-5800 jls519f@smsu.edu, Southwest Missouri State Univ – Oo
Skinner, Randall, 801-422-6083 Rskinner@geology.byu.edu, Brigham Young Univ – Gg
Skinner, Wilbert R., (313) 747-3960 Univ of Michigan – Oa
Skinner, William R., (440) 775-8351 bill.skinner@oberlin.edu, Oberlin College – Gx
Skippen, George B., (613) 520-2600 Carleton Univ – Cg
Skippen, George B.gskippen@ccs.carleton.ca, Carleton Univ – Cg
Sklar, Leonard, 415-338-1204 leonard@sfsu.edu, San Francisco State Univ – Gm
Sklarew, Debbie S., (509) 376-9760 debbie.sklarew@pnl.gov, Pacific Northwest National Lab – Og
Skowron, Gosia, (801) 585-9762 gskowron@egi.utah.edu, Univ of Utah – Gc
Skowron, Loretta M., (217) 333-4977 Illinois State Water Survey – Ca
Skubis, Steven T., (315) 312-2799 SUNY, Oswego – Ow
Skyllingstad, Eric, (541) 737-5697 skylling@coas.oregonstate.edu, Oregon State Univ – Oa
Sladek, Chris, (775) 784-6970 csladek@unr.edu, Univ of Nevada, Reno – Gg
Slater, Brian K., (614) 292-5891 slater.39@osu.edu, Ohio State Univ – Sd
Slater, Gregory F., (905) 525-9140 (Ext. 26388) gslater@mcmaster.ca, McMaster Univ – Cg
Slater, Lee S., 973-353-5109 Rutgers, The State Univ of New Jersey, Newark – Yg
Slatkin, Montgomery, (510) 642-6300 Univ of California, Berkeley – Po
Slatt, Roger M., (405) 325-3253 sgg@hoth.gcn.ou.edu, Univ of Oklahoma – Go
Slattery, Donna, 202-633-2479 Smithsonian Institution / National Air &

Space Museum – Oo
Slattery, William, 937 775-3455 william.slattery@wright.edu, Wright State Univ – Gr
Slaughter, Maynard, (970) 330-5893 mslaugh@unco.edu, Univ of Northern Colorado – Cg
Slaughter, Richard, 608/262-2399 rich@geology.wisc.edu, Univ of Wisconsin, Madison – Pg
Slavetskas, Carolcslavets@binghamton.edu, Binghamton Univ – Oo
Slawinski, Michael A., (709) 737-7541 mslawins@mun.ca, Memorial Univ of Newfoundland – Ys
Slaymaker, Susan Clark, (916) 278-6064 scs@csus.edu, California State Univ, Sacramento – Yg
Sledzinski, Grazyna, 585 3232313 msledzin@aol.com, Wayne State Univ – Gg
Sleep, Norman H., (650) 723-0882 Stanford Univ – Yg
Sleezer, Richard O.sleezerr@emporia.edu, Emporia State Univ – Os
Sletten, Ronald S., (206) 543-0571 Univ of Washington – So
Sletten, Ronald S., 206-543-0571 sletten@u.washington.edu, Univ of Washington – OsSb
Slingerland, Rudy L., (814) 865-6892 sling@geosc.psu.edu, Pennsylvania State Univ, Univ Park – Gs
Sloan, Doris, (510) 642-3703 dsloan@socrates.berkeley.edu, Univ of California, Berkeley – Ge
Sloan, Heather, 718-960-8008 Heather.Sloan@lehman.cuny.edu, Lehman College (CUNY) – Yr
Sloan, Heather, 718-960-8008 hsloan@lehman.cuny.edu, American Museum of Natural History – Yr
Sloan, Jon R., (818) 677-4880 jon.sloan@csun.edu, California State Univ, Northridge – Pm
Sloan, Lisa C., (831) 459-3693 lsloan@es.ucsc.edu, Univ of California, Santa Cruz – Pe
Sloan, Lisa C., (831) 459-3693 lsloan@pmc.ucsc.edu, Univ of California, Santa Cruz – Pe
Slobodkin, Lawrence B., (631) 632-8605 bzll@notes.cc.sunysb.edu, SUNY, Stony Brook – Ob
Slocum, Carol, (609) 652-4249 Stockton State College – Oo
Slowey, Niall C., (979) 845-8478 nslowey@ocean.tamu.edu, Texas A&M Univ – Ou
Slucher, Ernie R., (614) 265-6627 ernie.slucher@dnr.state.oh.us, Ohio Dept of Natural Resources – Gr
Small, Bryan, (303) 370-6404 BSmall@dmns.org, Denver Museum of Nature & Science – Pv
Small, Christopher, (845) 365-8354 Columbia Univ – Or
Small, Christopher, (845) 365-8354 small@ldeo.columbia.edu, Columbia Univ – Yr
Small, Eric E., 303-735-5033 eric.small@colorado.edu, Univ of Colorado – Hw
Small, Lawrence F., (541) 737-3960 smalll@coas.oregonstate.edu, Oregon State Univ – Ob
Smalley, Gabriela W., 609-896-5097 gsmalley@rider.edu, Rider Univ – ObcOp
Smalley, Glendon W., (615) 598-5714 Univ of the South – Sf
Smalley, Robert, Jr.,, (901) 678-4929 smalley@ceri.memphis.edu, Univ of Memphis – Ys
Smalley, Jr., Robert, (901) 678-4929 smalley@ceri.memphis.edu, Univ of Memphis – Yd
Smart, Kevin J., (405) 325-3157 kjsmart@ou.edu, Univ of Oklahoma – Gc
Smayda, Theodore J., (401) 874-6171 tsmayda@gso.uri.edu, Univ of Rhode Island – Ob
Smedes, Harry W., (541) 552-6479 Southern Oregon Univ – Gi
Smethie, William, (845) 365-8566 Columbia Univ – Cg
Smilnak, Roberta A., (303) 556-3144 Metropolitan State College of Denver – Og
Smirnov, Aleksey K., (906) 487-2365 asmirnov@mtu.edu, Michigan Technological Univ – Yv
Smirnov, Anna, (418) 654-3711 asmirnov@nrcan.gc.ca, Universite du Quebec – Cs
Smith, Alan L., (909) 537-5409 9095375112, California State Univ, San Bernardino – Gv
Smith, Albert C., (904) 227-1121 Univ of South Florida, St Petersburg – Ob
Smith, Alison J., (330) 672-3709 alisonjs@kent.edu, Kent State Univ – Gn
Smith, Arthur, (610) 436-3335 West Chester Univ – Oe
Smith, Betty E., 217-581-6340 cfbes@eiu.edu, Eastern Illinois Univ – Oi
Smith, C. Ken, (931) 598-3219 ksmith@sewanee.edu, Univ of the South – Sf
Smith, Carl, 304 594 2331 smith@geosrv.wvnet.edu, West Virginia Univ – Gg
Smith, Catherine H., (505) 667-0113 chsmith@lanl.gov, Los Alamos National Lab – Ca
Smith, Cathy, (831) 459-2504 csmith@pmc.ucsc.edu, Univ of California, Santa Cruz – Oo
Smith, Cathy, 573-368-2167 nrsmitc@mail.dnr.state.mo.us, Missouri Dept of Natural Resources – Oo
Smith, Charles C., 205/247-3609 ccsmith@gsa.state.al.us, Geological Survey of Alabama – Pg
Smith, Charles K., (210) 733-2840 csmith55@mail.accd.edu, San Antonio College – Oy
Smith, Charles T., (302) 831-0692 31438@udel.edu, Univ of Delaware – Gg
Smith, Craig R., (808) 956-8623 csmith@soest.hawaii.edu, Univ of Hawai'i, Manoa – Ob
Smith, David C., (702) 895-3971 Missouri Dept of Nat Resources – Gg
Smith, David C., (401) 874-6172 dcsmith@gsosun1.gso.uri.edu, Univ of Rhode Island – Ob
Smith, David E., (804) 982-3058 des3e@virginia.edu, Univ of Virginia – Ob
Smith, David R., (410) 293-6553 drsmith@usna.edu, United States Naval Academy – Ow
Smith, Debbie, 609-258-5809 debbie@princeton.edu, Princeton Univ – Gg
Smith, Deborah A., (609) 258-5809 debbie@princeton.edu, Princeton Univ – Oo
Smith, Deborah K., (508) 289-2472 dsmith@whoi.edu, Woods Hole Oceanographic Institution – Yr
Smith, Dena M., 303-735-2011 dena.smith@colorado.edu, Univ of Colorado – Pg
Smith, Dena M., 303-735-2011 dena@colorado.edu, Univ of Colorado – Po
Smith, Diane R., (210) 999-7656 dsmith@trinity.edu, Trinity Univ – Gi
Smith, Dina, 617-552-8300 dina.smith.1@bc.edu, Boston College – Ys
Smith, Donald R., (831) 459-5041 dsmith@biology.ucsc.edu, Univ of California, Santa Cruz – Oo
Smith, Dorian G. W., (780) 492-3955 dorian.smith@ualberta.ca, Univ of Alberta – Gz
Smith, Douglas, (512) 471-4261 doug@maestro.geo.utexas.edu, Univ of Texas at Austin, Jackson School of Geosciences – Gi
Smith, Douglas L., (210) 458-5751 Univ of Florida – Yh
Smith, Edward C., 217-244-2773 esmith@isgs.uiuc.edu, Illinois State Geological Survey – Hw
Smith, Eugene I., (702) 895-3971 gsmith@ccmail.nevada.edu, Univ of Nevada, Las Vegas – Gi
Smith, Florence P., (630) 252-7980 Argonne National Lab – Ca
Smith, G W., 724 938-4463 smith_gw@cup.edu, California Univ of Pennsylvania – Og
Smith, Gary A., (505) 277-2348 gsmith@unm.edu, Univ of New Mexico – Gs
Smith, Gerald J., (716) 645-6800 (Ext. 2470) stratigrapher@msn.com, SUNY, Buffalo – Gs
Smith, Gerald R., (734) 764-0491 grsmith@umich.edu, Univ of Michigan – Ps
Smith, Gordon W., (765) 494-7681 McGill Univ – Nx
Smith, H D., SmithHD@cf.ac.uk, Cardiff Univ – Oo
Smith, J. B., 205-247-3623 jbsmith@gsa.state.al.us, Geological Survey of Alabama – Oo
Smith, J. Leslie, (604) 822-4108 lsmith@eos.ubc.ca, Univ of British Columbia – Hw
Smith, Jacqueline A., 518-485-3756 smithj@strose.edu, The College of Saint Rose – GlmGe
Smith, James A., (434) 924-7991 jas9e@virginia.edu, Univ of Virginia – Hy
Smith, James A., (609) 258-4615 jsmith@radap.princeton.edu, Princeton Univ – Hg
Smith, James E., (905) 525-9140 (Ext. 24534) smithja@mcmaster.ca, Univ of Arizona – Hw
Smith, James E., (905) 525-9140 (Ext. 24534) smithja@mcmail.cis.mcmaster.ca, McMaster Univ – Hw
Smith, Janet S., (717) 477-1757 jssmit@ship.edu, Shippensburg Univ – Oi
Smith, Janice L., (330) 672-3709 jan@austin.cc.tx.us, Austin Community College – Gp
Smith, Jeffery L., (509) 335-7648 Washington State Univ – Sb
Smith, Jennifer R., 314-935-9451 jensmith@levee.wustl.edu,

Washington Univ – Ga
Smith, Jim, +353 21 4902151 j.smith@ucc.ie, Univ College Cork – Pl
Smith, Jim A., (702) 895-3971 Princeton Univ – Sp
Smith, Joseph V., (773) 702-8110 Univ of Chicago – Gz
Smith, K. L., (702) 895-3971 Wichita State Univ – Oo
Smith, Karen E., (330) 672-2680 geology@kent.edu, Kent State Univ – Oo
Smith, Karl A., (805) 756-2262 ksmith@tc.umn.edu, Univ of Minnesota, Twin Cities – Nx
Smith, Ken D., (775) 784-4218 ken@seismo.unr.edu, Univ of Nevada, Reno – Ys
Smith, Langhorne, 518-473-6262 lsmith@mail.nysed.gov, New York State Geological Survey – Go
Smith, Larry N., 406-496-4379 lsmith@mtech.edu, Montana Tech of The Univ of Montana – Hw
Smith, Lisa R., 217-244-2503 smith@isgs.uiuc.edu, Illinois State Geological Survey – Gg
Smith, Martha Gentry, (931) 372-3121 mgsmith@tntech.edu, Tennessee Technological Univ – Oo
Smith, Matthew, (352) 392-2106 msmith@geology.ufl.edu, Univ of Florida – Gi
Smith, Michael E., (707) 664-1976 mike.smith@sonoma.edu, Sonoma State Univ – Gd
Smith, Michael S., (910) 962-3496 smithms@uncw.edu, Univ of North Carolina Wilmington – Gz
Smith, Ned P., (407) 674-8096 Florida Institute of Technology – Op
Smith, Norman D., (402) 472-2663 nsmith3@unl.edu, Univ of Nebraska, Lincoln – Gs
Smith, Pamela, (702) 895-3971 Univ of Arizona – Oo
Smith, Paul E., (619) 546-7169 Univ of California, San Diego – Ob
Smith, Paul H., (505) 667-7494 Los Alamos National Lab – Oo
Smith, Paul L., (604) 822-6456 psmith@eos.ubc.ca, Univ of British Columbia – Pg
Smith, Paul L., 605-394-1990 Paul.Smith@sdsmt.edu, South Dakota School of Mines & Technology – Oo
Smith, Peter, (520) 621-2725 psmith@lpl.arizona.edu, Univ of Arizona – Or
Smith, Peter C., (902) 426-3474 smithp@mar.dfo-mpo.gc.ca, Dalhousie Univ – Op
Smith, Peter J., (780) 492-0343 peter.smith@ualberta.ca, Univ of Alberta – Oo
Smith, Phillip J., (765) 494-3286 pjsmith@purdue.edu, Purdue Univ – Oa
Smith, Richard P., (610) 436-3335 Lockheed Martin Idaho Tech – Gt
Smith, Richard S., (479) 575-3355 Univ of Arkansas, Fayetteville – Oy
Smith, Robert B., (801) 581-7129 rbsmith@mines.utah.edu, Univ of Utah – Ys
Smith, Robert C., 717.702.2021 roberts@state.pa.us, Pennsylvania Bureau of Topographic & Geologic Survey – Cg
Smith, Robert K., (210) 458-4455 robert.smith1@utsa.edu, The Univ of Texas at San Antonio – Gz
Smith, Robert L., (541) 737-2926 rsmith@coas.oregonstate.edu, Oregon State Univ – Op
Smith, Robert W., (210) 736-7656 Lockheed Martin Idaho Technologies – Cg
Smith, Ronald B., (203) 432-3129 ronald.smith@yale.edu, Yale Univ – Oa
Smith, Ronald M., (509) 376-5831 rmsmith@pnl.gov, Pacific Northwest National Lab – Hg
Smith, Shane V., 330-941-1752 svsmith@ysu.edu, Youngstown State Univ – Gs
Smith, Stephen V., (808) 956-8693 svsmith@soest.hawaii.edu, Centro de Inv Cientifica y de Educación Superior de Ensenada – Cm
Smith, Steven J., 319-273-6495 smithsj@uni.edu, Univ of Northern Iowa – Og
Smith, Stewart W., 425-823-1893 stew@ess.washington.edu, Univ of Washington – Ys
Smith, Susan M., 406-496-4173 ssmith@mtech.edu, Montana Tech of The Univ of Montana – Oo
Smith, Terence E.tsmith@uwindsor.ca, Univ of Windsor – Gi
Smith, Terry L., (805) 756-2262 tsmith@calpoly.edu, California Polytechnic State Univ – So
Smith, Theodore C., (210) 458-5751 California Geological Survey – Oo
Smith, Thomas M., (804) 924-3107 tms9a@virginia.edu, Univ of Virginia – Og
Smith, W. John, (215) 898-7889 wjsmith@sas.upenn.edu, Univ of Pennsylvania – Py
Smith, William G., 404-656-3214 Georgia Dept of Nat Resources – Hw
Smith, William H., (314) 935-5638 whsmith@dasi.wustl.edu, Washington Univ – Xc
Smith, William L., (757)864-8914 bill.l.smith@larc.nasa.gov, Univ of Wisconsin, Madison – Or
Smith, III, Thomas J., (305) 348-1267 Univ of South Florida, Saint Petersburg – Oo
Smith-Engle, Jennifer M., (361) 825-2436 jsengle@falcon.tamucc.edu, Texas A&M Univ, Corpus Christi – Gs
Smith-Engle, Jennifer M., (361) 825-2436 jsengle@tamucc.edu, Texas A&M Univ, Corpus Christi – Gs
Smith-Konter, Bridget R., (915) 747-6118 bkonter@geo.utep.edu, Univ of Texas, El Paso – YgOr
Smithson, Scott B., (307) 766-5280 sbs@uwyo.edu, Univ of Wyoming – Ys
Smosna, Richard A., (304) 293-5603 rsmosna@wvu.edu, West Virginia Univ – Gr
Smothers, Kent W., (217) 333-6167 Illinois State Water Survey – Cg
Smoyer, Karen E., (780) 492-3287 karen.smoyer@ualberta.ca, Univ of Alberta – Oo
Smrekar, Suzanne E., (818) 354-4192 Jet Propulsion Lab – Xy
Smylie, Douglas E., (416) 736-2100 (Ext. 66438) doug@core.yorku.ca, York Univ – Yg
Smyth, Joseph R., 303-492-5521 joseph.smyth@colorado.edu, Univ of Colorado – Gz
Smyth, Rebecca C., (512) 471-0232 rebecca.smyth@beg.utexas.edu, Univ of Texas at Austin, Jackson School of Geosciences – Hg
Smyth, Thomas J., (919) 515-2838 North Carolina State Univ – Sc
Smyth, William D., (541) 737-3029 smyth@coas.oregonstate.edu, Oregon State Univ – Op
Smythe, William, (310) 825-2434 Univ of California, Los Angeles – Ys
Snead, John I., (225) 578-3454 snead@lsu.edu, Louisiana State Univ – Gm
Snider, Henry I., (860) 228-9815 snider@easternct.edu, Eastern Connecticut State Univ – Ge
Snieder, Roel, 303-273-3456 rsnieder@mines.edu, Colorado School of Mines – Ye
Snodgrass, Eric R., (217) 333-3537 snodgrss@atmos.uiuc.edu, Univ of Illinois, Urbana-Champaign – Oa
Snoeckx, Hilde, 850-474-2746 snoeckx@uwf.edu, Univ of West Florida – Ou
Snoke, Arthur W., (307) 766-3386 snoke@uwyo.edu, Univ of Wyoming – Gc
Snoke, J. Arthur, (540) 231-6028 snoke@vt.edu, Virginia Polytechnic Institute & State Univ – Ys
Snow, Eleanour, (813) 974-0319 snow@chuma.cas.usf.edu, Univ of South Florida, Tampa – Gz
Snow, Jonathon, 713-743-3399 jsnow@uh.edu, Univ of Houston – Ca
Snow, Julie A., (724) 738-2503 julie.snow@sru.edu, Slippery Rock Univ – Oa
Snyder, George H., (561) 993-1500 ghs@gnv.ifas.ufl.edu, Univ of Florida – Sc
Snyder, Jeffrey A., (419) 372-0533 jasnyd@bgnet.bgsu.edu, Bowling Green State Univ – Gm
Snyder, Lori D., (715) 836-5086 snyderld@uwec.edu, Univ of Wisconsin, Eau Claire – Gx
Snyder, Peter K., (217) 244-0928 pksnyder@atmos.uiuc.edu, Univ of Illinois, Urbana-Champaign – Oa
Snyder, Victor, (809) 767-9705 Univ of Puerto Rico – Sp
Snyder, Walter S., (208) 426-3645 wsnyder@boisestate.edu, Boise State Univ – Gr
Sobel, Adam H., (212) 854-6587 Columbia Univ – Oa
Sobol, John A., (901) 682-4147 Univ of Memphis – Oo
Soja, Constance M., (315) 228-7200 csoja@mail.colgate.edu, Colgate Univ – Pi
Sokolova, Elena, (204) 747-8252 elena_sokolova@umanitoba.ca, Univ of Manitoba – Gz
Solar, Gary S.solargs@bscmail.buffalostate.edu, SUNY, Buffalo – Gc
Solar, Gary S., (716) 878-4900 solargs@buffalostate.edu, Buffalo State College – GcpGt
Solecki, William, (973) 655-5129 soleckiw@mail.montclair.edu, Montclair State Univ – Ou
Solecki, William, 212-7724536 wsolecki@hunter.cuny.edu, Hunter College (CUNY) – Ou
Soll, Wendy E., (505) 665-6930 weasel@lanl.gov, Los Alamos National Lab – Hw
Solomatov, Viatcheslav S.slava@nmsu.edu, New Mexico State Univ, Las Cruces – Yg
Solomatov, Viatcheslav S., 314-935-7882 slava@levee.wustl.edu,

Washington Univ – Yg
Solomon, Barry J., (801) 537-3388 nrugs.bsolomon@state.ut.us, Utah Geological Survey – Ng
Solomon, Douglas K., (801) 581-7231 ksolomon@mines.utah.edu, Univ of Utah – Hw
Solomon, Sean C., (202) 478-8850 scs@dtm.ciw.edu, Carnegie Institution of Washington – Yg
Solomos, Theophanes, (301) 405-4348 ts@umail.umd.edu, Univ of Maryland – Oo
Soloviev, Alexander V., (954) 262-3659 soloviev@ocean.nova.edu, Nova Southeastern Univ – Op
Somasundaran, Ponisseril, (212) 854-2926 ps24@columbia.edu, Columbia Univ – Nx
Somayazulu, Maddury, (202) 478-8911 zulu@gl.ciw.edu, Carnegie Institution of Washington – Yx
Somers, Lee H., (313) 936-0518 Univ of Michigan – Oo
Somers, Jr., Arnold E., (229) 333-5664 gsomers@valdosta.edu, Valdosta State Univ – Oo
Somerville, Richard C., (619) 534-4644 rsomerville@ucsd.edu, Univ of California, San Diego – Oa
Sommerfield, Christopher K., 302-645-4255 cs@udel.edu, Univ of Delaware – On
Sommerfield, Christopher K., 302-645-4255 cs@udel.edu, Univ of Delaware – Gs
Sonder, Leslie J., 603-646-2372 Leslie.Sonder@dartmouth.edu, Dartmouth College – Gq
Sondergeld, Carl H., (405) 325-6866 csondergeld@ou.edu, Univ of Oklahoma – Np
Sonel, Nurettinsonel@eng.ankara.edu.tr, Ankara Univ – Go
Sonett, Charles P., (520) 621-6935 sonett@dakotacom.net, Univ of Arizona – Yg
Song, Qingtao, (541) 737-6396 qsong@coas.oregonstate.edu, Oregon State Univ – Op
Song, Xiaodong, (217) 333-1841 xsong@uiuc.edu, Univ of Illinois, Urbana-Champaign – Ys
Sonzogni, William C., (608) 262-8062 Univ of Wisconsin, Madison – Oc
Soong, Yin S., (717) 871-2432 Yin.Soong@millersville.edu, Millersville Univ – Op
Sophocleous, Marios A., (785) 864-2113 marios@kgs.ku.edu, Univ of Kansas – Hw
Sorauf, James E., (607) 777-2561 Binghamton Univ – Pi
Soreghan, Gerilyn S., (405) 325-4482 lsoreg@ou.edu, Univ of Oklahoma – Gs
Soreghan, Michael J., (405) 325-3393 msoreg@ou.edu, Univ of Oklahoma – Gs
Sorensen, Sorena S., (202) 633-1820 sorensens@si.edu, Smithsonian Institution / National Museum of Natural History – Gp
Sorlien, Chris C.chris@crustal.ucsb.edu, Univ of Missouri – Yg
Sorlien, Christopher C., 845-359-7631 chris@crustal.ucsb.edu, Univ of California, Santa Barbara – Yr
Sorooshian, Soroosh, (520) 621-1661 soroosh@hwr.arizona.edu, Univ of California, Irvine – Hq
Soster, Frederick M., 765-658-4670 fsoster@depauw.edu, DePauw Univ – Gs
Sottile, Jr., Joseph, (606) 257-4616 Univ of Kentucky – Nm
Souch, Catherine J., (317) 274-1103 Indiana Univ / Purdue Univ, Indianapolis – Pe
Soufiane, Azzedine, (418) 654-3704 ess-ins-asoufiane@x1.nrcan.gc.ca, Universite du Quebec – Pl
Soule, Peter T., (828) 262-7056 soulept@appstate.edu, Appalachian State Univ – Oa
Soule, S. Adam, (508) 289-3213 ssoule@whoi.edu, Woods Hole Oceanographic Institution – Gu
Sousa, Wayne P., (510) 642-2435 Univ of California, Berkeley – Ob
Sousounis, Peter J., (313) 936-0488 Univ of Michigan – Oa
Southam, Gordon, (519) 661-3197 gsoutham@uwo.ca, Univ of Western Ontario – Py
Southam, John R., (305) 284-1898 Univ of Miami – Op
Southard, John B., (617) 253-3397 southard@mit.edu, Massachusetts Institute of Technology – Gs
Southard, Randal J., (530) 752-7041 rjsouthard@ucdavis.edu, Univ of California, Davis – So
Sowers, Todd, (814) 863-8093 sowers@geosc.psu.edu, Pennsylvania State Univ, Univ Park – Cs
Spalding, Brian P., bps@ornl.gov, Oak Ridge National Lab – Sc
Spane, Frank A., (509) 376-8329 frank.spane@pnl.gov, Pacific Northwest National Lab – Hg
Spang, John H., (979) 845-8972 spang@geo.tamu.edu, Texas A&M Univ – Gt
Spangler, Daniel P., (352) 392-2106 Univ of Florida – Hw
Spann, Louise, (801) 585-9689 lspann@egi.utah.edu, Univ of Utah – Gz
Spanos, J. Timothy, (780) 492-5892 tim@phys.ualberta.ca, Univ of Alberta – Ys
Sparkes, Sydney, 205/247-3554 ssparkes@gsa.state.al.us, Geological Survey of Alabama – Hw
Sparkman, Teri, 615-322-2976 teri.pugh@vanderbilt.edu, Vanderbilt Univ – Oo
Sparks, David, (979) 458-1051 sparks@geo.tamu.edu, Texas A&M Univ – Oo
Spayd, Steven E., (609) 984-6587 steve.spayd@dep.state.nj.us, New Jersey Geological Survey – Hw
Speakman, LuElla, 202-633-1790 speakmanl@si.edu, Smithsonian Institution / National Museum of Natural History – Gg
Spear, Frank S., (518) 276-6103 spearf@rpi.edu, Rensselaer Polytechnic Institute – Gp
Spear, Steven G., (760) 744-1150 (Ext. xt. 2513) sspear@palomar.edu, Palomar College – Gg
Speck, Angela, (573) 882-8371 speckan@missouri.edu, Univ of Missouri – Xc
Specker, Jennifer, (401) 874-6858 jspecker@gsosun1.gso.uri.edu, Univ of Rhode Island – Ob
Speece, Marvin A., (406) 496-4188 mspeece@mtech.edu, Montana Tech of the Univ of Montana – Ye
Speed, Don, (602) 285-7244 Phoenix College – Gg
Speer, James, (812) 237-2257 gespeer@isugw.indstate.edu, Indiana State Univ – Pe
Speer, Kevin G., (850) 645-4846 speer@ocean.fsu.edu, Florida State Univ – Op
Speer, Roberta D., (806) 651-2582 rspeer@mail.wtamu.edu, West Texas A&M Univ – Og
Speidel, David H., (718) 997-3323 Graduate School of the City Univ of New York – Cg
Speidel, David H., (718) 997-3323 speidel@qcvaxa.acc.qc.edu, Queens College (CUNY) – Cg
Spell, Terry L., (702) 895-1171 tspell@nevada.edu, Univ of Nevada, Las Vegas – Cc
Spence, Dan, 859-572-5115 spence@nku.edu, Northern Kentucky Univ – Oo
Spence, George D., (250) 721-6187 gspence@uvic.ca, Univ of Victoria – Ys
Spencer, Edgar W., (540) 463-8866 spencere@wlu.edu, Washington & Lee Univ – Gc
Spencer, Joel Q., (785) 532-2249 joelspen@ksu.edu, Kansas State Univ – CcGs
Spencer, Jon E., (520) 770-3500 Arizona Geological Survey – Gg
Spencer, Kathleen, Kathy_Spencer@mail.und.nodak.edu, Univ of North Dakota – Oo
Spencer, Larry T., (804) 683-5189 lts@oz.plymouth.edu, Plymouth State College – Gg
Spencer, Mary, (859) 257-5730 mary.spencer@uky.edu, Univ of Kentucky – Oo
Spencer, Matt, (906) 635-2085 mspencer@lssu.edu, Lake Superior State Univ – Gl
Spencer, Patrick K., (509) 527-5222 spencerp@whitman.edu, Whitman College – Pg
Spencer, Ronald J., (403) 220-6447 spencer@geo.ucalgary.ca, Univ of Calgary – Cg
Spencer, Steven M., (850) 488-9380 Florida Geological Survey – Eg
Spera, Frank J., (805) 893-4880 Univ of California, Santa Barbara – Gi
Spero, Howard J., (530) 752-3307 spero@geology.ucdavis.edu, Univ of California, Davis – Pe
Spetz, Dennis L., (502) 852-6844 Univ of Louisville – Oo
Spetzler, Hartmut A., 303-492-6715 spetzler@colorado.edu, Univ of Colorado – Ys
Spiegelman, Marc W., (845) 365-8425 mspieg@ldeo.columbia.edu, Columbia Univ – Yg
Spieler, Oliver, 089/2180 4221 spieler@min.uni-muenchen.de, LMU - Univ of Munich – Gz
Spieler, Richard C., (954) 262-3613 spielerr@nova.edu, Nova Southeastern Univ – Ob
Spiesberger, John L., (215) 573-5388 johnsr@sas.upenn.edu, Univ of Pennsylvania – Ob
Spiess, Fred N., (619) 534-1621 fns@mpl.ucsd.edu, Univ of California, San Diego – Yr
Spilde, Michael N., (505) 277-5430 mspilde@unm.edu, Univ of New Mexico – Gz

Spilker, Linda J., (818) 354-1647 linda.j.spilker@jpl.nasa.gov, Jet Propulsion Lab – Xy
Spindel, Robert C., (206) 543-1310 spindel@apl.washington.edu, Univ of Washington – Op
Spinelli, Glenn, 505.835.6512 spinellig@missouri.edu, New Mexico Institute of Mining and Technology – Hw
Spinosa, Claude, (208) 426-5905 cspinosa@boisestate.edu, Boise State Univ – Pi
Spittler, Thomas E., (707) 576-2949 California Geological Survey – Oo
Spitz, Yvette H., (541) 737-3227 yspitz@coas.oregonstate.edu, Oregon State Univ – Ob
Spivak, S. T., (807) 343-8797 sam.spivak@lakeheadu.ca, Lakehead Univ – Oo
Spongberg, Alison L., (419) 530-4091 alison.spongberg@utoledo.edu, Univ of Toledo – Co
Spooner, Edward T. C., (416) 978-3280 etcs@geology.utoronto.ca, Univ of Toronto – En
Spooner, Ian S., (902) 585-1312 ian.spooner@acadiau.ca, Acadia Univ – Ge
Spotila, James A., (540) 231-2109 spotila@vt.edu, Virginia Polytechnic Institute & State Univ – Gt
Spratt, Deborah A., (403) 220-6446 spratt@geo.ucalgary.ca, Univ of Calgary – Gc
Spray, John G., (506) 453-3550 jgs@unb.ca, Univ of New Brunswick – Gp
Sprenke, Kenneth F., (208) 885-5791 ksprenke@uidaho.edu, Univ of Idaho – Ye
Springer, Abraham E., (928) 523-7198 abe.springer@nau.edu, Northern Arizona Univ – Hw
Springer, Dale A., (570) 389-4747 dspringer@bloomu.edu, Bloomsburg Univ – PivOe
Springer, Everett P., (505) 667-0569 everetts@lanl.gov, Los Alamos National Lab – Hq
Springer, Gregory S., (740) 593-9436 springeg@ohio.edu, Ohio Univ, Lancaster – Gm
Springer, Kathleen B., (909) 307-2669 242 kspringer@sbcm.sbcounty.gov, San Bernardino County Museum – Og
Springer, Robert K., 204-727-9686 springer@brandonu.ca, Brandon Univ – Gi
Springer, Scott, (541) 737-2368 springer@coas.oregonstate.edu, Oregon State Univ – Op
Sprinkle, James T., (512) 471-4264 echino@mail.utexas.edu, Univ of Texas at Austin, Jackson School of Geosciences – Pi
Spruill, Richard K., (252) 328-4399 spruillr@mail.ecu.edu, East Carolina Univ – Hw
Spry, Paul G., (515) 294-9637 pgspry@iastate.edu, Iowa State Univ of Science & Technology – Em
Spudis, Paul D., (713) 486-2139 Rice Univ – Xg
Squelch, Andrew P., +61 8 6436 8725 A.Squelch@curtin.edu.au, Curtin Univ of Technology – Nmr
Squires, Richard L., (818) 677-2514 richard.squires@csun.edu, California State Univ, Northridge – Pg
Squyres, Steven W., (607) 255-3508 sws6@cornell.edu, Cornell Univ – Xg
Srimal, Neptune, (305) 919-5969 srimal@fiu.edu, Florida International Univ – Gt
Srinivasan, Gopalan, (416) 946-0278 srini@geology.utoronto.ca, Univ of Toronto – Ow
Sritharan, Subramania I., (937) 376-6275 Central State Univ – Hg
Srivastava, Ramesh C., (773) 702-8125 srivast@geosci.uchicago.edu, Univ of Chicago – Oa
Srogi, LeeAnn, (610) 436-2721 esrogi@wcupa.edu, West Chester Univ – Gp
St. Antoine, Mary, (906) 635-2075 mstantoine@lssu.edu, Lake Superior State Univ – Cg
St. Jean, Joseph, (919) 966-4516 jstjean@email.unc.edu, Univ of North Carolina, Chapel Hill – Pm
St. John, James C., (404) 894-1754 jim.stjohn@eas.gatech.edu, Georgia Institute of Technology – Oa
St. John, Kristen E., 540-568-6675 stjohnke@jmu.edu, James Madison Univ – Ou
Staff, George M., (512) 388-3682 gstaff@austin.cc.tx.us, Austin Community College – Pe
Stafford, C. Russell, (812) 237-3989 anstff@isugw.indstate.edu, Indiana State Univ – Ga
Stahl, Peter D., (307) 776-2179 Univ of Wyoming – Sb
Stahl, Susansusans, Athabasca Univ – Oo
Stahl, Terry L., (202) 478-8870 Carnegie Inst of Washington – Oo
Stahle, David W., (479) 575-3703 dstahle@comp.uark.edu, Univ of Arkansas, Fayetteville – Pb
Stalnaker, Jack, 337-482-6851 jstalnaker@louisiana.edu, Univ of Louisiana, Lafayette – Yg
Stamm, Alfred J., (315) 312-2806 stamm@oswego.edu, SUNY, Oswego – Ow
Stancyk, Stephen E., (803) 777-3944 Univ of South Carolina – Ob
Standlee, Larry, (817) 272-2987 standla@uta.edu, Univ of Texas, Arlington – Gc
Standridge, Debbiedas@canes.gsw.edu, Georgia Southwestern State Univ – Oo
Staneva, Marieta P., (814) 949-5258 mps5@psu.edu, Pennsylvania State Univ, Altoona – Oy
Stanford, Loudon R., (208) 885-7479 Univ of Idaho – Gm
Stanford, Scott D., 609-292-2576 scott.stanford@dep.state.nj.us, New Jersey Geological Survey – Gl
Stanley, Clifford R., (902) 585-1344 cliff.stanley@acadiau.ca, Acadia Univ – Cg
Stanley, Craig D., (941) 751-7636 cds@mail.ifas.ufl.edu, Univ of Florida – Sp
Stanley, Daniel J., (202) 633-1354 Smithsonian Institution / National Museum of Natural History – Ou
Stanley, Rolfe S., (802) 656-0247 rstanley@zoo.uvm.edu, Univ of Vermont – Gc
Stanley, Steven M., (410) 516-7042 stanley@jhu.edu, The Johns Hopkins Univ – Po
Stanley, Jr., George D., (406) 243-5693 fossil@selway.umt.edu, Univ of Montana – Pi
Stanton, Heather I., (970) 267-2159 Heather_Stanton@partner.nps.gov, Colorado State Univ – Gg
Stanton, Stephen, (573) 882-4860 Univ of Missouri – Oo
Stanton, Timothy P., (831) 656-3144 stanton@nps.navy.mil, Naval Postgraduate School – Op
Stapleton, Michael G., (724) 738-2495 micheal.stapleton@sru.edu, Slippery Rock Univ – So
Stapor, Jr., Frank W., (931) 372-3695 fstapor@tntech.edu, Tennessee Technological Univ – Gs
Stark, Colin, (845) 365-8742 Columbia Univ – Gg
Stark, Robert, 089/2180 4329 stark@lrz.uni-muenchen.de, LMU - Univ of Munich – Gz
Starkes, Geraldine, (709) 737-8142 gstarkes@esd.mun.ca, Memorial Univ of Newfoundland – Oo
Starks, Tammy D., (501) 569-3546 tdstarks@ualr.edu, Univ of Arkansas, Little Rock – Oo
Starnes, James E., (601) 961-5714 Mississippi Office of Geology – Gg
Starrs, Paul F., (775) 784-6930 starrs@unr.edu, Univ of Nevada, Reno – Oy
Stauffer, David R., (814) 863-3932 stauffer@essc.psu.edu, The Pennsylvania State Univ, Univ Park – Ow
Stauffer, Mel R., (306) 966-5708 mel.stauffer@usask.ca, Univ of Saskatchewan – Gc
Stead, Douglas, (604) 291-5387 Simon Fraser Univ – Ng
Steadman, David W., (352) 392-1721 (Ext. 464) steadman@flmnh.ufl.edu, Univ of Florida – Pq
Stearley, Ralph F., (616) 526-6370 rstearle@calvin.edu, Calvin College – PvOgGh
Stearns, Charles R., (608)262-0780 chucks@ssec.wisc.edu, Univ of Wisconsin, Madison – Ow
Stearns, David W., (405) 325-3253 Univ of Oklahoma – Gc
Stearns, Richard G., 615-322-2976 stearnrg@ctrvax.vanderbilt.edu, Vanderbilt Univ – Yg
Stebbins, Jonathan F., (650) 723-1140 stebbins@pangea.stanford.edu, Stanford Univ – Cg
Stebbins, Robert H., (804) 371-3242 J. Sargeant Reynolds Community College – Oo
Steck, Lee, (505) 665-3528 Los Alamos National Lab – Gg
Steckler, Michael, (845) 365-8479 Columbia Univ – Yv
Steed, G M., steed@cf.ac.uk, Cardiff Univ – Gg
Steel, Ronald J., (512) 471-0954 rsteel@mail.utexas.edu, Univ of Texas at Austin, Jackson School of Geosciences – Gs
Steele, Andrew, 202 478 8974 a.steele@gl.ciw.edu, Carnegie Institution of Washington – Xy
Steele, Carol S., (727) 553-3975 csteele@marine.usf.edu, Univ of South Florida, Saint Petersburg – Oo
Steele, Kenneth F., (479) 575-7937 ksteele@uark.edu, Univ of Arkansas, Fayetteville – Cg
Steele, Michael A., (570) 408-4763 Wilkes Univ – Py
Steenhuis, Tammo S., (607) 255-2489 tss1@cornell.edu, Cornell Univ – So
Steeples, Donald W., (785) 864-2730 don@ku.edu,

Univ of Kansas – Ys
Steer, David N., (330) 972-2099 dsteer@uakron.edu, Univ of Akron – Ys
Stefanova, Ivanka, 651-646-0665 stefa014@umn.edu, Univ of Minnesota, Twin Cities – Pl
Steffem, Konrad, (303) 492-4524 Univ of Colorado – Or
Steffens, Katja, 089/2180 6533 katja.steffens@lmu.de, LMU - Univ of Munich – Gg
Steidinger, Karen A., (727) 896-8626 Univ of South Florida, Saint Petersburg – Ob
Steidl, Gregg M., (609) 984-6587 gregg.steidl@dep.state.nj.us, New Jersey Geological Survey – Ge
Steidl, Jamison H., (805) 893-4905 steidl@crustal.ucsb.edu, Univ of California, Santa Barbara – Ys
Steidtmann, James R., (307) 766-2739 steidt@uwyo.edu, Univ of Wyoming – Gr
Steig, Eric J., 206-685-3715 steig@u.washington.edu, Univ of Washington – CsOl
Steiger, Scott, 315 312 2802 steiger@oswego.edu, SUNY, Oswego – Ow
Stein, Carol A., (312) 996-9349 carol.a.stein@uic.edu, Univ of Illinois, Chicago – Yg
Stein, Holly J., (970) 491-4318 hstein@cnr.colostate.edu, Colorado State Univ – Em
Stein, Seth, (847) 491-5265 seth@earth.northwestern.edu, Northwestern Univ – Ys
Steinat, Rosemary, 202-633-2480 Smithsonian Institution / National Air & Space Museum – Oo
Steinberg, Roger T., (361) 698-1385 rsteinb@delmar.edu, Del Mar College – Gr
Steineck, Paul L., (914) 251-6646 SUNY, Purchase – Pm
Steiner, Jeffrey, (212) 650-6465 City College (CUNY) – Cp
Steiner, Jeffrey, (212) 650-6465 Graduate School of the City Univ of New York – Gx
Steiner, Maureen B., (307) 766-6584 magnetic@uwyo.edu, Univ of Wyoming – Ym
Steinfeldt, Joel A., 217-244-2761 steinfeldt@isgs.uiuc.edu, Illinois State Geological Survey – Gg
Steinker, Don C., (419) 372-7200 Bowling Green State Univ – Po
Steinmetz, John C., (812) 855-5067 jsteinm@indiana.edu, Indiana Geological Survey – Pm
Stelck, Charles R., (780) 492-3266 Univ of Alberta – Ps
Stelling, Pete, 907 786-6889 pstelling@uaa.alaska.edu, Univ of Alaska, Anchorage – Gv
Steltenpohl, Mark G., (334) 844-4893 steltmg@auburn.edu, Auburn Univ – Gt
Stelzer, Diane, (715) 346-2629 dstelzer@uwsp.edu, Univ of Wisconsin, Stevens Point – Oo
Steneck, Robert S., (207) 563-3146 Univ of Maine – Ob
Stenstrom, Richard C., (608) 363-2289 stenstrom@beloit.edu, Beloit College – Ge
Stephen, Ralph A., (508) 289-2583 rstephen@whoi.edu, Woods Hole Oceanographic Institution – Ys
Stephens, Beatriz R., (706) 542-2652 beatrice@gly.uga.edu, Univ of Georgia – Oo
Stephens, Daniel B., (505) 822-9400 dbsteph@dbstephens.com, New Mexico Institute of Mining and Technology – Hw
Stephens, Graeme L., (970) 491-8550 stephens@atmos.colostate.edu, Colorado State Univ – Oa
Stephens, James D., (208) 282-3365 Idaho State Univ – Gz
Stephens, Nicolenicostep@nova.edu, Nova Southeastern Univ – Oo
Stephens, Pam K., (859) 257-6933 pkstep0@email.uky.edu, Univ of Kentucky – Oo
Stephens, Pamela A., 940-397-4448 pamela.stephens@mwsu.edu, Midwestern State Univ – Pm
Stephenson, David A., (307) 734-4432 dave.stephenson@centergeo.org, Univ of Wyoming – Hy
Steppe, Cecily N., 410-293-6558 natunewi@usna.edu, United States Naval Academy – Ob
Stern, Charles R., 303-492-7170 Univ of Colorado – Gi
Stern, David A., 516-572-7275 Nassau Community College – Ge
Stern, Herschel I., (619) 757-2121 MiraCosta College – Oy
Stern, Melvin E.stern@ocean.fsu.edu, Florida State Univ – Op
Stern, Robert J., (972) 883-2442 rjstern@utdallas.edu, Univ of Texas, Dallas – Gt
Sternberg, Ben K., (520) 621-2439 bkslasi@u.arizona.edu, Univ of Arizona – Ye
Sternberg, Richard W., (206) 543-6487 rws@ocean.washington.edu, Univ of Washington – Ou
Sternberg, Robert S., (717) 291-4134 rob.sternberg@fandm.edu, Franklin and Marshall College – Ym
Sternberg, Rolf, (973) 655-7386 Montclair State Univ – Oo
Stetler, Larry D., (605) 394-2464 Larry.Stetler@sdsmt.edu, South Dakota School of Mines & Technology – Ng
Stevens, Bjorn B., (310) 825-1217 Univ of California, Los Angeles – Oa
Stevens, Calvin H., (408) 924-5029 stevens@geosun.sjsu.edu, San Jose State Univ – Ps
Stevens, James B., (409) 880-8233 Lamar Univ – Gr
Stevens, Joan M., (805) 756-2261 jstevens@calpoly.edu, California Polytechnic State Univ – Oo
Stevens, Liane M., 781-891-2978 lstevens@bentley.edu, Bentley College – Gp
Stevens, Margaret S., (409) 880-8233 Lamar Univ – Pv
Stevens, Philip S., (812) 855-0732 pstevens@indiana.edu, Indiana Univ, Bloomington – Oa
Stevens, Robert G., (509) 786-9231 stevensr@coopext.cahe.wsu.edu, Washington State Univ – Sb
Stevens, Stan, 413-545-0773 sstevens@geo.umass.edu, Univ of Massachusetts – Oo
Stevens (Landon), Lora R., 562-985-4817 lsteven2@csulb.edu, California State Univ, Long Beach – Gn
Stevenson, David J., (626) 395-6108 djs@gps.caltech.edu, California Institute of Technology – Xy
Stevenson, David L., 0191 3742516 Durham Univ – Oo
Stevenson, Ross K., (514) 987-3000 (Ext. xt 7835) stevenson.ross@uqam.ca, Universite du Quebec a Montreal – Cg
Steward, Grieg F., 808-956-6775 grieg@hawaii.edu, Univ of Hawai'i, Manoa – Ob
Steward, Robert G., (727) 553-3954 rsteward@marine.usf.edu, Univ of South Florida, Saint Petersburg – Op
Stewart, Brian W., (412) 624-8883 bstewart@pitt.edu, Univ of Pittsburgh – Cc
Stewart, David P., (513) 529-3216 Miami Univ – Gl
Stewart, Dion, 770-274-5068 dstewart@gpc.edu, Georgia Perimeter College – Gi
Stewart, Dona J., (404) 413-5779 djstewart@gsu.edu, Georgia State Univ – Oo
Stewart, Dorothy L., (509) 376-5056 dot.stewart@pnl.gov, Pacific Northwest National Lab – Oo
Stewart, Gary, 405-372-6063 Oklahoma State Univ – Go
Stewart, James D., (206) 543-8336 Vincennes Univ Junior College – Oe
Stewart, Joe D., (213) 763-3318 jdstewar@nhm.org, Los Angeles County Museum of Natural History – Pv
Stewart, Kevin G., (919) 962-0683 kgstewar@email.unc.edu, Univ of North Carolina, Chapel Hill – Gc
Stewart, Mark T., (813) 974-8749 mark@chuma1.cas.usf.edu, Univ of South Florida, Tampa – Hw
Stewart, Meg E., (845) 437-7708 mestewart@vassar.edu, Vassar College – Oi
Stewart, Robert H., (979) 845-2995 rstewart@ocean.tamu.edu, Texas A&M Univ – Op
Stewart, Robert R., (403) 220-3265 stewart@ucalgary.ca, Univ of Calgary – Ye
Stewart, Robert W., (250) 472-4060 rstewart@islandnet.com, Univ of Victoria – Op
Stewart, Ronald E., (703) 993-3187 rstewart@gmu.edu, George Mason Univ – Oo
Stewart, Ronald E., 514-398-1380 ronald.stewart@mcgill.ca, McGill Univ – Oa
Stewart-Mukhopadhyay, Sarah T.sstewart@eps.harvard.edu, Harvard Univ – Xg
Steyn, Douw G., (604) 822-8995 dsteyn@eos.ubc.ca, Univ of British Columbia – Oa
Stickel, Robert, (404) 894-0153 Georgia Institute of Technology – Oa
Stickney, Dale M., (916) 322-4207 California Geological Survey – Gi
Stickney, Michael C., 406-496-4332 mstickney@mtech.edu, Montana Tech of The Univ of Montana – Ys
Stickney, Robert R., (979) 845-3854 stickney@tamu.edu, Texas A&M Univ – Ob
Stidham, Christiane W., 631 632-8059 christiane.stidham@sunysb.edu, SUNY, Stony Brook – Yg
Stieglitz, Ronald D., 920-465-2711/2128 stieglir@uwgb.edu, Univ of Wisconsin, Green Bay – Gr
Stierle, Andrea, (406) 496-4117 astierle@mtech.edu, Montana Tech of the Univ of Montana – Oo
Stierle, Donald, (406) 496-4117 dstierle@mtech.edu, Montana Tech of the Univ of Montana – Co
Stierman, Donald J., (419) 530-2860 donald.stierman@utoledo.edu,

Univ of Toledo – Yg
Stifel, Peter B., stifel@friend.ly.net,
Univ of Maryland – Pg
Stigall, Alycia L., (740) 593-0393 stigall@ohio.edu,
Ohio Univ, Lancaster – Po
Stiles, Lynn F., (609) 652-4677 stilesl@pollux.stockton.edu,
Richard Stockton College of New Jersey – Oo
Stillings, Lisa, (775) 784-5803 Univ of Nevada, Reno – Cm
Stillings, Lisa, 775-784-5803 stilling@usgs.gov,
Univ of Nevada, Reno – Hw
Stimac, John P., (217) 581-7016 Eastern Illinois Univ – Gc
Stimpson, Brian, (204) 474-9821 stimpsn@cc.umanitoba.ca,
Univ of Manitoba – Nr
Stimpson, Ian G., +44 01782 583182 i.g.stimpson@esci.keele.ac.uk,
Keele Univ – Gc
Stinchcomb, Bruce L., (314) 595-2380 bstinchcomb@stlcc.cc.mo.us,
Saint Louis Community College, Florissant Valley – Pg
Stine, Scott W., 510-885-3159 scott.stine@csueastbay.edu,
California State Univ, East Bay – Gm
Stinson, Amy L., 949-451-5622 astinson@ivc.edu,
Irvine Valley College – Gc
Stinson, Amy L., (949) 361-1260 mesiem@aol.com,
Santiago Canyon College – Gc
Stix, John, (514) 343-6832 Universite de Montreal – Gi
Stix, John, 514-398-5391/6768 stix@eps.mcgill.ca,
McGill Univ – Gv
Stixrude, Lars P., (734) 647-9071 stixrude@umich.edu,
Univ of Michigan – Ys
Stock, Carl W., 205-348-1883 cstock@geo.ua.edu,
Univ of Alabama – Pi
Stock, Joann M., (626) 395-6938 jstock@gps.caltech.edu,
California Institute of Technology – Gt
Stockli, Daniel F., (785) 864-4995 stockli@ku.edu,
Univ of Kansas – Gc
Stockton, Charles W., (520) 621-7680 stockton@ltrr.arizona.edu,
Univ of Arizona – Hq
Stoddard, Edward F., (919) 515-7939 skip_stoddard@ncsu.edu,
North Carolina State Univ – Gp
Stoddard, Paul R., (815) 753-7929 prs@geol.niu.edu,
Northern Illinois Univ – Gt
Stodt, John A., (801) 585-9690 jstodt@egi.utah.edu,
Univ of Utah – Yg
Stoecker, Diane, (410) 221-8407 stoecker@hpl.umces.edu,
Univ of Maryland – Ob
Stoertz, Mary W., (740) 593-0665 stoertz@ohio.edu,
Ohio Univ, Lancaster – Hw
Stoessel, Achim, (979) 862-4170 achim@advect.tamu.edu,
Texas A&M Univ – Op
Stoessell, Ronald K., (504) 280-6795 rstoesse@uno.edu,
Univ of New Orleans – Cl
Stofan, Ellen R., (818) 354-2076 ellen.r.stofan@jpl.nasa.gov,
Jet Propulsion Lab – Xg
Stoffa, Paul L., (512) 471-6405 pauls@utig.ig.utexas.edu,
Univ of Texas at Austin, Jackson School of Geosciences – Ye
Stohr, Christopher J., 217-244-2186 stohr@isgs.uiuc.edu,
Illinois State Geological Survey – Ng
Stokes, Patricia, (801) 537-3320 nrugs.pstokes@state.ut.us,
Utah Geological Survey – Oo
Stokes, Rodger T., (727) 553-1647 Univ of South Florida, Saint
Petersburg – Ob
Stokes, Sharon, (215) 895-2341 Drexel Univ – Oo
Stoklosa, Michele, (208)426-3645 mstoklosa@boisestate.edu,
Boise State Univ – Gs
Stolar, John, (610) 436-2788 jstolar@wcupa.edu,
West Chester Univ – Og
Stoll, Heather M., (413) 597-4541 heather.m.stoll@williams.edu,
Williams College – Cl
Stolper, Edward M., (626) 395-6504 ems@gps.caltech.edu,
California Institute of Technology – Cp
Stone, Alan T., (410) 516-8476 astone@jhu.edu,
The Johns Hopkins Univ – Cl
Stone, Charles G., 501 683-0126 agc@arkansas.gov,
Arkansas Geological Commission – Gc
Stone, George, (414) 297-7427 Milwaukee Area Technical College – Gl
Stone, Greg W., (225) 388-6188 Louisiana State Univ – Og
Stone, John O., 206-221-6332 stone@ess.washington.edu,
Univ of Washington – Cca
Stone, John O. H., (206) 685-9514 Univ of Washington – Cc
Stone, Loyd, (913) 532-5732 Kansas State Univ – Sp
Stone, Peter H., (617) 253-2443 phstone@mit.edu,
Massachusetts Institute of Technology – Oa
Stone, William J., (505) 665-8340 wstone@lanl.gov,
Los Alamos National Lab – Gg
Stoner, Joseph, (541) 737-9002 jstoner@coas.oregonstate.edu,
Oregon State Univ – Yrm
Stoner, Lester B., (217) 581-7270 cflbs@ux1.eiu.edu,
Eastern Illinois Univ – Og
Stooksbury, David E., (402) 472-8765 Univ of Nebraska, Lincoln – Oa
Storey, Brad J., 410-293-6563 storey@usna.edu,
United States Naval Academy – Og
Storey, Neil D., (801) 537-3386 nrugs.nstorey@state.ut.us,
Utah Geological Survey – Oo
Stork, Allen L., (970) 943-3044 astork@western.edu,
Western State College of Colorado – Giv
Storm, Lara, 207 859 5808 lstorm@colby.edu,
Colby College – Gpz
Stormer, Jr., John C., (904) 456-7884 Rice Univ – Gi
Storrs, Glenn W., (513) 345-8500 Univ of Cincinnati – Pv
Storrs, Glenn W., (513) 455-7141 storrsg@cincymuseum.org,
Cincinnati Museum Center – Pv
Stott, Lowell D., (213) 740-5120 stott@usc.edu,
Univ of Southern California – Pm
Stotz, Angela C., (606) 257-8026 Univ of Kentucky – Oo
Stoudt, Emily L., 432 552-2244 stoudt_e@utpb.edu,
Univ of Texas, Permian Basin – Gd
Stout, James H., 612-624-4344 jstout@umn.edu,
Univ of Minnesota, Twin Cities – Gp
Stout, Nathan, (702) 895-4302 stout@ccmail.nevada.edu,
Univ of Nevada, Las Vegas – Oo
Stowell, Harold H., 205-348-5098 hstowell@geo.ua.edu,
Univ of Alabama – Gp
Stowell Gale, Juliajulia.stowell@beg.utexas.edu,
Univ of Texas at Austin, Jackson School of Geosciences – Gc
Stracher, Glenn B., (478) 289-2073 stracher@ega.edu,
East Georgia College – Om
Strack, Kurt M., (713) 743-3399 Univ of Houston – Ye
Strack, Otto D., (612) 625-3009 strac001@tc.umn.edu,
Univ of Minnesota, Twin Cities – Hw
Stradford, Todd, (608) 342-1674 Univ of Wisconsin, Platteville – Or
Straffin, Eric, (814) 732-1574 estraffin@edinboro.edu,
Edinboro Univ of Pennsylvania – Gm
Strahler, Alan H., (617) 353-5984 alan@bu.edu,
Boston Univ – Or
Strain, Priscilla L., 202-633-2481 Smithsonian Institution / National Air
& Space Museum – Or
Stramski, Dariusz, (858) 534-3353 dstramski@ucsd.edu,
Univ of California, San Diego – Op
Strangeway, Robert J., (310) 206-6247 strange@igpp.ucla.edu,
Univ of California, Los Angeles – Oo
Strasser, Jeffrey C., (309) 794-7218 glstrasser@augustana.edu,
Augustana College – Gm
Strasser, Stefan, 089/2180 4340 stefan.strasser@vr-web.de,
LMU - Univ of Munich – Gz
Stratton, James F., (217) 581-6376 cfjfs@ux1.eiu.edu,
Eastern Illinois Univ – Pg
Stratton, Loretta A., (585) 245-5291 loretta@geneseo.edu,
SUNY, Geneseo – Oo
Straub, Christopher, (520) 626-6782 cstraub22@hotmail.com,
Univ of Arizona – So
Straub, David, (514) 398-8995 david.straub@mcgill.ca,
McGill Univ – Op
Straub, Derek J., 570-372-4767 straubd@susqu.edu,
Susquehanna Univ – Oa
Straub, John, (202) 478-8903 j.straub@gl.ciw.edu,
Carnegie Institution of Washington – Oo
Straub, Katherine H., 570-372-4318 straubk@susqu.edu,
Susquehanna Univ – Ow
Stravers, Jay A., (815) 753-7927 jay@geol.niu.edu,
Northern Illinois Univ – Gl
Strawn, Daniel G., (208) 885-2713 dgstrawn@uidaho.edu,
Univ of Idaho – Sc
Strayer, Luther M., (510) 885-3083 luther.strayer@csueastbay.edu,
California State Univ, East Bay – Gc
Streck, Martin J., (503) 725-3379 streckm@pdx.edu,
Portland State Univ – Gi
Streepey, Meg, (765) 973-2168 streeme@earlham.edu,
Earlham College – Gt
Streett, Pamela S., 520-621-6954 pam@lpl.arizona.edu,
Univ of Arizona – Oo
Streile, Gary P., (509) 372-6042 gary.streile@pnl.gov,

Pacific Northwest National Lab – Hq
Strick, James E., (717) 291-3856 james.strick@fandm.edu, Franklin and Marshall College – Ghe
Strickland, Cassandracstrickland@yvcc.edu, Yakima Valley College – Gg
Strickland, Richard M., (206) 543-3131 strix@ocean.washington.edu, Univ of Washington – Ob
Stringer, Carolyn F., (850) 487-2219 Florida Geological Survey – Oo
Stringer, Gary L., (318) 347-1893 Univ of Louisiana, Monroe – Oe
Strobel, Darrell F., (410) 516-7829 strobel@jhu.edu, The Johns Hopkins Univ – Oa
Strock, Jeffrey S., (507) 752-7372 stroc001@tc.umn.edu, Univ of Minnesota, Twin Cities – So
Strole, Torie L., 217-244-2392 strole@isgs.uiuc.edu, Illinois State Geological Survey – Gg
Strom, JanJan.Strom@usd.edu, Univ of South Dakota – Oo
Strom, Robert G., 520- 621-2720 rstrom@LPL.arizona.edu, Univ of Arizona – Xg
Strong, David F., (250) 721-7002 dstrong@uvvm.uvic.ca, Univ of Victoria – Em
Strother, Paul, 617-552-8395 paul.strother.1@bc.edu, Boston College – Pb
Strother, Paul K., (617) 552-8395 strother@bc.edu, Boston College – Pb
Strout, Charles R., (509) 359-2789 cstrout@mail.ewu.edu, Eastern Washington Univ – Gg
Strub, P. Ted, (541) 737-3015 tstrub@coas.oregonstate.edu, Oregon State Univ – Op
Strutton, Peter, (541) 737-2065 strutton@coas.oregonstate.edu, Oregon State Univ – Ob
Struzhkin, Viktor, 202 478 8952 v.struzhkin@gl.ciw.edu, Carnegie Institution of Washington – Gy
Stuart-Gerlach, Dana R., (508) 289-3369 dgerlach@whoi.edu, Woods Hole Oceanographic Institution – Ou
Stubler, Craig, (805) 756-2188 cstubler@calpoly.edu, California Polytechnic State Univ – So
Stucki, Joseph W., (217) 333-9636 Univ of Illinois, Urbana-Champaign – Sc
Stucky, Richard, (303) 370-6434 RStucky@dmns.org, Denver Museum of Nature & Science – Pv
Student, James J., 989-774-2295 stude1jj@cmich.edu, Central Michigan Univ – Gg
Studinger, Michael, (845) 365-8598 Columbia Univ – Yr
Stuiver, Minze, 206-685-1735 minze@u.washington.edu, Univ of Washington – Cc
Stull, Robert J., (323) 343-2408 rstull@calstatela.edu, California State Univ, Los Angeles – Gi
Stull, Roland B., 604-822-5901 rstull@eos.ubc.ca, Univ of British Columbia – Oa
Stump, Brian W., (214) 768-1223 Southern Methodist Univ – Ys
Stump, Brian W., (214) 768-1223 stump@lanl.gov, Los Alamos National Lab – Ys
Stump, Edmund, (480) 965-3971 ed.stump@asu.edu, Arizona State Univ – Gc
Stumpf, Andrew J., 217-244-6462 stumpf@isgs.uiuc.edu, Illinois State Geological Survey – Gl
Stumpf, Rick, (301) 713-3000 Univ of South Florida, St Petersburg – Og
Stupazzini, Marco, 089/2180 4143 stupa@geophysik.uni-muenchen.de, LMU - Univ of Munich – Yg
Sturchio, Neil C., (630) 252-3986 Argonne National Lab – Cg
Sturchio, Neil C., (312) 355-1182 sturchio@uic.edu, Univ of Illinois, Chicago – Cg
Sturges, Wiltonsturges@ocean.fsu.edu, Florida State Univ – Op
Sturgess, Steven W., 573-368-2149 steve.sturgess@dnr.mo.gov, Missouri Dept of Natural Resources – Oo
Sturman, Darko B., (416) 586-5872 Royal Ontario Museum – Gz
Sturz, Anne A., (619) 260-4600 asturz@acusd.edu, Univ of San Diego – Cm
Stutz, Jochen P., (310) 825-1217 jochen@atmos.ucla.edu, Univ of California, Los Angeles – Oa
Styles, Peter, +44 01782 584116 p.styles@esci.keele.ac.uk, Keele Univ – Ge
Su, Wen-June, 217-244-2185 wenjune@isgs.uiuc.edu, Illinois State Geological Survey – Ng
Su, Xiaobo, (541) 346-4568 xiaobo@uoregon.edu, Univ of Oregon – Oo
Suarez, Francisco, fsuarez@cicese.mx, Centro de Inv Científica y de Educación Superior de Ensenada – Gc
Sublette, Kerry, 918-631-3085 kerry-sublette@utulsa.edu, Univ of Tulsa – Ge
Suchy, Daniel R., (785) 864-2160 dsuchy@kgs.ku.edu, Univ of Kansas – Gg
Suczek, Christopher A., (360) 650-3590 suczekca@cc.wwu.edu, Western Washington Univ – Gs
Suen, C. J., (559) 278-3086 john_suen@csufresno.edu, California State Univ, Fresno – Hw
Sues, Hans-Dieter, (416) 586-5855 Royal Ontario Museum – Pv
Sugarman, Peter J., (609) 292-2576 pete.sugarman@dep.state.nj.us, New Jersey Geological Survey – Gr
Sugarman, Peter P., (609) 292-6842 Rutgers, The State Univ of New Jersey – Gr
Sugihara, George, (619) 534-5582 gsugihara@ucsd.edu, Univ of California, San Diego – Ob
Sui, Daniel Z., (409) 845-7154 d-sui@tamu.edu, Texas A&M Univ – Or
Sukop, Michael, (305) 348-3117 sukopm@fiu.edu, Florida International Univ – Hq
Sulanowska, Margaret, (508) 289-2306 msulanowska@whoi.edu, Woods Hole Oceanographic Institution – Cm
Sulanowski, Jacek K., jsulanowski@bridgew.edu, Bridgewater State College – Pe
Sullivan, Daniel A., (713) 974-3704 Houston Community College System – Oo
Sullivan, Jack B., (301) 405-0106 jsullivan@umail.umd.edu, Univ of Maryland – Ou
Sullivan, James D., (213) 265-8837 East Los Angeles College – Oy
Sullivan, Raymond, (415) 338-2061 sullivan@sfsu.edu, San Francisco State Univ – Go
Sullivan, Robert M., (717) 783-9897 rsullivan@state.pa.us, The State Museum of Pennsylvania – PvgPs
Sullivan, Walter "Bill", 207 859 5803 wasulliv@colby.edu, Colby College – Gct
Sullivan-Watts, Barbara K., (401) 874-6659 bsull@gso.uri.edu, Univ of Rhode Island – Og
Sullwold, Naomi E., (805) 965-0581 (Ext. 2315) Santa Barbara City College – Oe
Sultan, Mohamed, (630) 252-1929 Argonne National Lab – Cg
Sultan, Mohamed, 269-387-5487 mohamed.sultan@wmich.edu, Western Michigan Univ – Or
Sumida, Stuart S., (909) 537-5346 California State Univ, San Bernardino – Pv
Summa, Catherine L., 507-457-5269 csumma@winona.edu, Winona State Univ – Gt
Summons, Roger, (617) 452-2791 rsummons@mit.edu, Massachusetts Institute of Technology – Py
Sumner, Dawn Y., (530) 752-5353 sumner@geology.ucdavis.edu, Univ of California, Davis – Gs
Sumrall, Colin, (865) 974-2366 csumrall@utk.edu, Univ of Tennessee, Knoxville – Pi
Sun, Hongbing, 609-895-5185 hsun@rider.edu, Rider Univ – HwCgSc
Sun, Wen-Yih, (765) 494-7681 wysun@purdue.edu, Purdue Univ – Oa
Sun, Yifei, (818) 677-3532 yifei.sun@csun.edu, California State Univ, Northridge – Oi
Sundareshwar, P. V., 605-394-2492 pv.sundareshwar@sdsmt.edu, South Dakota School of Mines & Technology – Oo
Sundbeck, Carl, (805) 965-0581 (Ext. 2317) sundbeck@sbcc.net, Santa Barbara City College – Oy
Sundby, Bjorn, (418) 723-1986 (Ext. 1767) bjorn_sundby@uqar.uquebec.ca, Universite du Quebec a Rimouski – Oc
Sundby, Bjorn, 514-398-4883 sundby@eps.mcgill.ca, McGill Univ – Cm
Sundell, Kent A., (307) 268-2498 ksundell@caspercollege.edu, Casper College – Gg
Sundell, Ronald C., 906-227-1359 rsundell@nmu.edu, Northern Michigan Univ – Oo
Sunder, Joseph H., (412) 392-3892 Point Park College – Oo
Sunderman, Jack A., (260) 481-6249 Indiana Univ / Purdue Univ, Fort Wayne – Gr
Suneson, Neil H., (405) 325-3031 nsuneson@ou.edu, Univ of Oklahoma – Gc
Superchi-Culver, Tonia, 303-492-5211 toni.culver@colorado.edu, Univ of Colorado – Pg
Suppe, John, 609-258-4119 suppe@princeton.edu, Princeton Univ – Gc
Surdam, Ronald C., 307-766-2286x223 rsurdam@uwyo.edu, Wyoming State Geological Survey – Go
Surge, Donna M., 919-843-1994 donna64@unc.edu,

Univ of North Carolina, Chapel Hill – Pe
Susak, Nicholas J., (506) 453-4803 nsusak@unb.ca, Univ of New Brunswick – Cg
Suszek, Thomas J., (920) 424-2268 suszek@uwosh.edu, Univ of Wisconsin, Oshkosh – Gg
Sutherland, Bruce R., (780) 492-0573 bruce.sutherland@ualberta.ca, Univ of Alberta – Oo
Sutherland, Stuart, (604) 822-0176 ssutherland@eos.ubc.ca, Univ of British Columbia – Pg
Sutphin, Octavia L., (609) 292-2576 octavia.sutphin@dep.state.nj.us, New Jersey Geological Survey – Oo
Suttle, Curtis, (604) 822-8610 csuttle@eos.ubc.ca, Univ of British Columbia – Ob
Suttner, Lee J., (812) 855-4957 suttner@indiana.edu, Indiana Univ, Bloomington – Gd
Sutton, Chris, (360) 650-3581 geology@cc.wwu.edu, Western Washington Univ – Oo
Sutton, Erin F., (865) 594-5598 Erin.F.Sutton@state.tn.us, Tennessee Division of Geology – GgOi
Sutton, Paul C., (303) 871-2513 psutton@du.edu, Univ of Denver – Oi
Sutton, Sally J., (970) 491-5995 sallys@cnr.colostate.edu, Colorado State Univ – Gd
Sverdrup, Keith A., 414-229-4017 sverdrup@uwm.edu, Univ of Wisconsin, Milwaukee – Ys
Sverjensky, Dimitri A., (410) 516-8568 sver@jhu.edu, The Johns Hopkins Univ – Cl
Svitra, Zita V., (505) 667-7616 Los Alamos National Lab – Oo
Swain, Frederick M., 612-624-5028 swain002@umn.edu, Univ of Minnesota, Twin Cities – Ps
Swain, Geoffrey W., (321) 674-8096 Florida Institute of Tech – Ob
Swaminathan, Sudha, 508-929-8605 sswaminathan@worcester.edu, Worcester State College – Yg
Swan, Andy, 0044 (0)20 8547 2000 a.swan@kingston.ac.uk, Kingston Univ – Ggb
Swanson, Basil I., (505) 667-5814 Los Alamos National Lab – Oo
Swanson, David K., (907) 479-2657 Univ of Alaska, Fairbanks – Sd
Swanson, Donald A., (808) 967-8863 donswan@usgs.gov, Univ of Hawai'i, Manoa – Gv
Swanson, Eric R., 210-458-5458 Eric.Swanson@utsa.edu, Univ of Texas, San Antonio – Gv
Swanson, Frederick J., (541) 750-7355 Oregon State Univ – Gm
Swanson, Karen, (201) 595-2589 William Paterson College – Cc
Swanson, Mark T., (207) 780-5024 mswanson@usm.maine.edu, Univ of Southern Maine – Gc
Swanson, R. L., (631) 632-8704 lswanson@notes.cc.sunysb.edu, SUNY, Stony Brook – Og
Swanson, Samuel E., (706) 542-2415 sswanson@uga.edu, Univ of Georgia – Gi
Swanson, Sherman, (775) 784-4057 sswanson@agnt1.ag.unr.edu, Univ of Nevada, Reno – Hs
Swanson, Susan K., (608) 363-2132 swansons@beloit.edu, Beloit College – Hw
Swanson, Terry W., 206-543-1923 tswanson@u.washington.edu, Univ of Washington – CcGe
Swapp, Susan M., (307) 766-2513 swapp@uwyo.edu, Univ of Wyoming – Gp
Swarbrick, Richard E., 0191 3743334 r.e.swarbrick@durham.ac.uk, Durham Univ – Go
Swart, Peter K., 305 421-4103 pswart@rsmas.miami.edu, Univ of Miami – Cg
Swartwood, Jade L., (570) 389-4108 jswartwo@bloomu.edu, Bloomsburg Univ – Oo
Swartz, Harry J., (301) 405-4337 hs28@umail.umd.edu, Univ of Maryland – Oo
Swaters, Gordon E., (780) 492-7159 gordon.swaters@ualberta.ca, Univ of Alberta – Op
Sweeney, Jack, jsweeney@salemstate.edu, Salem State College – Yv
Sweeney, Mark D., (509) 373-0703 mark.sweeney@pnl.gov, Pacific Northwest National Lab – Yg
Sweeney, Mark R., (605) 677-6142 Mark.Sweeney@usd.edu, Univ of South Dakota – Gs
Sweet, Walter C., (614) 292-2326 Ohio State Univ – Ps
Sweigard, Richard J., (606) 257-1173 Univ of Kentucky – Nm
Swenson, John B., 218-726-6844 jswenso2@d.umn.edu, Univ of Minnesota, Duluth – Gr
Swetnam, Thomas W., (520) 621-2112 tswetnam@ltrr.arizona.edu, Univ of Arizona – Pe
Swett, Keene, (319) 351-4644 keene-swett@uiowa.edu, Univ of Iowa – Gs
Swift, Donald J. P., (757) 683-4937 dswift@odu.edu, Old Dominion Univ – Ou
Swift, Elijah V., (401) 874-6146 lige@gso.uri.edu, Univ of Rhode Island – Ob
Swift, Robert P., (505) 665-7871 bswift@lanl.gov, Los Alamos National Lab – Nr
Swift, Stephen A., (508) 289-2626 sswift@whoi.edu, Woods Hole Oceanographic Institution – Yr
Swihart, George H., (901) 678-2606 gswihart@memphis.edu, Univ of Memphis – Gx
Swindle, Timothy D., (520) 621-4128 tswindle@lpl.arizona.edu, Univ of Arizona – Xm
Swinehart, James B., (402) 472-7529 jswinehart1@unl.edu, Univ of Nebraska, Lincoln – Gr
Swinford, Edward M., (614) 265-6473 mac.swinford@dnr.state.oh.us, Ohio Dept of Natural Resources – Gr
Swisher III, Carl C., 732 445 5363 cswish@rci.rutgers.edu, Rutgers, The State Univ of New Jersey – Cc
Switzer, Paul, (650) 723-2879 switzer@stanford.edu, Stanford Univ – Gq
Swope, R. J., (317) 278-0132 rjswope@iupui.edu, Indiana Univ / Purdue Univ, Indianapolis – Gz
Sydnor, Robert H., (916) 327-0807 California Geological Survey – Ng
Sydora, Richard D., (780) 492-3624 rsydora@phys.ualberta.ca, Univ of Alberta – Xy
Sykes, Lynn R., (845) 365-8880 sykes@ldeo.columbia.edu, Columbia Univ – Ys
Sykes, Michael P., +61 8 9266 3521 M.Sykes@curtin.edu.au, Curtin Univ of Technology – Yve
Sylva, Sean, 508-289-3546 ssylva@whoi.edu, Woods Hole Oceanographic Institution – Oc
Sylvester, Arthur G., (805) 893-3156 Univ of California, Santa Barbara – Gt
Sylvester, Paul J., (709) 737-4736 sylvester@sparky2.esd.mun.ca, Memorial Univ of Newfoundland – Ca
Sylvester, Steven, (717) 291-3821 steve.sylvester@fandm.edu, Franklin and Marshall College – Ca
Symbalisty, E.M.D., (505) 667-9670 esymbalisty@lanl.gov, Los Alamos National Lab – Oo
Symes, William S., (713) 348-5997 symes@caam.rice.edu, Rice Univ – Oo
Symons, David T., (519) -25-3-30 (Ext. ext. 2493) dsymons@uwindsor.ca, Univ of Windsor – Ym
Syrett, William, (814) 865-6172 The Pennsylvania State Univ, Univ Park – Ow
Syverson, Kent M., (715) 836-3676 syverskm@uwec.edu, Univ of Wisconsin, Eau Claire – Gl
Syvitski, James P., 303-492-7909 james.syvitski@colorado.edu, Univ of Colorado – Gs
Szabo, John P., (330) 972-8039 jpszabo@uakron.edu, Univ of Akron – Gl
Szecsody, James E., (509) 372-6080 jim.szecsody@pnl.gov, Pacific Northwest National Lab – So
Szekielda, Karl H., (212) 772-4019 szekielda@aol.com, Hunter College (CUNY) – Or
Szeto, Anthony M. K., (416) 736-2100 (Ext. 77703) szeto@yorku.ca, York Univ – Yg
Szidarovszky, Ferenc, (520) 621-6557 ferenc@email.arizona.edu, Univ of Arizona – Hq
Szilagyi, Jozsefjszilagyi@unl.edu, University of Nebraska - Lincoln – Hq
Szlavecz, Katalin, (410) 516-8947 szlavecz@jhu.edu, The Johns Hopkins Univ – Pi
Szpunar, Jerzy, (514) 398-4755 McGill Univ – Oo
Söllner, Frank, 089/2180 6519 fank.soellner@iaag.geo.uni-muenchen.de, LMU - Univ of Munich – Gg

T

Tabet, David E., (801) 537-3373 nrugs.dtabet@state.ut.us, Utah Geological Survey – Ec
Tabidian, M. Ali, (818) 677-2536 ali.tabidian@csun.edu, California State Univ, Northridge – Hw
Taboada Castro, María T., 00 34 981 167000 teresat@udc.es, Coruna Univ – Sd
Tabrizi, Azam, (804) 427-7189 Tidewater Community College – Pm
Tabrum, Alan R., (412) 622-3265 Carnegie Museum of Natural History – Pv
Taggart, Christopher T., (902) 494-7144 chris.taggart@dal.ca, Dalhousie Univ – Ob
Taggart, Ralph E., 517-353-5175 taggart@msu.edu,

Michigan State Univ – Pb
Tahnk, William R., (541) 737-5703 tahnk@coas.oregonstate.edu, Oregon State Univ – Oa
Taib, Samsudin Hj., 03-79674235 samsudin@um.edu.my, Univ of Malaya – Yg
Taillefert, Martial, (404) 894-6043 mtaillef@eas.gatech.edu, Georgia Institute of Technology – Cm
Taira, Randal Y., (509) 372-4866 randal.taira@pnl.gov, Pacific Northwest National Lab – Hg
Taira, Taka'aki, (202) 478-8836 taka@dtm.ciw.edu, Carnegie Institution of Washington – Ys
Tait, C. Drew, (505) 667-7603 Los Alamos National Lab – Cp
Takahashi, Taro, (845) 365-8537 taka@ldeo.columbia.edu, Columbia Univ – Oc
Takle, Eugene S., (515) 294-9871 Iowa State Univ of Science & Technology – Oa
Takle, Eugene S., (515) 294-9871 gstakle@iastate.edu, Iowa State Univ of Science & Technology – Oa
Talbot, James L., (360) 733-4282 talbot@cc.wwu.edu, Western Washington Univ – Gc
Talbot, Robert W., (603) 862-1792 robert.talbot@unh.edu, Univ of New Hampshire – Oa
Talley, John H., (302) 831-8258 waterman@udel.edu, Univ of Delaware – Hw
Talley, John H., 302-831-2833 waterman@udel.edu, Univ of Delaware – Eg
Talley, Lynne D., (619) 534-6610 ltalley@ucsd.edu, Univ of California, San Diego – Op
Talwani, Manik, (713) 348-6067 manik@rice.edu, Rice Univ – Yr
Tan, Denis N.denistan@um.edu.my, Univ of Malaya – Go
Tan, Siang S., (213) 620-4793 California Geological Survey – Gg
Tandon, Kush, (541) 737-9658 kush@coas.oregonstate.edu, Oregon State Univ – Yr
Tanguay, Marc G., (514) 340-4556 Ecole Polytechnique – Or
Tanimoto, Toshiro, (805) 893-8375 toshiro@quake.crustal.ucsb.edu, Univ of California, Santa Barbara – Ys
Tanji, Kenneth K., (530) 752-6540 kktanji@ucdavis.edu, Univ of California, Davis – So
Tanji, Kenneth K., (530) 752-6540 kktanji@ucdavis.edu, Univ of California, Davis – Sc
Tao, Vincent C., (416)736-5221 tao@yorku.ca, York Univ – Oi
Tapanila, Leif, 208-282-3871 tapaleif@isu.edu, Idaho State Univ – Pi
Tapanila, Lori, 208-282-5024 tapalori@isu.edu, Idaho State Univ – Gg
Tapp, J. B., 918-631-3018 jbt@utulsa.edu, Univ of Tulsa – Gc
Tappen, Christine M., 212-769-5377 ctappen@amnh.org, American Museum of Natural History – Gi
Taranik, James V., (775) 682-8735 jtaranik@mines.unr.edu, Univ of Nevada, Reno – Ye
Tarduno, John A., (585) 275-5713 john@earth.rochester.edu, Univ of Rochester – Ym
Tarman, Donald W.dwtarman@csupomona.edu, California State Polytechnic Univ, Pomona – Gc
Tary, Anna K., (781) 891-2236 atary@bentley.edu, Bentley College – Gl
Taskey, Ronald D., (805) 756-1160 rtaskey@calpoly.edu, California Polytechnic State Univ – Sf
Tasse, Gilles, (514) 987-3984 tasse.gilles@uqam.ca, Universite du Quebec a Montreal – Oo
Tasse, Normand, (418) 654-2609 ntasse@inrs.uquebec.ca, Universite du Quebec – Ge
Tatabatai, M. A., (515) 294-7848 malit@iastate.edu, Iowa State Univ of Science & Technology – Sc
Tatham, Robert H., (512) 471-9129 tatham@mail.utexas.edu, Univ of Texas at Austin, Jackson School of Geosciences – Yg
Taubeneck, William H., (541) 737-1228 Oregon State Univ – Gx
Tauxe, John D., (423) 574-5348 Oak Ridge National Lab – Hw
Tauxe, Lisa, (858) 534-6084 ltauxe@ucsd.edu, Univ of California, San Diego – Ym
Taylor, Alfred R., (276) 676-5577 alfred.taylor@dmme.virginia.gov, Division of Mineral Resources – Gr
Taylor, Brian, (808) 956-6649 taylorb@hawaii.edu, Univ of Hawai'i, Manoa – Gt
Taylor, Danny L., (775) 784-6922 dtaylor@mines.unr.edu, Univ of Nevada, Reno – Nm
Taylor, Donavon, (715) 425-3395 donavon.h.taylor@uwrf.edu, Univ of Wisconsin, River Falls – Sp
Taylor, Edith, 785-864-3621 etaylor@ku.edu, Univ of Kansas – Pb
Taylor, Edward M., (541) 737-1232 taylore@geo.oregonstate.edu, Oregon State Univ – Gv
Taylor, Elwynn, (808) 956-3899 setaylor@iastate.edu, Iowa State Univ of Science & Technology – Oa
Taylor, Frank J. R., (604) 822-4587 mtaylor@eos.ubc.ca, Univ of British Columbia – Ob
Taylor, G. Jeffrey, (808) 956-3899 Univ of Hawai'i, Manoa – Xm
Taylor, Gary, (702) 895-4615 California Geological Survey – Em
Taylor, Geoff R., 61293858251 g.taylor@unsw.edu.au, Univ of New South Wales – Oi
Taylor, Gordon T., (631) 632-8688 gtaylor@notes.cc.sunysb.edu, SUNY, Stony Brook – Ob
Taylor, Gregory R., (530) 898-6369 taylgr@shasta.csuchico.edu, California State Univ, Chico – Oa
Taylor, Hugh P., (626) 395-6116 hptaylor@gps.caltech.edu, California Institute of Technology – Cs
Taylor, John F., (724) 357-4469 jftaylor@iup.edu, Indiana Univ of Pennsylvania – Ps
Taylor, Lawrence A., (865) 974-6013 lataylor@utk.edu, Univ of Tennessee, Knoxville – Cp
Taylor, Lawrence D., (517) 629-0308 ltaylor@albion.edu, Albion College – Gl
Taylor, Matthew, 303-871-2656 mtaylor7@du.edu, Univ of Denver – Og
Taylor, Michael H., 785-864-5828 mht@ku.edu, Univ of Kansas – Gt
Taylor, Penny M., 413-538-3236 pmtaylor@mtholyoke.edu, Mount Holyoke College – Gg
Taylor, Peter A., (416) 736-2100 (Ext. 77707) pat@yorku.ca, York Univ – Oa
Taylor, Richard P.richard_taylor@carleton.ca, Carleton Univ – Cg
Taylor, Robert W., (973) 655-4129 Montclair State Univ – Oo
Taylor, Robert W., (205) 851-5462 rtaylor@aamu.edu, Alabama A&M Univ – Sc
Taylor, Sid, (902) 867-2299 staylor@stfx.ca, Saint Francis Xavier Univ – Gg
Taylor, Stephen B., 503-838-8398 taylors@wou.edu, Western Oregon Univ – Gm
Taylor, Steven R., (505) 667-1007 taylor@lanl.gov, Los Alamos National Lab – Ys
Taylor, Thomas N., (913) 864-3625 Ohio State Univ – Pb
Taylor, Thomas N., 785-864-3625 tntaylor@ku.edu, Univ of Kansas – Pb
Taylor, Wanda J., (702) 895-4615 wjt@nevada.edu, Univ of Nevada, Las Vegas – Gc
Taylor, Wayne A., (505) 667-4253 Los Alamos National Lab – Cl
Tchakerian, Vatche P., (979) 845-3651 v-tchakerian@tamu.edu, Texas A&M Univ – Gm
Teasdale, Rachelrteasdale@csuchico.edu, California State Univ, Chico – Gv
Tebbens, Sarah F., (727) 553-1538 tebbens@marine.usf.edu, Univ of South Florida, Saint Petersburg – Yr
Tedesco, Lenore P., (317) 274-7154 ltedesco@iupui.edu, Indiana Univ / Purdue Univ, Indianapolis – Gs
Tedeski, Joseph R., 412.442.4295 jtedeski@state.pa.us, Pennsylvania Bureau of Topographic & Geologic Survey – Eo
Tedford, Richard H., (212) 769-5809 American Museum of Natural History – Pv
Teissere, Ron, (360) 902-1450 ron.teissere@wadnr.gov, Washington Division of Geology & Earth Resources – Gg
Teller, James T., (204) 474-9270 tellerjt@ms.umanitoba.ca, Univ of Manitoba – Gs
Telles, Franky, 909-941-2385 astrobiology@sbcglobal.net, Chaffey College – Gn
Telmer, Kevin, (250) 472-4182 Univ of Victoria – Cl
Tempel, Gina, (775) 784-4706 gina@mines.unr.edu, Univ of Nevada, Reno – Cl
Tempel, Regina, (775) 784-4706 gina@hydro.unr.edu, Univ of Nevada, Reno – Cl
Templeton, Alexis, 303-492-6069 alexis.templeton@colorado.edu, Univ of Colorado – ClGe
Templeton, Jeffrey H., 503-838-8858 templej@wou.edu, Western Oregon Univ – Gv
Ten Brink, Norman W.tenbrinn@gvsu.edu, Grand Valley State Univ – Gm

Faculty Index -T

Tencate, James, (505) 665-6667 tencate@lanl.gov, Los Alamos National Lab – Oo
Tenenbaum, Joel, (914) 251-6643 SUNY, Purchase – Oa
Teng, Ta-liang, (213) 740-5838 lteng@usc.edu, Univ of Southern California – Ys
Tepley, Frank, (541) 737-8199 ftepley@coas.oregonstate.edu, Oregon State Univ – CmGi
Ter-Simonian, Vardui, (213) 740-6106 tersimon@usc.edu, Univ of Southern California – Oo
Tera, Fouad, (202) 478-8472 tera@dtm.ciw.edu, Carnegie Institution of Washington – Cc
TerBerg, Robert A., (510) 495-2841 Rterberg@lbl.gov, Lawrence Berkeley National Lab – Hw
Terrell, Kenneth J., (404) 413-5780 geokjt@langate.gsu.edu, Georgia State Univ – Pv
Terry, Dennis O., (215) 204-8226 doterry@temple.edu, Temple Univ – Gr
Terry, Michael P., 605-394-5325 Michael.Terry@sdsmt.edu, South Dakota School of Mines & Technology – Gc
Terry, Richard E., (801) 378-2283 Brigham Young Univ – Sb
Tesfaye, Samson, (573) 681-5586 tesfayes@lincoln.edu, Univ of Missouri – GtcOr
Tetrault, Denis, (519) -25-3-30 (Ext. ext. 2495) deniskt@uwindsor.ca, Univ of Windsor – Gg
Tettenhorst, Rodney T., 614 247-4246 tettenhorst.2@osu.edu, Ohio State Univ – Gz
Tevesz, Michael J. S., (216) 687-3508 m.tevesz@csuohio.edu, Cleveland State Univ – Pi
Tew, Berry H., 205/247-3679 ntew@gsa.state.al.us, Geological Survey of Alabama – Gg
Tewksbury, Barbara J., (315) 859-4713 btewksbu@hamilton.edu, Hamilton College – Gc
Textoris, Daniel A., (919) 962-0690 dtextori@email.unc.edu, Univ of North Carolina, Chapel Hill – Gd
Teyssier, Christian P., 612-624-6801 teyssier@umn.edu, Univ of Minnesota, Twin Cities – Gc
Tezcan, Levent, +90 (312) 2977750 tezcan@hacettepe.edu.tr, Hacettepe Univ – Hw
Thackray, Glenn D., (208) 282-3560 thacglen@isu.edu, Idaho State Univ – Gl
Thale, Paul, 406-496-4653 pthale@mtech.edu, Montana Tech of The Univ of Montana – Oi
Tharp, Thomas M., (765) 494-8678 ttharp1@purdue.edu, Purdue Univ – Nm
Thayer, Paul A., (910) 962-3780 thayer@uncw.edu, Univ of North Carolina Wilmington – Gd
Themelis, Nickolas J., (212) 854-2138 njt1@columbia.edu, Columbia Univ – Nx
Theo, Lisa J., 715-346-4455 ltheo@uwsp.edu, Univ of Wisconsin, Stevens Point – Ou
Theokritoff, George, 973-353-5100 Rutgers, The State Univ of New Jersey, Newark – Pi
Therrien, Pierretherrien@ggl.ulaval.ca, Universite Laval – Gq
Therrien, Rene, (418) 656-5400 rene.therrien@ggl.ulaval.ca, Universite Laval – Hw
Thibault, Yvesythibault@stny.rr.com, Univ of Western Ontario – Gi
Thieling, Stanley C., (601) 961-5519 stan_thieling@deq.state.ms.us, Mississippi Office of Geology – Go
Thien, Steve J., (913) 532-7207 Kansas State Univ – Sc
Thiruvathukal, John V., (973) 655-4417 Montclair State Univ – Yg
Thissen, Joann, 516-572-7026 thissej@ncc.edu, Nassau Community College – Gg
Thistle, Davidthistle@ocean.fsu.edu, Florida State Univ – Ob
Thiuri, Philip, (201) 595-2213 William Paterson College – Oo
Thole, Jeffrey T., (651) 696-6426 thole@macalester.edu, Macalester College – Gg
Thomas, David J., (315) 312-3065 thomas@oswego.edu, SUNY, Oswego – Pi
Thomas, Debbie, 979 862 7742 dthomas@ocean.tamu.edu, Texas A&M Univ – Ou
Thomas, Donald M., (808) 956-6482 dthomas@soest.hawaii.edu, Univ of Hawai'i, Manoa – Ca
Thomas, Ellen, (860) 685-2238 Yale Univ – Pm
Thomas, Ellen, (860) 685-2238 ethomas@wesleyan.edu, Wesleyan Univ – Pm
Thomas, Gerald A.gthomas@clarion.edu, Clarion Univ – Oo
Thomas, Helmuth, (902) 494-7177 helmuth.thomas@dal.ca, Dalhousie Univ – Oc
Thomas, James D., (954) 262-3661 thomasjd@ocean.nova.edu, Nova Southeastern Univ – Ob
Thomas, Jay, (518) 276-3247 thomasj2@rpi.edu, Rensselaer Polytechnic Institute – Ce
Thomas, Jim, (775) 887-7648 jmthomas@usgs.gov, Univ of Nevada, Reno – Cg
Thomas, Kimberly W., (505) 667-4379 Los Alamos National Lab – Ct
Thomas, Margaret A., (860) 424-3583 margaret.thomas@po.state.ct.us, Geological and Natural History Survey of Connecticut – Gg
Thomas, Peter, 361749-6768 rumbles@mail.utexas.edu, Univ of Texas at Austin – Ob
Thomas, Ray G., (352) 392-7984 rgthomas@geology.ufl.edu, Univ of Florida – Og
Thomas, Robert C., (406) 683-7615 r_thomas@umwestern.edu, Univ of Montana, Western – Gs
Thomas, Roger D. K., (717) 291-4135 roger.thomas@fandm.edu, Franklin and Marshall College – PoGh
Thomas, Sharon M., (413) 664-4511 North Adams State College – Oy
Thomas, William A., (859) 257-6222 geowat@uky.edu, Univ of Kentucky – Gt
Thomasson, Joseph R., (785) 628-5665 Fort Hays State Univ – Pb
Thompson, Allan M., (302) 831-2585 thompson@udel.edu, Univ of Delaware – Gx
Thompson, Carol A., (254) 968-9739 cthompson@tarleton.edu, Tarleton State Univ – Hy
Thompson, Christopher J., (509) 376-6602 chris.thompson@pnl.gov, Pacific Northwest National Lab – Ca
Thompson, David E., (601) 354-6328 Mississippi Office of Geology – Gr
Thompson, David W. J., (970) 491-8360 davet@atmos.colostate.edu, Colorado State Univ – Yr
Thompson, Geoffrey, (508) 289-2397 gthompson@whoi.edu, Woods Hole Oceanographic Institution – Cm
Thompson, George A., (650) 723-3714 Stanford Univ – Yg
Thompson, Glennis, (510) 642-7025 Univ of California, Berkeley – Oo
Thompson, J. Robert, (623) 845-3679 bob.thompson@gcmail.maricopa.edu, Glendale Community College – Ge
Thompson, Jann W. M., (202) 633-1357 Smithsonian Institution / National Museum of Natural History – Pg
Thompson, Joel B., (727) 864-8991 thompsjb@eckerd.edu, Eckerd College – Py
Thompson, John F. H., (604) 687-1117 Univ of British Columbia – Em
Thompson, Joseph L., (505) 667-4559 Los Alamos National Lab – Ct
Thompson, Keith R., (902) 494-3491 keith.thompson@dal.ca, Dalhousie Univ – Op
Thompson, Kenneth W., 620-341-5985 thompsok@emporia.edu, Emporia State Univ – Oe
Thompson, Lonnie G., (614) 292-6652 thompson.3@osu.edu, Ohio State Univ – Gl
Thompson, Lorraine K., (989) 774-2992 thompl1@cmich.edu, Central Michigan Univ – Oo
Thompson, LuAnne, (206) 543-9965 luanne@ocean.washington.edu, Univ of Washington – Op
Thompson, Margaret D., (781) 283-3029 mthompson@wellesley.edu, Wellesley College – Gc
Thompson, Michael D., (630) 252-9269 Argonne National Lab – Yg
Thompson, Michael L., (515) 294-2415 Iowa State Univ of Science & Technology – Sd
Thompson, Peter J., (603) 862-1718 pjt3@cisunix.unh.edu, Univ of New Hampshire – Gc
Thompson, Robert N., (+44) 191 3342331 r.n.thompson@durham.ac.uk, Durham Univ – Gi
Thompson, Terrie, (520) 621-3131 programs@hwr.arizona.edu, Univ of Arizona – Oo
Thompson, Thomas L., (520) 621-3670 thompson@ag.arizona.edu, Univ of Arizona – So
Thompson, Todd, (207) 287-2801 Indiana Geological Survey – Gs
Thompson, Tommy B., (702) 327-5146 Colorado State Univ – Em
Thompson, Tommy B., (775) 327-5146 tommyt@mines.unr.edu, Univ of Nevada, Reno – Eg
Thompson, Warren C., (408) 656-2552 Naval Postgraduate School – Op
Thompson, William T., (417) 836-5228 thompson-w@rmc.ca, McGill Univ – Nx
Thompson, Woodrow B., (207) 287-2801 woodrow.b.thompson@state.me.us, Univ of Maine – Gl
Thompson, Jr., James B., (617) 495-1672 thompson@eps.harvard.edu, Harvard Univ – Gp
Thoms, Richard E., (503) 725-3379 dick@ch1.ch.pdx.edu, Portland State Univ – Ps

Thomsen, Charles E., (916) 484-8184 thomsec@arc.losrios.edu, American River College – Oy
Thomsen, Laurenz A.l.thomsen@iu-bremen.de, Univ of Washington – Ob
Thomson, Colin J., (613) 545-6178 Queen's Univ – Ys
Thomson, Dennis W., (814) 863-1585 thomson@ems.psu.edu, The Pennsylvania State Univ, Univ Park – Ow
Thomson, Jennifer A., (509) 359-7478 jthomson@mail.ewu.edu, Eastern Washington Univ – Gp
Thomson, Kenneth C., (417) 836-4306 kct548f@smsu.edu, Southwest Missouri State Univ – Em
Thomson, Richard E., (250) 363-6555 ThomsonR@pac.dfo-mpo.gc.ca, Univ of British Columbia – Op
Thonnard, Norbert, (865) 974-9702 nthonnar@utk.edu, Univ of Tennessee, Knoxville – Cc
Thorbjarnarson, Kathryn W., (619) 594-5586 kthorbjarnarson@geology.sdsu.edu, San Diego State Univ – Hq
Thorkelson, Derek J., (604) 291-5390 dthorkel@sfu.ca, Simon Fraser Univ – Gt
Thorleifson, Harvey, (612) -62-7-47 (Ext. ext 224) thorleif@umn.edu, Univ of Minnesota, Twin Cities – Gg
Thorleifson, Leonard H., (612) 627-4780 Ext. 224 thorleif@umn.edu, Univ of Minnesota, Twin Cities – Gl
Thornberry-Ehrlich, Trista L., (970) 225-3584 trista@cnr.colostate.edu, Colorado State Univ – Gg
Thorncroft, Christopher D., (518) 442-4555 SUNY, Albany – Oa
Thorne, Kathryn F., (570) 662-4612 kthorne@mnsfld.edu, Mansfield Univ – Oo
Thorne, Paul D., (509) 372-4482 paul.thorne@pnl.gov, Pacific Northwest National Lab – Hg
Thorne, Richard M., (310) 825-5974 rmt@atmos.ucla.edu, Univ of California, Los Angeles – Oa
Thornton, Charles P., (814) 865-3836 cpthorn@gamewood.net, Pennsylvania State Univ, Univ Park – Gv
Thornton, Daniel C., 979 845 4092 dthornton@ocean.tamu.edu, Texas A&M Univ – Ob
Thornton, Edward B., (408) 656-2847 thornton@oc.nps.navy.mil, Naval Postgraduate School – On
Thornton, Edward C., (509) 373-0358 edward.thornton@pnl.gov, Pacific Northwest National Lab – Cg
Thorson, Robert M., (860) 486-1396 robert.thorson@uconn.edu, Univ of Connecticut – Gm
Thosteson, Eric D., (321) 674-8096 thosteson@fit.edu, Florida Institute of Technology – On
Thren, Robert C., (540) 458-8868 threnr@wlu.edu, Washington & Lee Univ – Oo
Thunell, Robert C., (803) 777-7593 Univ of South Carolina – Pe
Thurber, Clifford H., (608) 262-6027 clifft@geology.wisc.edu, Univ of Wisconsin, Madison – Ys
Thurber, David L., (718) 997-3300 Graduate School of the City Univ of New York – Cl
Thurber, David L., (718) 997-3325 dlthumper@aol.com, Queens College (CUNY) – Cl
Thurnau, Ellen, (202) 633-1821 thurnaue@si.edu, Smithsonian Institution / National Museum of Natural History – Oo
Thurtell, George W., (519) 824-4120 (Ext. 52453) gthurtel@lrs.uoguelph.ca, Univ of Guelph – Oa
Thwaites, Robin N., 61 7 3138 2400 r.thwaites@qut.edu.au, Queensland Univ of Technology – Os
Thy, PeterThy@geology.ucdavis.edu, Univ of California, Davis – Gx
Tiab, Djebbar, (405) 325-6777 dtiab@ou.edu, Univ of Oklahoma – Np
Tiampo, Kristy F., 519-661-3188 ktiampo@uwo.ca, Univ of Western Ontario – Yd
Tibert, Neil E., (540) 654-1423 ntibert@mwc.edu, Univ of Mary Washington – Pm
Tibi, Rigobert, (314) 935-7965 tibi@seismo.wustl.edu, Washington Univ – Oo
Tibuleac, Ileana, (775) 784-6256 ileana@seismo.unr.edu, Univ of Nevada, Reno – Ys
Tick, Geoffreygtick@geo.ua.edu, Univ of Alabama – Hw
Tidwell, Vincent C., 505.844.6025 vctidwe@sandia.gov, New Mexico Institute of Mining and Technology – Hw
Tiffney, Bruce H., (805) 893-2959 Univ of California, Santa Barbara – Pb
Tikoff, Basil, 608-262-4678 basil@geology.wisc.edu, Univ of Wisconsin, Madison – Gc
Tilbury, Graham, (727) 553-3989 gtilbury@seas.marine.usf.edu, Univ of South Florida, Saint Petersburg – Oo
Tilton, George R., (805) 893-3155 Univ of California, Santa Barbara – Cc
Tilton, Terry, (956) 721-5195 Laredo Community College – Gp
Timmons, Michael, (505) 835-5237 mtimmons@gis.nmt.edu, New Mexico Institute of Mining & Technology – Gg
Timmons, Stacy, (505) 835-6951 stacyt@gis.nmt.edu, New Mexico Institute of Mining & Technology – Gg
Timofeeff, Michael N., 607-777-2323 mtimofee@binghamton.edu, Binghamton Univ – Cl
Tindall, Sarah E., (610) 683-4446 tindall@kutztown.edu, Kutztown Univ of Pennsylvania – Gc
Ting, Mingfang, (217) 333-7105 ting@atmos.uiuc.edu, Columbia Univ – Oa
Tingey, David G., 801-422-7752 david_tingey@byu.edu, Brigham Young Univ – Oo
Tingley, Joseph V., (775) 784-6691 (Ext. 139) jtingley@unr.edu, Univ of Nevada – Em
Tingley, Susan L., (775) 784-6691 (Ext. 142) stingley@unr.edu, Univ of Nevada – Oy
Tinker, Scott W., (512) 471-0209 scott.tinker@beg.utexas.edu, Univ of Texas at Austin, Jackson School of Geosciences – Go
Tinkham, Doug, (705) 675-1151 (Ext. ext 2270) dtinkham@laurentian.ca, Laurentian Univ, Sudbury – Gp
Tinkler, Keith J., (905) 688-5550 (Ext. 3486) ktinkler@brocku.ca, Brock Univ – Gm
Tipping, Robert G., (612)627-4780 Ext. 226 tippi001@umn.edu, Univ of Minnesota, Twin Cities – Hw
Tircuit, Frances A., (225) 578-5320 ftirc@lsu.edu, Louisiana State Univ – Oo
Titley, Spencer R., (520) 621-6018 stitley@geo.arizona.edu, Univ of Arizona – Eg
Titus, Robert C., (607) 431-4733 titusr@hartwick.edu, Hartwick College – Ps
Titus, Sarah J., (507) 646-4419 stitus@carleton.edu, Carleton College – Gc
Tivey, Margaret K., (508) 289-3362 mtivey@whoi.edu, Woods Hole Oceanographic Institution – Cm
Tivey, Maurice A., (508) 289-2265 mtivey@whoi.edu, Woods Hole Oceanographic Institution – Gt
Toal, Gerard O., (540) 231-8047 Virginia Polytechnic Institute & State Univ – Oo
Tobia, Maria C., (916) 278-4683 maria@csus.edu, California State Univ, Sacramento – Gg
Tobias, Craig R., 910-962-3490 tobiasr@uncw.edu, Univ of North Carolina Wilmington – Cs
Tobin, Harold, 608-265-5796 htobin@geology.wisc.edu, Univ of Wisconsin, Madison – Yg
Tobin, Harold J., (505) 835-5920 tobin@griffy.nmt.edu, New Mexico Institute of Mining and Technology – Ys
Tobisch, Othmar T., (831) 459-2777 otobisch@pmc.ucsc.edu, Univ of California, Santa Cruz – Gc
Todd, Brenda, 402-554-2662 btodd@mail.unomaha.edu, Univ of Nebraska at Omaha – Oo
Todhunter, Paul, (701) 777-4593 Univ of North Dakota – Oy
Tokmakian, Robin T., (408) 656-3255 robint@ucar.edu, Naval Postgraduate School – Op
Toksoz, M N., (617) 253-7852 toksoz@mit.edu, Massachusetts Institute of Technology – Ys
Tolstoy, Maria, (845) 365-8791 Columbia Univ – Yr
Tomas, Carmelo R., (727) 896-8626 Univ of South Florida, Saint Petersburg – Ob
Tomascak, Paul B., 315 312 2786 tomascak@oswego.edu, SUNY, Oswego – Cg
Tomasko, Martin G., (520) 621-6969 mtomasko@lpl.arizona.edu, Univ of Arizona – Oa
Tomaso, Matthew S., (973) 655-7990 tomasom@mail.montclair.edu, Montclair State Univ – Ga
Tomhave, Dennis W., (605) 677-5227 dtomhave@usd.edu, South Dakota Dept of Env & Natural Resources – Gl
Tomkin, Jonathan, (225) 578-1984 jonathan@geol.lsu.edu, Louisiana State Univ – Gm
Tompson, Andrew F., (925) 422-6348 tompson1@llnl.gov, Lawrence Livermore National Lab – Ng
Tomson, Janice H., (562) 938-4448 jtomson@lbcc.cc.ca.us, Long Beach City College – Gs
Tomson, Mason B., (713) 527-6048 mtomson@rice.edu, Rice Univ – Cg
Tonon, Fulvio, 801-585-0491 tonon@mines.utah.edu, Univ of Utah – Nr
Toogood, Landria J., 404-656-3214 Georgia Dept of Nat Resources – Oo
Toole, Dierdre, (508) 289-3552 dtoole@whoi.edu,

Woods Hole Oceanographic Institution – Cm
Toomey, Douglas R., (541) 346-5576 drt@newberry.uoregon.edu, Univ of Oregon – Ys
Toppozada, Tousson R., (916) 322-9309 California Geological Survey – Ys
Toran, Laura, (215) 204-2352 ltoran@temple.edu, Temple Univ – Hw
Torgersen, Thomas, (860) 445-3441 tom.torgersen@mailgw.er.doe.gov, Univ of Connecticut – Cl
Torlaschi, Enrico, (514) 987-3000 enrico@maia.sca.uqam.ca, Universite du Quebec a Montreal – Oo
Tormey, Brian B., (814) 949-5272 bbt1@psu.edu, Pennsylvania State Univ, Altoona – Gm
Toro, Jaime, (304) 293-5603 (Ext. xt. 4327) jtoro@wvu.edu, West Virginia Univ – Gc
Torreano, Scott, (615) 598-1271 storrean@sewanee.edu, Univ of the South – Sf
Torrens, Hugh S., +44 01782 583754 h.s.torrens@keele.ac.uk, Keele Univ – Gh
Torrents, Alba, (301) 405-1979 alba@eng.umd.edu, Univ of Maryland – Sb
Torres, Joseph J., (727) 553-1169 jtorres@marine.usf.edu, Univ of South Florida, Saint Petersburg – Ob
Torres, Marta, (541) 737-2902 mtorres@coas.oregonstate.edu, Oregon State Univ – OcCa
Torres, Michelle, 517-353-3275 torres@msu.edu, Michigan State Univ – Oo
Torres, Raymond, (803) 777-4506 torres@geol.sc.edu, Univ of South Carolina – Hy
Torrez, Betsy D., (409) 294-1560 Sam Houston State Univ – Gc
Tortell, Philippe, (604) 822-4728 ptortell@eos.ubc.ca, Univ of British Columbia – Ob
Toscano, Marguerite A., (727) 893-3100 Univ of South Florida, Saint Petersburg – Gu
Tosdal, Richard, (604) 822-5149 rtosdal@eos.ubc.ca, Univ of British Columbia – Em
Tóth, József, (780) 492-5740 joe.toth@ualberta.ca, Univ of Alberta – Hw
Totten, Iris M., (785) 532-2251 itotten@ksu.edu, Kansas State Univ – Oe
Totten, Matthew W., (785) 532-2227 mtotten@ksu.edu, Kansas State Univ – Gso
Totten, Stanley M., (812) 866-7245 totten@hanover.edu, Hanover College – Gl
Touchet, Arville, (318) 643-1644 Univ of Louisiana, Lafayette – Sd
Touchton, Joseph T., (334) 844-3952 jtouchto@acesag.auburn.edu, Auburn Univ – Sc
Towell, David G., (812) 855-0352 towell@indiana.edu, Indiana Univ, Bloomington – Cg
Towers, George W., (304) 384-6040 Concord Univ – Oi
Towery, Brooke L., (904) 484-1151 Pensacola Junior College – Gg
Townsend, David W., (207) 581-4381 Univ of Maine – Ob
Townsend, Margaret A., (785) 864-2111 townsend@kgs.ku.edu, Univ of Kansas – Hw
Townsend, Philip A., 301-689-7124 townsend@al.umces.edu, Univ of Maryland – Or
Tracy, Robert J., (540) 231-5980 rtracy@vt.edu, Virginia Polytechnic Institute & State Univ – Gp
Traganza, Eugene D., (408) 656-2552 Naval Postgraduate School – Oc
Tramer, Elliot J., (419) 530-2451 elliot.tramer@utoledo.edu, Univ of Toledo – Oo
Tran, Linda C., 281-765-7865 Linda.C.Tran@nhmccd.edu, North Harris College – Oi
Trapasso, Louis M., (270) 745-5986 Western Kentucky Univ – Oa
Trapp, R. J., (765) 496-6661 jtrapp@purdue.edu, Purdue Univ – Oa
Trapp, Richard A., (520) 770-3500 Arizona Geological Survey – Gg
Trask, C. Brian, 217-244-2421 trask@isgs.uiuc.edu, Illinois State Geological Survey – Ge
Trautz, Robert C., (510) 486-7954 rctrautz@lbl.gov, Lawrence Berkeley National Lab – Hw
Traverse, Alfred, (814) 863-3419 traverse@ems.psu.edu, Pennsylvania State Univ, Univ Park – Pl
Travis, Bryan J., (505) 667-1254 bjtravis@lanl.gov, Los Alamos National Lab – Hq
Treadwell-Steitz, Carol, (518) 564-2028 Plattsburgh State Univ (SUNY) – Gg
Trease, Lynn, (505) 667-0140 llt@lanl.gov, Los Alamos National Lab – Oo
Trefry, John H., (321) 674-8096 jtrefry@fit.edu, Florida Institute of Technology – Cm
Trefry, John H., (321) 674-7305 jtrefry@fit.edu, Florida Institute of Technology – Cm
Trehu, Anne M., (541) 737-2655 trehu@coas.oregonstate.edu, Oregon State Univ – Ygs
Treiman, Jerome A., (213) 620-3560 California Geological Survey – Ng
Trembanis, Arthur C., 302-831-2498 art@udel.edu, Univ of Delaware – On
Tremblay, Alain, (418) 654-2568 atremblay@inrs.uquebec.ca, Universite du Quebec – Gc
Tremblay, Bruno, (845) 365-8767 Columbia Univ – Op
Tremblay, Bruno, (514) 398-4369 bruno.tremblay@mcgill.ca, McGill Univ – Op
Tremblay, Michel, (418) 545-5011 Univ du Quebec a Chicoutimi – Oo
Tremblay, Thomas A., (512) 475-9537 tom.tremblay@beg.utexas.edu, Univ of Texas at Austin, Jackson School of Geosciences – Oy
Trenhaile, Alan S., (519) 253-2184 tren@uwindsor.ca, Univ of Windsor – Gm
Trent, Victoria P., (404) 656-3214 Georgia Dept of Nat Resources – Oi
Trentham, Robert C., 432 552-2432 trentham_r@utpb.edu, Univ of Texas, Permian Basin – Go
Treves, Samuel B., (402) 472-0872 streves1@unl.edu, Univ of Nebraska, Lincoln – Gi
Trevino, Ramon H., (512) 471-8920 ramon.trevino@beg.utexas.edu, Univ of Texas at Austin, Jackson School of Geosciences – Go
Treworgy, Janis D., (618) 374-5294 janis.treworgy@prin.edu, Principia College – GgPv
Trexler, James H., (775) 784-1504 trexler@mines.unr.edu, Univ of Nevada, Reno – Gs
Triay, Ines, (505) 665-1755 Los Alamos National Lab – Hg
Tribble, Gordon, (808) 522-9282 Univ of Hawai'i, Manoa – Og
Trifonoff, Karen M., (570) 389-4569 trifonof@bloomu.edu, Bloomsburg Univ – Oi
Triplehorn, Donald M., (907) 474-6891 Univ of Alaska, Fairbanks – Gs
Tripoli, Gregory J., (608)262-3700 tripoli@aos.wisc.edu, Univ of Wisconsin, Madison – Ow
Tripp, Bryce T., (801) 537-3317 nrugs.btripp@state.ut.us, Utah Geological Survey – En
Trixler, Frank, 089/2180 4317 Trixler@lrz.uni-muenchen.de, LMU - Univ of Munich – Gz
Trnka, Maureentrnka@nova.edu, Nova Southeastern Univ – Oo
Troch, Peter A., (520) 626-1277 patroch@hwr.arizona.edu, Univ of Arizona – Hs
Trocine, Robert P., (407) 674-8096 Florida Institute of Tech – Cm
Troedsson, Nils, 406-683-5563 Univ of Montana, Western – Oo
Troell, Arthur R., (210) 733-2829 atroell@mail.accd.edu, San Antonio College – GgOg
Tromp, Jeroen, (626) 395-8117 jtromp@gps.caltech.edu, California Institute of Technology – Ys
Trop, Jeffrey M., (570) 577-3027 jtrop@bucknell.edu, Bucknell Univ – Gs
Trost, G K., (501) 575-7317 garrett@dicksonstreet.com, Univ of Arkansas, Fayetteville – Oo
Trowbridge, Leslie W., (970) 35-1086 lesliewtrowbridge@msn.com, Univ of Northern Colorado – Oe
Troy, Marleen, (570) 408-4615 Wilkes Univ – Hw
Trudell, Leslie, (920) 832-6739 leslie.r.trudell@lawrence.edu, Lawrence Univ – Oo
Trudgill, Bruce D., (303) 273-3883 btrudgil@mines.edu, Colorado School of Mines – Gc
Trujillo, Alan P., (760) 744-1150 (Ext. xt. 2734) atrujillo@palomar.edu, Palomar College – Og
Trumbore, Susan E., (949) 824-6142 setrumbo@uci.edu, Univ of California, Irvine – Sc
Trupe, Charles H., (912) 681-0337 CHTrupe@GeorgiaSouthern.edu, Georgia Southern Univ – Gc
Trust, Michael, (617) 287-5287 Univ of Massachusetts, Boston/Harbor Campus – Or
Trzcienski, Jr., Walter E., (514) 343-5977 trzwalt@fas.umontreal.ca, Universite de Montreal – Gx
Tsang, Chin-Fu, (510) 486-5782 cftsang@lbl.gov, Lawrence Berkeley National Lab – Hw
Tsang, Monica, (513) 529-3216 Miami Univ – Hw
Tsang, Yvonne Y., (510) 486-7047 ytsang@lbl.gov, Lawrence Berkeley National Lab – Hw
Tschudi, Marktschudi@car.edu, Fort Hays State Univ – Or
Tsegaye, Teferi D., (205) 851-5462 tlsegaye@aamu.edu, Alabama A&M Univ – Sp
Tseng, Peng-Hsiang, (505) 667-5518 tseng@lanl.gov,

Los Alamos National Lab – Ng
Tshudy, Dale, (814) 732-2453 dtshudy@edinboro.edu, Edinboro Univ of Pennsylvania – Pi
Tso, Jonathan L., (540) 831-5638 jtso@radford.edu, Radford Univ – Gp
Tsoflias, George P., 785-864-4584 tsoflias@ku.edu, Univ of Kansas – Yg
Tsongas, Theodora A., (509) 335-6422 tsongas@wsunix.wsu.edu, Washington State Univ – Oo
Tsujita, Cameron J., (519) 661-2111 (Ext. 86740) ctsujita@uwo.ca, Univ of Western Ontario – Pi
Tsvankin, Ilya D., (303) 273-3060 ilya@dix.mines.edu, Colorado School of Mines – Ye
Tu, Wei, 912 681-5233 WTu@GeorgiaSouthern.edu, Georgia Southern Univ – Oi
Tucholke, Brian E., (508) 289-2494 btucholke@whoi.edu, Woods Hole Oceanographic Institution – Gu
Tucker, Greg E., 303-492-6985 Univ of Colorado – Gm
Tucker, Maurice E., 0191 37425240 m.e.tucker@durham.ac.uk, Durham Univ – Gs
Tucker, Robert D., (314) 935-6892 tucker@levee.wustl.edu, Washington Univ – Cc
Tucker, Stevens P., (408) 656-3162 Naval Postgraduate School – Og
Tuit, Caroline B., (609) 258-7428 ctuit@princeton.edu, Princeton Univ – Ob
Tulaczyk, Slawek, (831) 459-5207 tulaczyk@pmc.ucsc.edu, Univ of California, Santa Cruz – Gl
Tull, James F., (850) 644-1448 tull@gly.fsu.edu, Florida State Univ – Gc
Tullis, Jan A., (401) 863-1921 jan_tullis@brown.edu, Brown Univ – Gc
Tullis, Jason A., (479) 575-4770 jatullis@uark.edu, Univ of Arkansas, Fayetteville – Or
Tullis, Terry E., (401) 863-3829 terry_tullis@brown.edu, Brown Univ – Yx
Tuman, Vladmir S., (209) 667-3466 California State Univ, Stanislaus – Yv
Tung, Wen-wen, (765) 494-0272 wwtung@purdue.edu, Purdue Univ – Oa
Tunnicliffe, Verena, (250) 721-7135 Univ of Victoria – Ob
Tupas, Luis M., (808) 956-2449 ltupas@soest.hawaii.edu, Univ of Hawai'i, Manoa – Ob
Turco, Richard, (310) 825-6936 turco@atmos.ucla.edu, Univ of California, Los Angeles – Oa
Turcotte, Donald L., (607) 255-7282 turcotte@geology.cornell.edu, Cornell Univ – Yg
Turcotte, Donald L., 530-752-6808 Turcotte@geology.ucdavis.edu, Univ of California, Davis – Yg
Turekian, Karl K., (203) 432-3188 kkt@hess.geology.yale.edu, Yale Univ – Xm
Turnbull, Margaret, (202) 478-8863 turnbull@dtm.ciw.edu, Carnegie Institution of Washington – Oo
Turnbull, William D., (312) 665-7092 turnbull@fmnh.org, Field Museum of Natural History – Pv
Turner, A. Keith, (303) 273-3802 kturner@mines.edu, Colorado School of Mines – Ng
Turner, Brian R., 0191 3742539 b.r.turner@durham.ac.uk, Durham Univ – Gs
Turner, Elizabeth, (705) 675-1151 (Ext. ext. 2267) eturner@laurentian.ca, Laurentian Univ, Sudbury – Pg
Turner, Eugene, (818) 677-3532 eugene.turner@csun.edu, California State Univ, Northridge – Or
Turner, Lana, (314) 935-5480 Saint Petersburg Junior College – Oo
Turner, Mort D., (314) 935-5480 Univ of Colorado, Denver – Eg
Turner, Raymond M., (520) 629-9455 Univ of Arizona – Po
Turner, Robert E., (225) 388-6454 Louisiana State Univ – Ob
Turner, Robert S., rtt@ornl.gov, Oak Ridge National Lab – Cl
Turnock, Allan C., (204) 474-6911 ac_turnock@umanitoba.ca, Univ of Manitoba – Gp
Turpening, Roger M., (906) 487-1784 roger@mtu.edu, Michigan Technological Univ – Ye
Turski, Mark P., (603) 535-2749 markt@psc.plymouth.edu, Plymouth State College – Og
Tweddle, Jacqueline, (541) 737-6560 jtweddle@coas.oregonstate.edu, Oregon State Univ – Ob
Twine, Tracy E., (217) 333-7105 twine@atmos.uiuc.edu, Univ of Illinois, Urbana-Champaign – Oa
Twiss, Robert J., (530) 752-1860 twiss@geology.ucdavis.edu, Univ of California, Davis – Gc
Twohy, Cynthia, (541) 737-5690 twohy@coas.oregonstate.edu, Oregon State Univ – Oa
Tyburczy, James A., (480) 965-2637 jim.tyburczy@asu.edu, Arizona State Univ – Yx
Tyce, Robert, (401) 874-6879 tyce@oce.uri.edu, Univ of Rhode Island – Ou
Tyler, Dan, (541) 737-5692 dtyler@coas.oregonstate.edu, Oregon State Univ – Oa
Tyler, E. Jerry, (608) 262-0853 ejtyler@wisc.edu, Univ of Wisconsin, Madison – Sd
Tyler, Scott W., 775-784-6250 tyler@unr.edu, Univ of Nevada, Reno – Hg
Tyler, Stanley C., (949) 824-2685 styler@uci.edu, Univ of California, Irvine – Oa
Tynan, Cynthia T., (206) 860-6793 Cynthia.Tynan@noaa.gov, Univ of Washington – Ob
Tziperman, Eli, (617) 384-8381 Harvard Univ – Og
Törnqvist, Torbjörn E., 504-314-2221 tor@tulane.edu, Tulane Univ – Gs

U

Uchupi, Elazar, (508) 289-2830 Woods Hole Oceanographic Institution – Ou
Uddin, Ashraf, (334) 844-4885 uddinas@auburn.edu, Auburn Univ – Gt
Ufnar, David F., 601 266-5423 David.Ufnar@usm.edu, Univ of Southern Mississippi – Gs
Ugland, Richard, 541-552.6479 ugland@sou.edu, Southern Oregon Univ – Gg
Uhle, Maria, (865) 974-2366 Univ of Tennessee, Knoxville – Cl
Ulanski, Stanley L., (540) 568-6130 ulanski@jmu.edu, James Madison Univ – Og
Uliana, Matthew M., 570-372-4215 uliana@susqu.edu, Susquehanna Univ – Hw
Ullman, William J., (302) 645-4302 ullman@udel.edu, Univ of Delaware – Cm
Ulmer, Gene C., (215) 204-7171 gulmer@temple.edu, Temple Univ – Cp
Ulmer-Scholle, Dana S., (505) 835-5673 dulmer@nmt.edu, New Mexico Institute of Mining and Technology – Gs
Ulrych, Tadeusz J., (604) 822-3100 tulrych@eos.ubc.ca, Univ of British Columbia – Yg
Umhoefer, Paul J., (928) 523-6464 paul.umhoefer@nau.edu, Northern Arizona Univ – Gt
Underwood, Jeffrey, (775) 784-6999 jeffu@unr.edu, Univ of Nevada, Reno – Ow
Underwood, Michael B., (573) 882-4685 UnderwoodM@missouri.edu, Univ of Missouri – Gs
Unkefer, Clifford J., (505) 665-2560 Los Alamos National Lab – Oo
Unrug, Kot F., (606) 257-1883 Univ of Kentucky – Nr
Unruh, Jeffrey R., (530) 752-0352 unruh@geology.ucdavis.edu, Univ of California, Davis – Gc
Unsworth, Martyn, (780) 492-3041 unsworth@phys.ualberta.ca, Univ of Alberta – Oo
Unsworth, Michael H., (541) 737-5428 unswortm@coas.oregonstate.edu, Oregon State Univ – Oa
Untersteiner, Norbert, 206-543-1367 norbert@atmos.washington.edu, Univ of Washington – Oo
Upadhyay, Hans D., (773) 442-6055 h-upadhyay@neiu.edu, Northeastern Illinois Univ – Gc
Upchurch, Michael L., (434) 951-6363 michael.upchurch@dmme.virginia.gov, Division of Mineral Resources – Gd
Uptegrove, Jane, 609-292-2576 jane.uptegrove@dep.state.nj.us, New Jersey Geological Survey – Gr
Urban, Edward, (410) 516-4239 scor@jhu.edu, The Johns Hopkins Univ – Og
Urbanczyk, Kevin, (915) 837-8110 kevinu@sulross.edu, Sul Ross State Univ – Gi
Urbano, Lensyllurbano@memphis.edu, Univ of Memphis – Hw
Uren, Norman F., +61 8 9266-3565 uren@geophy.curtin.edu.au, Curtin Univ of Technology – Ye
Urosevic, Milovan, +61 8 9266-2296 M.Urosevic@curtin.edu.au, Curtin Univ of Technology – YerYx
Urquhart, Joannejurquhar@bowdoin.edu, Bowdoin College – Ge
Urquhart, Mary, 972-883-2496 urquhart@utdallas.ede, Univ of Texas, Dallas – Xy
Urzua, Alfredo, 617-552-8339 alfredo.urzua.1@bc.edu, Boston College – Hq

Urzua, Alfredo, (617) 552-8339 urzua@bc.edu, Boston College – Hq
Ustin, Susan, (530) 752-0621 slustin@ucdavis.edu, Univ of California, Davis – Or
Ustin, Susan L., (530) 752-0621 slustin@ucdavis.edu, Univ of California, Davis – Or
Utgaard, John E., (618) 453-3351 utgaard@geo.siu.edu, Southern Illinois Univ Carbondale – Pi
Utgard, Russell O., 614 247-4246 rutgard@regents.state.oh.us, Ohio State Univ – Ge
Utley, Tom, 321-674-8120 tutley@fit.edu, Florida Institute of Technology – Ow
Utter, James M., (914) 251-6642 SUNY, Purchase – Oo
Uzunlar, Nuri, 605-394-2494 Nuri.Uzunlar@sdsmt.edu, South Dakota School of Mines & Technology – Gg

V

Vacher, H. Leonard, (813) 974-5267 vacher@chuma.cas.usf.edu, Univ of South Florida, Tampa – Hw
Vacquier, Victor D., (619) 534-4803 vvacquier@ucsd.edu, Univ of California, San Diego – Ob
Vadas, Robert L., (207) 581-2974 Univ of Maine – Ob
Vaiden, Robert C., 217-244-4299 vaiden@isgs.uiuc.edu, Illinois State Geological Survey – Hw
Vail, Lance W., (509) 372-6237 lance.vailv, Pacific Northwest National Lab – Hg
Vail, Peter R., (713) 348-4888 vail@rice.edu, Rice Univ – Gr
Vail, Steve, (201) 595-2545 William Paterson College – Oo
Vajoczki, Susan, (905) 525-9140 (Ext. 23520) vajoczki@mcmaster.ca, McMaster Univ – Yr
Valastro, Salvadore, (512) 471-6600 Austin Community College – Cc
Valdes, Juan B., (520) 621-2266 jvaldes@email.arizona.edu, Univ of Arizona – Hs
Valenti, Christine, (973) 655-4448 Montclair State Univ – Gg
Valentine, Gregory A., (505) 665-0259 Los Alamos National Lab – Yg
Valentine, James W., (510) 643-5791 Univ of California, Berkeley – Po
Valentine, Michael J., (253) 879-3129 mvalentine@ups.edu, Univ of Puget Sound – Ym
Valentino, David W., (315) 312-2798 dvalenti@oswego.edu, SUNY, Oswego – GtcGp
Valentino, Karen R., (315) 312-5506 kvalent1@oswego.edu, SUNY, Oswego – GgOm
Vali, H., 514-398-3025 vali@eps.mcgill.ca, McGill Univ – Om
Vallance, James W., (514) 398-6860 Michigan Technological Univ – Gv
Valle-Levinson, Arnoldo, 75768305578 arnoldo@ccpo.odu.edu, Old Dominion Univ – Op
Valley, John W., (608) 263-5659 valley@geology.wisc.edu, Univ of Wisconsin, Madison – Gp
Van Alstine, James B., (320) 589-6313 vanalstj@mrs.umn.edu, Univ of Minnesota, Morris – Pg
Van Arsdale, Roy B., (901) 678-4356 rvanrsdl@memphis.edu, Univ of Memphis – Gc
Van Avendonk, Harm, 512-471-0429 harm@ig.utexas.edu, Univ of Texas at Austin, Jackson School of Geosciences – Yg
Van Besien, Alphonse C., (662) 236-1048 Univ of Mississippi – Ng
Van Biersel, Thomas P., 225/578-5320 tvanbier@lsu.edu, Louisiana State Univ – Hw
Van Buren, Fatemah, (510) 643-2684 Univ of California, Berkeley – Oo
Van Burgh, Dana P., (307) 268-2536 vbbison@aol.com, Casper College – Oe
Van De Poll, Henk W., (506) 453-4804 Univ of New Brunswick – Gs
Van de Ven, Christopher M., 517-629-0633 cvandeven@albion.edu, Albion College – Oi
Van Der Flier-Keller, Eileen, (250) 472-4019 fkeller@uvic.ca, Univ of Victoria – Cl
van der Hilst, Robert, (617) 253-6977 hilst@mit.edu, Massachusetts Institute of Technology – Ys
Van der Hoven, Stephen, (309) 438-3493 sjvande2@ilstu.edu, Illinois State Univ – Hg
van der Lee, Suzan, 847-491-8183 suzan@earth.northwestern.edu, Northwestern Univ – Ys
van der Pluijm, Ben A., (734) 763-0373 vdpluijm@umich.edu, Univ of Michigan – Gc
van der Vink, Gregorygvdv@princeton.edu, Princeton Univ – Gg
van der Voo, Rob, (734) 764-8322 voo@umich.edu, Univ of Michigan – Ym
Van Devender, Thomas R., (520) 883-3029 Univ of Arizona – Gg
van Dijk, Deanna, (616) 526-6510 dvandijk@calvin.edu, Calvin College – GmOny
Van Diver, Bradford B. Bravandi@cs.com, SUNY, Potsdam – Gp
Van Doren, Lisa, (614) 265-6589 lisa.vandoren@dnr.state.oh.us, Ohio Dept of Natural Resources – Oo
Van Geen, Alexander, (845) 365-8644 Columbia Univ – Cg
van Hees, Edmond H., (313) 577-9436 midas@wayne.edu, Wayne State Univ – Em
Van Horn, Stephen R., (740) 826-8306 svanhorn@muskingum.edu, Muskingum College – Ge
Van Houten, Franklyn B., (609) 258-4115 fbvan@princeton.edu, Princeton Univ – Gs
Van Iten, Heyo, (812) 866-7303 vaniten@hanover.edu, Hanover College – Pg
van Keken, Peter J., (734) 764-1497 keken@umich.edu, Univ of Michigan – Yg
Van Kooten, Gerald K., (616) 526-6374 gkv2@calvin.edu, Calvin College – EgCePi
Van Leuven, Judy, (831) 459-4478 judy@es.ucsc.edu, Univ of California, Santa Cruz – Oo
Van Mooy, Benjamin, (508) 289-2740 bvanmooy@whoi.edu, Woods Hole Oceanographic Institution – Oc
Van Niewenhuise, Donald, 713-743-3399 Univ of Houston – Ps
van Oss, Carel J., 716-829-2900 SUNY, Buffalo – Gy
Van Roosendaal, Susan, (801) 581-8218 Univ of Utah – Oo
Van Ryswick, Stephen, 410-554-5544 svanryswick@dnr.state.md.us, Maryland Geological Survey – Ge
Van Schmus, W. Randall, (785) 864-2727 rvschmus@ku.edu, Univ of Kansas – Cc
Van Staal, Cees, (613) 995-4333 Cees.Van_Staal@nrcan-rncan.gc.ca, Univ of Ottawa – Gt
Van Tassell, Jay, (541) 962-3351 jvantass@eou.edu, Eastern Oregon Univ – GsrPg
Van Vleet, Edward S., (727) 553-1165 evanvleet@marine.usf.edu, Univ of South Florida, Saint Petersburg – Co
VanAller-Hernick, Linda A., (518) 486-3699 New York State Geological Survey – Pg
VanBrocklin, Matt, (315) 229-5197 mvanbrocklin, St. Lawrence Univ – Oo
Vance, Joseph A., 206-685-0118 Univ of Washington – Gi
Vance, R. K., (912) 681-5640 RKvance@GeorgiaSouthern.edu, Georgia Southern Univ – Eg
Vandemark, Douglas C., 603-862-0195 doug.vandemark@unh.edu, Univ of New Hampshire – Or
Vander Auwera, Jacqueline, 32 4 366 22 53 jvdauwera@ulg.ac.be, Universite de Liege – Gx
Vanderlip, Richard L., (913) 532-7249 Kansas State Univ – So
Vandewege, Reynold J., 662-352-8431 rjv16@msstate.edu, Mississippi State Univ – Ow
VanGundy, Robert D., (540) 328-0201 Univ of Virginia College, Wise – Gs
Vanhoff, Barry, (541) 737-6840 bvanhoff@coas.oregonstate.edu, Oregon State Univ – Op
VanHorn, Jason, (616) 526-7623 jev35@calvin.edu, Calvin College – Oi
Vaniman, David T., (505) 667-1863 vaniman@lanl.gov, Los Alamos National Lab – Gx
Vanko, David A., (410) 704-2220 dvanko@towson.edu, Towson Univ – Gx
Vann, David R., (215) 898-4906 drvann@sas.upenn.edu, Univ of Pennsylvania – Py
Vaos, Stephen, 954-201-6674 svaos@broward.edu, Broward Community College, Central Campus – Gu
Varadi, Ferenc D., (310) 206-6484 Univ of California, Los Angeles – Oo
Varady, Robert G., (520) 626-4393 rvarady@email.arizona.edu, Univ of Arizona – Oo
Varanasi, Prasad, (631) 632-8313 pvaranasi@notes.cc.sunysb.edu, SUNY, Stony Brook – Oa
Varekamp, Johan C., (860) 685-2248 jvarekamp@wesleyan.edu, Wesleyan Univ – CgGzv
Varga, Robert J., (330) 263-2202 rvarga@wooster.edu, The College of Wooster – Gc
Vargo, Gabriel A., (727) 553-1167 gvargo@marine.usf.edu, Univ of South Florida, Saint Petersburg – Ob
Varley, Deborah, (604) 822-2034 dvarley@eos.ubc.ca, Univ of British Columbia – Oo
Varma, Madan M., (610) 683-4441 varma@kutztown.edu, Kutztown Univ of Pennsylvania – Yg
Varol, Bakivarol@eng.ankara.edu.tr, Ankara Univ – Gd

Faculty Index -V

Varricchio, David J., 406-994-6907 djv@montana.edu,
Montana State Univ – Pv
Vasquez, Cecile I., 413-538-2278 cvasquez@mtholyoke.edu,
Mount Holyoke College – Oo
Vassiliou, Andreas H., 973-353-5100 ahvass@andromeda.rutgers.edu,
Rutgers, The State Univ of New Jersey, Newark – Gz
Vastano, Andrew C., (979) 845-9826 dvastano@ocean.tamu.edu,
Texas A&M Univ – Op
Vaughan, Michael T., 631 632-8030 michael.vaughan@sunysb.edu,
SUNY, Stony Brook – Yx
Vazquez, Jorge A., (818) 677-2564 jvazquez@csun.edu,
California State Univ, Northridge – Gi
Vazquez, Rogeliorvazquez@cicese.mx,
Centro de Inv Científica y de Educación Superior de Ensenada – Ye
Veblen, David R., (410) 516-8487 dveblen@jhu.edu,
The Johns Hopkins Univ – Gz
Veblen, Linda, (410) 516-4654 lav@nrc.gov,
The Johns Hopkins Univ – Gi
Veblen, Thomas T., (303) 492-8528 veblen@colorado.edu,
Univ of Colorado – Og
Vedder-Greene, Jocelyn, (801) 585-9762 Univ of Utah – Ca
Veeder, Glenn J., (818) 354-7388 Jet Propulsion Lab – Oo
Veeger, Anne I., (401) 792-2187 veeger@uri.edu,
Univ of Rhode Island – Hw
Vega, Anthony J., (814) 226-2023 avega@clarion.edu,
Clarion Univ – Oo
Veisze, Paul M., 916-563-3276 veiszep@arc.losrios.edu,
American River College – Oi
Veizer, Jan, (613)562-5800 Ext 6461 Veizer@science.uottawa.ca,
Univ of Ottawa – Cl
Velasco, Aaron A., (505) 665-8037 velasco@lanl.gov,
Los Alamos National Lab – Yg
Velasco, Aaron A., (915) 747-5101 velasco@geo.utep.edu,
Univ of Texas, El Paso – Yx
Velbel, Michael A., 517-355-4626 velbel@msu.edu,
Michigan State Univ – Gd
Vendeville, Bruno C., (512) 471-8334 bvdv@mail.utexas.edu,
Univ of Texas at Austin, Jackson School of Geosciences – Gc
Vengosh, Avner, 919 684 5847 vengosh@duke.edu,
Duke Univ – Hw
Venkatesan, M I., (310) 206-2561 Univ of California, Los Angeles – Co
Venn, Cynthia, (570) 389-4141 cvenn@bloomu.edu,
Bloomsburg Univ – OgClOc
Vennum, Walt, (707) 664-2942 walt.vennum@sonoma.edu,
Sonoma State Univ – Gi
Venteris, Erik R., 614-265-6459 erik.venteris@dnr.state.oh.us,
Ohio Dept of Natural Resources – Sc
Vento, Frank J.fvento@clarion.edu,
Clarion Univ – Ga
Ventura, Stephen J., (608) 262-6416 sventura@wisc.edu,
Univ of Wisconsin, Madison – Or
Venzke, Edward, (202) 633-1822 venzkee@si.edu,
Smithsonian Institution / National Museum of Natural History – Gv
Vepraskas, Michael J., (919) 515-1458 North Carolina State Univ – Sd
Ver Ploeg, Alan J., (307)766-2286e230 verploeg@uwyo.edu,
Wyoming State Geological Survey – Gr
Ver Straeten, Charles, 486-2004 cverstrae@mail.nysed.gov,
New York State Geological Survey – Gr
Verardo, Stacey, (703) 993-1045 sverardo@gmu.edu,
George Mason Univ – Pe
Verburg, Paul S. J., (775) 673-7425 pverburg@dri.edu,
Desert Research Institute – Sb
Verhoef, Annea.verhoef@reading.ac.uk,
Univ of Reading – Sp
Verlinde, Johannes, (814) 863-9711 verlinde@essc.psu.edu,
The Pennsylvania State Univ, Univ Park – Ow
Verma, Shashi B., (402) 472-6702 agme009@unlvm.unl.edu,
Univ of Nebraska, Lincoln – Oa
Vermeij, Gareet, (916) 752-2234 Univ of California, Berkeley – Po
Vermeij, Geerat J., (530) 752-2234 vermeij@geology.ucdavis.edu,
Univ of California, Davis – Pe
Vermeul, Vincent R., (509) 376-8316 vince.vermeul@pnl.gov,
Pacific Northwest National Lab – Hg
Vermillion, Nicole, 770-274-5050 ndecorte@gpc.edu,
Georgia Perimeter College – Gg
Veronis, George, (203) 432-3148 george.veronis@yale.edu,
Yale Univ – Op
Verosub, Kenneth L., (530) 752-6911 verosub@geology.ucdavis.edu,
Univ of California, Davis – Ym
Versteeg, Roelofversteeg@ldeo.columbia.edu,
Columbia Univ – Yg
Vervoort, Jeffrey D., 509-335-5597 vervoort@wsu.edu,
Washington State Univ – Ca
Vervoort, Jeffrey D., 509-335-5597 vervoort@wsu.edu,
Washington State Univ – Cc
Vesecky, John F., (313) 764-5151 Univ of Michigan – Or
Vesper, Dorothy, 304 293 5603 dorothy.vesper@mail.wvu.edu,
West Virginia Univ – Hw
Vespucci, Paul D., (703) 323-3113 nvvespp@nv.cc.va.us,
Northern Virginia Community College - Loudoun Campus – Ca
Vetter, Eric W., (808) 544-0200 evetter@hpu.edu,
Univ of Hawai'i, Manoa – Oo
Vetter, Scott K., (318) 869-5055 svetter@centenary.edu,
Centenary College of Louisiana – Gi
Veverka, Joseph, (607) 255-3507 veverka@cuspif.tn.cornell.edu,
Cornell Univ – Xg
Vezina, Alain F., (902) 426-7706 vezinaa@dfo-mpo.gc.ca,
Dalhousie Univ – Ob
Viani, Brian E., (925) 423-2001 viani@llnl.gov,
Lawrence Livermore National Lab – Cg
Vice, Daniel H., (570) 450-3112 dhv1@psu.edu,
Pennsylvania State Univ, Hazleton – Eg
Vice, Mari A., (608) 342-1055 Univ of Wisconsin, Platteville – Gd
Vicenzi, Edward P., (202) 633-1824 vicenzi@si.edu,
Smithsonian Inst / National Museum of Natural History – CaXc
Vick, Timothy D., 507-646-4401 tvick@carleton.edu,
Carleton College – Oe
Vidal, Antoniovidalv@cicese.mx,
Centro de Inv Científica y de Educación Superior de Ensenada – Ys
Vidal Romaní, Juan Ramon, 00 34 981 167000 xemoncho@udc.es,
Coruna Univ – Gm
Vidale, John E., 206-543-6790 vidale@ess.washington.edu,
Univ of Washington – YsGt
Videtich, Patricia E., 616-331-3887 videticp@gvsu.edu,
Grand Valley State Univ – Gd
Vieira, David J., (505) 667-7231 Los Alamos National Lab – Oo
Vieira, Mario E. C., (410) 293-6567 United States Naval Academy – Op
Vierrether, Chris, (573) 368-2143 Missouri Dept of Nat Resources – Oo
Vignona, Laine C., (715) 425-3729 laine.vignona@uwrf.edu,
Univ of Wisconsin, River Falls – Cl
Vigrass, Laurence W., (306) 757-7800 Univ of Regina – Eo
Viguerie, Ramona, (337) 482-6468 geology@louisiana.edu,
Univ of Louisiana, Lafayette – Oo
Villalard-Bohnsack, Martine, (401) 254-3243 Roger Williams Univ – Ob
Villareal, Tracy A., (361) 749-6732 t.villareal@mail.utexas.edu,
Univ of Texas at Austin – Ob
Villarrubia, Jose, (809) 265-3851 Univ of Puerto Rico – Sf
Vimont, Daniel J., (608)262-2828 dvimont@wisc.edu,
Univ of Wisconsin, Madison – Oa
Vincent, Dayton G., (765) 494-3290 dvincent@purdue.edu,
Purdue Univ – Oa
Vincent, Paul, (229) 333-5752 Valdosta State Univ – Oi
Vincent, Paul, (541) 737-9176 pvincent@coas.oregonstate.edu,
Oregon State Univ – Gu
Vincent, Robert K., (419) 372-0160 rvincen@bgnet.bgsu.edu,
Bowling Green State Univ – Yg
Vinton, Bonita L., (724) 738-2048 bonita.vinton@sru.edu,
Slippery Rock Univ – Gg
Violet, Regina G., (865) 574-4971 violetrg@ornl.gov,
Oak Ridge National Lab – Oo
Viskupic, Karen, (208)426-3658 karenviskupic@boisestate.edu,
Boise State Univ – Gg
Vislova, Tatiana, (607) 436-3066 SUNY, Oneonta – Gx
Visscher, Pieter T., (860) 445-3462 visscher@uconnvm.uconn.edu,
Univ of Connecticut – Co
Vitek, John D., 405-744-6358 jvitek@okstate.edu,
Oklahoma State Univ – Gm
Vitton, Stanley J., (906) 487-2527 vitton@mtu.edu,
Michigan Technological Univ – Ng
Vivoni, Enrique R., 505-835-5611 vivoni@nmt.edu,
New Mexico Institute of Mining and Technology – Hs
Voegeli, Henry E., (203) 932-7106 Univ of New Haven – Oo
Vogel, Stefan W., (815) 753-7948 svogel@geol.niu.edu,
Northern Illinois Univ – Gl
Vogel, Thomas A., 517-355-4626 vogel@msu.edu,
Michigan State Univ – Gi
Vogl, Jim, (352) 392-6987 jvogl@geology.ufl.edu,
Univ of Florida – Gct
Vogt, Kelli, 614-265-6588 kelli.vogt@dnr.state.oh.us,
Ohio Dept of Natural Resources – OiiOi

Vogt, Steven, (831) 459-2151 vogt@ucolick.org,
Univ of California, Santa Cruz – Xc
Voight, Barry, (814) 865-9993 voight@ems.psu.edu,
Pennsylvania State Univ, Univ Park – Gv
Volborth, Alexis, (406) 496-4134 Montana Tech of the Univ of Montana – Ca
Volkert, Richard A., (609) 292-2576 rich.volkert@dep.state.nj.us,
New Jersey Geological Survey – Gx
Voller, Vaughan R., (612) 625-0764 volle001@tc.umn.edu,
Univ of Minnesota, Twin Cities – Om
Vollmer, Frederick W., 845-257-3760 vollmerf@newpaltz.edu,
SUNY New Paltz – Gct
von Bitter, Peter H., (416) 586-5592 peterv@rom.on.ca,
Royal Ontario Museum – Pm
von Bitter, Peter H., 416-586-5591 peterv@rom.on.ca,
Univ of Toronto – Pm
Von Damm, Karen L., 603-862-0142 kvd@eos.sr.unh.edu,
Univ of New Hampshire – Cl
von Frese, Ralph R., (614) 292-5635 vonfrese@geology.ohio-state.edu,
Ohio State Univ – Ye
Von Herzen, Richard P., (508) 289-2465 rvonh@cliff.whoi.edu,
Woods Hole Oceanographic Institution – Yh
Von Hillebrandt, Christa, 787 2653845 christa@midas.uprm.edu,
Univ of Puerto Rico – Ys
Von Reden, Karl F., (508) 289-3384 kvonreden@whoi.edu,
Woods Hole Oceanographic Institution – Ct
von Salzen, Knut, (250) 363-8287 Knut.vonSalzen@ec.gc.ca,
Univ of British Columbia – Oa
Von Schwind, Joseph J., (408) 656-2552 Naval Postgraduate School – Op
von Seggern, David, (775) 784-4242 vonseg@seismo.unr.edu,
Univ of Nevada, Reno – Ys
Vonder Haar, Thomas H., (970) 491-8448 vonderhaar@cira.colostate.edu,
Colorado State Univ – Oa
Vondra, Carl F., (515) 294-5867 cfvondra@iastate.edu,
Iowa State Univ of Science & Technology – Gr
Voner, Frederick R., 740-376-4845 vonerf@marietta.edu,
Marietta College – Gx
Vong, Richard J., (541) 737-5693 vong@coas.oregonstate.edu,
Oregon State Univ – Oa
Voorhees, Kent J.kvoorhee@mines.edu,
Colorado School of Mines – Co
Voroney, R. Paul, (519) 824-4120 (Ext. 53057) pvoroney@lrs.uoguelph.ca,
Univ of Guelph – Sb
Vorosmarty, Charles J., (603) 862-3978 charles.vorosmarty@unh.edu,
Univ of New Hampshire – Hq
Vose, Margaret E., (207) 780-5186 vose@usm.maine.edu,
Univ of Southern Maine – Og
Voss, Regis D., (515) 294-1923 Iowa State Univ of Science & Tech – So
Voulgaris, George, (803) 777-2549 voulgaris@geol.sc.edu,
Univ of South Carolina – On
Vrba, Elisabeth S., (203) 432-5008 eilsabeth.vrba@yale.edu,
Yale Univ – Pv
Vuille, Mathias, (413) 545-0659 mathias@geo.umass.edu,
Univ of Massachusetts – Ow
Vuke, Susan M., 406-496-4326 svuke@mtech.edu,
Montana Tech of The Univ of Montana – Gg
Vulava, Vijay M., (843) 953-1922 vulavav@cofc.edu,
College of Charleston – CgHwSc

W

Waag, Charles J., (208) 426-3658 cwaag@boisestate.edu,
Boise State Univ – Gc
Wach, Grant D., 902-494-8019 Grant.Wach@Dal.Ca,
Dalhousie Univ – Go
Waday, Erika, (775) 784-6115 ewaday@unr.edu,
Univ of Nevada, Reno – Oo
Waddington, David, (604) 527-5236 Douglas College – Gg
Waddington, Edwin D., 206-543-4585 edw@ess.washington.edu,
Univ of Washington – Ol
Waddington, Janet B., (416) 586-5593 janetw@rom.on.ca,
Royal Ontario Museum – Pi
Waddington, Michael, (905) 525-9140 (Ext.23217) wadding@mcmail.
cis.mcmaster.ca, McMaster Univ – Pe
Wade, Bridget, (732) 445-2044 bwade@rci.rutgers.edu,
Rutgers, The State Univ of New Jersey – Pm
Wade, Terry L., (979) 862-2323 Texas A&M Univ – Oc
Wadhwa, Meenakshi, (773) 702-8164 Univ of Chicago – Og
Wadhwa, Meenakshi, (312) 665-7639 mwadhwa@fmnh.org,
Field Museum of Natural History – Xm
Wadleigh, Moire A., (709) 737-2331 moire@sparky2.esd.mun.ca,
Memorial Univ of Newfoundland – Cs
Wadsworth, William B., (562) 907-4200 Whittier College – Gi
Waff, Harve S., (541) 346-5577 waff@oregon.uoregon.edu,
Univ of Oregon – Yg
Wagar, Timothy, (507) 280-2866 Univ of Minnesota, Twin Cities – Sc
Wagger, Michael G., (919) 515-4269 North Carolina State Univ – Sb
Wagner, David L., (916) 324-7380 California Geological Survey – Gg
Wagner, George H., (479) 575-3355 Univ of Arkansas, Fayetteville – Cg
Wagner, John R., (864) 656-5024 jrwgnr@clemson.edu,
Clemson Univ – Oe
Wagner, Lara S., (202) 478-8841 wagner@dtm.ciw.edu,
Carnegie Institution of Washington – Oo
Wagner, Mary E., (215) 898-5724 mwagner@sas.upenn.edu,
Univ of Pennsylvania – Gp
Wagner, Peter J., 312-665-7634 pwagner@fmnh.org,
Field Museum of Natural History – Pi
Wagner Riddle, Claudia, (519) 824-4120 (Ext. 2787) criddle@lrs.
uoguelph.ca, Univ of Guelph – Ow
Wagoner, Jeffrey L., (925) 422-1374 wagoner1@llnl.gov,
Lawrence Livermore National Lab – Gs
Wahl, Timothy E., (612)627-4780 Ext. 227 tewahl@umn.edu,
Univ of Minnesota, Twin Cities – Oo
Wahlen, Martin, (619) 534-0828 mwahlen@ucsd.edu,
Univ of California, San Diego – Cs
Waines, Russell H., (914) 257-3768 SUNY New Paltz – Pg
Waite, Cynthia, (213) 740-6109 waite@usc.edu,
Univ of Southern California – Oo
Waite, Gregory P., (906) 487-3554 gpwaite@mtu.edu,
Michigan Technological Univ – Ys
Wake, Cameron P., (603) 862-2329 cameron.wake@unh.edu,
Univ of New Hampshire – Gl
Wake, David B., (510) 642-3059 wakelab@uclink.berkeley.edu,
Univ of California, Berkeley – Po
Wake, Marvalee H., (510) 642-4743 Univ of California, Berkeley – Po
Wakefield, S J.wakefield@cf.ac.uk,
Cardiff Univ – Cm
Wala, Andrew M., (606) 257-2959 Univ of Kentucky – Nm
Walawender, Michael J., (619) 594-6543 San Diego State Univ – Gi
Walcek, Christopher J., (518) 437-8720 SUNY, Albany – Oa
Walczak, Joanna, walczak@nova.edu,
Nova Southeastern Univ – Oo
Waldinger, Rachel, 512-475-9551 rachel.waldinger@beg.utexas.edu,
Univ of Texas at Austin, Jackson School of Geosciences – Or
Waldner, Jeffrey S., (609) 984-6587 jeff.waldner@dep.state.nj.us,
New Jersey Geological Survey – Yr
Waldron, John, (902) 420-5749 Dalhousie Univ – Gc
Waldron, John, 850-474-2746 jwaldron@uwf.edu,
Univ of West Florida – Oy
Waliser, Duane E., (631) 632-8647 waliser@notes.cc.sunysb.edu,
SUNY, Stony Brook – Oa
Walker, Brian K., (954)262-3675 walkerb@nova.edu,
Nova Southeastern Univ – Oo
Walker, Charles S., (404) 657-6132 Georgia Dept of Nat Resources – Hw
Walker, Charles T., (562) 985-4818 California State Univ, Long Beach – Cl
Walker, David, dwalker@ldeo.columbia.edu,
American Museum of Natural History – Gi
Walker, David, (845) 365-8658 dwalker@ldeo.columbia.edu,
Columbia Univ – Cp
Walker, Ernest G., (306) 966-4181 Univ of Saskatchewan – Pv
Walker, Gayle, (707) 664-2334 gayle.walker@sonoma.edu,
Sonoma State Univ – Oo
Walker, Grahamgtwalker@bu.edu,
Boston Univ – Oe
Walker, J. D., (785) 864-7711 jdwalker@ku.edu,
Univ of Kansas – Gc
Walker, James A., (815) 753-7936 jim@geol.niu.edu,
Northern Illinois Univ – Gi
Walker, James C. G., (734) 764-2466 Univ of Michigan – Oa
Walker, Jeffrey R., (845) 437-5546 jewalker@vassar.edu,
Vassar College – Sc
Walker, Joan, 617-353-5743 joanw@bu.edu,
Boston Univ – Oo
Walker, John L., (630) 252-6803 jlwalker@anl.gov,
Argonne National Lab – Cg
Walker, Kenneth R., (865) 974-6017 kwalker5@utk.edu,
Univ of Tennessee, Knoxville – Pe
Walker, Mark, (775) 784-1938 mwalker@equinox.unr.edu,
Univ of Nevada, Reno – Oo
Walker, Nan D., (225) 388-5331 Louisiana State Univ – Og
Walker, Peter A., (541) 346-4541 pwalker@oregon.uoregon.edu,
Univ of Oregon – Oo

California State Univ, Fullerton – Go
Warger, Jane, 909-941-2385 jane.warger@chaffey.edu, Chaffey College – Cg
Wark, David A., (518) 276-2674 warkd@rpi.edu, Rensselaer Polytechnic Institute – Gi
Warland, Jon, (519) 824-4120 (Ext. 6374) jwarland@lrs.uoguelph.ca, Univ of Guelph – Oa
Warme, John E.jwarme@mines.edu, Colorado School of Mines – Pi
Warner, Don L., 573-341-4876 Univ of Missouri, Rolla – Hw
Warner, Jeffrey M., (814) 865-3197 jmw139@psu.edu, The Pennsylvania State Univ, Univ Park – Ow
Warner, Kristin J., 717.702.2029 kwarner@state.pa.us, Pennsylvania Bureau of Topographic & Geologic Survey – Oo
Warner, Mark J., (206) 543-0765 warner@u.washington.edu, Univ of Washington – Op
Warner, Richard D., (864) 656-5023 wrichar@ces.clemson.edu, Clemson Univ – Gz
Warner, Timothy A., (304) 293-5603 (Ext. 328) tim.warner@mail.wvu.edu, West Virginia Univ – Or
Warnke, Detlef A., (510) 885-4716 dietz.warnke@csueastbay.edu, California State Univ, East Bay – Ou
Warren, JosephJoe.Warren@stonybrook.edu, SUNY, Stony Brook – Ob
Warren, Lesley A., (905) 525-9140 (Ext. 27347) warrenl@mcmaster.ca, McMaster Univ – Yr
Warren, Linda M., 202 478 8842 warren@dtm.ciw.edu, Carnegie Institution of Washington – Ys
Warren, Paul, (310) 825-2015 Univ of California, Los Angeles – Xm
Warren, Richard G., (505) 667-7063 rgw@lanl.gov, Los Alamos National Lab – Gi
Warren, Richard L., (520) 621-2320 Univ of Arizona – Oo
Warren, Sarita N., (912) 681-5361 SNWarren@GeorgiaSouthern.edu, Georgia Southern Univ – Oo
Warren, Stephen G., 206-543-7230 sgw@atmos.washington.edu, Univ of Washington – Oal
Warrick, Arthur W., (520) 621-1516 aww@ag.arizona.edu, Univ of Arizona – Sp
Warrick, Beth, 573-368-2193 nrwarrb@mail.dnr.state.mo.us, Missouri Dept of Natural Resources – Hy
Warwick, John, (775) 673-7379 john.warwick@dri.edu, Univ of Nevada, Reno – Hs
Washburn, Albert Lincoln, (206) 543-8140 Univ of Washington – Gm
Washburn, Robert H., (814) 641-3600 Juniata College – Gs
Washburne, James C., (520) 626-4107 jwash@hwr.arizona.edu, Univ of Arizona – Hs
Washington, Paul A., (318) 342-1898 Univ of Louisiana, Monroe – Hw
Wasklewicz, Thad, (901) 678-4452 twsklwcz@postoffice.memphis.edu, Univ of Memphis – Oy
Wassenaar, Len, (306) 239-2270 Univ of Saskatchewan – Hw
Wasserburg, Gerald J., (626) 395-6139 isotopes@gps.caltech.edu, California Institute of Technology – Cc
Wassermann, Joachim, 08141/5346762 jowa@geophysik.uni-muenchen.de, LMU - Univ of Munich – Yg
Wasson, John T., (310) 825-1986 jtwasson@ucla.edu, Univ of California, Los Angeles – Xm
Waters, Michael R., (979) 845-5246 mwaters@tamu.edu, Texas A&M Univ – Ga
Waterstone, Marvin, (520) 621-1478 marvinw@u.arizona.edu, Univ of Arizona – Oo
Watkins, David K., (402) 472-2177 dwatkins1@unl.edu, Univ of Nebraska, Lincoln – Pm
Watkins, Joel, 979-845-1300 watkins@geo.tamu.edu, Texas A&M Univ – Gg
Watkins, Rodney, (414) 278-2706 rw@mpm.edu, Milwaukee Public Museum – Pi
Watkinson, A. John, (509) 335-2470 watkinso@mail.wsu.edu, Washington State Univ – Gc
Watkinson, David H.dwatkson@ccs.carleton.ca, Carleton Univ – Em
Watkinson, David H., (613) 520-2600 Carleton Univ – Em
Watling, Leslie D., (207) 563-3146 Univ of Maine – Ob
Watney, Lynn W., (785) 864-2184 lwatney@kgs.ku.edu, Univ of Kansas – Gr
Watney, W. L., (785) 864-2184 lwatney@kgs.ku.edu, Univ of Kansas – Gr
Watson, David B., (865) 241-4749 v6i@ornl.gov, Oak Ridge National Lab – Hw
Watson, E. Bruce, (518) 276-8838 watsoe@rpi.edu, Rensselaer Polytechnic Institute – Cp
Watson, Gerald F., (919) 515-7076 jerry_watson@ncsu.edu, North Carolina State Univ – Oa
Watson, Iain M., (906) 487-2045 watson@mtu.edu, Michigan Technological Univ – Gv
Watson, Kenneth M., (619) 534-6620 Univ of California, San Diego – Op
Watson, Kimberly, (518) 276-6474 kellek@rpi.edu, Rensselaer Polytechnic Institute – Oo
Watson, Mary E., (919) 733-7353 North Carolina Geological Survey – Gn
Watters, Brian R., (306) 585-4663 Univ of Regina – Gi
Watters, Robert J., (775) 784-6069 watters@mines.unr.edu, Univ of Nevada, Reno – Nr
Watters, Thomas R., 202-633-2483 Smithsonian Institution / National Air & Space Museum – Xg
Wattrus, Nigel J., (218) 726-7154 nwattrus@d.umn.edu, Univ of Minnesota, Duluth – Yr
Watts, Chester F., (540) 831-5637 cwatts@radford.edu, Radford Univ – Ng
Watts, Chester F., 540-831-5637 cwatts@vt.edu, Virginia Polytechnic Institute & State Univ – Ngg
Watts, D. Randolph, (401) 874-6507 rwatts@gso.uri.edu, Univ of Rhode Island – Op
Watts, Doyle, 937 775-3455 doyle.watts@wright.edu, Wright State Univ – Or
Watts, Pam, (601) 266-4526 Patricia.Watts@usm.edu, Univ of Southern Mississippi – Oo
Waugh, Darryn W., (410) 516-8344 waugh@jhu.edu, The Johns Hopkins Univ – Oa
Waugh, Richard A., (608) 342-1386 Univ of Wisconsin, Platteville – Gc
Waugh, Truman, 785-864-2119 twaugh@ku.edu, Univ of Kansas – Ca
Wawrzyniec, Timothy, 505-277-4204 tfw@unm.edu, Univ of New Mexico – Gt
Wax, Charles L., (662) 325-3915 wax@geosci.msstate.edu, Mississippi State Univ – Oa
Waychunas, Glenn A., (510) 495-2224 GAWaychunas@lbl.gov, Lawrence Berkeley National Lab – Gz
Wayne, Williamwwayne3@unl.edu, Univ of Nebraska, Lincoln – Gm
Wdowinski, Shimon, 305 421-4730 swdowinski@rsmas.miami.edu, Univ of Miami – Yd
Weatherly, Georges L.weatherly@ocean.fsu.edu, Florida State Univ – Op
Weathers, Maura S., (607) 255-2295 weathers@geology.cornell.edu, Cornell Univ – Oo
Weaver, Andrew J., (250) 472-4001 weaver@ocean.seos.uvic.ca, Univ of Victoria – Op
Weaver, Barry L., (405) 325-4492 bweaver@ou.edu, Univ of Oklahoma – Ct
Weaver, Douglas J., (702) 295-5916 douglas_weaver@lanl.gov, Los Alamos National Lab – Oo
Weaver, J. Scott, (607) 821-2208 Alfred Univ – Gz
Weaver, John T., (250) 721-6155 weaver@phys.uvic.ca, Univ of Victoria – Ym
Weaver, Kelli D., 217-265-8211 weaver@isgs.uiuc.edu, Illinois State Geological Survey – Ge
Weaver, Michael J., (208) 885-4351 mjweaver@uidaho.edu, Univ of Idaho – Eg
Weaver, Stephen G., (719) 389-6954 sweaver@coloradocollege.edu, Colorado College – Gx
Weaver, Thomas A., (505) 667-8464 tweaver@lanl.gov, Los Alamos National Lab – Yg
Weaver-Bowman, Kristin, (714) 278- 3882 kweaverbowman@fullerton.edu, California State Univ, Fullerton – Ys
Webb, Bruce L., (801) 378-2147 Brigham Young Univ – Sc
Webb, Elizabeth, (519) 661-2111 (Ext. xt. 80208) ewebb5@uwo.ca, Univ of Western Ontario – Cs
Webb, Fred, (828) 262-2166 webbfj@appstate.edu, Appalachian State Univ – Gr
Webb, Gregory E., 61 7 3138 2804 ge.webb@qut.edu.au, Queensland Univ of Technology – Gs
Webb, Jamieson D., ((85) 0) -769- (Ext. 51 ext. 2858) jwebb@gulf-coast.edu, Gulf Coast Community College – Gg
Webb, John A., +61 3 9479 1273 john.webb@latrobe.edu.au, Latrobe Univ – Ge
Webb, Peter N., (614) 292-7285 webb.3@osu.edu, Ohio State Univ – Pm
Webb, Robert H., (520) 626-3293 rhwebb@usgs.gov, Univ of Arizona – Gm
Webb, Spahr C., (845) 365-8439 scw@ldeo.columbia.edu, Columbia Univ – Yr

Faculty Index -W

Webb, III, Thompson, (401) 863-3128 thompson_webb_iii@brown.edu, Brown Univ – Pe
Webber, Karen L., (504) 280-6791 kwebber@uno.edu, Univ of New Orleans – Gv
Webber, Karen L., (504) 280-7395 kwebber@uno.edu, Univ of Michigan – Gv
Webber, William D., (509) 372-1413 wd.webber@pnl.gov, Pacific Northwest National Lab – Gg
Weber, Gerald E., (831) 459-5429 gweber@pmc.ucsc.edu, Univ of California, Santa Cruz – Ng
Weber, John C., 616-331-3191 weberj@gvsu.edu, Grand Valley State Univ – Gc
Weber, Keith, (208) 282-2757 webekeit@isu.edu, Idaho State Univ – Oi
Weber, Licia A., (812) 855-2687 weber@indiana.edu, Indiana Geological Survey – Ec
Weber, Neil V., (270) 809-2888 neil.weber@murraystate.edu, Murray State Univ – Oy
Weber, RobertNone, New Mexico Institute of Mining & Technology – Oo
Weber, Rodney J., (404) 894-1750 rweber@eas.gatech.edu, Georgia Institute of Technology – Oa
Weber-Diefenbach, Klaus, 089/2180 6549 klaus.diefenbach@iaag.geo.uni-muenchen.de, LMU - Univ of Munich – Gg
Webers, Gerald F., (651) 696-6288 webers@macalester.edu, Macalester College – Pg
Weborg-Benson, Kimberly, (716) 673-3293 kim.weborg-benson@fredonia.edu, SUNY, Fredonia – GgOaPy
Webre, Cherri B., (225) 388-8328 cherri@lgs.bri.lsu.edu, Louisiana State Univ – Oo
Webster, Clyde L., (909) 824-4548 Loma Linda Univ – Ct
Webster, Ferris, (302) 645-4266 Univ of Delaware – Op
Webster, Gary D., (509) 335-4369 webster@wsu.edu, Washington State Univ – Pi
Webster, James D., 212-769-5401 jdw@amnh.org, American Museum of Natural History – Eg
Webster, John R., (701) 858-3873 webster@minotstateu.edu, Minot State Univ – Gi
Weddle, Thomas K., (207) 287-2801 thomas.k.weddle@state.me.us, Univ of Maine – Hy
Wedel, Brett, 785-864-2051 bwedel@kgs.ku.edu, Univ of Kansas – Oo
Weeraratne, Dayanthie S., (202) 478-8837 dayanthie@dtm.ciw.edu, Carnegie Institution of Washington – Ys
Weertman, Johannes, (847) 491-3197 j_weertman2@northwestern.edu, Northwestern Univ – Om
Weglein, Arthur, (713) 743-3399 aweglein@mail.arco.com, Univ of Houston – Ye
Wegner, John, 404 727 4206 jwegner@emory.edu, Emory Univ – Sf
Wehling, Ed, (763) 433-1174 ed.wehling@anokaramsey.edu, Anoka-Ramsey Community College – Oa
Wehmiller, John F., (302) 831-2926 jwehm@udel.edu, Univ of Delaware – Cl
Wehmiller, John F.jwehm@udel.edu, Univ of Delaware – Gu
Wehner, Peter J.pewehner@dbcity.com, Austin Community College – Gv
Wehrmann, Laura, 089/2180 LMU - Univ of Munich – Pg
Wei, Mingzhen, 573-341-4221 weim@umr.edu, Univ of Missouri, Rolla – Np
Weibel, C. Pius, 217-333-5108 weibel@isgs.uiuc.edu, Illinois State Geological Survey – Gr
Weiblen, Paul W., 612-625-3477 pweib@umn.edu, Univ of Minnesota, Twin Cities – Gi
Weide, David L., (702) 895-3118 weide@nevada.edu, Univ of Nevada, Las Vegas – Gm
Weidner, Donald J., (631) 632-8211 donald.weidner@sunysb.edu, SUNY, Stony Brook – Gy
Weight, Willis D., (406) 496-4329 wweight@mtech.edu, Montana Tech of the Univ of Montana – HwGq
Weihaupt, John G., (303) 556-2823 Univ of Colorado, Denver – Ou
Weil, Arlo B., (610) 526-5113 aweil.brynmawr.edu, Bryn Mawr College – Gc
Weil, Raymond R., (301) 405-1314 rw17@umail.umd.edu, Univ of Maryland – Sb
Weiland, Thomas J.tjw@canes.gsw.edu, Georgia Southwestern State Univ – Gi
Weimer, Paul, 303-492-3809 paul.weimer@colorado.edu, Univ of Colorado – Go
Weinbeck, Robert S.weinbeck@dc.ametsoc.org, SUNY College at Brockport – Oa
Weinbeck, Stephen W.sweinbec@brockport.edu, SUNY College at Brockport – Ow
Weinberg, Joan M., (520) 621-2828 jweinber@u.arizona.edu, Univ of Arizona – Oo
Weinberger, Alycia J., (202) 478-8820 weinberger@dtm.ciw.edu, Carnegie Institution of Washington – Oo
Weinle, Art, (313) 882-2569 Wayne State Univ – Gg
Weirich, Frank H., (319) 335-0156 frank-weirich@uiowa.edu, Univ of Iowa – Gm
Weis, Dominique, (604) 822-1697 dweis@eos.ubc.ca, Univ of British Columbia – Cc
Weisberg, Michael K., 718-368-5769 mweisberg@kbcc.cuny.edu, American Museum of Natural History – Xm
Weisberg, Robert H., (727) 553-1568 rweisberg@marine.usf.edu, Univ of South Florida, Saint Petersburg – Op
Weisenfluh, Gerald A., (859) 257-5500 jerryw@uky.edu, Univ of Kentucky – Ec
Weislogel, Amy L.aweislogel@geo.ua.edu, Univ of Alabama – Gs
Weisman, Robert A.scsweisman@stcloudstate.edu, Saint Cloud State Univ – Ow
Weismiller, Richard, (301) 405-1306 Univ of Maryland – Or
Weiss, Albert, (402) 472-6761 agme011@unlvm.unl.edu, Univ of Nebraska, Lincoln – Oa
Weiss, Benjamin, (626) 395-6863 bpweiss@mit.edu, Massachusetts Institute of Technology – Ym
Weiss, Chester J., 540-231-6521 cjweiss@vt.edu, Virginia Polytechnic Institute & State Univ – Ye
Weiss, Christopher C., 806-742-4712 chris.weiss@ttu.edu, Texas Tech Univ – OaaOa
Weiss, Dale S., (617) 552-3653 weissdd@bc.edu, Boston College – Hw
Weiss, Dennis, (212) 650-6849 Graduate School of the City Univ of New York – Pg
Weiss, Ray F., (619) 534-2598 rfweiss@ucsd.edu, Univ of California, San Diego – Cm
Weissel, Jeffrey K., (845) 365-8533 jeffw@ldeo.columbia.edu, Columbia Univ – Yr
Weissmann, Gary, 505-277-4204 weissman@unm.edu, Univ of New Mexico – Hw
Weiterman, Sandra, (804) 427-7189 Tidewater Comm College – Pm
Welby, Charles W., (919) 515-7158 cwwelby@unity.ncsu.edu, North Carolina State Univ – Hw
Welch, Steve M., (913) 532-7236 Kansas State Univ – Sp
Weldon, Elise M., (541) 346-4647 lili@uoregon.edu, Univ of Oregon – Gl
Weldon, Ray J., (541) 346-4584 ray@newberry.uoregon.edu, Univ of Oregon – Gc
Welford, Mark R., (912) 681-5943 MWelfGeog@GeorgiaSouthern.edu, Georgia Southern Univ – Oy
Welhan, John A., (208) 282-4254 welhjohn@isu.edu, Idaho State Univ – Hw
Welhan, John A., (208) 282-4254 welhjohn@isu.edu, Univ of Idaho – Hw
Welle, Patrick G., 218-755-4103 pwelle@bemidjistate.edu, Bemidji State Univ – Oo
Wells, Darlene V., 410-554-5518 dwells@dnr.state.md.us, Maryland Geological Survey – Gs
Wells, Joseph G., (614) 265-1030 joseph.wells@dnr.state.oh.us, Ohio Dept of Natural Resources – Oy
Wells, Michael L., (702) 895-0828 mlwells@unlv.nevada.edu, Univ of Nevada, Las Vegas – Gc
Wells, Mike, 573-751-1134 nrwellm@mail.dnr.state.mo.us, Missouri Dept of Natural Resources – Oo
Wells, Neil A., (330) 672-2951 nwells@kent.edu, Kent State Univ – Gs
Wells, Randy, (941) 388-4441 Univ of South Florida, St Petersburg – Ob
Wells, Sam H., (918) 540-6266 Northeastern Oklahoma A&M College – Gg
Wells, Steve G., (775) 673-7470 sgwells@dri.edu, Univ of Nevada, Reno – Oo
Welsch, Daniel L., 301-687-9720 dwelsch@frostburg.edu, Frostburg State Univ – HgCgSb
Welsh, James L., (507) 933-7335 welsh@gac.edu, Gustavus Adolphus College – Gx
Welzenbach, Linda, (202) 633-1825 welzenbl@si.edu, Smithsonian Institution / National Museum of Natural History – Xm
Wen, Lianxing, (631) 632-1726 lianxing.wen@sunysb.edu, SUNY, Stony Brook – Ys

Wendlandt, Richard F., (303) 273-3809 rwendlan@mines.edu, Colorado School of Mines – Gx
Weng, Qihao, (812) 237-2255 geweng@isugw.indstate.edu, Indiana State Univ – Or
Wenk, Hans-Rudolf, (510) 642-7431 wenk@seismo.berkeley.edu, Univ of California, Berkeley – Gz
Wennberg, Paul O., (626) 395-2447 wennberg@gps.caltech.edu, California Institute of Technology – Oa
Wenner, David B., (706) 542-2393 dwenner@uga.edu, Univ of Georgia – Cs
Wenner, Jennifer M., (920) 424-7003 wenner@uwosh.edu, Univ of Wisconsin, Oshkosh – Gi
Wenz, Helmut C., (731) 587-7438 hwenz@utm.edu, Univ of Tennessee, Martin – Oo
Wenz, Zachary J., 785-864-4909 zwenz@kgs.ku.edu, Univ of Kansas – Gg
Werner, Alan, 413-538-2134 awerner@mtholyoke.edu, Mount Holyoke College – GleGm
Werner, Bradley T., (619) 534-0583 Univ of California, San Diego – Gm
Wernicke, Brian P., (626) 395-6192 brian@gps.caltech.edu, California Institute of Technology – Gc
Wesely, Marvin L., (630) 252-5827 mlwesely@anl.gov, Argonne National Lab – Oa
Wesler, Kit W., (270) 809-3457 kit.wesler@murraystate.edu, Murray State Univ – Oo
Wesnousky, Steve, (775) 784-6067 stevew@seismo.unr.edu, Univ of Nevada, Reno – Ys
Wesnousky, Steven G., (775) 784-6067 stevew@seismo.unr.edu, Univ of Nevada, Reno – Ys
Wesolowski, David J., (865) 574-6903 wesolowskid@ornl.gov, Oak Ridge National Lab – Cg
Wessel, Paul, (808) 956-4778 pwessel@hawaii.edu, Univ of Hawai'i, Manoa – Yr
West, David P., (802) 443-3476 dwest@middlebury.edu, Middlebury College – Gc
West, Gordon F., (416) 978-3155 west@physics.utoronto.ca, Univ of Toronto – Ye
West, Larry T., (706) 542-0906 Univ of Georgia – Sd
West, Olivia M., (615) 576-0505 qm5@ornl.gov, Oak Ridge National Lab – Ng
West, Ronald R., (785) 532-2248 rrwest@ksu.edu, Kansas State Univ – Po
West, Terry R., (765) 494-3296 trwest@purdue.edu, Purdue Univ – Ng
Westerik, Jan, (304) 384-6025 Concord Univ – Oy
Westerink, Joannes J., (219) 631-6475 Univ of Notre Dame – On
Westerman, David S., (802) 485-2337 Norwich Univ – Gx
Westermann, Gerd, (905)525-9140 (Ext. 27694) McMaster Univ – Yr
Westgate, James W., (409) 880-8237 Lamar Univ – Pe
Westgate, John A., (416) 287-7235 Univ of Toronto – Gl
Westjohn, David B., 517-887-8921 Michigan State Univ – Hy
Weston, Donald, (510) 231-5626 dweston@uclink.berkeley.edu, Univ of California, Berkeley – Ob
Westrop, Stephen R., (405) 325-0542 swestrop@ou.edu, Univ of Oklahoma – Pi
Wetherill, George W., (202) 244-3435 bravenewworld@starpower.net, Carnegie Institution of Washington – Og
Wethey, David S., (803) 777-3936 Univ of South Carolina – Ob
Wetzel, Dan L., (907) 488-3746 Univ of Alaska, Fairbanks – Oo
Wetzel, Laura R., (727) 864-8484 wetzellr@eckerd.edu, Eckerd College – Yr
Whalen, Michael T., 907-474-5302 ffmtw@uaf.edu, Univ of Alaska, Fairbanks – Gs
Whalen, Patricia A., (479) 575-3370 Univ of Arkansas, Fayetteville – Ps
Whaley, Peter W., (270) 809-4275 peter.whaley@murraystate.edu, Murray State Univ – Gr
Wheatcraft, Stephen W., 775-784-1973 steve@hydro.unr.edu, Univ of Nevada, Reno – Hq
Wheatcraft, Steve, (775) 784-1973 steve@mines.unr.edu, Univ of Nevada, Reno – Hw
Wheatcroft, Robert A., (541) 737-3891 raw@coas.oregonstate.edu, Oregon State Univ – Ob
Wheaton, John R., 406-657-2629 jwheaton@mtech.edu, Montana Tech of The Univ of Montana – Hw
Wheeler, Andy J., +353 21 4903951 a.wheeler@ucc.ie, Univ College Cork – Gu
Wheeler, George L., (203) 932-7171 Univ of New Haven – Cl
Wheeler, Patricia A., (541) 737-0558 pwheeler@coas.oregonstate.edu, Oregon State Univ – Ob
Wheeler, Richard, (931) 221-1004 wheelerr@apsu.edu, Austin Peay State Univ – Oo
Wheeler, William A., (806) 894-9611 wwheeler@spc.cc.tx.us, South Plains College – Gc
Whelan, Gene, (509) 372-6098 gene.whelan@pnl.gov, Pacific Northwest National Lab – Hg
Whelan, Jean K., (508) 289-2819 jwhelan@whoi.edu, Woods Hole Oceanographic Institution – Co
Whelan, Peter M., (320) 589-6315 whelan@mrs.umn.edu, Univ of Minnesota, Morris – Gx
Whipkey, Charles, (540) 654-1428 cwhipkey@mwc.edu, Univ of Mary Washington – Cg
Whisenhant, Michelle K., 410-293-6563 whisenha@usna.edu, United States Naval Academy – Ow
Whisonant, Robert C., 540 831-5224 rwhisona@radford.edu, Radford Univ – Ga
Whistler, David P., (213) 763-3310 dwhistle@nhm.org, Los Angeles County Museum of Natural History – Pv
Whitaker, Rodney W., (505) 667-7672 rww@lanl.gov, Los Alamos National Lab – Oa
White, Angelique, (541) 737-6397 awhite@coas.oregonstate.edu, Oregon State Univ – Ob
White, Bekki, 501 296-1880 bekki.white@arkansas.gov, Arkansas Geological Commission – Go
White, Craig M., (208) 426-3633 cwhite@boisestate.edu, Boise State Univ – Gi
White, Donald J., (415) 338-1209 Ecole Polytechnique – Ye
White, George W., (301) 687-4264 gwhite@frostburg.edu, Frostburg State Univ – Oo
White, James D., +64 3 479-9009 james.white@otago.ac.nz, Univ of Otago – Gv
White, James W. C., 303-492-5494 jwhite@colorado.edu, Univ of Colorado – Cs
White, Jeffrey, (812) 855-0731 whitej@indiana.edu, Indiana Univ, Bloomington – Hs
White, Jeffrey G., (919) 515-2389 Jeffrey_White@ncsu.edu, North Carolina State Univ – Or
White, Jeffrey R., (812) 855-0731 Indiana Univ, Bloomington – Hg
White, John C., 859-622-1275 John.White@eku.edu, Eastern Kentucky Univ – Gp
White, John R., (352) 392-1804 (Ext. 349) jrwh@gnv.ifas.ufl.edu, Univ of Florida – Sf
White, Joseph C., (506) 453-4804 clancy@unb.ca, Univ of New Brunswick – Gc
White, Joseph D., (254) 710-2141 Joseph_D_White@baylor.edu, Baylor Univ – Or
White, Lisa D., (415) 338-1963 lwhite@sfsu.edu, San Francisco State Univ – Pg
White, Mark D., (509) 372-6070 mark.white@pnl.gov, Pacific Northwest National Lab – Hy
White, Paul G., (828) 227-7367 Western Carolina Univ – Gm
White, Scott, (970) 247-7475 white_s@fortlewis.edu, Fort Lewis College – Oy
White, Scott M.swhite@geol.sc.edu, Univ of South Carolina – Yr
White, Scott W., (785) 864-2073 whites@kgs.ku.edu, Univ of Kansas – Oo
White, William A., (512) 471-0338 bill.white@beg.utexas.edu, Univ of Texas at Austin, Jackson School of Geosciences – Ge
White, William B., (814) 865-1152 wbw2@psu.edu, Pennsylvania State Univ, Univ Park – Cg
White, William M., (607) 255-7466 white@geology.cornell.edu, Cornell Univ – Cc
Whitehead, James, (506) 453-4593 Univ of New Brunswick – Gg
Whitehead, Robert E., (705) 675-1151 (Ext. ext. 2298) rwhitehead@laurentian.ca, Laurentian Univ, Sudbury – Ce
Whiteman, C. David, (509) 376-7859 Washington State Univ – Oo
Whiteway, James, (416) 736-2100 (Ext. 22310) whiteway@yorku.ca, York Univ – Oa
Whitfield, Thomas G., 717.702.2023 twhitfield@state.pa.us, Pennsylvania Bureau of Topographic & Geologic Survey – Gg
Whitford-Stark, James L., (915) 837-8083 Sul Ross State Univ – Xg
Whiticar, Michael J., (250) 721-6514 whiticar@uvic.ca, Univ of Victoria – Co
Whiting, Peter J., (216) 368-3989 peter.whiting.case.edu, Case Western Reserve Univ – Gm
Whitlatch, Robert B., (860) 445-3467 whitlatc@uconnvm.uconn.edu, Univ of Connecticut – Ob
Whitlock, Cathy, 406-994-6910 whitlock@montana.edu, Montana State Univ – Pey
Whitlock, Cathy, (541) 346-4555 whitlock@oregon.uoregon.edu,

Univ of Oregon – Pe
Whitlock, Linda D., (260) 481-6249 Whitlock@ipfw.edu, Indiana Univ / Purdue Univ, Fort Wayne – Oo
Whitlock, William W., (276) 676-5829 william.whitlock@dmme.virginia.gov, Division of Mineral Resources – Gr
Whitman, Brian E., (570) 408-4882 Wilkes Univ – Hw
Whitman, Dean, (305) 348-3089 whitmand@fiu.edu, Florida International Univ – Yg
Whitman, Jill M., (253) 535-8720 whitmaj@plu.edu, Pacific Lutheran Univ – Ou
Whitmeyer, Steven J.whitmesj@jmu.edu, James Madison Univ – Gc
Whitney, Donna L., 612-626-7582 dwhitney@umn.edu, Univ of Minnesota, Twin Cities – Gp
Whitney, Earl M., (505) 667-3595 whitney@lanl.gov, Los Alamos National Lab – Oo
Whitney, James A., (706) 548-6894 jamesawhitney@hotmail.com, Univ of Georgia – Gi
Whitney, Sandra, (706) 542-3240 swhitney@arches.uga.edu, Univ of Georgia – Ga
Whitson, Kristin B., (423) 425-4546 kristin-whitson@utc.edu, Univ of Tennessee, Chattanooga – OooOo
Whittecar, Jr., G. Richard, (757) 683-5197 rwhittec@odu.edu, Old Dominion Univ – Ou
Whittemore, Donald O., (785) 864-3965 donwhitt@kgs.ku.edu, Univ of Kansas – Cl
Whittington, Alan, 573/884-7625 WhittingtonA@missouri.edu, Univ of Missouri – Gv
Whittington, Melanie, (314) 977-2236 melanie@eas.slu.edu, Saint Louis Univ – Yr
Whitworth, Joan M., (606) 793-2965 Morehead State Univ – Oe
Whitworth, Thomas M., 573-341-6234 mikew@umr.edu, Univ of Missouri, Rolla – Hy
Wiberg, Patricia L., (434) 924-7546 pw3c@virginia.edu, Univ of Virginia – Og
Wicander, Reed, (989) 774-3179 wican1r@cmich.edu, Central Michigan Univ – Pl
Wicke, Jana, 705-566-8101 Ext. 7798 Cambrian College of Applied Arts & Technology – Nm
Wickens, David J., (601) 961-5512 david_wickens@deq.state.ms.us, Mississippi Office of Geology – Oo
Wickham, Jacob B., (408) 656-2552 Naval Postgraduate School – Op
Wickham, John S., (817) 272-2987 wickham@uta.edu, Univ of Texas, Arlington – Gc
Wickham, Thomas, 724 938-4180 wickham@cup.edu, California Univ of Pennsylvania – Oy
Wicks, Carol M., (573) 882-3231 WicksC@missouri.edu, Univ of Missouri – Hw
Wicks, Frederick J., (416) 978-5395 Royal Ontario Museum – Gz
Wicks, Frederick J., (416) 978-5395 fredw@rom.on.ca, Univ of Toronto – Gz
Wicks, Frederick J., (416) 978-5395 fredw@rom.on.ca, Royal Ontario Museum – Gz
Wickstrom, Lawrence H., (614) 265-6598 larry.wickstrom@dnr.state.oh.us, Ohio Dept of Natural Resources – Go
Widdowson, Mark A., (703) 231-7153 Virginia Polytechnic Institute & State Univ – Hw
Widom, Elisabeth, (513) 529-3216 widome@muohio.edu, Miami Univ – Cc
Wiebe, Robert A., (717) 291-3280 bob.wiebe@fandm.edu, Franklin and Marshall College – Gi
Wieded, Ted, (925) 685-1230 twieden@dvc.edu, Diablo Valley College – Oo
Wiedenfeld, Robert, 956-968-5585 bwiedenf@ag.tamu.edu, Texas A&M Univ – Sc
Wiederspahn, Mark, (512) 471-0406 markw@ig.utexas.edu, Univ of Texas at Austin, Jackson School of Geosciences – Oo
Wiens, Douglas A., (314) 935-6517 doug@kermadec.wustl.edu, Washington Univ – Ys
Wiens, Kyle C., 806-742-3120 kyle.wiens@ttu.edu, Texas Tech Univ – OaaOa
Wierenga, Peter J., (520) 621-1899 wierenga@ag.arizona.edu, Univ of Arizona – Sp
Wiese, Katryn, (415) 452-5061 kwiese@ccsf.edu, City College of San Francisco – Gg
Wiese, Jr., Robert G., (330) 823-3614 wieserg@muc.edu, Mount Union College – Eg
Wiesner, Mark R., (713) 348-5129 wiesner@rice.edu, Rice Univ – Og
Wigand, Peter, (775) 673-7387 pwigand@dri.edu, Univ of Nevada, Reno – Pl
Wiggert, Jerry, 228-688-3491 Jerry.Wiggert@usm.edu, Univ of Southern Mississippi – Og
Wiggins, Michael, (405) 325-6781 mwiggins@ou.edu, Univ of Oklahoma – Np
Wigmosta, Mark S., (509) 372-6238 mark.wigmosta@pnl.gov, Pacific Northwest National Lab – Hs
Wijesekera, Hemantha, (541) 737-2568 hemantha@coas.oregonstate.edu, Oregon State Univ – Op
Wijesinghe, Ananda M., (925) 423-0605 Lawrence Livermore Nat Lab – Nr
Wilbur, Henry M.hmw3q@virginia.edu, Univ of Virginia – Oo
Wilch, Thomas I., (517) 629-0759 twilch@albion.edu, Albion College – Gl
Wilcock, Peter W., (410) 516-5421 The Johns Hopkins Univ – Gm
Wilcock, William S., 206-543-6043 wilcock@u.washington.edu, Univ of Washington – Gu
Wildeman, Thomas R.twildema@mines.edu, Colorado School of Mines – Ct
Wilder, Gary F., (276) 676-5831 gary.wilder@dmme.virginia.gov, Division of Mineral Resources – Gg
Wilder, Lee, (603) 271-1976 lwilder@des.state.nh.us, New Hampshire Geological Survey – Gg
Wilder, Margaret, 520-626-7231 mwilder@email.arizona.edu, Univ of Arizona – Oo
Wilding, Larry P., (979) 845-3604 Texas A&M Univ – Sd
Wiles, Gregory C., (330) 263-2298 gwiles@wooster.edu, The College of Wooster – Gl
Wiley, Robert, 713-743-3399 Univ of Houston – Ys
Wiley, Robert C., (301) 405-4352 rw29@umail.umd.edu, Univ of Maryland – Oo
Wiley, Tomtom.wiley@state.or.us, Oregon Dept of Geology & Mineral Industries – Gg
Wilf, Peter D., (814) 865-6721 pwilf@geosc.psu.edu, Pennsylvania State Univ, Univ Park – Py
Wilheit, Thomas T., (409) 845-0176 wilheit@ariel.tamu.edu, Texas A&M Univ – Ow
Wilhelmson, Robert B., (217) 333-8651 bw@ncsa.uiuc.edu, Univ of Illinois, Urbana-Champaign – Oa
Wilhelmy, Jerry B., (505) 665-3188 Los Alamos National Lab – Oo
Wilhite, Donald A., (402) 472-4270 dwilhite@enso.unl.edu, Univ of Nebraska, Lincoln – Oa
Wilkens, Allen, (202) 282-7374 Univ of the District of Columbia – So
Wilkerson, Christine M., (801) 537-3332 nrugs.cwilkers@state.ut.us, Utah Geological Survey – Gg
Wilkerson, Forrest, (507) 389-2617 forrest.wilkerson@mnsu.edu, Minnesota State Univ – Oy
Wilkerson, M. S., (765) 658-4666 mswilke@depauw.edu, Univ of Illinois, Urbana-Champaign – Gc
Wilkerson, M. Scott, 765-658-4666 mswilke@depauw.edu, DePauw Univ – Gct
Wilkes, Gerald P., (434) 951-6364 gerry.wilkes@dmme.virginia.gov, Division of Mineral Resources – Gg
Wilkes, Paul G., +61 8 9266 2330 P.Wilkes@curtin.edu.au, Curtin Univ of Technology – Ye
Wilkey, Patrick L., (630) 252-6258 Argonne National Lab – Ng
Wilkie, Ann C., (352) 392-8699 acwi@gnv.ifas.ufl.edu, Univ of Florida – Sb
Wilkie, Richard W., (413) 545-2078 rwilkie@geo.umass.edu, Univ of Massachusetts – Oo
Wilkins, David E., (208) 426-2390 dwilkins@boisestate.edu, Boise State Univ – Oy
Wilkinson, Bruce H., (734) 764-1435 eustasy@umich.edu, Univ of Michigan – Gs
Wilks, Daniel, (607) 255-1750 dsw5@cornell.edu, Cornell Univ – Oa
Wilks, Maureen, (505) 835-5322 mwilks@gis.nmt.edu, New Mexico Institute of Mining & Technology – Gp
Williams, Alan E., (951) 827-4611 alan.williams@ucr.edu, Univ of California, Riverside – Cs
Williams, Allison M., (905) 525-9140 (Ext. 24334) awill@mcmaster.ca, McMaster Univ – Oo
Williams, Bruce A., (509) 372-3799 bruce.williams@pnl.gov, Pacific Northwest National Lab – Ng
Williams, Bryan, (520) 626-3406 Univ of Arizona – So
Williams, Carolyn S., (540) 231-6894 wilcar@vt.edu, Virginia Polytechnic Institute & State Univ – Oo
Williams, Charles A., (518) 276-3369 willic3@rpi.edu, Rensselaer Polytechnic Institute – Yg

Williams, Christopher J., 717 291-3814 chris.williams@fandm.edu, Franklin and Marshall College – Pe
Williams, Curtis J., (714) 484-7000 (Ext. 48181) Cypress College – Gg
Williams, Dave E.williams.david@interchange.ubc.ca, Univ of British Columbia – Oc
Williams, David A., (270) 827-3414 williams@uky.edu, Univ of Kentucky – Eg
Williams, Douglas F., (803) 777-7525 Univ of South Carolina – Cs
Williams, Floyd J., (909) 888-6511 (Ext. 1403) San Bernardino Valley College – Gc
Williams, G. Dent, (706) 569-3028 Columbus State Univ – Pv
Williams, Graham, (902) 426-5657 Dalhousie Univ – Pl
Williams, Graham D., +44 01782 583187 g.d.williams@esci.keele.ac.uk, Keele Univ – Gc
Williams, Gus T., (630) 252-4609 gpwilliams@anl.gov, Argonne National Lab – Hq
Williams, Harold, (709) 737-8396 hank@sparky2.esd.mun.ca, Memorial Univ of Newfoundland – Gt
Williams, Harris, (423) 974-6169 hew@sci.nccu.edu, North Carolina Central Univ – Oy
Williams, Harry F. L., (940) 565-3317 williams@unt.edu, Univ of North Texas – Gm
Williams, Ian S., (709) 737-8395 ian.williams@uwrf.edu, Univ of Wisconsin, River Falls – Yg
Williams, Jewell B., 615-322-2976 jewell.williams@vanderbilt.edu, Vanderbilt Univ – Oo
Williams, John W., (408) 924-5050 williams@geosun.sjsu.edu, San Jose State Univ – Ng
Williams, Kay R., (717) 477-1602 krwill@ship.edu, Shippensburg Univ – Oa
Williams, Kevin K., 716-878-5116 williakk@buffalostate.edu, Buffalo State College – GmXg
Williams, Kevin L., (206) 543-3949 williams@apl.washington.edu, Univ of Washington – Op
Williams, Kim R., (303) 273-3245 krwillia@mines.edu, Colorado School of Mines – Ca
Williams, Knox, (970) 482-0421 knoxsuz@aol.com, Colorado Geological Survey – Oa
Williams, Lisa A., (314) 984-7722 lwilliams@stlcc.cc.mo.us, Meramec Community College – Og
Williams, Mark D., (509) 372-6069 mark.d.williams@pnl.gov, Pacific Northwest National Lab – Hw
Williams, Mark W., (303) 492-8830 Univ of Colorado – Og
Williams, Michael L., (413) 545-0745 mlw@geo.umass.edu, Univ of Massachusetts – Gc
Williams, Nancy, (716) 878-6731 ncullis@buffalostate.edu, Buffalo State College – Oo
Williams, Paul F., (506) 453-5185 pfw@unb.ca, Univ of New Brunswick – Gc
Williams, Quentin C., (831) 459-3132 qwilliams@pmc.ucsc.edu, Univ of California, Santa Cruz – Gy
Williams, Scott T., (434) 951-6360 scott.williams@dmme.virginia.gov, Division of Mineral Resources – Gc
Williams, Stanley N., (480) 965-1438 stanley.williams@asu.edu, Arizona State Univ – Gv
Williams, Steve E., (307) 766-3597 Univ of Wyoming – Sb
Williams, Tom, (707) 253-3161 Napa Valley College – Ge
Williams, Trevor, (845)-3658626 Columbia Univ – Gu
Williams, Vincent P., (404) 657-6145 Georgia Dept of Natural Resources – Oo
Williams, William M., (514) 398-4755 McGill Univ – Nx
Williams, II, Richard T., (865) 974-6169 rwilliams@utk.edu, Univ of Tennessee, Knoxville – Yg
Williams-Jones, Anthony E., (514) 398-1676 willyj@eps.mcgill.ca, McGill Univ – Ce
Williams-Jones, Glyn, (604) 291-3306 glynwj@sfu.ca, Simon Fraser Univ – Gv
Williamson, Douglas, 212-772-5265 douglas.williamson@hunter.cuny.edu, Hunter College (CUNY) – Oi
Willis, Grant C., (801) 537-3355 nrugs.gwillis@state.ut.us, Utah Geological Survey – Gg
Willis, Julie B., 208-496-1905 willisj@byui.edu, Brigham Young Univ, Idaho – GtOri
Willoughby, Hugh E., 305-348-0243 hugh.willoughby@fiu.edu, Florida International Univ – Oa
Willoughby, Ralph H., (803) 896-7708 South Carolina Geological Survey – Eg
Wills, Christopher J., (415) 557-1668 California Geological Survey – Ng
Willsey, Shawn P., (208) 732-6421 swillsey@csi.edu, College of Southern Idaho – GgcGv
Wilmut, Michael, (250) 472-4343 Univ of Victoria – Oo
Wilson, Alexander T., (520) 298-5446 atwilson@u.arizona.edu, Univ of Arizona – Cs
Wilson, Alicia M., 803 777-1240 awilson@geol.sc.edu, Univ of South Carolina – Hw
Wilson, Blake, (785) 864-2118 bwilson@kgs.ku.edu, Univ of Kansas – Oi
Wilson, Charles A., (225) 388-6283 cwilson@lsu.edu, Louisiana State Univ – Ob
Wilson, Chester H., (313) 232-6291 Charles Stewart Mott Community College – Ye
Wilson, Clark R., (512) 471-5008 clarkw@maestro.geo.utexas.edu, Univ of Texas at Austin, Jackson School of Geosciences – Yg
Wilson, Elizabeth A., (207)-581-2152 wilsoneliz@earthlink.net, Univ of Maine – Eo
Wilson, Gary S., +64 3 479-7519 gary.wilson@otago.ac.nz, Univ of Otago – Gl
Wilson, Greg C., 616-331-2392 wilsong@gvsu.edu, Grand Valley State Univ – Gm
Wilson, James H., (408) 656-2690 wilson@oc.nps.navy.mil, Naval Postgraduate School – Op
Wilson, James R., (801) 626-6208 jwilson@weber.edu, Weber State Univ – Gz
Wilson, Jeffey S., (317) 274-1128 jeswilso@iupui.edu, Indiana Univ, Indianapolis – Oy
Wilson, Jeffrey A., (734) 647-7461 wilsonja@umich.edu, Univ of Michigan – Gg
Wilson, John A., (512) 471-6087 Univ of Texas at Austin, Jackson School of Geosciences – Pv
Wilson, John D., (780) 492-0353 jaydee.uu@ualberta.ca, Univ of Alberta – Oa
Wilson, John L., (505) 835-5634 jwilson@nmt.edu, New Mexico Institute of Mining and Technology – Hw
Wilson, John M., 410-554-5553 jmwilson@dnr.state.md.us, Maryland Geological Survey – Gg
Wilson, Lorne G., (520) 621-9108 lorne@email.arizona.edu, Univ of Arizona – Hw
Wilson, Lucy A., (506) 648-5607 lwilson@unbsj.ca, Univ of New Brunswick, Saint John – Ga
Wilson, Mark A., (330) 263-2247 mwilson@wooster.edu, The College of Wooster – Pi
Wilson, Mark R., (709) 737-8386 mwilson@sparky2.esd.mun.ca, Memorial Univ of Newfoundland – Cs
Wilson, Michael C., (604) 527-5236 wilsonmi@douglas.bc.ca, Douglas College – Ga
Wilson, Michael P., (716) 673-3843 wilson@fredonia.edu, SUNY, Fredonia – GmHgYg
Wilson, Moyra E., 0191 3742501 moyra.wilson@durham.ac.uk, Durham Univ – Gs
Wilson, Nicholas, (702) 895-3250 nwilson@nevada.edu, Univ of Nevada, Las Vegas – Og
Wilson, P. Christopher, (561) 468-3922 cwilson@gnv.ifas.ufl.edu, Univ of Florida – So
Wilson, Rick I., (709) 737-8386 California Geological Survey – Ge
Wilson, Roy R., (860) 465-4370 wilsonr@easternct.edu, Eastern Connecticut State Univ – Oi
Wilson, Steven D., (217) 333-0956 Illinois State Water Survey – Hw
Wilson, T. Desmond, (604) 527-5222 wilsond@douglas.bc.ca, Douglas College – Eg
Wilson, Terril E., (907) 474-7388 fftew@uaf.edu, Univ of Alaska, Fairbanks – Nm
Wilson, Terril (Ted), (520) 621-4689 wilsont@email.arizona.edu, Univ of Arizona – Nm
Wilson, Terry J., (614) 292-0723 twilson@mps.ohio-state.edu, Ohio State Univ – Gc
Wilson, Thomas H., 304 293 6431 tom.wilson@mail.wvu.edu, West Virginia Univ – Ye
Wilton, Derek H., (709) 737-8389 dwilton@sparky2.esd.mun.ca, Memorial Univ of Newfoundland – Eg
Wilton, Robert D., (905) 525-9140 (Ext. 24536) wiltonr@mcmaster.ca, McMaster Univ – Oo
Wiltschko, David V., (979) 845-9680 wiltschko@geo.tamu.edu, Texas A&M Univ – Gc
Wiltshire, John C., (808) 956-6042 Univ of Hawai'i, Manoa – Gu
Wimbush, Mark, (401) 874-6515 m.wimbush@gso.uri.edu, Univ of Rhode Island – Op
Winant, Clinton D., (619) 534-2067 cwinant@ucsd.edu, Univ of California, San Diego – On
Winchell, Robert E., (562) 985-4920 California State Univ, Long Beach – Gz

Winchester, John A., j.a.winchester@esci.keele.ac.uk,
Keele Univ – Gp
Winchester, John W., 850-576-0954 winchester@ocean.fsu.edu,
Florida State Univ – Oc
Winckler, Gisela, 845-365-8756 Columbia Univ – Cg
Winder, C. G., (519) -66-1-21 (Ext. x 82253) cwinder@uwo.ca,
Univ of Western Ontario – Gg
Windom, Kenneth E., (515) 294-2430 kewindom@iastate.edu,
Iowa State Univ of Science & Technology – Cp
Windsor, Jr., John G., 321-674-8096 jwindsor@fit.edu,
Florida Institute of Technology – Oc
Wine, Paul H., (404) 894-3425 pw7@prism.gatech.edu,
Georgia Institute of Technology – Oa
Winebrenner, Dale P., 206-543-1393 dpw@apl.washington.edu,
Univ of Washington – Olr
Wing, Boswell, 514-398-6772 wing@eps.mcgill.ca,
McGill Univ – Cs
Wing, Boswell A.wing@essic.umd.edu,
Univ of Maryland – Cg
Wing, Scott L., (202) 357-2649 Smithsonian Institution / National Museum of Natural History – Pb
Wingeart, Karen M., 410-293-6566 wingeart@usna.edu,
United States Naval Academy – Ow
Winglee, Robert M., 206-685-8160 winglee@ess.washington.edu,
Univ of Washington – XyYx
Winguth, Arne M., 817 272 2987 awinguth@uta.edu,
Univ of Texas, Arlington – Opa
Winguth, Cornelia, 817 272 2987 cwinguth@uta.edu,
Univ of Texas, Arlington – GugOm
Winhusen, Carol A., 937 775-3455 Wright State Univ – Og
Winkler, Dale A., (214) 768-2750 Southern Methodist Univ – Pv
Winklhofer, Michael, 089/2180 4143 michael@geophysik.uni-muenchen.de, LMU - Univ of Munich – Yg
Winn, Christopher D., (808) 956-7625 cwinn@triton.ucsd.edu,
Univ of Hawai'i, Manoa – Ob
Winslow, Margaret S., (212) 650-6984 Graduate School of the City Univ of New York – Gc
Winsor, Roger A., (828) 262-7053 winsorra@appstate.edu,
Appalachian State Univ – Oo
Winstead, Nathaniel S., (814) 867-1695 nsw@psu.edu,
Pennsylvania State Univ, Altoona – Ow
Winter, John D., (509) 527-5113 winter@whitman.edu,
Whitman College – Gp
Winterer, Edward ., (858) 534-2360 jwinterer@ucsd.edu,
Univ of California, San Diego – Gd
Winterkamp, Judith L., (505) 667-1264 judyw@lanl.gov,
Los Alamos National Lab – Oo
Wintsch, Robert P., (812) 855-4018 wintsch@indiana.edu,
Indiana Univ, Bloomington – Gp
Wirth, Karl R., (651) 696-6449 wirth@macalester.edu,
Macalester College – Gi
Wisdom, Jack, (617) 253-7730 wisdom@mit.edu,
Massachusetts Institute of Technology – Og
Wise, James A., (850) 644-6265 Washington State Univ – Oo
Wise, Michael A., (202) 633-1826 wisem@si.edu,
Smithsonian Institution / National Museum of Natural History – Gz
Wise, William S., (805) 893-3429 Univ of California, Santa Barbara – Gz
Wise, Jr, Sherwood W., (850) 644-6265 wise@gly.fsu.edu,
Florida State Univ – Pm
Wiseman, Alan, 615-322-2976 alan.wiseman@vanderbilt.edu,
Vanderbilt Univ – Oo
Wiseman, Jr., William J., (225) 388-2955 Louisiana State Univ – Op
Wishner, Karen, (401) 874-6402 kwishner@gso.uri.edu,
Univ of Rhode Island – Ob
Wiswall, C. Gil, (610) 436-2570 cwiswall@wcupa.edu,
West Chester Univ – Gc
Withers, Mitchell M., (901) 678-2007 withers@ceri.memphis.edu,
Univ of Memphis – Ys
Withers, Tony, 612-624-3336 withe012@umn.edu,
Univ of Minnesota, Twin Cities – Gi
Withjack, Martha O., (732) 445-3445 drmeow3@yahoo.com,
Rutgers, The State Univ of New Jersey – Gc
Witte, Ronald W., (609) 292-2576 ron.witte@dep.state.nj.us,
New Jersey Geological Survey – Gl
Witte, William, (907) 474-7809 fnwkw@uaf.edu,
Univ of Alaska, Fairbanks – Gg
Witten, Alan J., (405) 325-1563 awitten@ou.edu,
Univ of Oklahoma – Ye
Witter, Donna L., 330-672-7002 dwitter@kent.edu,
Kent State Univ – Op
Witter, Rob, 541-574-7969 Rob.Witter@dogami.state.or.us,
Oregon Dept of Geology & Mineral Industries – Og
Wittke, James, 9285239565/9044 james.wittke@nau.edu,
Northern Arizona Univ – Gi
Witzke, Brian J., (319) 335-1590 bwitzke@igsb.uiowa.edu,
Univ of Iowa – Ps
Wixman, Ronald, (541) 346-4568 rwixman@uoregon.edu,
Univ of Oregon – Oo
Wizevich, Michael, (607) 255-2844 mcw26@cornell.edu,
Cornell Univ – Eg
Wobus, Reinhard A., (413) 597-2470 reinhard.a.wobus@williams.edu,
Williams College – Gi
Woessner, William W., (406) 243-5698 gl_www@selway.umt.edu,
Univ of Montana – Hw
Wofsy, Steven C., (617) 495-4566 scw@io.harvard.edu,
Harvard Univ – Oa
Wohl, Ellen E., (970) 491-5298 ellenw@cnr.colostate.edu,
Colorado State Univ – Gm
Wohletz, Kenneth H., (505) 667-9202 wohletz@lanl.gov,
Los Alamos National Lab – Gv
Wohlford, Duane D., (607) 436-3066 wohlfodd@oneonta.edu,
SUNY, Oneonta – Gp
Wojtal, Steven F., (440) 775-8352 steven.wojtal@oberlin.edu,
Oberlin College – Gc
Wolcott, Donna L., (919) 515-7866 donna_wolcott@ncsu.edu,
North Carolina State Univ – Ob
Wolcott, Thomas G., (919) 515-7866 tom_wolcott@ncsu.edu,
North Carolina State Univ – Ob
WoldeGabriel, Giday, 505-667-8749 wgiday@lanl.gov,
Los Alamos National Lab – Gx
Wolf, Aaron T., (541) 737-2722 wolfa@geo.oregonstate.edu,
Oregon State Univ – Hg
Wolf, Ann M., (814) 863-0841 amw2@psu.edu,
Pennsylvania State Univ, Univ Park – Sc
Wolf, Lorraine W., (334) 844-4878 wolflor@auburn.edu,
Auburn Univ – Ys
Wolf, Michael B., (309) 794-7304 glwolf@augustana.edu,
Augustana College – Gi
Wolfe, Jack A., (520) 629-9455 jwolfe@geo.arizona.edu,
Univ of Arizona – Pb
Wolfe, Mark E., (614) 265-6595 mark.wolfe@dnr.state.oh.us,
Ohio Dept of Natural Resources – En
Wolfe, Paul J., 937 775-3455 paul.wolfe@wright.edu,
Wright State Univ – Ye
Wolff, John A., 509-335-2825 jawolff@mail.wsu.edu,
Washington State Univ – Gi
Wolfgram, Diane, (406) 496-4353 dwolfgram@mtech.edu,
Montana Tech of the Univ of Montana – Emo
Wolfsberg, Andrew V., (505) 667-3599 awolf@lanl.gov,
Los Alamos National Lab – Hw
Wolfsberg, Kurt, (505) 667-4464 Los Alamos National Lab – Cc
Wolman, M. Gordon, (410) 516-7090 wolman@jhu.edu,
The Johns Hopkins Univ – Gm
Wolny, Dave, (970) 248-1154 dwolmy@mesastate.edu,
Mesa State College – Ys
Wolosz, Thomas H., (518) 564-4031 woloszth@plattsburgh.edu,
Plattsburgh State Univ (SUNY) – Pe
Woltemade, Christopher J., (717) 477-1143 cjwolt@ship.edu,
Shippensburg Univ – Hg
Wong, C. S., (250) 363-6407 Univ of Victoria – Oc
Wong, Chi S., (250) 363-6407 WongCS@pac.dfo-mpo.gc.ca,
Univ of British Columbia – Cm
Wong, George T. F., (757) 683-4932 gwong@odu.edu,
Old Dominion Univ – Oc
Wong, Martin, (315) 228-7203 mewong@mail.colgate.edu,
Colgate Univ – Gtc
Wong, Teng-fong, 631-632-8212 teng-fong.wong@stonybrook.edu,
SUNY, Stony Brook – Yx
Wong, Victor, (516) 632-8212 vwong@cicese.mx,
Centro de Inv Científica y de Educación Superior de Ensenada – Ys
Woo, David, 510-885-3160 david.woo@csueastbay.edu,
California State Univ, East Bay – Or
Woo, Ming-ko, (905) 525-9140 (Ext.23526) woo@mcmail.cis.mcmaster.ca, McMaster Univ – Hg
Wood, Charles W., (334) 844-3997 cwood@acesag.auburn.edu,
Auburn Univ – Sc
Wood, David A., (210) 733-2840 dwood30@mail.accd.edu,
San Antonio College – Oo
Wood, Eric F., (609) 258-4675 efwood@pucc.princeton.edu,
Princeton Univ – Hg

Wood, Harold R., (671) 735-2688 hrwood@uog.edu,
Univ of Guam – Oo
Wood, James R., (906) 487-2894 jrw@mtu.edu,
Michigan Technological Univ – Cl
Wood, Lesli J., (512) 471-0328 leslie.wood@beg.utexas.edu,
Univ of Texas at Austin, Jackson School of Geosciences – Gr
Wood, Michael M., (510) 885-3486 California State Univ, East Bay – Gz
Wood, Roger C., (609) 652-4252 Richard Stockton College of New Jersey – Pv
Wood, Scott A., (208) 885-5966 swood@uidaho.edu,
Univ of Idaho – Cg
Wood, Spencer H., (208) 426-3629 swood@boisestate.edu,
Boise State Univ – Ge
Wood, Stephen L., (321) 674-8096 swood@fit.edu,
Florida Institute of Technology – Oo
Wood, Wendywendyw@nova.edu,
Nova Southeastern Univ – Oo
Woodard, Gary C., (520) 621-5399 gwoodard@sahra.arizona.edu,
Univ of Arizona – Oo
Woodard, Howard J., (605) 688-4774 Howard.Woodard@sdstate.edu,
South Dakota State Univ – ScoGg
Woodburne, Michael O., (909) 787-5028 michael.woodburne@ucr.edu,
Univ of California, Riverside – Pv
Woodbury, Allan D., (204) 474-9137 Univ of Manitoba – Hw
Woodcock, Curtis E., (617) 353-5746 curtis@bu.edu,
Boston Univ – Or
Wooden, Eric, 757.727.5710 eric.wooden@hamptonu.edu,
Hampton Univ – Ob
Woodgate, Rebecca A., (206) 221-3268 woodgate@apl.washington.edu,
Univ of Washington – Op
Woodhead, James A., (323) 259-2556 woodhead@oxy.edu,
Occidental College – Gz
Woodhead, Peter M. J., (631) 632-8731 pwoodhead@notes.cc.sunysb.edu,
SUNY, Stony Brook – Ob
Woodhouse, Connieconnie.woodhouse@noaa.gov,
Univ of Arizona – Pe
Woodin, Sarah A., (803) 777-4254 Univ of South Carolina – Ob
Wooding, Frank B., (508) 289-3334 Woods Hole Oceanographic Inst – Yr
Woodland, Bertram G., (312) 665-7648 Field Museum of Natural History – Gp
Woodland, Mary, 573-751-2867 nrwoodm@mail.dnr.state.mo.us,
Missouri Dept of Natural Resources – Oo
Woodley, Teresa, (604) 822-3146 twoodley@eos.ubc.ca,
Univ of British Columbia – Oo
Woodruff, William H., (505) 665-2557 Los Alamos National Lab – Oo
Woods, Adam D., (714) 278-2921 awoods@fullerton.edu,
California State Univ, Fullerton – Gs
Woods, Alan J., (541) 758-0713 alan.woods@science.oregonstate.edu,
Oregon State Univ – Gm
Woods, Terri L., (252) 328-6014 woodst@mail.ecu.edu,
East Carolina Univ – Cl
Woodward, Lee A., (505) 277-5309 Univ of New Mexico – Gc
Woodward, Mac B., 501 683-0113 Arkansas Geological Commission – Go
Woodwell, Grant R., (540) 654-1427 gwoodwel@mwc.edu,
Univ of Mary Washington – Gc
Wooldridge, C F.wooldridge@cf.ac.uk,
Cardiff Univ – Oo
Woolery, Edward W., (859) 257-3016 woolery@uky.edu,
Univ of Kentucky – Ys
Woolsey, James R., (662) 915-7320 inst@olemiss.edu,
Univ of Mississippi – Oo
Wooten, Richard M., (919) 733-2423 North Carolina Geological Survey – Ng
Wopenka, Brigitte, (314) 935-4771 bwopenka@wustl.edu,
Washington Univ – Ca
Worcester, Peter A., (812) 866-7306 worcestr@hanover.edu,
Hanover College – Gi
Wormuth, John H., (979) 845-7132 jwormuth@ocean.tamu.edu,
Texas A&M Univ – Ob
Worrall, Fred, 0191 3742525 fred.worrall@durham.ac.uk,
Durham Univ – Cg
Worsley, Thomas R., (740) 593-1101 Ohio Univ, Lancaster – Gu
Wortel, Matthew J., (319) 335-3992 matthew-wortel@uiowa.edu,
Univ of Iowa – Oo
Worthington, Galen, 785-864-2043 galen@kgs.ku.edu,
Univ of Kansas – Ge
Wortmann, Ulrich B., 416-978-2084 na,
Univ of Toronto – Gg
Woussen, Gerard, (418) 545-5011 gwoussen@uqac.uquebec.ca,
Universite du Quebec a Chicoutimi – Gi
Wraith, Albert E., (091) 222-7907 McGill Univ – Nx
Wraith, Jon M., (406) 994-1997 jwraith@gemini.oscs.montana.edu,
Montana State Univ – Sp
Wright, Carrie L., (812) 465-1145 clwright@usieagles.org,
Univ of Southern Indiana – Oe
Wright, Daniel G., 902 426 3147 wrigtdan@mar.dfo-mpo.gc.ca or dan.wright@dal.ca, Dalhousie Univ – Op
Wright, Dawn J., (541) 737-1229 dawn@dusk.geo.orst.edu,
Oregon State Univ – Ou
Wright, Elaine, (215) 898-7009 elwright@sas.upenn.edu,
Univ of Pennsylvania – Oo
Wright, James A., (709) 737-2334 jim.wright@mun.ca,
Memorial Univ of Newfoundland – Yh
Wright, James D., (732) 445-5722 Rutgers, The State Univ of New Jersey – Cs
Wright, James (Jim) E., (706) 542-4394 jwright@gly.uga.edu,
Univ of Georgia – Gt
Wright, Judith V., (509) 376-2138 Washington State Univ – Oo
Wright, Kathyrn, (269) 387-5486 kathyrn.wright@wmich.edu,
Western Michigan Univ – Oo
Wright, L. Donelson, (804) 684-7102 wright@vims.edu,
College of William & Mary – Ou
Wright, Scott, (727) 896-8626 Univ of South Florida, St Petersburg – Oo
Wright, Stephen F., (802) 656-4479 swright@uvm.edu,
Univ of Vermont – Gc
Wright, V P., wrightVP@cf.ac.uk,
Cardiff Univ – Gs
Wright, William H., (707) 887-7816 terryw100@gmail.com,
Sonoma State Univ – Gc
Wright, Jr., Herbert E., 612-624-5215 hew@umn.edu,
Univ of Minnesota, Twin Cities – Gl
Wronkiewicz, David J., (630) 252-4385 Argonne National Lab – Cg
Wronkiewicz, David J., (573) 341-4679 wronk@umr.edu,
Univ of Missouri, Rolla – Ce
Wu, Charles T., (519) 661-3791 ctwu@uwo.ca,
Univ of Western Ontario – Oo
Wu, Chin, 608-263-3078 chinwu@engr.wisc.edu,
Univ of Wisconsin, Madison – Hs
Wu, David T., (303) 273-2066 dwu@mines.edu,
Colorado School of Mines – Oo
Wu, Francis T., (607) 777-2512 Binghamton Univ – Ys
Wu, Mandy, (604) 822-2736 mwu@eos.ubc.ca,
Univ of British Columbia – Oo
Wu, Patrick, (403) 220-7855 ppwu@ucalgary.ca,
Univ of Calgary – Ys
Wu, Qian, (313) 747-3475 Univ of Michigan – Og
Wu, Ru-shan, (831) 459-5135 wrs@es.ucsc.edu,
Univ of California, Santa Cruz – Ys
Wu, Shuang-Ye, (937) 229-1720 Shuang-Ye.Wu@notes.udayton.edu,
Univ of Dayton – Oy
Wu, Xiao-chun, (403) 823-7707 Royal Tyrrell Museum of Palaeo – Pv
Wu, Xiaoqing, (515) 294-9872 wuxq@iastate.edu,
Iowa State Univ of Science & Technology – Oa
Wu, Yi-Hwa, (660) 562-1869 ywu@nwmissouri.edu,
Northwest Missouri State Univ – Oi
Wuebbles, Donald J., (217) 244-1568 wuebbles@atmos.uiuc.edu,
Univ of Illinois, Urbana-Champaign – Oa
Wuerthele, Norman, (412) 622-3265 Carnegie Museum of Natural History – Pv
Wulff, Andrew, (270) 745-5976 andrew.wulff@wku.edu,
Western Kentucky Univ – Gv
Wulff, Andrew H., (562) 907-4220 Whittier College – Gi
Wunderman, Richard, (202) 633-1827 wunderma@si.edu,
Smithsonian Institution / National Museum of Natural History – Gv
Wunsch, Carl I., (617) 253-5937 cwunsch@mit.edu,
Massachusetts Institute of Technology – Op
Wunsch, David R., (603) 271-6482 dwunsch@des.state.nh.us,
New Hampshire Geological Survey – Hw
Wurstner, Signe K., (509) 372-6115 signe.wurstner@pnl.gov,
Pacific Northwest National Lab – Hw
Wurth, Robert, (314) 977-3137 rwurth@eas.slu.edu,
Saint Louis Univ – Yr
Wyckoff, John W., (303) 556-2590 jwyckoff@spatial.cudenver.edu,
Univ of Colorado, Denver – Oy
Wyckoff, William K., (406) 994-6914 uesww@montana.edu,
Montana State Univ – Oo
Wyld, Sandra J., (706) 542-9908 swyld@gly.uga.edu,
Univ of Georgia – Gc
Wylie, Albert, (906) 487-2531 aswylie@mtu.edu,

Michigan Technological Univ – Go
Wylie, Ann G., (301) 405-4079 awylie@umd.edu,
Univ of Maryland – Gz
Wyllie, Peter J., (626) 395-6461 wyllie@gps.caltech.edu,
California Institute of Technology – Cp
Wyman, Derek A., (618) 938-0117 Univ of Saskatchewan – Cg
Wymer, Douglas, (205) 652-3414 dwymer@uwa.edu,
Univ of West Alabama –
Wyngaard, John C., (814) 863-7714 wyngaard@ems.psu.edu,
The Pennsylvania State Univ, Univ Park – Ow
Wynn, Thomas C., (570) 484-2081 twynn@lhup.edu,
Lock Haven Univ – GsPiEo
Wyrtki, Klaus, (808) 956-7729 ocean@soest.hawaii.edu,
Univ of Hawai'i, Manoa – Op
Wysession, Michael E., (314) 935-5625 michael@wucore.wustl.edu,
Washington Univ – Ys
Wysocki, Mark, (607) 255-2568 mww3@cornell.edu,
Cornell Univ – Oa
Wyss, Andre R., (805) 893-8628 Univ of California, Santa Barbara – Pv

X

Xia, Jianghai, (785) 864-2057 jxia@kgs.ku.edu,
Univ of Kansas – Ye
Xia, Renjie, (217) 244-6166 Illinois State Water Survey – Hs
Xia, Zong-Guo, (617) 287-5286 Univ of Massachusetts, Boston/Harbor Campus – Or
Xia, Zong-Guo, 718-960-8683 zongguo.xia@lehman.cuny.edu,
Lehman College (CUNY) – Oi
Xiao, Shuhai, 540-231-1366 xiao@vt.edu,
Virginia Polytechnic Institute & State Univ – Pi
Xie, Lian, (919) 515-1435 lian_xie@nscu.edu,
North Carolina State Univ – Oa
Xie, Xiao-bi, (831) 459-5094 xie@es.ucsc.edu,
Univ of California, Santa Cruz – Ys
Xiu, Dongbin, (765) 496-2846 dxiu@purdue.edu,
Purdue Univ – Gq
Xu, Huifang, 608/265-5587 hfxu@geology.wisc.edu,
Univ of Wisconsin, Madison – Gz
Xu, Li, (508)289-3673 lxu@whoi.edu,
Woods Hole Oceanographic Institution – Yr
Xu, Shangping, 414-229-6148 xus@uwm.edu,
Univ of Wisconsin, Milwaukee – Hg

Y

Yabusaki, Steven B., (509) 372-6095 steve.yabusaki@pnl.gov,
Pacific Northwest National Lab – Hg
Yacobucci, Peg M., (419) 372-7982 mmyacob@bgsu.edu,
Bowling Green State Univ – Po
Yacucci, Mark A., 217-265-0747 yacucci@isgs.uiuc.edu,
Illinois State Geological Survey – Ge
Yalcin, Kaplan, (541) 737-1230 yalcink@geo.oregonstate.edu,
Oregon State Univ – Og
Yalda, Sepideh, 717-872-3293 Sepi.Yalda@millersville.edu,
Millersville Univ – Oa
Yamazaki, Hidekatsu, (250) 721-8918 Univ of Victoria – Op
Yan, Jun, (270) 745-8952 junu.yan@wku.edu,
Western Kentucky Univ – Oi
Yan, Xiao-Hai, (302) 831-3694 Univ of Delaware – Or
Yan, Y E., (630) 252-6322 eyan@anl.gov,
Argonne National Lab – Gg
Yanai, Michio, (310) 825-1653 yanai@atmos.ucla.edu,
Univ of California, Los Angeles – Oa
Yancey, Thomas E., (979) 845-0643 yancey@geo.tamu.edu,
Texas A&M Univ – Ps
Yandle, Tracy, 404 727 5652 tyandle@emory.edu,
Emory Univ – Ob
Yang, Huijun, (727) 553-1639 Univ of South Florida, St Petersburg – Op
Yang, Panseok, (204) 474-9452 yangp@cc.umanitoba.ca,
Univ of Manitoba – Ca
Yang, Ping, (979) 845-7671 pyang@ariel.met.tamu.edu,
Texas A&M Univ – Oa
Yang, Wan, 316 -978-3140 Wichita State Univ – Gs
Yang, Y, YangY6@cf.ac.uk,
Cardiff Univ – Hg
Yang, Youqing, yangyo@missouri.edu,
Univ of Missouri – Yg
Yang, Zong-Liang, (512) 471-3824 liang@mail.utexas.edu,
Univ of Texas at Austin, Jackson School of Geosciences – Ow
Yao, Wensheng, (727) 553-3922 wyao@marine.usf.edu,
Univ of South Florida, Saint Petersburg – Oc
Yapp, Crayton J., (214) 768-3897 Southern Methodist Univ – Cs
Yarar, Baki, (303) 273-3768 byarar@mines.edu,
Colorado School of Mines – Nx
Yarbrought, Robert A., 912-681-0846 RYarbrough@GeorgiaSouthern.edu,
Georgia Southern Univ – Oo
Yarger, Douglas N.doug@iastate.edu,
Iowa State Univ of Science & Technology – Oa
Yarger, Douglas N., (515) 294-9872 doug@iastate.edu,
Iowa State Univ of Science & Technology – Oa
Yassin, Barbara E., (601) 961-5571 barbara_yassin@deq.state.ms.us,
Mississippi Office of Geology – Oy
Yates, Kay, (541) 737-5193 kyates@coas.oregonstate.edu,
Oregon State Univ – Oo
Yates, Martin G., (207) 581-2154 yates@maine.maine.edu,
Univ of Maine – Em
Yates, Mary Anne, (505) 667-7090 Los Alamos National Lab – Oo
Yau, Man Kong, (514) 398-3719 peter.yau@mcgill.ca,
McGill Univ – Oa
Yayanos, A. A., (858) 534-2935 ayayanos@ucsd.edu,
Univ of California, San Diego – Ob
Yeager, Kevin, 228-688-7098 Kevin.Yeager@usm.edu,
Univ of Southern Mississippi – On
Yeats, Robert S., (541) 737-1226 yeatsr@geo.oregonstate.edu,
Oregon State Univ – Gt
Yee, Nathan, (732) 932-1105 nyee@envsci.rutgers.edu,
Rutgers, The State Univ of New Jersey – Py
Yegulalp, Tuncel M., (212) 854-2984 yegulalp@columbia.edu,
Columbia Univ – Nm
Yeh, Jim, (520) 621-5943 yeh@hwr.arizona.edu,
Univ of Arizona – Sp
Yeh, Joseph S., (512) 471-3323 joseph.yeh@beg.utexas.edu,
Univ of Texas at Austin, Jackson School of Geosciences – Gm
Yeh, Tian-Chyi J., (520) 621-5943 ybiem@hwr.arizona.edu,
Univ of Arizona – Hq
Yelderman, Jr., Joe C., (254) 710-2185 joe_yelderman@baylor.edu,
Baylor Univ – Hw
Yelle, Roger, (520) 621-6243 yelle@lpl.arizona.edu,
Univ of Arizona – Oo
Yett, Jan R., 714-432-5647 jyett@occ.cccd.edu,
Orange Coast College – Pm
Yin, An, (310) 825-8752 yin@ess.ucla.edu,
Univ of California, Los Angeles – Gt
Yin, Qing-zhu, 530-752-0934 Yin@geology.ucdavis.edu,
Univ of California, Davis – Cc
Yingling, Jon C., (606) 257-1105 Univ of Kentucky – Nm
Yoakum, Barbara, (410) 293-6928 yoakum@usna.edu,
United States Naval Academy – Oo
Yoder, James A.jyoder@whoi.edu,
Woods Hole Oceanographic Institution – Cm
Yogodzinkski, Gene M., 803-777-9524 gyogodzin@geol.sc.edu,
Univ of South Carolina – Gi
Yoh, Shing, (908) 737-3692 syoh@kean.edu,
Kean Univ – Ow
Yokel, Lee S., (334) 460-6381 lestockng@aol.com,
Univ of South Alabama – Ge
Yokoyama, Tetsuya, (301) 405-6630 yokoyama@geol.umd.edu,
Univ of Maryland – Cg
Yonkee, William A., (801) 626-7419 ayonkee@weber.edu,
Weber State Univ – Gc
Yoo, Kyungsoo, 302-831-1376 kyoo@udel.edu,
Univ of Delaware – Os
Yool, Stephen R., 520-621-8549 yools@email.arizona.edu,
Univ of Arizona – Or
York, Derek, (416) 978-3157 york@physics.utoronto.ca,
Univ of Toronto – Cc
Young, Charles T., (906) 487-2072 ctyoung@mtu.edu,
Michigan Technological Univ – Ye
Young, David P., (785) 864-2096 young@kgs.ku.edu,
Univ of Kansas – Oo
Young, Davis A.youn@calvin.edu,
Calvin College – Giz
Young, Donald W., (602) 542-5025 Univ of Arizona – Oo
Young, Edward, (310) 267-4930 eyoung@ess.ucla.edu,
Univ of California, Los Angeles – Cs
Young, George S., (814) 863-4228 young@ems.psu.edu,
The Pennsylvania State Univ, Univ Park – Ow
Young, Graham A., (204) 988-0648 gyoung@cc.umanitoba.ca,
Univ of Manitoba – Pi
Young, Grant M., (519) 661-3193 gyoung@uwo.ca,

Univ of Western Ontario – Gc
Young, Harvey R., 204-727-9798 young@brandonu.ca, Brandon Univ – Gd
Young, James E., (828) 262-8482 youngje@appstate.edu, Appalachian State Univ – Oe
Young, Jeffrey, 204-474-8863 jyoung@cc.umanitoba.ca, Univ of Manitoba – Gg
Young, John A., (608)262-5963 jayoung@wisc.edu, Univ of Wisconsin, Madison – Oa
Young, Leonard M., (318) 342-1888 Univ of Louisiana, Monroe – Gs
Young, Michael, (702) 895-0400 michael@dri.edu, Univ of Nevada, Reno – Hg
Young, Patricia, (303) 866-3872 patricia.young@state.co.us, Colorado Geological Survey – Oo
Young, Richard A., (585) 245-5296 young@geneseo.edu, SUNY, Geneseo – Gm
Young, Richard E., (808) 956-7024 ryoung@hawaii.edu, Univ of Hawai'i, Manoa – Ob
Young, Robert S., (828) 227-3822 ryoung@wcu.edu, Western Carolina Univ – Gs
Young, Roger A., (405) 325-5753 Univ of Oklahoma – Ye
Young, Susan L., 501 296-1877 susan.young@arkansas.gov, Arkansas Geological Commission – Oo
Young, William R., (858) 534-1380 wryoung@ucsd.edu, Univ of California, San Diego – Op
Youngs, Leslie G., (916) 322-8078 California Geological Survey – Yg
Yu, Jaehyung, 361-593-3586 kfjy000@tamuk.edu, Texas A&M Univ, Kingsville – Oi
Yu, Qian, 413-545-2095 qyu@geo.umass.edu, Univ of Massachusetts – Oi
Yu, Zicheng, 610-758-6751 ziy2@lehigh.edu, Lehigh Univ – Pe
Yuan, Fei, (507) 389-2617 fei.yuan@mnsu.edu, Minnesota State Univ – Or
Yuan, Xiaojun, (845) 365-8820 Columbia Univ – Op
Yue, Stephen, (514) 398-4755 McGill Univ – Nx
Yuen, Cheong-yip R., (630) 252-4869 yuenr@anl.gov, Argonne National Lab – Ge
Yuen, David A., 612-624-1868 davey@umn.edu, Univ of Minnesota, Twin Cities – Yg
Yule, J. Douglas, (818) 677-6238 j.d.yule@csun.edu, California State Univ, Northridge – Gt
Yund, Richard A., (401) 863-1931 richard_yund@brown.edu, Brown Univ – Gy
Yung, Yuk L., (626) 395-6940 yly@gps.caltech.edu, California Institute of Technology – Oa
Yunwei, Sun, (925)422-1587 sun4@llnl.gov, Lawrence Livermore National Lab – Ng
Yuretich, Richard F., (413) 545-0538 yuretich@geo.umass.edu, Univ of Massachusetts – Cl
Yurkovich, Steven P., (828) 227-7367 yurkovich@wcu.edu, Western Carolina Univ – Gx
Yurtsever, Ayhan, 089/2180 4346 ayhan.yurtsever@ltz.uni-uemchen.de, LMU - Univ of Munich – Gz
Yusoff, Ismail, 03-79674153 ismaily70@um.edu.my, Univ of Malaya – Hg
Yuter, Sandra, 919-513-7963 sandra_yuter@ncsu.edu, North Carolina State Univ – Oa
Yutzy, Gale, (301) 687-4369 gyutzy@frostburg.edu, Frostburg State Univ – Oo
Yvon-Lewis, Shari A., 979 458 1816 syvonlewis@ocean.tamu.edu, Texas A&M Univ – Oc

Z

Zachos, James C., (831) 459-4644 jzachos@es.ucsc.edu, Univ of California, Santa Cruz – Cm
Zachry, Doy L., (479) 575-2785 dzachry@uark.edu, Univ of Arkansas, Fayetteville – Gr
Zadra, Ayrtonayrton.zadra@ec.gc.ca, McGill Univ – Oa
Zaengle, John F., (715) 394-8311 jzaengle@staff.uwsuper.edu, Univ of Wisconsin, Superior – Gs
Zafiriou, Oliver C., (508) 289-2342 ozafiriou@whoi.edu, Woods Hole Oceanographic Institution – Oc
Zahm, Laura, 713-896-8560 laura.zahm@beg.utexas.edu, Univ of Texas at Austin, Jackson School of Geosciences – Gg
Zajac, Roman N., (203) 932-7114 zajacrn@charger.newhaven.edu, Univ of New Haven – Oo
Zakardjian, Bruno, (418) 723-1986 (Ext. 1570) bruno_zakardjian@uqar.qc.ca, Universite du Quebec a Rimouski – Ob
Zakrzewski, Richard J., (785) 628-4286 rzakrzew@fhsu.edu, Fort Hays State Univ – Pv
Zaleha, Michael J., (937) 327-7331 mzaleha, Wittenberg Univ – Gs
Zamzow, Craig E.zamzow@clarion.edu, Clarion Univ – Gi
Zanda, Brigitte M., (732) 445-3232 zandahew@rci.rutgers.edu, Rutgers, The State Univ of New Jersey – Xm
Zandt, George, (520) 621-2273 zandt@geo.arizona.edu, Univ of Arizona – Ys
Zanetti, Kathleen, (702) 895-4789 zanettik@nevada.edu, Univ of Nevada, Las Vegas – Gg
Zaniewski, Kasimierz J., (414) 424-7109 Univ of Wisconsin, Oshkosh – Oo
Zaporozec, Alexanderazaporoz@wisc.edu, Univ of Wisconsin - Extension – Hw
Zarillo, Gary A.zarillo@fit.edu, Florida Institute of Technology – Ou
Zaspel, Craig E., (406) 683-7366 c_zaspel@umwestern.edu, Univ of Montana, Western – Yg
Zawadzki, Isztar I., (514) 398-1034 isztar.zawadzki@mcgill.ca, McGill Univ – Oa
Zawiskie, John M., (248) 645-3252 Wayne State Univ – Gg
Zborowski, Andrewzborowsk@fit.edu, Florida Institute of Technology – Oo
Zebker, Howard A., (650) 723-8067 zebker@pangea.stanford.edu, Stanford Univ – Yx
Zeh, Janet, (860) 424-3680 Geological and Natural History Survey of Connecticut – Oo
Zehner, Richard, (775) 784-7055 zehner@unr.edu, Univ of Nevada, Reno – Gg
Zehr, Jonathan P., (831) 459-4009 zehrj@cats.ucsc.edu, Univ of California, Santa Cruz – Ob
Zeiger, Elaine, (312) 665-7627 Field Museum of Natural History – Oo
Zeigler, E. Lynn, 678-891-3767 ezeigler@gpc.edu, Georgia Perimeter College – Gd
Zeitler, Peter K., (610) 758-3671 pkz0@lehigh.edu, Lehigh Univ – Cc
Zeitlhöfler, Matthias, 089/2180 6570 matthias@zeitlhoefler.de, LMU - Univ of Munich – Gg
Zelazny, Lucian W., (540) 231-9781 lzelazny@vt.edu, Virginia Polytechnic Institute & State Univ – So
Zelazny, Lucian W., (540) 231-9781 zelazny@vt.edu, Virginia Polytechnic Institute & State Univ – Sc
Zell, Paul D.pdz1@psu.edu, Pennsylvania State Univ, Altoona – Pi
Zellers, Sarah D., 660-543-8930 szellers@ucmo.edu, Univ of Central Missouri – Gg
Zelt, Colin A., (713) 348-4757 czelt@rice.edu, Rice Univ – Yg
Zen, E-An, (301) 405-4081 ezen@geol.umd.edu, Univ of Maryland – Og
Zeng, Hongliu, (512) 475-6382 hongliu.zeng@beg.utexas.edu, Univ of Texas at Austin, Jackson School of Geosciences – Gs
Zeng, Zhen, (604) 822-3932 zzeng@eos.ubc.ca, Univ of British Columbia – Oa
Zenger, Donald H., (909) 607-2422 dzenger@pomona.edu, Pomona College – Gr
Zentilli, Marcos, (902) 494-3873 Dalhousie Univ – Eg
Zentner, Nick, (509) 963-2828 nick@geology.cwu.edu, Central Washington Univ – Gg
Zhan, Hongbin, (979) 862-7961 zhan@geo.tamu.edu, Texas A&M Univ – Hw
Zhang, Don, 405-325-5928 donzhang@ou.edu, Univ of Oklahoma – Np
Zhang, Dongxiao, (505) 667-3541 dongzhang@lanl.gov, Los Alamos National Lab – Hg
Zhang, Fuqing, (979) 845-7671 fzhang@tamu.edu, Texas A&M Univ – Oa
Zhang, Minghua, (631) 632-8318 mzhang@notes.cc.sunysb.edu, SUNY, Stony Brook – Oa
Zhang, Qiaofeng (Robin), (270) 809-6760 robin.zhang@murraystate.edu, Murray State Univ – Oi
Zhang, Qing, (319) 335-0405 qing-zhang@uiowa.edu, Univ of Iowa – Gc
Zhang, Renduo, (417) 836-5301 Univ of Wyoming – Sp
Zhang, Renyi, (979) 845-7671 zhang@ariel.met.tamu.edu, Texas A&M Univ – Oa
Zhang, Yang, 919-515-9688 yang_zhang@ncsu.edu, North Carolina State Univ – Oa
Zhang, Ye, (307) 766-3386 Univ of Wyoming – Hy

Zhang, Yongwei, (417) 836-5301 Southwest Missouri State Univ – Ou
Zhang, You-Kuan, (319) 335-1806 you-kuan-zhang@uiowa.edu, Univ of Iowa – Hw
Zhang, Youxue, (734) 763-0947 youxue@umich.edu, Univ of Michigan – Cp
Zhang, Yuxia, (408) 656-2690 zhang@nps.navy.mil, Naval Postgraduate School – Op
Zhang, Zhong, 612-624-9593 zhang125@umn.edu, Univ of Minnesota, Twin Cities – Cl
Zhao, Jing-Zhi, (404) 894-3634 jing.zhao@eas.gatech.edu, Georgia Institute of Technology – Oa
Zhao, Xixi, (831) 459-4847 xzhao@es.ucsc.edu, Univ of California, Santa Cruz – Ym
Zhdanov, Michael S., (801) 581-7750 mzhdanov@mines.utah.edu, Univ of Utah – Ye
Zhen, Yan, (718) 997-3329 Queens College (CUNY) – Cl
Zheng, Chunmiao, 205-348-0579 czheng@ua.edu, Univ of Alabama – Hw
Zheng, Yan, (718) 997-3329 yzeng.sees.qc, Graduate School of the City Univ of New York – Cm
Zhou, Hua-Wei, (713) 743-3424 hzhou@uh.edu, Univ of Houston – Ys
Zhou, Xiaobing, (406) 496-4350 xzhou@mtech.edu, Montana Tech of the Univ of Montana – Or
Zhou, Ying, 540-231-6521 yingz@vt.edu, Virginia Polytechnic Institute & State Univ – Ys
Zhou, Yu, (845) 437-5543 Vassar College – Ou
Zhu, Chen, (812) 856-1884 chenzhu@indiana.edu, Indiana Univ, Bloomington – Oo
Zhu, Lupei, (314) 977-3131 lupei@eas.slu.edu, Saint Louis Univ – Ys
Zhu, Ping, 305-348-7096 zhup@fiu.edu, Florida International Univ – Oa
Zhuang, Qianlai, 765-494-9610 qzhuang@purdue.edu, Purdue Univ – Oo
Zieg, Michael J., (724) 738-2501 michael.zieg@sru.edu, Slippery Rock Univ – Gi
Ziegler, Alfred M., (773) 702-8146 Univ of Chicago – Gg
Zieman, Joseph C., (804) 924-0570 jcz@virginia.edu, Univ of Virginia – Ob
Zierenberg, Robert A., (530) 752-1863 zierenberg@geology.ucdavis.edu, Univ of California, Davis – Cs
Zimbelman, James R., 202-633-2471 Smithsonian Institution / National Air & Space Museum – Xg
Zimmerman, Andrew, 352-392-0070 azimmer@geology.ufl.edu, Univ of Florida – Co
Zimmerman, Brian S., (814) 732-2529 bzimmerman@edinboro.edu, Edinboro Univ of Pennsylvania – Em
Zimmerman, Mark, 612-626-0572 zimme030@umn.edu, Univ of Minnesota, Twin Cities – Yx
Zimmerman, Patrick R., (605) 394-2291 Patrick.Zimmerman@sdsmt.edu, South Dakota School of Mines & Technology – Oa
Zimmerman, R. Eric, (630) 252-6816 Argonne National Lab – Ng
Zimmerman, Ronald K., (225) 388-8302 ron@lgs.bri.lsu.edu, Louisiana State Univ – Go
Zimmerman, Jr., Jay, (618) 453-3351 zimmerman@geo.siu.edu, Southern Illinois Univ Carbondale – Gc
Zimmermann, George, (609) 652-4308 Richard Stockton College of New Jersey – Oo
Zingmark, Richard G., (803) 777-4873 Univ of South Carolina – Ob
Zink, Albert, 089/2180 4340 LMU - Univ of Munich – Gz
Zinner, Ernst K., (314) 935-6276 ekz@howdy.wustl.edu, Washington Univ – Xm
Zinsmeister, William J., (765) 494-0279 wjzins@purdue.edu, Purdue Univ – Ps
Zipper, Carl E., (540) 231-9782 czip@vt.edu, Virginia Polytechnic Institute & State Univ – Oe
Ziskin, Michael H., (203) 932-7107 Univ of New Haven – Oo
Zitzer, Stephen F., 702-895-0436 stephen@dri.edu, Desert Research Institute – Sf
Zizzo, Gayle, (520) 621-6025 gzizzo@geo.arizona.edu, Univ of Arizona – Oo
Zlotkin, Howard, 12012003161 hzlotkin, New Jersey City Univ – Oe
Zlotnik, Vitaly A., (402) 472-2495 vzlotnik1@unl.edu, Univ of Nebraska, Lincoln – Hq
Zoback, Mark D., (650) 725-9295 zoback@pangea.stanford.edu, Stanford Univ – Yg
Zois, Constantine S., (908) 737-3693 Kean Univ – Ow
Zollweg, James, (208) 426-3250 Boise State Univ – Ys
Zollweg, James A., (585) 395-2352 jzollweg@brockport.edu, SUNY College at Brockport – Hs
Zolotov, Mikhail Y., (480) 965-4739 zolotov@asu.edu, Arizona State Univ – Cg
Zou, Xiaolei, (850) 644-6025 zou@met.fsu.edu, Florida State Univ – Ow
Zreda, Marek G., (520) 621-4072 marek@hwr.arizona.edu, Univ of Arizona – Hw
Zuber, Maria T., 617-253-6397 zuber@mit.edu, Massachusetts Institute of Technology – Xy
Zuberer, David A., (979) 845-8322 d-zuberer@tamu.edu, Texas A&M Univ – Sb
Zumwalt, Gary S.gzumwalt@coes.latech.edu, Louisiana Tech Univ – Pi
Zuppann, Charles W., 812 855 9199 czuppann@indiana.edu, Indiana Geological Survey – Go
Zurawski, Ronald P., (615) 532-1502 Ronald.Zurawski@state.tn.us, Tennessee Division of Geology – GgEog
Zurick, David N., (859) 622-1427 david.zurick@eku.edu, Eastern Kentucky Univ – Oo
Zwack, Peter, (514) 987-3000 zwack.peter@uqam.ca, Universite du Quebec a Montreal – Ow
Zwettler, Kathie M., (608) 262-9418 kmzwettl@wisc.edu, Univ of Wisconsin - Extension – Oo
Zwiers, Francis, (250) 472-7304 Francis.Ziers@ec.gc.ca, Univ of Victoria – Oo
Zyko, Karen L., (860) 424-3378 karen.zyko@po.state.ct.us, Geological and Natural History Survey of Connecticut – Oi
Zyvoloski, George A., (505) 667-1581 Los Alamos National Lab – Ng